HEAT TRANSFER SCIENCE AND TECHNOLOGY

Edited by

Bu-Xuan Wang
Institute for Thermal Science and Engineering
Tsinghua University, Beijing, The People's Republic of China

HEMISPHERE PUBLISHING CORPORATION
A member of the Taylor & Francis Group

New York Washington Philadelphia London

HEAT TRANSFER SCIENCE AND TECHNOLOGY

2 3 4 5 6 7 8 9 0 BCBC 8 9

Library of Congress Cataloging-in-Publication Data

Heat transfer science and technology.

Bibliography: p.
Includes index.
Proceedings of the International Symposium on Heat Transfer, held in Tsinghua University, Peking, Oct. 15–18, 1985.
1. Heat—Transmission—Congresses. I. Wang, Pu-hsüan. II. International Symposium on Heat Transfer (1985 : Tsinghua University)
QC319.8.H46 1987 536′.2 86-33535
ISBN 0-89116-571-1 Hemisphere Publishing Corporation

Contents

Natural Heat Convection

Forced Heat Convection

Boiling Heat Transfer

Condensation Heat Transfer

Thermal Radiation

Heat and Mass Transfer in Porous Media

Nuclear Reaction Heat Transfer

Preface

The International Symposium on Heat Transfer (ISHT), Beijing 1985 was held in Tsinghua University during October 15–18, 1985. The Symposium was sponsored by the (Research) Institute for Thermal Science and Engineering, Tsinghua University, and supported by the China State Commission of Education.

The theory and practice of heat transfer had made vast contributions to production activities and daily life of mankind in the past decades, and is still a very active field with motivations and applications in various frontiers including life and biology, material, energy, aerospace, environmental and information sciences, and related high technologies. The scope of the ISHT, Beijing 1985 reflects the increasing importance of international exchange of various information about new ideas and recent achievements in expanding fields of heat transfer.

The Symposium had a registered attendance of 151 scholars and scientists from different countries including Australia, Canada, Federal Republic of Germany, Italy, Japan, People's Republic of China, United Kingdom, the United States of America, and the Soviet Union. The Chinese attendants were limited to 80, and nearly 47% of the total came from overseas. A total of 112 scientific papers on different topics, including 4 invited keynote addresses, were accepted to be presented at the Symposium, of which 3 papers were withdrawn and 52 papers were submitted by Chinese scholars and scientists.

This Proceedings includes 109 papers presented at the Symposium. All of the papers were reviewed for their main technical contents as stated in the preliminary abstracts and were printed from the camera-ready manuscripts supplied by the authors. So the editor assumes no responsibility for the accuracy and completeness of the typing itself.

It is a great pleasure for me to take this opportunity to express our gratitude and warm thanks to members of the International Committee and Organizing Committee, to the authors and the invited speakers, and to all the persons who had contributed to the Symposium and made this Proceedings possible.

Bu-Xuang Wang
Professor, Tsinghua University
Chairman, ISHT, Beijing '85

Organizing Committee and Staff

SPONSORED BY

Institute for Thermal Science and Engineering,
Tsinghua University, Beijing, People's Republic of China

The China State Commission of Education

INTERNATIONAL COMMITTEE

Bu-Xuan Wang (Chairman)
Tsinghua University, Beijing, PRC

Jun-kai Feng
Institute of Engineering Thermophysics, Chinese Academy of Sciences, Beijing, PRC

F. Mayinger
Technische University of Munich, Germany

T. Mizushina
Kyoto University, Japan

Y. Mori
University of Electro-Communications, Tokyo, Japan

C. L. Tien
University of California at Berkeley, USA

K. T. Yang
University of Notre Dame, USA

Shi-Ming Yang
Shanghai Jiaotong University, China

W. J. Yang
University of Michigan, USA

ORGANIZING COMMITTEE

Xiaowen Zhang (Chairman)
Vice-Pressident of Tsinghua University

Caiquan Liu (Secretary-General)

STAFF

Chongchen Fang
Zepei Ren
Dajun Ye
Zengyuan Guo
Wensheng Rong
Lianzhong Zhang

KEYNOTE PAPERS

Introductory Remarks on Current Research Activities of Heat Transfer in China

BU-XUAN WANG, ZENGYUAN GUO, and ZEPEI REN
Tsinghua University
Beijing 100084, PRC

ABSTRACT

The importance of heat transfer researches to sciences and technologies in China is discussed. The academic activities and the present stage of concurrent heat transfer researches in China are described briefly. Some aspects of theoretical and methodological studies are given, and examples of heat transfer applications are listed.

I. IMPORTANT ROLE OF HEAT TRANSFER RESEARCHES IN CHINA

Heat transfer research is of first importance in improving energy conservation system in China. For the past 36 years, China had been putting stress on the development of fuel power and has made an universally accepted progress. However, the exploitation rate in China still cannot meet the urgent needs of her economic development. The current project is well-known to the world that, by the year of 2000, the energy production is only expected to be two times that of the present, which is grossly insufficient to support the projected increase of industrial and agricultural annual production by a factor of 4 over the same time frame. At the meanwhile, the efficiency of energy production and utilization in China is quite low as compared to those of the developed countries now. It is clear, China must pay attention to energy saving and to variety in energy exploitation and utilization. Great attraction is thus paid to heat transfer researches in China.

Futhermore, heat transfer researches may contribute much more to advance material science and engineering, to develop thermal environment science and engineering, to improve high technologies in modern information systems, to push for ward the bioengineering, and therefore to speed up the industrial and agricultural production in China.

Heat transfer is one of the very active discipline in technical sciences during the past five years in China.

II. ACADEMIC ACTIVITIES ON HEAT TRANSFER IN CHINA

There were some national conferences and symposiums on heat and mass transfer in the early sixties, sponsored by colleges and universities and by the Chinese Mechanical Engineering Society. Such activities were discontinued for more than ten years as the result of unexpected reasons. The Chinese Society of Engineering Thermophysics (CSETP) was founded in 1979 , and was subdivided into three Divisions: Heat and Mass Transfer, Combustion, and Aerothermodynamics in 1980. A quarterly "Journal of Engineering Thermophysics" (JETP), published in Chinese and with English abstracts, started publication from February 1980. At the same time, the degree system (BS, MS and Dr.), carried out in universities and instiutes, initiated also from January 1981 in China. All these backgrounds benifit the activities on heat transfer researches, and the actual demands spur much more persons into action to work on heat transfer researches.

Since 1980, we have annual nation-wide heat transfer conferences once every year, sponsored by Division on Heat and Mass Transfer, CSETP. The 1st National Conference on Engineering Thermophysics in Colleges and Universities was held in Nanking, September 1983. Moreover, there are also a few papers in heat transfer presented in symposiums held by the Chinese Mechanical Engineering Society (CMES), the Chinese Society of Chemical Engineering and Technology (CSChET), the Chinese Society of Aeronautics and Astronautics (CSAA), the Chinese Metals Society (CMS), the Chinese Society of Measurement (CSM), or the Chinese Solar Energy Society (CSES) etc. Such a large number of papers were written in Chinese without exception, and only a few of them was accepted for publication in journals or to be collected in bounded volumes. It was therefore seldom known to the foreign heat transfer communities. China opens to the world recently. Most of Chinese scholars worked in the field of heat and mass transfer lack chance and experience to show their achievements in international conferences or related journals. So, we combined with our colleagues try to take this opportunity to introduce and summarize here some concurrent heat transfer researches in China.

III. RESEARCHES ON HEAT TRANSFER IN CHINA

It is quite difficult to make a comprehensive report on all of the research activities on heat and mass transfer in China, especially in such a short description. The authors are devoted themselves to give only a brief introduction to some aspects as follows.

1. Heat Conduction A new rigorous method on the conventional variable separation technique together with the finite element principle, was proposed to solve the transient heat conduction problem with random geometry and arbitrary time-dependent boundary condition [1,2], and was used successfully to predict the complex passive transient thermal environment in/around the underground cave for fruit preservation [3]. In order to get the actual thermophysical properties of soil for taking care of moisture migration, a new "heating and cooling" method for measuring thermal diffusivity and conductivity of dispersed media in the scene with a probe was developed in detail [4].

Similarity criteria for thermal scale modeling of spacecraft with radiative-conduction system was discussed briefly, and design, construction and test for thermal scale models of three satallites with passive thermal control were summalized [5]. Thermal conductivity measurement of materials during ablation with treatise on the moving boundary problems had been presented [6], and the measurement of the thermal conductivity for a charring ablation thermal protection material under transient heating was repoted recently [7]. A comprehensive analysis algorithm for ablation, temperature fields and thermal stresses of the carbon-base nose-tips had been made public [8].

The Lighthill's technique was applied to the solution of heat conduction in a cylindrical body undergoing solidification [9]. A semi-infinite body inbeded by a thermocouple was simulated by a two-dimensional RC network containing 19x25 modes, so as to analyze the distortion of temperature field due to the thermocouple inbeded [10]. As an inverse problem of heat conduction, the thermal conductivity and its dependence on temperature had been calculated numerically according to the temperature history measured at several points in an ingot mold [11]. For solving the unsteady heat conduction problems with finite element method, the functional responding to a parabolic partial differential equation is still absent today, so the $\partial T/\partial t$ had to be expanded in accordance with the finite difference technique, and three-points backward difference scheme and Galerkin scheme were recommended [12].

Based on the theory of unsteady heat conduction in a semiinfinite solid with plane heat sourse of constant heat rate, a method of simultaneous measurement of the thermal diffusivity and conductivity of insulating materials was developed, and the measuring techniques were well discussed, so that the maximum cumulative errors of the measuring results could be controlled as both the thermal diffusivity and conductivity within ±6%, [13, 14]. The transient process after changing the heating rate in a conventional twin-plate device for measuring thermal conductivity of building materials were analyzed theoretically, and as the result, a method for determining the thermal diffusivity of insulating materials with the guarded twin-plate device was proposed and realized [15,16], so as to extend the function of the twin-plate device. The physical model for measurement of thermal diffusivity by pulsed laser technique was discussed, and a computerized laser thermal diffusivity measurement apparatus were reported, which features automatic and high speed measurement, and is recommended to be operated in the range of 300-1800°C. [17]. A theoretical analysis of heat conduction factors and an experimental research on heat conduction properties for composite materials were later reported [18].

2. Single-phase Heat Convection Application of a transient method to the investigation on natural heat convection across air layers at various angles of inclination was reported [19]. The laser double-mirror interferometer was used to measure the temperature field for natural convection near a flat plate of uniform temperature [20], while a holographic interferometer was used to study the free convection heat transfer for horizontal cylinder [21], the results obtained are well compared with those reported in literature. Measurement of gaseous temperature field by holographic interferometer was also reported [22].

The mechanism of the thermal drag for flow of thermofluids was investigated both analytically and experimentally [23]. A new technique—"double block correction" was suggested to speed up convergence of numerical calculation of heat transfer and fluid flow [24].

A numerical heat transfer solution was obtained for laminar flow in entrance region of a circular tube with uniform heat flux for either μ=const. and μ=f(T), [25]. The mass and heat transfer in film flow of non-Newtonian power-law fluids were studied analytically and experimently [26,27]. The laminarization in the combined convective heat transfer process of the turbulent-natural co-direction flow was affirmed experimentally, and the effect of Re and Gr/Re^2 was analyzed, [28]. Turbulent heat transfer to air in parallel plates and annuli for both single and double side heating with varying heat flux were studied experimentally [29,30], and the effect of sinusoidal heat flux distribution and also the effect of double-side heating on heat transfer coefficients were analyzed. Experimental data for heat transfer to liquid sodium flowing in circular tube and concentric annuli were re-ported [31,32], and it was emphasized that the bilateral heating has an obvious effect on heat transfer for liquid metals flowing in an annuli[32].

To verify the effectiveness of heat and mass transfer analogy, naphthalene sublimation experiments of external flow across single cylinder and internal duct flow were carried out on a suction-type wind tunnel [33], the results show that the maximum deviation of mass transfer data from the corresponding heat transfer correlation is within ±5.5% The heat transfer and draft loss performance of air flowing across staggered banks of oval tubes fitted with rectangular fins were studied experimently [34]. Flow visulization in a complex duct was to be reported [35].

3. Two-phase Flow and Heat Transfer An analytical model was presented, possibly for the first time, to cover the effect of a pressure gradient caused by the increasing vapor film thickness along the flow direction [36-38]. It was found that this effect will be prominent for low-velocity laminar film boiling on a horizontal plate surface, especially for a large ratio between the density of the liquid and that of the vapor. A physical model and a semi-empirical theory had been proposed for the turbulent flow of subcooled liquid along a horizontal plate, the empirical values of the constants are determined through experimental investigation with subcooled water, and the equation thus obtained is expected to be adaptable for practical use in engineering analysis of high-temperature

uenching process, [38,39].

An empirical correlation had been developed for CHF (critical heat flux) with ressure ranging from 80 to 150 atm. and mass velocity ranging $(4-12)x10^6$ kg/m^2hr, nd local quality at CHF point from -0.20 to +0.15, [40]. Rewetting heat transer during bottom flowing of tubular test section was also investigated [41].

The microstructure of flow and Benard cells inside liquid drops on horizontal lat plate was recorded, and a qualitative analysis for the mechanism of triggerng Benard cells in liquid drops is inducted, [42]. The experimental results nd its discussion on the deterioration in the bend of a 25x2 mm vertical U-haped tube with radius of 700mm under the conditions of p=45-144 bars, G=800-000 kg/m^2s and q=80-330 kW/m^2 was reported [43].

The conductance-probe technique was used to measure the time-averaged void raction in the two-phase flow of air-water mixture through horizontal and vertial tubes [44], and the research on characteristics of two-phase flow and heat ransfer in helically-coiled tubes were repoted recently [45].

A general correlation for predicting average heat transfer coefficient of ilm condensation was suggested and the results calculated by this correlation vere reported to be well in agreement with the experimental data in literature[46].

Based on statistical physics and thermodynamics, a physical model of the fornation and growth of ice nuclei had been presented [47].

4. Thermal Radiation Method for measuring and checking the normal thermal emissivity of the metals and coatings were discussed [48], and apparatus for measuring the hemispherical total emissivity of the coatings and metals in the range of 1000-1800 K was described with discussion on possible errors of measurement [49]. Method for measuring multiple thermophysical properties, including thermal conductivity, hemispherical total emmittance and normal spectral emittance (0.65μ) of metals and alloys at high temperature, had been developed [50]. Measurement of minute radiometric quantities in the spectral range from visible to near infra-red had been reported also [51].

The application of conformal mapping and orthogonal curvilinear coordinate system in the analytical calculation of shape factor was described [52]. A method for calculating view factors by a computer programme was presented [53]. Radiative heat transfer had been analyzed by the net heat-flux method [54].

The Monte-Carlo method was used to solve the radiative heat transfer problems of a cylindrical furnace with taking into account the presence of convective heat transfer [55]. The problems of Monte-Carlo solution of radiative heat transfer in furnace were compared and discussed, and a revised method was developed to reduce the computing time and to raise the accuracy of computation [56].

Experiments were conducted in a fluidized twin-bed combustion boiler with steam capacity 10 tons/hr to determine the influence of fluidizing velocity, bed particle size and bed temperature, and on the radiative heat transfer coefficient [57].

The cause of instability in laser beam was analyzed by using radiation combined with conduction and convective heat transfer model [58]. The infra-red multi-spectral radiative pyrometry, which is based on the principle of the multicolor technique and on the principle of equalization of two spectral radiance, was investigated, and it was shown that this method can diminish the effect of emissivity on the accuracy of measurement [59].

5. Heat and Mass Transfer in Porous Media To meet the needs for energy saving with adaption of thermal insulation, the thermal conductivity and diffusivity of wet porous building materials were studied experimently [60]. It was found that, the effect of moisture migration on thermal diffusivity will be somewhat strange in the range of small moisture content and a point of inflection appears on the curve of thermal diffusity of aerocrete vs moisture content [60]. The complex phenomena of heat and mass transfer in wet porous building materials had been treated theoretically [61].

The capillary pressure and permeability of heat pipe screen wicks had been measured, and the Happel's theory was used to derive the coefficients of permeability of screen wicks [62]. The results of experimental research in pool boiling and boiling of the liquid layer from porous surfaces, which were formed by sintering screens of different meshes on copper blocks, were reported [63,64]. Experiments on pool boiling heat transfer of machined porous surface tubes had also been presented [65].

An analytical model was presented for the calculation of the heat transfer coefficient between a cell and an aqueous medium, and the analysis estimates the temperature difference across the membrane during freezing and its effect on the attendant water transport via thermoosmosis and other mechanisms [66].

6. High Temperature Heat Transfer A lot of work had been done on the film cooling for protection of the material surface exposed to high temperature gas flow [67-70], and was recently summarized in a monograph "Film cooling" [71]. Experimental investigation of impingement cooling of concave surfaces of turbine airfoils had also been reported [72,73]. The finite element method had been applied to predict the temperature field of a fir-tree turbine blade rotor [74].

A new calculation method was presented, so that the gas and electron temperature distribution of arc plasma can be obtained by measuring the line intensity only and then by solving a system of equations, the experimental work can therefore be reduced much more [75]. The effect of pressure on heat transfer to a particle exposed to a thermal plasma was shown to be caused mainly by the presence of the Knudsen effect for small particles with radii typical for applications in plasma chemistry and plasma processing [76]. The effects of combustion on the laminar boundary layer heat transfer of a blunt body stagnation point was investigated by using methane gas as an injecting media [77]. A method utilizing a flat plate specimen placed at a small angle to a supersonic arc heated flow had been developed [78].

7. Enhanced Heat Transfer It was reported that the axial groove is a suitable wick for cryogenic heat pipe and an axial groove neon heat pipe was used successfully for cooling a 1x0.5 m^2 aluminum plate to 30K [79]. The researches and applications of heat pipes, including the performance studies of axially grooved heat pipes, variable conductance heat pipes, thermosyphons and their various applications had been carried out at the Chinese Academy of Space Technology and at Chongqing Univ. for more than ten years [80,81]. The influence of inclination angle on limit to heat transport and film coefficients had been analyzed [82].

In order to find out the better types of tube inserts for augmentation of heat transfer of gas flow in lower Re range, flow visualization by means of hydrogen bubble technique and the comparative tests of friction drag and heat transfer of air flow in vertical tube with uniform temperature were made [83]. Heat transfer coefficient and power required for helical ribbon mixers were reported [84].

Experimental observations were made on enhanced heat transfer in fluids subjected to high frequency oscillations, which may be developed as a new effective technique [85].

Heat transfer and friction correlations are provided for turbulent flow in concentric annuli having different rib roughness [86,87]. Heat transfer augmentation and flow friction of perforated fin surfaces in conductive-convective heat transfer system were studied [88]. Methods of augmentation of convective heat transfer in cooling channels of the turbine vane with rough surface and dynamic coating on airfoil surface in virtue of silicon additives in fuel were suggested [89], it could be expected to raise the turbine inlet gas temperature to 1600K when the coolant to gas flow ratio equals 0.03-0.04, as concluded by the authors themselves.

The experimental results with R_{11} and R_{113} showed that, condensation heat transfer coefficient for saw-teeth-shape finned tube will be ten times as large as that of smooth tube [90]. It was reported that, under the experimental conditions, the boiling heat transfer coefficient increases by 1-3 times as compared with that of an usual thermosyphon, and the condensation heat transfer coefficient will be 67-100% higher than that of an usual thermosyphon [91].

8. Heat Exchangers Theoretical analysis and numerical calculation of transient and periodic steady-state temperature distribution of rotor and fluid in rotary regenerator are reported [92, 93], the results obtained were ckecked well with actual experimental measurement.

A generalized effectiveness expression of heat exchangers, which holds for crossflow heat exchangers with non-uniform inlet fluid temperature as well as with uniform one, was proposed for the better understanding of the heat transfer process in crossflow heat exchangers [94]. Three evaluation criteria for heat exchanger performance were suggested [95].

It was proposed that the heat pipe heat exchanger can be treated as a conventional shell-and-tube heat exchanger so far as the arrangement of tube bundles and calculation methods are concerned, some design problems such as calculation model, selection of parameters and design configuration etc. were summarized [96].

A mathematical model, the so-called "discret flow model" used for calculating the temperature of fluid when it flows from the exit of a heat exchanger, was established [97].

The method for experimental research on industrial equipments under production conditions and the method for processing the data obtained were discussed in detail [98]. The assessment for effects of the channel geometry on heat transfer performance and flow resistance in plate heat exchangers were reported [99].

9. Industrial Heat Transfer Heat flux meters may be used extensively in China as to indicate and control the heat loss through shell of heat equipments or from heat piping. The heat transfer analysis had been made on a high speed calibration device suggested for heat flux meters [100]. The possible disturbance of the temperature field due to heat flux meter and its error for heat flux measured were analyzed [101]. The transition time for the surface-installed heat flux meters was also estimated analytically [102].

Heat transfer problems with high value of Gr, consequent low Re/Gr, that occur in the cement industry, were investigated experimentally, and the conditions for which the effect of natural convection can be neglected were found[103].

The artificial freezing procedure of stratum and the temperature distribution of the freezing-wall were analyzed by finite element method, a simple and effective method was presented for calculating transient heat conduction with release of latent heat [104].

The climatological methods for estimating ultraviolet global radiation and diffusive radiation under various sky conditions were suggested according to a series of radiation measurement at ten cities in different regions [105].

The mathematical modelling of three-dimensional flame heat transfer with computer code was used to calculate the furnace of 100MW bituminous coal fired boiler, the results was in close accordance with test data [106]. The heating and evaporation processes of a single coal-water slury droplet under forced convection had been studied, and a calculation model was established [107]. The temperature and velocity fields of flowing slurry at various tube section were obtained in terms of numerical approach of its momentum and energy equations, and the drag and heat transfer coefficients of COM (coal-oil mixture) flowing in tube with constant wall-temperature were determined both analytically and experimentally [108]. A packet model was derived for heat transfer between the fluidized bed and the immersed tube surface, with the consideration of presence of surface resistance and its effect on local voidage by introducing a physically justified property boundary layer [109]. As to prevent sulfuric corrosion at high temperature, a model on the temperature field of the front part of boiler and a corresponding method for calculation with emphasis on treating upon the nonlinear and conjugate boundary conditions was proposed [110].

A model for computer simulation of heat transfer process in the rotary combustion engine had been presented, which performs a leakage mass balance, accounts for sweep flow, boiling on water side and other ignition and combustion characteristics, [111]. Measurements and analysis of the transient wall temperature of the rotary combustion engine was also reported, and a rapid responsible surface thermocouple was developed for suiting the measurement [112].

A numerical method proposed for calculating temperature rise of the friction brake will provide a favorable condition for designing mechanical press [113]. Temperature measurement and calculation of 4.5 tons ingot mold of vermicular cast iron during its usage was presented [114]. The temperature field in rectangular and circular planar was under consideration with uneven arbitrary distributed heat sources [115].

A set of empirical correlations covering a wide range of spacing, obtained by using transient technique for natural convection heat transfer across air layers inclined at various angles, were given. Based on these correlations, it was shown that there are two ranges of the optimum air layer spacing with small convective heat loss from the view of variation of convective heat loss with air layer spacing [116].

IV. CONCLUDING REMARK

The papers reviewed above are only those published in Chinese literature, mostly in JETP (J. Engineering Thermophysics). These are certainly very small part of the work done by Chinese scholars and scientists. There are also many research reports in the field of heat and mass transfer published in journals of Chinese universities and institutes. It is really pity, however, that we cannot cover all of those here because of the limitation of space. Even though, it may be clear, the research activities of heat transfer in China are quite widespread and profound.

REFERENCES

[1] B.X. Wang and Y. Jiang, "An eigenvalue method for the analysis of long-term transient heat conduction", JETP, 5(3):284-287, 1984.

[2] Y. Jiang, "Basic research on thermophysics for the utilization of natural temperature difference in subsurface spass", Dissertation for Dr.-Eng., Tsinghua University, Beijing, Feb. 1985.

[3] B.X. Wang and Y. Jiang, "The thermal analysis of passive underground cold store for fruit preservation", pp.49-54. Proceedings, the National Heat & Mass Transfer Coference, held in Wuhan, China, Nov. 1984, Science Press.

[4] B.X. Wang and Y. Jiang, "The heating-cooling method for measuring thermal diffusivity and conductivity of dispersed medium in the scene with a probe", JEPT, 6(3):249-254, 1985.

[5] W.H. Tian, Q.F. Ma and M.Z. Zheng, "Thermal scale modeling of spacecraft", JEPT, 1(1):88-94, 1980

[6] B.L. Zhou, Z. Wei and J.H. Lin, "Thermal conductivity measurement of materials during ablation--a treatment on the moving boundary problems", JETP,2(2):166-172, 1981.

[7] W.C. Sun and G.L. Wu, "Measuring the thermal conductivity of a charring ablation thermal protection material under transient heating", JETP 6(1):76-78, 1985.

[8] Z.Z. Huang, "A comprehensive analysis algorithm for ablation, temperature fields and thermal stresses of the carbon-base nose-tips", J. Chinese Society of Astronautics, p.30, No.3 1984.

[9] Y.W. Song, "On the solution of heat conduction in a cylindrical body undergoing solidification by the method of singular perturbation", JETP, 2(4):359-365, 1981.

[10] M.K. Yue, "A simulation analysis of temperature field distortion by a inbeded thermocouple", JETP, 2(4):346-352, 1981.

[11] C.M. Yu, "The inverse problem of heat conduction in the calculation of thermophysical properties", JETP, 3(4):372-378, 1982.

[12] X.Q. Kong, "Time finite difference scheme and its variable step size calculation in solution of unsteady conduction prolems by finite element method", JETP, 3(3):263-269, 1982.

[13] B.X. Wang et al., "A plane heat source method for simultaneous measurement of the thermal diffusivity and conductivity of insulating materials with constant heat rate", JETP, 1(1):80-87, 1980.

[14] B.X. Wang et al., "On the effect of heat capacity of the heater for plane heat source method with constant heat rate to determine simultaneously the thermal diffusivity and conductivity of

insulating materials", JETP,4(1):83-89, 1983.
[15] B.X. Wang and Z.P. Ren, "Research on the application of twin-platedevice for measurement of the thermal conductivity to determine the thermal diffusivity of insulating materials", JETP, 2(3): 262-268, 1981.
[16] B.X. Wang, Z.P. Ren and Z.H. Fang, "Meassuring method and error analysis for determining the thermal diffusivity of low-conductivity materials with a conventional twin-plate device", JETP, 3(1):52-59, 1982.
[17] T.G. Xi et al., "Computerized laser thermal diffusivity measurement apparatus", JETP, 1(2): 147-155, 1980.
[18] T.G. Xi et al., "A theoretical analysis of heat conduction factors and an experimental research and prediction of heat conduction properties for composite materials", JETP, 4(2):153-158, 1983.
[19] Z.S. Chen et al., "Application of the transient method to the investigation of natural convection heat transfer across air layers at various angles of inclination", JETP 1984--Special Issue for the US-China Binational Heat Transfer Workshop, Oct. 4-6, 1983.
[20] Y.Q. Yao and S.Y. Ko, "The measurement of temperature field of natural convection near the flat plate with the laser double-mirror interferometer (LDMI)", JETP, 6(1):72-75, 1985.
[21] S.Y. Huang, "Study of the local heat transfer coefficients for horizontal cylinder by free convection with holographic interferometer", JETP, 5(3):294-296, 1984.
[22] S.P. He, X.P. Wu and J.S. Cheng, "Measurement of gaseous temperature field by holographic interferometery", JETP, 5(4):364-370, 1984.
[23] Z.Y. Guo et al., "The thermal drag and thermal round about flow in fluid flow systems", JETP, 6(2):160-165, 1985.
[24] Z. Zhang, "Double block correction--a new technique to speed up convergence of numerical calculation of heat transfer and fluid flow", JETP, 5(4):364-370, 1984.
[25] W.D. Wu and L.J. Gao, "Numerical calculation of laminar flow and heat transfer in an entrance region of a circular tube with uniform heat flux", JETP, 2(1):60-66, 1981.
[26] T.Q. Jiang et al., "A theoretical study of mass and heat transfer in film flow of non-Newtonian power law fluids", J. Chinese Society of Chemical Industry and Technology, No.2 1982.
[27] T.Q. Jiang et al., "Experimental study on mass transfer in film flow of non-Newtonian fluids", J. Chinese Society of Chem. Industry and Technology, pp.368-373, No.4 1984.
[28] Q.Y. Zheng and Y.S. Wang, "Experimental research on combined convective heat transfer process of turbulent-natural co-direction flow occuring in vertical annular tube", JETP, 2(1):67-69, 1981.
[29] D.M. Yan et al., "Turbulent heat transfer in annular tubes with varying heat flux", JETP, 3(3): 249-255, 1982.
[30] D.M. Yan and G.H.Xu, "Effect of sinusoidal heat flux distribution and double-side heating on heat transfer coefficient in annuli", JETP, 2(1):53-59, 1981.
[31] S.K. Shi et al., "Experimental study of heat transfer to liquid metal sodium flowing in circular tube and annuli", JETP, 2(2):173-180, 1981.
[32] Y.J. Zhang et al., "The effect of bilateral heating on heat transfer coefficient to sodium flowing in an annlus", JETP, 3(4):386-388, 1982.
[33] H.H. Zhang et al., "Investigation of forced convective heat transfer using naphthalene sublimation technique", JETP, 6(1):49-55, 1985.
[34] B.X. Liu and Z.H. Cai, "Heat transfer and draft loss performance of air flowing across staggered banks of oval tubes fitted with rectangular fins", JETP, 3(4):365-371, 1982.
[35] Z. W. Ni et al., "Studies on the flow resistance, heat transfer and flow visulization in a complex duct", JETP, 6(4), 1985.
[36] B.X. Wang and D.H. Shi, "Investigation of the saturated laminar-flow boiling on a horizontal plate surface", JETP 1984--Special Issue for the US-China Binational Heat Transfer Workshop, Oct.4-6, 1983.
[37] B.X. Wang and D.H. Shi, "Film boiling in a forced-convection flow along a horizontal flat plate", JETP, 5(1):55-62, 1984.
[38] D.H. Shi, "Film boiling heat transfer for forced flow of subcooled liquid", Dissertation for Dr.-Eng., Tsinghua University, Beijing, CHINA, November 1984.
[39] B.X. Wang and D.H. Shi, "Film boiling heat transfer for turbulent flow of subcooled liquid along a horizontal plate", JETP, 6(2):148-153, 1985.
[40] G.X. Luo, Z.M. Shi and P.F. Wang, "Experimental investigation on pressure effects on critical heat flux in a rod bundle of square array", JETP, 5(1):82-84, 1984.

[41] D.M. Yan et al., "Rewetting heat transfer during bottom flowing of turbulent test section", JETP, 6(1):63-65, 1985.
[42] N.L. Zhang and Y.R. Xu, "Microstructure of flow and Benard cells in evaporating drops on horizontal plate", JETP, 6(4), 1985.
[43] T.K. Chen et al., "An investigation on heat transfer deterioration in vertical U-shaped boiling tubes", JETP, 6(2):154-159, 1985.
[44] H.H. Gao, F.T. Zhou and X.J. Chen, "The void fraction measurement by using conductance-probe technique in air-water two-phase flow", JETP, 6(1):56-59, 1985.
[45] F.T. Chou, "Research on the characteristics of two-phase flow and heat transfer in helically-coiled tube", Dissertation for Dr.-Eng., Xi'an Jiaotong University, Xi'an, CHINA, May 1985.
[46] J.F. Lin et al., "A general correlation for predicting average heat transfer coefficient of film condensation", JChIT (J. Chemical Industry & Technology), pp.353-362, No.4 1983.
[47] Y.B. Li and Y.L. Pan, "A physical model of forming ice nuclei on vessel surfaces from vapor in moist air through sublimation", JEPT, 3(3):176-182, 1982.
[48] X.S. Ge and S.X. Cheng, "On the measuring and checking of the normal thermal emissivity of the metals and coatings", JETP, 1(3):273-279, 1980.
[49] X.S. Ge et al., "An apparatus for measuring the hemispherical total emissivity of the high temperature coatings and metals", JETP, 3(2):169-175, 1982.
[50] L.Q. Yao et al., "A method for measuring multiple thermaphysical properties of metallic materials at high temperature", JETP, 2(2):185-187, 1981.
[51] J.S. Zheng, "Measurement of minute radiometric quantities in the spectral range from visible to near infra-red", Acta Physica Sinica, 29(3):286, 1980.
[52] H.Z. Zheng, "Two analytical calculation methods on the shape factor", JEPT, 3(4):389-393, 1982.
[53] W.L. Li and Z.X. Fu, "A method for calculating view factors by a computer program", JETP, 3(1):76-83, 1982.
[54] S.S. Xie and C.G. Bei, "Radiative heat transfer in multi-faces system Method of net heat-flux equations", J. Chemical Industry & Technology, pp.275-286, No.3. 1983.
[55] Y.F. Zhao and X.C. Xu, "The mathematical simulation of heat transfer in a cylindrical furnace", JETP, 4(3):275-280, 1983.
[56] W.D. Yao et al., "The problems on Monte-Carlo solution of radiant heat transfer in furnace of a large boiler", JETP, 5(3):288-290, 1984.
[57] H.S. Zhang et al., "The radiative heat transfer of the immersed tube in fluidized bed combustion boiler", J. Fuel Chemistry & Technology, pp.63-68, No.1. 1984.
[58] B.H. Bian, "The effect of unsymmetrical heat sources on the surface temperature field", J. Tsinghua University, 25(1):72-79, 1985.
[59] D.Z. Zhu, "An infra-red multi-spectral radiation pyrometry", JETP, 6(4), 1985.
[60] B.X. Wang and R. Wang, "On the thermal conductivity of moist porous building materials", JEPT, 4(2):146-154, 1983.
[61] B.X. Wang and Z.H.Fang, "A theoretical study of the heat and mass transfer in wet porous building materials", JETP, 6(1):60-62, 1985.
[62] T.Z. Ma et al., "A study of capillary and permeable characterstics of heat pipe screen wicks", JETP, 1(2):156-164, 1980.
[63] T.Z. Ma et al., "Nucleate boiling heat transfer from porous surface", JEPT 1984--Special Issue for the US-China binational Heat Transfer Workshop, Oct.4-6, 1983.
[64] T.Z. Ma et al., "Boiling heat transfer from porous surface by sintering screens", JETP, 5(2):164-171, 1984.
[65] L.X. Zhang et al., "Experiments of pool boiling heat transfer of machined porous surface tubes", JETP, 3(3):242-248, 1982.
[66] Z.Z. Hua, E.G. Cravalho and L.M. Jiang, "The temperature difference across the cell membrane during freezing and its effect on water transport", J. Shanghai Institute of Mechanical Engineering, pp.1-11, No.1 1984.
[67] S.Y. Ko et al., "Experimental investigation on the turbulent film cooling effectiveness over an adiabatic flat plate", JETP, 1(1):72-79, 1980.
[68] D.Y. Liu et al., "Experimental investigation on the coefficients of heat transfer of film cooling", JEPT, 1(4):371-377, 1980.
[69] S.Y. Ko et al., "Investigation of film cooling near the exits of discrete holes on a convex surface", JETP 1984--Special Issue for the US-China Binational Heat Transfer Workshop, Oct. 4-6, 1983.

[70] J.Z. Xu et al., "Experimental investigation on the near-field characteristics of film cooling with 30° injection from a row of holes on a convex surface", JETP, 5(2):182-186, 1984.
[71] S.Y. Ko et al., "Film cooling", (in Chinese), Science Press, 1985.
[72] J.R. Cheng and B.G. Wang, "Experimental investigation of simulating impingement cooling of concave surface of turbine airfoils", JETP, 1(2):165-175, 1980.
[73] B.G. Wang and J.R. Cheng, "Experimental investigation of impingement cooling by round jets of multiple rows", JETP, 3(3):253-241, 1982.
[74] H.T. Lin, "Air cooling in turbine rotors and calculation of the temperature field of a fir-tree blade root assembly by the finite element method", JETP, 2(1):70-72, 1981.
[75] Z.Y. Guo. "A new calculation method of the temperature of the arc plasma in non-LTE", JETP, 3(1):60-66 1982.
[76] X. Chen, "Effect of pressure on heat transfer to a particle in a thermal plasma", JETP, 5(1): 69-74, 1984.
[77] L.H. Yan and X.T. Yu, "Calculation of stagnation point heat transfer with methane gas injected in dissociated air boundary layer", JETP, 3(4):379-385, 1982.
[78] Y.D. Han and K.X. Wang, "A primary research on ablation experimental technique of the turbulent flow over a flat plats specimen", JETP, 2(1):46-52, 1981.
[79] C.S. Hua, "Performance investigation of nitrogen axial groove heat pipes and application of neon cryogenic heat pipe", JETP, 2(4):353-358, 1981.
[80] Z.Q. Hou and Y.P. Wen, "Research and application of heat pipe", JETP 1984--Special Issue for the US-China Bi-national Heat Transfer Workshop, Oct.4-6, 1983.
[81] M.D. Xin and Y.G. Chen, "Heat pipe researches in Chongqing University", J. Chongqing University, pp.1-11, No.2 1984.
[82] Z.F. Zhang et al., "Experimental investigation of performance and thermal resistance of gravity assisted heat pipes", JETP, 3(1):67-75, 1982.
[83] G.G. Huang et al., "An investigation on augmentation of single phase heat transfer in tube by means of inserts", J. Chemical Industry & Technology, pp.23-25, No.1 1983.
[84] H.Z. Wang, Heat transfer coefficient and power requirement in helical ribbon mixers", J. Chem. industry & Technology, pp.375-380, No.4 1984.
[85] U.H. Kurzweg and L.D. Zhao, "Experimental observations on enhanced heat transfer in fluids subjected to high frequency oscillations", J. Tsinghua University, 25(1):52-57, 1985.
[86] W.Z. Gu, "Calculation of heat transfer and friction in concentric annuli with various form of roughness", JETP, 1(3):280-289, 1980.
[87] T.Z. Ma et al., "Heat transfer and friction for turbulent flow in annuli with repeated-rib roughness", JETP, 2(3):269-275, 1981.
[88] J.R. Shen et al., "An investigation on the heat transfer augmentation and friction loss performance of perforated fin surfaces", JETP, 6(1):174-177, 1985.
[89] W.Z. Gu et al., "A new method of cooling turbine vane", JETP, 1(3):280-289, 1980.
[90] S.P. Wang et al., "Experiments of saw-teeth-shape finned tube enhancing condensation and its dimensionless correlations", JETP, 5(4):374-377, 1984.
[91] M.W. Tong et al., "Heat transfer enhancement in a two-phase closed thermosyphon", JETP, 5(4): 370-373, 1984.
[92] Z.P. Ren and S.Y. Wang, "Analysis of the transient and steadystate heat transfer in rotary regenerative heat exchanger", JETP, 5(3):269-274, 1984.
[93] Z.P. Ren et al., "Heat transfer performance of rotary regenerative heat exchanger", JETP, 6(4), 1985.
[94] Z.Y. Guo et al., "Thermal analysis of the crossflow heat exchangers with non-uniform inlet fluid temperature", JETP, 6(1):66-68, 1985.
[95] Z.W. Ni et al., "Three evaluation criteria for heat exchanger performance", JETP, 5(4):387-389, 1984.
[96] Z.Y. Jiang and T.Z. Ma, "Some problems in the design of gas-gas heat pipe heat exchangers",JETP, 6(3):255-262, 1985.
[97] Z.F. Dong and J. Wu, "A new method for determining heat exchanger--discret fluid flow transient method", JETP, 6(3):275-278, 1985.
[98] X.C. Xu and R.Y. Lin, "The experimental research method under production conditions for heat exchanger", JETP, 1(4):384-392, 1980.
[99] Z.W. Ni and Z.L. Jiao, "The assessment for effects of the channel geometry on the performance

of heat transfer and flow drag in the plate heat exchangers", pp.364-371, Proceedings of the Nationwide Heat & Mass Transfer Conference, held in Wuhan, CHINA, Nov. 1984, Science Press, Beijing.
[100] B.X. Wang, L.Z. Han and Z.H. Fang, "Heat transfer analysis of a high speed calibration device for heat flux meters", Acta Metrologica Sinica, 5(3):171-179, 1984.
[101] B.X. Wang et al., "Disturbance of the temperature field due to a heat flux meter and its error estimation", pp.411-416, Proceedings of the Nationwide Heat & Mass Transfer Conference, held in Wuhan CHINA, Nov. 1984, Science Press, Beijing.
[102] B.X. Wang et al., "The transition time for the surface-installed heat flux meter", pp.417-419, Proceedings of the Nationwide Heat & Mass Transfer Conference, held in Wuhan CHINA, Nov. 1984, Science Press, Beijing.
[103] X.M. Zhang et al., "Experimental research about convection heat transfer from a horizontal rotating cylinder", JETP, 6(3):268-271, 1985.
[104] Y. Zhang and S.S. Gan, "Using finite element method to analyze the formation and temperature distribution of the freezing-wall in the artifial freezing method", JETP, 5(2):175-181, 1984.
[105] Y.H. Zhou, "Climatological study of ultraviolet radiation", Acta Energiae Solaris Sinica, 5(1):1-11, 1984.
[106] X.C. Xu, "Application of mathematical modelling of three-dimensional flame heat transfer in utility boiler", JETP, 3(2):161-168, 1982.
[107] W.B. Fu and Y.H. Li, "A study of heating and evaporation processes of single coal-water slurry droplet under forced convection", JETP, 6(3):279-282, 1985.
[108] K.F. Cen et al., "Investigation of flow properties and heat transfer process of coal slurry inside the pipeline", JETP, 4(1):46-52, 1983.
[109] H.S. Zhang et al., "The heat transfer calculation and experimental study of the immersed tube in the fluidized bed combustion boiler with low grade coal", JETP, 5(1):63-68, 1984.
[110] H.J. Zhang, "Analysis on the temperature field of the front part of the boiler", JETP, 5(3):281-283, 1984.
[111] C.F. Ma and Z.M. Qi, "Analysis and computer simulation of heat transfer process in the rotary commbustion engine", JETP, 1(4):378-381, 1980.
[112] C.F. Ma et al., "Measurement and analysis of the transient wall temperature of the rotary combustion engine", JETP, 3(1):84-88, 1982.
[113] C.M. Yu et al., "A numerical method for calculating temperature rise of the friction brake", Chinese J. Mechanical Engineering, 20(4):43-52, 1984.
[114] C.M. Yu et al., "Temperature measurement and calculation of 4.5 tons ingot mold of vermicular cast iron during its usage", Scientific Report of Tsinghua University, QH81024, July 1981.
[115] S.B. Lin and G.D. Ma, "The analytic calculation method of thermal field in a rectangular and circular planar under periodic heat flow", JETP, 1(3):256-262, 1982.
[116] Z.S. Chen and X.S. Ge, "Theoretical and experimental investigation on determination of the optimum air layer spacing of the flat-plate solar collector with small convective heat loss", Acta Energiae Solaris Sinica, 6(3):287-296, 1985.

Heat Transfer in High Technology Problems

WEN-JEI YANG and K. KUDO
Department of Mechanical Engineering and Applied Mechanics
The University of Michigan
Ann Arbor, Michigan 48109 USA

ABSTRACT

Recent progress in transport phenomena research related to the problems of electronic equipment cooling and materials processing is reviewed. Part I deals with the cooling of single chips and electronic packages. In Part II on materials processing, efforts are focussed on transport phenomena in the fabrication of new materials, welding and cladding, steel processing, polymer processing and composite materials. While the well-known heat transfer mechanisms and techniques are involved in the heat removal from electronic devices, transport phenomena related to materials processing provide a challenging dimension for research.

INTRODUCTION

In the 1950's and 1960's, a significant achievement was made in heat transfer research for the sake of enhancing conventional and nuclear power generation. The main thrust was in the high heat flux problems such as boiling and condensation. The 70's saw the emphasis of heat transfer research involving energy problems, particularly in the utilization of nonpolluting, renewable energy resources. The principal consideration rested on the economical utilization of very low energy density in these energy resources. The era of high technology in the 80's has created the heat transfer problems in computers, communication systems and manufacturing processes in space. The main features include high energy density in very compact space and Marangoni (surface tension) effects in space processing.

HEAT REMOVAL FROM ELECTRONIC EQUIPMENT

The advent of the Integrated Microelectronics Revolution was motivated by technology that can simultaneously generate several microelectronic circuit elements on a single chip of semicon-

---

K. Kudo is an Associate Professor of Mechanical Engineering on leave from Hokkaido University, Sapporo, Japan.

ductor material. There continues to be a technology trend to greatly increase the number and speed of circuits and circuit elements integrated on a single chip while drastically reducing the size of the elements. This miniaturization trend brings advantages as well as drawbacks. The advantages include: (i) prodigious advancements in system reliability due to the continuous planar interconnection of conductors, (ii) tremendous increase in circuit speed due to the short electron transport distances, and (iii) enormous increase in system capability and compactness due to high complexity, functional density and functional miniaturization possible on a single chip. The drawbacks are sacrifices in functionality and life durability of microelectronics components at elevated temperature. The main problems in thermal control for microelectronics are caused by: (i) nonuniformity in thermal conductance within the volume of a chip and (ii) nonuniform distribution of heat sources over the surface of a chip. The former is due to the presence of poor thermally conducting materials and a large number of material interfaces, while the latter is a result of the energy barriers presented to the flow of electrical charges at the P/N junctions or interface between different impurity doped regions of the semiconductor chip. A brief history of cooling electronics equipment is available in reference [1]. Heat transfer approaches evolute with progressively increasing chip power dissipation density: via simple natural convection in an enclosed package, enhanced natural convection by mounting an extended surface sink and by perforated or open covers, forced convection by fan-driven air, immersion cooling, and boiling.

A clear understanding of state-of-the-art heat removal and thermal control techniques as well as the thermal potential and limits of more advanced concepts is of great importance to the design, development and production of electronic equipment. This

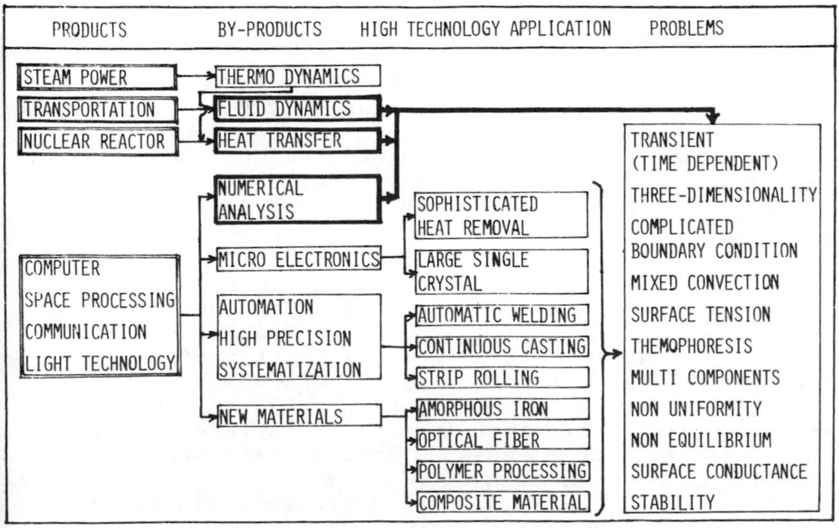

FIGURE 1. Heat transfer in high technology problem

section focuses on advanced thermal techniques and recent development of thermal-fluid mechanics in electronic packages.

Single Chip Cooling

Recently, two advanced thermal techniques have been proposed for direct heat removal from a high heat-flux single chip: boiling jet impingement cooling [2] and thermal contact conductance enhancement [3].

Ma and Bergles [2] employed pool boiling in the presence of impinging liquid jets to stabilize the temperature of microelectronic chips at power levels well above those expected during normal operation. In general, special cooling devices are required for a chip operated at power levels above 1 to 5 W or at relatively high heat fluxes exceeding 2×10^5 W/m^2. A conventional solution is direct immersion with pool boiling. However, the chip junction temperature may still be excessive. The superposition of forced convection on pool boiling will elevate the heat flux to counter the excess chip temperature. Jet impingement cooling of individual chips provides this convection. The boiling curves with forced convection approximately coincided with the extrapolation of the pool boiling curve. Burnout heat fluxes varied as the cube root (1/3) of jet velocity, but were only weakly dependent on subcooling at higher values of subcooling. Heat fluxes up to 10^6 W/m^2 were recorded. It was concluded that the use of subcooled jet impingement can achieve high chip powers: powers in excess of 20 W/chip can be accommodated within the usual 85°C limit on junction temperature.

Heat transfer across an aluminum junction operating in a vacuum or in a gaseous environment is a common problem in electronics packaging. Antonetti and Yovanovich [3] studied the use of metallic coatings to enhance thermal contact conductance. A new thermo-mechanical model for coated contacts was employed to predict the thermal contact conductance using the data obtained from experiments performed on nominally flat, rough, nickel specimens in contact with nominally flat, smooth, nickel specimens coated with a silver layer. It was disclosed that a thin coating, about 1 to 10 \m in thickness, of soft metals such as lead, tin and silver in an aluminum to aluminum joint resulted in an order of magnitude improvement in the thermal contact conductance.

Electronic Package Cooling

Electronic packaging such as enhanced feature telephones and modems forms a rectangular enclosure. High power densities in these packages require both local and global heat transfer enhancements. Recent trends in user-oriented electronic packaging favor a horizontal configuration. A characteristic of horizontal enclosures is that radiation, conduction and convection are small and of the same order of magnitude. Natural convection in vertical enclosures can produce heat transfer coefficients of an order of magnitude greater than the horizontal case. However, this configuration is often unacceptable for customer desk top applications. Forced convection is another alternative, as it can deliver a heat dissipation capability ten times that of natural convection between vertically mounted printed wiring boards (PWB). Design disadvantages of this method include fan cost, decreased reliability, noise, filter cost and maintenance.

Inclined. A tilted enclosure offers a viable thermal enhancement alternative. It offers a better user interface for desk top equipment, i.e., the plane of user interface approximately 310° perpendicular to the line of sight [4].

Air cooling of integrated circuit packages in natural convection was investigated experimentally and theoretically by Kennedy and Kanehl [4] and Torok [5], respectively. For an open-ended channel with a height of 2.65 cm, test results indicated that local temperature rises can be reduced by as much as 20 percent and 25 percent for angles of inclination of 10 and 20 degrees, respectively. Increasing the tilt angle beyond 20 degrees for a power dissipation of 10W resulted in no more thermal enhancement. By means of a general application code, FIDAP (Fluid Dynamics Analysis Program), with finite elements, Torok [5] obtained flow and thermal field prediction in an electronic package. Figure 2 shows the results for a tilt angle of 7 degrees. Venting from the package enhanced the heat transfer coefficient by as much as 40 percent at a higher Rayleigh number of 10^6. At a Rayleigh number lower than 10^5, however, no appreciable effect on heat transfer was detected through venting. In the latter case,

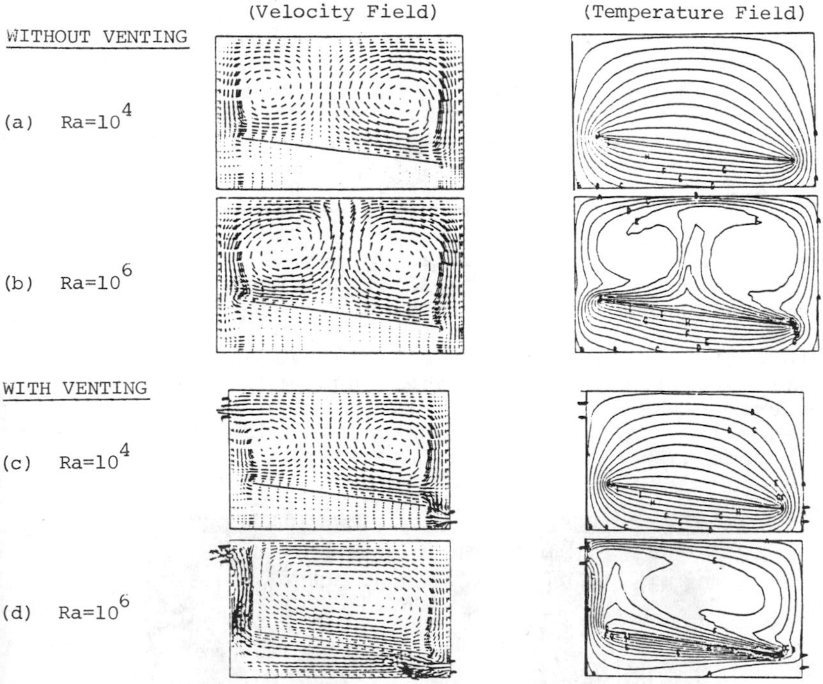

FIGURE 2. Velocity and temperature field in rectangular package with a tilted plate

convection was very weak and the temperature profile was attributed to conduction as shown in Figure 2(a). A comparison of Figures 2(a) and 2(c) leads to the conclusion that venting contributes little change in the flow and temperature profiles at lower Rayleigh numbers. Its effect on heat transfer becomes important only at higher Rayleigh numbers.

Horizontal. Forced air convection cooling of LSI packages mounted on a printed wiring card is common in high speed computers. For example, the neighboring cards form a narrow channel for air flow. Ashiwake, et al [6] studied the effects of package arrangement on heat transfer resistance and pressure drop in an air-cooled array of wiring cards. Figure 3 shows the flow patterns around the packages obtained by flow visualization. As expected, it was concluded that the air temperature rise is higher in the in-line arrangement than in the staggered case and vice versa for the pressure drop across the air flow channel.

A numerical analysis was conducted by Braaten and Patankar [7] to determine the effect in buoyancy of the fully developed flow and heat transfer in an in-line array of rectangular blocks. It was found that the secondary flow induced by buoyancy led to a significant enhancement to heat transfer over the forced convection results, along with a smaller increase in pressure drop. Mounting electronics packages facing upward, the use of highly conductive circuit boards, and the use of high Prandtl number fluids all lead to enhanced heat transfer.

Vertical. Buoyancy-induced convection is often a convenient and inexpensive mode of heat transfer which is commonly employed in cooling of electronic equipment. Figure 4 is a common geometry consisting of vertical fins (electronics packages) attached to the base (a circuit board). A shroud (the neighboring circuit board) is often placed near the fins so that the resulting chimney effect induces increased flow over the fins resulting in higher heat transfer. Numerical analyses on three-dimensional flow and thermal fields were made on the cooling of a vertical shrouded fin array by natural convection [8], forced convection [9] and mixed convection in [10] for a horizontal orientation and in [11] for a vertical one. Both the flow and thermal fields were three-dimensional and laminar in nature.

Figure 5 illustrates the development of axial velocity (W) and temperature (T) profiles for the Prandtle number of 0.7. Z denotes the dimensionless axial coordinate. In the inter-fin region, buoyancy force is very strong near the inlet, decreases axially and becomes very weak towards the exit. In general, the

(a) 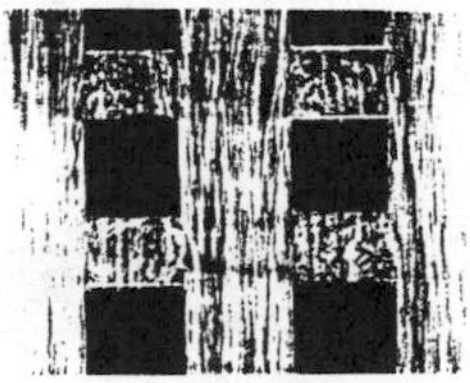

(b) 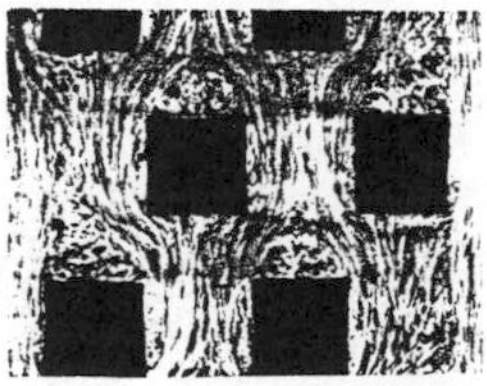

FIGURE 3. Flow patterns arround chips

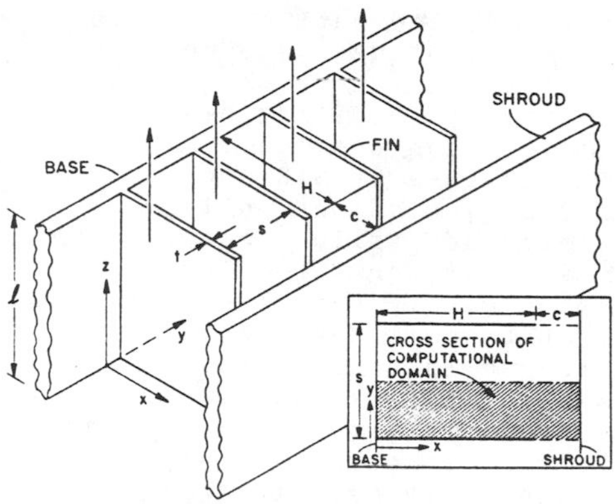

FIGURE 4. Shrouded board array

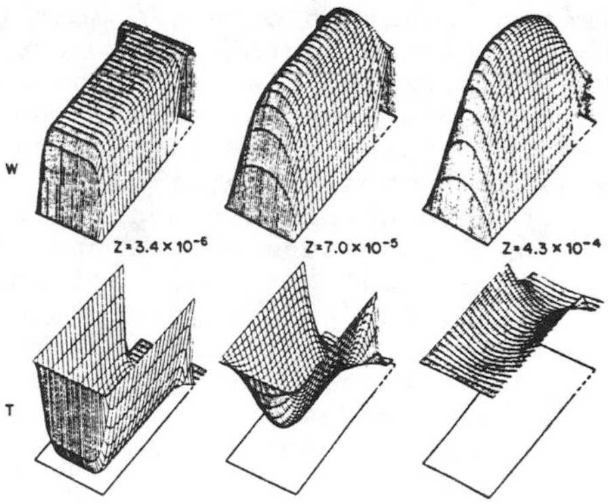

FIGURE .5 Axial velocity and temperature distribution

flow rate increases with increasing tip clearance or inter-fin spacing.

Oosthuizen [12] dealt with two-dimensional natural convection induced by the chimney effect. He determined the increase in the heat transfer rate from the heated portion of the wall that was induced by adding a length of adiabatic wall above the heated section of the wall. For a Prandtl number of 0.7, results indicated an approximately 50 percent increase in the heat transfer performance can be achieved by the addition of a long adiabatic wall (10 times the heated section).

Coyne [13] presented a technique to predict temperature distributions in air channels between printed circuit boards (PCB) in equipment frames with both horizontal and vertical variations in natural-convective heat dissipation. The study is an improvement over the existing analyses which assumed (i) sealed air channel, (ii) uniform frame heat flux and (iii) equally dissipating PCBs.

Natural convection wakes generated by isolated thermal sources in an extensive ambient medium are of importance in the natural-convection cooling of electronic equipment. In the absence of an externally induced flow, the heat released from an electronic component results in a buoyancy-driven flow that rises above the energy source as a wake or plume. The interaction of such a wake with the flows arising from other thermal sources is crucial to the positioning of components, because the heat transfer from an element is strongly influenced by the flow generated adjacent to it by wakes in its neighborhood. Jaluria [14] conducted a numerical study of the interaction of natural convection wakes arising from isolated, finite-sized thermal sources located on a vertical adiabatic surface, with a horizontal, isothermal surface at the leading edge. It was found that low values of the Grashof number $Gr \leq 10^4$, led to low velocity and temperature near the leading edge due to axial conduction, nonboundary-layer effects.

Immersion cooling. Complete immersion of electronic assemblies in fluids of appropriately high dielectric strength and low dielectric constant offers a most promising alternative to conventional thermal control measures. This technique has been successfully implemented in high power radar equipment, logic and memory devices and the Cray-2 supercomputer. Boiling heat transfer at the heat dissipating surfaces can provide a mild local environment for microelectronic devices (VHSIC and VLSI) and an accommodate substantial heat flux variations while minimizing temperature excursions and component failure rates. Many references deal with pool boiling heat transfer from single integrated circuit packages or transistor cans. Bar-Cohen and Schweitzer [15] investigated ebullient thermal transport from a pair of vertical, isoflux plates immersed in a saturated fluid. The wall superheat at constant imposed heat flux was found to decrease as the channel narrowed as shown in Figure 6, a thermosyphonic boiling behavior. The boiling thermosyphon analysis yielded the mass flux through the channel. Its result and a formulation for the convective component of heat transfer formed the basis for correlating the heat transfer performance from the channel walls.

Aakalu [16] determined the influence of air on the condensation of dielectric vapor. A closed-loop cooling system with a chemically inert low boiling temperature dielectric çoolant (FC-87 by 3M Company) in contact with the heat generating electrical components for large frame computers. The loop was operated at a near atmospheric pressure so as to maintain stresses in the liquid-cooled printed circuit boards within reasonable levels. Air was let into the vapor space at each of machine shutdowns to keep the internal pressure within a narrow band. Although the air in the system was expelled every time the machine was turned on, some air remained in the condenser space. This resulted in a reduction in the heat removal capability. Aakalu disclosed that a sudden drop in the heat transfer co-efficient (from 225 to 28 $W/m^2.^oC$) would occur when the mass of air in the system exceeded a critical amount. The critical mass is a strong function of spacing between condenser fins.

Fusion cooling. The cooling of an inflight electronic device in

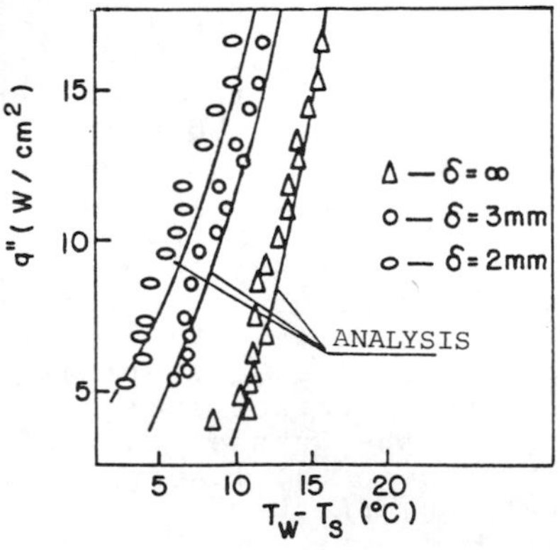

FIGURE 6. Pool boiling curves for different channel width in water

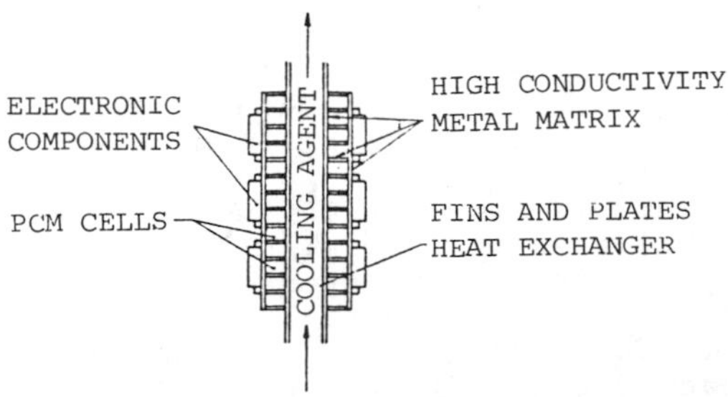

FIGURE 7. Phase change material heat sink

modern high performance aircraft involves special problems because of high and nonuniform heat dissipation and constraints imposed by the flight conditions. Light, compact, minimum energy input and relative independence of the different electronic units are also important consideration. The fusion cooling by a phase change material (PCM) is an attractive candidate to meet the requirements. Witzman et al [17] studied the potential of using paraffin hydrocarbons as a latent heat reservoir for stabilizing the temperature of electronic devices. Figure 7 shows a PCM heat sink that can dissipate up to 100 W of heat within 10 minutes. It consists of two panels partially filled with PCM. The electronic components are mounted on the exterior surfaces of the panels. Over very short times when the coolant is cut off, the discharged heat is removed through the PCM layers and carried away by the coolant that flows through the compact heat exchanger. Under normal operating conditions, however, the heat dissipated from the electronic elements is removed through the high-conductivity metal matrix (fins) which is placed parallel to the PCM

Summary

No new heat transfer mechanism and technique is involved in the thermal stabilization of microelectronics components. One only observes the applications of heat transfer technology in common practice to remove high heat fluxes from highly compact space.

TRANSPORT PHENOMENA IN MATERIALS PROCESSING

Materials processing transforms basic substances into useful elements through various operations such as forming, joining and modification. Most materials processing operations involve the transfer of heat, mass and momentum, accompanied by melting and solidification. This section deals with transport phenomena involved in the manufacturing processes of some materials for high-tech applications, welding and cladding, steel processing, polymer processing and composite materials.

Fabrication of New Materials

The following materials processing operations are considered: high-purity single crystals [18,19], optical fiber preforms [20] and amorphous iron [21].

High-purity single crystals. The floating-zone method is a noncrucible process that has been popularly employed as a crystal growth operation for high purity single crystals. Figure 8 shows the basic arrangement of floating-zone crystal growth. The upper rod is a polycrystalline feed material and the lower rod is a pure single crystal, with a melting zone suspended between them by surface tension. An induction heating coil supplies the heat required for melting. A continuous growth of the single crystal on the bottom rod is obtained by either moving the coil upward or moving the rods downward. The top or the bottom crystal is rotated at a constant speed to produce a uniform cylindrical crystal of better quality. However, the size of the crystal grown by the floating-zone method is limited by gravity. Materials processing in space has an advantage in the

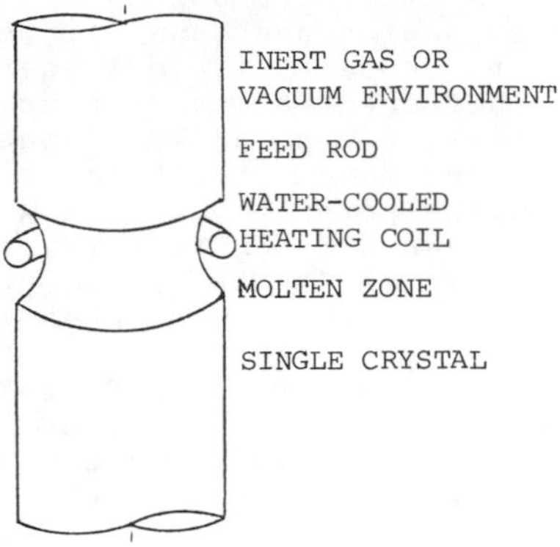

FIGURE 8. Floating zone crystal growth

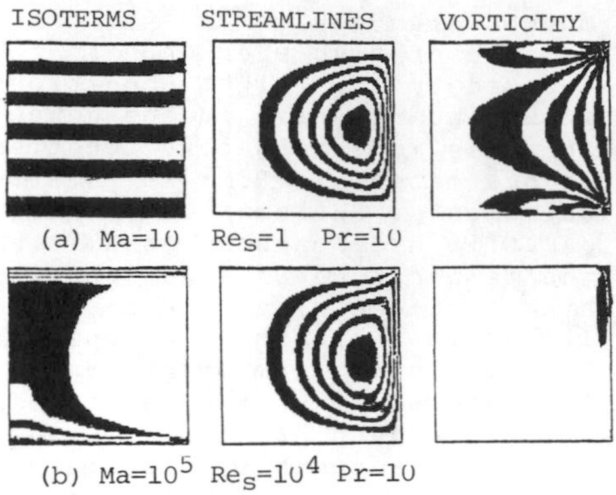

FIGURE 9. Temperature, flow and vorticity contour in molten zone

control of fluid motions and their influence on the crystal morphology. In a reduced-gravity environment, the main driving forces of the flow in the molten zone are the surface-tension gradients along the free surface, the buoyancy force and the electromagnetic force from the induction heating coil. Fu and Ostrach [18] investigated the flow caused by surface-tension gradients in a simulated floating-zone configuration. An up-wind finite difference scheme was used to determine the axisymmetric flow and temperature fields. The flows were found to form essentially a single-cell pattern with the core located nearer to the free surface. The axial velocity near the free surface was two to four times larger than that near the axis. While the Marangoni number Ma determines the relative importance of thermal convection to thermal diffusion, the surface-tension Reynolds number Re_s decides the relative importance of the vorticity convection to the vorticity diffusion. For Marangoni numbers less than 10^3, heat transfer is dominated by thermal diffusion, as illustrated in Figure 9(a). When $Ma>10^3$, Figure 9(b), the thermal convection becomes dominant over most of the floating zone. For Re_s less or equal to 10, the vorticity is distributed by diffusion. When Re_s reaches 10^4, it confines the vorticity diffusion to the region close to its source. Two peak surface velocities appear in the flow with $Pr>1$ and not in the flow with $Pr<1$. In conclusion, the buoyancy force generates negative vorticity which weakens the positive vorticity induced by the surface tension force.

Glicksman et al [19] investigated the stability of the crystal-melt interface surrounding a vertical, coaxial melt annulus undergoing stable convection flow. A long cylindrical sample of succinonitrile was heated by a coaxial, electrical wire (at the sample center) so that a vertical melt annulus formed between the wire and the surrounding crystal-melt interface (at a temperature below the melting point). A helical crystal-melt interface was formed, Figure 10, when the Grashof number exceeded certain critical values, in the range of 140 to 180. These critical Grashof numbers are more than an order of magnitude smaller than those corresponding to the convectively induced in-

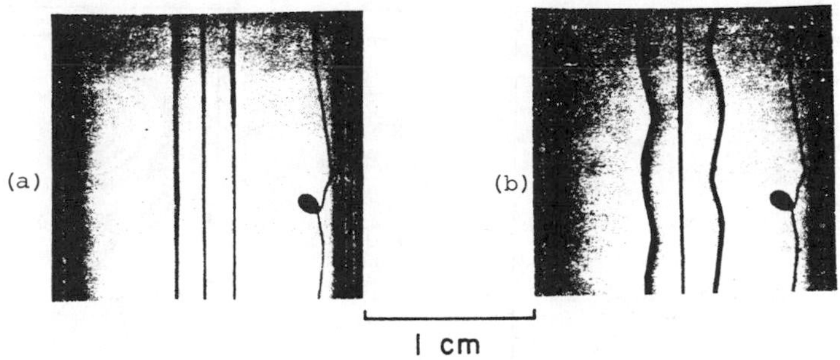

FIGURE 10. Crystal-melt interface (a vertical wire in the center is an electric heater)

stabilities in a similar chamber having non-deformable boundaries. The helical waves of the interface rotated with either a right-or-left-handed sense and with periods ranging from a few minutes to more than twelve hours. The dependence of convective instabilities on gravity is manifested through the Grashof number.

Optical fibers. Optical fibers, fine silica rods, with graded refractive index are produced from thicker silica rods called preforms which have a similar sectional refractive-index profile. Currently, the modified chemical vapor deposition (MCVD) is used in the production of preforms. In this process, Figure 11, a mixture of oxygen and the submicron-sized chlorides (aerosols) of silicon ($SiCl_4$), germanium ($GeCl_4$) and other dopants flows through a hollow quartz tube which is mounted and rotating in a lathe. An oxy-hydrogen flame heats the gaseous mixture to temperatures sufficiently high to cause a complete reaction of the chlorides into the oxides, SiO_2 and GeO_2 which then rapidly coagulate into particles of 0.1 to 0.3 μm diameter and collect on the interior of the tube. Subsequently, they are made fine so they can form a vitreous deposit as the flame traverses along the tube. The strong radial temperature gradient existing between the hot aerosol and the cold wall causes the thermophoretic displacement of the aerosol toward the wall, the main role in the deposition process. Because of a nearly exponential decay in the radial temperature gradient with axial distance along the tube, the process by torch heating converts about 50 percent of the products of reaction into waste.

Gerardin et al [20] improved the deposition rate by means of

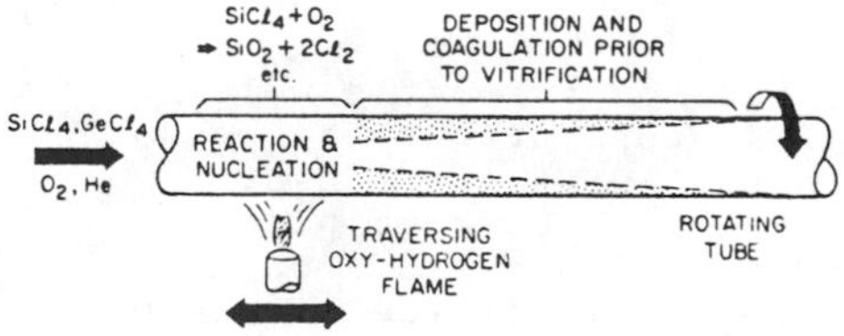

FIGURE 11. Schematic of the MCVD process

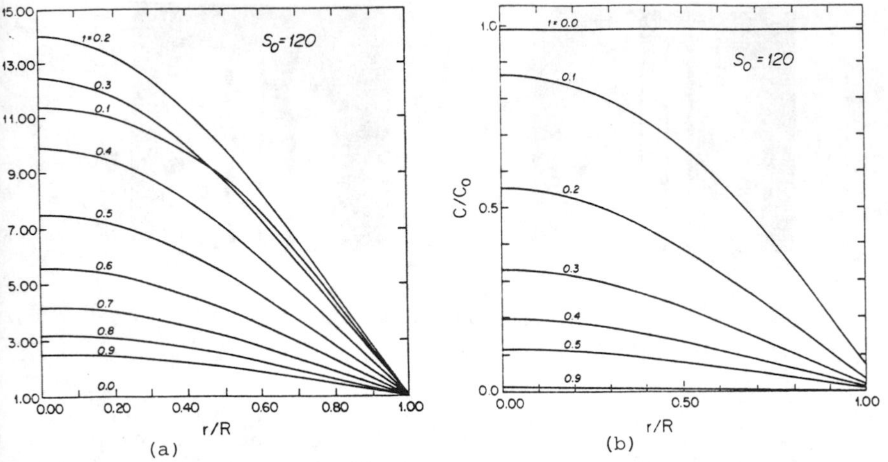

FIGURE 12. Temperature and aerosol concentration vs radial position for several non-dimensioned times (P=300W)

laser radiation. They studied the unsteady, one-dimensional (radial) thermophoretic motion in a closed isothermal cylinder subject to a sudden steady interaction with a laser. Figure 12 shows a typical aerosol response: the radial temperature gradient as the thermophoretic force for the aerosols to move toward the tube wall, Figure 12(a) and the resulting decrease in the aerosol concentration, Figure 12(b). For a tube with 20mm inner diameter, most of the aerosols can be deposited within 40 seconds by a 2.5-W laser and within 5 seconds by a 300-W laser.

Amorphous iron. An amorphous, or non-crystalline, iron structure is useful in the electric power generation field. It can be produced by rapidly quenching from the molten state to below 400°C at a rate of 10^6 °C/sec or greater. To achieve this rapid cooling rate, the iron is accelerated and heated by a plasma arc in the railgun so that it arrives at a prechilled target (generally, a copper surface at 20°C to -173°C) in a semi-molten state. The material melts upon impact, at a temperature between 1650 and 2000°C. Hayes et al [21] used a transient finite-element program to model the plasma spray coating process for building up a layer of amorphous iron on the base metal. It was disclosed that thin layers of amorphous iron (0.025mm) can be produced on either a room-temperature base or a prechilled base. Thicker layers can be built up as a series of thin layers.

Welding and Cladding

To improve productivity, more use is made of automated welding. In the welding process, three thermal factors affect the weld quality: convection flow in the weld pool, transient heat conduction in the workpieces and cooling by a back-up plate through the interface conductance. Convection plays a major role in arc, laser and electron beam weld pool. It is the single most important factor influencing the geometry of the pool, including pool shape, undercut and ripple. Therefore, if the welding process is to be automated with a closed-loop control of the weld zone size, the mechanisms of heat and fluid flow in the molten weld pool must be understood.

Chan et al [22] developed a two-dimensional transient model for convective heat transfer and surface tension driven fluid flow in laser melted pools. The surface tension gradient induced a recirculating flow in the pool at a velocity of one or two orders of magnitude higher than that of the scanning speed. The process was convection dominated, which determined both the molten pool shape and the cooling rate change. When the Marangoni number exceeded a critical value, the return velocity from the edge of the pool was so high that it formed a jet. It caused the counter rotating vortex to heat the lower center part of the pool, which deepened the pool.
The formation of weld defects such as humping, undercutting and tunnel bead are common in high travel speed and high current GTAW (gas tungusten arc welding) welding. Lin and Eagar [23] proposed that the transition from shallow to deep weld penetration, which caused a dramatic change in surface depression, was caused by a self-stabilizing vortex that was created by the circumferential rotation of the weld pool. In case of traveling GTA welds, the deep vortex penetration and the surface-tension driven reverse fluid flow caused weld defect formation.

Pulsing of the welding power on a real-time scale during welding is a means to minimize the weld defect formation and thermal stress problems. Hou and Tsai [24] performed a theoretical analysis of a three-dimensional heat flow behavior in the pulsed current GTAW process. It was observed that the weld penetration is almost linear and increases with increasing peak or average current. The welding speed is not a critical parameter.

Currently, Rolls Royce uses laser cladding (to clad material A to material B). The use of the cladding material in the form of a powder is preferred as an instantaneous layer of powder on or near the surface appears to enhance the laser energy coupling efficiency. Two methods of laser cladding are by gas borne powder injection and by melting a pre-placed power layer. Currently, Rolls Royce uses the powder injection process: The cladding material, a fine powder of 70-micron average size, is injected into the laser-generated melt pool of the moving substrate. Weerasinghe and Steen [25] developed a computer simulation model to determine the clad bead dimensions and the temperature distribution in the heated zone (bead and substrate).

Steel Processing

Two kinds of steel processing are treated: strip rolling and continuous casting.

Strip rolling. The metal rolling process involves extremely high pressures and velocities that create a large amount of heat generated by plastic deformation and friction. The adequate cooling of the roll and the rolled product is essential for a better control of the material properties and the surface condition (such as the lubricant behavior). Tseng [26] developed

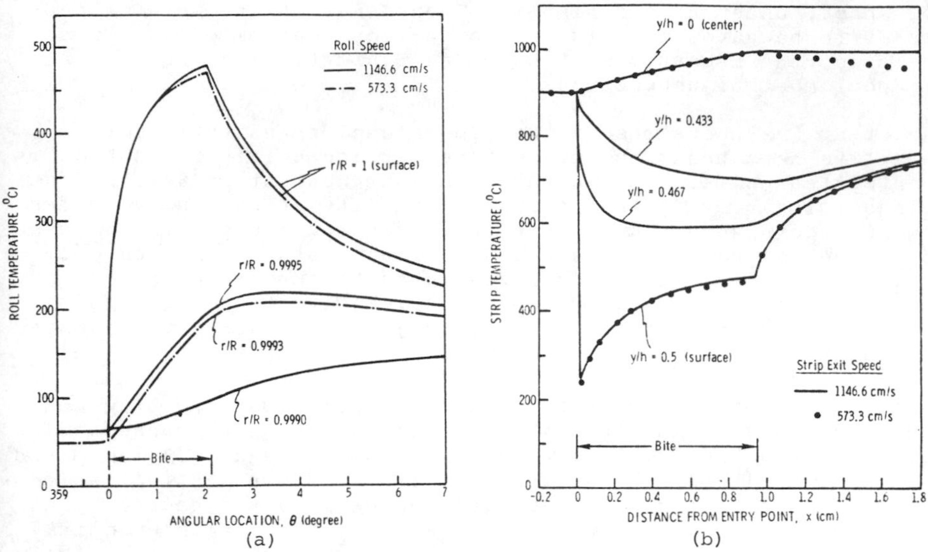

FIGURE 13. Roll and strip temperatures for hot rolling case

an effective finite-difference model to evaluate the temperature profiles of the work-roll and the strip during both cold and hot rollings. Very large variations in the roll temperature were found within a very thin layer (less than 1 percent of the roll radius) near the bite, (interface) region as shown in Figure 13(a). Such a high temperature variation creates very large thermal stress within the thin layer and, in turn, controls the rate of roll wear or roll failure. As the roll leaves the bite, the surface temperature immediately decreases due to heat convection to the ambient and heat conduction into the immediate subsurface layer. Figure 13(b) shows the strip temperature which also exhibits very large variations near the bite region.

Continuous casting. The problem of alloy solidification has important applications in crystal growth, ablation, welding and metal casting. It involves the combined effects of the fluid flow field with heat and mass transfer processes. Continuous casting of alloys is widely used as an energy saving and fully automated process. Prantil and Dawson [27] applied a mixture theory to examine the problems involving solidification of a binary alloy during continuous casting (conceptually represented in terms of solidification of a multiphase mixture). Nonequilibrium solute segregation and suppression of grain refinement during rapid solidification of alloys were taken into account. The mixture theory model which uses the energy, mass and momentum equations can be a useful tool for a two-dimensional analysis of continuous casting.

Polymer Processing

Flow and mixing are two essential requirements in polymer processing. The coupling of creeping flow and continuous mixing is a special feature in polymer processing equipment. Arimond and Erwin [28] decoupled axial and transverse flow problems in analyzing the mixing performance of a Kenics static mixer. The Kenics is a motionless mixer driven not by shear but by an applied pressure gradient. It consists of a circular pipe divided into two semicircular regions by a twisted ribbon, see Figure 14. A stable iterative scheme was employed to solve this three-dimensional flow problem on a two-dimensional finite-difference mesh.

Shen and Kwon [29] used boundary integral formulations to simulate two-dimensional creeping flow in polymer processing: the gate and juncture problems in injection molding.

Composite Materials

Many varieties of composites are fabricated by combining a reinforcing fiber (rigid rod-like particles or short fibers) with a polymeric or metallic matrix. These composites are molded, extruded or formed into the desired shape while the matrix is in a liquid state. During this process, flow causes the fibers to become oriented. Then the matrix is solidified by cooling or chemical reaction and the orientation pattern is retained in the finished product. The orientation pattern controls the mechanical properties of the product. The composite is stronger and stiffer in the direction in which most fibers are oriented, and weaker and more compliant in the direction of least orientation. By relating the orientation pattern to processing conditions, one can design and control the process to obtain the most desirable mechanical properties in the final product.

The orientation behavior varies with fiber concentration, C. There are three regimes of concentration: dilute, semi-concentrated and highly concentrated. In the dilute regime, $C<(d/l)^2$, fibers are free to rotate and interactions between fibers are rare. d and l denote the diameter and length of fibers, respectively. Interactions between fibers are frequent in the semi-concentrated regime with $(d/l)^2<C<(d/l)$. In the

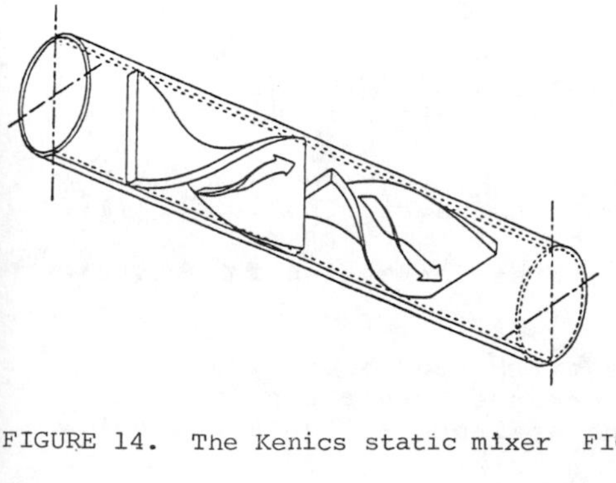

FIGURE 14. The Kenics static mixer

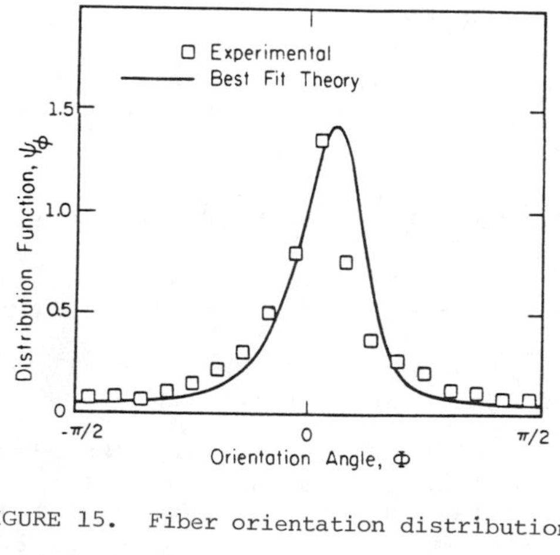

FIGURE 15. Fiber orientation distribution

highly concentrated regime, C>(d/l), the spacing between fibers is on the order of d and the maximum volume fraction C decreases with increasing aspect ratio.

The Jeffery model [30] assumed that the solution of the flow field is uncoupled from that of the fiber orientation state in the dilute concentration. The flow field around a single rigid ellipsoidal particle rotating in a viscous Newtonian liquid was solved by Jeffery in 1923. Recently, Gilver [31] developed a model to predict the short fiber orientation on the dilute suspension theory. First, the velocity field for a Newtonian solvent (i.e. matrix) was determined by the finite element method. Subsequently, fiber orientations within the domain were calculated in terms of the fluid kinematics by solving the Jeffery's equations. Results were obtained for the fiber orientation in plane Poiseuille flow past a square slot. Folger and Tucker [32] investigated the orientation behavior of fibers in concentrated suspensions. Using the Jeffery model for the viscous force and the measured interaction coefficient, the theory correctly predicted a disperse steady-state distribution in simple shear flow, Figure 15.

Summary

Some recent studies on the transport phenomena in materials processing are reviewed. This research field is of great practical importance and requires a cooperative effort among researchers with a diversity of background ranging from materials scientists to heat transfer and fluid flow specialists. It is envisioned that such a cooperative research will become popular with the advancement of materials engineering.

In sharp contrast to microelectronic cooling, which relies on the well-known conventional flow mechanisms, materials processing provides an opportunity to explore new transport phenomena that can challenge the researchers. This is particularly true in the case of space processing where the Marangoni effects may play an important role in transport phenomena.

REFERENCES

1. Mayer, A. N., Heat Transfer Technology in the Service of the Integrated Microelectronic Revolution, ASME HTD-vol. 28, pp. 1-3, 1983.

2. Ma, C. F., and Bergles, A. E., Boiling Jet Impingement Cooling of Simulated Microelectronic Chips, ibid., pp. 5-12.

3. Antonetti, V. W., Using Metallic Coating to Enhance Thermal Contact Conductance of Electronic Packages, ibid., pp. 71-77.

4. Kennedy, K. J., and Kanehl, J., Free Convection in Tilted Enclosures, ibid., pp. 43-47.

5. Torok, D., Augmenting Experimental Methods for Flow Visualization and Thermal Performance Prediction in Electronic Packaging Using Finite Elements. ASME HTD-vol.

32, pp. 49-57, 1984.

6. Ashiwake, N., Nakayama, W., Daikoku, T., and Kobayashi, F., Forced Convective Heat Transfer from LSI Packages in an Air-Cooled Wiring Card Array, ASME HTD-vol. 28, pp. 35-42, 1983.

7. Braaten, M. E., and Patankar, S. V., Analysis of Laminar Mixed Convection in Shrouded Arrays of Heated Rectangular Blocks, ASME HTD-vol.32, pp. 77-84, 1984.

8. Karki, K. C., and Patankar, S. V., Cooling of a Vertical Shrouded Fin Array by Natural Convection: A Numerical Study, ibid., pp. 33-40.

9. Sparrow, E. M., Baliga, B. R., and Patankar, S. V., Forced Convection Heat Transfer from a Shrouded Fin Array with and without Tip Clearance, Journal of Heat Transfer, vol. 100, pp. 572- 579, 1978.

10. Acharya, S., and Patankar, S. V., Laminar Mixed Convection in a Shrouded Fin Array, Journal of Heat Transfer, vol. 103, pp. 559-565, 1981.

11. Zhang, Z., and Patankar, S. V., Influence of Buoyancy on the Vertical Flow and Heat Transfer in a Shrouded Fin Array, International Journal of Heat and Mass Transfer, vol. 27, pp. 137-140, 1984.

12. Oosthuizen, P. H., A Numerical Study of Laminar Free Convective Flow Through a Vertical Open Partially Heated Plane Duct, ASME HTD-vol. 32, pp. 41-47, 1984.

13. Coyne, J. C., An Analysis of Circuit Board Temperatures in Electronic Equipment Frames Cooled by Natural Convection, ibid., pp. 59-65.

14. Jaluria, Y., Interaction of Natural Convection Wakes Arising from Thermal Sources on a Vertical Surface, ibid,. pp. 67-76.

15. Bar-Cohen, A., and Schweitzer, H., Thermosyphon Boiling in Vertical Channels, ASME HTD-vol. 28, pp. 13-20, 1983.

16. Aakala, N. G., Condensation of Heat Transfer of Dielectric Vapors in Presence of Air, ibid., pp. 21-27.

17. Witzman, S., Shitzewr, A., and Zvirin, Y., Simplified Calculation Procedure of a Latent Heat Reservoir for Stabilizing the Temperature of Electronic Devices, ibid., pp. 29-34.

18. Fu, B-I, and Ostrach, S., Numerical Solutions of Thermo-capillary Flows in Floating Zones, ASME PED-vol. 10, HTD-vol. 29, pp. 1-9, 1983.

19. Glicksman, M. E., Fang, Q. T., Coriell, S. R., McFadden, G. B., and Boisvert, R. F., Convectively Induced Crystal-Melt Instabilities-Influence of Gravity and Rotation, ibid., pp. 11-13.

20. Gerardin, D., Streesing, N., Cipolla Jr., J. W., and Morse, T. F., One Dimensional Unsteady Thermophoretic Motion, ibid., pp. 31-37.

21. Hayes, L. J., Devloo, P., and Spann, M. L., Finite Element Model of a Plasma Spray Coating Process, ibid., pp. 25-29.

22. Chan, C., Mazumder, J., and Chen, M. M., A two-Dimensional Transient Model for Convection in Laser Melted Pools, ibid., pp. 71-79.

23. Lin, M. L., and Eager, T. W., Influence of Surface Depression and Convection on Arc Weld Pool Geometry, idid., pp. 63-69.

24. Hou, C. A., and Tsai, C. L., Theoretical Analysis of Weld Pool Behavior in the Pulsed Current GTAW Process, ibid., pp. 117-127.

25. Weerasinghe, V. M., and Steen, W. M., Computer Simulation Model for Laser Cladding, ibid., pp. 15-23.

26. Tseng, A. A., A Numerical Heat-Transfer Analysis of Strip Rolling, ibid., pp. 55-62.

27. Prantil, V. C., and Dawson, P. R., Application of a Mixture Theory to Continuous Casting, ibid., pp. 47-54.

28. Arimond, J., and Erwin, L., Modelling of Continuous Mixers in Polymer Processing, ibid., pp. 91-98.

29. Shen, S. F., and Kwon, T. H., Boundary Integral Formulation for Creeping Flows with Application to Polymer Processing, ibid., pp. 81-89.

30. Givler, R. C., A Numerical Technique for the Prediction of Short-Fiber Orientation Resulting from the Suspension Flow, ibid., pp. 99-103.

31. Jeffrey, G. B., The Motion of Ellipsoidal Particles Immersed in a Viscous Fluid, Proceedings of Royal Society, Series A, vol. 102, pp. 161, 1923.

32. Folger, F., and Tucker III, C. L., Orientation Behavior of Fibers in Concentrated Suspensions, ASME PED-vol. 10, HTD-vol. 29, pp. 105-115, 1983.

New Developments in Two-Phase Flow Heat Transfer with Emphasis on Nuclear Safety Research

F. MAYINGER
Technical University Munich
FRG

INTRODUCTION

The literature on two-phase flow - with and without heat transfer shows an explosive-like growth of published papers within the last ten years. Many of these papers were published as a result of nuclear safety research. It is impossible to deal with all new developments reported in this extensive literature. So one has to ask: Are there trends of special interest, where this report could be concentrated on? Looking over the situation, there seem to be three very promising fields of research having high practical actuality, especially for nuclear safety, namely:

- fluiddynamic and thermodynamic nonequilibrium in steady state,
- transient conditions,
- scaling.

The discussion on new developments in two-phase flow heat transfer, therefore, will be limited on these subjects.

FLUIDDYNAMIC AND THERMODYNAMIC NONEQUILIBRIUM

One-component vapour-liquid flow is - strictly speaking - always in a thermo- and fluiddynamic nonequilibrium condition, even if there is no heat transfer, because pressure drop on the flow path generat continuous evaporation which changes vapour content and slip ratio between the phases. Also in adiabatic two-component gas-liquid flow with no heat and mass transfer between the phases there are many situations where fluiddynamic nonequilibrium - changes of flow pattern and of the velocity ratio between the phases - exists. Usually this disequilibrium does not affect heat transfer very much. Howeve there is a special situation where it can govern the heat transport from a wall. Such a condition is given with anular flow, where the liquid film cooling the wall is strongly dependent on the droplet entrainment in the vapour core. A reliable knowledge of the amount of liquid entrained in the vapour core is not only of great benefit for a better analysis of different phenomena like momentum- and hea transfer at the phase interfaces, but also is an urgent need for th design of heat exchanger components. Only by knowing the thickness of the liquid film at the wall it is possible to predict the dryout reliably.

The reason for separating droplets out of the liquid film at the wall are surface waves, as unanimously assumed in the literature.

n addition, vapour bubbles generated at the wall and penetrating he liquid film promote entrainment. Liquid droplets, however, are ot only separated out of the liquid film and entrained into the apour core, they are also replaced again - called de-entrainment - nto the liquid film. Only if the amount of entrained and de-enrained droplets is equal, the fluiddynamic conditions are in equiibrium. In reality, there are three different regions along the low path of the two-phase mixture:

A first in which the droplet separation out of the liquid film into the vapour core prevails,

a second in which the de-entrainment - replacing of droplets onto the liquid film - is larger, and finally,

a third one where is equilibrium between entrainment and de-entrainment.

here is an extensive literature on entrainment measurements and redictions (see, for example, /1-5/). To calculate the entrainment ne has to use a momentum balance between the droplet-enriched vaour core and the liquid anulus, which is rather difficult to formlate due to the fact that the spectrum of the droplet diameter, the elocity difference between droplet and vapour and the shear stress t the liquid film surface are not known. A simplified calculation as, for example, performed by Langner /5/ which predicts measured esults quite well.

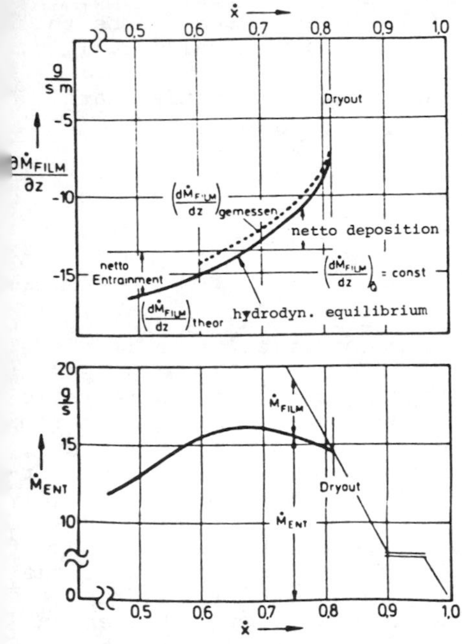

FIGURE 1. Entrainment as function of quality $\dot{x}$ (Fluid: R12, $\dot{m}$ = 500 kg/m²s, p = 12,8 bar, Δh_{in} = 3,03 kJ/kg, $\dot{q} = \dot{q}_{Bo,5m}$ = 3,88 W/cm²; L = 5 m, D = 0,014 m.)

In Figure 1, as an example, the liquid mass carried in the core is shown as it was measured in a tube of 5 m length and 14 mm inner diameter. Flowing substance was the refrigerant R12 at a pressure of 12,8 bar. In this figure one can distinguish two regions: one, where the droplet separation and another, where the droplet deposition prevails. In the upper diagram of Fig.1 the derivative of the film mass flow rate is shown, which is the diminution of the film, and the lower diagram gives the entrainment of droplets in the vapour core. Both diagrams are plotted versus the quality $\dot{x}$. The liquid is evaporating on its way through the tube by heat addi ion and, therefore, the increase of the quality $\dot{x}$ is proportional to the length of way the two-phase mixture has passed through the tube. The abscissa in Fig.1, therefore, can also be regarded as th coordinate of the flow path along the tube axis. In the upper part of Fig.1 two curves are shown; one representing the calculated values by Langner /5/ and the other, the dotted one, shows measured results. If the decrease of the liquid film would be only caused by the evaporation due to heat addition, the gradient of the liqui film thickness versus quality and also versus the tube length woul be constant, because the tube had constant heat flux density. The heat flux density was, during this experiment, adjusted in a way that at the end of the 5 m long tube dryout occurred. This had the advantage that for calculating the film thickness and the entrainment a well-defined boundary condition existed.

At low quality - that is in the lower part of the tube - the decrease of the liquid film thickness is larger than it would correspond to the evaporation. This means that the separation of liquid particles prevails the deposition. This is the entrainment-governe region. At high qualities the decrease of the liquid film is small than it would correspond to the amount evaporated. Here, therefore the evaporation is compensated by droplet re-deposition. Fluiddynamic equilibrium exists in this diabatic two-phase flow at this p sition, at which the entrainment curve is crossing the horizontal line of constant rate of evaporation (see upper diagram of Fig.1). Transfering this crossing point from the upper diagram in Fig.1 to the lower one, one realizes that this is the position where the en trainment shows a maximum. Philosophically one could say that natu tries to stabilize the situation, because a liquid film which is t thin would cause the danger of dryout, which is compensated for a while by increasing deposition. This effect can be explained with the flow pattern and the velocity distribution in the vapour core. At low quality and high entrainment the velocity profile is flat, whereas at high quality it takes the form of a parabola. The force rectangular to the flow direction onto the droplets are larger in the latter flow situation than with a flat profile having almost constant velocity.

Thermodynamic nonequilibrium exists in two regions of a boiling channel which are far away from each other; the first one is the subcooled boiling zone at the entrance of the channel, and the second one is the post-dryout region. Subcooled boiling was frequent studied in the literature, especially for local voids in water-cooled reactors. This void is a function of the growing and the co densing of the vapour bubbles. The condensing velocity of vapour bubbles in a subcooled liquid was studied in the last years more i detail /6,7/. It was found that up to Jakob-numbers $Ja = \rho_F \cdot c \cdot \Delta T/\rho_L \cdot \Delta h_V = 100$ the condensation of the vapour bubbles and the subcoole liquid is governed by the heat transfer at the phase interface. The temperature difference ΔT in the Jakob-number is formed with the subcooling of the liquid. The heat transfer coefficient at the phas boundary is mainly a function of the Reynolds-number, which is form with the relative velocity between bubble and liquid, and of the Prandtl-number, as Chen /7/ showed:

$$Nu = 0{,}185\ Re^{0{,}7} Pr^{0{,}5}; \qquad (1)$$

The influence of the Prandtl-number, as measured by Chen /7/ with different substances, is demonstrated in Figure 2. The holographic interferograms presented there are taken at approximately identical values of Reynolds- and Jakob-numbers. The Prandtl-number influences the heat transfer at the phase interface mainly via the thickness of the boundary layer on the liquid side.

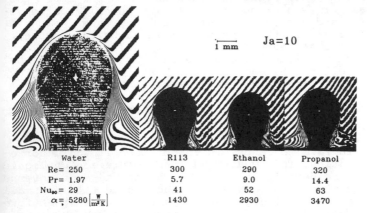

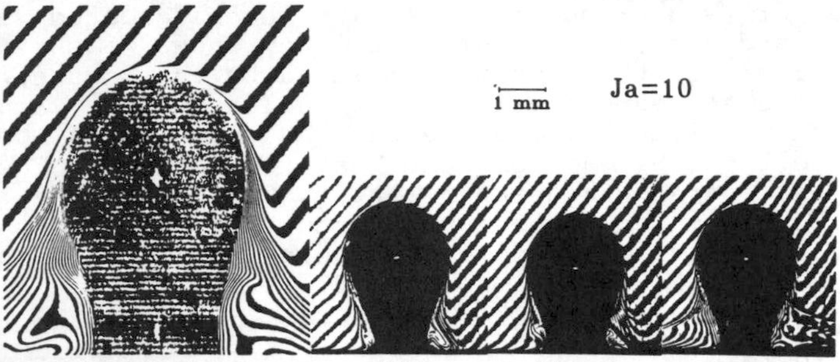

	Water	R113	Ethanol	Propanol
Re=	250	300	290	320
Pr=	1.97	5.7	9.0	14.4
Nu_{∞}=	29	41	52	63
α=	5280 $\left[\frac{W}{m^2 K}\right]$	1430	2930	3470

FIGURE 2. Comparison of boundary layer conditions around a condensing bubble for different Pr-numbers.

Chen /7/ presented also a correlation for the time-dependent volume of a condensing bubble with the Fourier-number $Fo = a_F \cdot t/D_{Bl,Dep}^2$ as dimensionless time.

$$D_{Bub}/D_{Bub,max} = (1 - 0{,}56\ Re^{0,7} \cdot Pr^{0,5} \cdot Ja \cdot Fo)^{0,9} \qquad (2)$$

Post-dryout heat transfer is the other situation where large thermodynamic nonequilibrium exists between the phases. These heat transfer conditions exist in fossil-fired boilers of the Benson-type and during certain accident situations in the core of nuclear reactors. Older physical models predicting the heat transfer under post-dryout conditions only take in account the fluiddynamic influence of the droplets, but neglect the high thermodynamic disequilibrium between the superheated vapour and the droplets being on saturation temperature. Newer models, as - for example - presented by Iloeje /8/, Plummer /9/, Ganic /10/ or Groeneveld /11/, split the heat transport from the wall to the fluid in different parts to describe the disequilibrium between the phases. They distinguish between

- convective heat transport from the wall to the vapour,
- convective heat transport from the wall to droplets being for a moment in a non-wetting contact with the wall,
- heat transport to droplets wetting the wall,
- convective heat transport from the superheated vapour to the droplets entrained in the vapour,
- radiation from the wall to the vapour,
- radiation from the wall to the droplets.

To get good agreement between the predictions by these models and measurements, the spectrum of the droplet diameters, the variable concentration of the droplets over the flow cross section, the relative velocity between droplet and vapour and, last not least, the boundary conditions at the wall, have to be known. Schnittger /12/ checked the predictions of different models (8-11) with own measurements and found that an additional physical phenomenon is, probably, missing. He, finally, came up with the idea that the complicated heat transfer behaviour along the flow path during post-dryout can only be explained if a droplet fragmentation is assumed. Iloeje /8/ mentioned such a fragmentation, and he explained this phenomenon with a very brief contact of the droplets on the wall without wetti it. Schnittger /12/ sees another possibility for dispersing a large droplet in a number of small particles. He assumes that the high superheating of the vapour boundary layer at the wall does not allo the droplet to contact the wall and makes large velocity difference in this boundary layer, and shear stresses resulting from it, responsible for droplet fragmentation. Based on his measurements and on the balance equations for energy and momentum he developed a new model /12/ for predicting post-dryout heat transfer. His theoretica predictions agree well with measured data in vertical upflow, as well as in horizontal flow, as Figure 3 demonstrates. In the case o: horizontal flow the model gives only a mean temperature between the upper and lower part of the tube, which slightly overestimates the situation at the lower part.

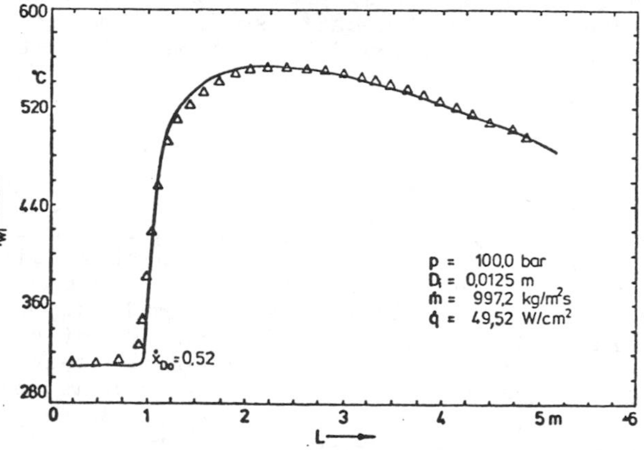

FIGURE 3. Post-dryout wall temperatures (Fluid water, measure data by KWU Δ, calculated by Schnittger ——

The thermo- and fluiddynamic conditions become even more interesting if not a straight tube but a bend or coil is studied. Measurements in such geometrical configurations are presently performed by Lautenschlager /13/ and show that the bend produces a strong secondary flow of the vapour-droplet mixture.

Figure 4 presents the wall temperature and, by this, the heat transfer conditions circumferentially and longitudinally to the bended tube. The influence of the bend can even be realized some inches upstream before the bending starts.

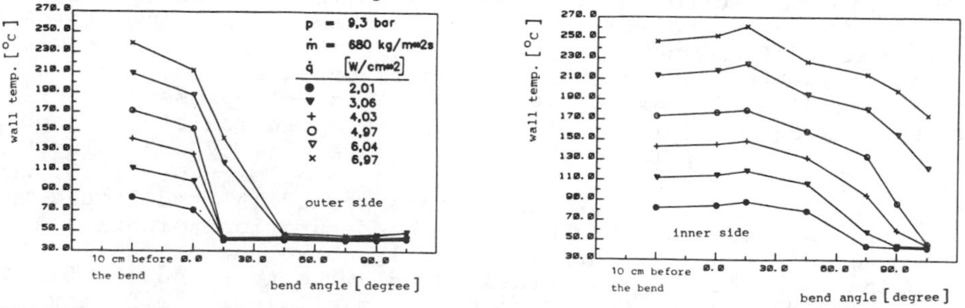

FIGURE 4. Heat transfer conditions along and around the wall of a bend.

From Fig.4 it can be seen that at first the heat transfer at the outer side of the bend improves, which easily can be explained by centrifugal forces enlarging the droplet concentration in this area However, after a short distance also the heat transfer at the inner side of the bend is increasing which can be seen by the decreasing wall temperatures.

What is the reason for this unexpected phenomenon? Centrifugal forces produce a pressure drop over the cross section of the bended tube. This generates a secondary flow in such a way that in the center of the tube cross section there is an outward radial velocit due to centrifugal forces and at the tube wall, where friction prevails, there is an inward flow due to the mentioned pressure difference. This movement of the secondary flow improves the heat transfe conditions so that - compared to a straight tube - the overall heat transfer coefficient in a bended tube is always better. The seconda flow can even act to such an extent that the improved spray cooling allows a wetting of the wall also on the inner side of the bend. Post-dryout heat transfer or spray cooling, therefore, is not only a problem of thermodynamic nonequilibrium but also involves very complicated fluiddynamic processes. It would be worthwhile to study these processes and also the thermodynamic behaviour more in detail

TRANSIENT CONDITIONS

Strong transient conditions occur with a pressure release which, for a water-cooled nuclear reactor, would be the case during a loss of coolant by accident. Pressure release in a vessel which is fille with liquid or vapour or both, which undergo the saturation conditi always causes a two-phase mixture. In the saturated liquid vapour bubbles are produced by flashing and, during the expansion of saturated vapour, liquid droplets are condensing. Both phase changes are combined with high energy transport between the phases and need a finite driving force to be started, which is well-known as boilin delay or condensation delay. Pressure release in fluids being origi ally in thermodynamic saturation condition, therefore, always cause at first and for a short period unstable conditions.

By measuring the time-depending pressure- and temperature-course during the depressurization - also called blowdown - we realize occurrences as demonstrated in Figure 5. The experimental results shown there were gained by Viecenz /14,15/ with a vessel of 0,5 m height and under the conditions of fast pressure release. The vess was filled up to 1/3 with saturated liquid - refrigerant R12 - before the release started. The temporal course of the pressure release is expected to be influenced by foaming. By destroying a burs disk at the upper nozzle of the vessel the pressure in the vessel was released from approximately 10 bar to ambient pressure within about 15 s. After starting the pressure release it took a few tent of a second until flashing started (A') and the pressure decrease during the period A is mainly due to the expansion of the vapour only in the vessel. The boiling delay more clearly can be seen fro the temperature/pressure diagram in the lower part of Fig.5, which gives the information that during this period (A) the liquid temperature is remarkably higher than the saturation temperature, and also the vapour temperature is slightly above the saturation line. At point (A') flashing evaporation and, with this, bubble formatio starts and during the period A'-C' the swell level is rising in th vessel. Due to its volumetric increase the flashing evaporation

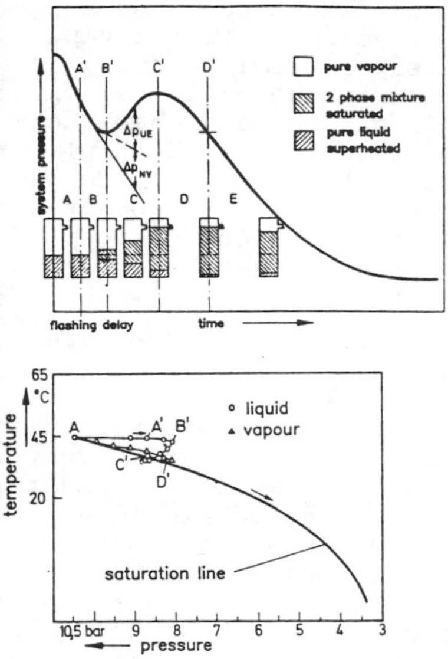

FIGURE 5. Pressure and temperature course during depressurization.
Fluid: R12

during the period A'-C' is acting against the pressure decrease by loss of coolant out of the leakage. In the first part of the period B'-C even more vapour is produced by flashing out of the superheated liquid than can flow out from the leakage. At the moment C' the swell level reaches the nozzle of the vessel. In the period B'-C' the superheating of the liquid is reduced strongly, as can be seen from the lower diagram in Fig.5. When the swell level reaches the nozzle (C') a two-phase mixture is now leaving the vessel and its mass flow rate is higher than that of pure vapour, which is the reason - together with the reduction in flash evaporation - that the pressure decreases again. This behaviour is mainly governed by the delayed vapour production on one side, and the phase separation on the other. For a better understanding and a more precise description of the blowdown we, therefore, need more information about phase separation.

Viecenz /14,15/ performed a detailed experimental and theoretical analysis of phase separation, based on the information available in the literature. He developed a correlation for the void fraction and also for the drift flux velocities of the phases. His void fraction correlation is presented in Figure 6.

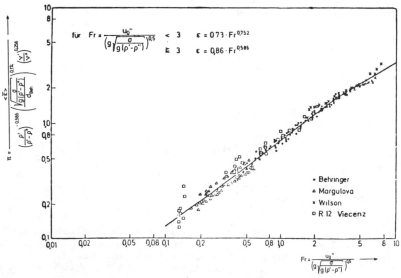

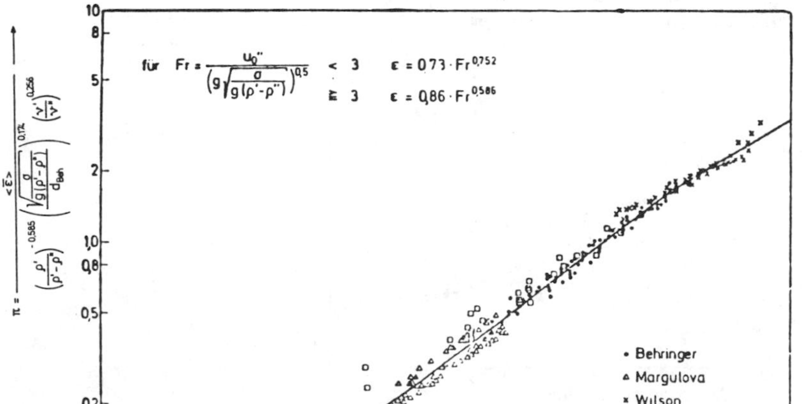

FIGURE 6. Void fraction as a function of Froude number.

The equation for the void fraction is mainly determined by a modified Froude-number in which the superficial velocity of the vapour, as characteristic velocity, is used. This superficial velocity is the velocity of the vapour, if it would flow in the empty vessel. In addition, the diameter of the bubbles is chosen as a characteristic length, and this bubble diameter is expressed by the Laplace-constant. At a Froude-number of 3 a change in the slope of the separation curve occurs. Therefore, the constant and the exponent witl the Froude-number is changed at fluiddynamic conditions correspondi to a Froude-number 3. This discontinuous behaviour can be explained by the change in flow pattern at this Froude-number.

The correlation in Fig.6 at first was developed for steady-state flow conditions only. Using a computer program which predicts the time-depending evaporation rate in the liquid, this correlation, however, is also good for calculating the transient behaviour of th void fraction and the swell level during depressurization. A very careful and complicated energy balance, however, has to be used to get good results. Energy is stored in the superheated liquid and in the superheated vapour as long as thermodynamic equilibrium is not yet reached, and energy is also carried out of the vessel through the nozzle. This energy balance is linked with the continuity equation for mass and, therefore, in addition one needs a reliable correlation for critical mass flow rate. The detailed correlation procedure is explained in /14/.

Results gained with these correlation methods are presented in Figure 7, for an example with a large leakage opening, i.e. a fast and violent blowdown where the foaming mixture reaches the nozzle within about half a second. For comparison sake also other phase se paration models - as for example presented by Wilson /16/ or Zuber /17/ - were used in the mentioned calculation procedure and the results are shown in Fig.7, too.

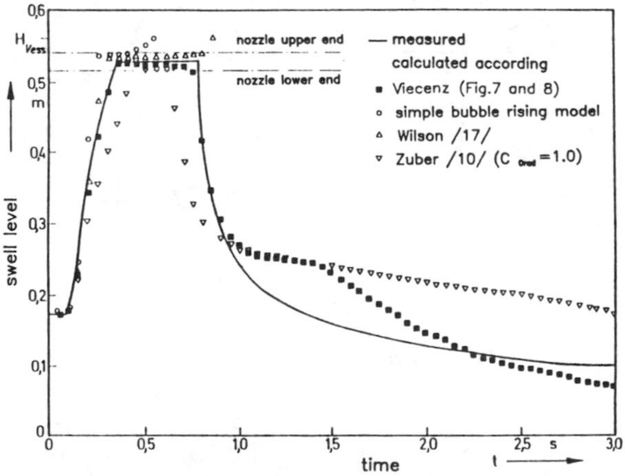

FIGURE 7. Comparison of measured and calculated swell level during a blowdown in a vessel.

Another type of transient conditions occur in a nuclear reactor, if the primary fluid circulating pumps fail due to a station blackout, or if a control rod would be suddenly withdrawn. Boiling crisis could occur in such a situation on the fuel rods, which could endanger the cladding.

The inertia of the rotating parts of the pumps and also of the liquid mass in the loop is so large that it would take seconds or even minutes until the mass flow rate is substantially reduced. Hein /18/ and Moxon /19/ measured the critical heat flux under the transient conditions of reducing the driving heat of the pump. They found that with a failure of the pumps the critical heat flux can be predicted fairly well by assuming quasi-steady conditions. Quasi-steady state has to be interpreted here in such a way that the calculation always assumes thermodynamic equilibrium by using one of the burn-out equations well-known from the literature /20/. Results calculated by this method and measured data are compared in Figure 8.

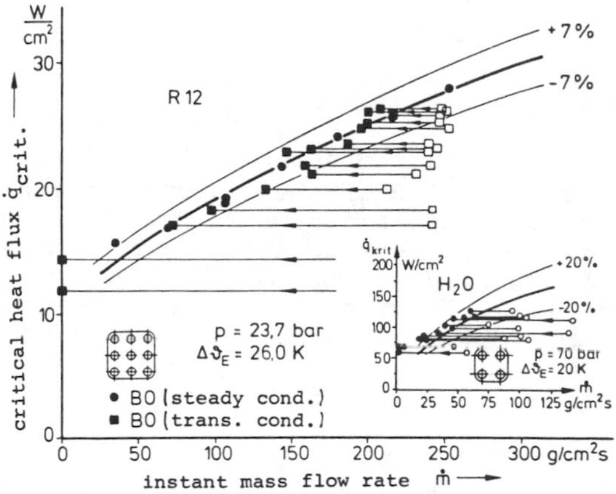

FIGURE 8. Critical heat flux with loss of flow.

This figure has to be interpreted in the following way:
Starting from a steady-state condition (empty symbols) the mass flow rate was lowered along the horizontal lines, marked in the figure with arrows, that is under constant heat flux conditions. This mass flow reduction was continued until boiling crisis occurred, which is marked with the full symbols. For comparison, critical heat flux as it would occur under steady-state conditions is given in the figure as full-line curve, which is calculated from an equation of the literature. The figure gives results for water and for the refrigerant R12. One can see that the transient data fall well together with the steady-state predictions. Only at very low mass flow rates there are deviations. Here, however, one has to realize that buoyancy forces influence the flow behaviour.

One can also imagine incidents in which the heat addition to a boiling channel is suddenly increased and then the question raises, which criteria are valid for predicting the onset of the boiling crisis. Hein and Kastner /21/ found that very rapid power excursions allow much higher heat flux densities until boiling crisis occurs than it is the case under steady-state conditions. Figure 9 shows that with these power excursions critical heat flux densities can be reached, which are by a factor of 1,5 - 3 higher than those in steady state situation.

Arguing why steep power excursions allow such high heat fluxes, one has to take in account different phenomena, well known from fluid-dynamics and heat transfer. At first it needs a certain time until nucleate boiling changes in film boiling, that is until the dry patch at the wall can be formed by evaporating the liquid film there.

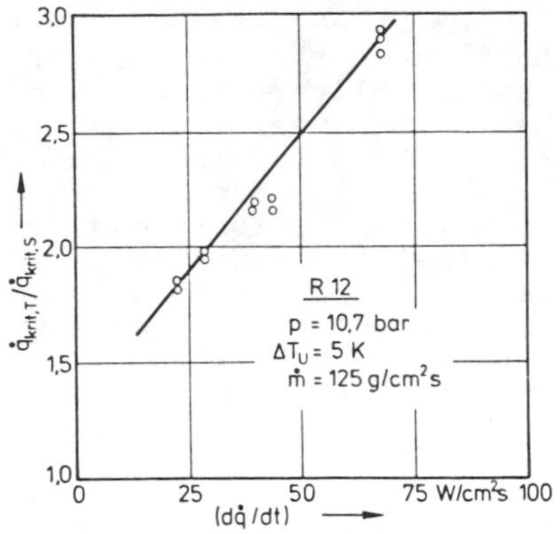

FIGURE 9. Critical heat flux with sudden power excursions (index T = transient conditions; S = steady-state conditions.)

By this, automatically a progression of the critical heat flux density is given with decreasing time period of the power transient. By evaporating liquid upstream of the position where the boiling crisis in the channel will occur, a violent acceleration of the fl is generated during this short power excursion, which improves the heat transfer remarkably. The liquid layer at the wall shows, with accelerated flow, a steeper temperature profile than with steady-state conditions.

SCALING

Transducing results from experimental models to original condition of large power plants or chemical installations is an old engineer problem. The scaled-down model has to be similar to the original with respect to its geometrical design, its fluiddynamic phenomena and its thermodynamic conditions. In two-phase flow scaling is usu ly not practiced via reducing the dimensions but by using another substance, a so-called modelling fluid, with lower latent heat of evaporization. So scaling the thermodynamic properties becomes an portant problem. Scaling criteria for two-phase flow, especially a with respect to thermodynamic similarity, are discussed in detail in /22-24/.

In the literature thermodynamic properties usually are scaled by using the same density ratio between liquid and vapour for the modelling experience as under original conditions. If there would exist a universal equation of state for all substances, one could show that scaling of the properties could be successfully performe via the thermodynamic consistency using the theorem of correspondi states. In /22,24/ is shown that already the Van der Waals-equatic can serve for a first step in this direction by reducing the data with the critical values. By applying the data to the reduced pres ure, not only transport properties like viscosity and thermal conductivity but also caloric properties and even their derivatives show similar behaviour for different fluids over a wide range of thermodynamic conditions. In Figure 10 examples of viscosity and t derivative of the latent heat of evaporization with respect to pressure are shown for the substances water and refrigerant R12. By applying a constant multiplier for the relevant property almos full agreement can be reached in the course of the curves over the

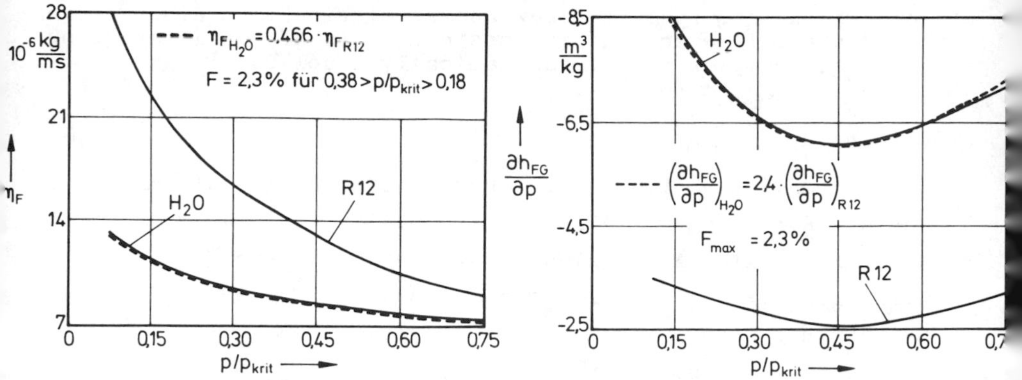

FIGURE 10. Scaling of thermodynamic properties, a) viscosity of liquid, b) derivative of latent heat of evaporization with respect to pressure.

pressure along the saturation line, as shown in Fig.10. This means that the thermodynamic condition can be advantageously scaled by using the same reduced pressure in the modelling experiment as in the original. The multiplier then resulting from the properties adjustment has to be taken in account with defining the fluiddynamic parameter, for example the Reynolds-number, by choosing a velocity in the modelling experiments which balances the multiplier in the viscosity. Scaling complicated situations in two-phase flow apparatus is difficult, however possible by this method, as demonstrated in /20,22,23/. Even the fluiddynamic and thermodynamic conditions during the blowdown of a nuclear reactor can be scaled by this metho with the modelling fluid R12, as Figure 11 shows.

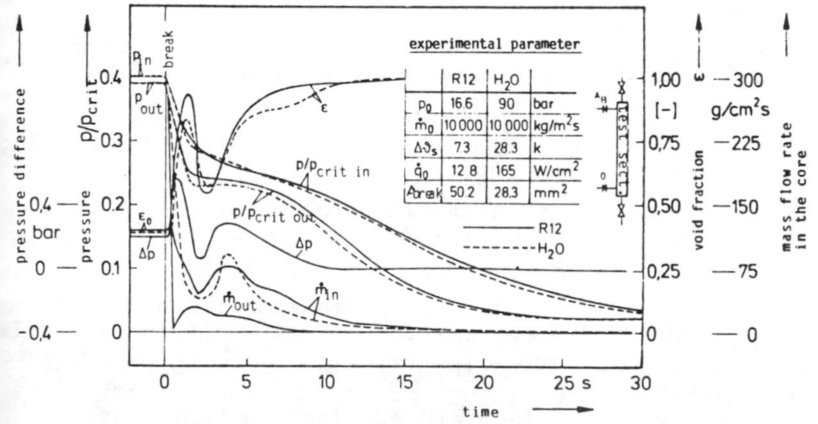

FIGURE 11. Experimental data during a blowdown (loss of coolant accident); comparison of tests performed in a water facility and in a R12-loop.

Two different experiments performed at two different institutions - one using water, the other one R12 - are compared there. The geometrical dimensions in both experiments were extendingly identical.

The comparison encourages to draw the conclusion that also in transient two-phase flow scaling is possible, even if the interesting phenomena are complex and fluiddynamically coupled.

LITERATURE

1. Hutchinson,P.; Whalley,P.B. Chem.Eng.Sci.28 (1973),26.
2. Keeys,R.K.; Ralph,R. AERE-Report No.6294 (1970).
3. Wicks,M.; Dukler,A.E. AICHE Journal 6(1960) 463.
4. Collier,J.G.; Hewitt,G.F. Trans.Inst.Chem.Eng.39 (1961),127.
5. Langner,H.; Diss., Techn.Univ.Hannover, 1978.
6. Nordmann,D.; Diss., Univ.Hannover, 1980.
7. Chen,Y.M.; Diss.Techn.Univ.München, 1985.
8. Iloeje,O.C.; Plummer,D.N.; Rohsenow,W.M.; Griffith,P.; MIT Dept.of Mech.Engng., Report 72718-92, Sept.1974.
9. Plummer,D.N.; Ph.D.Thesis,Massachusetts Inst.of Techn.,May 1974.
10. Ganic,E.N.; Rohsenow,W.M. Int.J.Heat/Mass Transf.20,855 (1977).
11. Groeneveld,D.C.; Delorme,G.G.I.; Nuclear Engineering and Design 36, 17-26, North Holland Publishing Company, 1976.
12. Schnittger,R.B.; Diss.Univ.Hannover, 1982.
13. Lautenschlager,G.; Unpublished information.
14. Viecenz,H.J.; Diss.,Univ.Hannover, 1980.
15. Viecenz,H.J.; Mayinger,F.; Two-Phase Flows and Heat Transfer, Hemisphere Publ.Corporation, New York, 1348-1404, 1978.
16. Wilson,J.F. et al. Trans.Am.Nuc.Soc.4 No.2, 1961.
17. Zuber,N.; Findley,J.A.; Transaction of the ASME, Journal of Heat and Mass Transfer, Vol.87 (1965), 453-468.
18. Hein,D.; Kastner,W.; M.A.N.- Report No.45.03.02.
19. Moxon,D.; Edwards,P.A.; European Two-phase Heat Transfer Meeting Bournemouth, and AEEW-R 553, 1967.
20. Mayinger,F.; Strömung und Wärmeübergang in Gas/Flüssigkeits-Gemischen, Springer-Verlag, Wien - New York, 1982.
21. Mayinger,F., et al. Brennst.-Wärme-Kraft 18(1966)No.6,288-294.
22. Ishii,M.; Jones,O.C.jr.; Derivation and application of scaling criteria for two-phase flows, Hemisphere Publ.Corp.,163-185,1976
23. Mayinger,F.; Scaling and modelling laws in two-phase flow and boiling heat transfer, Proceedings of NATO Advanced Study Inst. "Two-Phase-Flows and Heat Transfer", Hemisphere Publ.Corp., Vol.1, 129-161, Washington, 1976.

Dropwise Condensation—Progress toward Practical Applications

ICHIRO TANASAWA
Institute of Industrial Science
University of Tokyo
Tokyo, Japan

1. INTRODUCTION

At the Sixth International Heat Transfer Conference held at Toronto, Canada in 1978, the author was given an opportunity to deliver a keynote lecture entitled "Dropwise Condensation — The Way to Practical Applications[89]." In this lecture the author presented a state-of-the-art review on the study on dropwise condensation and showed the author's personal outlook upon the future investigation aiming at practical applications. In addition to this, the author published a monograph[77] and a couple of reviews[75,92] on dropwise condensation almost at the same time. The monograph[77], which is of about one hundred pages, has very detailed contents, the results of early researches and the studies on the microscopic mechanisms of dropwise condensation being extensively introduced.

Since then, a considerable progress has been made in the research on dropwise condensation. A main objective of this paper is to supplement the four articles cited above, reviewing the results of studies on dropwise condensation carried out during the last decade since 1975 and reconsidering about the problems to be solved before the practical application of dropwise condensation results in success. However, some of the results of researches published prior to 1975 are also quoted for the convenience of readers.

2. MEASUREMENT OF HEAT TRANSFER COEFFICIENT

The experimental studies on the heat transfer by dropwise condensation have been carried out only under very limited conditions. It is only recently that the data on the heat transfer coefficient of dropwise condensation of steam at atmospheric pressure have been put in order. Those data for other vapors and the measurements under different thermal conditions are still scarce. Therefore, the description which follows is mostly concentrated on dropwise condensation of steam at atmospheric pressure.

2.1 Results of measurements on steam at 1 atm

Since the first observation by Schmidt et al.[2] the heat transfer coefficient of dropwise condensation of steam near 1 atm has been measured by a lot of researchers. The results have been summerized by Tanner et al.[26], Graham[45], and Tanasawa[77,89,92], showing a considerably wide scattering. However, since the experimental techniques have made a great progress during these fifteen years, the reproducibility and reliability of the experimental data have been established to some extent. We owe much to several investigators such as Le Fevre and Rose[21,25], and Citakoglu and Rose[40]. At present, we can say that the heat transfer coefficient of dropwise condensation of steam at 1 atm, under the normal gravitational acceleration and on a vertical copper surface is about 250±50 kW/(m^2K), provided that there is no effect of noncondensable gases, the height of the condensing surface is less than about 50 cm, the heat flux is of the order of 0.1~1 MW/m^2, and the steam velocity is less than 10 m/s. The variation of the magnitude of the heat transfer coefficient mentioned above is mostly due to the effect of steam velocity and the effect of surface characteristics. The reasons why the most of the past measurements reported much lower heat transfer coefficients may be, as several authors have pointed out, (i) inaccuracy in the measurement of surface temperature and heat flux, and (ii) carelessness for the build-up of noncondensable gases on the condensing surface. [Of these two reasons, see Wilcox and Rohsenow[51] and Ochiai et al.[83] for (i), and see Citakoglu and Rose[40], Tanner et al.[26], Hampson[20], Takeyama and Shimizu[70], Sahde and Mikić[73], and Utaka and Tanasawa[95] for (ii).]

2.2 Measurements on low pressure steam

Few results are available on the heat transfer coefficient of dropwise condensation of steam under the pressure lower than 1 atm.

Reports have been published by Gnam[4], Brown and Thomas[27], Tanner et al.[38], Graham[45], Wilmshurst and Rose[48], and Tsuruta and Tanaka[101]. Among these, the first two papers seem subject to doubt because of the narrow pressure range and the inaccuracy of measurement.

Figure 1 shows the results obtained by the remaining four groups of researchers. Although there is a little difference between researchers, every result shows a tendency of decreasing heat transfer coefficient along with reducing pressure.

Very recently Hatamiya[106] has carried out extensive measurements down to 0.1 kPa using five different types of condensing surfaces. The surfaces used are two chromium-plated, two gold-plated and one thin gold surfaces. Although the results are not shown here, almost similar tendency as shown in Fig.1 has been recognized.

2.3 Effect of drop size

The heat transfer coefficient of dropwise condensation is dependent upon the distribution of drop size on the condensing surface. And what determines largely the drop size distribution is the maximum drop diameter or the departing drop diameter, together with the fraction of area covered by sliding drops. Thus, the authors would like to emphasize that the relationship between the drop size and the heat transfer rate should be studied more intensively. Some of the factors, which have influence upon the heat transfer coefficient of dropwise condensation, are closely related to the

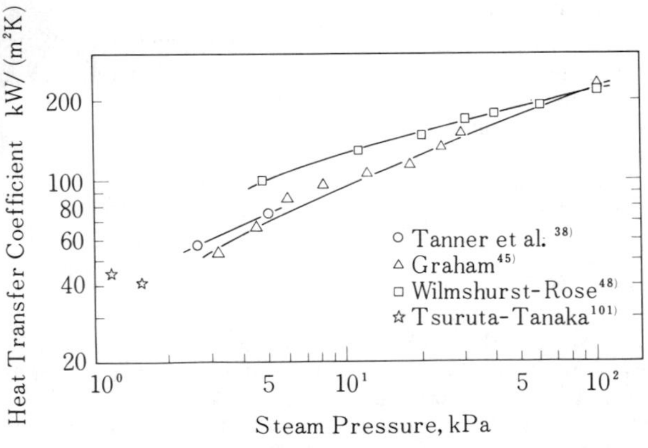

Figure 1. Dependence of heat transfer coefficient on steam pressure

maximum drop diameter. They are, for example, the vapor velocity, the physical and chemical conditions of the surface such as roughness and wettability, and the inclination and the height of the surface.

A number of reports have been published in which the effect of vapor velocity [6,13,15,26,36,45], the effect of the height[6,8,21] or the inclination[6,9,42, 49,57] of the condensing surface are measured. The drop size measurement has nc been carried out in most of them, however.

Tanasawa et al.[82] have measured the dependence of the heat transfer coefficier on the departing drop diameter, using the gravitaional, the centrifugal and the steam shear forces to change the departing drop diameter. The result is shown in Fig.2. It has been found that, no matter what kind of force may be used to prompt drop departure, the heat transfer coefficient is proportional to the departing drop diameter to the power of about -0.3. It is very interesting to not that Tanaka[74,76] and Rose[81] have theoretically derived results quite similar to the above.

2.4 Condensation curve

In boiling it is well known that the mode of heat transfer shifts from convectio to nucleate boiling, then to transition boiling, and finally to film boiling, along with the increase of surface superheat. The heat flux also varies with the degree of superheat. The curve which represents the change of heat flux wit the superheat is called boiling curve and was first obtained by Nukiyama[3].

In dropwise condensation, a similar curve, which may be called "condensation curve", must be obtained. In the case of condensation curve, however, it is very difficult to realize large surface subcooling, and hence no measurement has been done until recently.

The first measurement on steam at atmospheric pressure has been carried out by Takeyama and Shimizu[70]. [For organic vapors (nitrobenzene, aniline and ethanediol) a paper by Wilmshurst and Rose[69] has been published.] However, as Westwater[71] and Ochiai et al.[83] have pointed out, there remains some doubt in the accuracy of their measurement. Utaka and Tanasawa[95,97] have made measurement with much higher accuracy using a condensing surface having a special configuration of concave sphere. The results is shown in Fig.3, with the result obtained by Takeyama and Shimizu being drawn by a thin solid line. It is seen that the shape of condensation curves are almost similar to that of the boiling curve, while the maximun heat fluxes are very high, exceeding 10 MW/m^2. Dropwise condensation is maintained from a lower degree of subcooling up to the maximum heat flux. At much larger surface subcooling, the condensate begins to accumulate on a part of the condensing surface and a state of pseudo-film condensation

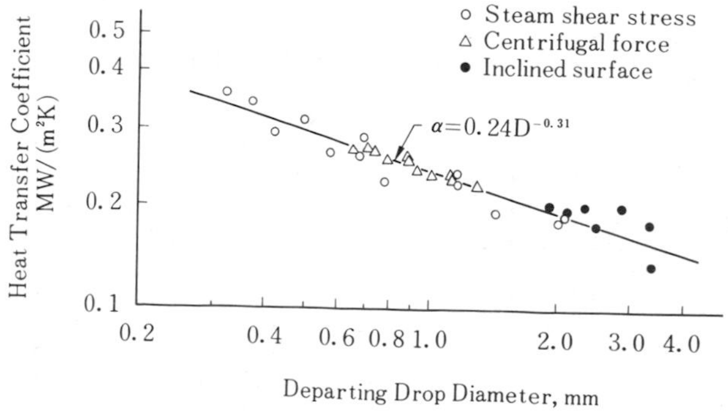

Figure 2. Dependence of heat transfer coefficient on departing drop diameter

is observed. When the surface temperature decreases below 0°C, a layer of ice is formed on the surface and film condensation occurs on it (on-ice condensation).

One of the important conclusions derived from this experiment is that dropwise condensation continues up to considerably larger heat flux. This means that the heat transfer performance on lower tubes among a vertical tube array is not deteriorated much by impingement of liquid dripping from the upper tubes[90,94].

Very recently, some other experimental results have been published on the condensation curves of steam[102,105] and propylene glycol[104].

3. HEAT TRANSFER THEORY

The theory of heat transfer by film condensation was established by Nusselt[1] more than 70 years ago, while no complete theory has yet been proposed for dropwise condensation. This is due to the very complicated nature of dropwise condensation phenomenon. On top of the randomness of the local and instantaneous distribution of drop size, the location of drops change very frequently due to coalescence and detachment of drops, resulting in fluctuation and non-uniformity of the surface temperature. Therefore, how to handle such random processes becomes the most important problem to be solved.

Of the theoretical treatments proposed up to present, the one proposed by Tanaka [64,72,74] seems to be the nearest to completion. Tanaka has derived governing equations for the processes of drop formation, growth, coalescence and vanish from a certain geometrical and statistical consideration, and succeeded in obtaining the apparent drop growth rate, the drop size distribution and the heat transfer coefficient. However, there are three points that the present author feels unsatisfactory; the first is that the fluctuation and non-uniformity of the surface temperature is not taken into consideration, the second is that the theory requires an assumption on the distance between nucleation sites, and the third is that the rate of drop growth(not accompanying coalescence) is assumed to obey the result derived from a quasi-steady conduction model (such as of Fatical and Katz[5]). Even so, it is very interesting indeed that the drop size distribution and the dependence of the heat transfer coefficient on the maximum drop size derived by the Tanaka's theory agree fairly well with experimental results.

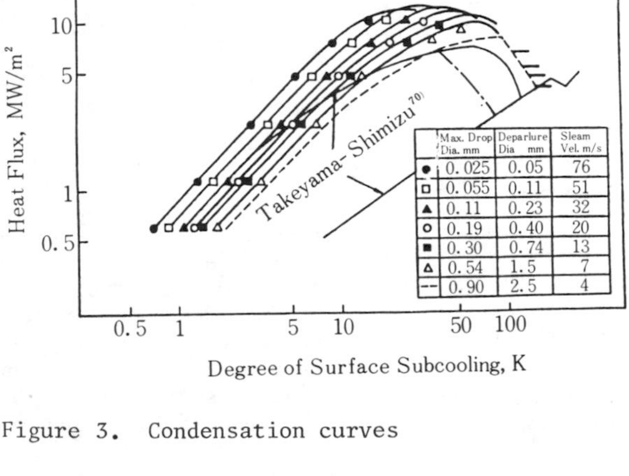

Figure 3. Condensation curves

Another theoretical approach, proposed by Le Fevre and Rose [28] and Rose[34], attempts to derive the heat transfer coefficient of dropwise condensation from the heat transfer resistances attributed to a single drop and the distribution of drops on the surface. Agreement with the experimental results[28,69,88] seems very well. However, this theory is not self-contained, since four constants, which are to be determined experimentally, are included in the theory. In addition, the fluctuation and non-uniformity of the surface temperature are not considered, either.

As a substitute for those analytical approaches as mentioned above, attempts have been made to handle the randomness, which is characteristic of dropwise condensation, by computer simulation[32,47,53,59]. However, the attempts have not obtained much success, since the handling of so many droplets on the surface demands so much computing time and expense.

It might be interesting to note that, regarding the drop size distribution on the surface, fair agreement has been obtained between theories[61,72,76] and experiments[57,60].

4. FUTURE PROBLEMS IN HEAT TRANSFER MEASUREMENTS

4.1 Effect of material thermal properties

The surface temperature of condensing surface has been assumed tacitly (or carelessly) to be uniform and constant in measurements of dropwise condensation heat transfer. However, since there are drops of various different sizes on the surface, and since these drops are moving randomly due to coalescence and detachment, the temperature and the heat flux on the condensing surface must be non-uniform and fluctuated. As a matter of fact, such a temperature fluctuation has been observed by several investigators[50,56,70,87].

Fluctuation or non-uniformity of the surface temperature is caused by the finite thermal conductivity of the surface material. Its possible effect on the heat transfer coefficient was first indicated by Mikić[43]. However, the definite support by experimental results has not yet been obtained in spite of efforts by several researchers during these twenty years. Two quite different tendencies are found among the experimental results, if omitting the one[33] which seems less reliable.

The results obtained by four groups[26,62,66,79] of researchers, measuring the dependency of the heat transfer coefficient on the material thermal properties, are shown in Fig.4. The thermal conductivity is taken on the horizontal axis as a representative of the material thermal properties.

The results show two different tendencies. The one obtained by Aksan and Rose [62](broken line A in Fig.4) makes us consider that the heat transfer coefficient is substantially independent of the surface thermal properties. The results of experiments done afterwards by Rose[86] and Stylianou and Rose[96](in which the overall heat transfer rates are compared) have shown similar tendencies. On the contrary, Tanner et al.[26], Wilkins and Bromley [66] and Hannemann and Mikić[79](lines B,C and D in Fig.4) insist that the heat transfer coefficients on the poor conductive materials are lower.

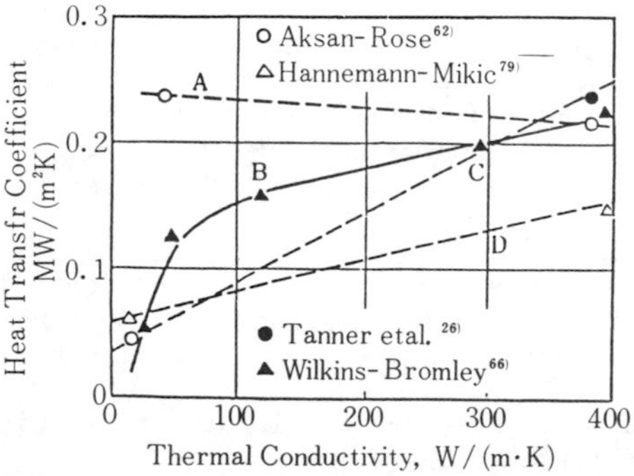

Figure 4. Effect of material thermal properties on heat transfer coefficient

Regarding such a discrepancy between the results, discussions have been continued for more than ten years. However, judging from a view point of a third party, it seems still early at present to decide which one prevails over the other. The constriction resistance theory proposed by Mikić and his coworkers [43,68,78] seems quite well-founded, while, as Rose[81,85] has been insisting, the non-uniformity and fluctuation of the surface temperature and heat flux may be homogenized by frequent coalescence between drops and hence the heat transfer remains almost unaffected.

The conclusion here is that the more reliable and accurate experimental verification is necessary to proceed on the further discussion. From the result of measurement on dropwise condensation on a glass surface[93], the author guesses that there certainly is an effect of the surface thermal properties, at least in the case of low heat flux condensation.

4.2 Dropwise condensation at small vapor-to-surface temperature difference

In section 2.4, the change of the heat flux along with the increase in the surface subcooling ΔT, i.e., the condensation curve, is discussed. On the contrary, the phenomena which possibly occur when one reduces ΔT are also of interest. At smaller ΔT (e.g. less than 0.2 K) the rate of drop nucleation may be reduced because the critical radius of drop formation is larger. Since the measurement under smaller ΔT accompanies technical difficulties, the most of the results [26, 45,93,100] reported up to present involve, more or less, some inaccuracy. More effort in the future is needed.

4.3 Measurement on high pressure steam

The heat transfer coefficients of dropwise condensation of steam at pressures higher than 1 atm have been reported by Wenzel[9](up to 4 atm), O'Bara et al[36] and Dolloff and Metzger[41](up to 9.5 atm). However, the heat transfer coefficient at 1 atm is in doubt in the last two reports. Although Wenzel has reported a reasonable value at 1 atm, his results at higher pressure region seem doubtful since the heat transfer coefficient decreases with the increase of pressure. Future study is necessary.

4.4 Dropwise condensation of vapors other than steam

(i) Liquid metals Reports have been published by Ivanovskii et al.[37], Gel'man[46], Kollera and Grigull[44,54] and Rose et al.[58,84,88] on dropwise condensation of mercury vapor, no reports having been published on metal vapors other than mercury. The results obtained by the first and the fourth authors seem reliable, but the range of experimental conditions remain rather restricted.

(ii) Organic compounds When considering possible applications to chemical plants and heat exchanging devices, dropwise condensation of organic vapors shoul be investigated more extensively. However, in reality, it is not the case, mainl because the method of surface treatment has not been established yet to maintain dropwise condensation of organic vapors.

Up to present, experiments have been carried out on vapors of ethylene glycol (ethanediol), aniline, nitrobenzene and propylene glycol by Topper and Baer[7], Mizushina et al.[35], Peterson and Westwater[29], Wilmshurst and Rose[69] and Utaka et al.[104]. The data are not sufficient as in the case of metal vapors.

Dropwise condensation of vapors of Freons which are the most important organic working substances, has been discussed by Iltsheff[55].

5. PROBLEMS FOR PRACTICAL APPLICATIONS

Although much remains still uncompleted regarding both the experimental and analytical studies on dropwise condensation, there might be no doubting that the heat transfer coefficient of dropwise condensation is remarkably high. If the presupposition of film condensation, which the design procedures of most of condensers are usually based on, could be shifted to that of dropwise condensation, substantial economization of materials, dimensions and, possibly, costs might be achieved. Of course, it is solely the vapor-side heat transfer that can be improved by dropwise condensation, since the overall heat transfer is dependent upon the thermal resistance of the solid wall and the coolant-side heat transfer And still, a rough estimation reveals that, if dropwise condensation is available on the vapor-side, the overall heat transfer coefficient is doubled and hence the heat transfer surface area is reduced to half, in the case of water vapor. This will result in a considerable reduction of the cost of condenser.

An objection which might be raised here most commonly is that the vapor-side thermal resistance can be reduced without employing dropwise condensation. As a matter of fact socalled high-performance condensing surfaces of a variety of types have been developed in these years. Among them are microstructured surfaces which make use of the effect of the capillary force on top of the effect of extended area. These high-performance condensing surfaces have indeed considerably small vapor-side heat transfer resistance at their optimum operating condition. However, such an excellent performance is usually limited within a relatively narrow range of operating condition. Especially, at higher condensing rate, cavities on the microstructured surface are flooded with condensate and the capillary force becomes ineffective.

On the other hand, in the case of dropwise condensation, a very high heat transfer rate is available even at a very high heat flux. In addition to this, there is a possibility that the low-energy surface for dropwise condensation can be manufactured less expensively, if an excellent technique for surface treatment is developed in the future. In reality, however, to find out techniques most effecfive in promoting dropwise condensation for a sufficiently long period of time is the most important and serious problem to be solved before the practical application of dropwise condensation ceases to be a mere dream.

The methods (both in laboratory scale and in industrial scale) employed up to present to maintain dropwise condensation are classified into following five categories.

(i) To apply an appropriate nonwetting agent, i.e., organic promoter, to the condenser surface before operation.

(ii) To inject the nonwetting agent intermittently (or constantly) into the vapor.

(iii) To coat the condensing surface with a thin layer of inorganic compound such as metal sulfide.

(iv) To plate the condensing surface with a thin layer of noble metal such as gold.

(v) To coat the condensing surface with a thin layer of organic polymer such as ptfe.

Of these five methods, the last two methods seem to be feasible when considering the application to industrial condensers. Since the reasons have been discussed in detail elsewhere[75,77,89,92], the author would not mention the methods (i), (ii) and (iii) in this writing.

To begin with the noble metal coating, the fact that the gold-plated surface produces dropwise condensation had been known from experience. Extensive investigation was made by Erb and Thelen[24,30] using silver, gold, rhodium, palladium chromium and platinum surfaces. Three of these surfaces (gold, palladium and rhodium surfaces) were reported to exhibit good dropwise condensation for more than 12,500 hours (1.43 years). Also it was found that, of the three, gold was the most reliable. A question was raised, however, whether or not the gold surface possessed genuine nonwettability. Wilkins, Bromley and Read[65] reported that the steam condensed as a film on a very carefully cleaned gold surface. Although Erb[67] believed that gold itself possessed nonwettability, the present author would like to agree personally to Westwater's remark[71] that the nonwettability of gold surface is due to some compounds adsorbed on the surface during the process of electroplating. Recent experiments by Woodruff and Westwater[91,98] have given some proof to this opinion. (More recently, similar experiment has been done by O'Neil and Westwater[103] for silver surface.) However, it remains still unclear which elements in the electroplating liquid are effective for promoting dropwise condensation.

A problem attributed to the use of gold-plated surface is in its cost. Not only the cost of gold itself, but also the cost required for plating process should be reduced. In this respects, the author is now seeking for the feasibility of utilizing chromium-plated surface which has been known to promote dropwise condensation, and which may cost less when compared with the noble metal plating.

The method, which is considered to be most promising to maintain dropwise condensation for a sufficiently long period of time is to coat the condensing surface with a certain organic polymer. Among a variety of organic high polymers developed after the World War II, there are some which have very low surface energy and are nonwettable with liquids. The most typical of them is polytetrafluoroethylene (ptfe, or Teflon). Attempts have been made to produce longterm dropwise condensation using polymer-coated surfaces[14,22,30,31,35,39,105].

There are two difficulties to be cleared away before the polymer coating approach to the practical use of dropwise condensation succeeds. The first is to form a film with good adhesion to the metal substrate, with few voids or pinholes, and with sufficiently high mechanical strength. The second is to make the film thin enough, because most of the high polymers have very low thermal conductivity. The required thickness of the film is less than a few micron in the case of ptfe, if the increase in the thermal resistance due to the coating should not nullify the gain by dropwise condensation. (In these respects, Westwater[99] seems to consider that the gold-plating is more feasible.) These two difficulties are interrelated each other, since reducing the thickness of the film, for example, will result in the reduction of mechanical strength of the film. However, several new processes of forming very thin and strong polymer coating on solid surface are being developed.

At any rate, the use of polymer coatings have the advantage of lower cost when compared with the noble metal plating method. The author's prospect is that the polymer coating will be the first to be used as the method of promoting dropwise condensation in industrial scale.

REFERENCES [Asterisked references are written in Japanese]

Abbreviations:
IJHMT: *Int. J. Heat Mass Transfer*
Mth IHTC: *Proc. Mth Int. Heat Transfer Conf.*
JHT: *Trans. ASME, J. Heat Transfer*

[1] Nusselt, W., *Z.VDI*, 60(1916), 541, 569.
[2] Schmidt, E., Schurig, W. and Sellschopp, W., *Techn. Mechan. u. Thermodyn.*, 1(1930), 53.
[3]*Nukiyama, S., *J. JSME*, 37(1934), 37.
[4] Gnam, E., *VDI-Fosch.-h.*, 382(1937), 17.
[5] Fatica, N. and Katz, D.L., *Chem. Engng. Progr.*, 45(1949), 661.
[6] Hampson H. and Özişik, N., *Proc. IMechE*, IB(1952), 282.
[7] Topper, L. and Baer, E., *J. Colloid Sci.*, 10(1955), 225.
[8] Gregorig, R., *Chemie-Ing.-Techn.*, 28(1956), 551.
[9] Wenzel, H., *Allg. Wärmetechnik*, 8(1957), 53.
[10] Blackman, L.C.F., Dewar, M.J.S. and Hampson, H., *J. Appl. Chem.*, 7(1957), 160.
[11] Blackman, L.C.F. and Dewar, M.J.S., *J. Chem. Soc.*, (1957), 162.
[12] Bobco, R.P and Gosman, A.L., *ASME Paper*, No.57-S-2(1957).
[13] Furman, T. and Hampson, H., *Proc. IMechE*, 173(1959), 147.
[14] Kulberg, G.K. and Kendall, H.B., *Chem. Engngn. Progr.*, 56(1960), 82.
[15]*Michiyoshi, I., *J. JSME*, 65(1962), 1483.
[16] Watson, R.G.H., Birt, D.C.P., Honour, C.W. and Ash, B.W., *J. Appl. Chem.*, 12(1962), 539.
[17] Tanner, D.W., Poll, A., Potter, W., Pope, D. and West. D., *J. Appl. Chem.*, 12(1962), 547.
[18] Osment, B.D.J., Tudor, D., Speirs, R.M.M. and Rugman, W., *Trans., IChE*, 40(1962), 152.
[19] Welch, J.F. and Westwater, J.W., *International Developments in Heat Transfer* (1963), 302.
[20] Hampson, H., *International Developments in Heat Transfer*(1963), 310.
[21] Le Fevre, E.J. and Rose, J.W., *IJHMT*, 7(1964), 272.
[22] Depew, C.A. and Reisbig, R.L., *I & EC, Process Design and Development*, 3 (1964), 365.
[23] Edwards, J.A. and Doolittle, J.S., *IJHMT*, 8(1965), 663.
[24] Erb, R.A. and Thelen, E., *I & EC*, 57(1965), 49.
[25] Le Fevre, E.J. and Rose, J.W., *IJHMT*, 8(1965), 1117.
[26] Tanner, D.W., Potter, C.J., Pope, D. and West, D., *IJHMT*, 8(1965), 419.
[27] Brown, A.R. and Thomas, M.A., *3rd IHTC*, Vol.2(1966), 300.
[28] Le Fevre, E.J. and Rose, J.W., *3rd IHTC*, Vol.2(1966), 362.
[29] Peterson, A.C. and Westwater, J.W., *Chem. Engng. Progr., Symposium Series*, 62, No.64(1966), 135.
[30] Erb, R.A. and Thelen, E., *U.S. Dept. of Interior, Office of Saline Water, R & D Report*, No.184(1966).
[31] Butcher, D.W. and Honour, C.W., *IJHMT*, 9(1966), 835.
[32] Gose, E.E., Mucciardi, A.N. and Baer, E., *IJHMT*, 10(1967), 15.
[33] Griffith, P. and Lee, M.S., *IJHMT*, 10(1967), 697.
[34] Rose, J.W., *IJHMT*, 10(1967), 755.
[35] Mizushina, T., Kamimura, H. and Kuriwaki, Y., *IJHMT*, 10(1967), 1015.
[36] O'Bara, J.T., Killian, E.S. and Roblee, L.H.S., *Chem. Engng. Sci.*, 22(1967), 1305.
[37] Ivanovskii, M.N., Subbotin, V.I. and Milovanov, Y.V., *Teploenergetika*, 14 (1967), 81.
[38] Tanner, D.W., Pope, C.J., Potter, D. and West. D., *IJHMT*, 11(1968), 181.
[39] Kosky, P.G., *IJHMT*, 11(1968), 374.
[40] Citakoglu, E. and Rose, J.W., *IJHMT*, 11(1968), 523.
[41] Dolloff, J.B. and Metzger, N.H., *ONR Technical Report*, No.2(1968).
[42] Citakoglu, E. and Rose, J.W., *IJHMT*, 12(1969), 645.
[43] Mikić, B.B., *IJHMT*, 12(1969), 1311.
[44] Kollera, M. and Grigull, U., *Wärme-und Stofübertragung*, 2(1969), 31.
[45] Graham, C., *Ph.D. Thesis, Massachusetts Institute of Technology*(1969).

46] Gel'man, L.I., *"Problems of Heat Transfer and Hydraulics of Two-phase Media"*(ed. Kutateladze, S.S., translated by Blunn, O.M.), Pergamon Press (1969), 184.
[47] Tanasawa, I. and Tachibana, F., *4th IHTC,* Vol.6(1970), Cs.1.3.
[48] Wilmshurst, R. and Rose, J.W., *4th IHTC,* Vol.6(1970), Cs.1.4.
[49] Tower, R.E. and Westwater, J.W., *Chem. Engng. Progr., Symp. Ser.,* 66, No.102(1970), 21.
[50] Abdelmessih, A.H. and Nijaguna, B.T., *Proc. 1970 Heat Transfer and Fluid Mechanics Institute* (ed. Surpkaya, T.), Stanford Univ. Press (1970), 74.
[51] Wilcox, S.J. and Rohsenow, W.M., *JHT,* 92(1970), 359.
[52] Graham, C. and Aerni, W.F., *Report Presented at 7th Annual Technical Symp., Association for Senior Engineers* (1970).
[53] Tanasawa, I. and Tachibana, F., *Proc. AICA Symp., Simulation of Complex Systems* (1971), G-6/1.
[54] Kollera, M. and Grigull, U., *Wärme-und Stoffübertragung,* 4(1971), 244.
[55] Iltscheff, S., *Kältetechnik-klimatisierung,* 23(1971), 237.
[56]*Chiba, Y., Ohwaki, M. and Ohtani, S., *J. Chem. Engng., Japan,* 36(1972), 78.
[57]*Tanasawa, I. and Ochiai, J., *Trans. JSME,* 38(1972), 3193.
[58] Rose, J.W., *IJHMT,* 15(1972), 1431.
[59] Glicksman, L.R. and Hunt, Jr., A.W., *IJHMT,* 15(1972), 2251.
[60] Graham, C. and Griffith, P., *IJHMT,* 16(1973), 337.
[61] Rose, J.W. and Glicksman, L.R., *IJHMT,* 16(1973), 411.
[62] Aksan, S.N. and Rose, J.W., *IJHMT,* 16(1973), 461.
[63]*Tanasawa, I. Tachibana, F. and Ochiai, J., *Trans. JSME,* 39(1973), 278.
[64]*Tanaka, H., *Trans. JSME,* 39(1973), 3099.
[65] Willkins, D.G., Bromley, L.A. and Read, S.M., *AIChE J.,* 19(1973), 119.
[66] Willkins, D.G. and Bromley L.A., *AIChE J.,* 19(1973), 839.
[67] Erb, R.A., *Gold Bulletin,* 6(1973), 2.
[68] Horowitz, J.S. and Mikić, B.B., *5th IHTC,* Vol.3(1974), 259.
[69] Wilmshurst, R. and Rose, J.W., *5th IHTC,* Vol.3(1974), 269.
[70] Takeyama, T. and Shimizu, S., *5th IHTC,* Vol.3(1974), 274.
[71] Westwater, J.W., *5th IHTC,* Vol.6(1974), 234.
[72]*Tanaka, H., *Trans. JSME,* 40(1974), 2283.
[73] Sahde, Jr., R.L. and Mikić, B.B., *AIChE Annual Meeting,* Paper 67B(1974).
[74] Tanaka, H., *JHT,* 97(1975), 72.
[75]*Tanasawa, I., *J. JSME,* 78(1975), 439.
[76] Tanaka, H., *JHT,* 97(1975), 341.
[77]*Tanasawa, I., *Progress in Heat Transfer,* Vol.4, Yokendo (1976), 229.
[78] Hannemann, R.J. and Mikić, B.B., *IJHMT,* 19(1976), 1299.
[79] Hannemann, R.J. and Mikić, B.B., *IJHMT,* 19(1976), 1309.
[80] Detz, C.M and Vermesh, R.J., *AIChE J.,* 22(1976), 87.
[81] Rose, J.W., *IJHMT,* 19(1976), 1363.
[82]*Tanasawa, I., Ochiai, J., Utaka, Y. and Enya, S., *Trans. JSME,* 42(1976), 2846.
[83]*Ochiai, J., Tanasawa, I. and Utaka, Y., *Trans. JSME,* 43(1977), 2261.
[84] Necmi, S. and Rose, J.W., *IJHMT,* 20(1977), 877.
[85] Rose, J.W., *IJHMT,* 21(1978), 80.
[86] Rose, J.W., *IJHMT,* 21(1978), 835.
[87] Tanasawa, I., Ochiai, J. and Funawatashi, Y., *6th IHTC,* Vol.2(1978), 477.
[88] Niknejad, J. and Rose, J.W., *6th IHTC,* Vol.2(1978), 483.
[89] Tanasawa, I., *6th IHTC,* Vol. 6(1978), 393.
[90] Tanasawa, I., *6th IHTC,* Discussions (1978), 35.
[91] Woodruff, D.W. and Westwater, J.W., *IJHMT,* 22(1979), 629.
[92]*Tanasawa, I., *Science of Machine,* 31(1979), 99.
[93] Tanasawa, I. and Shibata, Y., *Condensation Heat Transfer, Proc. 18th ASME-AIChE Heat Transfer Conf.,* (1979), 79.
[94] Marto, P.J. and Nunn, R.H. (ed.)*"Power Condenser Heat Transfer Technology"* (1980), 368.

[95]*Utaka, Y. and Tanasawa, I., *Trans. JSME,* 46B(1980), 1844.
[96] Stylianou, S.A. and Rose, J.W., *JHT,* 102(1980), 477.
[97]*Utaka, Y. and Tanasawa, I., *Trans. JSME,* 47B(1981), 1620.
[98] Woodruff D.W. and Westwater, J.W., *JHT,* 103(1981), 685.
[99] Westwater, J.W., *Gold Bull.,* 14(1981), 951.
[100]*Kaino, K., Izumi, M., Shimizu, S. and Takeyama, T., *Trans. JSME,* 48B(1982), 2181.
[101]*Tsuruta, T. and Tanaka, H., *Trans. JSME,* 50B(1984), 1600.
[102]*Izumi, M., Susuki, U. and Takeyama T., *Trans. JSME,* 50B(1984), 1600.
[103] O'Neil, G.A. and Westwater, J.W., *IJHMT,* 27(1984), 1539.
[104]*Utaka, Y., Saito, A., Tani, T., Shibuya, H. and Katayama, K., *Trans. JSME,* 50B(1984), 2418.
[105] Holden, K.M., Wanniarachchi, A.S., Marto, P.J., Boone, D.H. and Rose, J.W., *Paper presented at 1984 ASME Winter Annual Meeting* (1984).
[106]*Hatamiya, S., *Ph.D. Thesis, Univ. Tokyo* (1985).
[107]*Yamaguchi, A., Kumagai, S. and Takeyama, T., *Trans. JSME,* (1985), to be published.

GENERAL PAPERS

HEAT CONDUCTION

Heat Transfer Problems in the Freezing Process of Biomaterial Cryopreservation

TSE CHAO HUA
Cryobiological Engineering Laboratory
Shanghai Institute of Mechanical Engineering
Shanghai 200093, PRC

ABSTRACT

A thermal analysis is presented for the cryopreservation procedure of biomaterials. Theoretical and experimental study is shwon to freezing process, especially to heat transfer equations and thermal control of freezing process. It is important to the study and design of the cryopreservation protocal.

PRESENT STATE AND BASIC PROBLEMS OF CRYOPRESERVATION

Since Polge discovered accidentally in 1949 that glycerol protected the spermatozoa during freezing, [1] cryobiology has had 36 years history. Up to now, the important cells which can be preserved at low temperature are erythrocytes (red blood cells), lymphocytes, platelets, barrow stem cells, spermatozoa, embryos and some plant cells. The important cryopreserved organs and tissues are cornea, skin, and pancreas. The preservation of these biological materials has brought great practical interest to the agriculture, medicine, biology, fishery, and animal husbandry.

Nevertheless, the biomaterials which can be cryopreserved are very limited. Even for those cryopreserved cells which we just mentioned, the survival probability is not high in many cases. The reason is partly the lack of the thermal study in the procedures of cryopreservation.

As we know, the basic effects which affect the survival of cryopreservation are:

1) the cooling procedures and their cooling rates;
2) the storage temperature;
3) the warming procedures and their warming rates;
4) the cryoprotective agents (CPA), their concentrations, and the addition and removal procedures.

<u>Storage temperature</u>: Recent studies find it would easily injure the cells if the storage temperature is above -40°C. The lower the storage temperature, the better would be the cryopreservation. It seems there is no lower limit to storage temperature. Cravalho successfully preserved the erthrocytes at -272.29°C with high recovery.[2]

The practical storage temperature ranges between -80°C (dry ice) and -196°C (liquid nitrogen). Rowe had successfully preserved erythrocytes for 12 years at -196°C with no observable biochemical and functional deterioration.[3] It is likel that storage at -196°C would enable storage period for decades.

<u>Cryoprotective agents (CPA)</u>: In most cases, almost no survival of cells can be

obtained unless a cryoprotective agent is mixed with the cells during the cryopreservation procedure. Some common CPA are glycerol, dimethyl-sulfoxide (DMSO), polyvinylpyrollidone (PVP), hydroxyethyl starch (HES), etc. Although th mechanism of cryoprotective agents is still unclear, the following conditions ar evident.

1) Different kind of cells favors different kind and concentration of CPA. For example, bull spermatozoa favor 4-9% glycerol, and cow embryos favor generally 1.5M DMSO.[4]
2) Different freezing procedure needs different concentration of CPA. For example, the rapid cooling rate (approximately -100°C/min) for red cells uses 10-20% glycerol, [3] and the slow cooling rate (about -1°C/min) uses 40-80% glycerol.[5]

As we know, most difficulties of cryopreservation would appear in the temperature range between 0°C and -40°C (or -60°C) which could be called "the risky temperature range". The basic problem of cryopreservation is to select the appropriate CPA, the optimal cooling processes and warming processes in order that the biomaterial could pass through "the risky temperature range" without injury.

The cooling and warming processes which Cravalho, Cosman and the author developed with a cryomicroscope in M.I.T. for cow embryos is shown in Fig.1 as an example. The temperature procedure is controlled by a microcomputer. It is quite evident that the procedure in Fig. 1 is complex since embryo is a complex cell. Even if the cells are simple, there are still some complex processes during the cryopreservation. They are freezing, thawing, concentrate and dilute of solution, water transport, etc.

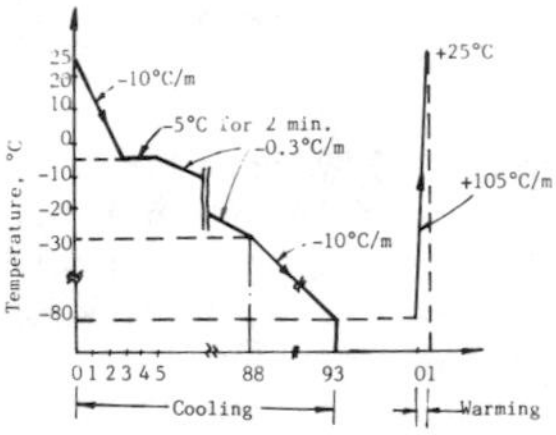

Fig.1 A temperature procedure for cow embryes

HEAT & MASS TRANSFER EQUATIONS FOR FREEZING PROCESS

Cooling process of pure water. The author did some experiment with a programmab freezer, and the result is partly shown in Fig.2.[6] The dash line represents the temperature in the chamber of the freezer, its cooling rate is about 1°C/min in this case. The solid line represents the center temperature of a tube, havin a diameter of 10 mm and filled with 3 ml pure water, which is placed in the central region of the freezer chamber.

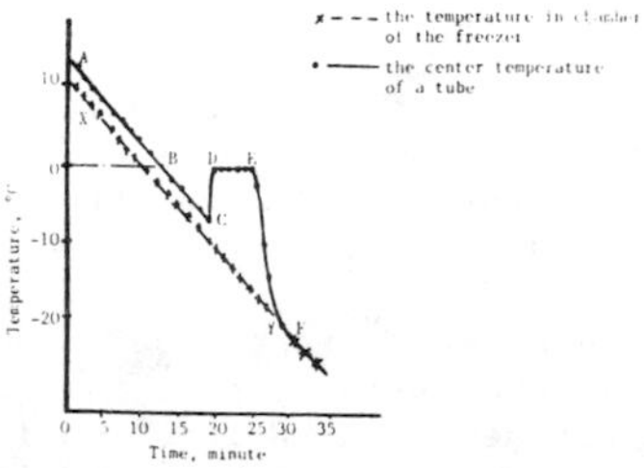

Fig.2 A cooling curve of pure water in a programmable freezer[6]

During the first part of cooling process, A-B, sensible heat is removed from the sample, and the cooling rate of the sample is very close to the cooling rate of the chamber. Nucleation happens at point C, which is called supercooling, an release heat of fusion is at D-E, the freezing temperature of water. After the whole sample is frozen, sensible heat is extracted from the solid state, E-F. Please note that the cooling rate of the sample is substantially higher than tha of the chamber and is very difficult to control. In most cases, this section is associated with cell injury.

Cooling process of biological solution

Biological samples, which are much more complex than pure water, contain usually ternary or more components. The following phenomena would occur.

1) Freezing point depression $\theta = T_o - T_f$ (1)

T_o - freezing point of pure water, 273.15K;

T_f - freezing point of biological solution.

The Gibbs - Helmholtz relationship is written as

$$\theta = K_f \cdot \Omega \tag{2}$$

and

$$\Omega = \sum_i \phi_i \gamma_i m_i \tag{3}$$

where K_f = 1.86°C/osmolality

Ω - the osmolality of the solution

i - the ith solute of the solution

ϕ_i - the osmolality coefficient of the ith solute

γ_i - the dissociation constant of the ith solute

m_i - the solution molality of the ith solute in the solution

2) Osmosis during freezing

The biological cell membrane is usually semi-permeable - that is permeable to water and non-permeable to the solutes. In the living body, cells are in the osmotic equilibrium with its environment. During the freezing process, the extracellular solution (the environment of the cells) freezes first, so that the concentration and the osmotic pressure of the extracellular solution become highe than that of the intracellular solution, and water transports from the inside of cell to the outside. This is a complex heat and mass transport process.

Mazur's equations for freezing process

P.Mazur is the first scientist who studied successfully the freezing process and proposed the following equations for the kinetics of water transport from the cell at subzero temperature.[7]

$$\frac{dV}{dT} = \frac{V(V+nV_w)}{nV_w}\left[\frac{L_f}{RT_f^2} - \frac{d\ln(Pout/Pin)}{dT}\right] \tag{4}$$

$$\frac{dV}{dt} = \frac{K_w ART}{V_w} \ln(Pout/Pin) \tag{5}$$

$$\frac{dT}{dt} = -B \tag{6}$$

$$\frac{dV}{dT} = -\frac{K_wART}{BV_w} \ln(Pout/Pin) \tag{7}$$

Initial conditions are:

$$V = Vi, \frac{dv}{dT} = 0 \quad \text{as } T = T_f$$

where
- V - volume of water inside the cell;
- Vi - initial volume of water inside the cell at $T=T_f$;
- T - temperature inside and outside the cell;
- T_f - freezing point of aqueous medium;
- n - osmoles of solute inside the cell, moles;
- A - surface area of cell;
- K_w - water permeability of cell membrane;
- B $= -\frac{dT}{dt}$ -- rate of temperature change;
- R - gas constant, $82.057\text{x}10^{12}\mu^3$.atm/(mole.K);
- V_w - molar volume of pure water, $18\text{x}10^{12}\mu^3$/mole;
- L_f - molar heat fusion of ice, $5.95\text{x}10^{16}\mu^3$atm/mole;
- Pout, Pin - the vapor pressures of water outside and inside the cell respectively.

The temperature differnce across the cell membrane and its effect on water transport

Mazur's equations assume that during cooling the temperature is uniform throughout the inside and outside of the cell at any time. The temperature distribution inside a cell during freezing has been studied analytically by Mansoori [8] who assumed that the temperature of the membrane is the same as the surrounding medium, i.e., no temperature difference exist across the membrane. Mansoori then calculated the temperature profile inside the cell by means of the heat conduction equation and found that the effect of the non-uniform temperature inside the cell on water transport is negligible for practical cases.

However, many papers [9,10] have discussed the significance of the thermo-osmosis phenomenon due to the temperature difference across the membrane and these efforts have shown that very small temperature differences on the order of 0.01°C could generate pressures on the order of 1 atm. In order to solve this problem, the author and others proposed a model. [11] The analysis based on this model can calculate the heat transfer coefficient between a cell and its surrounding meduim, estimate the temperature difference across the cell membrane during freezing and its effects on the water transport via thermo-osmosis and other mechanism. And the Mazur's equations have been modified by the author as follows:

$$\frac{dV}{dt} = \frac{KwART}{BVw(1-e^{-t/\tau})}\left[\ln\frac{X_i^o}{X_i} + \frac{L_f}{R}\left(\frac{1}{T_f} - \frac{1}{T}\right) + \frac{Ls}{RT^2}B\,\tau(1-e^{-t/\tau})\right] \tag{8}$$

$$t = -\frac{1}{B}(T - T_f) + \tau(1 - e^{-t/\tau}) \tag{9}$$

Initial conditions are

$$V = V_i, \quad T = T_f \quad \text{as} \quad t=0 \tag{10}$$

where

$\tau = \rho cV/(hA)$ -- time constant of temperature lag;
$X_i = V/(v+nVw)$, $X_i{}^o = V_i/(V_i+nVw)$;
X_i -- the molar fraction of intracellular water;
Ls -- the molar sublimation heat of ice;
T -- the temperature inside the cell.

CELL INJURY DURING FREEZING

The recovery of frozen-thawed biological cell is affected by a number of factors such as cooling rate, storage temperature, warming rate, concentration of CPA and the protocols of introducing and removing cryoprotectants, etc.

Mazur[12] studied the effect of cooling rate on the recovery shown in Fig.3. There exists an "optimal" cooling rate for each cell type in some definite cases. The figure illustrates that due to the differences in cell characteristics the "optimal" cooling rate for one cell type may differ by several order of magnitude when compared with other cell type.

If the cooling rate is much lower than the "optimal" one which is called "slow cooling", the intracellular solution always remains in the thermodynamic equilibrium with the extracellular solution. In most cases this means the cell has large surface area-to-volume ratio and high water permeability for this given cooling rate. The "slow cooling" process is a mass transfer dominated process in which there is enough time for water to be transported across the cell membran

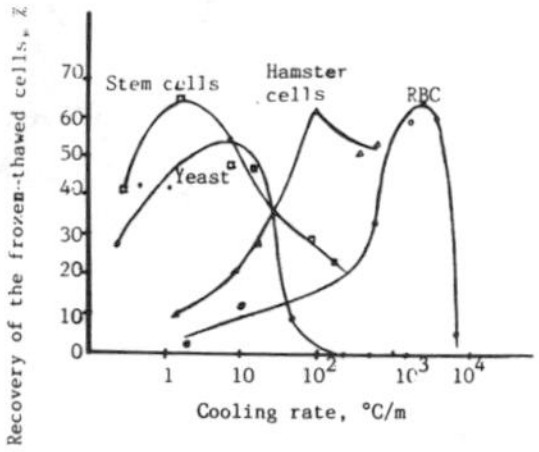

Fig.3 Effects of cooling rates on the recovery for several cell types [12]

Mazur and others [13] argued that "slow cooling" extends the exposure time of the cell to concentrated solution which results in the damage of cell. It is called "solute damage".

If the cooling rate is much higher than the "optimal" one, it is called "fast cooling". In most cases this means the cell has small surface area-to-volur ratio and low water permeability of the cell membrane. There is not enough time for water to be transported from the inside to the outside of the cell. There exists non-equilibrium between intracellular solution and extracellular solution. The fast cooling is a heat transfer-dominated process.

Since the cooling rate is high and water inside the cell could not transport adequately through the membrane to the outside of the cell, the intracellular water becomes supercooled and forms the intracellular ice which would recrystallizate during thawing. The recrystallization of the intracellular ice is the mechanism of cell damage in fast cooling.[14]

THE THERMAL CONTROL OF FREEZING

Cryomicroscope

The desire to view and record the significant events of cryogenic injury to living cell has led directly to attempts to create a variable low temperature environment for cell on a microscope, i.e., cryomicroscope system. Diller and Cravalho built the first satisfying cryomicroscope in M.I.T. in 1976.[15] We have just built an inverted cryomicroscope in Shanghai Institute of Mechanical Engineering (SIME) this year. Fig.4 and Fig.5 show the block diagram and total view of the inverted cryomicroscope system in our laboratory (the details will be published).

The cryomicroscope system could observe directly the cell during freezing and thawing, and could provide a phenomenological basis for interpreting experimental trials.

A primary design consideration for the low temperature stage in this system is satisfying the good heat transfer performance and optical performance. Since the heat capacities of the low temperature stage and the specimen are small enough for the cooling and warming rates required, the system works very well for both freezing and thawing processes.

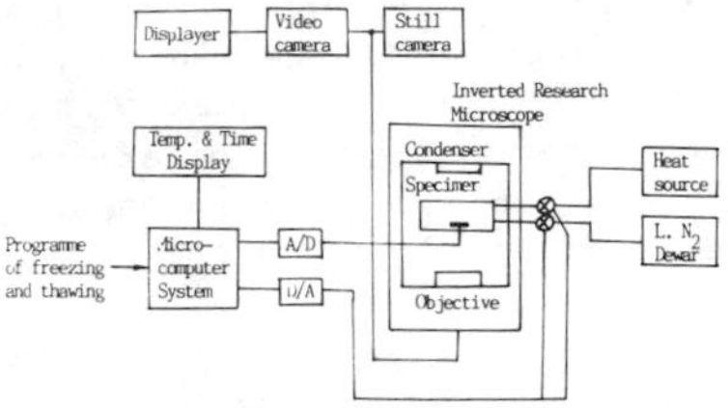

Fig.4 Block diagram of the cryomicroscope system in SIME

Fig.5 Total view of the cryomicroscope in SIME

Programable freezer

A programable freezer has been built in Shanghai Institute of Mechanical Engineering as shown in Fig.6.

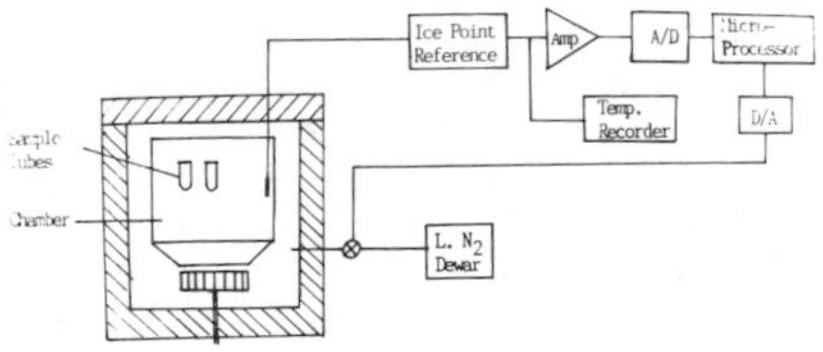

Fig.6 Schematic representation of programable freezer

A microprocessor is used in the cooling rate controller unit. The protocol of cooling and the low temperature limit of the chamber can be controlled easily by the microprocessor.

Some biologists might think the temperature history of the biological system (inside the sample tubes) would be the same as the protocol which is inputted to the microprocessor. But it is not true in most cases. Since the thermal capacities of the sample tubes are not small enough, the heat transfer coefficient inside the chamber is not large enough, and especially since the sample solution would release latent heat during freezing, the temperature of the biological sampl usually can not follow the protocol which is put in the microprocessor. Often there is large deviation between them. Fig.7 illustrates the deviation in our experimental trial. A plastic freezing bag of biological material, containing 16.7ml human serum, 16.6ml DMSO and 106.7ml 1640, is clamped by two aluminium plates. The thickness of the sample bag is about 15mm. The temperature protocol inputted to the microprocessor is: (1) -5°C/min, from room temperature to 5°C; (2) keeping 5°C for 1 minute; (3) -1°C/min, from 5°C to -30°C; (4) keeping -30°C for 1 minute; (5) -4°C/min, from -30°C to -60°C; (6) keeping -60°C for 1 minute; (7) -7°C/min, from -60°C to -100°C. The solid line in Fig.7 represents the temperature in the center of the bag; and the dash line represents the temperature in the chamber of the freezer. The deviation between these two temperature curves is very large. For example, the temperature of the biological material is 90°C higher than that in the chamber at pcint C; and also the cooling rate of CD is much higher than that we desire.

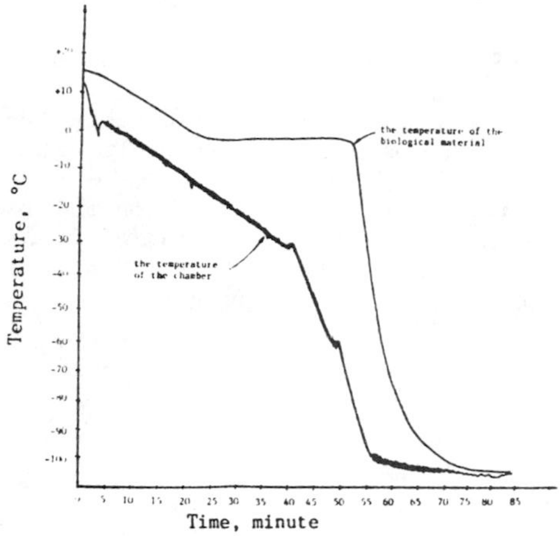

Fig. 7 The temperature derivation and cooling rate derivation in the programable freezer.

It is evident that keeping the practical temperature process of biological material always consistent with the protocol desired is a quite difficult task. We are going to develop some special software to overcome these difficulties.

REFERENCES

[1] Polge,C., Smith,A.U., and Parkes,A.S., "Revival of spermatozoa after vitrification and dehydration at low temperatures", Nature , 164,666, 1949.

[2] Cravalho,E.G., Huggins,C.E., Diller,K.R., and Watson, W.W., "Blood freezing to nearly absolute zero temperature: -272.29°C", J.Biomech.Eng. , 103. 24-26. 1981.

[3] Rowe,A.W., Lenny,L.L., and Mannoni,P.,"Cryopreservation of red cells and platelets", in Low temperature presservation in medicine & biology ,ed. by M.J.Ashwood-Smith and J.Farrant, University Park Press, Baltimore, 285-310, 1980.

[4] Polge,C.,"Freezing of spermatoza", in Low temperature preservation in medicine & biology , ed.by M.J.Ashwood-Smith and J.Farrant, University Park Press, Baltimore, 45-64, 1980.

[5] Huggins,C.E.,"A general system for the preservation of blood by freezing", in Long-Term Preservation of Red Blood Cells , National Academy of Sciences-National Research Council Bull., 160-180, 1965.

[6] Hua,T.C.,"The analysis of freezing and thawing in cryopreservation of biomaterials", Chinese Cryogenic Engineering p23-30 No.1, 1985.

[7] Mazur, P.,"Kinetics of water loss from cells at subzero temperatures and the likelihood of intracellular freezing", J.General Physiology , 47,347-369, 1963.

[8] Mansoori, G.Ali,"Kinetics of water loss from cells at subzero centigrade temperature", Cryobiblology 12, 34-45, 1975.

[9] Katchalsky,A.,Curran,P.F., Nonequilibrium Thermodynamics in Biophysics , Harvard University Press, Cambridge, MA. 1965.

[10] House,C.R., Water transport in cells & tissues Edward Arnold (Publishers) Ltd.,London, 1974.

[11] Hua,T C., Cravalho,E.G.,Jiang,L.,"The temperature difference across the cel membrane during freezing and its effect on water transport", Cryo-Letters 3,255-264, 1982.

[12] Mazur,P., Leibo,S.P., Farrant,J., Chu,E.H.Y.,"Interactions of cooling rates warming rates, and protective additive on the survival of frozen mammalian cells", in The frozen cell ed.by G.E.W.Wolstenholme, Churchill, London, 69-85, 1970.

[13] Mazur,P., Leibo,A.P., and Chu,E.H.Y., "A two-factor hypothesis of freezing injury evidence from Chinese hamster tissue culture cells", Exp.Cell.Res. 71,345-355, 1972.

[14] Mazur,P.,"The role of intracellular freezing in the death of cells cooled at supra-optimal rates", Cryobiology , 14,251-271, 1977.

[15] Diller,K.R., Cravalho,E.G., and Huggins,C.E., "An experimental study of freezing in erythrocytes", Med.Biol.Engng. , 14,321, 1976.

The Varying Conductivity Problem Solutions by Variable Substitution Combined with Boundary Element Method of Submerged Potential Distribution

XIANG-QIAN KONG
Harbin Shipbuilding Engineering Institute
PRC

DE-MING WANG
Shanghai 711 Research Institute
PRC

ABSTRACT

An effective formulation is presented in this paper to the solution of steady temperature field with varying conductivities by the variable substitution combined with boundary element method.

In numerical computation, the source-sink and /or dipole distributions on the boundary, and the technique of submergence are used. The comparison with conventional boundary element method shows that the application of submerging technique further improve the accuracy of the whole solution. In particular, the improvement of accuracy near the boundary is of great significance.

INTRODUCTION

Through generalized Green's theorem in linear PDE's, the solution of unknown variables in the region is substituted into that on the boundary of the region. This is the essentiality of boundary element method (BEM) or boundary integration method (BIE). The main merits of BEM, compared with conventional regional methods such as finite element method (FEM) and finite difference method (FDM) etc., can be concluded as follows:

(1) While the region problems are transformed into the boundary problems, essentially speaking, the dimension of the system is reduced by one. Therefore save the computer time and storage greatly;

(2) BIE solutions have higher accuracies than that of FDM and FEM under the same division and interpolation function;

(3) In dealing with the external problems, since the infinite boundary conditions are satisfied automatically, BEM appears to be very powerful.

References [2], [3] and [4] have validated these merits by applying BEM in solving heat conduction problems of constant thermal properties.

While the domain with sharp varying of temperatures and the practical problems in the engineering of metallurgy and low-temperature techniques, the thermal conductivities of the materials often vary greatly. In these problems, the control differential equations are nonlinearity. Thus, a new thesis has been mentioned out. An iteration scheme to solve varying conductivity heat conduction problems by BEM was introduced in reference [5]. This method assumes that the temperature distribution in the domain is as T^*, then one gains $k^* = k(T^*)$ from varying thermal conductivity $k(T)$, and

$$\nabla(k \nabla T) = 0 \tag{1}$$

become a Poisson equation

$$\nabla^2(kT) = \nabla(T^* \nabla k^*) \qquad (2)$$

Solving equation (2) by the conventional BEM, then comparing the last results as do in the iteration techniques.

Thus, besides iteration, the volume integration of distribution function in the domain must be performed. So, the calculating working are increased greatly. In fact, it losses the performance of reducing dimension. Thus, it can not be recognized as a effective method.

In this paper the origenal function of thermal conductivity

$$K(T) = \int k(T)\, dT \quad [10],[11] \qquad (3)$$

is applied as the to be solved variable. The nonlinear equation (1) is restored into Laplace equation of K. At the same time the boundary conditions of T are transformed accordingly into the equation of K. We can apply BEM to solve K, then gain the temperature field T from equation (3).

FORMULATION OF VARIABLE SUBSTITUTION

In the practice of solving, equation (3) can be written as follows:

$$K(T) = \int_{To}^{T} k(T)\, dT \qquad (3')$$

where T_o is some reference temperature, depending on the working range of k(T). For special problems we can extend the range of k(T), so T_o can be selected arbitrarily.

After the transformation is introduced into equation (1), it can be written to

$$\nabla^2 K = 0 \qquad (M \in \Omega) \qquad (4)$$

The boundary conditions can also be expressed in terms of K:

(1) The 1st kind of boundary condition:

$$T = f(M) \qquad (M \in \partial\Omega) \qquad (5)$$

Substituting equation (5) into (3'), we have

$$K(M) = \int_{T_o}^{f(M)} k(T)\, dT \qquad (M \in \partial\Omega) \qquad (6)$$

(2) The 2nd kind of boundary condition:

$$-k\,(\partial T/\partial n) = q(M), \quad (M \in \partial\Omega) \qquad (7)$$

Since dK/dT = k, so

$$-\partial K/\partial n = q(M) \qquad (M \in \partial\Omega) \qquad (8)$$

(3) The 3rd kind or nonlinear boundary condition:

$$k(\partial T/\partial n) = H(T), \qquad (M \in \partial\Omega) \qquad (9)$$

After substituting into equation (3'), we have

$$\partial K / \partial n = G(K) \qquad (M \in \partial\Omega) \tag{10}$$

Where G(K) is a nonlinear function generally. If G(K) can be expressed explicitly, then we can deal equation (10) as the 2nd kind of boundary condition. Its iteration scheme may be as follows:

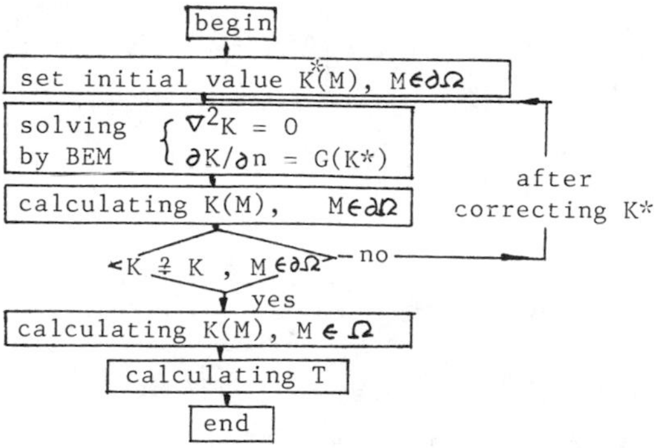

Fig.1 Iteration scheme of solving G(K) expressed by explicit formulation

If G(K) can not be expressed by explicit formulation, then the iteration scheme is slight different. Its iteration loop is as follows:

Set initial value T^* for $M \in \partial\Omega$:
- from $\partial K/\partial n = H(T)$ gain $(\frac{\partial K}{\partial n})^*$ through $\nabla^2 K = 0$ → gain K^*
- from $K = \int_{T_0}^{T} k(T)\, dT$ → gain K^{**} (11)

Some adequate method is applied to use the difference of K^* and K^{**} in modifying T^*.

Summarizing above discussion, after transforming equation (3), the solutions of steady temperature field with variable thermal conductivity are concluded to solve following problems:

$$\nabla^2 K = 0 \tag{12-a}$$

$$K \big|_{\partial\Omega_1} = b \tag{12-b}$$

$$\partial K/\partial n \big|_{\partial\Omega_2} = g \tag{12-c}$$

For the 1st and 2nd kind of boundary conditions, b and g, in the equations (5), (7) and (12) are known functions. For the 3rd or nonlinear boundary conditions, g is a function of K. According to the performance of BEM, only need to do the iteration solution of the unknown value on the boundary.

SINGULARITY DISTRIBUTIONS ON THE BOUNDARY

Generally speaking, the calculation formulas of BEM can be derived from Green's Formula or Galerkin's Method [1]:

$$C_j K_{Mj} = \int_\Gamma \left(\frac{\partial K}{\partial n}\varphi - K\frac{\partial \varphi}{\partial n}\right) d\Gamma \qquad (13)$$

where Γ is the boundary of the domain Ω ; n is the outer normal; φ is the fundamental solution of the governing equation (12-a). For the two-dimensional conditions, take

$$\varphi = \ln \frac{1}{r}, \text{ (r is the distance from the moving point on the boundary to } M_j) \qquad (14)$$

$$c_j = \begin{cases} 2\pi, & (M_j \in \Omega) \\ \pi, & (M_j \in \Gamma) \\ 0, & (M_j \bar{\in} \bar{\Omega} = \Omega \cup \Gamma) \end{cases} \qquad (15)$$

When $M_j \in \Gamma$, equation (13) is just the integration equation of to be solving unknown function $\partial K/\partial n$ or K on the boundary.

From the point of view of potential, to solve equation (13) directly is impliedly that the single-layer-potential is applied for the 1st kind of boundary value problems and double-layer-potential is applied for the 2nd kind of boundary value problems in fact. For the 1st kind of boundary value problems, its integration equation is of the 1st kind of Fredholm Type. Reference [6] had pointed out, since the coefficient matrix after discretization has a bad characteristics, its solution is often presented impulsive or with lower accuracies.

In this paper, it starts from the theory of singularity and applies the singular point distributions, i.e. the source-sink and/or dipole distributions, on the boundary to solve Laplace's Boundary Value Problems. For the 1st kind of boundary value problems the double-layer-potential is applied, and for the 2nd kind of boundary value problems the single-layer-potential is applied.

Noting R^2 as the whole plane, $\Omega^{-1} \cup \Omega = R^2$, $\Omega^{-1} \cap \Omega = \Gamma$, assuming K_e as a harmonics in Ω^{-1}, apply Green's Formula, we have

$$\iint_{\Omega^{-1}} \left(\ln\frac{1}{r}\cdot\nabla^2 K_e - K_e\cdot\nabla^2 \ln\frac{1}{r}\right) d\Omega = \oint_\Gamma \left(\ln\frac{1}{r}\cdot\frac{\partial K_e}{\partial n} - K_e\frac{\partial \ln\frac{1}{r}}{\partial n}\right) d\Gamma = 0 \qquad (16)$$

Where the definition of r is equal to equation (13). If Ω^{-1} is an infinite domain, it is in addition to assume $K_e \to 0$ $(r \to \infty)$.

For the 1st kind of boundary value problems noting $\partial K_e/\partial n = \partial K/\partial n|_\Gamma$, from equations (13) to (16), we have

$$K_{M_j} = \int_\Gamma \varepsilon \left(\partial \ln\frac{1}{r} / \partial n\right) d\Gamma \qquad (17)$$

Where $\varepsilon = (K_e - K)/2\pi|_\Gamma$ denotes the distribution intensity of the double-layer-potential, i.e. dipole. It is determined by following integration equation:

$$b_{P_j} = \int_{\Gamma} \varepsilon (\partial \ln \frac{1}{r} / \partial n) \, d\Gamma + \pi \varepsilon_j \quad (M_j \rightarrow P_j \in \Gamma) \tag{18}$$

Equally, for the 2nd kind of boundary value problems we have

$$K_{M_j} = \int_{\Gamma} \sigma \ln \frac{1}{r} \, d\Gamma \tag{19}$$

The single-layer-potential, i.e. the source-sink distribution intensity, is determined by following integration equation:

$$\partial K_j / \partial n_j = g_j = \int_{\Gamma} \sigma (\partial \ln \frac{1}{r} / \partial n_j) \, d\Gamma + \pi \sigma_j \tag{20}$$

For the problems as equation (12), distributing dipole on the 1st kind of boundary and source-sink on the 2nd kind of boundary, we have (omitting derivation):

$$K_{M_j} = \int_{\Gamma_1} \varepsilon (\partial \ln \frac{1}{r} / \partial n) \, d\Gamma + \int_{\Gamma_2} \sigma \ln \frac{1}{r} \, d\Gamma \tag{21}$$

$$b_j = \int_{\Gamma_1} \varepsilon (\partial \ln \frac{1}{r} / \partial n) d\Gamma + \int_{\Gamma_2} \sigma \ln \frac{1}{r} \, d\Gamma + \pi \varepsilon_j, \quad (M_j \in \Gamma_1) \tag{22}$$

$$g_j = \frac{\partial}{\partial n_j} \int_{\Gamma_1} \varepsilon \frac{\partial \ln \frac{1}{r}}{\partial n} \, d\Gamma + \int_{\Gamma_2} \sigma \frac{\partial \ln \frac{1}{r}}{\partial n_j} \, d\Gamma + \pi \sigma_j, (M_j \in \Gamma_2) \tag{23}$$

Integration equation (18) and (20) are of the 2nd kind of Fredholm Type. For such kind of integration equations, there are more complete analysis theoretically [7],[8].

In the cases of numerical calculation, dividing the boundary into N elements, we assume that the intensity of the singular point is a constant in every element. The position of the element is pushed out of the domain (submerged) slightly. Such submergence technique of singular point has been applied in the speciality fields of aviation and naval architecture to calculate the potential flow, since it can give the higher accuracies with the same condition of element division.

EXAMPLES AND ANALYSIS

Example 1. The 1st kind of boundary value external problem. As is shown in fig.2, there is a semi-circle hole in the semi-infinite plane. Assuming the thermal conductivity

$$k = k_o \, (1 + \beta T) \tag{24}$$

Where $\beta = 100$ and $k_o = 1$ can be used in the numerical calculation.
Its boundary conditions are

$$\begin{aligned} &y = 0, \quad T = 0; \quad r \rightarrow \infty, \quad T \rightarrow 0; \\ &r = 1, \quad T = (\sqrt{1 + 2\beta \sin\theta} - 1)/\beta \end{aligned} \tag{25}$$

The transformed relation of this example is

$$K(T) = \int_0^T (1 + \beta T)\, dT = T + (\beta T^2/2) \tag{26}$$

Then we have

$$T = (1 + 2\beta K - 1)/\beta \tag{27}$$

Its analytical solution is

$$T = \frac{\sqrt{1 + 2\beta y/(x^2 + y^2)} - 1}{\beta} = \frac{\sqrt{1 + 2\beta \sin\theta / r} - 1}{\beta} \tag{28}$$

Distributing dipole on the semi-circle surface, then apply the image technique to satisfy the boundary conditions on the plane wall y = 0 its fundamental solution is as follows:

$$\varphi = \ln (1/r_{MM'}) - \ln (1 / r_{MM''}) \tag{29}$$

Where M'' is the symmetric point of M' about X axis.

Table 1 lists the comparison of the results of surface flux solving from the conventional BEM with dipole distribution having different submerged depth. Table 2 lists the comparison of the results solving from with increasing of r under that the accuracies of surface heat flux is improved obviously. With the increasing of the number of element to be divided, the depth submerged should be increased.

Example 2. The 2nd kind of boundary value external problem. Changing the boundary condition on the semi-circle surface from example 1 to the 2nd kind

$$-k(\partial T / \partial n)_{r = 1} = \sin\theta \tag{30}$$

and the other parts keep the same with example 1. Its analytical solution is still equation (28).

Distributing source-sink on the semi-circle surface, use the image technique to satisfy the boundary conditions on the plane wall, then solve the problem by BEM.

Table 3 lists the comparison of the temperatures solving from the conventional BEM with singularity techniques having different submerged depth. Table 4 lists the comparison of the results solving from with increasing of r under that the angle θ is given.

The analytical solution of this example is the same as example 1, only the way that introducing the problem is different, so here take the source-sink distribution on the boundary. Table3 shows that it is generally required to submerge deeper for the source-sink distribution conditions. After taking the technique of submergence, even if we divide the domain into less element numbers (N=8), the relative errors of whole points are still less than the conventional BEM with more element division (N=16).

Example 3. Mixing boundary value internal problem of the 1st and 2nd kind. Fig.3 shows a plate rectangular domain. Both of the top and bottom sides are thermal-isolation boundary conditions. Both of the left and right sides are of the 1st kind of boundary conditions. The analytical solutions of this problem is as follows:

$$T = (\sqrt{1 + 2\beta(\beta/2 + 1)(1 - x)} - 1) / \beta \tag{31}$$

We take formulas (21) to (23) to solve this example, Fig.4 shows that the heat fluxes on the left and right sides are calculated with different schemes. Fig.5 shows the comparison of section temperatures calculated with different methods. Similarly, "submergence" has gotten the good results.

CONCLUSIONS

(1) This paper applies transforming equation (3) and combines with BIE to solve the problems of varying conductivity. It makes the nonlinearity of the governing differential equations transfering into the algebraic equations, those are simple essentially. Such ideas of simplifying formulation have general significance;

(2) The source-sink and/or dipole distributions are not limited on the boundary of the domain, but have more mobilities. Especially, the improvement of accuracy makes BEM more practical adaptabilities;

(3) While solving complicated engineering problems, such fully improved BEM would be expected more effective than FDM and FEM.

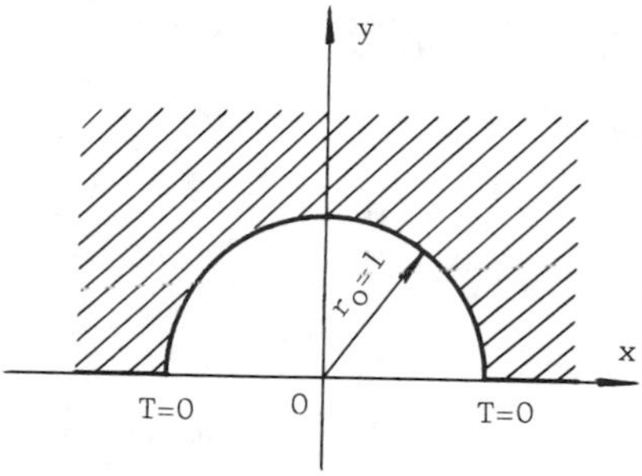

Fig.2 Plane semi-circle hole external problem

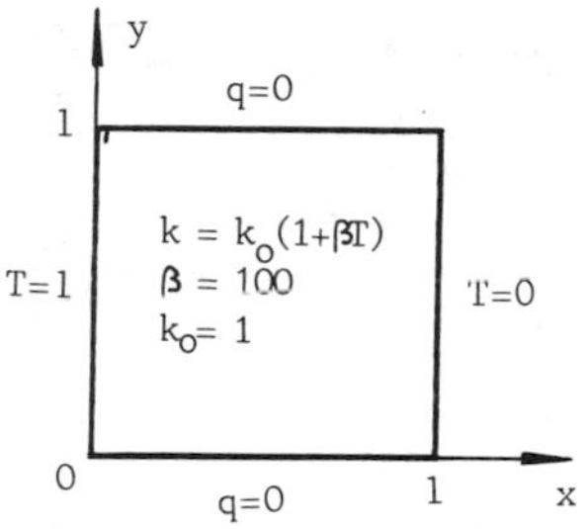

Fig.3 Plane rectangular domain internal problem

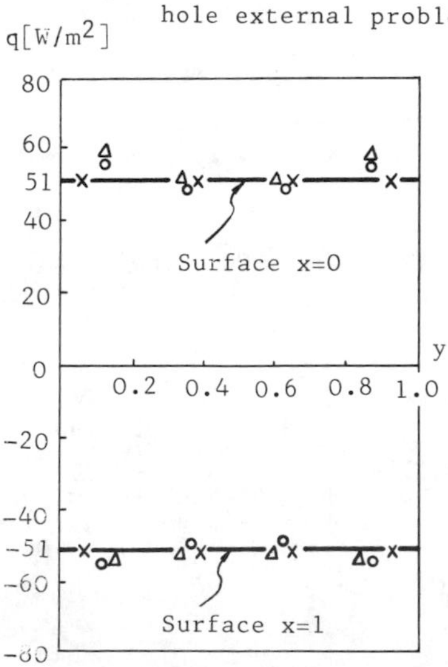

Fig.4 Heat flux q on both side of the plane

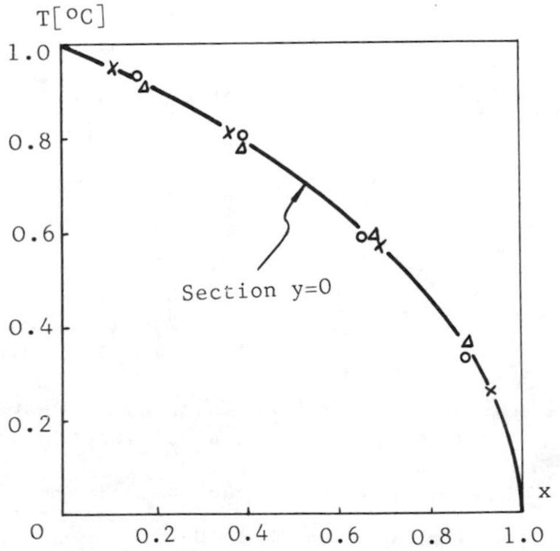

Fig.5 Section temperature distribution

- ∘ conventional BEM
- ▲ $\delta_n/r_o = 0$
- × $\delta_n/r_o = 0.075$
- — analytical solution

Table 1. Comparison with calculating results for surface heat flux $q = k(\partial T/\partial n)$

δ_n/r_o Dimensionless normal distance of surface submergence with double-layer potential distribution;

N Element numbers on the semi-circle.

(a) $N = 8$, $r_o = 1$, q [W/m^2]

θ (Rad.)	Conventional BEM	$\delta_n/r_o = 0$	$\delta_n/r_o = 0.01$	Analytical solution
0.19635	0.19066	0.19384	0.19565	0.19509
0.98175	0.81258	0.82614	0.83384	0.83147
1.76715	0.95850	0.97450	0.98358	0.98078
2.55254	0.54295	0.55201	0.55715	0.55557

(b) $N = 16$, $r_o = 1$, q [W/m^2]

θ (Rad.)	Conventional BEM	$\delta_n/r_o = 0$	$\delta_n/r_o = 0.02$	Analytical solution
0.09817	0.09642	0.09786	0.09805	0.09802
0.49087	0.46379	0.47064	0.47155	0.47140
0.88357	0.76040	0.77177	0.77327	0.77301
1.27627	0.94133	0.95540	0.95726	0.95694

Table 2. Temperature variation with r under that the angle θ is given (N=16).

$\theta = 1.47262$ (rad.), $r_o = 1$ T [^{o}C]

r/r_o	Conventional BEM	$\delta_n/r_o = 0$	$\delta_n/r_o = 0.002$	Analytical solution
1.0	0.13145	0.13143	0.13143	0.13143
1.5	0.10585	0.10553	0.10553	0.10562
3.0	0.07222	0.07200	0.07200	0.07206
12.0	0.03201	0.03190	0.03190	0.03193

Table 3. Comparison with calculating results for surface temperature

δ_n/r_o Dimensionless normal distance of surface submergence with single-layer potential distribution;

N Element numbers on the semi-circle surface.

$N = 8$, $r_o = 1$, T[^{o}C]

θ (Rad.)	Conventional BEM	$\delta_n/r_o = 0$	$\delta_n/r_o = 0.15$	Analytical solution
0.19635	0.05231	0.05556	0.05349	0.05326
0.98175	0.11736	0.12412	0.11982	0.11934
1.76715	0.12826	0.13561	0.13094	0.13041
2.55284	0.09426	0.09979	0.09628	0.09588

Table 4. Temperature variation with r under that the angle θ is given (N=16).
$\theta = 1.47262$ (rad.), $r_o = 1$ T [°C]

r/r_o	Conventional BEM	$\delta_n/r_o = 0$	$\delta_n/r_o = 0.14$	Analytical solution
1.0	0.13088	0.13426	0.13147	0.13143
1.5	0.10517	0.10801	0.10565	0.10562
3.0	0.07174	0.07374	0.07208	0.07206
12.0	0.03178	0.03276	0.03194	0.03194

REFERENCES

[1] C.A. Brebbia, "Progress in Boundary Element Method", V.1 (1981).

[2] L.C.Wrobel & C.A. Brebbia. "The Boundary Element Method for Steady State and Transient Heat Conduction". 1st Int. Conf. on Numerical Methods in Thermal Probs. (1979).

[3] C.A. Brebbia &L.C. Wrobel, "Steady and Unsteady Potential Problems Using The Boundary Element Method", Recent Advances in Numerical Methods in Fluids", V.1. (1980).

[4] Cheng Li-ren & Guo Kuan-lian, "The Boundary Element Method For Heat Conduction Problems", Journal of Engineering Thermophysics of China, Vol. 6, No.1 (1985).

[5] K. Onishi & T. Kuroki, "Boundary Element Method in Singular and Nonlinear Heat Transfer", Boundary Element Methods in Engineering, Proceedings of the 4th Int. Seminar (1982).

[6] J.L. Hess & A.M.O. Smith, "Calculation of Potential Flow about Arbitrary Bodies". Progress in Aeronautical Sciences V.8. pp 1-138 (1967).

[9] W.C. Webster, "The Flow about Arbitrary Three Dimensional Smooth Bodies", J. of Ship Research V. 19, No. 4 (1975)

[10] Khader M.S. & Hama M.C. "An iterative Boundary Integral Numerical Solution for General Steady Heat Conduction Problems", J. Heat Transfer V.103. pp26-31 (1981)

[11] R.K.McMordie, "Steady-state Conduction with Variable Thermal Conductivity", Trans. ASME, J. Heat Transfer 84(1), 92(1962)

[7] Fudan University Math. Department, "Mathematical Physics Equations", Shanghai Science and Technology Pub. House, China. (1961)

[8] Smilnof, "The Course of Advanced Mathematics", V.4, No. 1,2. People's Education Pub. House, China (1958), Translated from Russian.

Improved Calculation Method of Fin Efficiency

CHANGWEN MA
Institute Nuclear Energy Technology
Tsinghua University
Beijing, PRC

ABSTRACT

Common assumptions used in the existing calculation method for fin efficiency, are discussed. An analysis of the influence of various parameters on the fin efficiency is made. It covers the parameter range, which is of interest in practice.

An improved method based on the analysis is presented.

INTRODUCTION

The improvement of heat transfer caused by finned surfaces has been studied for a long time. Much of the development work for such a surface is carried out in the laboratory. For using the experimental data obtained from the laboratory to practice, with different materials and geometries, the following formula is usually applied [1]-[3]

$$h=h_o[(A_1/A)+\eta(A_2/A)] \tag{1}$$

where
- h_o = measured mean heat transfer coefficient
- h = equivalent heat transfer coefficient
- A_1 = heat transfer surface of the gully area between the fins, i.e. area with fins absent (see Fig.1)
- A_2 = heat transfer surface of the fins (see Fig.1)
- A = calculating (base) heat transfer surface (see Fig.1)
- η = fin efficiency

In equation (1), h_o only considers the influence of flow conditions (it can be given by experiment), and all effects of inner thermal resistance are incorporated in the fin efficiency . The fin efficiency is usually calculated by using Gardner's method[4], and is defined by

$$\eta = \frac{\left(\int_0^{A_2} (t-t_f)\, dA\right)}{(t_o-t_f)A_2} \tag{2}$$

where
- t = temperature on the fin surface.
- t_o = wall temperature at the gully surface between the fins
- t_f = coolant temperature
- dA = elemental area of the fin surface
- A_2 = surface area of the fin.

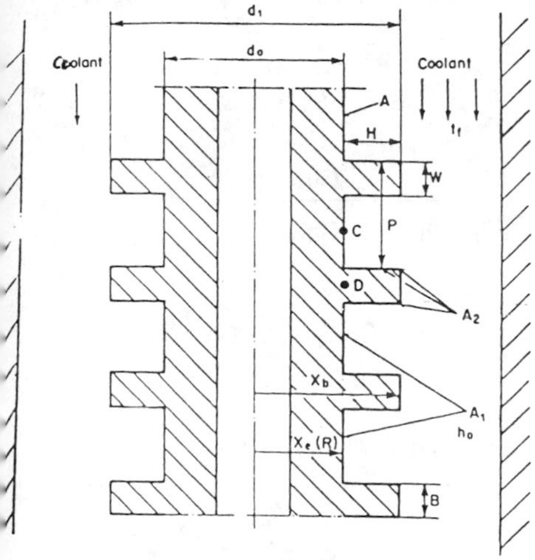

Fig.1 Sketch of the Finned Surface

$A = P \pi d_o$

$A_1 = (P - B)\pi d_o$

$A_2 = B\pi d_1 + 2 \times \frac{\pi}{4}(d_1^2 - d_o^2)$

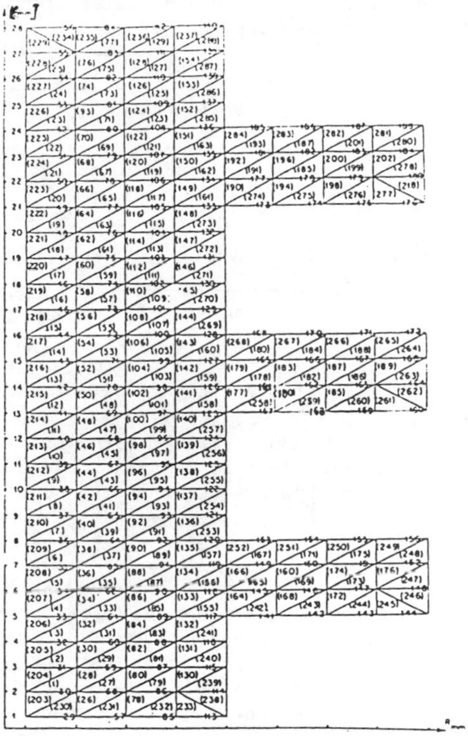

Fig.2 Cross Section of the Finned Surface.

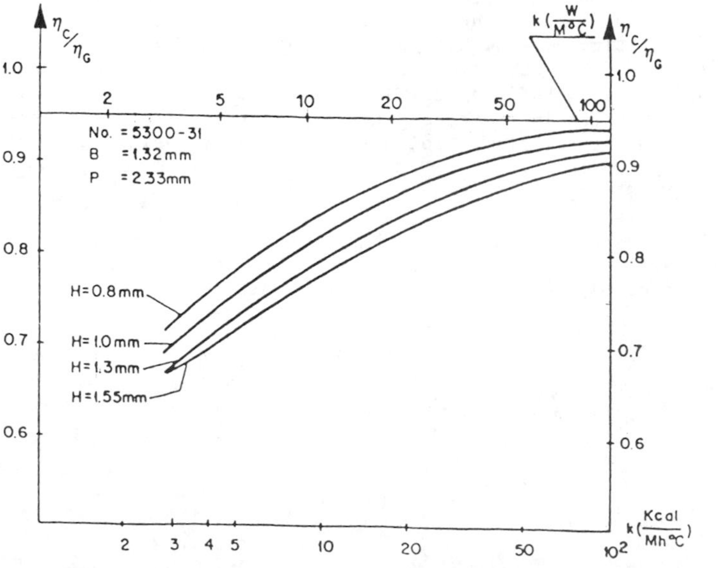

Fig.3 Effect of Conductivity and Fin Height on Fin Effeciency

K. Gardner considered the effects of geometry, material and convective heat transfer on the temperature distribution inside the fin (but only inside it), and gave the following formulae

$$\eta_{Gard} = \frac{\tanh\,(H\sqrt{2h_o/Kw})^{0.5}}{(H\sqrt{2h_o/Kw})}, \quad \text{(for a straight fin)} \tag{3}$$

$$\eta_{Gard} = \frac{2}{U_b(1-(U_e/U_b))} \cdot \frac{(I_1(U_b) - K_1\,(U_b)}{I_o\,(U_b) - K_o(U_b)} \tag{4}$$

where $U_b = \dfrac{H\ 2h_o/Kw}{((X_e/X_b)-1)}$, $U_e=U_b(X_e/X_b)$, $\beta = I_1(U_e)/K_1(U_e)$,

(for a cylindrical fin) (5)

where η_{Gard}= fin efficiency calculated by using the Gardner's method

H = fin height

K = conductivity of the fin

W = width of the fin

h_o = average convective heat transfer coefficient over the whole heat transfer surface

X_b, X_e= radius of the fin top and fin root, respectively.

In these formulae Gardner did not consider the influence of extended surface on temperature distribution in the base metal. He assumed that temperatures on the gully surface between the fins (Fig.1, point C) and on the fin root (Fig.1 point D) were the same. Hudina[5], Sparrow[6]-[7] Mantle and others[8] have shown that this assumption is doubtable. However, until now this problem has not been discussed in detail in the published literature.

Practically, for a metal finned surface, the temperature at the surface of the gully between the fins (point C in Fig.1) is always higher than that at the fin root centre (point D in Fig.1). Because all heat transferred from the fin surface to the coolant must go through the root section of the fin, and as fin surface is uaually much greater than the root section, so the heat flux at the fin root (point D in Fig.1) is higher than that at the gully surface (point C in Fig.1). This means that the temperature at the fin root is lower than that at the gully surface, i.e. the fin efficiency calculated by using Gardner's method will be higher than in practice. So it appears that it is necessary to consider a correction to the fin efficiency calculated by using Gardner's method

THEORECTICAL ANALYSIS

Theoretical analysis can show us which parameters are important in the process of heat transfer from finned surface to coolant. For this purpose the following continuity, momentum, energy and heat transfer equations and the boundary conditions are used

$$\partial\rho/\partial\tau + \nabla(\rho V) = 0 \tag{6}$$

$$\rho\, DV/D\tau = \rho\partial V/\partial\tau + (V\nabla)V = -\nabla P + \mu\nabla^2 V + F \tag{7}$$

$$DT/D\tau = a\nabla^2 T \tag{8}$$

$$h(T_f - T_w) = -\,K_f\,\partial T/\partial n \tag{9}$$

$$DT_s/Dx = a_s \nabla^2 T_s \tag{8'}$$

$$K\partial T/\partial n = K_f \partial T/\partial n \tag{10}$$

where ρ = density, τ =time, V =velocity,
p = pressure, F = body force, μ =viscosity,
a = Temperature diffusivity, K = conductivity of wall,
K_f= conductivity of fluid, T = temperature of fluid,
h = heat transfer coefficient, T_s= Temperature of wall,
$\partial T/\partial n$ = temperature gradient, T_f= main temperature of fluid,
T_w= temperature of wall, a_s= temperature diffusivity of wall

From equations (6)-(10), after some simplifications, the following criterion system can be obtained

$$Nu= f(Re,\ Pr,\ Bi,\ L_1/L_1',\ L_2/L_2',\ L_3/L_3') \tag{11}$$

where Nu=Nusselt Number, Re=Reynolds Number,
Pr=Prandtl Number, Bi=Biot Number,
L_1/L_1', L_2/L_2', L_3/L_3' = geometrical criteria

On the right hand side of equation (11), the Re and Pr criteria characterise the effects of flow situations and type of coolant, and the Bi, L_1/L_1', L_2/L_2', L_3/L_3' criteria characterise the effects of inner thermal resistance and geometry. Obviously, the correction factor of the fin efficiency must be a function of the Bi, L_1/L_1', L_2/L_2', L_3/L_3' criteria. For the finned surface the important geometrical parameters are fin height, H, fin width, w, fin pitch, P, and fin root radius, R. So it is reasonable to analyse the effects of the dimensionless criteria Bi, W/H, P/H and R/H on the fin efficiency (or to analyse the effects of the dimensional parameters concuctivity, K, fin height, H, convective heat transfer coefficient, h, fin width, w, fin pitch, P, and fin root radius, R).

CALCULATIONS AND RESULTS

Several groups of calculations have been made to determine the effects on fin efficiency of conductivity, K, fin height, H, convective heat transfer coefficient, h, fin width, w, fin pitch. P, and fin root radius, R. The finite element two-dimensional program CONDU[11] has been used in these calculations.

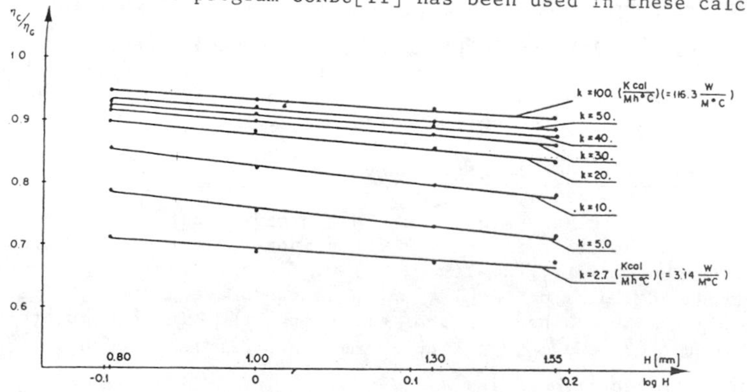

Fig.4 Effect of Conductivity and Fin Height on Fin Efficiency

CONDU solves the temperature distribution within the finned object and the heat flux distribution on the heat exchange surface. Calculations were mainly made for a finned heating element with straight fins, with a few calculations also made for cylindrical fins. A cross-section of the finned element is shown in Fig.2. Calculation was performed for three fins and only the data from the central fin was used, to avoid the effects of end conditions[12]. The section was divided into 288 elements. On all convective heat transfer surfaces the local heat transfer coefficient, h, remained unchanged. For all calculations the total heating power was kept constant. Calculation results are shown in Figs.3, 4, 5 and 6. The vertical ordinate of these figures is the ratio (η_C/η_G). η_C is the corrected fin efficiency, which considers the effects of extended surface on the temperature distribution in the base of the fin, and was calculated by the program CONDU in accordance with the following equation

$$\eta_C = \frac{(\Sigma(T-T_f)\Delta F_2)/F_2}{(\Sigma(T-T_f)\Delta F_1)/F_1} \tag{12}$$

η_G is the fin efficiency, calculated by using Gardner's method, and in accordan with equation (3) (for a straight fin) or equations (4) and (5) (for a cylindri fin). In equation (12),

T = local wall temperature, T_f = coolant temperature,
ΔF_2= elemental area of fin surface, F_2 = total surface area of a fin,
ΔF_1= elemental area of the gully surface between the fins,
F_1= total surface area of a gully between the fins.

(η_C/η_G)shows the relative difference between the fin efficiencies with and with out consideration of the effects of extended surface on temperature distributic in the bas of the fin. In practice it is a correction factor of the fin efficiency, calculated by using Gardner's method.

a) Effect of conductivity on fin efficiency

Several sets of calculations with different thermal conductivity of a finne heating element have been done. Conductivity was varied from 2.7 to 100.1 [Kcal/mh°C]. In every group fin height, H, fin width, w, fin pitch, P, and tot power, Q, were kept constant. Results are given in Figs.3 and 4. From these figures it can be seen that, when the conductivity, K, changed from 2.7 to 100. Kcal/mh°C (η_C/η_G) increased by more than 20%; the higher the conductivity, the higher is (η_C/η_G). This means that, conductivity is not high enough, using Ga dner's method to calculate fin efficiency may introduce a considerable error. From Figs.3 and 4 it can also be seen, that for the interesting practical rang of conductivity (K = 5-50 Kcal/mh°C) at a given geometry the curve of η_C/η_G= f(is almost a straight line in log-lin corrdinates.

b) Effect of fin height on fin efficiency

Four sets of ccalculations have been performed to determine the effect of fin height, H, on the fin efficiency. Fin height was varied from 0.80 to 1.55 In every group conductivity, K, fin width, w, fin pitch, P, total power, Q, re ined unaltered. Calculation results and parameters are given in Figs.3 and 4. From these figures it can be seen that the higher the fin, the greater the dif ference between the η_C and η_G. In logarithmic coordinates the curve (η_C/η_G)=f(is close to a straight line (Fig.4).

c) Effect of convective heat transfer coefficient,h, on fin efficiency

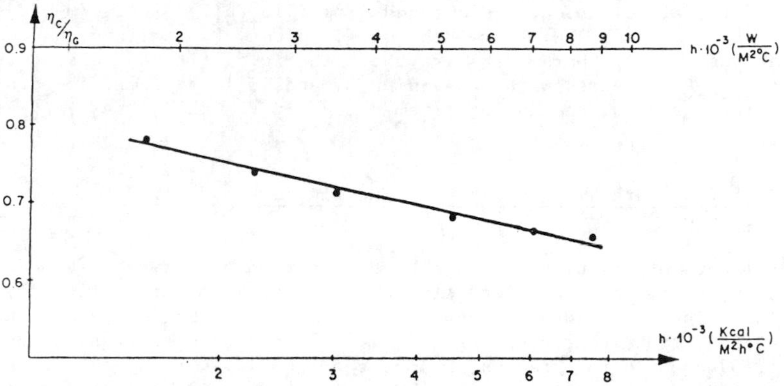

Fig.5 Effect of Convective Heat Transfer Coef. on Fin Efficiency

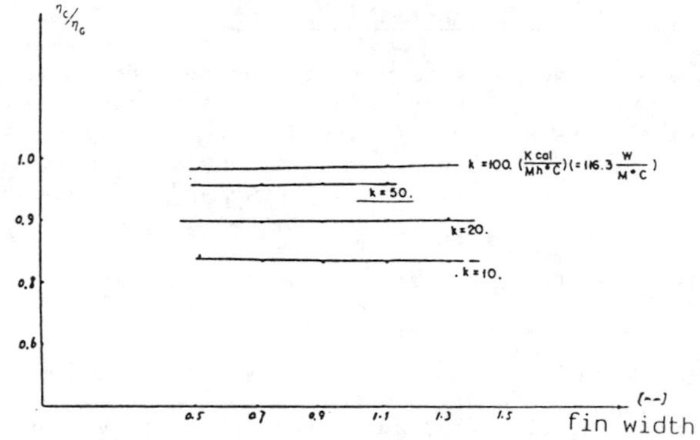

Fig.6 Effect of Fin Width on Fin Effeciency

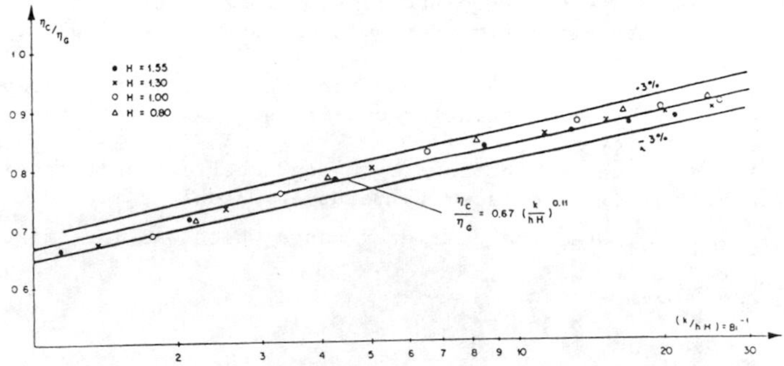

Fig.7 The Relationship $\eta_C/\eta_G = F(Bi)$

Six calculations have been performed to determine the effect of convective heat transfer coefficient, h, on the fin efficiency. In these calculations heat transfer coefficient, h, was changed from 1518 to 7590 [$Kcal/m^2h^oC$] (conductivity, K, fin height, H, fin width, w, fin pitch, P, remained unaltered). results and calculation parameters are shown in Fig.5. It can be seen from Fig.5 that η_C/η_G decreases with increasing h, and in logarithmic coordinates the curve increasing h, and in logarithmic coordinates the curve $(\eta_C/\eta_G)=f(h)$ is close to a straight line.

d) Effects of fin width, w, fin pitch, P and fin root radius, R, on fin efficiency

Several sets of calculations have been performed to determine the effects of fin width, w, fin pitch, P, and fin root radius, R, on fin efficiency. In these calculations fin width changed from 0.52mm to 1.32mm (corresponding to w/H=0.34-0.85), fin pitch, P, changed from 2.33 to 3.03mm (corresponding to P/H=1.5-1.95) and fin root radius, R, changed from 12.74 to 32.74mm (corresponding to R/H=8.21-21.12). Results show that (η_C/η_G) appears not to change with w/H, P/H and R/H (typical result was shown in Fig.6).

e) Effect of Biot criterion, Bi, on fin efficiency

As mentioned above, the correction factor of fin efficiency must be a function of the criteria Bi, R/H, P/H and w/H; and the calculation analysis above has shown that the effects of R/H, P/H and w/H are not significant. At the same time the effects of H, K. h on fin efficiency are considerable. So it seems that there must be a relationship between η_C/η_G and the criterion Bi=hH/K Calculation data were evaluated in the form $\eta_C/\eta_G=f(Bi)$, and results are shown in Fig.7. As expected, it was found that the data were best represented by the equation

$$\eta_C/\eta_G = 0.67\ Bi^{-0.11} \qquad (13)$$

The agreement between equation (13) and the calculation values is rather good. In the range H=0.8-1.55mm, K=5-50 [$Kcal/mh^oC$], h=1518-6000 [$Kcal/m^2h^oC$] the scatter is less than ± 3% (see Fig.7).

CONCLUSIONS

a) For a finned surface, the assumptions of Gardner's method may not be satisfied because of the existence of temperature differences between fin root and gully surface (between the fins). Therefore using Gardner's method to calculate fin efficiency, without any correction, may introdu a considerable error.
b) Calculation analysis shows that the error increases with increasing fin height and convective heat transfer coefficient, and decrease with increasing conductivity.
c) The fin efficiency can be obtained by using a correction factor applied to the fin efficiency calculated by means of Gardner's method, i.e. = $(\eta_C/\eta_G)\cdot\eta_G$ Over the parameter range which was investigated the correction factor

$$(\eta_C/\eta_G) = 0.67\ Bi^{-0.11}$$

REFERENCES

[1] Tatsuhiro Ueda and others: Trans. Japan, Soc. Mech. ENGRS, 30 (1964)
[2] Obermeier E., Schaber A.:
Heat Transfer Conf. (1978) Smooth Ins. Paper FC(a) - 22
[3] Kikkawa K. and others:
Bull. Sci. Eng. RES, Lab. Waseda Univ. Tokyo, (1977) No. 76-32-34
[4] Gardner K.A. and others:
Efficiency of extended surface, Trans. ASME, Vol.67,P. 621-631 (Nov. 1945)
[5] Hudina M. and others: EIR-Report TM-IN-572, Wurenlingen
[6] Sparrow E.M. and others: Trans. ASME, Vol. 103 (Feb. 1981)
[7] Sparrow E.M. and others: ASME, J. of Heat Transfer, Vol. 92 (1970)
[8] Mantle P.L.and others: Heat Mass Transfer, Vol. 14, p. 1825-1834 (1971)
[9] Rohsenow W.: Handbook of heat transfer (1973)
[10] C.W.Ma: EIR-Report, TM-23-82-9 (1982), Würenlingen
[11] C.W.Ma: EIR-Report, TM-24-83-4 (1983), Würenlingen
[12] C.W.Ma: EIR-Report, No. TM-238307 (1983), Würenlingen

Heat Transfer Study of Fluidized Bed Coating

CHAO-YANG WANG and CHUAN-JING TU
Department of Thermophysics Engineering
Zhejiang University
Hangzhou, PRC

ABSTRACT

The paper studies the heat transfer characteristics of fluidized bed coating, and presents a simple theoretical relationship between coating thickness on an object and the physical properties of the system, which is easily applied for industrial purposes.

The theory of fluidized bed coating was based on a new model and developed using the series-expansion method, found in authors' other papers, in solving the problem of heat conduction with a moving boundary as applied for the coating film. The approximate solution has been compared with the exact, numerical solution and experimental data given in the literature, the agreement is good.

NOMENCLATURE

a	thermal diffusivity (m^2/s)	T	temperature (^oC)
c	specific heat of coating material ($J/kg.^oC$)	X	coating thickness (cm)
		ρ	density of coating material (kg/m^3)
h	heat-transfer coefficient ($w/m^2.^oC$)	θ	dimensionless temperature $\frac{T_w-T_f}{T_f-T_\infty}$
H	dimensionless thickness X/X_f		
k	thermal conductivity ($w/m.^oC$)	τ	dimensionless time at/X_f^2
t	time (s)		

INTROCUCTION

Since Knapsack Grieshein Company invented the fluidized bed coating technique, it has been widely used in a routine production process of coating metals with plastics. Fluidized bed coating offers many advantages in applying plastic coatings for decorative purposes, electrical applications, or functiona protection aginst weathering, corrosion and friction.[1]

In coating plastics by means of the fluidized bed technique, a fusible polymeric resin in powder form is applied to the surface of an object that is immersed in a bed or chamble of powder through which a current of gas is passec The gas serves to levitate the resin powder in such a manner that it resembles a boiling liquid in appearance. In the usual thermal form, the object is heatec to a temperature enough above the melting or sintering range of the resin so that, after the object is removed from the heat source, it remains enough heat on its surface to melt or sinter the resin powder particles, which then form ar adherent coating.

In spite of the widespread use of the fluidized bed coating and many experiments done, the theoretical analysis has not been fully done owing to the complexity of the process. Gutfinger and Chen [2] presented a model of the process and its solution by the heat-balance integral, but unfortunately, accuracy of their model and solution is poor. In present paper, we attempt to find a simpler and more accurate theoretical relationship between coating thickness on an object immersed in a fluidized bed of coating material and the physical properties as well as operating conditions of the system.

THE MATHEMATICAL MODEL OF THE PROBLEM

The discussion presented in this paper deals with the growth of coating films on vertical plates in a fluidized bed. We consider one-dimensional heat conduction in a coating film that extends from x=0 to x=X(t). The face x=0 is the object surface. If the surface temperature, Tw, is at or above the melting or softening point, Tf, the coating commences. The thickness of the coating film X(t) as a function of time is the quantity we wish to find.

In the present paper the discussion is limited to cases where the following assumptions made:

1. The thermal properties of material, ,c,k, are constant.
2. The temperature within the fluidized bed is uniform throughout and constant.
3. The temperature of the particles and the fluid is the same.
4. The object temperature, T_w, is constant and equals the average value of initial and final object temperatures during coating process, that is,

$$\bar{T}_w = \frac{T_{wi} + T_{wf}}{2} .$$

5. The surface temperature of the coating film is constant and equals the melting or softening point of the material, Tf.
6. The thickness of films does not depend on orientation of the coated object in the fluidized bed.
7. The heat-transfer coefficient between the object and fluidized bed is constant during coating.

The most questionable assumption is No.5. In fact the temperature on the surface of the coating must remain higher than the softening point of the polymer if the coating is to continue to build up. So, we predict that the theoretical results will be a bit higher than experimental data owing to the assumption No.5. In addition, assumption No. 4 would cause some errors in theoretical results.

Under the assumptions made above, the equations describing the process are:

$$\frac{\partial T}{\partial t} = a \frac{\partial^2 T}{\partial x^2} \qquad (1)$$

with the boundary conditions

$$T(0,t) = T_w \qquad (2)$$

$$T(X(t), t) = T_f \tag{3}$$

$$-k \left.\frac{\partial T}{\partial x}\right|_{x=X(t)} = h(T_f - T_\infty) + \rho c(T_f - T_\infty)\frac{dX(t)}{dt} \tag{4}$$

APPROXIMATE SOLUTION OF THE PROBLEM

The essential difficulty in equations (1) - (4) is in the determination of the unknown moving boundary, X(t). This is a nonlinear problem because it involves a moving boundary whose location is unknown a priori. We are unable to solve it in an exact analytical manner. Thus, numerical or approximate methods have to be used. In recent years, although various approximate solutions have been developed in the literature, the accuracy of most solutions is superior only for small Ste number, and steeply lowered for large Ste number. In our problem, Ste number can be represented as following:

$$Ste = \frac{T_w - T_f}{T_f - T_\infty} .$$

Obviously, Ste number is generally large in fluidized bed coating. In order to ensure the validity of the theoretical solution, here we apply the series-expansion method introduced in [3] for the problem. We may also note that our main interest is the determination of film thickness as a function of time, rather than the temperature distribution in the film.

The temperature distribution in the film can be expressed in the following series form:

$$T = T_f + \sum_{n=1}^{\infty} a_n \frac{(x-X(t))^n}{n!}$$

[3] has deduced the recurrence formula of a_n. If we take only first two terms in the series, the approximate expression of temperature will be:

$$T = T_f + \left.\frac{\partial T}{\partial x}\right|_{x=X(t)}(x-X(t)) - \frac{1}{a}\left.\frac{\partial T}{\partial x}\right|_{x=X(t)} \cdot \frac{dX(t)}{dt} \cdot \frac{(x-X(t))^2}{2!} \tag{5}$$

Substituting equation (4) into (5), and applying equation (5) for equation (2), we obtain:

$$T_w = T_f + \frac{1}{k}\left[h(T_f - T\ ) + \rho c(T_f - T\ )\frac{dX(t)}{dt}\right]X(t)$$

$$+ \frac{1}{2ak}\left[h(T_f - T_\infty) + \rho c(T_f - T_\infty)\frac{dX(t)}{dt}\right] \cdot \frac{dX(t)}{dt} \cdot X^2(t) \tag{6}$$

When the growth of coating film stops and final coating thickness, X_f, is reached, heat balance equation (4) becomes:

$$K \cdot \frac{T_w - T_f}{X_f} = h(T_f - T_\infty)$$

or

$$X_f = \frac{k}{h} \cdot \frac{T_w - T_f}{T_f - T_\infty} \tag{7}$$

Defining the following dimensionless variables by the introduction of the reference length x_f,

$$H = \frac{X(t)}{X_f} \tag{8}$$

$$\tau = \frac{at}{X_f^2} \quad ; \qquad \theta = Ste = \frac{T_w - T_f}{T_f - T_\infty} \tag{10}$$

Equation (6) becomes:

$$(H \frac{dH}{d\tau})^2 + H \frac{dH}{d\tau} (2 + \theta H) + 2\theta (H - 1) = 0 \tag{11}$$

Solving equation (11), we obtain:

$$2\theta = (1 + \theta)(1 - H) - \frac{1}{4}\theta (1 - H)^2 - (1 + \frac{1}{2}\theta) \ln(1 - H) + (\frac{1}{\theta} - 1 - \frac{1}{2}H)Y$$

$$+ 2\ln(1 - \frac{1}{2}\theta H + Y) + (1 + \frac{1}{2}\theta) \ln \left[\frac{Y + 1 + \frac{1}{2}\theta}{1 - H} + \frac{\theta}{2} \cdot \frac{1 - \frac{1}{2}\theta}{1 + \frac{1}{2}\theta} \right]$$

$$+ (\frac{1}{2}\theta - 1) \ln(2\theta - \frac{1}{2}\theta^2 - \theta H + \theta Y) - A(\theta) \tag{12}$$

there

$$Y = \sqrt{(1 + \frac{1}{2}\theta)^2 + \theta(1 - \frac{1}{2}\theta)(1 - H) + \frac{1}{4}\theta^2 (1 - H)^2}$$

$$A(\theta) = 1 + \frac{3}{4}\theta + (\frac{1}{0} - 1) \cdot \sqrt{1 + 2\theta} + 2\ln(1 + \sqrt{1 + 2\theta}) + (1 + \frac{1}{2}\theta)\ln$$

$$(\sqrt{1 + 2\theta} + 1 + \frac{1}{2}\theta + \frac{\theta}{2} \frac{1 - \frac{1}{2}\theta}{1 + \frac{1}{2}\theta}) + (\frac{1}{2}\theta - 1)\ln(2\theta - \frac{1}{2}\theta^2 + \theta\sqrt{1 + 2\theta})$$

For a given θ , the coating thickness is easily calculated from equation (12) during the process.

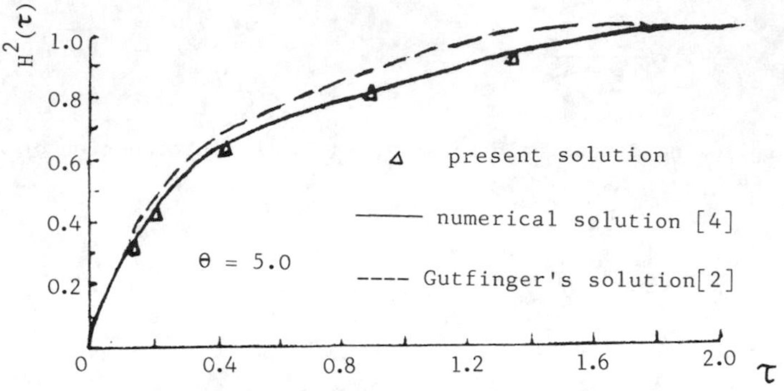

Fig.1 Comparison of the present solution, numerical solution [4] & Gutfinger's result

In Fig.1, the solution (12) and Gutfinger's result are compared with the exact, numerical solution [4] when Ste number equals 5.0. Apparently, our approximate solution well agrees with the exact numerical solution, but Gutfinger's solution deviates from the exact numerical curve with the maximum 12% error. In addition, the present solution is extensively compared with the exact numerical solution in authors' another paper [5], it is showed that good agreement is reached in the range of Ste number from 0.05 to 5. Therefore, it eliminates error in method to study heat transfer of fluidized bed coating using the expression (12).

Fig.2 presents dimensionless film thickness hX/k versus dimensionless time $h^2t/\rho ck$ with the dimensionless temperature as a parameter. It can be seen from Fig.2 that the larger θ, the larger the film thickness during the same immersion time, that is, the needed coating thickness can be obtained by raising either object preheat temperature or bed temperature. In cases where some objects cannot permit too high preheat temperature (for example, electrical appliance wiresand electronic components), the latter measure should been taken, that is, to increase properly the temperature of fluidized powder (to preheat the fluidizing gas). Furthermore, it can be concluded that increases in bed temperature are more effective than corresponding increases in object temperature (preheat temperature).

The coating rate is moderate in middle values of $\frac{h^2t}{ck\rho}$, so, in order to control the coating thickness precisely in ceating practice, it is best to choose the operating conditions of the fluidized bed coating system in the dark region of Fig.2.

Fig.3 presents coating thickness vs. immersion time curves for various object temperature. The parameters come from Pettigrew's [6] experiments. Obviously, the object temperature has a more pronounced influence on coating thickness. So, it is crucial for the theoretical results to choose an equivalent T_w to replace the actual object temperature varied from T_{wi} to T_{wf}. Our assumption that T_w is equal to $(T_{wi} + T_{wf})/2$ is first-order reasonable.

From equation (12) we can conclude that the parameters that affect the coating thickness are the object temperature, fluidized-bed temperature, and

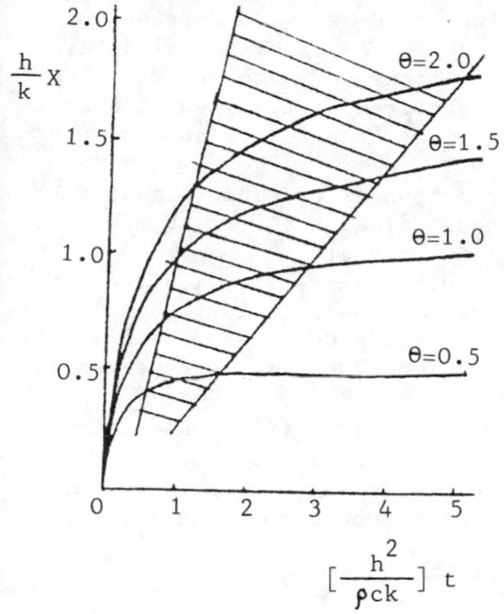

Fig.2 Dimensionless thickness as a function dimensionless time with 0 as a parameter

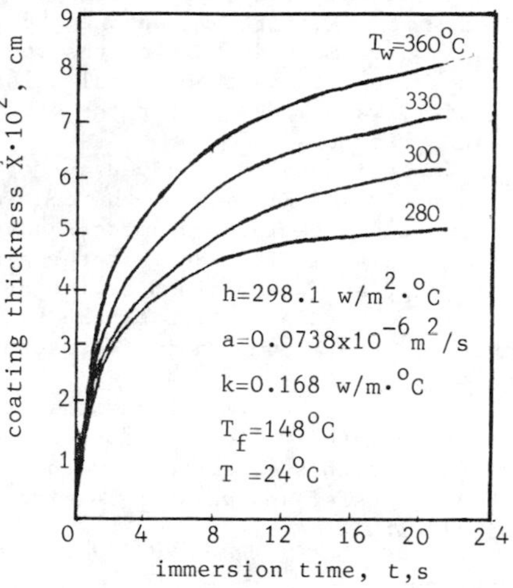

Fig.3 Effect of object temperature on coating thickness during a typical coating process

the properties of the coating material.

COMPARISON OF THE THEORETICAL SOLUTION WITH EXPERIMENTAL DATA

In this section we compared the simplified theoretical solution with some experimental data from the literature. This comparison will show us whether or not the present model can provide the correct guidance for parctical coating problems.

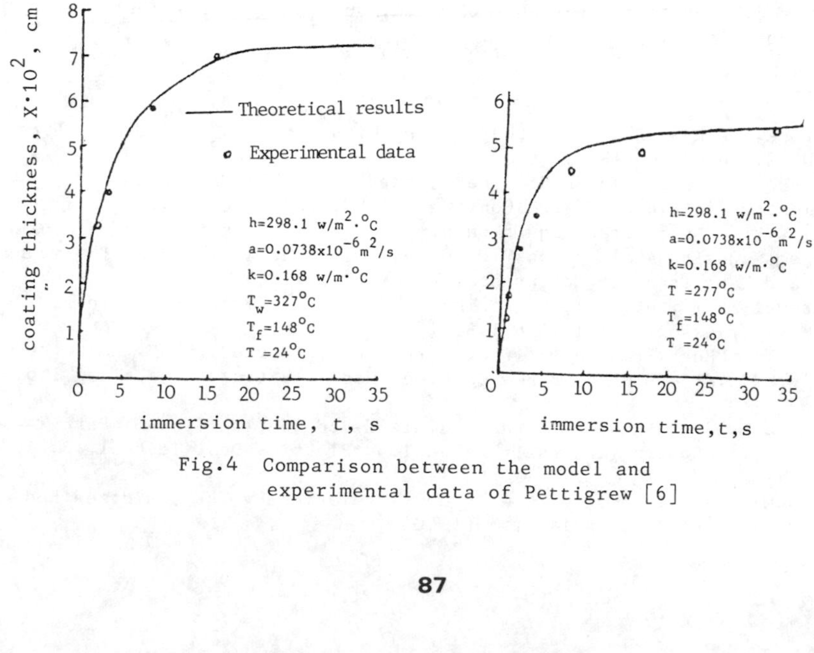

Fig.4 Comparison between the model and experimental data of Pettigrew [6]

Experimental studies of fluidized-bed coating processes were carried out by Pettigrew [6], Richart [7] and Lee [8]. In reporting the experimental data, Pettigrew gave more details on the operating conditions than the others. Thus, the comparison of the theoretical solution and Pettigrew's experimental data is straight-forward. While, for the other data, one has to estimate some of the coating parameters.

The experimental data given by Pettigrew are shown as coating thickness vs. immersion time in Fig.4 with the coating thickness calculated from theoretical equation (12). Various Parameters have been given in Fig.4.

From Fig.4, it is seen that the predicted results well agree with the experimental data, and the maximum deviation in coating thickness is less than 15 percent. The error in coating thickness by the model, as predicted before, is attributed to the assumptions of constant object and coating film surface temperatures.

If we relax the limitations that the object and coating film temperatures are constant, the validity and precision of the theoretical solution would be improved. In this case, the present approximate method is also suitable for the heat transfer analysis of coating process, but it would do so at the expense of simplicity.

CONCLUSIONS

The following conclusions may be drawn from the present study:

1. The series-expansion method is very successful to the heat conduction with a moving boundary, especially when Ste number is larger than 1.
2. The theoretical solution obtained is accurate enough and simple enough to guide the practical coating process.
3. The coating thickness is a strong function of object temperature and the object temperature has a more pronounced influence on coating rate in the case of long immersion times.
4. The control of coating thickness can be realized by changing preheat temperature of the object, bed temperature and heat transfer coefficient, and these three parameters can be adjusted by the fluidized-bed design.
5. The predicted coating thickness is higher than experimental data with the maximum error of 15 percent.
6. The series-expansion method is also suitable for the case that object temperature is treated as the variable of time, but the analysis result is not as simple as equation (12).

REFERENCES

[1] Landrock, A.H: "Fluidized Bed Coating with Plastics" Chem. Eng. Prog. Vol. 63, No.2, pp 67-74 (1967)

[2] Gutfinger, C. and Chen, W.H: "Heat Transfer with A Moving Boundary —— Application to Fluidized-Bed Coating" AIchE. Meeting, Los Angles, Dec. 1968

[3] Wang. C.Y: "Approximate Equation of Temperature Distribution In Stefan's Problems and Its Applications" Chinese Society of Engineering Thermophysics, Heat and Mass Transfer Conference, Wuhan, Dec. 1984

[4] Beaubeuef, R.T., et. al: "Freezing of Fluids In Forced Flow" Inter. J. Heat and Mass Transfer, Vol.10, pp 1581-1587 (1968)

[5] Wang,C.Y: "Heat Transfer of Fluid Freezing In Forced Flow" (to be submitted)

[6] Pettigrew,C.K: "Fluidized-Bed Coating" Mod. Plastics. 44, pp 150-156, August (1966)

[7] Richart, D.S: "A Report on the Fluidized-Bed Coating System. Part 2—— Plastics for Coating and Their Selection" Plastics Des. Technol. 2, pp 26-34 (July 1962)

[8] Lee, M.M: "Application of Electrical Insulation By The Fluidized Bed Process Electro-Technol. 66. pp 149-153 (1960)

An Eigenvalue Method for the Analysis of Long-term Transient Heat Conduction

BU-XUAN WANG and YI JIANG
Thermal Engineering Department
Tsinghua University
Beijing, PRC

ABSTRACT

The basic formulation of the eigenvalue method proposed for analysing the long - term transient heat conduction problem[1] is presented with discussions. It is developed from original so - called "State - space method"[4] in more logic and rigid foundation, and is quite different from that reported recently[5]. The reliability and feasibility of this eigenvalue method had been checked with illustrative example, and its applications in engineering practices are summarized briefly.

INTRODUCTION

It will be neccessary to treat the long - term transient heat conduction problems in arbitrary geometry with random time - dependent boundary conditions in thermal design and analysis, such as the prediction and design of an underground system which stores heat to surrounding soil or rock in summer and absorbes heat from soil or rock in winter, or vice versa, stores cool to soil or rock in winter and retains it in summer, so that the natural temperature difference between winter and summer or day and night can be used to provide useful energy for environment conditioning. A typical profile of such system is shown in Fig 1, i.e, a cave with a ventilation set. The temperature both of the ground surface and in the cave may be known while the heat flow rate from the wall of the cave is to be solved. There will be some difficults in solving such a problem. First, the geometrical size of the region in which the problem deals with is quite small as compared with its surroundings, on which the tempeature and the heat flow are most interested. So that a lot of much smaller elements should have to be taken for the conventional numerical methods. Consequently, as temperature of ground surface and in the cave changes in random way, the time step, $\Delta\tau$, taking for calculation, should be quite small too, as to compared with the preoid of time the problem deals with, T. For example, it is usually to limit to be a hour or a day while T is in several years. Moreover, it is difficult to

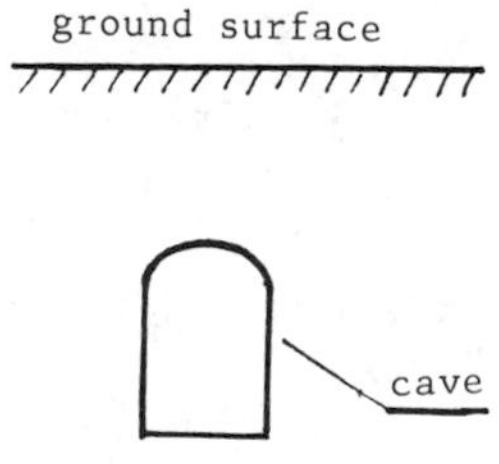

Fig.1 A typical long-term transient heat conduction to be treated

give the initial temperature distribution in the region to be analyzed and large amount of extra calculation work should be involved in determining the initial conditions. In this way, the calculation work would be extreme large and could be hardly accomplished with too much computertime.

Delsante et al[2] had developed a method with Fourier transforms to solve such problem, but it may be confused when the geometry of the region analyzed is not regular and the thermophysical properties in the region are not uniform. Akasaka used a method called "responce factors" based on the finite different principle, which can solve such problem better but also needs to spend much more computer time. We have developed a new way to deal with such problems, which had been reported first in 1982[4] as to simulate thermal environment of a building. Some transient heat conduction problems in subsurface spaces were recently solved by this way successfully. It is now called the "eigenvalue method", simply due to the fact that its main calculating work is realy to find the eigenvalues of the system to be dealed with. It is quite different in formulation from that reporte in a resent paper[5].

BASIC FORMULATION

According to Green's formulation, the temperature at any point r_o in the region Ω at time τ_o is given as:

$$t(r_o,\tau_o) = -\int_{-\infty}^{\tau_o}\int_{\Sigma_1} \lambda t_1 \partial G(r_o, r, \tau_o,\tau)/\partial n d\Sigma d\tau - \int_{-\infty}^{\tau_o}\int_{\Sigma_2} q_2 G(r_o,r,\tau_o,\tau) d\Sigma d\tau + \int_{-\infty}^{\tau_o}\int_{\Sigma_3} h t_{a3} G(r_o,r,\tau_o,\tau) d\Sigma d\tau \quad (1)$$

where Σ_1, Σ_2, Σ_3 are the boundaries with 1st, 2nd, and 3rd kind of boundary conditions respectively; t_1, q_2 and t_{a3} are the corresponding known states on the relative boundaries. $G(r_o,r,\tau_o,\tau)$ is the Green's function and can be found out from following equations:

$$\left.\begin{array}{llll} -\rho c \partial G/\partial \tau = \nabla(\cdot\lambda\nabla G) & , & r \in \Omega & \\ \lambda \partial G/\partial n = hG & , & r \in \Sigma_3 & \\ \partial G/\partial n = 0 & , & r \in \Sigma_2 & \\ G = 0 & , & r \in \Sigma_1 & \\ G = \delta(r,r_o)/\rho c & , & \tau = \tau_0 , & r \in \Omega \end{array}\right\} \quad (2)$$

in which, $\delta(r,r_o)$ is the Dirac's function. It is clear from equation (2) that, the Green's function G is depended only on the geometry of region Ω and the thermal physical properties. So, the problem can be solved in two steps: calculates G from Eq. (2) in advance, and then, integrates Eq. (1) with G and boundary states being known.

The variable separation technique is used to get the function G, that is, let:

$$G(r_o, r, \tau_o,\tau) = X_i(r) T_i(\tau) \quad (3)$$

and $X_i (i=1,2,......,)$ is satisfied by

$$\left.\begin{array}{lll} -\nabla(\lambda\cdot\nabla X_i) = \mu_i \rho c X_i & , & \\ -\lambda \partial X_i/\partial n = h X_i & , & r \in \Sigma_3 \\ \partial X_i/\partial n = 0 & , & r \in \Sigma_1 \\ X_i = 0 & , & r \in \Sigma_1 \end{array}\right\} \quad (4)$$

while T is satisfied by

$$\partial T_i/\partial\tau = \mu_i T_i \qquad \tau = \tau_0, \quad T_i = X_i(r_0) \tag{5}$$

Eq. (4) is an eigenvalue problem. According to Sturm - Liourille's law, there exist orthogonal complete functions set (X_i), i=1,2,....., and corresponding eigenvalue μ_i, satisfying Eq. (4) in the region by the weight ρc. Whenever X_i and μ_i were found out, the solution of Eq. (5) could be easely formulated as:

$$T_i = X_i(r_0)\exp(-\mu_i(\tau_0-\tau)) \qquad ,i=1,2,......., \tag{6}$$

Hence, the key for getting G is to solve the equation (4). Exact solution can be found in its analytical form by traditional method only for some kinds of geometric form of the region with uniform thermophysical properties. In general, Eq. (4) cannot be solved analysticaly, and so, we have to solve it in numerical way. Based on the idea of finit elements method, Eq.(4) can be transformed into a generalized eigenvalue problem as follows:

$$AX = CX\Lambda \tag{7}$$

where

$$A = \int_\Omega \lambda\nabla N(r)\cdot\nabla N(r)\,d\Omega + \int_{\Sigma_3} h\,N^T(r)\,N(r)\,d\Sigma \qquad C = \int_\Omega \rho c\,N(r)^T N(r)\,d\Omega \tag{8}$$

while N(r) is the base function for interpolation, and the diagonal matrix Λ consisting of n elements approximated to the μ_i, i=1.2,.... respectivity in Eq(4) which the matrix X is the values of the function $X_i(r)$ at the points $r=r_1, r_2, ..., r_n$, such as:

$$X_i(r) = N(r)\,X_i \qquad X = (x_1, x_2, ..., x_n) \tag{9}$$

With the X and Λ solved from Eq. (7), the approximate G can be found out as:

$$G(r_0, r, \tau_0, \tau) = N(r_0)\,X\exp(-\mu_i(\tau_0-\tau))\,X^T N(r)^T \tag{10}$$

The basic solution of transient heat conduction problem in an arbitrary geometry with random time - dependent boundary inputs will be

$$t(\tau_0, r_0) = N(r_0)X\int_{-\infty}^{\tau_0}\exp[-\Lambda(\tau_0-\tau)]\,X^T\Big[-\int_{\Sigma_1}\lambda\partial N/\partial n\,t_1(\tau,r)\,d\Sigma - \int_{\Sigma_2}N(r)q_2(\tau,r)\,d\Sigma + \int_{\Sigma_3}h\,N(r)\,t_{a3}(\tau,r)\,d\Sigma\Big]\,d\tau \tag{11}$$

If τ_0 extends to infinite, the steady solution will be

$$t(r_0) = N(r_0)\,X\Lambda^{-1}X^T\Big[-\int_{\Sigma_1}\lambda\partial N(r)/\partial n\,t_1(r)\,d\Sigma - \int_{\Sigma_2}N(r)\,q_2(r)\,d\Sigma + \int_{\Sigma_3}h\,N(r)\,t_{a3}(r)\,d\Sigma\Big] \tag{12}$$

In this way, all the work is to do on the generalized eigenvalue problem, Eq. (7), and the integration of Eq. (11). We have thus developed a special programme SG83.3 to do so.

PRELIMINARY CHECK

To test the reliability and the feasibility of the eigenvalue method suggested above, a sample of two - dimensional problem, as shown in Fig.2,

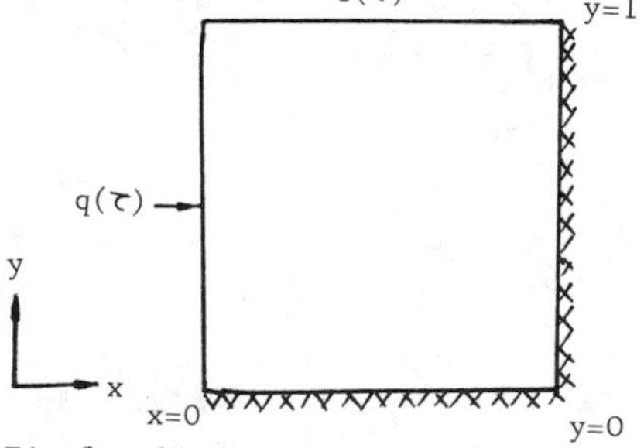

Fig.2 Illustrative example.

is chosen for illustration. At the boundary y=1, an uniform time-dependent distributed temperature $t(\tau)$, is given; at the boundary x=0, an uniform distributed heat flow time-dependent, $q(\tau)$, is given. By the traditional analytical method, the explicit solution for the mean temperature at the boundary x=0, $t_1(\tau)$, and the integrated heat flow over the boundary y=1, $Q_1(\tau)$, will be respectively as:

$$t_1(\tau_0)=\sum_{m=0}^{\infty}\sum_{n=0}^{\infty}\int_{-\infty}^{\tau_0}[W_{1,m,n,1}\,t(\tau)+W_{2,m,n,1}\,q(\tau)/\lambda]\exp[-\mu_{m,n}a(\tau_0-\tau)/l^2]\frac{a}{l^2}d\tau \tag{13}$$

$$Q_1(\tau_0)=\sum_{m=0}^{\infty}\{\sum_{n=0}^{\infty}\int_{-\infty}^{\tau_0}[W_{1,n,m,2}\,t(\tau)+W_{1,m,n1}\,q(\tau)/\lambda]\exp(-\mu_{m,n}\,a(\tau_0-\tau)/l^2)\frac{a}{l^2}d\tau + W_{1,m,0,2}\,t(\tau_0)\} \tag{14}$$

where

$$\mu_{m,n}=n\pi^2+(2m+1)^2\pi^2/4$$

$$W_{1,m,n,1}=\begin{cases} & n=0\\ & n>0\end{cases}$$

$$W_{2,m,n,1}=\begin{cases}8/\pi^2/(2m+1)^2 & n=0\\ 16/\pi^2/(2m+1)^2 & n>0\end{cases} \tag{15}$$

$$W_{1,m,n,2}=\begin{cases}(2m+1)^2\pi^2/2 & n=0\\ 0 & n>0\end{cases}$$

$$W_{1,m,0.2}=-2$$

while a is the thermal diffusivity.

By the eigenvalue method, the corresponding solutions can be formulated as:

$$t_1(\tau_0)=\sum_{i=1}^{n}\int_{-\infty}^{\tau_0}[W_{1,i,1}\,t(\tau)+W_{2,i}\,q(\tau)/\lambda]\exp(-\mu_i a(\tau_0-\tau)/l^2d\tau+ + W_{1,o,1}\,t(\tau_0)+W_{2,o}\,q(\tau_0)$$

$$Q_1(\tau_0)=\sum_{i=1}^{n}\int_{-\infty}^{\tau_0}[W_{1,i,2}\,t(\tau)+W_{1,i,1}q(\tau)/\lambda]\exp(-\mu_i a(\tau_0-\tau))/l^2d\tau + + W_{1,o,2}\,t(\tau_0)+W_{1,o,1}\,q(\tau_0)$$

The calculating results for the values of μ and W by Eq. (15) and by spacial computing program SG83.3, developed for eigenvalue method, are collecte in Table I. As $\exp(-\mu\tau)$ goes to zero with incresing τ when μ is larger, only a few terms of the series play roles actually. So, the results by eigenvalue method will be quite agree with the explicit analyzcal results.

Let $t(\tau)$ equal to zero, and $q(\tau)$ is a step function, that is

$$q(\tau)=\begin{cases}0 & \tau<0\\ \lambda/L & \tau\geq 0\end{cases}$$

We compute $t_1(\tau)$ from equation (13) to get the exact solution or from equati (16) to get the eigenvalue solution. Table II listed the results. The solutio by conventional finit elements method with different time steps, $\Delta\tau$, and also t CPU spended are listed in Table II for comparison. It is found that, the solut of the eigenvalue method is quite agree with the exact solution and spends much shorter computer time, only about 1/3 to 1/4 required by the conventional finit elements method for achieving the same accuracy.

Table I The H and W calculated by analytical method and eigenvalue method

No	Numerical Solution				Analytical Solution					
i	μ_i	$W_{1,i,1}$	$W_{2,i,1}$	$W_{1,i,2}$	n	m	$\mu_{n,m}$	$W_{1,n,m,1}$	$W_{2,n,m,1}$	$W_{1,n,m,2}$
1	2.4673	2.0	0.8206	4.934	0	0	2.4674	2.0	0.8106	4.935
2	12.353	.0	1.626	0	1	0	12.337	.0	1.6210	0
3	22.374	2.0	0.0914	43.84	0	1	22.2066	2.0	0.090	44.41
4	32,285	.0	0.1857	0	1	1	32.076	0	0.1801	0
5	42.817	.0	1.662	0	2	0	41.946	0	1.621	0
6	63.290	.0	.1979	0	2	1	61.685	0	0.1801	0
7	64.599	2.133	.0377	120.7	0	2	61.685	2.0	.0324	123.37
8	74.795	0	.084	0	1	2	71.560	0	.0648	0
9	109.77	0	.1044	0	2	2	101. 6	0	.0648	0
0		0.0556	- 0233	-7.0				0	0	-2(M+1)

Table II History of $t_1(\tau)$ computed with three schems: Exact, Eigenvalue and Finite elements by M-150 computer

$\frac{a}{l^2}$	Exact	Eigen value	Finite Elements			
			$\Delta\tau=.0003\ \frac{l^2}{a}$	$\Delta\tau=.0001\ \frac{l^2}{a}$	$\Delta\tau=.003\ \frac{l^2}{a}$	$\Delta\tau=.01\ \frac{l^2}{a}$
.001	.0133	-.0026				
.003	.0344	0.0266	0.0256	0.0236	0.0193	
.05	.0502	0.0467	0.0470	0.0439		
0.01	.0799	0.0799	0.0793	0.0780	0.0741	0.0639
0.03	.1498	0.1502	0.1505	0.150	0.1482	0.1424
0.1	.2666	0.2672	0.2668	0.2667	0.2658	0.2627
0.3	.4028	0.4034	0.4031	0.4030	0.4027	0.4012
1.0	.5349	0.5355	0.5353	0.5350	0.5352	0.5346
3.0	.5626	0.5632	0.5631	0.5630	0.5629	0.5630
10	.5626	0.5632				
CPU		27 sec	336 sec	101 sec	34 sec	15 sec

APPLICATIONS IN ENGINEERING PRACTICES

It is very often in engineering practices that, the temperature distribution at a given instant, τ_0 , is not required, the object is to find out the responce of temperature or boundary heat flow at certain point. In such cases, the solution can be formulated conveniently as Eg. (16) or (17) respectively, and the very practical problem is the option in the manners to do the integration. There are two different ways in option referred to whether the problem is with periodic or non-periodic conditions as follows:

. Problem with Periodic Boundary States.

Taking the system shown in Fig. 1 as an example. Suppose the periodical change of air temperature in the cave and the ground surface temperature can be considered as in the same manner every year, so that

$$t(\tau) = t(\tau + kp) \quad , \quad k=1,2,\ldots\ldots \tag{18}$$

where the period, P, equals to 8760 hours or 1 year; $t(\tau)$ is the air temperature in cave or the ground surface temperature. In this way, the integration of heat flow over the internal surfuce of the cave, $q(\tau)$, can be written as:

$$q(\tau) = \sum_{k=0}^{\infty} \sum_{i=1}^{n} \int_{\tau_0-p}^{\tau_0} \{W_{1,i}\, t_a(\tau) + W_{2,i}\, t_s(\tau)\} \exp\{-\mu_i(\tau_0 - \tau + kp)\}\, d\tau \tag{19}$$

or

$$q(\tau_0) = \sum_{i=1}^{n} \left[\int_{\tau_0-p}^{\tau_0} \{W_{1,i}\, t_a(\tau) + W_{2,i}\, t_s(\tau)\} \exp\{-\mu_i(\tau_0 - \tau)\} d\tau / \{1 - \exp(-\mu_i p)\}\right]$$

where $t_a(\tau)$ and $t_s(\tau)$ are air temperature in cave and the ground surface temperture respectively. Introducing an interpotation function $N_t(\tau)$, $t_a(\tau)$ and $t_s(\tau)$ can then be expressed as:

$$t_a(\tau) = N_t(\tau)\, T_a$$
$$t_s(\tau) = N_t(\tau)\, T_s \tag{20}$$

in which

$$T_a = (t_a(\tau_1),\ t_a(\tau_2),\ t_a(\tau_3), \ldots\ldots,\ t_a(\tau_m))^T$$
$$T_s = (t_s(\tau_1),\ t_s(\tau_2),\ t_s(\tau_3), \ldots\ldots,\ t_s(\tau_m))^T \tag{21}$$

$\tau_1, \tau_2, \ldots, \tau_m$ are the time interpolation points. Introduce Eq. (20) or (21) to Eq. (19), and so do the $q(\tau)$ to Q, it will be:

$$q(\tau) = N_t(\tau) Q = N_t(\tau)(\Phi T_a + \Theta T_s) \tag{22}$$

where Φ and Θ are m x m matrix, in which the i th row ϕ_i and θ_i are:

$$\phi_i = \sum_{k=1}^{n} \left[\int_{\tau_i-p}^{\tau_i} W_{1,k} N_t(\tau) \exp\{-\mu_k(\tau_i - \tau)\} d\tau\right] / (1 - \exp(-\mu_k p))]$$

$$\theta_i = \sum_{k=1}^{n} \left[\int_{\tau_i-p}^{\tau_i} W_{2,k} N_t(\tau) \exp(-\mu_k(\tau_i - \tau)) d\tau\right] / [1 - \exp(-\mu_k p)]$$

In quite a few practial egineering problems, $t_a(\tau)$ for part of time in 1 year, τ_1 to τ_l, for example, is known, which $q(\tau)$ is to be solved; for the remainder, τ_{l+1} to τ_m, the heat flows $q(\tau)$ is known and $t_a(\tau)$ is unknown. Let T_{a1} represents t_a between τ_1 and τ_l, and T_{a2} between τ_{l+1} to τ_m, so that $T_a = (T_a\ , T_{az})^T$; then,

$$\begin{pmatrix} Q_1 \\ Q_2 \end{pmatrix} = \begin{pmatrix} \Phi_{11} & \Phi_{12} \\ \Phi_{21} & \Phi_{22} \end{pmatrix} \begin{pmatrix} T_{a1} \\ T_{a2} \end{pmatrix} + \begin{pmatrix} \theta_1 \\ \theta_2 \end{pmatrix} T_s \tag{23}$$

As Q_1 and T_{a2} are unknown, equation (23) can be changed to

$$\begin{pmatrix} Q_1 \\ T_{a2} \end{pmatrix} = \begin{pmatrix} \tilde{\Phi}_{11} & \tilde{\Phi}_{12} \\ \tilde{\Phi}_{21} & \tilde{\Phi}_{22} \end{pmatrix} \begin{pmatrix} T_{a1} \\ Q_2 \end{pmatrix} + \begin{pmatrix} \tilde{\theta}_1 \\ \tilde{\theta}_2 \end{pmatrix} T_s \tag{24}$$

where $\tilde{\Phi}_{11} = \Phi_{11} - \Phi_{12}\Phi_{22}^{-1}\Phi_{21}$, $\tilde{\Phi}_{12} = \Phi_{12}\Phi_{22}^{-1}$. $\tilde{\Phi}_{21} = \Phi_{22}^{-1}\Phi_{21}$,

$\tilde{\Phi}_{22} = \Phi_{22}^{-1}$, $\tilde{\theta}_1 = -\Phi_{22}^{-1}\theta_2$ and $\tilde{\theta}_2 = \theta_1 - \Phi_{12}\Phi_{22}^{-1}\theta_2$

The final solution will be

$$q(\tau) = N_t(\tau)\left(\begin{pmatrix}\tilde{\Phi}_{11} & \tilde{\Phi}_{21} \\ 0 & I\end{pmatrix}\begin{pmatrix}T_{a1} \\ Q_2\end{pmatrix} + \begin{pmatrix}\tilde{\theta}_1 \\ 0\end{pmatrix} T_s\right) \tag{25}$$

$$t_a(\tau) = N_t(\tau)\left(\begin{pmatrix}I & 0 \\ \tilde{\Phi}_{21} & \tilde{\Phi}_{22}\end{pmatrix}\begin{pmatrix}T_{a1} \\ Q_2\end{pmatrix} + \begin{pmatrix}0 \\ \tilde{\theta}_2\end{pmatrix} T_s\right) \tag{26}$$

The sample calculations get quite good results as compared with the actual measurements in caves for store apples ets.[1]

2. Problems with Nonperiodic Boundary States

When the equation (18) is not satisfyed, including the analysis for the transient process in a short time, it is very often to simulated the real process hour by hour or day by day, another way should have to do the integration (11). For illustration, still take the system shown in Fig. 1 as an example, the first order spline function can be used to interpolate $t_a(\tau)$ and $t_s(\tau)$ such as:

$$q(\tau_0) = \sum_{i=1}^{n} \left\{ \sum_{k=1}^{\infty} \varphi_i \exp(-\mu_i k \Delta\tau) t_a(\tau_0 - k\Delta\tau) + \theta_i \exp(-\mu_i k \Delta\tau) t_s(\tau_0 - k\Delta\tau) \right\} + \varphi_0 t_a(\tau_0) + \theta_0 t_s(\tau_0) \tag{27}$$

where, $\varphi_i = W_{1,i}\Delta_i$, $\theta_i = W_{2,i}\Delta_i$, $i=1,2,\ldots\ldots, n$

$$\Delta_i = \{\exp(\mu_i \Delta\tau) - 2 + \exp(-\mu_i \Delta\tau)\} / \mu_i^2 / \Delta\tau$$

$$\varphi_0 = W_0 + \sum_{k=0}^{n} W_{1,i}\,[1 - \{\exp(\mu_i \Delta\tau) - 1\} / \mu_i \Delta\tau] / \mu_i$$

$$= K_0 - \sum_{i=0}^{n} W_{1,i}\,\{\exp(\mu_i \Delta\tau) - 1\} / \mu_i^2 / \Delta\tau$$

$$\theta_0 = \sum_{i=0}^{n} W_{2,i}\,[1 - \{\exp(\mu_i \Delta\tau) - 1\} / \mu_i \Delta\tau] / \mu_i$$

$$= -K_0 - \sum_{i=1}^{n} W_{2,i}\,\{\exp(\mu_i \Delta\tau) - 1\} / \mu_i^2 / \Delta\tau$$

in which $1/K_0$ is the thermal resistance from the air in cave to the ground surface. From equation (27), it is easy to calculated by:

$$\left.\begin{aligned} q(\tau) &= \sum_{i=1}^{n} q_i(\tau) + \varphi_0 t_a(\tau) + \theta_0 t_s(\tau) \\ \text{where} \quad q_i(\tau) &= C_i q_i(\tau - \Delta\tau) + \varphi_i t_a(\tau) + \theta_i t_s(\tau) \end{aligned}\right\} \tag{28}$$

in which $C_i = \exp(-\mu_i \Delta\tau)$.

Only a few terms in the series may be taken with n less than 20, to assure very good approximation.

With this scheme, a computer programme STES have be developed to simulate the thermal environment of subway system for which the temperature at the station and along the tunnel can be computed hour by hour over 1 year. The optimization of the ventilation plan in Beijing subway had been made by using this programme.

CONCLUSION

The basic formulation of this eigenvalue method for solving long - term transient heat conduction problems has been developed, which is quite different from that reported by Shih and Skladeng[5] and is in more logic and rigid foundation. As shown by the illustrative example and engineering applications with discussions, this eigenvelue method takes advantages in feasibility over other numerical methods presented so far. It is therfore to be expected for practical use in engineering analysis, such as to predict the thermal environment of underground systems.

REFERENCES

[1] Yi Jiang, The basic therophysical research on the utilization of natural temperature difference in subsurface space, Dissertation for the degree of Dr.- Engineering, Tsinghua Univ., Beijing. (1985)

[2] A.E.Delsante, A.N.Stokes, and P.J.Walsh,IJHMT,Vol 26,No 1, pp121-132. (1983)

[3] Hiroshi Akasaka: A method of calculation for air conditioning load from floors adjoining to the ground and the basement walls and floors, 4 th I.S.U.C.E.E.R.B, (1983)

[4] Yi Jiang,ASHRAE, Transactions, Vol 88,Part II, pp122-132. (1982)

[5] T.M.Shih and J.T.Skladang, Numerical heat transfer, Vol 6, pp.409-422, (1983)

The Rapid Extrapolation Method of Surface Temperature and Heat Flux and Its Application

DUQIANG WU, HUANAN JIAO, HANZHANG GUAN, and PUFA WANG
P.O. Box 1, Xianyang
Shanxi, PRC

ABSTRACT

In this paper, a rapid extrapolation method of surface temperature and heat flux and its application is reported. A great of evaluation has been accomplished with various weigh matrixes. This algorithm is applied to the temperature measuring. The result of restore evaluation is shown in this paper.

INTRODUCTION

It is important to search for an extrapolation method about transient surface temperature and heat flux in many aspects. However many experimental difficulties arise in implanting a probe at the surface for heat transfer measurement, for example, involving the motion of a projectile over a barrel surface, the sliding of a piston in the combustion chamber and the high temperature exhaustion of a rocket engine. Therefore, it is hopeful in these cases that the extrapolation of surface temperature and heat flux be accomplished by inverting the temperature as measured by a probe located interior to the surface of the object measured. It is obvious that this is a indirect method for measuring surface temperature. It is only the exact and rapid extrapolation method that appropriates the techniques in the current measuring temperature.

Murray Imber and Jamal Khan[1] developed an analytical method for the inverse heat conduction which is used in the temperature at two positions. Based upon these interior thermocouples readings, a closed form solution is obtained via Laplace transfer techniques for the transient temperatures beyond the two positions. Ching Jen Chen and Darrel M Thomsen[2] reported a simple method to determine a short time transient surface temperature and heat flux for the case of a hollow cylinder based on the inversion of the temperature profile measured only by one interior probe. Sparrow[3] treated the inverse problem in a different manner. In these methods the prediction temperature and heat flux formulas were expressed as repeated integral of error complementary function. Unfortunately the repeated integral was prolix and inconvenient to be calculated.

This paper develops a temperature extrapolation method. After two important formulas are testified, the Repeated Integral of Integeralization Error Complementary Function (RIIE) is presented. RIIE is expressed by a polynomial which includs only the error function. Moreover authors have compiled the numeral table about RIIE, worked out the rapid algorithm for the extrapolation.

EXTRAPOLATION EQUATIONS EXPRESSED BY ERROR FUNCTION

The material of the slab is considered to be homogeneous and isotropic with constant thermal diffusivity and the temperature of the material is initially uniform at T_o. For one-dimensional transient conduction under the semi-infinit boundary condition, it is assumed that the surface temperature is expressed as follow:

$$\theta(0,t) = \sum_{n=0}^{\infty} b_n t^n \tag{1}$$

where $\theta = T - T_o$. The inner temperature response is then

$$\theta(x,t) = \sum_{n=0}^{\infty} b_n (4t)^n \Gamma(n+1) i^{2n} \operatorname{erfc}(x/2\sqrt{t}) \tag{2}$$

The temperature response of the thermocouples measured at $x=x_p$ is

$$\theta(x_p,t) = \sum_{n=0}^{\infty} b_n (4t)^n \Gamma(n+1) i^{2n} \operatorname{erfc}(x_p/2\sqrt{t}) \tag{3}$$

where t, x is dimensionless time, coordinate, respectively $t=a\tau/L^2$ $x=X/L$. Where τ, X, L, a is dimension time, coordinate, the real thick of slab, thermal diffusivity, respectively.

It is not only prolix, but also many a great of calculation that the extrapolation with certain accuracy is directly made by the above-mentioned formulas which include the repeated integral of error complemental function. For convenience, we introduce a function

$$\Psi_n(s) = 2^n \Gamma(n/2+1) i^n \operatorname{erfc}(s) \tag{4}$$

Authors have certificated that two relations of $\Psi_n(s)$ exist

$$1.\quad \Psi_n(0) = 1 \tag{5}$$

$$2.\quad \Psi_{n+1}(s) = \Psi_{n-1}(s) - s\frac{\Gamma((n+1)/2)}{\Gamma(n/2+1)}\Psi_n(s) \tag{6}$$

Therefore, we call $\Psi_n(s)$ "the Repeated Integral of Integeralization Error Complemental function."

From EQ(6), $\Psi_n(s)$ is expressed by $\Psi_i(s)$ $(i=n-1, n-2, \ldots -1)$

$$\Psi_n(s) = C_i(n,s)\,\Psi_{i-2}(s) - Di(n,s)\,\Psi_{i-1}(s)$$
$$(i=n, n-1, \ldots 1) \tag{7}$$

Apparently $C_n(n,s) = 1$, $D_n(n,s) = -s\dfrac{\Gamma(n/2)}{\Gamma(n+1)/2)}$. It is assumed that $C_{n+1}(n,s) = 0$, $D_{n+1}(s) = 1$, the recurrence formulas of $C_i(n,s)$ & $D_i(n,s)$ are therefore

$$C_i(n,s) = D_{i+1}(n,s) \tag{8}$$

$$D_i(n,s) = C_{i+1}(n,s) - s\frac{\Gamma(i/2)}{\Gamma((i+1)/2)} D_{i+1}(n,s) \tag{9}$$

$C_1(n,s)$ & $D_1(n,s)$ can be obtained from above equations. Substituting

$C_1(n,s)$ & $D1(n,s)$ to Eq(2), the inner temperature solutions is then

$$\theta(x,t)=\sum_{n=o}^{\infty} b_n t^n [C_1(2n,s)e^{-x^2/4t}+D_1(2n,s)(1-\mathrm{erf}(x/2\sqrt{t})] \tag{10}$$

It shows that the extrapolation formulas can be expressed alone by error function, where $C_1(2n,s)$ & $D_1(2n,s)$ are the "s" polynomials

$$C_1(2n,s)=\sum_{i=1}^{n} A_{2i-1}(2n)s^{2i-1} \tag{11}$$

$$D_1(2n,s)=\sum_{i=1}^{n} A_{2i}(2n)s^{2i} \tag{12}$$

The following equations can be certificated as:

$$\begin{aligned}
&A_o(2n)=1\\
&A_1(2n)=-\left(\frac{\Gamma(n)}{\Gamma(n+\frac{1}{2})}+\frac{\Gamma(n-1)}{\Gamma(n-\frac{1}{2})}+\cdots+\frac{\Gamma(1)}{\Gamma(3/2)}\right)\\
&\cdots\cdots\\
&A_{2n}(2n)=\frac{\Gamma(\frac{1}{2})}{\Gamma(n+\frac{1}{2})}
\end{aligned} \tag{13}$$

For s<1, the above formulas are convenient to be calculated approximately. If "s" is close to one (or more than one), the results with some errors would arise. With similar principle, the prediction formulas under cylinder coordinate are experessed as

$$\theta(r_1,t)=\sum_{n=o}^{\infty} b_n . t^n [C_1(2n,s_1)e^{-\frac{(r-1)^2}{4t}}+D_1(2n,s_1)\mathrm{erfc}((r_1-1)/2\sqrt{t})] \tag{14}$$

$$\theta(1,t)=\sum_{n=o}^{\infty} b_n \cdot G_n(1,t) \tag{15}$$

where

$$G_n(1,t)=\sqrt{r_1}\Gamma(n+1)t^n \sum_{m=o}^{\infty} a_m(1)t^{m/2}/(\Gamma(n+m/2+1))$$

$$s_1=(r_1-1)/2\sqrt{t}$$

$$q(1,t)=\frac{K}{R_i}\sum_{n=o}^{\infty} b_n H_n(1,t) \tag{16}$$

$$H_n(1,t)=\tfrac{1}{2}G_n(1,t)+\sqrt{r_1}\Gamma(n+1)t^n\sum_{m=o}^{\infty} C_m(1)t^{(m-1)/2}/(\Gamma(n+(m+1)/2))$$

$$a_o(r)=1; \quad a_1(r)=(r-r_1)/8rr_1;$$

$$a_2(r)=\frac{(9r^2-2r_1r-7r^2)}{128r_1^2r^2} \quad ;a_3(r)=\frac{-75r_1^3+9r_1^2r+7r_1r^2+59r^3}{1024\,r_1^3r^3}$$

$$a_4(r)=\frac{3675r_1^4+300r_1^3r-198r_1^2r^2-292r_1r^3-3485r^4}{32768r_1^4r^4}$$

$\cdots\cdots$

$$C_o(r)=1 \quad ;C_1(r)=a_1(r) \quad ;C_m(r)=a_m(r)-\frac{d}{dr}(a_{m-1}(r))$$

Where r is the dimensionless radius ($r=R/R_i$, R_i, R_o respectively, the inner and outer radii, R_1, the radius at the probe). K is the thermal conductivity of the material.

Authors can certificate that $\{t^{n/2}\Psi(s)\}$ $(n=o,1,2,\cdots)$ are the solutions of differential equation of one-dimension transient heat conduction. The common solution can be expressed as

$$\theta(x,t)=\sum_{n=o}^{\infty} b_n \cdot t^{n/2} \cdot \Psi_n(s) \tag{17}$$

$$q(x,t)=\sum_{n=o}^{\infty} b_n \cdot t^{(n-1)/2}\,\Psi_{n-1}(s) \cdot \frac{\Gamma(n/2+1)}{\Gamma((n+1)/2)} \tag{18}$$

The surface temperature and heat flux are therefore as:

$$\theta(o,t)=\sum_{n=o}^{\infty} b_n \cdot t^{n/2} \tag{19}$$

$$q(o,t)=K \cdot \sum_{n=o}^{\infty} b_n t^{(n-1)/2} \cdot \frac{\Gamma(n/2+1)}{\Gamma((n+1)/2)} \tag{20}$$

For the constant surface temperature, the coefficients b_n are:

$$b_o=\theta_o;\, b_i=0 \qquad (i=1,2,\cdots) \tag{21}$$

For the constant surface heat flux, the coefficients b_n are:

$$b_1=2q_w/\sqrt{\pi}\,k \;;\; b_i=0 \qquad (i=0,2,3,\cdots) \tag{22}$$

For the boundary condition of the surface convection, the coefficients b_n satisfy the following relations:

$$-k \cdot \left.\frac{\partial\theta}{\partial x}\right|_{x=0}=h(\theta_o-\theta(x,t)|_{x=0}) \tag{23}$$

$$\begin{cases} b_o=0; \quad b_1=\theta_o \cdot \dfrac{2h}{\sqrt{\pi}\,k} \\ b_i=-\dfrac{h}{k}\dfrac{\Gamma((i+1)/2}{\Gamma(i/2+1)} \cdot b_{i-1} \qquad (i=2,3,\cdots) \end{cases} \tag{24}$$

$$\theta(x,t)=\theta_o\sum_{n=1}^{\infty}(-1)^{n-1}\frac{1}{\Gamma(n/2+1)}\left(\frac{h}{k}\sqrt{t}\right)^n\Psi_n(x/2\sqrt{t}) \tag{25}$$

$$\theta(0,t)=\theta_o\sum_{n=1}^{\infty}(-1)^{n-1}\frac{1}{\Gamma(n/2+1)}\left(\frac{h}{k}\sqrt{t}\right)^n$$

$$=\theta_o(1-e^{(h/k)^2t} \cdot \mathrm{erfc}(\frac{h}{k}\sqrt{t})) \tag{26}$$

ALGORITHM FOR RAPID EXTRAPOLATION

The extrapolation formulas Eq(2) &Eq(10) expressed by $\Psi_{2n}(s)$ or erfc (s) hav been obtained. If the numerical evaluation of $\Psi_{2n}(s)$ is not considered yet, Eq(is convenient. Authors have computed out the $\Psi_n(s)$ numerical table. This brir about the much more convenience and shorter evaluation time for extrapolation. W the table solution, the inverse problem that takes several hours with finite di ference algorithm has been computed only for twety minutes.

Assume that the temperature measured at x_p be f(t), then

$$f(t)=\sum_{n=o}^{\infty} b_n \cdot t^n \cdot \Psi_{2n}(x_p/2\sqrt{t}) \tag{27}$$

or

$$f(t)=\sum_{n=o}^{\infty} b_n \cdot t^n[C_1(2n,s_p)e^{-s_p^2}+D_1(2n,sp)\mathrm{erfc}(s_p)] \tag{28}$$

where $s_p=X_p/2\sqrt{t}$.

The extrapolation procedure is that the coefficiente b_n is determined by Eq(27) or Eq(28), the surface temperature and heat flux are obtained by substituting b_n to Eq(1) or Eq(15). In the concrete, the sampling series of f(t) is $f(t_1), f(t_2), \cdots$. The f(t) will be approximated by an N term. The b_n is the solution of the following linear equations:

$$(\Phi^T W^2 \Phi)B = \Phi^T W^2 F \tag{29}$$

where W is a weigh function diagonal matrix

$$F=[f(t_1), f(t_2) \cdots f(t_m)]^T$$

$$B=[b_o, b_1, \cdots b_N]^T$$

$$\Phi = \begin{bmatrix} \Psi_o(Sp_1) & t\,\Psi_2(Sp_1) & \cdots\cdots & t^N\,\Psi_{2N}(Sp_1) \\ \Psi_o(Sp_2) & t\,\Psi_2(Sp_2) & \cdots\cdots & t^N\,\Psi_{2N}(Sp_2) \\ \cdots & \cdots & \cdots & \cdots \\ \Psi_o(Sp_m) & t\,\Psi_2(Sp_m) & \cdots\cdots & t^N\,\Psi_{2N}(Sp_m) \end{bmatrix}_{mx(N+1)}$$

$$W= \begin{bmatrix} w(t_1) & & 0 \\ & w(t_2) & \\ & & \ddots \\ 0 & & w(t_m) \end{bmatrix}_{mxm}$$

$S_{pi}=X_p/2\sqrt{t_i}$

Calculating Ψ_n (s_p) is a key step in this method. However, $\Psi_n(s)$ numerial table has been prepared for users in Table 1.

The accuracy of this table can be demonstrated as following:

1. $\Psi_n(0)=1$
2. The numerial values of the table satisfy Eq(5)
3. The $\Psi_o(s)$ values of this table is as same as the datum given in mathematical handbooks.

Table 1 $\Psi_n(s)$ numerial table

n \ s	0	0.4	1	2	4
0	1	0.57161	0.15730	0.46777E-2	0.15417E-7
1	1	0.44688	0.89074E-1	0.17335E-2	0.32297E-8
2	1.000000002	0.36991	0.56790E-1	0 76564F-3	0.84012E-9
⋮	⋮	⋮	⋮	⋮	⋮
28	1.00000039	0.44947E-1	0.31068E-3	0.30902E-7	0.62034E-17
29	1.00000043	0.42650E-1	0.27260E-3	0.23849E-7	0.37560E-17
30	1.00000047	0.40505E-1	0.23970E-3	0.18483E-7	0.22918E-17

DISCUSSION

A great of evaluation has been accomplished with various weigh matrixes. This algorithm is applied to the temperature measuring. The result of restore evaluation is shown in Table 2.Assume that $\Theta(0,t)=1000\sin(2\pi t/3)$;$a=0.124cm^2/s$;

x_p=0.0129 cm, $\theta'(x_p,t)$ at X_p is obtained by the direct solution. $\theta'(0,t)$ is obtained by inverting the temperature $\theta(x_p,t)$ at x_p to surface.

The schematic drawing of experimental device is shown in Fig.1. The body of heat conduction is a brass bar around which heat-insulting material is full of. The diameter of the bar is 2 cm. The diffusivity is 0.3334 cm^2/s. A top surface is heated. Both temperatures at location X_1, X_2 are measured as $\theta(x_1,t)$, $\theta(x_2,t)$. $\theta'(x_1,t)$, $\theta'(x_2,t)$ are the results calculated from $\theta(x_1,t)$, $\theta(x_2,t)$, respectively. They are shown in Fig.2.

Table 2 Results of "restore evaluation"

t(ms)	$\theta(0,t)$	$\theta(x_p,t)$	$\theta'(0,t)$
0.15	309.2	2.42	312.4
0.45	809.0	82.0	806.3
0.75	1000	224.6	1000.34
1.05	809.0	337.0	810.7
1.35	309.0	358.4	307.1

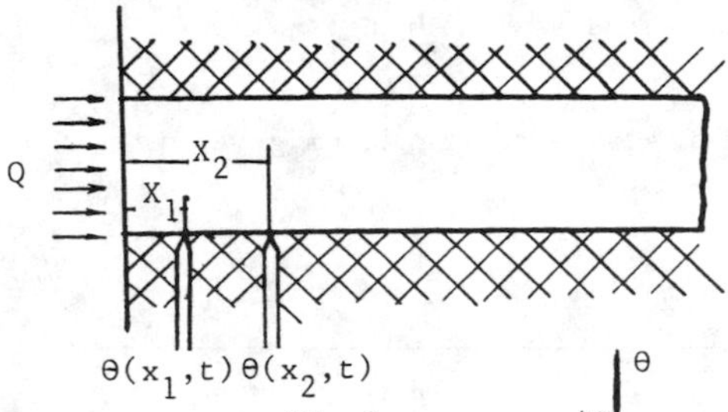

Fig.1

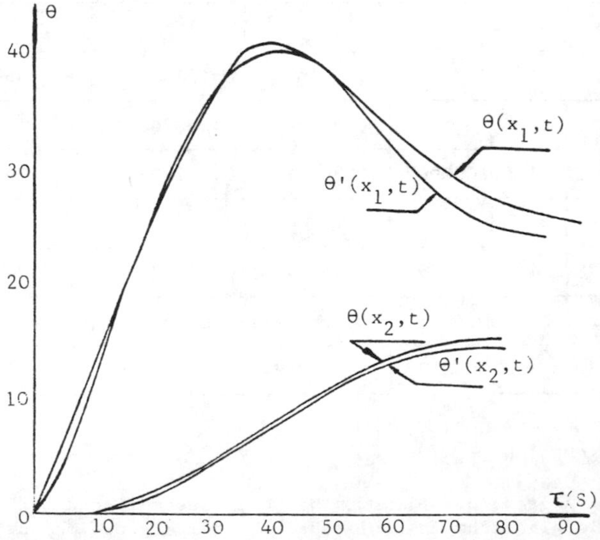

Fig.2

REFERENCE

1. Murray Imber and Jamal Khan. "Prodiction of Transient temperature Distributions with Embedded Thermocouples" AIAA Journal Vol 10.No 5 1972 pp 784-789
2. Ching Jen Chen and Darrel M.Thomsen "On Transient Cylindrical Surface Heat Flux Predicted from Interior Temperature Response" AIAA Journal Vol 13 No 5 1975 pp 697-699
3. Sparraw E.M, Hadji-Sheikh A and Lundgren T.S "The Inverse Problem in Transient Heat conduction" Journal of Applied Mechanics Vol 86 1964 pp 369-375
4. Murray Imber "A Temperature Extrapolation Method for Hollow Cylinders" AIAA Journal Vol 11.No 1 1973 pp 117-118

Enthalpy Method for the Solution of the Temperature Field During the Alloy Solidification Process

YIQIANG ZHANG
Shaanxi Institute of Mechanical Engineering
Xian, PRC

INTRODUCTION

The heat transfer problems involving melting or solidification generally re ferred to as "moving - boundary" or "phase - change" problems have numerous applications in various branches of science and engineering, so they have receive considerable attention from scholars at home and abroad and have been regarded as one of the basic subjects of heat transfer for a long time. To date, there are many methods for solving such probloms; but only by using the numerical method, the "moving - boundary" problems of multidimensions and multiphases may b solved. Among these numerical methods, the enthalpy method is the simplest to handle [1]; however, most of the papers which have been published on this subje refer to using the enthalpy method for solving the heat conduction problem involving phase - change which takes place at a discrete temperature. Using the enthalpy method for solving the alloy solidification problem which is significa in metallurgy, foundry and crystal growth is little seen. Some papers have pro posed that the enthalpy method may be used for solving the alloy solidification problem, but they do not present an enthalpy model which can be adapted to the physical change during the alloy solidification process [2]. Thus, this paper makes an attempt to use the enthalpy method for solving the alloy solidificatic problem.

During the alloy solidification which takes place over a temperature range, there exists a two - phase zone (also called a mushy region) consisting of liqu and solid between the purely solid and purely liquid phase, and there exist twc interfaces: the solid interface and the liquid interface which move with time. Furthermore, The release of the latent heat follows a certain law and occurs i the whole two - phase zone. Thus, this paper first starts from the alloy phase map, and according to the lever theorem of the alloy phase - map, deduces a rel tionship between the solid fraction within the two - phase zone and the temperature, then, sets up a mathematical model of the alloy solidification. Based o this analyisis, this paper presents an enthalpy model which reflects the law w the release of the latent heat follows, which is applicable for both eutectic solid solution alloys. Next, this paper demonstrates that the enthalpy equati corresponding to such an enthalpy model is equivalent to the governing set of differential equations which describes the heat conduction during the alloy solidification; consequently, the solution of a set of differential equations in region whose boundary moves can be transformed into the solution of an enthalp equation in a fixed region, so that the computed work can be much simplified. Finally taking the solidification of the Aluminum - silicon alloy in an infini long cylinder, this paper uses the enthalpy model in conjunction with a fully implicit finite difference scheme for solving the enthalpy equation. The nume solutions are in good agreement with the experimental data. Accompanied by the distribution of the temperature during the solidification being obtained, the location of the phase - change interfaces as a function of time can be determined, and so can the time required to complete the solidification.

(1) Physics Model and its Mathematical Description

During the alloy - solidification which takes place over a temperature range, there exists a two - phase zone, and the latent heat is released in the whole two - phase region; furthermore, the release of latent heat may be treated as the internal heat generation [4], so the internal heat generation rate (per unit volume) is

$$q = \rho r \frac{df_s}{dt} \qquad (1\text{-}1)$$

where ρ —— density, r —— latent heat, f_s —— the solid fraction within the two-phase zone. Hence, the heat conduction equation in the two - phase zone is

$$\nabla\cdot(K_t \nabla T_t) + \rho r \frac{df_s}{dt} = C_t \rho \frac{\partial T_t}{\partial t} \qquad (1\text{-}2)$$

where Tt —— the Temperature of the alloy in the two - phase zone;
Ct —— the specific heat of the alloy in the two - phase zone
Kt —— the thermal conductivity of the alloy in the two - phase zone

In the alloy phase - map shown in Fig.1, using the lever theorem , We can deduce a relationship between the solid fraction within the two - phase zone and the temperature.

For the alloy of which the constituent is c_o, when its tomperature is at Tt, it follows from the geometric relationship of Fig. 1 that

$$ON = \frac{T_2 - T_t}{m} \qquad (1\text{-}3)$$

$$BE = \frac{T_2 - T_1}{m} \qquad (1\text{-}4)$$

$$\frac{MN}{CE} = \frac{mc_o + T_2 - T_t}{mc_D} \qquad (1\text{-}5)$$

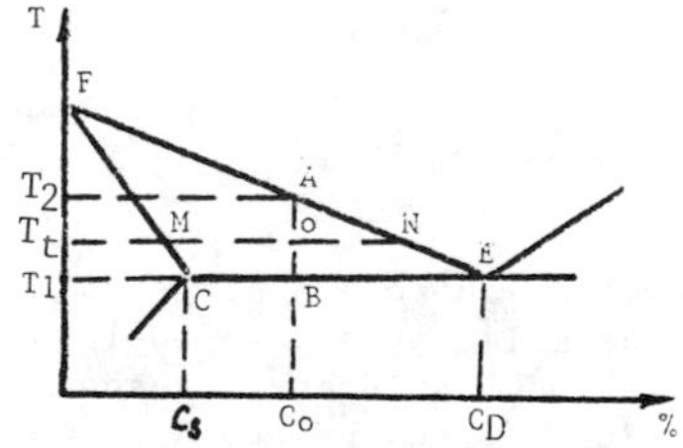

where m --- the slope of the liquid phase line
T2 -- the temperature of the alloy when the solidification begins
T1 -- the temperature of the alloy when the solidification is completed

It also follows from the lever theorem that

$$f_s = \frac{ON}{MN}, \quad f_{su} = \frac{BE}{CE} \qquad (1\text{-}6)$$

where f_{su} — solid fraction at the solid interface, for the solid solution alloy, $f_{su} = 1$; for the eutectic alloy, $f_{su} < 1$.

Hence, $$f_s = f_{su} \frac{CE}{BE} \cdot \frac{ON}{MN} \qquad (1\text{-}7)$$

Substitution of (1-3), (1-4) and (1-5) into (1-7) yields

$$f_s = f_{su} \cdot \frac{mC_D}{(T_2 - T_1)} \cdot \frac{(T_2 - T_t)}{(mc_o + T_2 - T_t)} \qquad (1\text{-}8)$$

$$\text{and} \qquad \frac{df_s}{dt} = f_{su} \cdot \frac{mC_D}{(T_2-T_1)} \cdot \frac{(-mc_o)}{(mc_o+T_2-T_t)^2} \cdot \frac{\partial T_t}{\partial t} \tag{1-9}$$

Substituting (1-9) into (1-2), we obtain

$$\nabla\cdot(K_t\nabla T_t) = [C_t+f_{su}\cdot\frac{mC_D}{(T_2-T_1)} \cdot \frac{mC_o r}{(mc_o+T_2-T_t)^2}] \cdot \rho\frac{\partial T_t}{\partial t} \tag{1-10}$$

Let $\overline{C}_t = C_t+f_{su} \cdot \frac{mC_D}{(T_2-T_1)} \cdot \frac{mc_o r}{(mc_o+T_2-T_t)^2}$, and call $\overline{C}_t$ the equivalent specific heat of the alloy in the two - phase zone; consequently, (1-10) becomes

$$\nabla\cdot(K_t\nabla T_t) = \overline{C}_t\rho\frac{\partial T_t}{\partial t} \tag{1-11}$$

The above equation (1-11) is the mathematical model and equation in the two - phase zone.

If the effects of the free convection in melting metal are not included in the analysis, that is, the different density for each phase is eliminated, the differential energy equations describing the alloy solidification can be stated as

$$\nabla\cdot(K_s\nabla T_s) = C_s\rho\frac{\partial T_s}{\partial t} \tag{1-12}$$

$$\nabla\cdot(K_t\nabla T_t) = \overline{C}_t\rho\frac{\partial T_t}{\partial t} \tag{1-13}$$

$$\nabla\cdot(K_1\nabla T_1) = C_1\rho\frac{\partial T_1}{\partial t} \tag{1-14}$$

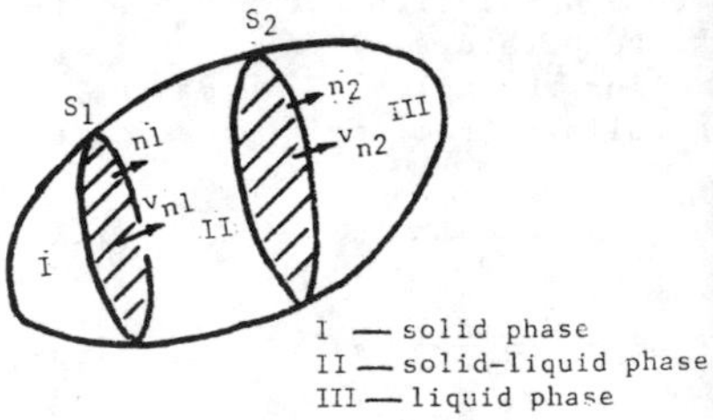

Fig.2

where the subscipts s,t and l refer to the solid phase, solid-liquid phase and liquid phase respectively.

Subject to the boundary conditions

at the outside boundary, $\qquad T(x,y,z,t) = f(t) \qquad$ (1-15)

at $z = s_1(x,y,t)$, $\qquad T_s(x,y,z,t) = T_t(x,y,z,t) = T_1 \qquad$ (1-16)

$$K_s\frac{\partial T_s}{\partial n_1} - K_t\frac{\partial T_t}{\partial n_1} = \rho r(1-f_{su})V_{n1} \tag{1-17}$$

at $z = s_2(x,y,t)$, $\qquad T_t(x,y,z,t) = T_1(x,y,z,t) = T_2 \qquad$ (1-18)

$$K_t\frac{\partial T_t}{\partial n_2} - K_1\frac{\partial T_1}{\partial n_2} = 0 \tag{1-19}$$

where $\vec{n_1}$, $\vec{n_2}$ denotes the unit normal vector at the phase - change interface S_1, S_2 respectively, their directions are shown in Fig.2; v_{n1}, v_{n2}, denotes the velocity of the interface S_1, S_2 in the $\vec{n_1}$, $\vec{n_2}$ direction respectively.

Subject to the initial condition

$$\text{at } t = 0, \quad T_1(x,y,z) = T_o \tag{1-20}$$

(2) Enthalpy Model and its Mathematical Verification

Among the numerical methods for solving the above set of equations, the enthalpy method is the simplest to handle. Thus an enthalpy function H(T) is introduced. It is defined as the sum of the sensible heat and the latent heat content. For the eutectic and solid solution alloys which are widely applied to engineering, their enthalpy model is as follows.

$$\text{at } T_s < T_1, \qquad H = \int_o^{T_s} \rho C_s T\, dT = PC_s T_s$$

$$\text{at } T_s = T_t = T_1, \quad \rho C_s T_1 \leq H \leq \rho C_s T_1 + (1 - f_{su})\rho r$$

$$\text{at } T_1 < T_t < T_2, \qquad H = \int_o^{T_1} \rho C_s T d\,T + (1 - f_{su})\rho r + \int_{T_1}^{T_t} \rho \bar{C}_t T d\,T \tag{2-1}$$

$$= \rho C_s T_1 + \rho C_t (T_t - T_1) + [1 - f_{su} \frac{mC_D}{(T_2 - T_1)} \cdot \frac{(T_2 - T_t)}{(mC_o + T_2 - T_t)}]\rho r$$

$$\text{at } T_2 < T1, \quad H - \int_o^{T_1} \rho C_s T d\,T + \int_{T_1}^{T_2} \rho C_t T d\,T + \rho r + \int_{T_2}^{T_1} \rho\, C_1 T\, dT$$

$$= \rho C_s T_1 + \rho C_t (T_2 - T_1) + \rho r + \rho C_1 (T_1 - T_2)$$

Obviously, at $f_{su} = 1$, the enthalpy model is applicable for the solid solution alloy; at $f_{su} < 1$, it is applicable for the euteotic alloy. According to the alloy phase - map and using the level theorem, we deduce the equivalant specific heat of the alloy in the two - phase zone, so the formula of the enthalpy model (2-1) reflects the law which the release of the latent heat follows during the alloy solidification.

Fig.3 shows a relationship between the enthalpy of an Aluminum - silicon alloy, whose silicon consitiuent is 7%, and the temperature. It may be seen from Fig.3 that there is a discontinuous jump in H.

In the following analyses, this paper demonstrates that the enthalpy equation in integral form together with the formula of the enthalpy model (2-1) is equivalent to a set of heat conduction equations describing the alloy solidification.

For a system of which the volume is v and the surface area is A, it may be derived from an energy balance that the enthalpy equation in integral form is

$$\frac{d}{dt}\int_V H dV = \int_A K \operatorname{grad} T \cdot \vec{n}\, dA \tag{2-2}$$

The above equation (2-2) holds true for an arbitrary region within the system. If both (2-1) and (2-2) are applied to each phase, then (1-12), (1-13) and (1-14) can be derived. Therefore what is left to prove is only that (2-1) and (2-2) satisfy the interface energy balance equations (1-17) and (1-19). Their verification is as follows:

Take a system as shown in Fig.4, which covers three phase zones. The shaded portions of the solid-lines in Fig.4 refer to the location of the interface S_1, S_2 at instant t respectively. Their area is Σ_1, Σ_2 respectively. The dash lines in Fig.4 refer to the location of S_1, S_2 at instant $t + \Delta t$ respectively. At instant t, the enthalpv value of the system is

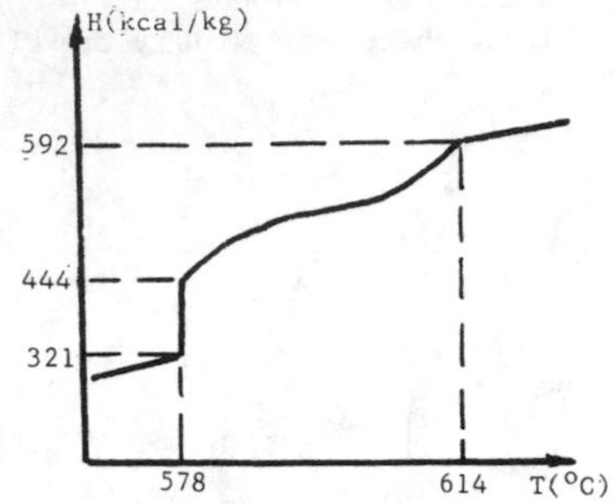

Fig.3

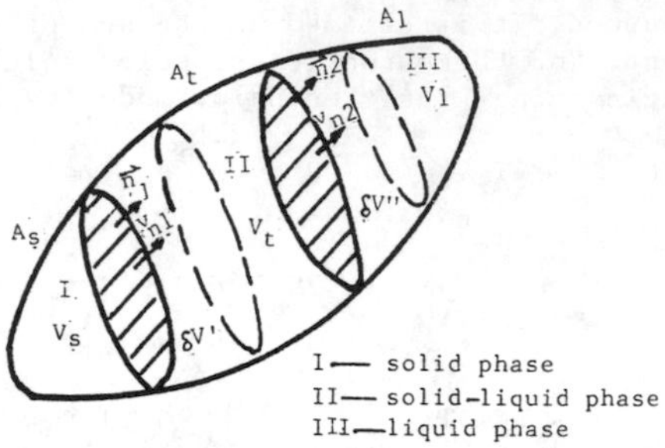

Fig.4

$$\int_V HdV = \int_{Vs} H_s dV + \int_{V_t-\delta V'} H_t dV + \int_{\delta V'} H_t dV + \int_{V_l-\delta V''} H_l dV + \int_{V''} H_l dV \quad (2\text{-}3)$$

At instant $t + \Delta t$, the enthalpy vàlue of the system is

$$\int_V HdV = \int_{V_s+\delta V'} H_s dV + \int_{V_t-\delta V'+\delta V''} H_t dV + \int_{V_l-\delta V''} H_l dV$$

$$= \int_{Vs} H_s dV + \int_{\delta V'} H_s dV + \int_{V_t-\delta V'} H_t dV + \int_{\delta V''} H_t dV + \int_{V_l-\delta V''} H_l dV \quad (2\text{-}4)$$

in addition, according to the definition of the limit of function, we can write the following formula

$$\frac{d}{dt}\int_V HdV = \lim_{\delta t \to 0} \frac{(\int_V HdV)_{t+\delta t} - (\int_V HdV)_t}{\delta t} \quad (2\text{-}5)$$

Then, substituting (2-3) and (2-4) into (2-5), and noticing that $\delta t \to 0$, $(V_t \to \delta V') \to V_t$, and $(V_l - \delta V'') \to V_l$, we obtain

$$\frac{d}{dt}\int_V Hd\,V = \frac{d}{dt}\int_{Vs} H_s dV + \frac{d}{dt}\int_{Vt} H_t dt + \frac{d}{dt}\int_{Vl} H_l dV +$$

$$+\lim_{\delta t \to 0}\int_{\delta V'} \frac{(H_{s.t+\delta t} - H_{t,t})}{\delta t} dV + \lim_{\delta t \to 0}\int_{\delta V''} \frac{(H_{t,t+\delta t} - H_{l,t})}{\delta t} dV \quad (2\text{-}6)$$

Because at $\delta t \to 0$. $\delta V' \to \Sigma_1, \frac{dV}{\delta t} \to v_{n1} d\Sigma_1$ and $(H_{t,t} - H_{s,t+\delta t})$

$\rightarrow (1-f_{su})\rho r$, the fourth term in the right - hand side of equation (2-6) becomes

$$\lim_{\delta t \to 0} \int_{\delta V'} \frac{(H_{s.t+\delta t} - H_{t,t})}{\delta t} dV = - \int_{\Sigma_1} (1-f_{su}) \rho r v_{n1} d\Sigma \qquad (2\text{-}7)$$

likewise, we also obtain

$$\lim_{\delta t \to 0} \int_{\delta V''} \frac{(H_{t,t+\delta t} - H_{1,t})}{\delta t} dV = 0 \qquad (2\text{-}8)$$

Thus, by introducing (2-7) and (2-8) into (2-6), equation (2-6) becomes

$$\frac{d}{dt}\int_V H dV = \frac{d}{dt}\int_{Vs} H_s dV + \frac{d}{dt}\int_{Vt} H_t dt + \frac{d}{dt}\int_{V1} H_1 dV - \int_{\Sigma_1} (1-f_{su}) \rho r v_{n1} d\Sigma \qquad (2\text{-}9)$$

Next, applying the first law of thermodynamics to each phase respectively, we may also write equation (2-9) as

$$\frac{d}{dt}\int_V H dV = \int_{A_s+\Sigma_1} (K\nabla T)_s \vec{n} dA + \int_{A_t+\Sigma_1+\Sigma_2} (K\nabla T)_t . \vec{n} dA + \int_{A_1+\Sigma_2} (K\nabla T)_1 . \vec{n} dA - \int_{\Sigma_1} (1-f_{su}) \rho r v_{n1} d\Sigma \qquad (2\text{-}10)$$

Noticing that at equation (2-10), $\vec{n}$ is the outward unit normal vector in each phase zone, we conclude that (1-10) can be written as

$$\frac{d}{dt}\int_V H dV = \int_{A_s+A_t+A_1} (K\nabla T) \vec{n} dA + \int_{\Sigma_1} [(K\frac{\partial T}{\partial n_1})_s - (K\frac{\partial T}{\partial n_1})_t - (1-f_{su}) \rho r v_{n1}] d\Sigma$$

$$+ \int_{\Sigma_2} [(K\frac{\partial T}{\partial n_2})_t - (K\frac{\partial T}{\partial n_2})_1] d\Sigma \qquad (2\text{-}11)$$

Then subtracting (2-11) from (2-2), we deduce

$$\int_{\Sigma_1} [(K\frac{\partial T}{\partial n_1})_s - (K\frac{\partial T}{\partial n_1})_T - (1-f_{su}) \rho r v_{n_1}] d\Sigma + \int_{\Sigma_2} [(K\frac{\partial T}{\partial n_2})_t - (K\frac{\partial T}{\partial n_2})_l] d\Sigma = 0 \qquad (2\text{-}12)$$

Because $(K\frac{\partial T}{\partial n})$ is continuous at the interface S_2, we have

$$(K\frac{\partial T}{\partial n_2})_t - (K\frac{\partial T}{\partial n_2})_1 = 0 \qquad (1\text{-}19)$$

Consequently, it follows that

$$(K\frac{\partial T}{\partial n_1})_s - (K\frac{\partial T}{\partial n_1})_t - (1-f_{su}) \rho r v_{n1} = 0 \qquad (1\text{-}17)$$

It may be seen from the above verification that the solution of a set of differential equations in a region whose boundary moves can be transformed into the solution of an equation in a fixed region.

(3) Numerical Solution of the Enthalpy Equation

Atthey proved that the numerical solution of a differential enthalpy equation converges, as the steps tend to zero, to a weak solution of the integral enthalpy equation [2]; Consequently, for the solidification of the Aluminum - silicon alloy in an infinitely long cylinder, this paper uses the finite difference method to solve the differential enthalpy equation.

First, substituting the values of the physical properties of the Aluminum - silicon alloy into (2-1), we obtain its enthalpy farmula

$$H=\begin{cases} 0.5565T & T < 578^{\circ}C \\ (321.7,\ 444.5) & T = 578^{\circ}C \\ 0.5565T + \left[250.7 - \dfrac{369.8 \times (614.0 - T)}{(659.4 - T)}\right] & 578^{\circ}C < 614^{\circ}C \\ 250.7 + 0.5565T & 614^{\circ}C \leq T \end{cases} \quad (3\text{-}1)$$

At the same time, the solidification of the Aluminum - Silicon alloy in an infinitely long cylinder satisfies the following enthalpy equation and its boundary and initial conditions

$$\begin{cases} \dfrac{\partial H}{\partial t} = \dfrac{1}{r} \dfrac{\partial}{\partial r} \left(Kr \dfrac{\partial T}{\partial r}\right) \\ t = 0,\ T = T_o \\ r = R,\ \ T = f(t) \end{cases} \quad (3\text{-}2)$$

The boundary and initial conditions can be determined according to measurements obtained by experiment. Next, using a fully implicit finite difference scheme, we carry out the discretization of the enthalpy equation and obtain its discretization equations:

$$\begin{cases} H_i^{n+1} + \dfrac{2K\Delta t}{(\Delta r)^2} \cdot T_i^{n+1} = H_i^n + K\Delta t\left[\left(\dfrac{1}{(\Delta r)^2} - \dfrac{1}{2r_i\Delta r}\right)T_{i-1}^{n+1} + \left(\dfrac{1}{(\Delta r)^2} + \dfrac{1}{2r_i\Delta r}\right)T_{i+1}^{n+1}\right] \quad i=1,2,\ldots n \\ H_o^{n+1} + \dfrac{4K\Delta t}{(\Delta r)^2} T_o^{n+1} = H_o^n + \dfrac{4K\Delta t}{(\Delta r)^2} T_1^{n+1} \end{cases} \quad (3\text{-}3)$$

Then, we use the step - marching method for solving in time coordinate, and the Gauss - siedel iterative method or overrelaxation iterative method for solving in the space coordinates. The numerical computation is performed on the microcomputer. The above form of the discretization equations ensures that the diagonal elements of the coefficient matrix have dominance over other elements, so that the iteration converges. The comparison of the numerical solutions with the experimental data is as follows.

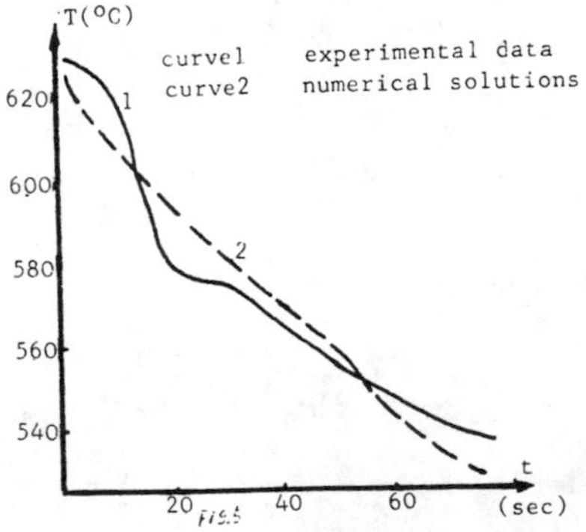

Fig.5

This experiment was performed in the Foundry laboratory. In the experiment, the Aluminum - silicon alloy (of which the silicon consitituent is 7% and the superheated temperature is 628°C) is poured into the cylindrical metal mould whose diameter is 45mm and length is 400mm. At two ends of the mould, an insulation is added to ensure that the heat transfer, in the middle region of the casting relative to the axial direction, is performed along the direction of radius. The thermocouples are embeded at the location of $r = 27$mm and $r = 45$mm respectively. The values of the temperature history measured at the location of $r = 45$mm, are treated as the boundary condition of this example; the ones at the location of $r = 27$mm are used to compare with the numerical solutions. Fig.5 shows their comparison. Obviously, the numerical solutions are in good agreement with the experimental data. Accompanied by the distribution of temperature during the solidification being obtained, the location of the phase - change interfaces at any instant can be determined, and so can the time required for the solidification to be completed.

(4) Conclusions

1. Among the methods for solving the alloy solidification problem the enthalpy method is the simplest to handle.

2. The enthalpy model deduced by this paper reflects the law which the release of the latent heat follows, and is applicable for both eutectic and solid solution alloys; furthermore, the numerical solutions based on this enthalpy model are in good agreement with the experimental data. Consequently, this mathematical model discribing the alloy solidifidation and this enthalpy model used for numerical calculation provide a reference for mathematical simulation and computer control of the alloy solidificaiton process in the areas of metallurgy, foundry and crystal growth.

3. This calculation supposes that the properties of thermophysics of an alloy are constant and adopts their valaes at a low temperature; for the values of the properties of thermophysics of an alloy at a high tomperature are not referred to in various engineering brochures. If we can determine their values at a high temperature, we could make the numerical solutions more satisfactory.

Finally, I wish to express here my thanks to Mr. Shen Chengli for his assistance in performing this experiment.

REFERENCES

[1] R.W.Lewis, Numerical methods in Heat Transfer, 1981

[2] D.R.Atthey, A finite difference scheme for melting problems J. Institute of mathematics and its application pp353-366, 1974.

[3] M.C. Flemings, Solidification Processing 1974.

[4] M.N.Ozisik Exact solution for freezing in cylindrical symmetry with Extend Freezing temperature range, J. Heat Transfer, pp 331-334, 1979.

[5] A.B.Crowley, Numerical Solution of Stefan Problems Int J. Heat mass Transfer, pp215-218, 1978.

[6] N.Shamsundar, Analyses of multidimensional Conduction phase change Via the enthalpy model, J. Heat Transfer, pp 333-340, 1975.

NATURAL HEAT CONVECTION

Heat Transfer by Natural Convection of Compressible Fluids with Temperature-Dependent Properties inside Quadratic Enclosures

H. D. BAEHR
Institut für Thermodynamik
Universität Hanover
FRD

M. THIELEMANN
Baubehörde der Freien und Hansestadt Hamburg
FRD

INTRODUCTION

Problems in buoyancy-driven laminar flow and heat transfer are usually solved by using the well-known Boussinesq-approximation (BA). Here, the fluid is assumed as incompressible and only a linear temperature-dependence of density is taken into account in the buoyancy-term of the momentum equation. The aim of the present work is to test the validity of the BA in the case of laminar natural convection inside enclosures with quadratic cross-sections (square cavities). This problem comprises the solution of the nonlinear partial differential equations representing the balances of mass, momentum and energy for compressible laminar flow considering density and also viscosity and heat conductivity as temperature-dependent properties of the fluid.

Solutions making use of the BA are much easier to achieve and have been given by several authors [1, 2]. The buoyancy-driven flow in a square cavity with vertical sides which are differentially heated has also been considered as a suitable case for testing and validating numerical methods and computer codes [2, 3]. Solutions of the problem without invoking the BA have been given by only a few authors [4, 5]. These investigations are restricted to gases and show appreciable deviations from the results found by BA only for very large temperature differences across the cavity.

THE PROBLEM AND ITS MATHEMATICAL FORMULATION

We consider the two-dimensional flow of a fluid in an upright square cavity, fig. 1. The vertical walls are kept at constant temperatures T_h and $T_c < T_h$. The horizont walls are adiabatic. The horizontal component u and the vertical component v of the velocity $\vec{v}$ are zero on all the boundaries.

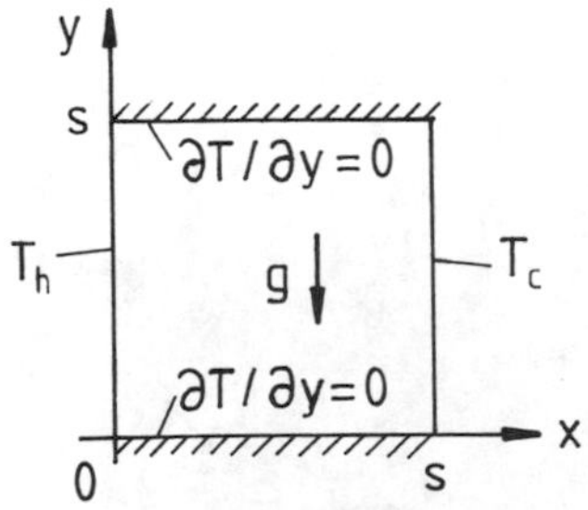

FIGURE 1. Square cavity of width s with vertical walls at different temperatures T_h and T_c

In the balance-equations for mass, momentum and energy, the thermophysical properties of the fluid, density ρ, viscosity μ and heat conductivity λ are considered as functions of temperature T, but the pressure dependence of these quantities will be neglected. The form of the governing equations taking full account of the temperature-dependence of the fluid properties is given in [6].

We use the stream function-vorticity formulation of the equations. We choose the quantities s, μ_o, $\rho_o s/\mu_o$ and $\rho_o s^2/\mu_o$ as scale factors for length, stream function Ψ, velocities u,v and vorticity ω respectively. Density and viscosity are scaled by their values ρ_o and μ_o at the reference temperature

$$T_o = \frac{1}{2}(T_h + T_c) \tag{1}$$

We then get the following equations with nondimensional variables marked by an asterisk:

$$-\omega^* = \frac{\partial}{\partial x^*}\left(\frac{1}{\rho^*}\frac{\partial \Psi^*}{\partial x^*}\right) + \frac{\partial}{\partial y^*}\left(\frac{1}{\rho^*}\frac{\partial \Psi^*}{\partial y^*}\right) \tag{2}$$

and

$$\frac{\partial}{\partial x^*}(\rho^* u^* \omega^*) + \frac{\partial}{\partial y^*}(\rho^* v^* \omega^*) + \frac{\partial \rho^*}{\partial x^*}\left(u^*\frac{\partial v^*}{\partial x^*} + v^*\frac{\partial v^*}{\partial y^*}\right) - \frac{\partial \rho^*}{\partial y^*}\left(u^*\frac{\partial u^*}{\partial x^*} + v^*\frac{\partial u^*}{\partial y^*}\right) =$$

$$- Ga\frac{\partial \rho^*}{\partial x^*} + \frac{\partial}{\partial x^*}\left(\mu^*\frac{\partial \omega^*}{\partial x^*}\right) + \frac{\partial}{\partial y^*}\left(\mu^*\frac{\partial \omega^*}{\partial y^*}\right) + \frac{\partial \mu^*}{\partial x^*}\left(\frac{\partial^2 v^*}{\partial x^{*2}} + 2\frac{\partial^2 v^*}{\partial y^{*2}} + \frac{\partial^2 u^*}{\partial x^*\partial y^*}\right)$$

$$- \frac{\partial \mu^*}{\partial y^*}\left(\frac{\partial^2 u^*}{\partial x^{*2}} + 2\frac{\partial^2 u^*}{\partial y^{*2}} + \frac{\partial^2 v^*}{\partial x^*\partial y^*}\right) + 2\frac{\partial^2 \mu^*}{\partial x^*\,\partial y^*}\left(\frac{\partial v^*}{\partial y^*} - \frac{\partial u^*}{\partial x^*}\right)$$

$$+ \left(\frac{\partial^2 \mu^*}{\partial x^{*2}} - \frac{\partial^2 \mu^*}{\partial y^{*2}}\right)\left(\frac{\partial u^*}{\partial y^*} + \frac{\partial v^*}{\partial x^*}\right) \tag{3}$$

with

$$u^* = \frac{1}{\rho^*}\frac{\partial \rho^*}{\partial y^*} \quad \text{and} \quad v^* = -\frac{1}{\rho^*}\frac{\partial \rho^*}{\partial x^*}$$

and $Ga = g\rho_o^2\, s^3/\mu_o^2$ as Galilei-number.

In the energy equation a linear temperature-dependence of specific enthalpy h is assumed:

$$h = h_o + c_p\,(T - T_o). \tag{4}$$

We neglect the dissipation-function and the term $\vec{v}$ grad p (with p as pressure) which are much smaller than the other terms. With these simplifying assumptions one gets for the non-dimensional temperature

$$\theta = (T - T_c)/(T_h - T_c) \tag{5}$$

the differential equation

$$\frac{\partial}{\partial x^*}(\rho^* u^* \theta) + \frac{\partial}{\partial y^*}(\rho^* u^* \theta^*) = \frac{1}{Pr}\left[\frac{\partial}{\partial x^*}\left(\lambda^*\frac{\partial \theta}{\partial x^*}\right) + \frac{\partial}{\partial y^*}\left(\lambda^*\frac{\partial \theta}{\partial y^*}\right)\right] \tag{6}$$

Here λ^* is the non-dimensional heat conductivity scaled by λ_o, and $Pr = \mu_o c_p/\lambda_o$ is the Prandtl number.

The boundary condition are

$\Psi^* = 0$ on all boundaries,

$\omega^* = -\frac{1}{\rho^*}\frac{\partial^2 \Psi^*}{\partial y^{*2}}$ on $y^* = 0$ and $y^* = 1$,

$\omega^* = -\frac{1}{\rho^*}\frac{\partial^2 \Psi^*}{\partial x^{*2}}$ on $x^* = 0$ and $x^* = 1$,

$\theta = 1$ on $x^* = 0$,

$\theta = 0$ on $x^* = 1$,

$\partial\theta/\partial y^* = 0$ on $y^* = 0$ and $y^* = 1$.

The three partial differential equations (2), (3) and (6) must be solved simultaneously with ρ^*, μ^* and λ^* as temperature functions characteristic of the fluid.

THE NUSSELT-NUMBER

From the viewpoint of an engineer the most important characteristic of the problem is the rate of heat transfer across the cavity. For the heat transfer rate divided by the depth of the cavity, one gets

$$\dot{Q}' = -\lambda(T_h)\int_0^s (\partial T/\partial x)_{x=0}\, dy = -\lambda(T_c)\int_0^s (\partial T/\partial x)_{x=s}\, dy \qquad (7).$$

Defining the average Nusselt-number by

$$Nu = \frac{\dot{Q}'}{\lambda_o(T_h - T_c)} \qquad (8)$$

we get from (7)

$$Nu = -\lambda^*(\theta=1)\int_0^1 (\partial\theta/\partial x^*)_{x^*=0}\, dy^* = -\lambda^*(\theta=0)\int_0^1 (\partial\theta/\partial x^*)_{x^*=1}\, dy^* \qquad (9)$$

In the limit of pure heat conduction, Nu = 1 holds. The equality of the heat transfer rates at both the hot and cold walls is a sensitive criterion for testing the accuracy of the numerical solution of the combined flow and heat transfer problem.

Within the scope of the BA, the Nu number depends on two other quantities only:

$$Nu = F_{BA}(Gr, Pr) \qquad (10)$$

with $Gr = Ga\, \beta_o(T_h - T_c)$ as Grashof number where β_o denotes the coefficient of thermal expansion at the temperature T_o. In most cases and for a relatively wide range of Pr numbers, Gr und Pr appeare in (10) only as product $Ra = Gr \cdot Pr$, the Rayleigh number.

In the general case of variable properties treated here, Nu will not only depend on Ra or on Gr and Pr, but a dependence on the non-dimensional temperature difference $\beta_o(T_h - T_c)$ and on further parameters appearing in the property-functions of the fluid is to be expected:

$$Nu = F_{VP}(Gr,\ Pr,\ \beta_o(T_h - T_c),\ \text{param. of property-funct.}) \qquad (11).$$

Moreover, the functions F_{BA} in (10) and F_{VP} in (11) will assume different forms.

THE SOLUTION

The partial differential equations (2), (3) and (6) have been solved by finite differences methods. A 44 x 25 grid has been used with the finer mesh size in the x-direction. Second upwind differences were employed and the difference equations have been solved by a successive overrelaxation (SOR) method. Details of the methods used and the calculations performed can be found in [7].

Calculations have been performed for air, water and oil. The thermal properties of these fluids have been represented by temperature polynominals. The reference temperature T_o was 293 K for air and water and 333 K for oil. For 27 values of the Gr number, first the solution applying the BA was calculated. At each value of Gr, solutions of the equations for compressible flow with variable properties were calculated using different values of $(T_h - T_c)$ between 0.5 K and 64 K. In order to keep the value of the Gr number constant, the cavity width s and $(T_h - T_c)$ have been selected in such a way that $s^3 (T_h - T_c)$ remained unchanged.

RESULTS

The results for the average Nu number using the BA could be represented by

$$F_{BA}(Gr,\ Pr) = F_{BA}(Ra) = 0.124\ Ra^{0.314} \qquad (12)$$

with an average deviation of 1,2 % and a maximum deviation not exceeding 3,2 %. Equ. (12) agrees well with the bench mark solution [3] except for $Ra = 10^6$ wherer the deviation is about 8 %.

The results of our calculations for compressible flow with variable properties deviate considerable from the BA especially at low Ra numbers. It is interesting to note that the Nu numbers calculated at the same Gr number with different $(T_h - T_c)$ differ only slightly, in all cases much less than 1 %. Therefore, in equ. (11) the explicit dependence of Nu on $\beta_o(T_h - T_c)$ can be neglected for all practical purposes. It was also possible to correlate the results for the three fluids with different Pr numbers by an equation with the Ra number as the only variable:

$$F_{VP}(Gr,\ Pr,\ \ldots) = F_{VP}(Ra) = 0.075\ Ra^{0.35} \qquad (13)$$

The average devation from (13) is 1.4 %, the maximum deviation is less than 4 %. In fig. 2 the results for the average Nu number are shown graphically.

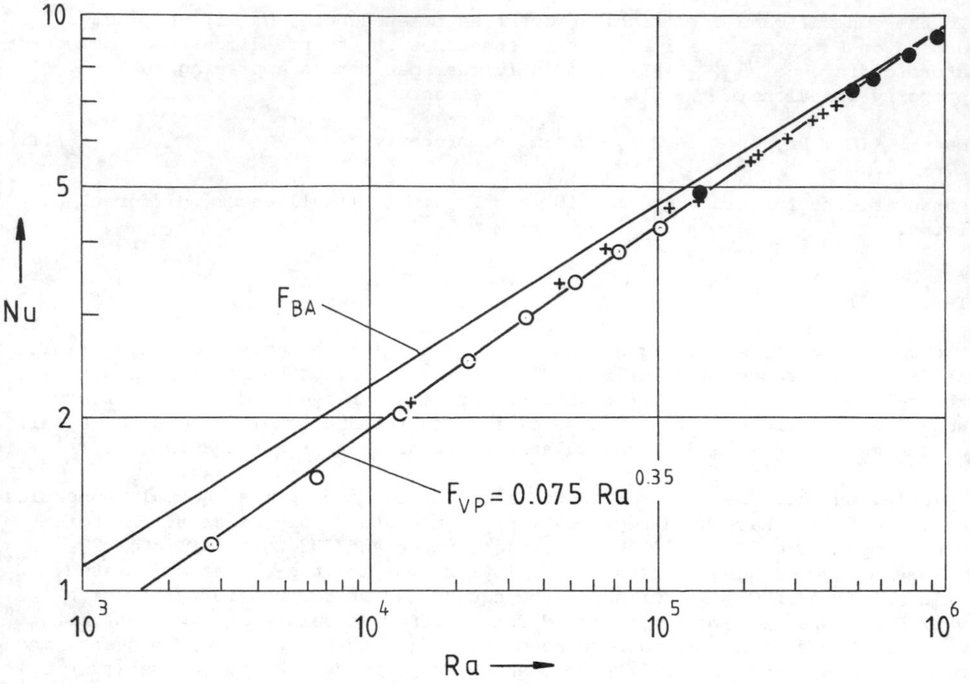

FIGURE 2. Average Nu number defined by equ.(8) as function of Ra number. Results for variable properties: o air with Pr = 0.70, + water with Pr = 6.9, • oil with Pr = 59.4

The main conclusion to be drawn from our study is that the BA gives Nu numbers which are to high at smaller Ra numbers compared with the solution of the complete equations for compressible flow with variable fluid properties. The consideration of the temperature dependence of the properties seems to be of minor influence. This has been shown by additional calculations with combinations of constant and variable viscosity and heat conductivity. The average Nu numbers obtained in all these cases at constant Ra differed less them 1 % from the Nu number calculated with the correct temperature dependence of all properties. But there always existed a marked difference to the Nu number obtained by the BA.

At smaller Ra numbers the BA exaggerates the buoyancy-effects on the velocity and temperature fields. This is shown in fig. 3 where for air at Gr = 3954 temperature profiles are compared. The profiles calculated by the BA are steeper at the verticals walls than the profiles calculated for variable properties thus giving higher Nu number. In fig. 4 the velocity for y/s = 0.5 are shown. Here again the BA overestimates the vertical velocities.

The results of this work do not agree with results of other authors [4, 5] who report only small deviations of the Nu numbers from those obtained by using the BA. The cause of this discrepancy is not known. Unfortunately, experimental results for this problem are not available.

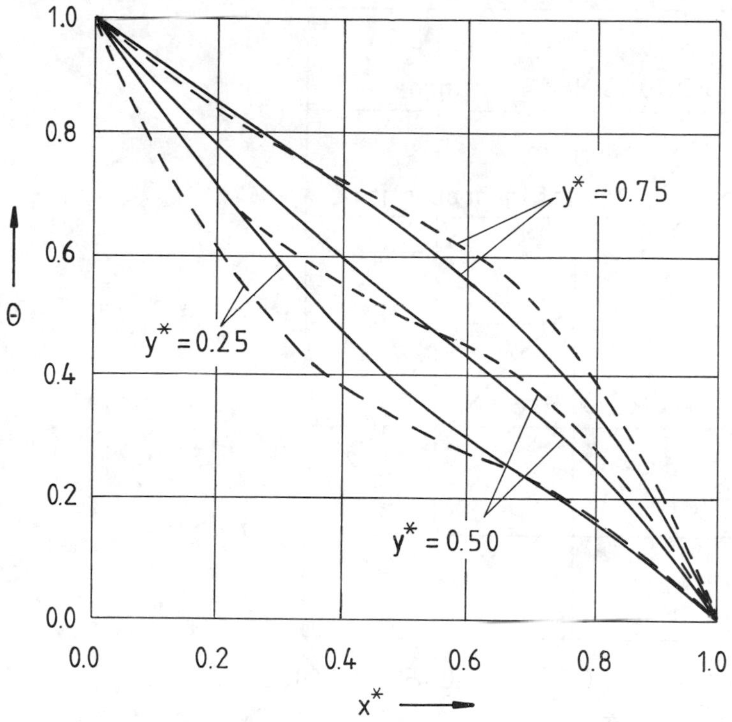

FIGURE 3. Temperature profiles for air at Gr = 3954 and $T_h - T_c = 64$ K.
—— variable fluid properties, --- Boussinesq-approximation

REFERENCES

1. Ostrach, S., Natural Convection Heat Transfer in Cavities and Cells, Proc.Int.Heat Transfer Conf., pp. 365 - 379, Hemisphere, Washington, D.C. 1982

2. De Vahl Davis, G., Jones, I.P., Natural Convection in a Square Cavity: a Comparison Exercise, Int. J. Num. Methods Fluids, vol. 3, pp. 227 - 248, 1983

3. De Vahl Davis, G., Natural Convection of Air in a Square Cavity: a Benchmark Numerical Solution, Int. J. Num. Methods Fluids, vol. 3, pp. 249 - 264, 1983

4. Leonardi, E., Reizes, J.A., Convective Flows in Closed Cavities with Variable Fluid Properties, Numerical Methods in Heat Transfer, ed. R.W. Lewis, K. Morgan and O.C. Zienkiewicz, J. Wiley and Sons, New York, 1981

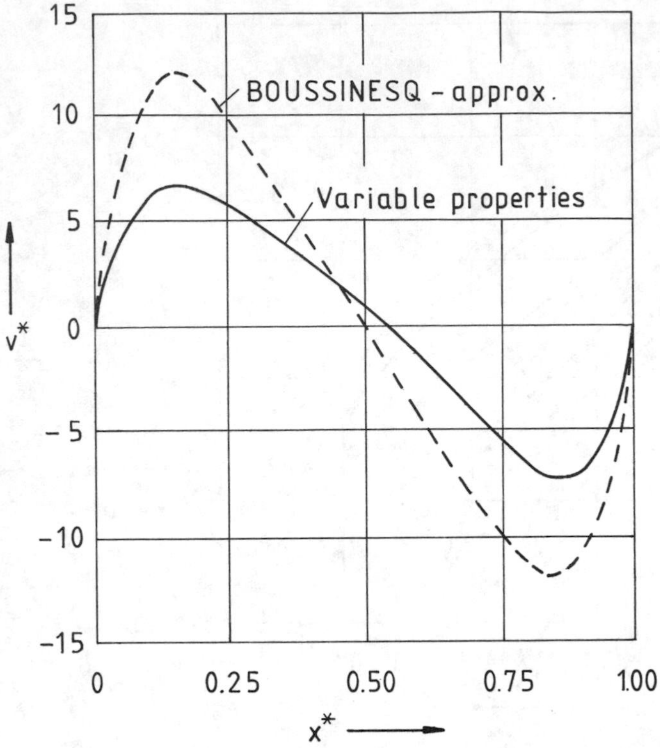

FIGURE 4. Dimensionless vertical velocity v* at y* = 0.5 for air at Gr = 3954 and $T_h - T_c = 64$ K

5. Markatos, N.C., Percleous, K.A., Laminar and Turbulent Natural Convection in an Enclosed Cavity, Int. J. Heat Mass Transfer, vol. 27, no. 5, pp. 755 - 772, 1984

6. Bird, R.B., Stewart, W.E., Lightfoot, E.N., Transport Phenomena, John Wiley and Sons, New York, 1960

7. Thielemann, M., Berechnung der freien Konvektion in Hohlräumen mit quadratischem Querschnitt unter Berücksichtigung temperaturabhängiger Stoffwerte, Dissertation Hochschule d. Bundeswehr Hamburg, 1985.

Studies on Natural Convection Heat Transfer from Horizontal Cylinders in a Vertical Array

ISHIHARA ISAO and KATSUTA KATSUTARO
Department of Mechanical Engineering
Kansai University
Osaka, Japan

1. INTRODUCTION

Experimental studies on natural convection heat transfer from heated horizontal cylinders in a vertical array were carried out by Marsters[1] and other investigators. However, the theoretical analysis of the heat transfer from the cylinders immersed in the buoyant plume has not been reported, since the thermal and hydrodynamic characteristics of the plume above the cylinder are not well known except the case with a horizontal line heat source[2,3].

In this paper, both analytical and experimental results were presented. The cylinders used were made of copper, 500 mm long and 50.8 mm diameter, and heated isothermally. The maximum number of cylinder was six in an array.

In the case of two-cylinder array, local and average Nusselt numbers for the bottom and top cylinders were measured for variety of spacings by means of Schlieren and calorimetric methods respectively. In the ones of more than two cylinders, the average Nusselt numbers at close spacing were obtained.

In the analytical study, Levy's integral method[4] was applied for the boundary layers of cylinders in a closely spaced array. The theoretical values for the top cylinder in the array agreed well with the experiment.

2. THEORETICAL ANALYSIS

To solve the bounbary layer equation for cylinders in a closely spaced array by the Levy's integral method, following assumptions were made:

(1) Heat transfer characteristics of the cylinder is not influenced by the presence of cylinder above it.
(2) Hot buoyant plume separated from the cylinder smoothly joins into the boundary layer flow of the neighbouring upper cylinder at its stagnant point keeping to hold the plume temperature.
(3) When joining, the momentum of rising plume is not transfered to the above boundary layer flow.
(4) Boussinesq approximation can be applied to calculation.

2.1 Basic Equations

Fig.1 shows the coordinate system used in a vertical array. According to the boundary layer approximations, the momentum and energy equations for the horizontal cylinder with constant temperature yield integral forms [4] as follows:

$$[\nu\frac{\partial^2 u}{\partial y^2}]_0^{\delta} + g\beta\cos\phi\int_0^{\delta}\frac{\partial(T-Ta)}{\partial x}dy - g\beta(Ts-Ta)\sin\phi - g\beta\sin\phi\frac{d\phi}{dx}\int_0^{\delta}(T-Ta)dy = 0 \quad (1)$$

$$\frac{\partial}{\partial x}\int_0^{\delta t}u(T-Ta)dy - a[\frac{\partial(T-Ta)}{\partial y}]_0^{\delta t} = 0 \quad (2)$$

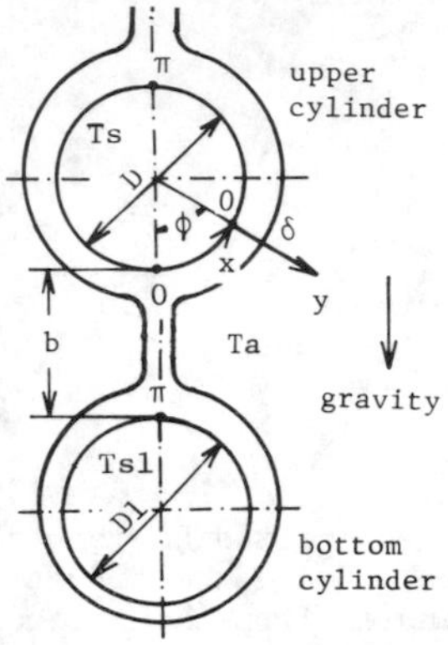

FIGURE 1. Coordinate system in vertical array

Herein, the velocity and temperature profiles propos[ed] by Eckert are employed.

$$u = u_x\frac{y}{\delta}(1-\frac{y}{\delta})^2 \quad (3)$$

$$\frac{T-Ta}{Ts-Ta} = (1-\frac{y}{\delta_t})^2 \quad (4)$$

Substituting equations (3) and (4) into the integral equations (1) and (2), the following equation is obtained from assuming $\delta=\delta_t$.

$$\frac{d}{d\phi}(\frac{\delta}{D})^4 + \frac{4}{3}\cot\phi(\frac{\delta}{D})^4 - \frac{240}{GrPr}\frac{1}{\sin\phi} = 0 \quad (5)$$

where $Gr = g\beta(Ts-Ta)D^3/\nu^2$ and $\phi = 2x/D$.

Integration of equation (5) yields

$$(\frac{\delta}{D})^4 = \frac{1}{\sin^{4/3}\phi}[\frac{240}{GrPr}\int_0^{\phi}\sin^{1/3}\phi d\phi + C] \quad (6)$$

2.2 Heat Transfer Coefficients

For the bottom cylinder. In accordance with the assumption (1), the bottom cylinder has the same heat transfer characteristics as in the case of a single horizo[n]tal cylinder, hence integral constant C in equation (6) is equal to zero. Consequently the local heat transfer coefficient for the bottom cylinder becomes

$$\frac{Nu_1}{\sqrt[4]{GrPr}} = \frac{2}{\sqrt[4]{240}}\sin^{1/3}\phi[\int_0^{\phi}\sin^{1/3}\phi d\phi]^{-1/4} \quad (7)$$

and numerical integration gives an average value

$$\frac{Nu_{m1}}{\sqrt[4]{GrPr}} = 0.44 \quad (8)$$

For the upper cylinder. The heat transfer coefficient for the upper cylinder (i.e. the second cylinder) can be determined by estimating the integral constant. Denoting W as the mass flow rate of boundary layer departed from the top of bott[om] cylinder ($\phi=\pi$),

$$\frac{W}{2} = [\int_0^{\delta}\rho u dy]_{1,\phi\to\pi} = (\frac{\rho u_x\delta}{12})_{1,\phi\to\pi} \quad (9)$$

is given. Heat balance at the bottom cylinder results in

$$\pi D_1\alpha_{m1}(Ts_1-Ta) = [2\int_0^{\delta}\rho C_p u(T-Ta)dy]_{1,\phi\to\pi} = [\frac{1}{15}\rho Cp\delta u_x(Ts_1-Ta)]_{1,\phi\to\pi} \quad (10)$$

Therefore

$$\frac{W}{2} = 5\pi D_1\alpha_{m1}/(4Cp) \quad (11)$$

Furthermore, W is equivalent to the mass flow rate in the bounbary layer at the stagnant point of the upper cylinder (ϕ=0). Then

$$\frac{W}{2} = (\rho u_m \delta)_{2,\phi\to 0} \tag{12}$$

is presented.

The velocity profile along a heated and horizontal plate facing downward [5] is given by

$$u = \frac{g\beta(Ts-Ta)}{4\nu}\sin\phi\delta^2(\frac{y}{\delta})(1-\frac{y}{\delta})^2 \tag{13}$$

and integration of equation (13) yields an average velocity

$$u_m = g\beta(Ts-Ta)\sin\phi\delta^2/(48\nu) \tag{14}$$

Substituting equation (14) into equation (12) leads to

$$(\frac{\delta}{D})^4_{2,\phi\to 0} = [\frac{W}{2}\frac{48\nu}{g\beta(Ts-Ta)\rho\sin\phi}\frac{1}{D^3}]^{4/3} \tag{15}$$

Equation (6) can be rewritten as follows:

$$(\frac{\delta}{D})^4_{2,\phi\to 0} = \frac{180}{GrPr} + \frac{C}{\sin^{4/3}\phi} \simeq \frac{C}{\sin^{4/3}\phi} \tag{16}$$

Correspondence between equation (15) and (16) leads to

$$C = (24W/\mu)^{4/3}Gr^{-4/3} \tag{17}$$

The constant C is obtained from substituting equation (11) into the above.

$$C = (60\pi\ Nu_{m1})^{4/3}(GrPr)^{-4/3} \tag{18}$$

where Nu_{m1} is the average Nusselt number for the bottom cylinder and is defined by $Nu_{m1} = \alpha_{m1}D_1/\lambda$.

Hence the boundary layer thickness on the upper cylinder is given by equation (6).

$$(\frac{\delta}{D})^4 = \frac{1}{\sin^{4/3}\phi}[\frac{240}{GrPr}\int_0^{\phi}\sin^{1/3}\phi d\phi + (60\pi\ Nu_{m1})^{4/3}(GrPr)^{-4/3}] \tag{19}$$

When the Grashof number for the upper cylinder is equal to that for the bottom cylinder, the local heat transfer coefficient results in

$$\frac{Nu_2}{\sqrt[4]{GrPr}} = 2\sin^{1/3}\phi[240\int_0^{\phi}\sin^{1/3}\phi d\phi + (60\pi\ K_1)^{4/3}]^{-1/4} \tag{20}$$

where $K_1 = Nu_{m1}/\sqrt[4]{GrPr}$. When K_1= 0.44, numerical integration leads to

$$K_2 = Nu_{m2}/\sqrt[4]{GrPr} = 0.3272 \tag{21}$$

For n-cylinder array. The heat transfer coefficient for the top cylinder (the n-th cylinder) in n-cylinder array of a close spacing (n>2) can be obtained in the same manner, and when all cylinders in the array have a same value of Grashof number, the coefficient for the top cylinder becomes

$$\frac{Nu_n}{\sqrt[4]{GrPr}} = 2\sin^{1/3}\phi[240\int_0^{\phi}\sin^{1/3}\phi d\phi + (60\pi\sum_{i=1}^{n-1}K_i)^{4/3}]^{-1/4} \tag{22}$$

where $K_i = Nu_{mi}/\sqrt[4]{GrPr}$. And hence

$$240\int_0^{\phi}\sin^{1/3}\phi d\phi \ - 180\sin^{1/3}\phi \tan\phi + (60\pi \sum_{i=1}^{n-1} K_i)^{4/3} = 0 \qquad (23)$$

is deduced and ϕ_{max} is obtained from solving equation (23) with respect to angle ϕ. As the sum of K_i increases towards infinity, ϕ_{max} approaches 90°. For the upper cylinder in two-cylinder array, ϕ_{max} is 74°.

3. ANALYTICAL RESULTS

Fig.2 shows the local heat transfer coefficient derived from the theoretical sol tion for the n-th cylinder (i.e., the top cylinder) in n-cylinder array, in whic each cylinder has the same value of Grashof number. $Nu/\sqrt[4]{GrPr}$ curve at n=1 is fo the bottom cylinder, which is identical to Levy's solution for a single cylinder The curves at n≧2 have a similar tendency each other and indicate to be zero at ϕ=0. $Nu/\sqrt[4]{GrPr}$ decreases and ϕ_{max} (arrow marks and numerical values above the curves) moves towards 90° with increasing elevation in the array.

The average heat transfer coefficient $Nu_{mn}/\sqrt[4]{GrPr}$ for the top cylinder is shown as a solid curve in Fig.6. The coefficient decreases with increasing n and is inversely propotional to $\sqrt[4]{n}$ at a large value of n.

4. COMPARISON OF EXPERIMENTAL RESULTS WITH ANALYTICAL VALUES

4.1 Two-Cylinder Array

Photograph 1 illustrates Schlieren photographs which were taken for two-cylinder array at spacing b=1 and 50.8 mm. Figs.3 and 4 show the local heat transfer coefficients of the upper and bottom cylinder respectively which are obtained from the photographs at various spacings.

As shown in Fig.3, the experimental results of the upper cylinder in a closer sp ing array agree with the analytical values. However, the discrepancy between the theoretical and experimental distribution of $Nu/\sqrt[4]{GrPr}$ at spacing of b=1 or 5 mm

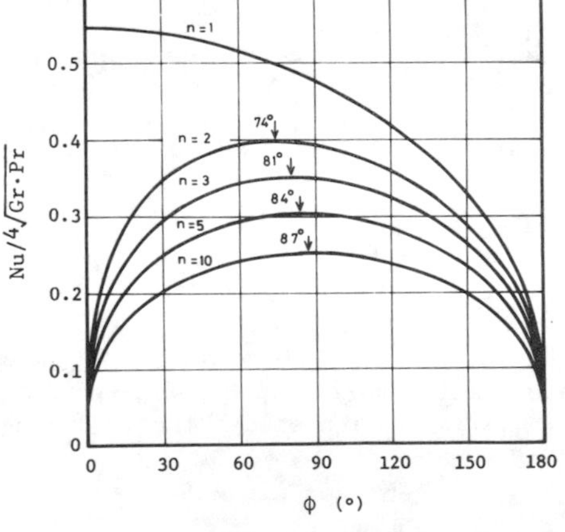

FIGURE 2. Local heat transfer coefficient for top cylinder in array

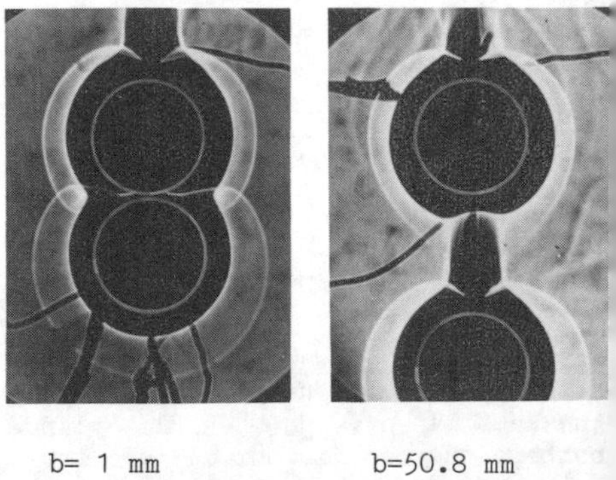

b= 1 mm b=50.8 mm

($GrPr=5\cdot10^5$, D=50.8 mm)

PHOTOGRAPH 1. Schlieren photographs for two-cylinder array

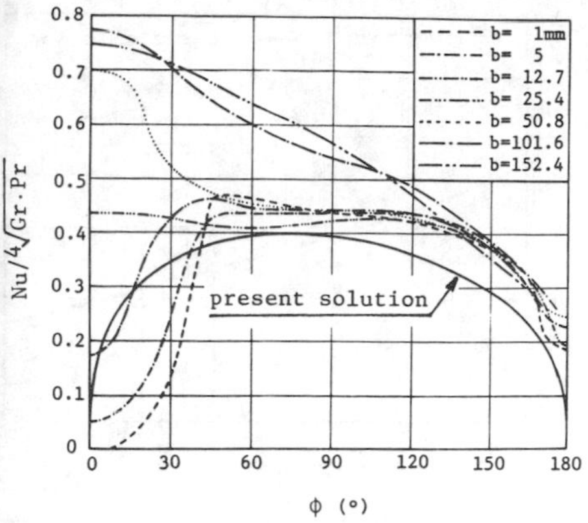

FIGURE 3. $Nu/\sqrt[4]{GrPr}$ distribution for the upper cylinder in two-cylinder array

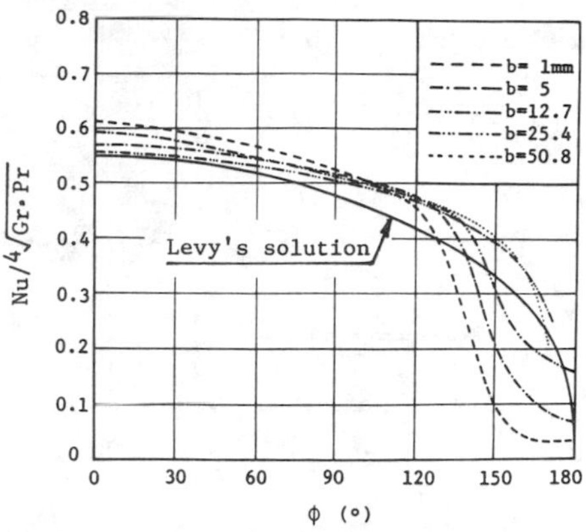

FIGURE 4. $Nu/\sqrt[4]{GrPr}$ distribution for the bottom cylinder in two-cylinder array

appears at angle ranging 0-45°. This is based on the reason that bounbary layer flow joined with that from the preceding cylinder actually occurs at some wide angle centered at the stagnant point as shown in the photograph, while in the analysis, it is assumed to occur concentrically at this point.

As a rising plume has a faster velocity with increasing distance from the top of cylinder and its temperature approaches to ambient temperature, the plume gives a forced convection like effect to the upper cylinder. This effect becomes large at spacing of b=50.8 mm and $Nu/\sqrt[4]{GrPr}$ at $\phi<30°$ in the upper cylinder is greater than that of the bottom. At wider spacing, $Nu/\sqrt[4]{GrPr}$ increases over a wide range of angle.

In Fig.4, the distribution of $Nu/\sqrt[4]{GrPr}$ based on experimental results is found to be similar to the predicted ones for wider spacings. With decreasing spacing, $Nu/\sqrt[4]{GrPr}$ at the top parts of the bottom cylinder decreases due to the presence of the upper cylinder and heat transfer from the two cylinders to ambient air in the clearance region between these cylinders tends to depend on heat conduction. Therefore $Nu/\sqrt[4]{GrPr}$ distributions of the two cylinders in this region result in symmetry with respect to the horizontal centerline of these cylinders respectively.

Fig.5 presents the relationship between the average heat transfer coefficients $Nu_m/\sqrt[4]{GrPr}$ of the two cylinders and the spacing b. The average heat transfer coefficients of the both cylinders reduce with decreasing b and approach assymptotically to constant values. The constant value for the upper cylinder agreed well with the predicted one and that of the bottom cylinder accidentally coincides with the Levy's solution.

On the other hand, the heat transfer characteristic of the bottom cylinder becomes equal to that of a single cylinder with increasing b, and $Nu_m/\sqrt[4]{GrPr}$ of the upper cylinder displays to be maintained higher than that of the bottom one. Even at spacing wider than 400 mm, the influence of buoyant plume separated from the bottom cylinder on the upper one still remains.

4.2 Array with Many Cylinders

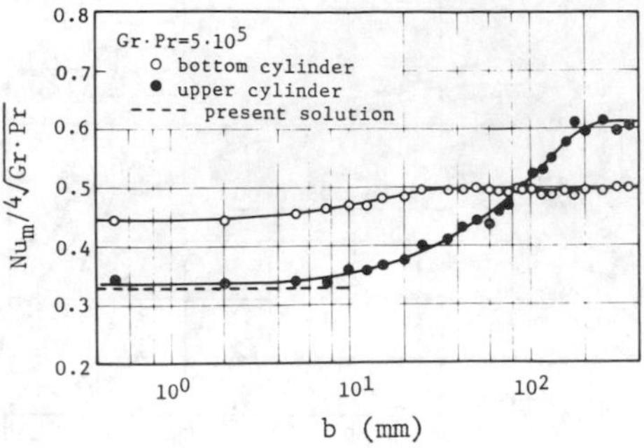

FIGURE 5. Average heat transfer coefficients for two-cylinder array in various spacings

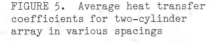

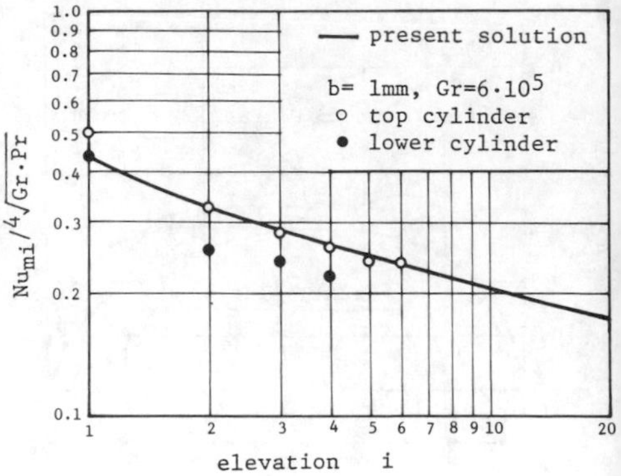

FIGURE 6. Relationship between average heat transfer coefficients and elevation in array

Fig.6 illustrates the relationship between the average heat transfer coefficient $Nu_{mi}/\sqrt[4]{GrPr}$ and the elevation i in the array where spacing is maintained 1mm. In this experiment, the vertical array was made by adding a cylinder one by one above a single cylinder and the maximum number of cylinder was six.

As shown in Fig.6, $Nu_{mi}/\sqrt[4]{GrPr}$ of the i-th cylinder in the array has two values at a given i, which are indicated by open and closed circles respectively. The open circle at i=1 denotes the value for a single cylinder and the closed circle does that of the bottom one. At i larger than 1, the open circle means the value of the top cylinder in the array with i cylinders.

As mentioned previously, further addition of one or more cylinders above the top cylinder causes to reduce the heat transfer coefficient of this cylinder to the value indicated by the closed circle. Therefore, if the cylinder is not influenced by the one above it, it remains the value at the open circle.

As no influence due to upper cylinders is assumed in the present analysis, the curve shown in Fig.6 which is obtained by integration of equation (22) must be compared with the values of the top cylinder indicated by open circles.

As the result, a good agreement is found between the theoretical and the experimental values of $Nu_m/\sqrt[4]{GrPr}$.

5. CONCLUSION

Both analytical and experimental studies were carried out on natural convection heat transfer from horizontal cylinders in the vertical array with various number of cylinder. In the case where the diameter and Grashof number for each cylinder in the array have the same values respectively, the following conclusions were drawn.

(1) In two-cylinder array, the heat transfer coefficients of these two cylinders approach to asymtotic values, which are smaller than the value for a single cylinder, with decreasing spacing.

(2) In the array spaced closcly, the cylinder taken position between them was affected by both the adjacent upper cylinder and the lower ones.

(3) The theoretical solution for Nusselt number of the top cylinder in the array spaced very closely was obtained and its values agreed well with experimental results.

NOMENCLATURE

a : thermal diffusivity
b : spacing
Cp : specific heat
D : cylinder diameter
g : gravitational acceleration
Gr : Grashof number
Pr : Prandtl number
T : temperature
u : velocity in x direction
v : velocity in y direction
W : mass flow rate of boundary layer
x : circumferential distance from the botton of cylinder
y : radial distance from cylinder surface
α : heat transfer coefficient
β : coefficient of volumetric expansion
δ : velocity bounbary layer thickness
δ_t : thermal boundary layer thickness
λ : thermal conductivity
μ : dynamic viscosity
ν : kinematic viscosity
ρ : density
ϕ : angle between vertical line and y direction
ϕ_{max} : angle corresponding to the maximum heat transfer coefficient

Subscripts

a : ambient fluid
i : i-th cylinder or elevation in array
m : average
n : n-th cylinder i.e., top cylinder or number of cylinder
s : cylinder surface
1 : first cylinder i.e., bottom cylinder
2 : second cylinder or upper cylinder

REFERENCES

1. Marsters, G. F., Arrays of Heated Horizontal Cylinders in Natural Convection, Int. J. Heat Mass Transfer, vol. 15, pp. 921-933, 1972.

2. Brodowicz, K. and Kierkus, W. T., Experimental Investigation of Laminar Free Convection Flow in Air above Horizontal Wire with Constant Heat Flux, Int. J. Heat Mass Transfer, vol. 9, pp. 81-94, 1966.

3. Fujii, T., Morioka, I. and Uehara, H., Buoyant Plume above a Horizontal Line Heat Source, Int. J. Heat Mass Transfer, vol. 16, pp. 755-768, 1973.

4. Levy, S., Integral Methods in Natural Convection Flow, J. Appl. Mech., vol.

22, pp. 515-522, 1955.

Sugawara, S. and Michiyoshi, I., Heat Transfer from a Horizontal Plate by Natural Convection (in Japanese), Trans. JSME., vol. 21, no. 109, pp. 651-657, 1955.

Analysis of Natural Convection Heat Transfer in Enclosures Divided by a Vertical Partition Plate

TATSUO NISHIMURA, MITSUHIRO SHIRAISHI, and YUJI KAWAMURA
Department of Chemical Engineering
Hiroshima University
Higashi-Hiroshima, Japan

INTRODUCTION

Natural convection heat transfer in an enclosed space filled with a fluid, with one vertical wall heated and the opposing vertical wall cooled has been receiving considerable attention. Numerous experimental and numerical computational studies reported the heat transfer mechanism and the correlations of heat transfer rate. Excellent reviews [1,2] are available and there is no need to repeat them. The interest in the fundamental topic is fueled by applications in many real-life situations such as the enclosed air layer in double glazed windows, buildings or solar collectors. Under some circumstances, it is desirable to suppress natural convection, if possible. So recently natural convection in a rectangular enclosure with internal partitions as shown in Fig. 1 [3,4] has been studied to determine the effect of heat transfer reduction. However, this procedure in which many partition plates are inserted obliquely into an enclosure might be complex in the structure and also expensive in the cost. In this study, another procedure in which a single partition is inserted vertically into an enclosure is considered. It is similar to an efficient procedure of heat radiation reduction. If this procedure for natural convection has the heat transfer reduction comparable to that for the above procedure, the procedure proposed here is superior from an engineering standpoint.

Natural convection in enclosures divided by a single partition is relatively unknown, in fact we are aware of only two fundamental studies [5,6]. Duxbury [5] experimentally investigated air-filled enclosures divided by a vertical heat conducting plate as shown in Fig. 2 for aspect ratios between 5/8 and 5 for Rayleigh numbers between 10^3 and 5×10^6. The side walls of the enclosure are isothermal but are maintained at different temperatures, while the upper and lower walls are not completely insulated, but have a linear temperature variation. He indicated that the heat transfer rate through the divided enclosure of height H, width W, across which a temperature difference ΔT was maintained, corresponded roughly to that across an undivided enclosure of height H, width W/2 and a temperature difference $\Delta T/2$. Nakamura et al. [6] performed numerical computational and experimental studies including the effect of heat radiation for the same configuration as that of Duxbury. A good agreement was found between the numerical and experimental Nusselt numbers.

From the above literatures, it appears that there is a great lack of information on the heat transfer mechanism through the partition and the correlation of heat transfer rate. Therefore, in this paper the heat transfer mechanism was studied in details by a numerical calculation, and the boundary layer solution derived on the basis of the numerical results gave the heat transfer correlation.

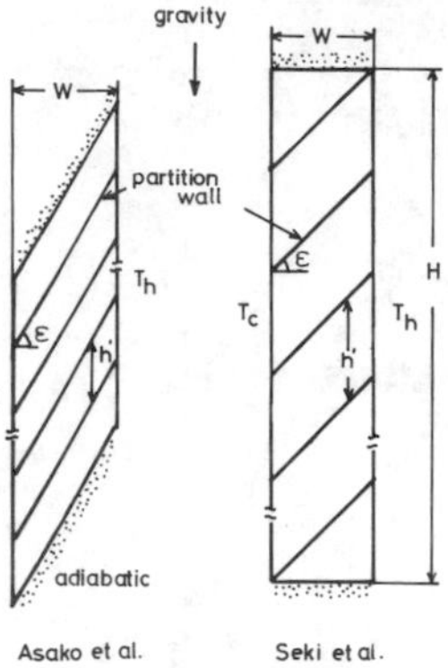

FIGURE 1. Schematic diagram of enclosure with oblique partition plates

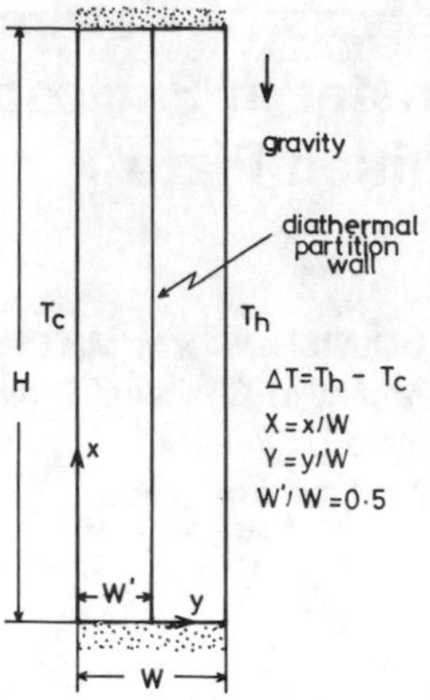

FIGURE 2. Schematic diagram of enclosure divided by a vertical partition plate

Furthermore, the experiments were carried out to validate the boundary laye approximation at high Rayleigh numbers.

1. ANALYSIS

We consider a two-dimensional rectangular enclosure of height H and width W divided by a vertical partition plate, which is fixed on the center of th enclosure as shown in Fig. 2. The analysis postulates laminar motion and th Boussinesq approximation of negligible variation in physical properties, excep for the density in the bouyancy term. The upper and lower walls of the enclosur are insulated, while the vertical walls are isothermal: the right-hand side wal at temperature T_h and the left-hand side at T_c, where $T_h > T_c$. With th assumptions stated above, the governing equations for this problem are expresse in the following non-dimensional form in terms of stream function, vorticity an temperature:

$$U\partial\Omega/\partial X + V\partial\Omega/\partial Y = Pr\nabla^2\Omega - RaPr\partial\Theta/\partial Y \qquad (1)$$

$$\Omega = -\nabla^2\Psi \qquad (2)$$

$$U\partial\Theta/\partial X + V\partial\Theta/\partial Y = \nabla^2\Theta \qquad (3)$$

where $Ra = g\beta(T_h - T_c)W^3/\alpha\nu$, $Pr = \nu/\alpha$, $\Theta = (T - T_c)/(T_h - T_c)$

The non-dimensional boundary conditions are given as

at $Y=0$, $X=0 - H/W$: $\Psi=0$, $\Omega=-\nabla^2\Psi$, $\Theta=0$

at $Y=1$, $X=0 - H/W$: $\Psi=0$, $\Omega=-\nabla^2\Psi$, $\Theta=1$

at $Y=0 - 1$, $X=H/W$: $\Psi=0$, $\Omega=-\nabla^2\Psi$, $\partial\Theta/\partial X=0$ (4)

at $Y=0 - 1$, $X=0$: $\Psi=0$, $\Omega=-\nabla^2\Psi$, $\partial\Theta/\partial X=0$

at $Y=0.5$, $X=0 - H/W$: $\Psi_{0.5}\pm=0$, $\Omega_{0.5}\pm=-\nabla^2\Psi_{0.5}\pm$, $\Theta_{0.5-}=\Theta_{0.5+}$, $\partial\Theta/\partial Y_{0.5-}=\partial\Theta/\partial Y_{0.5}$

umerical solutions of Equations (1) to (4) were obtained with the use of the inite element method in the Rayleigh number range 10^3 - 10^6 for Pr=10 and aspect atios H/W=4 and 8.

. NUMERICAL RESULTS AND DISCUSSION

treamlines and isotherms at H/W=4 are shown in Fig. 3 for various Rayleigh umbers. Streamlines and isotherms have the centro-symmetry property with respect o the center of the partition plate. At Ra=10^3, streamlines form a single irculation in each cell constructed by the partition. Its center is not located n the middle part of each cell, but in the upper part for the cold cell and in ne lower part for the hot cell respectively. Corresponding isotherms are almost arallel to the side walls, indicating that most of the heat transfer is by heat onduction. As the Rayleigh number increases, the center of circulation shifts oward the middle part of the cell and the effect of convection is more ronounced in the isotherms. At Ra=10^6, the density of isotherms is more severe ear the side walls and the partition plate, but diminishes in the core region ar away from the side walls, indicating a thermal boundary layer formation. ereafter we focus on the temperature field in the boundary layer regime.

igure 4 shows the horizontal temperature profiles for Ra=10^6 at different evels, which clearly indicates the existance of thermal boundary layers along ne side walls and the partition plate. In the cold and hot cells, the horizontal emperature gradient at the side wall varies in the vertical direction, while the emperature gradient at the partition is almost constant. These results suggest nat two thermal boundary layers differ in the chracteristic. Figure 5 shows the ertical temperature profiles. As the Rayleigh number increases, the variation of ne temperature with the Rayleigh number becomes smaller. The partition emperature at Y=0.5 varies linearly in the vertical direction except near the pper and lower walls. The temperature profiles at Y=0.25 and 0.75, whose ocations are regarded as the core region in the cold and hot cells respectively s shown in Fig. 4, vary linearly at nearly the same rate as the partition emperature at high Rayleigh numbers.

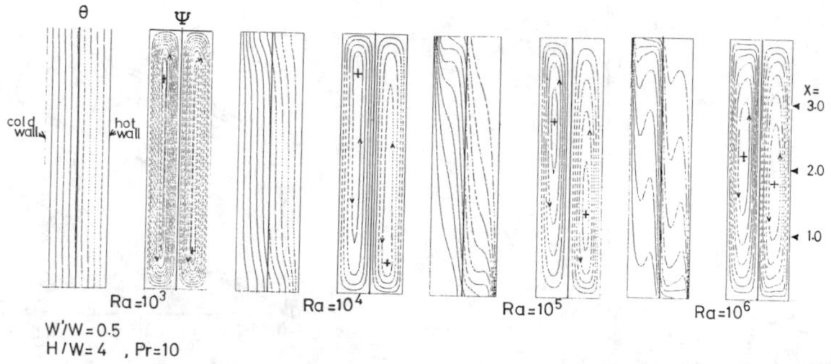

FIGURE 3. Isotherms and streamlines for H/W=4

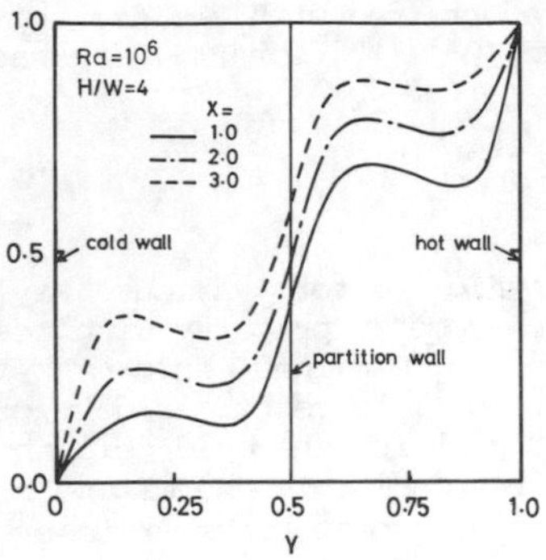

FIGURE 4. Temperature profiles at horizontal section

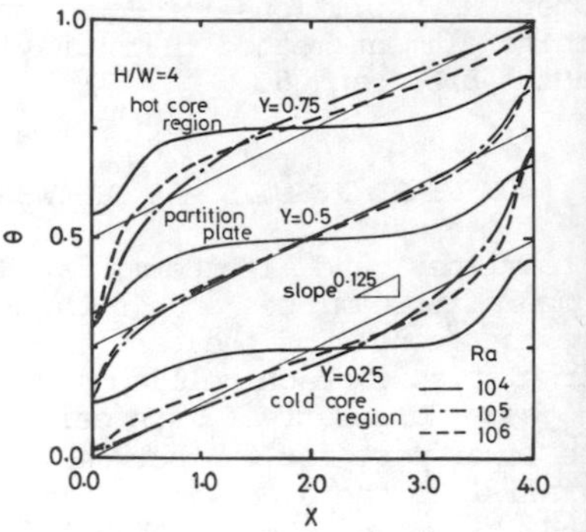

FIGURE 5. Temperature profiles at vertical section

Figure 6 shows the heat flux distributions corresponding to local Nusselt numbe at Ra=10^6. The heat fluxes along the cold and hot side walls vary in the vertic direction. The maximum heat flux appears near the starting point of the flow a the minimum near the departure point for the cold and hot side walls. On t other hand, the heat flux along the partition is nearly uniform except near t upper and lower walls. These results mean that the boundary layer at the si walls become thicker in the flow direction, while the boundary layer thickness the partition is almost constant. The reason for uniform heat flux is that t temperatures at the core region and the partition vary linearly with the sa rate as shown in Fig. 5.

The streamlines and isotherms for H/W=8 are similar to those for H/W=4, but a not shown in this paper. Figure 7 represents the vertical temperature profiles the partition plate (Y=0.5) and both the cold and hot core regions (Y=0.25 a Y=0.75) at Ra=10^6. The temperature variation is linear and also the temperatu gradient at the partition is almost equal to that at the core region.

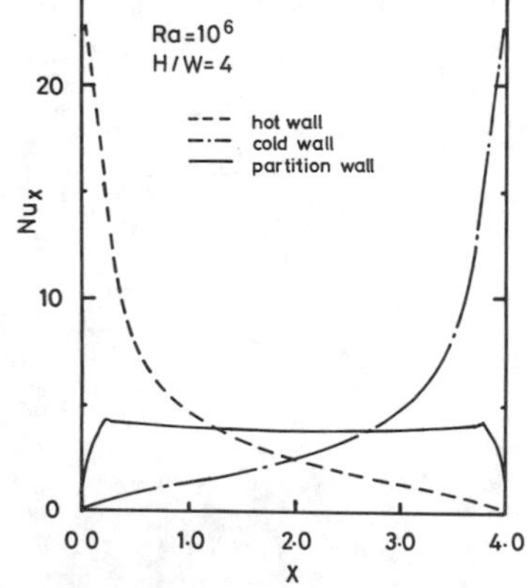

FIGURE 6. Local Nusselt number profiles

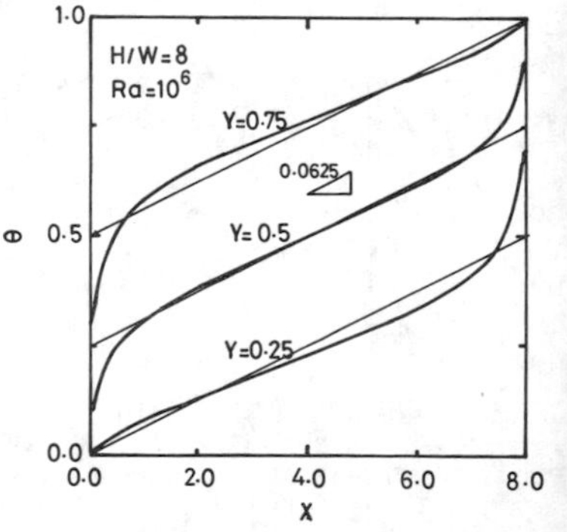

FIGURE 7. Temperature profiles at vertical section for H/W=8

From the above results, the temperature gradient at the partition and the temperature difference between the partition and the core region give $\Delta T/2H$ and $\Delta T/4$ respectively for H/W=4 and 8. Thus it is deduced that these relations are independent of Rayleigh number and aspect ratio at high Rayleigh numbers.

3. BOUNDARY LAYER MODEL

It is found from the present numerical calculations that the thermal boundary layer with constant thickness is developed along the partition plate at high Rayleigh numbers as shown in Fig. 8. So we try to estimate the heat transfer rate for this system from the boundary layer model. The following assumptions are given on the basis of the numerical results. The temperature at the partition varies linearly in the vertical direction, a constant temperature gradient being positive. The temperature gradient at the core region is the same as that at the partition, but the temperature is different. These temperature profiles are represented using the coordinates s and z shown in Fig. 8.

partition plate (s=0): $T=B + Gz$ (5)

core region (s→∞): $T=Gz$ (6)

where B is the temperature difference between the partition and the core region and G is the temperature gradient.

For this system, since the fluid motion is vertical everywhere and the temperature gradient is the same everywhere, a solution exists of the simple form:

$$T=\Phi(s) + Gz \tag{7}$$

$$u=u(s) \tag{8}$$

The energy and momentum equations are given as

$$Gu=\alpha d^2\Phi/ds^2 \tag{9}$$

$$\nu d^2u/ds^2 + \beta g\Phi=0 \tag{10}$$

Equations (9) and (10), along with the boundary conditions satisfied by Φ and u (Φ =B, u=0, at s=0 and Φ=u=0 at s→∞), give the solution as

$$\Phi=B \exp(-s/l) \cos (s/l) \tag{11}$$

where $l=(4\alpha\nu/\beta gG)^{1/4}$

Furthermore the Nusselt number is obtained from Equation (11) by the simple form:

$$Nu=QW/(\lambda\Delta T)=BW/(l\Delta T) \tag{12}$$

Since B and G are given as $\Delta T/4$ and $\Delta T/2H$ respectively, from the numerical results, the Nusselt number correlation is finally represented as follows

$$Nu=0.149\ Ra^{1/4}(H/W)^{-1/4} \tag{13}$$

Figure 9 shows the comparison of numerical and boundary layer solutions of the Nusselt number. The boundary layer solutions for H/W=4 and 8 agree well with the numerical results, which validates the boundary layer approximation.

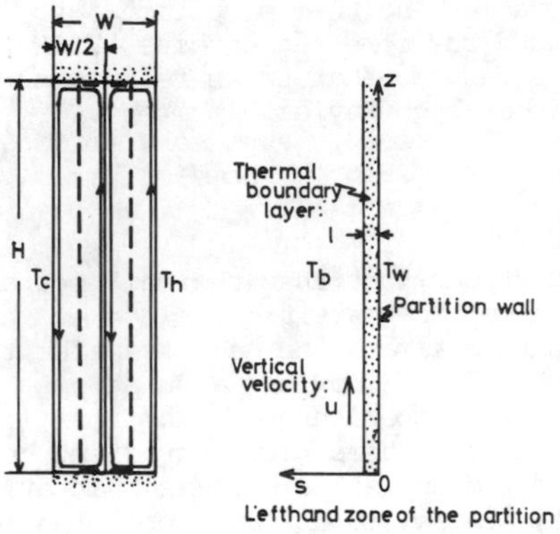

FIGURE 8. Schematic diagram of boundary layer model at the partition plate

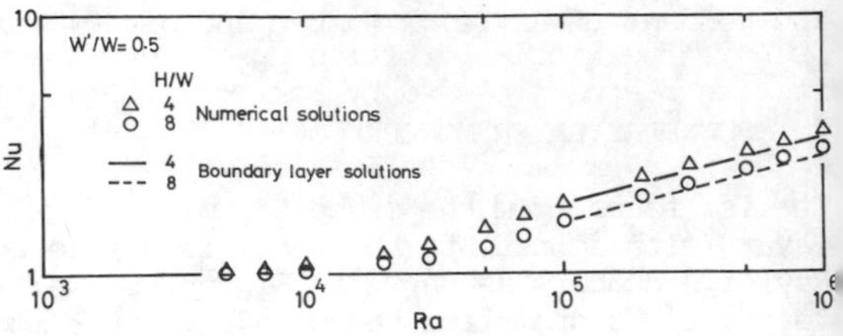

FIGURE 9. Comparison of numerical and boundary layer solutions in average Nusselt number

4. HEAT TRANSFER REDUCTION BY PARTITION PLATE

Returning to the engineering application which motivated this fundamental investigation, we now assess the thermal insulation capability of the vertical partition. The heat transfer reduction η is defined as

$$\eta = (Nu_o - Nu_p)/Nu_p \quad (1\cdot$$

where Nu_p is the Nusselt number with the partition, and Nu_o is the Nusselt number without the partition which is obtained from the correlation by Churchill [2 Figure 10 shows the relation between the heat transfer reduction and the Rayleigh number. At high Rayleigh numbers ($Ra > 10^5$), η scarcely depends on the Rayleigh number and the aspect ratio, and it has a constant value. Thus the partition pla

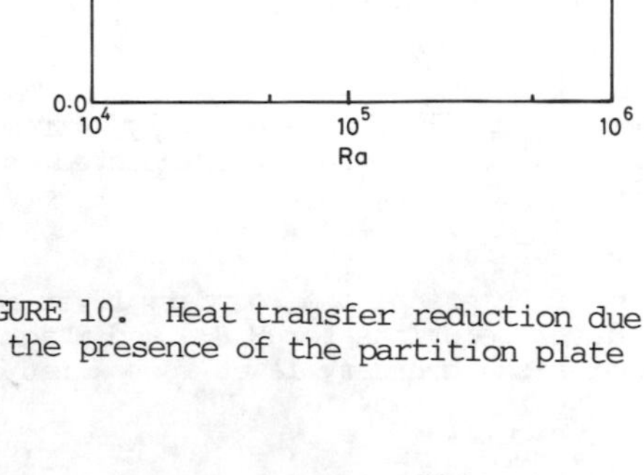

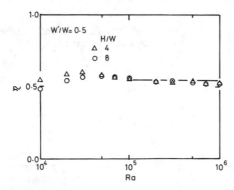

FIGURE 10. Heat transfer reduction due to the presence of the partition plate

ıs the effect of reducing the heat transfer by 55% in the high Rayleigh number ınge. From an engineering standpoint this is a sizeable reduction. The ɔmparison of heat transfer reduction is also performed between this procedure by single vertical partition plate and the other procedure by many oblique ırtition plates proposed by Seki et al.[4], as shown in Fig. 1. However, they do ɔt represent the relation between the heat reduction and the Rayleigh number. So ıe heat transfer reduction is roughly estimated under a condition (H/W=13.3, /H=0.1, Pr=0.7 and Ra=220000) from their results, and therefore η is -0.37, 0.0 ıd 0.56 for the oblique angle ε=0, 50 and 70 deg., respectively. Although a .rect comparison is difficult because of different conditions, it is expected ıat the thermal insulation capability for the present procedure is almost equal superior to that proposed by Seki et al.

EXPERIMENTS

e experiments were carried out to validate the boundary layer approximation at yleigh numbers higher than those considered in the numerical calculation. perimental apparatus is the same as that used in the previous study [7]. The in parts of experimental apparatus consist of the test section, heating and oling parts. The test section is a lucite rectangular enclosure of height 300mm and width W=75mm (H/W=4), whose two opposing vertical walls constructed copper plates are maintained at different temperatures, while the upper and wer walls are insulated. The vertical partition plate is made of aluminum foil ım thick to satisfy the condtions of the above analysis and it is fixed on the nter of the enclosure. Water is used as the working fluid. The experiment was rried out in the range $10^7 < Ra < 7x10^8$, and the temperature profiles and the erage Nusselt numbers were measured.

gure 11 shows the experimental temperature profiles at the partition plate =0.5) and both the cold and hot core regions (Y=0.25 and 0.75) at Ra=5.7x10^7. ese temperatures satisfactorily agree with those assumed by Equations (5) and) in the boundary layer approximation described above. The comparison between e experiment and the boundary layer solution in the Nusselt number is shown in g. 12. The Nusselt numbers without the partition also shown in this figure are good agreement with the correlation by Churchill [2], validating this perimental technique. The Nusselt numbers are smaller than those without the rtition and also agree with the boundary layer solution of Equation (13). Thus boundary layer approximation is useful in the high Rayleigh number range.

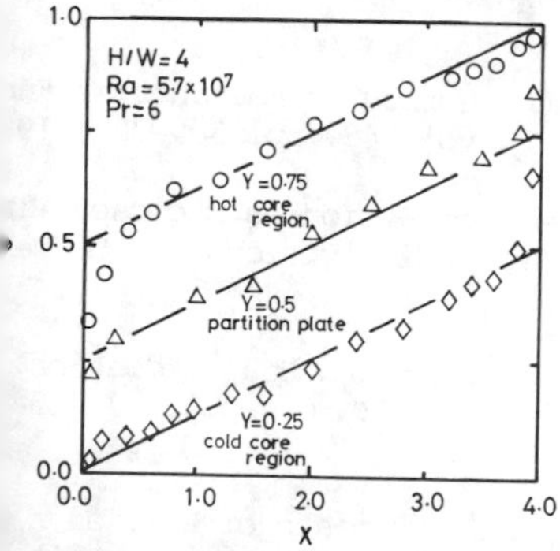

GURE 11. Temperature profiles vertical section

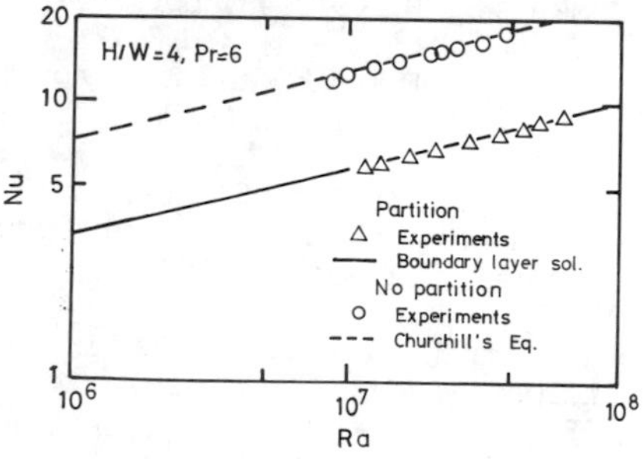

FIGURE 12. Comparison of experiment and boundary layer solution in average Nusselt number

CONCLUSIONS

Laminar natural convection in rectangular enclosures divided by a vertic partition plate was investigated by the numerical calculation, the boundary lay approximation and experiments. The enclosure was bounded by isothermal vertic walls at different temperatures and adiabatic horizontal walls. The partiti plate was fixed on the center of the enclosure and its thickness was neglected.
1) At high Rayleigh numbers, the thermal boundary layer with a constant thickne is developed along the partition plate. The partition temperature varies linear in the vertical direction. The core temperature in the cold and hot cel increases linearly at nearly the same rate as the partition temperature. The relations are confirmed in both numerical calculation and experiment.
2) The boundary layer solution is derived on the basis of the numerical resul and then the Nusselt number correlation is presented as

$$Nu=0.149\ Ra^{1/4}\,(H/W)^{-1/4}$$

This correlation predicts well the experimental data in the high Rayleigh numb range.
3) The heat transfer rate through the enclosure with the partition is compar with that without the partition to assess the thermal isulation capability of t partition. The partition has the effect of reducing the heat transfer rate about 55% in the high Rayleigh number range. From an engineering standpoint th is a sizeable reduction.

Acknowledgement- This study was supported by Computer Center of Hirosh: University.

REFERENCES

1. Catton,I., Natural Convection in Enclosures, Proceedings of the 6th Int. H Transfer Conf., vol. 6, pp. 13-43, 1978.

2. Churchill,S.W., Free Convection in Layers and Enclosure, ch. 2.5.8, Heat Exchanger Design Handbook, Hemisphere Publishing Co., 1983.

3. Asako,Y. and H.Nakamura, Heat Transfer in a Parallelogrammic Enclosure, Tr JSME, Ser. B, vol. 48, pp. 105-113, 1982.

4. Seki,N., S.Fukusako and A.Yamaguchi, Natural Convection in Rectangular Enc sures with Partition Plates, Trans JSME, Ser. B, vol. 48, pp. 300-307, 19

5. Duxbury,D., An Interferometric Study of Natural Convection in Enclosed Pl Air Layers with Complete and Partial Central Vertical Divisions, Ph.D.The University of Salford 1979.

6. Nakamura,H., Y.Asako and T.Hirata, Natural Convection and Heat Radiation Enclosures with Partition Plate, Trans JSME Ser B, vol. 50, pp. 2647-2654 1984.

7. Nishimura,T., T.Takumi, Y.Kawamura and H. Ozoe, Experiments of Natural Convection Heat Transfer in Rectangular Enclosures Partially Filled with Particles, forthcoming Kagaku Kogaku Ronbunshu 1985.

Numerical Calculation of Three-Dimensional Turbulent Natural Convection in a Cubic Enclosure Heated from below and Cooled from a Portion of Two Adjacent Vertical Walls

HIROYUKI OZOE, MASASHI HIRAMITSU and TOSHIHO MATSUI
Department of Industrial and Mechanical Engineering
Okayama University
Okayama 700, Japan

NTRODUCTION

'urbulent natural convection occurs in many circumstances including energy and nvironmental problems. The primary objective of this paper is to develop a imulation technique to study turbulent natural convection in large enclosures uch as passive solar rooms in which it is intended to use the minimum amount f expensive electric energy for heating or cooling. The method is expected to e applicable to turbulent natural convection in other systems such as for emoval of the heat generated within the casing of electronic equipment by ntegrated circuits, for turbulent convection in a room with a strong heat ource such as a fire, and for turbulent convection of reacting gas in a high emperature reactor vessel. The followings are short review on the numerical omputation of turbulent natural convection. All reports employed k-ε model xcept as noted. Fujii and Fujii [1] computed a free convection along a vertical ot plate using a Glushko model, Plumb and Kennedy [2] and Lin and Churchill 3] computed the same problem . Farouk and Guceri [4]computed a free convection bout a horizontal cylinder. Two-dimensional turbulent natural convection was omputed by Fraikin et al.[5] for a rectangular enclosure. Farouk and Guceri[6] omputed a turbulent convection in a horizontal annular space. Markatos and ericleous [7] reported solutions up to $Ra=10^{16}$ but the solution at $Ra=10^{8}$ ppears to differ from the experimental data by Ozoe et al.[8]. Ozoe et al.[9] omputed two-dimensional turbulent natural convection in a long rectangular hannel heated on one vertical wall and cooled on the opposing one and hermally insulated along upper and lower horizontal boundaries. They obtained a onvergence for Rayleigh number up to 10^{11} for water and found a good agreement f the time-averaged vertical velocity in a laminar flow at middle height for $a = 1.52 \times 10^{8}$ and $Pr = 9.17$ with their previous experimental data [8]. They lso compared the turbulent time-averaged velocity at $Ra = 6.3 \times 10^{10}$ and $Pr =$ .7 and found that a combination of turbulent Prandtl number $\sigma_t = 4$ and $C_1 =$ .296 of k-ε model equation gives much better agreement in the time-averaged elocity, time-averaged temperature and the average Nusselt number than those iven by Jones and Launder[10] for forced convection. The maximum eddy iffusivity obtained was 14 times the molecular kinematic viscosity.

hree-dimensional turbulent natural convection was computed by Markatos & Cox 11] for the flow in an L-shaped room of 3 m by 16 m by 14 m long. A heat source f 3.2 MW was presumed. The grid was 9 by 12 by 16. They obtained 1000 Kelvin emperature on the ceiling for the 293 K ambient temperature and 5 m/s for the aximum velocity. However, the computing scheme was the same as the previous ne and their result is to be tested with the experiments. Ozoe et al. [12] eveloped a computing scheme using a k-ε model for three-dimensional turbulent atural convection in a cubical enclosure heated from below and cooled from a art of one of a vertical wall. The computation was stable for Rayleigh numbers f 10^6 and 10^7. The maximum eddy diffusivity was 5.8 times the molecular inematic viscosity for $Ra = 10^7$ and $Pr = 0.7$. Thermal boundary condition of heir calculation was the one to give a quasi two-dimensional flow.

he present study is to extend their computing scheme for more general hree-dimensional thermal boundary conditions. The model regime is the one as hown in Fig.1. The floor of a cubic room is heated isothermally and a part of two adjacent vertical walls are isothermally cooled and all other walls are thermally insulated.

MATHEMATICAL EQUATIONS FOR THREE-DIMENSIONAL TURBULENT FLOW

The model equation is the same as the one developed by Ozoe et al.[12] and is a result of natural extension of the two-dimensional model equation employed by Fraikin et al.[5] Farouk & Guceri [6] and Ozoe et al.[9]. Boussinesq approximation was presumed that all physical properties except the density in the buoyant term are not functions of temperature. Three momentum equations were partially differentiated by x, y and z and subtracted to eliminate the pressure terms to give vorticity equation for three components as shown in Eqs. (1) to (3). In these equations an angle ϕ of the heated floor from the horizontal plane is included but sample calculation in this work is for a horizontal orientation.

$$\frac{\partial\Omega_1}{\partial\tau} + U\frac{\partial\Omega_1}{\partial X} + V\frac{\partial\Omega_1}{\partial Y} + W\frac{\partial\Omega_1}{\partial Z} - \Omega_1\frac{\partial U}{\partial X} - \Omega_2\frac{\partial U}{\partial Y} - \Omega_3\frac{\partial U}{\partial Z} = (\sigma + \nu_t^*)\left(\frac{\partial^2\Omega_1}{\partial X^2} + \frac{\partial^2\Omega_1}{\partial Y^2} + \frac{\partial^2\Omega_1}{\partial Z^2}\right) + \frac{\partial\nu_t^*}{\partial X}\frac{\partial\Omega_1}{\partial X}$$
$$+ 2\frac{\partial\nu_t^*}{\partial Y}\frac{\partial\Omega_1}{\partial Y} + 2\frac{\partial\nu_t^*}{\partial Z}\frac{\partial\Omega_1}{\partial Z} - \frac{\partial\nu_t^*}{\partial Y}\frac{\partial\Omega_2}{\partial X} - \frac{\partial\nu_t^*}{\partial Z}\frac{\partial\Omega_3}{\partial X} - \left(\frac{\partial^2\nu_t^*}{\partial Y^2} + \frac{\partial^2\nu_t^*}{\partial Z^2}\right)\Omega_1 + \frac{\partial^2\nu_t^*}{\partial X\partial Y}\Omega_2 + \frac{\partial^2\nu_t^*}{\partial X\partial Z}\Omega_3$$
$$+ 2\left[\frac{\partial^2\nu_t^*}{\partial X\partial Y}\frac{\partial W}{\partial X} + \frac{\partial^2\nu_t^*}{\partial Y^2}\frac{\partial W}{\partial Y} + \frac{\partial^2\nu_t^*}{\partial Y\partial Z}\frac{\partial W}{\partial Z} - \left(\frac{\partial^2\nu_t^*}{\partial X\partial Z}\frac{\partial V}{\partial X} + \frac{\partial^2\nu_t^*}{\partial Y\partial Z}\frac{\partial V}{\partial Y} + \frac{\partial^2\nu_t^*}{\partial Z^2}\frac{\partial V}{\partial Z}\right)\right] - \sigma\frac{\partial T}{\partial Y}\cos\phi \quad (1)$$

$$\frac{\partial\Omega_2}{\partial\tau} + U\frac{\partial\Omega_2}{\partial X} + V\frac{\partial\Omega_2}{\partial Y} + W\frac{\partial\Omega_2}{\partial Z} - \Omega_1\frac{\partial V}{\partial X} - \Omega_2\frac{\partial V}{\partial Y} - \Omega_3\frac{\partial V}{\partial Z} = (\sigma + \nu_t^*)\left(\frac{\partial^2\Omega_2}{\partial X^2} + \frac{\partial^2\Omega_2}{\partial Y^2} + \frac{\partial^2\Omega_2}{\partial Z^2}\right) + 2\frac{\partial\nu_t^*}{\partial X}\frac{\partial\Omega_2}{\partial X}$$
$$+ \frac{\partial\nu_t^*}{\partial Y}\frac{\partial\Omega_2}{\partial Y} + 2\frac{\partial\nu_t^*}{\partial Z}\frac{\partial\Omega_2}{\partial Z} - \frac{\partial\nu_t^*}{\partial X}\frac{\partial\Omega_1}{\partial Y} - \frac{\partial\nu_t^*}{\partial Z}\frac{\partial\Omega_3}{\partial Y} + \frac{\partial^2\nu_t^*}{\partial X\partial Y}\Omega_1 - \left(\frac{\partial^2\nu_t^*}{\partial X^2} + \frac{\partial^2\nu_t^*}{\partial Z^2}\right)\Omega_2 + \frac{\partial^2\nu_t^*}{\partial Y\partial Z}\Omega_3$$
$$+ 2\left[\frac{\partial^2\nu_t^*}{\partial X\partial Z}\frac{\partial U}{\partial X} + \frac{\partial^2\nu_t^*}{\partial Y\partial Z}\frac{\partial U}{\partial Y} + \frac{\partial^2\nu_t^*}{\partial Z^2}\frac{\partial U}{\partial Z} - \left(\frac{\partial^2\nu_t^*}{\partial X^2}\frac{\partial W}{\partial X} + \frac{\partial^2\nu_t^*}{\partial X\partial Y}\frac{\partial W}{\partial Y} + \frac{\partial^2\nu_t^*}{\partial X\partial Z}\frac{\partial W}{\partial Z}\right)\right]$$
$$+ \sigma\left(\frac{\partial T}{\partial Z}\sin\phi + \frac{\partial T}{\partial X}\cos\phi\right) \quad (2)$$

$$\frac{\partial\Omega_3}{\partial\tau} + U\frac{\partial\Omega_3}{\partial X} + V\frac{\partial\Omega_3}{\partial Y} + W\frac{\partial\Omega_3}{\partial Z} - \Omega_1\frac{\partial W}{\partial X} - \Omega_2\frac{\partial W}{\partial Y} - \Omega_3\frac{\partial W}{\partial Z} = (\sigma + \nu_t^*)\left(\frac{\partial^2\Omega_3}{\partial X^2} + \frac{\partial^2\Omega_3}{\partial Y^2} + \frac{\partial^2\Omega_3}{\partial Z^2}\right) + 2\frac{\partial\nu_t^*}{\partial X}\frac{\partial\Omega_3}{\partial X}$$
$$+ 2\frac{\partial\nu_t^*}{\partial Y}\frac{\partial\Omega_3}{\partial Y} + \frac{\partial\nu_t^*}{\partial Z}\frac{\partial\Omega_3}{\partial Z} - \frac{\partial\nu_t^*}{\partial X}\frac{\partial\Omega_1}{\partial Z} - \frac{\partial\nu_t^*}{\partial Y}\frac{\partial\Omega_2}{\partial Z} + \frac{\partial^2\nu_t^*}{\partial X\partial Z}\Omega_1 + \frac{\partial^2\nu_t^*}{\partial Y\partial Z}\Omega_2 - \left(\frac{\partial^2\nu_t^*}{\partial X^2} + \frac{\partial^2\nu_t^*}{\partial Y^2}\right)\Omega_3$$
$$+ 2\left[\frac{\partial^2\nu_t^*}{\partial X^2}\frac{\partial V}{\partial X} + \frac{\partial^2\nu_t^*}{\partial X\partial Y}\frac{\partial V}{\partial Y} + \frac{\partial^2\nu_t^*}{\partial X\partial Z}\frac{\partial V}{\partial Z} - \left(\frac{\partial^2\nu_t^*}{\partial X\partial Y}\frac{\partial U}{\partial X} + \frac{\partial^2\nu_t^*}{\partial Y^2}\frac{\partial U}{\partial Y} + \frac{\partial^2\nu_t^*}{\partial Y\partial Z}\frac{\partial U}{\partial Z}\right)\right] - \sigma\frac{\partial T}{\partial Y}\sin\phi \quad (3)$$

The vorticity is defined as a curl of the velocity and the velocity is given as a curl of the vector potential and, as a result, the Poisson equation is given for vorticity and vector potential as follows.

$$\vec{\Omega} = \nabla \times \vec{V}, \quad \vec{V} = \nabla \times \vec{\psi}, \quad \vec{\Omega} = \nabla \times \vec{V} = -\nabla^2\vec{\psi} \quad (4), (5), (6)$$

The energy equation is given as follows with turbulent Prandtl number.

$$\frac{\partial T}{\partial\tau} + U\frac{\partial T}{\partial X} + V\frac{\partial T}{\partial Y} + W\frac{\partial T}{\partial Z} = \frac{\partial}{\partial X}\left[\left(1 + \frac{\nu_t^*}{\sigma_t}\right)\frac{\partial T}{\partial X}\right] + \frac{\partial}{\partial Y}\left[\left(1 + \frac{\nu_t^*}{\sigma_t}\right)\frac{\partial T}{\partial Y}\right] + \frac{\partial}{\partial Z}\left[\left(1 + \frac{\nu_t^*}{\sigma_t}\right)\frac{\partial T}{\partial Z}\right] \quad (7)$$

The following two equations represent in dimensionless form the conservation of the turbulent kinetic energy and of the rate of dissipation of turbulent kinetic energy. This is the three-dimensional extension of the equations employed by Fraikin et al. for two-dimensional field.

$$\frac{\partial K}{\partial\tau} + U\frac{\partial K}{\partial X} + V\frac{\partial K}{\partial Y} + W\frac{\partial K}{\partial Z} = \left(\sigma + \frac{\nu_t^*}{\sigma_k}\right)\left(\frac{\partial^2 K}{\partial X^2} + \frac{\partial^2 K}{\partial Y^2} + \frac{\partial^2 K}{\partial Z^2}\right) + \frac{1}{\sigma_k}\left(\frac{\partial\nu_t^*}{\partial X}\frac{\partial K}{\partial X} + \frac{\partial\nu_t^*}{\partial Y}\frac{\partial K}{\partial Y}\right.$$
$$\left. + \frac{\partial\nu_t^*}{\partial Z}\frac{\partial K}{\partial Z}\right) + \nu_t\left[\left(\frac{\partial U}{\partial Y} + \frac{\partial V}{\partial X}\right)^2 + \left(\frac{\partial V}{\partial Z} + \frac{\partial W}{\partial Y}\right)^2 + \left(\frac{\partial W}{\partial X} + \frac{\partial U}{\partial Z}\right)^2 + 2\left(\frac{\partial U}{\partial X}\right)^2 + 2\left(\frac{\partial V}{\partial Y}\right)^2 + 2\left(\frac{\partial W}{\partial Z}\right)^2\right.$$
$$- E - \sigma\frac{\nu_t^*}{\sigma_t}\left(\frac{\partial T}{\partial X}\sin\phi - \frac{\partial T}{\partial Z}\cos\phi\right) \quad (8)$$

$$\frac{\partial E}{\partial \tau} + U\frac{\partial E}{\partial X} + V\frac{\partial E}{\partial Y} + W\frac{\partial E}{\partial Z} = (\sigma + \frac{\nu_t^*}{\sigma_\varepsilon})(\frac{\partial^2 E}{\partial X^2} + \frac{\partial^2 E}{\partial Y^2} + \frac{\partial^2 E}{\partial Z^2}) + \frac{1}{\sigma_\varepsilon}(\frac{\partial \nu_t^*}{\partial X}\frac{\partial E}{\partial X} + \frac{\partial \nu_t^*}{\partial Y}\frac{\partial E}{\partial Y} + \frac{\partial \nu_t^*}{\partial Z}\frac{\partial E}{\partial Z})$$

$$+ C_1\frac{E}{K}\nu_t^*[(\frac{\partial U}{\partial Y} + \frac{\partial V}{\partial X})^2 + (\frac{\partial V}{\partial Z} + \frac{\partial W}{\partial Y})^2 + (\frac{\partial W}{\partial X} + \frac{\partial U}{\partial Z})^2 + 2(\frac{\partial U}{\partial X})^2 + 2(\frac{\partial V}{\partial Y})^2 + 2(\frac{\partial W}{\partial Z})^2].$$

$$- C_2\frac{E^2}{K} - C_\varepsilon\sigma\frac{E}{K}(\frac{\partial T}{\partial X}\sin\phi - \frac{\partial T}{\partial Z}\cos\phi)\frac{\nu_t^*}{\sigma_t} \ (7) \qquad (9)$$

The dimensionless time-averaged eddy diffusivity is related as follows, per Jones and Launder, to the dimensionless time-averaged turbulent kinetic energy and the dimensionless time-averaged rate of dissipation of turbulent kinetic energy.

$$\nu_t^* = C_\mu\frac{K^2}{E} \qquad (10)$$

The dimensionless time-averaged variables in the above equations are defined as

$X = x/x_0$, $Y = y/y_0$, $Z = z/z_0$, $U = u/u_0$, $V = v/v_0$, $W = w/w_0$, $\tau = t/t_0$, $K = k/k_0$, $E = \varepsilon/\varepsilon_0$, $T = (\theta - \theta_0)/(\theta_h - \theta_c)$, $\nu_t^* = \nu_t/\nu_{t_0}$, $x_0 = y_0 = z_0 = [g\beta(\theta_h - \theta_c)/(\alpha\nu)]^{-1/3} = H/Ra^{1/3}$, $u_0 = v_0 = w_0 = \alpha/x_0$, $k_0 = (\alpha/x_0)^2$, $\varepsilon_0 = \alpha^3/x_0^4$, $\nu_{t_0} = \alpha$ and $t_0 = x_0/u_0$.

The following empirical constants recommended by Launder and Spalding [13] were used, except for C_ε in the buoyant term of the E-equation, which was adopted from Fraikin et al.

$C_\mu = 0.09$, $C_1 = 1.44$, $C_2 = 1.92$, $C_\varepsilon = 0.7$, $\sigma_K = 1$ $\sigma_\varepsilon = 1.3$ and $\sigma_t = 1$.

The boundary conditions for the cubic room shown in Fig.1 are summarized as follows:

. Temperature $T = 0.5$ at $Z = Ra^{1/3}(=\bar{H})$ $T = -0.5$ from $Z = 0.09\ \bar{H}$ to $0.66\bar{H}$ at $X = 0$ and $Y = 0$. $\partial T/\partial n = 0$ on all other walls where n is a coordinate normal to each wall.

. Velocity $U = V = W = 0$ at $X = 0, \bar{H}$, at $Y = 0, \bar{H}$ and $Z = 0, \bar{H}$.

. Vorticity The components of vorticity are extrapolated from the fluid velocity at one time step earlier. (This approximation holds rigorously at the final steady state.) Thus:

$\Omega_1 = 0$, $\Omega_2 = -\partial W/\partial X$, $\Omega_3 = \partial V/\partial X$ at $X = 0$, H.

$\Omega_1 = \partial W/\partial Y$, $\Omega_2 = 0$, $\Omega_3 = -\partial U/\partial Y$ at $Y=0$, H. $\Omega_1 = -\partial V/\partial Z$, $\Omega_2 = \partial U/\partial Z$, $\Omega_3 = 0$, at $Z=0$, H

. Vector potential The boundary conditions derived by Hirasaki & Hellums [14] for a rigid wall were adopted: $\partial\psi_1/\partial X = \psi_2 = \psi_3 = 0$ at $X = 0, \bar{H}$.

$\psi_1 = \partial\psi_2/\partial Y = \psi_3 = 0$ at $Y = 0, \bar{H}$. $\psi_1 = \psi_2 = \partial\psi_3/\partial Z$ at $Z = 0, \bar{H}$.

. Dimensionless time-averaged kinetic energy K. The dimensionless time-averaged kinetic energy K was set to zero on all rigid walls. $K = 0$ at $X = 0, \bar{H}$, at $Y = 0, \bar{H}$ and at $Z = 0, \bar{H}$.

. Dimensionless time-averaged dissipation rate of turbulent kinetic energy E. The rate of dissipation of turbulent kinetic energy is proportional to $k^{3/2}/\ell$ where ℓ is a characteristic length expressing the scale of the turbulence. Since both k and ℓ approaches zero on the wall, the value of ε is indecisive. However, the rate of dissipation ε at a short distance from the wall can be given following the mixing length theory by Prandtl as employed by Fraikin et al. and Ozoe et al. The dimensionless rate of dissipation at ΔY from the wall is given by the next equation.

$$E_{\Delta Y} = C_\mu^{3/4}K^{1.5}/(\kappa\,\Delta Y) \qquad (11)$$

COMPUTED RESULTS

The turbulent free convection has been known to occur at $Gr = 10^9$ over a vertical isolated plate at constant temperature in an unconfined fluid. On the other hand, it occurs at $Ra = 2 \times 10^4$ for the fluid heated from below and cooled from above. There is no definitive agreement on the critical Rayleigh number for the convection in a confined regime heated from side and/or from below. Sample calculation was carried out for Ra = 10^6 and Pr = 0.7. The transient computation was carried out for the convergence of the average Nusselt number over the floor . The average Nusselt number can be obtained as the ratio of the heat flux at convection state versus that at conduction state for the same thermal boundary conditions. The average Nusselt number was 5.37. On the other hand, the average Nusselt number for a room with a single cold wall was 6.04. This is because the heat flux at conduction state for two cold windows was 1.3[illegible] times that for a single cold wall and the relative increase of convective heat flux due to the two cold windows was only 1.18 times that for a single col[illegible] wall. This relatively small increase in the convective heat flux is due to the complicated triangular flow regime resulted from the symmetrical two col[illegible] windows located side by side and the fluid suffers relatively large frictiona[illegible] loss with side walls. The flow characteristics are explained in detail graphically as follows.

Since the thermal boundary condition is symmetrical in terms of the diagona[illegible] plane Y = X for this case , the vectors were projected to this diagonal plan[illegible] or another diagonal plane perpendicular to this. These are shown in Figs.2 an[illegible] 3. Plane A - A' is a symmetrical plane. The maximum vertical velocit[illegible] components 3.17 exists at $(X,Y,Z)=(0.91\bar{H},0.91\bar{H},0.5\bar{H})$. The velocity vector[illegible] along the curved planes as indicated by B - B', C - C' and D - D' wer[illegible] projected over A - A' plane. In Fig.3, only those vectors on Planes A-A' an[illegible] G-G' are indicated. The velocity vectors in G-G'plane are symmetrical in term[illegible] of a central plane. The main component of flow is that in A-A' but almost th[illegible] same magnitude of spiral velocity components appear near both sides in G-G plane.

These complicated flow characteristics are rather simple in a top view as see[illegible] in Fig.4. (a) is that at Z = 0.02H near the top plate. Main flow is toward th[illegible] two cold walls. It is much stronger in (b) at Z = 0.147H. At the middl[illegible] height(not shown), velocity components are mostly in vertical directions an[illegible] almost zero in a horizontal plane. At Z = 0.853H(not shown), main components o[illegible] velocity are in horizontal and in (c) at Z = 0.98H the stagnation point o[illegible] downward velocity appears at X = Y = 0.15H. The main component of the flow i[illegible] the circulating flow as seen in Fig.2 but the flow itself becomes quit[illegible] complicated near the corner due to the drag on the walls.

The temperature contour maps were drawn on the plane of constant Y due to th[illegible] difficulty to draw it for the curved plane. Fig.5 (a), (b) and (c) are for Y [illegible] 0.049H, 0.5H and 0.951H respectively. In (a) isothermal lines are dense ove[illegible] the floor since cold fluid comes down to the floor. The temperatur[illegible] stratification at higher than middle height is also apparent . In (c), th[illegible] temperature gradient over the floor becomes small because the fluid is heate[illegible] up when it comes to this plane. The temperature stratification is rather weak In (b) at middle, the profile is in between (a) and (c). In Fig.6, isotherma[illegible] lines are drawn for the horizontal planes at Z = 0.5H and 0.951H in (a) and (b[illegible] respectively. The isotherms are symmetrical in terms of X = Y plane. At th[illegible] middle height, over the two cold walls, dense isothermal lines indicate a stro[illegible] temperature gradient on the cold windows. Interesting isothermal two circles [illegible] T = 0.1 appeared near the cold walls as seen in (a). This indicates a stro[illegible] reverse gradient of isotherms of T = 0.1 at Z = 0.5H in the plane at Y = 0.[illegible] as seen in Fig.5 (b). In Fig.6 (b), the isotherms at Z = 0.951H correspon[illegible] well to the velocity vectors of Fig.4 (c). The minimum temperature contour T [illegible] 0.0 in this plane appears to correspond to the stagnation flow and temperatu[illegible] increases in all horizontal directions in this plane.

The contour maps of dimensionless eddy diffusivity are shown in Fig.7 The maximum dimensionless eddy diffusivity 0.943 (1.35 times the molecul[illegible] kinematic viscosity) occurs at Z = 0.773H. The contour maps at four differe[illegible] levels are shown in (a), (b), (c) & (d) for Z = 0.049H, 0.227H, 0.773H a[illegible] 0.951H, respectively. The contour maps indicate the value of dimensionless ed[illegible] diffusivity. The maximum value in the plane shown in (a) corresponds to t[illegible] location near the stagnation point of Fig.4(a) where no horizontal velocit[illegible] exists. In Fig.7 (b), at Z = 0.227H four peaks appeared near four corners d[illegible] to three downwards velocity at three corners and one upward. In Fig.7 (c), at [illegible] = 0.773H, the maximum value of the eddy diffusivity appeared near the corn[illegible] against to the cold walls and where the horizontal main flow turns to go [illegible] from the floor. This high value of the eddy diffusivity is caused by t[illegible] concentrated upward flow along the corner near X = Y = H. In (d) most of t[illegible]

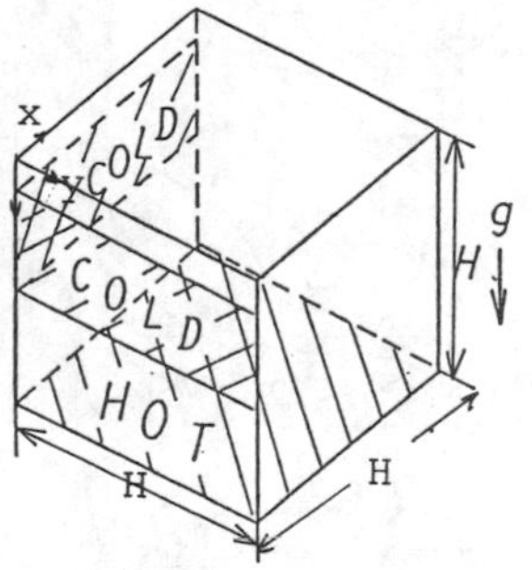

Fig.1 Schematics of the system.

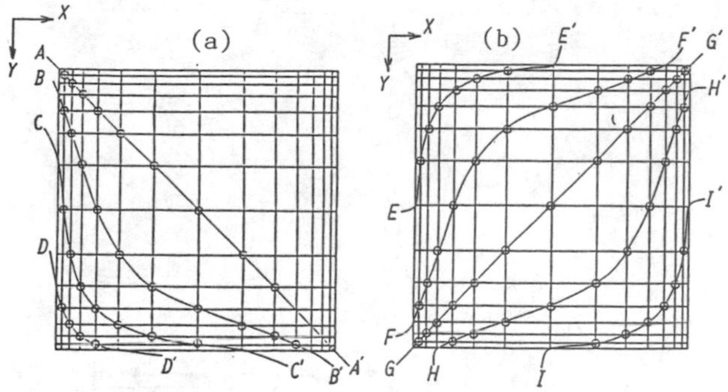

Fig.2 Top view of the system with various vertical planes.

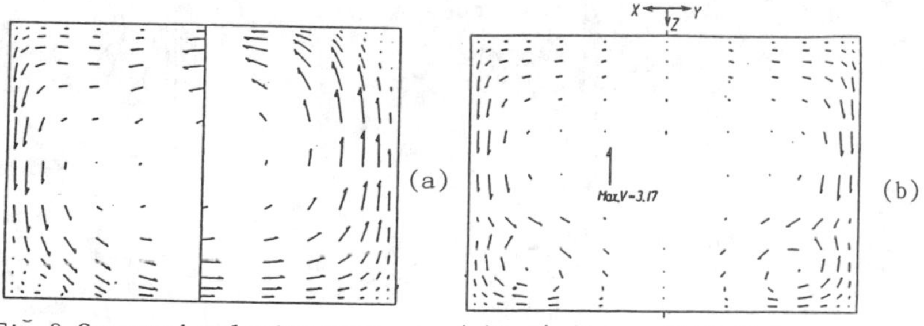

Fig.3 Computed velocity vectors. (a)A–A' plane. (b)G–G'plane.

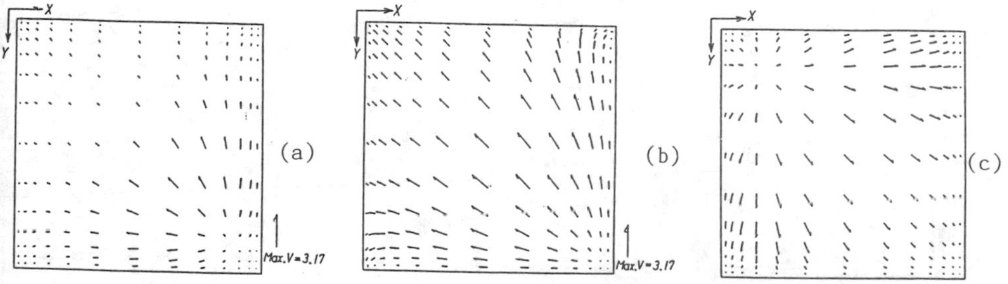

Fig.4 Computed velocity vectors in horizontal planes. (a) Z=0.02H (b) Z=0.147H (c) Z=0.98H.

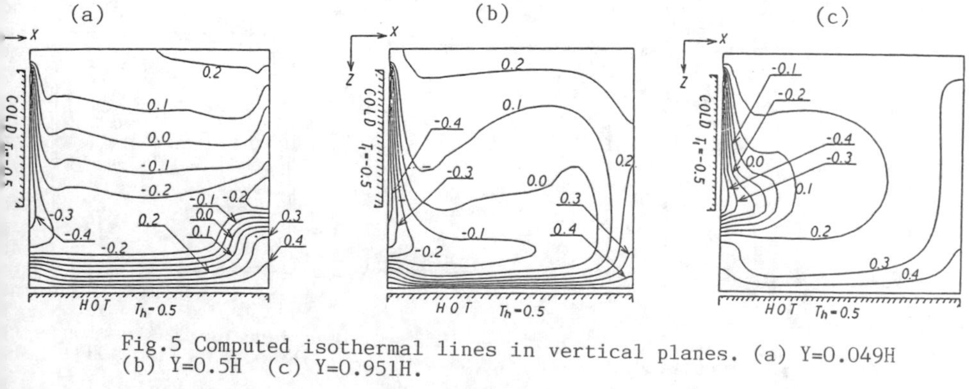

Fig.5 Computed isothermal lines in vertical planes. (a) Y=0.049H (b) Y=0.5H (c) Y=0.951H.

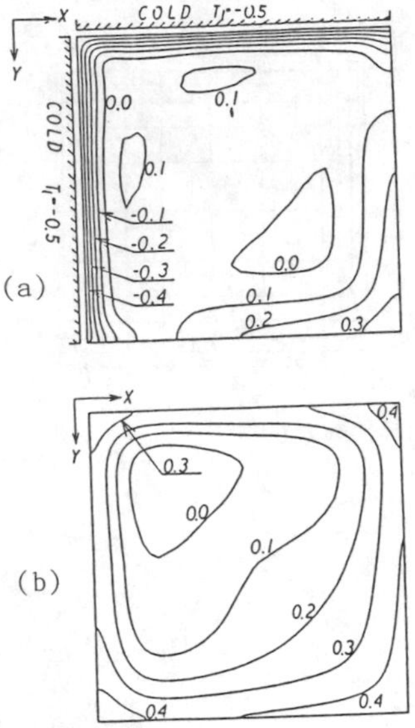

Fig.6 Computed isothermal lines in horizontal planes. (a) Z=0.5H, (b) Z=0.951H.

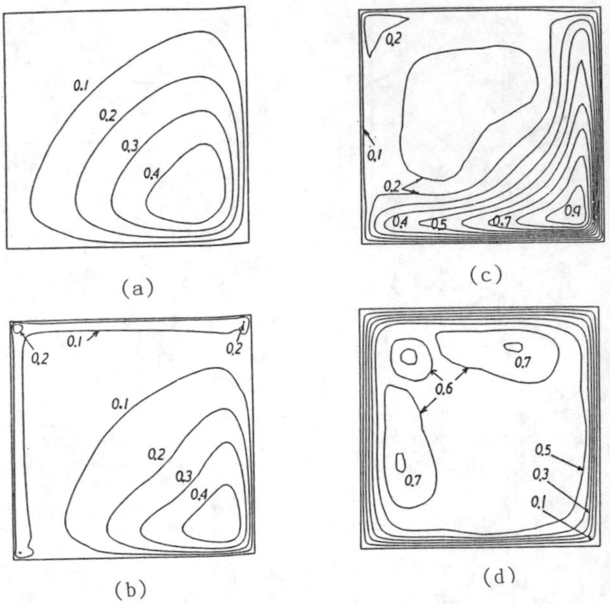

Fig.7 Computed contours of the dimensionless eddy diffusivity in horizontal planes. (a)Z=0.049H, (b)Z=0.227H (c)Z=0.773H (d)Z=0.95H

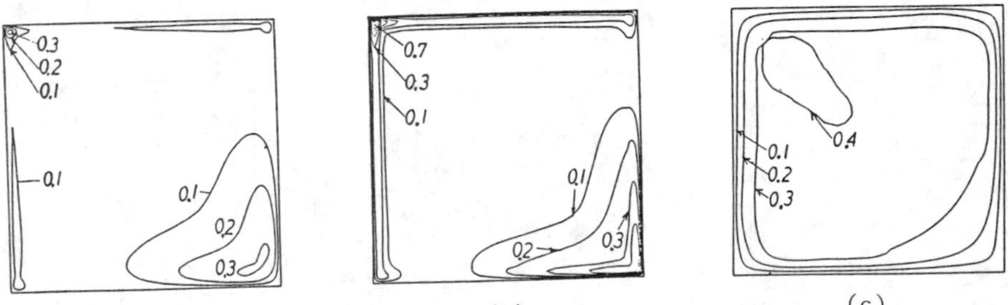

Fig.8 Computed contours of the dimensionless turbulent kinetic energy. (a)Z=0.227H (b)Z=0.5H (c)Z=0.95H

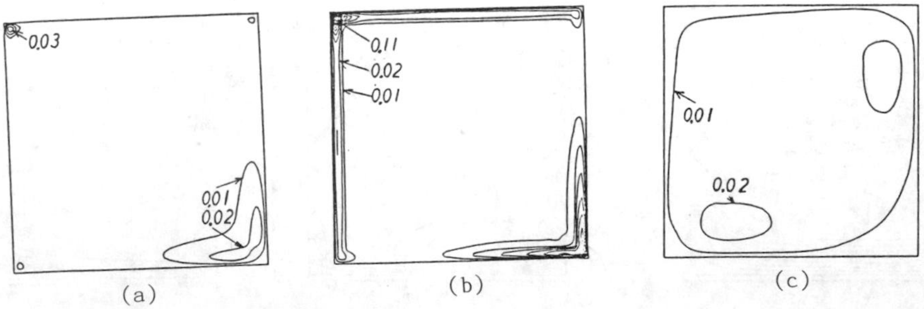

Fig.9 Computed contours of the dimensionless rate of dissipation of turbulent kinetic energy. (a)Z=0.227H (b)Z=0.5H (C)Z=0.95H

regime has the value larger than 0.5.This characteristic is quite different from those on other horizontal planes. Three peak points appear to correspond to the places where the fluid turns the direction of flow from vertical downward to horizontal and these produce wide area of high eddy diffusivity.

The dimensionless time-averaged turbulent kinetic energy K and the dimensionless time-averaged rate of dissipation of turbulent kinetic energy E are also shown in Figures 8 and 9 in horizontal contour maps. These maps are complicated and the eddy diffusivity profile as seen in Fig.7 are resulted as a combination of these two profiles via Equation (10).

Computations were further tested for Rayleigh number of 10^7 and Pr = 0.7. However, it required more refinement in grid size and dimensionless time step width for a convergence and the converged solution has not yet been obtained.

SUMMARY

Three-dimensional turbulent natural convection in a cubic enclosure heated from a floor and cooled from a portion of two vertical adjacent walls was numerically computed at $Ra=10^6$ and Pr=0.7. Stable solution was obtained. All computed physical properties were symmetrical in terms of the diagonal vertical plane due to the symmetrical thermal and geometrical boundary conditions and this assures to some extent the reliability of this computational code and scheme. The maximum dimensionless eddy diffusivity was 0.943 which is only 1.35 times the molecular kinematic viscosity. The similar computational scheme is expected to be applicable for other turbulent convection.

NOMENCLATURE

E dimensionless time-averaged rate of dissipation of turbulent kinetic energy
g acceleration due to gravity, m/s^2
H height of the enclosure, m $\overline{H}$ = dimensionless height, $Ra^{1/3}$
K dimensionless time-averaged turbulent kinetic energy
k turbulent kinetic energy = $(\overline{u'^2} + \overline{v'^2} + \overline{w'^2})/2$, m^2/s^2
Nu average Nusselt number $=qH/[\lambda(\theta_h - \theta_c)]$
q average heat flux, $J/(m^2\ s)$
Pr Prandtl number $=\nu/\alpha$
Ra Rayleigh number = $g\beta(\theta_h - \theta_c)H^3/(\alpha\nu)$
T dimensionless time-averaged temperature
t time, s
u,v,w velocity components, m/s
x,y,z coordinates, m

GREEK LETTERS

α thermal diffusivity, m^2/s
α_t eddy diffusivity for heat transfer, m^2/s
β volumetric coefficient of expansion, K^{-1}
ε time-averaged rate of dissipation of turbulent kinetic energy, m^2/s^3
θ temperature, K
θ_0 average temperature $=(\theta_h + \theta_c)/2$, K
κ von Karman's constant =0.42
ℓ length scale, m
μ viscosity, Pa s
λ thermal conductivity, J/(m s K)
ν kinematic viscosity = μ/ρ, m^2/s
ν_t eddy diffusivity, m^2/s
ν_t^* dimensinless time-averaged eddy diffusivity $=\nu_t/\alpha$
ρ density
σ Prandtl number $=\nu/\alpha$
σ_k Prandtl number for the turbulent kinetic energy
σ_t turbulent Prandtl number $=\nu_t/\alpha_t$

σ_ε Prandtl number for the rate of dissipation of turbulent kinetic energy
τ dimensionless time
ϕ angle of inclination of the cube from horizontal plane
ψ_i dimensionless time-averaged vector potential
Ω_i dimensionless time-averaged vorticity

REFERENCES

1. Fujii,M. and Fujii,T., Numerical calculation of turbulent free convectior along a vertical plate, Trans. JSME, Vol.43, No.384,pp.2797-2807, 1978.

2. Plumb,O.A. and Kennedy,L.A.,Application of a k-ε turbulence model to natural convection from a vertical isothermal surface. J.Heat Transfer, Trans. ASME, Vol.99C, pp.79-85, 1977.

3. Lin,S.J. and Churchill,S.W.,Turbulent free convection from a vertical isothermal plate,Num. Heat Transfer, Vol.1, pp.129-145, 1978.

4. Farouk,B. and Guceri,S.I.,Natural convection from a horizontal cylinder - turbulent regime. J.Heat Transfer, Trans. ASME, Vol.104c,PP.228-235, 1982.

5. Fraikin,M.P., Portier, J.J. and Fraikin,C.J,. Application of a k-ε turbulence model to an enclosed buoyance driven recirculating flow.19th ASME-AIChE Nat. Heat Transfer Conf. Paper, 80-HT-68, 1980.

6. Farouk,B. and Guceri,S.J., Laminar and turbulent natural convection in the annulus between horizontal concentric cylinders. J.Heat Transfer. Trans. ASME. Vol.104, pp.631-636, 1982.

7. Markatos.N.C. and Pericleous, K.A., Laminar and turbulent natural convection in an enclosed cavity, Int.J.Heat Mass Transfer, Vol.27, No.5 pp.755-772, 1984.

8. Ozoe,H.,Ohmuro,M.,Mouri,A.,Mishima,S.,Sayama,H. and Churchill,S.W. Laser-Doppler measurements of the velocity along a heated vertical wall of a rectangular enclosure. J.Heat Transfer,Vol.105,pp782-788,1983.

9. Ozoe,H., Mouri,A., Ohmuro,M., Churchill, S.W. and Lior,N., Numerical calculations of laminar and turbulent natural convection in water in rectangular channels heated and cooled isothermally on the opposing vertical walls. Int.J.Heat Mass Transfer,vol.28,pp125-138,1985.

10. Jones,W.P. and Launder,B.E., The prediction of laminarization with a two-equation model of turbulence. Int.J.Heat Mass Transfer Vol.15,pp.301-314,1972.

11. Markatos,N.C. and Cox,G., Turbulent buoyant heat transfer in enclosure containing a fire source, Proceedings of the Seventh International Hea Transfer Conference, München, Vol.6, IH4, pp.373-379 1982.

12. Ozoe,H., Mouri,A., Hiramitsu, M., Churchill. S.W. and Lior.M., Numerica calculation of three-dimensional turbulent natural convection in a cubica enclosure using a two-equation model, The 22nd Natural Heat Transfe Conference, Niagara Falls, HTD-vol.32, pp.25-32. Aug., 1984.

13. Launder,A.E. and Spalding,D.B., The numerical computation of turbulent flows. Comp. Meth. Appl. Mech. and Eng., vol.3, pp.269-289, 1974.

14. Hirasaki,G.J. and Hellums,J.D., A general formulation of the boundar conditions on the vector potential in three-dimensional hydrodynamics Quart.Appl.Math., vol.26, pp.331-342, 1968.

A Study of Natural Convective Heat Transfer in a Vertical Rectangular Enclosure

RENQIA SUN (JEN-HSIA) and MING-YU WANG
Power Engineering Department
Nanjing Institute of Technology
Nanjing, Jiangsu, PRC

ABSTRACT

The natural convective heat transfer in a vertical rectangular enclosure with hot and cold walls isothermal and other walls adiabatic is investigated experimently with a Mach-Zehnder interferometer. The medium enclosed in the cavity is air under normal pressure. The Rayleigh number, Ra, varies from 400 to 4×10^5, and the aspect ration, A_s, ranges from 4 to 100. Demarcation lines between the conduction regime and transition regime, and between the transition regime and the boundary layer regime are determined in terms of Ra and A_s. An overall correlation of heat transfer in laminar flow is also obtained.

NOMENCLATURE

A_s	aspect ratio, H/W
D	depth of enclosure
g	gravitational acceleration
Gr	$\beta g \Delta T\, W^3/\nu^2$, Grashof number
h	heat transfer coefficient
H	height of enclosure
k	thermal conductivity
Nu	Nusselt number, hW/k
Pr	Prandtl number, ν/α
Ra	Rayleigh number, PrGr = $\beta g \Delta T W^3/\alpha\nu$
T	local temperature of fluid in enclosure
T_h, T_c	hot and cold wall temperatures
ΔT	temperature difference beween the hot and cold wall, $T_h - T_c$
W	width of enclosure
x,y	coordinates, see Figure 1.
x_o, y_o	dimensionless corrdinates, x/W and y/W respectively
α	thermal diffusivity
β	coefficient of volumetric thermal expansion
θ	dimensionless temperature, $(T-T_c)/(T_h-T_c)$
ν	kinematic viscosity
ρ	density

INTRODUCTION

M.-Y. Wang is now at the Department of Thermoscience Engineering, Zhejiang University, Hangzhou, Zhejiang, The People's Republic of China.

Natural convection heat transfer in vertical rectangular enclosures is of considerable practical importance such as in the field of thermal insulation, solar energy collection and cooling of electronic equipments, and has thus recei much attention in the literature. Batchelor (1), Poots (2) and Gill (3) are among the first to investigate the buoyancy induced flow in vertical enclosures analytically. Early attempts to solve this problem by numerical methods are mad by Wilkes and Churchill (4), Elder (5), de Vahl Davis (6), MacGregor and Emery (and Newell and Schmidt (8). For aspect ratio A_s greater than one, a number of experimental work have been done. We might mention Mull and Reiher (11), Eckert and Carlson (12), Elder (13), MacGregor and Emery (7) and Yin, Wung and Chen (14 Other work published relevant the present subject can be found from the excellen review by Ostrach (15,17) and Catton (16).

The objective of the present study is to determine experimentally, using a Mach-Zehnder interferometer, the limits of laminar flow regimes in rectangular enclosures with aspect ratio greater than 1. Two vertical isothermal walls of t rectangular enclosure, opposite to each other, are maintained at hot wall temper ature T_h and cold wall temperature T_c respectively, and the rest walls are all adiabatic. In respect of this subject, Batchelor (1), in his analysis, consider the limiting cases for laminar flow of air in rectangulr enclosures, and deduced a relation for the limit of the conduction regime. He noted that several differ flow regimes occur and could be determined by the parameters, Rayleigh number Ra and aspect ratio A_s which takes values between 5 and 200 for thermal insulation of buildings. Eckert and Carlson (12) conducted an experimental study with an Mach-Zehnder interferometer; defined the laminar flow regimes by the characteris temperature fields in the core as conduction regime, transition regime and boundary layer regime; and obtained the approximate limits of these regimes on a Ra-A_s plot. The aspect ratio employed in their experiments ranged from 2.33 to 46. Obtaining temperature profiles in enclosures of A_s from 4.9 to 78.7 at various R with thermocouples, Yin, Wung and Chen (14), in their investigation, found that the slopes of limiting lines of regimes are lower than those obtained by Eckert and Carlson (12). As the heat transfer mechanisms of these regimes are quite different to one another, so will their correlations in Nu, Ra and A_s differ, it is felt that an investigation on these limits is worthwhile. And it is also useful to formulate, based on the pertinent data of the present study, an overall correlation of natural convective heat transfer in laminar flow inside the vertical rectangular enclosures with an aspect ratio greater than one.

TESTING MODEL AND EXPERIMENTAL SET-UP

As shown in Figure 1, the vertical rectangular enclosure is a hexagonal cavi with both the hot wall acge and the cold wall bdhf vertical and the other walls insulated thermally. With a temperature difference between the hot and cold wal ΔT, if the dimensionless ratio $D/W \geqslant 5$, the fluid flow and the heat transport in the enclosure could be treated as a two-dimensional problem (13,18). For the present study, the fluid employed is air under normal pressure.

The hot wall is made of an aluminum plate, 300 mm wide x 1,000 mm high x 12 thick, on the back of which are three main sets of horizontal nichrome ribbon strip heaters arranged along the height of the plate and two sets of guard heater placed along the back edges of the plate. The voltage and current of each set o heaters can be independently measured and adjusted. Mica and asbestos are used for insulation. The hot junctions of 15 pairs of copper-constantan thermocouple calibrated beforehand are inserted into the bottoms of respective holes dirilled to a depth of 10 mm on the back for monitoring the surface temperature distribution of the hot wall. In such a construction, the uniformity of hot wall temperature attained is within 1.3^oC at a testing temperature difference of 50^oC.

The cold wall is made of a copper plate, of the same size and thickness as the hot plate. As shown in Figure 2, the back of the cold wall forms one of the surfaces of a water jacket with horizontal crossings inside arranged at regular intervals from top to bottom to enhance the heat transfer between the cold wall and the water pumped from a thermostatic water bath. There are also 15 pairs of copper-constantan thermocopules installed on its back in a similar fashion as did on the hot wall. The monitored temperature difference on the cold wall is less than 0.5°C.

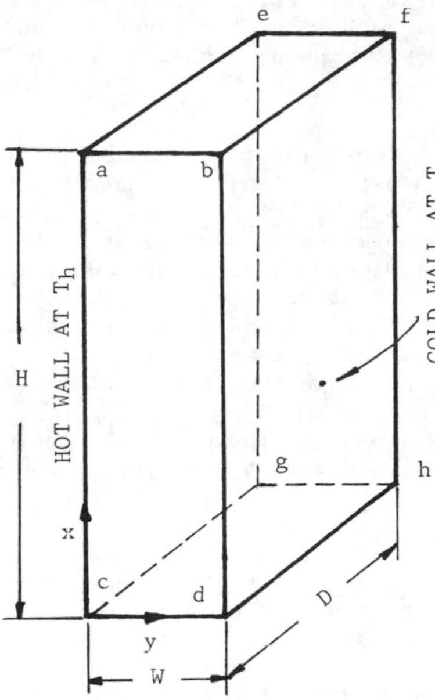

Figure 1 Vertical Rectangular enclosure

The horizontal top and bottom walls are made of pressed bakelite board with smooth flat surfaces and good insulating property. On its back is covered with plastic foam material of sufficient thickness.

The vertical front and back wall are of similar construction. Each wall is made from a pressed bakelite board, 8 mm thick and 2,000 mm high, with a circular hole of 120 mm in diameter located near its mid-height for the mounting of optical window glass of the Mach-Zehnder interferometer.

Height of the enclosure is fixed at 1,000 mm. The distance between the hot and cold wall can be adjusted and then fastened to the top and bottom walls with sealing gaskets attached to those edges of the hot and cold walls facing top and bottom to make the seams between the contacting surfaces air-

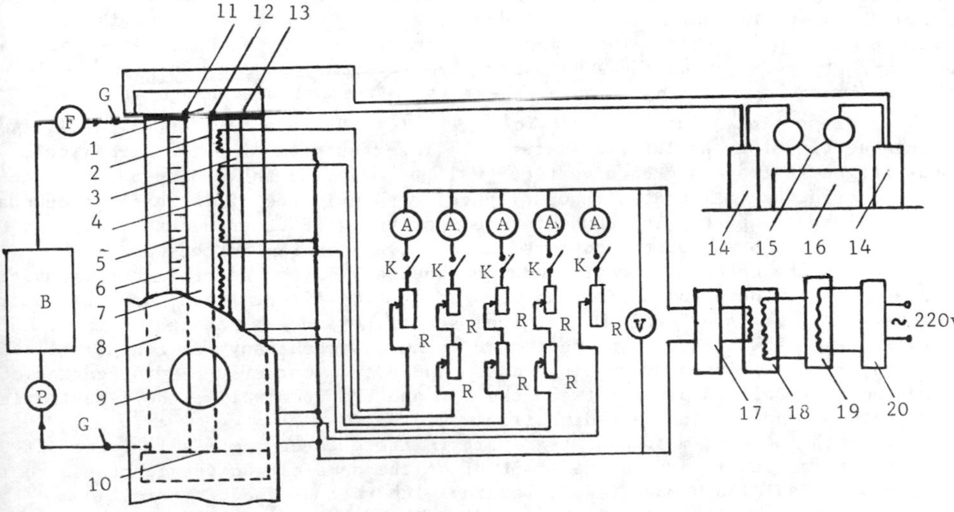

Figure 2

Figure 2. Schematic of experimental set up. 1.top wall, 2.hot wall, 3.thermal i sulation, 4.heaters, 5.cold wall, 6.water jacket, 7.air cavity, 8.front wall, 9 optical glass window, 10.bottom wall, 11,12.thermocouples, 13.gasket, 14.ice-water cell, 15.rotary swith, 16.dvm, 17.bridge rectifier, 18.transformer, 19.au transformer, 20.A.C.stabilizer, A-D.C.ammeter, B-thermostatic bath, F-flow mete G-glass thermometer, K-keyboard, P-circulating water pump, R-rheostat, V-D.C.vol meter.

tight. The front and back walls are clamped in position to the assembly of the hot, cold and horizontal walls, using foamed plastic strips glued on the edges of the above four walls facing back and front to seal off. Release the clamps, then the front and back walls could be easily moved up or down relative to the assembly of the four walls, so the the optical window glass mounted on these wa could be adjusted to any position of the rectangular section formed by the asse of the four walls. The assembled enclosure is then suspended on a frame and ca be moved up and down, and rotated a small angle for alignment. In this way, th air layer within the enclosure could be viewed through the interferometer in it whole extent.

A schematic of the experimental set-up is shown in Figure 2. The optical glasses of the Mach-Zehnder interferometer employed have a diameter of 100 mm.

EXTENT OF VARIOUS HEAT TRANSFER REGIMES

According to Eckert and Carlson (12), the natural convective heat transfer for a fluid in laminar flow within a vertical rectangular enclosure would show three different regimes: conduction regime, transition regime and boundary lay regime.

In the conduction regime, heat transfer is mainly through the thermal condu tion of fluid inside the enclosure while the Rayleigh number is small. The tem perature distribution within the enclosure would be linear with dimensionless t perature gradient, $\partial\theta/\partial y_o = -1$. Each fringe in the interferogram represents a temperature distribution function at a certain horizontal section.

In the boundary layer regime, the heat transfer from the hot to the cold wa is mainly through convection. Thermal boundary layer is built up on each verti cal wall. Rayleigh number is usually greater than 10^5, but the flow is still laminar (Ra $\leq 10^7 \sim 10^9$). The temperature gradients are concentrated within th boundary layers. Beyond these boundary layers, nearly on the mid-plane between the hot and cold walls in the core, there is no temperature change along any horizontal section, that is, $(\partial\theta/\partial y_o)_{y_o=\frac{1}{2}} = 0$. Using a horizontal parallel fringe setting for the interferometer, the interferogram of this regime would show frigne shifts along each vertical wall, similar to those displayed on the hot and cold vertical plates in open space, with evidence of increase in bounda layer thickness upwards and downwards respectively. Each fringe sufficiently away from the top and bottom corners shows a flat portion in the core.

Between the conduction regime and the boundary layer regime is the transiti regime in which both conduction and convection take effect in the heat trasnpor process. In the interferogram, original set of horizontal straight fringes in parallel will shift into slanting curved fringes without any horizontally flat portion, if the fringes curved in the top and bottom corners are disregarded. Near the vertical mid-plane between the hot and the cold walls, the dimensionle temperature gradient in the y-direction would be $-1 < (\partial\theta/\partial y_o)_{y_o=\frac{1}{2}} < 0$.

Since the mechanism of heat transport in these three regimes is quite different, it is useful to find the limit of or the demarcation between these regi in terms of Rayleigh number for enclosures with various aspect ratios.

In the present investigation, we define the demarcation between the conduction regime and the transition regime as a curve in the Ra – A_s diagram with Ra

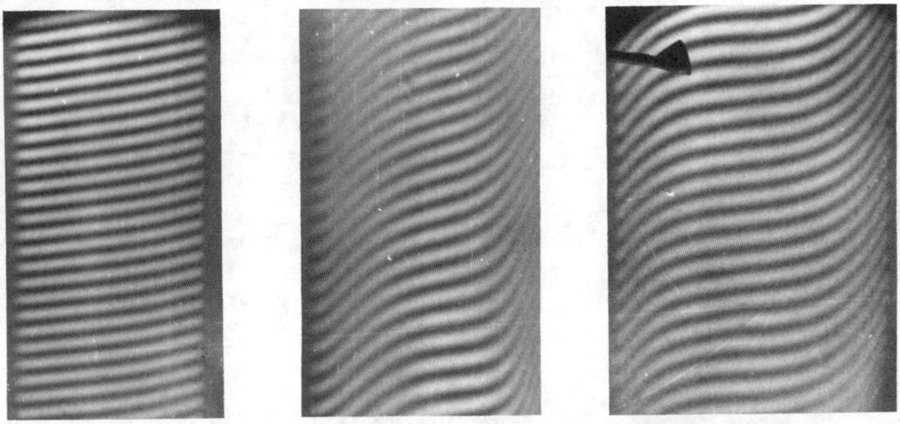

Figure 3 Figure 4 Figure 5

Figure 3-6 Flow regime characteristics

as abscissa increasing in the direction to the right that the points to the left of the curve represent a regime in which the temperature field of the fluid, besides these near the corners, possesses a dimensionless temperature gradient, $\partial\theta/\partial y_o=-1$, and points just to its right have to show a temperature distribution of $\partial\theta/\partial y_o \succ -1$ at $y_o=\frac{1}{2}$. Similarly, the demarcation between the transition regime and the boundary layer regime is defined as a curve in the Ra-A_s diagram, to the left of which the Ra-A_s points show a temperature distribution of $(\partial\theta/\partial y_o) \succ -1$ at $y_o=\frac{1}{2}$, and to the right of which any Ra-A_s point must have not only an evidence of thermal boundary layer showing on both the hot and the cold wall, but also in the core, besides the top and bottom corner regions, a temperature distribution of $\partial\theta/\partial y_o \geqslant 0$ at $y_o = \frac{1}{2}$.

As an illustration, Figure 3-6 are appended. The interferogram as shown in Figure 3 exhibits the feature that a series of straight fringes starts to bend. However, some straight fringes in the upmost portion, near to the mid height of the air layer, are still showing. Therefore, the heat transfer is mainly of conduction. Figure 4 displays an interferogram in the transition regime since there is no single fringe with flat portion in the core, although the fringes becomes much inclined and curved and the thermal boundary layers are forming on both walls. Figure 5 shows that the boundary layer are almost separated with each other but fringes with flat portion in the core is still rare. The core portion of some fringes shown in Figure 6 displayes flatness with an indication that the boundary layers are going to be completely seperated from each other. It is in the situation approaching the boundary layer regime. The lower portion of this interferogram is close to the bottom corner of the enclosure where the fringes become more distorted.

According to the above definitions, Ra-A_s points on these demarcation curves are determined in the following way. Set the interferometer to a series of horizontal parallel fringes when the testing rays pass through an uniform density field, such as when the enclosure is at $\Delta T = 0$. For a given aspect ratio, start the test by heating up the hot wall to increase ΔT from zero upward in steps to a series of testing values of ΔT. To cross a demarcation line, the increment of ΔT is set to be 1°-2^oC. For each value of ΔT, after the steady state has been reached, interferograms are photographed.

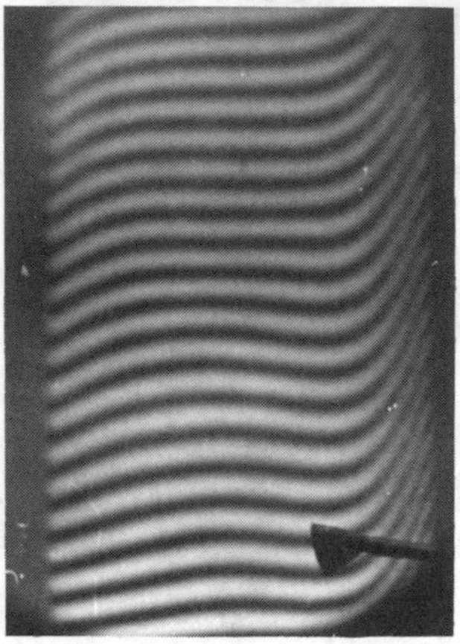

Figure 6

From these tests, we found the following points which fall on the demarcation line between the conduction regime and the transition regime:

A_s:	8	10	20	25	33.33	50	80
Ra:	2199	2392	4398	6408	7583	9301	13194

Employing the least-squares technique, the following equation is found to represent this demarcation line,

$$Ra = 403.68\ A_s^{0.81} \tag{1}$$

The data obtained for the demarcation line between the transition regime and boundary lay regime are:

As:	4	16.67	20	25	50	70
Ra:	21990	68114	87375	103096	146600	219900

and the corresponding equation is:

$$Ra = 7694.7\ A_s^{0.78} \tag{2}$$

As shown in Figure 7, the demarcation lines defined and obtained in the present investigation located close to those found by Eckert and Carlson (12) and S.H.Yin et. al. (14) and have slopes intermediate between them.

HEAT TRANSFER CORRELATION

Natural convective heat transfer results in vertical rectangular enclosure could be correlated by dimensionless parameters in the following form:

$$Nu = f(Ra,\ Pr,\ A_s) \tag{3}$$

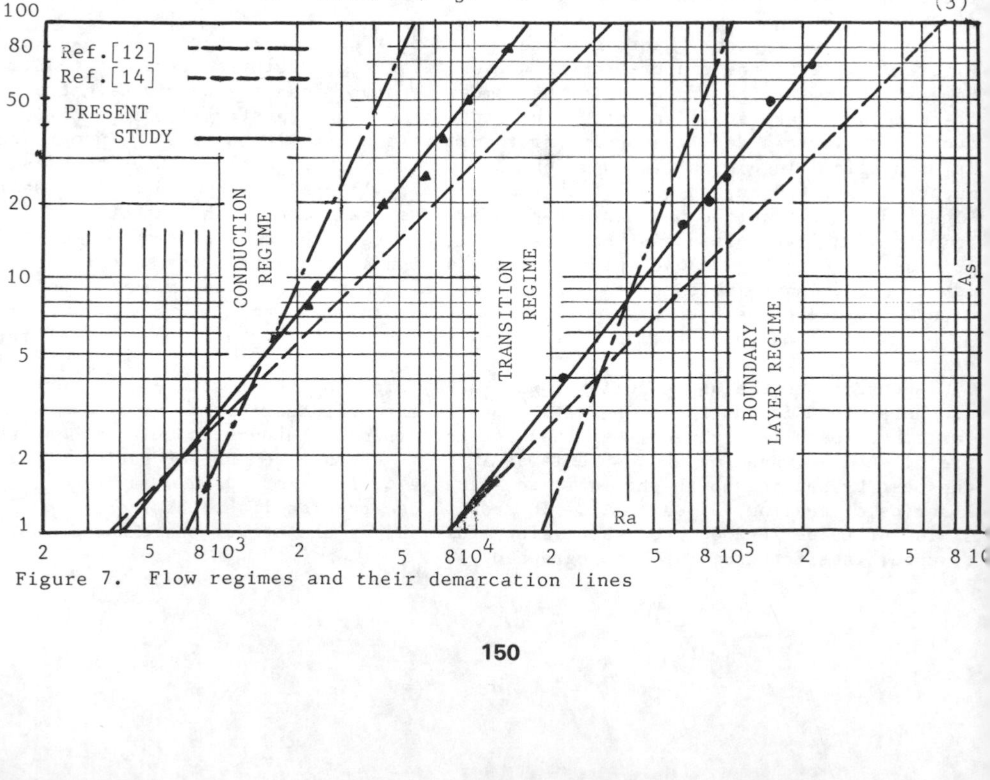

Figure 7. Flow regimes and their demarcation lines

For air layers at a temperature range not far from atmospheric, Pr is approximately constant. Equation (3) could then be simplified to:

$$Nu = f(Ra, A_s) \tag{4}$$

Experimental results of natural convection heat transfer in vertical rectangular enclosure were reported in the past. Jacob (9,10) correlated the test data of Mull and Reiher (11) and derived an equation for laminar flow at a range of $Gr = 2 \times 10^4 \sim 2 \times 10^5$ and $A_s = 10 \sim 42$:

$$Nu = 0.18\ Gr^{\frac{1}{4}}\ A_s^{-1/9} \tag{5}$$

Eckert and Carlson (12) gave the following correlation for the boundary layer regime with property values based on the wall temperature (either T_h or T_c):

$$Nu = 0.119\ Gr^{0.3}\ A_s^{-0.1} \tag{6}$$

A rough estimation from this paper, the range of Gr is about 10^4 to 5×10^6. MacGregor and Emery (7) obtained the following equation, evaluated from their data from experiments with a constant heat flux wall,

$$Nu = 0.420\ Ra^{\frac{1}{4}}\ Pr^{0.012}\ A_s^{-0.30} \tag{7}$$

which holds for $Pr = 1 \sim 2 \times 10^4$, $Ra = 10^4 \sim 10^7$, $A_s = 10 \sim 40$.
Yin et al. (14) with their experimental data obtained an overall correlation for $A_s = 4.9 \sim 78.7$ and $Gr = 1.5 \times 10^3 \sim 7 \times 10^6$:

$$Nu = 0.210\ Gr^{0.269} A_s^{-0.131} \tag{8}$$

Randall, Mitchell and El-Wakil (19) found that the effect of A_s negligible over their test ranges: $A_s = 9 \sim 36$, $Gr = 4 \times 10^3 \sim 3.1 \times 10^5$, $\phi = 45^o \sim 90^o$ and reported the equation:

$$Nu = 0.118\ [Gr\ Pr\ \cos^2\ (\phi\ 45^o)]^{0.29} \tag{9}$$

where ϕ is the enclosure tilt angle from the horizontal. With $\phi = 90^o$ for vertical rectangular enclosures, Equation (9) reduces to,

$$Nu = 0.0965\ Ra^{0.29} \tag{10}$$

In the present investigation, interferograms and other test data are evaluated with all the property values based on the average temperature of air, $(T_h+T_c)/2$, for Nu and Ra at various aspect ratios. Employing the least-squares technique, an overall correlation equation is obtained:

$$Nu = 0.191\ Ra^{0.30}\ A_s^{-0.19} \tag{11}$$

for the range of: $Ra < 4.24 \times 10^5$, and $16.67 \leq A_s \leq 100$. For comparison purposes, take $Pr = 0.713$ according to the temperature range in present work, Equation (11) is transformed into the following form:

$$Nu = 0.173\ Gr^{0.30}\ A_s^{-0.19} \tag{12}$$

After comparing Equations (11) and (12) with the previously reported Equations (5~10), it is seen that the experimental data obtained from present investigation are reasonably in agreement with the previously reported results.

CONCLUSION

An experimental study of natural convective heat transfer of air enclosed in vertical rectangular enclosures with large aspect ratios by the use of interferometry is reported. The demarcation lines in laminar flow, between conduction regime and transition regime and between transition regime and boundary layer regime, are determined in a Ra-A_s diagram by examination and evaluation of interferograms and other test data. Equations of demarcation lines are also obtained

An overall correlation for laminar flow in this type of heat transfer in ter of Nu, Ra and A_s is obtained which is in tolerable agreement with the previously reported experimental results.

REFERENCES

1. G.K.Batchelor, Heat transfer by free convection across a closed cavity betwe vertical boundaries at different temperatures, Q.Appl.Math. 12(3): 209-233, (1954)
2. G.Poots, Heat transfer by laminar free convection in enclosed plane layers, Q.J.Mechanics and Appl.Math. 2:257-273, (1958)
3. A.E.Gill, The boundary-layer regime for convection in a rectangular cavity, J.Fluid Mech. 26(3):515-536, (1966)
4. J.O.Wilkes and S.W.Churchill, The finite-difference computation of natural convection in a rectangular enclosure, A.I.Ch.E.Jl. 12(1):161-166, (1966)
5. J.W.Elder, Numerical experiments with free convection in a vertical slot, J.Fluid Mech. 24(4):823-843, (1966)
6. G.de Vahl Davis, Laminar natural convection in an enclosed rectangular cavit Int.J.Heat Mass Transfer 11:1675-1693, (1968)
7. R.K.MacGregor and A.F.Emery, Free convection through vertical plane layers-moderate and high Prandtl number fluids, J.Heat Transfer 91: 39-402, (1969)
8. M.E.Newell and F.W.Schmidt, Heat transfer by laminar natural convection with in rectangular enclosures, J.Heat Transfer 92:159-168, (1970)
9. M.Jacob, Free heat convection through enclosed plane gas layers, Trans. ASME 68:189-193, (1946)
10. M.Jacob: Heat Transfer, vol.1, p.534-539, J.Wiley, New Jork, (1949)
11. W.Mull and H.Reiher, Der Wärmeschutz von Luftschichten, Beih.Gesundh-Ing. Reihe 1, Heft 28, (1930), (as reported in (9,10))
12. E.R.G.Eckert and W.O.Carlson, Natural convection in a layer enclosed between two vertical plates with different temperatures, Int.J.Heat Mass Transfer 2: 106-120, (1961)
13. J.W.Elder, Lamina free convection in a vertical slot, J.Fluid Mech. 23(1):7 -98, (1965)
14. S.H.Yin, T.Y.Wung and K.Chen, Natural convection in an air layer enclosed within rectangular cavities, Int.J.Heat Mass Transfer 21:307-315, (1978)
15. S.Ostrach, Natural convection in enclosures, Advances in Heat Transfer, eds. J.P.Hartnett and T.F.Irvine, vol.8, p.161-227, (1972)
16. I.Catton, Natural convection in enclosures, Proc. 6th Int. Heat Transfer Conf. vol.6, p.13-31, (1978)
17. S.Ostrach, Natural convection heat transfer in cavities and cells, Proc.7th Int. Heat Transfer Conf. vol.1, p.365-379, (1982)
18. Ming-Yu Wang, Natural convection heat transfer inside rectangular cavity wi large aspect ratios, M.S.Thesis, Nanjing Institute of Technology, Nanjing, (1984)
19. K.R.Randall, J.W.Mitchell and M.M.EL-Wakil, Natural convection heat transfe characteristics of flat plate enclosures, J.Heat Transfer 101:120-125, (19

General Correlating Equations for Free Convection Heat Transfer from a Vertical Cylinder

S. M. YANG
Shanghai Jiaotong University
Shanghai, PRC

ABSTRACT

Analysis and experimental evidences indicate that for $Ra_L^{1/4} < 32$, physical laws for free convection heat transfer from vertical cylinder deviate from that of the vertical flat plate. Therefore, different general correlating equations are necessary. Up to now, no such general correlating equations for vertical cylinder are available. It is the purpose of this paper to fill this gap. The practical importance of such general correlating equations are evident because these equations are particularly useful in modern computer calculations. The method proposed by Churchill and Usagi is extended to the present case. In the whole span of independent variable, four differnt regimes may be identified: the diffusion, the curvature significant, laminar and turbulent regimes. Satisfactory general correlating equations for the first three regimes and for all four regimes are developed. The degenerating case of $D = \infty$ and the application of these general correlating equations to free convection mass transfer are discussed.

NOMENCLATURE

D	diameter of cylinder [m]	k	thermal conductivity [J/m.s.k]
g	acceleration due to gravity [m^2/s]	N_u	hz/k, local Nusselt number at z
		$\overline{N_u}$	$\bar{h}$z/k, mean Nusselt number over 0-z
h	local heat transfer coefficient [$J/m^2.s.K$]	P_r	ν/α, Prandtl number
		L	height of cylinder, [m]
$\bar{h}$	mean heat transfer coefficient over 0-z [$J/m^2.s.K$]	R_a	$g\beta(T_s - T_b)z^3/\nu\alpha$, Rayleigh number
		z	linear dimension [m]

Greek Symbols

α thermal diffusivity [m^2/s]

β thermal coefficient of expansion [K^{-1}]

ν kinematic viscosity [m^2/s]

Subscripts

b bulk

D with cylinder diameter D as characteristic dimension

L with cylinder height L as characteristic dimension

s surface

O limiting behavior for small z

∞ limiting behavior for large z

INTRODUCTION

General correlating equations are available in the literature of free convection heat transfer for the cases of vertical flat plate and horizontal cylinder. To compare with the usual experimental simple power law relationships between dimensionless groups, these general correlating equations have distinct advantages. The usual simple power law relationships are applicable only in a restricted range of the independent dimensionless group. To apply these simple power law relationships it is necessary to calculate the value of the independent dimensionless group first in order to select the proper correlation. By virture of their general applicability, the general correlating equations apply to the whole range of independent dimensionless group and prior confirmation of the application range is unnecessary. Therefore, the general correlating equations are particularly useful in computer calculations.

As has been shown in a previous paper by the author [1], that the dimensionless equation relevant to the free convection heat transfer of a vertical cylinder is

$$Nu_D = f_1(Ra_D D/L, Pr) \qquad (1)$$

or its equivalent

$$Nu_{L}/Ra_L^{\frac{1}{4}} = f_2(Ra_L^{\frac{1}{4}} \cdot D/L, Pr) \qquad (2)$$

In the whole range of $Ra_D D/L$, four different regimes may be identified. In addition to the usual turbulent and laminar regimes, for sufficient small values of $Ra_D D/L$, there is a boundary layer regime where the curvature effect is significant, for even smaller values of $Ra_D D/L$, there is a diffusion regime where the boundary layer analysis no longer valid. In the two regimes with small values of $Ra_D D/L$, experimental correlations differ from that of the vertical flat plate, so the general correlating equations for free convection heat transfer of the vertical flat plate proposed by Churchill and Chu [2] will not be expected to apply to the whole range of $Ra_D D/L$ for the present case. New general correlating equations are needed. To the best knowledge of the author, no general correlating equations are available for free convection heat transfe of the typical geometric case of a vertical cylinder. It is the purpose of this paper to fill this gap.

CORRELATING EQUATION FOR DIFFUSION, CURVATURE SIGNIFICANT AND LAMINAR REGIMES

The method developed in terms of the model of Churchill and Usagi [3] in establishing the general correlating equations reveals its great power in developing full range formulation of intrisic physical laws of a physical phenomenon as indicated in the cases of free convection heat transfer from vertical flat plate and that of horizontal cylinder. This method will be extended to the present case to present simple but general correlations for space-mean value of the heat transfer rate for free convection heat transfer from a vertical cylinder.

In terms of the model of Churchill and Usagi [3], the general correlating equations take the following form:

$$y^n\{x\} = y_o^n\{x\} + y_\infty^n\{x\} \qquad (3)$$

where x is the independent variable and y is the dependent variable. To develop correlating equations by this method, appropriate expressions for limiting behavior for both small and large values of the independent varible x, i.e. $y_o(x)$ and $y_\infty(x)$, are required.

In this section, general correlating equation for the range including diffusion, curvature significant boundary layer and laminar regimes will be considered. The expression for limiting behavior for large value of $Ra_D D/L$ will be proposed in the first place. Next, that for the small value of $Ra_D D/L$ will also be proposed.

The criterion for the difference of heat transfer rate of vertical cylinder within 5% of that of the vertical flat plate has been established by Sparrow and Gregg [7]:

$$Ra_L^{\frac{1}{4}} \cdot \frac{D}{L} \geqslant 32 \qquad (4)$$

In terms of $Ra_D D/L$:

$$Ra_D D/L \geqslant 1.05 \times 10^6 \qquad (5)$$

For even larger $Ra_D D/L$, the curvature effect is negligible and correlating equation for vertical flat plate may also be used for vertical cylinder. The expression developed by Churchill and Chu [2] for free convection of vertical flat plate in the laminar regime $10^5 < Ra_L < 10^9$ apparently is applicable also for large values of $Ra_D D/L$ for vertical cylinder in the regimes considered in this section. Their expression derived from the laminar boundary layer theory is of the following form:

$$Nu_L = Ra_L^{\frac{1}{4}} \cdot f\{Pr\} \qquad (6)$$

Multiply both sides of the equation by D/L converts equation (6) to the form of equation (1):

$$Nu_D = (Ra_D D/L)^{\frac{1}{4}} \cdot f\{Pr\} \qquad (7)$$

where

$$f\{Pr\} = 0.670/[1 + (0.492/Pr)^{9/16}]^{4/9} \qquad (8)$$

Equation (8) represents the various computed values of $f\{Pr\}$ within 1% from $Pr = 0$ to $Pr = \infty$ and is in general agreement in the laminar regime with the widely scattered experimental data compiled by Ede [4].

For $Ra_D D/L \rightarrow 0$, a general accepted solution has not been derived. As proposed by Elenbaas [5], the physical mode of heat transfer for very small $Ra_D D/L$ is pure conduction through a cylindrical fluid film. For pure conduction form an infinitely small wire, $\overline{Nu} = 0$, but for a wire of finite dimension $\overline{Nu}$ has a finite value. Since for pure conduction, the orientation of the cylinder does not make any difference in the heat transfer rate, the limiting value for horizontal cylinder may also be used in the case of vertical cylinder. Therefore the empirical limiting Nu value of 0.36 proposed by Tsubouchi and Masuda [6] based on the analysis of several sets of data is adopted in the present case as the limiting $\overline{Nu}$ value for $Ra_D D/L \rightarrow 0$.

Combining the latter value with equation (7) in the form of the model of Churchill and Usagi results in the following test expression for $Ra_L < 10^9$:

$$\overline{Nu}_D^n = (0.36)^n + \left(\frac{0.670[Ra_D D/L]^{\frac{1}{4}}}{[1 + (0.492/Pr)^{9/16}]^{4/9}}\right)^n \tag{9}$$

Test plots of experimental data in this form indicate that n = 1 is a reasonable choice, yielding the following correlation:

$$\overline{Nu}_D = 0.36 + \frac{0.670[Ra_D D/L]^{\frac{1}{4}}}{[1 + (0.492/Pr)^{9/16}]^{4/9}} \tag{10}$$

Equation (10) is shown in Fig.1 as the solid line curve. This curve is seen to provide a good representation of representative data from various sources [10, 12-15] for all Pr. The dotted line in Fig. 1 represents the experimental correlation of the heat transfer of vertical flat plate in the laminar regime. It represents an equation similar to equation (10) but without the 0.36 term on the right hand side.

The boundary condition of the majority of the data plotted in Fig. 1 is constant wall heat flux. However, the data of Carne [10], which was obtained under the constant wall temperature boundary condition, are also plotted in Fig.1. No significant difference is evident between these two different boundary conditions. Therefore, the same correlating equation is recommended for both constant heat flux and constant wall temperature boundary conditions.

CORRELATING EQUATION INCLUDING ALSO THE TURBULENT REGIME

It is well established that in the turbulent regime the correlating equation of vertical flat plate may be used for vertical cylinder as well. Therefore, it is possible to take advantage of the analysis of Churchill and Chu [2] for vertical flat plate on the limiting value of Nu number for large Ra numbers. No asymptotic solution is available for $Ra \rightarrow \infty$, but Churchill and Chu [2] recommended the following expression for large Ra numbers:

$$\overline{Nu}_L \longrightarrow 0.150\, Ra_L^{1/3}\, \varphi\{Pr\} \tag{11}$$

where $\varphi\{Pr\}$ is a function which approaches unity for $Pr \rightarrow \infty$ and is proportional to $Pr^{1/3}$ for $Pr \rightarrow 0$. $\varphi\{Pr\}$ takes the following specific form:

$$\varphi\{Pr\} = [1 + (0.492/Pr)^{9/16}]^{-16/27} \tag{12}$$

On the other extreme, however, the limiting value $\overline{Nu}_D = 0.36$, which is different from the limiting value for the vertical flat plate, must be used. Transformation of characteristic dimension yields

$$\overline{Nu}_L \rightarrow 0.36 \frac{L}{D} \tag{13}$$

The resulting test expresion in terms of Churchill and Usagi is therefore of the following form:

$$\overline{Nu}_L^n = (0.36\frac{L}{D})^n + [0.150Ra_L^{1/3}\varphi\{Pr\}]^n \tag{14}$$

Trial plots indicate that n = 1/2 is a reasonable choice. The straight line drawn in Fig.2 represents the following relationship

$$\overline{Nu}^{\frac{1}{2}} - 0.60(\frac{L}{D})^{\frac{1}{2}} = 0.387(\frac{Ra_L}{[1 + (0.492/Pr)^{9/16}]^{16/9}})^{1/6} \tag{15}$$

Representative experimental data from various sources [11,12,15] are also plotted in the same Figure. It is apparent from Fig.2 that the expermental data are in good agreement with the general correlating equation (15).

DISCUSSIONS

It is interesting to note the degenerating case of $D \rightarrow \infty$. For $D \rightarrow \infty$, physically the problem degenerates to the vertical flat plate case. Therefore, the present case may be considered as a general case which includes the vertical flat plate as a special case. With $D \rightarrow \infty$, the second term on the left side of equation (15) drops out, yielding the following reduced equation:

$$\overline{Nu}_L^{\frac{1}{2}} = 0.387(\frac{Ra_L}{[1 + (0.492/Pr)^{9/16}]^{16/9}})^{1/6} \tag{16}$$

In comparison with the correlating equation recommended by Churchill and Chu for vertical flat plate [2], equation (16) displays a slight reduction of the discrepancies between the experimental data and the correlating line for $Ra_L > 10^4$. In reference [2] all data fall below their correlating line. Now nearly all experimental data still fall below the correlating line similar to the situation in reference [2]. The principal uncertainy in the correlations proposed herein arises from the uncertainty in the limiting solution and experimental data for $Ra \rightarrow \infty$. Further works on the limiting solution and experimental data of very high Ra numbers are highly desirable.

Mass transfer data of Wilke, Tobias and Eisenberg [9] are also plotted in Fig.2 and reasonalbe agreement with equation (15) is indicated. Experimental evidence, therefore, supports the following extension. With $\overline{sh}$ substituted for $\overline{Nu}$, Sc for Pr and Ra' ($Ra' = g\gamma(\omega_s - \omega_b)z^3/\nu\mathcal{D}$ is the Rayleigh number for mass transfer, where ω_s and ω_b are surface mass fraction and bulk mass fraction and D is the diffusivity) for Ra, equation (15) and (10) are expected to hold for mass transfer as long as the net rate of mass transfer is not so high as to affect the velocity field significantly. On the basis of the results of Saville and Churchill [8] another extension can be made. For the special case of Sc = Pr in simultaneous heat and mass transfer, $\overline{Nu}$ and $\overline{Sh}$ can be calculated from equation (15) merely by substituting Ra + Ra' for Ra.

CONCLUSIONS

1. To compare with the usual simple power law relationships, the general correlation have distinct advantages. Unlike the application of simple power law relationships, prior check of the specific application range is unnecessary in the application of the general correlating equations. Therefore, the general correlating equations are particularly useful in computer calculations.

2. Equation (15) provides a good representation of the mean heat transfer rate for free convection from a vertical cylinder over a complete range of Ra and Pr from 0 to ∞. For $Ra_L > 10^4$, equation (16) shows a slight improvement over previous recommended correlating equation for the vertical flat plate, which is a special case of a cylinder with $D \rightarrow \infty$.

3. For the heat transfer of diffusion, curvature significant and laminar regimes, i.e. for $Ra_L < 10^9$, the simpler general correlating equation (10) rather than equation (15) is recommended for convenience.

REFERENCES

[1] Yang, S.M.,Free Convection of Heat Outside Slender Vertical Cylinders and Inside Vertical Tubes, Journal of Xian Jiaotong University, Vol. 14, No. 3, 115-131, (1980).(in Chinese).

[2] Churchill, S.W. and Chu, H.H.S., Correlating Equations for Laminar and Turbulent Free Convection From a Vertical Plate, Int. J. Heat Mass Transfer, Vil. 18, 1323-1329, (1975).

[3] Churchill, S.W. and Usagi, R., A General Expression for the Correlation of Rate of Transfer and Other Phenomena, A.I.Ch.E Jl., Vol. 18, 1121-1128 (1972)

[4] Ede. A.J., Advances in Free Convection, in "Advances in Heat Transfer", edited by Hartnett, J.P. and Irvine, T.F. Jr., Vol 4, pp. 1-64, Academic Press, New York (1967).

[5] Elenbaas, W., Dissipation of Heat by Free Convection from Vertical and Horizontal Cylinders, J.Appl. Phys., Vol. 19, 1148 (1948).

[6] Tsubouchi, T. and Masuda H., Heat Transfer by Natural Convection from Horizontal Cylinders at Low Rayleigh Numbers, Report Inst. High Speed Mech., Japan, Vol. 19, 205-219 (1967/1968).

[7] Sparrow, E.M. and Gregg, J.L., Laminar Free Convection Heat Transfer from the Outer Surface of a Vertical Circular Cylinder, Trans. ASME, Vol. 78 1823-1829 (1956).

[8] Saville, D.A. and Churchill, S.W., Simultaneous Heat and Mass Transfer in Free Convection Boundary Layers, A.I.Ch.E.Jl., Vol. 16 268-273 (1970).

[9] Wilke, C.R., Tobias, C.W. and Eisenberg, M., Free Convection Mass Transfer at Vertical Plates, Chem. Engng., Vol. 49, 663-674 (1953).

[10] Carne, J.B.,Heat Loss by Natural Convection from Vertical Cylinders. Phil. Mag., Journal of Sci., Ser. 7, Vol.24, 635-653 (1937).

[11] Fujii, T., Experimental Studies of Free Convection Heat Transfer, Bull J.S.M.E., Vol. 2, No. 8, 555-558 (1959).

[12] Pchelkin, I.M., Heat Transfer of Vertical Tube under Natural Convection, in "Convective and Radiative Heat Transfer", pp. 56-64, AH USSR, Moscow (1960). (in Russian).

[13] Hama, F.R. et. al., The Axisymetric Free Convection Temperature Field along a Vertical Thin Cylinder, J. Aero. Space Sci., Vol. 26, No. 6, 335 (1959).

[14] Mueller, A.C., Heat Transfer from Wires to Air in Parallel Flow, Transaction American Institute of Chemical Engineers, Vol. 38, 613-627 (1942).

[15] Ren, H.Z., Laminar Free Convection Heat Transfer of Vertical Cylinder in Air, Journal of Xian Jiaotong University, (1964). (in Chinese).

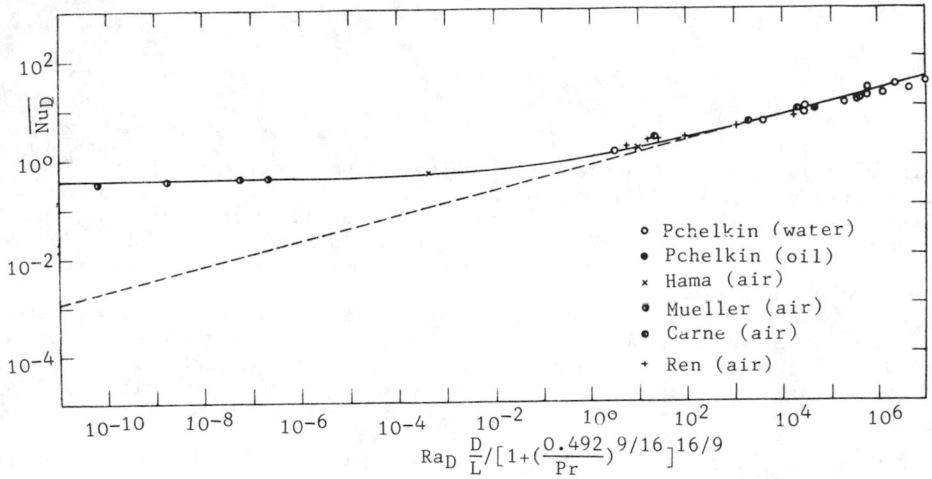

Fig.1 Comparison of correlating equation (10) with experimental data of vertical cylinder for $Ra_L < 10^9$

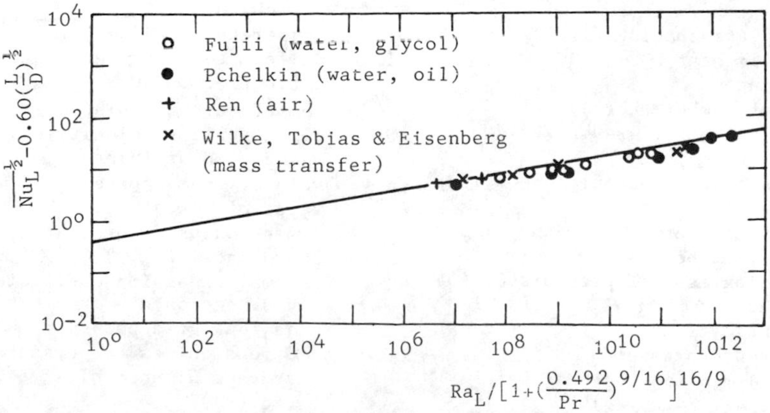

Fig.2 Comparison of Correlating equation (15) with experimental data for experimental data for vertical cylinder

Directly Solving a Conjugate Flow and Heat Transfer Problem—Natural Convection in a Vertical Plane-Walled Channel and External Natural Boundary Layer Flow

ZHENG ZHANG
Beijing Institute of Chemical Technology
Beijing, PRC

ABSTRACT

A conjugate flow and heat transfer problem can be met in many cases. There are some papers to show the numerical solution of this kind of problems, but all of them use iteration method to match the thermal boundary conditions on the common wall. It will take much computer time to finish the iterations. For more effective calculation, a directly numerical method to solve this kind of problem is derived in this paper.

An interesting conjugate problem, the effect of the external natural boundar layer flow on the natural convection inside the channel, is solved by our direct method, as an example. Some discussion of the calculation results is shown up i the paper, which shows that heat rate passing through the common wall should be considered as an important effect, especially in small Gr region.

NOMENCLATURE

a	average surface heat conductance
an,ap,as	coefficients of the discretization eq.(16)
b_{ij}	constant term in eq.(16)
b	interplate spacing(fig.1)
c	specific heat
d	constant
Gr	Grashof number(eq.(5))
g	accelaration of gravity
k	heat conductivity
L	dimensionless channel length(1/Gr)
l	channel length(fig.1)
Nu	Nusselt number
Pr	Prandtl number
P	dimensionless dif.pressure(eq.(5))
p	pressure in channel
p_o	ambient pressure
p'	difference pressure
Q', Q_1', Q_2'	dimensionless heat rate (eq.(24,26,28))
Q, Q_1, Q_2	value of Q', Q_1', Q_2' at exit of the channel
$Q_{in}=Q$	total dimensionless heat rate entered from wall 1
$Q_{out}=Q_1$	dimensionless heat rate brought out by internal flow at exi
q', q_1', q_2'	heat rate(eqs.(23),(25),(27))
q, q_1, q_2	value of q', q_1', q_2' at exit of th channel
T	temperature
T_1	uniform temperature of wall 1
T_o	ambient temperature
	temperature of inlet air
U,V	dimensionless velocity(eq.(5))
U_o	value of U at inlet
u,v	velocity components
u_o	value of u at inlet
X,Y	dimensionless coordinates(eq.(5))
x,y	coordinates(fig.1)
β	thermal expansion coefficient
$\delta y_1, \delta y_2$	y direction distance between two adjcent grid points
θ	dimensionless temperature(eq.(5)
$\bar{\theta}$	average dimensionless temperatur (eq.(5))
ν	kinematic viscosity
ρ	density
ρ_o	ambient density

Subscripts

1 internal flow
2 external flow
i number of calculation domain
j grid location in y direction
w2 wall(2)

Superscripts

* refer to a reference case

INTRODUCTION

The nature convection flow in a vertical, plane-wall channel has been studied by numerical method in many papers since 1962. Among them, Bodoic and Osterle[1] first described the numerical investigation of developing natural convection flow and heat transfer between two heated vertical plates with same uniform temp. T_w. Carpenta et al.[2] and Sparrow et al.[3] discussed the effect of radiation on developing laminar convection flow and heat transfer between two vertical flat plates Sparrow and Tao[4] numerically investigated a buoyancy-driven fluid flow and heat transfer in a pair of interacting vertical parallel channels. This is a conjugate flow and heat transfer problem. An iteration scheme had to be used to match heat flux and temperature continuity at the common wall. There were several other papers[5,6] which discussed different conjugate flow and heat transfer problems. There was no exception without adapting an iteration scheme to match the heat flux and temperature continuity at the common wall. Normally it visits two flows alternatively and transfered thermal information from one flow to the other across the common boundary many times.

In this paper, a similar conjugate flow and heat transfer problem, which describes interaction between the nature convection in a vertical plane-walled channel and natural convection boundary layer flow external to the channel, is discussed, but iteration is discarded. For directly solving the problem a discretization energy equation, which connects two flows, is derived to match the thermal boundary conditions at the common wall to instead of iteration method in the paper.

ANALYSIS

The problem we are interested in is sketched on fig.1.

A vertical channel is formed by two parallel flat walls of length l and infinite width, separated by a distance b.Wall(1) is maintained at uniform temperature T_1. Wall(2) is a very thin plate, so it has no resistance to heat flux and the wall is directly exposed to the ambient air with temperature T_o. Fluid rises in the channel by natural convection and is assumed to enter the channel at T_o with a uniform velocity profile u_o. Since wall(2) will be heated by rising fluid. The temperature on the wall(2) will be graduately increased along the length. The natural convection boundary layer flow will be caused by the tempeature difference between the wall and the ambient air. It will bring heat out from the channel, and give the effect on heat transfer and flow inside of the channel. This is a conjugate flow and heat transfer problem.

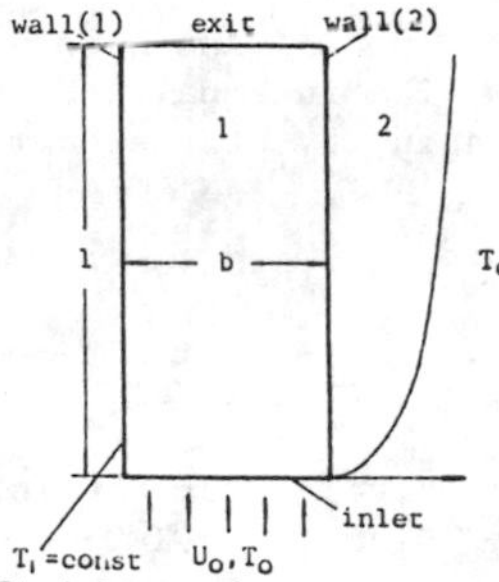

Fig.1 Scheme of the vertical channel-conjugate free internal flow and external boundary layer flow

EQUATIONS

For moderate differences between T_1 and T_o, both internal and external flows are governed by[7]

$$\frac{\partial u}{\partial x} + \frac{\partial v}{\partial y} = 0 \tag{1}$$

$$u\frac{\partial u}{\partial x} + v\frac{\partial u}{\partial y} = \nu\frac{\partial^2 u}{\partial y2} - \frac{1}{\rho}\frac{dp}{dx} - g \tag{2}$$

$$u\frac{\partial T}{\partial x} + v\frac{\partial T}{\partial Y} = \frac{k\partial^2 T}{\rho c\partial y2} \tag{3}$$

By defining $p' = p - p_o$, there p_o is the pressure at a particular elevation outside the channel and $dp_o/dx = -\rho_o g$, equation(2) can be written as

$$u\frac{\partial u}{\partial x} + v\frac{\partial u}{\partial y} = \nu\frac{\partial^2 u}{\partial y2} - \frac{1}{\rho}\frac{dp'}{dx} + \beta g\,(T - T_o) \tag{4}$$

where β is the expansivity of the fluid defined by $\beta = -1/\rho(\partial\rho/\partial T)_p$.

To start it, dimensionless variables for both flows are introduced as following:

$$Gr = \frac{\beta g(T_1 - To)}{l\nu^2}b^4, \quad P = \frac{p'b^4}{\rho l^2\nu^2 Gr^2}, \quad \theta = \frac{T - To}{Tl - To}$$
$$X = \frac{x}{lGr}, \quad Y = \frac{y}{b}, \quad U = \frac{b^2 u}{l\nu Gr}, \quad V = \frac{bv}{\nu} \tag{5}$$

Under the assumption of constant physical properties and exactly same fluids in both flows except the density difference to drive buoyancy. The dimensionles conservation equations for both internal and external flows are same as:

$$\frac{\partial U}{\partial X} + \frac{\partial V}{\partial Y} = 0 \tag{6}$$

$$U\frac{\partial U}{\partial X} + V\frac{\partial U}{\partial Y} = -\frac{dP}{dX} + \frac{\partial^2 U}{\partial Y^2} + \theta \tag{7}$$

$$U\frac{\partial \theta}{\partial X} + V\frac{\partial \theta}{\partial Y} = \frac{1}{Pr}\frac{\partial^2 \theta}{\partial Y^2} \tag{8}$$

Initial and boundary conditions:

The internal and external problems are coupled by temperature and heat flux continuity at the common wall of two flows.

$$\theta_{1,Y=1} = \theta_{2,Y=0} \tag{9}$$

and $$\left(\frac{\partial \theta}{\partial Y}\right)_{1,Y=1} = \left(\frac{\partial \theta}{\partial Y}\right)_{2,Y=0} \tag{10}$$

additively, for the internal flow

$\theta_1 = 0$ at $X=0$ and $\theta_1 = 1$ at $Y_1 = 0$ (11)

$U_1 = U_o$ at $X=0$ and $U_1 = 0$ at $Y_1 = 0$ and $Y_1 = 1$ (12)

$P_1 = 0$ at both $X=0$ and $X = 1/Gr$ (13)

for the external flow

θ_2, $U_2 \longrightarrow 0$ as $Y \longrightarrow \infty$ and $X \longrightarrow 0$ (14)

and $U=0$ at $Y=0$ (15)

It is obvious that the flow and heat transfer in two flows are tightly couple by the thermal boundary conditions at the common wall(eqs.(9,10)). Since an iteration scheme would spend too much computer time to reach convergence, it is prefered to use a direct method to solve a conjugate problem in one procedure. Here the prefered method is derived as following.

SOLUTION METHODOLOGY

At first, the equations(6) and (7) for both internal and external flows can be solved independently, since velocity boundary condition has no interconnection between two flows at the common wall.

For solving the coupled heat transfer problem, the numerical solutions were carried out by adapting the Patankar-Spalding(P.-S.)method to match the thermal boundary condition in one connection. A marching procedure is adapted to solve a set of linear algebraic eqs. in Y direction at each forward step[6].

$$ap_{i,j}\ \theta_{i,j} = as_{i,j}\ \theta_{i,j-1} + an_{i,j}\ \theta_{i,j+1} + b_{i,j} \qquad (16)$$

$$i \in [1,2,]\ , \qquad j \in [2, M_i - 1]$$

which include two sets of discretization eqs, for two flows respectively, in which i=1 is for internal flow and i=2 is for external boundary layer flow.

As shown on fig.2, the common wall plays a dual rule in two calculation domains. At first it is the last point($j=M_1$) of the internal domain(i=1). Then it is the first point(j=1) in the domain of external flow(i=2).

To match the thermal boundary conditions of (9) and (10) at the common wall, the following formulars may be obtained:

$$\theta_{1,M_1} = \theta_{2,1} \qquad (17)$$

and
$$\frac{(\theta_{1,M_1} - \theta_{1,M_1-1})}{\delta Y_1} = \frac{(\theta_{2,2} - \theta_{2,1})}{\delta Y_2} \qquad (18)$$

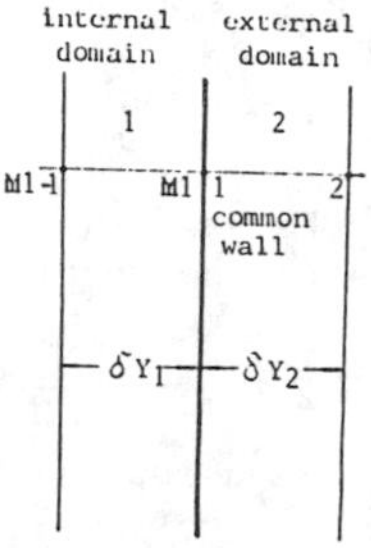

Fig.2 Grid in two sides of the common wall

Since
$$an_{1,M_1-1} = \left(\frac{1}{Pr}\right)\frac{1}{\delta Y_1} \qquad (19)[9]$$

and
$$as_{2,2} = \left(\frac{1}{Pr}\right)\frac{1}{\delta Y_2} \qquad (20)[9]$$

then
$$an_{1,M_1-1}\ (\theta_{1,M_1} - \theta_{1,M_1-1}) = as_{2,2}(\theta_{2,2} - \theta_{2,1})$$

$$ap_{1,M_1}\ \theta_{1,M_1} = ap_{2,1}\ \theta_{2,1} =$$

$$= an_{1,M_1-1}\ \theta_{1,M_1-1} + as_{2,2}\ \theta_{2,2} + b_{1,M_1} \qquad (21)$$

in which

$$ap_{2,1} = ap_{1,M_1} = an_{1,M_1-1} + as_{2,2} \quad \text{and} \quad b_{1,M_1} = 0 \qquad (22)$$

By combining eqs. (16),(21) and (22), a same type of numerical formulas for solving the temperature fields in both internal and external flows is obtained in one series, which can be solved at one procedure. There is no need to iterate at all. During the calculation procedure, the thermal boundary conditions(17,18) are automatically matched.

There is one more thing to say for the internal flow only. One pressure boundary condition is overspecified, since there are two pressure boundary conditions (13) for only the first derivative of P respective to X in differential eq.(17). The overspecified one (P=0 at X=1/Gr) would serve as a replacement for an unknown boundary condition of U at X=0 demanded by the pressence of ∂U/∂X in the conservation eqs. In general, some iteration is needed to do the replacement. To assume U_o across the inlet and to calculate the pressure at exit, and if it is not equal to zero, then to assume another value of velocity at inlet until zero pressure is obtained at exit. It is also prefered to discard the iteration procedure in this paper. To match the pressure boundary condition at exit, the only thing to do is to take the value of X at which P returns to zero as L.Once L is known, Gr can be evaluated from the identity LGr=1.

To sum up, the calculation procedures in our program are described as follows

(1) To start our program at X=0 by setting P=θ=0, U=a given value of U_o and the nominal Grashof number Gr. Of course Pr=0.7 is set in the program;

(2) Excuting the marching procedure. When the temperature eq.(8) is being solved, both internal and external flows should be connected together by using eqs.(16), (21),(22), but the velocity profile may be obtained separately in each flow, and dimensionless pressure P of internalflow is obtained at every stage;

(3) Recording the value of X as P is reached to zero and setting it as the dimensionless channel length L;

(4) Other heat transfer presentation parameters, such as Nusselt number, dimensionless heat flux, temperature along the common wall, are calculated at a certain given stages. The formulation to calculate these parameters will be shown in the next section.

FORMULATION OF PRESENTATION PARAMETERS

The heat rate absorbed by the fluid rising inside the vertical channel from entrance to a particular elevation X is

$$q_1' = \rho c \int_0^b u(T-T_o)\, dy \tag{23}$$

Dimensionless heat rate of q_1' is

$$Q_1' = \frac{q_1'\, b}{\rho c \nu\, l Gr (T_1 - T_o)} = \int_0^1 U\theta\, dy \tag{24}$$

The heat rate absorbed by the boundary layer flow rising along the common wall outside the vertical channel from entrance to a particular elevation X is

$$q_2' = \rho c \int_0^{y_{M1}} u(T-T_o)\, dy \tag{25}$$

Dimensionless heat rate of q_2' is

$$Q_2' = \frac{q_2'\, b}{\rho c \nu l Gr\, (T_1 - T_o)} = \int_0^{Y_{M1}} u\, \theta\, dY \tag{26}$$

Total dimensionless heat rate emitted from wall 1(fig.1) for the length of 0 to X should be equal to the sum of aforementioned two heat rates, i.e.:

$$q' = q_1' + q_2' \tag{27}$$

and

$$Q' = Q_1' + Q_2' = \frac{qb}{\rho c \nu\, l Gr\, (T_1 - T_o)} \tag{28}$$

An average surface heat conductance over the channel length can be defined by:

$$a = \frac{q}{l\, (T_1 - T_o)} \tag{29}$$

where q is the value of q' at X=L, Nusselt number is defined by Nu=a b/k. By substituting from eqs.(28) and (29), and the definition of Prandtl number $Pr=\rho c\nu/k$, it is to be given by

$$Nu = Q\ Pr\ Gr \tag{30}$$

where Q is the value of Q' at the top of the channel (X = L).

The flow in the entire channel is given by

$$f = bu_o = \int_0^b u dy \tag{31}$$

Its dimensionless form can be written as

$$\frac{fb}{l\nu Gr} = \int_0^b U dY = U_o \tag{32}$$

Total dimensionless heat rate brought out by the internal flow which leaves from the top of the channel is:

$$Q_{out} = Q_1 = \frac{q_1\ \ b}{\rho c\nu lGr\ (T_1 - T_o)} \tag{33}$$

Total dimensionless heat rate transfered into the channel from the whole wall(1). is:

$$Q_{in} = Q = \frac{qb}{\rho c\ \nu lGr(T_1 - T_o)} \tag{34}$$

The residue value of dimensionless heat rate $Q_2 = Q - Q_1 = q_2 b/[c\rho\nu lGr(T_1 - T_o)]$ represents the heat rate part which transfers through the common wall and is then brought away by the boundary layer flow rising along the common wall outside the vertical channel.

RESULTS AND DISCUSSION

Since Pr = 0.7 is more common to happen in the life, this parameter is fixed in all calculation cases. During the calculation, there are 60-102 grid-points for the internal flow and only 22 grid-points spread over the cross section of the external flow in Y direction. 600-1000 points are covered along the length. Special grid spacing is designed to make much more grid-points to be adjacent to the channel walls in our program.

For comparing, the calculation is not only done to the conjugate flow case which we are interested in, but also done to a reference case shown on fig.3. It is a vertical plate-channel consisting of two flat plates. One of them is kept at temperature of T_1 as the conjugate case aforementioned, but the other one is kept adiabatic which is different from the conjugate flow.

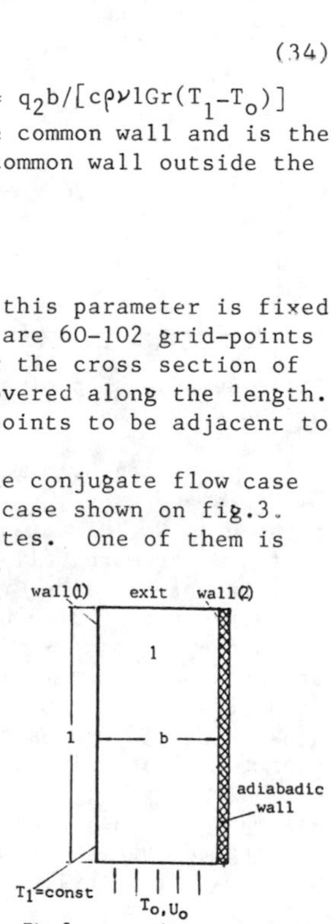

Fig.3 a comparing vertical channel

For both cases, dimensionless heat rate Q_{in} and Q_{out} versus Gr is compared in fig.4, Nusselt number Nu versus Gr is in fig.5, dimensionless average temperature $\bar{\theta}$ across the channel section and dimensionless common wall temperature θ_{w2} (both temperature values refer to the place of the channel exit) versus Gr is in fig. 7, and dimensionless internal flow rate U_o versus Gr is in fig.6. For clearity, all of the values in the reference case are shown with a superscript *.For directly showing the differences between two cases, relative values of foregoing parameters, such as Nu/Nu^*, Q_{in}/Q^*_{in}, $1-U_o/U^*_o$, $\theta^*_{w2}-\theta_{w2}$ and $\bar{\theta}^*-\bar{\theta}$ versus Gr are shown on fig.8.

From the results, some discusion might be described in the following:

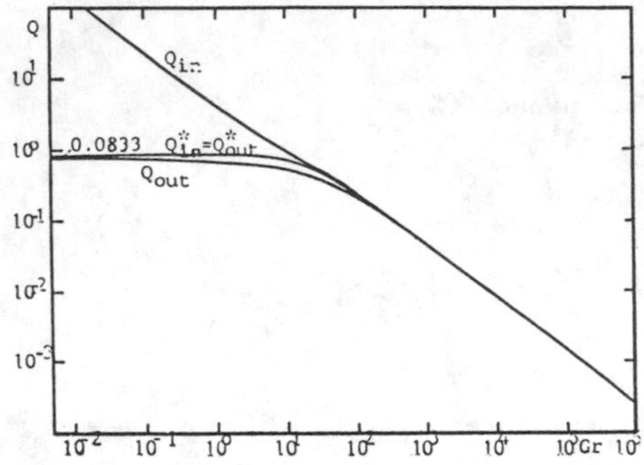

Fig.4 Relationship of dimensionless heat flux Q_{in}, Q_{out}, Q_{out}^* versus Gr

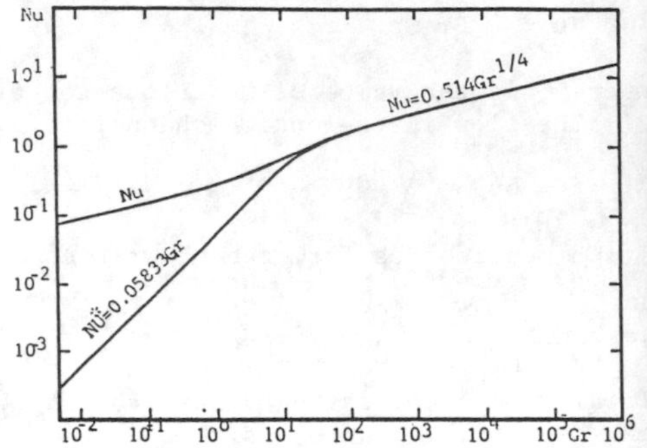

Fig.5 Relationship of Nusselt number Nu and Nu^* versus Gr

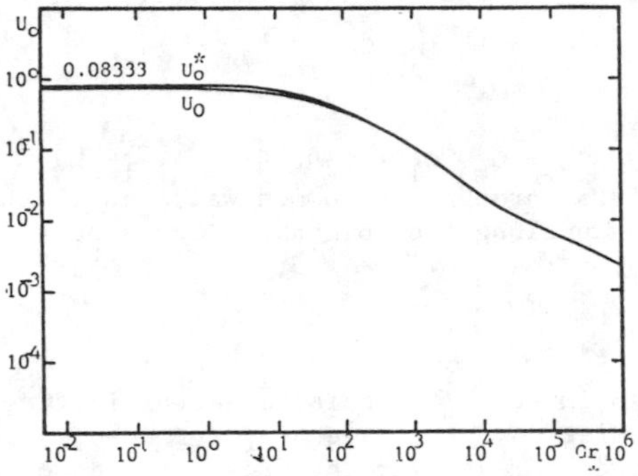

Fig.7 Relationship of average velocity U_o and U_o^* versus Gr

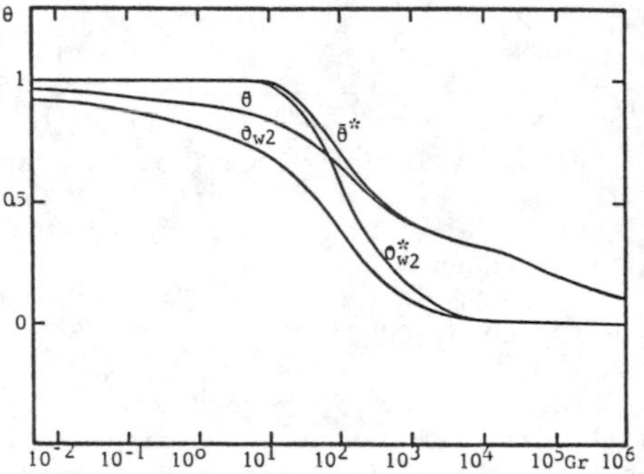

Fig.6 Relationship of dimensionless average temperature $\bar{\theta}$,$\bar{\theta}^*$ and temperature on the wall 2 θ_{w2}^*,θ_{w2}^* versus Gr (all of the temperature are refer to the exit of the channel)

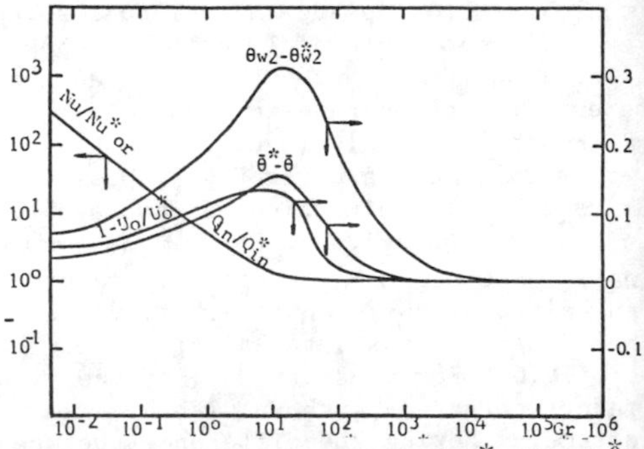

Fig.8 Relationship of Nu/Nu^*, Qin/Q_{in}^*, $1-U_o/U_o^*$, $\bar{\theta}^*-\bar{\theta}$ and $\theta_{w2}^* - \theta w2$ versus Gr

As shown on fig.4, dimensionless heat rate Q_{in} and Q_{out} in both cases will go down as Gr increases. When Gr is up to 10^3 or more, all values of Q_s' overslap at the same Gr number, and show that the (-3/4) power relation is to the Gr no. ($Q \propto Gr^{-3/4}$). The proportional constant is 0.734. As Gr decreases, in the reference case, $Q_{in}^* = Q_{out}^*$ increases first and then goes to a flat value 0.08333=1/12(it can be varified by fully developed assumption). As there is no heat passing through the common wall, $Q_{in}^* = Q_{out}^*$ is obvious at any Gr. Since $\bar{\theta}$ is always smaller than $\bar{\theta}^*$(fig.7) and U_o is also smaller than U_o^* (fig.6), Q_{out} is always less than Q^*_{out}, but Q_{out} will become much closer to Q_{out} when Gr is very small. Since the external boundary layer flow brings heat out passing through the common wall, Q_{in} is always larger than Q^*_{in}, but the difference is small when Gr is large. The difference becomes obvious only after Gr is smaller than 10. As Gr decreases further, Q_{in} is still going up when Q^*_{in} tends to a constant value, since heat keeps going out through the common wall in the conjugate problem. This makes Q_{in} much higher than Q^*_{in}. (this result can also be found from fig.8).

When Gr is greater than 10^2, Nu is almost equal to Nu^* and goes up with increasing Grashof no., lastly Nu is proportional to $Gr^{1/4}$, which is similar to the relationship for the natural boundary layer flow along a vertical wall. The proportional constant 0.514(see fig.5) is a little bit different from the results in[3], which may be caused by the calculation error with coarse grid in the boundary layer when the width of the channel b is too large with respect to a large Gr.

When Gr is less than 10^2, Nu^* goes down with decreasing Gr faster and faster, at last Nu^* is proportional to Gr with the proportional constant 0.05833. Since more heat transfers through the common wall as the channel length increases(Gr decreases),Nu will go much higher than Nu^* just as heat rate relations in fig.4. Similar results can be seen on fig.8 by the curves of Nu/Nu^* versus Gr.

The results above can be explained by the internal temperature changing as Gr changes. As shown on fig.6, when Gr is high both θ_{w2} and θ^*_{w2} are almost zero. Thus no heat flux would pass through the common wall at high Gr. In this way, the common wall plays a rule of an adiabatic wall. As a result, $\bar{\theta} = \bar{\theta}^*$, $Nu=Nu^*$, $Q_{in}=Q^*_{in}$ and U=U as Gr is high. After Gr is less than 10^4,θ^*_{w2} starts to leave zero and goes up with Gr decreasing gradually. Since heat flux passes through the common wall to the external cold air, for same Gr,θ_{w2} should always be less than θ^*_{w2} as θ^*_{w2} is not equal to zero. As Gr further decreases, θ^*_{w2} increases, and reaches to 1 at last, and keeps it constant(1) when Gr is small(about 8 or less). It shows that all of the internal flow is heated up to the temperature of wall(1) as wall(2) is adiabatic. For a conjugate flow, there is a certain of heat passing through the common wall, so θ_{w2} will not reachs to 1 even if Gr is less than 10^{-2} .

Since dimensionless average temperature is a somewhat average of the dimensionless temperature in the channel, $\bar{\theta}^*$ and $\bar{\theta}$ is somewhere between θ^*_{w2},θ_{w2} and 1 as the result shown on fig.6.

For the same reason, U_o is equal to U^*_o when Gr is high. As Gr decreases, longer channel will give larger driven forces to bring more air flow up and then both U_o & U^*_o increase. When Gr is less than 10^3, $\bar{\theta}$ less than $\bar{\theta}^*$ makes U_o less than U^*_o. The largest relative velocity difference $1-U_o/U^*_o$ is about 14%(at about 10 of Gr) and the largest relative average temperature difference $\bar{\theta}^*- \bar{\theta}$ is about 16-17%(at about 15 of Gr). U^*_o has a limit value of 1/12, when Gr is small(which can be verified by setting $\bar{\theta}$=1, dp/dx=0 for a long vertical channel with one uniform temperature wall and one adiabadic wall). This makes obvious result of Q^*_{in}, lim=1/12, and $Nu^*_{in,lim} = Q^*_{in,lim}PrGr=0.05833Gr$ as shown on fig.4 and 5. This proves that our results are accuracy enough.

The results on fig.4,5 and 8 show that the assumption of adiabadic wall can be adapted only for large Gr. For a small Gr, such as Gr is less than 10^2, heat rate passing through the common wall by the natural convection outside the channel should be considered, if the common wall is not made by an isolate material. Nu/Nu^* and Q_{in}/Q^*_{in} would become very large when Gr is very small.

Fig.9 and 10 show the development of dimensionless tempersture $\bar{\theta}$,$\bar{\theta}^*$ and dimensionless velocity profile U^*/U^*_o, U/U_o at a particular Grashof number(Gr=1). There is some difference of temperature profile and velocity profile between two cases as showing on the figures at some stages in the development. It can be explained in similar manner as aforementioned.

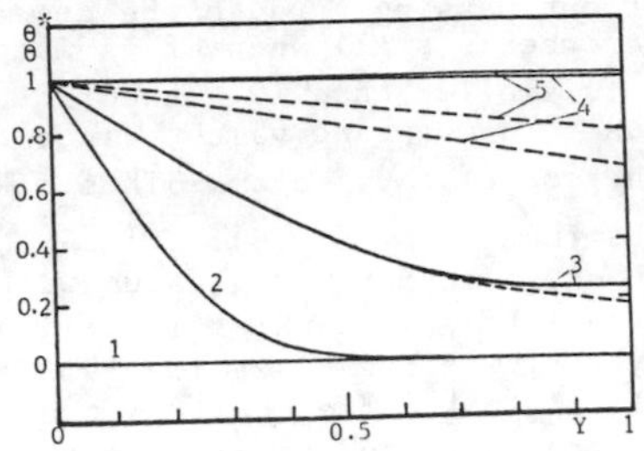

Fig.9 development of dimensionless temp. profile 1.X/L=0 2.X/L=10^{-3} 3.X/L=10^{-2} 4. X/L=10^{-1} 5.X/L=1 Gr=1,Pr=0.7
- - - - θ, ——— θ*

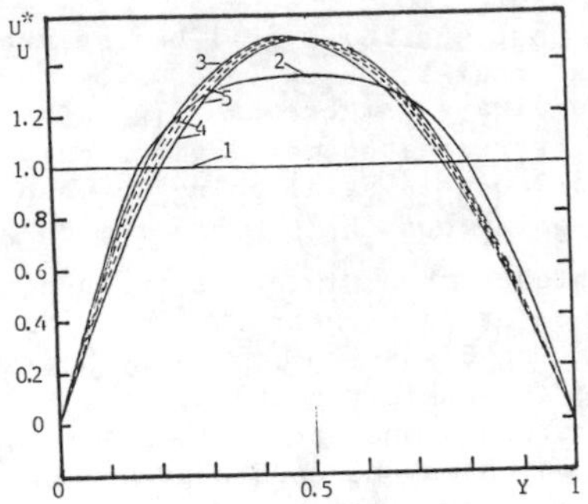

Fig.10 development of dimensionless vel. profile 1.X/L=0 2.X/L=10^{-3} 3.X/L=10^{-2} 4.X/L=10^{-1} 5.X/L=1 Gr=1,Pr=0.7
- - - - U/U_o ——— U^*/U_o^*

CONCLUSION

(1) Derived numerical formula can be used to solve a conjugate flow and hea transfer problem directively and makes the Patankar-Spalding method to treat a conjugate flow and heat transfer problem more effectively. As an example, a problem of conjugate flow and heat transfer of natural convection in a vertical plane-walled channel and external to the channel has been investigated in this paper. Another example will be presented recently[10];

(2) For the solved problem, large influence of the external boundary layer flow on the internal flow and heat transfer has been found in small Gr region. For large Gr, the common wall plays as an adiabatic wall, but the influence wi increase with decreasing Gr. It should be considered when Gr is less than abou 100;

(3) The development of velocity and temperature profile for the solved pro shows the influence of conjugate heat flux is to be seen only after θ_{w2} is grea than 0.

REFERENCES

1 Bodoia,J.R., and Osterle,J.F., Journal of Heat Transfer, Trans. of ASME, Vol. 84, pp.40-44,1962.
2 Carpenter,J.R.,Briggs,D.G., and Sernas,V., Int. J. of Heat and Mass Transfer Vol.15,pp.2293-2309,1972.
3 Sparrow,W.M.,Shah,S., and Prakash,C., Numerical Heat Transfer,Vol.3, pp.297-314,1980.
4 Sparrow,E.M., and Tao,W.Q., Numerical Heat Transfer, Vol.5, pp.39-58,1982.
5 Prakash,C., and Sparrow,E.M., Int. J. of Heat Transfer, Vol.24,No.5,pp.895-9 1981.
6 Sparrow,E.M., and Faghri, J. of Heat Trans, Trans. of ASME, Vol.102,No.3, pp.402-407,1980.
7 Schlichting,H., Boundary Layer Theory, Fourth Ed., p.302, McGraw-Hill Book, 1960.
8 Patankar,S.V., and Spalding,D.B., Heat & Mass Transfer in Boundary Layers, Second Ed., Intertext Books, London, 1970.
9 Patankar,S.V., Computation of Boundary Layer Flow, Course for Graduated Students, 8353, Univ. of Minnesota, 1981.
10 Zhang,Zheng, Parallel Flow and Heat Transfer in Double Tube Heat Exchanger, (to be published)
11 Patankar.S.V., Proc. 6th Int. Heat Transfer Conf., Toronto,Vol.3,p.297-,197

FORCED HEAT CONVECTION

Mixed Convection in the Laminar Entry Flow of Curved Circular Pipes

MITSUNOBU AKIYAMA, MASAMI SUZUKI,
MICHIYOSHI SUZUKI, and ICHIRO NISHIWAKI
Department of Mechanical Engineering,
Faculty of Engineering,
Utsunomiya University
Utsunomiya 321, Japan

K. C. CHENG
Department of Mechanical Engineering,
Faculty of Engineering,
The University of Alberta
Edmonton, Alberta, Canada T6G 2G8

ABSTRACT

A numerical and experimental study has been made on laminar convective heat-transfer in the thermal entrance region of curved tubes where special attentions have been paid to the interaction of centrifugal and buoyant body forces. Numerical calculation of the set of three dimensional parabolic type equations is extended over the ranges of Dean number to be from 20 to 1000,and of the product of Reynolds number and Rayleigh number to be 0 through $5x10^6$ for prandtl numbers of 0.7 and 5. Flow behavior of the results reveals the special characteristic of hydrodynamic and thermal entrance regions which can be categorized into three distinct regimes; namely centrifugal convection regime, fully mixed convection regime and natural convection regime. In fully developed region, a set of heat transfer correlations suitable for the three indivisual heat transfer regimes is proposed on the basis of the flow behavior. Experimental data were provided by flow visualization using a combination of hydrogen bubble and particle tracer methods, and by heat transfer experiment of axially uniform heat flux and constant temperature peripherally through which direct comparisons with numerical results were made satisfactory.

NOMENCLATURE

a=tube radius, De= $Re(a/R)^{1/2}$ Dean number, Gr= $\beta g a^4 (dT_b/dZ)/\nu^2$ Grashof number, Gz= 2RePr.a/Z Graetz number, fRe=$2(-\partial p/\partial z)/\overline{w}$, Nu=$Prw(\partial\theta/\partial z)/(\theta_w-\theta_b)$ Nusselt number, P= $(\rho\nu^2/a^2)p$, Ra= GrPr Rayleigh number, R_c=radius of cuveture, R=(a)r length in radial direction, Re= $2\overline{w}$ Reynolds number, W= $(\nu/a)w$ main flow, U =$(\nu/a)u$ flow in radial direction, V=$(\nu/a)v$ flow in peripheral direction, Z =$R_c\Omega$ main axis length, θ= $(T-T_b/a(dT_b/dZ))$ non dimensional temperature ,Sub.s straight tube.

1. INTRODUCTION

Curved tubes are used in various applications such as heat exchanger and cooling equipment, and have important role in the industrial field. Due to such importance, numerous theroretical as well as experimental works have been conducted on flow and heat transfer in curved pipes covering a wide range of aspects. In spite of the large number of works , the main results of theoretical work were more or less confined to the fields where the temperature and velocity fields were fully developed and the flow regimes were pure forced convection with centrifugal forces, and experimental works leave some ambiguities behind. It is hard to summarize then. The entry flow in curved channel with thermally-induced buoyant flow is obtaining attention recently by Patankar et al (1), Akiyama, et al (2,3,4) and Chilukuri and Humphrey(5). Yet,knowledge on thermal entry flow and buoyant flow in curved tube is quite insufficient. In this paper, a numerial solution was obtained for the thermal and associated hydrodynamic entry region in a curved circular tube placed on horizontal plane under the effect of natural convection developed by the acceleration of gravity acting vertically on the laminar forced convection. As the means for verifying numerical results, distributions of the main flow and secondary flow were obtained experimentally using the hydrogen bubble and particle tracer methods. Upon comparison with the numerical solution from the set of parabolic type equations of motion confirmed was their agreement. Thereafter, Two heat transfer experiments were conducted under the conditions of the no buoyant effect and buoyant effect. The accuracy of the present method of solving the problem with the parabolic type energy equation was confirmed.

Based on these varification, solutions for the set of three dimensional parabolic type partial differential equations adaptable to the system accomodating the two body forces were seeked. A detailed flow and heat transfer results were,then obtained in order to understand the overall phenomena in the entrance region affected by the two types of convections. It was hoped that these results and interpretation would facilitate the correlation of experimental data available in the present and future. Finally, three correlation equations were proposed for three distinct regimes of the mixed convetive heat transfer in fully developed region of curved pipe flow.

2. GOVERNING EQUATIONS and NUMERICAL COMPUTATION

Assumptions and conditions applied to the governing equatios are as followes: (1) Taking the effect of buoyancy into consideration, Boussinesq approximation is used and other property values were assumed to be constant. Transport equations of momentum and energy were coupled. (2) Viscosity and thermal dissipations in the main flow direction were disregarded, therefore parabolic type partial differential equations were adopted for the transport equations (3) The curvature ratio, the value of radius of curvature to the pipe radius R_c/a was comparatively mild and the value being $Rc/a \geqq 9$ could be assumed where no flow separation was allowed. (4) Incompressible Newtonian fluid with no heat generation was assumed. (5) Fully developed flow in curved pipe at the thermal entry was applied. (6) For thermal boundary condition, uniform heat flux in axial flow direction and constant temperature along the peripheral direction in curved section were assumed. Upon satisfying the above assamptions and conditions and by using the toroidal coordinate system, a set of governing equations and conditions which constitute formal mathematical statment of this problem is obtained. These were thereupon nondimensionalized by nomenclature listed below. The equations were rewritten by donor-cell type finite difference equation with uniform mesh. For the pressure term Poisson's type equation was introduced and set of finite difference equations were solved towards the downstream by marching procedure. The solution procedure was similar to those of Patanker and Spalding(6) and an earlier work of Harlow and Welch(7). Since the energy equation was coupled with the equation of motion, iterations were required in the present analysis.

3. EXPERIMENT

Three different experiments were carried out for the purpose of verifying the parabolic type partial differential equations which were the approximation used in the diffusion terms of motion and energy equations.

The development in the velocity field initially having a fully developed velocity profile in straight tube and entering smoothly to curved tube was sought for the axial flow and secondary flow using the hydrogen bubble and particle tracer methods. It is noted that in obtaining both velocity distributions of the secondary flow as well as the main flow, we have used photographs which would appeal directly to the visual sense and in that there should be some significance . The inside diameter of the tube is 48mm and water is drawn into the test sectin having various lengths of the curved sections. In order to avoid the water flow from the buoyancy effect, due to the temperature difference between the surrounding room temperature and the flow fluid itself,it was necessary to construct the tube with a double wall and water having the sam temperature as the test water is introduced between the inner and outer tubes for the thermal compensation. In the experiment, the flow stream was kept with in the condition of Dean number to be 0 to 1000. However, an average flow velocity of 1 to 3 cm/sec was used to avoid possible effects of the buoyant current of the hydrogen bubble. A stream containing a line of hydrogen bubbles ejected from a fine wire intermittently and tracer particles introduced at the front part of the test section, was photographed. A slit ray of illumination parallel to the main flow is used and photograph of the axial flow was taken from either horizontal or vertical position. For secondary flow, a slit ray perpendicular to the main flow was flashed on the stream at the desired cross section from the entrance of the curved tube and the secondary flow was photographed from the rear. The results of respective velocity distributions were compared directly with the results of the numerical solutions.

First heat transfer experiment was conducted under the thermal boundary condition of the uniform heat flux in the main flow direction and constant wall temperature in the peripheral direction of the cross section. It is noted that numerical solution under the same conditions as the experiment where the thermal field was in the developing stage, has not been found to this date.

In this experiment, compensation heaters were placed in between thick thermal insulator materials, and thermopiles were inserted between the main heater and compensation heater whereby heat loss was prevented therough keeping the temperature gradient between the main and guard heater to be neglegible. A 1mm thick copper tube of 20mm inner diameter was used in order to avoid temperature variation in the peripheral direction and to make the heat flux uniform in the main axial direction. The total load at the heated area was 1150 watt or less, and the mixed mean temperature difference being 2 ℃ or at most 4 °c. As a consequeuce, the parameter ReRa was also $1.5x10^4$ or less and was within the neglegible range of the buoyant force effect when compared with Dean number of 200 to 2000.

Second heat transfer experiment was carried out using a larger pipe diameter of 48.6mm. Keeping Dean number to be 100, the values of the parameter ReRa were set at $5x10^4$, $1x10^5$, $3x10^5$, $5x10^5$, $7x10^5$, and $1x10^6$. A strong buoyancy was expected.

4 RESULTS AND DISCUSSIONS

4.1 Comparison of Solution for Parabolic Type Equation of Motion With Visualization Experiment

The secondary flow development at Z/a=3.54 (Ω=22.5°) and Z/a=28.26 (Ω=180°) from the entrance to the curved pipe for De=333 is shown in Fig.1. Both the center and strength of the vortex change in conjunction with the development. It can be seen that the experimental and theoretical solutions are in good agreement. Fig.2 shows the time lines for the main flow velocity distributions taken by the hydrogen bubble method. The results can be converted to the velocity distribution, as shown in Fig.3. A very satisfactory agreement with the theoretical solutions was obtained. It may be concluded that approximation made on the equation of motion to be of a parabolic type partial differential equation can be applied in sofar to an entry flow of the mildly curved tube with only exception to the very near the entry portion of Z/(2a)=1.5, say.

4.2 Comparison of The Thermal Entrance Region Solution With Heat Transfer Experiments

In Fig.4, the results of the average Nusselt number variations along the main axis direction obtained from numerical calculation are indicated with solid lines. Calculation has been made under the assumption of no buoyant force effcect. The Dean numbers used were De=0, 50, 100, 200, 300, 400, 800, and 1000. The Dean numbers realized in the experiment were 106.6, 195.8, 396.4, and 398.8. No buoyant force effect is observed in present experiment where the value of product of Reynolds number and Rayleigh number ReRa, which represents the buoyant force effect appropriately is kept lower than $1.5x10^4$. The second heat transfer results are shown in Fig.5, where strong buoyant force effects are observed in both analysis and experiment. The values for ReRa used in the experiments were $1x10^5$, $3x10^5$, $5x10^5$, $7x10^5$ and $1x10^6$.

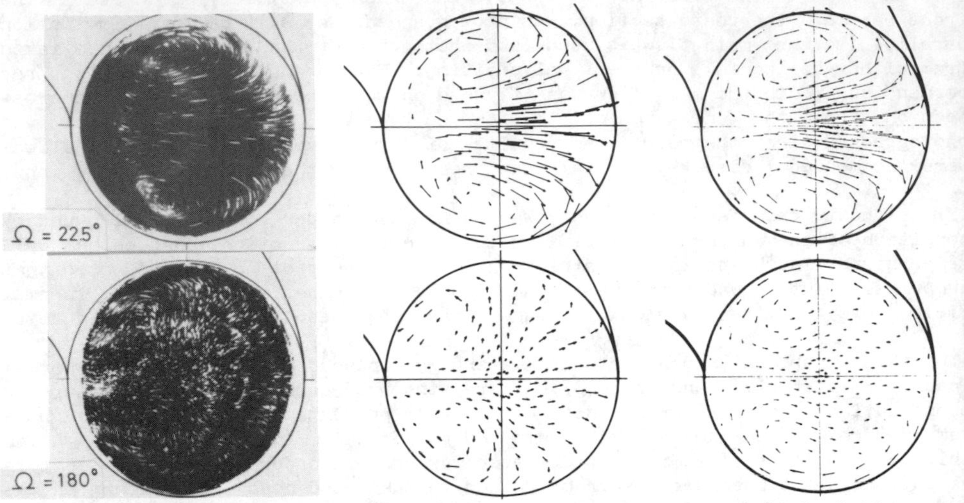

(a)Photograph (b)Measurement (c)Calculation

Fig.1 Secondary Flow (De=333)

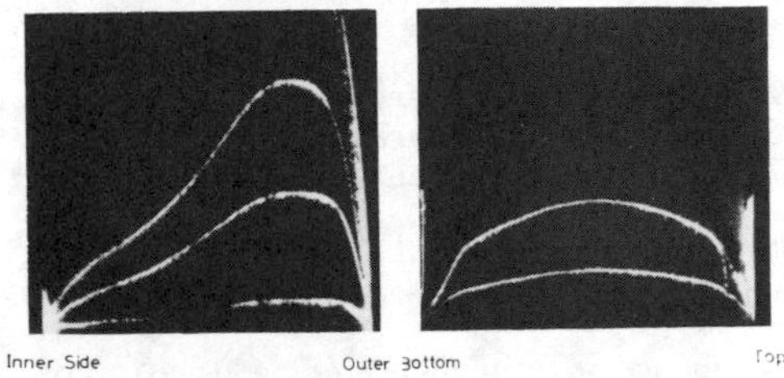

(a) Horizontal (b) Vertical

Fig.2 Hydrogen Bubble Tracer (=22.5)

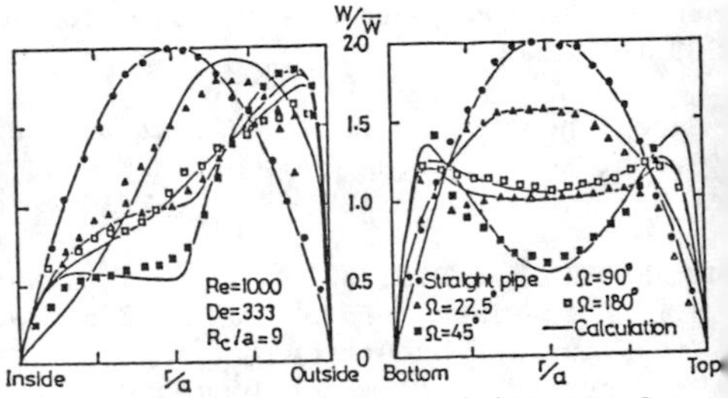

(a)Horizontal (b)Vertical

Fig.3 Main Flow Distribution

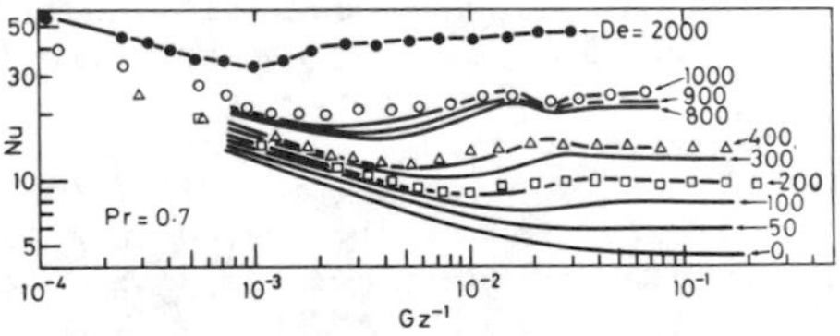

Effect of Dean number on Nusselt number develoment for non-buoyant effect field with Pr=0.7

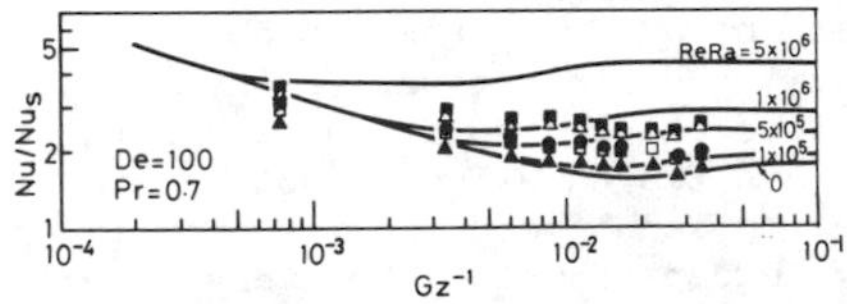

Fig.5 Effect of buoyant force on Nussélt number development

4.3 Characteristics of Mixed Convection in The Thermal Entrance Region

As an example, the details of structural change in distribution pattern of the main velocity contour, secondary flow vector,peripheral velocity distribution, peripheral friction coefficient, temperature contour and peripheral Nusslt number are sequentially shown in Fig.6. Parameters used are De=100, Pr=0.7 and ReRa=$1x10^6$.

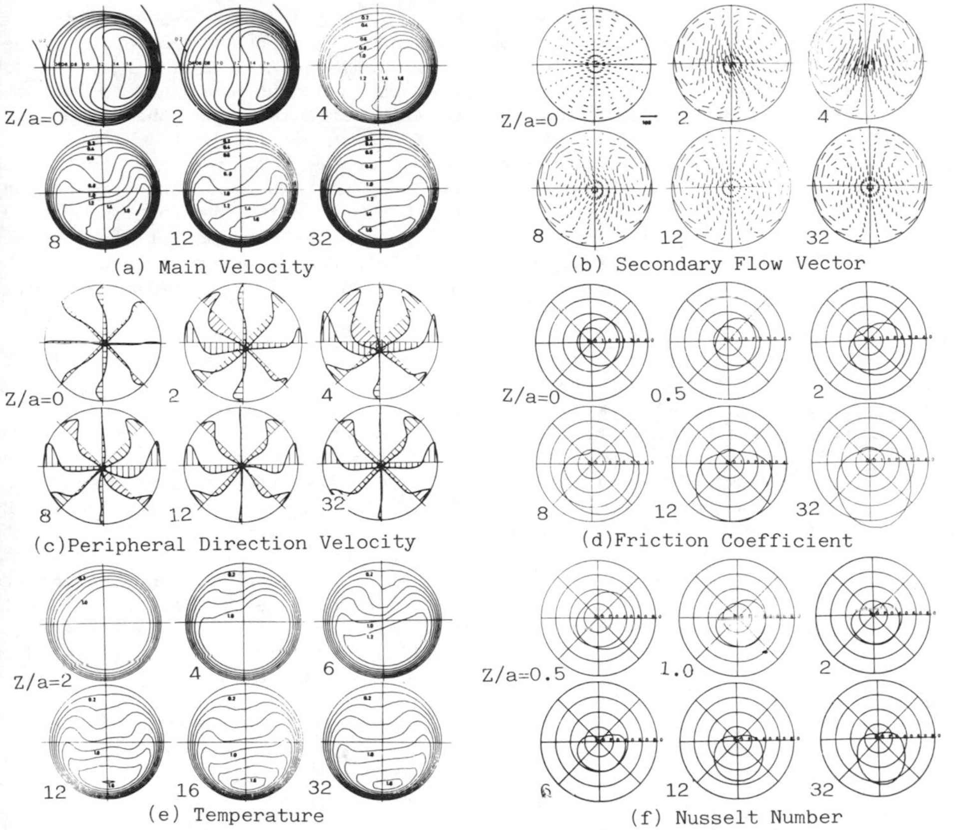

Fig.6 Development of Mixed Convection (De=100,ReRa=1x10^6)

(1) <u>Changing mechanism of the velocity and temperature distributions</u>

At the commencement of heating, there is a typical velocity distribution of fully developed region in a curved tube. At the position Z/a=0 shown in the top left hand of the figure, one can see such distribution patterns; the maximum velocity point of the main flow is located at the outer side, accompanies a secondary flow facing outward at the center of the cross section and friction factor keeping high in outer side. All values are symmetrical with respect horizontal center line. When the heating is started, a thermal boundary layer with a sharp temperature gradient developes in the vicinity of the tube wall. As early as when Z/a=2, one diameter from the entry, the effect of buoyant force shows up first in the temperature distribution and consequently on the heat transfer. Nusselt number changes in the peripheral direction is shown in detail for non-dimensinonal axis length of Z/a=0.5, 1.0, and 2.0. At Z/a=0.5, departure from Leveque solution domain is already observed owing to the mixed secondary flow and its value decreases gradually. At Z/a=2, the value becomes close to the minimum, while the axis of symmetry rapidly revolves close to 90° in the clockwise direction from the horizontal direction to the vertical direction and the basic distribution pattern also settles down. The secondary flow gains its strength herewith the change in the thermal field which naturally changes the symmetrical axis of a plain of vortex circulation from horizontal to almost vertical condition. The buoyant force will act on the main flow velocity distribution very slowly and a slight distortion due to rising convection appears first in the vicinity of the confluence point of the secondary flow where very weak axial velocity field exists. Coincidentally with the nominal amount of the axial flow rate, the buoyancy induced secondary flow also becomes stronger in this inner side of the curvature. Subsequently, as the Z/a proceeds along the tube axis, the natural convection which has been dominating over the thermal field and secondary flow, gradually effects onto the main flow distribution and consequently onto the friction coefficient. Finnally it reaches to the fully developed mixed convection region. In the convection region where the effect of buoyancy is as strong as this case, there is not any large difference between the entry length of velocity and of temperature. However it is certain to say that during the process of development, the thermal field is clearly the motive force of change. This regime can be regarded as the natural convection regime near to the fully mixed convection dominance regime.

(2) Development of local friction coefficient and Nusselt number

The development of local friction coefficient averaged over peripherally is shown in Fig.7 with ReRa as a parameter. With increase of ReRa, the $fRe/(fRe)_S$ becomes larger. Up to ReRa=$5x10^5$, the value of $fRe/(fRe)_S$ continues to approach a certain constant value without taking the maximum value after its rise in the vicinity of the entrance. When ReRa=$1x10^6$ or more, however, the value rises sharply. After taking the maximum value, it decreases again and reaches to the constant value. Due to actuate high heat flux, the secondary flow created by the natural convection is strengthend which in turn accompanies an overshooting phenomenon in flow and heat transfer. Fig.5(f) shows the change in the local Nusselt number averaged in the peripheral direction for the parameters corresponding to that in Fig.7. As for the abscissa, reciprocal of the Graetz number Gz^{-1} is used. If the ReRa=0, Leveqe solution region will come close to Gz =10^{-2}. Thereafter it reaches to the intermediate region. While the secondary flow due to the bend continues to develope, the Nusselt number will take a mimimum value at this point. Rising will continiue up to the point of approximately Gz^{-1} =$8x10^{-2}$ and gradually approaches the constant value leading to the state of the fully development. This point corresponds to Z/2a=32. Incidentally, the entrance lenght for a straight tube with the same value of Reynolds number is calculated to be Z/2a=17 or 39 depend on definition of fully development. It means that if there is no buoyant force effect the thermal entry length for mildly curved pipe with Prandtl number 0.7 is almost same as the hydrodynamic entry length of a straight tube under the same Reynolds number. The pattern of Nusselt number behavior does not change much with the increase in ReRa, but value itself increases. Notable fact is that the length of thermal entrance region is shortend conspicuously. For example at ReRa=$5x10^6$, fully development will be reached at the order of Gz =$1.4x10^{-2}$ and this corresponds to Z/2a=6.

(3) Three distinct regimes of entry length behavior

When the fully developed laminar flow in a curved tube horizontally placed is led into the thermally heated or cooled section, the flow starts and continues to develope with the effect of buoyant force and it was found that the thermal entrance length and associated hydrodynamic entry length show distinctly different features in three regimes. Concept of the categorization is illustrated in Fig.8. (i) Centrifugal convection regime When ReRa is small, the centrifugal force dominates and the length of thermal entrance region is relatively long and its value depends on Dean number and Prandtl number. The entrance length of flow is sufficientlly short, or is not effected by the thermal field. (ii) Fully mixed convection regime In a fully mixed convection regime both of the centrifugal and buoyancy force could not be neglected. The length of thermal entrance region is shortened as the ReRa becomes larger. The length of hydrodynamic entrance region, in contradiction, continues to grow until the buoyant force shows the approximately equivalent effect as that of the centrifugal force, however it is shorter than the length of thermal entrance region. When the ReRa becomes much larger, the thermal entrance length becomes shorter. With higher buoyant force effect, the length of hydrodynamic entrance will become shorter. However, it will be shightly longer than that of the thermal entrance length. <u>(iii) Natural convection regime</u> In a region where the natural convection is predominant with sufficiently large ReRa, the development of the thermal entry length is completed within a short length and the development of the hydrodynamic entrance length will follow thereafter. The detailed example shown in Fig.6. represents this state.

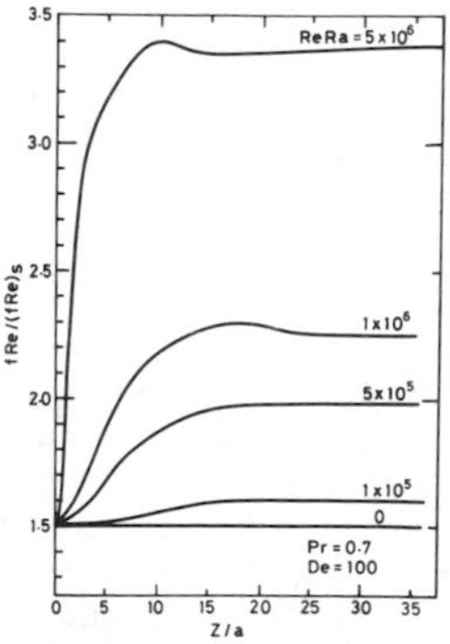

Fig.7 Development of Friction Coefficient

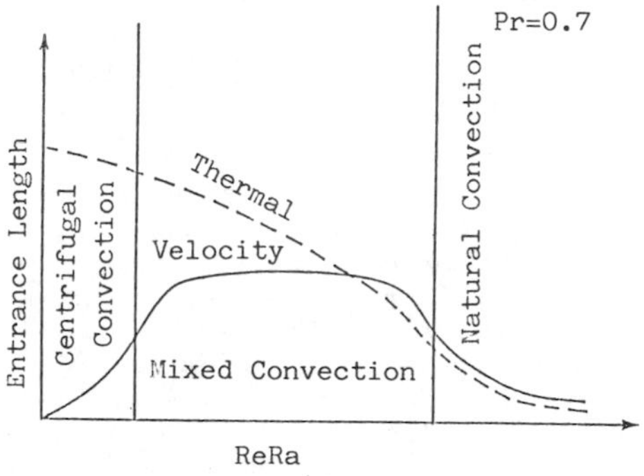

Fig.8 Entranth Lengths Map

(4) Fully developed regime

Inasmuch as the samc to the entry region, depending on the intensity of the buoyant force and centrifugal force, there are three distinct regimes in fully developed field. To understand the three distinct regimes clearly, three dimensional mapping is shown for friction coefficient and Nusselt number variations with De and ReRa in Fig.9 and Fig.10, respectively.

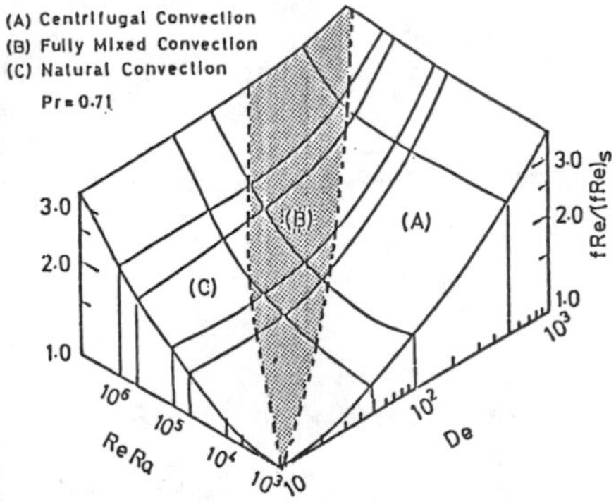

Fig.9 Friction Coefficient in Fully Developed Field

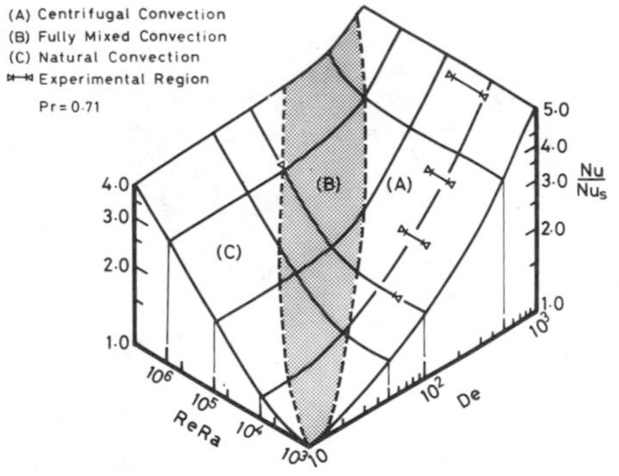

Fig.10 Nusselt Number in Fully Developed Field

The correlation equations for the Nusselt number corresponding to the above three regimes are proposed with the limitations of applicability. The equations are adopted considering rigorous interrelationship among them and error is kept within 4%.

(i) Centrifugal convection region

$$\frac{Nu}{Nu_S} = 1.00 + 0.0183 De^{0.825} Pr^{0.521}$$

(ii) Mixed convection region

$$\frac{Nu}{Nu_S} = (1.00+0.0183De^{0.825}Pr^{0.521}) \times [1.00+0.00614\left(\frac{ReRa}{De^{1.80}Pr^{1.537}}\right)^{0.7849}]$$

(iii)Natural convection region

$$\frac{Nu}{Nu_S} = 1.00+0.00167\,(ReRa)^{0.339}$$

The limitations for eq.(2)

$$\frac{ReRa}{De^{2.434}Pr^{1.537}} \geqq 0.8 \quad \text{and} \quad 1.57 \geqq \frac{ReRa}{De^{2.825}Pr^{1.537}}$$

CONCLUSIONS

A numerical solution was explored to demonstrate the interaction between centrifugal force and buoyant force in developing curved-circular-tube flows flowing in the horizontal plane subjected to the vertical gravitational force. The ranges of parameters treated were wide being De=0~1000, ReRa=0~5x10^6 , Pr=0.7 and 5, and thermal boundarry condition is of axially uniform heat flux and constant temperature peripherally. Though the solutions were obtained under the restriction of parabolic type Navier-Stokes equations, the applicability of the results is believed to be wide and general with only exception at the very near entrance region. (1) The respective relationship of the thermal and associated hydrodynamic entrance length and their characteristics are different in three distinct regimes; (i) centrifugal convection (ii) fully mixed convection (iii) natural convection . (2) Categorization of flow and heat transfer into three regimes in developing region can be extended to the fully developed region. Characteristics of friction coefficient and Nusselt number are differentiated in to three regimes, and the correlation equations for Nusselt number with clarified limitation of applicability for these three regimes are presented. (3) Flow visualization was performed on the distributions of the main flow and secondary flow develoment process with experiments using the hydrogen bubble and particle tracer methods and these distributions were quantified from the pictures obtained. The results obtained therefrom agreed well with the numerical solution. (4) Heat transfer experiment was conducted in the thermal entrance region with the thermal boundary condition of uniform heat flux in the main flow direction and constant temperature peripherally and was compared with the numerical solution for the regime where the centrifugal convection was the dominating factor and a good agreement was observed.

The authors are pleased to acknowledge Mr. Urai's assistance on printing the present manuscript.

REFERENCES

(1) Patankar,S.V.,Pratap,V.S. and Spalding,D.B.,Int.J.Heat.Mass Transfer, Vol.62(1974)pp.539-551
(2) Akiyama,M.,et al, 17th Japan Heat transfer Symp. Paper,(1980),pp.16-18
(3) Akiyama,M.,et al, Trans. Japan Soci. Mech.Eng.,Vol.50,(1984),pp.1197-1204
(4) Akiyama,M.,et al,ASME-JSME Themal Eng.Joint Conf.Vol.3(1983)pp.27-33
(5) Chilukuri,R. and Humphrey,J.A.C.,Int.J.Heat.Mass Transfer, Vol.24(1980)pp.305-314
(6) Patankar,S.V.and Spalding,D.B.,Int.J.Heat.Mass Transfer, Vol.15(1972)pp.1787-1806
(7) Harlow,F.H.,and Welch,J.F.,Phys.Fluids,Vol.8,(1965)pp.2182-2189

he Effects of Parameters of Cast-Iron lliptical Tubes with Rectangular Fins n the Performances of Heat Transfer nd Flow Resistance

ANG-MO CHENG and LO-YI TAO
azhong University of Science and Technology
azhong, PRC

BSTRACT

The effects of parameters of cast-iron elliptical tubes with rectangular ins on the performances of heat transfer and flow resistance are investigated. he empirical dimensionless expression of convective heat transfer is given. he application result of elliptical tubes with rectangular fin also is given.

OMENCLATURE

major axis of ellipse
fin length along major axis of ellipse
minor axis of ellipse
fin width along minor axis of ellipse
ɔ specfic heat of air
characteristic dimension of transverse single row of tubes
u Euler number, $\Delta P/\rho u_m^2$
fin pitch
u Nusselt number, $\alpha d/\lambda$
Pressure of air
e Reynolds number, $u_m d/\nu$
1 transverse tube pitch of transverse single row of tubes
w tube outside convective heat transfer area

t fin thickness
T temperature of tube outside surface
T_f average temperature of air
$u_i(i=1,2,3)$ velocities of air
u_m air velocity at minimum cross section
$x_i(i=1,2,3)$ coordinates
ρ density of air
ϵ turbulent viscosity
ϵ_H turbulent thermal diffusivity
λ thermal conductivity of air
ν viscosity of air
α mean outside convective heat transfer coefficient

NTRODUCTION

Finned elliptical tubes have better heat transfer performance and lower flow resistance than finned circular tubes. Some investigations have been conducted in the heat transfer and flow resistance characteristics of finned elliptical tube banks made of steel and aluminium [1-4], but there have not been any investigations in the characteristics of cast-iron finned elliptical tube banks. In [5], we presented the experimental results of convective heat transfer of air flowing across a single cast-iron elliptical tube with rectangular fins (CETRF). Now, this paper investigates the effects of geometrical dimensions of CETRF and transverse tube pitch on the convective heat transfer and flow resistance characteristics of a single transverse CETRF row. According to the exper mental results, a new kind of economizer has been designed. The industrial experiments show that the performance of CETRF economizer is superior to that of cast-iron finned circular tube economizer.

EXPERIMENTAL EQUIPMENT AND MEASUREMENTS

The experiments were performed in an open-circuit wind tunnel (Fig.1)

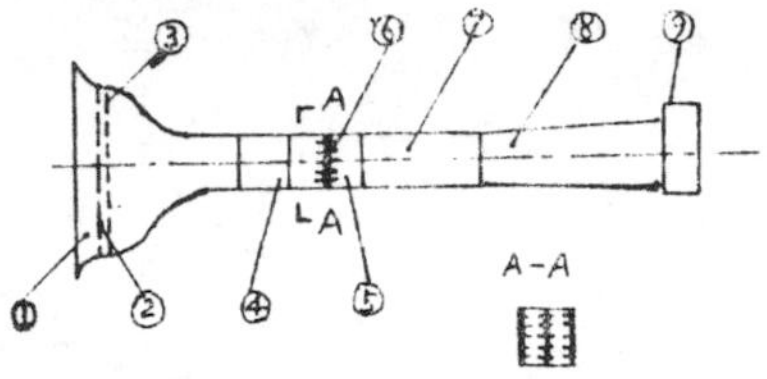

Fig.1 Schematic diagram of the wind tunnel

1. contract section 2. honeycomb
3. damp net 4. section for velocity measurement
5. test sectio
6. CETRF 7. section for pressure measurement
8. diffuser
9. blower

Before test, air velocity in front of test-section was checked, the resul shew that the velocity was uniform. A pitot tube was installed in the sectio to measure the air velocity. The pressure drop cross the single row of tubes was measured by a tilting micromanometer attached to the static pressure hole The inlet air temperature was measured by thermometers.

Fig.1 shows the tube layout, two half tubes were located beside a tube, e tric heater was used to heat in the tube, the tube ends were insulated, tempe tures of the tube surface were measured by chromel-alumel thermocouples.

ANALYSIS AND PROCESS OF EXPERIMENTAL DATA

Differential equations:

$$\frac{\partial u_i}{\partial x_i} = 0 \qquad (1$$

$$u_i \frac{\partial u_j}{\partial x_i} = - \frac{1}{\rho} \frac{\partial P}{\partial x_j} + \frac{\partial}{\partial x_i} \left[(\nu + \epsilon) \left(\frac{\partial u_j}{\partial x_i} + \frac{\partial u_i}{\partial x_j} \right) \right] \qquad (2$$

$$\rho c_p u_i \frac{\partial T}{\partial x_i} = \frac{\partial}{\partial x_i} \left[(\lambda + \epsilon_H) \frac{\partial T}{\partial x_i} \right] \qquad ($$

$$\alpha_1 (T - T_f) \Big|_w = -\lambda \frac{\partial T}{\partial n} \Big|_w \qquad (4$$

$$\alpha = \frac{1}{S_w} \iint_w \alpha_1 \, d\sigma \qquad (5$$

Boundary conditions:

(1) $u_1 = u_o$, $u_2 = u_3 = 0$, $P = P_o$, $T = T_o$

$x_1 < - \frac{A}{2}$

(2) $u_i = 0$, $T = T_w$ (i=1,2,3)

$$|x_1| \leqslant \frac{A}{2}, \quad \frac{s_1-B}{2} \leqslant |x_2| \leqslant \frac{s_1}{2}, \quad \frac{x_1^2}{(a/2)^2} + \frac{(x_2-s_1/2)^2}{(b/2)^2} \geqslant 1, \quad x_3 = \pm \frac{L-t}{2};$$

$$x_1 = \pm \frac{A}{2}, \quad \frac{s_1-B}{2} \leqslant |x_2| \leqslant \frac{s_1}{2}, \quad \frac{L-t}{2} \leqslant |x_3| \leqslant \frac{L}{2};$$

$$|x_1| \leqslant \frac{A}{2}, \quad |x_2| = \frac{s_1-B}{2}, \quad \frac{L-t}{2} \leqslant |x_3| \leqslant \frac{L}{2};$$

$$\frac{x_1^2}{(a/2)^2} + \frac{(x_2-s_1/2)^2}{(b/2)^2} = 1, \quad |x_2| \leqslant \frac{s_1}{2}, \quad |x_3| \leqslant \frac{L-t}{2}.$$

(3) $\frac{\partial u_i}{\partial x_3} = 0$, $\frac{\partial T}{\partial x_3} = 0$ (i=1,2,3)

$$|x_2| < \frac{s_1-B}{2}, \quad x_3 = \pm \frac{L}{2}.$$

(4) $\frac{\partial u_i}{\partial x_2} = 0$, $\frac{\partial T}{\partial x_2} = 0$ (i=1,2,3)

$$|x_1| > \frac{a}{2}, \quad x_2 = \pm \frac{s_1}{2}, \quad |x_3| < \frac{L-t}{2}.$$

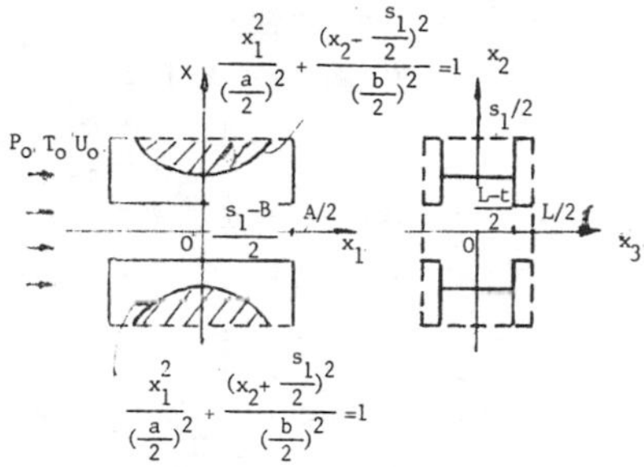

Fig.2 Schematic diagram of tube outside flow

Using nondimensional transform to analyze Eqs.(1)-(5) and the boundary conditions, we get

$$Nu = F(Re, Pr, \frac{a}{d}, \frac{b}{d}, \frac{A}{d}, \frac{B}{d}, \frac{t}{d}, \frac{L}{d}, \frac{s_1}{d}) \quad (6)$$

$$Eu = G(Re, \frac{a}{d}, \frac{b}{d}, \frac{A}{d}, \frac{B}{d}, \frac{t}{d}, \frac{L}{d}, \frac{s_1}{d}) \quad (7)$$

where $Nu = \alpha d/\lambda$, $Eu = \Delta P/\rho u_m^2$, $Re = u_m d/\nu$.

Table Dimensions of CETRF

Tube	a	b	A	B	L	t
1	80	40	112	68	25	5
2	100	40	140	68	25	5
3	120	40	168	68	25	5
4	100	40	120	68	25	5
5	100	40	160	68	25	5
6	100	40	140	56	25	5
7	100	40	140	80	25	5

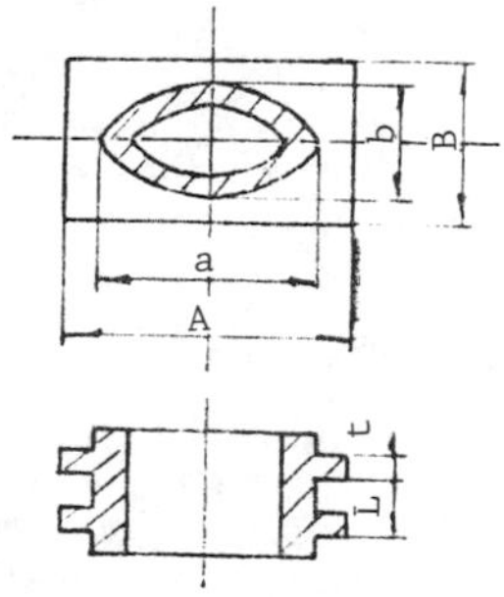

Fig.3 Schematic diagram of CETRF

Eqs.(6),(7) indicate that Nu and Eu are all the functions of Reynolds numbe Re and nondimensional groups a/d , b/d , A/d , B/d , t/d , L/d , s_1/d or some combinations of these groups.

Seven kinds of CETRF which had different geometrical dimensions constituted different single transverse rows of tubes separetly. The shape and dimensio of CETRF are shown in Fig.3 and the table. The experimental purpose is to dete mine Eqs.(6) and (7).

A. Effect of geometric dimensions on the convective heat transfer of single ro of tubes

1) Effect of s_1

Fig.4 shows the experimental results of mean outside heat transfer coeffi cients for Tube 2 at s_1=75mm, 100mm and 200mm as a function of the maximum air velocity.

The figure indicates that α increases as s_1 increases under the same velocit u_m . This phenomenon may be explained as follows: when u_m maintains constant increase of s_1 means increase of air face velocity of the single row of tubes, the increase of air face velocity intensifies the outside convective heat tran fer and makes the mean outside convective heat transfer coefficients increase. But the figure indicates also that the effect of s_1 on the heat transfer coeff cient is very weak as s_1 increases to a certain value. This critical value of is different for tubes having different dimensions.

2) Effect of A

Fig.5 presents the experimental results of mean outside heat transfer coef cient for Tubes 2,4 and 5 as a function of u_m at s_1=75 mm.

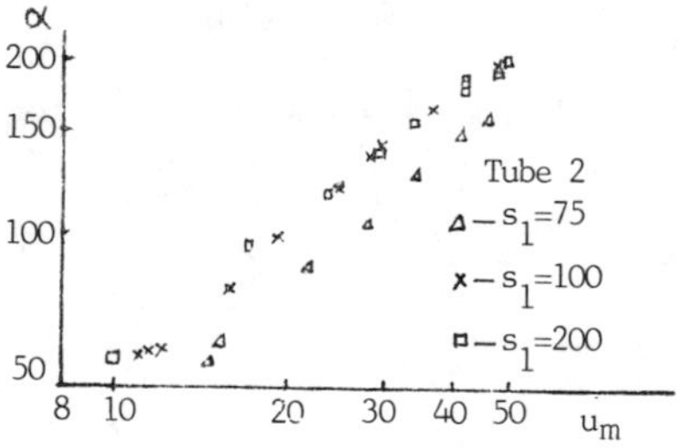

Fig.4 Effect of s_1 on α

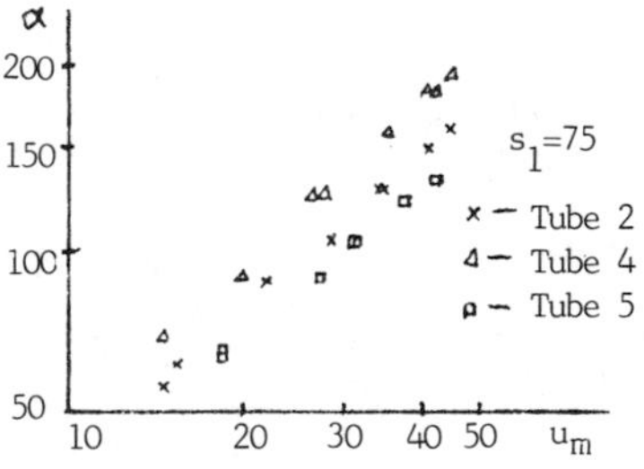

Fig.5 Effect of A on α

The values of A for Tubes 2,4 and 5 are 140 mm, 120 mm and 160 mm separetly the other dimensions of these tubes are identical.

The figure shows that α decreases with an increase of A.

3) Effect of B

The mean outside heat transfer coefficients for Tuabes 2, 6 and 7 are shown in Fig.6 (s_1=200mm) and in Fig.7 (s_1=75mm) as a function of maximum velocity u_m

The values of B for Tubes 2,6 and 7 are 68 mm, 56 mm and 80 mm separetly, the other dimensions are identical.

Fig.6 indicates that the change of B has little effect on heat transfer coefficient when s_1 equals 200 mm. Fig.7 indicates that B has a greater effect on α when s_1 equals 75 mm. α decreases with an increase of B.

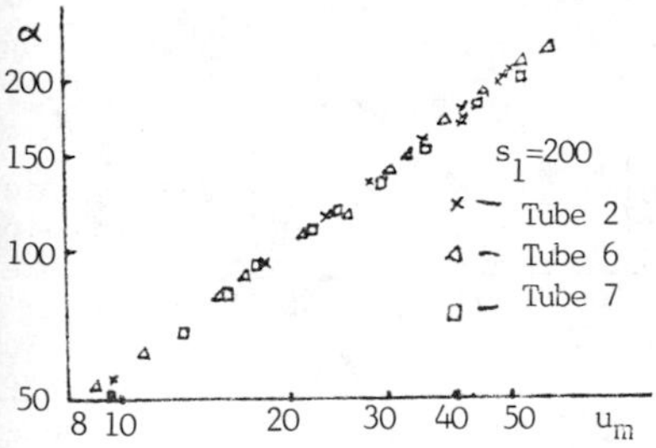

Fig.6 Effect of B on α

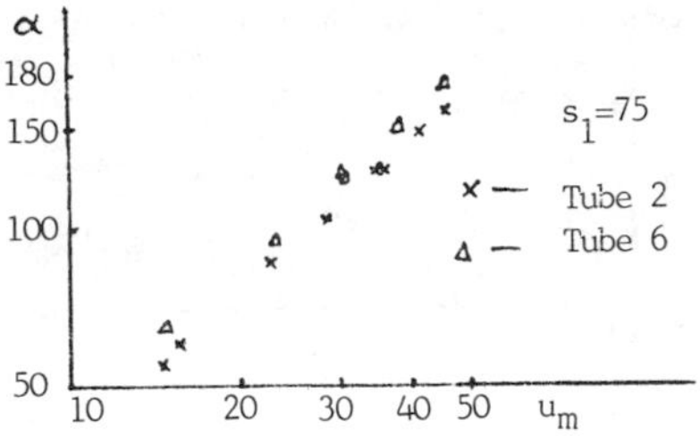

Fig.7 Effect of B on α

4) Effect of a

Fig.8 shows the experimental results of mean outside heat transfer coefficients for Tubes 1, 2 and 3. The transverse tube pitch s_1 equals 75 mm.

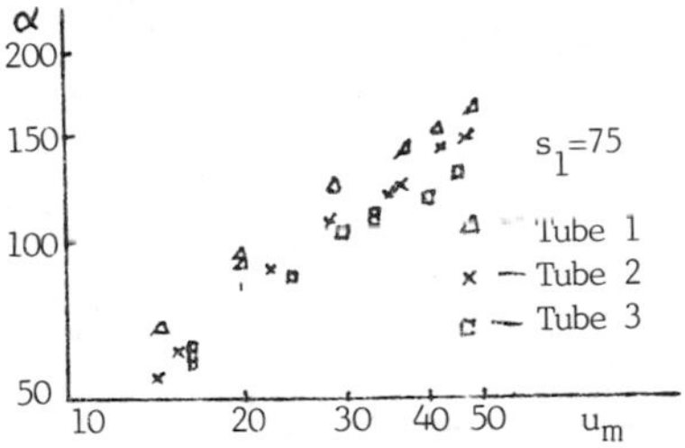

Fig.8 Effect of a on α

The values of a for Tubes 1, 2 and 3 are 80 mm, 100 mm and 120 mm separetly.

The figure indicates α decreases as a increases. But the values of A for Tubes 1, 2 and 3 are 112 mm, 140 mm and 168 mm separetly, therefore the figure shows the total effects of a and A on α. Hence, the relation of α with a is not gotten here. The relation may be derived from analysis of total experimental data.

B.Effects of geometric dimensions on flaw rsistance of single row of tubes

1) Effect of s_1

Fig.9 presents the experimental results of Eu for Tube 3 as a function of u_m. The transverse tube pitches equal 75 mm and 100 mm separetly. The air velocity at minimum cross section was used to process the experimental data.

The figure shows that Eu increases as s_1 increases, and Eu has little relation to u_m, that is, the pressure drop Δp across the single row of tubes is linear with u_m^2.

2) Effect of A

Fig.10 indicates that Eu increases as A increases, and the effect of u_m on Eu is very weak.

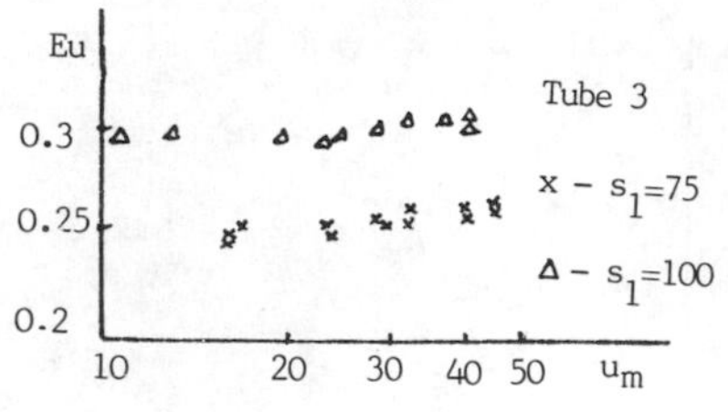

Fig.9 Effect of s_1 on Eu

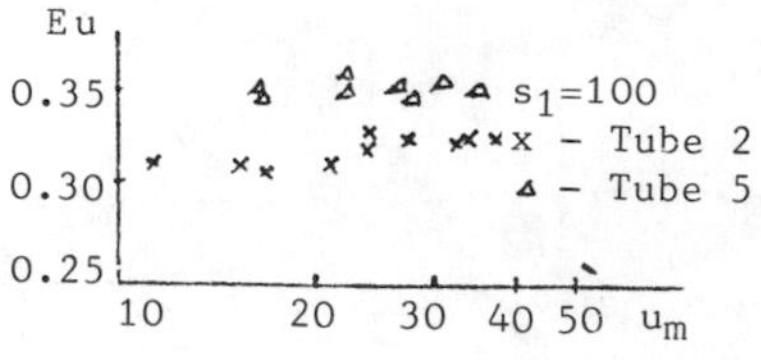

Fig.10 Effect of A on Eu

3) Effect of B

Fig.11 presents the experimental results of Eu for Tubes 2 and 6 at s_1=75mm
It may be derived from the figure that Eu is directly proportional to B.

4) Effect of a

Fig.12 shows the experimental results of Eu for Tubes 1, 2 and 3 as a functi of u_m at s_1=100 mm.

The values of a of Tubes 1,2 and 3 are 80 mm, 100 mm and 120 mm separetly, the values of A are 112 mm, 140 mm and 168 mm separetly, the other dimensions a identical. Therefore, the figure shows the total effects of a and A on Eu.

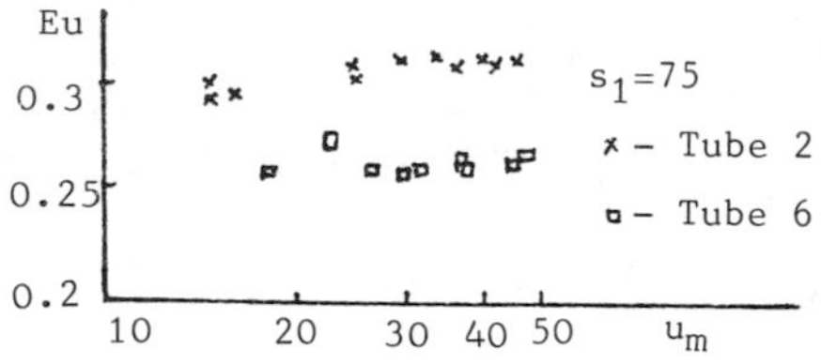

Fig.11 Effect of B on Eu

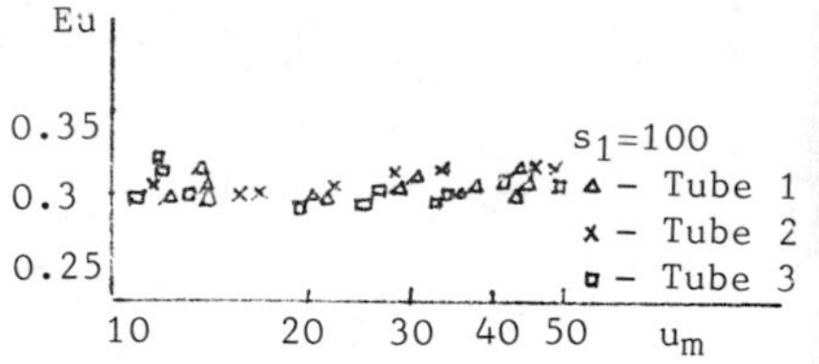

Fig.12 Effect of a on Eu

Process the total convective heat transfer data to get an empiric equation. Write Eq.(6) in following form:

$$Nu = m_o Re^{m1} (\frac{a}{b})^{m2} (\frac{A}{a})^{m3} (\frac{B}{b})^{m4} (\frac{s_1 - b}{s_1})^{m5} (\frac{s_1 - b}{L - t})^{m6} \quad (8$$

where $Nu = \alpha b/\lambda$, $Re = u_m b/\nu$.

A step-by-step linear regression analysis method was used to determine the constants m_o, m_1, m_2, m_3, m_4, m_5 and m_6 of Eq.(8). In order to complete the re gression analysis for the total convective heat transfer data, Eq.(8) was rewri ten as

$$\ln Nu = \ln m_o + m_1 \ln Re + m_2 \ln(\frac{a}{b}) + m_3 \ln(\frac{A}{a}) + m_4 \ln(\frac{B}{b}) + m_5 \ln(\frac{s_1 - b}{s_1}) + m_6 \ln(\frac{s_1 - b}{L - t}) \quad (9$$

In the process of the step-by-step linear regression, the signi-ficance level was 0.05, an F level of 3.84 was used for introucing any parameter which was significant and an F level of 3.835 was used for removing any parameter which was not significant. F-examinations of parameters lnRe, $\ln(^a/b)$, $\ln(^A/a)$, $\ln(B/b)$, $\ln[(s_1-b)/s_1]$ and $\ln[(s_1-b)/(L-t)]$ were made one by one. If a paramet had an F level greater than 3.84, the parameter was in-drodiced into the corre tion; if a parameter's F level was less than 3.835, the parameter was removed he correlation. The regression analysis process was carried out on a digital omputer. The analysis results showed that the F level of the parameter $\ln[(s_1-b)/$ $L-t)]$ was less than 3.84, therefore, that parameter was not introduced into the orrelation.

The final results of the data process were as follows: $\ln m_0=-2.734$, $m_1=0.79$, $m_2=-0.33$, $m_3=-0.77$, $m_4=-0.13$, $m_5=0.43$, that was

$$\ln Nu=-2.734+0.79\ln Re-0.33\ln(\frac{a}{b})-0.77\ln(\frac{A}{a})-0.13\ln(\frac{B}{b})+0.43\ln(\frac{s_1-b}{s_1}) \qquad (10)$$

he standard deviations of the regression coefficients in Eq.(10) were m_1-0.64%, m_2-2.67%, m_3-4.07%. m_4-3.3%, m_5-1.38% separe ly. Eq.(10) had standard deviation f 4.06% and multi-correlation coefficient of 0.99.

The above results indicate that the parameters Re, a/b, A/a, B/b, $(s_1-b)/s_1$ ave significant effects on the convective heat transfer characteristic of the ingle CETRF row.

It may be derived from Eq.(10) that

$$Nu = 0.065Re^{0.79}(\frac{a}{b})^{-0.33}(\frac{A}{a})^{-0.77}(\frac{B}{b})^{-0.13}(\frac{s_1-b}{s_1})^{0.43} \qquad (11)$$

The applicable conditions for Eq.(11) are $0.21\times10^5 \leqslant Re \leqslant 1.4\times10^5$, $2 \leqslant (a/b) \leqslant 3$, $.2 \leqslant (A/a) \leqslant 1.6$, $1.4 \leqslant (B/b) \leqslant 2$, $0.45 \leqslant [(s_1-b)/s_1] \leqslant 0.8$.

ONCLUSIONS

1) Within the experimental reange, the parameters Re, a/b, A/a, B/b, $(s_1-b)/s_1$ have significant effects on Nu. Nu increases with the increases of Re, $(s_1-b)/s_1$, and decreases with the increases of a/b, A/b and B/b.
2) Within the experimental reange, Eu has little relation to u_m.
3) An economizer designed on the basis of the above experimental results has been used after end of an industrial boiler 4T/h to recover heat from chimney gases. The industrial experimental results show that the CETRF-economizer's overall coefficient of heat transfer is 20% greater, heat transfer rate per unit volume 40% greater and pressure dron 60% less than the cast-iron finned circular tube economizer's separetly.

EFERENCES

1] F.J. Schulenberg: Finned Elliptical Tubes and Their Application in Air-Cooled Heat Exchangers, ASME J. Eng. for Industry, vol.88, No.2

2] H.Braucer: Compact Heat Exchangers, Chemical and Process Eng. 1964

3] Li Bin, Experimental Investigation of Heat Trausfer and Flow Resistance of a staggered Finned tube Bank, Paper Presented in Conference on Heat and Mass Transfer of CSET, Huanshan, China, Oct. 1981.

4] Liu Baoxing, Cai Zuhui, Heat Trausfer and Draft Loss performance of Air Flowing Across Staggered Banks of Oval Tubes Fitted with Rectaugular Fins, Journal of Engineering Thermophysics (in China), vol.3, No4, 1982.

5] Cheng Shaugmo, Research on Convective Heat Transfer performance of Cast-Iron Elliptical Finned Tubes, Paper Presented in Couference on Heat and Mass Trausfer of CSET, Wuhan, China, Nov. 1984.

Temperature Distributions in the Laminar Wake behind a Slender Streamlined Heated Body of Revolution

XIAN ZHI DU and BO YI CHEN
Naval Academy of Engineering
PRC

ABSTRACT

An empirical expression of temperature distributions in the steady laminar wake behind a slender streamlined heated body of revolution was derived in terms of Independent Constants Method. A semi-analytical solution of the wake radius and the temperature elevation of the wake centerline was obtained. The temperature measurement system of high resolution was designed with a digital tour measurement unit and an analogue drawing equipment. The temperature distributions of the wake behind the body from $\frac{x}{d}$ = 0 to 110 were fully explored in a recirculating water channel. The channel-velocity profiles and fairly well laminar flow patterns in the wake were visualized with the aid of H_2 bubbles.

INTRODUCTION

The foundational problems of wakes are their velocity, pressure and temperature distributions. Some theoretical and experimental investigations had been carried out already. An early searching test of distributions of temperature and velocity in the wake behind a heated obstacle was made by Fage and Falkner (1932)(Ref.1). In 1938, Hall and Hislop (Ref.2) made the same measurements behind a heated cylindrical body. But in their measurements, the width of wake behind the heated body was somewhat indefinite, it was not explicit to distinguish between a radius of the wake temperature field and that of the wake velocity field, and Hall and Hislop just observed only two traverses at x/d = 9.7 and 17 behind the body. Furthermore, they didn't concern the effect of natural convection in the wake. Heinrich and Eckstrom (1963)(Ref.3) obtained an empirical solution of velocity field in turbulent wakes behind bodies of revolution, Hama and Peterson (1976)(Ref.4) established semi-empirical equations of velocity field in a laminar wake behind a slender streamlined body of revolution, but they provided only the velocity distributions in the wake much closer behind the body at a distance 4 to 20 diameters or 0 to 5 lengths of the body downstream. No applicable expression had thus far been gotten for the temperature field in transient process from the trailing edge to fully developed wake behind a slender streamlined heated body of revolution. However, in problems of aerodynamic deceleration of aircraft, drag reduction and heat transfer of underwater moving body and recently especially in the problem of dispersion and control pollutants from aircraft and water vehicle propulsion units, the velocity and temperature distributions must be known much closer as well as farther behind the body producing wake. Prompted by these, a semianalytical and experimental consideration of temperature distributions in the wake behind the body had been researched in the present paper.

AN EMPIRICAL SOLUTION WITH INDEPENDENT CONSTANTS

Until recently, no mathematically accurate solution has been obtained for overall region of the flow field in a wake, and the classical assumptions cause their results to agree with experiments only relatively far downstream from the body. An expression of the temperature distribution, which is applicable for transient process from the trailing edge to fuly developed wake behind a slender streamlined heated body of revolution, was derived in terms of Independent Constants Method which is very promising and is thus further pursued in the present study of wake temperature field because the classical assumptions usually considered valid far downstream may be placed by this method. The main procedure to obtain the empirical solution may be described as follows:

It was assumed that the temperature profiles in different cross sections of the wake are similar, and that the wake radius b_t and the temperature elevation on the centerline of the wake t_{1m} vary with axial length of the wake x according to some power law. These assumptions can be expressed as

$$\theta = \frac{t_1}{t_{1m}} = \exp(S\eta_i^2) \tag{1}$$

$$\frac{b_t}{R} = K_1 \left(\frac{x}{d} \right)^n \tag{2}$$

$$\frac{t_{1m}}{t_\infty} = K_2 \left(\frac{x}{d} \right)^{-m} \tag{3}$$

where $\eta_i = r/x^n$, r——radial coordinate, d——mamimum diameter of the body, $R = d/2$, t_1——difference betwwen wake temperature t_w and free-stream temperature t_∞, and K_1, K_2, m, n are unknown constants which are to be determined from measured temperature distributions in the wake.

Introducding (1), (2) and (3) into the differential equation of energy in the laminar far-wake behind the body of revolution

$$u_\infty \frac{\partial t_1}{\partial x} = a\left(\frac{1}{r} \frac{\partial t_1}{\partial r} + \frac{\partial^2 t_1}{\partial r^2}\right) \tag{4}$$

yield the rather complicated quadratic solution for

$$4a\eta_i^2 \, x^{1-2n} S^2 + (4ax^{1-2n} + 2nu_\infty\eta_i^2)S + mu_\infty = 0 \tag{4a}$$

Where u_∞ is velocity of the free-stream.

Employing the quadratic formula, assuming temporarily m = 2n and applying the boundary conditions

a) $t_1 = 0$ for $r = \infty$, b) $t_1 = t_{1m}$ for $r = 0$

we can obtain much simpler solution

$$\theta = \frac{t_1}{t_{1m}} = \exp(-nu_\infty \eta_i^2 / 2ax^{1-2n}) \tag{5}$$

In order to insure the agreement of the theoretical and experimental temperature profiles both at the wake centerline and at the arbitrary boundary of the wake, Eq.(5) whould be modified with the definition (Ref.5) of thermal boundary layer because when Eq.(5) was derived the wake boundary was defined as the points where $t_1 = 0$ for $r = \infty$, and the positions of $t_1 = 0$ for $r = b_t$ are very difficultly to be measured in experiments. Introducing a coefficient of modification M. into Eq.(5), then we have

$$\theta_m = \frac{t_1}{t_{1m}} = \exp(-M \cdot nu_\infty \eta_i^2 / 2ax^{1-2n}) \tag{5a}$$

Let the thermal wake boundary be defined as that points where temperature elevation is ξ of the local maximum temperature elevation, i.e.,

$$(t_1/t_{1m})_{r=b_t} = \xi \quad , (0 < \xi < 0.1) \tag{6}$$

Now combining Eqns.(5a) and (6) yields

$$M. = -2a\ln\xi \,/ nu_\infty (K_1 Rd^{-n})^2 x^{2n-1}$$

Thus the empirical solution of the temperature distributions in the laminar wake behind the body of revolution is expressed by the equation

$$\theta_m = \frac{t_1}{t_{1m}} = \exp[(\frac{r}{b_t})^2 \ln\xi\,] \tag{7}$$

The power law expressions of Eqns. (2) and (3) indicated that the wake radius b_t and the temperature elevation on the centerline of the wake t_{1m} should plot as straight lines on log-log coordinate paper from which the constants K_1,K_2,m and n are determined. Obviously, it requires a large number of experimental data including b_t and t_{1m}, which want not only a tremendous amount of the experimental work, but also cost very much. Thus we employ another way to determine the b_t and t_{1m} by utilizing directly the semi-empirical expressions of the wake velocity distribution obtained by Hama and Peterson (Ref.4), since their experimental conditions were in a considerably similar case with ours,and may derive the semi-analytical solution of b_t and t_{1m} as follows:

Based on the principle of heat balance, the difference between the enthalpy of the wake and of the free-stream should be equal to the heat Q. emited by the slender streamlined heated body of revolution in arbitrary cross sections of the wake behind the body, i.e.,

$$2\pi\rho c_p \int_0^{b_t} t_1 urdr = Q. \tag{8}$$

Where u was given by the equations in Ref. (4)

$$\frac{u}{u_\infty} = 1 - (1 - \varphi)\exp\left[-\left(\frac{r}{b}\right)^2 \ln 2\right] \tag{9}$$

$$Kx = \frac{2}{1-\varphi} + \ln\frac{1-\varphi}{1+\varphi} - 2 \tag{10}$$

$$b = [c_d\, d^2 \ln 2/4(1-\varphi^2)]^{1/2} \tag{11}$$

Where $\varphi = u|_{r=0}/u_\infty$, $K = 64\nu/c_d u_\infty d^2$, $c_d = 0.553$, and ν is kinematic viscosity of the free-stream.

Assuming that the ratio of the thermal wake radius b_t to the halfwake radius b of the wake velocity profile varies rectilinearly with the axial length x of the wake, i.e.,

$$\frac{b_t}{b} = \psi(x) = \varepsilon + kx \tag{12}$$

Where ε = constant, it was determined as $\varepsilon = 3$ from the boundary condition $t_{1m} = t_0$ for $x = x_0$ at the initial section of the present experimental region.

Introducing Eqns.(7) and (9) to (12) into Eqn.(8), we can obtain the semi-analytical solution for the calculation of the wake radius b_t and its maximum temperature elevation as follows

$$b_t = [\psi^2 c_d d^2 \ln 2/4(1-\varphi^2)]^{\frac{1}{2}} \tag{13}$$

$$t_{1m} = 4Q.(1-\varphi^2)/\pi\rho c_p u_\infty \left[\frac{\xi-1}{\ln\xi} - (1-\varphi)/(\psi^2\ln 2 - \ln\xi)\right]\psi^2 c_d d^2 \ln 2 \tag{14}$$

For the convenience of application Eqns.(13) and (14) may be equivalently changed into

$$f_b = \frac{b_t}{\sqrt{c_d d^2}} = \frac{0.769}{J.}\frac{x}{d} + 1.245\ , \quad (0 \leqslant \frac{x}{d} \leqslant 5J.) \tag{15}$$

$$f_t = \frac{\dfrac{t_{1m}}{Q.}}{2\pi\rho c_p u_\infty c_d d^2} = \begin{cases} 2\ J.\left(\frac{x}{d}\right)^{-1/2}, (0.2J. \leqslant \frac{x}{d} \leqslant 1.3J.) \\ 2.5J.\sqrt{J.}\left(\frac{x}{d}\right)^{-3/2}, (1.3J. \leqslant \frac{x}{d} \leqslant 6J.) \end{cases} \tag{16}$$

Where $J. = c_d u_\infty d/64\nu$ is non-dimensional, J. = 21.9-28.3 in the present experiment. Evidently, it is able to predict the values of b_t and t_{1m} merely from the known parameters of the flow field.

EXPERIMENT

Analogue Apparatus of the Wake

The experiment was carried out in a recirculating water channel 20 cm. wide, 25 cm. deep and 328 cm. long, which was made of 5mm. thick plexiglass plate. The mean-flow velocity profiles (Fig.3(a)) were fairly uniform by the insertion of a series of fine-mech screens and small-bore phenolic resin honeycombs. The free-stream velocity u_∞ was maintained at 6-18 cm./s. and the water temperature was 30°C.

The wake to be studied was created by a slender streamlined heated body of revolution of length 192mm. and diameter 15mm. which had a parabolic nose, a cylindrical middle body and a pointed stern. The body was supported by a vertic streamlined strut, whose cross section was a NACA64_A-021 axisymmetric wing profi The wake of the strut had no obvious influence on the main wake of the body. Th electrical heating elements within the body was capable of dissipating up to 176 w.

Temperature measurement systems

Temperature measurements (Fig.1) were made both by the digital tour measurement system (40μv/°C) which contains nine pairs of copper-constantan couple each of which has a hot probe 0.5mm. in diameter and lower output, and by the analogu drawing system (2.2x$10^4$$\mu$v/°C) which includes a high sensitive thermistor model MF51 with a probe 1.5mm. in diameter which has rapid respond to heat and whose time-constant is 0.2 second in water. The probes were mounted on a traversing mechanism, which enabled us to locate them within 0.1mm. accuracy in both the horizontal and the vertical direction in a plane normal to the free-stream direc tion . The miniature motor forced the probes to move along horizontal lines at a speed of 1mm./s. When the volts created by displacement of the probes and the temperature elevation of the wake were fed into a model LZ3 pen recorder the analogue charts of the wake temperature distributions were drawn as shown in Fig 6.

Visualization of the flow field

The tracer of H_2 bubbles visualizes smooth rectilinearing laminar flow patte in the wake as shown in Fig.3(b). In order to make the visualization clear the Na_2SO_4 solution of 0.5N was selected as the electrolyte. The photographs of visualization indicate that the thermal wake is axisymmetric (Fig.3(b)) in horizontal planes, and that the heating produces natural convection and causes the thermal wake to buoy up in vertical planes (Fig.3(c)) when the free-stream velocity was lower.

RUSULTS AND DISCUSSION

Wake radius and its maximum temperature elevation

The temperature distributions of the wake were measured with close spacings behind the body from x/d=0 to 110, and Fig.6 is a sample of the distributions. With the wake boundary defined as Eq.(6) (choosing ξ = 0.05), the experimental wake radius b_t and its maxiumu temperature elevation t_{1m} are indicated in Fig The constant K_1,K_2, m and n are determined from these experimental results as follows:

$$K_1=0.94 \quad , \qquad n=0.293$$
$$K_2=0.155 \quad , \qquad m=0.66 \qquad (17)$$

It may be seen that the assumption m=2n used in the solution of Eq.(4a) is qui accurate, and that the relative error between the assumption and the present experimental results is less than 12 percent. Furthermore, the present experimental results agree well with the empirical solution (Eq.(7)) and the semi-analytical solution (Eqs.(15) and (16)) as shown in Fig.6. The maximum relative errors of b_t and t_{1m} are respectively less than 16 percent and 12 percent.

Temperature profiles in crosss sections of the wake

From the observed traverses, contours of equal temperature were plotted, and typical diagram is shown in Fig.2. With the aid of this diagram the mean position of the wake center was estimated. The contours of equal temperature shows that any cross section of the wake presents an upright ellipse, and that there exists the regions of asymmetry at the top and bottom. These may be mainly due to the action of the buyant force. As shown in Fig.5, the experimental temperature profiles in different cross sections of the wake are similar and very approximate to a Gaussian distribution.

The buoyant effect of the thermal wake

The present experimental results indicated that the thermal wake centerline rose up to a certain height and rotated at an angle ($\beta \approx 2.3^{o}$) relative to the horizontal plane as shown in Fig.7. Of course, if the free-stream velocity is higher, natural convection created by the buoyant force may be neglected so that the axis of the thermal wake will be identical with that of the slender streamlined heated body of revolution.

CONCLUSION

(1) The temperature profiles in a steady laminar wake behind a slender streamlined heated body of revolution from x/d=0 to 110 were fully explored. The experimental results and the semi-empirical analysis indicated that the temperature profiles in all cross sections of the wake obey a Gaussian distribution (Figs.5 and 6), and that the wake radius b_t and the temperature elevation on its centerline t_{1m} vary with its axial length x according to some power law (Fig.4).
(2) The contours of equal temperature in any cross section of the thermal wake present upright ellipses (Fig.2). The centerline of the thermal wake rose up to a certain height and rotated at an angle ($\beta = 2.3^{o}$) relative to the horizontal plane(Fig.7).
(3) The visualization of flow field with the aid of H_2 bubbles can vividly reveal the flow mechanism. The results of surveys given by the analogue drawing system can be seen at a mere glance, which will be suitable for a rapid judgement.

Thus the theoretical and experimental work of this paper lays down a primary fundation for a practical application of the wake.

REFERENCES

[1] FAGE, A. and FALKNER,V.M., 1932, Proc.Roy.Soc.A 135,702.

[2] HALL,A.A. and HISLOP,G.S., 1938, Proc. Cambridge Phil.Soc.34.48-67.

[3] HEINRICH,H.G. and ECKSTROM,D.J., 1963, Technical Documentary Report SAD-TDR-62-1103.

[4] HAMA,F.R. and PETERSON,L.F., 1976, J.Fluid Mech.76,1.

[5] WANG BU XUAN, 1982, Engineering heat and Mass Transfer.

[6] MERRITT,G.E., 1972, Wake Laboratory Experiment (Cornell Aeronautical Laboratory).

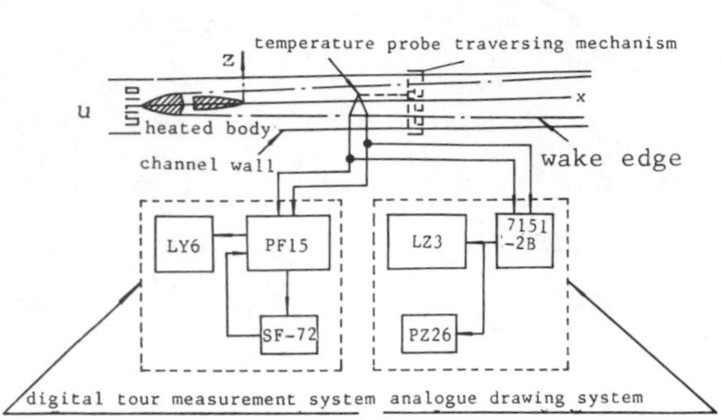

LY6——digital type printer; PF15,PZ26——digital voltmeter;
SF-72——amplifier; LZ3——pen recorder; 7151-2B——bridge.

Fig.1 Scheme of the temperature measurement system.

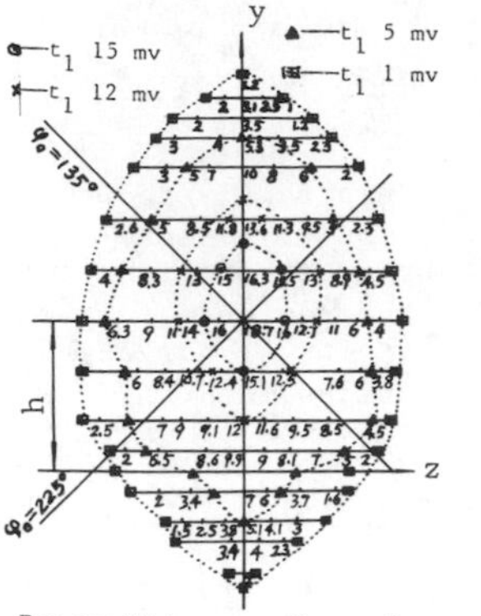

For conditions see Figure 7.

Fig.2 Contours of equal temperature in the section of the wake.

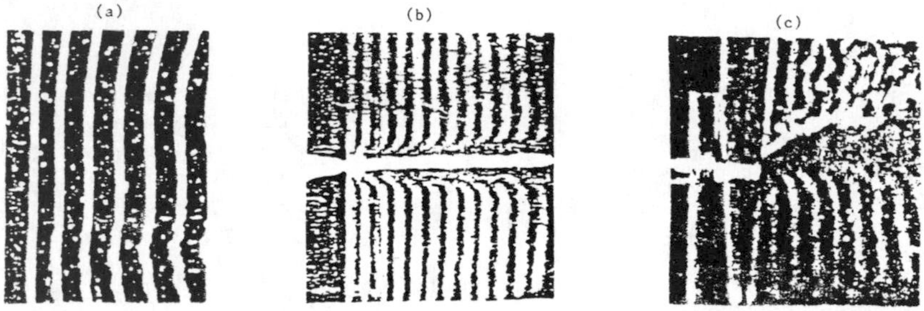

(a) Water channel-velocity profiles; (b) Flow patterns of the thermalwake in the horizontal plane; (c) Flow patterns of the wake in the vertical plane.

Fig.3 Photographs of visualization with the aid of H_2 bubbles (Red=1500)

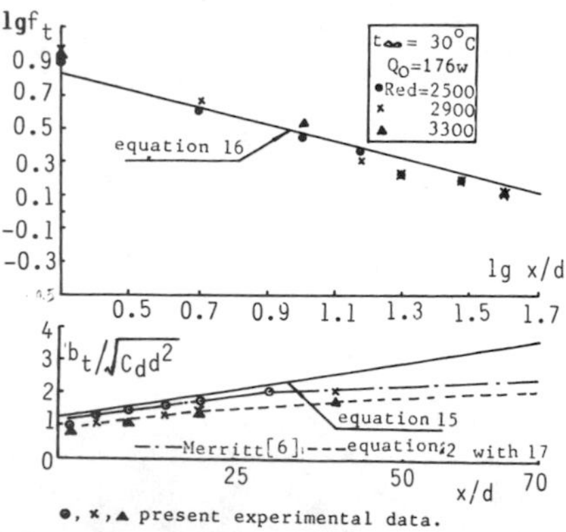

●, ×, ▲ present experimental data.

Fig.4 Maximum temperature elevation within the wake and wake radius VS downstream distance.

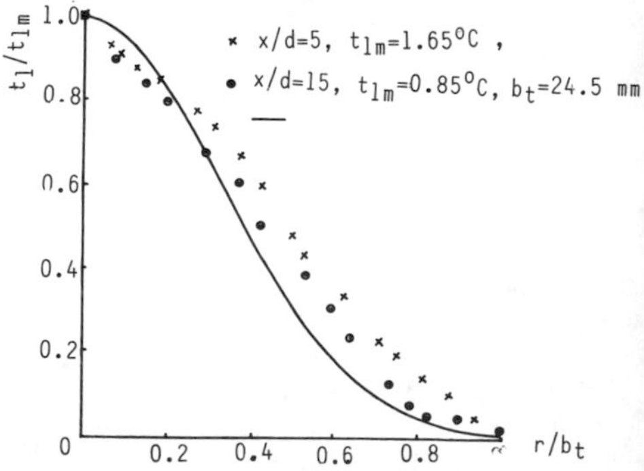

Fig.5 Non-dimensional temperature elevation VS. normalized radial coordinate.

(1.25mv/mm)

$(0.045°C/mv)$

$x/d=2$ $x/d=5$ $x/d=10$ $x/d=15$ $x/d=20$ $x/d=30$ $x/d=40$

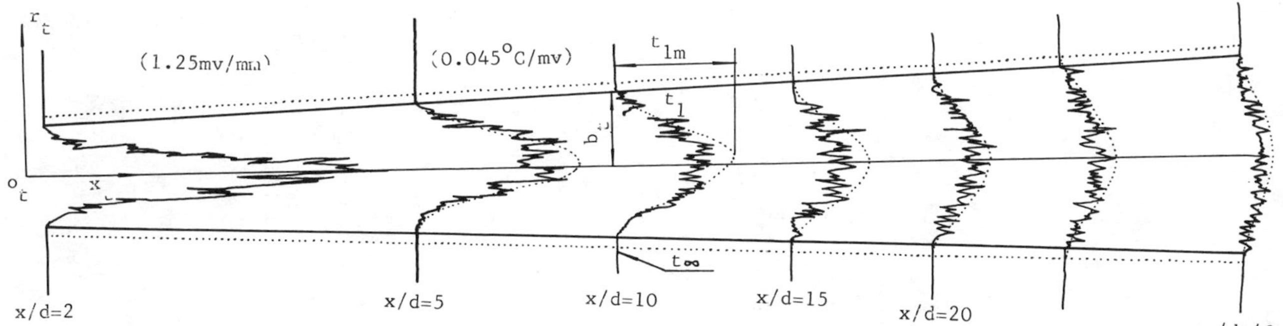

For conditions see figure 7.

...... Equation 7 with 15 and 16 ; —— present experimental curves.

Fig.6 Experimental and theoretical temperature distributions in the laminar wake behind a streamlined heated body of revolution.

$\beta = 2.3°$

$x/d=2$ 10 20 30

$u_\infty=13.6\text{cm/s},$
$t_\infty=30°C$

$Re_d=2500, \ \bar{d}=15\text{mm}$

$Q_0=176\text{W}$

——heating, ----not heating

a slender streamlined
heated body of revolution

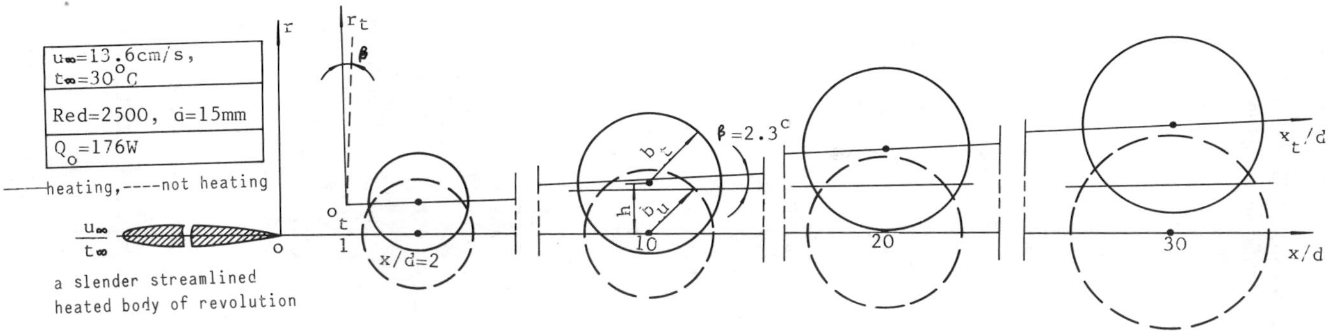

Fig.7 Buoying effect on the thermal wake behind the slender streamlined heated body of revolution

Thermal Drag in Fluid Flow Systems

ZENGYUAN GUO, WEIHONG BU, and GUANZHONG CHANG
Department of Engineering Mechanics
Tsinghua University
Beijing, PRC

INTRODUCTION

In most of heat transfer equipments used in power, chemical and metallurgical industry the fluid flow is usually associated with the heating or cooling process. Consequently, several attempts were made to investigate the interaction of fluid flow and heat transfer. Абрамович[1] studied the gas flow in constant-area ducts with heating and referred to the pressure drop in duct resulting from the gas acceleration due to heating as the thermal drag. Shapiro[2] summarized the choking phenomenen in duct flow caused by various thermal effects in the case where the upstream conditions were kept unchanged. Cebeci and Brashaw[3] present some analytical results for the flows that are coupled to heat transfer. With the development of the computer in capacity and function, it is now principlly possible to obtain the 3-D flow field and temperature field for various problems of the combined heat transfer and fluid flow[4] by solving their governing equations numerically. It is, however, still important to gain an insight into the physical mechanism of effects of the heating or cooling process on the fluid flow in order to have a deeper understanding of the coupling of heat transfer and fluid flow process on the one hand, and to facilitate the engineering analysis concerned on the other hand.

In this paper the physical mechanism of the thermal drag is investigated under some simplifications both analytically and experimently. Defining the thermal drag coefficient and the dimensionless heating number, an analytical relation between them which can be utilized to predict the pressure drop(such as pressure drop in combustors, acceleration drag in gas-liquid two phase flow and the clogging effect in gas-blast circuit breakers) due to heating has been obtained based on the solution of the governing equations for 1-D duct flow with simple heating. A phenomenon of the thermal roundabout flow which may occur in miscellaneous problems of practical interest(such as particles penetration into high temperature jets, interaction of gas flow and electric arc etc.) is revealed and discussed.

THERMAL DRAG IN 1-D DUCT FLOWS

1. Mechanism of the thermal drag

As shown in Fig. 1, we consider a case where a constant-area duct is fed by a infinite vessel at a pressure, p_o , which is greater than the back pressure, p_2 . Under the idealizations that (i) the flow is steady and one-dimensional; (ii) there is no heat or work exchange between the system and the environment; (iii) the heat source is homogeneously distributed in the constant-area duct;

(iv) the viscous dissipation produces negligible effect on the duct flow, the gas flow in the constant-area duct can be then described by the following 1-D, steady, invisid conservation equations:

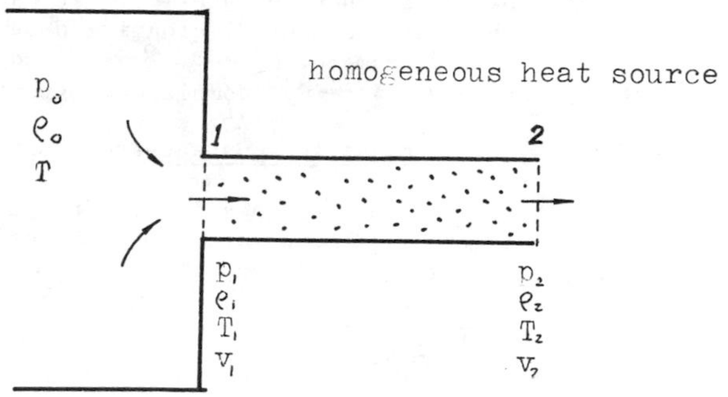

Fig.1 1-D thermal drag in the constant-area duct

continuity equation

$$\rho_1 V_1 = \rho_2 V_2 \qquad (1)$$

momentum equation

$$p_1 + \rho_1 V_1^2 = p_2 + \rho_2 V_2^2 \qquad (2)$$

$$p_o = p_1 + \rho_1 V_1^2 \qquad (3)$$

energy equation

$$C_p T_1 + \frac{V_1^2}{2} + s = C_p T_2 + \frac{V_2^2}{2} \qquad (4)$$

$$C_p T_o = C_p T_1 + \frac{V_1^2}{2} \qquad (5)$$

where s=S/Am, S is the overall heat generation, A is the cross-sectional area and m is the mass flux dencity.

The substitution of the equation (1) to equation (2) yields

$$p_1 - p_2 = \rho_1 V_1 (V_2 - V_1).$$

Here p_1 must be greater than p_2 for $V_2 > V_1$ due to gas heating. This implies that the heat addition to the flowing gas inevitably leads to gas acceleration and the consequent pressure drop in duct. This is just the physical mechanism of thermal drag.

Equations (1) to (5) are supplemented by the state equations:

$$p_1 = \rho_1 R T_1 \quad , \quad p_2 = \rho_2 R T_2 \qquad (6,7)$$

and their solution results in the ratio, ξ, of the mass flux density with heating, m_h, to that without heating, m_c:

$$\xi = \frac{m_h}{m_c} = \frac{1}{\sqrt{1+s/C_p T_o}} \qquad (8)$$

If $s > o$, $\xi < 1$. This means that the heat addition to the flowing gas reduces the flow rate in the duct. In this regard, this effect may be called thermal clogging.

Let $s/c_p T_o = H_e$ and it can be defined as the dimensionless heating number which reflects the ratio of the heat addition absorbed by per unit mass of flowing gas to its initial stagnation enthalpy. It can be easily seen from equation (8) that the strength of the thermal clogging depends merely on the dimensionles heating number H_e (for example, $H_e \rightarrow 0$, $\xi \rightarrow 1$: no clogging; $H_e = 3, \xi = 1/2$: half-clogging; $H_e \rightarrow \infty$, $\xi \rightarrow 0$: complete clogging).

2. Thermal drag coefficient in constant-area duct flows

In view of the fact that the heating of the gas flow produces an additional pressure drop, as friction factor or form drag coefficient defined in viscous flows, we may define a thermal drag coefficient:

$$C_t = \frac{p_1 - p_2}{\frac{1}{2}\rho_2 V_2^2} \tag{9}$$

If the thermal drag is regarded as an equivalent drag force, the momentum equations for the flow system as shown in Fig.1 can be written as

$$p_1 + \rho_1 V_1^2 = p_2 + \rho_2 V_2^2 + \frac{F}{A} \tag{10}$$

$$p_o \cong p_1 + \rho_1 V_1^2 \tag{11}$$

and gives the mass flux density

$$\dot{m}_f = V_2 = \frac{p_o - p_2 - \frac{F}{A}}{RT_{2,c}} \tag{12}$$

where F is the equivalent drag force, $T_{2,c}$ represents the outlet temperature of gas flow without heating which is identical to the inlet temperature, T_1, for the incompressible, constant-area duct flow.

Equalizing $\dot{m}_f$ to $\dot{m}_h$ and doing some operations, the analytical relation betwe the thermal drag coefficient and the heating number is given by

$$C_t = \frac{2H_e}{1+H_e} \tag{13}$$

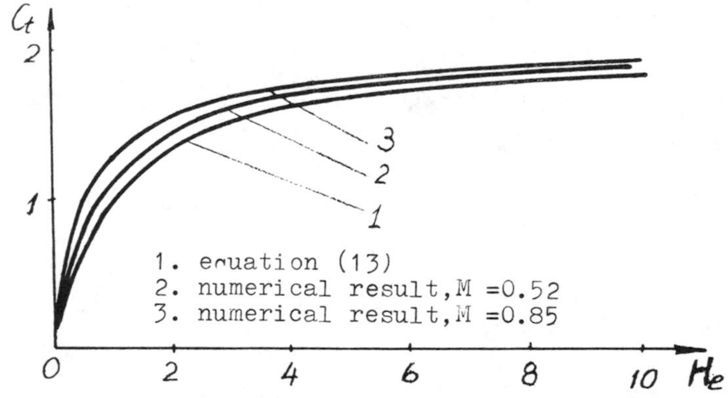

Fig.2 dependence of the thermal drag coefficient on the dimensionless heating number

It can be found that the thermal drag coefficient increases monotonically with the heating number. In the limiling case of very large H_e , C_t approaches to a value of 2. It means that the pressure drop caused by thermal clogging is twice as the dynamic head at duct exit. The graphic illustration of equation (13) is given in Fig.2.

The analytical relation (13) provides a simple way to predict the pressure drop in duct flows due to heating. The procedure for evaluating the pressure drop is similar to that for evaluating the pressure drop due to friction. Their analogy is given as follows:

$$R_e \rightarrow C_f \rightarrow \Delta p_f$$

$$H_e \rightarrow C_t \rightarrow \Delta p_t$$

It is worth noting that the approximate expression (3) no longer holds and should be replaced by the relation for the adiabatic and isentropic flow when the velocity (or Mach number) at duct inlet is not small:

$$p_o = p_1 \left(1+ \frac{k-1}{2} M_1^2\right)^{\frac{k}{k-1}} \qquad (14)$$

Here M_1 is the Mach number at section 1 (duct inlet). In this case the thermal drag coefficient, C_t , is dependent not only on the heating number, H_e , but also on the Mach number, M_1. The C_t—H_e curves with M_1 as parameter obtained by means of the numerical solution of governing equations are also plotted in Fig.2. These results confirm that equation (13) may be used with a good degree of approximation unless M_1 exceeds 0.2.

3. Thermal drag coefficient in varying-area duct flows

Consider a case of a converging duct with the homogeneous heat source in it connected by a infinite tank at a pressure, p_o , which is higher than the back pressure, p_2 , as shown in Fig.3. The gas flow is accelerated in this type of duct both due to area variation and the heating effect. The governing equations for this case can be written as:

continuity equation

$$\rho VA = \text{const} \qquad (15)$$

momentum equation

$$V\frac{dV}{dx} = -\frac{1}{\rho}\frac{dp}{dx} \qquad (16)$$

energy equation

$$ds = C_p dT + d\left(\frac{V^2}{2}\right) \qquad (17)$$

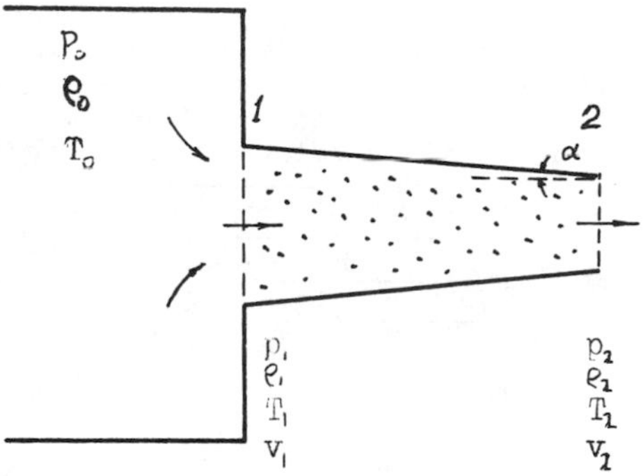

Fig.3 1-D thermal drag in the converging duct

The numerical approach of the above conservation equations with p_o, p_2 given yields the results illustrated in Fig.4. There total drag coefficient C_{total} covers contributions of the cross-sectional area variation and the thermal drag. From Fig.4 we find that (i) the total drag coefficient C_{total} , in the converging duct is greater than the thermal drag coefficient, C_t, in the constant-area duct when the heating number is not large. On the contrary, C_{total} becomes less than C_t when the heating number is sufficient large; (ii) the total drag coefficient hardly ever varies with the heating number if the half angle of convergence, α, exceeds 8°. This feature largly simplifies the prediction of the total pressure drop in the converging ducts.

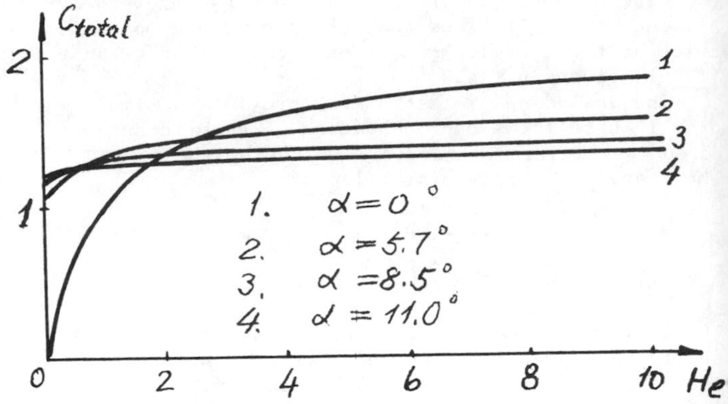

Fig.4 thermal drag coefficient in the converging ducts

THERMAL DRAG IN 2-D PROBLEMS AND THERMAL ROUNDABOUT FLOW

For all real problems the assumption of one-dimensionality for the entire flow is at best an approximation. For brevity we start with 2-D problems resulting from nonuniform heat supply in radial direction. Consider a case in which the heat supply is concentrated on the central part of duct cross section, as shown in Fig.5. In order to reveal the main feature of 2-D thermal drag phenomenon, we make the seemingly rather rough assumption that there is no heat, mass or momentum transfer between and inside the hot zone(with heat supply) and the cold zone(without heat supply). As a result, this 2-D problem can be reduced to two 1-D problems which may be treated seperately except the connection condition between two zones----radial pressure balance on the cross section. One of characteristic results obtained based on the above assumption is the ratio of mass flux density in the cold zone to that in the hot zone:

$$\frac{\dot{m}_c}{\dot{m}_h} = \sqrt{\frac{T_{2,n}}{T_{2,c}}} \tag{18}$$

where $\hat{m}_c = \sqrt{\frac{(P_o-P_2)P_2}{R\,T_{2,c}}}$, mass flux density in the cold zone,

$$\hat{m}_h = \sqrt{\frac{(P_o-P_2)P_2}{R\,T_{2,h}}},$$

mass flux density in the hot zone, $T_{2,c}$, $T_{2,h}$ is the temperature at outlet of the cold and hot zone respectively. Evidently, the higher the temperature of the hot zone is, the more the mass flux density decreases

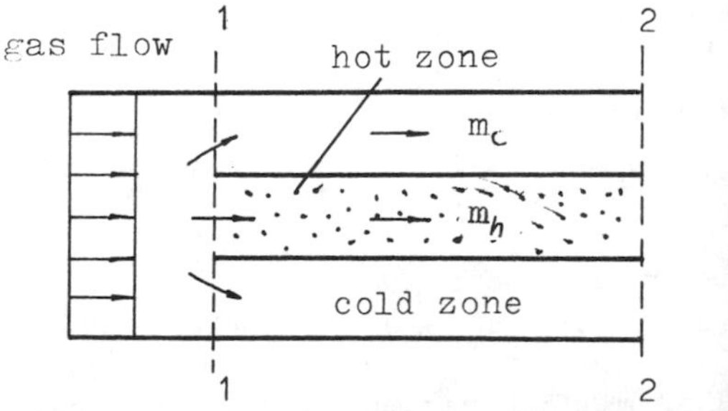

Fig.5 2-D thermal drag in the constant-area duct

(for instance, $\dot{m}_c/\dot{m}_h = 1.76$, when $T_{2,c}$ =300K and $T_{2,h}$ =1000K; m_c/m_h=3.16, when $T_{2,c}$=300K and $T_{2,h}$=3000K).

The difference of the mass flux densities between the cold and hot zone may be understood in the following manner: the flowing gas through the hot channel undergoes a thermal drag, namely, it must be against a pressue drop while the flowing gas through the cold channel is not this case since the viscous force has been here neglected. The oncoming stream, as common knowlege, prefers to pass through the channel (the cold zone in our case) in which lower resistance is needed to overcome. As a result, the flow rate will be redistributed among different flow channels with different flow resistance. The oncoming flow goes roundabout the hot zone (or hot channel), as shown in Fig.5, here referred to as thermal roundabout flow. The hot zone acts as a porous solid body which forces the oncoming flow in part to go round and in part go through at the same time. Therefore, the phenomenon of the thermal roundabout is attributed to the thermal drag in 2-D heating problems.

The flow rate through the duct consists of the flow rate in the cold zone and the hot zone:

$$\dot{m}_{un}A = \dot{m}_c A_c + \dot{m}_h A_h \tag{19}$$

where $\dot{m}_{un}$ is the mean mass flux density in the duct with nonuniform heat supply distribution on the cross section, A_c, A_h and A is the cross section area of the cold channel, hot channel and the duct respectively.

Substituting the equation (8) into the expression (19) leads to

$$\frac{\dot{m}_{uni}}{\dot{m}_{un}} = \frac{1+A_c/A_h}{1+\frac{A_c}{A_h}\sqrt{1+\frac{s}{C_pT_o}}} \tag{20}$$

From equation (20) the following remarks can be drawn:

(i) $m_{uni} < m_{un}$, when $s > 0$. This indicates greater strength of the thermal clogging with uniform heat supply in comparison with the case with nonuniform heat supply under equal overall amount of heating

(ii) $m_{un} \rightarrow m_{uni}$, when $A_h \gg A_c$. This shows that the expression for the thermal clogging in the case of uniform heat supply can be used to approximate to that in the case of nonuniform heat supply as long as the cross-sectional area of the hot channel is sufficient large.

EXPERIMENT

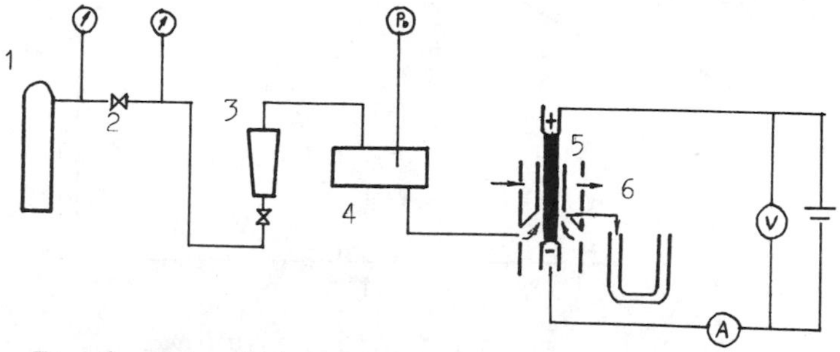

Fig.6 test flow system
1. gas reservoir 2. throttle valve 3. flowmeter
4. tank 5. electric arc 6. water

The experiments for the purpose of demonstrating the thermal drag(or thermal clogging)were carried in the test flow system, as schematically shown in Fig.6. The test section is the constant-area ducts of 37 mm in length, 6mm and 8mm inner diameter which is connected with a tank. the supplied gas(argon) comes from the gas reservoir. The D.C. electric arc burning inside the test duct was taken as the heat supply for the gas heating. The strength of the heat supply was changable by controlling the arc current. The stagnation pressure, p_o , in the tank was maintained at the value desired during the experiment. The measure parameters cover the tank pressure, p_o , inlet and outlet pressure of the test duct, p_1 and p_2 , gas flow rate, G, arc voltage and current, U and I, the amount of heat taken away by the cooling water, Q, and so on. After some arrangements of the measured data the thermal drag and the thermal clogging (flow rate variation) caused by arc heating are presented in Fig.7. It can be found that the pressure drop in test ducts and the reduction in the flow rate become more evident. These results were also rearranged in the relation between the thermal drag coefficient and the dimensionless heating number, then plotted in Fig.8 for checking the analytical result, equation (13). The qualitative agreement between the experimental results and the analytical expression (13) shows the avaliability of the latter though further work is needed in order to achieve more accurate experimental data.

ACKNOWLEDGEMENT:

The authors wish to acknowledge the Foundation of Chinese Academy of Science for the financial support of the Research described in this paper

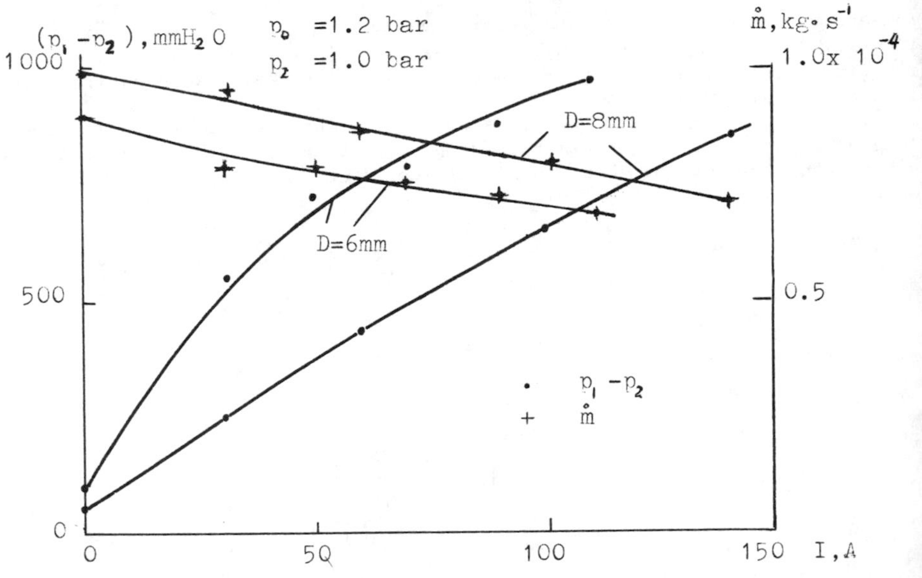

Fig.7 pressure drop and flow rate variation in ducts under arc heating

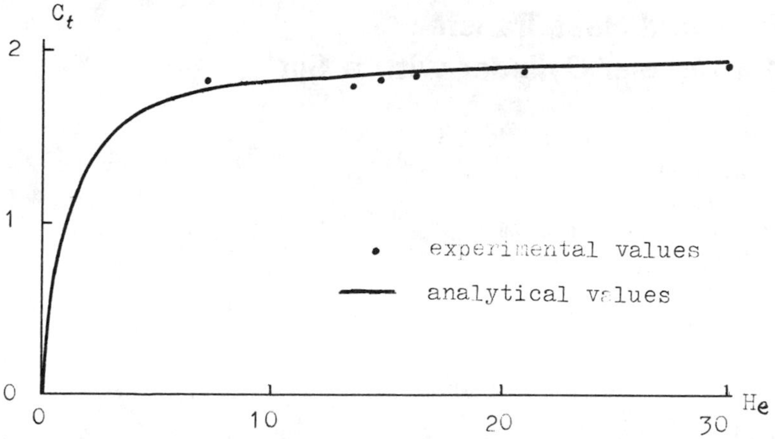

Fig. 8 comparison between experimental values and analytical values

REFERENCES

1. Абрамович, Г.Н., прикладная газовая динамика, Гостехиздат, 1953
2. Shapiro,A.H.,The Dynamics and Thermodynamics of Compressible Fluid Flow, Newyork 1953
3. Cebeci,T.and Bradshaw,P.,Physical and Computational Aspects of Convective Heat Transfer, Springer-Verlage, 1984
4. Patanka,S.V.,Numerical Heat Transfer and Fluid Flow, Newyork, 1980

Fluid Flow and Heat Transfer around a Circular Cylinder with a Slit

TAMOTSU IGARASHI
The National Defense Academy
1-10-20 Hashirimizu, Yokosuka, Japan

INTRODUCTION

The author[1,2] has reported that the flow around a circular cylinder with a two-dimentional slit placed along the diamter presents three typical flow patterns depending upon the inclination angle β of the slit. The first is the type of self-injection ($0° \leq \beta \leq 40°$), the second is the type of intermittent boundary layer suction ($60° \leq \beta \leq 75°$) and the third is the type of alternating boundary layer suction and blowing ($78° \leq \beta \leq 90°$). Particularly, the mechanism of the boundary layer suction and fluid flow in the slit as well as the mechanism of the vortex formation in the wake and fluid flow on the rear surface of the cylinder were clarified in ref.[2].

Such a circular cylinder with a slit has been studied from industrial use and also received practical application. For example, by installing a slit in parallel with the main flow, the cylinder is utilized as a bluff-body flame-holder[3,4] for stabilizing flame. Since in the case of the slit installed at right angles, it shed vortices at regular intervals, the cylinder is available as a Karman vortex flowmeter[5∿7]. Neither of these studies, however, have been carried out from the standpoint of fluid or thermal engineerings.

The purpose of this study is to clarify the fluid flow and heat transfer around a circular cylinder with a slit. This heat transfer is related to the heat transfer in separated region controlled by the boundary layer suction and blowing or by the jet into the wake. This model is of interest in connection with the heat transfers from walls with an unheated leading edge or with a break down in the thermal boundary layer and also film cooling.

EXPERIMENTAL APPARATUS AND PROCEDURE

The configuration of a circular cylinder with a slit is shown in Fig. 1. The experiments were carried out in the same wind tunnel as that in the previous reports[1,2]. The free stream velocity, u_o, ranged from 6 to 20 m/s, and the turbulence intensity was 0.5% in this range. The diameter of the cylinder was 34 mm, the Reynolds numbers were, therefore, in the range $1.3\times10^4 \leq Re \leq 4.5\times10^4$. The value of the slit ratio used for the flow characteristics measurements was $s/d=0.080$ ($\alpha=4°35'$) and that for heat transfer $s/d=0.087$ ($\alpha=5°$). The outside of an acrylic resin tube having a slit was wrapped with a stainless steel sheet of 0.03 mm in thickness. This sheet was heated directly by ac current, the measurements of heat transfer were performed under a constant heat flux. The temperature of the heating surface was measured with sixteen copper-constantan thermocuples of 0.07 mm in diameter.

LASSIFICATION OF FLOW PATTERNS

ccording to the previous reports[1,2], the flow patterns can be classified by he inclination angle β of the slit and the slit angle α as follows:

a) $0° \leqq \beta \leqq 40° - \alpha$: self-injection into the wake (pattern A),
b) $40° \leqq \beta + \alpha \leqq 55°$: region with a little effect of a slit,
c) $56° \leqq \beta + \alpha \leqq 62°$: transition region,
d) $62° - \alpha \leqq \beta \leqq 75°$: intermittent boundary layer suction (pattern B),
e) $75° \leqq \beta \leqq 78°$: transition region,
f) $78° \leqq \beta \leqq 90°$: alternating boundary layer suction and blowing (pattern C).

oreover, Reynolds numbers have no influence on the flow patterns.

LOW CHARACTERISTICS

he flow characteristics obtained are summarized below. Typical pressure istributions are shown in Fig. 2. For the pattern A, the base pressure rises onsiderably depending upon the influence of the self-injection. At β=15°, the

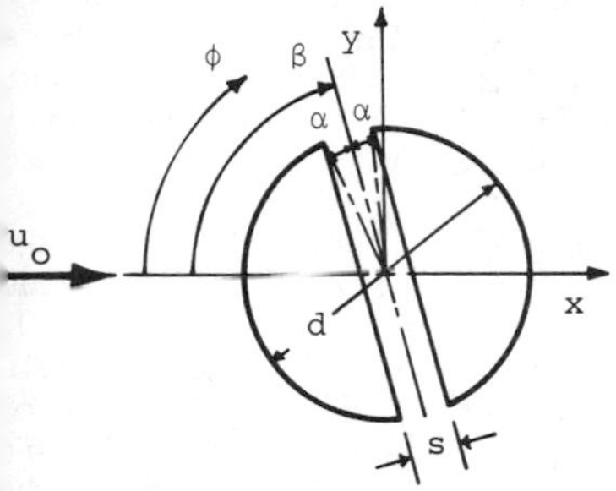

IGURE 1. Coordinate system

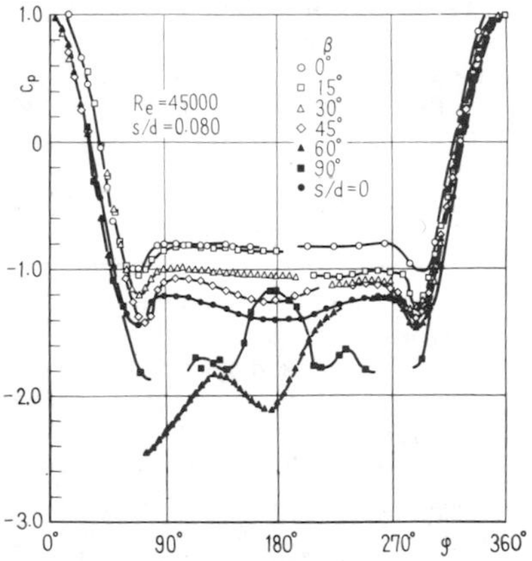

FIGURE 2. Pressure distribution

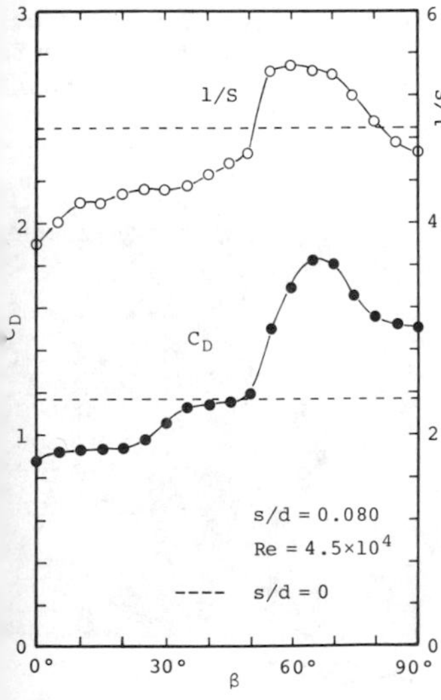

IGURE 3. Drag coefficient
d Strouhal number

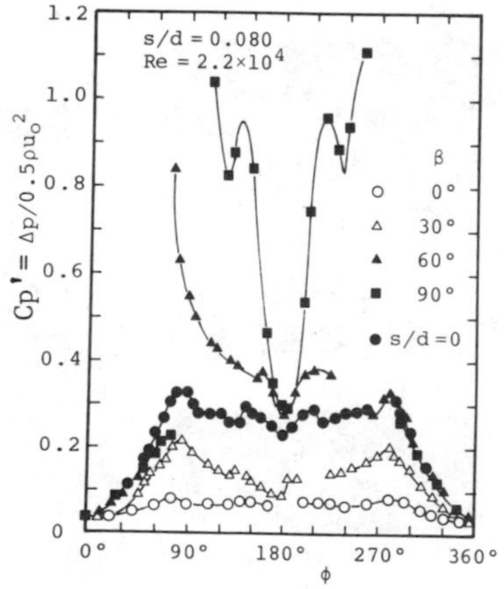

FIGURE 4. Distributions of
R.M.S. fluctuating pressure

pressure difference between both sides of the outlet of the slit becomes apparent, this was caused by the deflection of the jet. The effect of the jet decreases with increasing β, then at β=45° the distribution approaches that of the cylinder without a slit. At β=60° for the pattern B, the flow on the upper side of the cylinder becomes a potential flow owing to the intermittent boundary layer suction. Therefore, behind the slit the pressure reduces remarkably. The pressure distribution on the rear surface comes to be non-symmetric and has two peaks. At β=90° for the pattern C, suction and blowing take place alternately at both ends of the slit with the period of vortex shedding behind the cylinder. The time-averaged flow is, therefore, symmetric and the profile of the pressure distribution on the rear suface has three peaks produced by the temporary separation and the reattachment of the flow along the rear surface of the cylinder. These peculiar pressure distributions for the patterns B and C are caused by the adhered vortex on the rear surface and the flow of suction side. Figure 3 shows the variations in the drag coefficient and in the reciprocal of Strouhal number with β. The two profiles show a similar tendency to each other. The drag coefficient for the pattern A is reduced by 20 to 30% as compared with that of a cylinder without a slit, while those for the patterns B and C increase about 50 and 20%, respectively. The Strouhal number for the pattern A is larger than that of the cylinder without a slit, whereas for the pattern B it is smaller. On the other hand, the value for the pattern C is slightly larger than that of the cylinder without a slit. As being pointed out in the previous report[2], this is closely related to the fact that the vortex streets for the pattern C are identical to that of an oscillating aerofoil.

For typical three patterns the distributions of the R.M.S. fluctuating pressure on the surface of the cylinder are shown in Fig. 4. At β=0 and 30°, the value of Cp' on whole surface is relatively small in comparison with that of s/d=0. The drop in heat transfer on rear surface of the cylinder should be predictable from the previous report[8]. At β=60°, the value of Cp' on the upper side of the cylinder, increases drastically behind the slit, then decreases rapidly and takes the minimum value at the rear stagnation point. On the lower side, the value of Cp' agrees approximately that of s/d=0. At β=90°, the value of Cp' on the front surface is slightly lower than that of s/d=0, while the value on the rear surface increases remarkably. Especially the value just behind the slit exceeds the dynamic pressure. The values decrease rapidly near the rear stagnation point. The porfile is in contrast to that of the pressure distribution: the maximum and minimum values in the profile of Cp correspond to

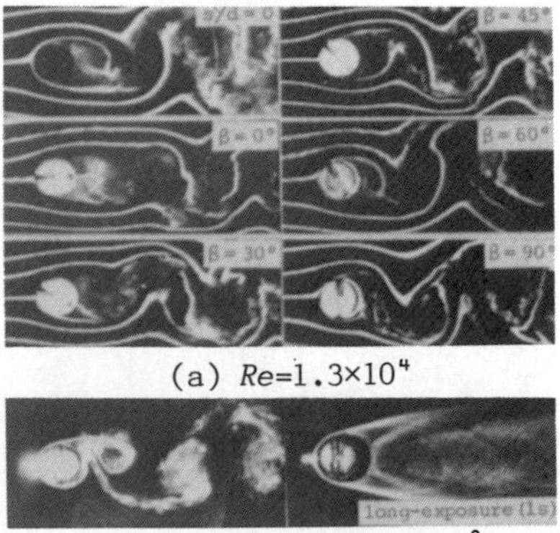

(a) Re=1.3×10^4

(b) β=90°, Re=4.4×10^3

FIGURE 5. Visualization of the flow around a circular cylinder with a slit

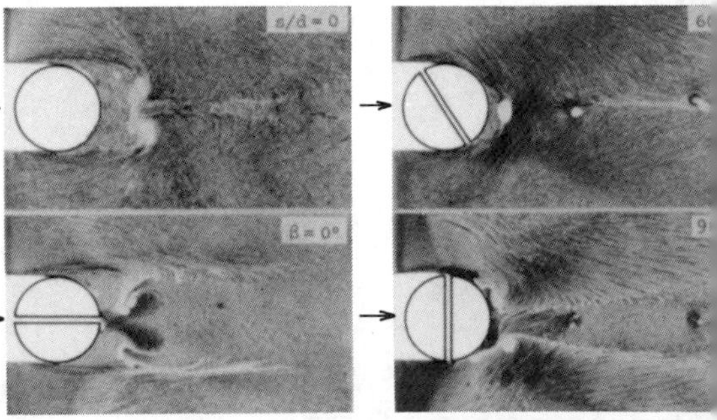

Re=4.5×10^4

FIGURE 6. Visualization of wake flow by oil-film method

the minimum and maximum in the profile of *Cp'*, respectively.

The flow around the cylinder was visualized using a smoke wind tunnel. The instantaneous photographs obtained are shown in Fig. 5. The wake flow was visualized by oil-film method on a thin plate inserted into the center of the axial section of the cylinder. The patterns obtained are shown in Fig. 6. For s/d=0, two concentrating parts of oil-film appears at about x/d=1.0. These parts can be correspond to the centers of the vortices formed by the upper and lower shear layers separated from the cylinder. In the case of self-injection, the shear layers are parallel to the main flow and the vortex formation region moves downstream. The jet blowing into the wake occurs a flip-flap phenomenon and the jet reaches to a distance of 0.7d. For the pattern B, the stagnant region behind the cylinder narrows and the two parts mentioned above combined in one. For the pattern C, a vortex adhered to the surface behind the slit of blowing side is formed and the flow of suction side moves along the rear surface and is caught in the adhered vortex. The locus of the upper and lower vortex streets overlap each other and spread out the downstream.

LOCAL HEAT TRANSFER

To examine the accuracy of the measurement of heat transfer, a heat insulator is packed closely in the slit. The local heat transfer coefficient on the front surface is in agreement with that obtained by the previous experiments, and the coefficient at the stagnation point agrees closely with the theoretical value. The variations in distributions of the local heat transfer coefficient around the cylinder with β are shown in Figs. 7 (a) and (b). At β=0, 15 and 30°, the local heat transfer coefficient near the slit on the front surface increases 1.5 to 2.0 times as high as that of the front stagnation point of the cylinder without a slit. On the upper side of the rear surface, the value decreases considerably as is expected. In addition, it is independent of the angle of β. On the contrary, the value on lower side increases with increasing β because of the deflected jet. At β=45°, there is no effect of the slit. In particular, on the lower side of the front surface and on the upper side of the rear surface, the heat transfer coefficients almost agree with those of the cylinder without a slit. At β=60, 70 and 75° for the pattern B, the coefficients increase

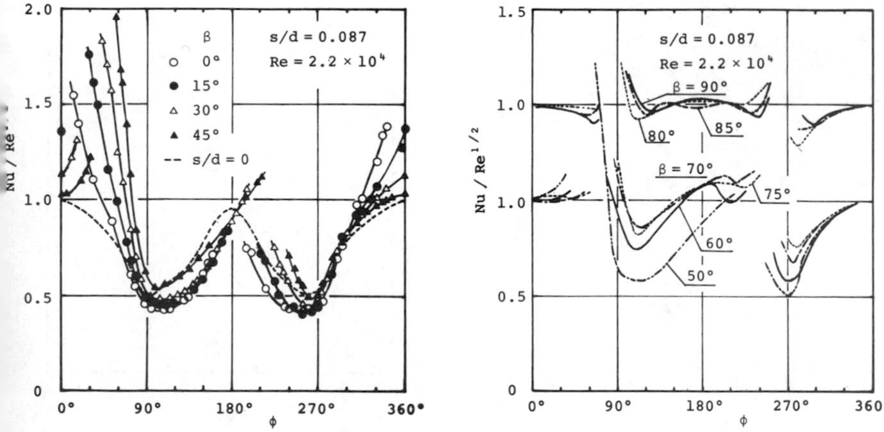

(a) (b)

FIGURE 7. Local heat transfer around a circular cylinder with a slit remarkably on the circumference of the cylinder. As a result of the intermittent boundary layer suction, the coefficient on the lower side of the front surface is equivalent to that of the cylinder without a slit. The values in the region from near the separation point to the rear surface increase considerably with increasing β. This fact are attributable to the separated vortex adhered on the lower side of the rear surface of the cylinder. On the upper side of the front surface, the coefficients increase slightly toward downstream. At β=80, 85 and 90° for the pattern C, the profiles become symmetric because of the symmetric flow and is roughly uniform circumference except just behind the slit. The heat transfer coefficients are nearly equal to that at the front stagnation point in the case of the cylinder without a slit. The coefficient on the rear surface have two minima at ϕ=125 and 235° near the second peak on the pressure distribution.

DEPENDENCE OF REYNOLDS NUMBERS

At β=0, 60 and 90° for typical three flow patterns, the local heat transfer coefficients with three different Reynolds numbers are shown in Figs. 8(a)∿(c). On the front side of the cylinder, the boundary layer is always laminar and the Nusselt number increases with the square root of the Reynolds number, except

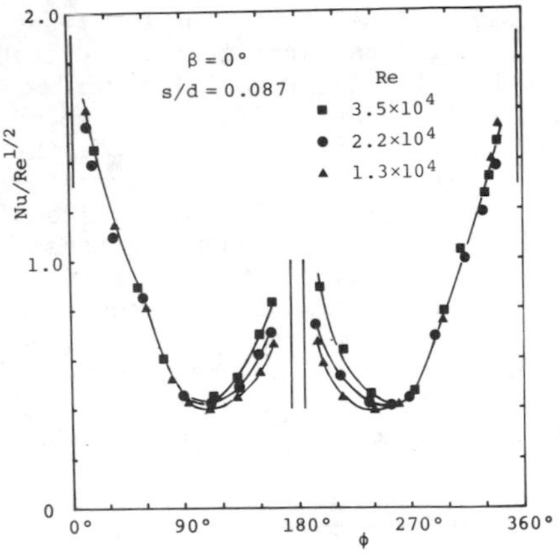

(a)

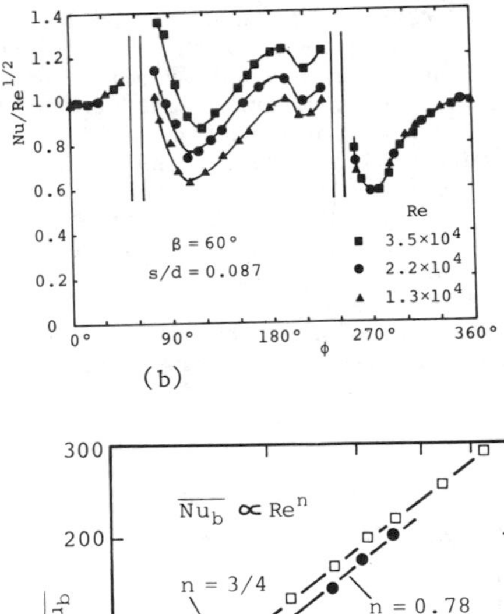

(b)

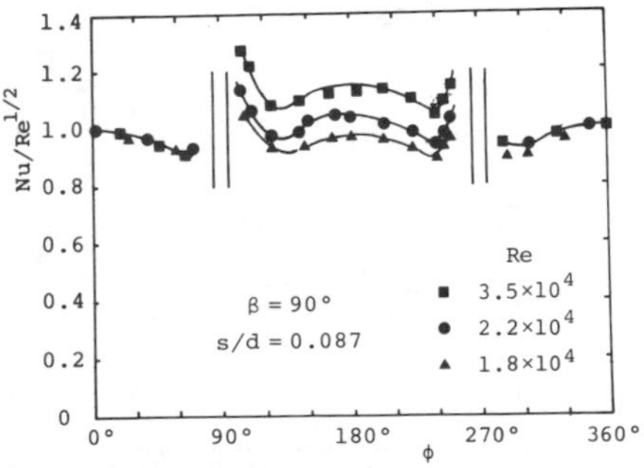

(c)

FIGURE 8. Local heat transfer at varying Reynolds number

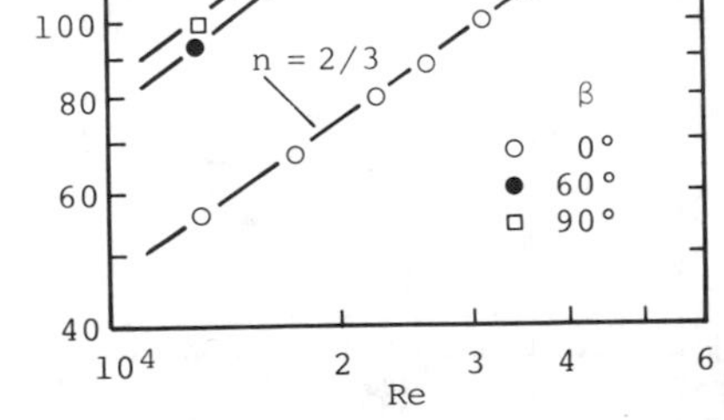

FIGURE 9. Dependence of Reynolds number on average heat transfer

for just behind the slit at β=60°. On the rear surface or behind the slit, the values of $Nu/Re^{1/2}$ increase with increasing Re. At β=60°, the coefficient behind the slit presents the same feature as that in the separated region, while on the lower side, the correlation of $Nu \propto Re^{1/2}$ holds up to the slit. On the rear surface, the location of the maximum heat transfer, ϕ=180~190°, corresponds to that of the minimum pressure. This fact was caused by the vortices formed on the rear surface of the upper and lower sides of the cylinder. Namely, as is evident from Fig. 6, the location is closer to the region overlapping the two centers of vortices reattaching on the rear surface, which are fromed by roll up of the upper and lower shear layers, and hence it is always exposed to either vortex. The dependence of the average heat transfer coefficient on Reynolds numbers was examined. The results obtained at the rear surface responsible for typical three flow patterns are shown in Fig. 9. At β=0°, the dependence satisfies the basic relation $\overline{Nu_b} \propto Re^{2/3}$ in agreement with the previous results for bluff bodies: a circular cylinder, a flat plate, a triangular prism and a square prism [9~13]. At β=60°, the relation is given by $\overline{Nu_b} \propto Re^{0.78}$. At β=90°, it is presented by the relation $\overline{Nu_b} \propto Re^{3/4}$.

CORRELATION BETWEEN AVERAGE HEAT TRANSFER AND FLOW CHARACTERISTICS

The average heat transfer coefficients are shown in Fig. 10. The coefficient on the front half of the cylinder, $\overline{Nu_f}$, is a constant, regardless of the inclination angle β, and is equal to that at the front stagnation point on a cylinder without a slit. In the range of 0°≦β≦20°, the coefficient on the rear

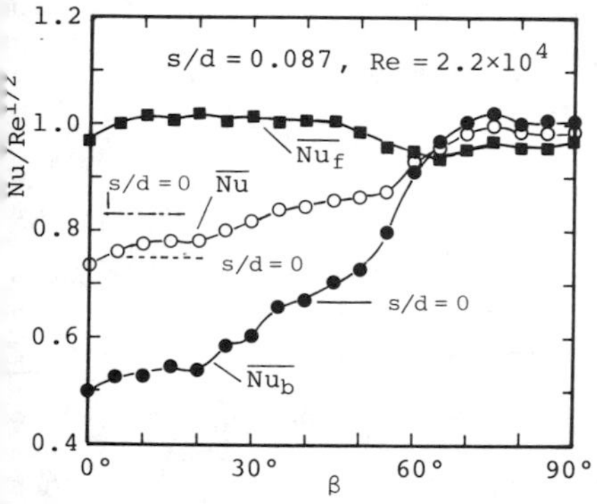

FIGURE 10. Average heat transfer

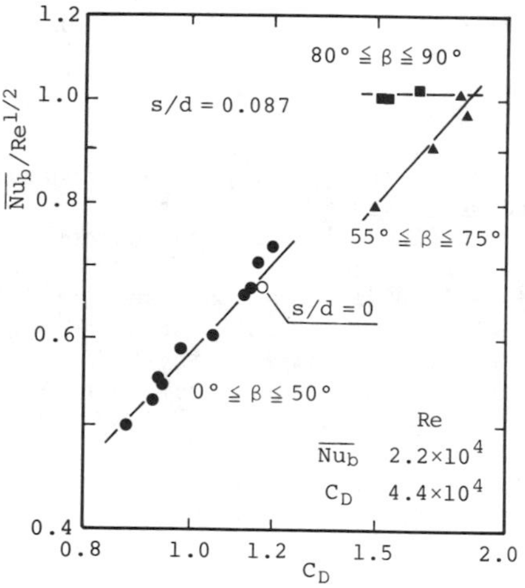

FIGURE 11. Correlations of average heat transfer on rear surface vs drag coefficient

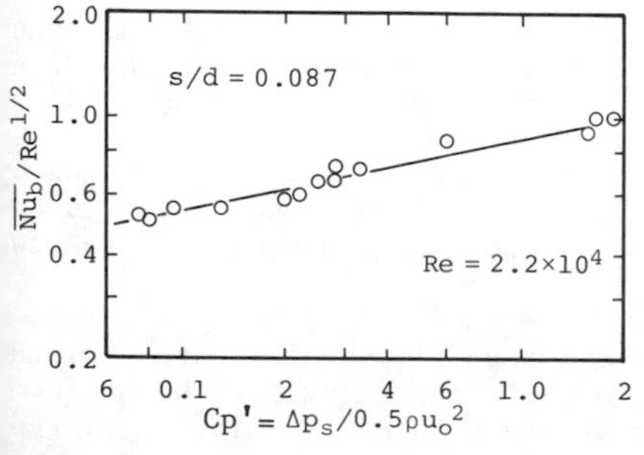

FIGURE 12. Correlations of average heat transfer on rear surface vs R.M.S. fluctuating pressure

half of the cylinder, $\overline{Nu_b}$, is 30% lower than that on the front half. The coefficient increases with increasing β, beyond β=60° it is slightly in excess of that on the front surface. At β=0∿5° the total heat transfer is nearly equal to that on a cylinder without a slit, and above β=75° the value is 30% larger.

From Figs. 10 and 3, it is seen that the heat transfer on the rear half of the cylinder is closely related to the drag coefficient. Though the measured Reynolds numbers are different, the relation between $\overline{Nu_b}$ and C_D is illustrated in Fig. 11. Up to β=75° the Nusselt number $\overline{Nu_b}$ increases with increasing in the drag coefficient C_D, beyond β=80° it becomes a constant. Especially, the data point of a cylinder without a slit lies in a straight line for the self-injection. As is obvious from this, on the rear surfaces of the two cylinders with and without a slit, the heat transfers are ascribed to the same mechanism. In the case of the pattern C, the characteristics of heat transfer differ from those for other patterns, because of the differences in the mechanism of the vortex formation above mentioned.

In the previous papers[8,14], the heat transfer coefficient in separated region of a circular cylinder can be given by the maximum R.M.S. value of fluctuating pressure at the separation point or at the rear stagnation point. In this experiment, the flow around the cylinder is unsymmetrical, and the flow on suction side becomes a dominant factor. Therefore, the R.M.S. value of fluctuating pressure is largest near the separation point on the suction side of the cylinder. In the range of 0°≦β≦90°, the correlation between the R.M.S. fluctuating pressure and the average heat transfer coefficient on the rear half of the cylinder for $Re=2.2\times10^4$ is shown in Fig. 12. The plots of data for every pattern give a straight line. As to the circular cylinder with a self-injection or a boundary layer suction, it was confirmed that the R.M.S. fluctuating pressure at the separation point was one of the most dominant factors of the heat transfer on the rear half of the cylinder.

CONCLUSIONS

Experiments were performed to investigate the fluid flow and heat transfer around a circular cylinder with a two-dimensional slit placed along the diameter. The following conclusions were obtained.

(1) Three typical flow patterns were observed depending upon the inclination angle of the slit as follows: self-injection type, intermittent boundary layer suction type and alternating boundary layer suction and blowing type. These characteristics of the local heat transfer were clarified in connection with the flow around the cylinder.

(2) For the pattern A, the drag coefficient C_D reduces to 0.9 and the Strouhal number S increases to 0.26. The local heat transfer coefficient on the rear surface reduces considerably, while it satisfies the basic relation $Nu \propto Re^{2/3}$.

(3) For the pattern B, the value S has nearly constant 0.18, and the value C_D increases to 1.80. In the pattern C, the value S increases to 0.21 and the value C_D decreases to 1.5. The Nusselt numbers on the rear surface of the two patterns are correlated with $Nu \propto Re^{3/4}$.

(4) The average heat transfer on front half is nearly equal to that of the front stagnation point on a cylinder without a slit, regardless of the angle of β. The total heat transfer coefficient for the pattern A is slightly larger than that of a cylinder without a slit, and those for the patterns B and C are as large as 30%.

(5) The average heat transfer on the rear surface of the cylinder is closely related to not only the drag coefficient but also the maximum values of the R.M.S. fluctuating pressure around the separation point.

NOMENCLATURE

C_D	drag coefficients
Cp	pressure coefficient
Cp'	R.M.S. coefficient of fluctuating pressure, $\Delta p/0.5\rho u_o^2$
d	diameter of circular cylinder
f	vortex shedding frequency
h, h_m	local and average heat transfer coefficients
$Nu, \overline{Nu}$	local and average Nusselt numbers, hd/λ, $h_m d/\lambda$
Δp	R.M.S. of fluctuating pressure
Re	Reynolds number, $u_o d/\nu$
S	Strouhal numbers, fd/u_o
s	width of slit
u_o	free stream velocity

Greek symbols

α	slit angle, $\sin^{-1}(s/d)$
β	inclination angle of slit
λ	thermal conductivity of fluid
ν	kinematic viscosity of fluid
ρ	density of fluid

Subscripts

b	rear face
f	front face
s	separation point

REFERENCES

1. Igarashi, T., Flow Characteristics around a Circular Cylinder with a Slit (1st Report), *Bull. JSME*, 21-154, 656-664 (1978).
2. Igarashi, T., Flow Characteristics around a Circular Cylinder with a Slit (2nd Report), *Bull. JSME*, 25-207, 1389- 1397 (1982).
3. Filippi, F. and Fabrovich-Mazza, L., Control of Bluff-body Flameholder Stability Limits, *Proc. 8th Symp. on Comb.*, 956-963 (1961).
4. Tsuji, H., Flame Stabilization by a Bluff-body Flameholder in High-velocity Gas, *J. Jap. Soc. Aero. Space Sci.* (in Japanese), 10-99, 122-132 (1968).
5. Tsuchiya, K., Ogata, S. and Ueta, M., Karman Vortex Flow Meter, *Jour. Jap. Soc. Mech. Eng.* (in Japanese), 72-607, 46-55 (1969).
6. Yamasaki, H., Ishikawa, Y. and Kurita, Y., The Karman Vortex Flowmeter, (A New Flow Measurement on Regular Vortex Shedding), *J. Soc. Inst. and Contr. Engr.* (in Japanese), 10-3, 173-188 (1971).
7. Kurita, Y., Karman Vortex Flowmeter, *J. Soc. Inst. and Contr. Engr.* (in Japanese), 18-5, 407-412 (1979).
8. Igarashi, T., Correlation between Heat Transfer and Fluctuating Pressure in Separated Region of a Circular Cylinder, *Int. J. Heat Mass Transfer*, 27-6, 927-937 (1984).
9. Richardson, P. D., Estimation of the Heat Transfer from the Rear of an Immersed Body to the Region of Separated Flow, *ARL*, 62-423, Brown University (1960).
10. Igarashi, T., Hirata, M. and Nishiwaki,N., Heat Transfer in Separated Flows (Part 1), *Heat Transfer-Jap. Res.*, 4- 1, 11-32 (1975).
11. Igarashi, T. and Hirata, M., Heat Transfer in Separated Flows (Part 2), *Heat Transfer-Jap. Res.*, 6-3, 60-78 (1977).
12. Igarashi, T. and Hirata, M., Heat Transfer in Separated Flows (Part 3), *Heat Transfer-Jap. Res.*, 6-4, 13-39 (1977).
13. Igarashi, T., Heat Transfer from a Square Prism to an Air Stream, *Int. J. Heat Mass Transfer*, 28-1, 175-181 (1985).
14. Igarashi, T., Fluid Flow and Heat Transfer in Separated Region of a Circular Cylinder, *Proc. ASME-JSME Thermal Eng. Joint Conf.* Vol. 3, 87-93 (1983).

Empirical Correlations of Turbulent Friction Factors and Heat Transfer Coefficients of Aqueous Polyacrylamine Solutions

EUG Y. KWACK and JAMES P. HARTNETT
Energy Resources Center
University of Illinois at Chicago
Chicago, Illinois 60680, USA

1. INTRODUCTION

The friction factor of a viscoelastic fluid in turbulent channel flow is much lower than tha of a Newtonian fluid at a fixed Reynolds number. This behavior is related to the elasticity of the viscoelastic fluid. Moreover, there appears to be a lower asymptotic limit for the fully developed turbulent friction factor which is a function only of the Reynolds number [1].

If heat transfer accompanies the turbulent flow of a viscoelastic fluid then the dimensionless heat transfer j-factor will be lower than for a Newtonian fluid at the same Reynolds number. Furthermore, it will be found that the reduction in heat transfer will be equal to or greater than the reduction in friction factor [2-5]. In addition, there is an asymptotic value of the dimensionless heat transfer j-factor, but more fluid elasticity is required to reach the heat transfer asymptote than to reach the friction factor asymptote [5,6].

Dimensional analysis suggests that a new dimensionless parameter must be introduced to take care of the fluid elasticity [7]. The Weissenberg number, defined as $\lambda V/D$ where λ i the characteristic time of the fluid, V the mean velocity and D the tube diameter provides a quantitative measure of the elastic behavior of the viscoelastic fluid. It has been reported that the fully established friction factor and the fully established heat transfer j- factor for aqueous solutions of polyacrylamide can be successfully correlated as a functio of the Reynolds number and the Weissenberg number [7]. It was found that the friction factor reached its asymptotic lower limit at a Weissenberg number of approximately 10. Above this so-called critical Weissenberg number for friction the friction factor remained at the asymptotic value. In the case of the heat transfer j-factor, a Weissenberg number 100 was necessary to reach the asymptotic heat transfer limit.

Empirical correlations were developed for aqueous polyacrylamide solutions giving the friction factor and the heat transfer j-factor as a function of the Reynolds number and th Weissenberg number. The proposed correlations were limited to Reynolds numbers of 20,000 and 30,000 [8]. In the current study the empirical correlations are extended to cover the Reynolds number range from 10,000 to 80,000.

2. EXPERIMENTAL APPARATUS AND PROCEDURES

The polymer solutions used in the current study were 300 and 500 wppm polyacrylamide solutions (Separan AP-273 from Dow Chemical Company) dissolved in Chicago tap water. The chemistry of Chicago tap water can be found elsewhere [9]. The test fluids, 300 and 500 wppm Separan solutions, were continously circulated in the flow loop for 16 and 8 hours, respectively. At regular time intervals the friction factors and heat transfer oefficients were measured at several flow rates corresponding to a wide range of the Reynolds number from 10,000 to 80,000. Samples of the circulating test fluid were ollected at the same time as the friction and heat transfer measurements. The steady shear viscosity of these test fluids removed from the loop was measured with the Weissenberg rheogoniometer with Couette geometry and with a capillary tube viscometer 0.05334 cm I.D. and ℓ/d = 375). A detailed description of the flow loop, the viscometer, he related instrumentation and test procedures can be found elsewhere [7,10].

. RESULTS AND DISCUSSIONS

.1 Rheological Properties

The apparent viscosity of the degraded 300 and 500 wppm Separan solutions are shown as unction of the shear rate for various hours of shear in Figure 1. The viscosity continuously decreases with circulation time for the entire shear rate range, especially in the low shear ate range. In previous studies [7,11,12], the characteristic time of the fluid was determined by a simple generalized Newtonian model such as Powell-Eyring model [13] in onjuction with the steady shear viscosity measurements. However, it was found that the viscosities of degraded 300 and 500 wppm Separan solutions were almost constant with espect to the shear rate, making it impossible to determine the characteristic time by the Powell-Eyring model. Alternatively, the empirical relationship between the friction actor and the Weissenberg number at a Reynolds number of 30,000 which had been well-established earlier [8] was used. From the measured friction factor it was possible to determine the Weissenberg number $\lambda V/d$ which yielded the characteristic time since the mean velocity and the tube diameter were known. The calculated values are shown in Figure 1 where it may be seen that the characteristic time of the fluid continuously

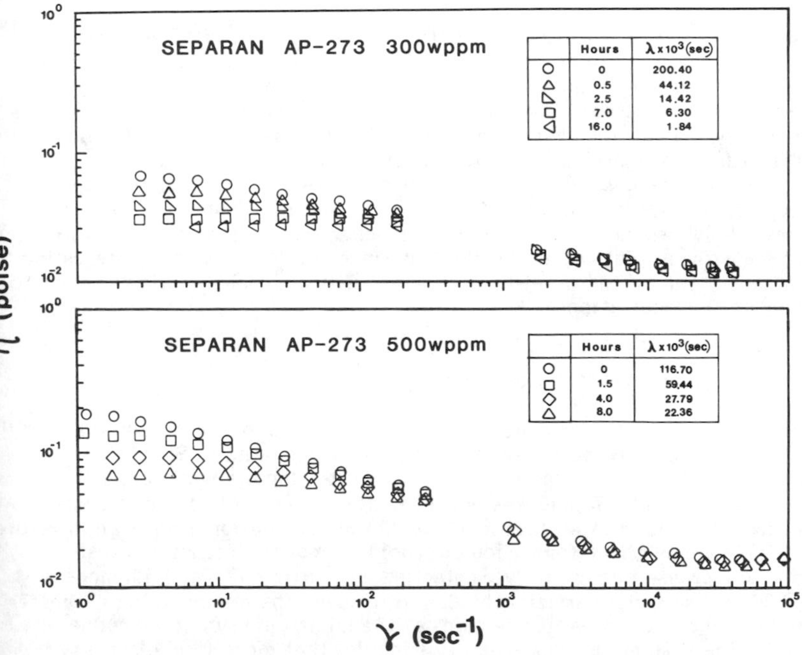

IGURE 1. Apparent viscosity vs. shear rate for 300 and 500 wppm Separan solutions.

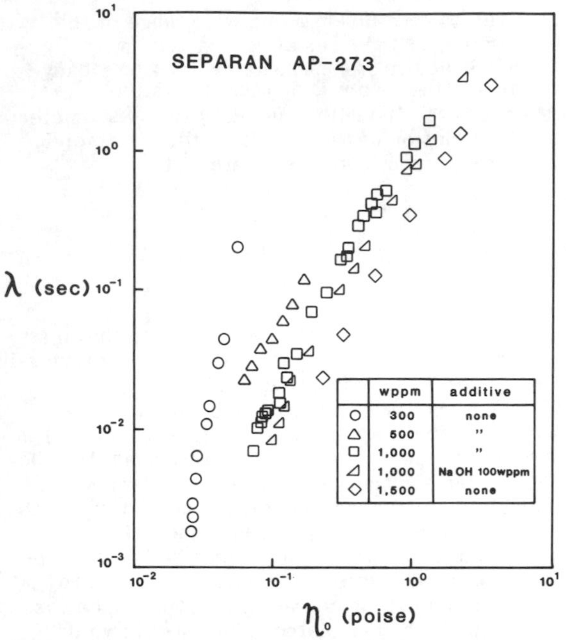

FIGURE 2. Characteristic time vs. zero shear rate viscosity for various Separan solutions.

decreases with shearing time, resulting from continuous loss of elasticity. Since the zero shear rate viscosity is known to be strongly related to fluid elasticity, the characteristic times are plotted as a function of the zero shear rate viscosity in Figure 2. For comparison, the results of 1,000 to 1,500 wppm Separan solutions used in previous studies [7,11,12] whose characteristic times were determined by the Powell-Eyring model are also included in this figure. This figure reveals that the characteristic time linearly decreases with decreasing zero shear rate viscosity on a logarithmic graph for a given polymer solution. The slope of 500 wppm solution is the same as those of 1,000 and 1,500 wppm solutions, whereas the 300 wppm solution has much steeper slope. It is seen that the slope becomes steeper for a given solution, as the fluid continues to degrade. It is interesting to note that for the same value of the zero shear rate viscosity the fluid elasticity seems to decrease with polymer concentration.

3.2 Friction factor and Heat Transfer Coefficient

The friction factors and heat transfer coefficients were simultaneously measured at x/d equal to 110 and 430 respectively, to obtain the fully developed values. Since the viscosities of 300 and 500 wppm Separan solutions are not too high, a wide range of Reynolds number from 10,000 to 80,000 was covered. The dimensionless heat transfer coefficient and corresponding friction factor of the 300 wppm solution are shown in Figure 3 as functions of the hours of shear for various Reynolds numbers. The heat transfer coefficient begins to increase immediately on starting circulation of the fluid, since a concentration of 300 wppm is not sufficiently elastic to reach the minimum heat transfer asymptote. However, the friction factor remains at the minimum asymptotic value for about a half hour before it starts to increase. This implies that more fluid elasticity is required to reach the minimum heat transfer asymptote. The rates of increase of friction factor and heat transfer coefficient are the highest at the time period when they start to

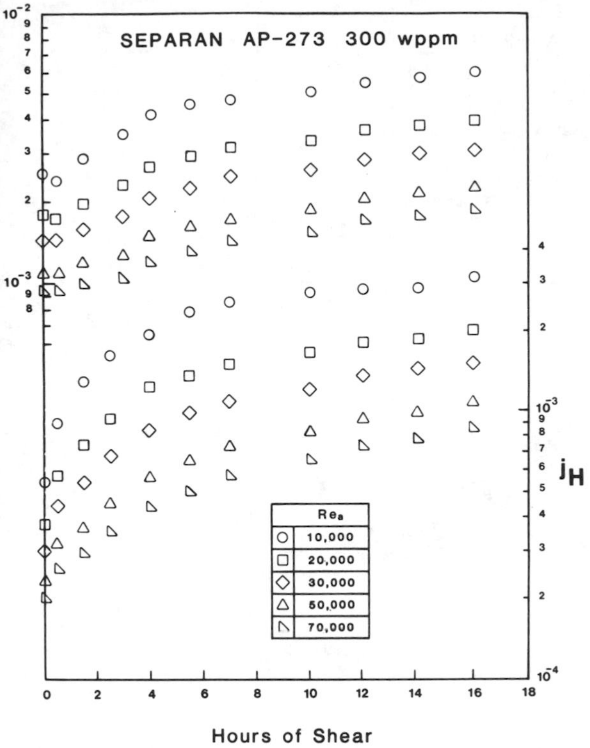

FIGURE 3. Fanning friction factor and heat transfer j-factor of 300 and 500 wppm Separan solutions vs. hours of shear for various Reynoldds numbers.

increase from their minimum asymptotes. It is seen that the rate of increase of the heat transfer coefficient is much larger than that of the friction factor. For example, the friction factor increases by about 2.5 times over the 16 hours of circulation, whereas the heat transfer coefficient increases by about 5 times during the same period. This is consistent with the earlier finding that the heat transfer reduction is greater than the drag reduction. Similar behavior was observed for the 500 wppm Separan solution.

The results of the 300 and 500 wppm solutions were replotted as functions of the Weissenberg number for various Reynolds numbers in Figure 4. The friction factor remains constant at a fixed Reynolds number if the Weissenberg number is large. However, as the Weissenberg number decreases, the critical Weissenberg number for friction Ws_{cf} is reached; below this value the friction factor increases with decreasing Weissenberg number. The friction factor approaches the Newtonian value, as the Weissenberg number becomes zero. Although similar behavior was observed for each Reynolds number the critical Weissenberg number for friction seems to increase with the Reynolds number. For example, the critical Weissenberg number for friction is about 4 at a Reynolds number of 10,000 and it is about 18 at a Reynolds number of 70,000. Similar behavior was observed for heat transfer. However, since the 300 and 500 wppm are not concentrated enough to reach the minimum heat transfer asymptote, the critical Weissenberg number for heat transfer Ws_{ch} cannot be determined.

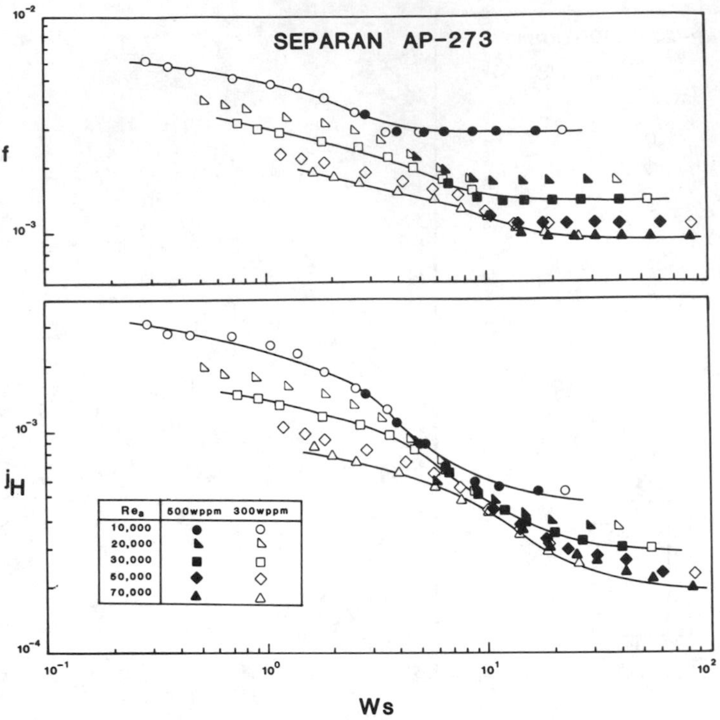

FIGURE 4. Fanning friction factor and heat transfer j-factor of 300 and 500 wppm Separan solutions vs. Weissenberg number for various Reynolds numbers.

3.3 Empirical Correlations

Based on the experimental results obtained with the 300 and 500 wppm Separan solutions the following correlations are proposed for fully developed friction factors and fully developed heat transfer coefficients of viscoelastic fluids in turbulent pipe flow. These correlations can be applied over the Reynolds number range from 10,000 to 80,000.

$$f = f_v + (f_n - f_v)\,[1.0 + C_1\,Ws + C_2\,Ws^5]^{0.50} \tag{1}$$

$$j_H = j_{H,v} + (j_{H,n} - j_{H,v})\,[1.0 + C_3\,Ws + C_4\,Ws^5]^{0.35} \tag{2}$$

where
$f_n = 0.079\,Re_a^{-0.25}$
$f_v = 0.20\,Re_a^{-0.48}$
$j_{H,n} = 0.0155\,Pr^{0.17}\,Re_a^{-0.17}$
$j_{H,v} = 0.03\,Re_a^{-0.45}$
$C_1 = 7.1 \times 10^9\,Re_a^{-2.6}$
$C_2 = 0.15\,Re_a^{-0.37}$
$C_3 = 3.5. \times 10^5\,Re_a^{-1.6}$
$C_4 = 7.0 \times 10^{-4}\,Re_a$

The functions f_n and $j_{H,n}$ are well known correlations for Newtonian friction factor and heat transfer coefficients. The functions f_v and $j_{H,v}$ are minimum drag and heat transfer asymptotes. Of course, these are functions of the Reynolds number.

Statistical analysis show that Equations (1) and (2) are in good agreement with the experimental data, with 126 of the friction factor predictions and 121 of the heat transfer predictions being within 5% of the 137 measurements. With two exceptions (12.7% and 10.2%) all of the remaining friction factors lie within 10% of the predictions. As for the heat transfer, only one point (11.2%) lies outside the 10% limit. Empirical correlations are shown in Figure 4 as solid lines for Reynolds numbers of 10,000, 30,000 and 70,000. These empirical correlations can be used to predict the fully developed friction factor and the fully developed heat transfer coefficient of aqueous polyacrylamide solutions if the physical properties including the rheological properties and characteristic time are known. Furthermore, the correlations of friction factor can be used to determine the characteristic time of dilute or degraded polymer solutions with a simple pressure drop measurements in a capillary tube [14]. Since these correlations are based on experimental results of aqueous Separan solutions, the general applicability of these correlations should be examined for other aqueous polymer solutions.

ACKNOWLEDGEMENTS

The authors wish to express appreciation to Mr. Moses L. Ng for his assistance in the experimental program.

NOMENCLATURE

C_1, C_2, C_3, C_4	coefficients defined by Equations (1) and (2)
c_p	specific heat of fluid
d	diameter of tube
f	Fanning friction factor, $\tau_w/(\rho V^2/2)$
f_n	Fanning friction factor of Newtonian fluid
f_v	minimum drag asymptote
h	convective heat transfer coefficient, $q_w/(T_w-T_b)$
k	thermal conductivity of fluid
j_H	heat transfer j-factor, St $Pr_a^{2/3}$
$j_{H,n}$	heat transfer j-factor of Newtonian fluid
$j_{H,v}$	minimum heat transfer asymptote
ℓ	total length of test section
Nu	Nusselt number, hd/k
Pr_a	Prandtl number based on viscosity at the wall, $\eta c_p/k$
q_w	heat flux at the wall

Re_a	Reynolds number based on apparent viscosity at the wall, $\rho Vd/\eta$
St	Stanton number, $Nu/(Re_a Pr_a)$
T	Temperature
T_b	bulk temperature of fluid
T_w	inside wall temperature
V	average velocity
Ws	Weissenberg number, $\lambda V/d$
Ws_{cf}	critical Weissenberg number for friction
Ws_{ch}	critical Weissenberg number for heat transfer
x	axial coordinate

Greek Symbols

$\dot{\gamma}$	shear rate
η	apparent viscosity evaluated at the wall
η_o	zero shear rate viscosity
λ	characteristic time of fluid
ρ	density of fluid
τ_w	wall shear stress

REFERENCES

1. Virk, P.S., Drag reduction fundamental, A.I.Ch.E.J., Vol. 21, p. 625 (1975).

2. Khabakhpasheva, E.M. and B.V. Perepelitsa, Turbulent heat transfer in weak polymeric solutions, Heat Transfer - Sov. Res., Vol. 5, p. 117 (1973).

3. Mizushina, T., H. Usui and T. Yamamoto, Turbulent heat transfer of viscoelastic fluids flow in pipe, Lett. Heat and Mass Transfer, Vol. 2, p. 19 (1975).

4. Ng, K. S., Y.I. Cho and J.P. Hartnett, Heat transfer performance of concentrated polyethylene oxide and polyacrylamide solutions, A.I.Ch.E. Symp. Ser. (19th National Heat Transfer Conf.), No. 199, Vol. 76, p. 250 (1980).

5. Kwack, E.Y., J.P. Hartnett and Y.I. Cho, Turbulent heat transfer in circular tube flows of viscoelastic fluids, Wärme-und Stoffübertragung, Vol. 16, p. 35 (1982).

6. Cho, Y.I. and J.P. Hartnett, Non-Newtonian fluids in circular pipe flow, in Advances in Heat Transfer (edited by T.F. Irvine, Jr. and J.P. Hartnett), Vol. 15, p. 59, Academic Press, New York (1982).

7. Kwack, E.Y., Y.I. Cho and J.P. Hartnett, Effect of Weissenberg number on turbulent heat transfer of aqueous polyacrylamide solutions, Proc. 7th Int. Heat Transfer Conf., Vol. 3, p. 63 (1982).

8. Kwack, E.Y. and J.P. Hartnett, Empirical correlations of turbulent friction factors and heat transfer coefficients for viscoelastic fluids, Int. Commun. Heat Mass Transfer, Vol. 10, p. 451 (1983).

9. Kwack, E.Y., Y.I. Cho and J.P. Hartnett, Solvent effect on drag reduction of Polyox solutions in square and capillary tube flows, J. Non-Newtonian Fluid Mech., Vol. 9, p. 79 (1981).

10. Kwack, E.Y., Effect of Weissenberg number on turbulent heat transfer and friction factor of viscoelastic fluids, Ph.D. thesis, University of Illinois, Chicago (1983).

11. Kwack, E.Y. and J.P. Hartnett, Effect of diameter on critical Weissenberg numbers for polyacrylamide solutions in turbulent pipe flow, Int. J. Heat Mass Transfer, Vol. 25, p. 797 (1982).

12. Kwack, E.Y. and J.P. Hartnett, Effect of solvent chemistry on critical Weissenberg numbers, Int. J. Heat Mass Transfer, Vol. 25, p. 1445 (1982).

13. Bird, R.B., Experimental test of generalized Newtonian models containing a zero shear viscosity and a characteristic time, Can. J. Chem. Eng., Vol. 43, p. 161 (1965).

14. Kwack, E.Y. and J.P. Hartnett, New method to determine characteristic time of viscoelastic fluids, Int. Commun. Heat Mass Transfer, Vol. 10, p. 77 (1983).

Numerical Study of the Friction and Heat Transfer Performances of Straight Plate-Fin Surfaces

XING LUO and ZU-HUI CAI
Shanghai Institute of Mechanical Engineering
Shanghai, PRC

ABSTRACT

Friction and heat transfer performances of the straight plate-fin surfaces with a more realistic thermal boundary condition were computed for both fully developed laminar and turbulent flows. The special feature of the condition co sidered here is the accounting of finite fin thermal resistances in both transv and longitudinal directions. It was concluded that the effect of the longitudi al conduction on the heat transfer can usually be neglected except in the case o a low Peclet number flow, while the effect of the transverse thermal resistance is much more significant. It was also found that the effect of the transverse thermal resistance is greater in turbulent flow than in laminar flow. The pred ed friction and heat transfer performances were shown to be in good agreement with experimental data.

INTRODUCTION

It is of importance to be able to predict friction and heat transfer performances of non-circular ducts accurately in the design of compact heat exchangers. Straight plate-fin surfaces with rectangular section (Fig.1) are frequently used in effective compact heat exchangers. In these rectangular ducts, two streams of fluid are separated from one another by the spacers of the ducts, and the flanks of the ducts are composed by fins. The friction and heat transfer performances of usual rectangular ducts have been analyzed by many investigators as shown in Table 1.

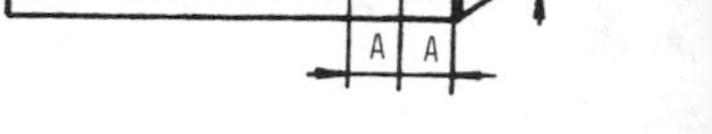

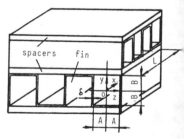

Fig.1 Schematic diagram of th straight plate-fin surf

Table 1 Boundary Conditions Considered in Open Literature

Laminar Flow		Turbulent Flow		Remark
Thermal Boundary Condition	Aspect Ratio	Thermal Boundary Condition	Aspect Ratio	T_n : Constant axial wal temperature with finite normal wall thermal resistance
T [1,2,3]	1,2,3,4,6,8	T [6,7]	1	H_p : Constant axial wal heat flux with finite peripheral wall heat conduction
H [1,2,3]	1,2,3,4,6,8	H [7]	1	
H_1 [2]	1,2,4,8			
T_n [4]	1,2,4,8			
H_p [5]	1			

However, the realistic condition of the straight plate-fin surfaces differs from those of usual rectangular ducts as shown in Table 1. The heat flow from the fluid to the fin surfaces must be transferred to the top and bottom spacers via heat conduction through the fins and, in turn, transferred to another fluid, as shown in Fig.1. Thus, the effect of the finite fin thermal resistances in both y and x-directions should be included in the thermal boundary condition of the straight plate-fin surfaces. This situation considered here is, to the authors' knowledge, not discussed in open literature.

The primary objective of this paper is to investigate the effect of the finite fin thermal resistances in y and x-directions on the laminar and turbulent convective heat transfer. For turbulent heat transfer, an approximate algebraic expression of the mixing length including an extended form of van Driest's damping factor was used to obtain the velocity distributions close to the experimental results. In order to examine the accuracy of the numerical calculations obtained in this paper, the heat transfer performances for the usual constant temperature (T) and constant heat flux (H) conditions were also computed and the performances of the stainless steel straight plate-fin surfaces with different aspect ratios were determined by the modified selected point matching technique [8].

NOMENCLATURE

A	duct half width
A^+	constant, eq.8
a	thermal diffusivity
a^*	dimensionless thermal diffusivity, $1+\varepsilon_H/a$
B	duct half height
de	equivatent diameter, $4AB/(A+B)$
f	friction factor, $2\tau_w/\rho w^2$
G,H	functions defined by eq.6
j	j-factor, $Nu/RePr^{1/3}$
L	duct length
L^+	dimensionless duct length, $2L/deRePr$
l	characteristic length
l^*	dimensionless characteristic length, l/A
Nu	Nusselt number, $\alpha de/\lambda_f$
P	pressure
Pe	Peclet number, $w_m de/a$
Pe^*	characteristic Peclet number, $w_s A^2/aL$
Pr	Prandtl number, ν/a
Pr_t	turbulent Prandtl number, $\varepsilon_M/\varepsilon_H$
Re	Reynolds number, $w_m de/\nu$
T	temperature
T_{in}	temperature at entrance
T_s	temperature at spacer
u_τ	friction velocity, $(\tau_w/\rho)^{\frac{1}{2}}$
W	dimensionless velocity, w/w_s
W_m	dimensionless average velocity, w_m/w_s
w	velocity
w_m	average velocity
w_s	characteristic velocity, $\lvert\partial P/\partial x\rvert A^2/\rho\nu$
x,y,z	coordinates
$X=x/L$, $Y=y/B$, $Z=z/A$	dimensionless coordinates
x^+	dimensionless coordinate, $2x/deRePr$
α	heat transfer coefficient; parameter, eq.6
γ	parameter, B/L
δ	fin thickness; damping factor
ε_H	turbulent thermal diffusivity
ε_M	turbulent kinematic viscosity
η	fin height-to-space ratio, B/A
θ	dimensionless temperature, $(T-T_s)/(T_{in}-T_s)$
θ_m	dimensionless bulk temperature
θ_w	dimensionless average wall temperature
λ_f	thermal conductivity of fluid
λ_w	thermal conductivity of fin
λ^*	dimensionless thermal conductivity, $\lambda_w\delta A/2\lambda_f B^2$
ν	kinematic viscosity
ν^*	dimensionless viscosity, $1+\varepsilon_M/\nu$
ρ	fluid density
τ_w	fluid shearing stress at wall

GOVERNING EQUATIONS AND NUMERICAL METHOD

It is known that the fin used in plate-fin surfaces is very thin so that the temperature across the thickness of the fin can be considered uniform. The temperature distribution of the spacers is simplified as constant temperature.

Besides, the assumptions made in the present analysis are as follows: (1) flui flow is steady and fully developed; (2) properties of both fluid and solid surfaces are constant; (3) heat conduction of fluid in the fluid flow direction an viscous dissipation of fluid are negligible; (4) the tips of the entrance and exit of surfaces are adiabatic; (5) transverse velocity components for turbulen flow are neglected.

According to above assumptions, the governing equations that take account o the finite transverse and longitudinal thermal resistances of the fins are as follows,

$$\frac{1}{\eta^2}\frac{\partial}{\partial Y}\left(\nu^*\frac{\partial W}{\partial Y}\right)+\frac{\partial}{\partial Z}\left(\nu^*\frac{\partial W}{\partial Z}\right)=-1 \tag{1}$$

$$W_{Y=1}=W_{Z=1}=0 \tag{2a}$$

$$\left(\frac{\partial W}{\partial Y}\right)_{Y=0}=\left(\frac{\partial W}{\partial Z}\right)_{Z=0}=0 \tag{2b}$$

$$Pe^*W\frac{\partial\theta}{\partial X}=\frac{1}{\eta^2}\frac{\partial}{\partial Y}\left(a^*\frac{\partial\theta}{\partial Y}\right)+\frac{\partial}{\partial Z}\left(a^*\frac{\partial\theta}{\partial Z}\right) \tag{3}$$

$$X=0,\ Y<1,\ Z<1:\quad \theta=1 \tag{4a}$$

$$Z=1,\ 0<X<1,\ Y<1:\quad a^*\frac{\partial\theta}{\partial Z}=\lambda^*\left(\frac{\partial^2\theta}{\partial Y^2}+\gamma^2\frac{\partial^2\theta}{\partial X^2}\right) \tag{4b}$$

$$Y=1:\quad \theta=0 \tag{4c}$$

$$Z=1:\quad \left(\frac{\partial\theta}{\partial X}\right)_{X=0}=\left(\frac{\partial\theta}{\partial X}\right)_{X=1}=0 \tag{4d}$$

$$\left(\frac{\partial\theta}{\partial Y}\right)_{Y=0}=\left(\frac{\partial\theta}{\partial Z}\right)_{Z=0}=0 \tag{4e}$$

where $\nu^*=a^*=1$ for laminar flow and for turbulent flow $\nu^*=1+\varepsilon_M/\nu$, $a^*=1+\varepsilon_M/a$. The thermal boundary condition expressed in equations (4) is designated as T_2 condition in which the temperature of the top and bottom spacers is constant an the finite fin thermal resistances in both transverse and longitudinal direction are considered. Two special cases of this boundary condition are:(i) when the longitudinal thermal resistance of the fin approaches infinity, that is the cas of $\gamma^2\approx 0$ in equation (4b), T_2 reduces to the T_1 boundary condition neglecting the longitudinal fin condition; (ii) when λ^* approaches infinity, T_2 reduces t the usual constant wall temperature condition designated as T . Equations (1- were solved numerically by using finite-difference method. An implicit scheme was used, in which the upwind scheme was applied to the convective term, and ce tral-difference scheme to the diffusive terms. The $(\partial\theta/\partial Z)$ term at the boundary surface (Z=1) was represented by the backward-difference scheme while the first order partial derivatives at the other surfaces were represented by the central difference scheme.

2.1 Laminar Flow

For laminar flow, the equidistant mesh spacings were used, and the mesh siz used was X=0.02, Y=0.05, Z=0.05. Equations (1-2) were first solved numerica to predict the fully developed velocity field. This solution was then coupled with equations (3-4) to predict the developing temperature field with over-relaxation Gauss-Seidel iteration.

T2 Boundary Condition Because of the longitudinal fin heat conduction to be a considered in this condition, the temperature distribution at each section is related to those at upstream and downstream adjacent-sections. Therefore a rep iteration is needed in the calculation. The step by step procedure is as follo

(i) according to the temperature distribution of fluid just obtained (at first it is assumed), determine the temperature distribution at the fin by matching that to the boundary condition at the fin surfaces (4b,4d); (ii) begin at the first section, calculate the temperature distribution of the fluid in the duct one after the other by solving the finite-difference form of equations (3) and (4); (iii) repeat the above steps until the temperature variations of the fluid and the fin are smaller than given precision (10^{-5}) after iterating.

T Boundary Condition Under this condition, the temperature distribution of fluid at each section is related only to the upstream adjacent-section, therefore the calculation is simplified greatly and is not limited by computer capacity, so that any number of sections can be calculated.

The predicted laminar temperature development in duct of η=4.39 with T_2, T_1 and T boundary conditions are shown in Fig.2a.

2.2 Turbulent Flow

From the point of view of the engineering purposes, the product of the modified Buleev's characteristic length and the damping factor of the rectangular duct was adopted as the mixing length in the present calculation. The main stream velocity distribution was then obtained by solving the momentum equation in the main stream direction.

Turbulent Viscocity and Thermal Diffusivity For turbulent flow in rectangular ducts, the Buleev's characteristic length 1 was obtained by using the modified Buleev's formula proposed by Gessner et al [9],

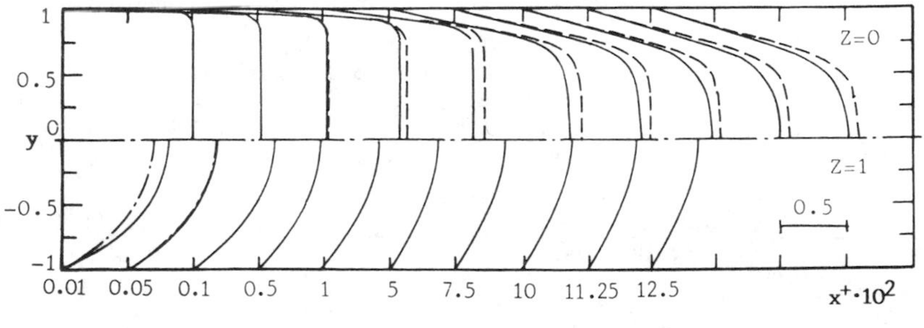

(a) Laminar flow, λ^*=2.03 γ=0.0154 (—— T_1, T_2;---T)

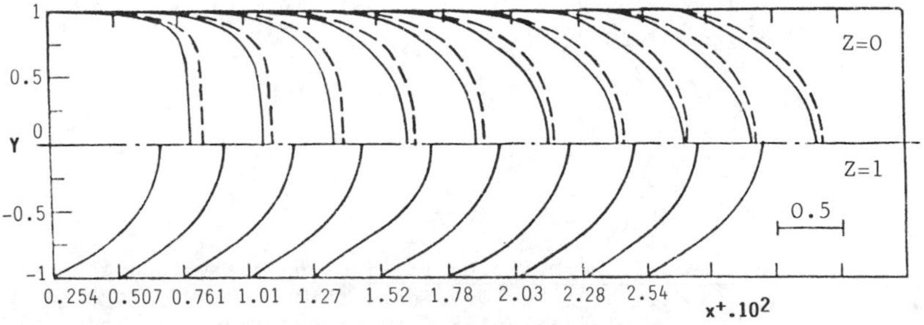

(b) Turbulent flow, λ^*=2.03, Re=10^4 (—— T_1, T_2; --- T)

Fig.2 Temperature developments in duct of η=4.39, profiles across wide duct dimension

$$L^* = \frac{1}{A}$$

$$= \frac{2}{\left[\frac{1}{\eta^2(1-Y')^2} + \frac{1}{(1-Z')^2}\right]^{\frac{1}{2}} + \left[\frac{1}{\eta^2(1-Y')^2} + \frac{1}{(1+Z')^2}\right]^{\frac{1}{2}} + \left[\frac{1}{\eta^2(1+Y')^2} + \frac{1}{(1-Z')^2}\right]^{\frac{1}{2}} + \left[\frac{1}{\eta^2(1+Y')^2} + \frac{1}{(1+Z')^2}\right]^{\frac{1}{2}}} \quad (5)$$

where $Y' = \begin{cases} Y, & Y > 1-1.6/\eta \\ 1-1.6/\eta, & Y \leq 1-1.6/\eta \end{cases}$, $Z' = \begin{cases} Z, & Z > 1-1.6\,\eta \\ 1-1.6\eta, & Z \leq 1-1.6\,\eta \end{cases}$.

The damping factor δ can be calculated with the procedure suggested by Barr(et al [10], of which the following partial differential equations

$$\left.\begin{aligned} &\frac{1}{\eta^2}\frac{\partial^2 G}{\partial Y^2} + \frac{\partial^2 G}{\partial Z^2} = \alpha(H-1) \\ &\frac{1}{\eta^2}\frac{\partial^2 H}{\partial Y^2} + \frac{\partial^2 H}{\partial Z^2} = -\alpha G \\ &Y = 1 \text{ or } Z = 1 : \quad G = H = 0 \\ &\left(\frac{\partial G}{\partial Y}\right)_{Y=0} = \left(\frac{\partial G}{\partial Z}\right)_{Z=0} = \left(\frac{\partial H}{\partial Y}\right)_{Y=0} = \left(\frac{\partial H}{\partial Z}\right)_{Z=0} = 0 \end{aligned}\right\} \quad (6$$

were solved numerically, where $\alpha=2(Au_\tau/A^+\nu)^2$. Then, the damping factor $\delta = [1-\sqrt{G^2+(1-H)^2}]$. But for high Reynolds number, the convergency becomes ver difficult due to the large value of α when equations(6) are solved using a fini difference method. To avoid this difficulty, a product of the damping factors of two sets of parallel plates was suggested as an approximate representation o the damping factor of rectangular duct in this paper, that was,

$$\delta = \left[1- \frac{\cosh[(1-Y)\sqrt{2\alpha}\,\eta]+\cos[(1-Y)\sqrt{2\alpha}\,\eta]}{\cosh(\sqrt{2\alpha}\,\eta)+\cos(\sqrt{2\alpha}\,\eta)}\right]\left[1- \frac{\cosh[(1-Z)\sqrt{2\alpha}\,]+\cos[(1-Z)\sqrt{2\alpha}}{\cosh(\sqrt{2\alpha}\,)+\cos(\sqrt{2\alpha})}\right] \quad (7)$$

Fig.3 compared the numerical and approximate damping factors obtained in presen work for a rectangular duct. It shows that the approximate solution adopted in this paper is very close to the numerical solution and can be regarded as reaso able for engineering purposes. Substituting equations(5) and (7) into the van Driest formula[11,eq.23.37a], we obtain

$$\nu^* = 1+\varepsilon_M/\nu = 1+\frac{1}{2}\alpha(1+1/\eta)A^{+2}\varkappa^2\,(L^*\delta)^2\sqrt{\frac{1}{\eta^2}\left(\frac{\partial W}{\partial Y}\right)^2+\left(\frac{\partial W}{\partial Z}\right)^2} \quad (8)$$

where $\varkappa$ and A^+ are constants, $\varkappa=0.4$, $A^+=26$. Thus, we have

$$a^* = 1 + \varepsilon_H/a = 1 + (\nu^*-1)Pr/Pr_t \quad (9)$$

<u>Turbulent Velocity and Temperature Distributions</u> Using the values of ν^* and a^* obtained from equations(8-9), the fully developed turbulent velocity and develo ing temperature distributions were calculated from equations(1-2) and(3-4) respectively by the method of finite-difference. These finite-difference equation were solved with over-relaxation alternate direction iteration. The unequidistant mesh spacings were used and the mesh scheme was shown in Fig.4. The spacings near the wall are small and increase progressively towards the duct centre line according to the arithmetical progressions, that is,

$$\Delta X_i = 1/Nx,\ \Delta Y_j = (Ny-j)/\sum_{l=1}^{Ny} 1\ ,\quad \Delta Z_k = (Nz-k)/\sum_{l=1}^{Nz} 1\ ,$$

where the indexes i,j and k indicate positions in the X,Y and Z-directions, respectively; the origin is designated by i=j=k=0 and the positions at X=1,Y=1 and Z=1 are designated by i=Nx, j=Ny and k=Nz, respectively.

Because there are unknown velocity gradients in the representation of ν^*(eq.8), the velocity distribution should be assumed first. An iterative approach was needed. According to the velocity distribution assumed, the values of ν^* were obtained by solving equation(8). Then, the equations(1-2) were solved for velocity distribution W(Y,Z). After that, this velocity distribution was used to initiate the second iteration. This procedure was continued until the requirement of accuracy was attained.

For turbulent heat transfer, the effect of longitudinal heat conduction of the fin is negligible because $(\gamma/Pe^*)^2 \ll 1$, and then T_2 reduces to the T_1 boundary condition. Using the information of ν^* and W(Y,Z) obtained, the temperature distribution for this condition was ready to be solved from equations(3, 4,9). In calculating, the turbulent Pr_t was taken as 0.9. The predicted turbulent temperature development in duct of η=4.39 with T_1 and T boundary conditions are shown in Fig.2b.

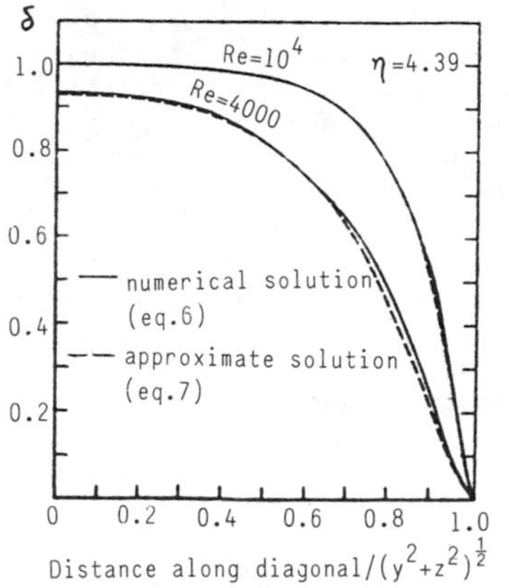

Fig.3 Comparison of approximate solutions of damping factors with numerical solutions

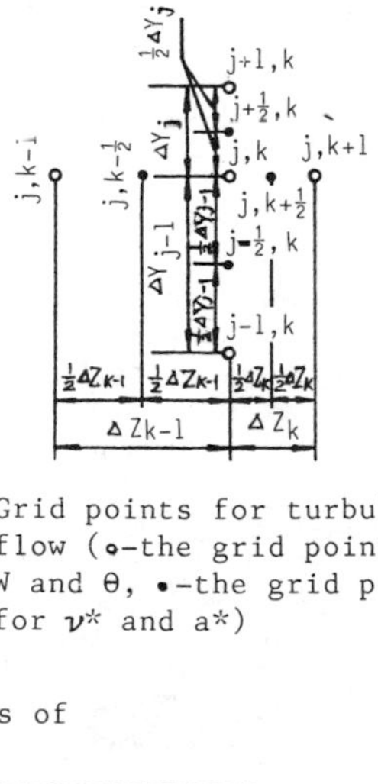

Fig.4 Grid points for turbulent flow (∘-the grid points for W and θ, •-the grid points for ν^* and a*)

Table 2 The geometric parameters of the stainless steel test cores

No	de (mm)	η = B/A	γ = B/L	$\frac{\lambda_w \delta}{2\lambda_f A}$	Fin Thickness	Spacer Thickness
1	2.83	1.07	0.0049	30.2	0.15mm	0.5mm
2	4.69	3.02	0.0156	26.4	0.15mm	0.5mm
3	3.43	4.39	0.0154	39.2	0.15mm	0.5mm

RESULTS AND DISCUSSION

3.1 Friction and Convective Heat Transfer Performances

The predicted friction factors, local and average Nusselt numbers for laminar flow in rectangular ducts with constant wall temperature are shown in Fig.5a and 5b, respectively, and compared with the numerical results under the same boundary condition noted by Kays and Crawford [2]. Further comparisons with the experimental data for both laminar and turbulent flow in a square duct presented by Kays and London [12] are shown in Fig.6, and very good agreement is also in evidence between them.

In Fig.7 are shown the experimental and predicted results of three stainless steel test cores (the geometric parameters were shown in Table 2) obtained in present work. In the region of laminar flow, j-factors obtained experimentally lie between the predictions calculated under the H and T_1 boundary conditions, respectively, and reveal that the experimental results of high η surfaces lie close to the predictions for T_1 condition, while for the surfaces of $\eta \leq 2$, the results are close to that for H condition. As mentioned in the introduction, the experiments in this work were performed with a transient technique, and the experimental boundary condition of the test core did not correspond, therefore, to those employed in the present predictions. In view of the combined effect of the low conductivity of stainless steel and the corner region, this trend of the experimental results seems to be reasonable. As already noted, the predicted friction factors for laminar flow agree well with the available numerical and experimental results [2,12]. They also agree well with the present experimental results in lower Reynolds number region. As Reynolds number increases, however, the deviations become evident gradually due to the effect of the undeveloped velocity field. Unfortunately, the experiments performed in present work were limited in the range of Reynolds numbers ($Re \leq 7000$) by the performance of the centrifugal blower of the test rig, and therefore, the quantitative comparisons in the region of the turbulent flow regime was very difficult to make. Fig.7 does indicate, however, that reasonable trends exist between the predictions and the experimental results in this work.

3.2 The Effect of the Thermal Boundary Condition on Heat Transfer

The effect of the longitudinal heat conduction of the fin on Nusselt number is shown in Fig.8. It reveals that the effect is significant only for the case of a low Peclet number flow except when the duct length of the surface is very short.

The predicted local and average Nusselt numbers for laminar flow of the straight plate-fin surfaces with T_1 and T boundary conditions were given in Table 3a. It indicates

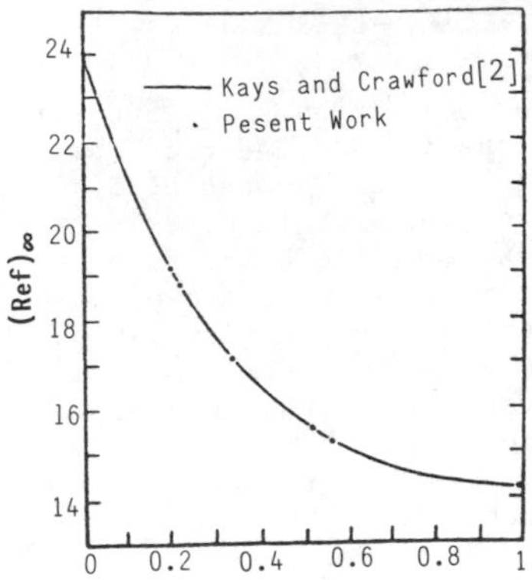

Fig.5a Friction Factors for laminal flow

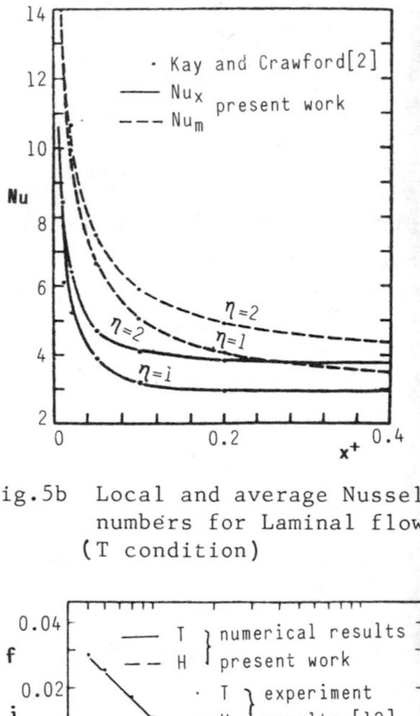

Fig.5b Local and average Nussel numbers for Laminal flow (T condition)

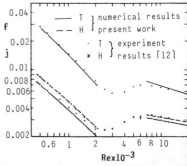

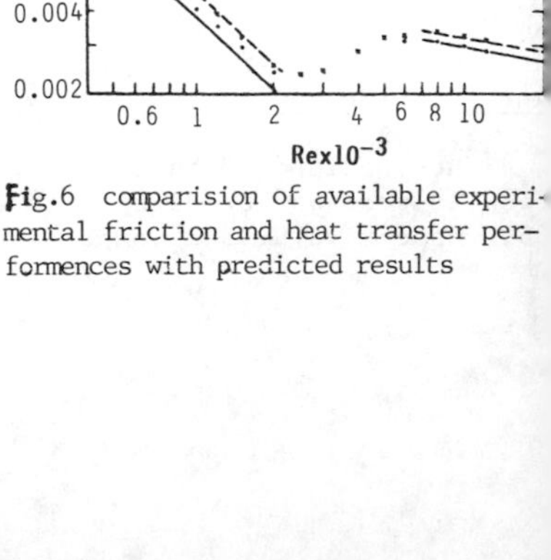

Fig.6 comparision of available experimental friction and heat transfer performences with predicted results

that there is a significant effect of η on the local and average Nusselt numbers. For the surfaces of $\eta \approx 1$ or $\eta < 1$, the effect of the transverse thermal resistance vanishes, while for the surfaces of $\eta \geq 3$, the effect becomes significant. As an example, for a stainless steel surface of $\eta = 5$, of which the value of $\lambda_w \delta / 2\lambda_f A$ is 20, the deviation of the average Nusselt number from that for T boundary condition may be greater than 10%. The other feature projected from this Table is the deviation of local or average Nusselt number from the corresponding value for T boundary condition increases with increasing downstream distance. This is because the peak of the local Nusselt number at the fin surface moves from the root to the middle of the fin with increasing downstream distance due to the development of the temperature field(See Fig.9).

For turbulent flow, because of the greatly enhanced convective heat transfer between fluid and the wall, the effect of the transverse fin thermal resistance is much more significant than that for laminar flow. One can see from Table 3b that for a stainless steel surface with high value of η, the local Nusselt number may be about 20% less than that for T condition. For the surfaces of lower η ($\eta \leq 1$), however, this effect can also be neglected.

CONCLUSION

To investigate the effect of the finite thermal resistances of the fins, consideration was given to the T_2, T_1 and T boundary conditions of the straight plate-fin surfaces. It has been shown that the effect of the longitudinal heat conduction of the fin is negligible except in the case of a low Peclet number flow, while the heat transfer performance of the surfaces is quite sensitive to the transverse thermal resistance of the fin. It has also been shown that the effect of the transverse theraml resistance in turbulent flows is much more substantial than that in laminar flows. It was concluded that for the surfaces having higher value of η, the effect of the transverse thermal resistance should be taken into consideration, especially in turbulent flows, or else substantial errors may occur.

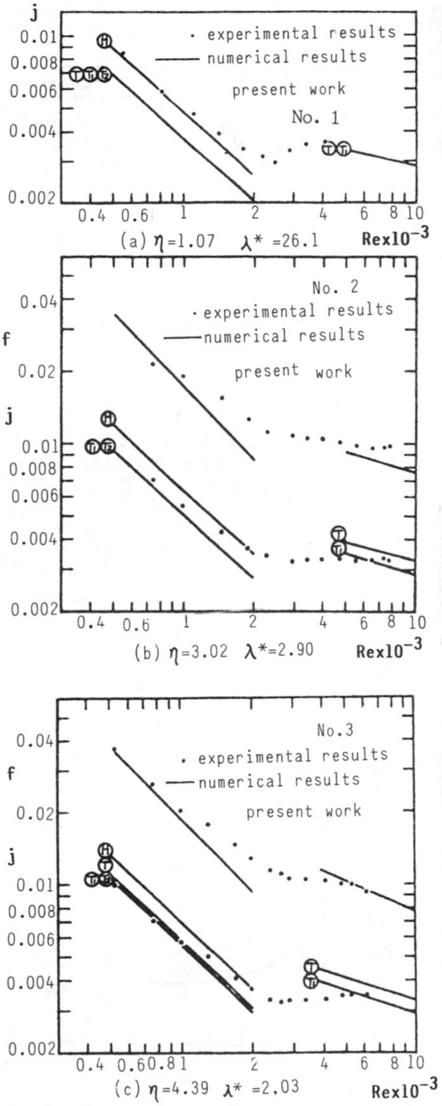

Fig.7 Comparison of present experimental friction and heat transfer performances with predicted results

ACKNOWLEDGEMENT

This research was performed under the auspices of a grant from the Science Fund of the Chinese Academy of Sciences.

Table 3 Local and average Nusselt numbers under T_1 and T boundary conditions
(a) Laminar Flow

x^+	Local Nusselt Number								Average Nusselt Number							
	$\eta = 1$		$\eta = 2$			$\eta = 5$			$\eta = 1$		$\eta = 2$			$\eta = 5$		
	T	$\lambda_w\delta/2\lambda_f A$	T	$\lambda_w\delta/2\lambda_f A$		T	$\lambda_w\delta/2\lambda_f A$		T	$\lambda_w\delta/2\lambda_f A$	T	$\lambda_w\delta/2\lambda_f A$		T	$\lambda_w\delta/2\lambda_f A$	
		20		40	20		40	20				40	20		40	2
0.02	4.47	4.48	4.84	4.84	4.70	5.97	5.76	5.65	6.46	6.46	6.86	6.84	6.61	8 02	7.68	7.5
0.04	3.66	3.66	4.06	4.04	3.91	5.35	5.04	4.93	5.20	5.21	5.60	5.58	5.40	6.79	6.48	6.3
0.10	3.09	3.09	3.57	3.54	3.38	5.07	4.64	4.50	4.05	4.05	4.48	4.45	4.29	5.81	5.45	5.3
0.20	2.98	2.98	3.43	3.40	3.23	4.96	4.49	4.34	3.53	3.53	3.98	3.95	3.79	5.41	5.00	4.8
0.40	2.98	2.98	3.39	3.36	3.20	4.88	4.41	4.26	3.25	3.25	3.69	3.66	3.50	5.16	4.72	4.5

(b) Turbulent Flow

x^+	Local Nusselt Number								Average Nusselt Number							
	$\eta=1$ Re=11000		$\eta=1.8$ Re=14000			$\eta=4.39$ Re=10000			$\eta=1$ Re=11000		$\eta=1.8$ Re=14000			$\eta=4.39$ Re=1000		
	T	$\lambda_w\delta/2\lambda_f A$	T	$\lambda_w\delta/2\lambda_f A$		T	$\lambda_w\delta/2\lambda_f A$		T	$\lambda_w\delta/2\lambda_f A$	T	$\lambda_w\delta/2\lambda_f A$		T	$\lambda_w\delta/2\lambda_f$	
		20		40	20		100	40		20		40	20		100	4
0.002	31.1	30.7	36.7	35.4	34.6	32.1	30.7	29.0	36.3	35.7	42.7	41.4	40.6	37.1	35.5	33.
0.004	28.5	28.0	34.6	33.0	32.1	30.0	28.3	26.4	32.7	32.2	39.0	37.6	36.8	33.9	32.2	30.
0.010	27.0	26.6	33.4	31.6	30.5	29.0	26.6	24.3	29.5	29.0	35.8	34.3	33.3	31.1	29.2	27.
0.015	26.9	26.4	33.2	31.3	30.2	28.7	26.1	23.6	28.7	28.2	34.9	33.2	32.3	30.0	28.2	26.
0.025	26.9	26.4	—	—	—	28.3	25.5	22.8	28.0	27.5	—	—	—	29.6	27.2	24.

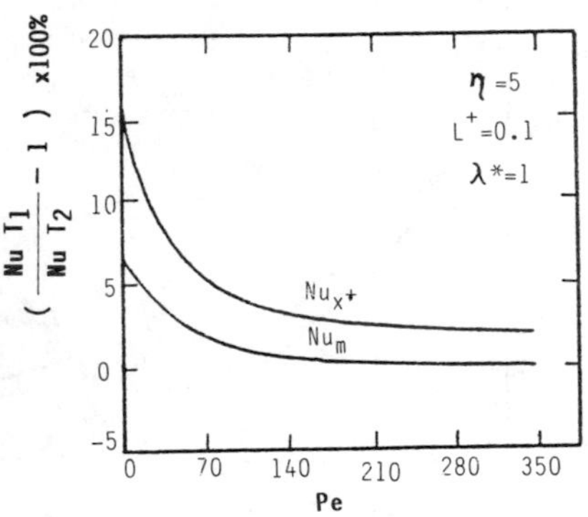

Fig.8 Comparison between the Nusselt numbers under T_1 and T_2 boundary conditions

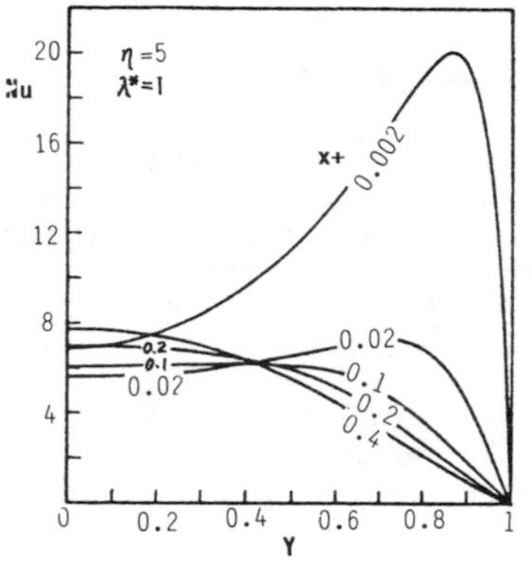

Fig.9 Local Nusselt number along the fin

REFERENCES

[1] Shan,R.K., and London,A.L.,Trans.ASME,J.Heat Transfer, Vol.96,pp. 159-165, 1974.
[2] Kays,W.M.,and Crawford,M.E.,"Convective Heat and Mass Transfer",2nd ed., McGraw-Hill, New York, 1980.
[3] Montgomery,S.R., and Wibulswas, P.,Appl.Scient.Res.Vol. 18,No.4, pp.247-259, 1967.
[4] Javeri, V., Int.J.Heat Mass Transfer, Vol.21,pp.1029-1034, 1978.
[5] Lyczkowski,R.W.,Institute of Gas Technology,Technical Information Center, File3229. Chicago,Ill., 1969.
[6] Launder, B.E.and Ying,W.M.,Proc.Instn.Mech.Engrs,Vol.187,37/73, pp.455-461, 1973.
[7] Emery,A.F.,Neighbors,P.K., and Gessner,F.B.,Trans.ASME,J.Heat Transfer,Vol. 102, pp.51-57,1980.
[8] Cai,Z.H.,Li,M.L.,Wu,Y.W.,and Ren,H.S.,Int.J.Heat Mass Transfer,Vol.27,pp. 971-978, 1984.
[9] Gessner,F.B.,and Emery,A.F.,Trans.ASME,J.Fluid Eng.,Vol.99,pp.347-356,1977.
[10] Barrow,H.,Hornby,R.P., and Mistry,J.,Proc.6th Int. Heat Transfer Conf., Toronto, Vol.2,pp.549-554,1978.
[11] Schlichting,H.,"Boundary Layer Theory", 2nd ed., McGraw-Hill,New York,pp. 722, 1979.
[12] Kays,W.M.,and London,A.L.,"Compact Heat Exchangers", 2nd ed.,McGraw-Hill, New York,1964.

Laminar Forced-Convection Heat Transfer in the Air Flowing through a Cascade of Parallel Plates: The Case of Uniform Heat Flux

E. NAITO
Department of Mechanical Engineering
Shiga Prefectural Junior College
Hikone 522, Japan

INTRODUCTION

It is the purpose of this study to clarify the characteristics of the fluid flow and heat transfer near the inlet between the flat plate-fins for intermediate Reynolds numbers and fluids below a Prandtl number of 1. In the previous report [1], numerical analyses are performed to investigate the effects of axial diffusions of vorticity and heat upon the velocity and temperature fields in the upstream and downstream regions of the leading edge for various Reynolds number. The physical system considered involve simultaneous development of the velocity and temperature distributions in the air flowing through a cascade of parallel plates with uniform wall temperature.

In several engineering devices, particularly a compact heat exchanger, there is wall-temperature change along the plate channel length. Hence, the present stud is an extension of the previous one and deals with laminar forced-convection he transfer in the parallel plates with uniform heat flux at the wall. As in the case of uniform wall temperature, the numerical solutions are obtained for the six different Reynolds numbers in the range of 30 to 1500. Comparisons are mad of the numerical solutions for this and the other inflow models, and with the approximate results using boundary layer theory. As independent of Reynolds nu ber, a close agreement as to local Nusselt numbers is found in the entrance region. The limitations for the application of approximate results in this model are investigated through this result. The relationship between the thermal entrance length and Reynolds number is compared with one for hydrodynamic and the mal entrance lengths. In addition, consideration has been given to the thermal entrance lengths for two types of heat transfer: one with a uniform heat flux a the wall and the other a uniform wall temperature reported previously.

GOVERNING EQUATIONS AND BOUNDARY CONDITIONS

The physical system considered here features simultaneous development of the velocity and temperature distributions, in the steady-state laminar flow of an incompressible Newtonian fluid with constant physical properties, at the entra region between parallel plates with uniform heat flux on the surface. The veloc ty profile is assumed to be uniform at a distance further upstream from the lead ing edge. However, a uniform heat flux is maintained at the plates, so the tem perature profile is assumed to be uniform up to the leading edge. The schematic representation of the physical model and coordinate systems is shown in Fig. 1.

The governing continuity, momentum and energy equations based on the dimensionless form are as follows:

TABLE 1. Parameters in Eq. (6)

	Parameters in ξ transformation, equation (6)	
Re	$\lvert C \rvert$	n
30	2	1
60	2	1
100	2	1
300	1	1
900	0.5	0.5
1 500	0.8	0.8

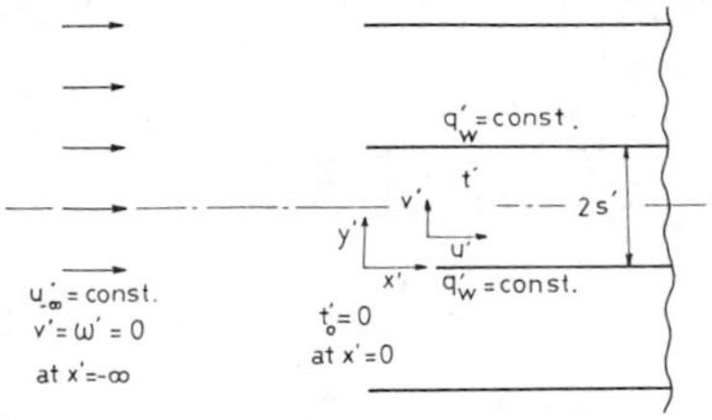

FIGURE 1. Physical model and coordinate.

$$\partial u/\partial x + \partial v/\partial y = 0 \tag{1}$$

$$u(\partial u/\partial x) + v(\partial u/\partial y) = -(\partial p/\partial x)/2 + 2\nabla^2 u/Re \tag{2}$$

$$u(\partial v/\partial x) + v(\partial v/\partial y) = -(\partial p/\partial y)/2 + 2\nabla^2 v/Re \tag{3}$$

$$u(8/Pe + \partial\theta/\partial x) + v(\partial\theta/\partial y) = 2\nabla^2\theta/Pe \tag{4}$$

where the dimensionless variables and parameters are defined by:

$$\left.\begin{array}{l} u = u'/u'_{-\infty}, \quad v = v'/u'_{-\infty}, \quad p = 2p'/\rho' u'^2_{-\infty}, \quad t' = t'_0 + q's'(8x/Pe + \theta)/\lambda' \\ x = x'/2s', \quad y = y'/2s', \quad Pe = Pr\ Re, \quad Pr = c'_p\,\rho'\,\nu'/\lambda', \quad Re = 4s'u'_{-\infty}/\nu' \\ u'_m/u'_{-\infty} = u'_0/u'_{-\infty} = 1 \end{array}\right\} \tag{5}$$

The axial coordinate transformation is given by:

$$\xi = \pm (Cx)^n / \{1 + (Cx)^n\} \tag{6}$$

where C and n are an arbitrary constant whose value depends on the Reynolds number, and are presented in Table 1. Here C is a positive value for $x > 0$ and a negative value for $x < 0$. Therefore, the infinite region $-\infty < x < \infty$ is transformed into the finite region $-1 \leq \xi \leq 1$ by the choice of sign in this equation (6).

Hence, the Laplacian operator $\nabla^2 \equiv \partial^2/\partial x^2 + \partial^2/\partial y^2$ is rewritten as follows:

$$\nabla^2 \equiv (d\xi/dx)^2 \partial^2/\partial\xi^2 + (d^2\xi/dx^2)\partial/\partial\xi + \partial^2/\partial y^2$$

By elimination of the pressure gradients from Eqs. (2) and (3), the momentum equations lead to the following vorticity transport equation:

$$(d\xi/dx)\{(\partial\psi/\partial y)(\partial\omega/\partial\xi) - (\partial\psi/\partial\xi)(\partial\omega/\partial y)\} = 2\nabla^2\omega/Re \tag{7}$$

where the stream function and vorticity are defined as follows:

$$u = \partial\psi/\partial y, \quad v = -(d\xi/dx)\,\partial\psi/\partial\xi \tag{8}$$

$$\omega = -\nabla^2\psi \tag{9}$$

With the transformation of Eq. (6), Eq. (4) is rewritten by:

$$(d\xi/dx)\{(\partial\psi/\partial y)(\partial\theta/\partial\xi) - (\partial\psi/\partial\xi)(\partial\theta/\partial y)\} = 2\{\nabla^2\theta - 4(\partial\psi/\partial y)\}/Pe \tag{10}$$

The boundary conditions for the governing equations are as follows:

$$u=1,\ v=0:\ \psi=y,\ \partial\psi/\partial\xi=\partial^2\psi/\partial\xi^2=0,\ \omega=1,\ \theta=0\ \text{for}\ x\to-\infty,\ \xi=-1,\ 0\leq y\leq 0.5 \quad (11$$

$$v=0,\ \partial u/\partial y=0:\ \psi=y,\ \partial\psi/\partial\xi=\partial^2\psi/\partial y^2=\omega=\partial^2\omega/\partial y^2=\theta=0$$
$$\text{for}\ -\infty<x\leq 0:\ -1\leq\xi\leq 0,\ y=0\ \text{and}\ 0.5 \quad (12$$

$$u=0,\ v=0:\ \psi=\partial\psi/\partial\xi=\partial\psi/\partial y=0,\ \partial\theta/\partial y=-2\ \ \text{for}\ 0\leq x<\infty:\ 0\leq\xi\leq 1,\ y=0 \quad (13$$

$$v=\partial u/\partial y=0:\ \psi=0.5,\ \partial\psi/\partial\xi=\partial^2\psi/\partial y^2=\omega=\partial^2\omega/\partial y^2=\partial\theta/\partial y=0$$
$$\text{for}\ 0\leq x<\infty:\ 0\leq\xi\leq 1,\ y=0.5 \quad (14$$

$$u=6y(1-y^2),\ v=0:\ \psi=6y^2(1/2-y/3),\ \omega=-6(1-2y),\ \theta=17/35-2y+4y^3-2y^4$$
$$\text{for}\ x\to\infty:\ \xi=1,\ 0\leq y\leq 0.5 \quad (15$$

NUMERICAL METHOD OF SOLUTION

Eqs. (7), (9) and (10) are approximated by the finite difference equations with the above second-order accuracy. When the centered-space difference forms are employed for the governing equations, somehow the vorticity distributions in the upstream region shows the very small and undesirable fluctuations in the x-direction. However, a phenomenon such as fluctuations did not occur with the u wind difference form adoped for the convection terms in the vorticity transport and energy equations. The second-order upwind difference form used here is as follows with the exception of the grids next to the boundary:

$$\dot{f}_0 = (3f_0-4f_{-1}+f_{-2})/(2\Delta)+\Delta^2\,\dddot{f}_0/3 \quad (16$$

On the other hand, the finite-difference forms on the neighbouring nodes of boundary are selected from the following equations:

$$\dot{f}_0 = (2f_0-2f_{-1}-\Delta\dot{f}_{-1})/\Delta-\Delta^2\,\dddot{f}_0/6 \quad (17$$

$$\dot{f}_0 = (2f_0-2f_{-1}+\Delta^2\ddot{f}_{-1})/(2\Delta)+\Delta^2\,\dddot{f}_0/3 \quad (18$$

where $f=\omega$ or θ, $\Delta=a$ or b and subscript numbers are mesh point numbers. The centered-space difference forms are used for all other terms in Eqs. (7) an (9), and for Eq. (10). Since the accuracy of these solutions will certainly be limited by the consideration of the boundary conditions, the finite difference forms for the derivatives of stream function on the neighbouring nodes of plate are derived from the Taylor's series expansions considering the boundary conditions: $\psi=\partial\psi/\partial y=0$ and $\partial^2\psi/\partial y^2=-\omega$. The following method are employed:

$$\dot{\psi}_{i,2}=(-2\psi_{i,4}+27\psi_{i,3}+162\psi_{i,2}-187\psi_{i,1}-102b\dot{\psi}_{i,1}-18b^2\ddot{\psi}_{i,1})/(108b)+b^5\psi^{(6)}_{i,2}/360 \quad (19$$

$$\ddot{\psi}_{i,2}=(-8\psi_{i,4}+135\psi_{i,3}-216\psi_{i,2}+89\psi_{i,1}-30b\dot{\psi}_{i,1}-18b^2\ddot{\psi}_{i,1})/(108b^2)+7b^4\psi^{(6)}_{i,2}/840 \quad (20$$

The values of vorticity at the plate are obtained by the following formula:

$$\omega_{i,1}=3(\psi_{i,1}-\psi_{i,2})/b^2-\omega_{i,2}/2 \quad (21$$

RESULTS AND DISCUSSION

The numerical analysis of this heat transfer problem in the air flow (Pr=0.71) was made for six different Reynolds numbers: 30, 60, 100, 300, 900 and 1500. The 51×26 $(\xi,\ y)$ grid was normally used, but to improve numerical accuracy near

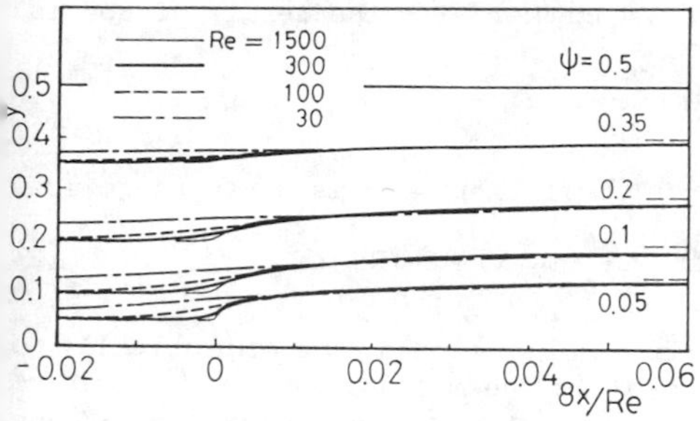

'IGURE 2. Development the stream-
.ines near the leading edge.

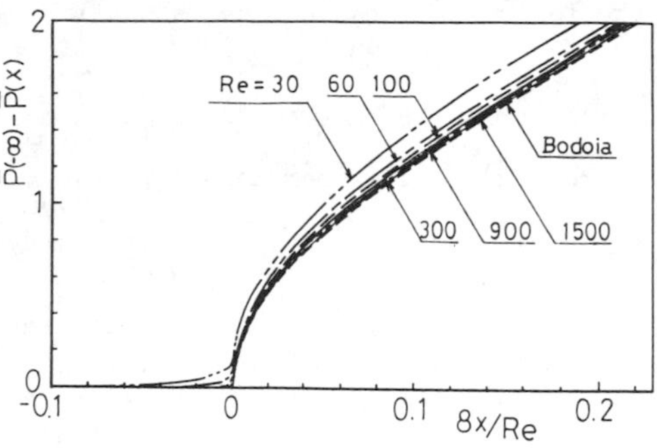

FIGURE 3. Average pressure drop.

'ABLE 2. Hydrodynamic entrance lengths and .ncremental pressure drop coefficients.

Researcher		L_h/Re (99%)	$\overline{K}$
Present Study[1]	Re= 30	0.021 7	0.87
	60	0.020 6	0.756
	100	0.020 5	0.717
	300	0.022 3	0.706
	900	0.022 3	0.697
	1 500	0.022 7	0.672
Bodoia and Osterle[3]		0.022	0.676
Collins and Schowalter[4]			0.676
Hwang and Fan[5]		0.021 1	0.625
Gary[6]	Case 1	0.019 06	0.531 16
	Case 2	0.020 93	0.596 75
Ishizawa[7]			0.66
Kiya, et al.[8]		0.022 25	0.666
McComas[9]		0.023 52	0.685 7
Morihara and Cheng[10]		0.021 15	
Roidt and Cess[11]		0.022 7	0.63
Schiller[12]			0.626
Schlichiting[13]		0.02	0.601
Schmidt and Zeldin[14]	Re= 100		0.748
	500		0.698
	10 000		0.669
Sparrow, et al.[15]			0.686

the plate, the mesh size for $0 \leq y \leq 0.2$ was again divided into one fourth. The iteration was continued until the following condition was satisfied:

$$|f^{k+1}_{i,j} - f^{k}_{i,j}|_{max} < 10^{-6} \tag{22}$$

in which f represents ψ or θ and k is an iterative number.

Flow Fields

Since the results of flow field are already reported in the previous report, the present study deals with a comparison between the incremental pressure drop coefficient or hydrodynamic entrance length in the present study and those by the other authors. To obtain an understanding of the fluidic movements in this model, the development of the streamlines near the eading edge for different Reynolds numbers is illustrated in Fig. 2. While the niform flow for Re=1500 closely approaches the leading edge, the effect of voricity-diffusion for Re=30 is extended from the leading edge to the far upstream egion. It is seen that the difficulty in the assumption of uniform influx at he inlet is increased with decreasing Reynolds number.

ncidentally, because the pressure drop is of special importance in many industril applications, the average pressure drop $\overline{p}(-\infty) - \overline{p}(x)$ which was obtained from the following equations was shown in Fig. 3 together with the result of Bodoia [2] for comparison.

$$\overline{p}(-\infty) - \overline{p}(x) = 48x/Pe + \overline{K}(x) \quad (23$$

where $\overline{K}(x)$ is the incremental pressure drop coefficient and is given as follows

$$K(x) = 2\int_0^{0.5}(u^2 + v^2 - 1)dy - 4\int_\infty^0\int_0^{0.5} v\,\omega\,dy\,dx + 2\int_0^x\{\tau_w(x) - 24/Re\}dx \quad (24$$

This result is higher than Bodoia's result, and approaches it asymptotically wi increasing Reynolds number. $\overline{K}(0)=\overline{p}(-\infty)-\overline{p}(0)$ is the mechanical energy loss whi is produced in the upstream region. $\overline{K}(0)$=0.119 for Re=30 comes about 13.6% whi $\overline{K}(L_h)$=0.87, but the ratio, $\overline{K}(0)/\overline{K}(L_h)$, for Re=1500 is within 0.71%.

Hence, L_h is the hydrodynamic entrance length, and is defined as the dimensionless axial position at which the center-line velocity reaches 99% of its fully developed value, that is u_c=1.485. In Table 2, the values of L_h/Re and $\overline{K}(L_h)$ ar compared with the results previously reported by the other authors. The value of L_h/Re for Re=1500 and the Roidt and Cess result agree very well with each other, though they may be obtained by a different method of solution.

Thermal Fields

The dimensionless temperature of fluid is rewritten from Eq. (5) as follows:

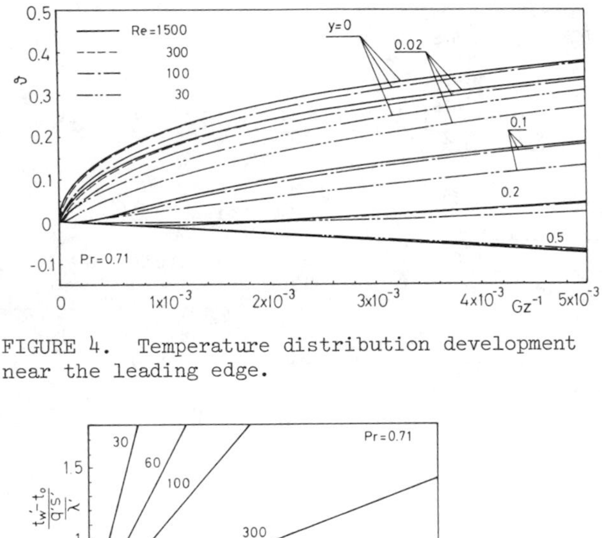

FIGURE 4. Temperature distribution development near the leading edge.

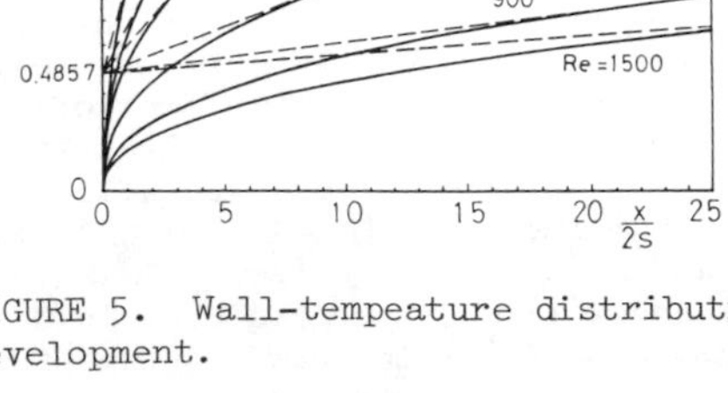

FIGURE 5. Wall-tempeature distribution development.

$$(t'-t'_0)/(q's'/\lambda')=8x/Pe + \theta(x,y) \quad (25$$

where the first term of th equation, 8x/Pe, is the di mensionless bulk temperatu described later, and the s cond term is the differenc of temperature from the bu temperature in a profile (this incremental temperature is only referred to a the "temperature"). The d velopment of the temperatu distribution near the lea ing edge is illustrated in Fig. 4. These temperature distributions show differences among Reynolds numbe and inflow models, but al profiles of fully develope temperature agree precise regardless of those numbe or models and are given by $\theta_f = 17/35 - 2y + 4y^3 - 2y^4$. Since the local Nusselt nu ber varies inversely with the degrees of the wall te perature, the developing wall temperature distribution is worthy of note.

Except near the leading edge, the result of Re=300 coincides well with one for Re=1500. Although the temperature distributions for Re=900 are omitted in this figure, it agrees very well with the one for Re=1500. With decreasing Reynolds number, the wall temperature becomes lower, and the difference between the temperature at the wall and one at the center-line decreases.

The bulk temperature is given by:

$$t_b' = \int_0^{0.5} u'\,t'\,dy' \Big/ \int_0^{0.5} u'\,dy' = t_0' + q's'(8x/Pe + \theta_b)/\lambda' \tag{26}$$

On the other hand, the heat balance as it advances dx a distance x from the leading edge is given as follows:

$$2q'\,dx' = \rho'\,c_p'\,u_m'\,2s'\,dt_b' \tag{27}$$

Integrating this equation from 0 to x with respect to x, the bulk temperature leads to as follows:

$$(t_b' - t')/(q's'/\lambda') = 8x/Pe \tag{28}$$

Consequently, it is clear from Eqs. (26) and (28) that the incremental dimensionless temperature must be zero: $\theta_b = 0$.
The wall temperature discussed previously was the incremental dimensionless temperature, so the wall temperature given by the following equation is shown in Fig. 5.

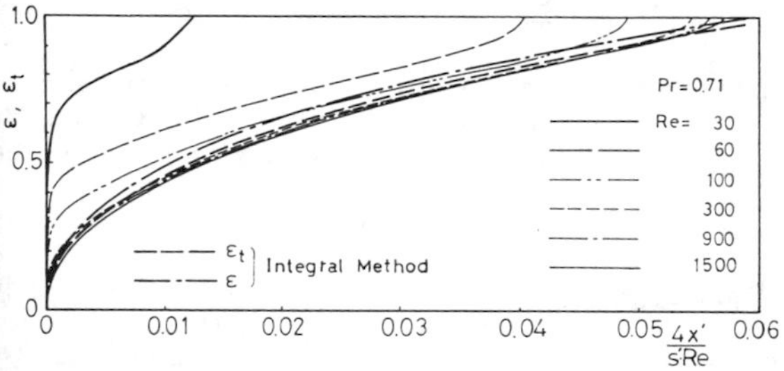

(a) Thermal boundary layer thicknesses for uniform heat flux.

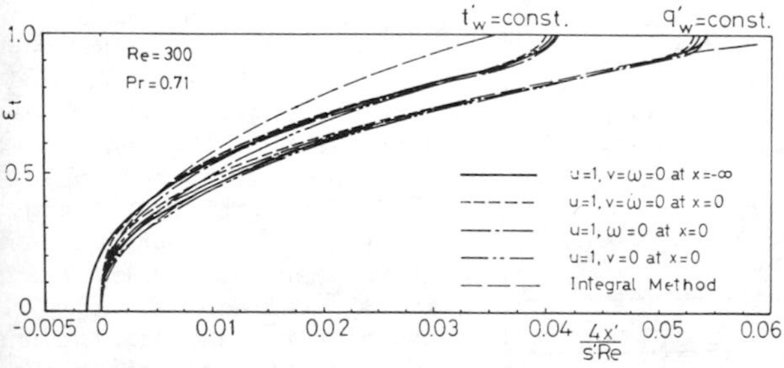

(b) Thermal boundary layer thicknesses for two types of heat transfer.

FIGURE 6. Thermal boundary layer thichness.

$$(t_w' - t_0')/(q's'/\lambda') = 8x/Pe + \theta_w \tag{29}$$

The broken lines in this figure are the asymptotic lines for each Reynolds number, the value of intercept is the incremental temperature of fully developed state, $\theta_w = 17/35$. As seen from Eqs. (28) and (29), the broken lines show also a similar tendency of bulk temperature for Reynolds numbers, respectively. The wall temperature is gradually approaching each asymptotes with increasing Reynolds number.

The thermal boundary layer thickness ε_t is defined as the distance from the plate to the point where the temperature of transverse section at the point at arbitrary axial position is $(t' - t_w')/(t_0' - t_w') = 0.99$ and is normalized by the half spacing s. Figure 6-(a) shows the state of growth of thermal boundary layer thickness for six different Reynolds numbers together with the hydrodynamic and thermal boundary layer thicknesses of the boundary layer equations using the integra method [16] for comparison. The thermal boundar layer thicknesses above Re=300 present no great difference and agree rather well with the result of the integra method. However, ones below Re=100 grow faster with decreasing Reynolds number Figure 6-(b) gives a comparison of variations for the thermal boundary layer thickness under the two coditions of the heat transfer and inflow at Re=300. T starting point of thermal boundary layer thicknesses is the leading edge, x=0, for the case of uniform heat flux, but one for uniform wall temperature under t same flow moves off for the upstream region. In contrast to the results of uniform heat flux, the thermal boundary layer thicknesses for uniform wall temperature do not agree with each other near the leading edge and with the result of the integral method [17]. Except the result using the integral method, however the other ones agree with each other for $\varepsilon_t = 0.8$ to 1. The intersection point the two thermal boundary layers on the plate facing each other with uniform hea flux is further downstream than one for uniform wall temperature.

The local Nusselt number is given by:

$$Nu_x = 4s'\, h'_x/\lambda' = 4/\theta_w \qquad (30$$

in which h_x is the local heat transfer coefficient. The variations of local Nusselt number for six Reynolds numbers are shown in Fig. 7-(a) together with t results of Hwang and Fan [18], Siegel and Sparrow [19], Bhatti and Savery [20], and the result with the integral method for comparison. The result for Re=30 a 60 individually approaches the value of a fully developed Nusselt number, Nu = 8.235. It is seen in this figure that a good agreement the local Nusselt numbers is present in the entrance regi for Reynolds number range fr 300 to 1500, except near the leading edge at Re=300. Thi similarity indicates good agreement with the results o Hwang and Fan, but is higher than the results of other au thors. For instance, the pr sent result for Re=1500 is about 7% higher than that of the integral method, about 1 higher at $Gz^{-1} = 3\times10^{-3}$ than the one of Siegel et al., an about 25% higher at $Gz^{-1} = 3$ 10^{-5} than one of Bhatti and Savery. Figure 7-(b) gives comparison of the behavior c local Nusselt numbers below Re=300 for various inflows. Re=300, the present result i lower than one for the unifc influx (u=1 and v=0 at x=0) and higher than ones for the uniform and irrotational influx (u=1, v=0 and ω=0 at x= or for the irrotational infl (u=1 and ω=0 at x=0). The r sult for the uniform and ir-

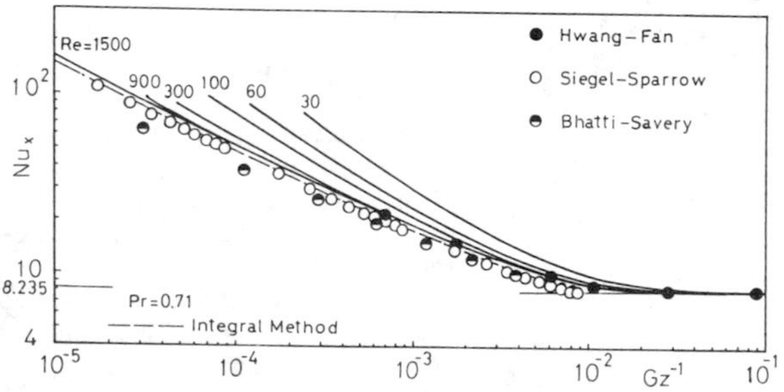

(a) Local Nusselt numbers for six different Reynolds numbers.

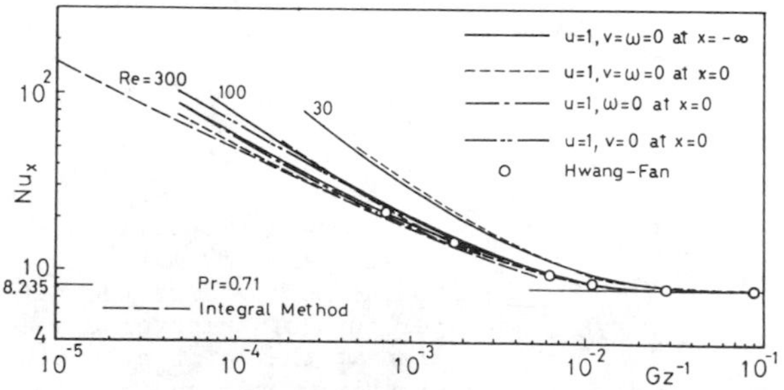

(b) Local Nusselt numbers for each inflow model.

FIGURE 7. Local Nusselt number.

TABLE 3. Thermal entrance lengths for two types of heat transfer

Re	Thermal entrance lengths L_t/Re, (99%)		Hydrodynamic entrance lengths, L_h/Re, (99%) 1)
	Uniform heat flux	Uniform wall temperature 1)	
30	0.141 6	0.03	0.021 7
60	0.064 4	0.020 7	0.020 6
100	0.032	0.018 3	0.020 5
300	0.026 1	0.018 3	0.022 3
900	0.026	0.018 4	0.022 3
1 500	0.025 8	0.018 4	0.022 7

rotational influx is in partial agreement with one of the integral method.

The thermal entrance length L_t for the combined entrance region is defined as the distance through which the local Nusselt number develops this one by no less than 101% of the asymptotic value, that is Nu=8.317. Table 3 shows the relationship between the thermal entrance length and Reynolds number together with the results of uniform wall temperature and flow for comparison. It is known from this table that the hydrodynamic entrance length is greater than the thermal entrance length of uniform wall temperature and is less than one of uniform heat flux for Reynolds numbers over Re=300.

CONCLUSIONS

A numerical analysis was conducted to investigate the characteristic of heat transfer in the air flowing through a cascade of parallel plates with uniform heat flux. The results can be summarized in the following conclusions:

1) A close agreement of the local Nusselt numbers is found in the entrance region with the Reynolds number range from 300 to 1500, except near the leading edge for Re=300. It is known that the axial diffusion of heat is not negligible near the leading edge below Re=300.

2) In comparison to the approximate results using boundary layer theory, the local Nusselt number result for Re=1500 is in agreement with that of Hwang and Fan, but is higher than the other author's results. Therefore, the result of Hwang et al. can be applied in this model for higher Reynolds number.

3) In the problem of combined hydrodynamic and thermal entrance regions, the hydrodynamic entrance length above Re=100 is greater than the thermal entrance length of uniform wall temperature and is shorter than that of uniform heat flux.

NOMENCLATURE

- = grid size in the x-direction
- = grid size in the y-direction
- = parameter in transformation, equation (6)
- p = specific heat [J/(kg K)]
- z = Greatz number =2 Pr Re/x
- = heat transfer coefficient $[W/(m^2 K)]$
- = incremental pressure drop coefficient, equation (24)
- = entrance length [m]
- u = Nusselt number, equation (30)
- = parameter in transformation, equation (6)
- Pr = Plandtl number, $=c_p' \rho' \nu'/\lambda'$
- Pe = Peclet number, = Pr Re
- p = pressure [Pa]
- q = heat flux $[W/m^2]$
- Re = Reynolds number = $4s'u'/\nu'$
- s = one half the spacing between parallel plates [m]
- t = temperature [K]
- u = velocity component in the x-direction [m/s]
- v = velocity component in the y-direction [m/s]
- x = axial coordinate [m]
- y = transverse coordinate [m]

∇^2 = Laplacian operator $\equiv \partial^2/\partial x^2 + \partial^2/\partial y^2$
δ = boundary layer thickness [m]
ε = dimensionless boundary layer thickness $=\delta/s$
θ = incremental temperature, equation (5)
λ = thermal conductivity [W/(m K)]
ν = knematic viscosity $[m^2/s]$
ξ = dimensionless axial coordinate, equation (6)
ρ = dencity $[kg/m^3]$
ψ = stream function, equation (8)
ω = vorticity, equation (9)

Subscripts
b = bulk mean value
c = center-line value
f = fully developed value
h = flow field
i = mesh point in the x-direction
j = mesh point in the y-direction
m = mean value
t = thermal field
x = local value
w = wall value
0 = value at leading edge
$-\infty$ = value at distance further upstream

Superscript
( )' = dimensional variables

REFERENCES

1. Naito, E., "Fluid Flow and Heat Transfer near the Leading Edge of Parallel Plates," Proceeding of ASME-JSME Thermal Engineering Joint Conferences (Honolulu, Hawaii, USA, March, 1983), Vol. 3, pp. 35-41.

2. Bodoia, J. R., "The Finite Difference Analysis of Confined Viscous Flow," Ph. D. Thesis, Carnegine Institute of Technology, Pittsburgh Pensylvania, (1959).

3. Bodoia, J. R. and Osterle, J. F., Appl. Sci. Res., Sec. A, Vol. 10, pp. 265-276, (1961).

4. Collins, M. and Schowalter, W. R., Phys. Fluid, Vol. 5, pp. 1122-1124, (1962).

5. Hwang, C. L. and Fan, L. T., Appl. Sci. Res., Sec. B, Vol. 10, pp. 329 -343, (1963).

6. Gary, V. K., J. Phys. Soc. Japan, Vol. 46, No. 1, pp. 300-302, (1979

7. Ishizawa, S., Trans. JSME, Vol. 32 No. 241, pp. 1373-1379, (1966).

8. Kiya, M., Fukusako, S. and Arie, M Trans. JSME, Vol. 37, No. 299, PP. 1325-1335, (1967).

9. McComas, S. T., Trans. ASME, Ser. Vol. 89, No. 4, pp. 847-850, (1967

10. Morihara, H. and Cheng, P. T., J. Comp. Phys., Vol. 11, No. 4, pp. 5 -572, (1973).

11. Roidt, M. and Cess, R. D., Trans. ASME, Ser. E, Vol. 29, No. 1, pp. 171-176, (1962).

12. Schiller, L., Z. Angew. Math. Mech Vol. 2, pp. 96-106, (1972).

13. Schlichting, H., Z. Angew. Math. Mech., Vol. 14, pp. 368-373, (1934

14. Schmidt, F. W. and Zeldin, B., AIC J., Vol. 15, No. 4, PP. 612-614, (1969).

15. Sparrow, E. M., Lin, S. H. and Lundgren, T. S., Phys. Fluid, Vol. 7, No.3, pp. 338-347, (1964).

16. Naito, E., Kagaku Kogaku, Vol. 38, No. 10, pp. 739-745, (1974); see also Heat Transfer Japanese Research, Vol. 4, No. 2, pp. 63-74, (1975).

17. Naito, E., Kagaku Kogaku Ronbunshu Vol. 10, No. 2, pp. 166-172, (1984 see also Heat Transfer Japanese Research, Vol. 13, No. 3, pp. 92-106 (1984).

18. Hwang, C. L. and Fan, L. T., Appl. Sci. Res., Sec. A, Vol. 13, pp. 40 - 422, (1964).

19. Siegel, R. and Sparrow, E. M., AIC J., Vol. 5, No. 1, pp. 73-75, (1959

20. Bhatti, M. S. and Savery, C. W., Tran ASME, J. Heat Transfer, Vol. 99, N 1, pp. 142-144, (1977).

Numerical Calculation of Separated Flows in Three-Dimensional Ducts of Wavy Wall Surfaces

AKIRA NAKAYAMA and HITOSHI KOYAMA
Department of Energy and Mechanical Engineering
Shizuoka University
Hamamatsu, 432 Japan

INTRODUCTION

Some detalied studies[1,2] on flows over wavy wall surfaces have been carried out in view of importance of determining wave generation in liquid surfaces, sediment transport and flow resistance. These periodically shaped surafces are also of great interest from the standpoint of the heat transfer augmentation. Corrugated walls[3] have been investigated by several workers as an effective heat transfer surface configuration. Although heat transfer can be augmented using such a surface configuration, the required pumping power increases considerablly. Periodically shaped smooth walls such as a sinusoidally shaped wall, on the other hand, may be expected to give high heat transfer rates without much loss in the pressure head. It is the purpose of this paper to introduce a general numerical prediction scheme, so as to investigate the hydrodynamic and heat transfer characteristics of laminar and turbulent flows within three-dimensional ducts of sinusoidally shaped wall surfaces. While some investigations have been already reported for two-dimensional[3] and axisymmetric[4] configurations, no theoretical works on the fully elliptic three-dimensional periodic flows of this class seem to have been reported elsewhere.

ANALYSIS

A general form common to all governing equations may be given in Cartesian form as

$$\frac{\partial}{\partial x_i}\left(u_i \phi - \Gamma \frac{\partial \phi}{\partial x_i} \right) = so \qquad (1)$$

where ϕ stands for any one of the dependent variables under consideration. The diffusion coefficient Γ and the source term so in Cartesian form are listed for the conservation equations for mass, momentum, turbulent kinetic energy k, its dissipation rate ε, and energy(u_h, the velocities; T, the temperature):

$$\phi = 1, \quad \Gamma = 0, \quad so = 0 \qquad (2)$$

$$\phi = u_h\ (h = 1,2,3), \quad \Gamma = \nu + \nu_t, \quad so = -\frac{1}{\rho}\frac{\partial p}{\partial x_h} + \frac{\partial}{\partial x_j}\Gamma\frac{\partial u_j}{\partial x_h} \qquad (3)$$

$$\phi = k, \quad \Gamma = \nu + \nu_t/\sigma_k, \quad so = P - \varepsilon \qquad (4)$$

$$\phi = \varepsilon, \quad \Gamma = \nu + \nu_t/\sigma_\varepsilon, \quad so = (c_1 P - c_2 \varepsilon)\, \varepsilon/k \qquad (5)$$

$$\phi = T, \quad \Gamma = \nu/Pr + \nu_t/\sigma_T, \quad so = 0 \qquad (6)$$

where $\nu_t = c_D k^2/\varepsilon$, and $P = \nu_t \frac{\partial u_j}{\partial x_i} \left(\frac{\partial u_i}{\partial x_j} + \frac{\partial u_j}{\partial x_i} \right)$ (7

Empirical constants and the Prandtl number Pr are given by

$c_D = 0.09,\ c_1 = 1.44,\ c_2 = 1.92,\ \sigma_k = 0.9,\ \sigma_\varepsilon = 1.3,\ \sigma_T = 0.9,\ Pr = 0.71$

It is possible to retain the Cartesian coordinate system even when dealing with irregular wall surface boundary. When using such a system, however, extensive in polative calculations must be carried out to satisfy required conditio To overcome complexities associated with irregular boundaries, the authors have proposed a general three-dimensional coordinate transformation procedure [5]. The procedure is quite different from the conventional general tensor approach the sense that the velocity base vectors $\bar{g}^*_i$ are arbitrary, and independent of t coordinate base vectors $\bar{g}_i$. ($\bar{g}^*_i$ may even be constant vectors, which, then, lea a considerable simplification of the momentum equations.) The resulting genera conservation equation in arbitrary velocity and coordinate systems is written a

$$\frac{\partial}{\partial x^i} J \left(D^i_j u^j \phi - \Gamma G^{ij} \frac{\partial \phi}{\partial x^j} \right) = J\, so \qquad (8$$

New expressions for the source term so in the scalar equations (4) and (5) readily be obtained by a straight-forward transformation based on $\bar{g}_i$ and $\bar{g}^*_i$. Fo the momentum (vector) equation, on the other hand, the following expression c be reduced through a scalar product operation [5]:

$$so = -\frac{1}{\rho} D^{jh} \frac{\partial p}{\partial x^i} - \gamma^h_{ik} D^i_j \left(u^j u^k - \tau^{jk}/\rho \right) + \frac{1}{J} \frac{\partial}{\partial x^i} J D^i_j \left(\tau^{jh}/\rho - \Gamma D^{kj} \frac{\partial u^h}{\partial x^k} \right) \qquad (9$$

where $\bar{g}_i = \frac{\partial \bar{r}}{\partial x^i}$, $\bar{u} = \bar{g}^*_i u^i$, $\bar{g}^i \cdot \bar{g}_j = \bar{g}^{*i} \cdot \bar{g}^*_j = \delta^i_j$, $J = \bar{g}_1 \cdot \bar{g}_2 \times \bar{g}_3$, $D^i_j = \bar{g}^i \cdot \bar{g}^*_j$

$$D^{ij} = \bar{g}^i \cdot \bar{g}^{*j}, \quad \gamma^k_{ij} = \bar{g}^{*k} \cdot \frac{\partial \bar{g}^*_j}{\partial x^i}, \quad \tau^{jk} = \rho\Gamma \left(\varepsilon^{jk} + \varepsilon^{kj} \right), \quad \varepsilon^{jk} = D^{ij} \left(\frac{\partial u^k}{\partial x^i} + \gamma^k_{im} u^m \right) \qquad (10$$

The general equation (9) may be appreciated by actually specifying $\bar{g}^*_i$. The us contravariant representation of the momentum equation, for example, may readily obtained by setting $\bar{g}^*_i = \bar{g}_i$. Naturally, D^{ij} and γ^k_{ij} for this case, reduce to the metric tensor G^{ij} and the Christoffel symbols of the second kind, respectively. the interpretation of the calculated results on the duct flows of the present c cern, it appears more convenient to retain the Cartesian velocity frame even wh using transformed coordinates. Thus, the velocity base vectors $\bar{g}^*_i$ are set to th Cartesian unit base vectors $\bar{e}_i$, while the coordinate base vectors $\bar{g}_i$ are specifi analytically following the family of grid systems (x^1, x^2, x^3) given by

$$\bar{r} = x^1 \bar{e}_1 + Y_{tb} x^2 \bar{e}_2 + x^3 \bar{e}_3 \qquad ($$

where Y_{tb} is the vertical distance between the upper wall boundary and the hori plane of symmetry such that x^2 varies from zero to unity. Discretization has be carried out by integrating the general conservation equation (8) within a fin volume element. Dirichlet and Neumann boundary conditions were imposed at the u stream inlet and downstream outlet of the calculation domain, respectively. Usua wall functions based on the constant stress layer assumption were applied to th grid nodes next to the wall to match the interior flow with the required wall c itions. Calculations start with solving the three momentum equations, and subse tly, the estimated velocity field is corrected by solving the pressure correcti equation [6] reformulated from the continuity equation (2) so that the velocity fulfils the continuity principle. Then, the scalar transport equations such as k, ε and T are solved. This sequence is repeated till the solution converges. T equirement for the convergence has been taken as satisfied when the maximum change each variable during an iteration becomes less than a prescribed value, 10^{-5}. l three dimensional duct flow calculations have been carried out using the grid des (38 x 17 x 11) with highly non-uniform grid spacing.

RBULENT FLOWS OVER WAVY WALL SURFACES

lker et al.[1,2] measured the wall pressure and wall shear distributions on a sinuso-ally shaped boundary. As schematically shown in Fig. 1, the wave sections are con-ructed such that the wave length λ equals to the mean channel height H. They made asurements changing the wave amplitude to length ratio a/λ. Calculations have en performed for the physical model shown in Fig. 1 with the same duct width 12 H in their experiment, using the aforementioned fully elliptic three-dimensional de. The calculated mean velocity vectors projected onto the cross-sectional plane x = 0 shown in Fig. 2, indicates that the two-dimensionality of the flow is main-ined nearly all over the cross-sectional plane. Figs. 3 present the velocity vec-rs plotted onto the z = 0 plane for three different wave amplitude to length ratios λ = 0.0254, 0.0625 and 0.1, at the Reynolds number based on the bulk velocity u_B d the mean channel height H , $Re = u_B H/\nu$ = 60,000. It is seen that the periodic ow essentially becomes fully-developed after the second crest of the wave. The ll pressure distribution plotted in Fig. 4 confirms the establishment of the peri-ic flow pattern. Thus, the comparison of the present prediction and the experiment Zilker et al. will be made focussing on the third wave section, using the dimen-onless coordinate $x^* = x/\lambda - 2$. Figs. 5 show the fully developed pressure patterns r the three different a/λ. The predicted patterns for a/λ = 0.0254 and 0.0625 are good accord with the experiment, while some discrepancy is appreciable in the case

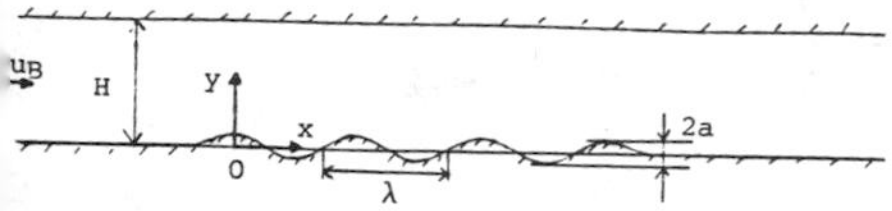

G. 1 Physical model and coordinates.

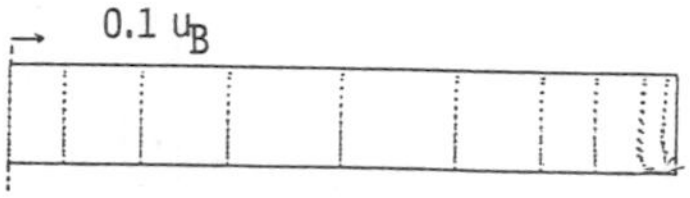

FIG. 2 Secondary flow velocity vectors.

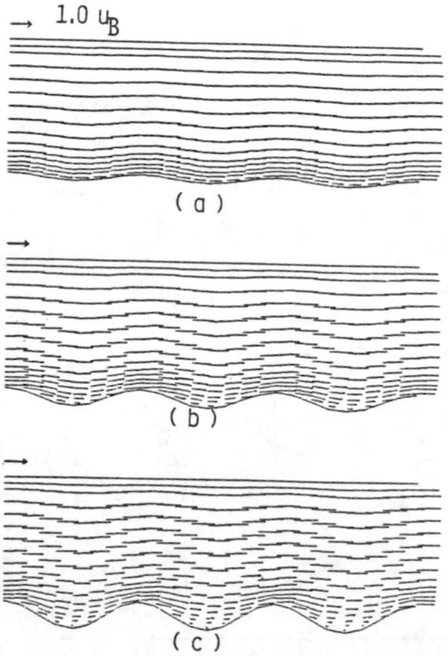

3. 3 Primary flow velocity vectors.

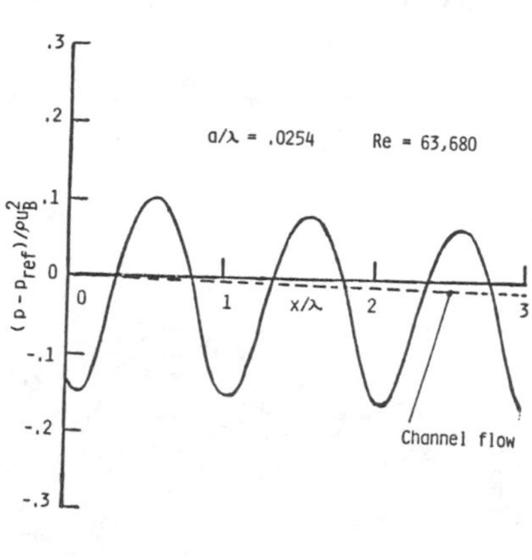

FIG. 4 Development of wall pressure.

of a/λ = 0.1, as a result of the underestimation of the recirculation zone wit wave trough. The wall shear distributions are compared in Figs. 6 for a/λ = 0. and 0.0625, where the ordinate variable is nondimensionalized by $\tau_{av} = \int_0^\lambda \tau\, dx/\lambda$ While the prediction yields profiles almost symmetric about the wave trough x* the measured wall shear varies over a wave length with a large amplitude, and ates significantly from the sinusoidal distribution, even when the measured wa pressure profile may well be approximated by a single harmonic. The discrepanc observed here for the wall shear distribution suggests that the direct employm of the conventional wall functions is inappropriate for turbulent flows in var pressure gradients. It is possible to use more elaborate low Reynolds number m to solve directly down to the wall without using any wall functions. However, a procedure would require extremely fine grids near the wall, hence, would not practical for the three dimensional calculations of this kind. Thus, more gene wall functions applicable even under adverse pressure gradients[7] should be deve and used for the better wall shear estimations. Although the present numerical scheme, when used with the usual wall functions, fails to provide accurate wal shear distributions over the wavy wall surfaces, the predicted mean velocity a pressure fields are found to be satisfactory. The turbulent kinetic energy pre ted by this scheme is also expected to be reasonable except at regions close t wall[5]. The three dimensional duct flow calculations discussed later, have been ied out using the wall functions, for the case of a relatively high wave ampli to length ratio, such that the form drag is large compared with the friction and the near-wall treatment with the wall functions would not significantly in nce the solutions.

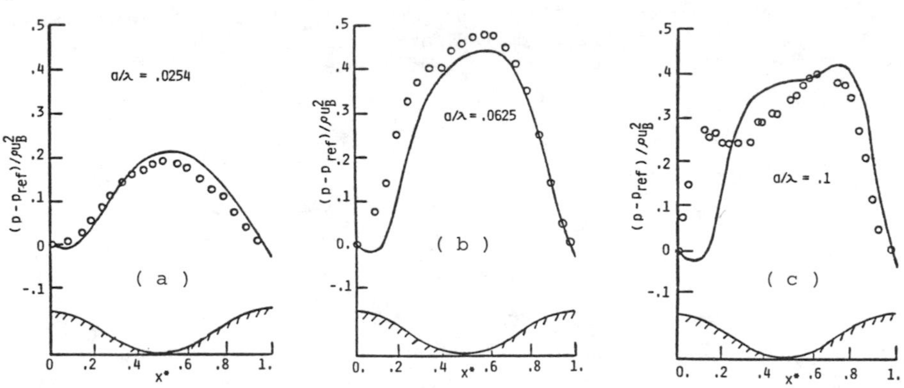

FIG. 5 Pressure distributions over a wave length.

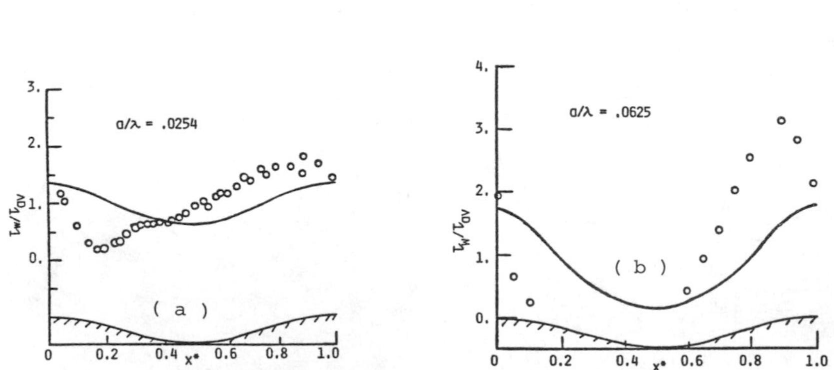

FIG. 6 Wall shear distributions over a wawe length.

AMINAR FLOW WITHIN A THREE-DIMENSIONAL DUCT OF WAVY WALL SURFACES

alculations were carried out for the three-dimensional duct of a square cross-sec-ion. The upper and lower walls are shaped sinusoidally with $a/\lambda = 0.2$. The calcula-ed velocity vector plots at Re = 300 are presented in Figs. 7 for the planes at /H = 0.48 (near the side wall), 0.34 and 0 (the plane of symmetry). No signif-cant change in the flow pattern can be discerned after the first wave. Figs. 8 how the periodic variation of the secondary velocity field over a half wave length, $\leqq x^* \leqq 0.5$. (Note, Figs. 8(a) to (e) correspond with the locations a to e ndicated in Fig. 7(c).) At $x^* = 0$, a pair of vortex rolls appears near the upper avy wall, and then disappears downstream. The observation on both Figs. 7 and 8 eveals that the fluid particles within the primary flow recirculation region are ontinuously drawn back upstream with complex spiral lateral motions. As may be xpected from the periodic nature of the flow, the secondary flow pattern shown in ig. 8(e), when turned upside down, becomes identical to that in Fig. 8(a). The sovels obtained at $x^* = 0.0625$ and 0.3125 are presented in Figs. 9. It is seen hat the recirculation region extends deep into the corner, where the axial momentum s low due to the viscous effect.

URBULENT FLOW WITHIN A THREE-DIMENSIONAL DUCT OF WAVY WALL SURFACES

he Reynolds number was increased to Re = 60,000 for the calculations of the turbu-

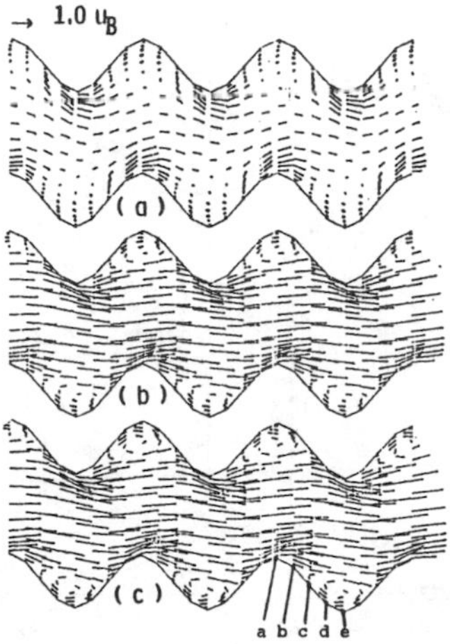

FIG. 7 Primary flow (laminar.)

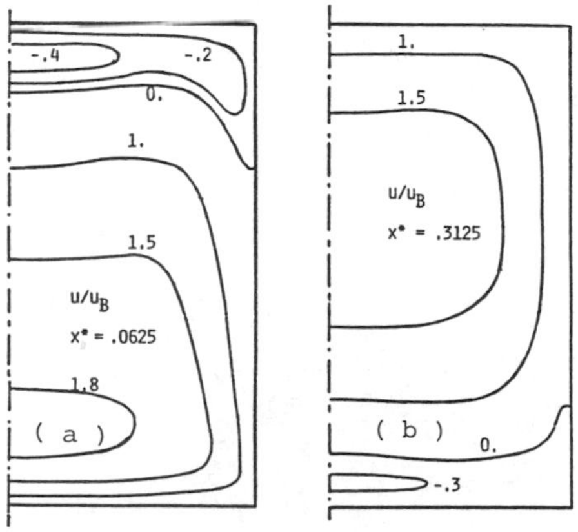

FIG. 9 Isovels (laminar.)

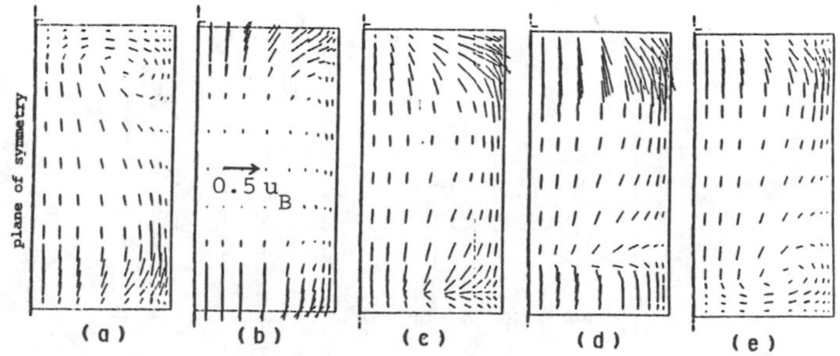

FIG. 8 Seconadry flow (laminar.)

lent flows in the same three-dimensional duct of wavy wall surfaces. The calcul mean velocity vectors are presented in Figs. 10 and 11 in a similar fashion. As seen in Figs. 10, the pattern of the primary flow velocity vectors changes sign cantly along the z direction. It is especially interesting to note that, at z/H 0.34 (Fig. 10(b)), almost all velocity vectors are pointing downstream. Thi fact may be appreciated in connection with the secondary flow activities in Fig 11 as follows. Fig. 11(a) indicates the existence of an additional pair of co rotating vortex rolls along the plane of symmetry, similar to what observed in ed rectangular channels for a certian Dean number range[8]. It is seen that the se ary flow current carries the fluid at the duct core of high axial momentum towa the upper wall. This secondary flow current energizes the fluid particles along boundary of the vortex rolls (say the plane at $z/H \cong 0.3$.) Thus, even the flui near the upper wall under an adverse pressure gradient, can proceed downstream, already observed in Fig. 10(b). Secondary flow activities associated with the reattachment may be seen in Figs. 11(c) and (d). Strong secondary flow curr is generated near the upper wall following the flow reattachment. The velocity of the secondary flow appears to be high especially near the corner of low axia momentum. The predicted isovels are shown in Figs. 12 which exhibit the contou quite different from those of the laminar flow. As seen in Fig. 12(a), the se ary flow current directed from the core to the upper wall distorts the contour considerably. The isovel pattern obtained downstream of the flow reattachment c upper wall (Fig. 12(b)) reveals that the region of high axial momentum shif

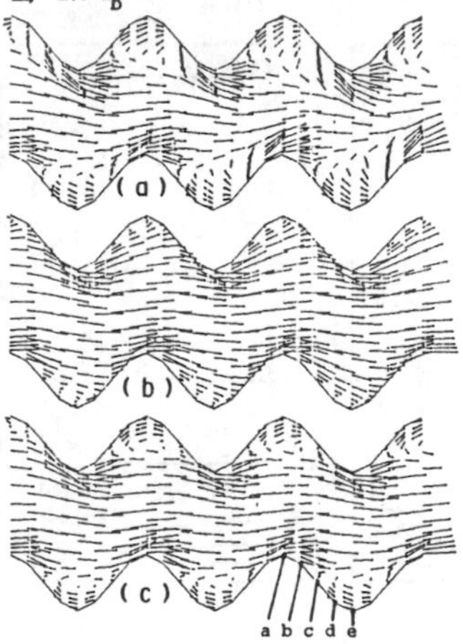

FIG. 10 Primary flow (turbulent.)

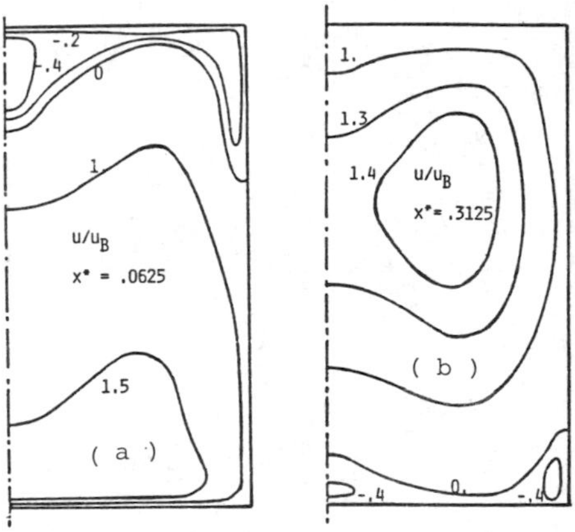

FIG. 12 Isovels (turbulent.)

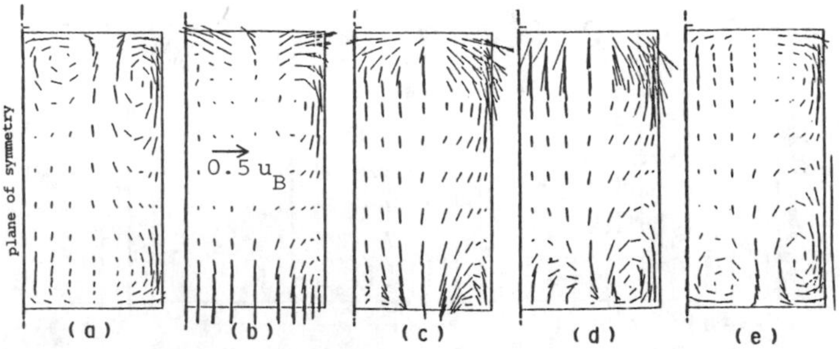

FIG. 11 Secondary flow (turbulent.)

ignificantly toward the side wall. The axial variation of the pressure is presen-ed in Fig. 13, where the predicted pressure drop in a straight square duct in the resence of the secondary flow of the second kind is also indicated for comparison. s going along the lower wall, the pressure minimum appears downstream of the crest, nd then the pressure rises downstream to cause the flow separation. As the flow eparates, the wall pressure becomes fairly uniform. After an appreciable increase ue to the flow reattachment, the wall pressure decreases drastically as the flow ccelerates over the upstream side of the wave. A similar pressure pattern is repea-ed along the upper wall. The pressure varies periodically also along the duct enter, but with a half wave length, as may be expected from the geometrical consi-eration. The predicted contours of the turbulent kinetic energy are shown in Figs. 4. While the contour lines near the lower wall at $x^* = 0$ (Fig. 14(a)) exhibit pattern characteristic of a constant stress layer, those near the upper wall ndicate a typical kinetic energy layer usually observed in the turbulent boundary ayers developed under adverse pressure gradients. The kinetic energy contours shown n Fig. 14(b) were obtained downstream of the flow reattachment on the upper wall. ne figure suggests the formation of the high kinetic energy layer within the fluid ocated away from the walls.

EAT TRANSFER RESULTS

ince the near-wall treatment based on the conventional wall functions fails to give easonable predictions on the wall shear distribution over the wavy wall surfaces, ne heat transfer coefficients obtained assuming the temperature law of the wall ould be seriously in error. Therefore, an attempt to predict the turbulent heat ransfer from the three-dimensional duct had to be abandoned. Instead, laminar flow alculations were carried out to investigate general heat transfer characteristics n the three-dimensional duct. The heat transfer results obtained for the three-imensional duct of a square cross-section with wave amplitude $a/\lambda = 0.1$ and at Re 300, are presented in Fig. 15 along with the results on a straight square duct, nere both the upper and lower surfaces are heated up to a constant temperature T_w, nile the side walls are maintained at a lower constant temperature T_s (note, h = /($T_w - T_s$); and k_t, the thermal conductivity of the fluid.) It is clearly seen that ne wavy wall surfaces are quite effective in view of the heat transfer enhancement. should be noted here that almost all previous studies on the periodic flows have ssumed two-dimensional flow configurations, and none of them considered the effect the three-dimensional fluid motions on the heat transfer characteristics, which ay become considerable in actual heat exchangers. Thus, calculations were made

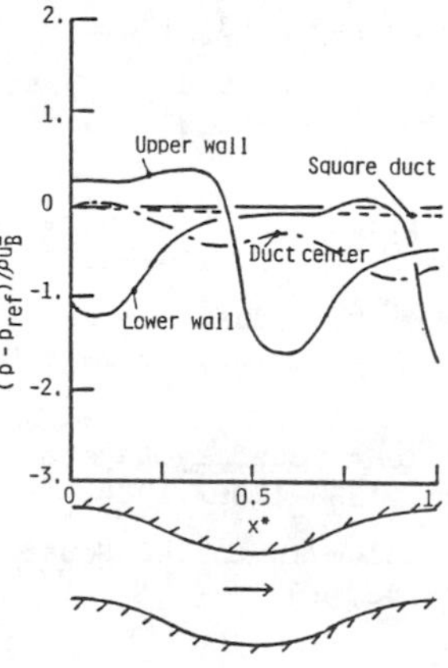

FIG. 13 Pressure (turbulent.)

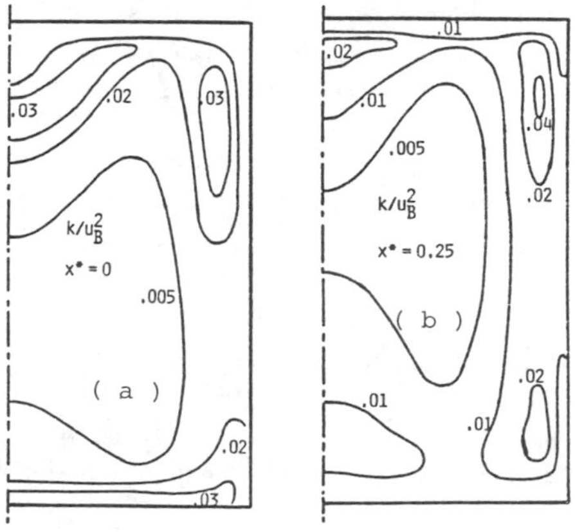

FIG. 14 Turbulent kinetic energy.

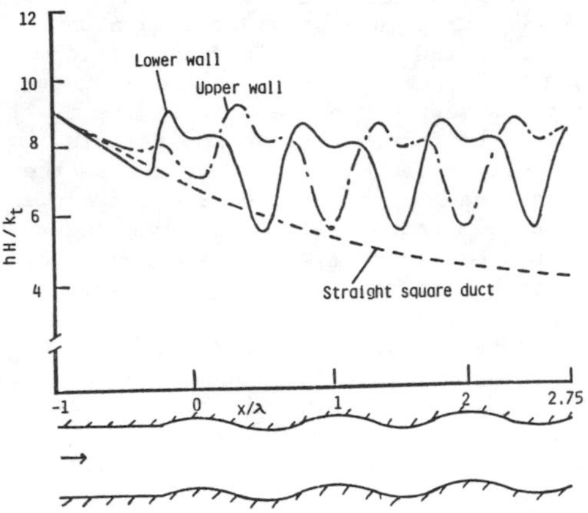

FIG. 15 Local heat transfer coefficient

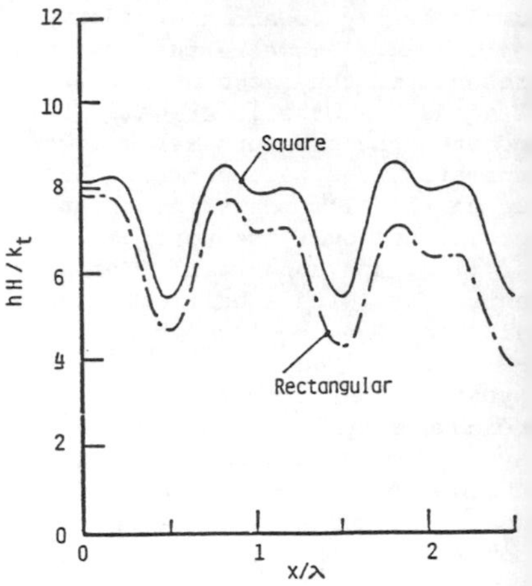

FIG. 16 Effect of secondary flo

also for a rectangular duct having the same wavy upper and lower surfaces, but a different duct width as wide as 6 H , so that the flow within the duct is ess ally two-dimensional. Comparison between the square duct and rectangular duct made in Fig. 16, which reveals that the secondary flow activities work not onl enhance the convective heat transfer , but also to shorten the running length for the establishment of the periodic pattern in the temperature field.

REFERENCES

1. Zilker, D. P., Cook, G. W. and Hanratty, T. J., Influence of the amplitude a Solid Wavy Wall on a Turbulent Flow, Part I, J. Fluid Mech., vol. 82, pp. 51, 1977.

2. Zilker, D. P. and Hanratty, T. J., Influence of the Amplitude of a Soild Wa Wall on a Turbulent Flow, Part II, J. Fluid Mech., vol. 90, pp. 257-271, 1

3. Izumi, R., Yamashita, H. and Oyakawa, K., Analysis on Channels Bent Many T Trans. Japan Soc. Mech. Engrs., vol. 48-435B, pp. 2245-2253, 1982.

4. Hijikata, K., Mori, Y. and Ishiguro, H., Turbulence Structure and Heat Tra of Pipe Flow with Cascade Smooth Turbulence Surface Promotors, Trans. Japa Mech. Engrs., vol. 50-458B, pp. 2555-2562, 1984.

5. Nakayama, A., Finite Difference Calculation Procedure for Three-Dimensiona lent Separated Flows, Int. J. Numer. Methods Eng., vol. 20, pp. 1247-1260, 1984

6. Patankar, S. V. and Spalding, D. B., A Calculation Procedure for Heat, Mas Momentum Transfer in Three-Dimensional Parabolic Flows, Int. J. Heat Mass fer, vol. 15, pp. 1787-1806, 1972.

7. Nakayama, A and Koyama, H., A Wall Law for Turbulent Boundary Layers in Ad Pressure Gradients, AIAA J., vol. 22, no. 10, pp. 1386-1389, 1984.

8. Cheng, K. C. Lin, R. C. and Ou, J. W., Graetz Problem in Curved Square Cha Trans. ASME, J. Heat Transfer, vol. 97, pp. 244-248, 1975.

Turbulent Transport Phenomena in Axisymmetric Jet with Negative Buoyancy

FUMIMARU OGINO, HIROMI TAKEUCHI, MASATSUGU TOKUDA, and TOKURO MIZUSHINA
Department of Chemical Engineering
Kyoto University
Kyoto, Japan

ABSTRACT

Measurements of the time-averaged and fluctuating velocity and temperature of the vertical round jet with negative buoyancy have been made. The measured velocities and temperatures of the up- and downflow are in fairly good agreement with the results obtained by a simple analysis. The values of the relative intensity of the vertical velocity fluctuation are nearly same as those of the momentum jet, whereas those of the radial velocity fluctuation of the upflow are much larger than those of the momentum jet. The values of the relative intensity of the temperature fluctuation of the upflow are smaller than those of the momentum jet, but those of the downflow are very large. A remarkable result was obtained from the measurements of the Reynolds stress and turbulent heat flux that the values of them were nearly zero in the downflow, indicating that the downflow transports neither momentum nor heat.

INTRODUCTION

One of the flow configurations important in the heat transfer processes with significant buoyancy effect is a vertical turbulent buoyant jet discharged into a stagnant environment. Measurements of the time-averaged and fluctuating velocity and temperature of the vertical round jet with positive buoyancy have been made by the authors [1, 2]. However, the state of the experimental knowledge is much less satisfactory for the negative buoyant jet.

Mizushina et al. [3] have presented experimental results on the spread of the upflow, the width of the downflow, the penetration distance, the decay of the centerline velocity and temperature, and the radial velocity and temperature profiles of the jet discharged upward into an ambient of higher temperature than that of the jet fluid.

The purposes of this paper are to compare the measured radial distributions of the time-averaged velocity and temperature with an analytical result based on the integral method and to present the results of the measurements of the turbulence quantities in the vertical round negative buoyant jet discharged into uniform temperature ambient.

EXPERIMENTAL APPARATUS AND PROCEDURE

The experimental apparatus is shown in Fig. 1 (a). The test tank was constructed of transparent acrylic plates with internal dimensions of 2 m long,

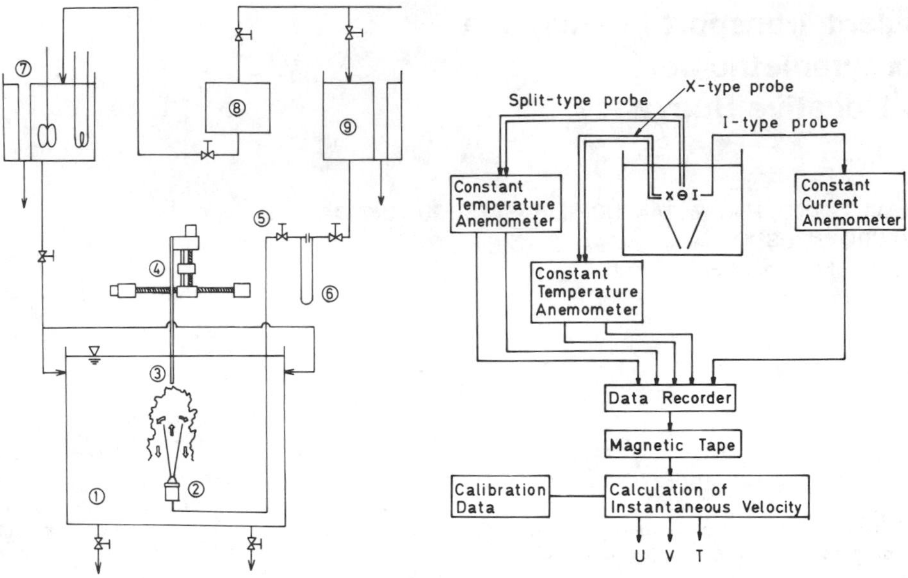

(a) (b)

FIGURE 1. Experimental apparatus and data reduction procedure. 1 test tank, 2 contraction nozzle, 3 probe support, 4 traverser, 5 needle valve, 6 manometer, 7,9 head tank, 8 heater

1 m wide and 1 m deep. Wall thickness was 2 cm. Filtered city water was used as ambient fluid. After the water was heated in a heater and its temperature was regulated in a head tank, the test tank was filled with it. For jet flow a contraction nozzle of diameter 5 mm was used. The nozzle was located at the center of the tank base, the level of the outlet being 0.3 m above the floor. The jet fluid was also city water and discharged vertically upward. Flow rate was measured by a calibrated orifice meter.

The radial and vertical components of the velocity fluctuation were measured with X-type film probe (TSI 1240-20W) and the temperature fluctuation with I-type film probe (TSI 1210-20W). The split-type film probe (TSI 1288W) was also used to distinguish the flow direction. The velocity signal was obtained by a constant temperature anemometer (DISA 55M10), while the temperature signal was obtained by a constant current anemometer (DISA 55M20) operating as a resistance bridge. The outputs from the anemometers were stored in a data recorder (TEAC DR-2000) as shown in Fig. 1 (b). The sampling frequency was 200 Hz and the sample size about 20 000. The velocity-temperature calibration curve was used for calculation of the actual velocity signal from the instantaneous constant temperature anemometer output.

Reynolds number at the outlet of the nozzle was 3 000 and Froude number 1 360. Measurements were made in the non-buoyant region near the discharging nozzle, $X^{\#} = 0.4$, and the buoyancy-dominating region, $X^{\#} = 0.9$.

ANALYSIS

If density variation is accounted for only in the buoyant force term, then the boundary layer forms of the momentum and energy equation are given by

$$U \frac{\partial U}{\partial X} + V \frac{\partial U}{\partial r} = \varepsilon_m \frac{1}{r} \frac{\partial}{\partial r} \left(r \frac{\partial U}{\partial r} \right) - \beta g (T - T_\infty) \tag{1}$$

$$U \frac{\partial T}{\partial X} + V \frac{\partial T}{\partial r} = \varepsilon_h \frac{1}{r} \frac{\partial}{\partial r} \left(r \frac{\partial T}{\partial r} \right) \tag{2}$$

The integration of Eqs. (1) and (2) gives

$$\frac{d}{dX} \int_0^\infty U^2 r dr = -\beta g \int_0^\infty (T - T_\infty) r dr \tag{3}$$

$$\frac{d}{dX} \int_0^\infty U T r dr = 0 \tag{4}$$

A uniform distribution is used for the profiles of velocity and temperature as,

$$U = U_u \; (0 \leqq r \leqq b_u), \quad U_d \; (b_u \leqq r \leqq b_d), \quad 0 \; (b_d \leqq r) \tag{5}$$

$$T = T_u \; (0 \leqq r \leqq b_u), \quad T_d \; (b_u \leqq r \leqq b_d), \quad 0 \; (b_d \leqq r) \tag{6}$$

The velocities U_u and U_d are respectively considered to be characteristic velocities of the upward and downward flows of the jet and the temperatures T_u and T_d characteristic temperatures of these two flows.

Choosing $b_u = 0.17X$ and $b_d = 0.488 d_o \sqrt{Fr}$, which have been obtained experimentally by Mizushina et al. [3], Eqs. (1) and (2) at r = 0 and the integrals of Eqs. (3) and (4) are specified in terms of four quantities U_u, U_d, T_u and T_d.

In the calculation the following assumptions are applied.

$$\varepsilon_m = \varepsilon_h = 0.17 b_u U_u \tag{7}$$

$$\frac{1}{r} \frac{\partial}{\partial r} \left(r \frac{\partial U}{\partial r} \right) \Big|_0 = (U_u - U_d)/b_u^2 \tag{8}$$

$$\frac{1}{r} \frac{\partial}{\partial r} \left(r \frac{\partial T}{\partial r} \right) \Big|_0 = (T_u - T_d)/b_u^2 \tag{9}$$

The boundary conditions are given by

$$X^{\#} = 1.66 d_o \sqrt{Fr}, \quad \left. \begin{array}{l} U_u = U_d = 0 \\ T_u - T_\infty = T_d - T_\infty = 3.8 (T_\infty - T_o)/\sqrt{Fr} \end{array} \right\} \tag{10}$$

Equation (10) is based on the experimental result of Mizushina et al. [3].

The calculated results for velocity and temperature are depicted in Fig. 2 along with the measured centerline velocity and temperature $U_m^{\#}$ and $T_m^{\#}$. Since the characteristic velocity and temperature of the upflow obtained by the present analysis are in fairly good agreement with $U_m^{\#}$ and $T_m^{\#}$, the results for the downflow are considered to be good approximation to the actual downward flow.

EXPERIMENTAL RESULTS

The time fractions of the upward and the downward flow obtained by the split-type film probe are shown in Fig. 3. There is no difference in the

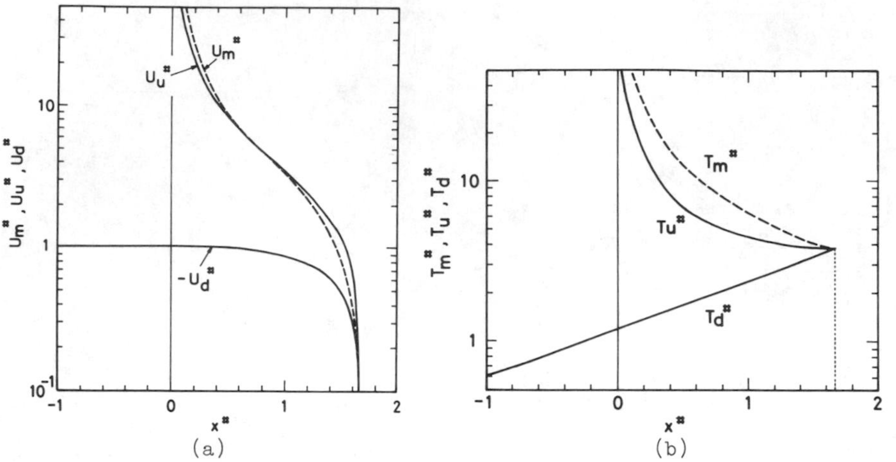

FIGURE 2. Characteristic velocities and temperatures of upflow and downflow.

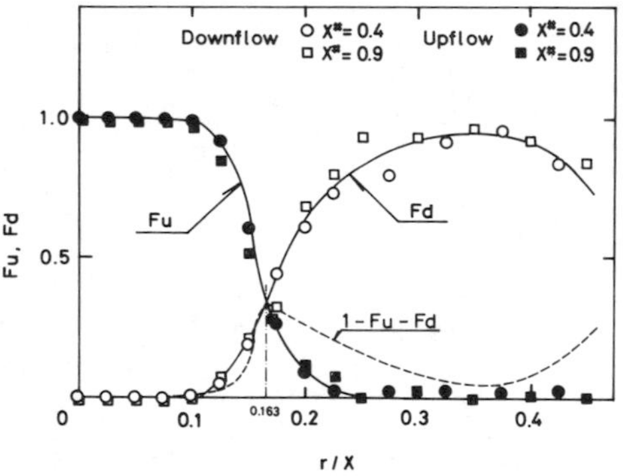

FIGURE 3. Time fractions of upflow and downflow.

measured values between at $X^\# = 0.4$ and 0.9. The sum of the time fractions of the upflow and the downflow is not unity because of the occurrence of the flow in the radial direction. The time fraction of the radial flow, 1 - Fu - Fd, is also depicted in Fig. 3.

In the region of $r/X < 0.1$ the time fraction of the upflow is unity and becomes zero at $r/X > 0.23$. On the other hand the time fraction of the downflow is zer at $r/X < 0.1$, increases gradually with r/X to a maximum at $r/X = 0.35$ and then decreases. Two time fractions are equal at $r/X = 0.163$, the values being 0.33. The location of $r/X = 0.163$ coincides with the spread of the upflow, $b_u = 0.17X$ obtained by the hydrogen bubble technique. The time fraction of the radial flo increases rapidly in the region of $0.1 < r/X < 0.163$, attains a maximum value of 0.33 at the boundary of the upflow and downflow and then decreases gradually at $r/X > 0.163$. It is interesting to note that the values of three time fractions are equal each other at $r/X = 0.163$.

The radial distributions of the axial and radial components of the time-average velocity are shown in Figs. 4 (a) and (b). The dot-dashed curves show

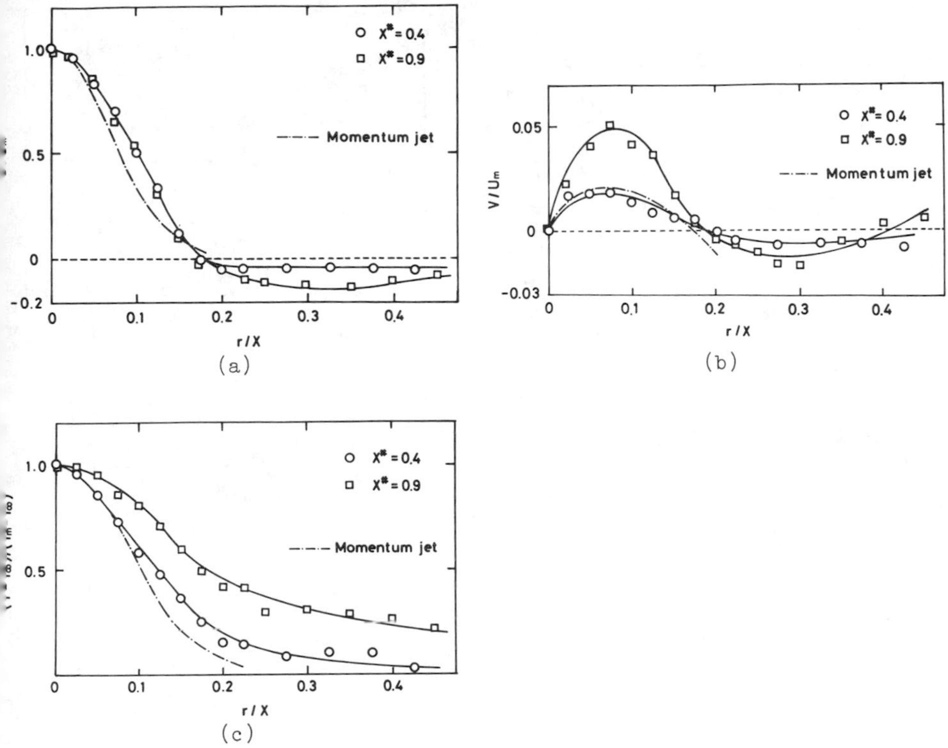

FIGURE 4. Radial distributions of time-averaged axial velocity, radial velocity and temperature.

the results of the momentum jet. The axial velocity profiles of the upflow at the two locations of X# = 0.4 and 0.9 show the similarity profiles. The velocity distributions in the downflow are nearly uniform, the average values of them being $-0.12U_m$ at $X^{\#} = 0.9$ and $-0.04U_m$ at $X^{\#} = 0.4$. This experimental result is in qualitative agreement with the analysis that the velocity of the downflow is $-0.22U_m$ at $X^{\#} = 0.9$ and $-0.09U_m$ at $X^{\#} = 0.4$ as shown in Fig. 2 (a).

From Fig. 4 (b) the radial component of the time-averaged velocity of the upflow is positive, the value at $X^{\#} = 0.9$ being larger than that at $X^{\#} = 0.4$ in accordance with the spread of the upflow. On the other hand the radial velocity in the downflow is negative, indicating that the fluid of the downflow is entrained by the upflow.

The radial distributions of the time-averaged temperature are shown in Fig. 4 (c), indicating that the similarity does not exist for the temperature profile of the negative buoyant jet. If the temperature at r/X = 0.3 is assumed to be the characteristic temperature T_d of the downflow, the measured values of $T_d^{\#}/T_m^{\#}$ is 0.08 at $X^{\#} = 0.4$ and 0.31 at $X^{\#} = 0.9$ as seen from Fig. 4 (c). This result is in good agreement with the analysis shown in Fig. 2 (b).

The radial distributions of the intensities of the axial and radial components of the velocity fluctuations are depicted in Figs. 5 (a) and (b). It can be seen from Fig. 5 (a) that the values of the intensity of axial velocity

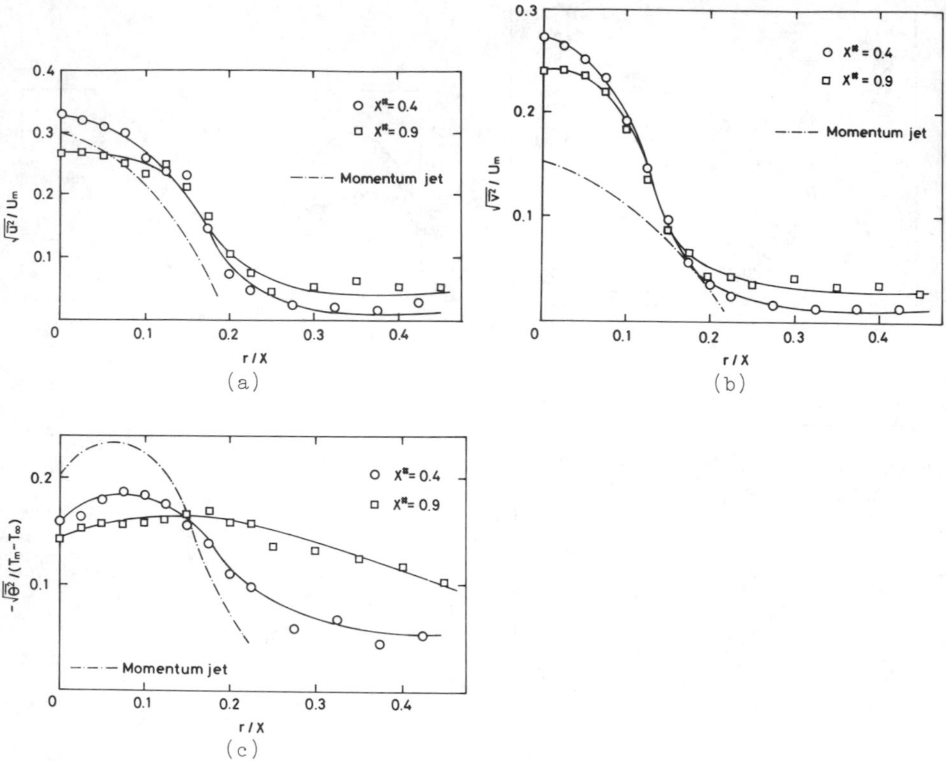

FIGURE 5. Radial distributions of intensity of axial velocity, radial velocity and temperature fluctuations.

fluctuation in the upflow region are nearly same as those of the momentum jet. On the other hand the intensity of the radial velocity fluctuation of the upflo is much larger than that of the momentum jet as shown in Fig. 5 (b). The reaso of the large values of the radial velocity fluctuations cannot be inferred so easily because of the intricate interlinkage of the turbulence quantities and buoyancy. The present result will serve a modelling of the transport equations for the turbulence correlations. The intensities of both components of velocit fluctuations of the downflow are smaller than those of the upflow and their radial distributions are nearly uniform.

The radial distributions of the normalized intensities of temperature fluctuations are depicted in Fig. 5 (c), showing that the values of the upflow are smaller than that of momentum jet. The production term of the temperature fluctuation is expressed by $2(-\overline{v\theta})(\partial T/\partial r)$. The turbulent heat flux and the temperature gradient become smaller than those of momentum jet as shown in Fig. 6 (b) and Fig. 4 (c), so that the value of the production term becomes smaller, causing smaller intensities of temperature fluctuations. The distributions of the intensities of the temperature fluctuations of the downflow are different between $X\# = 0.4$ and 0.9 as shown in Fig. 5 (c). Although the values of the normalized temperature fluctuation at $X^\# = 0.9$ are larger than those at $X^\# = 0.$ in the downflow region, the values of $\overline{\theta^2}$ itself at $X^\# = 0.4$ and 0.9 are nearly same each other at $r/X = 0.3$. Therefore the difference in the shape of the profiles of the relative intensities is said to be caused by the difference of

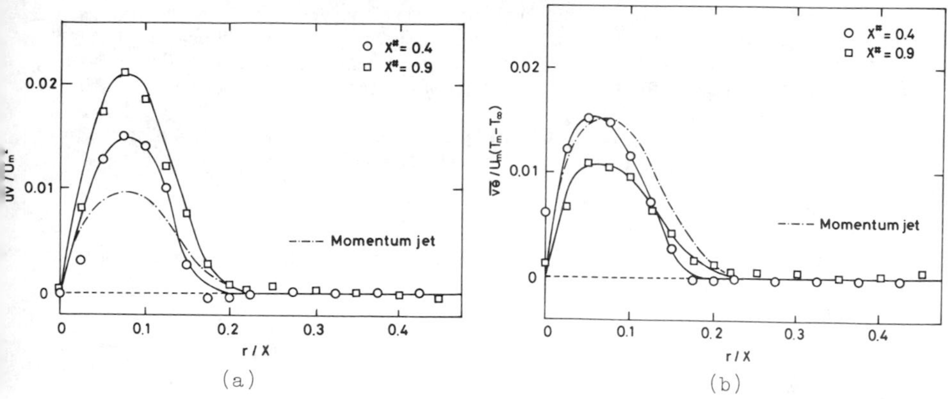

FIGURE 6. Radial distributions of Reynolds stress and turbulent heat flux.

the values of the centerline temperature.

The measured values of the Reynolds stress and the turbulent heat flux are shown in Figs. 6 (a) and (b). The values of the normalized Reynolds stress at $X^{\#}$ = 0.9 are larger than those at $X^{\#} = 0.4$, both still being larger than that of the momentum jet. In addition the radial distributions of them show peak values at r/X = 0.075. The production term of $\overline{uv}$ is $\overline{v^2}(-\partial U/\partial r)-\beta g(-\overline{v\theta})$. The value of the production by buoyancy is much smaller than that by stress and the value of $\overline{v^2}$ becomes very large in the upflow region as shown in Fig. 5 (b), so that the Reynolds stress of the upflow may be considered to become larger. The Reynolds stress of the downflow is seen to be nearly zero, indicating that the momentum is not transported by turbulence in the region of the downflow.

The radial distributions of the normalized turbulent heat flux are similar to those of the Reynolds stress, but the peak appears at r/X = 0.05. The peak value at $X^{\#} = 0.9$ is smaller than that at $X^{\#} = 0.4$, this result being different from that of the normalized Reynolds stress. The production term of $(-\overline{v\theta})$ is $\overline{v^2}(\partial T/\partial r)$. The small temperature gradient may cause the decrease of the turbulent heat flux at $X^{\#} = 0.9$, although the value of $\overline{v^2}$ is larger than that of momentum jet. A remarkable result is that the turbulence in the downflow does not transport heat as well as momentum, suggesting that the downflow seems to be an irrotational potential flow.

CONCLUSIONS

1. A simple analysis indicates that the velocities of the upflow and downflow are in fairly good agreement with the measured centerline velocity and that at r/X = 0.3, respectively. The calculated temperatures are also in fair agreement with the measured centerline temperature and that in the downflow region.

2. The boundary between the upflow and downflow is 0.163X.

3. The radial distributions of the time-averaged velocity of the upflow show the similarity profile, while the shape of the profiles of the time-averaged temperatures at two locations are different.

4. The values of the relative intensity of the vertical velocity fluctuation

are nearly same as those of the momentum jet, whereas those of the radial velocity fluctuation of the upflow are much larger than those of the momentum jet. The values of the relative intensity of the temperature fluctuation of the upflow are smaller than those of the momentum jet.

5. The turbulence of the downflow transports neither momentum nor heat.

ACKNOWLEDGMENT

This work was supported by the Ministry of Education, Science and Culture through a Grant in Aid for Scientific Research (No. 57045066).

NOMENCLATURE

b = width of the jet, m
d = diameter of the nozzle, m
Fr = $U_o^2/\beta g d_o(T_\infty - T_o)$, discharge Froude number
F = time fraction
g = gravitational acceleration, m/s^2
r = radial coordinate, m
T = time-averaged temperature, K
$T^\#$ = $\sqrt{Fr}\,(T - T_\infty)/(T_o - T_\infty)$
U = time-averaged axial velocity, m/s
$U^\#$ = $\sqrt{Fr}\,U/U_o$
u = axial velocity fluctuation, m/s
V = time-averaged radial velocity, m/s
v = radial velocity fluctuation, m/s
X = distance from the virtual source, m
$X^\#$ = $X/(d_o\sqrt{Fr})$
β = coefficient of volume expansion, 1/K
ε_h = eddy diffusivity of heat, m^2/s
ε_m = eddy diffusivity of momentum, m^2/s
θ = temperature fluctuation, K

Subscripts
d = downflow
m = centerline
u = upflow
o = discharge of the nozzle
∞ = ambient

REFERENCES

1. Ogino, F., Takeuchi, H., Kudo, I. and Mizushina, T., Heated Jet Discharged Vertically Into Ambients of Uniform and Linear Temperature Profiles, Int. J. Heat Mass Transfer, vol. 23, pp. 1581-1588, 1980.

2. Ogino, F., Takeuchi, H., Wada, H. and Mizushina, T., Buoyancy Effects on Turbulence in Axisymmetric Turbulent Buoyant Jet, Turbulence and Chaotic Phenomena in Fluids, ed. T. Tatsumi, pp. 495-500, Elsevier Science Publishers B. V. (North Holland), Amsterdam, 1984.

3. Mizushina, T., Ogino, F., Takeuchi, H. and Ikawa, H., An Experimental Study of Vertical Turbulent Jet with Negative Buoyancy, Wärme- und Stoffübertragung, vol. 16, pp. 15-21, 1982.

Heat Transfer and Pressure Drop of a Staggered Longitudinally-Finned Tube Bank

[illegible]. QIU and Z.-Q. CHEN
Department of Power Machinery Engineering
Xi'an Jiao-tong University
PRC

ABSTRACT

The heat transfer and pressure drop performances of a staggered tube bank, in which longitudinal plate fins were attached to the individual tube, were investigated. The heat transfer coefficients have been obtained via mass transfer measurements carried out with the naphthalene sublimation technique. Experimental results showed that the heat transfer was fully developed after the fourth row of the bank. For the thermally fully developed region, the heat transfer coefficient has been correlated as

$$Nu = 0.153\ Re^{0.696}\ Pr^{0.36}$$

The flow pattern on the individual tube has been revealed by the oil-lamp-black technique. On the rear fin, eddies appear in the wake downstream of the tube, and thus deteriorate the heat transfer of the plate fin affixed at the back of the tube.

NOMENCLATURE

- A heat transfer area
- d base tube diameter
- D naphthalene-air diffusion coefficient
- f pressure drop coefficient
- M total amount of mass transfer
- $\dot{M}$ mass transfer rate
- Nu Nusselt number
- Δp pressure drop
- Pr Prandtl number
- $\dot{Q}$ air flow rate
- Re Reynolds number
- s_1 transverse pitch of tubes
- s_2 longitudinal pitch of tubes
- Sc Schmidt number
- Sh Sherwood number
- T_f fluid temperature
- T_w surface temperature
- u_{max} maximun velocity
- β mass transfer coefficient
- ν kinematic viscosity
- ρ air density
- ρ_{nw} naphthalene vapor density at surface
- ρ_{no} naphthalene vapor density at inlet
- ρ_{ne} naphthalene vapor density at outlet
- τ time duration of a data run

*Project supported by the Science Fund of The Chinese Academy of Sciences

INTRODUCTION

The goal of this paper is to determine experimentally the cross flow heat transfer and pressure drop characteristics of a staggered tube bank in which longitudinal plate fins were attached to the individual tube (Fig.1). Though this kind of fin tube has a rather smaller extended surface, it can effectively prevent the ash deposition when the working fluid is ash-laden flue gas. So this kind of finned tube is expected to be widely used in the economizer of boiler and various kinds of industrial kiln or furnance.

In view of the wide-spread use of the longitudinally-finned tube, it is surprising that so little experimental result about their heat transfer and pressure drop characteristics has been published. Some experimental data have been published by Sharan [1] in 1966, but his experiments were performed in an industrial boiler, so the data were not accurate enough. For example, the heat transfer coefficient was obtained by deducting all other thermal resistances from the overall heat transfer coefficient, and the radiation heat transfer of the 300°C ash-laden flue gas has not been accounted in the data reduction. Baran [2] systematically studied the influence of the angle between the fin and the flow on the heat transfer and pressure drop performances, but only the in-line bank has been investigated and no data of the frequently used staggered tube bank has been presented. So it is clear that the heat transfer of a longitudinally-finned tube bank still remains unsolved.

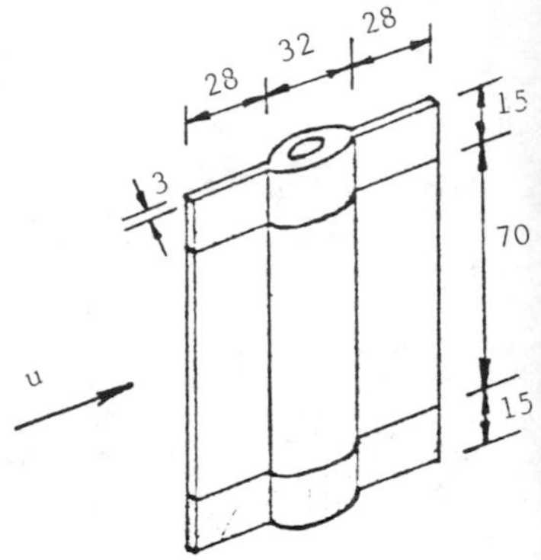

Fig.1

EXPERIMENTAL APPARATUS AND PROCEDURE

The convective heat transfer coefficient of the finned tube bank here was obtained by applying the analogy between heat and mass transfer. The mass transfer coefficient was measured via the naphthalene sublimation technique.

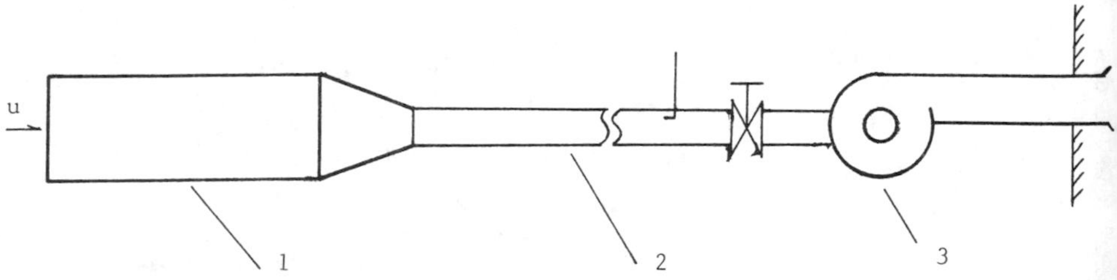

Fig.2

The significant advantages of this method are its higher accuracy and simplicity of the experimental apparatus. Futhermore, according to the heat and mass transfer analogy, a naphthalene surface corresponds to an isothermal surface in the counterpart heat transfer experiment. So no further correction for fin efficiency is needed as in the case of heat transfer experiment.

Experimental Apparatus

The test apparatus as shown in Fig.2, was composed of a rectangular duct 1 with cross sectional dimentions of 100x200mm, a contracted section 2 for measuring flow velocity, and a blower 3 working in the suction mode.

The finned tubes were seated in blind holes on the floor of the test section. In order to accommodate the need for frequent access to the test section for installing or removing the experimental naphthalene specimen, the upper wall of the test section was made removable, and the leak-proof seal was accomplished with the aid of O-rings. The flow velocity of the air was measured by a pitot tube with blockage ratio less than 1.5%.

The placement of the blower downstream of the test section, rather than upstream, was a purposeful decision intended to avoid preheating and disturbing of the air prior to its entry into the test section. The naphthalene vapor sublimated during the experimental runs was discharged to the atmosphere outside the laboratory in order to ensure the air entering the test section was naphthalene-free.

Fig.3 is the top view of the test section with the upper wall removed. In the streamwise direction nine rows of finned tubes and in the transverse direction two rows of finned tubes were deployed with pitch $s_1/d=100/32=3.13$, $s_2/d=50/32=1.56$. The clearance between the two adjacent fins is only 12mm, so

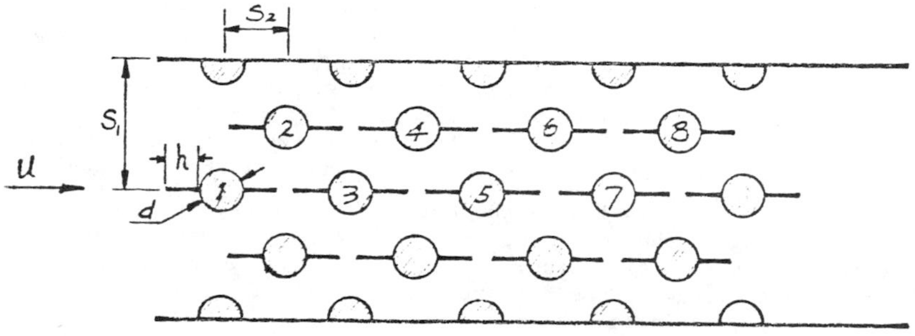

Fig.3

the flow appears to be four independent channel flows. For this reason, two rows of finned tubes in transverse are enough, this will be confirmed by the data of the mass transfer experiments (see Fig.6).

Naphthalene sublimation technique requires that the naphthalene surface should be very smooth, flat and very very clean. In addition, as the configuration of the finned tube is somewhat complex, it is difficult to fabricate the test specimen. We made it successfully with casting method. For each data run, a fresh naphthalene specimen was cast. The shadow parts of both ends, as shown in Fig.1, are metal frames and only the central part is naphthalene participating in the transfer process. The aluminium mold was first preheated to 70-80°C, so that the melting naphthalene could flow fluently to all parts of the mold, especially the 3 mm thin fin space. Once the casting process was finished, the test specimen was put into a sealed bag to minimize its extraneous sublimation and then placed in the laboratory for at least 12 hrs. to attain thermal equilibrium. All the handling of the specimen, including its unmolding was performed with particular care so as to avoid any contamination of the naphthalene surface. The detailed process of fabrication of the mold, casting and unmolding are available in [3].

Experimental Procedure

The reliability of our experimental technique and apparatus was confirmed in a preliminary experiment on mass transfer from a bare tube bank. Agreement of the data, when compared with the well-known Zukauskas correlation [4], was within 3.3% as shown in Fig.4.

Two experimental methods were used in the experiments. One was the one-by-one active method, i.e. in a data run, only one of the finned tube specimen was made of naphthalene, the others were made of plexiglass served to simulate the flow pattern, but did not participate in the transfer process. The second method was the all-active method, i.e. in a data run, all finned tube models participated in the mass transfer process. In our experiments, because the flow appears to be independent channel flow, so only eight naphthalene models (as shown in Fig.2) are necessary to simulate the all-active situation. The identity of these two methods has been verified by their experimental results shown in Fig.6.

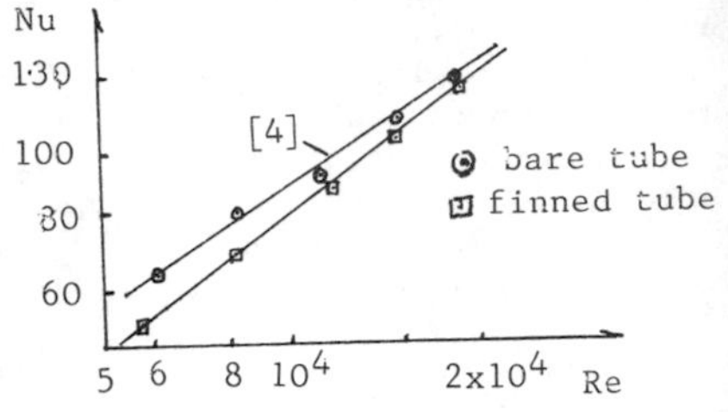

Fig.4

Before experiment, it is necessary to seat the specimen in the test section and blow it so that all the small naphthalene particles adhering to the naphthalene surface could be removed. At the same time, thermal equilibrium between the specimen and the air flow could be reached. After this, the test specimen was weighed (M_1). Then the specimen was seated in the appropriate location of the test section and a data run started. After a certain period of time, stop the blower and reweigh the test specimen (M_2), (M_1 $-M_2$) will be the total sublimed amount of naphthalene, including the extra sublimation during weighing, installing and removing of the test specimen. A correction must be made in order to deduct this extra sublimation. This can be done by repeating all the process described above, except the data run itself. Then weigh the specimen again (M_3). Obviously, the actual amount of sublimation must be

$$\Delta M = (M_1-M_2) - (M_2-M_3) \tag{1}$$

The time duration selected for a data run depends on the sublimation depth and the corresponding sublimation mass. If the sublimation depth is too large, the geometric configuration will be distored. But on the other hand, if the depth is too small, the mass sublimated will be also too small for weighing, then the relative accuracy of weighing will decrease sharply. In the present experiment, the average depth of sublimation was limited to 0.006 mm and the minimum amount of mass sublimated was 30 mg.

The mass of the specimen was measured with a precise analytical balance capable of being read to 0.1 mg. A stop watch capable of sense 0.1 sec. was used to measure the duration of a data run.

According to Ko [5], the temperature of the air flowing through the test section and that of the naphthalene surface are alsmot the same. The temperature of the air was sensed by three carefully calibrated thermocouples that could be read to 0.1^{o}C. The pressure drop and the velocity head were measured by two Rosemount 1151 capacitance-type differential pressure transdusers capabl of being read to 0.001 Torr and 0.002 Torr respectively. The flow rate was calculated from the mean velocity of air and the cross sectional area of the velocity measuring section. The mean velocity of the air was obtained according to the standard method ISO-3354. All the measurements were carried out in a seperated room so as to prevent the influence of human body to the air temperature of the test room.

The flow visualization on the finned tube surface was performed via the oil-lampblack technique [6] to facilitate the explanation and rationalization of the heat transfer results.

DATA REDUCTION

The average mass transfer coefficient is defined as follows:

$$\beta = \dot{M}/(A\Delta\rho_n) = \Delta M/(\tau A \Delta\rho_n) \qquad (2)$$

where $\Delta\rho_n$ is the logarithmic mean density difference given by

$$\Delta\rho_n = [(\rho_{nw}-\rho_{no})-(\rho_{nw}-\rho_{ne})]/[\ln(\rho_{nw}-\rho_{no})/(\rho_{nw}-\rho_{ne})] \qquad (3)$$

where ρ_{nw} can be obtained from the Sogin vapor pressure correlation [7] and the perfect gas law. Converting into SI system, we get

$$\rho_{nw} = 5.847\times10^{11}/\{[10\ \exp(3729/T_W)]\cdot T_W\} \qquad (4)$$

ρ_{no} is zero at the entrance of the test section because the entering air is free of naphthalene vapor. Then the naphthalene densities of the inner rows of the bank can be calculated as:

$$\rho_{ne} = \rho_{no} + \frac{\Sigma\dot{M}}{\dot{Q}}$$

the calculation was perfermed assuming the flow was a channel flow between two rows of finned tubes. The average Sherwood number is

$$Sh = \beta d/D = \beta d Sc/\nu$$

where Sc=2.5 [7]. The Reynolds number is defined as

$$Re = u_{max}\, d/\nu$$

The Sherwood number can be converted into Nusselt number as employing the analogy between heat and mass transfer

$$Nu = Sh\cdot(Pr/Sc)^m$$

the value m of the bare tube bank 0.36 [4] will be used here, then

$$Nu = Sh\ (0.7/2.5)^{0.36} = 0.632\ Sh \qquad (5)$$

The pressure drop coefficient is defined as

$$f = 2\Delta p/(n\rho\cdot u_{max}^2) \qquad (6)$$

RESULTS AND DISCUSSION

Flow visualization

The flow visualization was performed via the oil-lampblack technique. Once the thin film of the oil-lampblack mixture was brushed onto a surface to be observed, the surface was exposed to the air flow. Under the action of exerted shear stress, the mixture will move along the path of the fluid particles passing adjacent to the surface, thus the black and white streaks remained show the flow pattern of the air flow. Fig.5 is the photograph of the visualized flow pattern on the surface of the seventh row of the bank. The black zone parallel to the tube axis near the front generatrix of the base tube marks a seperation region. Emanating outward from the seperation region is a group of very fine lines which, except near the ends of the tube, are perpendicular to the tube axis. Closer inspection of the photograph reveals a slightly inclined flow near the ends of the base tube. This can be explained as follows: the stagnation of the on-coming flow against the base tube caused a pressure rise, since the flow velocity near the upper and bottom walls were relatively smaller than that in the central part of the tube. Therefore, the pressure near walls must be slightly smaller than the pressure in the central part. As a consequence, there was a pressure gradient towards the ends of the tube, thus the air flow would tend to the ends of the tube. In order to eliminate the end effect, 15 mm long metallic part, which did not participate in the transfer process, was made to meet this requirement. The black band on the rear fin showed that the flow at the rear fin was a recirculating one, which deteriorated the transfer process sharply, this would have been verified by the mass transfer experiments.

Fig.5

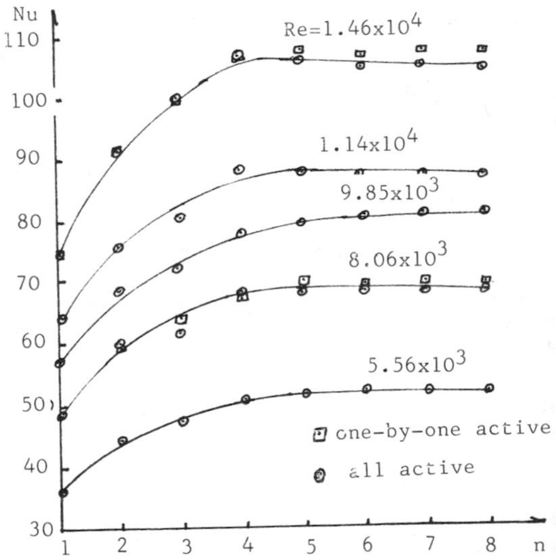

Fig.6

Heat Transfer Characteristics

The row-by-row heat transfer characteristics of the finned tube bank are shown in Fig.6. It can be seen that the heat transfer is fully developed after the fourt row of the bank. A fully developed Nusselt number can be correlated (see Fig.4) as follows:

$$Nu = 0.153\ Re^{0.696}\ Pr^{0.36} \qquad (Re = 5.5 \times 10^{3} \sim 1.8 \times 10^{4}) \qquad (7)$$

where the physical properties are evaluated at the average air temperature.

The uniformity of transfer coefficients of tubes No.5-No.8 (see also Fig.3) indicates that the sidewalls of the test section have no influence on the transfer process, thus our "indepedent channel flow assumption" is verified. In addition, from Fig.4, the heat transfer of the fin tube bank is lower than that of a bare tube bank. This result coincides with Sharan [1]. This fact can also be confirmed from the flow visualization pattern, the vortex prevailes the rear fin and thus decreases its heat transfer. As to the base tube, the seperation region exists in the front part of the base tube, which does not exist in the bare tube case. So it is expected that the heat transfer of the base tube will be worse than that of a bare tube. These expectations were verified by the measurements of the quasi-local transfer coefficients as shown in Fig.7. The quasi-local transfer coefficient was measured by exposing the surface to be investigated to the air flow while the remainder of the specimen was covered with an air-tight tape preventing it from participating in the transfer process. In addition, the transfer coefficient of the rear fin is about 13% lower than that of the front fin and the base tube. So it would be beneficial to have some artificial augmentation devices on the rear fin.

A tube bank in which only the front fin was attached to the base tube has been investigated. The average heat transfer coefficient of the fin tube did increase as expected, but owing to the decrease in heat transfer area, the total amount of heat transfered decreased. More fin tubes must be used when a definite amount of heat is to be transfered. Since the price of tube is higher than that of plate, so it is uneconomical to use this kind of finned tube in industry.

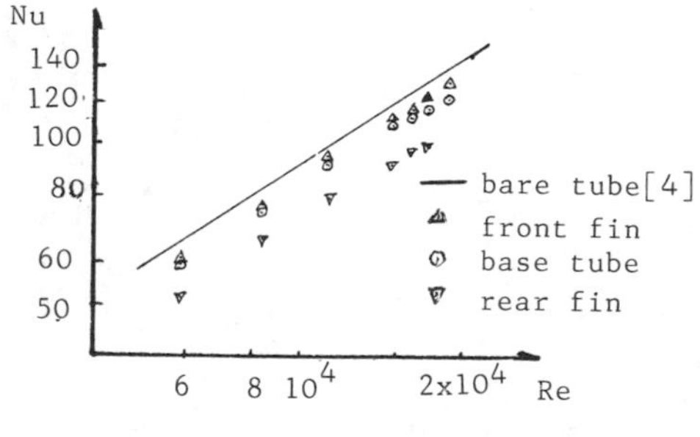

Fig.7

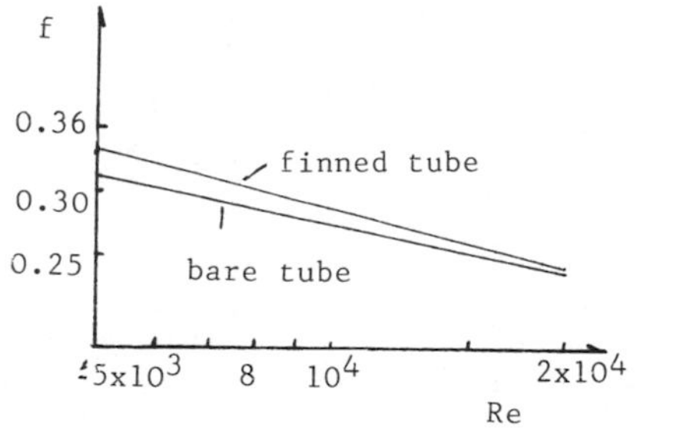

Fig.8

Pressure Drop

The pressure drop coefficients of the bare tube bank and the longitudinally finned tube bank are shown in Fig.8. The pressure drop coefficient of the finned tube bank is higher due to the additional seperation at the front part of the base tube.

CONCLUSIONS

(1) The heat transfer of a longitudinally finned tube bank is fully developed after the fourth row. Its heat transfer and pressure drop performances can be determined by eq. (7) and Fig.8 respectively.
(2) The quasi-local heat trasfer coefficient of the rear fin is about 13% lower than that of the front fin and the base tube, so it is advisable to provide some artificial augmentation devices on the rear fin.

REFERENCES

[1] H.N.Sharan, "Heat Transfer and Pressure Drop of Flue Gases in Cross Flow over Finned Tubes", Sulzer Tech. Rev., Research,No. 1966 pp.10-21, 1966.
[2] M.Baran, "Konvektiver Wärmeübergang und Druckverlust bei querangeströmten Flossenrohrbündeln", Wärme-und Stoffübertragung, v.18 pp. 149-156, 1984.
[3] J.Qiu, "Heat transfer and pressure drop of a Longitudinally-Finned Tube Bank", Master's Degree Thesis, Xi'an Jiao-tong University, 1985 (in Chinese)
[4] A.A.Zukauskas, "Heat transfer from Tubes in Crossflow" in "Advances in Heat transfer", v.8 pp.93 -160, 1972.
[5] S.Y.Ko, H.H. Sogin, "Laminar Mass and Heat Transfer from Ellipsoidal Surface of Fineness Ratio 4 in Axisymmetric Flow", Trans. ASME, v.80 pp.387-390, 1958.
[6] W.Merzkirch,"Flow Visualization", pp. 53-56, 1974.
[7] H.H.Sogin, "Sublimation from Disk to Air Streams Flowing Normal to Their Surfaces", Trans. ASME, v.80 pp. 61-71, 1958.

Liquid Metal Turbulent Heat Transfer in a Circular Tube

K. SUZUKI and A. TOHKAKU
Department of Mechanical Engineering
Kyoto University
Kyoto 606, Japan

INTRODUCTION

It is well-known that the turbulent Prandtl number, Pr_t, must be taken higher than unity in liquid metal turbulent heat transfer, but what is the correct relationship between the turbulent Prandtl number and the fluid Prandtl number, Pr, or the Peclet number, Pe, is still uncertain. In his review on the liquid metal turbulent heat transfer [1], Huetz commented on a calculation method of heat transfer employing the turbulence model of one-point closure and suggested the necessity of developing a calculation method which does not use the turbulent Prandtl number concept. To discard the usage of this concept, the local value of turbulent heat flux, $\rho C_p \overline{v\theta}$, must be calculated directly.

A simple closure of $\overline{v\theta}$ equation has been proposed recently by Suzuki [2]. The closed form of the equation was tested with a local equilibrium assumption and by substituting the experimental data by Lawn [3] into the hydrodynamic turbulent quantities appearing in the equation. The tested equation was concluded to be promising but this test cannot be counted as prediction because the experimental data were used. In this article, the hydrodynamic turbulent quantities are calculated employing a full Reynolds stress model. Three types of full Reynolds stress models are tested: the model by Launder, Reece and Rodi (Model I), the model by Pope and Whitelaw (Model II) and the one by Prud'homme and Elgobashi (Model III) [4-6]. All the governing equations are solved numerically with a finite difference scheme. The computation will be executed for a fully developed state of liquid metal flow in a circular tube heated at uniform heat flux.

CALCULATION METHOD

First, the following non-dimensional quantities are used throughout this text:

$$\eta=\frac{r}{R},\ \xi=\frac{x}{R},\ U^+=\frac{U}{U_\tau},\ (\overline{u^2})^+=\frac{\overline{u^2}}{U_\tau^2},\ (\overline{v^2})^+=\frac{\overline{v^2}}{U_\tau^2},\ (\overline{w^2})^+=\frac{\overline{w^2}}{U_\tau^2},\ k^+=\frac{k}{U_\tau^2},\ \tau_t^+=-\frac{\overline{uv}}{U_\tau^2},\ \varepsilon^+=\frac{R\varepsilon}{U_\tau^3}$$

$$R^+=\frac{U_\tau R}{\nu},\ P^+=\frac{P}{\rho U_\tau^2},\ \Theta^+=\frac{T-T_w}{\theta_\tau},\ q^+=\frac{q}{q_w},\ q_t^+=\frac{\overline{v\theta}}{U_\tau\theta_\tau},\ V^+=\int_\eta^1 U^+\eta d\eta,\ V_0^+=\int_0^1 U^+\eta d\eta$$

where r and x are respectively the radial and axial coordinates, R the radius of the tube, $\overline{U}$ and u the mean and fluctuating components of streamwise velocity, v and w the radial and peripheral fluctuating velocities, k the kinetic energy of turbulence, $\overline{uv}$ the Reynolds shear stress devided by $(-\rho)$, ρ the fluid density, ε the viscous dissipation rate of turbulence kinetic energy, P the pressure, U_τ the friction velocity, T and T_w the fluid and wall temperature, q the total heat flux at local position, q_w the wall heat flux assumed constant along x direction,

θ_τ the friction temperature and ν the kinematic viscosity.

Mometum equation, energy equation and the governing equation of $\overline{v\theta}$ are written in the following forms.

$$0=2+\frac{1}{R^+\eta}\frac{d}{d\eta}(\eta\frac{dU^+}{d\eta})+\frac{1}{\eta}\frac{d}{d\eta}(\eta\tau_t^+) \qquad \frac{\partial\Theta^+}{\partial\eta}=PrR^+[q_t^+-\frac{1}{\eta}(1-\frac{V^+}{V_b^+})] \qquad (1),(2$$

$$0=\frac{1}{\eta}\frac{d}{d\eta}\{C_q\frac{k^+}{\varepsilon^+}\eta(\overline{v^2})^+\frac{dq_t^+}{d\eta}\}-C_q\frac{k^+(\overline{w^2})^+}{\varepsilon^+\eta^2}q_t^+-\frac{\tau_t^+}{V_b^+}-(\overline{v^2})^+\frac{\partial\Theta^+}{\partial\eta}-3.2\frac{\varepsilon^+}{k^+}q_t^+-C(1+\frac{1}{Pr})\frac{\varepsilon^+}{k^+}q_t^+ \qquad (3$$

where the overall momentum balance for the fully developed flow, Eq. (4), has been introduced into Eq. (1). Equation (2) has been obtained by integrating or the energy equation with respect to η under the thermally fully developed state condition, Eq. (5), and accounting for the expression for the total heat flux, Eq. (6).

$$\frac{dP^+}{d\xi}=-2\ , \qquad \frac{\partial\Theta^+}{\partial\xi}=\frac{d\Theta_b^+}{d\xi}=\frac{1}{V_b^+}\ , \qquad q^+=-\frac{1}{PrR^+}\frac{\partial\Theta^+}{\partial\eta}+q_t^+ \qquad (4),(5),(6$$

The constant C_q in the diffusion term of Eq. (3) was set as follows: C_q=0.22 when the model I is employed, C_q=0.25 when the model II is employed and C_q=0.2 when the model III is employed. The function f_μ is given as follows.

$$f_\mu=\exp[-3.4(\frac{50\varepsilon^+}{50\varepsilon^++R^+k^{+2}})^2]$$

Re-optimization of the value of C has been performed in this study because the previously chosen value in reference [2] was obtained ignoring the diffusion term which is accounted for in the present study. The newly optimized value is 0.032. The molecular diffusion term of $\overline{v\theta}$ equation does not appear explicitly in Eq. (3) but has been taken into account when the dissipation term of $\overline{v\theta}$ was modelled [2].

The values of τ_t^+, k^+, $(\overline{v^2})^+$, $(\overline{w^2})^+$ and ε^+ appearing in Eqs. (1)-(3) must be sol simultaneously. A full Reynolds stress model must be employed for this purpose The three models mentioned above are tested. The model by Pope and Whitelaw (Model II) can be used directly because it is presented in the form suitable fo the cylindrical coordinates used in this study. Other two models are formulate with tensor notation for general use. Their forms specifically suitable for the cylindrical coordinates were derived by coordinate transformation but they are not presented here because of the lack of space.

Equations (1) through (3) together with other five equations for the five turbulence quantities mentioned above are solved with the boundary conditions both at η=0 and at the wall-side boundary. At η=0, the following boundary condition are adopted.

$$\frac{dU^+}{d\eta}=0,\ \frac{\partial\Theta^+}{\partial\eta}=0,\ \tau_t^+=0,\ q_t^+=0,\ \frac{d(\overline{u^2})^+}{d\eta}=\frac{d(\overline{v^2})^+}{d\eta}=\frac{d(\overline{w^2})^+}{d\eta}=\frac{d\varepsilon^+}{d\eta}=0$$

Since Eq. (1) is of the form integrated once with respect to η, it automatical satisfies the above boundary condition for Θ^+at η=0. Concerning the wall-side boundary, the following relationships are used at η=1 for the model by Prud'homme and Elgobashi (Model III).

$$U^+=0,\ \Theta^+=0,\ \tau_t^+=0,\ q_t^+=0,\ (\overline{u^2})^+=(\overline{v^2})^+=(\overline{w^2})^+=\tilde{\varepsilon}^+=0$$

where $\tilde{\varepsilon}^+$ is the isotropic part of ε^+ defined by the following equation.

$$\tilde{\varepsilon}^+=\varepsilon-\frac{2}{R^+}(\frac{dk^{+1/2}}{d\eta})^2 \qquad ($$

)ther two models by Launder, Reece and Rodi (Model I) and by Pope and Whitelaw (Model II) are for high Reynolds number turbulent flows and cannot be applied to :he near wall region. Thus, the wall-side boundary conditions are set at the :irst grid point from the wall using the following wall functions.

$$U^+=2.44\ln[1+0.41R^+(1-\eta^*)]+7.8\{1-\exp[-\frac{R^+(1-\eta^*)}{11}]-\frac{R^+(1-\eta^*)}{11}\exp[-\frac{R^+(1-\eta^*)}{3}]\} \quad (8)$$

$$\Theta^+=PrR^+(1-\eta^*)\ , \quad \tau_t^+=-\eta^*-\frac{1}{R^+}(\frac{dU^+}{d\eta}) \quad (9),(10)$$

$$(\overline{u^2})^+=-4.9\tau_t^+,\ (\overline{v^2})^+=-1.0\tau_t^+,\ (\overline{w^2})^+=-2.4\tau_t^+ \quad (11)$$

$$q_t^+=\frac{8.3Pr\tau_t^+[PrR^+\{V_0^+-V^+\}-\eta^*]}{V_0^+\eta^*[2\{C+Pr(C+3.2)\}(dU^+/d\eta)+8.3Pr^2R^+\tau_t^+]}\ , \qquad \varepsilon^+=\tau_t^+(\frac{dU^+}{d\eta}) \quad (12),(13)$$

where η^* is the value of η at the grid point where the above wall functions are applied. Eq. (8) is the approximated form of Reichardt profile [8] at η=1 and Eq. (9) is the solution of energy equation obtained by assuming that the thermal conduction is predominant. Equations (10) and (11) are the wall functions used by Launder, Reece and Rodi [4]. Equations (12) and (13) are the solutions obtained with the local equilibrium assumption [2]. $(dU^+/d\eta)$ in Eqs. (10), (12) and (13) is obtained by differentiating Eq. (8).

inite difference analogues of Eqs. (1) through (3) and other five equations for he turbulence quantities have been solved numerically with an iterative method. he total number of grid points alocated differs from one case to another depend- ng on the type of turbulence model adopted and on the Reynolds number concerned, ut it ranges between 40 at minimum and 55 at maximum. The grid points were llocated non-uniform fashion, being finer in the near wall region. At the start f the computation, the value of R^+ was assumed. The value of Reynolds number, e, was evaluated after the convergence of the numerical computation has been ttained, by making use of the following relationship.

$$Re=4V_0^+R^+ \quad (14)$$

he following convergence criterion was used in this study. At the end of each terative step, the residual of the finite differenced equation of each variable as calculated for every grid cell. The summed-up value of the calculated esidual over all grid cells was compared with the total flow rate of the variable oncerned inside the tube. The computation was terminated when the ratio just entioned fell below 10^{-4}.

he Fanning friction factor f, the Nusselt number Nu and the non-dimensional luid bulk mean temperature Θ_b^+ required in the calculation of Nu were calculated y the following relationships:

$$f=\frac{1}{2V_0^{+2}} \qquad Nu=\frac{2Pr\ R^+}{\Theta_b^+} \qquad \Theta_b^+=\frac{\int_0^1 U^+\Theta^+\eta d\eta}{\int_0^1 U^+\eta d\eta} \quad (15),(16),(17)$$

ne thermal eddy diffusivity, a_t, is not required at all in the present calcula- ion of Nusselt number but, only for the purpose to compare the present result ith available experimental data, its ratio to the fluid diffusivity, a, has been alculated from the following equation:

$$\frac{a_t}{a}=\frac{q_t^+}{q^+-q_t^+} \quad (18)$$

:SULTS AND DISCUSSIONS

.l of the three full Reynolds stress models stemmed out from the same principle of closure, but some differences still exist among them. The model by Prud'hom and Elgobashi (Model III) differs from others in a point that it includes a modification accounting for the low Reynolds number effect. The model by Launder, Reece and Rodi (Model I) differs from the one by Pope and Whitelaw (Model II) in the form of diffusion term. The full Reynolds stress model does not give satisfactory results in some flow situations [7]. Considering this, it may be worthwhile to see how large discrepancy can arise in the final result from the above mentioned differences among the models. Especially, it may be interesting to see how the wall functions used in the two models for high Reync number turbulence are effective. Therefore, in every of the following figures but one exception, all the results obtained with the three tested models are compared with each other.

First, the computed results of hydrodynamic quantities are discussed. Figure 1 compares the computed results of friction factor f with the Blasius' and Prandt formulae [9]. Figures 2 and 3 compare the calculated cross-sectional distributions of averaged velocity and fluctuating intensities ($u'=\sqrt{\overline{u^2}}$, $v'=\sqrt{\overline{v^2}}$, $w'=\sqrt{\overline{w^2}}$) with the logarithmic law of the wall and with the Lawn's data [3]. Figure 4 sh

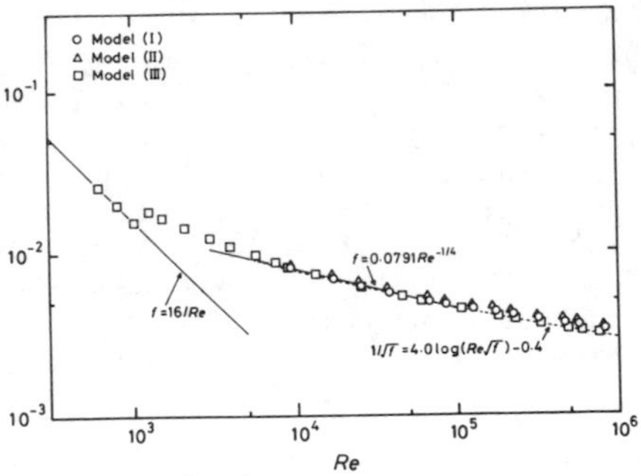

FIGURE 1 Friction factor

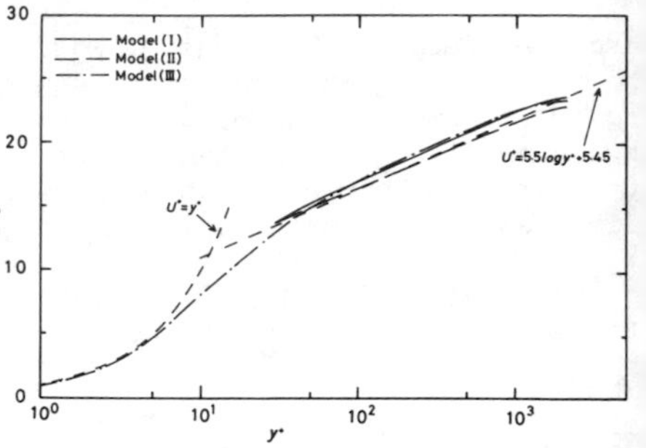

FIGURE 2 Velocity distribution

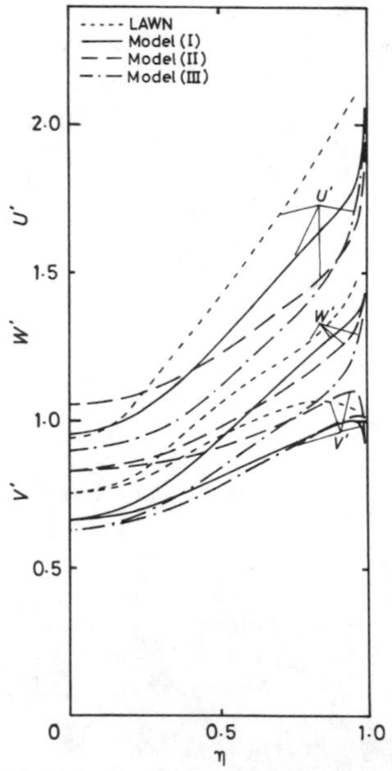

FIGURE 3 Turbulence intensities

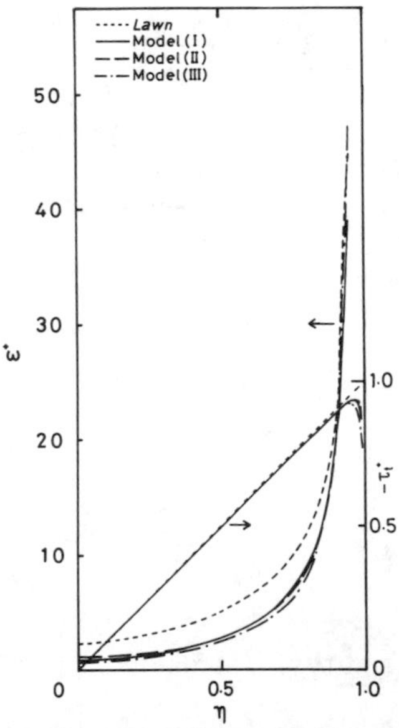

FIGURE 4 Distributions of τ_t^+ and ε

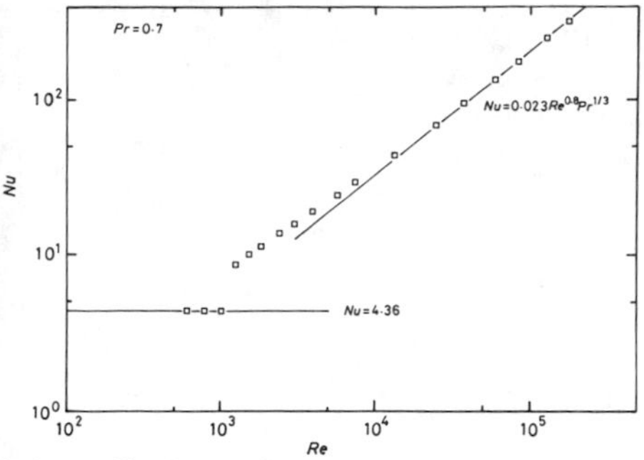

FIGURE 5 Plot of Nu versus Re (Pr=0.7)

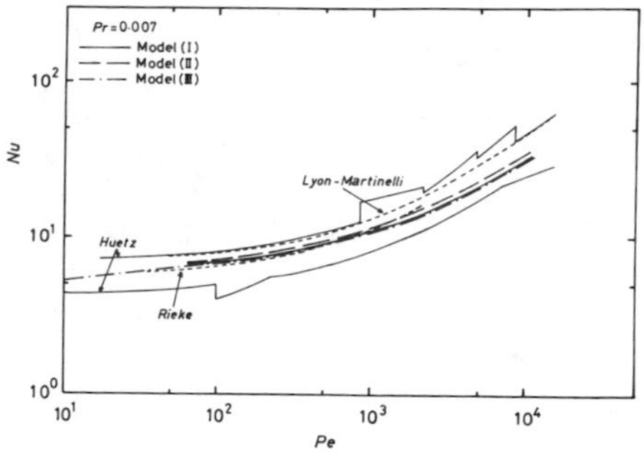

FIGURE 6 Plot of Nu versus Pe (Pr=0.007)

the present results of τ_t^+ and ε^+.

As for the value of f, the model by Prud'homme and Elgobashi (Model III) is superior to other two models. However, this model gives a little higher averaged velocity in the turbulent region as seen in Fig. 2. With this model, the transition Reynolds number is predicted lower and the calculated flow can remain in laminar state only at Re≦1000. Additionally, the calculated transition Reynolds number was observed to depend on the initial level of turbulence assumed at the initial stage of the computation. But it is not relevant to discuss this point because the capability to predict the transition Reynolds number is not embedded in the model so that it is out of the present scope. All the computed results of the three components of intensities show noticeable discrepancy from the experimental counterparts measured by Lawn. Similar discrepancy is also found if the present calculation is compared with Laufer's data [10]. Among the three results presently calculated, the one with the model by Launder, Reece and Rodi (Model I) looks to be closest to the Lawn's data. As for τ_t^+ and ε^+, the tested three models do not differ so much from each other. All the computed results of ε^+ lie below the Lawn's experimental data, but the results with the model of Prud'homme and Elgobashi (Model III) look to be best in a point that it is rather close to the Lawn's data in the near wall region. In conclusion, all the tested turbulence models give different results in every quantities discussed above but the difference is not essentially large. Each of them is better in one quantity but worse in another. Thus, on the whole, it is difficult to say which model is best. Probably, three models may be similar in applicability to the heat transfer calculation.

The computed heat transfer results are discussed in the next. To check the adequacy of the frame of the computation method, the present calculation was applied to the case of the fluid having the Prandtl number of 0.7 (Air). The obtained result of this preliminary study is shown in Fig. 5. In this particular calculation, the model of Prud'homme and Elgobashi (Model III) was used. The computed results show good agreement with the Colburn's empirical formula [11] in the Reynolds number range of $Re \geqq 10^4$. It also agrees with the established value of Nu=4.36 for laminar flow in the Reynolds number range $Re \leqq 10^3$. Too low transition Reynolds number was already discussed. The difference between the present computation and the Colburn's formula found in the Reynolds number range of $10^3 < Re < 10^4$ is related to the inability of the turbulence model of predicting the transitional flow. The flow in this Reynolds number range is sometimes laminar otherwise in intermittent nature having growing turbulent puffs, depending on the level of the disturbance existing in the inlet flow. This affects the heat transfer too but the turbulence model is not sufficient in accounting for such flow behaviour. If the used model is sufficient in this, the transition Reynolds number has not been too low. Additionally, the empirical formula, not only of Colburn but also of others, cannot describe the detailed

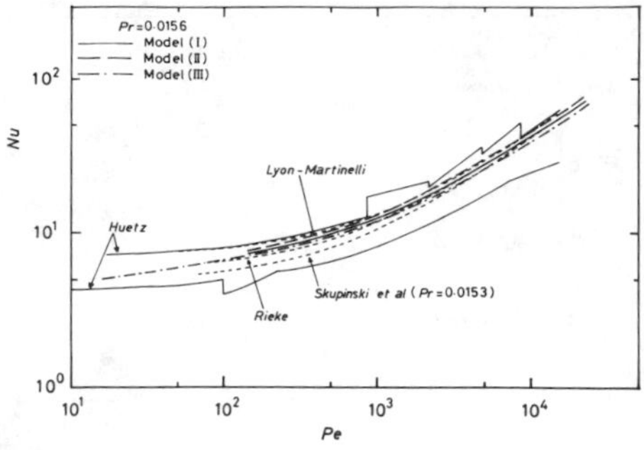

FIGURE 7 Plot of Nu versus Pe (Pr=0.0156)

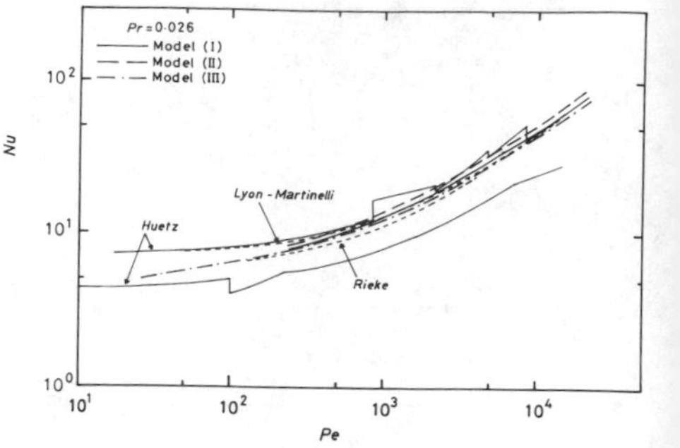

FIGURE 8 Plot of Nu versus Pe (Pr=0.026)

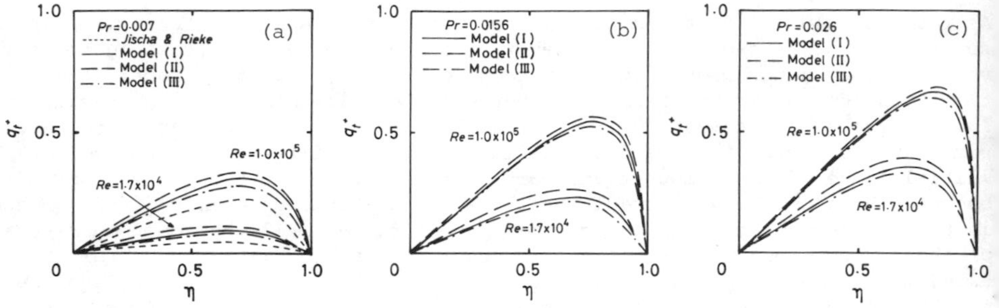

FIGURE 9 Distribution of q_t^+: (a) Pr=0.007, (b) Pr=0.0156, (c) Pr=0.026

dependence of Nu on such flow behaviour too. Therefore, it may be reasonable t conclude that the present method is adequate in principle as far as the turbule model can approximate the actual flow within a reasonable accuracy.

Figures 6 through 8 show the computed results of Nu for liquid metal plotted against the Peclet number, Pe. In every figure, the range of reliable experimental data of Nusselt number approved by Huetz [1] is also shown. No great difference is found among all the results obtained with three turbulence models I, II and III. But the models other than that of Prud'homme and Elgobashi (Model III) result in the Nu value beyond the range of experimental data at hig Peclet number. In this point, the model by Prud'homme and Elgobashi (Model III is a little better than other models, as far as the same value 0.032 is commonl used for C in all the calculation.

Apart from the well-recognized dependency of Nu on Pr through the Peclet number separate, weak dependency of Nu on Pr can be found in the computed results. Th computed value of Nu plotted against Pe decreases slightly with the decrease of Pr. Similar separate dependency of Nu on Pr has also been noticed in the previous study [2]. In this connection, it will be worthwhile to give detailed re-examination to the existing experimental data if such dependency can actuall be found among them.

Figures 9(a), (b) and (c) show the cross-sectional distribution of q_t^+ for two different Reynolds number and for three different Prandtl number. In any case, the difference among the results obtained with the three tested models is not large. For Pr=0.007, the present calculation gives higher value of q_t^+ compare to the result of Jischa and Rieke, which was obtained with the k-kL two-equatio

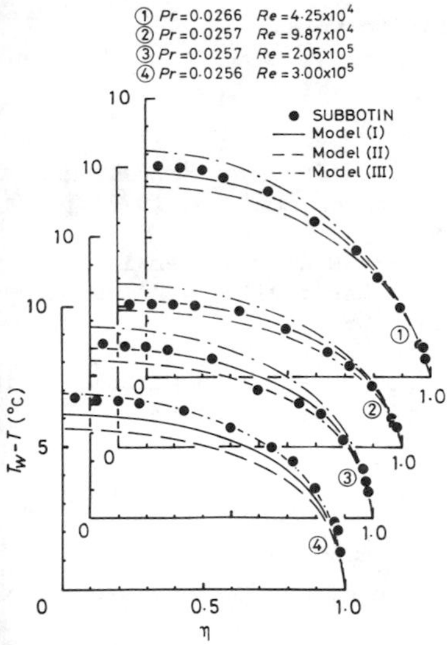

FIGURE 10 Temperature distribution

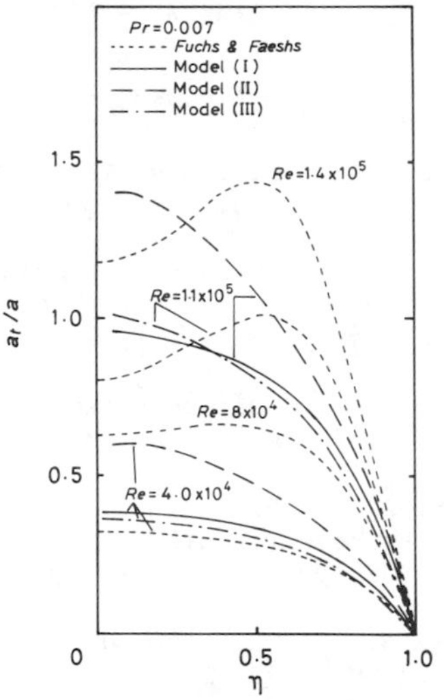

FIGURE 11 Distribution of a_t/a

model of turbulence, therefore, relying on the turbulent Prandtl number concept [12]. In the case of Re=10^5 and Pr=0.026, the value of $q_t{}'$ is computed rather closely to q^+. It indicates that the turbulent heat transport is the major mechanism in this high Peclet number case.

Presently computed cross-sectional distribution of temperature is compared with the experimental data of Subbotin et al [13] for four typical cases in Fig.10. The model by Pope and Whitelaw (Model II) gives a little lower temperature distribution but other results obtained with other two models agree fairly well with the experimental data.

In Fig. 11, the calculated value of a_t/a is compared with the experimental counterparts measured by Fuchs and Faesh [14].
In accordance with the temperature distribution just discussed, the results obtained with the model by Pope and Whitelaw (Model II) show higher value of a_t/a but other two results are comparatively good at low Reynolds number. But, in the core region, the computed results show the distribution different in shape from the experimental ones. This may imply the necessity of further refinement of Eq. (3). However, the discussion on the value of a_t must be done with reservation because the direct measurement of $\overline{v\theta}$ is difficult and scarce. In this connection, accumulation of accurate data of $\overline{v\theta}$ is highly desired.

CONCLUSIONS

The combined set of Eq. (3), momentum and energy equations and the full Reynolds stress model can be applied to the calculation of liquid metal turbulent heat transfer, at least, in the hydrodynamically and thermally fully developed state of the flow in a circular tube at relatively high Reynolds number. The results obtained on Nusselt number and temperature distribution are reasonably good. The full Reynolds stress models for high Reynolds number turbulence solved by making use of wall functions are found to give more or less the same results as the one obtained with the model accounting for the low Reynolds number effect. Minor refinement of Eq. (3) would still be necessary but the accurate experiment on $\overline{v\theta}$ distribution is highly desired for such refining work.

REFERENCES

[1] Huetz, J., Eddy Diffusivities in Liquid Metals, in *Progress in Heat and Ma Transfer*, Vol.7, ed. O. E. Dweyer, pp.3-23, Pergamon, Oxford, 1973.

[2] Suzuki, K., An Approach to Liquid Metal Turbulent Heat Transfer in A Circu Tube Solving $\overline{v\theta}$ Equation with Local Equilibrium Assumption, *Letters in Hea and Mass Transfer*, Vol. 9, No. 4, pp. 245-254, 1982.

[3] Lawn, C. J., The Determination of The Rate of Dissipation in Turbulent Pip Flow, *J. Fluid Mechanics*, Vol. 48, Part 3, pp.477-505, 1981.

[4] Launder, B. E., Reece, G. J. and Rodi, W., Progress in The Development of Reynolds-stress Turbulence Closure, *J. Fluid Mechanics*, Vol. 68, Part 3, pp. 537-566, 1975.

[5] Pope, S. B. and Whitelaw, J. H., The Calculation of Near-wake Flows, *J. Fl Mechanics*, Vol. 73, Part 1, pp.9-32, 1976.

[6] Prud'homme, M. and Elgobashi, S., Prediction of Wall-bounded Turbulent Flo with An Improved Version of A Reynolds-stress Model, *Proc. 4th Symp. on Turbulent Shear Flows*, pp.1.7-1.12, 1983.

[7] Yamada, J., Kawaguchi, Y., Suzuki, K. and Sato, T., Turbulent Boundary Lay Disturbed by A Cylinder - Test of 5-equation Model of Turbulence, *JSME Pre- print*, No. 814-5, pp. 44-47, 1981, (in Japanese)

[8] Reichardt, H., Vollständige Darstellung der Turbulenten Geschwindigkeits- verteilung in Glatten Leitungen, *Z. a. M. M.*, Bd. 31, pp. 208-219, 1951.

[9] Schlichting, H., *Boundary-Layer Theory*, 7th ed., McGraw-Hill, New York, p. 611, 1979.

[10] Laufer, J., The Structure of Turbulence in Fully Developed Pipe Flow, *NACA TR* - 1174, 1951.

[11] Colburn, A. P., A Method of Correlating Forced Convection Heat Transfer Da and A Comparison with Fluid Friction, *Trans. AIChE*, Vol. 29, pp. 174-210, 1933.

[12] Jischa, M. and Rieke, H. B., Modelling Assumptions for Turbulent Heat Transfer, in *Heat Transfer 1982*, ed. U. Grigull et al., Vol. 3, pp. 257-26 Hemisphere, Washington, D.C., 1982.

[13] Subbotin, B. I., Ibragimov, M. X and Nomofilov, E. B., Measurement of Temp erature Field in Turbulent Mercury Flows in A Tube, *Teploenergetika*, Vol. pp. 70-74, 1963. (in Russian)

[14] Fuchs, H. and Faesh, S., Measurement of Eddy Conductivity in Sodium, *Progress in Heat and Mass Transfer*, Vol. 7, ed. O. E. Dweyer, pp. 39-43, Pergamon, Oxford, 1973.

Laminar Heat Transfer and Flowfield Downstream of Backward-Facing Steps

FU-KANG TSOU and AMICHAI BARON
Department of Mechanical Engineering and Mechanics
Drexel University
Philadelphia, Pennsylvania 19104, USA

WIN AUNG
Division of Chemical, Biochemical and Thermal Engineering
National Science Foundation
Washington, D.C. 20550, USA

INTRODUCTION

In recent years, studies of separated flows and heat transfer has been of interest since the development in high performance heat exchanger and in the cooling of electronic equipment requires more quantitative information on heat transfer rates in separated flow regions. This paper deals with a numerical computation of separated flows and heat transfer associated with a plane flow past a backward-facing step. It is considered that the pressure field of the main flow is constant and the whole flow field is laminar, i.e., the flow approaching the step, prevailing in the recirculating region, reattaching to the wall and redeveloping into a boundary-layer type are all laminar. Furthermore, the plate upstream of the step is assumed to have a finite length so that the boundary-layer thickness at the step becomes a parameter that must be taken into account.

Details of the computational results of the above-mentioned problem are being reported elsewhere [1]. It is found that the reattachment distance is related to the system and flow parameters in a rather complicated way. The initial shear layer thickness, for example, can have opposing effects on the reattachment distance, depending on whether the velocity is held constant. The velocity and temperature profiles have been compared using the solutions for wall independent shear layer as well as data obtained from interferometric studies [2]. In this paper, supplementary information on the heat transfer results are given.

MATHEMATICAL TECHNIQUES

Differential Equations

Consider an incompressible, two-dimensional laminar flow. The conservation equations of the elliptic type which govern the stream function (ψ), the vorticity (ω), and the temperature (T) in Cartesian coordinates can be written as [3],

$$a\left[\frac{\partial}{\partial x}\left(\phi\,\frac{\partial \psi}{\partial y}\right) - \frac{\partial}{\partial y}\left(\phi\,\frac{\partial \psi}{\partial x}\right)\right] - \frac{\partial}{\partial x}\left[b\,\frac{\partial (c\phi)}{\partial x}\right] - \frac{\partial}{\partial y}\left[b\,\frac{\partial (c\phi)}{\partial y}\right] + S = 0 \qquad (1)$$

The views expressed in this article are the private opinions of the authors.

where ϕ stands for any one of the unknowns ψ, ω and T. The functions a, b, c and S, are given in Table 1.

Table 1. The functions a, b, c and S.

ϕ	a	b	c	S
ω	1	1	μ	0
ψ	0	$1/\rho$	1	$-\omega$
T	1	$\rho\alpha$	1	0

The definitions of ψ and ω are,

$$u = \frac{1}{\rho}\ \frac{\partial\psi}{\partial y}\ ,\ v = -\frac{1}{\rho}\frac{\partial\psi}{\partial x}\ . \tag{2}$$

$$\omega = \frac{\partial v}{\partial x} - \frac{\partial u}{\partial y}\ . \tag{3}$$

Eq. (1) through Eq. (3) and the boundary conditions described in the next paragraph are applied to the plane flow past a backward facing step (OABCD) shown in Fig. 1. The whole flow field is considered laminar and the grid distribution of the computational region (ABCDEF) is also indicated in the figure.

Boundary Conditions

It is assumed that heating starts at the location with the distance x_o from the leading edge (Fig. 1). The wall temperature (T_w) is kept constant for $x \geq x_o$. The sides of the computational region are chosen as,

$$AB = 10\ s,$$

$$CD = 40\ s,$$

$$DE = 22\ s,$$

where s is the step height. The region is sufficiently large to give computational results that are size independent.

On the west boundary AF, the Blasius velocity and temperature profiles [4] are prescribed. The approaching length upstream of the step shown in Fig. 1 is chosen to conform with the experimental condition in [2]: i.e., x_s = 45.5 cm, and x_o = 15.0 cm in most computational cases. However in some other cases, x_s is arbitrarily changed in order to obtain different values of δ/s at the step, for a fixed value of Re_s (= Us/ν). Since this approaching length is sufficiently long, one would expect that the streamlines above the surface AB (Fig. 1) will be bent toward it resulting in a thinner boundary layer at the upper corner (point B). The present method thus provides for the streamline curvature effect noted in [2], wherein the effect has been observed to give a heat transfer augmentation at the step of as much as 30% when compared with attached flow.

Along the north boundary FE, constant free stream velocity and temperature are prescribed. The east boundary DE is situated far away from the step. The property gradients along the streamwise direction are thus assumed to be negligible (i.e. $\partial/\partial x = 0$). On the solid boundary ABCD, the non-slip condition and onstant wall temperature apply. To obtain vorticity on the boundary, it is ssumed that, in the region near the wall: (1) the property gradients along he wall is zero; and (2) the velocity distribution normal to the wall is inear which leads to a linear variation of vorticity in this region. Integra-ion of the continuity equation,

$$= -\frac{1}{\rho}\nabla^2\psi, \tag{4}$$

ives [3],

$$_w = -\frac{3\psi_1}{(\Delta y)^2\rho} - 0.5\,\omega_1, \tag{5}$$

here the subscript "1" refers to the node next to the wall and Δy is the istance between this node and the wall node. The wall vorticity expressed in his equation is iterative since ω_1 varies in each iteration. Because of high radients of variables near the wall, very fine grid is utilized in the near all regions as is described in the paragraph to follow.

UMERICAL PROCEDURES

he system of governing differential equations with the boundary conditions is olved using the finite difference scheme that involves the iterative solution lgorithm described in [3] and [5]. In this scheme, the methods of upwind ifference and central difference are applied, respectively, to the convective erm and the diffusion term. Test of various non-uniform grid system is made nd a choice of the final grid distribution, 31 x 30, showing grid independency s adopted. The grids as shown in Fig. 1 are finer near the corner and the sur-ace. The finest grid, of dimension 0.025s, is located adjacent to the wall, nd the sizes of other grids are chosen such that each is within 150% of the ext grid [3] in order to avoid abrupt changes and to obtain convergence.

n the present algorithm, the Gauss-Seidel iteration scheme is applied for ob-aining convergence of each variable. The variable ϕ is said to be convergent f its residual is smaller than a pre-assigned value. The residual is defined s,

$$= \left|(\phi^n - \phi^{n-1})\Big/\phi^n\right|_{max} \tag{6}$$

here n refers to the n-th iteration. In the above expression, the residual β epresents the maximum value throughout the computational region. The pre-ssigned value is 10^{-3} for all cases except those for the smallest step (s = 0.38 m) for which a value of 10^{-4} is used.

esults

he results for the region ABCDEF in Fig. 1 presented in this section are for he step sizes s = 3.8 mm, 6.35 mm and 12.7 mm. The plate length upstream of he step is x_s = 455 mm unless otherwise stated. Experimental data from Ref. 2 s utilized for comparison with the results.

he contour plots of the dimensionless stream function, vorticity and temper-ture distribution for Re_s = 233 and s = 12.7 mm are shown in Figs. 2a, 2b, and c, respectively. They are representative of the results from the present study. n Fig. 2a, the streamlines are seen to be densely packed in the shear layer wing to a rapid velocity change. The streamline (ψ = 0) that divides the re-irculating region with the rest of the flow field is not shown, but the re-ttachment point (L) is indicated. The center of circulation is situated closer to the step than to the reattachment point. The streamlines outside of the re-circulating region are nearly parallel to each other.

Fig. 2b indicates that near the upper corner of the step strong vorticities are generated and swept downstream into the recirculation region. The vorticity changes sign in the region when a closed streamline is followed. The curve wit zero vorticity passes through the point of reattachmen (L). It is interesting note that an increase in Reynolds Number (Re_s) will lead to an increase in the vortex strength (not shown in the figure) for a fixed step size. This is becau an increase in the Reynolds number designates an increase in the velocity which in turn means a thinner initial shear layer thickness. A thinner initial shear layer thickness gives rise to higher vorticity values in the recirculation region. Larger vorticities lead to vortices of higher strengths and therefore to higher heat transfer. Hence, heat transfer increases with the Reynolds number.

Isotherms are plotted in Fig. 2c. In the proximity of the lower corner of the step the isotherms are further apart indicating lower heat transfer values. Away from the step in the region near the wall, the isotherms are parallel to i and are densely packed resulting in higher heat transfer values. Thus, heat transfer first increases in the streamwise direction and then tends to level of in the far downstream. The Stanton number results in Fig. 4 indicate such a behavior.

Reattachment Distance

The location of the reattachment point refers to the intersection of the dividing streamline and the downstream wall (Fig. 2a), the reattachment distance shown in the figure being 7.5 times the step height. Existing literature indicates that the reattachment distance is not only dependent on Reynolds number but related to the system and other flow parameters. Detailed treatment of the topic is given elsewhere [1]. Here the influence of the initial shear layer thickness on the reattachment distance will be assessed from our computational results. The reattachment distance effects the heat transfer in an indirect wa

Two methods are used to study the influence of the initial shear layer thicknes First, at a given step size, a Reynolds number (Re_s) is chosen and held fixed (i.e., the freestream velocity is fixed). Then by varying the hypothetical fla plate length upstream of the line AF in Fig. 1, velocity profiles with varying boundary-layer thicknesses at the step (the initial shear layer thickness) are specified on AF. The computational results using this method is shown in curve A and B (Fig. 3). It is seen that the reattachment distance (L) increases fairly fast as the displacement thickness (δ^*) at the step increases. Thus, an increase of the initial shear layer thickness elongates the streamwise extent of the separation bubble.

The second method utilizes a fixed length of the upstream plate (x_s) which is equal to 45.5 cm. The Reynolds number (i.e. the velocity) is varied to obtain the solution of the flow field. The results are shown in curves C and D (Fig. It is observed that the reattachment distance (L) decreases rather rapidly as t displacemnt thickness (δ^*) increases. Fig. 3 shows trends that are by and larg opposite, and suggest the need to exercise care while considering the influence of the upstream boundary layer. Clearly, changes in the boundary layer thickness cannot be divorced at all times from free stream velocity changes, and the effects of these two parameters are sometimes inter-mingled.

Heat Transfer Results

The computational results expressed in the form of local Stanton number are plotted in Fig. 4 for the cases of three Reynolds numbers (Re_s), 63, 110, and 132. The local heat transfer for each case grows rapidly from the corner of the step, eventually levelling off and then decreases. This situation is expected from the temperature distribution given in Fig. 2(c). At large distances from the step, the streamwise variation of the Stanton number appears to assume the trend of a hypothetical flat plate (shown in dash lines) obtained by setting the step size to zero. Comparison of the results with experimental data reported in [2] appears satisfactory except for the case of $Re_s = 110$ where the maximum difference may amount to 10%.

In all the above-mentioned representative cases, the trends for streamwise variation remain the same and the average heat transfer with the step is reduced from the flat plate value. A careful examination of the figure indicates that the amount of the reduction in the recirculation region depends on Reynolds number, being largest for $Re_s = 132$ and smallest for $Re_s = 63$. Since the Reynolds number is related to the boundary-layer thickness at the step or the initial shear layer thickness, the average Stanton number over the recirculating zone can be correlated with these two parameters. The correlation obtained from [2] indicates a strong dependence of the average Stanton number on the initial layer thickness.

An attempt is made to plot the heat transfer data in [2] and the present numerical results for all three step sizes using the functional relationship,

$$Nu/Re_s^{0.55} = f(\frac{\xi}{k}). \quad (7)$$

Such a plot is shown in Fig. 5 where both the experimental data and the computational results are nearly collapsed into a single curve. The correlation is better for the case of the smallest step (s = 3.8 mm). When the large and medium step sizes are considered, the spread among the curves and the data becomes more severe. Fig. 5, however, is an improved version of the plot obtained using the square root of Reynolds number ($\sqrt{Re_s}$) to replace the denominator $Re_s^{0.55}$ in the left hand side of Eq. (7). It is thus felt that a more complicated functional relation than that given in Eq. (7) has to be utilized.

Finally, the maximum heat transfer coefficient expressed in terms of the Nusselt number $(Nu)_{max}$ is found to vary linearly with the Reynolds number raised to the 0.55 power for a given slot. As shown in Fig. 6, the slopes of these curves are the same. The maximum Nusselt number for a fixed Reynold number is seen to be higher for larger steps. The following relation is obtained to represent the numberical results:

$$Nu_{max} = e^{b}\, Re_s^{0.55} \quad (8)$$

with $b = 0.66s - 4.05$

where the Reynolds number raised to the power of 0.55 is consistent with Eq. (7), s being measured in centimeters. This is a rather cumbersome expression, but is one that represents well the results of the present computation.

CONCLUSION

The stream function/vorticity method has been applied to obtain the numerical results of flow and heat transfer for laminar flow past a backward facing step. The method has the advantages that it gives directly the streamline distribution in the recirculation region and the vorticity transport for the entire flow field. The stream lines tend to bend at the step resulting in a thinner initial shear layer that in turn produces a stronger vorticity and hence a higher heat transfer in the region.

The influence of the initial shear layer thickness on the reattachment distance has been shown to depend on how this thickness is varied. For a fixed step size, the reattachment distance increases with the thickness for a fixed Reynolds number (Re_s); it decreases with the thickness for an upstream plate of fixed length.

The Stanton number results compare satisfactorily with available data. It starts from zero at the corner of the step and increases asymptotically to the flat plate value in the downstream direction. At any given streamwise position percentage reduction of the heat transfer from the flat plate value for a given step size increases with the Reynolds number. The maximum Stanton number for a given step size is found to be proportional to the Reynolds number raised to the 0.55 power.

REFERENCES

1. Aung, W., Baron, A. and F. K. Tsou, "Wall Independency and Effect of Initial Shear Layer Thickness in Separated Flow and Heat Transfer," to be published in IJHMT.

2. Aung, W., "An Experimental Study of Laminar Heat Transfer Downstream of Backsteps," J. Heat Transfer, v. 105, Nov. 1983 pp. 823-829.

3. Gosman, A.D., Pan, N. M., Runchal, A.K., Spalding, D.B. and Wolfstein, M., Heat and Mass Transfer in Recirculating Flows, Academic Press, 1969.

4. Eckert, E.R.G. and Drake, R.M., Analysis of Heat and Mass Transfer, McGraw-Hill, 1972.

5. Hwang, F. and Tsou, F.K., "Friction and Heat Transfer in Laminar Free Swirling Flow in Pipes," ASME Gas Turbine Heat Transfer, 1978. pp. 71-78.

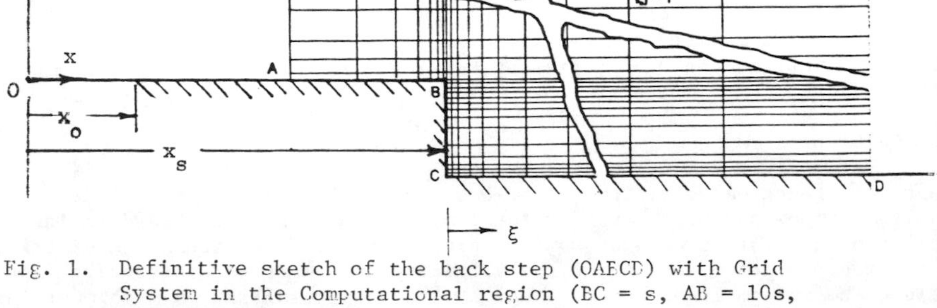

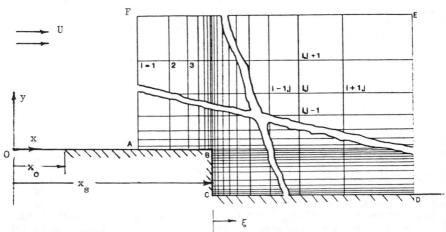

Fig. 1. Definitive sketch of the back step (OABCD) with Grid System in the Computational region (BC = s, AB = 10s, FE = 50s, ED = 22s.

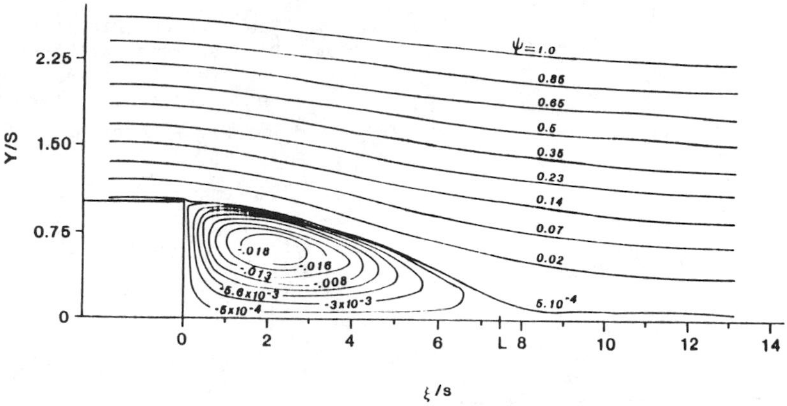

Fig. 2a. Stream Line Distribution for s = 12.7mm and Re_s = 233.

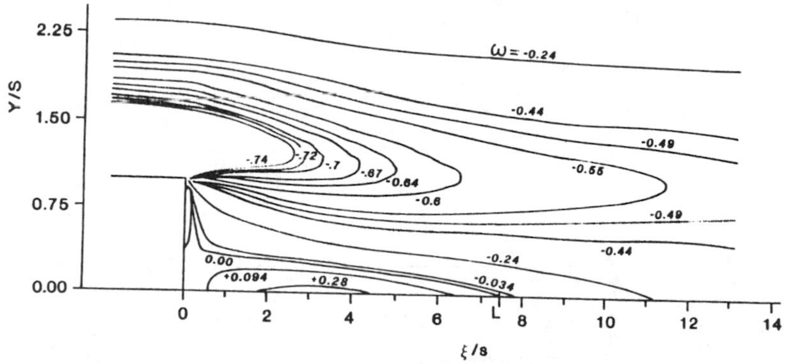

Fig. 2b. Vorticity Distribution for s = 12.7mm and Re_s = 233.

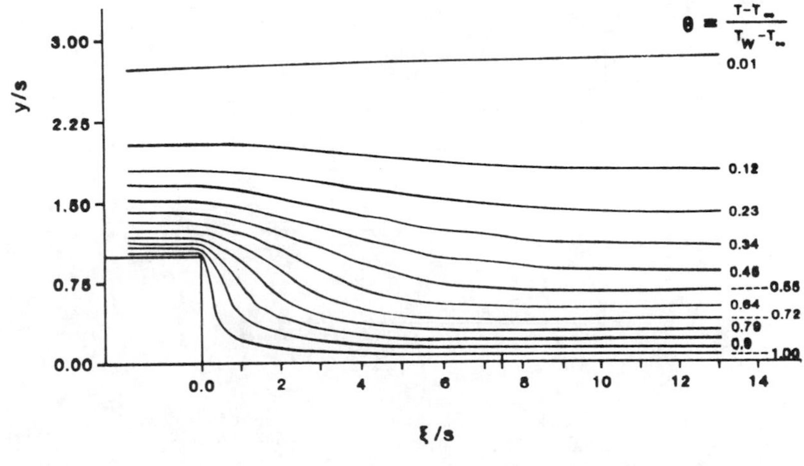

Fig. 2c. Temperature Distribution for s = 12.7mm and Re_s = 233.

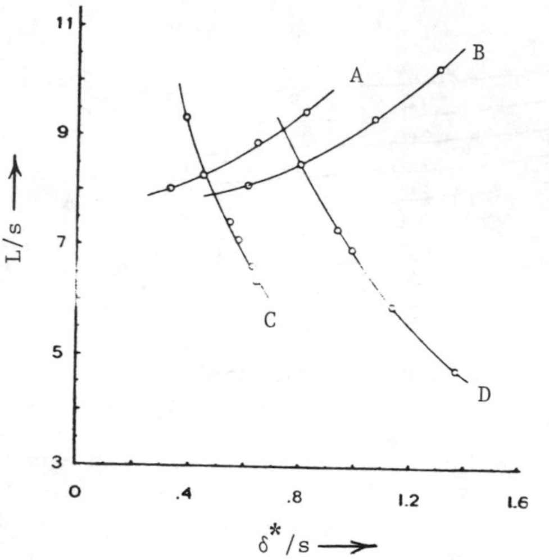

Fig. 3. Reattachment Distance Versu Displacement Thickness at the step (Curves A and B ~ Re_s = constant; curves C an D ~ $Re_s \neq$ constant).

A s = 12.7 mm, Re_s = 315.

B s = 6.35 mm, Re_s = 272.

C s = 12.7 mm } x_s = 0.455m.

D s = 6.35 mm }

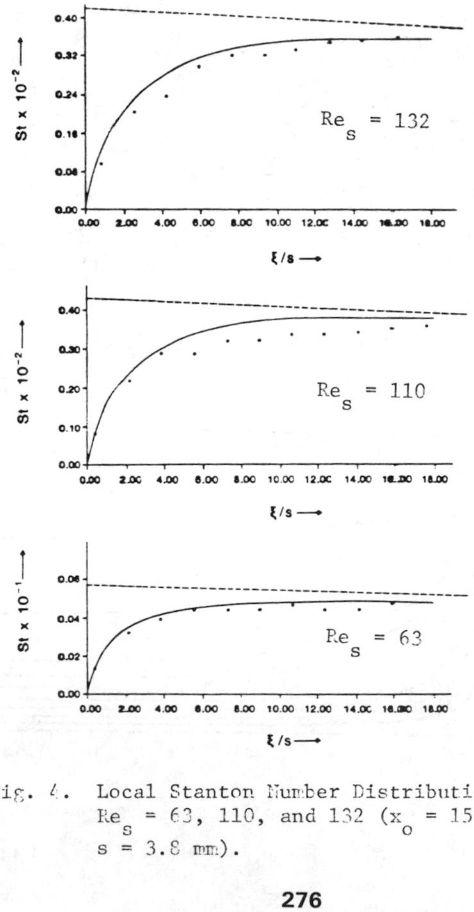

Fig. 4. Local Stanton Number Distribution for Re_s = 63, 110, and 132 (x_o = 150 mm, s = 3.8 mm).

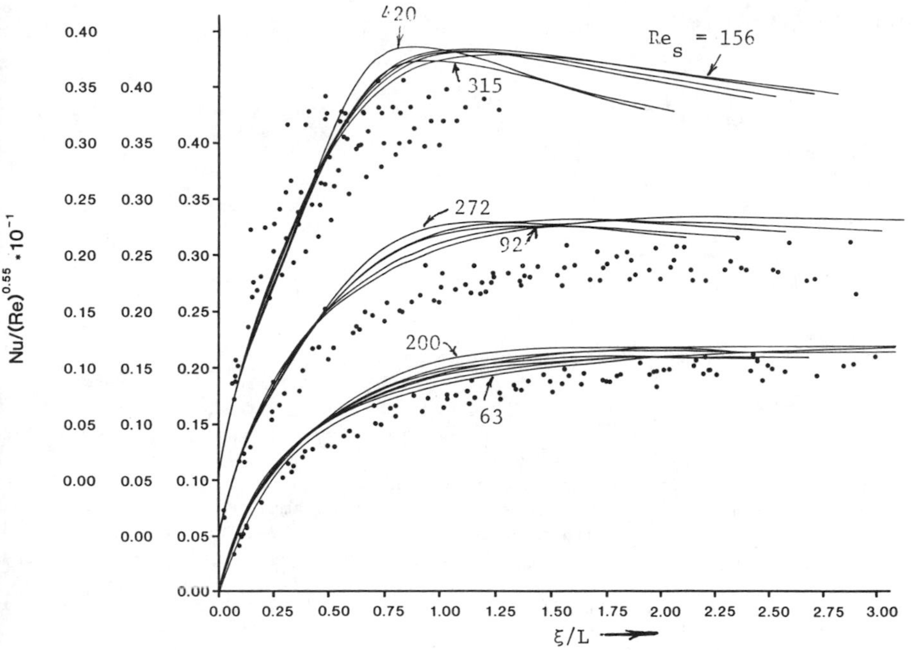

Fig. 5. Comparison of the Computed Nu/Re_s^{055} With The Experimental Values.

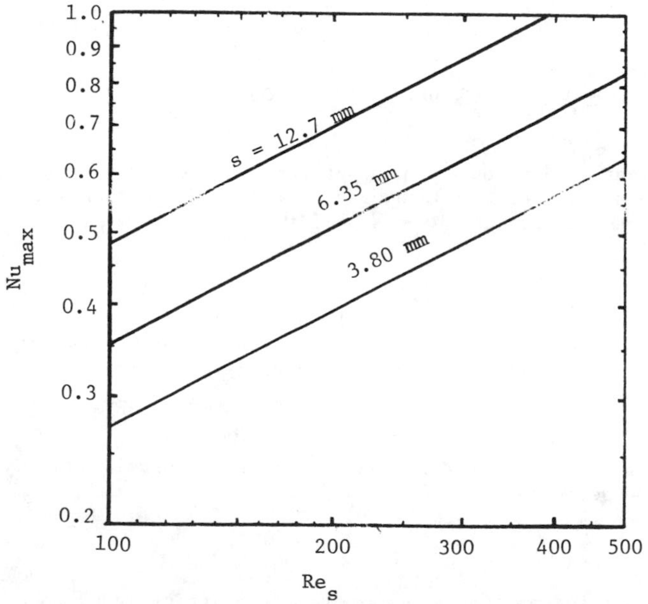

Fig. 6. Maximum Nusselt Number (Nu_{max}) Versus Reynolds Number (Re_s).

Convection Heat Transfer from a Horizontal Rotating Cylinder

XIMIN ZHANG and WEIYI LI
Tianjin University
Tianjin, PRC

ABSTRACT

There are some topics raised by engineering design about the calculation of heat transfer from a horizontal rotating cylinder to ambient air. These problems are usually with the higher values of Grashof number ($Gr \sim 10^8$) and the lower values of ratio $Re_r^2/Gr(\sim 10)$. This paper presents the results of experimental investigation.

1. The following empirical equation about the relationship of critical value of rotating Reynolds number $Re_{r,cri}$ with Grashof number for air was obtained:

$$Re_{r,cri} = 2.6 Gr^{0.456}$$

It can be used in a larger range of Gr from 10^4 to 10^9.

2. The experimental results of heat transfer at higher Gr number ($10^8 \sim 10^9$) and lower ratio, Re_r^2/Gr can be represented by the following equation:

$$Nu = 0.53\,[(0.0018 Re_r^{2.66} + Gr)\,Pr]^{0.25}$$

3. The investigators developed a new approach for measuring the rotating cylinder's surface temperature. Advantages of this new approach are as follows: simple structure, easy operation, high stability and accuracy.

NOMENCLATURE

The following nomenclature is used in this paper:
Nu-Nusselt number; D-diameter of the test cylinder, m; Gr-Grashof number; ν- Kinematic viscosity, m^2/s; Δt-Temperature difference = $t_w - t_f$, ^oC; t_w -Surface temperature of cylinder, ^oC; t_f-Temperature of ambient air, ^oC; Pr-Prandtl number.

Note: All physical properties in the experimental correlations have been evaluated at the temperature $t = (t_w + t_f)/2$.

INTRODUCTION

Convection heat transfer from a horizontal rotating cylinder to ambient air is a combination of free convection and revolving movement. Many experimental relations have been presented by some investigators [1]-[5], but their test apparatus usually consisted of small diameter rotating cylinder approximately 100 mm and the rotational speed was more than 10^4 rpm. so the test were carried out in lower Gr number and higher ratio of Re_r^2/Gr. The main purpose of this investigation was to obtain data on heat transfer of higher Gr number and lower ratio, Re_r^2/Gr to satisfy the requirement for calculating the heat transfer of a horizontal rotating cylinder with larger diameter. The following experimental range was covered in this paper:

Rotating Reynolds number Re_r: $4x10^3 - 5x10^4$; Grashof number: $2.7x10^8 - 9.8x10^8$; Nusselt number: 70-200; Cylinder surface temperature: 50-150°C; Rotational speed: 6-180 rpm.

EXPERIMENTAL EQUIPMENT

The schematic diagram of the test apparatus is shown in Fig.1. The rotating cylinder, 900mm in length and 500mm OD was made of steel sheet, 2mm in thickness. The mean deviation of OD is 0.4mm and the maximum deviation is +2mm.

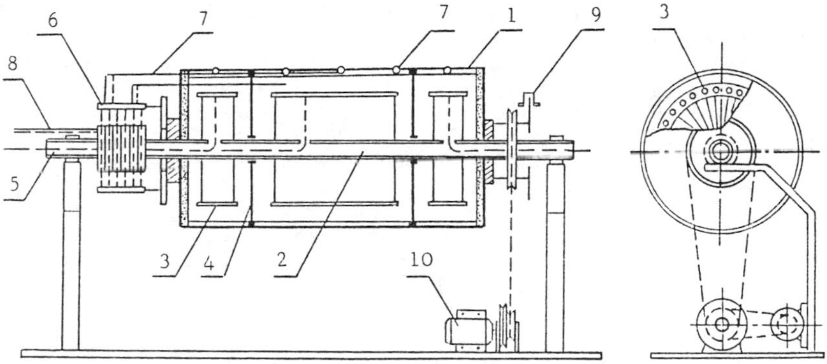

1. Cylinder, 2. shaft, 3. heaters, 4. sealing plate, 5. power mains,

6. slip rings assembly, 7. thermocouples, 8. Compensation lead-wire,

9. Speed counter, 10. Electric motor.

Fig.1 The schematic diagram of the test apparatus

The Cylinder casing consists of two layers in order to increase the strength of the cylinder and improve the uniformity of heat flow, the distance between the two layers is 18mm. On the surface of the inside layer there are 20 rectangular holes for improving the air convection. The cylinder cavity was divided into three parts. The middle part, 500mm, in which the main electric heater had been installed was the experimental section. The both ends are 200mm, each fixed with supplemental heaters for compensation the heat loss. The main and supplemental heaters are separated completely by a sealing plate. During operation, AC power was supplied to the heaters from three variable transformers, and the power mains were introduced through the center of shaft to heaters which are motionless. The cylinder was rotated by an adjustable-speed electric motor, the rotational speed can be adjusted, the minimum speed is 6 rpm. (The circumferential velocity is 0.16 m/s).

The emissivity of the cylinder surface was carefully determined by AGA 780 Thermovision and Normal emissivity meter, and found to be equal to 0.22.

MEASUREMENT

For each test, the surface temperature of the cylinder, the temperature of the ambient air, the rotational speed and the power input to the main heater were measured. The surface temperatures were measured by five 0.1mm Nickel-chrome Nickel-Aluminium thermocouples. These thermocouples were inserted in the small grooves milled on the surface. In order to estimate the temperatures of the main heater and the auxiliary heaters, another two thermocouples were directly installed in the cavity of cylinder.

The technique of measuring the voltage of the thermocouples located on a rotating body played the most important role in this experiment. A new approac was proposed for measuring a rotating body's temperature called the compensatio slip ring in which simple structure and easy operation are characterized. This system, as schematically indicated in Fig. 2, consist of many copper slip rings and sliping wire made by thermocouple compensation lead-wire around the slip rings. All thermocouples were connected to a potentiometer with an accuracy of class 0.1 to indicate the temperature. It was discovered that even at high rot tional speed the pointer of galvanometer fluctuates only 0.01mv. The repeatabi ty of experimental data led to a conclusion that measuring system was reliable, the accuracy are also satisfactory, so it can be used to mearsured the temperature of rotating bodies with low or middle rotaional speed.

Input power was measured by a class 0.2 mutual indictor and a class 0.2 power meter. For each series of runs, the electric power input to the main heater was steady. It was possible to maintain the same temperature inside the main heater and the auxiliary heaters by means of adjusting the power input to the auxiliary heaters for minimizing the end heat loss.

All data were taken in steady state conditions. It was assumed that steady state was attained when two consecutive readings taken at half an hour intervals gave almost the same results.

In order to provide a check on the adequacy of the experimental apparatus, a series of tests were run with the cylinder stationary, so that the results could be checked against available data on free convection from a horizontal cylinder. These results of tests were presented in Fig.5. The values of Nu number measured in natural heat transfer tests were in close agreement with equation of McAdams [6].

$$Nu = 0.53\ (Gr.\ Pr)^{0.25}$$

the average deviation was about 4%. It was concluded that the nonrotating test results convincingly demonstrated the adequacy of the test system. so the equation of McAdams was used as basis for processing the test data.

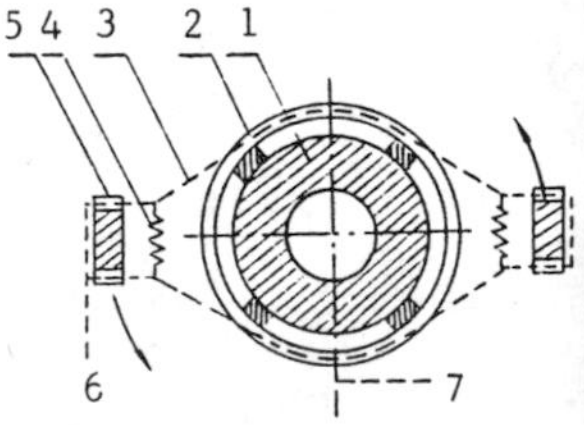

1. Shaft, 2. Slip ring, 3. Compensation lead-wire, 4. Spring, 5. Support, 6. & 7. Connection with potentiometer.

Fig.2 The slip rings assembly

In this research, fifty-two sets of experimental data were obtained.

THE CRITICAL REYNOLDS NUMBER

The important character of rotational cylinder heat transfer is that the lusselt number is independent of Re_r number up to a critical value, beyond which lu number increase with Re_r. It could be explained that the rotation leads to augmentation of the free convection velocity on the ascending side of the cyl-nder and to diminish it on the descending side, the heat transfer would thus e expected to be higher on the ascending side and lower on the descending side, s compared with the heat transfer when there is no rotation.

The critical number would be a function of Gr number and Pr number.

$$Re_{r,cri} = f(Gr,Pr) \qquad (1)$$

In order to investigate the critical Reynolds number three series of tests ere carried out within the Gr region 10^8. The variation of Nu with the Re_r was epresented in Fig.3. It showed that the coefficient of heat transfer would eep constant up to a critical value, but before the critical point the Nu are lightly higher than the values predicted by equation of free convection. Beyond he critical value Nu would increase with the 2/3 power of Re_r. The variation f $Re_{r,cri}$ with the Gr was plotted in Fig. 4, it may be expressed as follows:

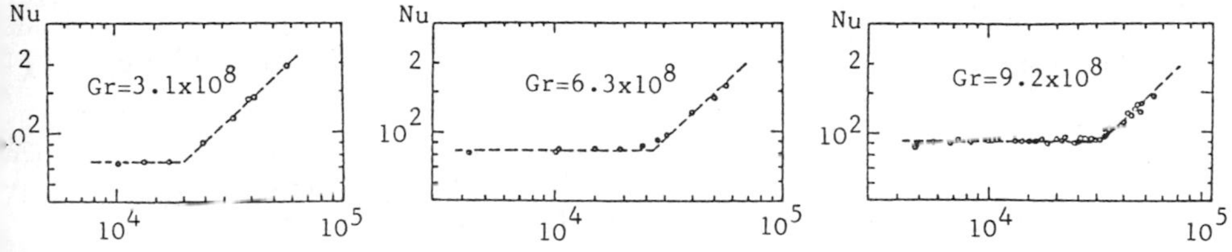

Fig.3 The variation of Nu with Re_r

$$e_{r,cri} = 2.6Gr^{0.456} \qquad (2a)$$

n consideration of the in-luence of properties the data an be correlated within 2% y the following equation:

$$e_{r,cri} = 3.05(Gr.Pr)^{0.456} \qquad (2b)$$

t was discovered that the re-ults given by the Eq. (2) are n close agreement with those f Kays and Anderson, 1], [4] the average deviation s lower than 5%. To compare ith Eq. (2b) the analytical q. (3) recommended by Anderson as also plotted in Fig. 4.

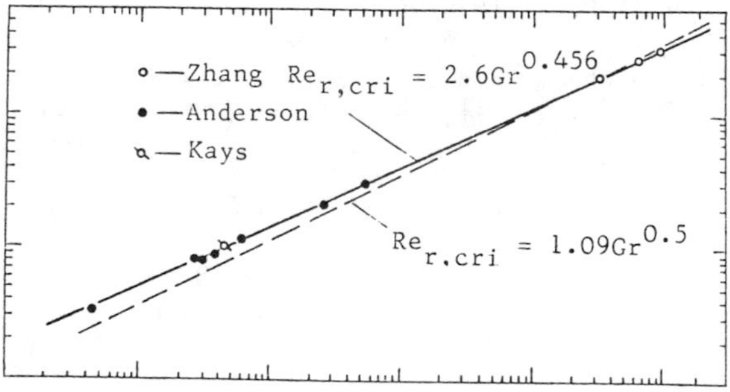

$$e_{r,cri} = 1.09Gr^{0.5} \qquad (3)$$

Fig.4 The variation $Re_{r,cri}$ with Gr

ig.4 illustrated that the data in this paper were very in agreement with Eq.(3) or higher Gr number, the average deviation was only 4%. However, Anderson ointed out that the theoretical values predicted by Eq. (3) were lower than the experimental ones within the region of lower Gr number. Table 1 listed the experimental data including those from reference [1].

Table 1

	Anderson[1]				Zhang		
Gr	4.8×10^4	3.2×10^5	2.7×10^6	5.4×10^6	3.1×10^8	6.3×10^8	9.2×10^8
Exp.	320	780	2100	3000	2×10^4	2.7×10^4	3.1×10^4
Eq. (3)	240	625	1800	2540	19200	27400	33000

The above analysis indicated that Eq. (2) which has been strongly supposted by experimental results can be recommended for practical use in a wide Gr range from 10^4 to 10^8.

According to the numerical solution of a velocity profile of free convectio laminar sublayer with constant surface temperature we might find the maximum velocity, U_{max}, of the sublayer at the side of the horizontal cylinder, [7], and furthermore we can obtain a circumferential linear velocity W_{cri} of the cylinder surface on the point of $Re_{r,cri}$ from the Eq. (2), thus, the ratio of W_{cri}/U_{max} could be calculated. In the Gr range from 10^4 to 10^9, its values would equal to 1.6–1.1 approximately. In other words, the linear circumference velocity of cylinder surface has been already beyond the maximum velocity of free convection sublayer when the Re_r equals to the $Re_{r,cri}$.

This phenomenon can be explained as follows: The rotation influences the heat transfer of a rotating cylinder only when the circumference velocity of air caused by rotation approximately equals to the maximum velocity U_{max} of free convection boundary layer. But the U_{max} is located at one third of the thickness of the boundary layer, the rotating flow of air is resulted from the transmission of momentum. So for producing a circumferential velocity of air located in the boundary layer to equal the velocity U_{max}, the circumferential velocity of cylinder W_{cri} must be larger than the maximum velocity of free convection boundary layer, U_{max}. The effect is decreased by increasing of Gr, because the Gr and the kinematic viscosity are increased with t. Therefore, the W_{cri}/U_{max} will decrease as Gr increases.

THE RELATIONSHIPS OF HEAT TRANSFER

At the condition of lower ratio, Re_r^2/Gr, the heat transfer of a horizontal rotating cylinder will be influenced by the natural convection and the forced rotational flow, its relation would be written in following function,

$$Nu = f(Gr, Re_r, Pr) \qquad (4)$$

when $Re_r > Re_{r,cri}$ Eq. (4) can be written in the form,

$$Nu = c[(a\,Re_r^b + Gr).Pr]^n \qquad (5)$$

Since our experimental conditions were still located in laminar flow regior the constants c & n of Eq. (5) were taken from the relationship of laminar free convection with constant surface temperature, which was recommended by McAdams, i. e. $c=0.53$, $n=0.25$, and the constants a & b must be determinated experimentally. Finally, we obtained $a=0.0018$, $b=2.66$. The data could be correlated within 7% by the following equation as shown in Fig.5.

$$Nu = 0.53\,[(0.0018\,Re_r^{2.66} + Gr).Pr]^{0.25} \tag{6a}$$

Fig. 5 also included the data of Kays [4], the deviation was less than 7%. Etmad [2] and Dropkin [3] also provided some correlations about the heat transfer of rotating cylinder as follows

$$Nu = 0.11\,[0.5Re_r^2 + Gr).Pr]^{0.35} \tag{7a}$$

$$Nu = 0.095[(0.5Re_r^2 + Gr).Pr]^{0.35} \tag{8a}$$

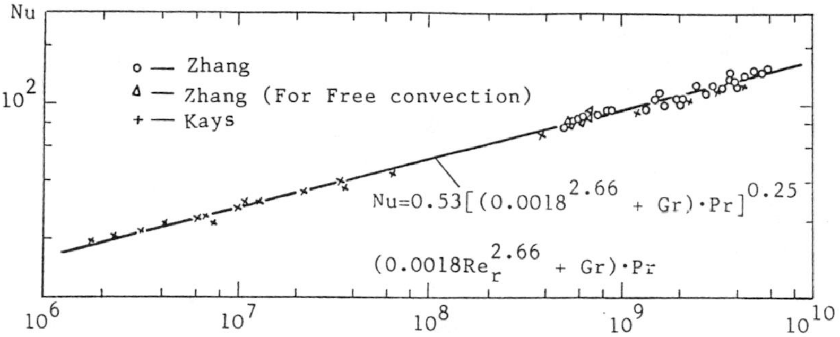

Fig.5 Nu versus $Pr.(0.0018Re_r^{2.66} + Gr)$ For Re_r above critical value

These equations compared with Eq. (6a) in Fig.6, in which there are two eries curves computed from two Gr numbers separately. for lower Gr number the alues of Eq. (6a) were between those from Eq. (7a) and Eq. (8a). But for igher Gr number results of Eq. (6a) will be lower than those of Eq. (7a) and 8a), these facts may be explained from two respects.

The influence of Re_r : In so far as concerns the influence of Re_r, the Gr an be omitted and equations can be reduced to the simple forms as follows:

$$Nu = 0.1Re_r^{0.666} \tag{6b}$$

$$Nu = 0.076Re_r^{0.7} \tag{7b}$$

$$Nu = 0.073Re_r^{0.7} \tag{8b}$$

Calculations indicated that in a larger region of Re_r from 10^3 to 10^5 the alues of Nu are in close agreement with deviation less than 9%. Therefore the nfluence of Re_r on the three equations mentioned above are vasically the same. t should be pointed out that the present investigation is in remarkably good greement with the results of Andenson [1] about the influence of Re_r.

The influence of Gr: Fig.6 the main difference between Eq. (6) and Eq. (7), 8) was in the region of lower ratio Re_r^2/Gr. From the analysis of rotating cylinder heat transfer we knew that when the ratio Re_r^2/Gr is low, the effect of free convection will be a dominative factor, when Re_r number approximately equals to $Re_{r,cri}$, the effect of Re_r might be neglected, as a consequence, the Eq. (6), (7), (8) will be reduced to following forms:

$$Nu = 0.53\ (Gr.Pr)^{0.25} \tag{6c}$$

$$Nu = 0.11\ (Gr.Pr)^{0.35} \tag{7c}$$

$$Nu = 0.095\ (Gr.Pr)^{0.35} \tag{8c}$$

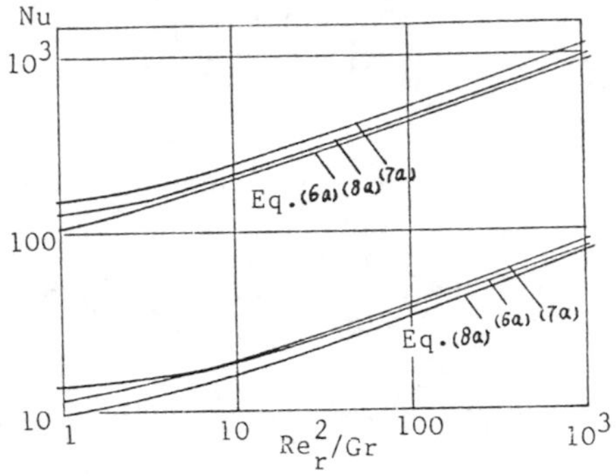

Fig.6 Comparision Eq.(6a) with Eq.(7a) & (8a)

The theory and practice have proved that the laminar free convection heat transfer is proportional to one-fourth power of (Gr.Pr) number. The Eq. (6) shows this characteristic, but the values of Nu obtained from Eq. (7c) (8c) give an overestimation in high Gr and an underestimation in low Gr. The condition of Eq. (7a)and (8a) were based on high Re_r^2/Gr, and hance they are not able to cover the rule of heat transfer on the lower ratio, Re_r^2/Gr and higher Gr number.

In view of the above mentioned analysis we recommended Eq. (6a) for calculaing the heat transfer of horizontal cylinder in the higher Gr and lower Re_r^2/Gr.

REFERENCES

[1] J.T. Anderson, O.A. Saunders, Convection from an isolated heated horizontal cylinder rotating about its axis, Proc. Roy. Soc. vol. 217, 1953. pp. 555-562.

[2] G.A. Etmad, Buffalo, N.Y., Free-convection heat transfer from a rotating horizontal cylinder to ambient air with interferometic study of flow, Tran ASME, vol. 77. 1955, pp. 1283-1289.

[3] D. Dropkin, Arieh Carmi, Natural-Convection heat transfer from a horizonta cylinder rotating in air, Trans. ASME vol. 73, 1951, pp. 741-749.

[4] W.M. Kays, I.S. Bjorklund, Heat transfer from a rotating cylinder with and without crossflow, Trans. ASME, No. 1, 1958, pp. 70-78.

[5] P.D. Richardson, Transition on a heated horizontal rotating, Trans. ASME, No. 8, 1961, pp. 386-387.

[6] W.H. McAdams, Heat transmmission, 3d-ed, McGram-Hill, 1954.

[7] J.H. Merkin, Free Convection boundary layer on an isothermal horizontal cylinder, ASME-AIchE Heat Transfer Conference. S.T. Louis. M., Aug. 9-11, 1976.

TWO–PHASE FLOWS AND VISUALIZATION

Experience with Two-Phase Flow Measurement Techniques

BERNHARD BRAND, VOLKER KEFER, HANS LIEBERT, and RAFAEL MANDL
Kraftwerk Union AG
Erlangen, FRG

ABSTRACT

In many cases the measurement of characteristic two-phase flow quantities is of great technical interest. In this paper different methods which were investigated and applied at KWU test facilities are described, namely

- full flow pipe instrumentation
- local velocity and void fraction measurements and
- liquid level detection within large volumes.

Measurement principles, accuracies and results are discussed and recommendations concerning the applicability of two-phase flow measurement techniques are given.

INTRODUCTION

In many technical applications and processes the measurement of two-phase flow quantities like density, local and average velocity, phase distribution or the overall mass flow rate is required. Especially in the field of

- determination of flow conditions to calculate the heat transfer in heated tubes and bundles
- performance of mass and energy balances for conventional and nuclear power generation systems
- analysis of thermohydraulic conditions during transients like Loss-of-Coolant-Accidents investigated experimentally for nuclear power plants and
- investigation of two-phase flows in various flow distribution systems or pipe networks

the authors of this paper are involved in design, testing and application of two-phase flow instruments. The following overview outlines the experience gained with some selected measurement techniques, explaning briefly the measurement principles and giving some up-to date results.

FULL-FLOW PIPE INSTRUMENTATION

The so-called "instrumented spool piece" (Figure 1) is a system designed to measure both single- and two-phase mass flow rates in pipes.

It mainly consists of a turbine, a drag disc and a three-beam gamma-densitometer

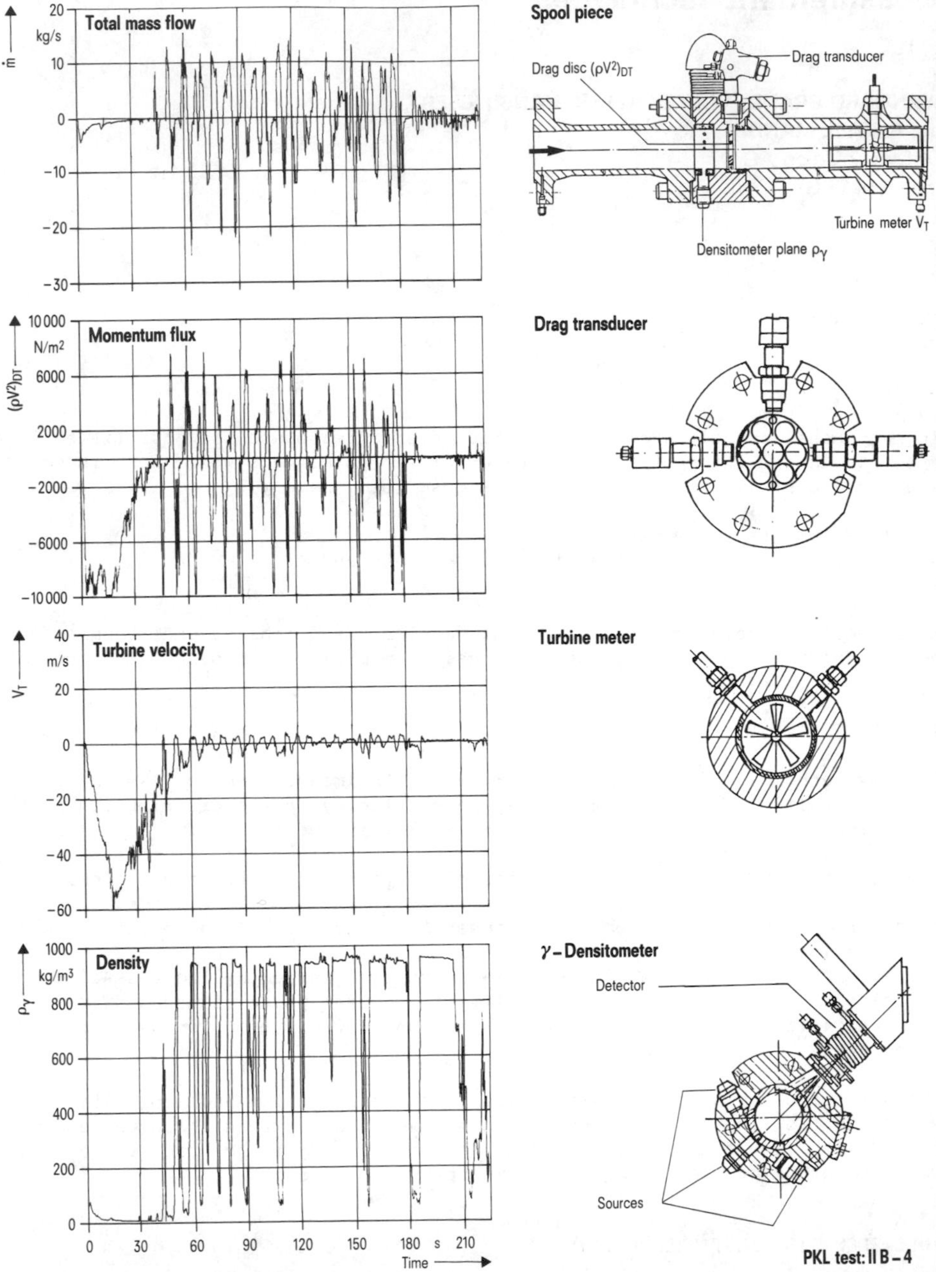

FIGURE 1. Arrangement of measuring devices within the "instrumented spool piece" and some measurement results

to determine velocity, momentum flux and density of the flow. Additionally it is instrumented with absolute and differential pressure transducers as well as thermocouples. The directly measured quantities, as mentioned above, can be combined to obtain void fraction, slip ratio, steam and water mass fraction as well as the overall mass flow rate.

Four of these systems are installed in the PKL test facility - a model (volume scale 1 : 145) designed to simulate Loss-of-Coolant Accidents in pressurized water reactors[1]. The results exemplarily shown in Figure 1 are taken from a PKL End-of-Blowdown experiment (hot leg break, combined injection), where in this case the signals of turbine, drag disc and gamma-densitometer are used to obtain the total two-phase mass flow rate passing through the pipe.

In general the results obtained under steady state conditions show good accuracy at high ($\dot{x} > 0.8$) and low ($\dot{x} < 0.05$) steam qualities, where the error is smaller than 5 %. However in extreme cases of transient flow or at qualities of about $\dot{x} \approx 0.3$ the error can increase up to 35 % and 50 % in forward and reverse flow directions respectively.

Since drag bodies and turbine meters are relatively expensive, alternative instruments like Venturi nozzles and Averaging Pitot Meters (APM) were tested in two-phase flow under steady state conditions covering a wide range of parameters. The evaluation of the signals of both instruments, Figure 2, is based on the "homogeneous model" which assumes equal velocities of the water and steam phase and homogeneous phase distribution.

---

[1] Brand, B.,Mandl, R., Watzinger, H., Safety Investigations of KWU ECC-Systems, Loca Thermohydraulic Phenomena Simulated in the PKL-Facility, International Symposium on Heat Transfer, 15 - 18 October, 1985, Beijing, China

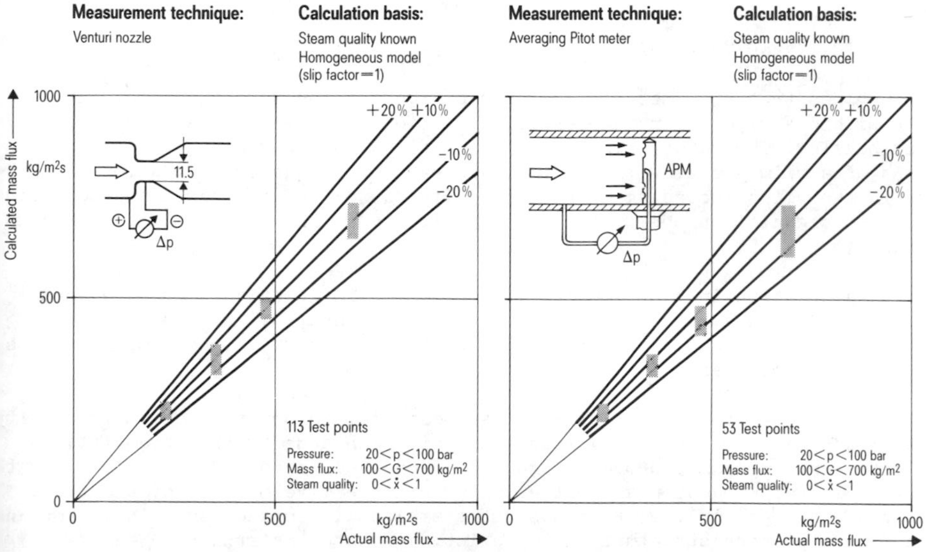

'IGURE 2. Application of Venturi nozzle and averaging Pitot meter in two-phase 'low. Plot of actual mass flux versus calculation from measurement.

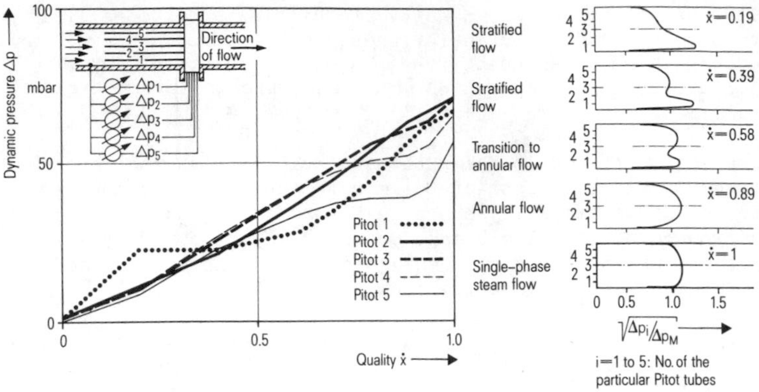

FIGURE 3. Flow pattern identification with Pitot tubes distributed over the pipe cross section.

The experiments resulted in good agreement between actual and calculated mass fluxes based on the Venturi or APM signals. Thus these simple techniques can also be recommended for application in two-phase flow, particularly in the range tested. It should be mentioned that the actual mass flux in these experiments was obtained by a single-phase orifice plate measurement, taken before the water was heated up and evaporated.

Additional tests with five single Pitot tubes distributed over the pipe cross section as illustrated in Figure 3, showed that flow pattern can be detected with a simple measurement technique as well.

LOCAL VELOCITY AND VOID FRACTION MEASUREMENT

Mini turbines (rotor diameter 11 mm) are designed to measure local velocities either within a pipe or in inaccessible locations such as fuel assembly top or bottom nozzles. The rotor, see Figure 4, is fixed in a capsule and located at the end of a slim stalk. Within this stalk carrier frequency modulation takes place in two coils according to the speed (frequency) of the passing rotor blad[e]. The output signal is proportional to the velocity of the medium passing through the turbine.

In-situ calibration in single-phase flow resulted in relatively small errors ($<$ 3 % of full scale). The turbines have fast response (0.1 seconds) and a good reliability. Figure 4 shows typical signals of local velocity measurements in a PKL reflooding experiment.

For the determination of mass flow rates additional information on fluid densit[y] is required. Besides using Δ p-cells to obtain mean density from the static head different impedance probes, Figure 5, were used in the PKL test facility t[o] detect local void fraction as well as water and steam velocities of the flow. Their basic design includes two sensors assembled within a ceramic insulator an[d] mounted on a support structure, e. g. existing reactor internals.

The void fraction is obtained by analysing magnitude and phase of the sensor impedance. A comparison between void fractions determined from impedance probes

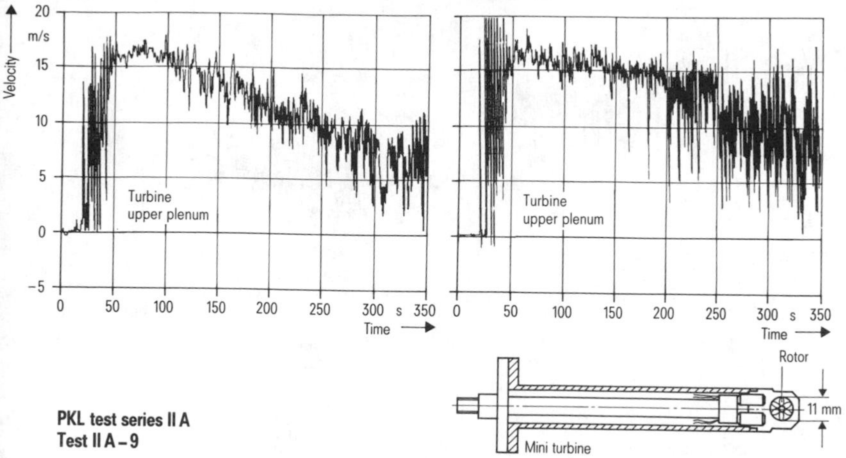

FIGURE 4. Local velocity measurement with mini turbines

and Δ p-measurements shows, Figure 6, that due to water bridging effects large differences can occur. Since, in addition, several sensors failed as a result of leakages in the ceramic-to-metal connection caused by strong thermal shocks, this method can be recommended only for certain applications.

Steam and water velocities can also be determined with impedance probes by cross correlating the impedance magnitude of two adjacent sensors. Results of this signal analysis are exemplarily shown and compared with turbine measurements in Figure 7. Caused by the data qualification procedure used to assess the impedance probe signals, only few acceptable values were obtained. These show certain scattering around the turbine signals.

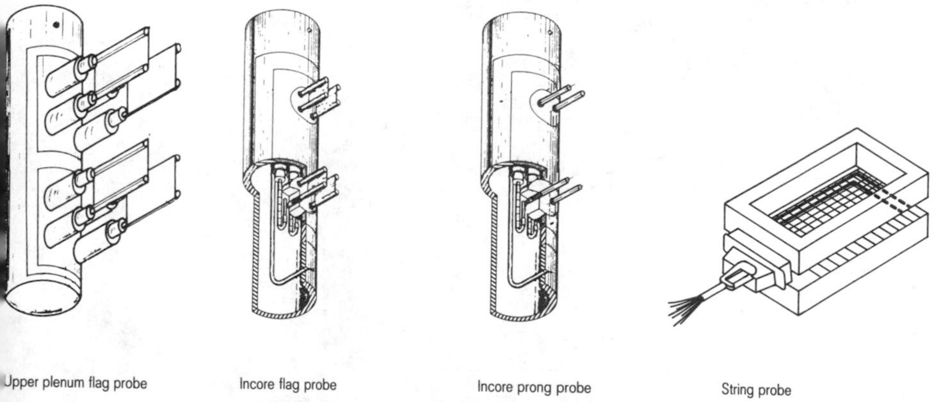

FIGURE 5. Different types of impedance probes for local velocity and void fraction measurements

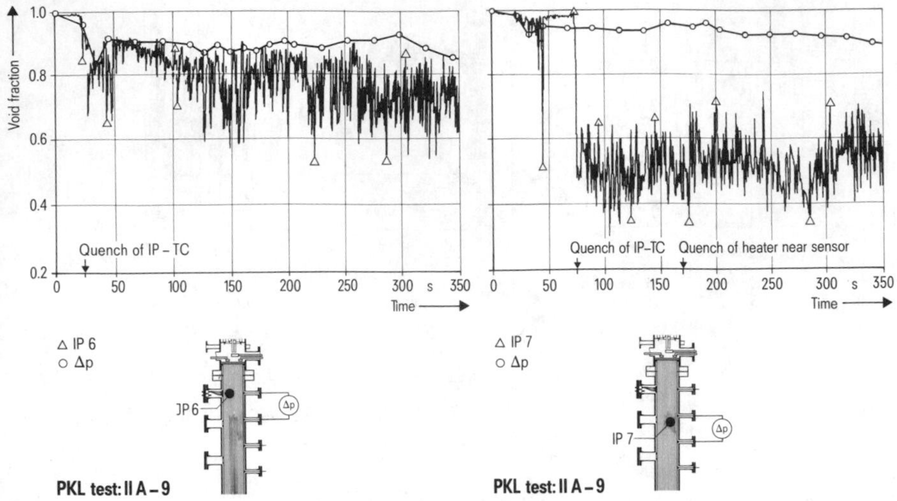

FIGURE 6. Comparison of void fractions determined with impedance probes and geodetic head (Δ p) measurements

LIQUID LEVEL DETECTION WITHIN LARGE VOLUMES

Very often inventory and distribution of liquids in large volumes are of great technical interest. Especially in case of LOCA experiments the distribution of emergency core cooling water in downcomer, core and upper plenum of the reactor pressure vessel has to be detected. This can be done by using a number of sensors of a conductivity liquid level detector (CLLD) system. Each of these sensors indicates, Figure 8, by measuring local conductivity, the presence of steam (low conductivity) or water (high conductivity).

Considering the extrem conditions (thermal shocks, pressure gradients) the hardware tested proved to be rather reliable, but the interpretation of the

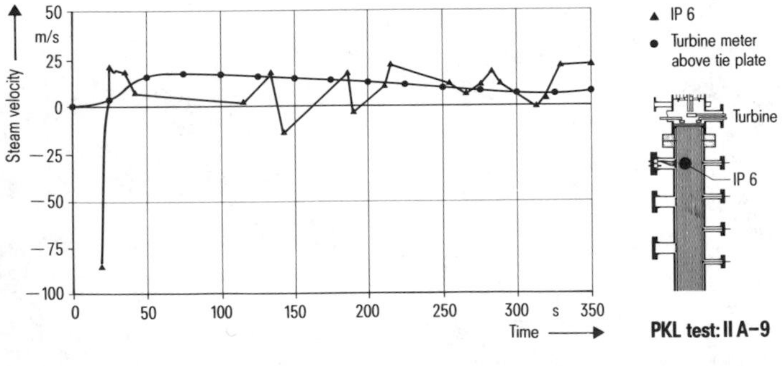

FIGURE 7. Comparison of velocities measured by impedance probe and turbine

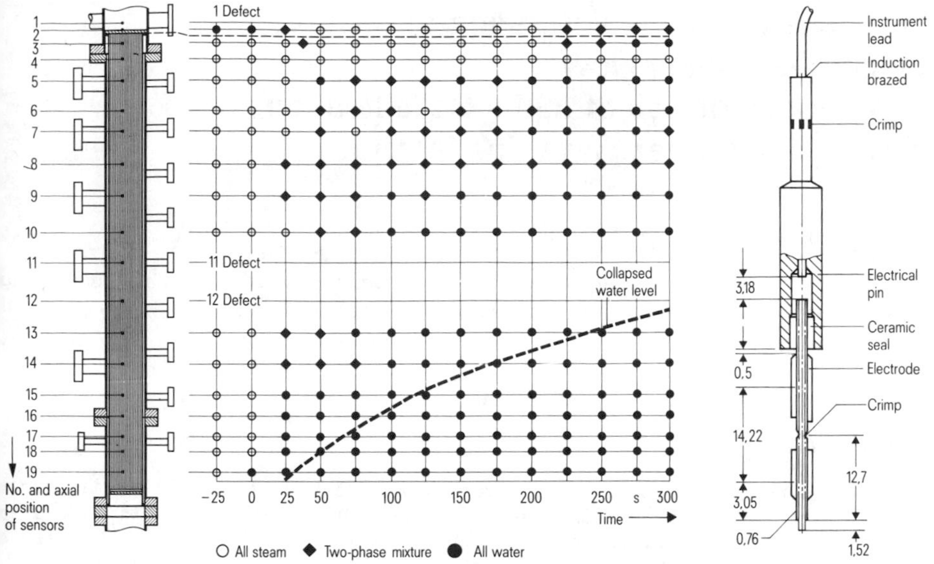

FIGURE 8. Detection of water distribution in a large volume using a conductivity liquid level detector system

results was made difficult by external interference like heat radiation or deposition of thin liquid films on the electrodes. In flooding experiments this system seems to overpredict water inventory, as illustrated in Figure 8.

Better results may be obtained by using special optical probes which consist of a 45-degree sapphire cone metallized into a housing. Continuous light generated by a high intensity lamp enters an input fiber and is reflected at the probe tip surface according to the ambient conditions (water, steam, resp.). These probes will be tested in the near future.

CONCLUSIONS

The requirement to determine two-phase flow quantities in many technical applications and processes led to the development and testing of various two-phase flow measurement techniques at KWU.

These investigations showed that for every technical application appropriate instruments must be selected and possibly adapted. Before choosing expensive two-phase flow instrumentation the testing of prototypes or less complicated and thus cheaper methods under similar conditions was found to be worthwhile.

It is imperative not to regard two-phase flow instrumentation as a "black box". Fundamental knowledge of the two-phase flow itself and the instrument capabilities is required for correct interpretation of measurement results.

ACKNOWLEDGMENT

The authors of this paper wish to express their thanks to USNRC which supported the experiments by supplying spool pieces, mini turbines and impedance probes for prototype testing in the PKL test facility. This was performed within the international 2D/3D contract.

Characteristics of Ice Formation in a Curved Channel Containing Flows Part 1: Unsteadiness of Initial Solidification and Heat Transfer at Water-Ice Interface

K. ICHIMIYA
Department of Mechanical Engineering
Yamanashi University
Kofu, Japan

R. SHIMOMURA
Department of Mechanical Engineering
Fukui Institute of Technology
Fukui, Japan

INTRODUCTION

Freezing problems in a channel containing flows relate practically to the block of water pipes and to the solidification of chemical process lines in cold area While these types of problems have been analyzed and experimented in the case c the straight channel [1∿6], they have not been studied in the curved channel since complicated and perplexing physical phenomena occur in the freezing proce due to curvature. For this reason, there is need to clarify experimentally the freezing situation in the curved channel containing flows.

At the first stage we tried to observe how the ice layer, which developed in th straight entrance section, behaved along the curved rectangular channel. The transient variation in the thickness of the ice layer and the heat transfer at water-ice interface were examined in the curved section.

EXPERIMENTAL FACILITY AND PROCEDURE

Experimental facility

The outline of the experimental facility is illustrated in Figure 1. Experime were performed with the apparatus in which water was used as the freezing medi The equipment consists of water and coolant circulation systems. Water flows from the constant temperature ba (reservoir) ① to the mixing chamber ⑥ via orifice ④ for meter. It runs through the stra entrance region, then the curved test section to the exit region back to the constant temperature bath. Flow rate is controlled b valve ③ . The 180° curved tes section is connected to the stra entrance region, and both are he horizontally. This section is c structed of rectangular channel, whose cross section is 40 (heigh x 25 (space) mm, with a curved radius 150 mm. The concave and convex walls (b=40 mm) are coole at a temperature through thin cc

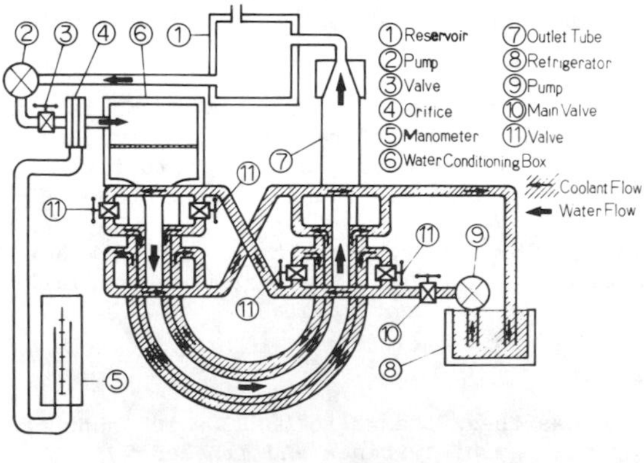

FIGURE 1. Experimental facility

TABLE 1. Experimental conditions

Cooling surface temperature °C	-5 , -10 , -15							
Water entrance temperature °C	2							
Water velocity cm/sec	1.2	2.3	3.5	4.7	6.1	8.9	11.9	18.2
Reynolds number	200	380	600	790	1020	1490	1990	3040
Dean number	64	122	192	253	326	477	637	973

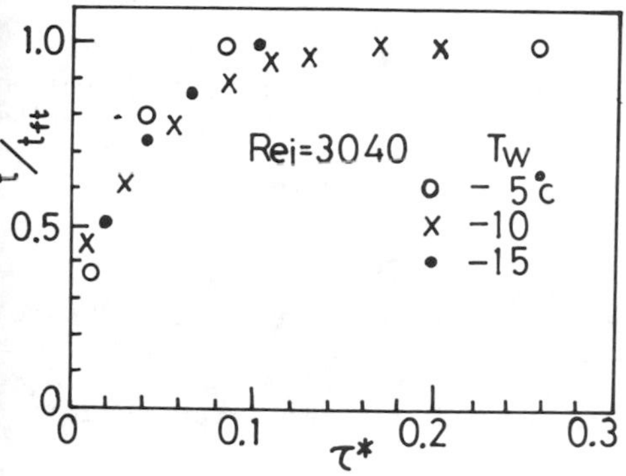

IGURE 2. Ice thickness at the entrance f curved section

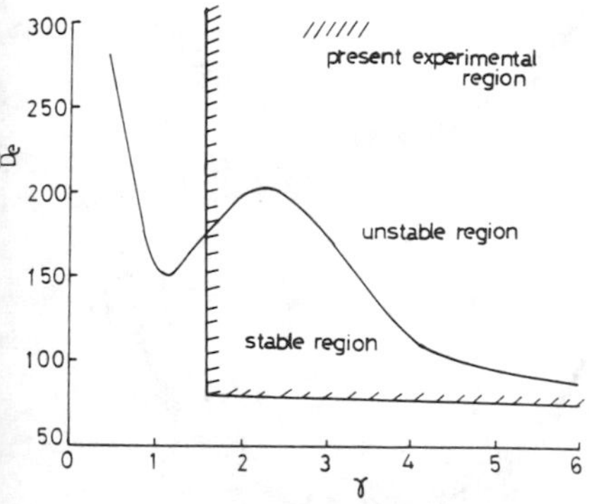

IGURE 3. Critical Dean number

plates, and the insulated upper and lower parts are made of transparent acrylic plate (a=25 mm). The coolant system is represented by the shaded portion in Figure 1. The counterflow configurations of the coolant are used to produce a uniform wall temperature along the flow direction. Copper-constantan thermocouples are set on the back surface of cooled walls.

Experimental procedure

After the flow rate and the water temperature were set at a given value, the coolant (ethylene-glycohol) circulation was adjusted, thus lowering the wall temperature below $0^{o}c$. The starting point of the experiment was determined when the ice began to generate at the cooling surface. The freezing situations were observed by measuring the thickness of the ice layer with a metric scale and by a photograph taken through the upper and lower transparent walls. The conditions adopted in the experiment are listed in Table 1.

Figure 2 represents the ice thickness at the entrance of the curved section with the cooled wall temperature Tw being the parameter. The ordinate is the dimensionless thickness t/t_{ft} and the abscissa the dimensionless time τ^* which is the product of Fourier number and Stefan number. The t_{ft} is the thickness of ice layer after full expiration. The thickness changes similarly for various cooled wall temperatures.

EXPERIMENTAL RESULTS AND DISCUSSIONS

Flow situation in the curved channel

n the curved channel, the centrifugal force-radial pressure gradient imbalance cting on the slow moving fluid near the side walls of the channel induces a otion of the fluid along the side walls and directed from the outer towards the nner curvature wall. The fluid in the core region of the flow moves along the enter plane of the channel, being directed from the inner to the outer curvature all. The position of the maximum velocity moves from the center to the concave ide, and the velocity gradient becomes stronger near the outer wall. The cross tream motion is refered to as a secondary flow. There are two kinds of secondary lows. The first is the primary secondary flow which flows fully in the whole ross section at the low Dean number [7]. The second is the additional secondary low which appears at the outer wall over the critical Dean number for the each spect ratio [8].

In Figure 3, the area of the additional secondary flow is shown as the unstable region in terms of the Dean number De for the aspect ratio γ [9,10]. In the pre experiment the aspect ratio changes from 0.625 to large value since water is fro and the width of the passage in which water flows, becomes narrow. The experimental region is, therefore, represented by the shadowed portion which extends from the stable region to the unstable one. It seems to be the transition stage in which the flow situation moves from the primary secondary flow to the generation of the additional secondary vortices [11].

Variation in the thickness of the ice layer

Figure 4(a) and (b) represent the freezing situation through the curved section for Rei=600 and 1990. Arrows show the direction of water flow, while the white and black portions represent the frozen and unfrozen layers, respectively. The difference of the freezing situation for Re can be found in these figures.

The quantitative variation of the ice layer are shown by time and position in Figure 5 and 6, respectively. For inlet velocity u=3.5 cm/sec, the interface becomes wavy at the exit of the curved section, while the thickness of the ice layer increases locally with time. This kind of situation could be found in th solidification through the straight tube [6]. For u=8.9 cm/sec, the ice thickn increases locally with an elapsed time up to about 15 minutes. After 20 minutes

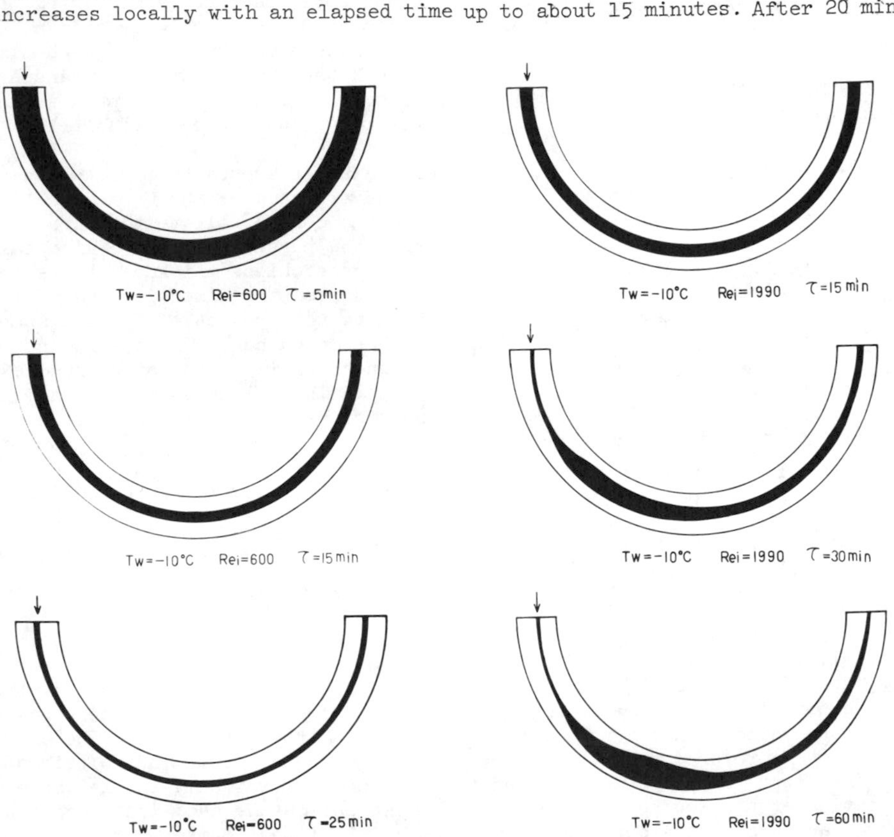

(a) (b)

FIGURE 4. Freezing situation

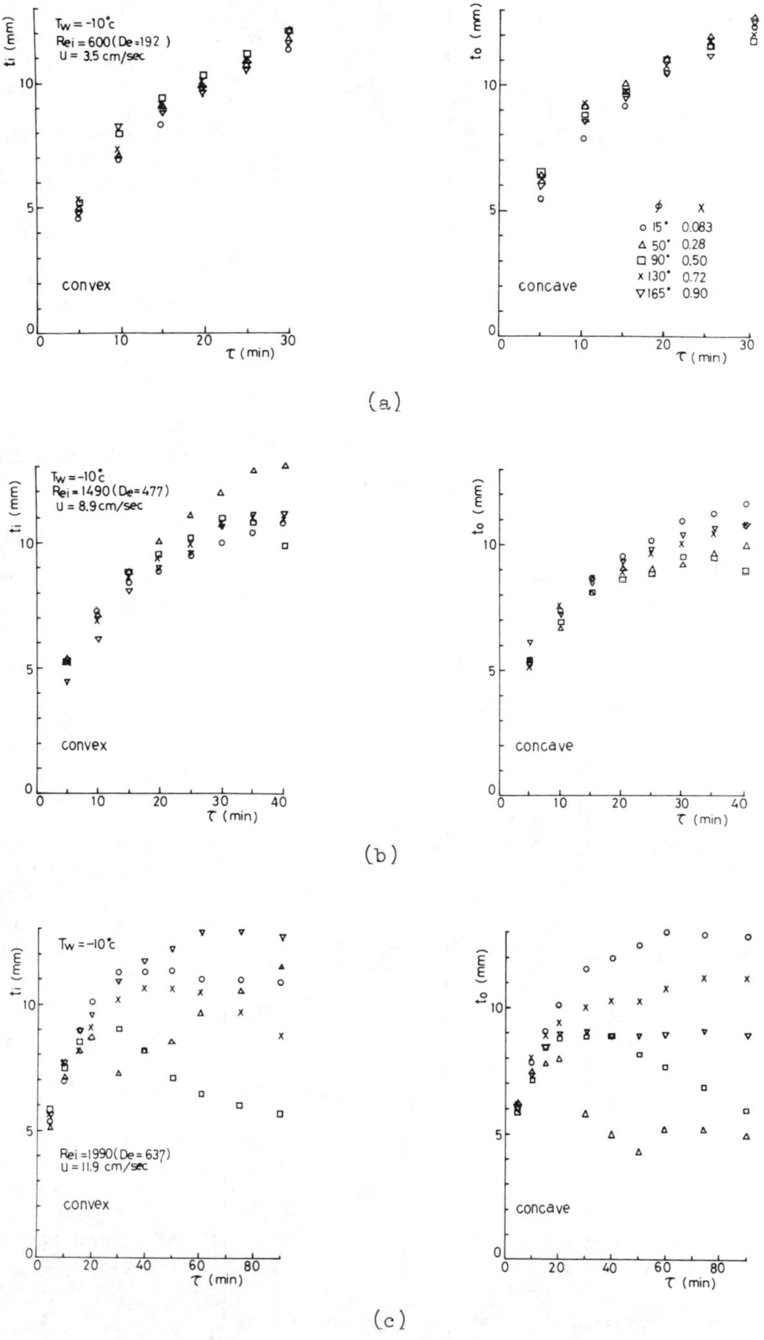

FIGURE 5. Thickness of ice layer (position as a parameter)

rate of increase of the ice thickness decreases and the thickness takes minimum at $\phi=90°$ and increases longitudinally (concave wall. After th thickness on the convex wall increases graduall from the entrance to $\phi=$ (x=0.28) and takes a pe it takes a minimum at ϕ 90°(x=0.50) and increas downstream. For u=11.9 after the thickness take minimum at $\phi=90°$ and inc with flow direction on t concave wall, it decreas toward downstream again. position of the minimum the maximum thickness m to $\phi=50°$ and $\phi=130°$(x=0. respectively after 20 mi The thickness on the cor wall which initially inc with time, decreases at entrance section after I minutes and increases ag with time in accordance the decrease of the ice thickness on the concave After it takes a peak at 50°, it reaches its mini at $\phi=90°$ and also increa to the flow direction. described above, the pos of the minimum thickness upstream with an increas the inlet velocity and i determined in terms of F the state approaches ste ness.

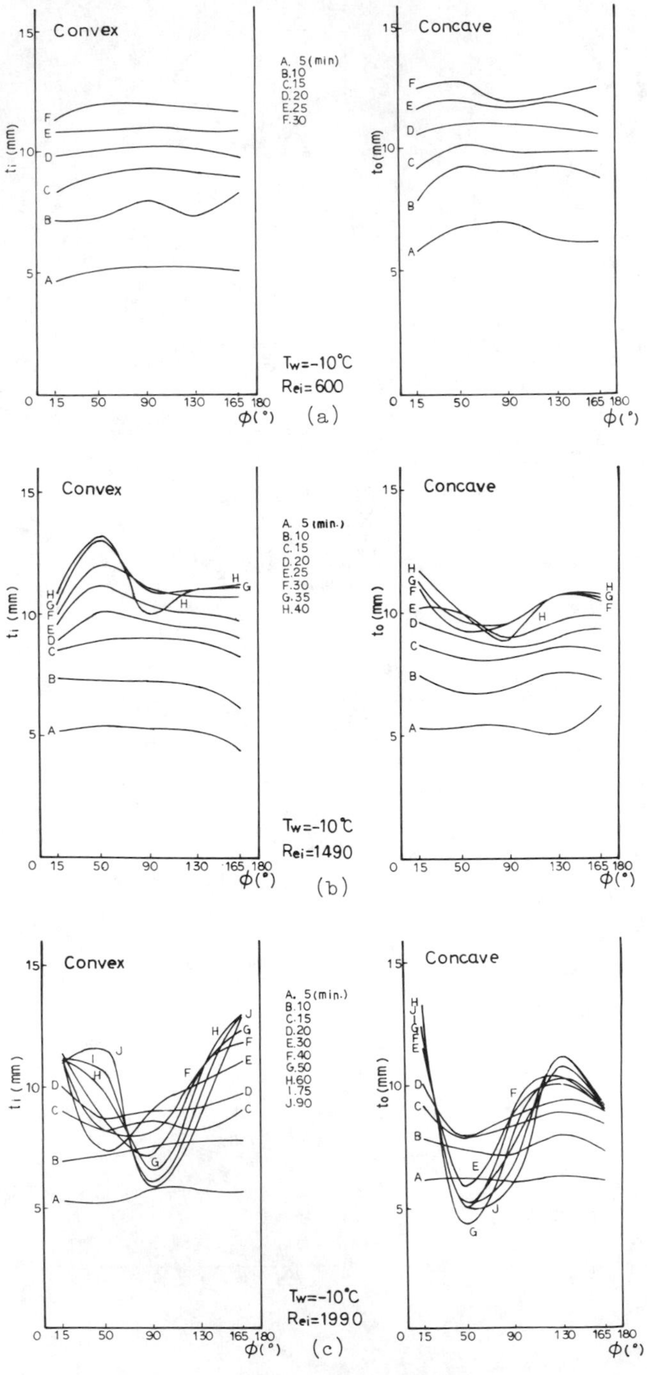

FIGURE 6. Thickness of ice layer (time as a parameter)

The relation between th position X_L and Rei is shown in Figure 7 as a parameter of cooled wal temperature. The posit X_L tends to move to the entrance as the wall te ature becomes lower. T relations are expressed as eq.(1) and (2).

For the concave wall

$$X_L = 12.1\ Re_i^{-0.48} \qquad ($$

For the convex wall

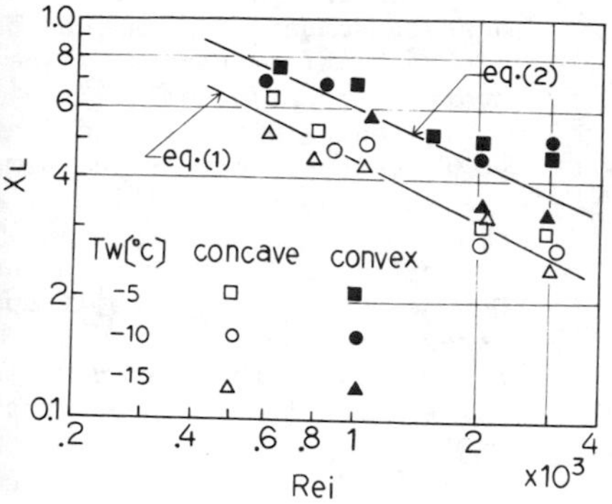

FIGURE 7. Position of the minimum thickness of ice layer

$$X_L = 11.4\ Rei^{-0.43} \quad (2)$$

It is estimated that the increase of heat transfer by acceleration with ice formation causes the unevenness of water-ice interface at comparatively low velocity. As the flow velocity increases, the effects of the secondary flow and the acceleration cause the heat transfer increase, and the situation on the convex wall develops corresponding to the decrease of the ice thickness on the concave wall.

Heat transfer on water-ice interface

Temperature in the ice layer is estimated by assuming quasi-steady state of freezing situation, as follows;

$$T = \frac{(Tf - Tw)\cdot \ln r + Tw\cdot \ln r_s - Tf\cdot \ln r_{i,o}}{\ln (r_s/r_{i,o})} \quad (3)$$

where r_s is the radius of the interface and $r_{i,o}$ the radius of the convex or concave wall, respectively.

Heat balance equation at the interface is

$$\pm L\rho \frac{dr_s}{d\tau} = \pm k_{i,o} \frac{dT}{dr} - q_c \quad (4)$$

The variation of the ice layer with time $(dr_s/d\tau)$ is approximated by the least square method.

Figure 8 (a) and (b) show the local Nusselt number of the water-ice interface and is compared with the average Nusselt number in the case of heating (Tw=const.)[12].

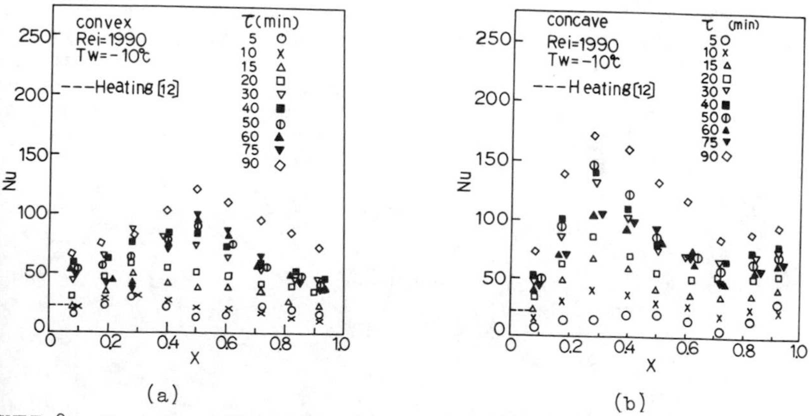

FIGURE 8. Heat transfer at water-ice interface

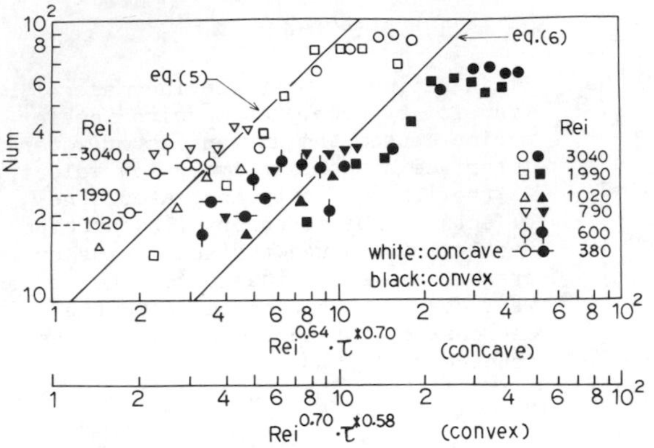

FIGURE 9. Average Nusselt number

While on the concave wall, the Nusselt number is less than th of the heating process at the starting point of solidificati it increases with time, and af 90 minutes it is over two time the value at the position of t minimum thickness than that at the entrance of the curved cha On the other hand, Nusselt num takes a peak at ϕ=90° on the convex wall, and this value is higher than that at the concav wall.

Figure 9 shows the average Nus number Num to grasp the whole situation of ice formation. T values of heating(T_w=const.) a also represented with dotted l in this figure. Abscissa is the function of Rei and dimensionless time τ^*. Ac cording to the figure, Num depends strongly on τ^* for the concave wall and on R for the convex wall, respectively. Experimental equations of average Nusselt number are expressed by eq. (5) and (6).

For the concave wall

$$Num = 8.0\ Rei^{0.64}\ \tau^{*0.70}$$

For the convex wall

$$Num = 3.2\ Rei^{0.70}\ \tau^{*0.58}$$

where the valid region of time is $\tau^* \geq 1.0$. The behavior becomes flat as the situation approaches its steady state.

CONCLUDING REMARKS

The transient process in the curved channel containing flows was experimentall examined

The variation in the thickness of the ice layer depends on the water velocity. The expansion of the flow migrates upstream with inlet velocity. The effects the acceleration due to freezing and the secondary flow by channel curvature interact with each other and affect the freezing process. These effects produ a great heat transfer through the thin ice layer, compared with heat transfer without solidification. The average Nusselt number around the wall at each ti expresses the characteristics of the freezing process.

ACKNOWLEDGEMENT

Support of this work by a Scientific Research Grant from the Ministry of Educa of Japan is gratefully acknowledged.

OMENCLATURE

. = width of channel, m, mm, b = height of channel, m, mm, c = specific heat, :J/kg K, D_H = hydraulic equivalent diameter (=2ab/(a+b)), k = thermal conduc-ivity, W/m K, L = latent heat, kJ/kg, q_c = heat flux, W/m^2, r = radial distance, : = radius of curvature, t = thickness of ice layer, mm, T = temperature, °c, K, 'f = freezing temperature, °c, K, Ti = inlet temperature, °c, K, Tw = cooling urface temperature, °c, K, u = inlet velocity, cm/sec, x = dimensionless length =X/πR), Rei = inlet Reynolds number $(=u \cdot D_H/\nu)$, De = Dean number $(=Re\sqrt{D_H/2R})$, u = Nusselt number $(=\alpha \cdot D_H/k)$, α = local heat transfer coefficient $(=q_c/(Tf-Tw))$ $'/m^2$ K, γ = aspect ratio, φ = angle from starting point of curvature, τ = time, * = dimensionless time $(=(c_s Tw \rho_s/\rho_L L)(\kappa L/a^2)\tau)$, ρ = density, kg/m^3, κ = thermal iffusivity, m^2/min,

uffix
 = liquid, s = solid, i = convex, o = concave,

EFERENCES

1. Hirschberg, H.G., Freezing of Piping Systems, Kaltetechnik, vol.14, pp.314 -321, 1962.
2. Zerkle, R.D. and Sunderland, J.E., The Effect of Liquid Solidification in a Tube upon Laminar Flow Heat Transfer and Pressure Drop, Trans. ASME. Ser. C, vol.90, pp.183-190, 1968.
3. Ozisik, M.N. and Mulligan, J.C., Transient Freezing of Liquids in Forced Flow inside Circular Tubes, Trans. ASME. Ser. C, vol.91, pp.385-389, 1969.
4. Mulligan, J.C. and Jones, D.D., Experiment on Heat Transfer and Pressure Drop in a Horizontal Tube with Internal Solidification, Int. J. Heat & Mass Transfer, vol.19, pp.213-219, 1976.
5. Thomason, S.B., Mulligan, J.C. and Everhart, J., The Effect of Internal Solidification on Turbulent Flow Heat Transfer and Pressure Drop in a Horizontal Tube, Trans. ASME. Ser. C, vol.100, pp387-394, 1978.
6. Gilpin, R.R., The Morphology of Ice Structure in a Pipe at or near Transition Reynolds Numbers, Proc. of 18th National Heat Transfer Conf., San Diego, PP.89-94, 1979.
7. Dean, W.R., Note on the Motion of Fluid in a Curved Pipe, Phil. Mag., vol.4 pp.208-223, 1927.
8. Cheng, K.C., Lin, R.C. and Ou, J.W., Fully Developed Laminar Flow in Curved Rectangular Channels, Trans. ASME. Ser. I, vol.98, pp.41-48, 1976.
9. Akiyama, M., Kikuchi, K., Nakayama, J., Suzuki, M., Nishiwaki, I. and Cheng, K.C., Two Stage Development of Entry Flow with an Interaction of Boundary-Wall Deans Instability Type Secondary Flows (Hydrodynamic Entry Region Problem of Laminar Flow in Curved Rectangular Channels), Trans. JSME., vol.47, no.421, pp.1705-1713, 1981.
10. Sugiyama, S., Hayashi, T. and Yamazaki, K., Flow Characteristics in the Curved Rectangular Channels (Flow Visualization of Secondary Flow), Trans. JSME., vol.48, no.434,pp.1870-1876, 1982.
11. Akiyama, M., Kikuchi, K., Suzuki, M., Nishiwaki, I., Cheng, K.C. and Nakayama, J., Numerical Analysis and Flow Visualization on the Hydrodynamic Entrance Region of Laminar Flow in Curved Square Channels, Trans. JSME., vol.47, no. 422, pp.1960-1970, 1982.
12. Akiyama, M. and Cheng, K.C., Laminar Forced Convection Heat Transfer in Curved Rectangular Channels, Int. J. Heat & Mass Transfer, vol.13,pp.471-490, 1970.

Experimental Study of Heat Transfer of Parallel Louvered Fins by Laser Holographic Interferometry

YASUO KUROSAKI, TAKAO KASHIWAGI, and HIROKI KOBAYASHI
Tokyo Institute of Technology
Oohkayama, Meguro-ku, Tokyo, Japan

HIDEO UZUHASHI
Tochigi Works, Hitachi Ltd.
Oohira-machi, Tochigi, Japan

SHIE-CHUNG TANG
Beijing University of Iron and Steel Technology
Beijing, PRC

1. INTRODUCTION

Recently many compact air-cooled heat exchanger begin to utilize the enhance fins to upgrade the performance of a heat exchanger. The study of heat transfe characteristics of these enhanced fins has been focused on[1]. Sparrow et al[2 calculated the heat transfer performances in two-dimensional plate arra neglecting the effects of tubes. Nakayama et al[3] proposed correlations of th Colburn j-facor and the friction factor for triangular tube arrangement t design the heat exchanger. Hatada et al[4] studied the effects of louver' geometry and their arrangement on heat transfer of louvered fins.

A number of informations are available concerning theoretical analyses ar global heat transfer performances in a heat exchanger with louverd fins however, little amount of experimental information to clarify the mechanism (heat transfer around louvers so far has prevailed[5].

The objectives of this paper are experimentally to study the mechanism of hea transfer in a louver-array and to develop the preferable geometrical arrangeme of louvers from the point of view of improving the performance of a hea exchanger. Our approach toward that goal was made via the following steps.

The first step treated in this paper is optically to visualize the temperatu field around louvers and to measure the heat transfer coefficients of t louver; the louver fins used in this experiment are simply parallel louver fins called offset strip fins shown in Fig.1. The tracer method by usi

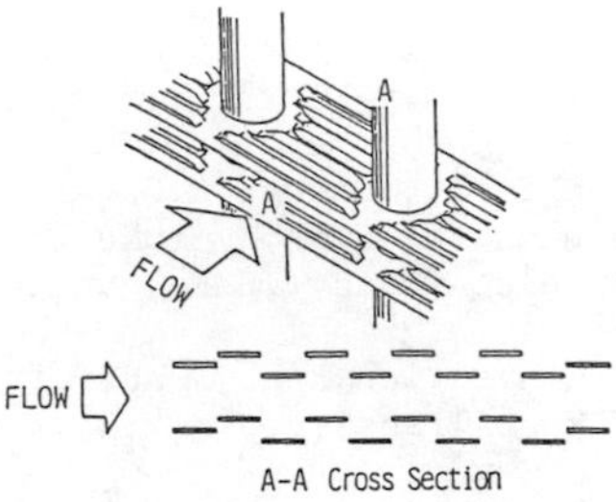

FIGURE 1. Parallel louvered fins.

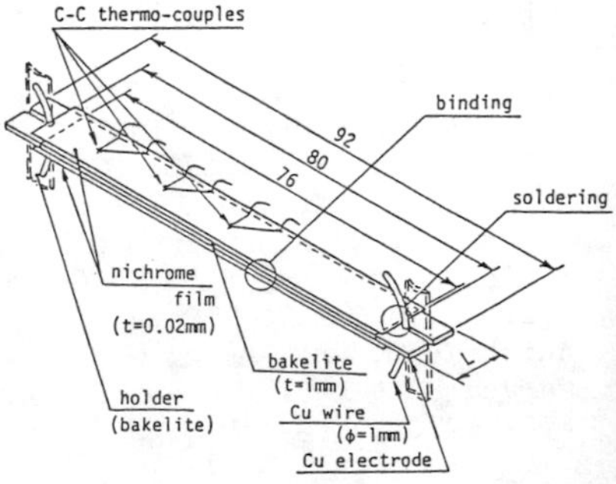

FIGURE 2. A model louver for experiment.

.uminum powder or hydrogen bubbles so far has been often applied to measure the ›locity profiles of an air flow through the array of louvers. In contrast with ıose, little has been reported concerning the visualization of temperature .eld around louver; our experiment has been done to visualize the isotherms ·ound louvers by employing the laser holographic interferometry. The thermal ›undary layers and wakes generated by an upstream louver were clearly observed › extend toward downstream louvers; the heat transfer coefficients obtained by ıe experiment were virtually affected by those boundary layers and wakes.

ıe second step is to examine plausible arrangement of louver for enhancing heat ·ansfer. The slight position shift of downstream louvers toward the direction ·oiding the influence of thermal wake was proposed from both the obsevation of ;otherms and the measurement of heat transfer coefficients in a staggered ›uver arrangement; its effectiveness was verified by the experiment. Applica-.on of the proposed minor rearrangement of louvers for enhanced fins is ·omising for improvement of the performance of a heat exchanger .

EXPERIMENT

1. Model Louver

ıe usual louvered fins used in a gas-cooled heat exchanger has a typical ›ometry as shown in Fig.1. The fin thickness is 0.1 - 0.25 mm ; the louver .dth is about 2 mm, and the louver pitch is 5 - 8 louvers/cm. Such a samll size ›uver is inadequate to use for visualization experiment of the temperature .elds around louvers by means of laser holographic interferometry. Therefore, ıe model louvers with the lengths of 5, 10, and 20 mm were used. The detail of model louver used is illustrated in Fig.2; the nichrome foils of 20 μm for .rect heating by current was pasted on both sides of a bakelite plate, and ıermocouples were mounted to measure the surface temperature. The thickness of ›uver at a test section is 1.18 mm. Therefore, the surface of louver is ›nsidered to have constant heat flux; the surface temperature attained 50 °C at ıe maximum.

.2. Apparatus for Heat Transfer Experiment

ıe experimental set up for heat tansfer is schematically demonstrated in Fig.3. ıe air flow from a blower passed through a flow meter and a series of screens ı a settling chamber prior to entering the test section. The test louvers were ıpported vertically in the test section of the duct; the air flow blew upward. ıe test cross section is 100 mm × 200 mm; both sides of which consist of ·ansparent plastic plates for optical measurement.

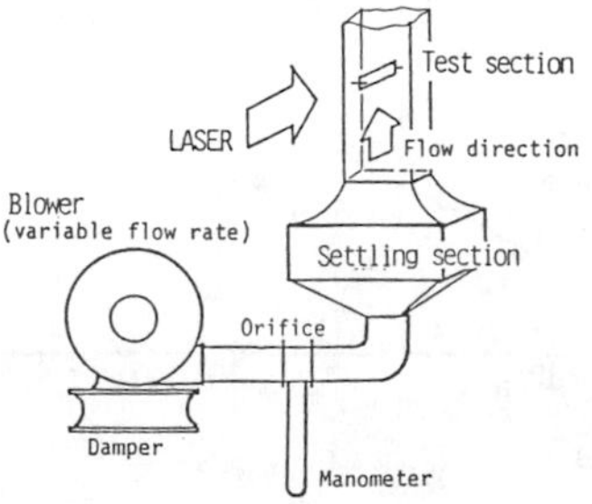

FIGURE 3. Experimental set-up.

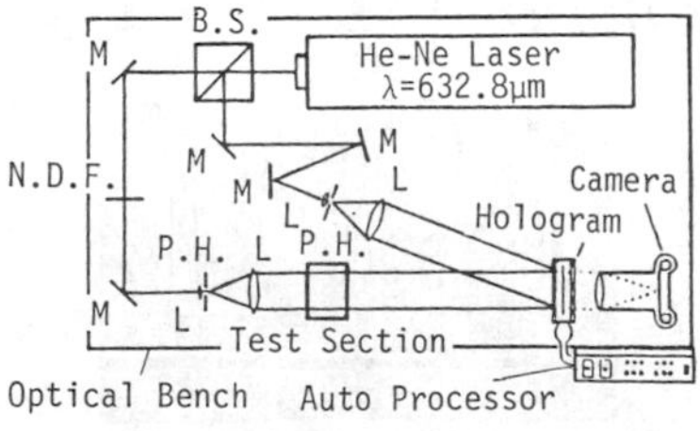

FIGURE 4. Optical system for laser holographic interferometry.

2.3. Optical Apparatus

A real time holographic interferometry was used for measurement of t temperture profile around louvers without disturbing any field(6). Fig.4 sho the optical system of the holographic interferometry used in this experimen The optical source is a 5 mW He-Ne gas laser; the beam emitted from the laser 6328 A) is divided into a test beam and a reference beam by a beam splitte The interference fringes obtained by the real time holographic method reveal t density contours. In this experiment, those fringes are identical with t isotherms since the pressure throughout the system is an atomospheric pressur Therefore, the temperature profile in air can be expressed as follows:

$$T^{-1}(x,y) = T_{\infty}^{-1} - [N(x,y)-1]\cdot\lambda /(273\cdot\beta\cdot L)$$

where $N(x,y)$: the fringe number , λ : the wavelengh of a beam, β : constant (2.91×10^{-4} for 6328 A in air), L : the path length of a beam through the te section, T_{∞} : the ambient temperature.

A line in fringes in this experimnt corresponds to the temperature difference 8 - 9 degrees.

2.4. Arrangement of Louvers

The following four arrangements of louvers were adopted in this study : (1) single louver, (2) two louvers aligned with the flow, (3) two parallel louver (4) a bank of staggered louvers and (5) a bank of modified staggered louvers .

3. Results

3.1 Heat Transfer Around a Single Louver

As the first step, the mean heat transfer coefficeint and the temperatu profile around a single model louver, which was located parallel to the strea were experimentally obtained to assess the facility dependence of the result The velocity of air flow was varied from 0 to 20 m/s: that is, the Reynol number ($= U_{\infty}\ell/\nu$) was from 0 to 1060. Practically, the Reynolds numbers bas on a louver length in the air-cooled heat exchanger are about 200 -500.

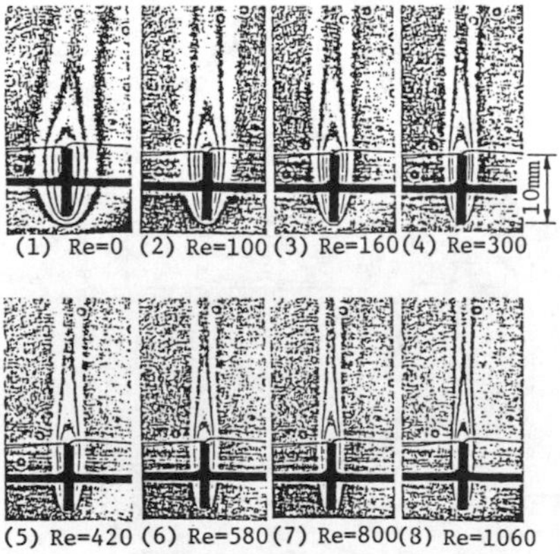

FIGURE 5. Isotherms around a single louver.

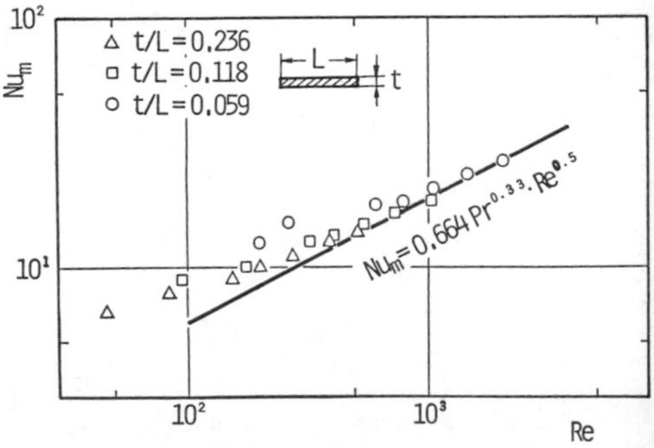

FIGURE 6. The mean Nusselt number for a single louver.

ie instantenous profiles around a louver was measured by means of the real time ıser holographic interferometry as shown in Fig.5.

ɔth the theoretical heat transfer results for a single plate and the xperimental ones obtained in the present study are shown in Fig.6. Both of them gree quite well in the range of the larger Reynolds number than 420. They epend little upon the aspect ratio of a louver (t/ℓ). Thus the heated plate sed in this experiment was assessed to be usable as a louver model.

.2 Heat Transfer of Two Louvers Aligned with the Flow

s the second step, the heat transfer around two model plates aligned with the .ow direction with the spacing (s) between louvers was studied. In order to xamine the effect of the wake from an upstream louver on the downstream ones ie visualization experiment was run for a series of different louver spacings. s an example, Figs.7(a) and (b) show two isotherms around louvers with the .ffernt spacings of 10 and 5 mm (s/ℓ = 1, 0.5).

.th increasing the flow velocity the effect of the thermal boundary layer ɜnerated from the upstream louver on the downstream one becomes evident specially in the case of the narrow spacing as shown in Fig.7(b).

ie mean Nusselt number for a flat plate due to simultaneous laminar natural and ɔrced convections is expressed in the following equation[7,8].

$$^3_{comb} = Nu^3_{for} + Nu^3_{nat} \quad (1)$$

iere Forced Conv.: $Nu_{for} Re_\ell^{-1/2} = 0.644 Pr^{1/3}$ (2)

Nat. Conv.: $Nu_{nat} Re_\ell^{-1/2} = 0.560 Pr^{1/4} (Gr_\ell / Re_\ell^2)^{1/4}$ (3)

ie mean heat transfer coefficient h_m for a flat plate with a uniform heat flux

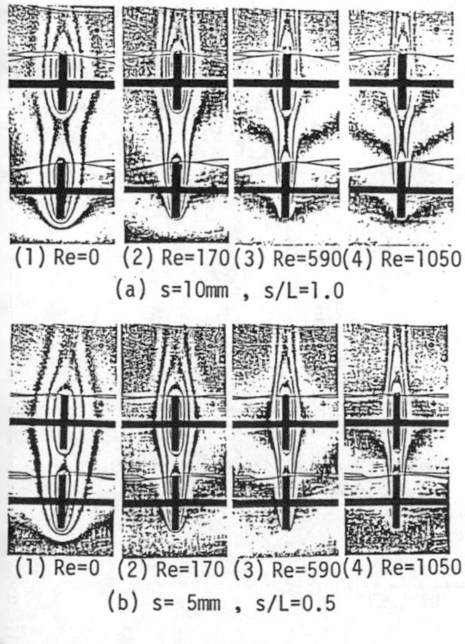

ɪGURE 7. Isotherms for two louvers ıligned with a flow.

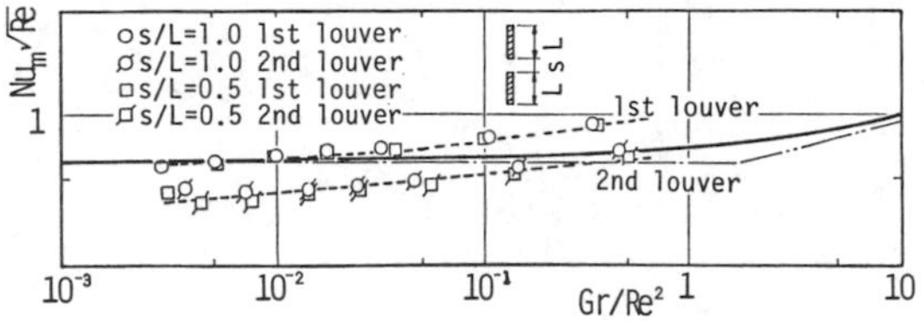

FIGURE 8. The mean Nusselt numbers for two alined louvers.

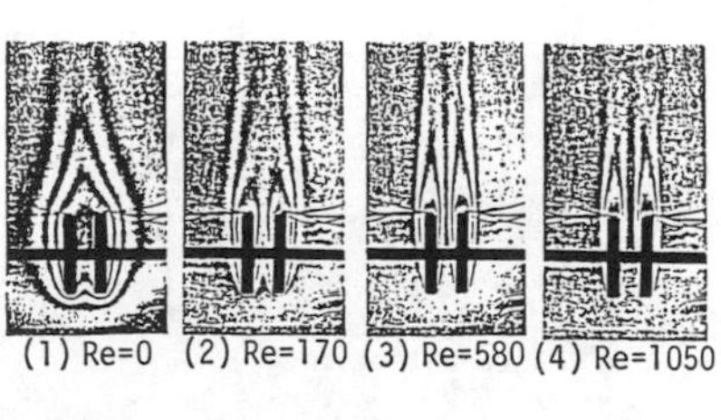

FIGURE 9. Isotherms for two parallel louvers.

was verified to be expressed theoretically as follows(9):

$$h_m = (Q/\ell)/(\Delta T)_{mean} = q/\{(Tw)_{x=4\ell/9} - T_\infty\} \qquad (4)$$

Instead of the wall temperature at x = 4ℓ/9, the measured wall temperature at = ℓ/2 was used here to obtain the mean heat transfer coefficient in Eq.(4). Th former temperature is equal to the latter within the experimental error. Fig. shows the results of heat transfer for each of two louvers obtained from Eq.(1) The evident difference between the heat transfers of the upstream louvers and c the downstream ones exsists; however, it is noted that little influence c louver separation distance on heat transfer appears except the range of th large Reynolds numbers.

3.3 Heat Transfer of Parallel Two Louvers

In order to evaluate the effect of other louvers placed parallel to each othe on heat transfer, an experiment was carried out to obtain the results about hea transfer of two parallel louvers as shown in Fig.9. The isotherms in Fig.8 show that the interaction of the boundary layers generated from two louvers is i connection with the velocity of a free stream; the boundary layers from bot louvers interact with each other in the range of the smaller Reynolds number than 500 . The larger Reynolds number becomes, the less interaction wa observed; then, each louver is able to be considered isolated.

The mean Nusselt number for each louver experimentally obtained in this case ar shown in Fig.10. This result indicates that the heat transfer coefficient i almost identical with the one for an isolated louver, provided that th dimensionless distance (d/ℓ) between two louvers is larger than 0.2. Disappear ance of the free stream between louvers for small louver separation (d/ℓ 0.1) yields lower heat transfer coefficients around louvers than the one with free stream.

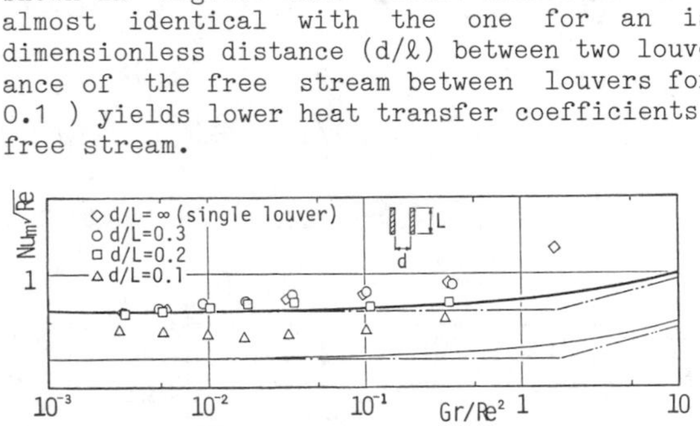

FIGURE 10. The mean Nusselt numbers for aligned louverers.

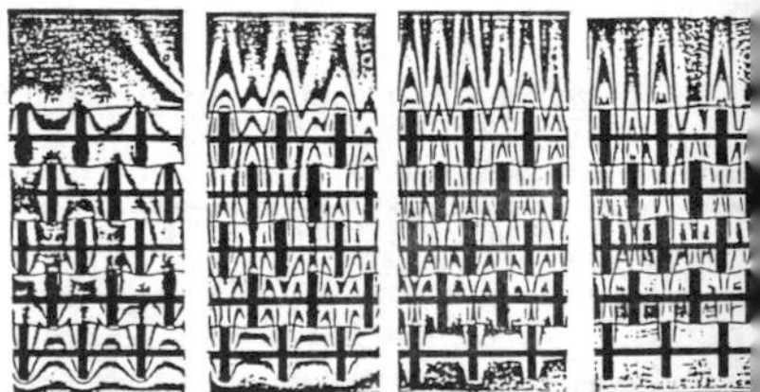

FIGURE 12. Isotherms for staggere louvers.

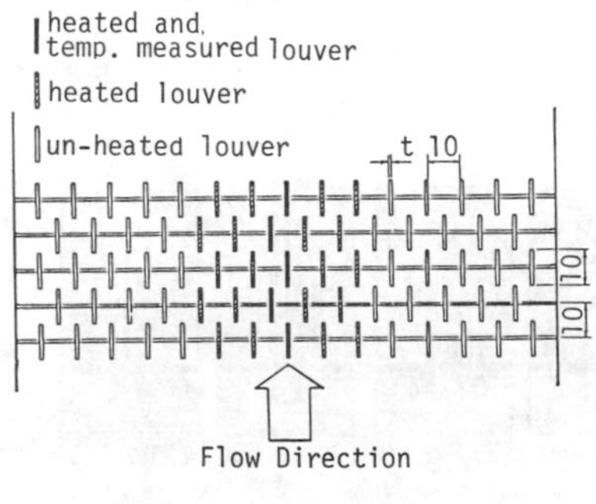

FIGURE 11. A bank of staggered model louvers.

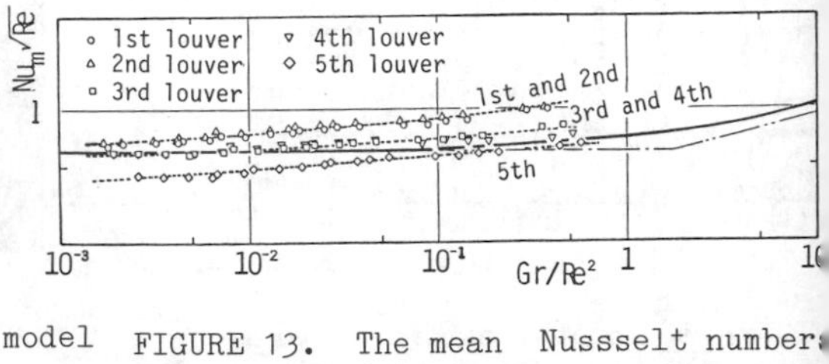

FIGURE 13. The mean Nussselt number for staggered louvers.

.4. Heat Transfer of Staggered Louver Arrangement

ost of practical heat exchanger with louvered fins have a staggered arrangement f louvers. The similar experiment was conducted for the staggered louver array hown in Fig.11. Heaters were provided to 20 louveres out of all in the middle ortion ; 5 of those in the center have themocouples. The distance between arallel louvers is 10 mm; the spacing between louvers aligned with the flow irection is 10 mm.

he isotherms for those louver arrangements obtained by the holographic nterferometry are shown in Fig.12. As predicted previously, this figure shows hat the heat transfer of downstream louvers is affected by the wake generated rom the upstream louver. Furthermore, the interactions among thermal boundary ayers from each louver in the same transverse row are seen to be less than hose in the same streamwise row.

he experimental results for the heat transfer characteristcs of a bank of taggered louvers are shown in Fig.13. It is noted that the mean heat transfer oefficient in the first transverse row is almost equal to the one in the second ne, which is shown by the dotted line in that figure; likewise, the one in the hird is almost equal to the one in the fourth whose value is lower than that in he first. These behaviors are attributed to the staggered arrrangement because he louver in the second transverse row is considered to be subject to the ncontaminated flow like the one in the first row.

.5. Modified Staggered Louver Arrangement

he results described above indicate the wake induced from the upstream louver vidently degrade the heat transfer of the downstream louver. The modified rrangement of louvers to lessen the degradation due to the wake is proposed ere. Its concept is to prevent the downstream louvers from submerging in hermal wakes convected from upstream louvers; practically, the louvers in the hird row are horizontally slided toward the center between louvers as shown in

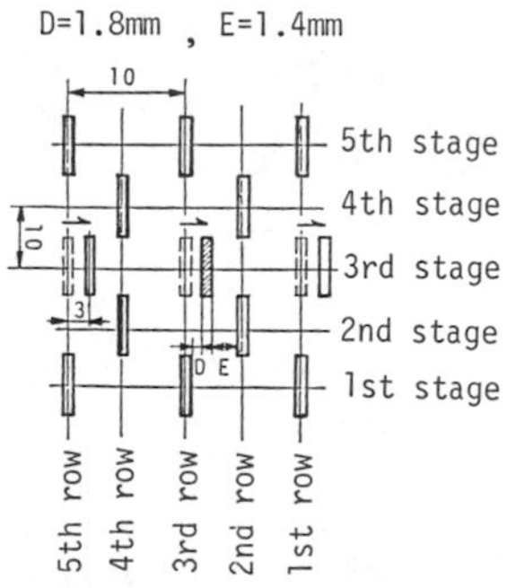

FIGURE 14. Modified arrangement of staggeed louvers.

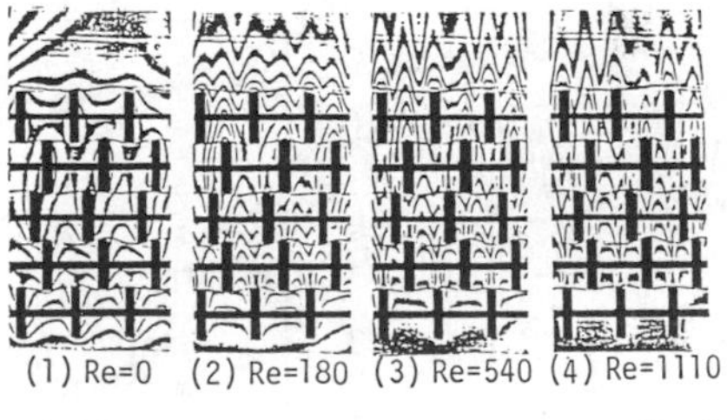

FIGURE 15. Isotherms for modified staggered louver.

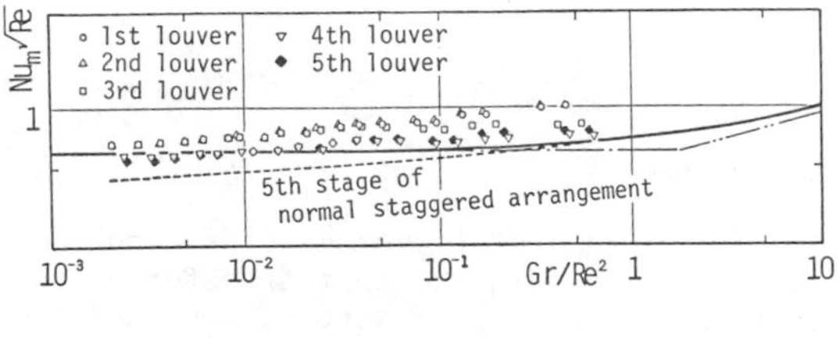

FIGURE 16. The mean Nusselt number for modified staggered louvers.

Fig.14. In this arrangement, where the louvers in the third row are slided, th effect of the louvers in the first row on the heat transfer of the fifth ro louvers may still remain. However, the spacing between the first and the fift louvers in this case is 3 times as wide as the one in regular staggere arrangement; therefore, the less influence of the first louvers' wakes on th fifth louvers are expected. The prediction that the proposed louver arrangemen will be promising for heat transfer enhancement was verified by the experiment The isotherms in this modified staggered array of louvers are shown in Fig.15 It is seen that influences of the wake from the first louver on the fifth one i this case are less than that in the former case. The heat transfer coefficient in this case were also shown in Fig.16. These results indicate that the hea transfer enhancement due to this arrangement is rather effective in the fift row louvers than in the third row louvers. The dimension of louver positic shift in this model, which is 3 mm, corresponds to 0.5 mm in the practica louvered fin. There are no problems in designing and manufacturing such modifie louvered fins. If the concept proposed here is developed to the other louve arrangement, further heat transfer enhancement can be expected.

3.6. Discussion About a Louvered Fin

It is important to discuss the comparison of a plate fin with a louvered one Imagine a flat plate the length of which is 5ℓ , and a 5 segmented strips length of which is ℓ, as shown in Fig.17(a). The correlations of heat transfe obtaind experimentally were referred to in driving the mean heat transfe coefficients based on the plate length of x. A solid line in Fig.17(b) shows th theoretical result for a flat plate. The mean heat transfer coefficient of th first louvers is in good agreement with the theoretical one; however, those fc segmented strips are quite higher than the one for flat plate. The triangula marks in Fig.17(b) indicate the mean heat transfer coefficients in the modifie louver arrangement proposed here. This arrangement is found to enhance the hea transfer 10 percent over the usual staggered one. Thus the use of the modifie array is encouraging for designing the high performance heat exchanger.

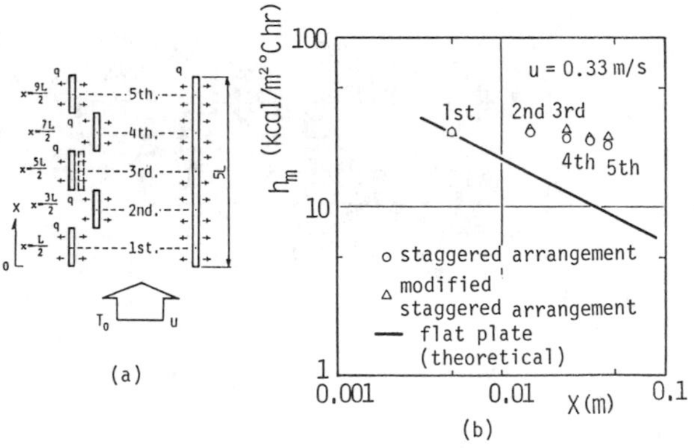

FIGURE 17. Comparison of a flat plate fin with a louvered fin.

4. CONCLUSION

(1) In a bank of louvered fins, the influence of the wake induced from upstream louver on the heat transfer around a downstream louver is greater th that of neighboring louvers in the same transverse row.

(2) In a staggered arrangement of louvers, the mean heat transfer coefficient

louver greatly depends upon both the number of louvers existing in the upstream and the separation distance between louvers in the fow direction.

3) In order to enhance heat transfer, the modified staggered arrangement of louver is proposed; the downstream-louvers submerging in the wake of an upstream should be slided out of it.

NOMENCLATURE

d	:	separation distance between two parallel louvers
Gr	:	Grashoff number
h_m	:	mean heat transfer coefficient of a louver/or louvers
ℓ	:	length of a louver
Nu_{comb}	:	mean Nusselt number due to natural and forced convection
Nu_{for}	:	mean Nusselt number due to laminar forced convection
Nu_{nat}	:	mean Nuseelt number due to laminar natural convection
Pr	:	Prandtl number of fluid
Re	:	Reynolds number ($= U_\infty \ell/\nu$)
s	:	separation distance between louvers aligned with a flow direction
t	:	thickness of a louver
U_∞	:	velocity of free stream

REFERENCE

1. Nakayama,W., Enhancement of Heat Transfer, Keynote RK13, Proc. the 7th Int.Heat Transfer Conf., vol. 1, pp.223-240, 1982.
2. Sparrow,E,M., Baliga,B.R. and Patankar,S.V., Heat Transfer and Fluid Flow Analysis of Interrupted Wall Channels, With Application to Heat Exchangers, J.Heat Transfer, vol. 99. no.1, pp.4-11, 1977.
3. Nakayama,W. and Xu,L.P., Enhanced Fins for Air-Cooled Heat Exchanger - Heat Transfer and Friction Factor Corrections, Proc. the ASME-JSME Thermal Eng. Joint Conf., vol.1, pp.495-492, 1983.
4. Hatada,T and Senshu,T., Heat Transfer Characteristics of Convex Louvered Fins for Air Conditioning Heat Exchangers, Trans. JSME, vol.50, no.453(B), pp.1415-1422, 1984.
5. Roadman,R.E. and Loehrke,R.I., Low Reynolds Number Flow Between Interrupted Flat Plates, J. Heat Transfer, vol.15, no.1, pp.166-171, 1983.
6. Kurosaki,Y., Isshiki,N., Ito,A. and Kashiwagi,T., Downward Flame Spread Along a Vertical Sheet of Thin Combustible Solid, J. Fire and Flammability, vol.10, pp.3-25, 1979.
7. Lloyd,J.R. and Sparrow,E.M., Combined Forced and Free Convection Flow on Vertical Surfaces, Int. J. Heat Mass Transfer, vol.13, pp.434, 1970.
8. Churchill,S.W., A Comprehensive Correlating Equation for Laminar, Assisting, Forced and Free Convection, AIChE J., vol.23, no.1, pp.10-16, 1977.
9. Mori,Y. and Kurosaki,Y., Laminar Forced Convective Heat Transfer With Uniform Surface Heat Flux, Proc. 11th Japan Nat. Cong. Appl. Mech., pp.253-257,1961.

Two-Phase Flow Instabilities in a Single Channel with Enhanced Heat Transfer, and Pressure-Drop Type Oscillation Thresholds

A. MENTEŞ, O. T. YILDIRIM, S. KAKAÇ, and T. N. VEZIROĞLU
Clean Energy Research Institute
University of Miami
Coral Gables, Florida 33124, USA

NOMENCLATURE

A	area, m
c	specific heat J/kgK
d	diameter,m
e	parameter, dimensionless
K	valve coefficient, dimensionless
L	length, m
m	mass flow rate, kg/s
n_1	parameter, dimensionless
n_2	parameter, dimensionless
P	static pressure, Pa (N/m^2)
Q	heat input, W
q	heat flux, W/m^2
T	temperature, K
t	time, s
s_1	nondimensional slope of pressure drop vs mass flow rate curve
s_2	parameter, dimensionless
V	volume, m^3

Greek Letters

α	heat transfer coefficient, W/m^2K
ρ	density, kg/m^3
τ	parameter, dimensionless

Subscripts

c	compressible, critical
e	equivalent, exit
N	nitrogen
o	steady-state
s	stable
tp	two-phase
v	vapor
w	wall

INTRODUCTION

The phenomenon of thermally induced two-phase flow instability is of intere for design and operation of many industrial systems and equipment, such steam generators, thermosiphon reboilers, refrigeration plants, and oth chemical process units.

cause of the economical reasons and space constraints, there is a tendency reduce the size of the heat exchanging systems by increasing the heat transfer efficient. This is being achieved by using fin-like protrudings, grooves the heat transfer surfaces, and also roughened and specially treated surfaces.

e understanding of two-phase flow instability is extremely important to the sign, control and performance prediction of any system having boiling flows. is necessary to determine the influence of pressure, flow rate, heat transfer efficient, temperature, etc., on two-phase flow oscillations. The effects various parameters, such as inlet and exit restrictions, inlet subcooling, at flux on two-phase flow instabilities in a single and parallel channel flow system with smooth tubes have been studied experimentally and eoretically [1-8].

sically there are three identifiable types of oscillations, namely, essure-drop type, densitywave type and thermal oscillations, which may be countered in two-phase flow systems. Extensive work has been carried out find the basic mechanisms and the predictions of these oscillations. A od review of the subject may be found in [1]* and [2].

recent years, there has been increased emphasis on techniques to enhance o-phase flow heat transfer; an excellent survey of the subject is presented Bergles and Joshi [9,10]. The resulting increase in the use of heat transfer gmentation and the engineering importance of the subject caused the present vestigation to be taken in order to study the effect of different heater rface configurations on two-phase flow instabilities.

PERIMENTAL INVESTIGATION

open-loop forced convection boiling upflow system operating between two nstant pressures has been designed and built to generate the main types of o-phase flow oscillations. Tests were carried out using Freon-11 as the st fluid. A schematic diagram, with the basic dimensions of the experimental t up, is shown Figure 1. More detailed information on the set up and the strumentation can be found in Refs. [1,11].

x different augmented heater surfaces, description of which is summarized Table 1 have been prepared. Experiments have been performed with each tube

BLE 1. Description of the heater tubes.

be	Description of the tube	d_e mm
	Bare	7.493
	Threaded, 7.938 mm - 16 threads per 25.4 mm	7.619
	With internal spring of 0.794 mm wire diameter and 19.05 mm pitch	7.446
	With internal spring of 0.432 mm wire diameter and 3.175 mm pitch	7.401
	With internal spring of 1.191 mm wire diameter and 6.350 mm pitch	7.192
	Coated with Union Carbide Linde High Flux Coating	7.073

umbers in square brackets refer to bibliography given at the end of the paper.

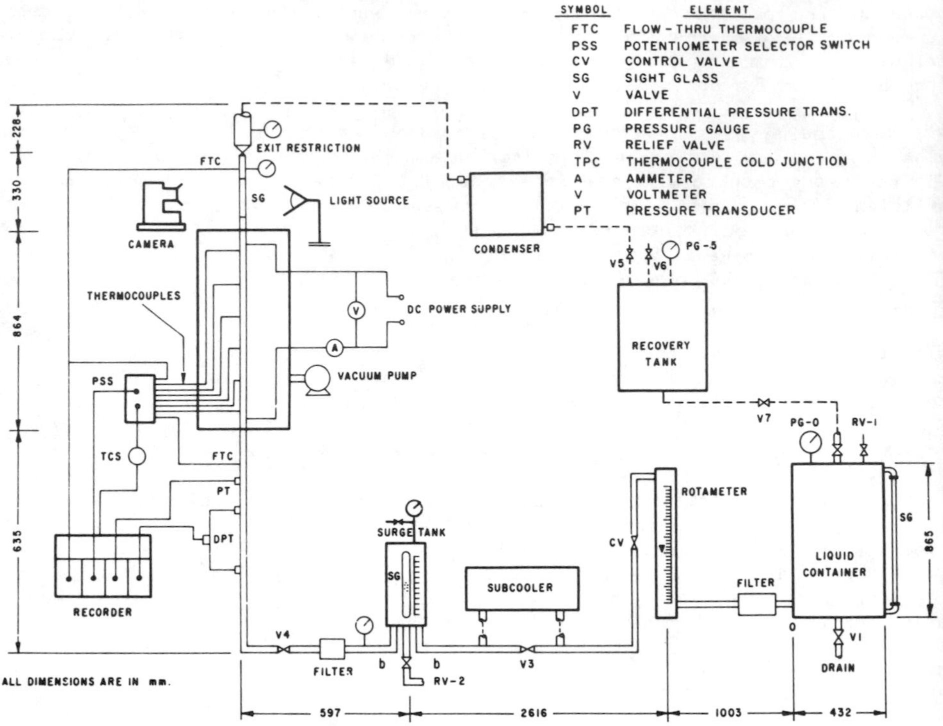

FIGURE 1. Schematic drawing of the experimental system.

at a constant inlet temperature and constant heat flux for various values of mass flow rate. This has been repeated for six different values of heat input and for various values of inlet temperatures. Each experiment was composed of sufficient number of tests to cover a wide range of boiling regimes.

Each heater tube was made of 7.493 mm inside and 9.525 mm outside diameter 605 mm long of Nichrome tubes. Tubes are classified according to their effective diameter which is defined as

$$d_e = \frac{4V}{\pi L} \tag{1}$$

where V is the net inside volume and L length of the heater tube.

For a given heater tube, several sets of experiments corresponding to various heat inputs and/or inlet temperatures were conducted. Stability boundaries were located for each case. Oscillations were identified by cyclic variations in pressures, flow rate and temperatures and, also by observing the transparent section of the set up, the pressure gauge pointers and the recorder. In defining the boundaries short-life transients were ignored, only the sustained oscillations were considered. Detailed information on the experimental procedure and experiments is given in reference [11].

Steady-State Characteristics

Pressure drop from the surge tank exit to the system exit is referred to

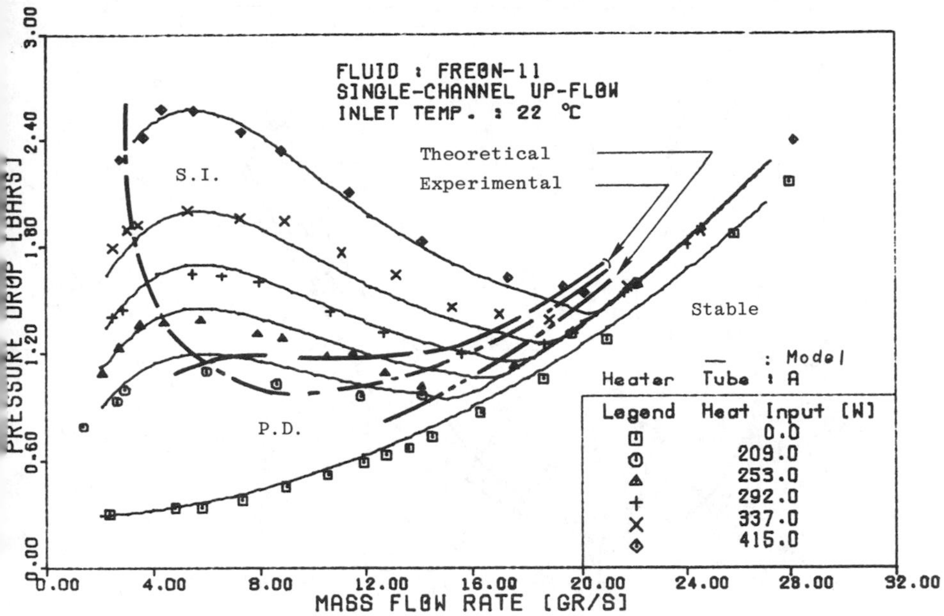

GURE 2. Comparison of theoretical and experimental steady-state aracteristics and oscillation boundaries for the system with heater tube at different heat inputs.

e system pressure drop; and experimental and theoretical data for the heater be A is presented on pressure drop versus mass flow rate coordinates in Fig. 2.

eady-state pressure drops for different heater tubes show the same type havior, which is to be expected since the pressure drop over the heater section the system is only about 4% of the total pressure drop. The flow was ngle-phase and steady along the right leg of the steady-state curve. The rst bubbles were observed when the pressure drop curve started to round up it approached its local minimum point. The number of bubbles and their ameters increased as the curve changed direction and pressure drop started increase with decreasing mass flow rate. The rate of pressure-drop increase higher for higher heat inputs to the system.

cillations

resholds for pressure-drop type oscillations and for density-wave type cillations superimposed on pressure-drop type oscillations were located during periments and indicated on the pressure-drop mass-flow rate coordinates ig. 2). The stability boundary maps are obtained by connecting these threshold ints. Boundaries for the pure pressure-drop type oscillations for different ater tubes are collected in Fig. 3. An examination of the figure shows that e curves for different heater tubes exhibit a similar pattern, with internally ringed tubes having narrower unstable regions. The same order of the tubes obtained for the superimposed oscillation boundaries.

ble 2 presents the values of system parameters during oscillations, and was epared to compare the system with different heater tubes. Mass flow rate ves the average value obtained after stabilizing the system, exit quality

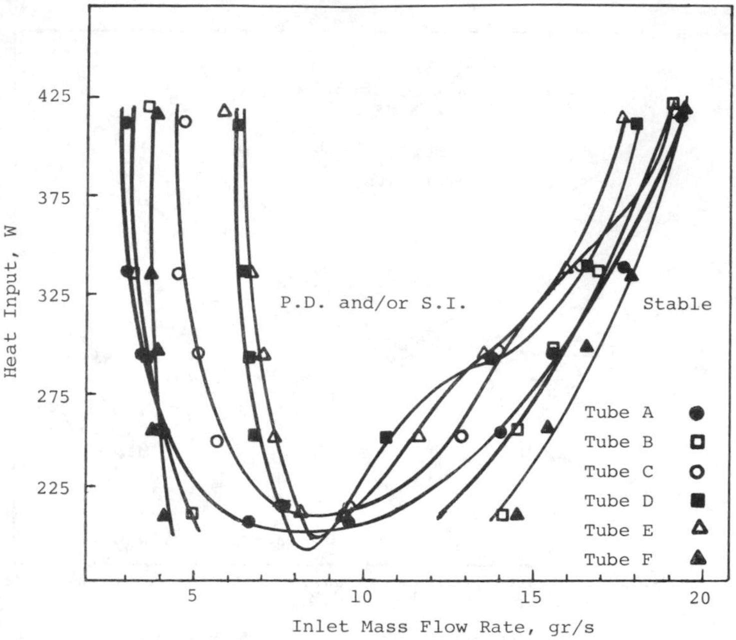

FIGURE 3. Boundaries for pressure-drop type oscillations.

is the calculated value at the orifice inlet. Temperature values are giv in two rows. Top row shows the steady-state values, while the bottom row giv the overall variation of heater wall temperature at the given location.

MATHEMATICAL MODELING

One-dimensional homogeneous equilibrium flow model is used to describe t flow in the system under consideration. Conservation equations are writt

TABLE 2. Sustained oscilation data for different tubes during superimpos oscillations.

Liquid Container Pressure : 6.53 Bar

Pc.Vc : 193.1 N.m (In the surge tank)

TUBE	Heat Input W	Mass Flow Gr/s	Exit x %	Temperatures Steady State & Amplitude [C]							Pressures [Bar]				Periods [Se	
				Inlet	Heater Wall					Exit	Surge	System	Amplitude*10		P.D.	D.W
					2	3	4	5	6		Tank	Exit	P.D.	D.W.		
A	425.0	8.32	9.1	22.1	66.7	N.A.	70.6	73.3	72.2	61.7	3.41	1.26	6.89	7.65	55.0	1.
				N.A.	N.A.	N.A.	N.A.	N.A.	27.7	10.0 2.5						
B	422.3	9.45	6.3	22.7	62.3	63.9	67.0	65.7	68.1	58.5	3.41	1.08	3.83	4.98	40.0	2.
				N.A.	10.0	7.0	7.0	7.5	15.0	7.5 2.5						
C	411.8	10.45	3.7	22.7	61.2	64.3	66.1	68.1	68.1	57.6	3.32	1.05	4.69	4.13	48.0	2.
				N.A.	N.A.	N.A.	N.A.	N.A.	8.8	8.8						
D	413.3	9.90	4.8	22.5	59.2	62.8	64.3	69.4	69.2	57.1	3.34	1.12	4.70	3.42	54.0	2.
				N.A.	5.0	N.A.	N.A.	N.A.	7.5	8.8						
E	416.8	8.72	8.2	22.5	56.2	63.4	65.9	68.5	67.4	58.0	3.38	1.10	8.16	2.98	88.0	4.
				2.5	17.6	17.8	12.5	12.5	8.8	8.8						
F	411.9	8.20	9.4	22.2	62.5	62.3	64.5	62.3	63.6	58.7	3.41	1.02	4.41	3.31	38.0	4.
				N.A.	N.A.	N.A.	5.0	N.A.	N.A.	7.5						

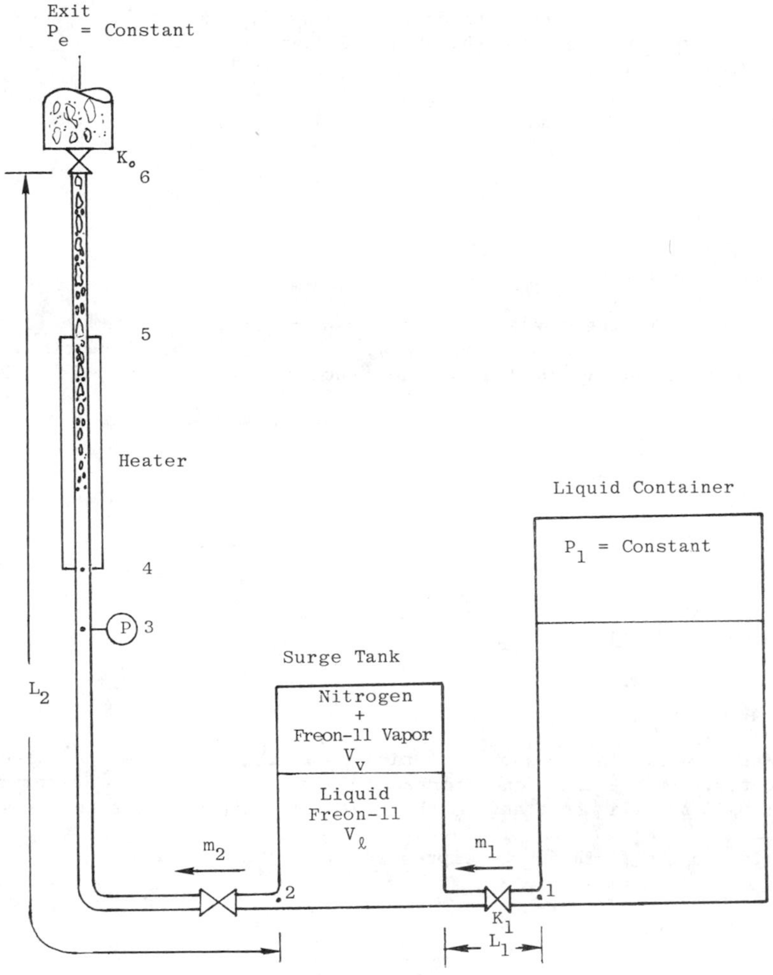

FIGURE 4. Schematic Diagram of the System for Mathematical Modeling.

using this model, and steady-state solution is obtained numerically using implicit, backward differencing scheme. Detailed information on formulati and solution can be found in reference [11].

Pressure-drop Oscillation Threshold

An approach introduced by Stenning et al. [12] is used to predict t pressure-drop type instability thresholds. Dynamic equations of the syst shown in Fig. 4 are written as:

$$P_1 - P_2 = K_1 m_1{}^2 + \frac{L_1}{A_1} \frac{dm_1}{dt} \tag{2}$$

$$P_2 - P_e = (P_2 - P_e)_s + \frac{L_2}{A_2} \frac{dm_2}{dt} \tag{3}$$

$$P_1 - P_e = \text{constant} \tag{4}$$

For the upstream of the surge tank, from surge tank to system exit and fr the liquid container to system exit respectively. Dynamic equations for t surge tank and the heater wall are also required for the analysis. These c be written as

$$m_1 - m_2 = -(\rho \ell - \rho_v) \frac{dV_v}{dt} \tag{5}$$

$$\frac{dV_N}{dt} = - \frac{(P_N V_N)_o}{P_N{}^2} \frac{dP_2}{dt} \tag{6}$$

for the surge tank, and

$$Q_e - Q_f = m_w C_w \frac{dT_w}{dt} \tag{7}$$

for the heater wall. In the above equation Q_e is the heat generated electrical in the heater, Q_f the heat transferred to the fluid, m_w the heater mass, the heater wall specific heat and T_w the wall temperature of the heater wall.

Heat transfer to the fluid can be expressed as:

$$Q_f = \alpha(T_w - T_f) \tag{8}$$

The equations 12 through 18 are linearized wind small perturbation analysi The characteristic equation of the linearized system of equations is fou to be

$$(s_1 + s_2 n_2)\tau_1 \tau_2 D^2 + [(\frac{s_1 s_2 n_2}{2e} + 1 + s_2 n_1)\tau_2 + s_1 \tau_1]D + 1 + \frac{s_1}{2e} = 0 \tag{9}$$

where

$$s_1 = \frac{m_o}{P_{2o}} [\frac{\partial(P_2 - P_e)}{\partial m_2}]_o \tag{10}$$

$$s_2 = \frac{Q_o}{P_{2o}} \left[\frac{\partial (P_2 - P_2)}{\partial Q} \right]_o \quad (11)$$

$$n_1 = \frac{P_{2o}}{T_{wo} - T_{fo}} \left(\frac{dT_f}{dP_2} \right)_o \quad (12)$$

$$n_2 = \frac{m_o}{\alpha_o} \left(\frac{d\alpha}{dm_2} \right)_o \quad (13)$$

$$\tau_1 = (\rho_\ell - \rho_v) \frac{V_{No} P_{2o}}{m_o P_{No}} \quad (14)$$

$$\tau_2 = \frac{m_w C_w}{Q_o} (T_{wo} - T_{fo}) \quad (15)$$

$$D = \frac{d}{dt} \text{, the differential operator} \quad (16)$$

$$e = \frac{P_{1o} - P_{2o}}{P_{2o}} \quad (17)$$

he equation 9 can be written in short form as

$$C_1 D^2 + C_2 D + C_3 = 0 \quad (18)$$

here

$$C_1 = (s_1 + s_2 n_2) \tau_1 \tau_2 \quad (19)$$

$$C_2 = (1 + s_2 n_1 + \frac{s_1}{2e} + \frac{s_2 n_2}{2e}) \tau_2 + s_1 \tau_1 \quad (20)$$

nd

$$C_3 = 1 + \frac{s_1}{2e} \quad (21)$$

or a second order system to be stable, all coefficients of the characteristic quation must be positive. Therefore, the conditions for the system to be stable an be set as:

$$s_1 > -s_2 n_2 \quad (22)$$

$$s_1 > -\tau_2 / (\tau_1 + \frac{\tau_2}{2e}) (1 + s_2 n_1 + \frac{s_2 n_2}{2e}) \quad (23)$$

$$s_1 > -2e \quad (24)$$

FORTRAN program has been developed to carry out the computations. The program ses the steady-state characteristics of the system to check the conditions or equations (32), (33) and (34). Detailed analysis is given in ref. [11].

onditions for stability were checked with the calculated values of parameters $_1$, n_2, s_1, s_2, τ_1, τ_2 and e. If any one of the conditions is not satisfied he system is said to be unstable.

The above analysis has been carried out for all of the heater tubes. The resul for the heater tube A are presented in the Table 3. Calculated threshol are also marked in Fig. 2.

TABLE 3. Pressure-Drop Type Oscillation Thresholds for the Heater Tube A.

Heat Input Threshold [Gr/s]	209 W	253 W	291337 W	415 W	 W
Experimental Value	9.0	13.5	16.0	17.8	19
Theory with Experimental Data	14.0	15.0	17.0	19.0	20
Theory with Theoretical Data	14.6	16.3	17.4	18.5	19

Table 3 indicates that the pressure-drop type oscillation thresholds can be located with linearized analysis and steady-state data. The predicti is within 1% for 415 W but error increases at lower heat inputs. For eve case theoretical values are conservative.

CONCLUSIONS

Results of experimental and theoretical study can be summarized as:

1. Instabilities are not affected by small changes in heat transfer.
2. Nucleate boiling increases sytem unstability.
3. Amplitude of the oscillations increases as the axial temperatu gradient over the heater increases.
4. For the same type heater surfaces system stability increases wi decreasing equivalent diameter.
5. Period of the oscillations depend on the heater surfaces.
6. Linearized analysis and steady-state data can be used to determi the oscillation thresholds.

ACKNOWLEDGEMENTS

The authors gratefully acknowledge the financial support of the National Scien Foundation, and wish to extend their thanks to Union Carbide Corporation, Lin Division for preparing special heat transfer augmented surfaces for t experimental study and the NATO Scientific Affairs Division. The secretari work of Ms. Donna Pressley is also acknowledged.

REFERENCES

1. Veziroglu, T.N., and Kakac, S., "Two-Phase Flow Instabilities", Fir Report, N.S.F. Project CME 79-20018, July 1983, Clean Energy Resear Institute, Coral Gables, Florida.
2. Kakac, S., and Veziroglu, T.N., "A Review of Two-Phase Flow Instabilities In Kakac, S., and Ishii, M. (eds.): Advances in Two-Phase Flow and He Transfer, Fundamentals and Applications, Martinus Nijhoff, The Hagu The Netherlands, Vol 2, 1983, pp. 577-669.
3. Stenning, A., and Veziroglu, T.N., "Flow Oscillation Modes in Forc Convection Boiling", Proceedings of the 1965 Heat Transfer and Flu Mechanics Institute, Stanford University Press, California, 1965, p 301-316.

. Saha, P., Ishii, M., and Zuber, N., "An Experimental Investigation of the Thermally Induced Flow Oscillations in Two-Phase Systems", Journal of Heat Transfer, Trans. ASME, Vol. 98, No. 4, 1976, pp. 616-622.

. Aritomi, M., Aoki, A., and Inoue, A., "Instabilities in Parallel Channel of Forced-Convection Boiling Upflow System, (III) System with Different Flow Conditions Between Channels", J. Nuclear. Sci. Tech., Vol. 16, 1979, pp. 343-355.

. Veziroglu, T.N., and Kakac, S., "Two-Phase Flow Instabilities and Effect of Inlet Subcooling", Final Report, N.S.F. Project ENG 75-16618, Feb. 1980, Clean Energy Research Institute, Coral Gables, Florida.

. Cumo, M., Plazzi, G., and Rinaldi, L., "An Experimental Study on Two-Phase Flow Instability in Parallel Channels with Different Heat Flux Profile", CNEN-RT/ING (81)1, 1981.

. Mentes, A., et al., "Effect of Heat Transfer Augmentation on Two-Phase Flow Instabilities in a Vertical Boiling Channel", Warme und Stoffubertragung, vol. 17, 1983, pp. 161-169.

. Bergles, A.E., "Principles of Heat transfer for Augmentation-II, Two-Phase Heat Transfer", in Kakac, S., Bergles, A.E. and Mayinger, F. (eds.), Heat Exchangers: Thermal Hydraulic Fundamentals and Design, McGraw Hill, 1981. pp. 857-881.

0. Bergles, A.E. and Joshi, S.D., "Augmentation Techniques for Low Reynolds Number in Tube Flow", in Kakac, S. and Shah, R.K. (eds.), Low Reynolds Number Flow Heat Exchangers, Hemisphere Publishing Corp., 1983. pp. 695-720.

1. Mentes, A., Two-Phase Flow Instabilities: Pressure-Drop Type Oscillation Thresholds, Ph.D. Thesis, University of Miami, 1985.

2. Stenning, A.H., Veziroglu, T.N. and Callahan, G.M.,"Pressure-Drop Oscillations in Forced Convection Flow with Boiling", Proc. of Symposium on Two-Phase Flow Dynamics, Eindhoven, the Netherlands, 1967.

Micro-Visualization of Fluidizing Behavior of Binary Particle Mixtures

YASHEN XIA and KWAUK MOOSON
Institute of Chemical Metallurgy
Academia Sinica
Beijing, PRC

ABSTRACT

The quality of fluidization affects directly heat transfer characteristics. Previous studies at this Institute demonstrated that a coarse Geldart Group-B powder and a fine Geldart Group-C powder could improve the otherwise poor fluidizing quality of either component, when mixed together in appropriate proportions.

To elucidate the above synergistic action, the present investigation was designed to visualize, by video recording under a microscope, the dynamic behavior of binary mixtures on a particle-size scale, that is, within a field of vision of the order of a few millimeters.

Two types of experiments were conducted:

1. Fluidization of the binary mixtures with different weight fractions of the components;
2. Flooding a single sessile coarse particle with a flowing dilute suspension of fine particles at different gas velocities.

It was found that synergistic action occurred only when the fine particles were capable of adhering to the coarse. The flooding experiments showed further that the fine particles had the preferred tendency of deposition, at first, on the south pole and then the north pole of the stationary coarse particle, increasing their coverage with increase in the suspension gas velocity, and then denuding the coarse particle around an expanding equatorial belt after certain maximum flow rate, until a sparsely covered north pole remanined as the only region with adhering fine particles.

Based on balance of forces on a fine particle at the surface of a coarse, for the actions of gravity, adhesion and hydrodynamics due to the adjacent flowing gas stream, a mathematical model was formulated to account for the shifting region of fine-particle coverage on the coarse.

NOMENCLATURE

d_p	average particle diameter, L
F_D	drag force on the particle, MLT^{-2}
F_H	adhering force on surface of particle, MLT^{-2}

F_G	gravity force of the particle, MLT^{-2}
u	superficial gas velocity, LT^{-1}
u_s	particle velocity, LT^{-1}
u_t	terminal velocity of particle, LT^{-1}
ρ_f	gas density, ML^{-3}
ρ_s	particle density, ML^{-3}
μ_f	viscosity of gas, $ML^{-1}T^{-1}$

Subscripts

1	component 1, or coarse particles
2	component 2, or fine particles
12	mixture of coarse and fines

INTRODUCTION

The quality of fluidization affects directly heat transfer characteristics in combustion and other chemical reaction systems. The wide disparity in behavior between the relatively quiescent fluidization of liquid-solid systems and bubbling, turbulent fluidization of gas-solid systems led early investigators to elaborate on criteria(1, 2, 3) for clearcut demarcation of the former, or particulate fluidization, from the latter, or aggregative fluidization. Subsequent investigations revealed, however, that not only liquid-solid systems may be made aggregative by judicious choice of parameters, but also for gas-solid systems, there may exist regions, though generally short, of operation predominantly particulate.

Previous studies at the Institute of Chemical Metallurgy on the mathematical modeling for the bed collapsing process(4) led to the formulation of dimensionless subsidence time Θ, correlating the physical properties of the solids and the fluid to the operating behavior of the particulate system. The larger the value of Θ, the smoother the fluidization. Especially, it has been demonstrated(5, 6) that for binary mixtures of Geldart Group-A and Geldart Group-B powders(7), fluidizing characteristics would be intermediate between those of the two constitutent components according to their relative proportion. When a Group-C powder was mixed with either a Group-A or a Group-B powder, however, the fluidizing characteristics of either component would be improved by the presence of the other, that is, the dimensionless time Θ often assumed some maximal value when plotted against composition of the mixture.

To elucidate the above synergistic action, it is needed to observe the dynamic behavior of binary mixtures on a particle-size scale, that is, within a field of the order of a few millimeters.

MATHEMATICAL MODELING AND COMPUTATION

Figure 1 presents an analysis of force balance in the vertical direction on a single fine particle adhering to a coarse while subject to the drag force of gas flowing upward:

$$\underbrace{F_{H12}\cos\theta}_{\text{adhesion}} + \underbrace{F_{D2}\sin\theta}_{\text{drag}} = \underbrace{F_{G2}}_{\text{gravity}} \qquad (1)$$

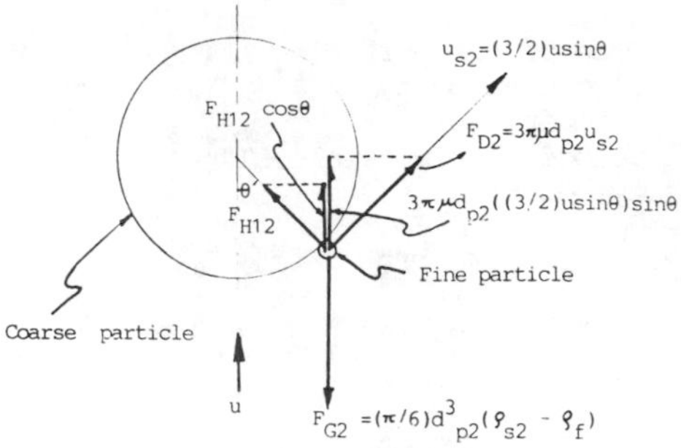

Figure 1. Force balance on a fine particle adhering to a coarse

Since the particles are small, flow is assumed to be laminar, in order to simplify the analysis. Thus the tangential component of the upward gas flow with velocity u is taken approximately as $u_{s2} = (3/2)u\sin\theta$, and the drag on the fine particle is Stokesian, $F_{D2} = 3\pi\mu d_{p2}u_{s2}$. Substitution of the weight of the fine particle $F_{G2} = (1/6)\pi d^3_{p2}(\rho_{s2} - \rho_f)$ gives

$$F_{H12}\cos\theta + (9/2)\pi\mu d_{p2}u\sin^2\theta = (1/6)\pi d^3_{p2}(\rho_{s2} - \rho_f) \qquad (2)$$

At the equator of the coarse particle, adhesion possesses no vertical component, and the fine particle is suspended at its terminal velocity u_{t2} by the gas stream, or

$$\theta = 90, \quad \cos\theta = 0, \quad \sin\theta = 1, \quad (9/2)\pi\mu d_{p2}u_{t2} = (1/6)\pi d^3_{p2}(\rho_{s2} - \rho_f) \qquad (3)$$

Normalize velocity u against u_{t2}, so that $u' = u/u_{t2}$, the above equation may be written alternatively as follows

$$F'_{H12}\cos\theta + F'_{D2}u'\sin^2\theta = 1 \qquad (4)$$

where F'_{H12} and F'_{D2} are respectively the adhesion force and drag force of the fine particle normalized to its weight F_{G2}.

Transposing gives

$$(F'_{D2}u') = \frac{1 - F'_{H12}\cos\theta}{\sin^2\theta}$$

$$= \frac{27\mu u}{d^2_{p2}(\rho_{s2} - \rho_f)} \qquad (5)$$

Equation (5) is presented graphically in Figure 2 as a family of θ-versus-($F'_{D2}u'$) curves with F'_{H12} as the parameter. These curves demonstrate the behavior of the fine particle on the surface of the coarse with increasing gas velocity. At the equator, θ=90°, where the vertical component of adhesion is zero, the fine particle is always balanced hydrodynamically by the upflowing gas at all gas velocities. The curves show also the case of no adhsion between the fine and the coarse particles, that is, $F'_{H12}=0$; the case when adhesion equals the fine particle weight, that is, $F'_{H12}=1$; and the case when adhesion exceeds the fine particle weight, that is, $F'_{H12}>1$. They clearly show the effect of upward gas

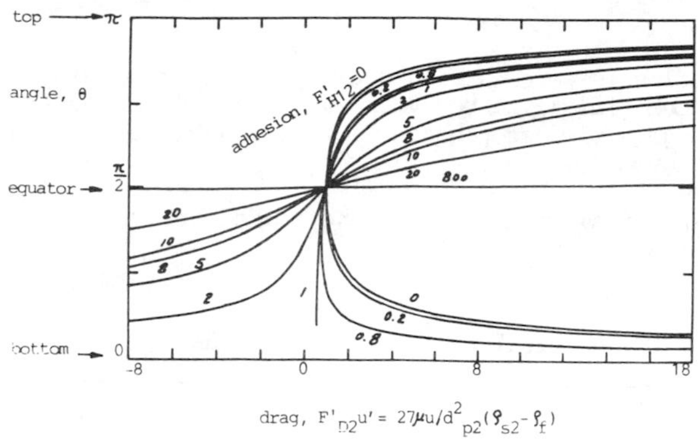

Figure 2. Graphical Representation of Interrelationship:

$$\theta \sim (F'_{D2}u') \sim F'_{H12}$$

velocity on the receding position of fines coverage towards the north pole in the northern hemisphere for all the cases. But for the southern hemisphere, the balancing position of fine adhesion varies as F'_{H12} varies. From these curves can be deduced the effect of gas velocity and the effect of particle size of the fines on their coverage on the coarse, as shown schematically in Figure 3.

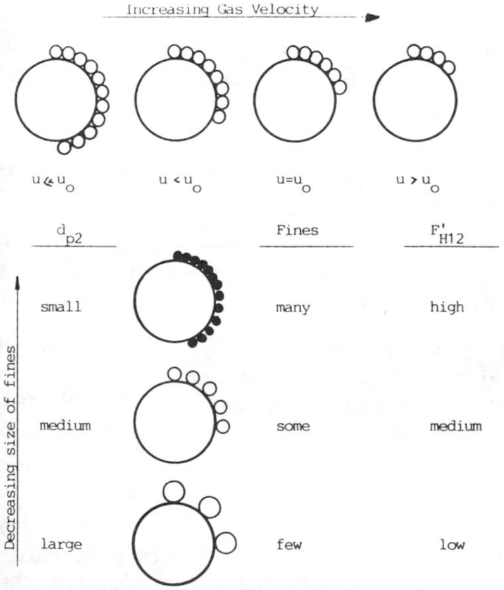

Figure 3. Factors affecting adhesion of a fine particle on coarse

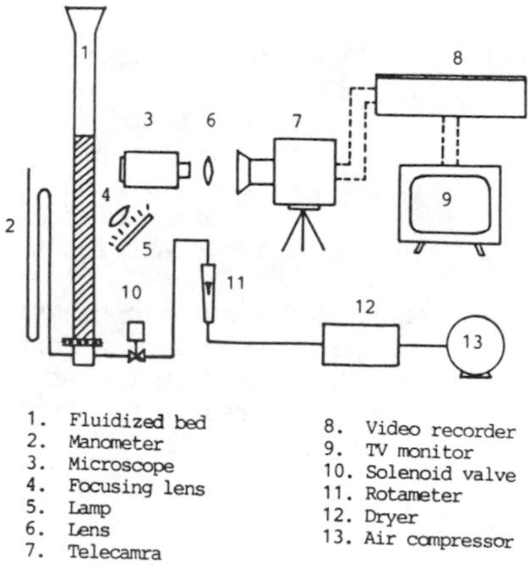

Figure 4. Schematic diagram of experimental apparatus

EXPERIMENTAL

Apparatus   Experiments were carried out in a 3-cm-square by 60-cm-high glass column, and the bed fluidizing behavior was recorded by means of a VO-2630CE videocorder through a microscope and a lens arrangement according to the schematic diagram of Figure 4.

Meterials   The physical properties of the solid particles used in this investigation are summarized in Table 1.

Table 1.   Physical Properties of Solids

	Materials	Code Number	Particle size		Density
			Mesh	$\bar{d}_p$, μ	ρ_s, g/cm^3
COARSE	Active carbon	C_{675}	20-40	675	0.920
	Microspherical silica gel	M_{650}	24-35	650	0.753
	Ion exchange resin	I_{300}	40-60	300	1.383
FINE	FCC catalyst	F_{122}	100-160	122	0.909
	Quartz powder	Q_6	-400	5.62	2.662

Procedure   Two types of experiments were conducted:

1. Fluidization of binary mixtures with different weight fractions of the components. The superficial gas velocity was chosen to be 5 cm/sec higher than that for minimum fluidization.
2. Flooding a single sessile coarse particle with a flowing dilute suspension of fine particles at different gas velocities, in an arrangement shown schematically in Figure 5.

The image of the dynamic behavior of binary mixtures was first enlarged under the microscope and the lens before being picked up by a telecamera, and then recorded by the videocorder, and by playing back frame by frame on the TV monitor, selected pictures were taken by a still camera, as shown in Figure 6.

RESULTS AND DISCUSSION

Figure 6 shows typically two sets of pictures for the fluidization of binary mixtures, one set for Geldart Group-B/A mixture and one for Group-B/C. For the Geldart Group-B/A mixture, that is, M_{650}/F_{122}, each component is shown to behave independently for all compositions, while for the Group-B/C mixture, the phenomena of fines adhesion on the coarse and of clustering of the fines themselves are shown to increase with fines fraction.

Figure 7 presents the behavior of the flow of a dilute suspension of fine

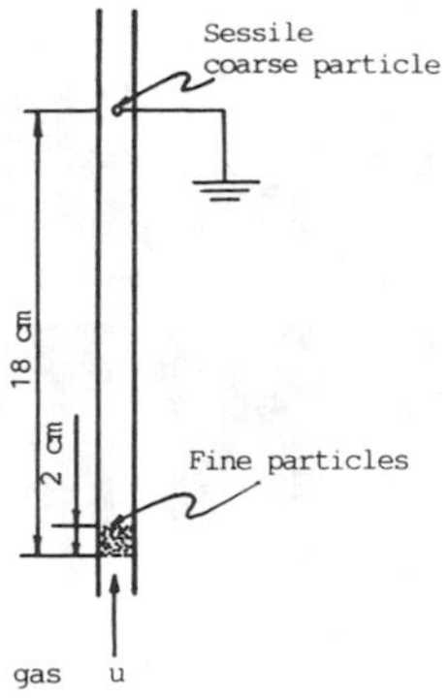

Figure 5. Schematic diagram of flooding experiments

particles of Geldart Group-C solid, Q_6, around a singly suspended sessile coarse particle of Geldart Group-B solid, M_{650}. Two sets of experiments are shown in which the coarse particle is fixed on a horizontal and on a vertical thin copper wire. Adhesion of fines starts from the lower or south pole of the coarse particle, increasing their coverage with gas velocity as more fines are thus carried upward. Then the north pole becomes populated with the fines, while an equatorial region remains essentially barren. With further increase in gas velocity, the lower side of the coarse particle becomes stripped of fines, followed by a gradual ascent of the denuded zone, until at sufficiently high gas velocity, the top, or north pole vicinity of the coarse remains the only region capped by fines.

The two significant findings on the clustering of fines with concentration both on the surface of the coarse and among themselves in the surrounding space, and on the upward receding of the region of fines adhesion on coarse with increasing gas velocity furnish certain insight into the mechanism of improvement in fluidizing characteristics of Geldart Group-A powder or Geldart Group-B powder with appropriate additions of Geldart Group-C powder. It was noted that adhesion of fine particles on coarse was enhanced by increasing the fines concentration. Moreover, the region of fines coverage of the surface of the coarse was found both experimentally and shown through mathematical deduction and computation, to recede upward with increasing gas velocity. With increasing fines concentration, cohesion among the fines, on the other hand, took place at the same time, leading to cluster formation. On account of their greater mass, these clusters clung less tenaciously to the coarse particle than when they were present as discrete individuals. Also, fines clusters in the surrounding space were noted from Figure 6 to bombard their kin on the coarse surface with greater momenta, thus tending to dislodge them, as compared to their un-agglomerated state, as shown in Figure 8.

CONCLUSIONS

1. Micro-visualization of fluidizing behavior of binary particle mixtures was realized by video recording under a microscope.

Figure 6. Fluidization of two binary mixtures

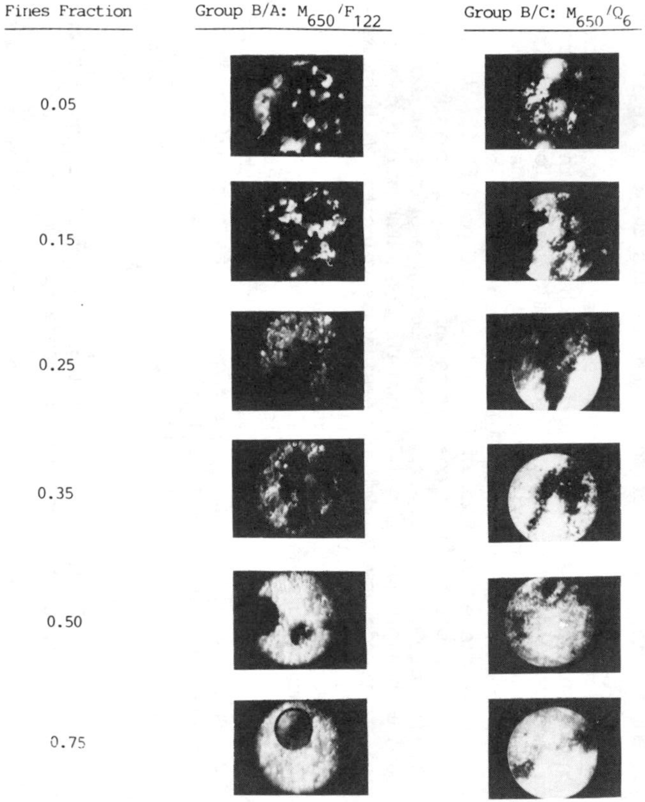

Figure 7. Pictures of flow behavior of fines adhesion on coarse with increasing gas velocity

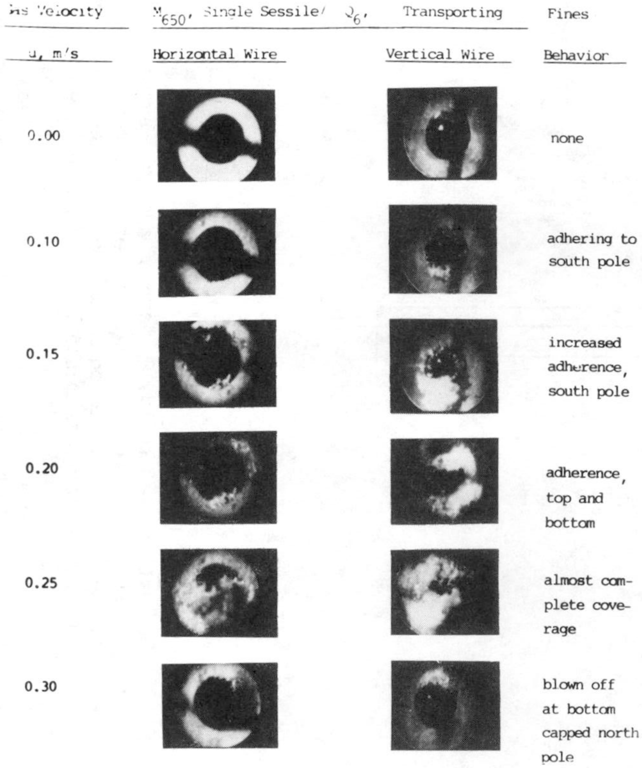

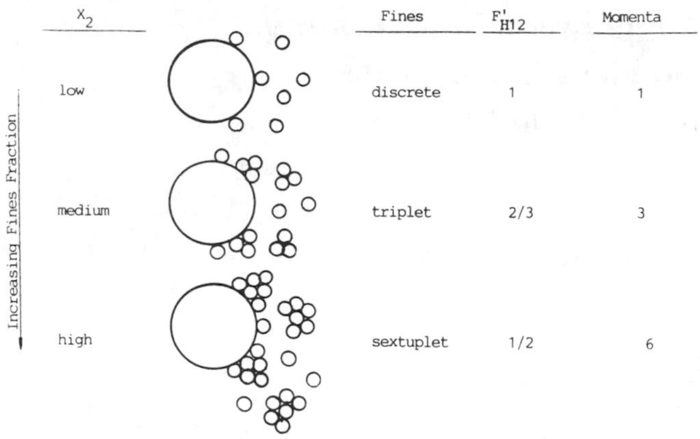

Figure 8. Schematic diagram of mechanism of the synergistic action

2. Synergistic action for binary mixtures was found to occur only when the fine particles were capable of adhering to the coarse.

ACKNOWLEDGEMENT

The authors wish to thank Mr. Ai Min for his support in the experiments.

REFERENCES

1. Wilhelm, R. H., and Kwauk, M., Chem. Eng. Prog. 44(2), 201(1948)
2. Romero, J. B., et al., Chem. Eng. Prog. Symp. Ser. 58, 28(1962)
3. Verloop, J., et al., Chem. Eng. Sci. 25(5), 825(1970)
4. Tung, Y., and Kwauk, M., in "Fluidization -- Science and Technology," Ed. M. Kwauk and D. Kunii, Science Press, Beijing, 1982, p.155
5. Yang, Z., Tung, Y., and Kwauk, M., paper to AIChE National Meeting, Denver, (1983)
6. Xia, Y., Yang, Z., Tung, Y., and Kwauk, M., abstract of paper to Fine Particle Society 16th Annual Meeting, Miami, April, 1985
7. Geldart, D., Powder Technol. 7(5) , 285(1973)

Recording Evaporation Time History of Droplet on Flat Plate by Laser Real-Time Interferometry

YOUREN XU
East China Technical University of Water Resources
Nanjing, PRC

NENGLI ZHANG
Department of Thermal Engineering
Tsinghua University
Beijing, PRC

ABSTRACT

An accurate optical method, based on holographic and shearing interferometry, was developed to record the profile changes of liquid droplet evaporating on a flat plate. Interfacial flow phenomena can be studied interferometrically in real time. Detailed profiles of evaporating droplet were calculate according to the time variant interference fringe patterns. Evaporation processes of analytically pure liquids and two-component liquids were tested and experimental results are shown in sequential photos.

INTRODUCTION

In studing the interfacial flow pattern of evaporating drops on a flat plate, the authors have used direct photography, laser shadowgraphy and holographic and shearing interferometry to reveal different aspects of the process [1,2,4]. A minute droplet which rests on a flat plate takes the form of a spherical segment due to the action of surface tension. As described in detail in reference [2], a collimated laser beam passing through the droplet forms a specific shadowgraph, which exhibiting two sets of rings: an inner rin reflects the diameter of the liquid-solid interface on the test plate and the outmost ring reflects the refraction of the lens like droplet. From the shadow graphic pattern volume of the droplet can be estimated by a simple geometric consideration. Volume-time histories of various liquids were experimentally obtained and three types of evaporation were recognized by the authors. Howeve for substable and unstable-type evaporations the droplet surface was rippled, resulting in irregular outer rings which made difficulties for determining the volume of the droplet. The direct photography was used to observe the morphology of droplets undergoing unstable-type evaporation. Although three stages: quiescent, vigorous and residual stages were recognized through the observation [1], the investigation was qualitative in nature and a quantitative procedure is strongly desirable in order to reveal the variety of the evaporation process. In reference [4] the authors developed a special laser interferrmeter which combined the holographic interferometry and shearing interferometry in determining the detail profile of a droplet resting on a horizontal glass plate. Because the method was based on a two exposures technique, however, only an instantaneous information could be recorded at a time.

For the investigation of the whole process of evaporation of a liquid droplet on a flat plate quantitatively it is important to record the sequentia changes with time for both of the surface shape and the size of the droplet when it is evaporating. The present study is a development of the work of reference [4]. The specially designed laser interferometer is modified to record the required information in real time. The information can be shown with time variant interference fringes. Quantitative computations can be performed to get accurate information about the detail profiles of the droplet according to the time variant interference patterns. Evidently, the present method recording the droplet profile interferometrically in real time has advantages over all other available procedures in acquiring the time history of an evaporating liquid droplet.

PRINCIPLE

Two plane waves of laser light which propagate with an angle of 2θ between them and which are oriented symmetrically about the normal to a high resolution photographic plate with the emulsion upwards as shown in Fig.1. An exposure of the photographic plate is made and after this plate is developed it contains a set of high spacial frequency interference fringes forming a holographic grating. The developed plate is then carefully replaced in its oringinal position and an optically flat high quality glass plate is put on over the grating emulsion. While a test droplet is placed on the glass plate the two laser beams passing through the liquid drop interfere each other form another set of interference grating like fringes at the emulsion plane. A set of moire fringes which indicate the difference of these two gratings caused by the optical effects of the droplet is formed at this plane. The moire fringe pattern changing its shape and size when the droplet undergoes evaporation is observed in real time which contains the information necessary for calculating the profile changes with time of the droplet. Two types of fringes occur. One appears in the peripheral region near the contact line of the liquid droplet with the solid and other appears over most of the droplet cross section. The former forms a holographic interferogram. The profile in this narrow peripheral region can approximately be determined using the relation

$$y \simeq N\lambda/(n_1\cos\theta' - \cos\theta) \qquad (1)$$

where y is the height of the liquid surface and $\cos\theta = \sqrt{1 - (\sin\theta/n_1)^2}$; θ is the half angle between the collimated laser beams; n_1 is defined as $\sin\theta/\sin\theta'$; N is the fringe order number which is assigned integer values at the center of each bright fringes sequentially from the outer edge of the pattern; and λ is the wavelength of laser light. The latter constitutes a shearing interferogram. An approximate equation can be used to evaluate the liquid profile as

$$dy/dx \simeq N\lambda/(2(y + d)(n_1\cos\theta' - \cos\theta)\mathrm{tg}\theta') \qquad (2)$$

where d denotes the glass plate thickness and x is the distance from the contact line at the direction toward the center of the pattern. A more accurate calculation has been made using the equations as following

$$y_N^2(n_1 - n\sin\theta \cdot \sin\beta_2)/\cos\beta_2 - y_N^1(n_1 - n\sin\theta \cdot \sin\beta_1)/\cos\beta_1$$
$$+ d(n_2 - n\sin\theta \cdot \sin\beta_2')/\cos\beta_2' - (n_2 - n\sin\theta \cdot \sin\beta_1')$$
$$- n(y_N^2 - y_N^1)\cos\theta = N\lambda \qquad (3)$$

and

$$x_N^2 = x_N + d\cdot tg\beta_2' + y_N^2 tg\beta_2 \quad (4)$$

$$x_N^1 = x_N - d\cdot tg\beta_1' - y_N^1 tg\beta_1 \quad (5)$$

where y_N^2 and y_N^1 are the height of the liquid surface at both sides of the N-th order of fringe respectively; x_N^2 and x_N^1 are the x-coordinates corresponding to y_N^2 and y_N^1; and x_N is the position of the N-th order of fringe. We have following relations to determine β_1, β_1', β_2 and β_2':

$$\beta_1 = \sin^{-1}(n \sin(\theta - y_N^{1\prime})/n_1) + y_N^{1\prime} \quad (6)$$

$$\beta_1' = \sin^{-1}(n_1 \sin\beta_1/n_2) \quad (7)$$

$$\beta_2 = \sin^{-1}(n \sin(\theta + y_N^{2\prime})/n_1) - y_N^{2\prime} \quad (8)$$

$$\beta_2' = \sin^{-1}(n_1 \sin\beta_2/n_2) \quad (9)$$

where $y_N^{2\prime}$ and $y_N^{1\prime}$ are the slopes at positions of x_N^2 and x_N^1 respectively. The changes of the pattern with time correspond to the evaporation process of the droplet and they can be recorded by a video cassette recorder for use as quantitative analyses and calculations.

EXPERIMENTAL SETUP AND PROCEDURE

The experimental setup is illustrated in Fig.2. A He-Ne laser 1 of 50 mw was used as a light source. The laser beam was spatially filtered, expanded and collimated by a collimator 2. The collimated light beam was then separated by a beam splitter 3 into two beams of light waves. Mirrors 4,5 and 6 were used to direct the two beams to the test section which consisted of a test droplet 7, a high quality glass plate 8 and a high resolution photographic plate 9. A mirror 10 and two lenses 11 and 12 cooperated with a video camera 13 were used to pick up the image of the interference pattern. The required information was displayed on a monitor 15 and recorded by a video cassette recorder 14 for late use as quantitative analyse and calculations with a microcomputer 16.

The photographic plate used was a holographic plate of Tianjin type-I. After the optical setup was adjusted correctly as required by the experimenta principle, an exposure was made with only the holographic plate in place. The plate was then developed and fixed and repositioned carefully. The glass substrate was thus placed on it as shown in Fig.1. A drop of test liquid was carefully placed on the glass plate by means of a 50 μl micro syringe. The needle tip of the micro syringe was slightly touching the substrate surface before and during the injection to permit the formation of a calm unsplashing test droplet. The substrate plates were cleaned and left overnight before use for the purpose of maitaining consistency and reproduction of the test results.

RESULTS AND DISCUSSION

The evaporation process of droplets of several analytically pure liquids and some two-component liquids were investigated using the experimental system. The equations mentioned above were used in calculations of the sizes and shapes of the droplets at sequential times during the evaporation process.

Fig.3 shows the sequential changes with time of the interference fringes of an analytically pure carbon tetrachloride droplet evaporating on a glass plate under room temperature in open air. The droplet under study experienced a typical stable-type evaporation. Fig.5 presents the droplet profile changes during the evaporation process, which were calculated according to the interference pattern of Fig.3.

A typical unstable-type evaporation process is presented in Fig.4, it shows the sequential changes of the interference fringes of an ethanol droplet evaporating on a glass plate. Three evaporation stages are evident: from (a) to (b) the interference pattern changed gradually and the process took quite a long time, which indicated the quiscent stage; photos (c) to (m) showed the violent changes of the droplet profile indicating the second stage of evaporation process, which experienced relatively short time compared to the first stage. The third stage of evaporation process, the residual stage, was displayed by the photos from (n) to (p). This stage took a very long period in which the drop size changed slowly while its shape remained essentially unchanged. Experimental resulte showed that the unstable-type evaporation process such as indicated in Fig. 4 had the characteristics of a typical evaporation pattern of two-component liquid droplets. The detailed discussion will be presented in a separate paper later.

In conclusion, we report an accurate optical technique in this paper, which is especially suitable to visualize and measure the profile changes of a liquid droplet undergoing evaporation on a flat plate. The detailed evaporation process can be studied interferometrically in real time through the profile changes of liquid droplets. Preliminary experimental results presented here indicate attractive interfacial flow phenomena in an evaporating droplet.

REFERENCES

1. Zhang, N. & Yang, W-J., J. Heat Transfer 104 (1982) p.656

2. Zhang, N. & Yang, W-J., Sci. Instrum. 54 (1983) p.93

3. Vest, C.M., Holographic Interferometry, John Wiley & Sons, (1979)

4. Xu,Y., Zhang, N., Yang, W-J. & Vest, C.M., Experiments in Fluids 2 (1984) p.142

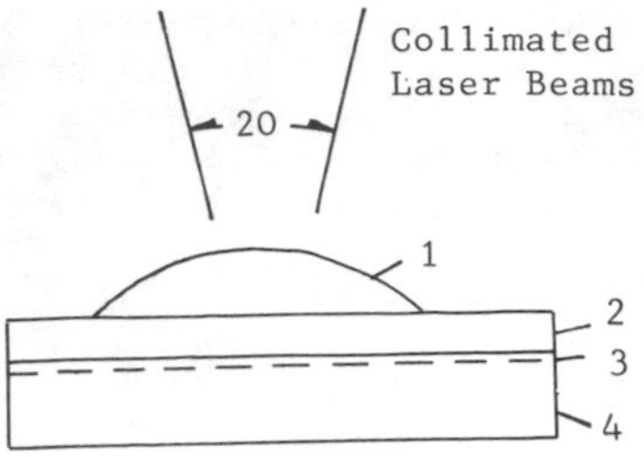

Fig.1 Schematic diagram of the test section
1— Droplet Sample
2— Glass Plate
3— Photographic Emulsion
4— Photographic Plate

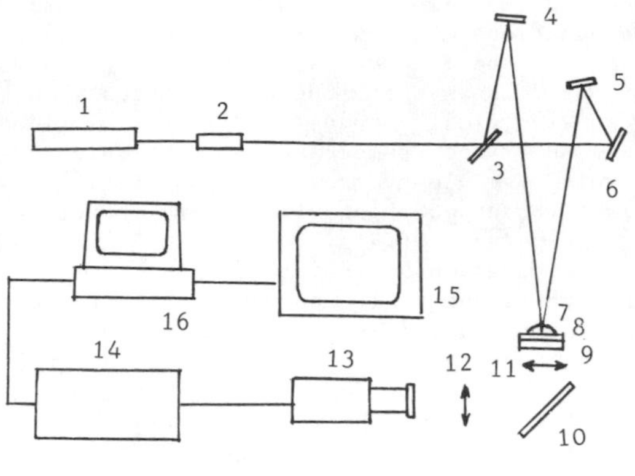

Fig.2 A schematic of experimental system

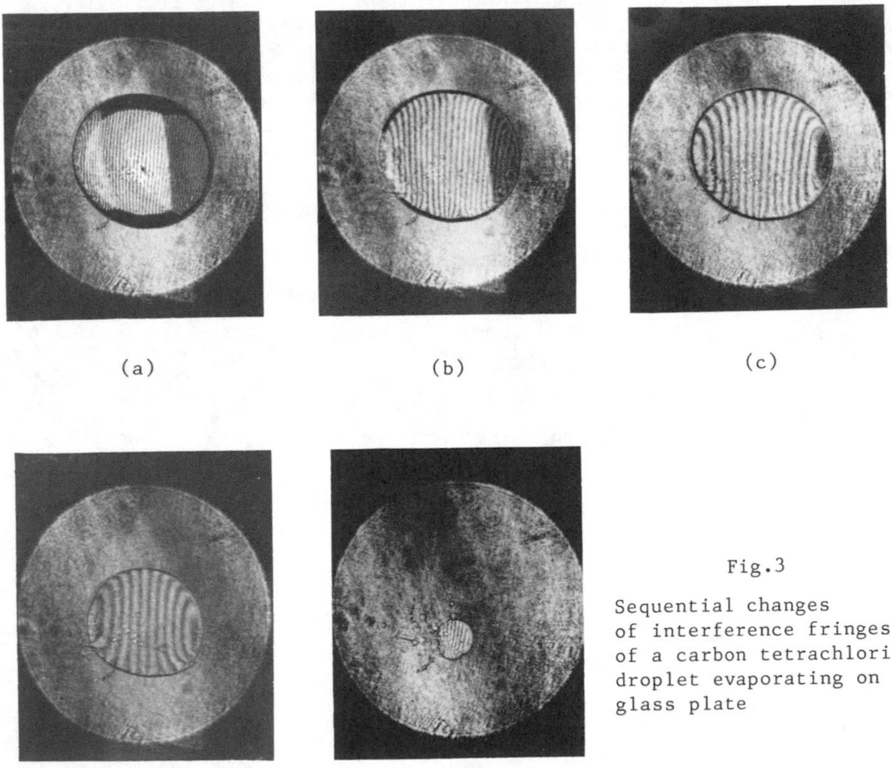

(a) (b) (c)

(d) (e)

Fig.3

Sequential changes of interference fringes of a carbon tetrachloride droplet evaporating on a glass plate

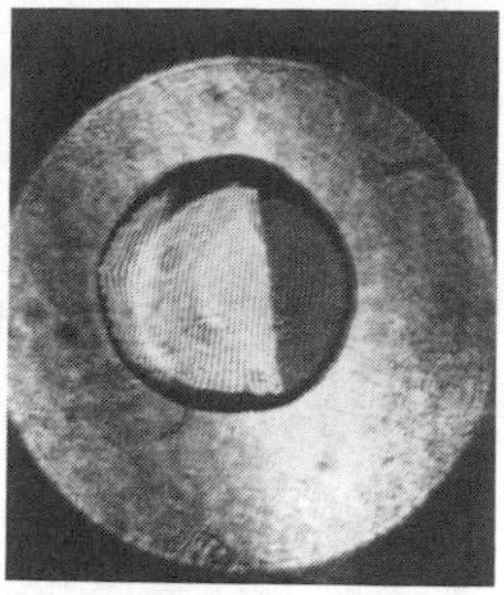

(a)

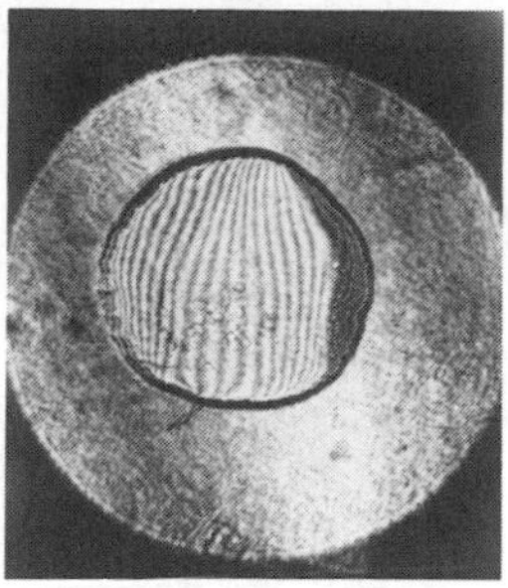

(b)

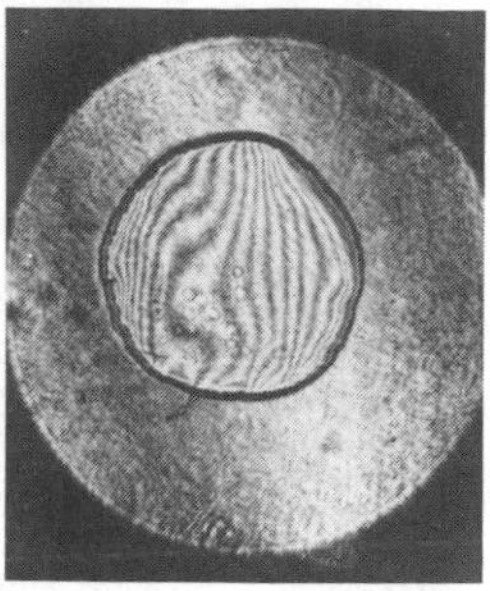

(c)

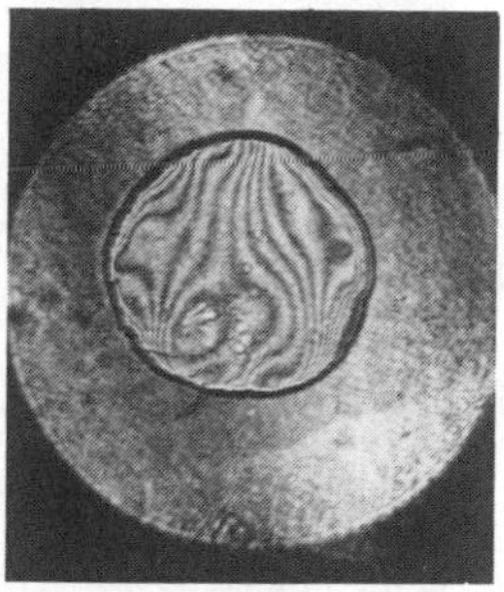

(d)

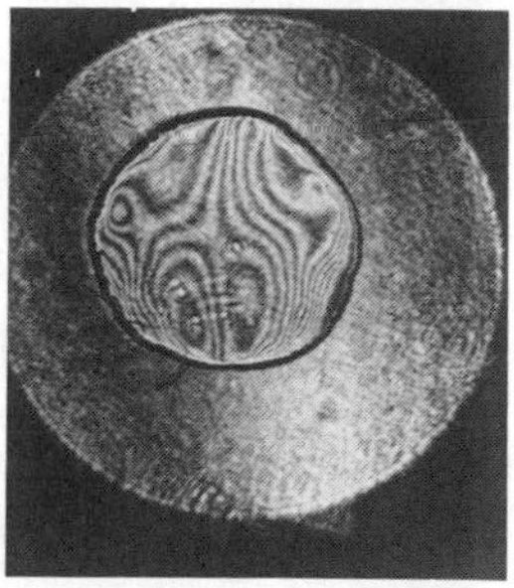

(e)

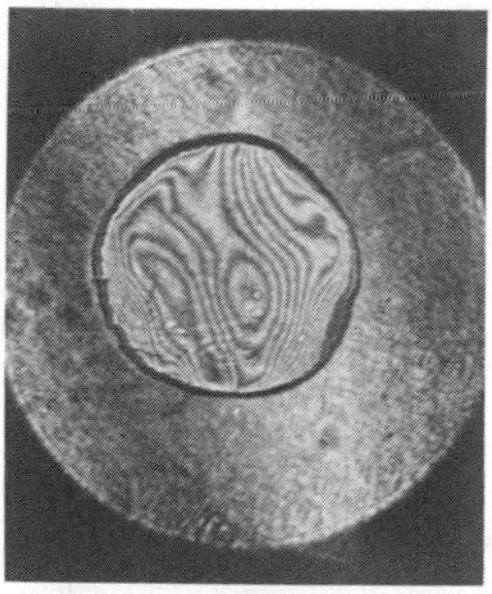

(f)

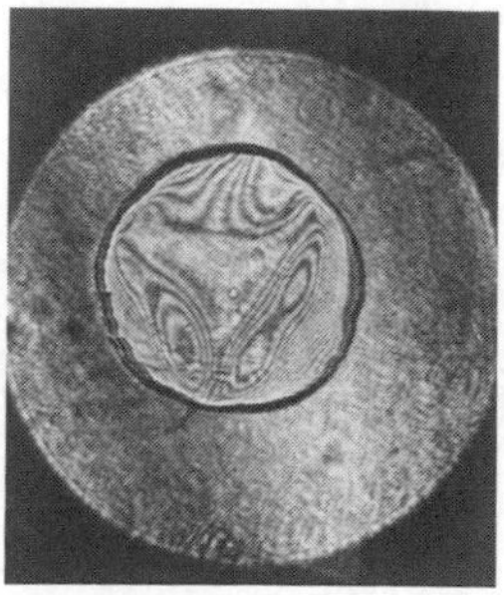

(g)

(h)

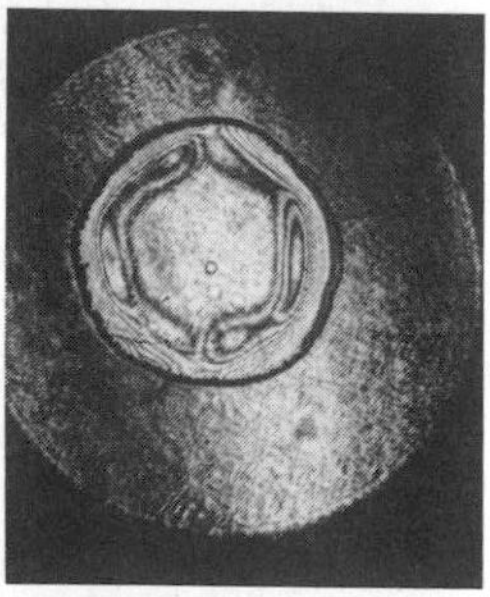

(i)

(j)

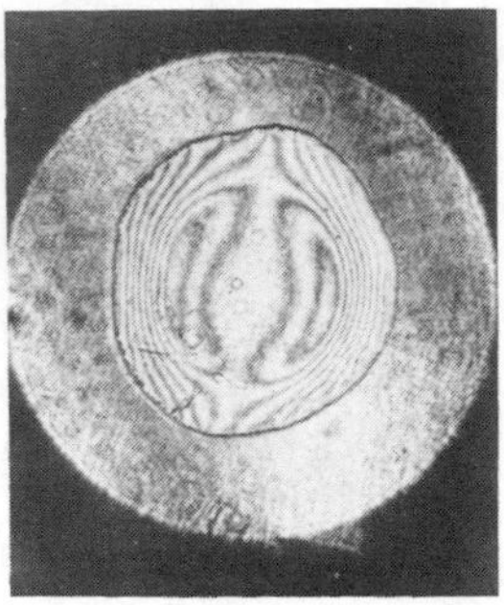

(k)

(l)

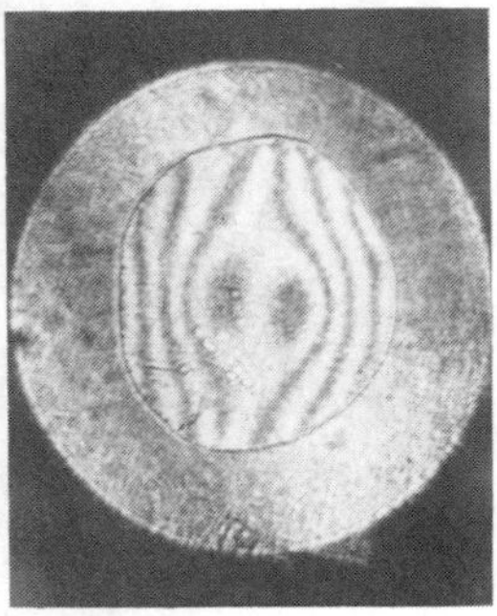

(m)

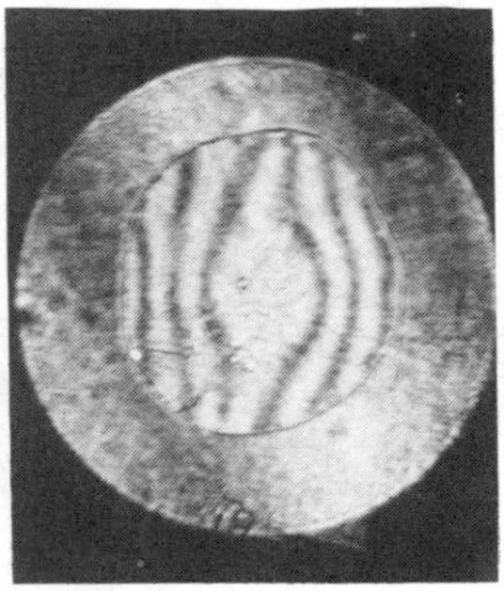

(n)

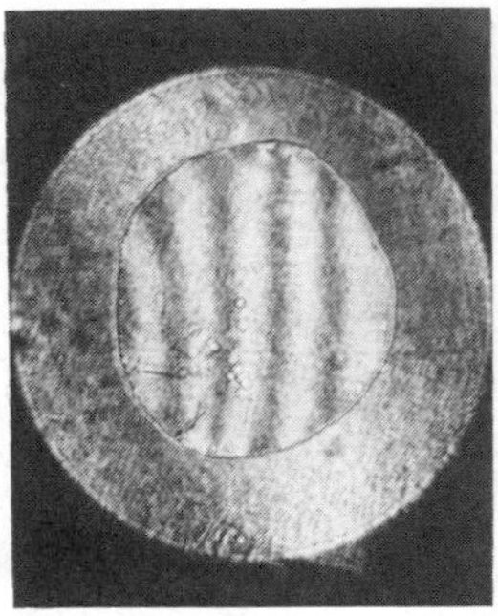

(o)

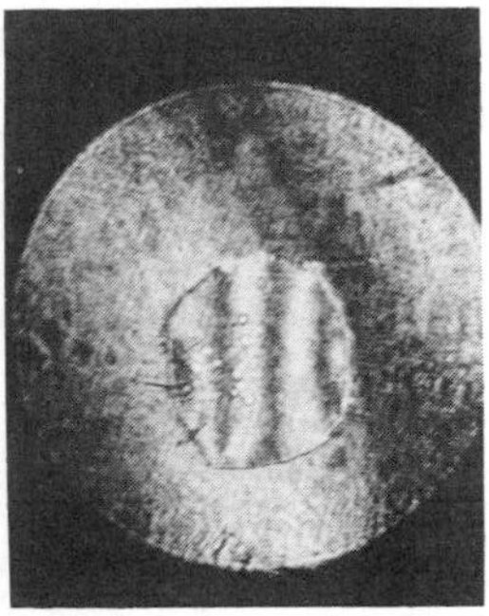

(p)

Fig.4 Sequential changes of interference fringes of an ethanol droplet evaporating on a glass plate (a typical evaporation process of two-component liquid droplet)

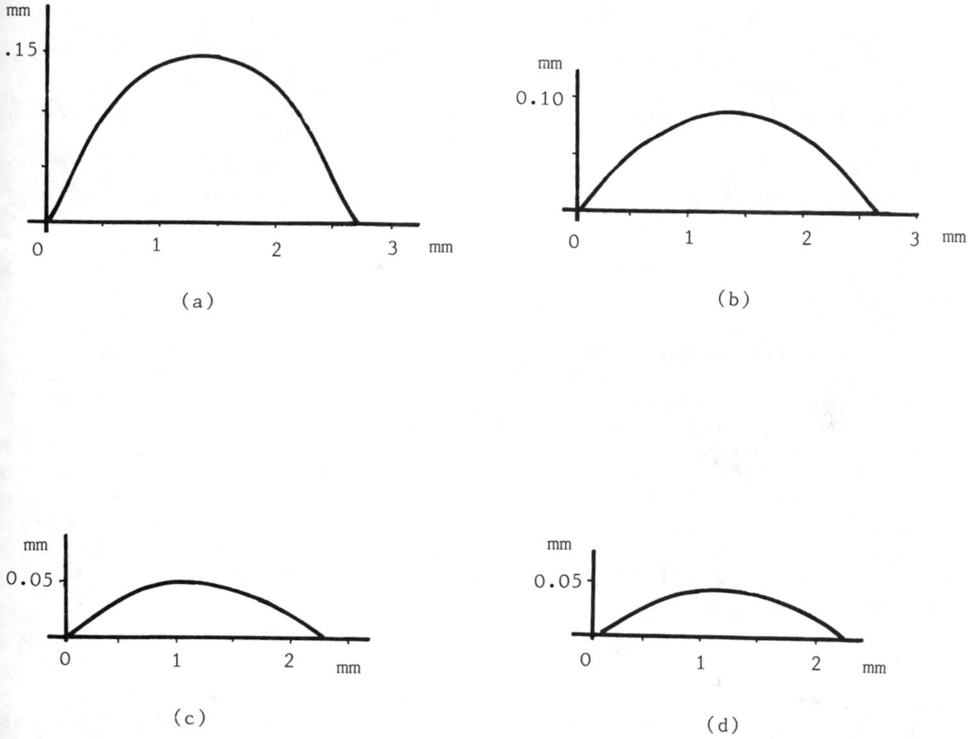

Fig.5 Droplet profiles calculated from interference pattern in Fig.3

Flow and Stability inside Evaporating Sessile Drops of Binary Liquid Mixtures

W. J. YANG
Department of Mechanical Engineering and Applied Mechanics
University of Michigan
Ann Arbor, Michigan 48109, USA

T. UEMURA
Faculty of Engineering Science
Osaka University
Toyonaka, Japan

C. L. CHAO
Chungchen Institute of Technolog
Taiwan

ABSTRACT

The double-diffusive convection induced by buoyancythermo-capillary action in minute drops of binary-liquid mixture evaporating on a horizontal plate is studied by holographic interferometry, shadowgraphy and direct photography. Water and silicon oil are selected as the solvents with various liquids of low boiling point as the solutes. Interfacial flow patterns and drop morphology are studied together with internal flow structures, cellular formation and heat transfer mechanisms during evaporation process. The mechanisms of interfacial turbulence, spontaneous emulsification and drop surface denture are explained.

INTRODUCTION

Natural convection in evaporating drops of pure liquids on a plate is summarized in reference 1. An unusual feature in thi thermal system is the effect of surface tension force on fluid motion which may be comparable or even more important than buoyancy-force effect. As a result of combined surface tension/buoyancy action, both the convective transport process and flow instability which determine the heat and mass transfer become very complex. Articles on Marangoni instabilities in reference 2 provide a good overview of such interfacial phenomenon. Another unusual feature is the everchanging shape and surface conditions (at microscopical scale) of evaporating droplets whose timewise varying size depend upon the plate temperature, liquid physical properties and ambient conditions. Reference 3 disclosed three distinct flow structures at the liquid-air interface: stable, substable and unstable. While th drops of stable and substable type maintain a lens shape as they diminish, the unstable type drop exhibits a crater shape during the course of evaporation. The evaporation process of unstable-interface type drop consists of three stages: quiescent, vigorous and residual stages which exhibit different drop shape. As a result of differences in shape and/or surface condition, the rate of evaporation and life time (the time required for a droplet to disappear) can be significantly different among droplets of different liquids. These two unusua features contribute to enormous difficulties in theoretical analysis of evaporating drops on a plate due to mathematical complication in the formulation of the problem. Because of the microscopical scale of fluid flow patterns in a minute liquid volume, the studies on the Marangoni instability and natural convection have been confined to a non-invasive experimental approach by optical methods such as shadowgraphy [3], interferometry [4], and holography [5] as well as by direct photography [3, 6]. The qualitative information thus far being obtained includes the time histories of droplet shape, size and flow patterns. Some experimental techniques and results are summarized in [7].

The present work deals with a double-diffusive convection in binary liquid drops evaporating on a horizontal plate by means of shadowgraphy, interferometry and direct photography. Water and index matching oil are used as the solvents with various volatile liquids of a relatively low boiling point as the solutes. Internal flow structures and stability phenomena in the evaporating drop are investigated.

EXPERIMENTAL APPARATUS AND PROCEDURE

Various experiments were conducted to record the microscopic flow patterns in binary-liquid droplets evaporating on a glass plate. The test setups for laser shadowgraphy, laser holographic interferometry and direct photography were identical with those employed in the previous studies [3, 4]. In brief, the laser shadowgraphy used the apparatus consisting of a laser light, a test plate, two aluminized mirrors, a screen and a camera (Fig. 1). The apparatus for direct photography consisted of a white light source, a glass plate coated with developed emulsion on one side, and a camera (Fig. 2). In the laser holographic interferometry, a drop was placed on a transparent substrate which was in turn placed on the emulsion of a high-resolution photographic plate. The droplet was then illuminated from above by two plane waves of laser light which propagated with an angle 2θ between them and which were oriented symmetrically about the normal to the plate (Fig. 3). Two separate exposures of the plate were made: one before the drop was placed on the substrate and one after it was so placed. After this plate was developed, it contained interference fringe patterns.

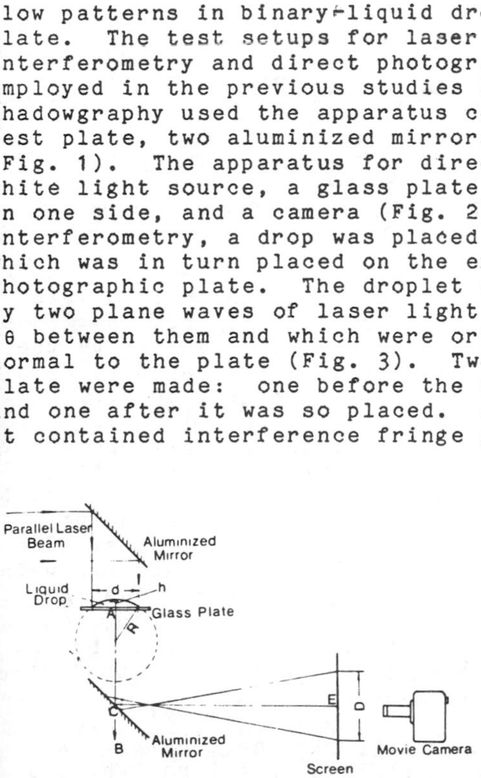

FIGURE 1. A schematic of laser shadowgraphic system

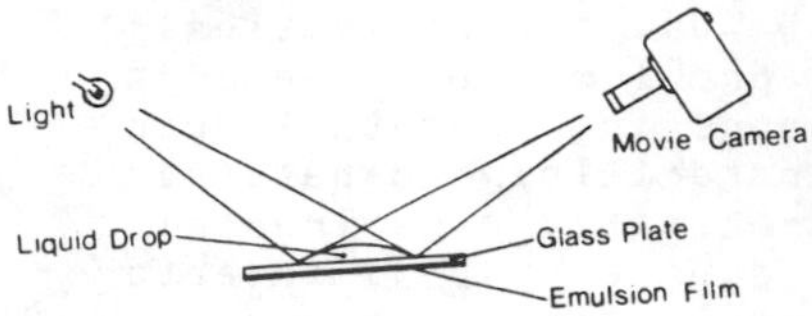

FIGURE 2. A schematic of apparatus for direct photography

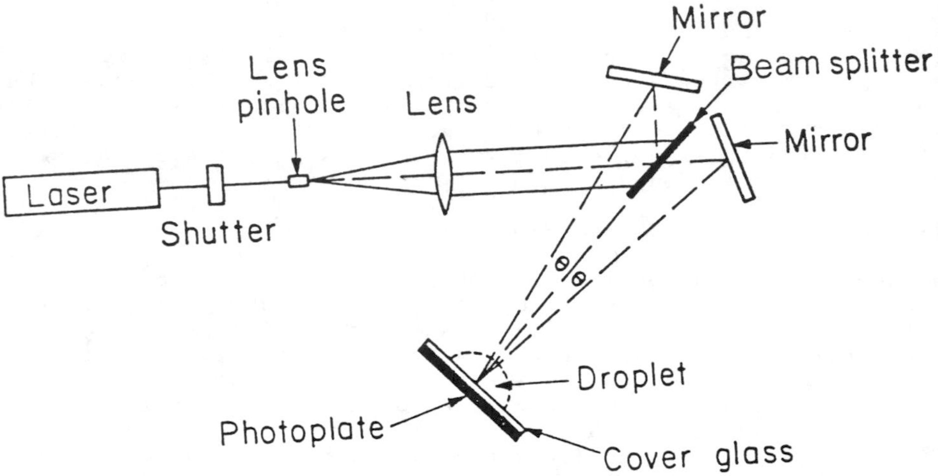

FIGURE 3. A schematic of laser holographic interferometric syste

A drop of 8 μℓ size was carefully placed on a glass plate using a 50 μℓ Monojet microsyringe. Details on the preparation of the test plate is available in reference 3. For observation by the naked eye, a tiny amount of aluminum pigment 1400 was spread on the test followed by the placement of a drop on the powder.

It is important to interpret the images in the shadowgraph and the interferogram: The shadowgraph is roughly sensitive to changes in the second derivative of density of the field. Due t temperature differences between the ascending and descending fluid streams, the convective pattern behaves like lenses. The descending stream of cold fluid whose density is higher (in whic the disturbance acts like a convex lens) produces a focusing effect on a parallel beam which traverses the fluid volume. Hence, it appears as a bright line. On the other hand, the ascending warm stream (in which the disturbance exerts the effec of a concave lens) appears as a dark area.

The interference fringes indicate changes in the density. A stream of ascending warm fluid appears as a dark line, while a cold stream is seen as a bright line.

XPERIMENTAL RESULTS AND DISCUSSION

ater and index matching oil were used as the solvents or base iquids. Both have high boiling points and thus evaporate very lowly. Whereas all the solutes at low boiling point evaporate ich faster, including acetone, benzene, carbon tetrachloride, hloroform, cyclohexane, ethanol, ethyl acetate, ethyl ether, ethanol and methylene chloride. The solvent to solute ratio was aried for each combination. The laser shadowgraphy, laser holo-raphic interferometry, direct photography and visual observation ere carried out on each test (using different drops of the same ize and concentration). For the same solute-solvent combina-ion, different ratios yielded different flow patterns. evertheless, both the interfacial and internal flow structures an be classified into certain regimes.

hree representative results are presented here: a 50% index atching oil - 50% cyclohexane mixture and a 50% water - 50% cetone mixture, as shown in Figs. 4, 5, and 6, 7, respec-ively. Each photo was taken on different drops of the same nitial size and composition. Finally, the proportion of oil-yclohexane was reduced to study the phenomenon of spontaneous mulsification in Fig. 8.

) Shape Varying Drop Evaporation - 50% Index Matching Oil/50% clohexane Mixture.

n Fig. 4, series A, B, and C indicate photos taken by holo-aphic interferometry, shadowgraphy and direct photography, espectively. When a droplet was placed on the plate, it took a early hemi-spherical shape, as shown in Figs. 4Ci and 5a. The orresponding holographic interferogram Fig. 4Ai, a finite nterferogram reconstruction, exhibited a circular image with arallel fringes. This is the quiescent stage of the evaporation ocess during which the top portion of the drop diminished due a high evaporation rate. Rigorous natural circulation started mediately following the placement of the drop, since the itical condition for the onset of natural convection was atisfied. Therefore, no cellular structure was formed in this age.

e rigorous evaporation stage was initiated when the peak of e droplet began to dent as a result of evaporation of the re volatile liquid, the interference fringes in the center gion became wider and curved, pushing other fringes closer ward the sides, Fig. 4Aii. The corresponding shadowgraph is en in Fig. 4Bii. As the droplet became dented, as sketched in g. 5b, the interferogram Fig. 4Aiii exhibited a fringe pattern at was practically symmetrical with respect to the equator of e circle. Two embryos, not visible in the photo, were formed the south and north poles with the neighboring fringes being nt center-ward. The condition for formation of cellular motion the liquid is on its way to be fulfilled. As these two bryos grew to form Benard cells, the second-generation embryos re nucleated, one on each side (45 degrees to the east and the st) of the nucleation sites of the first-generation embryos.

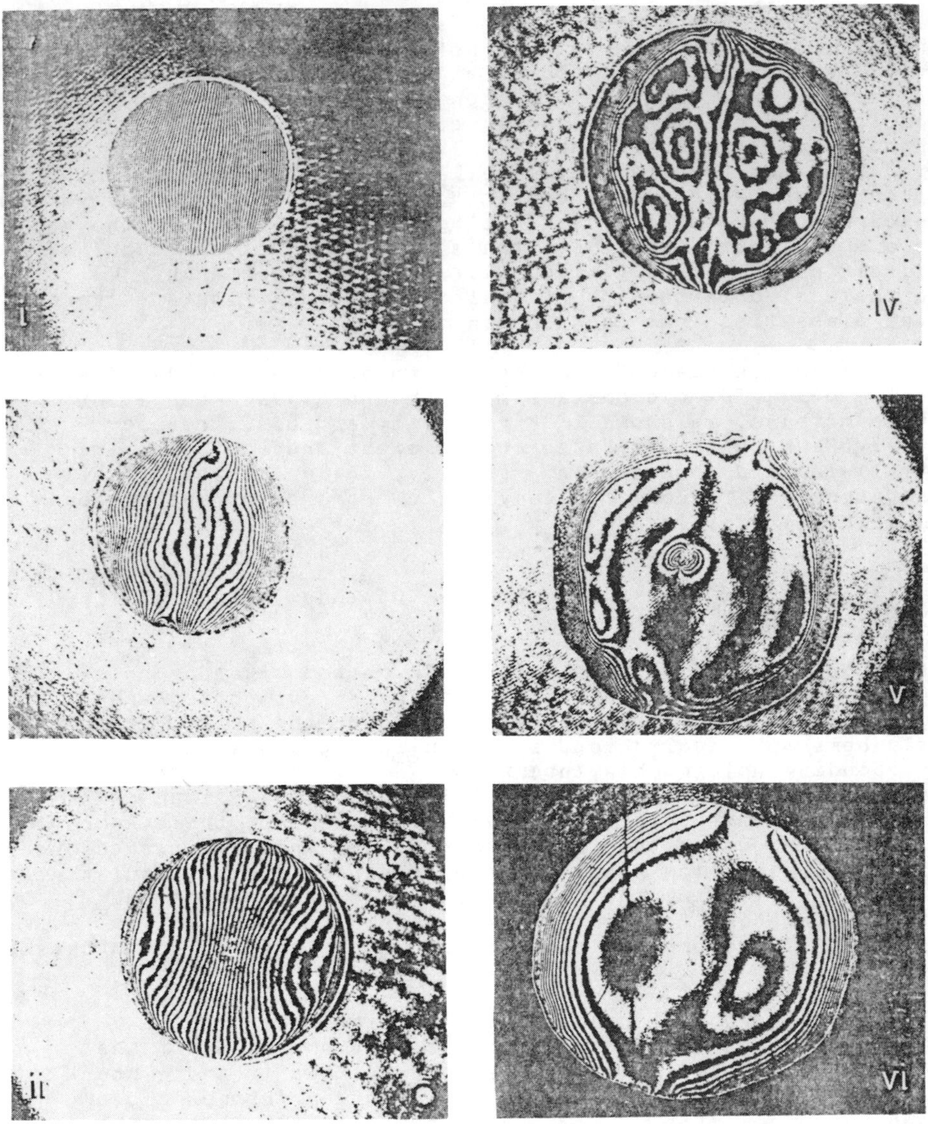

FIGURE 4A. Interferometric (series A), shadowgraphic (series B and direct photographic (series C) results of 50% oil - 50% cyclohexane drops during evaporation process

During the course of growth, these new generation embryos pushe the first pair of Benard cells toward the center, as depicted i Fig. 4Aiv. Notice that one Benard cell at the south-west position is somewhat disfigured. In each Benard cell, a stream of warm fluid (in dark) sprang up (upwelling) in the center lik a fountain jet, with colder stream (in bright) flowing downward along the edge of the cell. Presumably, the warm stream is hig

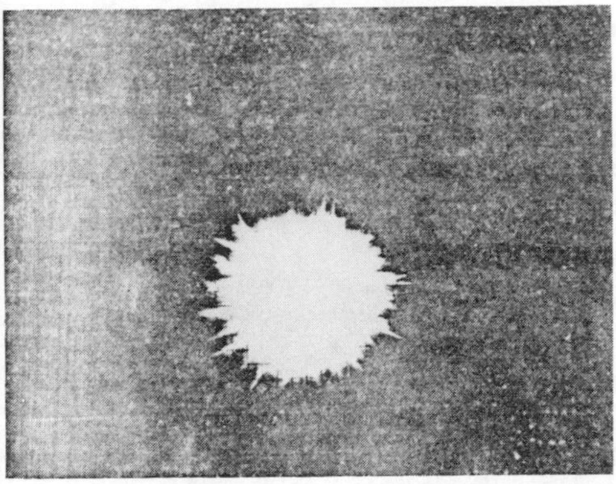

4Bii

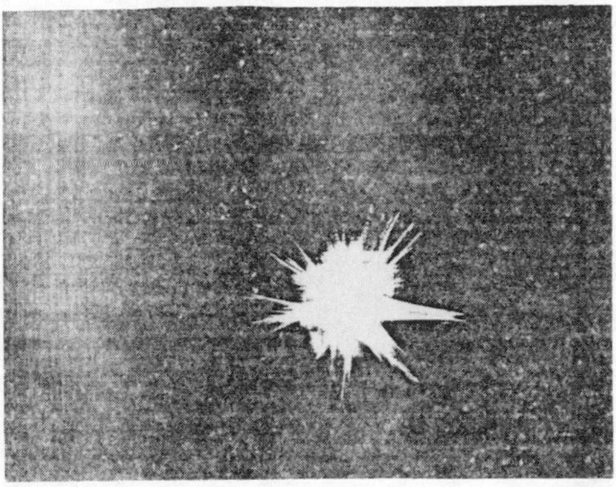

4Bvi

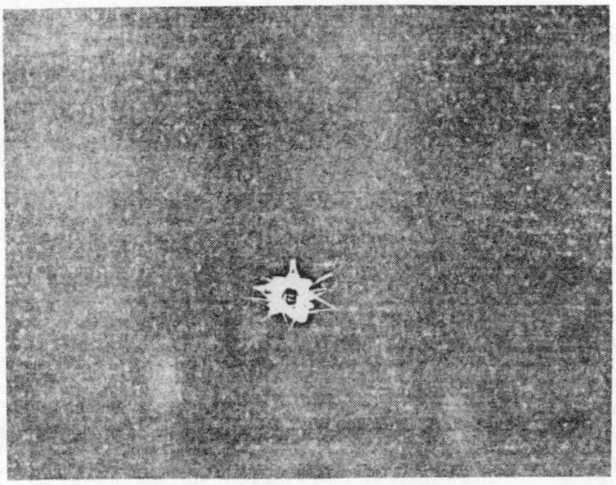

4Bvi

FIGURE 4B. Interferometric (series A), shadowgraphic (series B) and direct photographic (series C) results of 50% oil - 50% cyclohexane drops during evaporation process

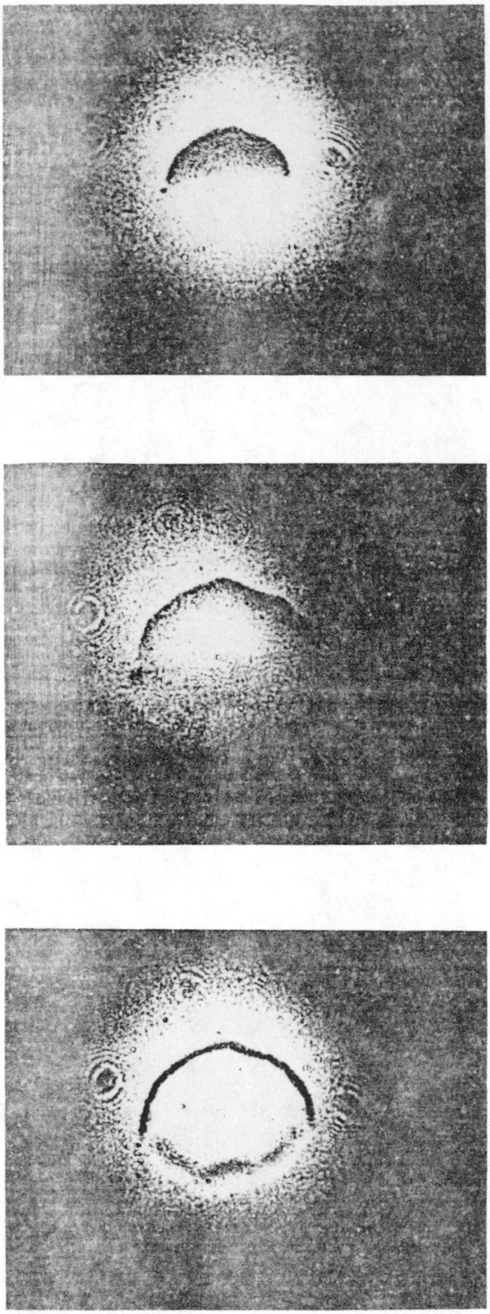

i

iv

vi

FIGURE 4C. Interferometric (series A), shadowgraphic (series B) and direct photographic (series C) results of 50% oil - 50% cyclohexane drops during evaporation process in the concentration of the more volatile component, while the colder stream is high in the concentration of the solvent. As the more volatile liquid was surfaced, it evaporated into the air. Therefore, high evaporation took place in the region with Benard cells. Some cellular structures were double-layered, others were triple-layered. The liquid layer must be thick enough to sustain the fluid circulation in a Benard cell. In other words, a Benard cell begins to disintegrate as soon as the thickness of the liquid layer is diminished to a certain level. Upon the disintegration of a Benard cell, the evaporation rate drops in the absence of flow circulation. In Fig. 4Aiv, the outer crust consisting of parallel fine fringes was the interference image of a ridge around the droplet, while the region with Benard cells formed a basin. The morphology of the droplet corresponding to Fig. 4Aiv is shown in Fig. 4Civ, or Fig. 5c. When the depth of the basin diminished to a certain level, which prohibited internal flow circulation, as a result of evaporation, the Benard cells started to disintegrate, beginning at the center region and proceeding outward. Figure 4Av shows the instant where four Benard cells (with two at left touching each other) remain along the western border. Notice the fringe pattern of dust at the center.

When all Benard cells had disappeared, the basin became a placid pool of liquid surrounded by a ridge as shown in Figs. 4Cvi and 5d. Heat transfer in the liquid film was by conduction. The evaporation process is in the residual stage. The placid liquid layer displayed a fringe pattern as depicted in Fig. 4Avi, while the corresponding shadowgraph exhibited a star-like pattern as seen in Fig. 4Bvi. Such a shadowgraphic pattern is typical of an unstable type liquid. With cyclohexane almost completely evaporated, eventaully surface tension acted to pull the surrounding ridge toward the basin and formed a flat plateau as sketched in Fig. 5e. Now, the evaporation of index matching oil is a very slow process, since heat is transferred by conduction alone.

The evaporation of cyclohexane is a stable type maintaining a lens shape in the entire course of phase change [3]. Oil forms a nearly hemispherical drop and evaporates very slowly [4]. However, when these two liquids are mixed, 50% each in this case, they exhibit an unstable-type evaporating behavior, as characterized by three stages during which the drop shape changes from a lens to a crater and finally to a flat top. The mechanism of forming a crater is as follows: When the drop is in a lens shape, its central region has the greatest height allowing a higher internal circulation. It results in the highest concentration of cyclohexane on the drop surface there, promoting high evaporation. The high departure rate of cyclohexane molecules coupled with the repulsive pressure of the adsorbed layer of cyclohexane lead to the buckling in of the drop surface and consequently the formation of a basin in the central region of the drop. The lowering of the drop surface continues until the local liquid depth becomes too low to sustain a natural circulation. Then, the heat transfer mechanism in the liquid film at the basin changes from convection to conduction, a slower process. The evaporation from the basin region drops.

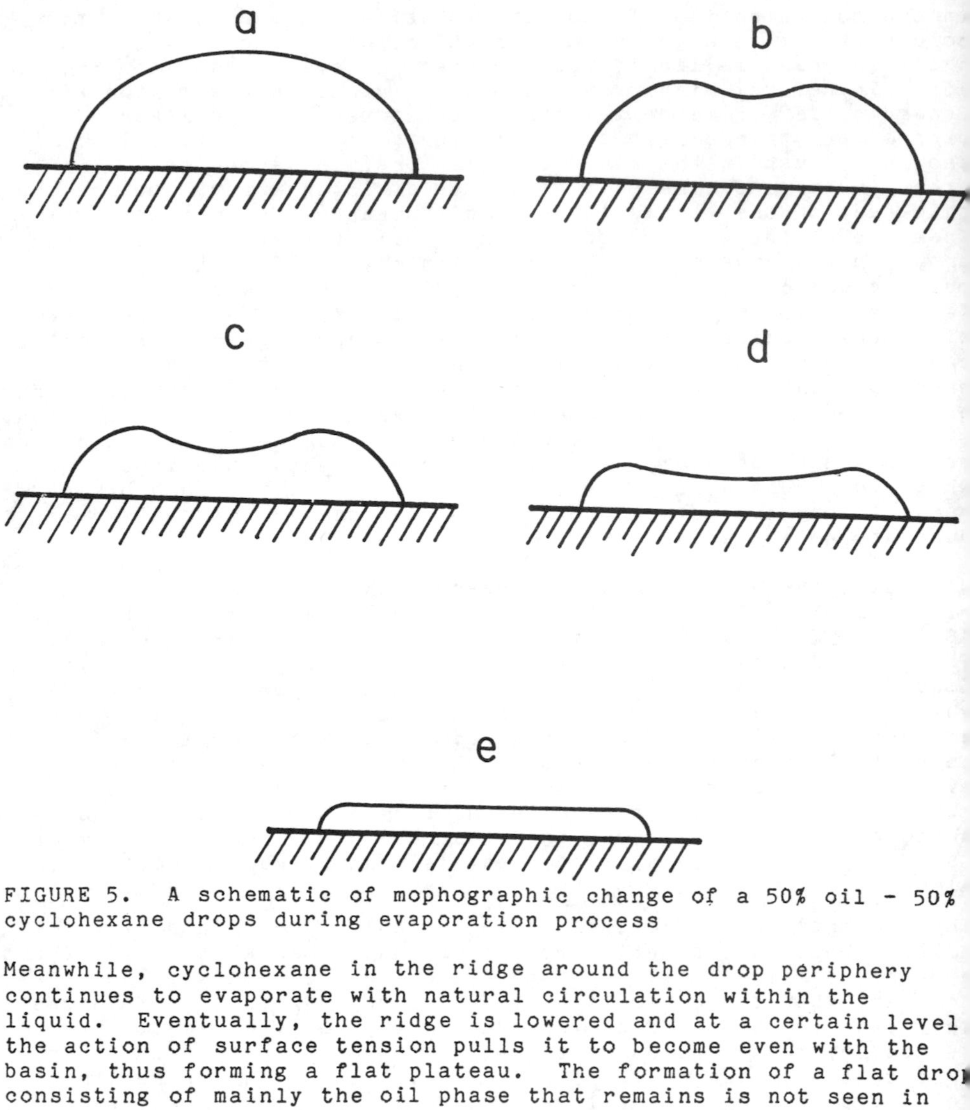

FIGURE 5. A schematic of mophographic change of a 50% oil - 50% cyclohexane drops during evaporation process

Meanwhile, cyclohexane in the ridge around the drop periphery continues to evaporate with natural circulation within the liquid. Eventually, the ridge is lowered and at a certain level the action of surface tension pulls it to become even with the basin, thus forming a flat plateau. The formation of a flat drop consisting of mainly the oil phase that remains is not seen in the residual stage during the evaporation of an unstable type drop of pure liquids. In the latter case, a new hemispherical drop is formed if the surface tension of the liquid film is strong enough to pull the torous toward the center. Otherwise, the torous will remain at the same spot until its complete evaporation. A reduction in the initial concentration of cyclohexane in the oil drop reduces the extent of denting and eventual flattening of the drop shape. At low initial cyclohexane concentration, the oil drop remains essentially a lens shape in the course of cyclohexane evaporation.

2) Lens-Shape Drop Evaporation - 50% Water/50% Acetone Mixture

The acetone-water mixtures represent an extreme case in convective patterns, since the evaporation of pure acetone droplet belongs to the unstable type [3]. The entire course of evaporation of an acetone water droplet (50% - 50% mixture for example) can be divided into four stages: initial, intermediate (or transition), asymptotic and equilibrated stages. The last two stages may be combined into one final stage. With very high solubility of acetone in water, the droplet retained a lens shape throughout the entire evaporation process. Series A and B in Fig. 6 correspond to the interferographic and shadowgraphic photos, respectively. During the initial stage immediately following the placement of a droplet on the plate, two sets of fringes were recorded in the photographs, for example Fig. 6-A-i. Those fringes which appear in the peripheral region form a holographic interferogram [4]. The second set of fringes which generally appear over most of the liquid cross section, constitute a shearing interferogram. Those holographic interference fringes resulted from convection in a region at the foot of the droplet. A flow visualization by suspension of fine aluminum particles revealed that the fluid was in a spiral (from inside and outward) motion and circulating around the droplet as schematically shown in Fig. 7-a. The remaining volume of the droplet from side to the top was stagnant.

The intermediate stage of evaporation process begins when natural convection takes place throughout the entire drop. A pair of fringe shifts in the shearing interferogram appeared in the droplet cross section, Fig. 6-A-ii. The peripheral convection continued as confirmed by flow visualization, Fig. 7-b. These two fringe shifts rapidly elongated in opposite directions, accompanied by a sequential reproduction of additional frings shifts as depicted in Fig. 6-A-iii which was reconstructed photographs from the same photographic plate. These fringe shifts, called "hook lines", resulted from constant concentrations. All "hook lines" were very active, moving around busily in the fluid cross section. In shadowgraph, these "hook lines" appeared like fine "roots" individually moving in the same direction in an uncoordinated manner, as illustrated in Fig. 6-B-iii. A leaf-shape outer edge of the image suggested that the evaporation is of a substable type with the surface of the droplet being a slightly rippled [3]. The bright spot at the image center was the image center was the image of the light beam. The rate of evaporation increased with time in this stage, resulting from rigorous internal circulation.

When the movements of the "hook lines" and "roots" in the holographic interferogram and shadowgraph respectively slowed down, the evaporation entered into the asymptotic stage with a steady rate of evaporation. The fringe shifts changed in appearance from a band of very shallow "U"-shaped fringes to a band of "N" shaped fringes, as seen in Fig. 6-A-v. In the shadowgraphic images, the "roots" broke down to form multiple-lined "ribs", as shown in Fig. 6-B-iv. The outer edge of the image had more "sawteeth", indicating more rippled droplet surface undergoing higher evaporation. As seen in Fig. 6-B-v, the multiple-lined

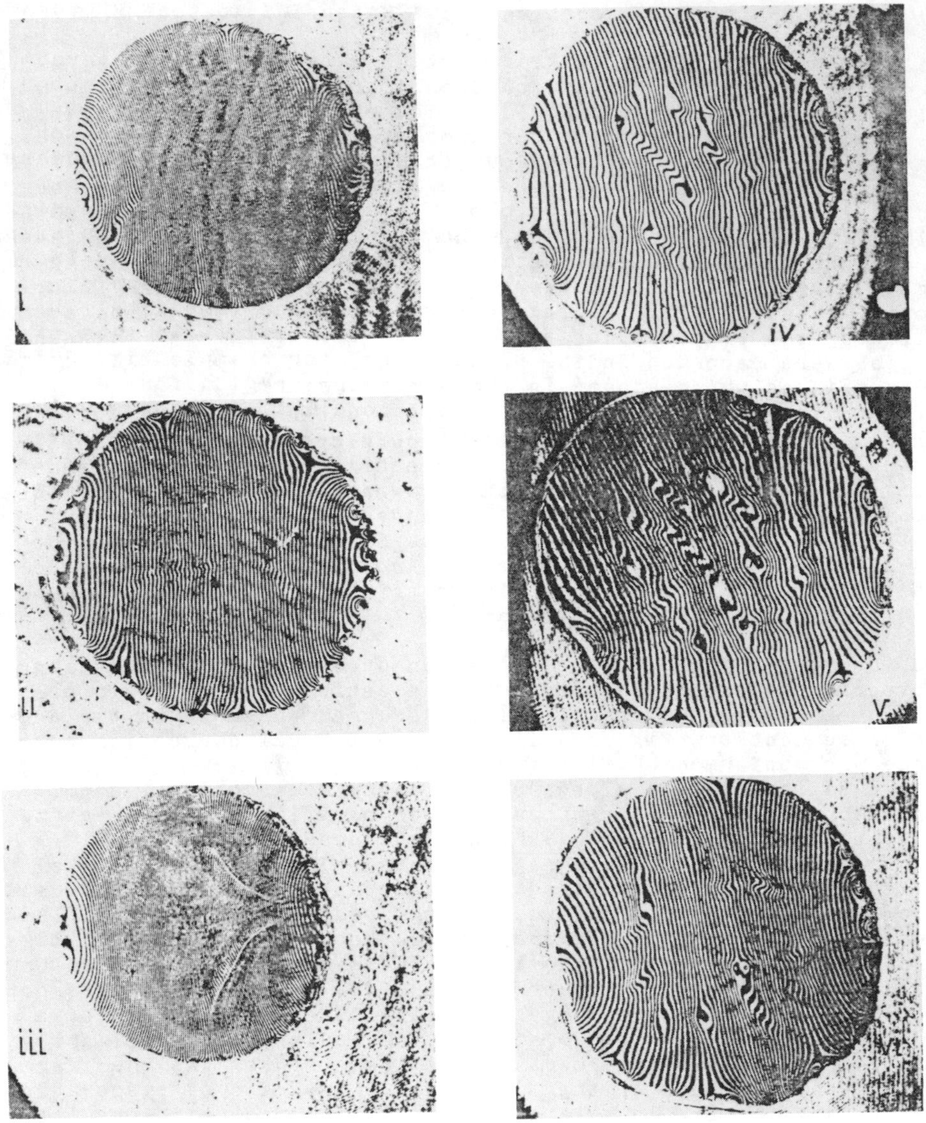

FIGURE 6A. Interferometric (series A) and shadowgraphic (series results of 50% water - 50% acetone drops during evaporation proc

"ribs" grew in width and length.

After a major fraction of the more volatile piece had evaporated the drop came in the final stage when the movements of all "hook lines" and "ribs" ceased. Heat transfer within the drop is by conduction alone. The image patterns in the equilibrated stage are depicted in Figs. 6-A-vi and 6-B-vi. Evaporation continued

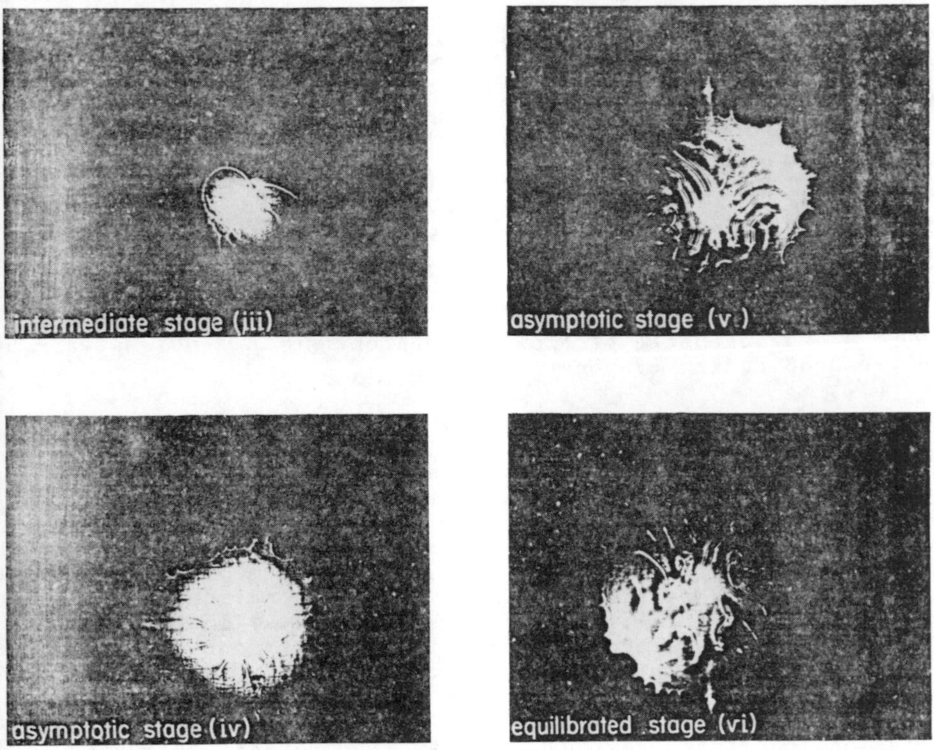

FIGURE 6B. Interferometric (series A) and shadowgraphic (series B) results of 50% water - 50% acetone drops during evaportion process

to take place under slow convection.

The molecular mechanism of interfacial turbulence can be explained as follows. At some point on the interface, an eddy (so-called "disturbance") of water brings up rather more acetone than is present at other points on the interface. Such eddies may originate in the formation of the drop, in thermal inequalities in the system, in an eddy of air, or in movements associated with previous occurrences once the process is started. At the interface, locally high concentration of acetone results in an increase in adsorption over a small region on the drop surface. An enhancement in the evaporation of acetone follows, causing a local reduction in the spreading pressure there. Spreading into this region then occurs from the adjacent film of liquid, causing liquid to be pushed into a bump at the point where the original evaporation occurred. These acetone molecules spreading from a locally high concentration in the surface carry some of the underlying water with them, bring up more acetone toward the same point, thus enlarging what was originally a small effect into a pronounced one. The net effect of surface turbuelnce is an appreciable depletion of volatile substance into the air, accompanied by strong circulation of the liquid inside the drop. The fringes show the temperature changes due to

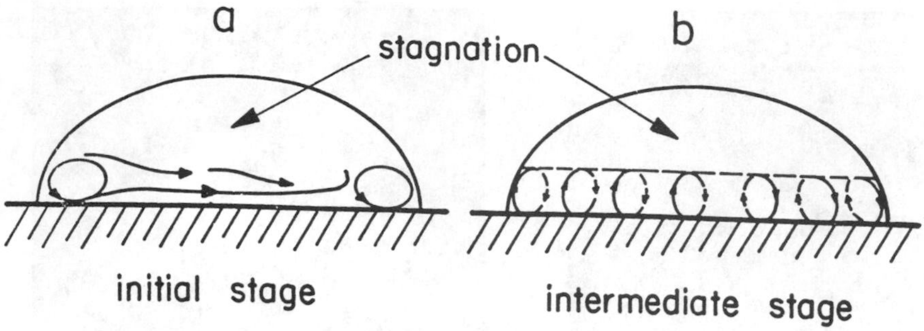

FIGURE 7. A schematic of flow-pattern change in a 50% water - 5C acetone drop during evaporation

redistribution of the solute in the drop.

(3) Spontaneous Emulsification - When Two Inmiscible Phases are Present, a Dispersion System is Formed.

Such a system is often thermodynamically unstable, because any fr energy associated with the interface between the disperse phase a the continuous phase can decrease by the aggregation or coalescen of the dispersed phase. Spontaneous emulsification is used to describe the process for the formation of emulsions that requires external mechanical work. Three mechanisms have been suggested t account for spontaneous emulsification in a system consisting of immiscible liquids with certain additives: interfacial turbulenc diffusion and stranding, and negative interfacial tension [8]. I the present work, spontaneous emulsification was observed in evaporating drops of cyclohexane mixed with a small amount of oil Figure 8 is a typical series of shadowgraphs showing the process spontaneous emulsification in a drop with 98.4% cyclohexane and 1 oil by volume. Figures 8i, ii and iii showed the flow patterns f three separate drops immediately following their contact with the plate. The dark spot in the photo center was a piece of black pa placed on the screen to block off the laser light in order to balance the illumination of the other faintly visible in Fig. 8i. They were of a concentric form. During the diffusion process, th outer rings of the shadowgraph were distorted from a circular sha and spiked. This implied that the drop surface was slightly rippl indicating a substable-type evaporation [3]. Figure 8ii illustra the onset of spontaneous emulsification at the top portion of the drop. The formation of fine emulsion drops resulted from the aggregation or coalescence of the dispersed phase. The spikes diminished and the outer rings appeared quite spherical. It indicated that interfacial turbulence had diminished and the drop surface became quite smooth. This is typical of a stable-type evaporation. The fine emulsion drops then migrated outward (mear from the top down toward the root along the drop surface) in concentric fronts, as seen in Fig. 8iii. Since the evaporation r of the volatile component is high at the drop top, there needs to a counter diffusion of cyclohexane toward the drop top. The fine emulsion drops continued to fall toward the liquid contact region and the white dots disappeared from the shadowgraph. While oil

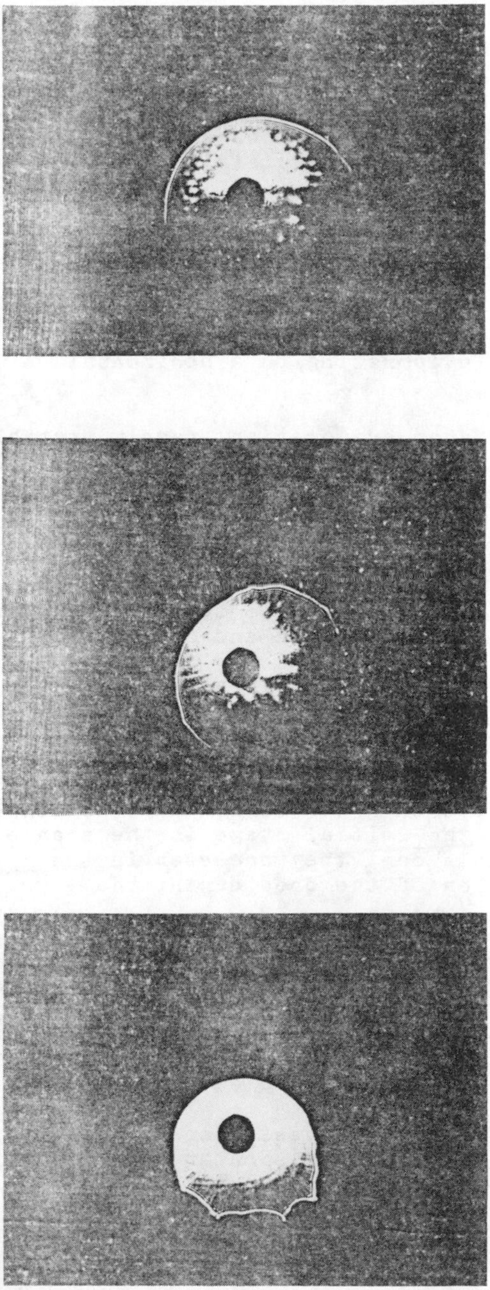

i

ii

iii

FIGURE 8. Shadowgraphs illustrating spontaneous emulsification in 98.4% cyclohexane -1.6% oil drops during evaporation process (heavier) remained on the plate around the drop periphery, cyclohexane contined to evaporate.

In conclusion, there was a transition of the evaporation process from a substable type to a stable one at the onset of spontaneous emulsification. While the fine emulsions of oil migrated downward from the top to the root along the drop surface, a counter diffusion of cyclohexane toward the drop top took place. The mechanism of spontaneous emulsification was "diffusion and stranding".

CONCLUSIONS

The optical methods, shadowgraphy and holographic interferometry, together with direct photography have been employed to study both natural convection and stability phenomena in minute drops of binary liquid mixtures evaporating on a horizontal plate. It is concluded that

(i) There are three types of drop evaporation, namely unstable substable and stable.

(ii) The surface of stable-type drops remain smooth and takes a lens shape, while that of unstable type drops is rippled and changes from a lens to a crater shape.

(iii) The history of both stable and substable type evaporation is characterized by the initial, intermediate and final stages.

The unstable-interface type drop undergoes four stages of the evaporation process, namely quiescent, rigorous, residual and equalized evaporation, the drop shape changes a lens shape to a torous one before its complete evaporation.

(iv) Benard cells are formed over a brief duration in all three type drops. They appear during the initial stage in the stable type drop, whereas in the unstable one, they are seen in the liquid film at the central portion of the drop during the rigorous evaporation stage.

(v) The mechanism of interfacial turbulence is that at the air-liquid interface, an increase in local evaporation (and thus a reduction in the interfacial tension) causes a momentary change in the local pressure inside the drop resulting in the liquid motion (stream of the more volatile substance) toward the interface.

(vi) A transition of the evaporation process from a substable type to a stable one occurs at the onset of spontaneous emulsification whose mechanism is diffusion and stranding.

ACKNOWLEDGEMENT

The work was supported by the National Science Foundation under the Grant Number MEA 83-04740 for which the authors are grateful. The second and third authors would like to express their appreciation for financial aid from their respective governments.

REFERENCES

1. Yang, Wen-Jei, "Natural Convection in Evaporating Droplets," Chapter 2 in Vol. I. Transport and Reaction Mechanisms, Handbook of Heat and Mass Transfer Operations (Ed.: N. P. Cheremisinoff), Gulf Publishing, West Orange, New Jersey (in press, 1984).

2. Zierep, J. and Oertel, H. Jr. (eds.), Convective Transport and Instability Phenomena, G. Braun GmbH. Karlsruhe, West Germany (1982).

3. Zhang, N. and Yang, Wen-Jei, "Natural Convection in Evaporating Minute Drops," Journal of Heat Transfer, vol. 104, pp. 656- 662, (1982).

4. Xu, Y., Zhang, N., Yang, Wen-Jei, and Vest, C. M., Optical Measurement of Profile and Contact Angle of Liquids on Transparent Substrates," Experiments in Fluids, vol. 2, pp. 142-144, (1984).

5. Xu, Y., Zhang, N., and Yang, Wen-Jei, "Direct-Recording Optical Methods for Visualizing Fluid Motion in Minute Liquid Drops," Flow Visualization III, Ed.: Wen-Jei Yang, Hemisphere, Washington, D.C., pp. 668-672 (1983).

6. Zhang, N. and Yang, Wen-Jei, "Evaporation and Explosion of Liquid Drops on a Heated Surface," Experiments in Fluids, vol. 1, pp. 101-111 (1983).

7. Yang, Wen-Jei, "Natural Convection in Evaporating Droplets," Handbook of Heat and Mass Transfer Operations, (ed. N. Cheremisinoff), vol. 1, Chap. 2, (1985).

8. Davis, J. T., and Rideal, E. K., Interfacial Phenomena, Chap. 8, Academic Press, New York, (1963).

Boiling Flow Instabilities in Parallel Channels with Enhanced Heat Transfer

O. T. YILDIRIM, A. MENTEŞ, S. KAKAÇ, and T. N. VEZIROĞLU
University of Miami
Coral Gables, Florida 33124, USA

INTRODUCTION

During the last two decades, the demand for higher heat transfers in the heat exchange has kept increasing. These high heat transfer rates can be obtained by using two-phas flows, or heat transfer augmented surfaces, or in some cases, both. Thus, the boilin flows and the associated two-phase flow instabilities have been studied by mar investigators.

Most of the work in boiling flows has been concentrated on single channel upflow system or on parallel channel systems with plain heat transfer surfaces. The boiling flo instabilities in multi-channel systems should be covered to a greater extent, studyir the effects of various parameters, such as inlet and outlet restrictions, mass flow rate the property variations, heat transfer coefficients, equal and unequal heat inputs on tl stability. Also, the mathematical modeling needs to be developed for the parallel chann system to take into account the above mentioned variables.

The research on two-phase flow instabilities in parallel channels have mainly bee concentrated in the United States, Japan and Russia. The state-ofthe-art of Japane research in this area has been presented by Nakanishi [1]. The extensive activities two-phase flow dynamics in two countries, the United States and Japan, are present in Two-Phase Flow Dynamics by Bergles and Ishigai [2]. The volume represents both revie and original contributions. On the other hand, the research in Russia on two-phase flc instabilities has been summarized by Veziroğlu and Kakaç [3]. Their study also includ the experimental results of investigation of two-phase flow instabilities in a single chann system with enhanced heat transfer.

References [4] through [7] give detailed information on the experiments carried out the authors regarding two-phase flow instabilities in multi-channel systems and sing channel systems with heat transfer augmentation.

GENERAL DESCRIPTION OF EXPERIMENTAL SYSTEM

The existing experimental setup, which had been built earlier and used in previous resear projects, was modified extensively for the purpose of the present project. Figure 1 a schematic diagram of the two-phase flow loop, showing basic dimensions a instrumentation. Test fluid Freon-11 is supplied from a main tank pressurized by nitrog gas. A thermostatically controlled immersion heater in the main tank and a cooling ur before the test section provide an inlet temperature range of -20°C to 90°C with a contr accuracy of ±10°C. Following the electrically heated test section is a recovery syste consisting of a condenser and a collector tank. Mixture of saturated liquid and vap is led through the condenser coil, which is cooled by refrigerated brine at 0°C. T condensed liquid is then stored in a recovery tank which is maintained at constant pressu This arrangement insures a constant level of exit pressure.

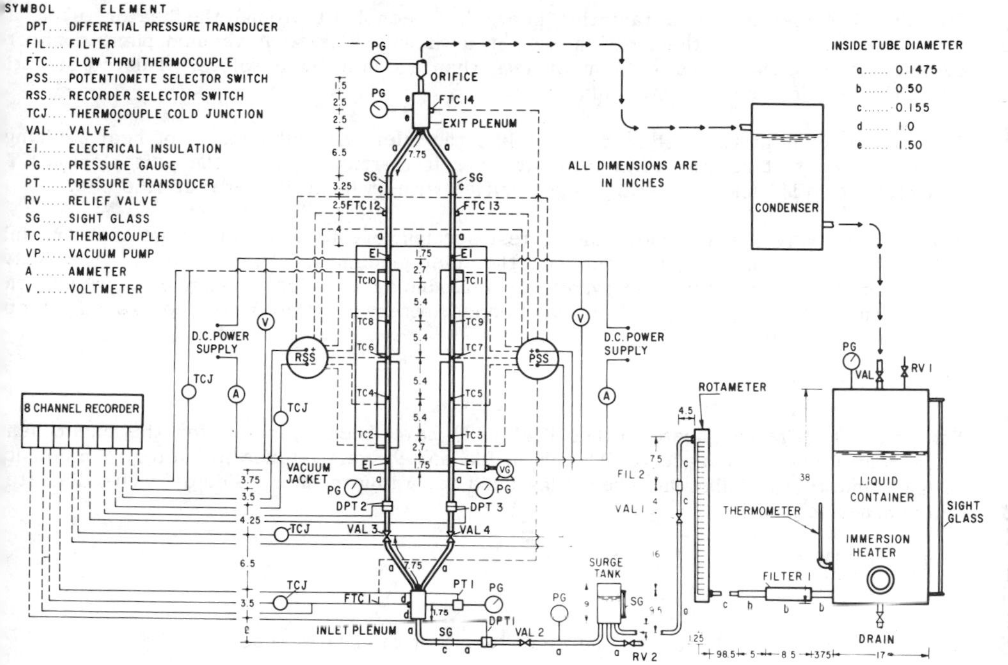

FIGURE 1. Two-parallel channel system.

All the tubing except the recovery section is made of 0.75 cm (0.295 in.) ID Nichrome pipe, and the connections are made by Swagelok-type joints. In the recovery section, tubing of much larger diameter (1.90 cm ID) is used to minimize pressure losses, thus maintaining a nearly constant level of exit pressure from the test section on. Two plenums of 6.35 cm (2.5 in.) ID are installed to simulate the inlet and exit headers of heat exchangers.

A 76 mm transparent piece of tubing is incorporated into the loop right after each heater. This part provides a very clear view of the two-phase flow behavior inside the tubes, and makes it possible to record the two-phase flow during the oscillations, including the various flow regimes, by means of a high speed camera.

A fine filter is installed at the main tank exit (Filterite LMO 48-378) to keep the test section and sensitive devices such as the rotameter and valves, free from entrained particles.

Appropriate instrumentation is installed to provide control and measurement of the test parameters; namely, the flow rate, temperature, pressure and the electrical heat input.

Basically, the experimental setup is an open-loop forced convection boiling flow system operating between constant inlet and exit pressures. It is specifically designed to generate pressure-drop, density-wave and thermal oscillations, and to investigate the effect of inlet subcooling, heat flux, flow rate and heater wall characteristics on these oscillations. The isolation of the vacuum assembly from the rest of the system has been achieved by two Teflon bulkhead unions at the two ends.

To minimize heat losses, a radiation guard has been built around the heater tubes, a the whole assembly has then been housed in a vacuum jacket. A vacuum pump connect to this jacket maintains a vacuum of less than 1.0 mm Hg absolute. This constructi keeps the heat losses to the minimum level.

The two plenums at the test section act like the inlet and exit headers of heat exchang systems. The test section terminates at an exit restriction after the exit plenum. T exit restriction is basically a sharp-edged orifice with an inner diameter of 1.66 mm.

Three needle valves are included in the test section, one between the surge tank and inl plenum, and the remaining two between the inlet plenum and heater inlets. These valv have been used to stabilize the system by introducing an inlet resistance. However, th have been used infrequently and only as needed; otherwise, they have been kept fully op while recording data.

HEATER TUBES

Six pairs of different heater tubes (Table 1) have been prepared for the experimen Each heater has been made of 7.5 mm inside and 9.5 mm outside diameter, 605 mm lo of Nichrome tubes. Tubes have been classified according to their inside effective diamete which is defined as,

$$d_e = \frac{4V}{\pi L}$$

where V is the net inside volume and L the length of the heater tube.

The tube A is the bare tube which has been used for comparison purposes. Descriptic of all of the tubes are given in Table 1.

TABLE 1. Description of the heater tubes.

TUBE	DESCRIPTION OF THE TUBE	d_e mm
A	Bare	7.4
B	Threaded, 7.938 mm -- 18 threads per 25.4 mm	7.6
C	Threaded, 7.938 mm -- 24 threads per 25.4 mm	7.6
D	With internal spring of 0.432 mm wire diameter and 3.175 mm pitch	7.
E	With internal spring of 1.191 mm wire diameter and 6.350 pitch	7.
F	Coated with: Union Carbide Linde High Flux Coating	7.

ΞXPERIMENTAL PROCEDURE

For sustained instability experiments, the following experimental procedure has been 'ollowed:

For a given heater test section, different sets of experiments corresponding to various ıeat inputs have been conducted. Each set has been composed of a sufficient number of tests to cover the available flow range. Stability boundaries have been determined n each case. Oscillations have been identified by the cyclic variations in pressures and low rates, and by observing the transparent section, the pressure gauge pointers and the ecorder. In defining the stability boundary, short life transients have been disregarded ınd only sustained oscillations have been considered.

A test preparation procedure was followed before each experiment: The test liquid was ransferred to the main tank, either from the recovery tank or from the original containers. 'or low temperature tests, the subcooling controller was set for the proper temperature. All the instrumentation and mechanical components were checked for faultless operation. An ice bath prepared in a thermos bottle served as the reference junction temperature or the thermocouples. The recorder was switched on half an hour before the experiment. 'his allowed the amplifiers and transducers to warm up, as suggested by the manufacturers. 'he amplifier settings were checked periodically as well, in order to insure the maintaining f proper calibrations.

or each heater tube, an initial experiment was conducted without heat input to find the ingle phase characteristics. Experiments with heat input were started with high mass low rate and continued to lower mass flow rates to cover the whole boiling region.

he procedure for the actual tests can be outlined as follows:

1. With enough liquid in the main tank, the tank was pressurized using nitrogen gas.
2. The surge tank was half filled with liquid Freon-11 and pressurized to a predetermined value by nitrogen gas.
3. Flow rate and heat inputs were increased gradually to the desired starting point, and the system was allowed to become steady, as indicated by recordings of system pressure, temperature and flow rate.
4. Measurements of temperature, pressure, flow rate and heat input were taken and critical observations were noted.
5. The mass flow rate was reduced by a small amount using the inlet control valve and the system was allowed to become steady, and the readings were taken. This procedure was repeated starting from Step 4 until sustained oscillations were observed. After reaching the unstable region, mass flow rate was first increased and then decreased very slowly to locate the boundaries.
6. While operating in the unstable region, first recordings of the heater inlet pressures and temperatures were made. Then the system was stabilized by closing the inlet throttling valve slowly and readings were taken. After taking the readings, the inlet throttling valve was brought into full open position and mass flow rate was reduced by a small amount.

ollowing each adjustment, the system was allowed to stabilize, and the procedure was ontinued, starting with Step 3. Experiments were stopped after reaching the dry-out.

ıe stability boundaries were approached several times during a test with smaller ljustments in flow rate. This eliminated the possibility of crossing directly into the ıstable region, which otherwise would cause misleading information regarding the stability oundary. After locating the stability boundary for pressure-drop type oscillations, tests ere continued as explained above towards lower mass flow rates to observe if the nsity-wave type oscillations existed.

The recorder kept a continuous history of flow variables during each test, thus providing detailed information about the instabilities for later analysis.

EXPERIMENTAL RESULTS

Experiments have been carried out for various heat inputs and heater tubes with Freon-1 as the working fluid. An experiment was also run with no heat input. The experiment are done in two series. The inlet temperature of the working fluid is kept constant aroun 22°C and identical heater tubes are used for both heaters. In the first series, variou heat inputs are given to one heater while keeping the heat input in the second heater th same (0 or 400 w). In the second series of experiments, various equal heat inputs are give to both heaters. A few of the runs have been repeated to confirm the repeatability o the results.

The raw data were processed by using a computer program. The program calculates th enthalpy, quality, pressure, outside and inside wall temperatures, the heat transfe coefficients, Nusselt numbers, Prandtl and Reynolds numbers, densities, density ratios velocities, conductivities, viscosities and specific heat constants at various points alon the tubes for each run. The program also draws a plot of total pressure-drop betwee the surge tank and the system exit and additional pressure drop to stabilize the syster against the mass flow rate. Some sample graphs are given in Figs. 3 and 7.

A typical recording of the oscillations observed in the experiments are presented in Fig. The oscillations were pressure-drop type and in phase for both tubes during all the tes runs. Their periods increased as the mass flow rate was reduced at the same heat inpu but the amplitudes stayed the same.

Plots of system pressure drop (from surge tank to system exit) versus mass flow rate ar used to present the results of the experiments. Figs. 3, 4 and 5 show such steady-stat characteristics of the system at various heat input combinations. Stability-instabilit boundaries are also indicated in these figures. These are typical sets of curves showin the total system pressure drop versus mass flow rate which is parabolic with zero he input and S-shaped for two-phase flows. Similarly Figs. 6 and 7 are the plots of addition pressure drop versus mass flow rate. The additional pressure drop is defined as th additional pressure drop needed to stabilize the system between the surge tank and inle plenum.

Up until the time this paper is written, the experiments on bare tubes, the tubes wi Union Carbide Linde High Flux Nucleate Boiling Coating, the tubes with 18 threads p inch and the tubes with 24 threads per inch have been tested. Among these, bare tub are found to be the most stable ones, whereas the coated tubes gave the most unstab results. This can be seen in Fig. 4 as a big unstable region which starts almost as so as the boiling starts. However, the oscillations recorded for the coated tubes were rath mild.

During the experiments with no heat input in one of the tubes, no oscillation was observe But a backflow at the cold tube was recorded at low total mass flow rates. This natur circulation between the tubes made it possible to continue the experiments to very lo total mass flow rates without any burn-out problem.

The experiments with unequal heat inputs gave surprising results for bare tubes and thread tubes with 24 threads per inch. Bare tubes showed an island of unstability on the syste pressure drop vs. mass flow rate graph as shown in Fig. 3, whereas threaded tubes h an island of stability on the graph (Fig. 4). This behavior was not observed during t equal heat input experiments.

An examination of additional pressure drop vs. mass flow rate also showed that t additional pressure drop required to stabilize the system for coated tubes was in the sa order for different heat inputs (Fig. 6), whereas, for other tubes, this additional pressu drop requirement varied considerably for various heat inputs (Fig. 7).

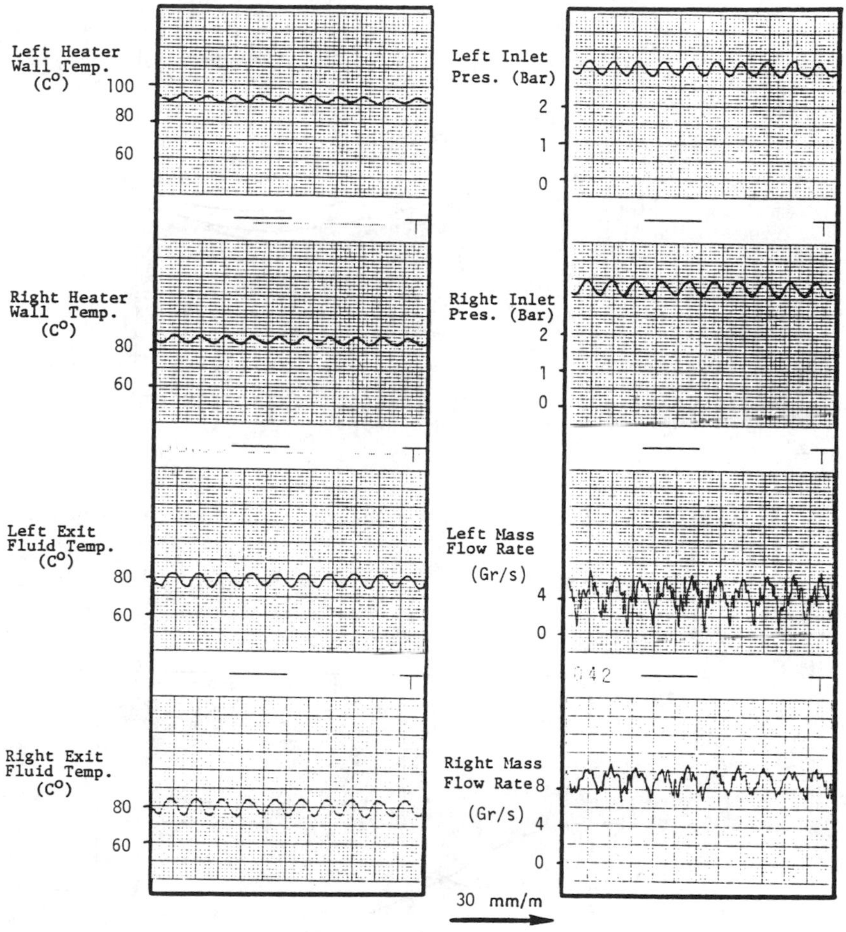

'IGURE 2. Typical pressure drop type oscillations. Mass flow rate = 12.0 Gr/s; right eat input = 400 w; left heat input = 300 w; inlet temperature = 22°C.

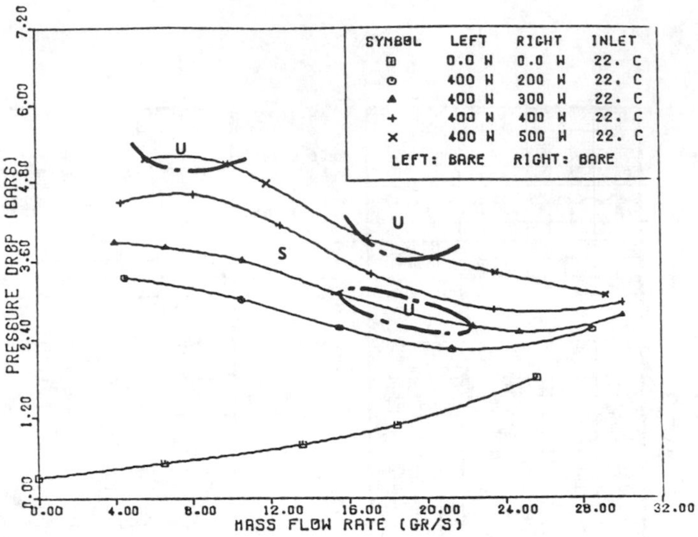

FIGURE 3. Steady-state characteristics of two-channel system with unequal heat input through the channels.

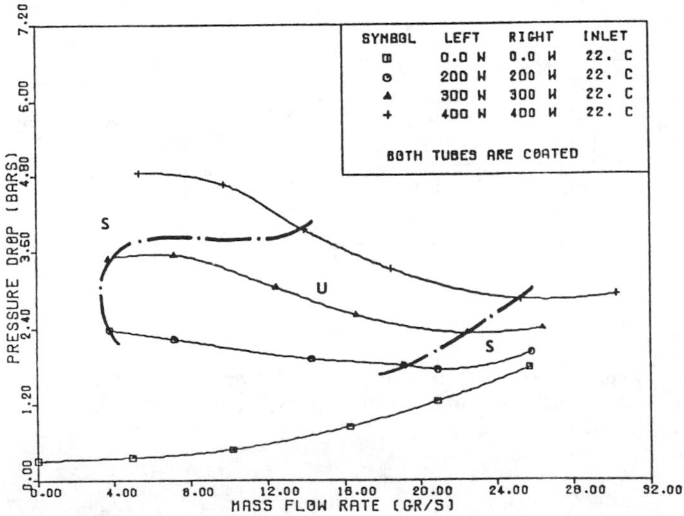

FIGURE 4. Steady-state characteristics of two-channel system with equal heat input through the channels.

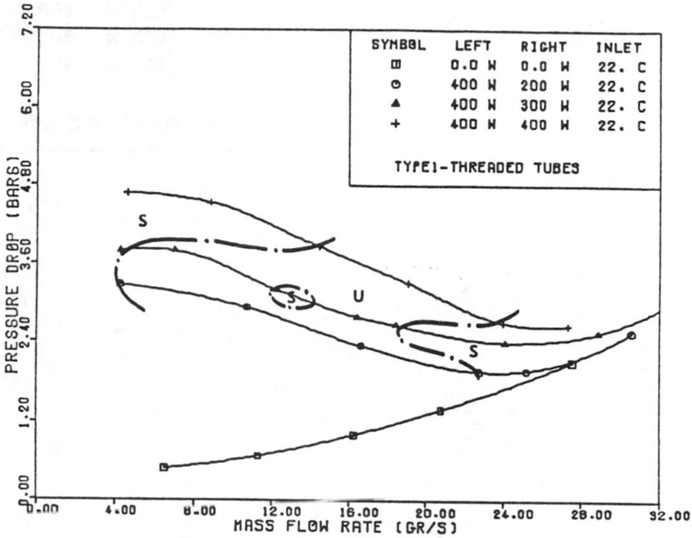

FIGURE 5. Steady-state characteristics of two-channel system with unequal heat input through the channels.

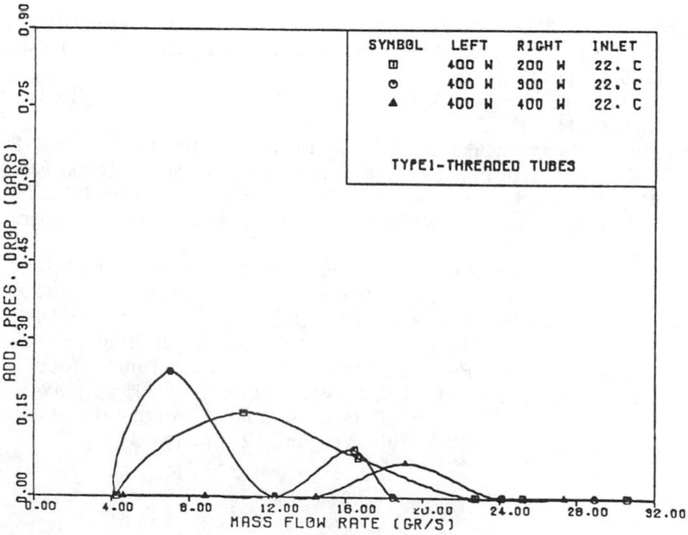

FIGURE 6. Additional pressure drop to stabilize the system.

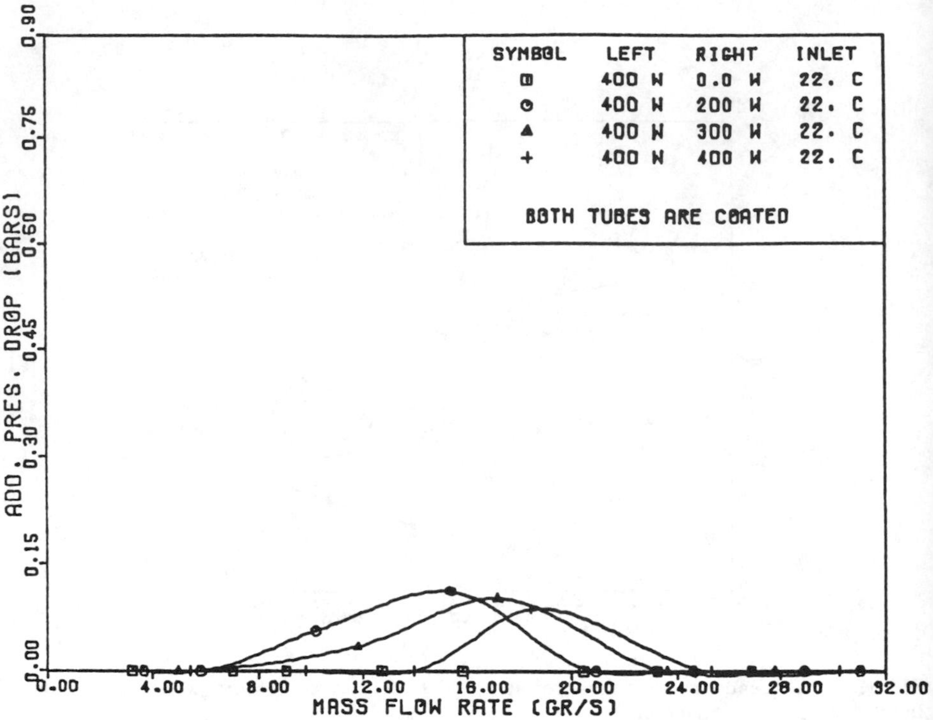

FIGURE 7. Additional pressure drop to stabilize the system.

REFERENCES

1. Nakanishi, S., "Recent Japanese Researchers on Two-Phase Flow Instabilitie Two-Phase Flow Dynamics, Bergles, A.E. and Ishigai, S. (eds.), Hemisphere Publish Corp., U.S.A. (1981).
2. Bergles, A.E. and Ishigai, S. (eds.), Two-Phase Flow Dynamics Hemisphere Publish Corp., U.S.A., (1981).
3. Veziroglu, T.N. and Kakaç, S., "Two-Phase Instabilities", Final Report NSF Proj CME 79-20018, Clean Energy Research Institute, Coral Gables, Florida (1983).
4. Veziroğlu, T.N. and Lee, S.S., "Boiling Flow Instabilities in a Two Parallel Chan Upflow System", AEC-Oak Ridge National Laboratory Subcontract No. 2975, Fi Report (1969).
5. Veziroğlu, T.N. and Lee, S.S., "Sustained and Transient Boiling Flow Instabilit in a CrossConnected Parallel Channel Upflow System", AECORNL-Union Carb Subcontract No. 2975, Final Report (1971).
6. Kakaç, S., et al., "Sustained and Transient Boiling Flow Instabilities in Cross-Connected Four Parallel Channel Upflow System", Heat Transfer 1974, Pr 5th Int. Heat Transfer Conf., Tokyo, Vol. 4, Paper B5.11, pp. 235-239, 1974.
7. Menteş, A., et al., "Effect of Heat Transfer Augmentation on Two-Phase Fl Instabilities," Warme-und Stoffubertragung 17, pp. 161-169.

Heat Transfer by Impinging Bubble Jet

HIDEO YOSHIDA and KUNIO HIJIKATA
Tokyo Institute of Technology
2-12-1 Ohokayama
Meguro-ku, Tokyo, Japan

YASUO MORI
University of Electro-Communications
1-5-1 Chofugaoka
Chofu, Tokyo, Japan

INTRODUCTION

It has been observed that the two-phase gas-liquid flow is usually associated with intense increase in heat transfer rate as compared with a single-phase flow under the similar flow rate condition [1]. Since this enhancement is considerable even in the absence of phase change, it is mainly attributed to heterogeneity and unsteadiness of the two-phase flow.

Heat transfer by an impinging bubble jet (a bubble-induced water jet) is considered to be a typical example of an effective heat transfer system using a two-phase flow. However, little work has been reported regarding impinging bubble jets. Yen [2] reported experimental results about average *Nu* number over a downward-fasing ice surface melted by a bubble-induced water jet, but the mechanism of the heat transfer was not discussed.

In the previous study [3], heat transfer problems of a regularly impinging bubble jet were treated; that is, uniform bubbles were periodically injected upward into a uniform upward liquid flow and impinged regularly on a heating plate fasing downward. The stagnation-point heat transfer rate and the liquid velocity induced by bubbles were measured by an electrochemical method based on the analogy between heat and mass transfer. These results indicate that enhancement of heat transfer at the stagnation point is due to rapid acceleration of fluid element between a bubble and the heating plate at the moment of bubble impingement. In other words, bubbles do not decelerate as much owing to the buoyancy force even in the pressure gradient of the stagnation flow, hence the liquid between the bubble and the plate is displaced intensely; by this accelerated fluid motion, the boundary layer is made thin and the heat transfer at the stagnation point is enhanced. In addition to the experimental study, observed heat transfer enhancement was quantitatively explained by a theoretical analysis which assumes a bubble as a rigid sphere and numerically calculates the unsteady flow field near the stagnation region.

In this paper, on the basis of the previous study, heat transfer performance near the stagnation point by an irregularly impinging bubble jet is investigated in the light of practical applications. In an actual impinging bubble jet, bubbles of various sizes are injected from a nozzle and spread widely and irregularly in the jet. Even if bubbles of uniform size are periodically injected upward from a nozzle, they rise irregularly in a stagnant liquid and coalesce with each other. Therefore, in the present study, effects of the gas flow rate and the nozzle-to-plate distance on heat transfer performance are extensively investigated. Firstly, flow characteristics of a bubble jet are discussed. In particular, the rate of bubble impingement on the plate is carefully measured, because it was proved to affect largely the heat transfer performance of an impinging jet in the previous study. Secondly, measurements of heat transfer rate along the impinging plate are made under the constant heat flux condition. Lastly, the measured heat transfer rates are compared with a prediction considering the statistical behavior of irregularly impinging bubbles.

EXPERIMENTAL APPARATUS

The schematic diagram of the experimental apparatus is shown in figure 1. A water container is 0.544m deep with 0.28m square cross section. Gas bubbles of N_2 gas were blown into a stagnant water from a nozzle of 0.8mm i.d. installed at the bottom of a container. Bubbles rose up and impinged irregularly on a downward-facing heating plate submerged in water as shown in figure 2(b), while the regularly impinging bubble jet as shown in figure 2(a) was also studied for reference; a liquid-flow loop shown in figure 1 was provided for the experiments of the regularly impinging bubble jet. The gas flow rate was measured by a flowmeter placed upstream the nozzle. The volume expansion of rising bubbles due to the decreasing static pressure in water was so small as to be neglected. At the bottom of the container, cooling pipes were fixed to keep the water temperature constant during experiments.

Figure 3 shows the details of the heating plate. The stainless steel foil 30μm thick was mounted on the plate of 15mm thickness with 150mm square surface and was electrically heated. In order to make the departure of impinged bubbles from the plate smooth, the heating plate has a slope of 6° on the periphery. Wall temperatures were measured by fifteen thermocouples embedded beneath the foil. Time-averaged heat transfer rates were obtained from the time-averaged temperature difference between the wall and the bulk liquid.

A void fraction, a rate of bubble impingement on the plate (a bubble impaction rate) and a gas velocity were measured by an electrical resistivity probe consisting of copper wire of 0.1mm diameter and 10mm long as shown in figure 4. The gas velocity was measured by the probe having two needles, whose vertical distance is about 3mm. On the other hand, an electrochemical method was adopted

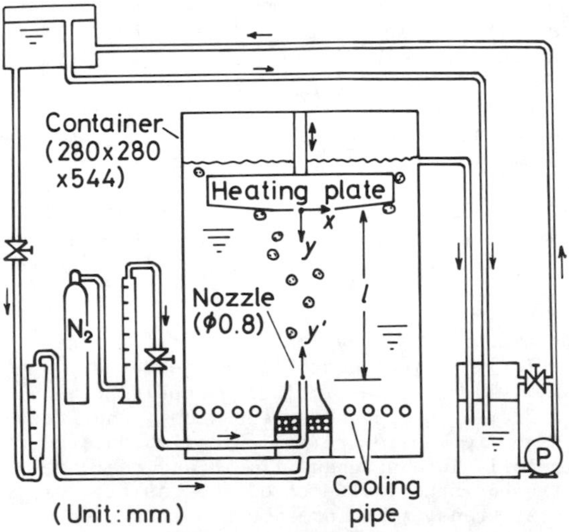

FIGURE 1. Scheme of experimental apparatus

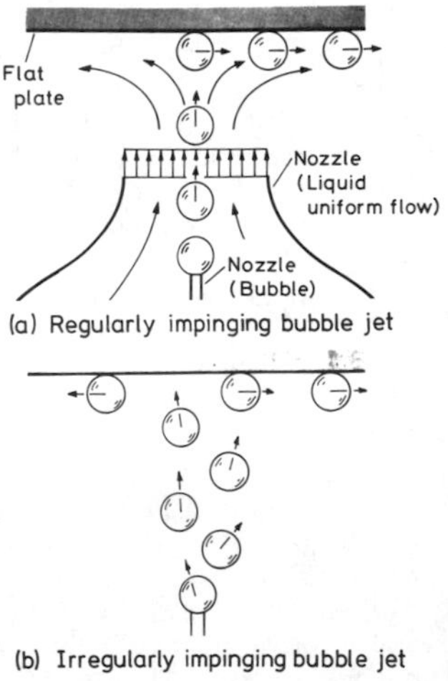

FIGURE 2. Regularly and irregularly impinging bubble jets

o measure the liquid velocity, using the electrolyte consisting of 0.01M-$_4$Fe(CN)$_6$, 0.01M-$K_3Fe(CN)_6$ and 2.0M-NaOH as a working fluid. Detailed informa-ion on these methods is given in references [4] and [5].

XPERIMENTAL RESULTS AND DISCUSSION

s indicated in figure 1, the stagnation point of the plate is taken as the rigin of the axisymmetric coordinate system; x and y denote distance parallel o and normal to the surface, respectively. The height above the nozzle is enoted by y', with the corresponding velocity component v. The nozzle-to-plate istance l was varied from 50 to 250mm, and the gas flow rate Q_g covered the ange from 1.25 to 20.0cm^3/s. Injected bubbles were uniform in size for the gas low rate less than about 2cm^3/s. For the larger gas flow rate, however, they ere not uniform as a result of coalescence.

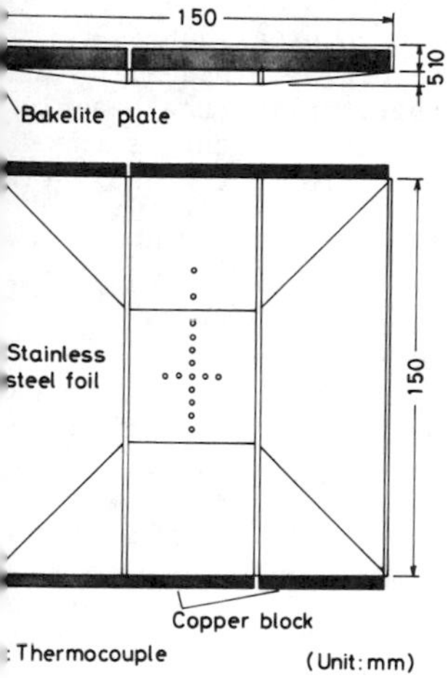

IGURE 3. Heating plate

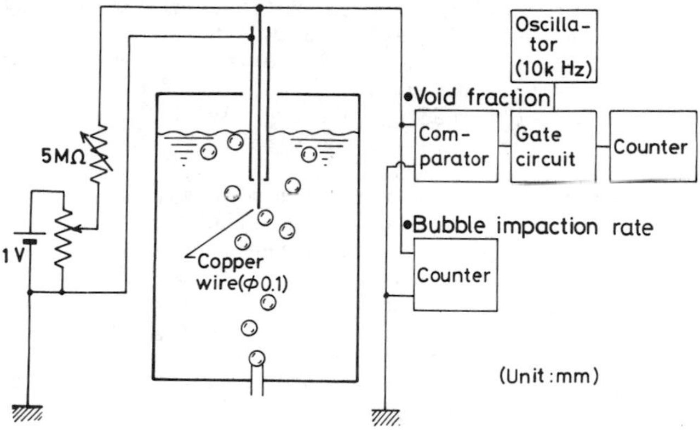

FIGURE 4. Electrical resistivity probe method

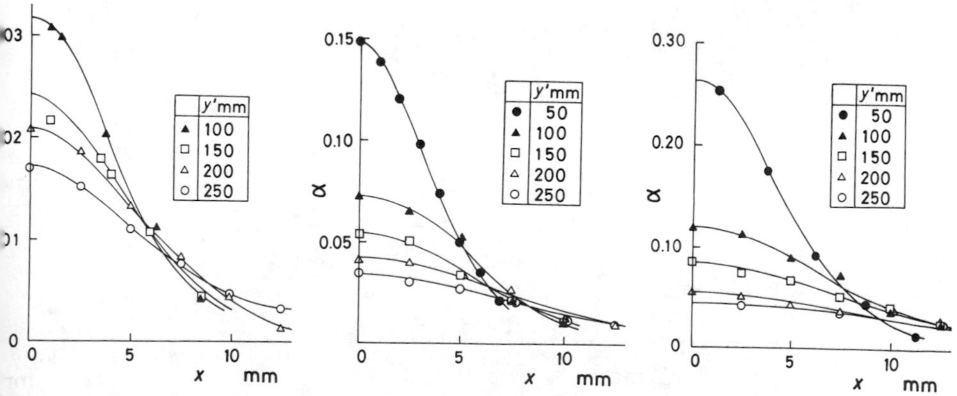

) Q_g = 1.25 cm^3/s (b) Q_g = 5.0 cm^3/s (c) Q_g =20.0 cm^3/s

GURE 5. Profiles of void fraction

Flow Characteristics

Firstly, the flow characteristics of a bubble jet without the impinging plate are discussed. Since bubble jets are driven by buoyancy force, the void fraction profile is very important. Figures 5(a-c) show the profiles of the void fraction for Q_g=1.25, 5.0 and 20.0cm^3/s, respectively. They are well represented by a Gaussian profile which is characteristic of the diffusion phenomena from a point source. In order to clarify the radial diffusion rate o bubbles, the half radii of void fraction profile $x_{h\alpha}$ are plotted against the distance from the nozzle y' in figure 6 by open symbols. It is seen that bubbl jets spread linearly with increasing y' and that diffusion of bubbles are enhanced with the increasing gas flow rate.

The profiles of gas and liquid velocities are depicted in figures 7(a-c). While the gas velocity is almost constant except in the region near the nozzle, the liquid velocity profiles agree well with the Gaussian. With the increasing gas flow rate, the liquid velocity v_l increases and reaches to about 50cm/s, which is a considerably large value compared with that in an usual natural convection case. The half radii of these profiles x_{hv} are also plotted in figure 6 by solid symbols. It is interesting to note that, unlike the behavior of $x_{h\alpha}$, the half radius of liquid velocity profile x_{hv} is independent of the ga flow rate; the spreading rate dx_{hv}/dy'=0.086 is nealy equal to that of a single phase jet reported in reference [6].

On the other hand, the rate of bubble impingement on the plate n was defined ar measured by counting the number of bubbles which impact on the electrical resistivity probe installed at the location 10mm upstream of the plate (i.e. y=10mm or y'=(l-10)mm). Figure 8 shows the measured distributions for the case of Q_g=1.25cm^3/s in which injected bubbles are uniform in size. The half radii of these distributions are plotted in figure 6. As clearly seen from figures 6 and 8, the axial variation of the x_{hn}-distribution coincides faithfully with that of $x_{h\alpha}$-profile. Thus it is concluded that the bubble behavior is unaffecte by the existence of the impinging plate except in the region very close to it.

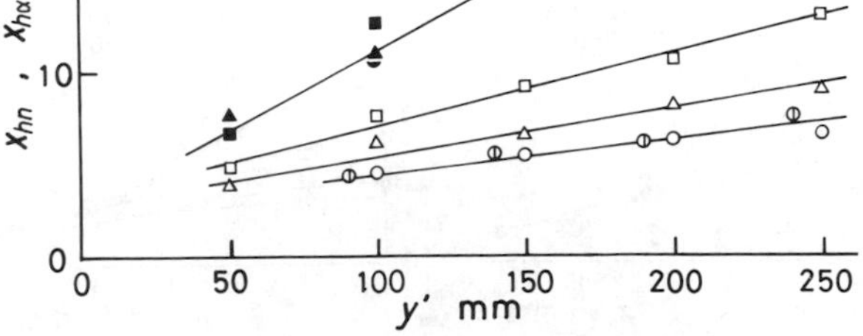

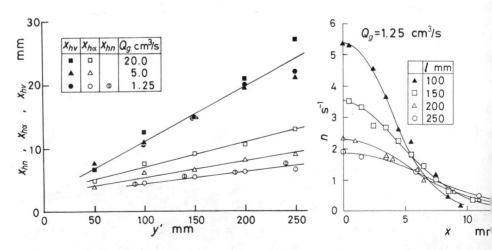

FIGURE 6. Axial variations of half radii

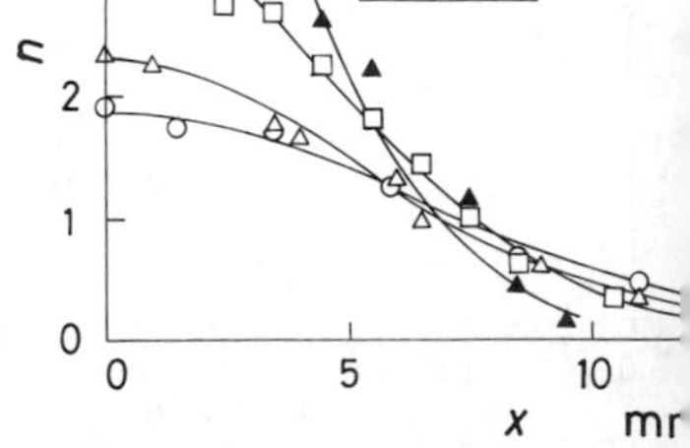

FIGURE 8. Distributions of ra of bubble impingement on plate

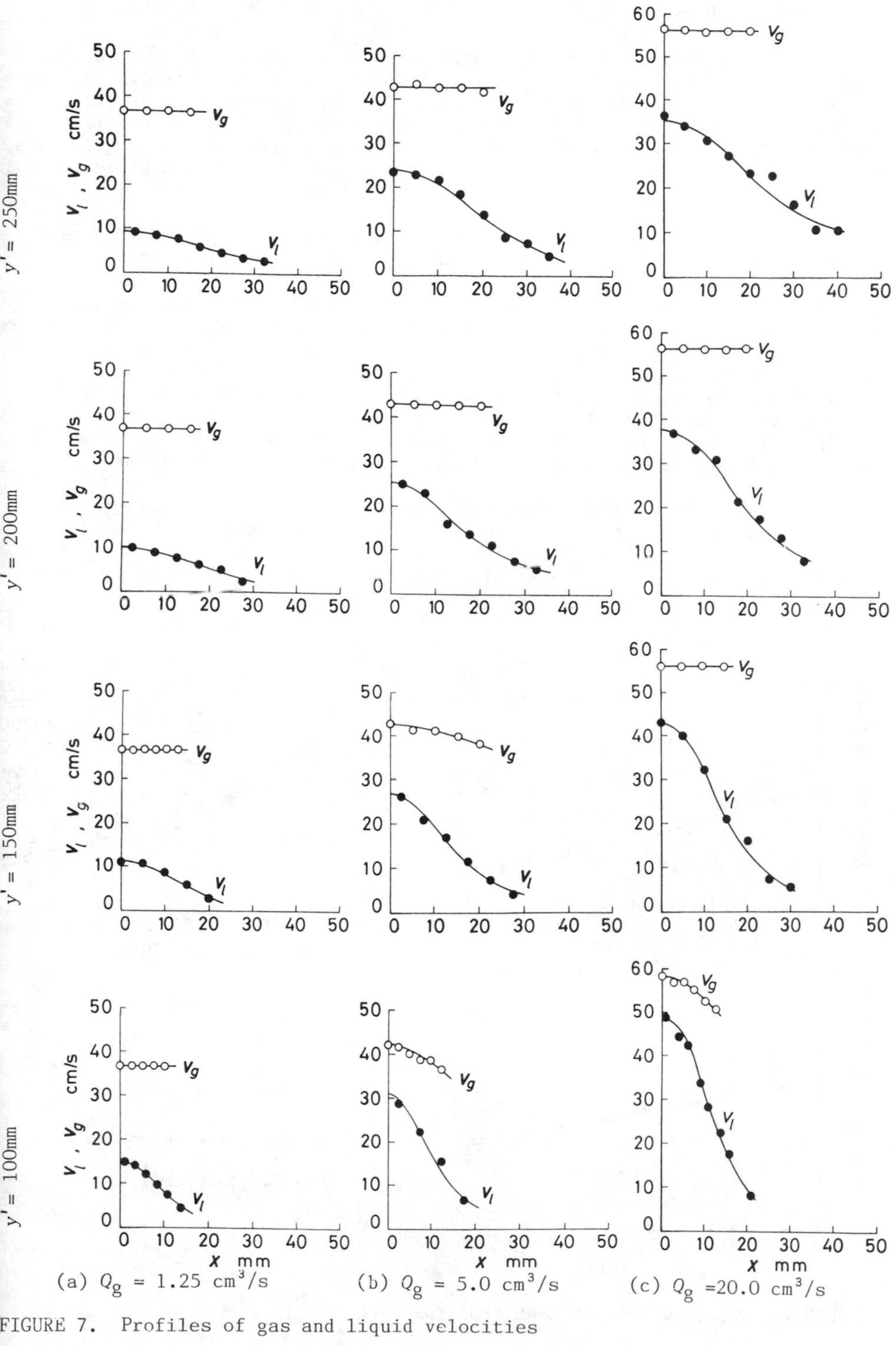

FIGURE 7. Profiles of gas and liquid velocities

Heat Transfer Characteristics

The distributions of the heat transfer rate along the surface are shown in figures 9(a-c) for Q_g=1.25, 5.0 and 20.0cm^3/s, respectively. For all the cases, the maximum heat transfer rates are attained at the stagnation point and decrease with increase of the nozzle-to-plate distance *l*. On the contrary, heat transfer rates far from the stagnation point increase with increaseing *l*. This tendency closely relates to the rate of bubble impingement on the plate shown in figure 8.

As the gas flow rate increases, the heat transfer rates increase substantially except in the region near the stagnation point. This fact is explained as follows:

(1) Along the stagnation stream line of the jet, the void fraction increases owing to the decrease of liquid velocity. Consequently, coalescence of bubbles frequently occurs, resulting in reduction of the rate of bubble impingement on the plate in the stagnation region.

(2) With increase of the gas flow rate, the consequent higher void fraction enhances the effect described above.

(3) Accordingly, the increment of heat transfer performance is suppressed for the high gas flow rate owing to the reduction of bubble impinging rate.

This non-linear nature of the heat transfer enhancement against the gas flow rate exhibits a striking contrast to the monotonic increase of the bubble-induced liquid velocity shown in figure 7. This fact indicates that the high heat transfer performance of impinging bubble jets is much affected by the behavior of impinging bubbles than by the macroscopic liquid motion induced by buoyancy force.

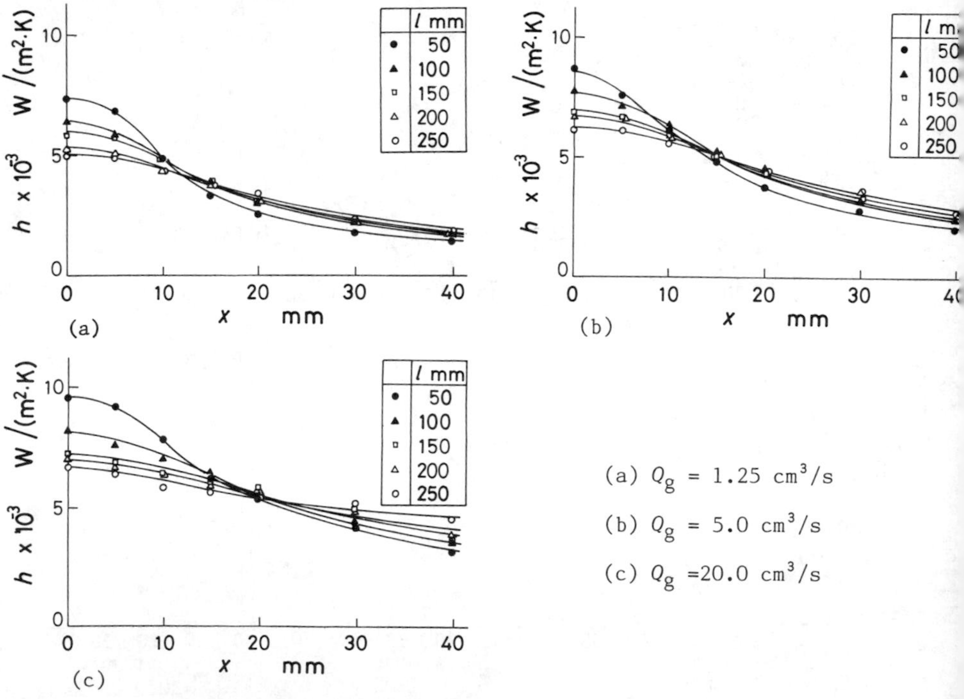

(a) Q_g = 1.25 cm^3/s

(b) Q_g = 5.0 cm^3/s

(c) Q_g =20.0 cm^3/s

FIGURE 9. Distributions of heat transfer rate

REDICTION OF STAGNATION-POINT HEAT TRANSFER RATE

he experimental results shown in the preceding section suggest that the heat ransfer rates of the bubble jet impinging irregularly and unsteadily on the urface can be explained by relating those of the regularly impinging bubble jet o the probability of bubble impingement. From now on, we distinguish between eat transfer rates of the regularly and irregularly impinging bubble jets by sing the nomenclature h_r and h_{ir}, respectively. The bubble injection frequency s designated by f. Bubbles are assumed to be uniform in size.

igure 10 shows a schematic outline of the statistical procedure to predict the tagnation-point heat transfer rate $h_{irs}(f)$ shown in figure 10(a) by using the istributions $n(x)$ and $h_r(f,x)$ shown in figures 10(b) and (d), respectively. We all the point at x=0 the stagnation point, which does not mean the instantane-us stagnation point but the time-averaged one.

ince we are concerned with the stagnation point, it is convenient to use the ne-dimensional probability density function for bubble impinging rate $p(x)$ nstead of the $n(x)$ on the basis of the axisymmetric nature of the phenomena. ne $p(x)$ can be obtained by normalizing the product of the $n(x)$ and a radial istance x as shown in figure 10(c). In order to facilitate understanding of ne procedure, a probability distribution function $P(x)$ for a discrete quantity is introduced as shown in figire 10(c'). Bubbles impinge at the point of x_1, $\cdots, x_i$ with the probability of $P(x_1)$, $\cdots$, $P(x_i)$, respectively. If the rate of eat transfer caused by each bubble impingement is assumed to be independent of ach other and to be only a function of f and x, the stagnation-point heat ansfer rate is calculated as follows:

$$_{scal}(f) = \sum_{j=1}^{i} [h_r(f,x_j) \cdot P(x_j)] \quad . \tag{1}$$

r the continuous probability density function, equation (1) is transformed to e integral form expressed as

$$_{scal}(f) = \int_0^\infty h_r(f,x) \cdot p(x) \mathrm{d}x \quad . \tag{2}$$

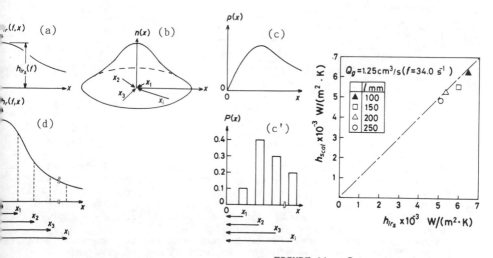

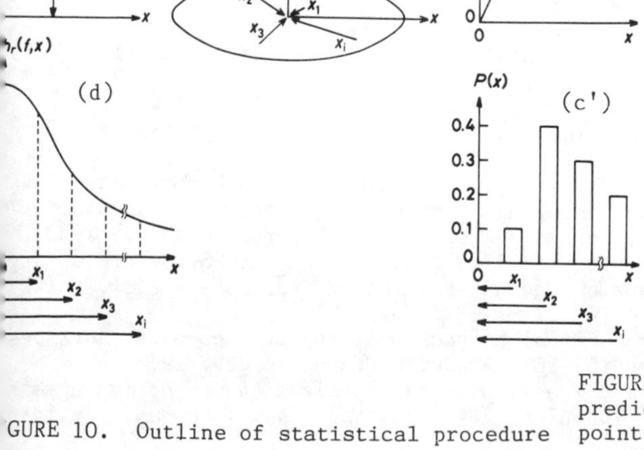

GURE 10. Outline of statistical procedure

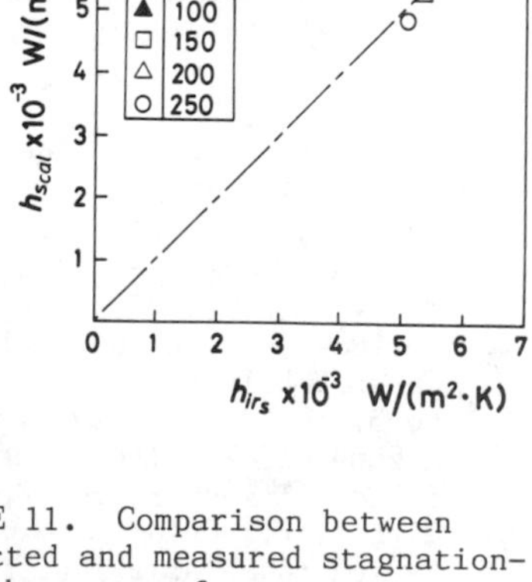

FIGURE 11. Comparison between predicted and measured stagnation-point heat transfer rate

Figure 11 compares the calculated results with the measured stagnation-point heat transfer rates for various nozzle-to-plate distances. The agreement between them is excellent. Thus we conclude that the high heat transfer performances of irregularly impinging bubble jets could be explained by the sum of the contribution by the individual bubble impingements.

CONCLUSIONS

(1) For the case of a bubble jet without an impinging plate, both the bubble diffusion rate and the liquid velocity increase with the increasing gas flow rate. On the other hand, the spreading behavior of liquid velocity is independent of the gas flow rate and is nearly the same as that of a single-phase flow. The distribution of the rate of bubble impingement on the plate varies in the manner similar to the void fraction profile without the impinging plate.
(2) Heat transfer rates increase far from the stagnation point with the increasing nozzle-to-plate distance, while decrease in the stagnation region. This tendency is corresponding to that of the rate of bubble impingement on the plate. On the other hand, heat transfer rates do not so increase with the increasing gas flow rate owing to the decrease of bubble impinging rate.
(3) The heat transfer performance of the irregularly impinging bubble jet is successfully explained by the statistical procedure taking account of the probability for the bubble impinging rate. This fact demonstrates that the hig heat transfer performances are ascribed to the individual bubble impingement process with the heterogeneity and the unsteadiness of the flow.

NOMENCLATURE

f = bubble injection frequency, s^{-1}
h = heat transfer rate, $W/(m^2\ K)$
l = nozzle-to-plate distance, m
n = rate of bubble impingement on plate, s^{-1}
P = probability distribution function
p = probability density function, m^{-1}
Qg = gas flow rate, m^3/s
v_g = gas velocity, m/s
v_l = liquid velocity, m/s
x, y, y' = coordinates, m
x_h = half radius, m
α = void fraction

Subscripts
cal = calculation
ir = irregularly impinging bubble je
n = rate of bubble impingement on plate
r = regularly impinging bubble jet
s = stagnation point
v = liquid velocity
α = void fraction

REFERENCES

1. Kubie, J., Bubble Induced Heat Transfer in Two Phase Gas-Liquid Flow, *Int. J. Heat Mass Transfer*, vol.18, no.4, pp.537-551, 1975.
2. Yen, Y.-C., Heat-Transfer Characteristics of a Bubble-Induced Water Jet Impinging on an Ice Surface, *Int. J. Heat Mass Transfer*, vol.18, no.7/8, pp.917-926, 1975.
3. Yoshida, H., Mori, Y., and Hijikata, K., Theoretical and Experimental Stud on Heat Transfer Mechanism of Impinging Bubble Jets, *Trans. Japan Soc. Mech. Engrs* (in Japanese), Series B, vol.49, no.445, pp.1904-1911, 1983.
4. Serizawa, A., Kataoka, I., and Michiyoshi, I., Turbulence Structure of Air Water Bubbly Flow-I. Measuring Techniques, *Int. J. Multiphase Flow*, vol.2, no.3, pp.221-233, 1975.
5. Mizushina, T., The Electrochemical Method in Transport Phenomena, *Advances in Heat Transfer*, vol.7, pp.87-161, Academic Press, London, 1971.
6. Giralt, F., Chia, C.-J., and Trass, O., Characterization of the Impingemen Region in an Axisymmetric Turbulent Jet, *Ind. Eng. Chem., Fundam.* , vol.16 no.1, pp.21-28, 1977.

BOILING HEAT TRANSFER

Post-CHF Heat Transfer and Pressure Loss in Once-through Boilers

BERNHARD BRAND, DIETMAR HEIN, WOLFGANG KASTNER, and WOLFGANG KÖHLER
Kraftwerk Union AG
Erlangen, FRG

ABSTRACT

To improve the design of fossil-fired steam generators, thermal-hydraulic studies were conducted on inside cooled tubes.

The experimental studies of which the parameter range covers part and full-load operation of the steam generators relate to

- post-CHF heat transfer
- the influence of tube orientation on heat transfer
- friction pressure loss in the wetted and unwetted regions

The experiments were analysed and made amenable to theoretical analysis by means of models which simulate the major physical phenomena.

INTRODUCTION

Kraftwerk Union constructs besides nuclear power plants also complete fossil-fired energy generating plants and is furthermore licenser for once-through boilers (BENSON principle). In the course of the work involved, extensive thermal-hydraulic studies on inside cooled tubes are conducted to improve the design. The emphasis lies on problems of heat transfer and pressure loss in evaporator tubes. The prime interest for the design of evaporator heat exchange surfaces is heat transfer in the unwetted region since the material temperatures in this area affect the lifetime of the evaporator tubes. Heat transfer is also influenced by the tube orientation since this affects the distribution of the liquid and vapour phases over the flow cross-section. Finally, knowledge of the pressure loss of two-phase flows is required for the fluid flow design of steam generators.

TEST FACILITY AND MATRIX

A high-pressure test facility is available for the experimental studies; its flowchart is shown in Figure 1.

The test facility essentially consists of a high pressure loop, 1000 kW DC power supply unit, cooling tower and a water treatment plant. The facility is suitable for performing tests with water, steam or two-phase mixture and its capability extends to supercritical pressures up to 330 bar. A detailed description of the test rig is given in (1).

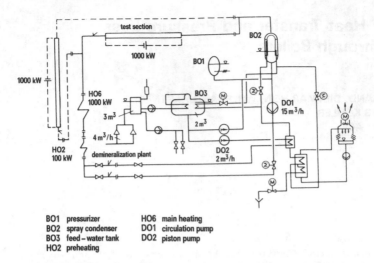

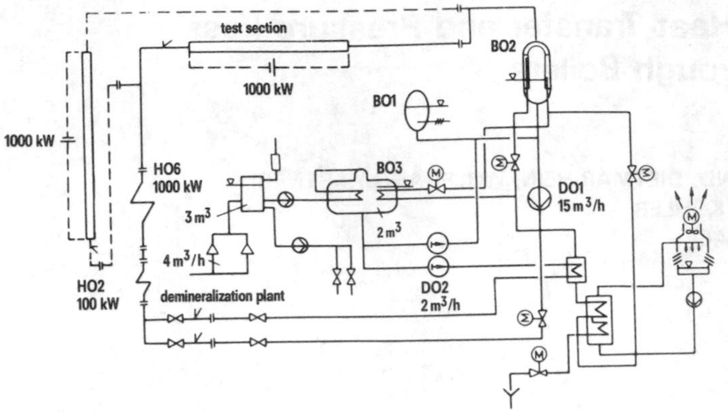

FIGURE 1. BENSON-Test rig

The studies on heat transfer and pressure loss were conducted on 6 m long internally cooled tubes. Flow through the tubes was vertically upwards or horizontal.

The test matrix comprised altogether 600 tests which covered the following parameters:

Tube diameter:	12.5, 14.0, 24.3	mm
Pressure :	50 - 250	bar
Mass velocity:	300 - 2500	kg/m² s
Heat flux :	up to 600	kW/m²

POST-CHF HEAT TRANSFER

The typical flow patterns and heat transfer regimes shown in Figure 2 occur in an evaporator tube in which flow is vertically upwards. In this process, the regimes of single-phase liquid flow, two-phase mixture and single-phase vapour flow are passed through as flow enthalpy increases. In the two-phase flo regime the location of the boiling crisis forms the boundary between the wetted and unwetted heat exchange regions. The wetted region features high heat transfer coefficients, whereas heat transfer is severely reduced in the unwetted (post-CHF) region. This gives rise to a sharp increase in the surface temperature in this region. On account of the operating parameters, the boiling crisis in fossil-fired steam generators is caused by dryout and therefore the unwetted region is termed post-dryout region.

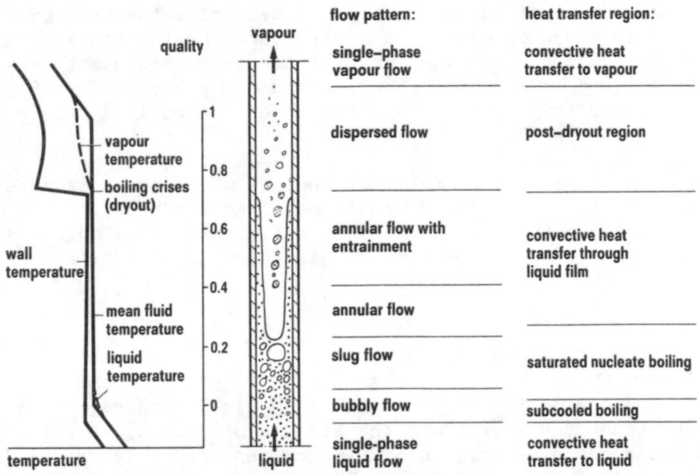

FIGURE 2. Typical wall temperature distribution in a vertical evaporator tube with uniform heating

On the basis of the experiments performed on the BENSON test rig and the results of literature studies a heat transfer model for the post-CHF regime has been developed with which heat transfer up to the critical pressure can be calculated (2), (3). The model allows for a representative thermal non-equilibrium between the vapour and liquid phases - which is dependent on the operating parameters - with the steam superheat being obtained with the aid of an energy balance.

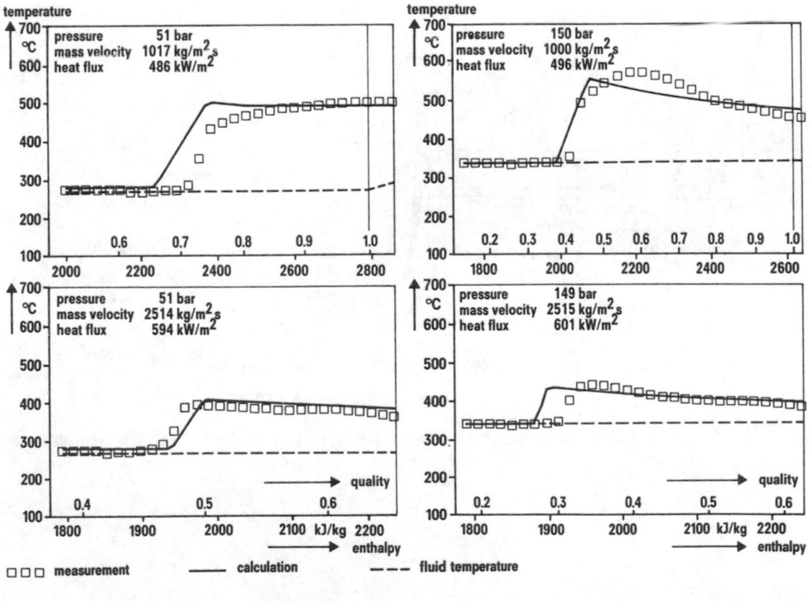

FIGURE 3. Calculated and measured wall temperature distribution (d=12.5 mm)

The steam superheat thus obtained can be used to determine the heat exchange surface temperature in conjunction with a conventional heat transfer equation for single-phase flow. Comparative analyses for the experiments and the described model show good agreement in the range under analysis from 50 bar up to the critical pressure (Figure 3). The location of the boiling crisis was calculated according to KON'KOV (4).

Figure 4 depicts in a three-dimensional view the calculated heat transfer coefficient as a function of the enthalpy of the fluid and pressure. In the region of nucleate boiling the heat transfer coefficient was calculated according to JENS and LOTTES (5), in single-phase flow GNIELINSKI's correlation (6) was used.

INFLUENCE OF TUBE LOCATION ON HEAT TRANSFER

While a symmetrical distribution of the vapour and liquid phases over the cross-section is present under vertical flow, the effects of gravity in a tube carrying horizontal flow cause phase separation of varying degrees of severity. The heavier liquid flows preferentially in the lower tube region while the lighter vapour flows above. Heat transfer in a horizontal tube is also affected by this separation phenomenon. Figure 5 shows a typical wall temperature pattern for a horizontal tube in which the effect of gravity on heat transfer is pronounced. Here, dryout occurs on the top side of the tube concomitant with a rise in wall temperature, whereas saturation conditions have not even been reached on the tube bottom. At this point stratified flow occurs which disappears at higher fluid enthalpy. Consequently, the top side of the tube can be rewetted. Finally, with a further heat increase the boiling crisis takes place a second time on the top side and at a far higher fluid enthalpy on the bottom of the tube as well.

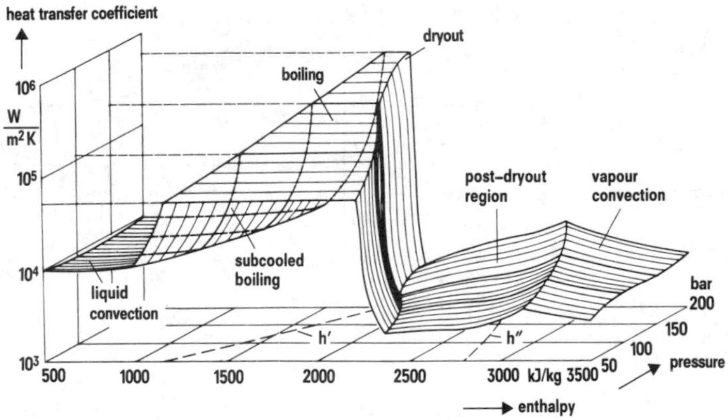

FIGURE 4. 3-Dimensional representation of heat transfer for forced flow conditions in tubes

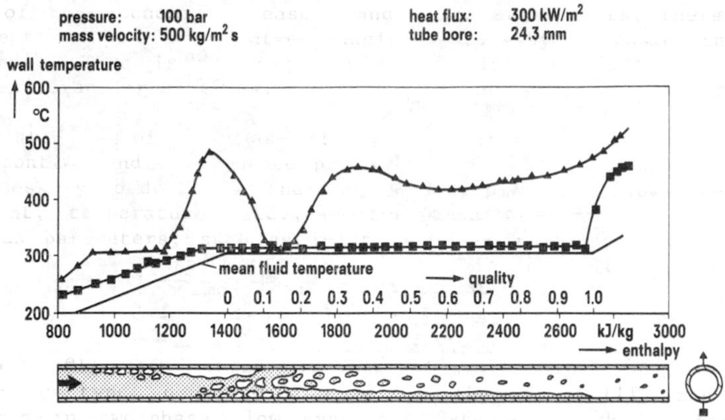

FIGURE 5. Influence of gravity on the heat transfer in a horizontal tube

Figure 6 shows as an example the pattern gained from the tests of the critical heat flux as a function of steam quality. As opposed to the pattern for the vertical tube where a monotone drop in the critical heat flux is to be expected for an increasing steam quality, the relationships are far more complicated in this case. As is apparent from the diagram, the boiling crisis on the top and bottom of the tube occurs at the same axial position for high mass fluxes. Here it is evident that a noticeable influence of tube orientation on heat transfer is no longer present. The Froude number derived with the aid of the superficial velocity of the vapour is a proven means of evaluating the influence of the tube orientation. As is to be seen from Figure 7, the difference in critical steam quality between the top and bottom of the tube at Froude numbers above 7 is less than 0.1. By contrast, pronounced stratification phenomena are to be anticipated at Froude numbers below 3.

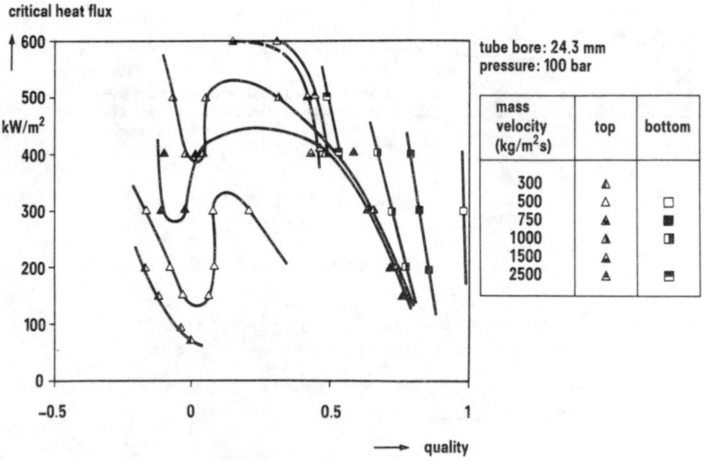

FIGURE 6. Critical heat flux in a horizontal tube

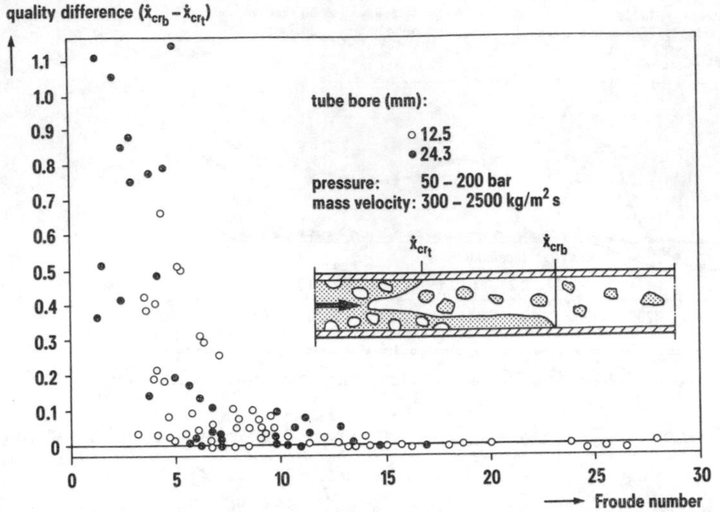

FIGURE 7. Difference of the critical quality at top and bottom of the tube as a function of the Froude number

FRICTIONAL PRESSURE LOSS IN THE WETTED AND UNWETTED REGIONS

In experiments investigating heat transfer in evaporator tubes the pressure drop was also measured. From these measurements the frictional pressure loss was calculated by eliminating the acceleration and gravitational components.

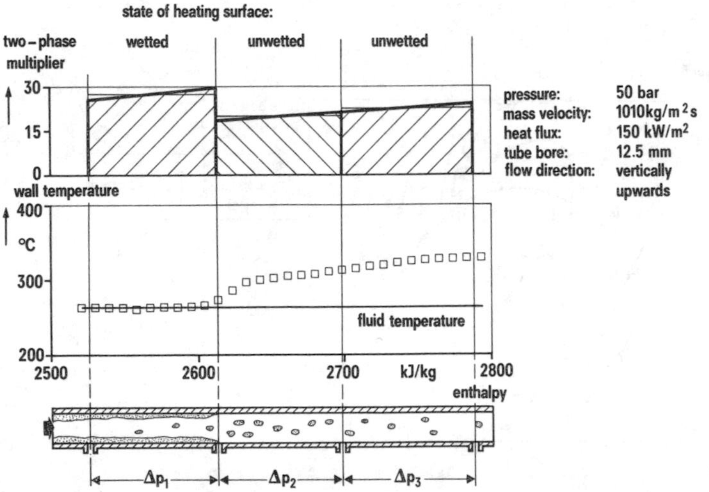

FIGURE 8. Two-phase multiplier in the wetted and unwetted zones of heated tube flow

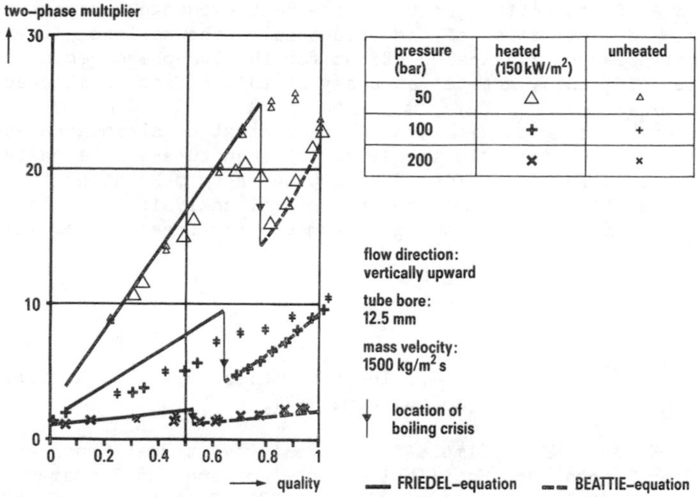

FIGURE 9. Influence of heating on the two-phase pressure drop

It became apparent that the state of the heat exchange surface (wetted/unwetted) has a strong influence on the frictional pressure loss which is considerably higher in the wetted than in the unwetted region. Figure 8 shows typical changes in two-phase flow multipliers as determined for wetted and unwetted regions of an evaporator tube.

Calculated pressure loss in the post-CHF region showed good agreement with experimental results when the BEATTIE model (7) was used. In the wetted region, the frictional pressure loss is influenced by a number of factors which do not appear in an exact form in commonly used equations. In the parameter range applicable to fossil-fired steam generators, the FRIEDEL equation (8) is sufficiently accurate for frictional pressure loss calculations. Figure 9 shows a typical comparision between calculated and measured two-phase flow multipliers in wetted and unwetted regions. In the same figure it can also be seen that the frictional pressure loss of non-heated flow is higher than that of heated flow in an unwetted region. This is to be attributed to the sharply reduced momentum exchange between liquid and heat exchange surface in the unwetted region.

SUMMARY AND PROSPECTS

Post-CHF heat transfer and the frictional pressure loss in the two-phase regime were studied for the parameter range of fossil-fired steam generators.

Proceeding from the experimental studies a heat transfer model was developed for the post-CHF region with which heat exchange surface temperature can be calculated with good accuracy.

The influence of the tube orientation on heat transfer was studied on tubes carrying horizontal and vertical flow. A procedure was developed with which the influence of the tube orientation can be taken into account.

A marked influence of the wetting state of the heat exchange surface was established during determination of the frictional pressure loss. It was demonstrated that pressure loss calculations for the two-phase regime can be performed with significantly greater accuracy if this effect is allowed for.

Studies are currently being carried out in the context of steam generator development on the BENSON test rig involving inclined tubes, internally rifled tubes and tubes heated from one side. The purpose is to make hitherto imprecisely modeled effects amenable to theoretical analysis to review new steam generator concepts and to improve design documents for fossil-fired steam generators.

REFERENCES

1. Hein, D., Keil, H. and Köhler, W., The BENSON Test Rig, VGB-Kraftwerkstechnik, 57th Annual Series, No. 6, June 1977.

2. Hein, D. and Köhler, W., A Simple-To-Use Post-Dryout Heat Transfer Model Accounting for Thermal Non-Equilibrium, Int. Workshop on Fundamental Aspects of Post-Dryout Heat Transfer, April 2 - 4, 1984, Salt Lake City, USA.

3. Köhler, W., Effect of the Wetting State of the Heat Exchange Surface on Heat Transfer and Pressure Loss in an Evaporator Tube, Dissertation, Munich University of Technology, 1984 (in German).

4. Kon'Kov, A. S., Experimental Study of the Conditions under which Heat Exchange Deteriorates when a Steam-Water Mixture Flows in Heated Tubes, Teploenergetika, Vol. 13 (1966) No. 12, P. 77.

5. Jens, W. H. and Lottes, R. A., Analysis of Heat Transfer, Burnout, Pressure Drop and Density Data for High Pressure Water, USAEC Report ANL-4627 (1951).

6. Gnielinski, V., New Equations for the Heat and Mass Transfer in Tubes and Channels Carrying Turbulent Flow, Forsch. Ing.-Wes. 41, No. 1 (1975), (in German).

7. Beattie, D. R. H., A Note on the Calculation of Two-Phase Pressure Losses, Nucl. Engng. and Design 25, pp. 395 - 402 (1973).

8. Mayinger, F., Flow and Heat Transfer in Gas-Liquid Mixtures, Vienna (1982) P. 54 (in German).

The Equivalent Model of Bubble Growth Rate at the Wall

YIDING CHAO and MINGDAO XIN

Department of Heat Power Engineering
Chongqing University
Chongqing, Sichuan, PRC

ABSTRACT

An equivalent model for bubble growth rate at the wall under general conditions is proposed. Unlike the existing models in the literatures, this model considers the effects of evaporation and relaxation microlayers on the bubble growth rate equal to those of a thermal layer, whose thickness and temperature change with time. Thereby a general relation for the bubble growth rate at the wall is derived. The comparison is made between the theoretical values and the experimental values published in the literatures with good agreement.

NOMENCLATURE

	a parameter, defined by Eq. (18)
	liquid thermal diffusibility $[m^2/s]$
	a parameter, defined by Eq. (18)
,b*	dimensionless bubble growth parameter during adherence
	liquid specific heat at constant pressure [J/kg.k]
1	bubble growth constant in pure liquid $[m/s^{\frac{1}{2}}k]$
(t)	thickness of equivalent boundary layer around bubble
(t)	equivalent coefficient of heat transfer
a	$=(\rho_1 c/\rho_2 L)\theta_o$, Jakob number for superheated pure liquid
	liquid thermal conductivity $[W/m^2k]$
	latent heat of vaporization at ambient pressure [J/kg]
	ambient pressure $[Pa=N/m^2]$
r	$=\nu/a$, liquid Prandtl number
(t)	equivalent spherical bubble radius [m]
o	equilibrium bubble radius [m]
*	radius of hemispherical bubble [m]
1	equivalent bubble radius according to modified Rayleish Solution [m]
R_2	equivalent bubble radius according to total diffusion [m]
R_d	departure radius of bubble [m]
R^+	$=\pi(\pi L\rho_2\theta_o)^{\frac{1}{2}}R/12aJa^2(7\rho_1 T)^{\frac{1}{2}}$
T	saturation temperature of liquid at ambient pressure [k]
$T_v(t)$	instantaneous temperature of vapor in the bubble [k]
t	bubble growth time during adherence [s]
t_d	bubble departure time [s]
t_w	waiting time between succeeding bubbles [s]
t^+	$=\pi^2 L\rho_2\theta_o t/84\rho_1 TaJa^2$
t_w^+	$=\pi^2 L\rho_2\theta_o t_w/84\rho_1 TaJa^2$
θ_o	superheating of heating surface [k]
$\theta(t)$	temperature difference of heat conduction [k]
$\Delta\theta_o$	superheating of bulk liquid [k]
ρ_1	liquid density $[kg/m^3]$
ρ_2	vapour density $[kg/m^3]$

INTRODUCTION

Bubble growth rates at the wall were extensively investigated in the last a few decads, and various bubble growth models were proposed. Some typical models are cited as follows:

(i) The "evaporation microlayer (beneath the bubble)" model, e.g. Cooper an Vijuk [1]. The derived relation is

$$R^*(t) = \frac{1}{(3\rho_1 T/2\rho_2 L\theta_o t^2)^{\frac{1}{2}} + (\pi Pr)^{\frac{1}{2}}/4Ja(at)^{\frac{1}{2}}} \tag{1}$$

(ii) The "relaxation microlayer" model, or model for bubble growth due to evaporation at the curved interface, e.g. Van Stralen [2] and Mikic et al.[3]. The relation derived by Van Stralen [2] is

$$R(t) = [bC_1\theta_o\exp(-t/t_d)^{\frac{1}{2}})]t^{\frac{1}{2}} \tag{2}$$

where $b = eR_d/C_1\theta_o t_d^{\frac{1}{2}}$, and the relation derived by Mikic et al. is

$$\frac{dR^+}{dt^+} = [t^+ + 1 - \frac{\theta_o - \Delta\theta_o}{\theta_o}(t^+/(t^+ + t_w^+))^{\frac{1}{2}}]^{\frac{1}{2}} - (t^+)^{\frac{1}{2}} \tag{3}$$

for $\Delta\theta_o = \theta_o$, $t_w^+ = \infty$, Eq.(3) is simplified as

$$\frac{dR^+}{dt^+} = (t^+ + 1)^{\frac{1}{2}} - (t^+)^{\frac{1}{2}} \tag{4}$$

(iii) The combination of the evaporation and the relaxation microlayer, e.g Van Stralen et al. [4], the derived relation is

$$R(t) = R_1(t)R_2(t)/[R_1(t) + R_2(t)] \tag{5}$$

where

$$R_1(t) = [\frac{2\rho_2 L}{3\rho_1 T}\theta_o\exp(-(t/t_d)^{\frac{1}{2}})]^{\frac{1}{2}}t$$

$$R_2(t) = 1.954[b^*\exp(-(t/t_d)^{\frac{1}{2}}) + \Delta\theta_o/\theta_o]Ja(at)^{\frac{1}{2}} + 0.373Pr^{-1/6}\exp[-(t/t_d)^{\frac{1}{2}}]Ja(at$$

$$b^* = 1.391\,[R_2(t_d)/(Ja(at_d)^{\frac{1}{2}})] - 0.191Pr^{-1/6}$$

$$R_2(t_d) = R_d R_1(t_d)/[R_1(t_d) - R_d]$$

Both t_d and R_d in above equations are determined experimentally.

Among the equations mentioned above, Eq. (2) is in quantitative agreement with the experimental data at atmospheric pressure [2], while Eq. (5) is in qua titative agreement at subatmospheric pressure [5,6]. The conditions on which E (5) is derived are at low pressure and with a hemispherical shape for the growi bubble. It should be pointed out that $b^* > 1$ for $p \geq 20.28$kPa[5,6], which seems be in contradiction with the physical background ($b^* \leq 1$[2,4]) of Eq. (5). The purpose of this paper is to derive a relation which could be applicable to all cases.

THE EQUIVALENT MODEL

We treat the real bubble as an equivalent spherical bubble with the equivalent radius R(t). As mentioned by Van Stralen et al. [4,5,6], the vaporization of the evaporation and the relaxation microlayer determines the bubble growth rate at the wall, and the superheat of bulk liquid has little effect on it. We consider the effects of both microlayers on the bubble growth equal to those of a liquid thermal layer, whose thickness and temperature change with time, as shown in Fig.1. Where d(t) denotes the thickness of the equivalent liquid thermal layer, θ(t) denotes the corresponding temperature difference of heat conduction.

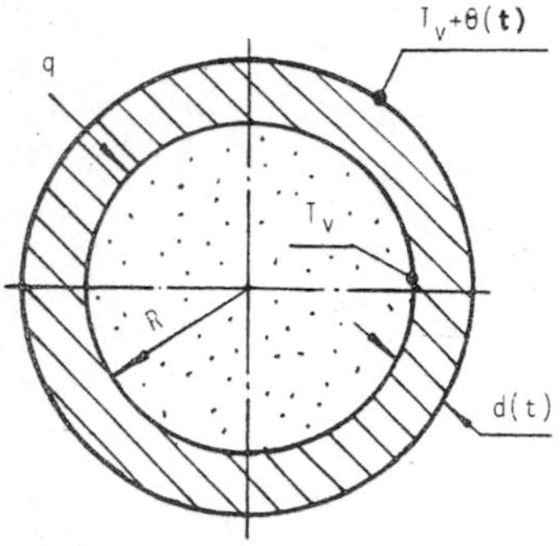

Fig. 1 Equivalent model

Before analysing, we make the following assumptions:

(i) When $t \to 0$, the growing bubble has the shape of a hemisphere, and satisfies the Rayleish solution for free bubbles [6], i.e.

$$R^* = (2\rho_2 L\theta_o/3\rho_1 T)^{\frac{1}{2}} t \tag{6}$$

(ii) $$\left.\frac{dR(t)}{dt}\right|_{t=t_d} = 0 \qquad \text{and} \qquad R(t_d) = R_d$$

The assumption that bubble initially has the shape of a hemisphere can be confirmed by many photographs [2,5,7]. Because at that time the bubble is very small, the pressure difference caused by the curved interface is so large that the wall resistance to the growing bubble is negligible, and because of the hemispherical shape of the bubble, its effect on liquid flow pattern is the same as that of the free bubble. Therefore assumption (i) is correct. In fact, Cooper and Vijuk [1] have used Eq. (6) as the initial growth relation in Eq.(1). Moreover, assumption (i) implies $R(0) = R_o = 0$ at $t = 0$, which is a standard assumption in many analyses because of its very small value.

At $t = t_d$, the bubble breaks away from the wall, the contact area between the bubble and the heating wall has vanished, so that heat supply due to the evaporation microlayer is stopped. On the other hand, the excess enthalpy of the relaxation microlayer has almost been consumed, its effects on the bubble growth are also small. Therefore, under saturated boiling condition, assumption (ii) is satisfied.

We may define the equivalent coefficient of heat transfer as

$$h(t) = k/d(t)$$

where k is liquid thermal conductivity.

By means of the energy balance for an equivalent spherical bubble, we get

$$4\pi R^2\theta(t)h(t) = \rho_2 L \frac{d}{dt}(4\pi R^3/3)$$

whence

$$h(t) = \frac{\rho_2 L}{\theta(t)} \frac{dR(t)}{dt} \tag{7}$$

When $t \to 0$, according to assumption (i)

$$R^* = (2\rho_2 L\theta_o/3\rho_1 T)^{\frac{1}{2}} t$$

where R^* is the radius of the hemispherical bubble, converting it into the equivalent spherical bubble radius as

$$R(t) = (1/2^{1/3})(2\rho_2 L/3\rho_1 T)^{\frac{1}{2}}\theta_o^{\frac{1}{2}}\, t \qquad (t \to 0)$$

whence

$$\frac{dR(t)}{dt} = (1/2)^{1/3}(2\rho_2 L/3\rho_1 T)^{\frac{1}{2}}\theta_o^{\frac{1}{2}} \qquad (t \to 0) \qquad (8)$$

During the whole growth period, the forms of relations for R(t) should be consistent. Therefore, we assume the validity of the following expression for th whole bubble growth period:

$$R(t) = (1/2)^{1/3}(2\rho_2 L/3\rho_1 T)^{\frac{1}{2}}[\theta(t)]^{\frac{1}{2}}t \qquad (0 \leq t \leq t_d) \qquad (9)$$

it can be seen from Eq.(8) and Eq.(9) that

$$\theta(0) = \theta_o \qquad (10)$$

through Eq.(7) And Eq.(8), we get the initial condition of h(t) as

$$h(0) = (\rho_2 L/2^{1/3})(2\rho_2 L/3\rho_1 T)^{\frac{1}{2}}\theta_o^{-\frac{1}{2}} \qquad (11)$$

the other condition at $t = t_d$ is obtained by mentioned assumption (ii) as

$$h(t_d) = 0 \qquad (12)$$

for $0 \leq t \leq t_d$, h(t) could be expressed with a function as the following:

$$h(t) = (1/2)^{1/3}\rho_2 L(2\rho_2 L/3\rho_1 T)^{\frac{1}{2}}[\theta(t)]^{-\frac{1}{2}}[1-(t/t_d)^n] \qquad (13)$$

the function above satisfies the two conditions, i.e. Eq.(11) and Eq. (12).

Substituting Eq.(13) and Eq.(9) into Eq.(7), we get an ordinary differentia equation of the first order for θ(t):

$$\frac{1}{\theta(t)}\frac{d\theta(t)}{dt} + 2\frac{t^{n-1}}{t_d^n} = 0 \qquad (14)$$

the general solution of this equation is

$$\theta(t) = C'\exp[(-2/n)(t/t_d)^n]$$

the constant of integration C' follows from Eq.(10), which yiels

$$\theta(t) = \theta_o \exp[-(2/n)(t/t_d)^n] \qquad (15)$$

Eq.(9) becomes

$$R(t) = \frac{1}{2^{1/3}}\left(\frac{2\rho_2 L\theta_o}{3\rho_1 T}\right)^{\frac{1}{2}} t \exp[-(1/n)(t/t_d)^n]$$

by taking $R(t_d) = R_d$, we obtain

$$n = \frac{1}{\ln\left[\frac{t_d}{R_d 2^{1/3}}\left(\frac{2\rho_2 L\theta_o}{3\rho_1 T}\right)^{\frac{1}{2}}\right]} \qquad (16)$$

therefore

$$R(t) = [B\ \exp(-A(t/t_d)^{1/A})]t \tag{17}$$

where

$$B = \frac{1}{2^{1/3}}\left(\frac{2\rho_2 L\ \theta_o}{3\rho_1 T}\right)^{\frac{1}{2}} \quad , \qquad A = \ln\left(\frac{Bt_d}{R_d}\right) \quad , \tag{18}$$

ρ_1, ρ_2 and T in B take the saturation values corresponding to ambient pressure. $(B\ t_d/R_d)$ in A is the ratio between the bubble departure radius according to the initial growth relation and the real departure radius R_d. It is worth pointin out that under certain boiling conditions, the values of $(B\ t_d/R_d)$ for different bubbles approach a constant provided the bubbles don't interfere each other, which has been confirmed by authors with some experimental data from the iiteratures[2, 5,7].

It can be verified that Eq.(17) satisfies Eq.(8), assumption (i) and assumptio (ii).

Rearranging Eq.(17), we get

$$R(t) = (Bt)\ \exp[-A(t/t_d)^{1/A}] \tag{17'}$$

where Bt denotes the bubble radius at any time t according to Eq.(6), $\exp[-A(t/t_d)^{1/A}]$ could be considered as a revising value.

For asymptotic isobaric bubble growth, Eq.(2) can be rewritten as

$$R(t) = (C_1\theta_o t^{\frac{1}{2}})\ b\ \exp[-(t/t_d)^{\frac{1}{2}}] \tag{2'}$$

where $C_1\theta_o t^{\frac{1}{2}}$ is identical with the asymptotic isobaric growth relation for free bubbles[2], $b\exp[-(t/t_d)^{\frac{1}{2}}]$ could also be considered as a revising value. The difference between Eq.(2') and Eq.(17') is that the former holds in the asymptotic isobaric period while the latter holds in the whole bubble growth period.

COMPARISON OF THEORETICAL PREDICTIONS AND EXPERIMENTAL DATA

The comparisons between the different relations for the bubble growth and the experimental data published in Ref.[2,5,6] are shown in Fig.2 - Fig.5.

It can be seen from Fig.2 - Fig.5 that well below atmosphere, Eq.(5) is in good agreement with the experimental data, and Eq.(2) in good agreement at atmosphere, while Wq.(17) is in good agreement with the experimental data in all case. Moreover, Eq.(17) is much simpler than Eq.(5).

REFERENCES

1. M.G.Cooper and R.M.Vijuk, Bubble growth in nucleate pool boiling, Proc. 4th Int. Heat Transfer Conf., Paris-Versailles, Vol. V, p. B2.1. Elsevier, Amsterdam (1970).
2. S.J.D.Van Stralen, The mechanism of nucleate boiling in pure liquids and in binary mixtures, Parts 1-2, Int. J. Heat Mass Transfer 9, 995-1020, 1020-1046 (1966).
3. B.B.Mikic, W.M.Rohsenow and P.Griffith, On bubble growth rates, Int. J. Heat Mass Transfer 13, 657-666 (1970).
4. S.J.D.Van Stralen, M.S.Sohal, R.Cole and W.M.Sluyter, Bubble growth rates in pure and binary systems: Combined effect of relaxation and evaporation micro-

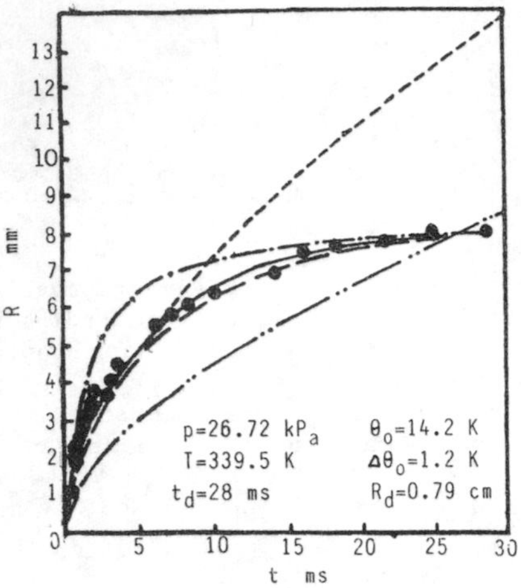

Fig.2 Comparison between theoretical predictions and experimental data of bubble growth radii [5] for water boiling.

——·· —— Eq.(1). ——·——·—— Eq.(2).
--------- Eq.(4). — — — — Eq.(5).
———— Authors Eq.(17).

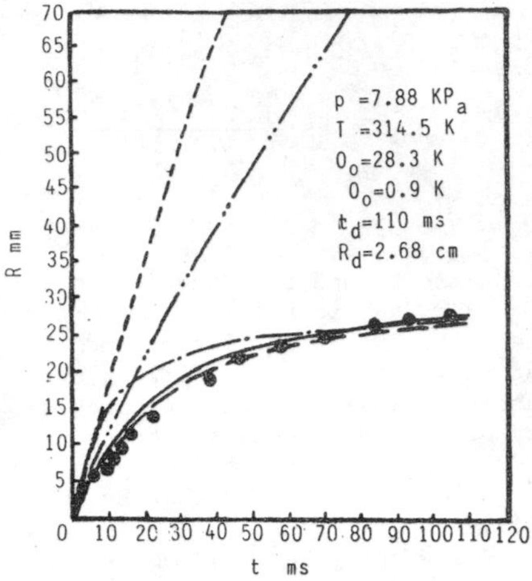

Fig.3 Comparison between theoretical predictions and experimental data of bubble growth radii [5] for water boiling.

——·· —— Eq.(1). ——·—— Eq.(2).
--------- Eq.(4). — — — — Eq.(5).
———— Authors Eq.(17).

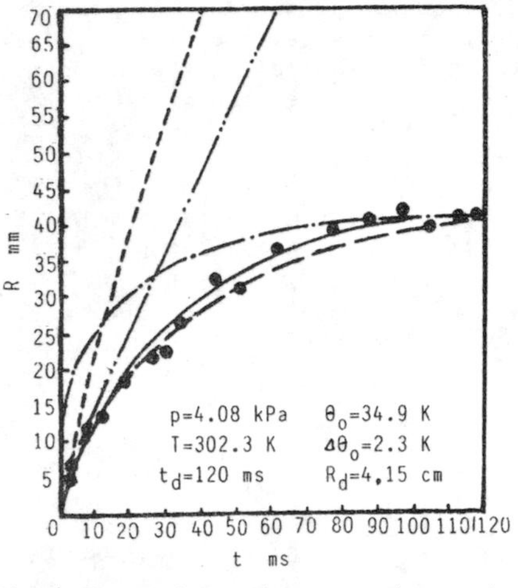

Fig.4 Comparison between theoretical predictions and experimental data of bubble growth radii [5] for water boiling.

——··——Eq.(1). ——·—— Eq.(2).
------- Eq.(4). — — — — Eq.(5).
———— Authors Eq.(17).

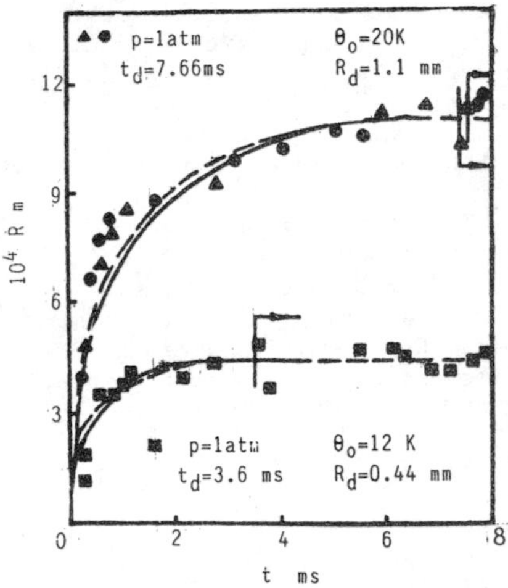

Fig.5 comparison between theoretica predictions and experimental data of bubble growth radii [2] for water bo ing.

— — — — Eq.(2).
———— Authors Eq.(17).

layers, Int. J. Heat Mass Transfer 18,453-467 (1975).
5. S.J.D.Van Stralen, R.Cole, W.M.Sluyter and M.S.Sohal, Bubble growth rate in nucleate boiling of water at subatmospheric pressure, Int. J. Heat Mass Transfer 18, 655-669 (1975).
6. S.J.D.Van Stralen, R.Cole, Boiling Phenomena, Vol. 1 andVol. 2, Hemisphere Publishing Corporation (1979).
7. C.Y. Han and P.Griffith, The mechanism of heat trausfer in nucleate pool boiling, Part I, Int. J. Heat Mass Transfer 8, 887-904 (1965).

Critical Heat Flux in Helical Coils

XUE-JUN CHEN and FANG-DE ZHOU
Engineering Thermophysics Research Institute
Xi'an Jiaotong University
Xi'an, PRC

ABSTRACT

A study of boiling water at high pressures in electrically heated helical coils of various diameters is reported. A concept about the influence region of liquid film near the location of CHF condition occured is presented. The correlations of the CHF at high quality region for straight horizontal tubes and helical coils are established. The experimental data are well agreement with the correlations.

INTRODUCTION

Heat exchangers consisting of helical coils have been widely used in many industrial fields. To investigate the characteristics of heat transfer in helica coils is an important subject. The critical heat flux (CHF) in forced convection boiling is a very important performance. However, few studies (1-3) have been done in the quality region under medium and high pressure.

This report based on the theoretical analysis and experimental results discusses the CHF in helical coils. Due to the close relation between the heat tran fer performances of straight tubes and helical coils, the investigating of CHF in straight tubes is also included.

EXPERIMENTAL APPARATUS

A shematic of the water flow loop is shown in Fig.1. Three helical coils and a straight horizontal tube (see table.1.) are made of 1 Cr 18 Ni 9 Ti stainless steel. All-temperatures were measured with calibrated NiCr-NiSi thermocouples. The test sections were installed 82 groups of thermocouples to determine the axial and circumferential temperature profiles. For the straight test section twenty thermocouples were spaced along the upper tube wall and eighteen thermocouple were spaced along circumference of the cross-sections at outlet. For the helical coils, twenty thermocouples were spaced along the inner of the tube wall and twnety-four thermocouples were spaced along the circumference of corss-sections of inlet, midst and outlet. The experimental range investigated were: pressure 6.5-10.5 Mpa. Mass velocity 500-2600 $Kg/m^2.S$, exit quality 0.01-0.95, heat flux 0.1-0.54 $Mw/_m{}^2$.

EXPERIMENTAL PROCEDURE

A series of single-phase flow and heat transfer tests were performed to establish the validity of the system and to check the testing exactness. The tes results were good agreement with the correlation found in the literatures (4).

After establish the required test section flow, inlet quality and outlet pressure were obtained, the test section power was slowly increased until thermocouple signal rose rapidly or large fluctuatin relative to the other wall thermocouples. At this time, all data were taken quickly. After this, the test section power was decreased until the thermocouple location had rewet. The test conditions were then readjusted, if necessary, the power was raised slowly until boiling crisis was again observed. Some tests were performed to determine the variation of the CHF along the axial position where the wall temperaure rose sharply during continuously increasing power.

THEORETICAL ANALYSIS

For flow boiling in a straight horizontal tube and helical coil tubes, the CHF condition was assumed to occur when the liquid consumption rate at a position exceeded the net liquid supply rate or the droplet deposition.

At high quality region, the flow pattern is annular flow. Liquid formed a film flow adjecent to the wall surface and some of which were entrained by vapour core, while the droplets in vapour core also deposited in liquid film. Due to the influence of the buoyancy force, the circumferential liquid film profile along wall surface was not homogeous. The thickness of liquid film decreased from tube bottom to upper. Assume that flow was steady, vapour core took a cylindrical form and the interface was smooth, see Fig.2. The thickness profile of liquid film was derived as following:

$$\frac{r}{\sin\theta} = \frac{oB}{\sin\beta} = \frac{e}{\sin\alpha} \tag{1}$$

Where e is the distance between the centres of vapour core and the tube.
r is the radius of the vapour core

The average thickness $\overline{m}$ of liquid near the dryout spot can be derived from eq. (1),

$$\overline{m} = \overline{m}_p\left(1 - \frac{\sin\Delta\theta}{\Delta\theta}\right) \tag{2}$$

Where $\overline{m}_p$ is the peripheral average thickness of liquid film.

In annular flow the mass balance equation for a small increment of length gives

$$\frac{\partial[(1-\alpha)\rho_L]}{\partial\tau} + \frac{\partial G_{Lf}}{\partial Z} = \frac{4}{d}\left(D - E - \frac{q}{i_{Lv}} - F_{Lf}\right) \tag{3}$$

Where $F_{L.f}$ is flashing mass flux of liquid film.

G_L is the film flow-rate/unit perimeter of heated surface.

α is the void fraction.

D is the rate of droplet deposition.

E is the rate of liquid entrainment.

Subscript f denotes the film flashing.

It has been shown (5) that as the film thickness and film flow rate in annular flow decreased, so the entrainment also decreased. In steady state for this particular condition,

$$\frac{q_c}{i_{Lv}} = D \tag{4}$$

and

$$D = KC \tag{5}$$

Where C is concentration of droplets in Vapour core
K is mass transfer coefficent,
which can be calculated from (8)

That is $$K = 87xu^*\left(\frac{\mu_1^2}{\sigma d \rho_1}\right)^{\frac{1}{2}} \qquad (6)$$

Where u^* is friction velocity, μ_1 is liquid viscosity.
Up to now, there is no suitable formulation of calculating the rates of droplet deposition and liquid entrainment for horizontal flow. We should revise the correlation suitable for the vertical tube. The average droplet concentration can be evaluated based on the Hutchinson and whalley's investigation (6),

$$C = f\left(\frac{\tau_i \bar{m}}{\sigma}\right) \qquad (7)$$

We can use the mean thickness of the liquid film in the influence region to evaluated the local droplet concentration.

The droplet concentration in vapour core was not uniform for a straight horizontal tube due to the effect of the force of gravity. The thickness of the liquid film depended on the circumferential locations.

That in the region near the top wall of the tube affected strongly on the droplet concentration of the deposition at the top wall of the tube and the CHF condition. The influence region of liquid film varied with the mass flowrate and pressure. Acoording to the analysis of the experimental result, the influence region was expressed with

$$\Delta\theta = \frac{\pi}{9}\left(\frac{G^*}{G}\right)^{1.5}\left(\frac{P^*}{P}\right)^{0.8} \qquad (8)$$

Where G is mass velocity $G^* = 1000\ kg/m^2 \cdot s$.
P is pressure $P^* = 10.5$ Mpa.

The droplet concentration of the deposition in influence region of the top wall of the tube can be evaluated with eq. (2), (7) and (8). Thus, the CHF correlation at high quality for a straight horizontal tube is

$$q_{c.s} = 87C\ i_{\ell.v}\ u^*\left(\frac{\mu_1^2}{d\sigma \rho_1}\right)^{\frac{1}{2}} \qquad (9)$$

On the basis of local, near wall condition, the performance of the CHF in helical coil and horizontal flow are probably quite similar. The differences of the CHF conditions between in coils and in straight tubes can be attributed to the formation of a secondary flow superimposed on the main flow. A pair of generally symmetrical vortices arises due to the centrifugal force, which occurs because of the coil geometry. When the CHF condition occured, the location where the temperature rose rapidly first would varied according to the balance of the buoyancy force and centrifugal force. Assuming that flow being steady, vapour core taking a Cylindrial form. Then, the location depends on φ. see Fig.3.

$$tg\varphi = \frac{(1-\alpha)\rho_1 U_1^2 - \alpha \rho_v u_v^2}{R\alpha g(\rho_1 - \rho_v)} \qquad (10)$$

When mass velocity was low, the centrifugal force was weak, $tg\varphi \to o$, $\varphi \doteq o$. The location where the temperature of the wall surface being the highest was at the top of the wall. When mass velocity was high, the buoyancy force could be negligible relative to the centrifugal force, that is $tg\varphi \to \infty$, $\varphi \doteq \frac{\pi}{2}$ When mass velocity was low and quality is high, the centrifugal force of vapour was large than that of liquid.

$$\alpha \rho_v u_v^2 > (1-\alpha)\rho_1 u_1^2 \qquad (11)$$

The liquid film inversion occured. The vapour tended to flow at the outside of the tube. The liquid film tended to flow to the top and bottom wall of the tube. In this condition the CHF would decrease with the decrease of the mass velocity.

According to the correlation of the CHF for horizontal tube, we considered the factors affecting on the CHF in helical coils, and obtained

$$\frac{q_{c.c}}{q_{c.s}} = 1+0.52\left(\frac{d}{D}\right)^{0.25}\left(\frac{x_e}{1-x_e}\right)^{2.2} \tag{12}$$

Where x_e is exit quality. D is the diameter of coil

In high quality and low mass velocity region the local droplet concentration at the influence region in coils became less than in horizontal tube due to the particular behavior of liquid film inversion.

$$\Delta\theta = \frac{\pi}{11}\left(\frac{G^*}{G}\right)^{1.5}\left(\frac{P^*}{P}\right)^{0.8} \tag{13}$$

EXPERIMENTAL RESULTS AND DISCUSSION

The CHF in the straight horizontal tube at quality region is plotted as a function of the exit quality in Fig.4.5.

At high quality the CHF increased with decreasing of pressure, decreased with increasing of exit quality and mass velocity. The calculating results of eq.(9) are excellent agree with the experimental data. The RMS-error for all the 172 data is 6.7 percent, with the maximum deviation being 18. percent. All CHF data were obtained in the case where there was a vapour-liquid mixture at the tube inlet.

In horizontal flow, boiling crisis was most often observed at the exit of the test section, only the thermocouple at the top of the tube indicated a rapidly temperature rose. This suggests that while the top surface of the tube was dry, the bottom surface of the tube still remained wet.

At low quality and low mass velocity the CHF conditions occured at both the extrance and outlet of the test section, with a wet region between them. This appears to be caused by the flow pattern transition from slug flow to annular flow.

The experimental data of the CHF in helical coils are also plotted as a function of the exit quality in Fig.6.7.8. The CHF increased with increasing of d/D at same mass velocity the CHF decreased with increasing of operating pressure. For a given quality the CHF initially increased with increasing of the mass velocity during mass velocity less than 1000 $kg/m^2 \cdot s$. Then, the CHF decreased with increasing mass velocity during mass velocity higher than 1000 $kg/m^2 \cdot s$. The literature (7) also found this behavior. The mass velocity had a strong effect on the location of the initiation of the CHF condition. In low mass velocity the buoyancy overcame the inertia and centrifugal forces resulting in vapour clotting at the top of the tube where the CHF condition occured first. In high mass velocity strong centrifugal force effected on the vapour core and led the vapour to the inner of the tube. At this case the CHF condition occured at the inner of the tube first. The location of the highest temperature of the tube wall varied along the test section. With increasing of the quality the vapour velocity increased. The influence of the centrifugal force was dominant. The location where the thickness of the liquid film was the thinnest, was removed to the inner of the tube. Both of the inner and outside of the tube were dryout under the CHF condition at low mass velocity and high quality. It was shown as above that if

$$u_v > \left[\frac{(1-\alpha)\rho_1}{\alpha\rho_v}\right]^{\frac{1}{2}}\rho_1$$

the liquid film inversion occured. The thickness of the outside wall varied thinner. The rate of droplet deposition at the outside wall of the tube decreased.

Hence the temperature of the tube wall of the outside also rose sharply. It ca that the CHF decreased with dereasing of the mass velocity in low mass velocity and high quality region. The eq. (12) is excellent agree with the experimental data. The RMS-error for all 128 data is 7.5 percent, with the maximum deviatio being 22.4 percent. The CHF in coil tubes is higher than in straight horizonta tube.

In mass velocity below 1000 $kg/m^2 \cdot s$. region the calculating results with eq (12),(13) are also good agreement with the experimental data. The RMS-error fo all 41 data is 7.2 percent, with the maximum deviation being 25.1 percent. As d/D is small, the condition of eq. (11) could not be satisfied. The CHF conditio always occured at the inner or the top wall of the tube first. The behavior o the CHF in coils of small d/D was quite simimlar to that in horzontal tube.

The temperature rise of the tube wall due to boiling crisis was not as seve in coils as in horizontal flow. This is due to the secondary flow spreading th liquid film over the perimeter and ensuring a wetted wall.

CONCLUSION

1. At high quality region, the CHF in a straight horizontal uniformly heated tube decreases with increasing operating pressure, mass velocity and exit quality. The CHF at high quality region can be evaluated with eq. (9).
2. The CHF at quality region in helical coils is higher than that in straight horizontal tube. The CHF increases with increasing d/D, decreases with increas ing exit quality and operating pressure. At a given quality pressure and d the CHF increases with increasing mass velocity during mass velocity less t 1000 $kg/m^2 \cdot s$ and decreases with increasing mass velocity during mass veloci higher than 1000 $kg/m^2 \cdot s$.
3. The location of the initiation CHF depends on the balance of centrifugal an buoyancy forces. The location of the highest temperature rise varies along the test section during the CHF condition occuring. At lowwer mass velocit and high quality region when the centrifugal force of vapour core is strong than that of liquid, the liquid film inversion occurs. Both of the inner a outside of the tube are dryout under the CHF condition.
4. The CHF at high quality in coils can be evaluated with eq. (12). At mass velocity below 1000 $kg/m^2 \cdot s$. the CHF can be evaluated with eq. (12) and (13

ACKNOWLEDGMENT

This work is supported by the Science Foundation of Chinese Academy of Science and National Commision of Science and Technology.

References

1. BaBaRiN. B.P. et al "Critical heat fluxes in Tubular Coils" Heat Transfer-Soviet Research Vol.3, No.4. 1971. pp 85-89.
2. Moribumi. K. et al "A study of Helically-Coiled Tube once-Through Steam Generator" Bulletin of the JSME. Vol.13. No.66. 1970. pp 1484-1494.
3. Unal. H.C. "Dryout and Two-Phase Flow Pressure Drop in Sodium Heated Helically Coiled Steam Generator Tubes at Elevated Pressure" Int. J Heat Mass Transfer. Vol.24 1981. pp 285-298.
4. Y.Mori and W. Nakayama. "study on Forced Convection Heat Transfer in Curvec Pipes" (Second report, turbulent region) Int. J. Heat Mass Transfer. Vol.1(1967. pp 37-59.
5. Whalley P.B. et al "The Calculation of Critical Heat Flux in Complex Situations Using an Annular Flow Model" 6th Int Heat Transfer Conference (Toront Vol.1. 1978. pp 64-69.
6. Hutchinson, P. and Whalley, P.B. "A Possible Characterisation of Entrainme

in Annular Flow". Chem. Engng. Sci Vol.28 1973.
7. Jensen, M.K. and Bergles, A.E. "Critical Heat Flux in Helically Coiled Tubes" Trans of the ASME. Heat Transfer. Vol.103 1981. pp 660-666.
8. Hewitt, G.F. and Hall-Taylor, N.S. "Annular Two-Phase Flow" pergamon press. Oxford. 1970.

Table 1. Test Section Dimensions

Test Section No	Inside Tube Diameter mm	Coil Diameter mm	d/D	Pitch mm	Total Coil Length mm	Inclination (0)	Tube Wall Thickness mm
1	15.95	195	0.0818	11.5	4865	3.72	2.0
2	16.02	405	0.0396	33.0	4815	1.65	2.0
3	16.05	800	0.0201	60.5	4775	0.86	2.0
4	16.05	Straight	-	-	2000	-	2.0

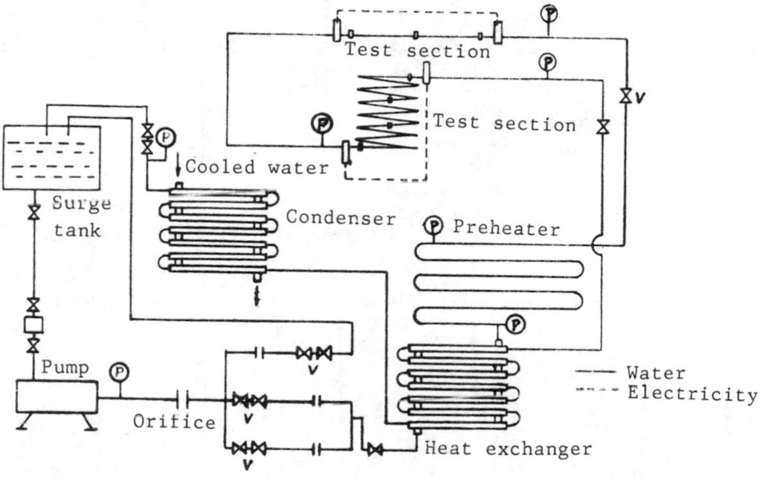

Fig.1 Schmatic layout of test loop

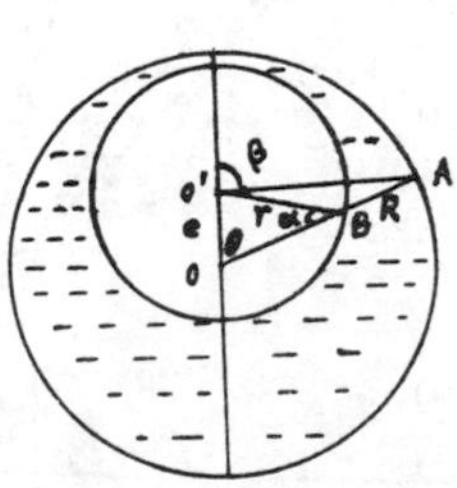

Fig.2 The thickness of liquid film

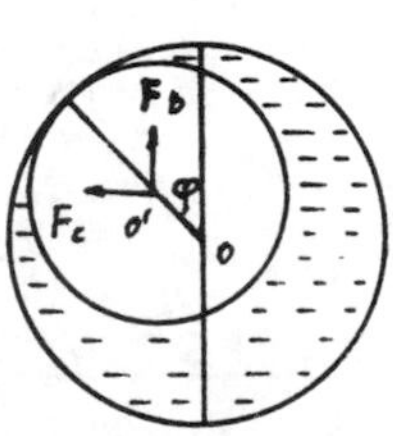

Fig.3 The location of the vapour core

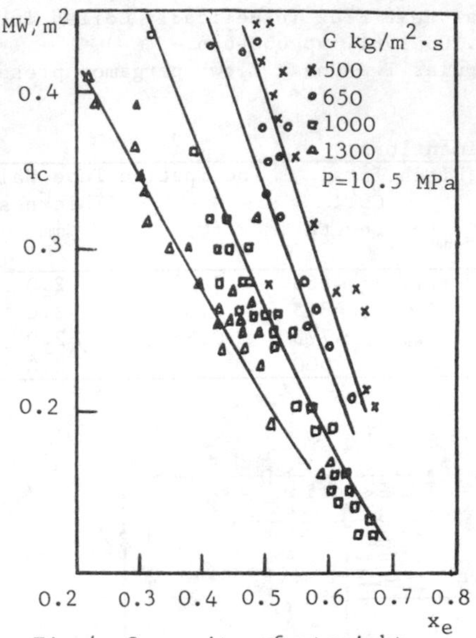

Fig.4 Composite of straight horizontal tube CHF data

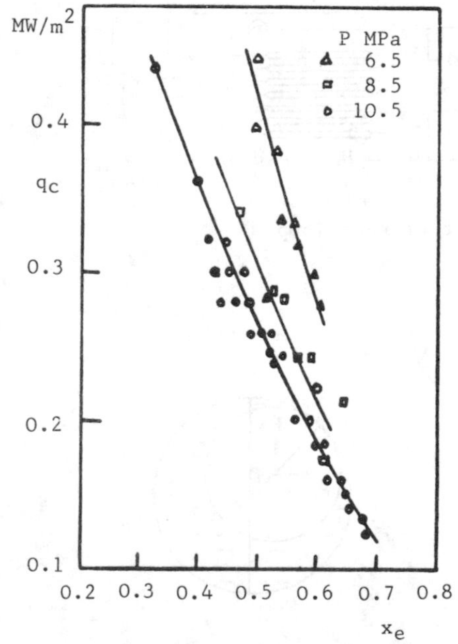

Fig.5 Effect of pressure on CHF in straight horizontal tube

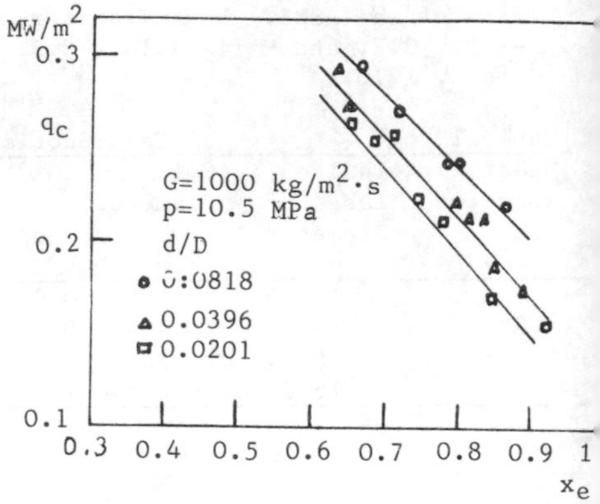

Fig.6 Influence of d/D on CHF in Coi

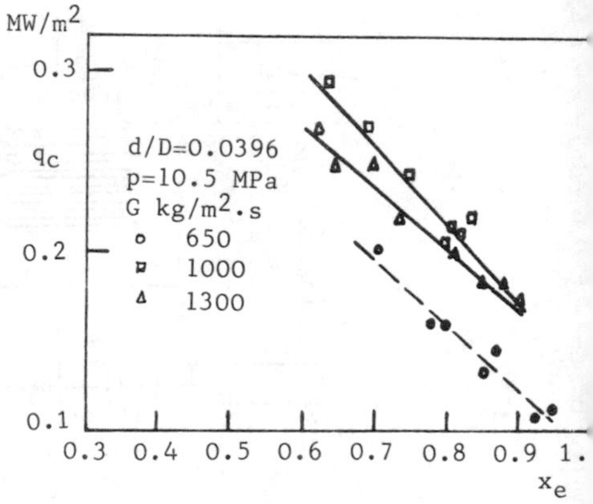

Fig.7 Dependence of CHF in Coil o mass velocity

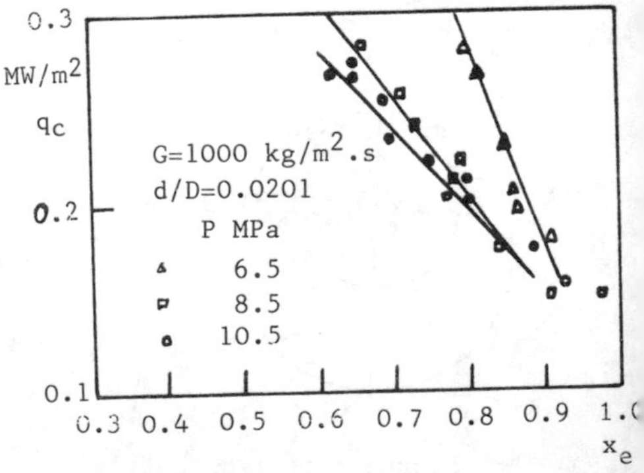

Fig.8 Effect of pressure on CHF in Coi

Boiling Heat Transfer on Horizontal Tube Bundles

YASUNOBU FUJITA, HARUHIKO OHTA, KEISUKE YOSHIDA,
SUMITOMO HIDAKA, and YUKIFUMI TOYAMA
Kyushu University
Fukuoka, Japan

KANEYASU NISHIKAWA
Kurume Technical College
Kurume, Japan

1. INTRODUCTION

Nucleate boiling heat transfer characteristics for a tube in a bundle differ from that for a single tube in a pool and this difference is known as 'tube bundle effect'. There exist two bundle effects, positive and negative. The positive bundle effect enhances heat transfer due to convective flow induced by rising bubbles generated from the lower tubes, while the negative bundle effect deteriorates heat transfer due to vapor blanketing caused by accumulation of bubbles.

Robinson and Katz[1] tested staggered tube bundles and found that the upper tubes in bundles have higher heat transfer coefficients than the lower tubes. Since then the effects of various parameters such as pressure, tube geometry and oil contamination on heat transfer have been examined[2-4]. Some workers attempted to clarify the mechanism of occurrence of 'bundle effect' by testing tube arrangements of small scale[5-8]. All of them reported only enhancement in heat transfer but the results by Nakajima and Morimoto[5] showed the symptom of heat transfer deterioration at higher heat fluxes.

As mentioned above, it has not been clarified so far even whether the 'tube bundle effect' should serve as enhancement or deterioration of heat transfer in nucleate boiling. In this study, experiments are performed in detail by using bundles of small scale, and effects of heat flux distribution, pressure and tube location are clarified. Furthermore, some consideration on the mechanisms of occurrence of 'tube bundle effect' is made and a method for prediction of heat transfer rate is proposed.

2. EXPERIMENTAL APPARATUS

FIGURE 1 shows the outline of experimental apparatus. The boiling vessel ① is a horizontal circular hollow cylinder with an inner diameter of 286mm and a length of 250mm. Two windows ④ for observation are installed on the vessel wall. Horizontal test tubes are cantilevered from one of flanges of the cylindrical vessel. FIGURE 2 shows details of the test tube. It is copper tube with an outer diameter of 25mm and a heated length of 120mm, whose surface is polished by No.0/4 emery paper. The tube is heated by electric cartridge heater inserted inside of it. Four thermocouples are axially inserted at the depth of 2.3mm from the tube surface at four circumferential locations.

Experiments were performed under the condition of saturated nucleate boiling of Freon 113($C_2F_3Cl_3$) of commercial grade. The system pressures tested are 0.1,

0.2, 0.5 and 1.0MPa. Heat flux is evaluated by dividing electric heat input by surface area. Surface temperatures are obtained by extrapolating the wall temperatures measured by thermocouples to the outer surface. The degree of superheat of heating surface is defined as the difference between the mean value of four surface temperatures and the saturation temperature corresponding to local liquid pressure at the same level as the tube concerned.

3. EXPERIMENTAL RESULTS

3.1 A Single Tube (Experiment A)

The relations between heat transfer coefficient α and heat flux q for a single tube at various pressures are shown in FIGURE 3. This gives the reference heat transfer characteristics for evaluating the tube bundle effect. There observed no appreciable difference in heat transfer characteristics between tubes of different positions in a bundle, which means no effect of liquid head above the tube concerned. Heat transfer coefficients in nucleate boiling obtained in this experiment agree well with those by Nishikawa et al.[9] for a horizontal flat plate.

3.2 A Vertical Row of Three Tubes (Experiments B and C)

FIGURE 4 (Experiment B) shows heat transfer coefficients at pressures of 0.1 and 1.0MPa of a vertical row of three tubes to which equal heat flux is supplied

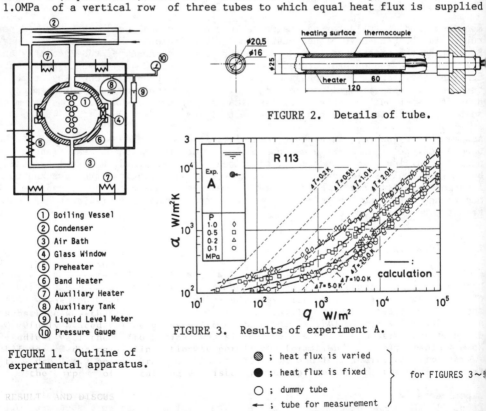

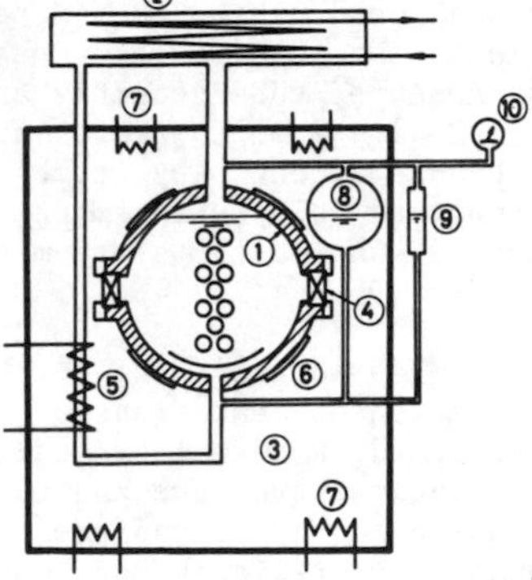

FIGURE 1. Outline of experimental apparatus.

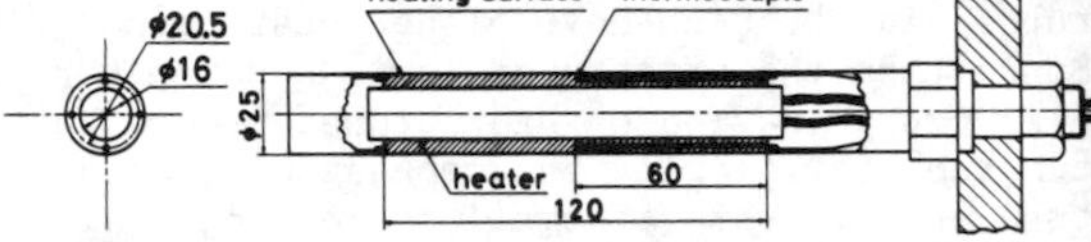

FIGURE 2. Details of tube.

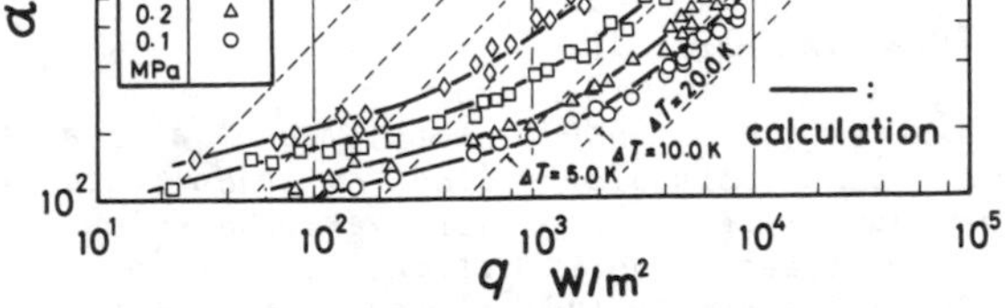

FIGURE 3. Results of experiment A.

◍ ; heat flux is varied
● ; heat flux is fixed
○ ; dummy tube
← ; tube for measurement

for FIGURES 3~5

The followings are observed. (1) Heat transfer characteristics of the bottom tube(No.1) are identical to those for a single tube over the whole range of heat flux. (2) Enhancement in heat transfer caused by the 'tube bundle effect' is remarkable in a low heat flux region, while it disappears in a high heat flux region. (3) Enhancement in heat transfer is greater for the top tube(No.3) than the middle one(No.2). (4) At high pressures, enhancement in heat transfer decreases. Furthermore, an additional result was obtained from another measurements where the vertical tube pitch varies from 32mm to 64mm. (5) No noticeable effect of tube pitch on enhancement in heat transfer was observed within the test range.

FIGURE 5 (Experiment C) shows heat transfer coefficients of the top tube(No.3) at pressures of 0.1 and 0.5MPa, in which heat transfer coefficient α_3 is plotted against its heat flux q_3, whereas the lower two tubes(No.2 and No.1) are held at the constant heat flux q_0. The following characteristics were observed. (1) Enhancement in heat transfer increases with increasing vapor generation rate from the lower two tubes. (2) In this case, the dependence of heat transfer coefficient α_3 of the top tube on heat flux q_3 is small. (3) When heat flux q_3 increases, enhancement in heat transfer disappears. (4) In contrast with the result of Experiment B (which is represented by a chain line in FIGURE 5), the remarkable enhancement is observed even at high pressures. Solid symbols in FIGURE 5 represent the results for similar measurements in which heat flux q_0 is given only to the middle tube. Comparison of data indicates that there exists little difference in the 'tube bundle effects' despite of considerable difference in vapor volume generated from the lower tubes. (5) Enhancement in heat

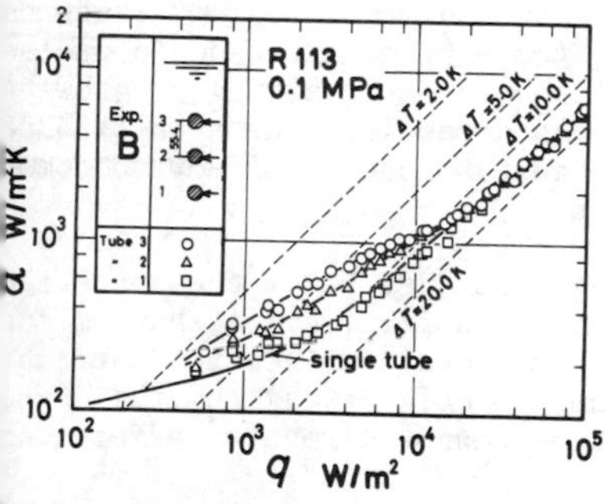

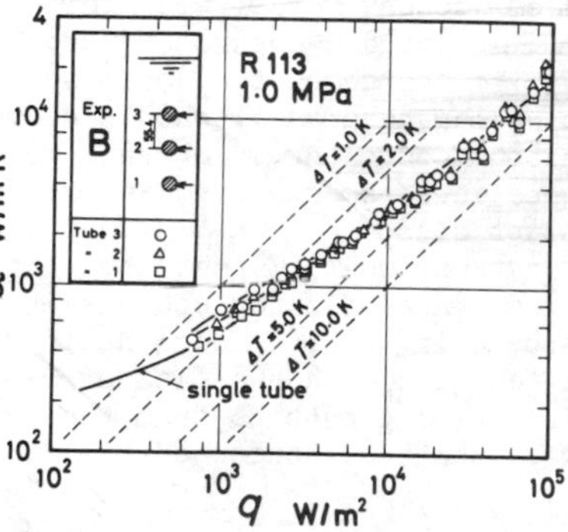

FIGURE 4. Results of Experiment B.

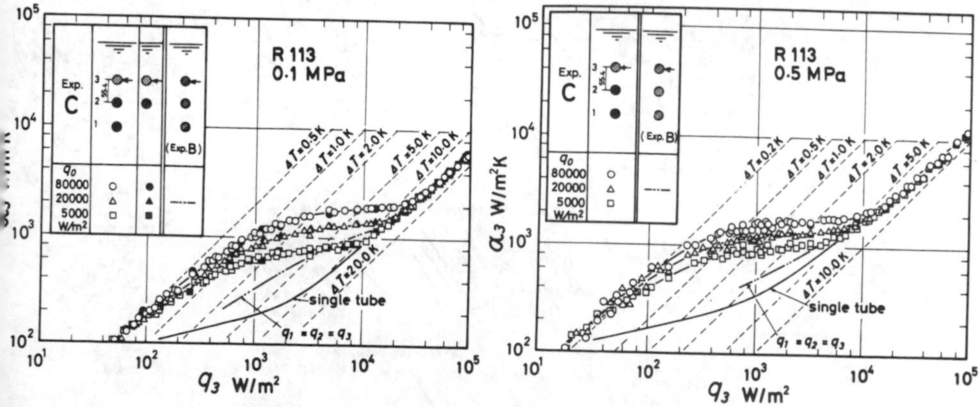

FIGURE 5. Results of Experiment C.

transfer is controlled by the neighboring tube just below the concerned tube, provided that heat flux added to the lower tubes is large.

4. DISCUSSIONS

We assume that boiling tube surface consists of two influence areas of different heat transfer mechanisms, the influence area by growing bubbles on the concerned tube, and the influence area by flowing bubbles from the lower tubes. In a low heat flux region (FIGURE 6(a)), since nucleation site density is low and also bubbles grow slowly owing to low wall superheat, the influence area by flowing bubbles is relatively large against the total area. On the other hand, in a high heat flux region (FIGURE 6(b)), increasing nucleation site density and large bubble growth rate cause the increase in influence area by growing bubbles and the simultaneous decrease in influence area exposed to flowing bubbles. Therefore, the tube bundle effect may be interpreted as the effect of rising bubbles which are originated from the lower tubes and act on a thermal boundary layer of the tube concerned. Thus, the intensity of this effect depends on both nucleation site density of the concerned tube and vapor generation rate from the lower tubes. For these reasons, enhancement in nucleate boiling heat transfer cannot be observed at higher heat fluxes where nucleation site density is very high.

The effect of pressure is considered as follows. At low pressures, the influence area by a single growing bubble becomes large corresponding to a large bubble diameter. However, this effect is completely compensated by low nucleation site density, which means relatively large area exposed to flowing bubbles (FIGURE 6(c)). At high pressures(FIGURE 6(d)), the extreme increase in nucleation site density overcomes the decrease in influence area by a single growing bubble corresponding to a small bubble diameter. As a result, the area exposed to rising bubbles decreases. For this reason, the tube bundle effect becomes weak at high pressures as observed in Experiment B.

We have seen in Experiments C that the neighboring tube just below the concerned one plays an important role for enhancement in heat transfer. This mechanism can be explained by using a flow model depicted in FIGURE 6(e). Since the top tube is completely covered with the flow of rising bubbles originated from the tube just below it, rising bubbles from the bottom tube cannot directly affect on heat transfer of the tube concerned.

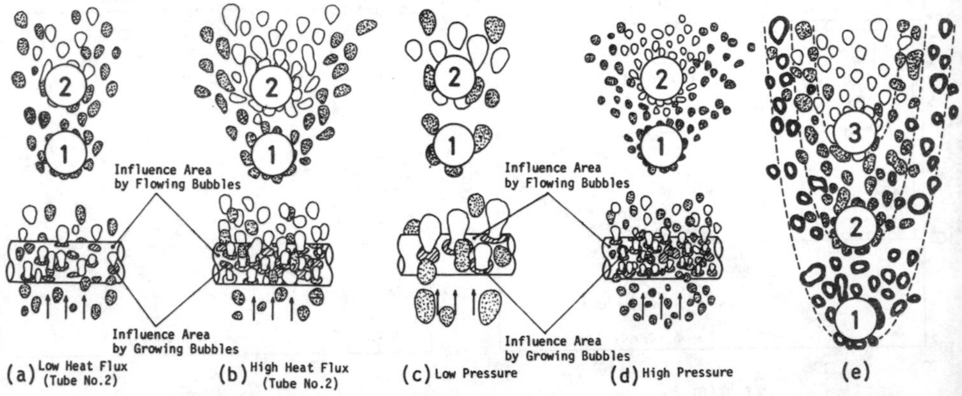

FIGURE 6. Effects of heat flux and pressure on behavior of flowing bubbles.

Experiment C shows that the enhanced heat transfer coefficient in a moderate heat flux region is insensible to the heat flux of itself. This fact means that the enhancement in this region is mainly governed by the forced convection induced by flowing bubbles. At lower heat fluxes, the enhanced heat transfer coefficient decreases along a constant superheat line with decreasing heat flux. Heat transfer coefficient by the forced convection would be remains constant regardless of heat flux provided that the surface superheat is reasonably defined. However, the temperature of liquid around the measured tube is slightly superheated. Thus, the degree of surface superheat defined as the temperature difference between the surface and saturated liquid is overestimated, which results in decrease in heat transfer coefficient.

In these measurements, the 'negative' tube bundle effect due to vapor blanketing has not been observed even at higher heat fluxes.

5. PREDICTION OF HEAT TRANSFER CHARACTERISTICS

5.1 A Single Tube

Mikic and Rohsenow[10] proposed a heat transfer model for an isolated bubble region in nucleate boiling. The model assumes that the heating surface under the influence area of growing bubbles around active sites is quenched by bulk liquid to which heat is transferred by unsteady heat conduction, while heat is removed from the remainder of surface by natural convection. The resulting heat flux is given in the following.

$$q_1 = 2\sqrt{\pi}\sqrt{\lambda_\ell \rho_\ell c_{p\ell}}\sqrt{f}\, D_b^2 (N/A)_1 \Delta T_1 + [1 - (N/A)_1 \pi D_b^2]\, \alpha_{NC1} \Delta T_1 \quad (1)$$

$$\alpha_1 = q_1 / \Delta T_1 \quad (2)$$

where, ΔT_1:degree of surface superheat[K], D_b:diameter of detaching bubble[m], $(N/A)_1$:active site density[1/m²], f:frequency of bubble emission[1/s], α_{NC1}:heat transfer coefficient due to natural convection[W/m² K], λ_ℓ, ρ_ℓ and $c_{p\ell}$:thermal conductivity[W/m K], density[kg/m³] and specific heat[J/kg K] of liquid, respectively. It is noteworthy that the application of eq.(1) can be extended to coalesced bubble region. In order to simplify the analysis, we also consider the bubble growth in uniform temperature field and assume that the cumulative number of cavities on the surface is inversely proportional to the n-th power of cavity radius. Thus, active site density $(N/A)_1$, can be related to surface superheat ΔT_1 as follows.

$$r_c = \frac{2T_{sat}\sigma}{L \rho_v \Delta T_1} \quad (3)$$

$$(N/A)_1 = C(1/r_c)^n = C\left(\frac{L \rho_v \Delta T_1}{2T_{sat}\sigma}\right)^n \quad (4)$$

where, r_c:radius of an active cavity[m], C:constant[m^{n-2}], Tsat, L and ρ_v :saturation temperature[K], letent heat of vaporization[J/kg] and density of vapor[kg/m³], respectively, σ:surface tension[N/m]. Heat transfer coefficient α_{NC1} due to natural convection is generally given as follows.

$$\alpha_{NC1} = C_{NC} \Delta T_1^{1/3} \quad (5)$$

Values of C_{NC} in TABLE 1, obtained based on the experimental data, have large dependence on pressure. The diameter of detaching bubble D_b and its product with frequency $f D_b$ are estimated from the correlations proposed by Cole and Rohsenow[11] and Zuber[12] respectively, and they are substituted into eq.(1) together with equations(4) and (5).

The relations between the active site density $(N/A)_1$ calculated by eq.(1) and the radius of cavity r_C by eq.(3) using the experimental data of q_1 and ΔT_1 are represented in FIGURE 7. Since the distribution of cavities on the heating surface is peculiar to the surface concerned, the relation between $(N/A)_1$ and r_C is represented by a single line irrespective of pressure provided that the idealized heat transfer model, the assumed criteria of activation and the simplified distribution of cavities are appropriate. In the present analysis, the relation between $(N/A)_1$ and r_C are approximated by the four representative lines through data for each pressure, and the corresponding values of n and C are given in TABLE 2. Bold solid lines in FIGURE 3 represent the calculated heat transfer coefficient. In a high heat flux region, the second term in the right hand of eq.(1) becomes negative corresponding to the increase in active site densities. In this case, the contribution of this term is ignored and the results are represented by broken lines.

5.2 A Vertical Row of Two Tubes

Heat transfer characteristics of a vertical row of two tubes, the most fundamental tube arrangement for consideration, are predicted as follows.

FIGURE 8 shows heat transfer characteristics of the upper tube at pressures of 0.1 and 0.5MPa when heat flux supplied to the lower tube is varied as a parameter. Heat transfer from the upper tube is assumed to be consist of heat transfer from the influence area by growing bubbles and that from the influence area by rising bubbles. Then heat flux of the upper tube can be given by the next form analogous to a single tube.

$$q_2 = 2\sqrt{\pi}\sqrt{\lambda_\ell \rho_\ell c_{p\ell}}\sqrt{f}\, D_b^2 (N/A)_2 \Delta T_2 + [1 - (N/A)_2 \pi D_b^2]\, \alpha_{2C} \Delta T_2 \tag{6}$$

$$(N/A)_2 = C\left(\frac{L\rho_v \Delta T_2}{2T_{sat}\sigma}\right)^n \tag{7}$$

$$\alpha_2 = q_2 / \Delta T_2 \tag{8}$$

where, α_{2C} indicates heat transfer coefficient due to the convection induced by rising bubbles generated from the lower tube(No.1).

FIGURE 9 shows the typical pattern of bubble behavior for this tube arrangement depicted based on visual observations. Since the rising bubbles from the lower

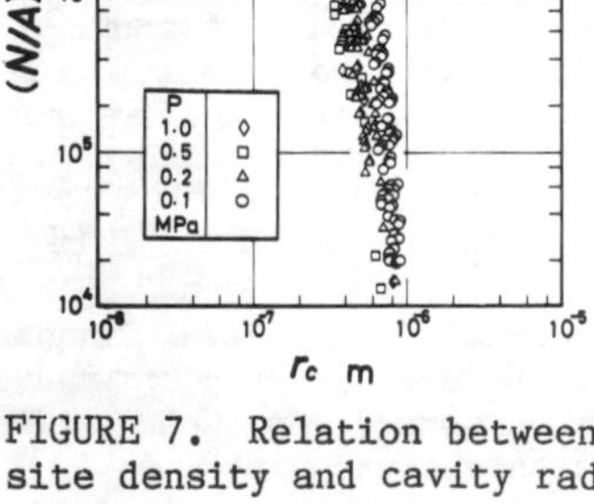

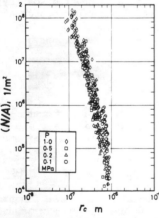

FIGURE 7. Relation between active site density and cavity radius.

TABLE 1. Values of C_{NC} in Equations (5) and (9).

Pressure, P, MPa	C_{NC}, $W/m^2 K^{4/3}$
0.1	97.4
0.2	128
0.5	194
1.0	251

TABLE 2. Values of n and C in Equations (4) and (7).

Pressure, P, MPa	n, -	C, m^{n-2}
0.1	6.7	1×10^{-36}
0.2	6.0	5×10^{-33}
0.5	5.0	6×10^{-27}
1.0	4.0	2×10^{-20}

ube(painted bubbles) pass along the bottom and side segment of the upper tube, ffect of these bubbles on heat transfer from the top segment may be small. hen, the surface area of upper tube is divided into four segments, which are ottom area A_B, side areas A_S and top area A_T, and we assume that convective eat transfer coefficients are equal for the bottom and side areas and given by $_{BS}$, while that for the top area is approximated with the heat transfer coeffi-ient for natural convection given by

$$_{NC2} = C_{NC}\,\Delta T_2^{1/3} \tag{9}$$

hen, eq.(6) is transformed as follows.

$$_2 = 2\sqrt{\pi}\sqrt{\lambda_\ell \rho_\ell c_{p\ell}}\,\sqrt{f}\,D_b^2\,(N/A)_2\Delta T_2 + [\,1 - (N/A)_2\pi D_b^2\,] \times [\,\alpha_{BS}(A_B/A + A_S/A) + \alpha_{NC2}(A_T/A)\,]\,\Delta T_2 \tag{10}$$

here, α_{BS} indicates heat transfer coefficient due to forced convection by ising bubbles and its value is prescribed by the bubble generation rate at the urface of the lower tube.

$$_1 = \frac{\pi}{6} D_b^3\,(N/A)_1 f \tag{11}$$

xperimental data of q_2 and ΔT_2 are substituted into eq.(10) and α_{BS} is esti-ated against q_1. On the other hand, $\dot{V}_1$ is calculated corresponding to q_1 by sing equations(1),(4),(5) and (11). The obtained relation between α_{BS} and $\dot{V}_1$ is hown in FIGURE 10. In this calculation, the ratio of each segment area to total rea was approximately estimated as $A_B/A = 1/4$, $A_S/A = 1/2$ and $A_T/A = 1/4$. It is lear from the figure that the relation between α_{BS} and $\dot{V}_1$ can be represented y a line irrespective of pressure.

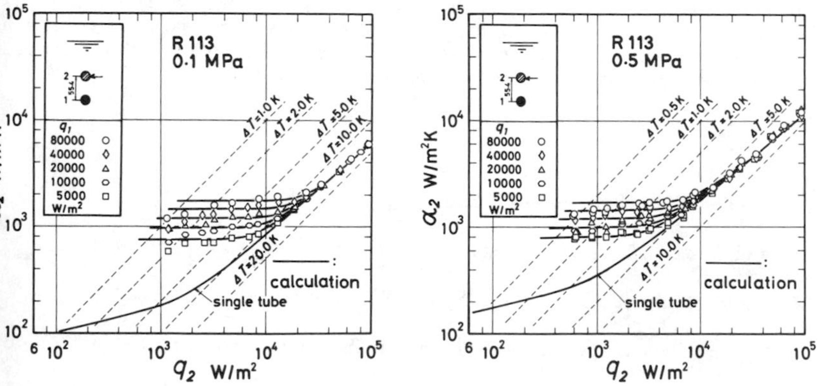

IGURE 8. Heat transfer characteristics of upper tube in a vertical row of wo tubes.

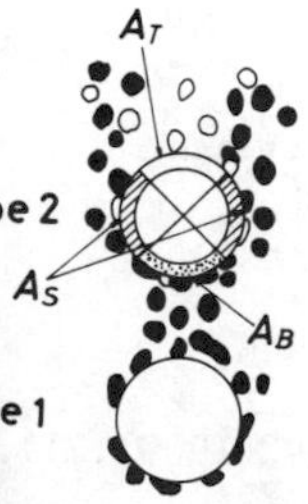

IGURE 9. Bubble behavior in vertical row of two tubes.

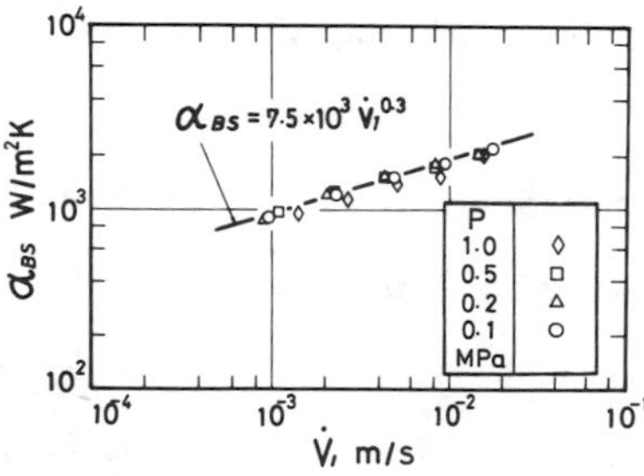

FIGURE 10. Relation between heat transfer coefficient α_{BS} and bubble production rate $\dot{V}_1$.

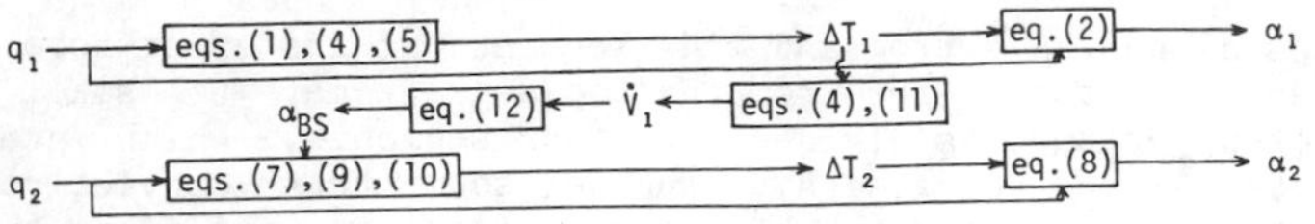

FIGURE 11. Flow chart for heat transfer prediction in a vertical row of two tubes.

$$\alpha_{BS} = 7.5 \times 10^3 \ \dot{V}_1^{0.3} \qquad (1$$

where units used are α_{BS} in [W/m^2 K], $\dot{V}_1$ in [m/s], and proportional constant i [(W/m^2 K) /(m/s)$^{0.3}$].

Consequently, heat transfer characteristics of the upper tube are predicted t the procedure shown in FIGURE 11. In the prediction, heat transfer coefficier of the lower tube is regarded as that of a single tube. The calculated charac teristics of the upper tube is represented by bold solid lines in FIGURE 8. In high heat flux region, the second term in eq.(10) is ignored because it become negative corresponding to the increase in active site densities $(N/A)_2$, and th results are represented by broken lines.

6. CONCLUSIONS

The fundamental characteristics for heat transfer from the small tube bundle were investigated in detail, and the effects of heat flux, pressure and positi of tube on heat transfer were clarified by both the experiment and the analysi Finally, more works for tube bundles of large scale are necessary to verify a refine the proposed model and to make clear the conditions under which decrea in enhancement or deterioration of heat transfer will appear.

REFERENCES

1. Robinson,D.B. and Katz,D.L., Chem. Eng. Prog., 47-6(1951),317.

2. Myers,J.E. and Katz,D.L., Chem. Eng. Prog. Symp. Ser., 49-5(1953),107.

3. Palen,J.W. et al., Chem. Eng. Prog. Symp. Ser., 68-118(1972),50.

4. Danilova,G.N. and Dyundin,V.A.,Heat Transfer-Soviet Research, 4-4(1972),48.

5. Nakajima,K. and Morimoto,K., Refrigeration, 44-495(1969),3. (in Japanese)

6. Güttinger,M., Proc. 4th Int. Heat transfer Conference(1970), HE 2.4.

7. Wallner,R., Proc. 5th. Int. Heat transfer Conference(1974), HE 2.4.

8. Hahne,E. and Müller,J., Int. J. heat Mass Transfer, 26-6(1983),849.

9. Nishikawa,K. et al., Refrigeration, 53-607(1978),9. (in Japanese)

10. Mikic,B.B. and Rohsenow,W.M., J. Heat Transfer, 91-2(1969),245.

11. Cole,R. and Rohsenow,W.M., Chem. Eng. Prog. Symp. Ser., 65-92(1969),211.

12. Zuber,N., USAEC Rep., AECU-4439(1959).

Convective Boiling Heat Transfer of Mixture of Immiscible Two-Liquids

KUNIO HIJIKATA
Tokyo Institute of Technology
2-12-1 Ohokayama
Meguroku, Tokyo, Japan

HIROKI ITO
Tokyo Institute of Technology
2-12-1 Ohokayama
Meguroku, Tokyo, Japan

YASUO MORI
University of Electro-Communications
1-5-1 Choufugaoka
Choufu City, Tokyo, Japan

INTRODUCTION

Thermal energy conversion of low or middle temperature difference to electric power is conventionally made by the Rankine cycle using the organic compound as a working fluid. However, the energy conversion efficiency from thermal energy to electric power is limited by the pinch point temperature difference in the high temperature side heat exchanging. In order to aviod the efficiency ceiling due to the pinch point temperature difference, utilization of mixture of miscible two liquids as the working fluid of the Rankine cycle has been proposed and its cycle efficiency has been calculated[1]. However, in the miscible mixture, mutual diffusion process is considered to greatly affect the thermo-fluid characteristics, but has not been clarified yet because of its complexity. On the other hand, thermo-fluid characteristics of the mixture of immiscible two liquids is simpler than that of miscible one, even though it has similar advantage in the cycle efficiency.

From these standpoints, the use of the mixture of immiscible two liquids as the working fluid of the Rankine cycle is proposed. The effective heat capacity of the mixture in the temperature range lower than the boiling point of the main working fluid is raised by use of latent heat of the other volatile material mixed with the main fluid as seen in Fig.1. In addition to that, the heat transfer coefficient in the liquid phase region is enhanced by turbulence caused by the boiling of the volatile liquid and by flow acceleration due to increase of the volume flow rate after the boiling. In this report, the boiling heat transfer performance of the mixture of immiscible two liquids and its effects on heat transfer augmentation are investigated.

Boiling of the mixture of immiscible liquids has scarcely been studied. Bragg [2] experimentally investigated film boiling of R113 located below water, which transfers heat by natural convection, and concluded that the heat transfer coefficient is decreased compared with pure liquid. Heat transfer problems from a dispersed liquid to other surrounding liquid were investigated by Tochitani et.al.[3] and Hara et.al[4] where the acceleration mechanism of liquid due to boiling of the volatile liquid was also clarified. In this study, a volatile liquid was premixed homogeneously with another immisible liquid having higher boiling temperature and heated in a hot channel. Some of fine droplets of volatile liquid struck the channel wall due to the turbulent motion and boiled there. These boiling performance and dependency of the volume fraction of the volatile liquid on heat transfer enhancement have been investigated.

PHASE EQUILIBRIUM OF MIXTURE OF IMMISCIBLE LIQUIDS

The saturation temperature of R113-water mixture at atmospheric pressure is shown against the mole fraction of steam in Fig.2. The boiling temperature has the minimum value at point C. On the left side of this point, only steam exists but no liquid-phase. On the contrary, R113 cannot exist in the liquid-phase on the right side of C. Therefore, only at the point C, the liquid R113 and water coexist. The total pressure P can be written by the saturation pressures $P_{satR}(T)$ and $P_{satW}(T)$ of each component, as follows:

$$P = (1-X_{gW}) \cdot P_{satR}(T) \text{ on the left side of C.} \quad P = P_{satR}(T) + P_{satW}(T) \text{ at C}$$
$$P = X_{gW} \cdot P_{satW}(T) \text{ on the right side of C.} \tag{1}$$

The boiling temperature of the mixture is uniquely given by the mole fraction of steam X_{gW}($= P_{gW}/P$) at a given total pressure. On the right side of point C, R113 vapor is in the superheated state. When the mixture of water in the continuous phase and R113 droplets is heated at atmospheric pressure, initiation of boiling of R113 droplets occurs at 47.5°C, the saturation temperature of pure R113 liquid. After the generation of pure R113 bubble, water in contact with this bubble evaporates into the bubble and the saturation temperature of R113 in the bubble decreases with decrease of its partial pressure. It enhances evaporation of R113 and equilibrium is finally established at the point A in Fig.2. With increase of the temperature, the mole fraction of steam changes along the curve CA in the figure.

As the partial pressure of steam increases after the perfect evaporation of R113 according as the temperature increase. the ratio of the partial volume V_{gR} of R113 vapor to that of the total volume V_{gT} can be witten as follows:

$$V_{gT}/V_{gR} = 1 + (P_W/P_R) = 1/(1 - X_{gW}) \tag{2}$$

This equation shows that the total volume of vapor at 80°C is estimated about twice as large as that without considering water evaporation. This water evaporation into R113 bubbles also reduces the limit of utilization of high temperature heat source due to the pinch point temperature difference, which can be clearly seen from the temperature-enthalpy diagram of the mixture.

EXPERIMENTAL APPARATUS

A schematic diagram of experimental apparatus is shown in Fig.3 which consists of mixing, test and condensing sections. R113 and water were used as a volatile

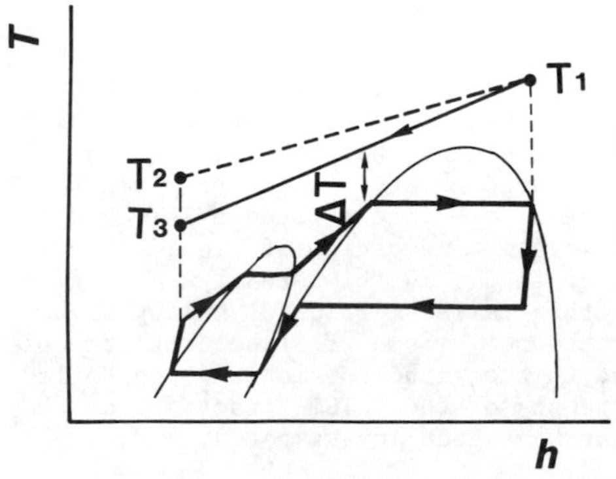

FIGURE 1. Cycle diagram

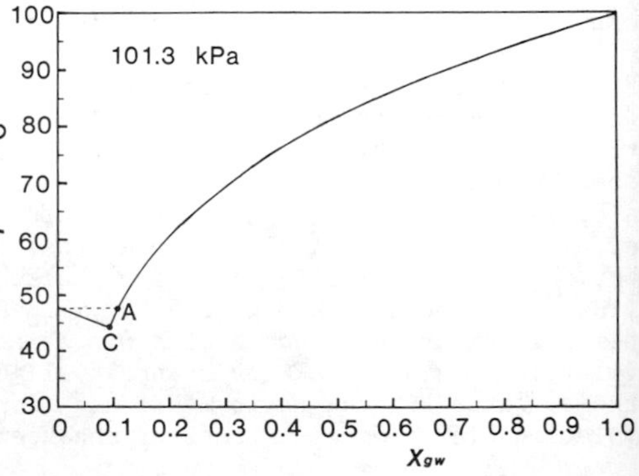

FIGURE 2. T-X diagram of R113-water

material and a main fluid, respectively. As R113 and water hardly mix with each other because of the density difference between them, dispersion of R113 droplets in water was performed by the following procedure. Water was injected into a mixing tank through a nozzle and R113 liquid at the bottom of the tank was sucked up to the throat of the nozzle and dispersed into water. Large R113 droplets quickly sank down and only small droplets of uniform size remained in the middle part of the mixing tank. The mixture of R113 droplets and water thus produced was led to a test section through a flowmeter and an inlet tank, where the bulk temperature and the pressure were measured.

The test section was a channel of 6mm×30mm in cross sectional area and 960mm in length. One side wall of 30mm in width of the section was covered by the copper thin film of 32 μm in thickness, which was directly heated by alternating current. Other walls were made of transparent acrylic resin for observation of bubble formation and flow pattern of the two-phase flow. The wall temperature distribution was measured by C-C thermocouples of 0.1mm diameter welded at the back of heating thin film. Temperature profiles in the cross section were measured by traversing C-C themocouples of 0.1mm in diameter. The void fraction distribution in the cross-section was measured by an electric capacitance.

The two-phase mixture of R113 and water flowed out through an outlet mixing tank for measuring the bulk temperature, condensed in a water cooled heat exchanger and returned to the mixing tank. Before experiments, the whole test loop was filled with water of 90°C and dissolved gas was evacuated for about one hour. After that, the water was cooled down below the boiling point of R113 and a suitable amount of R113 was led in the loop. The inlet temperature at the test section was adjusted by the cooling rate at the condenser. During the experiment, the behavior of boiling was observed.

In order to make clear effects of the volume fraction of R113 droplets and to check the accuracy of the measuring system, heat transfer performance in pure water was firstly examined. The experimental results of pure water well agreed with the theoretical values[5] and show that the flow does not fully develop even at the channel exit, when Re is below 5000. The volume fraction of R113 was measured by sampling the test fluid in the inlet tank. The distribution of the droplet size was measured by a microscope and an example is shown in Fig.4. The mean droplet diameter is about 15μm and the largest one is about 30μm. The

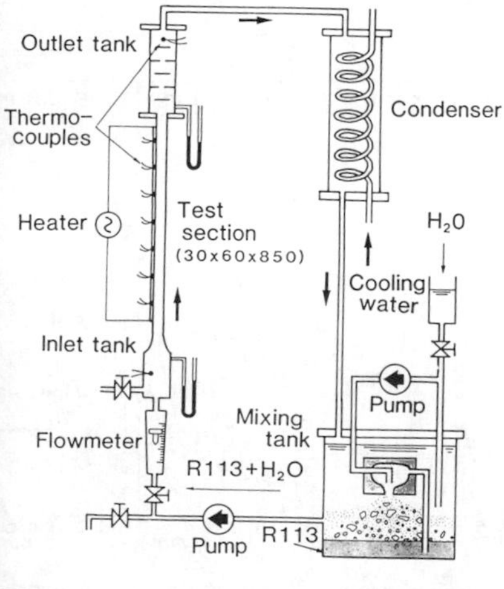

IGURE 3. Experimental apparatus

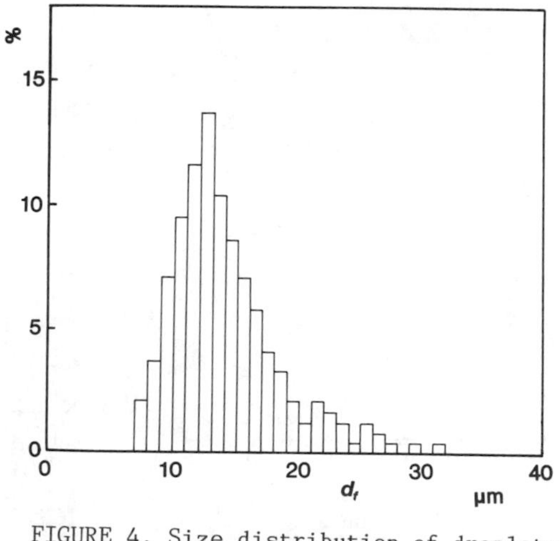

FIGURE 4. Size distribution of droplets

real distribution was assumed more uniform than the measured one, because the droplets coalesced with each other and large droplets quickly became larger during the sampling.

FLOW PATTERN AND HEAT TRANSFER

A distribution of wall temperature T_w along the channel heated by constant heat flux (q=113 kW/m^2) and temperature profiles in cross-sections are shown in Figs.5 and 6, respectively, where the mean velocity u is 0.287 m/s (Re=5700). The initial volumetric fraction ψ of R113 is taken as a parameter. The bulk temperature T_m shown by the solid line corresponds to the pure water flow case of ψ=0 and the outlet bulk temperature is 38.2°C which is lower than the boiling temperature of R113 at 101.3 kPa. Therefore, under this heating condition, the bulk temperature does not depend on the volumetric fraction of R113. The measured outlet temperature well agreed with the value calculated from the heat input and the flow rate. In case of ψ=0, the flow is fully developed in the downstream of x=40cm, where the temperature difference (T_w-T_m) becomes constant. On the contrary, when R113 droplets were premixed in water, the wall temperature became lower than that of the pure water case at x=10cm and decreased largely in the downstream of x=40cm.

The bubble diameter distribution at x=5cm, where the difference of wall temperature for cases with and without R113 droplets cannot be observed, is shown in Fig.7. The bubble diameter of R113 becomes about 6 times larger than the droplet diameter at atmospheric pressure due to evaporation, which explains the correlation between the distribution profile shown in Fig.4 and that of Fig.7. From this fact, it can be concluded that initial bubbles of R113 are generated from each droplet without coalescence. From the photographic observation, bubbles grew up from about 0.5mm at x=40cm to about 1mm at x=70cm. As the effect of steam pressure on the bubble size can be ignored in this temperature range, these large bubbles might be generated by coalescence of bubbles or from collision of bubble and R113 droplet. This machanism will be clarified in the near future.

In the downstream of x=40cm, the number of bubbles near the wall extremely increased and the bubbles covered almost the whole cross-section of the channel. Corresponding to these phenomena, the overshoot of the wall temperature appeared and the heat transfer coefficient increased drastically. However,

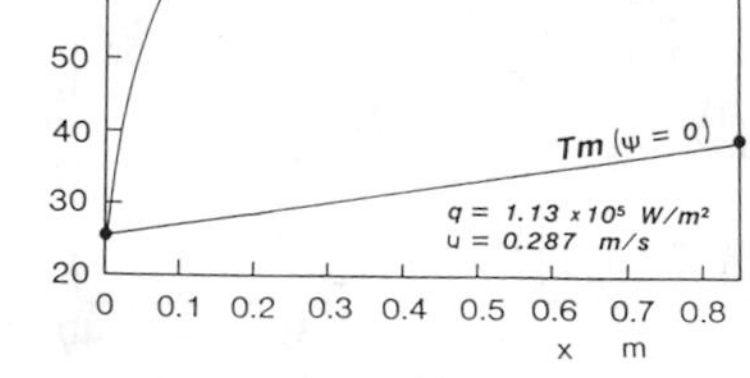

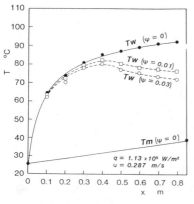

FIGURE 5. Distribution of T_w

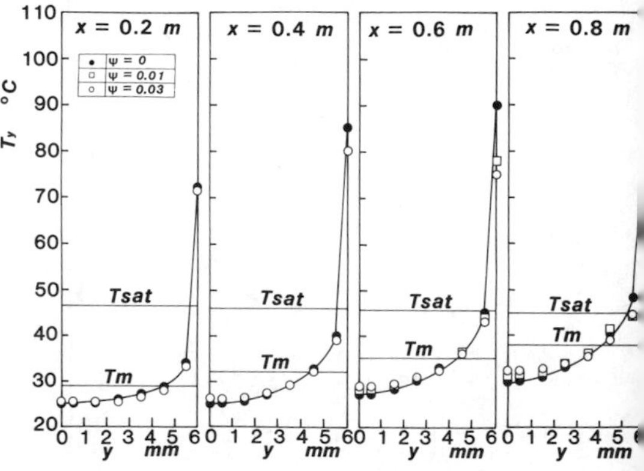

FIGURE 6. Profiles of T in cross-sections

the temperature in the cross section between x=60cm and 80cm is lower than the minimum saturation temperature shown by the point C in Fig.2 except the region near the wall. This fact means that bubbles cannot be generated or grow up except near the wall region.

The heat transfer augmentation by containing R113 droplets shown in Fig.5 is considered to be brought about by the following mechanisms.

(1). Boiling heat transfer of R113 droplets.
(2). Turblent mixing due to the motion of generated bubbles.
(3). Recondensation of R113 bubbles in the main flow.
(4). Flow acceleration by vapor generation.

Judging from the droplet number density and its heat transfer by phase change, the effects (1) and (3) are considered small. As dependency of heat transfer augmentation rate on the initial volumetric fraction of R113 droplets is not large, the effect (2) is considered most important.

HEAT TRANSFER AFTER R113 BOILING

In order to realize complete evaporation of R113 droplets, higher heat flux was applied compared with Fig.5. The heat flux q was 171 kW/m^2 and the mean velocity u was 0.199 m/s. The wall temperature distributions are shown in Fig.8 whose intial volumetric fractions of R113 ψ were 0, 0.01, 0.03 and 0.05, respectively. For ψ=0, boiling of water also occurred at x=30cm, where the wall temperature decreased and became constant in the downstream of it. The bulk temperature distribution for ψ=0 is linear in the flow direction. For ψ>0, it becomes flat at the minimum saturation temperature of the mixture shown by broken lines in the figure. In these conditions, the bulk temperature T_m can be calculated by the following equation.

$$Q = \{ h_{gR}(T_m) - h_{LR}(T_{in}) \}\times W_{LR} + \{ h_{gW}(T_m) - h_{gW}(T_{in}) \}\times W_{gW} + \{ h_{LW}(T_m) - h_{LW}(T_{in}) \}\times\{ W_{LW}-W_{gW} \} \qquad (3)$$

where T_{in} is the inlet temperature of the fluid and Q is the total heat input. W_{LR} and W_{LW} are the liquid mass flow rates for R113 and water, respectively, and W_{gR} and W_{gW} are the for gas mass flow rates. h_g and h_L are the enthalpies of gas and liquid phases, respectively. On the other hand, the mass flow rate can be obtained from the partial pressures of R113 and water as follows;

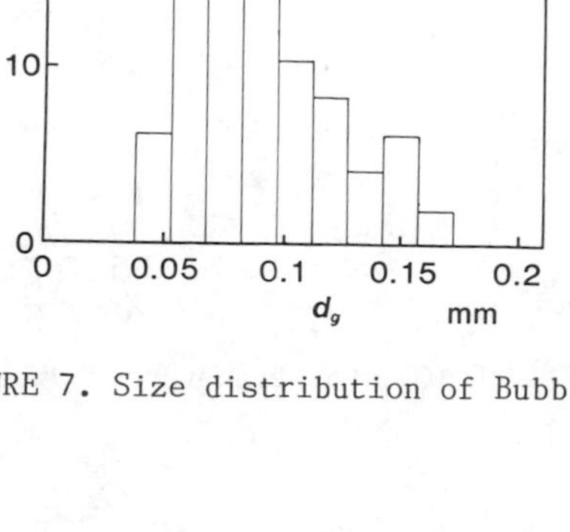

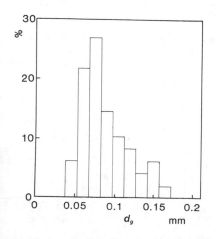

FIGURE 7. Size distribution of Bubble

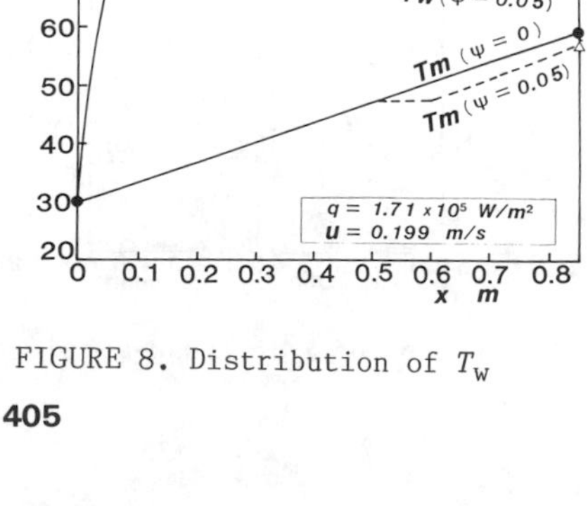

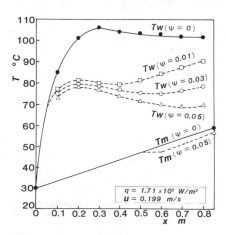

FIGURE 8. Distribution of T_W

$$W_{gW} = M_W \cdot P_W \cdot W_{gR}/(M_R \cdot P_R) = M_W \cdot X_{gW}(T_m) \cdot W_{gR}/\{M_R \cdot (1-X_{gW}(T_m))\} \quad (4)$$

where X_{gW} is the mole fraction of steam, which is only a function of T_m. M_W and M_R are the molecular weights and P_W and P_R are the partial pressures of water and R113, respectively. First, T_m is assumed, then W_{gW} is obtained from Eq.(4) and the total heat input Q can be calculated from Eq.(3). The distance x from the inlet is calculated from Q/qB, where B is the width of heating channel. The broken line in Fig.8 was obtained by this procedure. The calculated bulk temperature at the outlet of the channel for ψ=0.05 well agreed with the measured one as shown in the figure. The bulk temperature distribution for ψ=0.01 or 0.02 exists between the solid and broken lines.

The temperature profiles in the cross section are shown in Fig.9. The results at x=20 and 40cm for ψ=0.01 and 0.03 are not plotted, because they weekly depend on ψ. As the wall temperatures were also independent upon Ψ, the heat transfer coefficient was enhanced about twice as much by containing a small amount of R113. The temperature along the opposite side of the heating wall is higher than the saturation temperature at x=80cm as shown in Fig.9, therefore, all R113 droplets were considered to evaporate already. As the bulk temperature for ψ=0.01 increased with the wall temperature in the downstream of χ=60cm, it also shows that all R113 droplets have evaporated and the flow is fully developed. However, for ψ=0.05, the wall temperature is decreasing in the downstream of x=60cm, where the bulk temperature is alraedy higher than that of the boiling point, then turbulence generation by R113 boiling continues to affect on heat transfer.

These results show that the heat transfer coefficient at the outlet of the channel largely depends on the initial volumetric fraction of R113. It suggests that the heat transfer is controlled by the acceleration effect due to the phase change. From this standpoint, the ratio of the Nusselt numbers of R113-water mixture to the pure water at x=80cm is shown against χ_{tt} in Fig.10 where white and solid circles are experimental results. The solid line in the figure is an empirical relation proposed by Dengler[6] for evaporation in boiler tubes and given by the following equation.

$$Nu_{TP}/Nu_L = 3.5 \cdot (1/\chi_{tt})^{0.5} \quad (5)$$

$$\chi_{tt} = \{(W_{LW}+W_{LR})/(W_{gW}+W_{gR})\}^{0.9} \times (\rho_g/\rho_L)^{0.5} \times (\mu_g/\mu_L)^{0.1} \quad (6)$$

The volume change due to steam evaporation is considered and the results are

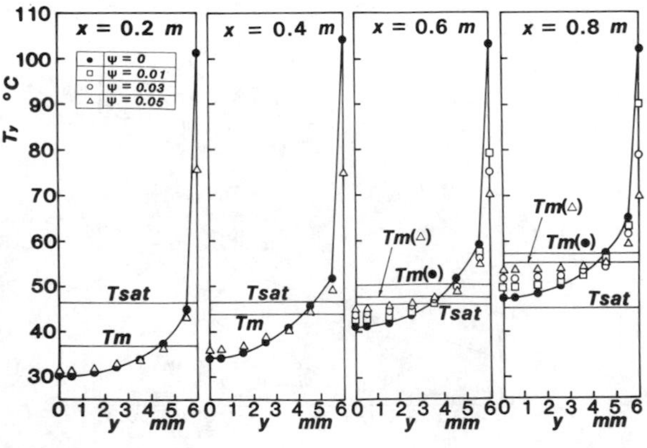

FIGURE 9. Profiles of T in cross-sections

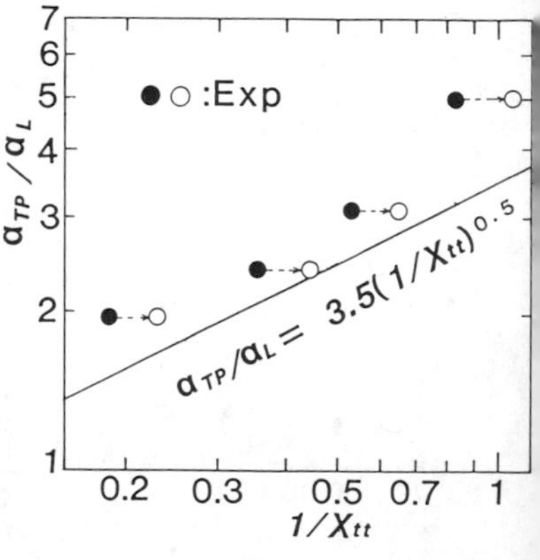

FIGURE 10. Heat transfer enhancement

shown by the white symbols, while the solid symbols neglect this effect. The temperature difference between the heating and adiabatic walls were used to calculated the Nusselt number. Experimental results considering thermal equilibrium between two components well agree with the empirical equation (6) except for ψ=0.05, which corresponds to the largest value of χ_{tt}. In case of ψ=0.05, the two phase flow of R113 vapor and water was not fully developed at the outlet of the channel.

LOCAL HEAT TRANSFER COEFFICIENT

The heat transfer coefficient largely depends on the void fraction α of vapor in a two phase region. Therefore, information of variation of the void fraction along the channel is required to discuss the heat transfer mechanism more precisely. From experimental results of static pressure and void fraction distributions along a channel, the fraction of volume flow rate of gas phase β was obtained. The ratio of Nusselt numbers of two-phase flow Nu_{TP} to that of water Nu_L is shown against β in Fig.11. If the gas flow only contributes to liquid acceleration and gives no effect on heat transfer mechanism, augmentation of heat transfer is expressed by $1/(1-\alpha)$. The experimental results are slightly higher than $1/(1-\alpha)$ in the foam and foam-slug flow regions. However, in the bubbly flow, they are larger than $1/(1-\alpha)$ and moreover larger than $1/(1-\beta)$, which is realized under no velocity slip condition between two phases.

Therefore, augmentation of heat transfer is established not only by liquid acceleration by the vapor void, but also by turbulence due to bubble generation near the wall. In order to clarify augmentation mechanism of heat transfer in the mixture of immiscible liquids, the relation between the pressure loss and the Nusselt number is shown in Fig.12, where Φ is the ratio of the pressure loss in the two-phase flow to that in the single water flow. In the figure, the experimental results are compared with those of the single phase water flow in the rough surface pipe by Dipprey[7](solid lines) and in the pipe having smooth cascade roughness reported by authers[8](broken lines). The results of the R113-water mixture are well correlated to the previous results of the single phase flow in the roughness pipe. Therefore, it can be concluded that the enhancement of heat transfer of the mixture of immiscible liquids in a small void fraction region is brought about by turbulence generation near the wall region due to the relative motion between bubbles and water.

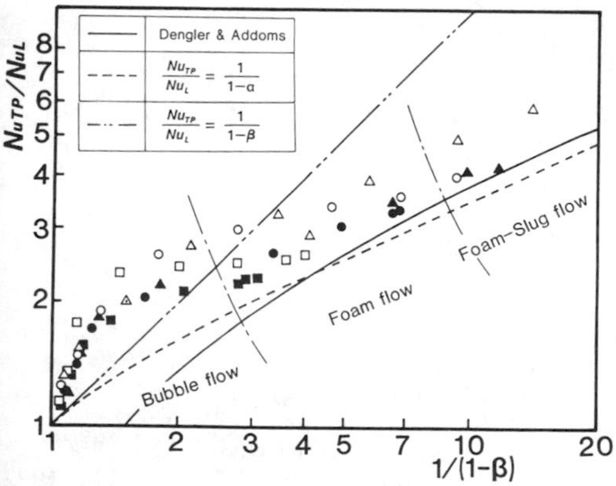

FIGURE 11. Increase of *Nu* number with increase of gas volumetric flow rate

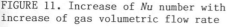

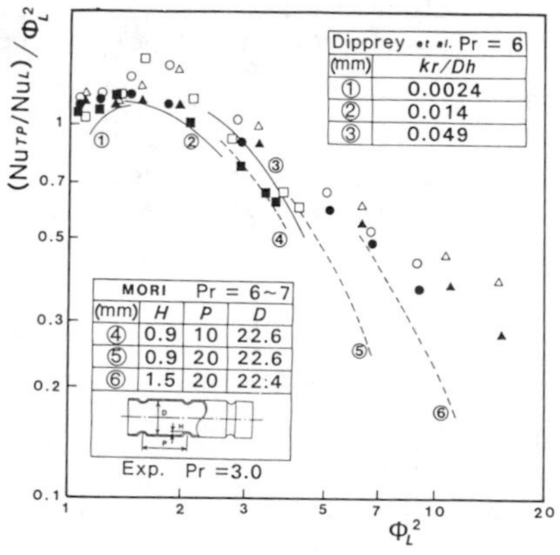

FIGURE 12. Relation between heat transfer rate and pressure loss

CONCLUSION

For the effective use of thermal energy of small or middle temperature difference, the use of a mixture of two immiscible liquids is proposed as the working fluid to increase the thermal efficiency of Rankine cycle. The fundamental mechanism of heat transfer of such liquids mixture has been experimentally investigated and the following conclusions are obtained.

1. Initial boiling of R113 droplets occures near the heating wall and the bubble size distribution corresponds to that obtained from boiling of the droplets without coalescence.
2. With successive boiling, the diameter and number density of bubbles increase and bubbles spread over the whole cross-section of the channel. The wall temperature decreases along the flow direction and enhancement of heat transfer is about twice as much as that of the water flow and weekly depends on the initial volumetric fraction of R113 droplets .
3. The heat transfer augmentation process can be explained by roughness effect caused by generated bubbles from comparison with results of the forced convective heat transfer performance in pipe flow with rough surface.
4. The heat transfer performance after complete evaporation of R113 droplts can be explained by acceleration effect of vapor void and well correlates with results reported so far about water boiler tubes.
5. The experimental results reported in this paper suggest that the use of the mixture of two immiscible liquids in the Rankine cycle has a great advantage not only for heat transfer performance but also for elevation of energy conversion efficiency.

This study was supported by grant in aid of Scientific Research (57550126) of the Ministry of Education, Science and Culture, Japan.

NOMENCLATURE

d	: droplet diameter	α	: void fraction
h	: enthalpy	β	: fraction of volumetric flow rate of gas phase
Nu	: Nusselt number		
P	: pressure	Φ	: pressure loss ratio of two-phase flow to liquid flow
Q	: volumetric flow rate		
T	: temperature	Ψ	: initial volumetric fraction of R113
V	: volume		
x, y	: flow direction and perpendicular direction to it		

Subscripts

l, g,:	liquid and gas phases	W, R :	water and R113
TP :	two-phase mixture	m, w :	bulk and wall

REFERENCES

[1]. Ototake, N., *Research on Effective Use of Thermal Energy*, SPEY14, p5, 1985
[2]. Bragg J. R. et.al., *4th. Int. Heat Transfer Conf.*, Vol.4, b7-1, 1970
[3]. Tohitani Y. et.al., *Warme und Stroffubertragung*, Vol.10, p71, 1977
[4]. Hara T. et.al., *Transaction of JSME*, Vol.41, No.349, p269, 1985
[5]. Kay, W. M., *Convective Heat and Mass Transfer*, p180, McGraw-Hill, 1966
[6]. Collier,J.R., *Convective Boiling and Condensation*, p209, McGraw-Hill, 1972
[7]. Dipprey, D. F., *Inter. J. Heat Mass Transfer*, Vol.6, p329, 1963
[8]. Mori, Y. et.al., *Trans. of JSME*, Vol.51, No.461, p160, 1984

Internal Thermosyphon Boiling Device

HUNG-TING HUANG, YUAN-MING CHEN, SONG-MIN ZHANG, and CHAO-WU FONG
Tianjin University
Tianjin, PRC

ABSTRACT

A flow model and design method of a newly developed internal thermosyphon boiling device were proposed. Results of calculation predicted the characteristics of two-phase flow and heat transfer inside this boiling equipment. Flow patterns, stability, rate of recirculation, fouling and etc. have been visually investigated qualitatively by means of a transparent heated tube. Water, sulfuric acid, triethanolamine, heavy oil and ethanol-water were used as working media. Experiments showed that this device operated very stably even at very high power density. High velocity two-phase annular flow dominated most parts of the boiling regions that results not only high heat transfer coefficient but also low rate of fouling. This internal thermosypyon boiling device has been tested and used as reboilers, evaporators and the boiling section of a heat pipe in laboratory and industrial processes.

INTRODUCTION

Two-phase flow and boiling of liquid in vertical tubes occur widely in nuclear, chemical and process equipments. The flow patterns and therefore mechanisms of two-phase flow and heat transfer in a pipe are quite different from those of single-phase flow.

Flow patterns and regions of heat transfer in a vertical heated tube have been described in detail by Collier (1). Among those regions, saturated nucleate boiling and forced convective heat transfer through liquid film are of the utmost importance. From the former to the latter, shear force of the vapor imposing on the interface is increasing progressively and the flow patterns will change from bubbly or plug flow to annular flow. A fundamental transition in the mechanism of heat transfer takes place. The process of saturated nucleate boiling is replaced by the process of "evaporation".

Chen (2) proposed a two-mechanisms approach which is assumed that both nucleation and convective mechanisms occur to some degree in the boiling heat transfer process. It has been found that in two-phase flow boiling, high vapor quality and liquid velocity will result high heat transfer coefficient. This can be obtained by increasing the heat flux or more exactly the power density (power per unit fluid volume) which can always be increased by reducing the tube diameter. But high power density may enhance two-phase flow instability or flow oscillation. Shellene et al (3), Lee et al (4) and some other investigators described the effects of the flow instability on the limit of operating conditions.

There are many sourses of disturbance which can stimulate instability. For example, poor nucleation can give rise to "bumping" in boiling liquid, intermittent flow patterns such as plug or slug flow have the unstable nature, the syste geomtrical parameters may be strongly related to hydrodynamic instabilities.

Different types of instability may have different mechanisms. In most industrial equipments, small diameter is preferred in order to obtain higher power density. Unfortunately, flow oscillation becomes more serious as diameter decreases. The principal advantages of the internal thermosyphon boiling device are: plug flow which is the main sourse of instability in most cases can be minimized or eliminated and high power density can be obtained even for large diameter.

MECHANISMS OF TWO-PHASE FLOW AND HEAT TRANSFER IN THE INTERNAL THERMOSYPHON BOILING DEVICE

Internal thermosyphon boiling device has a very simple configuration, merel inserting a tube into the heated tube as illustrated in figure 1.

Liquid boils and evaporated in the annulus and the liquid-vapor mixture flow upward into the separator at the top of the tube. Liquid phase returns to the inner tube and flows down to the bottom and then re-enters into the annulus at the lower end. Owing to the density difference of the fluids in the inner tube and the annulus, a thermosyphon recirculation occurs. Since the width of the annulus is quite narrow comparing to the diameter of the tube therefore bubbles cannot occupy the whole area of any cross-section along the tube, that means no air plugs exist as those in the "empty tubes". Annular flow plays the most important role in this situation. The two-phase mixture density in the annulus, the overall pressure drop, the rate of thermosyphon recirculation and power density can be adjusted easily by choosing a suitable diameter of the inner tube to obtain stable performance and high heat transfer coefficient.

FLOW MODEL AND METHOD OF DESIGN CALCULATION

As illustrated in figure 2, an equation can be established between the driving force and resistance as follows:

$$L\, g \rho_1 = \Delta P_{sp} + \Delta P_t + \Delta P_{spa} + \Delta P_{tp} \qquad (1)$$

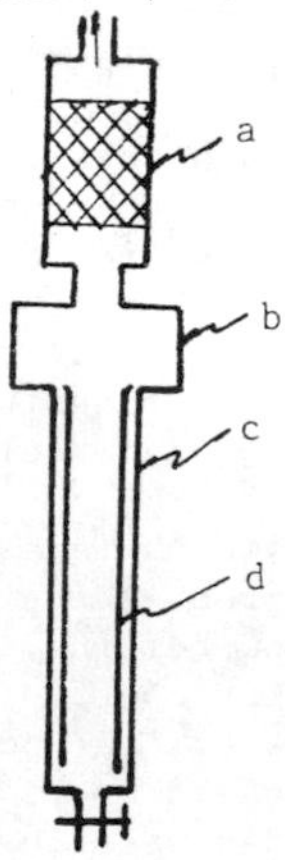

a) secondary separator or distillation column,
b) separator,
c) transparent heated tube,
d) internal thermosyphon device

Fig.1 Schematic diagram of internal thermosyphon boiling device

where L, length of the heated tube; ρ_l, density of liquid; ΔP_{sp}, single-phase frictional pressure drop in inner tube; ΔP_t, pressure drop at turnaround and entry; ΔP_{spa}, single-phase frictional pressure drop in annulus at the lower part where the temperature is below boiling point; ΔP_{tp}, two-phase flow pressure drop including losses due to frictional resistance, changes of momentum and changes of static head.

In most cases, two-phase flow pressure drop is much greater than the other terms on the right hand side of equation (1) and its values depends mainly on the vapor quality. Therefore calculation of fluid flow and heat transfer should be carried out simultaneously.

Model of Martinelli (5) (6) was used to predict two-phase flow pressure drop. Chen's two-mechanisms approach which was recommended for heat transfer calculations can be expressed as follows:

$$H_b = h_c + h_{nb}\ S \qquad (2)$$

where h_c, boiling heat transfer coefficient; h_{nb}, nucleate boiling heat transfer coefficient; S, nucleate boiling suppression factor; h_c, forced convective heat transfer coefficient; $h_c = F_{tp}\ h_1$; F_{tp}, is a function of Martinelli parameter X. In Chen correlation F_{tp} is given by

$$F_{tp} = 1 \qquad \text{for } 1/X \leqslant 0.1 \qquad (3)$$

$$F_{tp} = 2.35(1/X + 0.213)^{0.736} \qquad \text{for } 1/X \succ 0.1 \qquad (4)$$

$$X = \left(\frac{1-y}{y}\right)^{0.9} \left(\frac{\rho_g}{\rho_\ell}\right)^{0.5} \left(\frac{\mu_\ell}{\mu_g}\right)^{0.1} \qquad (5)$$

where y is vapor quality. Equations show that X decreases and F_{tp} becomes larger as y increases.

$$h_1 = 0.023\ (k_1/d_i)\ (Re_1)^{0.8}\ (Pr_1)^{0.4} \qquad (6)$$

h_1 is the heat transfer coefficient of liquid phase as if liquid flows alone in the same pipe. Several correlations for calculating h_{nb} can be found in literture. Among them Chen (2) and Mostinski (7) correlations are recommended: Chen correlation:

$$h_{nb} = 0.00122 \left(\frac{k_\ell^{0.79}\ C_{P\ell}^{0.45}\ \rho_\ell^{0.49}}{\sigma^{0.5}\ \mu_\ell^{0.29}\ \lambda^{0.24}\ \rho_g^{0.24}}\right) \Delta T_s^{0.24}\ \ \Delta P_s^{0.75} \qquad (7)$$

Mostinski correlation:

$$h_{nb} = 0.1011\ \ P_c^{0.69}\ (1.8\ P_r^{0.17})\ q^{0.7} \qquad (8)$$

where P_c, critical pressure; P_r, reduced pressure; q, heat flux. The suppression factor S is a function of Re_1 and F_{tp}.

Figures 3, 4 and 5 illustrate part of the calculation results.

EXPERIMENTAL OBSERVATION

This device has been tested in two types of loop: open loop, such as evaporators or thermosyphon reboilers and closed loop, as gravity-type heat pipe. In order to obtain better observation, electrically heated quartz tubes coated with transparent film of semi-conductor were used. Water, sulfuric acid, triethanolamine, heavy oil and ethanol-water were used as working media.Ranges of geometrical variables are:

Heated tube diameter, mm	15-50
Heated tube length, mm	400-2000
Width of annulus, mm	1.5-6.0

Problems of flow patterns, stability, rate of recirculation and fouling have been visually investigated. Experiments were also carried out in the "empty tube" (without inner tube) in order to make a comparison.

Flow Patterns and Stability Plug flow was the major flow pattern for tube diameter less than 20 mm in the case of boiling in "empty tube". Bumping phenomenon appears especially for the boiling of sulfuric acid. Sometimes, the Raschig rings were blown out from the secondary separator to the condenser or distillation column. For the gravity-type heat pipe, liquid recirculation was very weak or no recirculation existed at all because there was no downcomer in it. Liquid entrainment or flooding often occurred in the condensing section even at low heat flux.

As tube diameter increased instability could be improved but power density and therefore heat transfer coefficient would decrease significantly.

In the new device, annular flow predominated and operation showed quite smooth and stable with a wide range of heat flux. Geometrical variables were not quite sensitive to the flow instability, however larger bubbles might appear when the width of the annulus increased up to 5 mm.

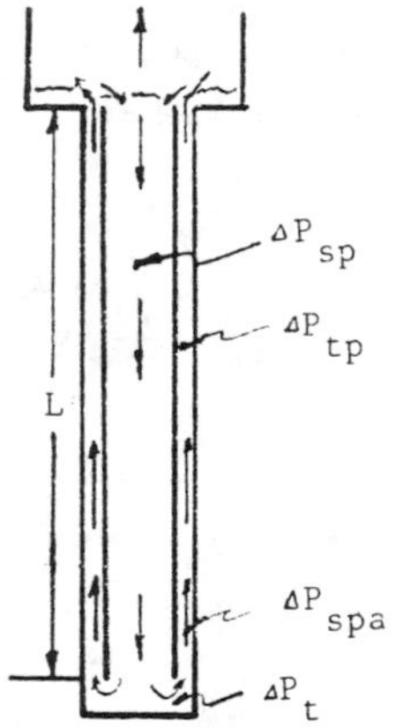

Fig.2 Physical model of fluid flow in the internal thermosyphon boiling device

Rate of Recirculation and Heat Transfer

Particle of polypropene which density is very close to water was used as the tracer in water system. However, settling velocity of the particle in a pool of quiescent water was measured and noted as u_s before putting into the tested loop where its moving velocity inside the inner tube was measured by means a stopwatch noted as u_m. The rate of liquid recirculation, G_T can be calculated readily since the diameter of the inner tube was known; the difference of u_s and u_m might be considered as the liquid velocity.

Circles and triangles plotted in figure 3 showed the measured values of G_T with annulus widths of 1.225 mm and 2.925 mm respectively and tube length of 1030 mm. They are in good agreement with the predicted ones.

Table 1 showed some data taken from previous work (9) on gravity-type heat pipe which consisted of a copper "empty tube". Most experimental values of h_b scatter from 3000 to 7000 W/m^2K; some of them might reach 10000 W/m^2k which were quite closed to the predicted values of the lower curve in Fig.5.

Table 1

D_i, mm	L, mm	Water filling, %	q, Kw/m^2	h_b, W/m^2k
29.5	819	17-72	8-25	3000-7000

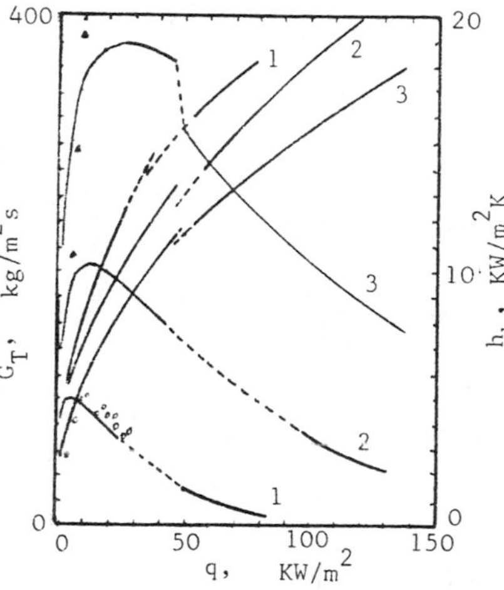

Fig.3. Curves of rate of recirculation and heat transfer coefficient for various widths of annulus: Tube length, 1030 mm; D_o, 28.15 mm; d_i, 21.37 mm;
curve 1 2 3
δ, mm 1.225 2.0 2.925

<u>Fouling</u> Serious fouling was found in the reboilers of tap water and sulfuric acid. For example, the quartz heated tube would become opaque in about two days of operating duration but it could maintain its transparence for a long time with an inner tube inserting into it. It is interesting to discover that deposit was found on the inside wall of the inner tube. Even though very little theoretical development has been done on the mechanism of deposit removal, this can be explained by the re-entrainment model (8) which described the "turbulent burst" phenomenon in which local high vorticity flows impinge on the surface. These vortices would provide a much higher particle removal rate than would the classical picture of turbulent flow. In the "empty tube" or the inner tube only liquid flows at very low velocity and the rate of removal of deposit must be very low. On the other hand, strong vortices in the two-phase flow regions could be observed by means of the tracer particles. This results high rate of fouling removal.

DISCUSSION

Fig.3 illustrates that when δ becomes narrower, power density and also y increase. Thus, rate of recirculation decreases because of the higher ΔP_{tp} in the annulus. This results lower h_l, but F_{tp} increases rapidly and consequently $F_{tp} \cdot h_l$ increases until y = 1 where dryout occurs. For "empty tubes", power density drops rapidly with the same heat flux, and everything mentioned above will go reversedly; $F_{tp} \rightarrow 1$; $F_{tp} \cdot h_l \rightarrow h_l$ and S drops to the level about 0.1-0.2 that means process approaches single-phase convective heat transfer. Especially, if larger diameter is selected in order to minimize flow oscillation, coefficient may fall dramatically as illustrated in Figs.4 and 5. The dash sections of the curves in those figures show the transition regions of Re_l and Re_g.

For the purpose of obtaining better recirculation long tubes are always selected in the design of vertical thermosyphon reboilers which cause too much trouble for installation of the distillation tower. Fig.4 predicts very high h_b for short heated tube with internal thermosyphon device. In fact, this type of reboiler with length of 400-2000 mm have been tested in laboratory and in industrial process.

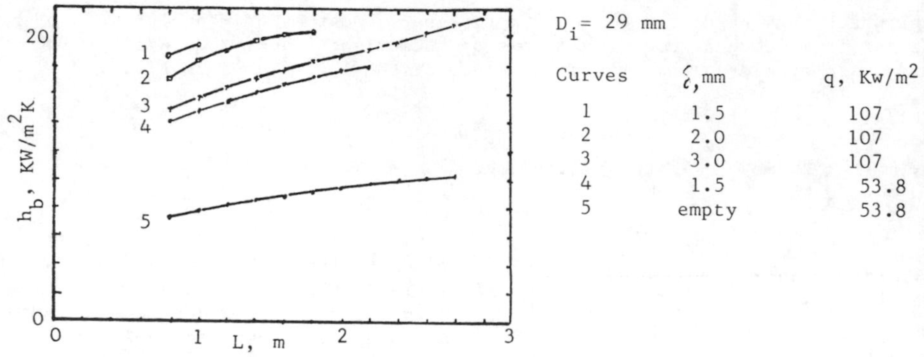

Fig.4 Comparison of heat transfer coefficients with various widths and "empty tube".

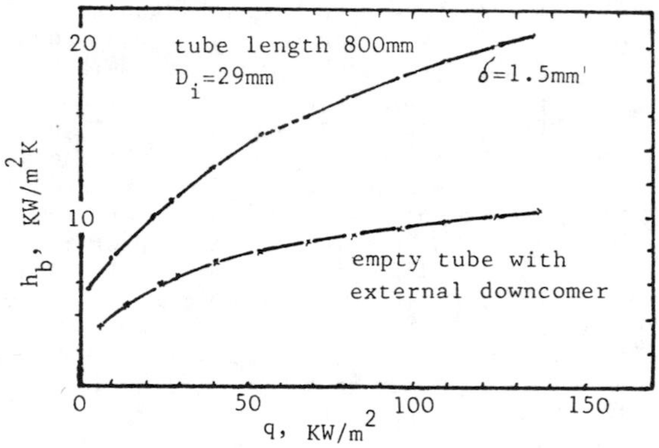

Fig.5 Relationship between heat transfer coefficient and heat flux for internal thermosyphon boiling device and "empty tube" with external downcomer.

REFERENCES

1. Collier, J.G., " Convective boiling and condensation", Mcgraw Hill Book Co., UK. 1972
2. Chen, J. C., Int. Eng. Chem. Process Design and Development, Vol. 5, p.322 1966
3. Shellene, K. B. et al, Chem. Eng. Progess Symp. Series, 64, no.82, p.102,1968
4. Lee, D. C. et al, Chem. Eng. Prog. 52, no. 4, p. 160, 1956
5. Martinelli, R. C. and Nelson, D. B., Trans. ASME Vol. 70, p. 695, 1984
6. Lockhart, R. W. and Martinelli, R. C., Chen. Eng. Prog. Vol.45, p.39, 1949
7. Mostinski, I. L., Teploenergetika, Vol. 4, no. 66, 1963
8. Cleaver, J. W. and Yates, B., Chem. Eng. Sci., 31,p.147, 1976
9. Huang, H. T., Chen, Y. M. and Fong, C. W., Chem. Ind. and Eng., Vol. 1, no. 3 and 4, p. 18, 1984

The Effect of a Thin Insulating Layer on Rapid Cooldown of Hot Metals in Saturated Water

Y. KIKUCHI, T. HORI, H. YANAGAWA, and I. MICHIYOSHI
Department of Nuclear Engineering
Kyoto University
Kyoto 606, Japan

INTRODUCTION

The process of cooling a high temperature material with boiling liquid is widely encountered in metallurgy, cryogenic engineering, nuclear technology and other industrial applications. One of the problems of interest in such a cooling process is the assessment of the minimum temperature to maintain the film boiling. At this temperature, usually termed the minimum film boiling temperature T_{min}, the vapor film which separates the hot surface from the liquid becomes unstable and a transition from the film to the transition boiling occurs.

Several investigators [1-6] have reported that, for rapid cooldown of metals coated with a thin insulating (low thermal conductivity) layer, an earlier onset of transition boiling occurs at significantly higher T_{min} than that predicted by the hydrodynamic model of Berenson [7] or the maximum liquid superheat theory of Spiegler et al. [8]. No theory, except the theoretical study of Kikuchi et al. [9], which explains these augmentations of T_{min} successfully, has yet been found. Kikuchi et al. assumed the occurrence of local and intermittent liquid-solid contacts in the film boiling regime for predicting the actual T_{min} for coated metals. The calculated results agreed well with the experimental data for Teflon-coated copper plates cooled in liquid nitrogen.

In order to apply the intermittent liquid-solid contact model to other liquids an experimental study has been conducted with a paint-coated silver cylinder cooled in saturated water under atmospheric pressure. This paper gives the experimental results of the effect of coating thickness on rapid cooldown processes. The measured values of T_{min} will also be compared with theoretical calculations by the intermittent liquid-solid contact model.

THEORETICAL ANALYSIS

On the basis of the experimental results published in the literature [10, 11], Kikuchi et al. [9] have proposed a theoretical model for explaining the rapid cooldown of insulated metals in boiling liquid. Fig. 1 shows the physical pictures for the film boiling. As the bubble departs from the antinode [Fig. 1(a)], the liquid rushes toward the solid surface, and then contacts the local portion of the surface [Fig. 1(b)]. The rapid evaporation of a residual microlayer of liquid, which is superheated on the surface, occurs [Fig. 1(c)]. To formulate a model describing these processes, the sequence of liquid-solid contact in the film boiling regime is divided into three periods: (a) dry period, (b) conduction period and (c) evaporation period.

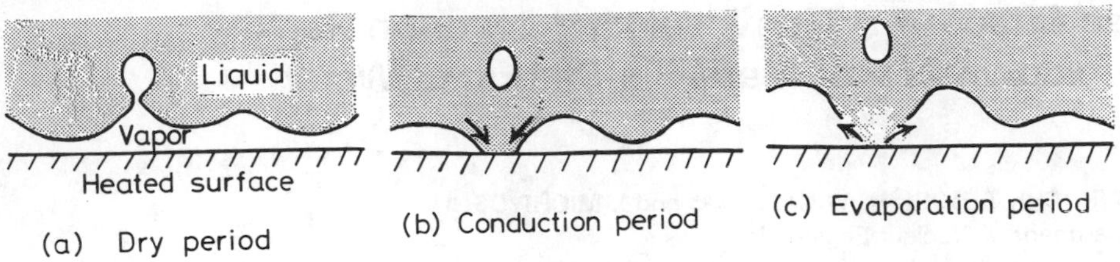

FIGURE 1. Intermittent liquid-solid contact model

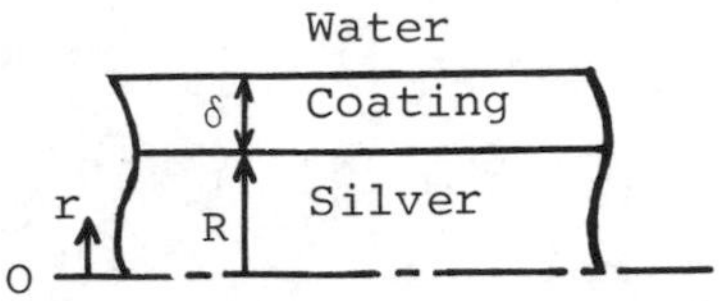

FIGURE 2. Coordinate system

Dry Period (t<0)

A dry condition is always maintained on the most areas of the solid surface, which are cooled with the film boiling heat transfer coefficient h_{fb}. The local portions, which will later experience liquid-solid contacts, are also cooled with h_{fb} during the dry period. With the coordinate system shown in Fig. 2, the flow of heat in both coating and silver sections obeys the one-dimensional equation of heat conduction

$$\frac{\partial T}{\partial t} = \frac{1}{r}\frac{\partial}{\partial r}\left(a\, r\frac{\partial T}{\partial r}\right) , \tag{1}$$

where a is the thermal diffusivity.

The boundary conditions are

$$r = 0 \quad : \quad \frac{\partial T}{\partial r} = 0 , \tag{2}$$

$$r = R \quad : \quad -k_o\frac{\partial T}{\partial r}\bigg|_{silver} = -k_w\frac{\partial T}{\partial r}\bigg|_{coating} = h_g(T_{silver} - T_{coating}) , \tag{3}$$

$$r = R+\delta \ : \quad -k_w\frac{\partial T}{\partial r} = h_{fb}(T - T_{sat}) , \tag{4}$$

where k is the thermal conductivity, and h_g is the contact coefficient between the coating and silver sections. The subscripts o and w refer to the silver and the coating, respectively.

According to some calculated results, the temperature transient is slow and a quasi-steady-state condition is maintained throughout the dry period except at the initial but short time when the coated cylinder is submerged into water. The temperature distribution through the coating layer is linear since the layer is very thin. In the silver region, however, a uniform temperature distribution is dominant because of its extremely high thermal conductivity. If the silver temperature T_o is assumed to be uniform, the temperature distribution in the coating layer is, with good accuracy, given by a simple equation

$$T-T_{sat}=\frac{(T_o-T_{sat})[1+h_{fb}(R+\delta-r)/k_w]}{1+h_{fb}/h_g+h_{fb}\delta/k_w} \quad , \quad R\le r\le R+\delta \ . \tag{5}$$

Conduction Period ($0\le t<\tau_c$)

Consider a cold liquid of uniform temperature T_{sat} suddenly coming into direct contact with the local portion of the hot surface of a paint-coated silver cylinder. During the short conduction period the flow of heat in each section, even in liquid, obeys the heat conduction. Although the contact portion is localized on the heated surface, the governing energy equation for the coating layer is simplified into the one-dimensional form of Eq. (1) since the thermal conductivity of coating is extremely low and the layer is so thin that two-dimensional (circumferential) conduction effects on the temperature transients can be neglected.

The boundary conditions are

$$r=R \quad : \quad -k_w\frac{\partial T}{\partial r}=h_g(T_o-T) \ , \tag{6}$$

$$r=R+\delta \ : \quad T-T_{sat}=\frac{T_o-T_{sat}}{(1+h_{fb}/h_g+h_{fb}\delta/k_w)[1+\sqrt{\rho_\ell c_\ell k_\ell/(\rho_w c_w k_w)}]} , \tag{7}$$

where T_o is the uniform silver temperature, ρ the density and c the specific heat. The subscript ℓ refers to the liquid. Eq. (7) was derived in Ref. 5.

Evaporation Period ($\tau_c\le t\le\tau_c+\tau_e$)

A thin film of cold (saturated) liquid has been superheated by conduction from the heated solid surface for the former conduction period. For the subsequent evaporation period, the superheated microlayer continues to cool locally the surface through rapid evaporation. For simplicity, during the evaporation period, the unknown time-dependent heat flux is represented by a time-averaged heat flux designated q_e [12]. The energy equation and the boundary condition at r=R become the same forms as Eqs. (1) and (6), respectively. The other boundary condition is

$$r=R+\delta \ : \quad -k_w\frac{\partial T}{\partial r}=q_e \ . \tag{8}$$

Calculation Procedure

Numerical methods are applied to transient conduction problems in each period. The relevant differential equations are reduced to finite-difference equations. An explicit method is used to calculate the unknown temperature distribution after a time interval.

The contribution of liquid-solid contact heat transfer to the overall film boiling heat transfer (h_{fb}) is neglected since the direct contact occurs on the local portions of the surface during a short time and the frequency of contacts is very low under the film boiling condition.

If the local surface temperature falls below the lowest value T_{iso}* of limit while the average wall temperature is quite high, the vapor generation rate will not be sufficient to maintain the film boiling and an onset of transition boiling will occur, resulting in an improvement of heat transfer from the bulk surface. On the other hand, if the local fall in the surface temperature is not too severe, a film of vapor is formed on the heated surface again and a dry condition is recovered.

EXPERIMENTAL APPARATUS AND PROCEDURES

The test specimen selected for the present experiment is a silver cylinder (20 mm in diameter and 70 mm long), whose heat transfer surface is coated with a thin refractory paint. The thickness of coating is varied from 5 to 156 μm. A chromel-alumel thermocouple (1 mm in diameter), which monitors the temperature T_o of the silver cylinder, is buried with solder into a hole in the center of the cylinder.

An electric furnace is used to heat the test cylinder to a desired temperature. The heated cylinder is vertically submerged in distilled water inside a stainless-steel vessel and is cooled down to the saturation temperature of water. The output signal from the thermocouple is recorded on a strip chart. During the runs the water is maintained at the saturation temperature under atmospheric pressure.

For the uncoated silver cylinder, the slope of the temperature-time trace of the center hole of the silver cylinder in the film boiling regime yields the heat transfer coefficient h_{fb}. Because of the high thermal conductivity of silver, the film boiling Biot number of the cylinder is very small ($\sim 7\times 10^{-3}$). Assuming the cylinder as a lumped-parameter system, i.e. an isothermal one, the heat transfer coefficient is calculated using the following equation

$$q = h_{fb}\Delta T_o = \rho_o c_o \frac{V_o}{S_o}\left|\frac{dT_o}{dt}\right| , \tag{9}$$

where V_o and S_o are the volume and the effective heat transfer area, respectively, of the cylinder. ΔT_o is defined as $\Delta T_o = T_o - T_{sat}$.

For the paint-coated cylinder, however, the transient conduction equation in the two-layer cylinder should be solved using a finite difference scheme since a discrepancy between the surface temperature T_w and the silver temperature T_o occurs.

The temperature at which the slope of the temperature-time trace suddenly increases is taken to be the minimum film boiling temperature $T_{min,o}$ of the silver cylinder with and without coating.

RESULTS AND DISCUSSION

Fig. 3 shows a comparison of theoretical and experimental results of temperature transients of a silver cylinder (20 mm in diameter) for four coating thicknesses

---

* T_{iso} is the minimum film boiling temperature measured for an isothermal surface which does not experience any temperature drop due to liquid-solid contacts. In the present study T_{iso} is assumed to be equal to the value measured for the uncoated silver.

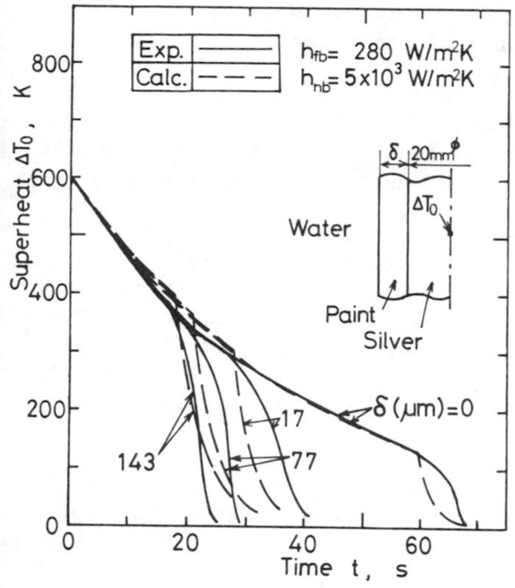

FIGURE 3. Effect of coating thickness on temperature change during cooldown

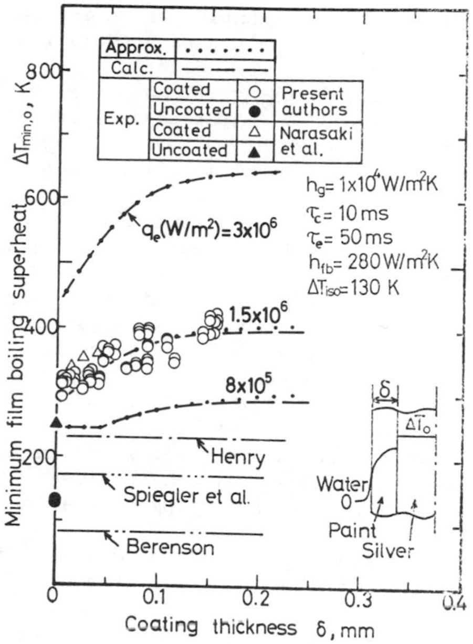

FIGURE 4. Effect of evaporation heat flux on minimum film boiling superheat

The abscissa is the elapse of time t and the ordinate is the superheat defined as $\Delta T_o = T_o - T_{sat}$. The paint coating gives a great improvement of heat transfer since the minimum film boiling temperature becomes higher with increasing the coating thickness and an earlier onset of transition boiling occurs. The experimental results are in fairly good agreement with the theoretical calculations, in which the film and the transition boiling heat transfer coefficients are assumed to be $h_{fb}=2.8\times10^2$ W/m²K and $h_{nb}=5\times10^3$ W/m²K, respectively. The isothermal minimum film boiling superheat ΔT_{iso} is taken as 130 K which is based on the experimental results of the uncoated silver cylinder.

In the above calculations several parameters concerning liquid-solid contact are also assumed that $q_e=1.5\times10^6$ W/m², $\tau_e=5\times10^{-2}$ s, $\tau_c=1\times10^{-2}$ s and $h_g=1\times10^4$ W/m²K. A way to determine these parameters will be described in the following.

The minimum film boiling temperatures are indicative of the general level of the enhancement of heat transfer in cooldown processes. The minimum film boiling superheats $\Delta T_{min,o}$ for the coated and the uncoated cylinders are brought together in Fig. 4. The abscissa is the thickness δ of paint coating. The figure also contains the measured values of Narasaki et al. [4] for a paint-coated stainless-steel cylinder of 20 mm in diameter. The $\Delta T_{min,o}$ is higher with thicker δ for the data of the present authors and Narasaki et al.

The experimental results are compared with the theoretical calculations by the intermittent liquid-solid contact model. Calculations were performed for three evaporation heat fluxes q_e. The calculated curve for q_e of 1.5×10^6 W/m² agrees well with the present measured values. The q_e of 1.5×10^6 W/m² for water is higher than that (2.5×10^5 W/m²) for liquid nitrogen, which was derived in the preceding paper [9]. The q_e is equal to or slightly higher than the critical heat flux, for example, which is predicted from a conventional correlation of Kutateladze [13] for saturated nucleate boiling of each liquid.

The dotted lines are the approximate solutions which were obtained under the assumption of a linear temperature distribution in coating layer, as described in the Appendix of Ref. 9. The approximate solutions are in good agreement with the exact numerical solutions, especially in the region of thin coating.

In this figure are also indicated the minimum film boiling temperatures T_{min} predicted by other investigators. The theoretical values given by the hydrodynamic model of Berenson [7] or the maximum liquid superheat theory of Spiegler et al. [8] are significantly different from the measured values for the paint-coated cylinder. This large discrepancy can not be explained by the bulk temperature drop ($=T_o-T_w$) across the coating layer since the temperature drop across a 0.1-mm-thick layer is of the order of only 18 K in the vicinity of the minimum film boiling point.

Henry's correlation [14], which takes into account the thermal properties of the heating surfaces, also gives lower values of T_{min} than the present data for various thicknesses of paint coating.

Fig. 5 shows the effect of the duration τ_e of the evaporation period on the minimum film boiling superheat $\Delta T_{min,o}$. The $\Delta T_{min,o}$ is higher with longer time of evaporation, especially in the thick coating region, because the local surface temperature drop due to liquid-solid contact becomes larger with increasing τ_e. The calculated results for $\tau_e=5\times10^{-2}$ s agree well with measured values.

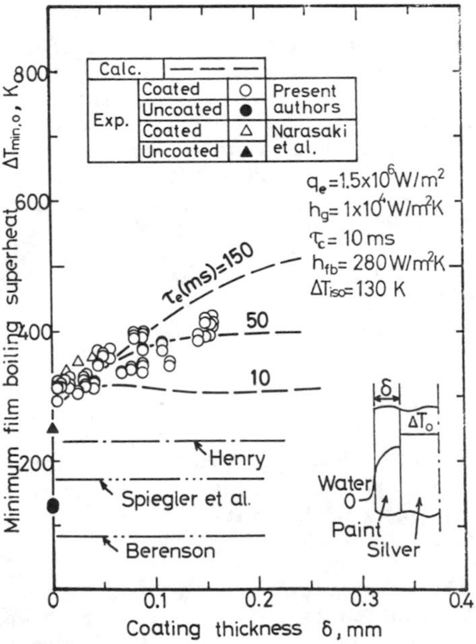

FIGURE 5. Effect of duration of evaporation period on minimum film boiling superheat

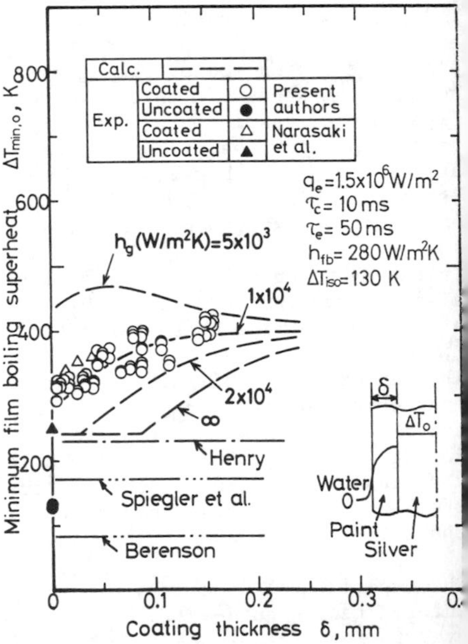

FIGURE 6. Effect of contact coefficient between coating and silve on minimum film boiling superheat

The τ_e of 5×10^{-2} s is near the contacting time for water, which was measured in the liquid pool experiments [11] as well as the liquid drop experiments [15] in the vicinity of the minimum film boiling (or Leidenfrost) point. The present value of 5×10^{-2} s is shorter than that (2.5×10^{-1} s) for the earlier study [9] of liquid nitrogen.

Fig. 6 shows a dependency of $\Delta T_{min,o}$ on the contact coefficient h_g between the coating and the silver. In the thin coating region there exists a strong effect of h_g on $\Delta T_{min,o}$ and then the $\Delta T_{min,o}$ is higher with lower h_g. No thermal resistance (complete contact) conditions correspond to $h_g=\infty$. The calculation for $h_g=1\times10^4$ W/m^{2}K agrees well with the experimental data. This value of 1×10^4 W/m^{2}K is equal to the earlier one [9] for Teflon-coated copper plates cooled in liquid nitrogen.

The weak effect of the duration τ_c of the conduction period on $\Delta T_{min,o}$ was confirmed by other calculations. The reference value of τ_c is, therefore, taken as 1×10^{-2} s.

Consequently, the parameters q_e, τ_e, τ_c and h_g have been determined well as constants. This suggests that the assumptions and formulations of the intermittent liquid-solid contact model is valid for the prediction of the effect of a thin insulating layer on the T_{min} of water.

CONCLUSIONS

The effect of a thin paint coating on the rapid cooldown of a silver cylinder in saturated water has been investigated experimentally. The paint coating gives a great improvement of heat transfer since the minimum film boiling temperataure becomes higher with increasing the coating thickness and then an earlier onset of transition boiling is caused. The experimental results agree well with theoretical calculations by the intermittent liquid-solid contact model. The enhancement of minimum film boiling temperature for insulated metals cooled in saturated water is consistent with the preceding results of liquid nitrogen.

This study was partially supported by the Special Project Research on Energy under Grant-in-Aid of Scientific Research of the Ministry of Education, Science and Culture.

REFERENCES

1. Cowley, C. W., Timson, W. J. and Sawdye, J. A., A Method for Improving Heat Transfer to a Boiling Fluid, Adv. Cryogenic Engng., vol. 7, pp. 385-390, 1962.

2. Butler, A. P., James, G. B., Maddock, B. J. and Norris, W.T., Improved Pool Boiling Heat Transfer to Helium from Treated Surfaces and Its Application to Superconducting Magnets, Int. J. Heat Mass Transfer, vol. 13, pp. 105-115, 1970.

3. Moreaux, F., Chevrier, J. C. and Beck, G., Destabilization of Film Boiling by Means of a Thermal Resistance, Int. J. Multiphase Flow, vol. 2, pp. 183-190, 1975.

4. Narasaki, M., Fuchizawa, S., Keino, T. and Takeda, N., On Water Quenching of Heated Metal Cylinder Coated with the Other Material, Proc. 18th Japan National Heat Transfer Symp., pp. 421-423, 1981.

5. Kikuchi, Y., Hori, T. and Michiyoshi, I., Effect of a Thin Insulating Layer on Rapid Cooldown of Metals in Liquid Nitrogen, Proc. 9th Int. Cryogenic Engng. Conf., pp. 77-80, 1982.

6. Nishio, S., Cooldown of Insulated Metal Plates, Proc. ASME-JSME Thermal Engng. Joint Conf., vol. 1, pp. 103-109, 1983.

7. Berenson, P. J., Film-Boiling Heat Transfer from a Horizontal Surface, J. Heat Transfer, vol. 83, pp. 351-358, 1961.

8. Spiegler, P., Hopenfeld, J., Silberberg, M., Bumpus Jr., C. F., and Norman, A., Onset of Stable Film Boiling and the Foam Limit, Int. J. Heat Mass Transfer, vol. 6, pp. 987-989, 1963.

9. Kikuchi, Y., Hori, T. and Michiyoshi, I., Minimum Film Boiling Temperature for Cooldown of Insulated Metals in Saturated Liquid, Int. J. Heat Mass Transfer, forthcoming.

10. Bradfield, W. S., Liquid-Solid Contact in Stable Film Boiling, I & EC Fundamentals, vol. 5, pp. 200-204, 1966.

11. Yao, S. C. and Henry, R. E., An Investigation of the Minimum Film Boiling Temperature on Horizontal Surfaces, J. Heat Transfer, vol. 100, pp. 260-267, 1978.

12. Kikuchi, Y., Hori, T. and Michiyoshi, I., Surface Temperature Transient upon Liquid-Solid Contacts, Proc. 1983 Annual Meeting of Atomic Energy Soc. Japan vol. 1, p. 21, 1983.

13. Kutateladze, S. S., Heat Transfer in Condensation and Boiling, AEC-tr-3770, 1959.

14. Henry, R. E., A Correlation for the Minimum Film Boiling Temperature, Chem. Engng. Prog. Symp. Ser., no. 138, vol. 70, pp. 81-90, 1974.

15. Makino, K. and Michiyoshi, I., The Behavior of a Water Droplet on Heated Surfaces, Int. J. Heat Mass Transfer, vol. 27, pp. 781-791, 1984.

Modelling Experimental Investigation of Two-Phase Gas-Liquid Flow Patterns in Horizontal Pipes

ZHONGQI LU and ENDE MA
Thermal Engineering Department
Tsinghua University
Beijing, PRC

NOMENCLATURE

d	pipe diameter	μ_G	viscocity of gas
G	mass rate	μ_L	viscocity of liquid
s	slip velocity	σ	surface tension
V	velocity	ρ_G	density of gas
V_G	superficial gas velocity	ρ_L	density of liquid
V_L	superficial liquid velocity		
V_M	mixture velocity	Fr_{TP}	Froud number for mixture
β	gas fraction based on volumetric flow	Re_{TP}	Renolds number for mixture
μ	average viscosity	We_{TP}	Weber number for mixture

INTRODUCTION

In two-phase flow systems the effect of flow patterns on the behavior of the dynamic flow and heat transfer is very important. The study of flow patterns belongs in the fundamental research that have strong practicality. Flow patterns and their transitions in horizontal pipes have been up to now experimentally studied by a number of investigators, and many flow pattern maps and their transition correlations have been proposed. However, there is probably any distance from practicability. Besides the complexity of flow patterns themselves, the following account for the occurrence of this impracticability:

1. Most of these studies were confined to air-water (i.e. two components) systems.
2. Most of these studies were confined to normal atmospheric condition.
3. Most of these studies were confined to adiabatic condition.
4. There have not common principles on the classification of flow patterns and common definitions of flow patterns yet.
5. The better coordinate parameters of flow pattern map have not been found as yet.

In order to reveal the effect of parameters and components on flow patterns and their transitions thoroughly, in this study the modelling experiment and analysis of horizontal heated channel flow patterns for single component fluid medium (F-12) at higher pressure (the pressures of F-12 were 10-15 bar, corresponding to water system pressures 65-95 bar) were made.

EXPERIMENTAL SYSTEM

The whole experimental system (shown in Fig. 1) was consisted of four parts: F-12 flow system, cooling water flow system, A.C. electric heated system, the measurement and monitered control of experimental parameters, and operating and control system of equipments.

For convenience of visual observation an outside tin oxide film plating transparent glass tube was used as experimental section. It's inner diameter was 1.9 cm and length was 100 cm. The plating film was divided in sections by copper hoops which was electrod collars too. The maximum heated power was 15 KW. In addition, there was a plexiglass casing in order to make vacuum space between glass tube and casing. There were two preheaters one 20 KW direct heated and the other 6.4 KW infrared electric indirect heated. The accuracy of voltmeters and amperemeters were 0.2%. Four copper-constantan thermal couples with 0.2% accuracy were used to measure medium temperature. The other measuring components for temperature were industrial thermocouples. There were pressure measuring points in the inlets and outlets of preheater and experimental section. Accuracy of pressure gauges was 0.25%. There was a turbine flowmeter in the main tube. It's display instrument was a milliameter with 0.5% accuracy. Another turbine flow meter for smaller flowrate was set in the bypass It's display instrument was a numerical frequency detector with 0.5% accuracy.

The flow patterns are fixed by the visualazation and photograph.

ANALYSIS AND COMPARISON OF EXPERIMENTAL RESULTS

The experimental medium was R-12, which working pressures was 10 bar and 15 bar. The flow rate range was 102-2021 $kg/m^2 \cdot s$. The outlet mass quality ranged 10^{-3}-0.53. About 300 experimental points were obtained.

1. The classification principles and definitions of flow patterns:

Up to now the flow pattern classification methods are not still unified; for the same flow pattern some difference in the definition still exist. It is quite evident that the classification of flow patterns should be futher defined. Usually, there are two methods of classification in theses; (1): stratified flow, wave flow, plug flow, slug flow, annular flow, dispersed flow (bubble or mist flow); (2): separeted flow (stratified and wave flow), intermittent flow (plug and slug flow), annular flow, dispersed flow(bubble or mist flow). In this paper it was held that following classification principles should be reasonable:

(1) The difference between various flow patterns is definite in the regime.

(2) Various flow patterns have evident performances both of flow and heat transfer.

(3) The classifation should be convenient to establish physical and mathema tical models.

In view of the above mentioned, the second classification method was used i this study.

The definitions of flow patterns in this paper are as follows:

Separeted flow--In the side view two phases are both continuous. The liqui phase flow at bottom and gas phase above, the interface between phases is clear discernible.

Intermittent flow--Liquid phase is continuous but gas phase is discontinuou In side view, the liquid plug filled in the cross section of the tube appears intermittently and moves foward. There is continuous or discontinuous liquid film at the tube wall occupied by gas space. The interface between phases is again clearly discernible.

Annular flow--Two phase are both continuous. Liquid phase flow around inside of tube and gas phase flow in center. In side view, the interface is undiscerning.

Bubble dispersed flow--Liquid phase is continuous and gas phase is discontinuous. Small gas bubbles disperse throughout liquid phase. Due to the dis-

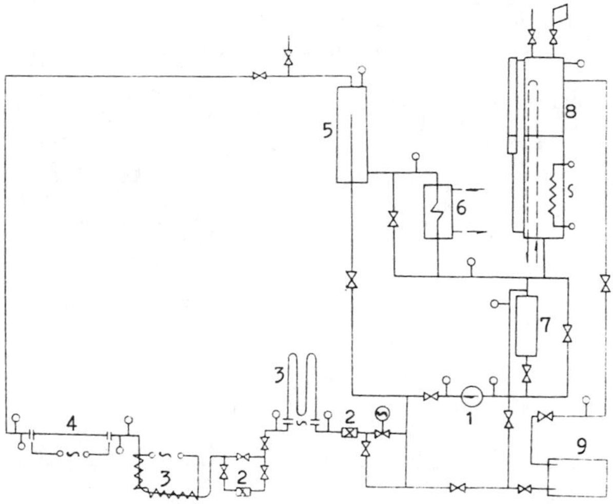

Fig.1 Schematic of test loop 1-pump 2-flowmeter 3-preheater 4-test section 5-mixter 6-cooler 7-fillter 8-regulator 9-tank

persed phase moving at nearly the same velocity as continuous phase with higher velocity. Some "gas-lines"--streamlines of bubble--can be seen by visual observation.

In brief, the critique of distinguishing flow patterns are whether or not interface is discernible; whether or not every phase is continuous, and the distribution of phases.

2. Analysis and comparison of results

The experimental results of this study were compared with prevalent flow pattern maps by converting them into same parametric coordinats.

Comparison with Hoogendoorn (1961) flow pattern map (Fig. 2): It is can be seen that in Hoogendoorn flow map the annular flow regime was not involved. Hoogendoorn flow map is based on the experimental data of R-11 system at lower pressure condition (1-2 bar). However the volume fraction of R-12 is lower than that of R-11 at the same mixture velocities V_M. This is mainly caused by that in the two experiments the pressure were unequel, so that the density difference of gas phases between R-11 and R-12 was larger. The liquid phase densities of R-11 at 2 bar and R-12 at 10 bar approach to each other (density ratio is 1:0.87), but the difference of densities between gas phases is larger (density ratio is about 1:6). Therefore the momentum of gas phase in R-12 is higher than that in R-11 for the same velocities of gas phases, and therefore the transition of R-12 toward intermittent and annular flow patterns will appeared at lower value of β, in other words, the transitions of R-12 will appeared at higher mixture velocity for the same β. In the bubble dispersed flow regime, because the bubble of R-12 has larger inertia and smaller buoyancy than the bubble of R-11 for the some β, the transition of R-12 toward dispersed flow pattern will be earlier than R-11.

Comparison with Mandhane (1974) flow pattern map (Fig. 3): Mandhane flow

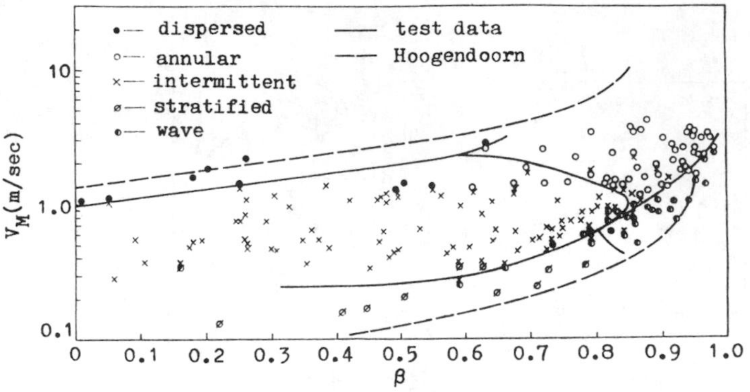

Fig.2 Comparison with Hoogendoorn map (1961)

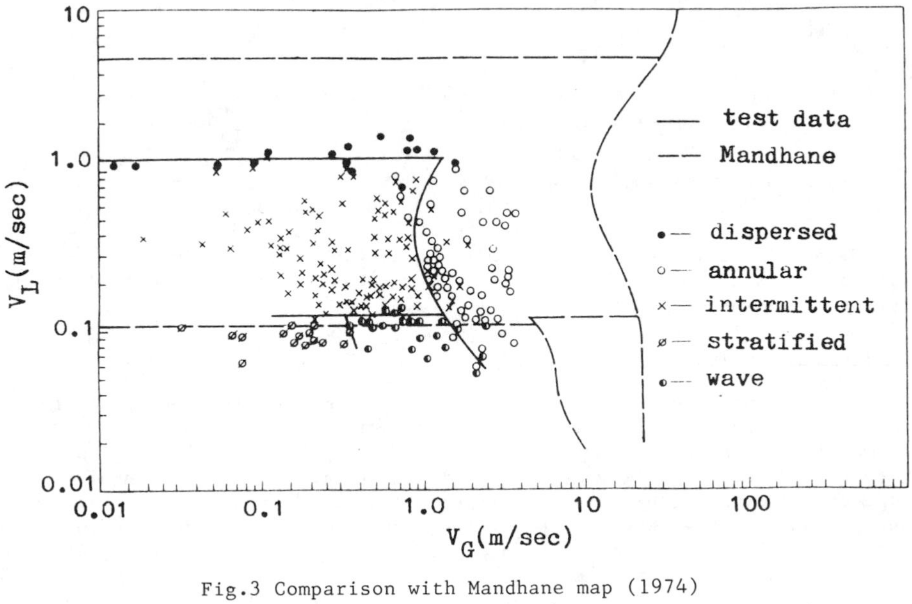

Fig.3 Comparison with Mandhane map (1974)

pattern map is based on the experimental data of air-water systems at normal atmospheric condition. Although it also involves some datum where the gas phase density is up to 50kg/m^3 , but only a few such data can not play an important role in the statistical result. The densities of liquid phase R-12 at 10 bar and water at normal condition approach to each other (density ratio is 1.2:1), but the difference of densities between their gas phases is very large (about 40:1). As will be readily seen there are large differences on the transition lines between wave--stratified flow patterns, and the transitions toward annular as well as bubble dispersed flow patterns, these differences can be also explain as Fig.2. In addition, for the transition toward bubble dispersed flow pattern there is also an effect of definition. The definition of Mandhane about this flow pattern is that the bubbles are fully developed and even distributed, but in this paper it merely considered that whether the dispersed bubbules appear or not. In the air-water flow pattern studied by Barnea et al (1980) the same definition are used, and the comparison of their experimental result with Mandhane map was also made. Results shown that the transition toward bubble dispersed flow pattern was earlier too. Due to the experimental medium and pressure were close to Mandhane's the departure is smaller than that of our condition. That is just as it should be.

Comparison with Taitel-Dukler (1976) flow pattern map (Fig.4): Taitel-Dukler flow map is a semitheoretical analysis result. The comparison show that the departures from each other are large except for the transition line between separated and intermittent flow regimes.

Weisman et al (1979) have proposed a overall flow pattern map. Similar result can be seen from this comparison(Fig. 5). There is a question that in the correlations of Weisman et al the transition between separated and intermittent flow regimes is only dependence on tube diameters or not.

These comparisons shown that the discrepancy of the flow pattern is large when the map is based on air-water experimental data at low pressure and existing semitheoretical flow map are used for predicting flow patterns, in particular for the transitions toward annular and bubble dispersed flow pattern.

PROPOSED FLOW PATTERN MAP

The modelling method is used for sorting out the data in this study. In the horizontal two-phase flow the main determinate criterions obtained by the continuity, mometum equations of each phase and the moment transfer equation on the interface of the two phases are

$$Fr,\quad Re,\quad We,\quad \rho_G/\rho_L\ ,\ \beta\ ,$$

In order to emphasise the effect of gas phase in Froude number a new velocity is defined by

$$V=V_G+\frac{\rho_L}{\rho_G}V_L$$

and then $$Fr_{TP}=V^2/(gd)=L_1$$

Fr_{TP} represents the ratio of inertial force of gas phase and gravitational attraction and involves the effect of density ratio of two phases. The remainder three criterions can be combined as

$$\frac{We_{TP}}{Re_{TP}^2}\beta^2=L_2$$

The definitions of We_{TP} and Re_{TP} are given by

$$Re_{TP}=\frac{G(1-\beta)^2}{d\mu}$$

$$We_{TP}=d\left[\frac{\rho_L V_L^2}{\sigma(1-\beta)^{\frac{1}{2}}}+\frac{\rho_G s^2\beta^{\frac{1}{2}}}{\sigma}\right]$$

and $$s=\frac{V_L}{\beta}-\frac{V_L}{1-\beta}$$

$$=\beta\mu_G+(1-\beta)\mu_L$$

which are rather similar to that derived by Eaton et al (1967). The experimental

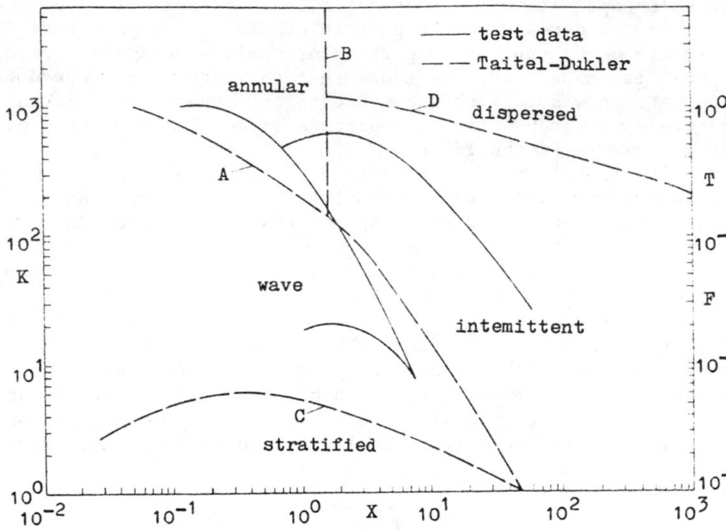

Fig.4 Comparison with Taitel-Dukler map (1976)

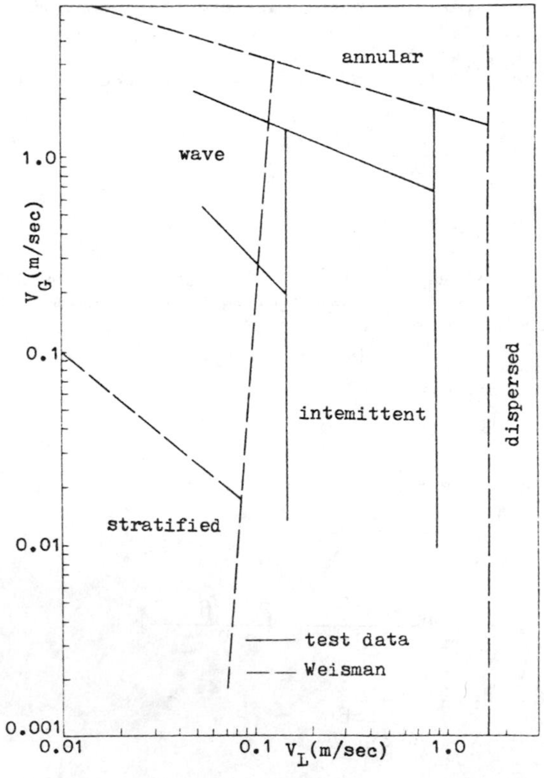

Fig.5 Comparison with Weisman map(1979)

data for R-12 at 10 bar and 15 bar have fairly correlated in L_1-L_2 coordinate. Then a new flow pattern map (Fig. 6) which covered about 300 data points was finually deduced. In this flow pattern map the transition between separated and non-separeted flow patterns is merely related to Fr_{TP} when $L_1>10^{-6}$. This is in agreement with the explanation of Fig. 2,3, i.e. this transition mainly depends on the ratio of momentum of gas phase and gravitational attraction. In addition, the transition between annular and bubble dispersed flow patterns happens at nearly $L_2=1.5\times10^{-6}$. It shows straightforwardly that this transition is independent on the gravitational field. Because the experimental points near by the dashed line are comparatively sparse when $L_2<10^{-6}$, perhaps this transition line somewhat is inaccurate quantitatively. But its trend is in the affirmative.

Up to now the map of horizontal two-phase flow patterns under heated conditions is not published yet. Usually the flow pattern maps deduced from adiabatic condition are used in predicting the two-phase flow patterns under non-adiabatic conditions. It is obviously unreasonable, because under non-adiabatic condition the flow is unsteady. The uninterrupted destruction and establishment of thermodynamic and hydrodynamic equilibrium make the problem more complex, so that the flow patterns and their transitions under non-adiabatic conditions considerably differ from that under adiabatic conditions.

In this study the flow patterns under heated conditions was also observed. Some experimetal points were obtained, but they are not enough for analysis and comparison. The transitions toward both intermittent and annular flow patterns had been observed. But from the experiment it is obvious that for the same pressure and total mass flow rate the transition power under heated conditions was more than that under adiabatic conditions. It clearly relates to the thermodynamic non-equilibrium.

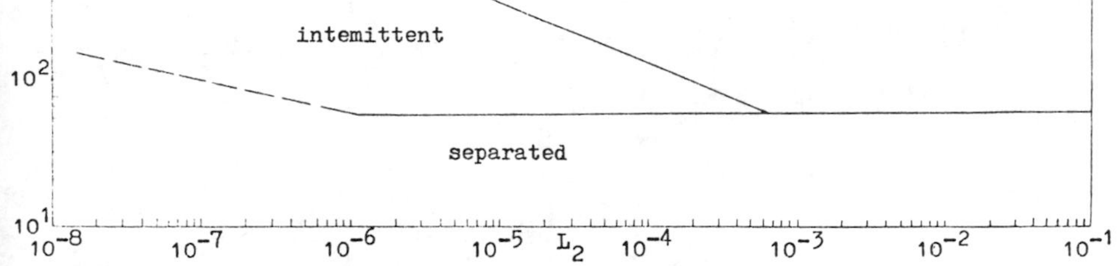

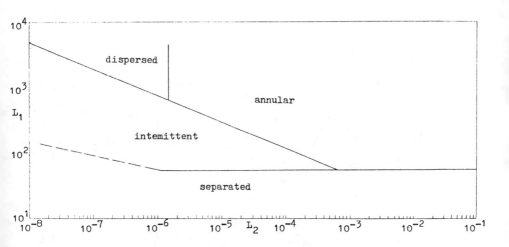

Fig.6 Proposed high pressure flow pattern map

CONCLUSIONS

1. The existing flow pattern maps of horizontal pipes are not satisfied for higher pressure, in particular, about the transitions toward to annular and bubble dispersed flow patterns.

2. Because the momentum transfer on interface (as well the heat and mass transfer under heated conditions), the existing flow pattern maps based on air-

water experiments are not satisfied for single component medium.

3. A new flow pattern map for single component medium R-12 at higher pressure 10-15 bar (Fig. 6) was proposed in this study (it is can model water systems at 65-95 bar).

4. Further study is necessary for the transitions of flow patterns and the flow pattern near by the bubble dispersed flow. This paper only presents partial results of our study program in progress.

REFERENCE

Baker, O. 1954 Simultaneous flow of oil and gas, Oil Gas J. 53, 185-195.

Barnea, D., Shohan, O., Taitel, Y., and Dukler, A.E. 1980 Flow pattern transition for horizontal and inclined pipes: experimental and comparison with theory, Int.J. Multiphase Flow, 6, 217-225.

Eaton, E.A. et al. 1967 The prediction of flow patterns, Liquid holdup and pressure losses occurring during continuous two-phase flow, J.Petrol. Techno June, 815-823.

Hetsroni, G. 1982 Handbook of Multiphase Systems, McGRAW-HILL, New York.

Hoogendoorn, C.J. and Beutelaar,A.A. 1961 Effects of gas density and gradual vaporization on gas-liquid flow in horizontal lines. Chem. Engng Sci. 16,208-215.

Mandhane, J.M., Gregory, G.A. and Aziz, K.A. 1974 A flow pattern map for gas-liquid flow in horizontal pipeline. Int.J.Multiphase Flow 1,537-554.

Taitel, Y. and Dukler, A.E. 1976 A model for predicting flow regime transition in horizontal and near horizontal gas liquid flow. A.I.Ch.E.J.22,47-55.

Weisman, J., Duncan,C., Gibson, J.and Crawford,T. 1979 Effects of fluid properties and pipe diameter on two-phase flow patterns in horizontal lines. Int.J. Multiphase flow 5,437-462.

orced-Flow Turbulent Film Boiling f Subcooled Liquid Flowing in Horizontal lat Duct

-XUAN WANG and DE-HUI SHI
ermal Engineering Department
inghua University
ijing 100084, PRC

STRACT

The analytical model for turbulent film boiling of subcooled liquid along a rizontal plate, as presented in a previous paper [1], is now extended to the se of subcooled liquid flowing in a horizontal duct. The corresponding mathe-tical description is given, and a semi-empirical heat transfer correlation thus rived is well verified by experimental data for subcooled water flow. The re-lts are expected to be adaptable for practical heat transfer calculation of gh-temperature quenching process for rolled metals.

MENCLATURE

a	thermal diffusivity
b	altitude of horizontal duct
c_p	specific heat at constant pressure
Nu	Nusselt number
Pr	Prandtl number
q	local heat transfer rate per unit area
Re	Reynolds number
t	temperature
u	velocity in x-direction

Greek symbols

α	local heat transfer coefficient
δ	thickness of vapor film
λ	thermal conductivity
ρ	density
μ	absolute viscosity
ν	kinematic viscosity

Subscripts

w	at surface
i	at entrance of duct
s	saturated condition
1	vapor
2	liquid

INTRODUCTION

As an important application to develop efficient cooling-control technology high-temperature quenching with water in metallurgical industries, the study forced turbulent film boiling of a subcooled liquid flowing in a duct has at-acted the attention of many researchers for a long time. An investigation of lm boiling in vertical tubes with subcooled nitrogen flow had been reported by linin et al [2], who took the well-known empirical correlation for single-phase turbulent flow to calculate the heat transfer rate for turbulent film boiling. A theoretical analysis of saturated film boiling for fluid of low vapor qualit flowing upward in a vertical tube had been reported even earlier by Dougall an Rohsenow [3]. Both experimental investigation and numerical calculation of su cooled and low vapor quality film boiling of water flowing in a vertical tube were recently reported by Fung [4], but the flow velocity was very low (0.5 m/ only). Although several industrial measurement data on the quenching of rolle metal have been reported in literature, the correlations of heat transfer data confining to local conditions, as recommended by different researchers, differ from each other so greatly that it is difficult to adopt them in practical use We have still known very little about the characteristic of such two-phase flo heat transfer, especially for the subcooled liquid flowing at high velocities More recently, an analysis of subcooled film boiling inside a horizontal duct concentric circular-tube annulus was reported by Viannay and Karian [5], but assumption may be untrue that all the heat flux transfers from the high-temper ture plate to the opposite cold plate of the flat duct or from the high-temper ture inside tube of the annulus to the outside cold tube, because the subcoole liquid flow will carry away actually most part of the heat flux taken from hig temperature surfaces.

A physical model was made to analyse film boiling heat transfer for turbul flow of subcooled liquid along a horizontal plate in our previous study [1]. This model is now effectively extended to the case for liquid flowing in a hor zontal flat duct. The results thus obtained will be adaptable to practical us in engineering analysis of high-temperature quenching process for rolled metal

II. THEORETICAL CONSIDERATION

The quenching process of high temperature rolled metal sheets is usually completed by highly subcooled water flowing in a horizontal flat duct as shown schematically in Fig.1. At y=b, the upside of the high-temperature rolled metal sheet acts as base of the horizontal flat duct.

Suppose the width of the duct (in z-direction) is so large that the influence of side walls of the duct can be neglected, the analytical model will be thence reduced to two-dimensional one. The velocity distribution within the liquid region should become very complex in nature due to the existance of both vapor-liquid interface and top cover plate. Fortunately, for the case of subcooled liquid flowing turbulentl with higher velocity, the velocity distribution within the liquid region can b considered as being uniform, taken as u_2=const.,[1].

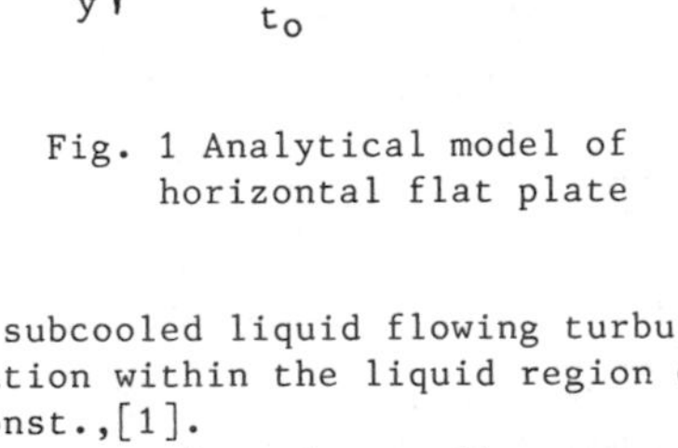

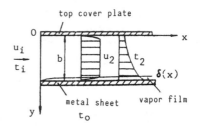

Fig. 1 Analytical model of horizontal flat plate

In actual engineering practice, the top cover plate is usually exposed to surrounding circumstance, so that heat loss should exist. However, such heat in its natural convective way is so small as compared with that transferred f the high-temperature metal surface, that it can really be neglected. So the cover plate could be regarded as a surface insulated thermally.

In short, the following assumptions can be made for further consideration

(1) Owing to the strong suppression of subcooled liquid flowing at high city, the thickness of the vapor film increases very slowly along the x-direc

1] and is thinner as compared with the altitude, b, of the horizontal flat duct, .e. $(b - \delta) \rightarrow b$, so that the heat flux transferred to the vapor film from high-emperature metal sheet surface can be assumed as completely transferred to the iquid flow. This is equivalent to assume the temperature of vapor dropped sharp-y from t_w to t_s, which results in very high heat transfer rate.

(2) The top wall of the duct is insulated thermally.

(3) The velocity distribution within liquid region can be considered as being uniform, u_2, and may be assumed as equal to mean velocity in duct, u_i.

(4) According to the previous analyses [1], the "turbulent viscosity for omentum exchange" ε_{M2} can be expressed as

$$\frac{\varepsilon_{M2}}{\nu_2} = k\left(\frac{u_i x}{\nu_2}\right)^m , \tag{1}$$

where k and m are empirical constants which can be determined experimentally.

From Fig.1, the energy equation within the liquid region will be [1]

$$u_i \frac{\partial t_2}{\partial x} = k\nu_2\left(\frac{u_i x}{\nu_2}\right)^m \frac{\partial^2 t_2}{\partial y^2} \tag{2}$$

with boundary conditions:

$$\left.\begin{array}{ll} x = 0, & t_2 = t_i; \\ y = 0, & \dfrac{\partial t_2}{\partial y} = 0; \\ y \rightarrow b, & t_2 \rightarrow t_s. \end{array}\right\} \tag{3}$$

By introducing a dimensionless subcooled temperature, $\theta = \dfrac{t_s - t_2}{t_s - t_i}$, equation (2) can be rewritten as

$$u_i \frac{\partial \theta}{\partial x} = k\nu_2\left(\frac{u_i x}{\nu_2}\right)^m \frac{\partial^2 \theta}{\partial y^2} \tag{4}$$

while the corresponding boundary conditions from Eq. (3) are:

$$\left.\begin{array}{ll} x = 0, & \theta = 1; \\ y = 0, & \dfrac{\partial \theta}{\partial y} = 0; \\ y \rightarrow b, & \theta \rightarrow 0. \end{array}\right\} \tag{5}$$

Let

$$\eta = \left(\frac{u_i x}{\nu_2}\right)^{m-1} x^2 = \left(\frac{u_i}{\nu_2}\right)^{m-1} x^{m+1} \tag{6}$$

or

$$\frac{\partial \theta}{\partial x} = \frac{\partial \theta}{\partial \eta}\frac{d\eta}{dx} = \frac{\partial \theta}{\partial \eta}(m+1)\left(\frac{u_i}{\nu_2}\right)^{m-1} x^m \tag{7}$$

Substituting Eq.(7) into Eq.(4), we get

$$\frac{\partial \theta}{\partial \eta} = \frac{k}{m+1}\frac{\partial^2 \theta}{\partial y^2} \tag{8}$$

with boundary conditions transformed from Eq.(5):

$$\left.\begin{array}{ll} \eta = 0, & \theta = 1; \\ y = 0, & \dfrac{\partial \theta}{\partial y} = 0; \\ y \rightarrow b, & \theta \rightarrow 0. \end{array}\right\} \tag{9}$$

Solving Eq.(8) combined with boundary conditions Eq.(9) by the method of separation of variables [6], we obtain the following subcooled temperature distribution:

$$\theta = \sum_{n=0}^{\infty} (-1)^n \frac{4}{(2n+1)\pi} \exp[-\frac{k}{m+1}(\frac{2n+1}{2b}\pi)^2\eta] \cos(\frac{2n+1}{2b}\pi y) \; . \qquad (10)$$

Let $Re_x = u_i x/\nu_2$. Introducing the definition of η, Eq.(6), we can rewrite Eq.(10) as

$$\theta = \sum_{n=0}^{\infty} (-1)^n \frac{4}{(2n+1)\pi} \exp[-\frac{k}{m+1}(\frac{2n+1}{2b}\pi)^2 Re_x^{m-1} x^2] \cos(\frac{2n+1}{2b}\pi y). \qquad (11)$$

From Fourier's law, the heat flux transferred from the vapor-liquid interface to the liquid would be:

$$\left.\begin{aligned} q_{1,x} &= -\rho_2 c_{p2} \varepsilon_{M2} \left.\frac{\partial t_2}{\partial y}\right|_{y=b} \\ \text{or} \quad q_{1,x} &= \rho_2 c_{p2} \varepsilon_{M2}(t_s-t_i) \left.\frac{\partial \theta}{\partial y}\right|_{y=b} \end{aligned}\right\} \qquad (12)$$

Substituting Eq.(11) into Eq.(12), we obtin

$$q_{1,x} = \rho_2 c_{p2} \varepsilon_{M2} (t_s-t_i) \sum_{n=0}^{\infty} \frac{2}{b} \exp[-\frac{k}{m+1}(\frac{2n+1}{2}\pi)^2 Re_x^{m-1} (\frac{x}{b})^2]. \qquad (13)$$

Define a "revised local Nusselt number" based on the subcooled temperature (t_s-t_i), $\widetilde{Nu}_b = \frac{q_{w,x} b}{(t_s-t_i)\lambda_2}$. For the highly subcooled liquid, $q_{w,x} \approx q_{1,x}$. So we get from Eq.(12)

$$\widetilde{Nu}_b = \frac{2\varepsilon_{M2}}{a_2} \sum_{n=0}^{\infty} \exp[-\frac{k}{m+1}(\frac{2n+1}{2}\pi)^2 Re_x^{m-1} (\frac{x}{b})^2]. \qquad (14)$$

Noticing $Re_x = \frac{u_i x}{\nu_2} = \frac{u_i b}{\nu_2} \cdot \frac{x}{b} = Re_b \frac{x}{b}$, and substituting ε_{M2} from Eq.(1) into Eq.(14), we obtain

$$\widetilde{Nu}_b = 2kRe_b^m Pr_2 (\frac{x}{b})^m \sum_{n=0}^{\infty} \exp[-\frac{k}{m+1}(\frac{2n+1}{2}\pi)^2 Re_b^{m-1} (\frac{x}{b})^{m+1}] \qquad (15)$$

$$\text{or} \quad \widetilde{Nu}_b = C_x Re_b^m Pr_2 \, , \qquad (16)$$

where

$$C_x = 2k(\frac{x}{b})^m \sum_{n=0}^{\infty} \exp[-\frac{k}{m+1}(\frac{2n+1}{2}\pi)^2 Re_b^{m-1} (\frac{x}{b})^{m+1}]. \qquad (17)$$

The coefficient C_x reflects the effects of x/b on local wall heat flux for giv subcooled liquid with known u_i and t_i . It is clear from Eqs.(16) and (17) th the wall heat flux will decrease gradually along the flowing direction.

III. EXPERIMENTAL STUDY

Experimental investigations were conducted on the turbulent film boiling f subcooled deionized water flowing at atmospheric pressure in a horizontal flat duct, with velocity ranging from 1 to 4.5 m/s and with temperature subcooled ranging from 22 to 72°C. The experimental installation with test section made of a short horizontal flat duct was the same as reported in previous paper [1]. As shown in Fig. 2, the flow passage of the test section measured 100 mm wide and 10 mm high with a 200 mm long, 80 mm wide and 8 mm thick copper plate as the bottom plate. When the copper plate had been heated below by an electric heater to reach 650 - 700 °C, the subcooled water flows suddenly through the test section. The surface temperatures of the copper plate in response of the cooling were measured by thermocouples at the distances of 50 mm and 150 mm respectively from the front edge of the duct, and recorded automatically, so that the heat flux $q_{w,x}$ can be calculated out. The measuring techniques and the reliability of the data taken from the experiments were discussed in detail, [7].

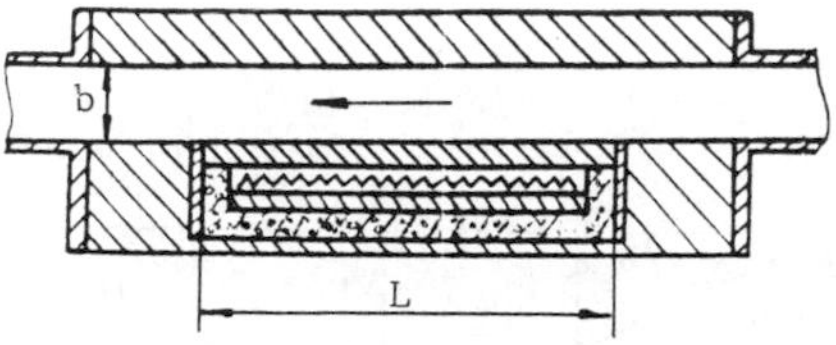

Fig.2 Construction of horizontal duct

The typical experimental data for x = 50 mm and x = 150 mm, (corresponding to x/b = 5 and 15 respectively), are plotted in Fig. 3 and all converge to a straight line, as indicated by the relation derived for forced-flow turbulent film boiling of subcooled liquid along a horizontal plate [1], with k = 0.0055 and m = 0.68. This is a rational result, the heat transfer relation for fluid flowing along a plate surface should be adaptable for the duct flow near the entrance [8]. The maximum deviation for all the data points, taken from x = 50 and x = 150 mm, did not exceed ± 25%. All the data points for x = 50 mm lie on the upside while those for x = 150 mm lie below the line

$$\widetilde{Nu}_x = 0.054\ Re_x^{0.84}\ Pr_2 \tag{18}$$

or

$$\widetilde{Nu}_x = \sqrt{\frac{k\ (m+1)}{\pi}}\ Re_x^{(m+1)/2}\ Pr_2 \tag{19}$$

with empirical constants k = 0.0055 and m = 0.68 [1].

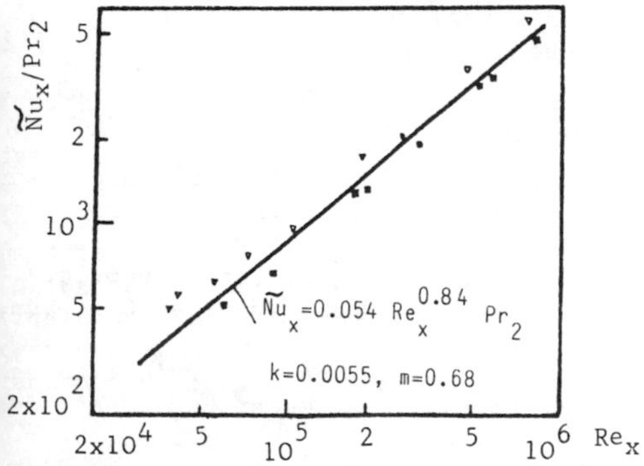

Fig.3 Plot of experimental data for turbulent film boiling of subcooled water

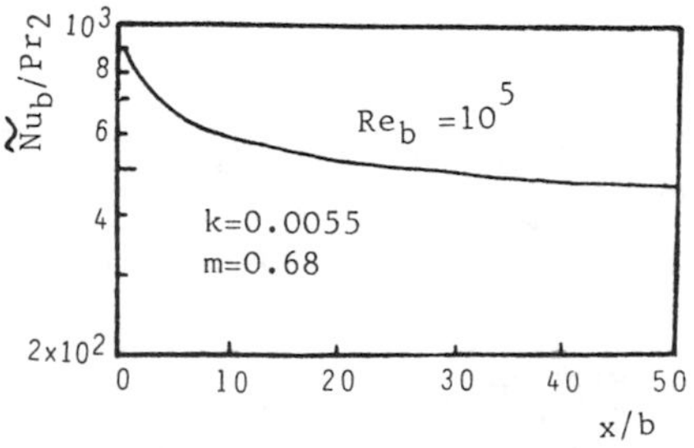

Fig.4 Plot of Eq.(15) for subcooled water flowing through horizontal flat duct

The calculating results from Eq.(15) for $Re_b = 10^5$, with empirical values of k=0.0055 and m=0.68, are plotted in Fig.4 and summarized briefly for the val of C_x from Eq.(17) in table 1. If x/b exceeds about 30, the first term of the progression in Eq.(15) will be much greater then the rest, and Eq.(15) can thus be simplified to

$$\widetilde{Nu}_b = 2k\, Re_b^m\, Pr_2 \left(\frac{x}{b}\right)^m e^{-\frac{k}{m+1}\left(\frac{\pi}{2}\right)^2 Re_b^{m-1}\left(\frac{x}{b}\right)^{m+1}}, \quad (20)$$

or for the case of water flowing with higher velocity, k=0.0055 and m=0.68,

$$\widetilde{Nu}_b = 0.011\, Re_b^{0.68}\, Pr_2 \left(\frac{x}{b}\right)^{0.68} e^{-0.008 Re_b^{-0.32}\left(\frac{x}{b}\right)^{1.68}} . \quad (21)$$

Table 1. Effect of x/b on $\widetilde{Nu}/(Re_b^m\, Pr_2)$ for subcooled water flowing through horizontal flat duct

Re_b \ x/b	1	5	10	20	30	40	50	60
10^5	0.342	0.264	0.237	0.212	0.198	0.190	0.183	0.178
$5X10^5$	0.437	0.342	0.306	0.274	0.257	0.245	0.237	0.230

IV. CONCLUDING REMARKS

An essential physical model to describe the turbulent film boiling for subcooled liquid flowing along a horizontal plate, suggested in the previous paper [1], is successfully extended to analyze the more universal case for subcooled liquid flowing in a horizontal flat duct. The results have been verified experimentally. It is reasonably to be expected that Eq.(15) with k=0.0055 and m=0.68 or Eq.(21) for x/b > 30 can be used to predict the heat transfer rate for high-temperature quenching of rolled metals with subcooled water flowing at much high velocity. However, it should be emphasized that k=0.0055 and m=0.68 are obtaine empirically from experiments with subcooled water only. Furthermore, the testir section was limited to x/b = 20, further experimental check for the theoretical considerations with larger values of x/b may be still needed.

REFERENCES

[1] B.X.Wang and D.H.Shi, "A semi-empirical theory for forced-flow turbulent fil boiling of subcooled liquid along a horizontal plate", Int.J.Heat Mass Trans fer, 28(9), (accepted for publication), 1985.

[2] E.K.Kalinin et al, "Investigation of film boiling in tubes with subcooled nitrogen flow", Proceedings of the 4th International Heat Transfer Conferen 1970.

[3] R.S.Dougall and W.M.Rohsenow, "Film boiling on the inside of vertical tubes with upward flow of the liquid with low qualities", Dep. Mech. Eng.,Mass. Ins Technol. Rep. No. 9097, p.29 1963.

[4] K.K.Fung, "Subcooled and low quality film boiling of water in vertical flow atmospheric pressure", NUREG/CR - 2461, 1981.

[5] S.Viannay and J.Karian, "Study of the accelerated cooling of a very hot wal with a forced flow of subcooled liquid in film boiling region", Proceedings

of the 7th International Heat Transfer Conference, 1982.
[6] V.S.Arpaci, Conduction Heat Transfer, Addison-Wesley, 1965.
[7] D.H.Shi, "Film boiling heat transfer for forced flow of fluid", Dissertation for degree of Dr.-Eng., Tsinghua University, Beijing, 1984.
[8] B.X.Wang, Engineering Heat and Mass Transfer, (in Chinese), Vol.1, Science Press, Beijing, 1982.

An Experimental Research to Prevent Heat Transfer Deterioration in a U-tube

RUI-CHANG YANG, ZHONG-QI LU, and DE-QIANG SHI
Department of Thermal Engineering
Tsinghua University
Beijing, PRC

ABSTRACT

This paper reports the results of experimental research for preventing heat transfer deterioration in a U-tube steam generator. The tests conducted include (1) eliminating the staying of the bubbles at the top of the U-tube by increasing the mass velocity and (2) preventing heat transfer deterioration at the top of the U-tube by installing swirlers in it. On the basis of present experimental research the practicability of the above mentioned methods in actual steam generators is discussed.

1. INTRODUCTION

This paper presents the second stage results of experimental research following "An Experimental Study of CHF in a U-tube by Fluid Modeling"[1]. The first stage results (see Ref.[1], illustrates that in a veritical U-tube shown in Fig.2 heat transfer deterioration occurs at the top in certain ranges of heat fluxes, mass velocities and inlet qualities, and often results in a burst of the tube. Present research is aimed at eliminating the staying of bubbles at the tube top and in turn preventing heat transfer deterioration. Tests with increased mass velocities in the tube and with installed swirlers have been carried out. According to this experimental research the practicability of these methods in actual U-tube steam generators is discussed.

2. EXPERIMENTAL RIG AND METHOD

The schematic diagram of the experimental rig is shown in Fig.1. The U-tube test section is shown in Fig.2. Modeling experiments with Freon 12 as working fluid in heated conditions have been conducted. For a U-type steam generating tube, when the mass velocity in the tube is lower than a certain value, the liquid-phase momentum becomes insufficient to carry away the bubbles appearing at the top of the tube due to buoyancy and centrifugal effects and then heat transfer deterioration occurs at the top of the tube which may results in a burst of it. Since the direction of buoyancy is opposite to the flow direction at the upstream part of the bend ($\vartheta < 90^\circ$, see Fig.2), the detachment of the bubbles from the wall becomes more difficult and heat transfer deterioration will occur first in this region. Therefore, the mass velocity G determined from equality of Fr Numbers of water and Freon-12 should be most suitable to reflect the influence of mass velocity on the heat flux under which heat transfer deteriorates. The experimental results of Ref.[1] justified this analysis. Hence, modeling method used in present research is as follows:

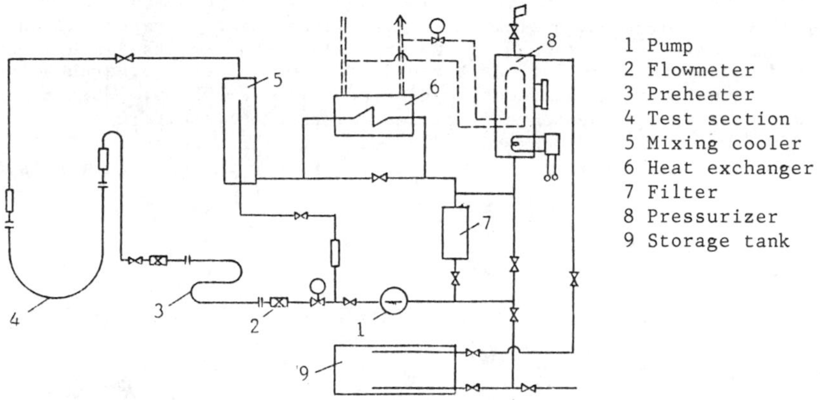

Fig. 1. Schematic Diagram of the Experimental Freon-12 Rig

$$(L/R)_F = (L/R)_W , \qquad (L/d)_F = (L/d)_W \tag{1}$$

$$(\rho_l/\rho_g)_F = (\rho_l/\rho_g)_W \tag{2}$$

$$(\Delta h_i/r)_F = (\Delta h_i/r)_W \tag{3}$$

$$(u_o^2/gd)_F = (u_o^2/gd)_W \tag{4}$$

$$(q/Gr)_F = (q/Gr)_W \tag{5}$$

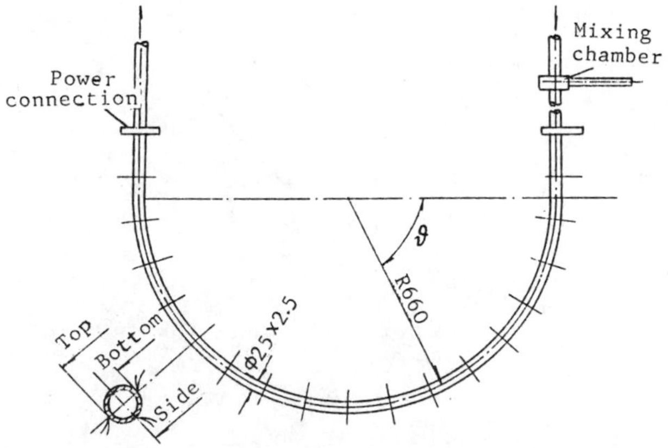

Fig. 2. U-tube Test Section

where L is the unfolded length, R is the curvature radius, d is the inner diameter of the tube, ρ_l and ρ_g are the liquid and the vapor phase densities respectively, Δh_i is the inlet quality, r is the latent heat of vaporization, u_o is the circulation velocity, and q is the heat flux. Subscripts F and W refer to Freon and water respectively. As the objective of present research we use a U-tube of an actual high pressure steam generator with R=693 mm. When a substantial difference in temperature is present along the circumference of a cross section of the tube, a two dimensional (radial r and circumferential φ) is used to solve for inner wall temperature t_i and heat flux q_i. The equation is as follows:

$$\frac{\partial^2 t}{\partial r^2} + \frac{1}{r}\frac{\partial t}{\partial r} + \frac{1}{r^2}\frac{\partial^2 t}{\partial \varphi^2} + \frac{H}{\lambda} = 0 \qquad (6)$$

The local heat transfer coefficient is obtained from

$$h_i = q_i/(t_i - t_s) \qquad (7)$$

where t_s is the saturation temperature of the fluid.

3. RESULTS AND ANALYSIS

(1) Experiments with increased mass velocity in the U-tube

As mentioned above, at the upstream part of the bend it is not easy for bubbles to detach themselves from the wall due to the direction of the buoyancy, so heat transfer deteriorates first in this region. Consider a bubble at the top of the U-tube (see Fig.3). Following forces act on the bubble: the drag of liquid phase $F_d=C_d \rho_l (r^3/D_h)\, u^2$, buoyancy force $F_b=C_b g(\rho_l - \rho_g)r^3$ and surface tension $F_s=C_s \sigma r$. Here r is the bubble radius, D_h is the hydrodynamic diameter, u is the actual velocity of the liquid phase, C_a , C_b, C_s are the coefficient of drag, bouyancy and surface tension respectively. As opposed to vertical tube

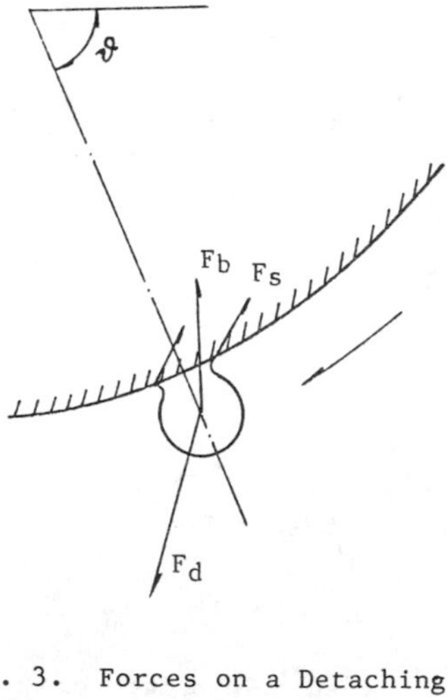

Fig. 3. Forces on a Detaching Bubble

condition, in a U-tube the only force which causes bubbles to leave the wall is the drag of the liquid phase. Therefore, increase of mass velocity appears to be one of the effective methods to prevent heat transfer deterioration. Experiments with increased mass velocity have been conducted under the following conditions: $P_F = 16.92$ kg/cm^2 , heat flux $q_F = (15\text{-}50) \times 10^3$ kcal/m^{2}h, inlet quality $x_{in} = 0.02\text{-}0.05$, mass velocity $G_F = 1450\text{-}3000$ kg/m^{2}s. The corresponding high pressure water parameters are $P_W = 106$ kg/cm^2, $q_W = (100\text{-}350) \times 10^3$ kcal/m^{2}h, $G_W = 1000\text{-}1800$ kg/m^{2}s, the inlet quality is the same. Typical experimental result is illustrated in Fig.4 where the abscissa indicates the mass velocity, while the ordinate indicates the maximum temperature excursion within the region of heat transfer deterioration. Besides the experimental parameters of Freon 12, the corresponding parameters of high pressure water are also listed in brackets in the figure. It can be seen that the value of temperature excursion at the tube top decreases as the mass velocity increases. And the temperature excursion vanishes when the mass velocity arrives at certain value. The mass velocity which causes the temperature excursion to vanish takes different values under different heat fluxes. A theoretical analysis of heat transfer deterioration in U-type tube was given in Ref.[1]. It is based on the balance of forces acting on a bubble detaching from the wall and is justified again by the present experiments. In order to increase the mass velocity the head of the circulating pump, of course, must be increased too. Consequently, the investment and maintenance expense of the pump will be higher.

(2) Experiments with installed swirlers

Different types of swirlers are used in steam generators such as boilers etc., to enhance the disturbance of the fluid, destroy the forming of vapor film and consequently to eliminate or to delay the heat transfer deterioration [2, 3, 4]. As mentioned above, the abnormal temperature rise at the top of the U-tube is caused by the staying of bubbles. Therefore, swirlers can be used to

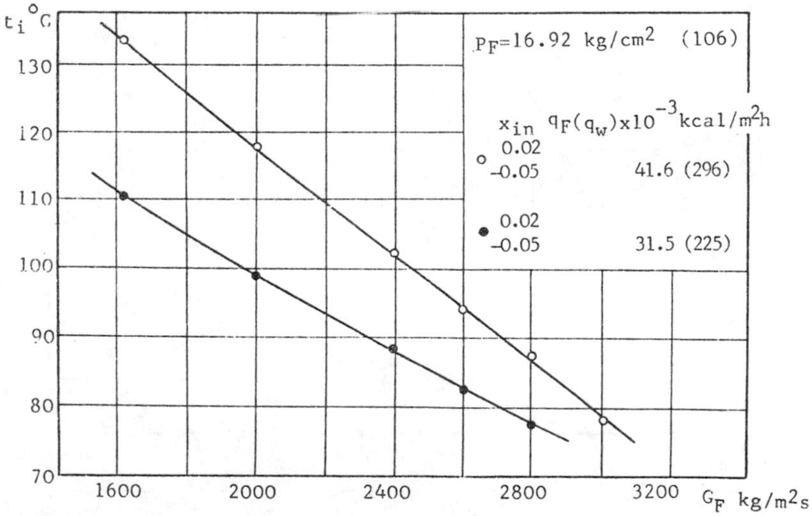

Fig.4. Maximum Wall Temperature vs Mass Velocity for Different Heat Fluxes

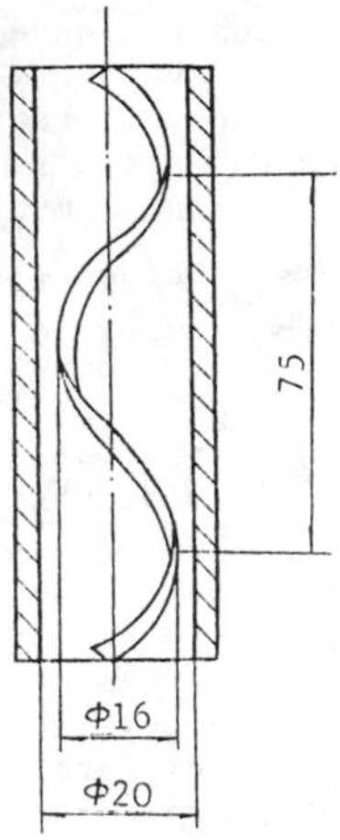

Fig. 5. Structure of the Tape-swirler

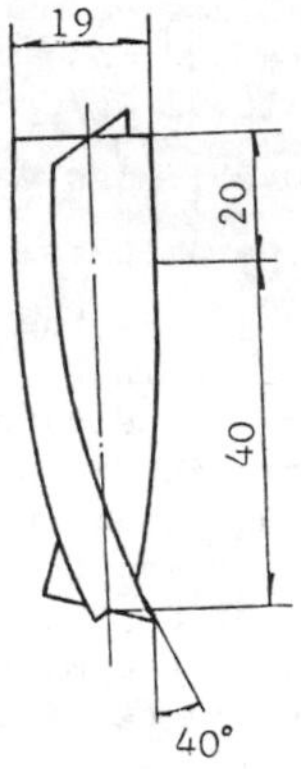

Fig. 6. Structure of the Fan-swirler

prevent this emergence. Two types of swirlers are tested in present research. The swirler shown in Fig.5 is a spiral tape of 0.5 mm thick and 6 mm wide made from stainless steel. The twist ratio S (defined as 360-deg twist length/i.d.) of the swirler is 4.6. Total length of the tape-swirler is 1145 mm. It is installed in the upstream part (ϑ=10-90°) of the bend. The fan-swirler shown in Fig.6 is made from stainless steel plate of 19 mm wide and 0.5 mm thick. Twist angle of its blade is 40°. Accordingly, its twist ratio S is 4.0 The fan-swirle is placed just at the inlet to the bend, i.e. in the straight section immediately adjecent to the bend. Tests have been carried out under different heat fluxes

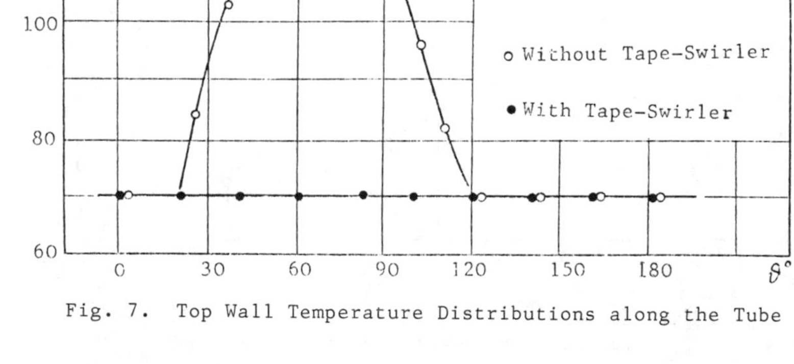

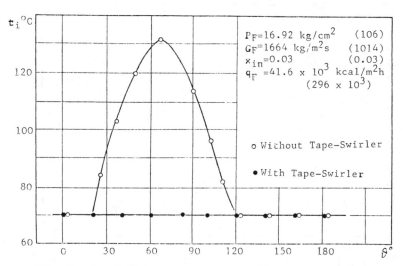

Fig. 7. Top Wall Temperature Distributions along the Tube

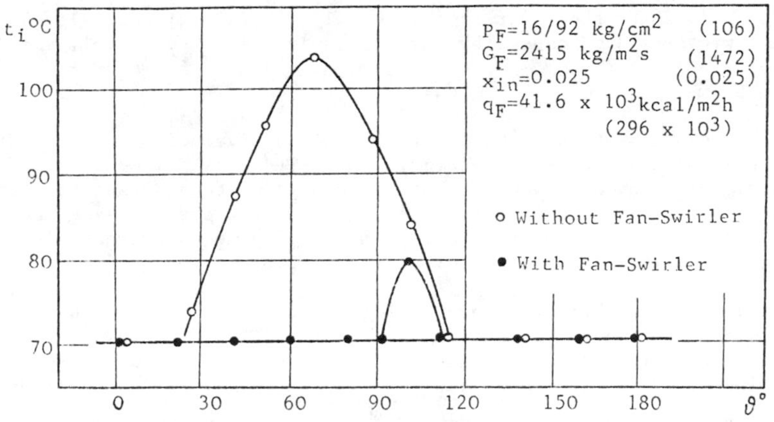

Fig. 8. Top Wall Temperature Distributions along the Tube

nd with different mass velocities. Typical results are shown in Fig.7 and Fig.8. ig.7 is the comparison of the top wall temperature distributions with and withut tape-swirler in the U-tube. The abscissa indicates the axial positions along he U-tube and the ordinate indicates the top wall temperature. Besides the tests nder heated conditions a visual observation of the effect of the swirler has also een conducted in a transperant tube with air-water mixture. It is shown that, wing to the existance of the tape-swirler, the liquid phase often flowing at the ottom of the tube now swirls along the tape-swirler, washing the top wall and estroying the vapor film formed there. So the tube wall is cooled and wetted nd prevented from the heat transfer deterioration. The centrifugal acceleration f the fluid caused by this type of swirlers a_c can be calculated from

$$a_c = 2\pi u_a / S^2 D_i \tag{8}$$

here u_a is the axial velocity of fluid in the condition without swirler, D_i is he inside diameter of the swirler. In the range of present experiments there s no temperature excursion at the top wall of the U-tube even when a_c falls down o 1.1 x 10^2m/s^2. Fig.8 is the comparison of the top wall temperature distribuions with and without using a fan-swirler. The action principle of a fan-swirler s similar to that of a tape-swirler. Having passed through the blades, the fluid wirls along the circumference of the tube, washes away the bubbles at the top nd prevents the top wall temperature to rise abnormally. Practice shows that he swirling intensity of the fan-swirler decreases quite rapidly. For example, ven when the centrifugal acceleration caused by the fan-swirler reaches 1.85x10^2 /s^2 some temperature excursion still exists at the downstream part (ϑ>90°) of he bend. This means that the influence region of a fan-swirler is limited.

Application of swirlers will increase the flow resistance of the U-tube. he pressure drop due to installing of swirlers has been measured. The resistance oefficient ζ is calculated from the following equation:

$$\Delta p = \zeta(\rho_1 u_o^2/2g)[1+x(\rho_1/\rho_g - 1)] \tag{9}$$

We find that ζ =1.72 for tape-swirler and ζ =1.15 for fan-swirler. It can be seen that the pressure drop caused by both the swirlers is relatively small.

4. CONCLUSIONS

(1) Present experiments show that increase of mass velocity can prevent U-tube from boiling heat transfer deterioration. The physical model based on the balance of forces on a detaching bubble and applied to analyze the emergenc of heat transfer deterioeation is justified again by the present experimental research. Of course, increase of circulation velocity means increase in expens of investment and maintenance of the circulating pump.

(2) Installation of swirlers shown in Fig.5 and Fig.6 in the U-tube is effective in respect to preventing of heat transfer deterioration. The pressur drop caused by the swirlers is not significant. Caution in manufacture and mount technology is necessary when these swirlers are applied to actual units.

ACKNOWLEDGEMENTS

The authors are grateful to Prof. Feng Jun-kai, Xu Xiu-qing, Zhen Qia-yu and Li Yao-zhu for their helpful suggestions and discussions in accomplishing this work.

REFERENCES

1. Lu Zhong-qi et al, An experimental study of CHF in a U-tube by fluid modeling, Proceedings of Condensed Papers, China-U.S. Seminar on Two-Phase Flows and Heat Transfer, May 1984.
2. Akagawa K. et al, Transactions of the Japan Society of Mechanical Engineers Vol. 49, No. 444 (1983).
3. Pai R. H. et al, Heat Engineering, Vol. 39. No. 6 (1964).
4. Brevi R. et al, Nuclear Science and Engineering, Vol. 46, No. 1 (1971).

The Effects of Peripheral Wall Conduction on Pool Boiling

. ZENG and Y. LEE
Department of Mechanical Engineering
University of Ottawa
Ottawa, Canada, K1N 6N5

INTRODUCTION:

Given uniform heat generation within a heater placed in an asymmetrical fluid boundary condition, as in the case of a horizontally placed cylindrical heater in pool boiling, heat flows by conduction within the wall of the heater and creates a non-uniform wall surface temperature distribution.

It is well known that when all other flow conditions are equal, the overall heat transfer rate is strongly affected by the heating boundary conditions. This is because the local heat transfer is a function of the local thermal conditions, which are affected by the physical dimensions and thermal properties of the heater [1]. Therefore, a study on the boiling heat transfer from a heater placed in an asymmetric heating condition must recognize the effect of the peripheral wall conduction on the overall heat transfer rate.

Pool boiling heat transfer has been studied extensively for many years. The effects of fluid and thermal properties, of surface finish and coating, of orientation and geometry of the heater(s), of agitation of the working fluid, of the force field, etc., have been investigated and a large number of correlations have been proposed. Many of the existing results on supposedly identical phenomena are inconsistent or differ widely with each other.

It is obvious that to compare the experimental results obtained by different investigators, all parameters governing the heat transfer process should be set equal. Seldom included is the effect of the variation of the surface temperature, which is dependent on the Biot number, and the specific heat generation rate of the heater. A few studies on boiling heat transfer indirectly recognize this variation of the surface temperature on the surface heat transfer coefficient.

Kovalov et al.[2], Bereson et al. [3] and Magrini et al. [4], all investigated the effects of the thickness and materials, used for the wall surface coating of the test heaters, on the overall heat transfer coefficient in nucleate pool boiling regions: none have measured the circumferential variation of the wall surface temperature.

Sauer et al. [5], and Jensen et al. [6], in their study on the effect of oil in nucleate pool boiling, have observed some significant variations of the circumferential wall temperature and attempted to explain the uncertainty in their experimental results via this variation. They concluded that the circumferential wall conduction could have a significant effect on both the local and average heat transfer coefficient in the nucleate region.

No studies on film pool boiling heat transfer which recognize the effect of this peripheral wall conduction on heat transfer seem to exist.

An analysis may be possible for the local boiling heat transfer coefficient in the film pool boiling region but in the regions of nucleate and transition pool boiling this is not feasible for the moment. Thus, an empirical approach to the problem seemed to be appropriate.

In the present study, a non-dimensional parameter, $K^* = K_f R / K_w b$ which was deduced from the governing energy differential equation [1], has been used to characterize the peripheral wall heat conduction of the heater.

EXPERIMENTAL:

Apparatus:

The experimental apparatus consisted primarily of a vertical boiling vessel of dimensions 420 mm x 280 mm x 380 mm containing an electrically heated horizontal test heater. The working fluid was 99% pure commercial grade freon-113.

Four test cylinders (direct electrical resistance heating) made of stainless steel 304 with three different wall thicknesses (1.24, two of 1.65 and 2.76 mm, respectively), one made of monel 400 (b = 2.82 mm)and the other inconel 600 (b = 2.28 mm), were used to vary the values of the parameter K* between 0.014 and 0.051. The outside diameters of the test cylinders were kept at about 25.4 mm to maintain a constant heater diameter and blockage ratio so that the hydrodynamic or fluid flow pattern around the test-section would be identical for each test heater.

A number of K-type (chromel-alumel) thermocouples were spot welded onto the center of the inner wall periphery of the test heaters. Several additional thermocouples were positioned on the inner wall 13 to 26 mm away from the center of the test heater to check the uniformity of the temperature distribution. One stainless steel 304 test heater had its thermocouples installed using the "split junction" method [1] to test the accuracy of the spot welded thermocouples. Two welding methods showed almost no difference in reading wall temperature. The circumferential wall temperature distribution was measured at every 30° interval for one half side and every 60° for the other half side. All thermocouples were calibrated in situ.

Great care was taken to maintain each test heater surface at the same roughness by polishing it with grit sizes 80, 180 and 320 silicon-carbide emery paper, respectively.

The test heaters were heated by a 15 KVA, 1200 Amp A.C. power supply. The electrical power input to the test heater was measured directly with two leads embedded in the cylinder 50.5 mm apart in the central portion by a digital voltmeter and an ammeter through the measuring circuits. This eliminated the uncertainty in estimating the electrical lead losses and thus little error was introduced in calculating the heat flux. The probable maximum error in the wall heat flux was estimated to be about 4%.

Experimental Procedure and Data Acquisition:

All tests reported here were conducted at atmospheric pressure.

For nucleate pool boiling tests, the liquid preheated to a pre-set temperature in the storage tank was charged to a level of about 70 mm above the top of test heaters. The power to the heater was then turned on at a relatively high level for about half an hour to let the liquid boil off the dissolved non-condensible gases. The power to the heater was then adjusted to a desired level to start the test.

For film pool boiling tests, the power to the test heater was turned on at the start of each test and the wall temperature of the test heater was brought up to about 350°C. Tests were performed by introducing the working fluid at or near the saturation temperature into the test vessel.

All the test data was acquired and reduced through a data acquisition system consisting of: a Hewlett-Packard 9835A computer; 3455A Digital Voltmeter; 3495A Scanner; 7245A Printer/Plotter and 98305 Real Time Clock. The raw and reduced data are stored on tape. Hard copy was also available.

Because of circumferential heat conduction, even with electrical resistance heating, the condition of constant heat flux is no longer applicable. Therefore, the local heat transfer coefficients should be deduced from a relation [1] which could be obtained from an energy balance made on the element of the test heater.

Since the thermocouples were spot-welded onto the inside tube wall, the solution of the "inverse" heat transfer problem was required in order to obtain the outside surface temperature. The difference was, in general, negligibly small.

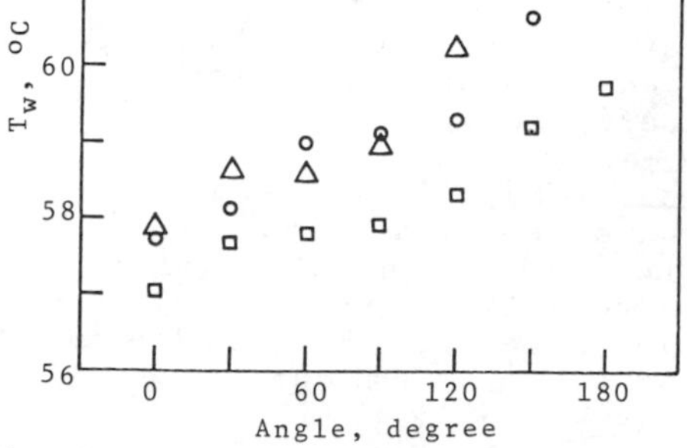

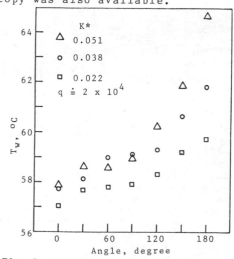

Fig. 1 Wall Temperature Distribution

RESULTS AND DISCUSSION:

An example of the circumferential surface temperature of the test cylinders made of s.s. 304, for three values of K*, is shown in Fig. 1. The heat flux was at about 2×10^4 $w/m^{2o}K$. Similar plots showed that there was no significant difference between temperatures at symmetric angles around the circumference.

A large value of K* implies a poor conductor and would add a large circumferential surface temperature variation. This was clearly seen in the figure.

Figs. 2(a) to 2(c) represent typical pool boiling test results obtained from three test cylinders, made of s.s.304, in the nucleate pool boiling region, while Figs. 3(a) to 3(c) show those in the film pool boiling region. As can been seen in the figures, the reproducibility of the test results is excellent.

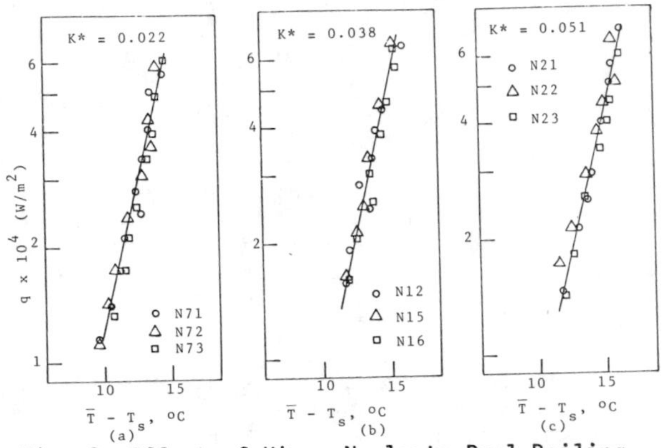

Fig. 2 Effect of K* on Nucleate Pool Boiling

The lines of the experimental results, correlated by least square regression analysis, are represented in Figs. 4(a) for the nucleate pool boiling region and in Fig. 4(b) for the film pool boiling region.

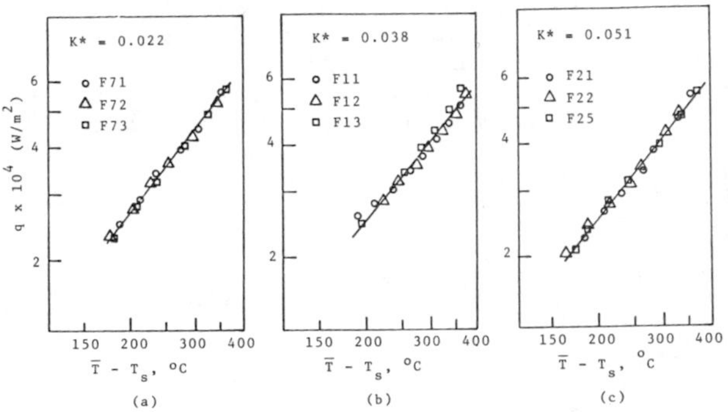

Fig. 3 Effect of K* on Film Pool Boiling

A large peripheral variation of the surface temperature is associated with a larger variation of the local heat transfer coefficient.

Therefore, if the value of the local heat transfer coefficient at the stagnation point for different values of K* are the same for a given test condition (which is a realistic assumption as can be deduced from Fig. 1), it may be concluded that a large value of the parameter K* would result in a smaller average heat transfer rate. This is confirmed by the present experiment as shown in Figs. 2 to 4.

The effect of the parameter K* seems to be less significant in the film pool boiling region as may be seen in Fig. 4(b). This is easy to understand as the Biot numbers involved are on the order of 10^{-2}.

The correlation of Stephan [7] is also plotted in Fig. 4(a) and those of Bromley [8], Sakurai et al. [9] and Breen and Westwater [10] are compared with the present results in Fig. 4(b). It can be seen that the Stephan correlation does not recognize the effect of K*. In the film pool boiling region, the empirical correlation proposed by Sakurai seems to correlate the present results the best.

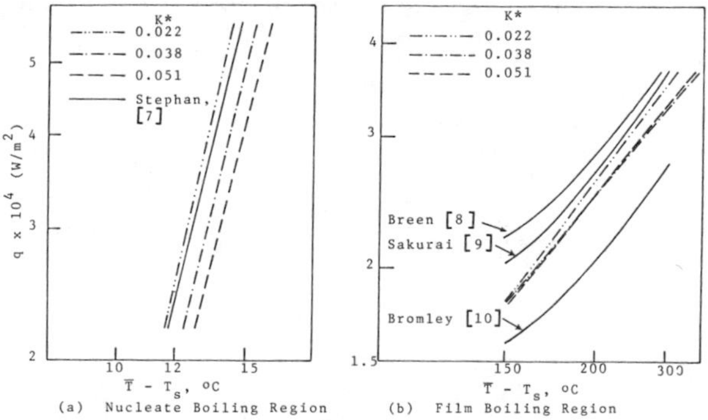

Fig. 4 Effect of K* on q

In Fig. 5(a), the results in the nucleate region were normalized using Stephan's correlation as the reference, whereas Sakurai's correlation for film pool boiling was used as the reference in Fig. 5(b).

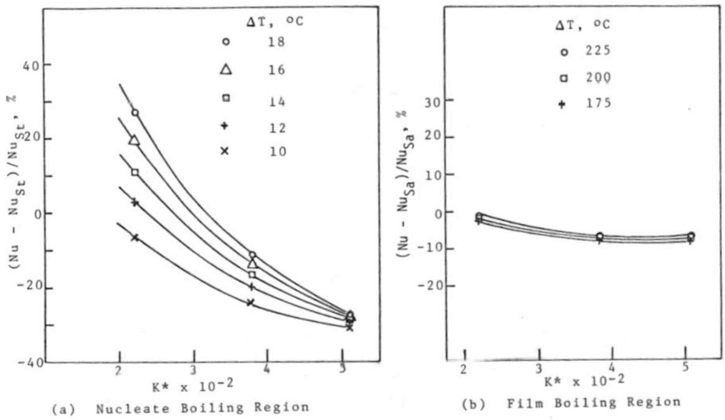

Fig. 5 Effect of K* on Nu

The added asymmetry of the thermal boundary conditions, due to the value of K*, affects the average heat transfer rate up to a maximum of about 60 to 70 percent for the ranges of K* and heat flux studied in the nucleate pool boiling region. This is illustrated in Fig. 5(a). However, as previously discussed the effect seemed to be negligible in the film pool boiling region; Fig. 5(b).

The finding is contrary to that which may be seen in forced convection heat transfer from a horizontal cylindrical heater in a single phase fluid cross flow. This may be because the heat transfer processes involved are not the same.

The present results are not compared with those of Kovalov et al. [2] and Magrini et al. [4] because our study was concerned with the Biot number of the test heater itself, while these investigators were interested in the thickness and material of the thin layer of the surface coating on a test cylinder.

The surface material has been known to affect the pool boiling heat transfer characteristics of many fluids. In the absence of adequate information on the interfacial energy conditions, an experimental approach is the only way to establish the relationship between the parameter K* and the surface material and their effect on heat transfer.

It can be seen in Fig. 6(a) and 6(b) that there is a significant difference between the heat transfer rates of the test heater made of s.s.304 and that of inconel 600, even though the values of K* are about the same. The effect is much more pronounced in the nucleate pool boiling region than in the film pool boiling region.

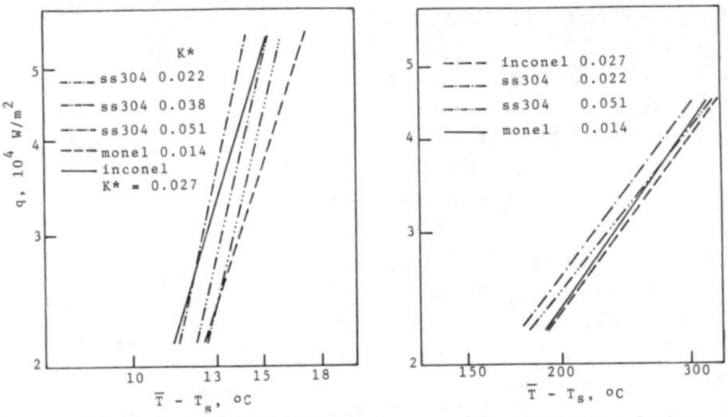

(a) Nucleate Boiling REgion (b) Film Boiling Region

Fig. 6 Effects of Surface Material and K* on q

The relationship between the parameter K* and the diameter of the heaters is to be investigated.

CONCLUSION:

The results from the present study lead to the conclusions that the parameter K* has a significant effect on the nucleate pool boiling heat transfer process, and that the many predictions available for nucleate pool boiling heat transfer cannot correlate pool boiling with different values of K*. In addition, the study also reconfirms that the relative magnitude of interfacial energies do significantly affect the pool boiling heat transfer process [11].

ACKNOWLEDGMENT:

The present study is financed by the Natural Science and Engineering Research Council of Canada under grant number A5175.

NOMENCLATURE:

b thickness of the test cylinder wall

D diameter of the test cylinder

F refers to Film boiling test run number in Fig. 3

K thermal conductivity

K* dimensionless parameter, $K_f R/K_w b$

N refers to Nucleate boiling test run number in Fig. 2

Nu Nusselt number, hD/K

q heat flux

R radius of of the test cylinder

T temperature

Subscripts:

ave average

f fluid

s saturation

St refers Stephan's correlation [7]

Sa refers Sakurai's correlation [9]

w wall

REFERENCES:

1. Lee, Y. and Kakade, S.G., Effect of Peripheral Wall Conduction on Heat Transfer from a Cylinder in Cross Flow, Int. Jour. of Heat and Mass Transfer, vol 19, pp. 1030-1038, 1976.

2. Kovalev, Zhukov, V.M., Kazakov, G.M. and Kuzmakichta, Effect of Coating with Low Thermal Conductivity upon Boiling Heat Transfer of Liquid on Isothermal and Non-isothermal Surface, 4th Int. Heat Transfer Conf., B.1.4., Paris, 1970.

3. Bereson, P.J., Experiments on Pool Boiling Heat Transfer, Int. Jour. Heat Mass Transfer, vol. 5, pp. 985-999, 1962.

4. Magrini, U. and Nannei, E., On Influence of the Thickness and Thermal Properties of Heating Walls on the Heat Transfer Coefficients in Nucleate Pool Boiling, Jour. of Heat Transfer, vol. 97, pp.173-178, 1975.

5. Sauer, H., Gibson, R.K. and Chongrungreong, S., Influence of Oil on the Nucleate Boiling of Refrigerants, 6th th Inter. Heat Transfer Conf., vol. 1, pp. 181-186, Toronto, 1978.

6. Jensen, M.K. and Jackman, D.L., Prediction of Nucleate Pool Boiling Heat Transfer Coefficients of Refrigerant-Oil Mixtures, Jour. of Heat Transfer, vol. 106, pp. 184-190, 1984.

7 Stephan, E. and Abdelsalam, M., Heat Transfer Correlations of Natural Convection Boiling, Int. Jour. Heat Mass Transfer, vol. 23, pp. 73-87, 1980.

8 Bromley, L.A.,Heat Transfer in Stable Film Boiling, Chem. Eng. Prof., vol. 48, p. 221, 1950.

9 Sakurai, A., Shiotsu, M. and Hata, K., Film Boiling Heat Transfer on Horizontal Cylinders (II), (in Japanese), 21st Nat.ional Heat Transfer Symposium of Japan, pp. 466-468, Kyoto, 1984.

10 Breen, B.P. and Westwater, J.W., Effect of Diameter of Horizontal Tubes on Film Boiling Heat Transfer, Chem. Eng. Prog., vol. 17, oo.367-377, 1974.

11 Lee, Y., Effect of Surface Materials on Pure Mercury Boiling Heat Transfer, Int. Jour. of Heat and Mass Transfer, vol. 17, pp. 376-377, 1974.

CONDENSATION HEAT TRANSFER

Effect of Drainage Strips on the Condensation Heat Transfer Performance of Horizontal Finned Tubes

H. HONDA and S. NOZU
Department of Mechanical Engineering
Okayama University
Okayama, Japan

INTRODUCTION

Experimental studies have shown that horizontal finned tubes realize condensation heat transfer enhancement in excess of the surface area increase due to finning [1∿9]. The enhancement is produced by combined gravity and surface tension forces which act to drive the condensate on the fin surface into the fin root. On the other hand, the surface tension acts to retain the condensate between fins on the lower parts of horizontal finned tubes [7,10,11], which results in a decrease in effective surface area. The latter effect is marked for fluids having a large value of surface tension to condensate density ratio σ/ρ.

Recently, attempts have been made to further enhance condensation heat transfer on horizontal finned tubes by use of drainage strips fitted to the tube bottom [7,12]. Honda et al. [7] showed that a porous drainage strip yields a marked increase in the average heat transfer coefficient due to a considerable decrease in the amount of retained condensate. Yau et al. [12] reported similar effects produced by a solid drainage strip. In these studies, however, no theoretical consideration was given to the effect of the drainage strip.

In the present work, the effects of solid and porous drainage strips with various heights on the condensation heat transfer performance of horizontal finned tubes are studied experimentally. An analytical model for film condensation on a horizontal finned tube reported in a previous work [13] is extended to include the effect of the porous drainage strip. The model is verified by comparison with the previous [7] and present experimental data.

ANALYSIS

Figure 1 shows the physical model and coordinates. A saturated vapor condenses on a horizontal low integral-fin tube fitted with a porous drainage strip at the tube bottom. The angular coordinate ϕ is measured from the tube top, and the coordinate x is measured downward from the tube bottom. The angle ϕ_f denotes the flooding angle below which the inter-fin space is almost filled with the retained condensate. The angular portitions $0 \leq \phi \leq \phi_f$ and $\phi_f \leq \phi \leq \pi$ are termed the u-region (unflooded region) and the f-region (flooded region), respectively.

In the u-region, the condensate on the fin surface is driven by combined gravity and surface tension forces into the fin root and is drained by gravity. In the f-region where only the fin tip is effective for condensation, the condensate on the fin tip is driven by surface tension force into the retained condensate. At the tube bottom, the condensate is drawn by combined gravity and capillary forces

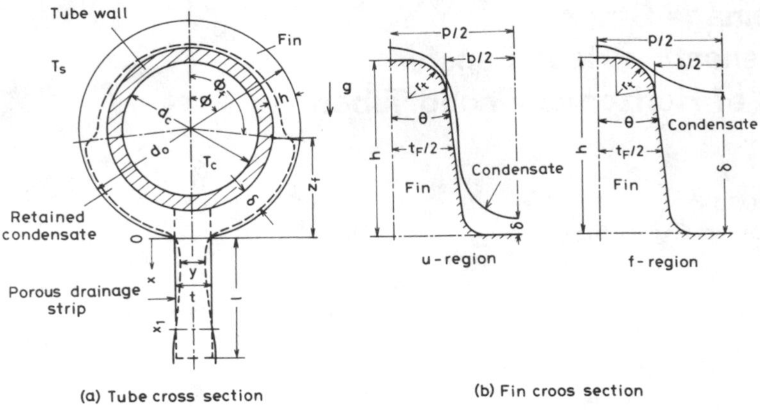

FIGURE 1. Physical model and coordinates.

in the porous drainage strip. The condensate flows through the porous drainage strip, overflows the strip at $x = x_1$ and finally drains off. The pressures of both phases below $x = x_1$ are assumed to be the same.

It is reasonable to assume that the mechanisms of condensate flow and heat tran fer in the u- and f-regions are basically the same as those for the case of a finned tube without the drainage strip. Consequently, following the theory developed in ref. [13], the average Nusselt number may be written as

$$Nu = \{Nu_u \eta_u (1 - \tilde{T}_{wu})\tilde{\phi}_f + Nu_f \eta_f (1 - \tilde{T}_{wf})(1 - \tilde{\phi}_f)\}/\{(1 - \tilde{T}_{wu})\tilde{\phi}_f + (1 - \tilde{T}_{wf})(1 - \tilde{\phi}_f)\}$$

where Nu_i, η_i and $\tilde{T}_{wi}$ are the Nusselt number, the fin efficiency and the dimensionless average wall temperature at the fin root for region i (i = u, f), resp tively. The expressions for Nu_i, η_i and $\tilde{T}_{wi}$ are given in ref. [13] (small modification is required for Nu_f). Thus, Nu is obtained from equation (1) when ar appropriate expression for $\tilde{\phi}_f$ is given.

According to the analysis of retained condensate presented in ref. [7], the cap lary pressure difference between the surrounding vapor and the retained condens at $\phi = \phi_f$ is given by

$$\Delta P = 2\sigma\cos\theta/b$$

Since the pressures of both phases at $x = x_1$ are the same, ΔP should be equal t the sum of the static pressure difference of the retained condensate between ϕ π and $\phi = \phi_f$, ΔP_1, and the pressure rise of the condensate flow in the porous drainage strip between x = 0 and $x = x_1$, ΔP_2, i.e.

$$\Delta P = \Delta P_1 + \Delta P_2$$

The pressure difference ΔP_1 is written as

$$\Delta P_1 = \rho g z_f = \rho g d_o (1 + \cos\phi_f)/2$$

Assumig the Darcy's law to hold, the basic equation for the condensate flow through the porous drainage strip is written as

$$dP/dx = \rho g - \rho\nu v/K$$

The heat transfer rate to the tube is related to the condensate flow rate by the following equation

$$\pi d_o q = \rho h_{fg} v y \quad (6)$$

Eliminating v from equations (5) and (6), integrating the resulting equation from $x = 0$ to $x = x_1$, and introducing the assumption of $y \simeq t$ yield

$$\Delta P_2 = \int_0^{x_1} \rho(g - \pi d_o q \nu/\rho h_{fg} K y)dx \simeq \rho g(1 - F/K)x_1 \quad (7)$$

where $F = \pi d_o q \nu/\rho g h_{fg} t$. Since ΔP_2 is equal to the vapor to condensate pressure difference at $x = 0$, its maximum value ΔP_{2max} is given by the capillary pressure as

$$\Delta P_{2max} = 2\sigma/r_p \quad (8)$$

Substituting equations (2) and (4) into equation (3) yields the expression for $\tilde{\Phi}_f$ as

$$\tilde{\Phi}_f = \cos^{-1}G/\pi \qquad \text{for } -1 \leq G \leq 1 \quad (9)$$

where $G = (4\sigma\cos\theta/b - 2\Delta P_2)/\rho g d_o - 1$. It should be mentioned here that $\tilde{\Phi}_f = 0$ for $1 \leq G$ and $\tilde{\Phi}_f = 1$ for $G \leq -1$.

APPARATUS AND PROCEDURE

The experimental apparatus is shown in Fig. 2. A test tube (5) was fitted concentrically through a test section consisting of a 102 mm ID, 300 mm long horizontal glass cylinder (3) and brass end heads (4). Vapor generated in a boiler (1) flowed through a superheater (2) and a vapor supply tube (6) into the test section. Most of the vapor condensed on the test tube and the condensate returned by gravity to the boiler through a condensate measuring tube (7). To photograph condensate menisci between fins at an arbitrary angular position, a 35 mm camera (15) was mounted on a movable stand which rotated around the glass cylinder.

The geometrical characteristics of the test tubes are listed in Table 1. All the tubes were made of copper. Tube A was a smooth tube. Tubes B, C and D had low integral-fins. Tube E (Thermoexcel C) had saw-tooth fins. Tubes BS to ES had a 1.9 mm thick, about 14 mm high solid drainage strip made of polyvinyl chrolide. Tubes BP to EP had 1.9 mm thick, 4 to 19 mm high porous drainage strips made of nickel (Nickel Cellmet #5). The drainage strips were fitted to a 2 mm wide, 1 to 1.5 mm deep longitudinal slot cut along the bottom of tubes B to E.

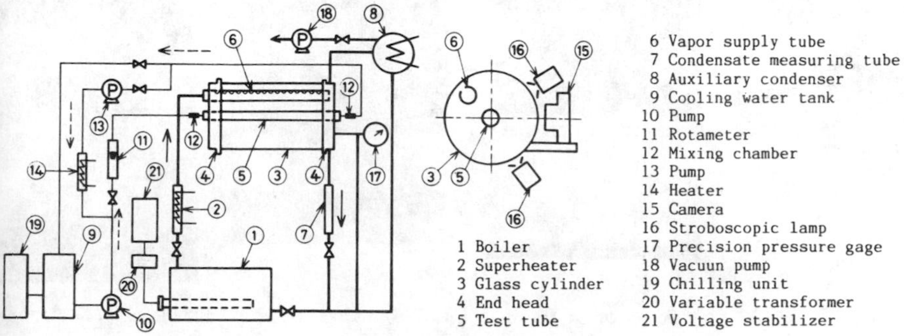

FIGURE 2. Schematic diagram of the experimental apparatus.

TABLE 1. Geometrical characteristics of test tubes.

Test series			A	B	C	D	E
Tube designation			A	B,BS,BP	C,CS,CP	D,DS,DP	E,ES,EP
Fin pitch	p	mm	---	0.98	0.64	0.50	0.72
Fin height	h	mm	---	1.46	0.92	1.13	1.03
Fin spacing	b	mm	---	0.74	0.46	0.39	---
Fin half tip angle	θ	rad	---	0.079	0.087	0.0	---
Outside diameter	d_o	mm	19.05	18.69	18.89	19.35	19.40
Inside diameter	d_c	mm	15.88	14.10	15.48	14.20	15.50
Actual area		m^2/m	0.060	0.190	0.197	0.294	---
Actual area/Nominal area			1.00	3.24	3.32	4.84	---
Effective length		mm	433	433	263	170	168

Methanol was used as a test fluid. Experiments were conducted at a constant saturation temperature of $T_s \simeq 343$ K. A more in-depth description of the apparatus and the test procedure is contained in ref. [7].

BEHAVIOR OF CONDENSATE

Figure 3 compares the side views of the series D tubes at $(T_s - T_{wm}) \simeq 8$ K. The figures suceeding to the tube designations denote the rounded values of l (in mr The light reflections at the condensate surfaces (indicated by arrows) show the positions where the condensate film thickness at the fin root increases abruptl The flooding point is defined as the mid-point of the two light reflections. I is apparent that the value of $\tilde{\phi}_f$ for tube DP is greater for larger l. The $\tilde{\phi}_f$ value for tube DS14 is a little greater than that for tube D.

The upper half of Fig. 4 shows the circumferential distributions of the normali condensate film thickness at the fin root δ/h for series B, D and E. The measu values of $\tilde{\phi}_f$ are also shown by vertical dotted lines. It is obvious for series B and D that $\tilde{\phi} = \tilde{\phi}_f$ corresponds approximately to the point at the maximum slope of the δ/h distribution. The value of $\tilde{\phi}_f$ increases as the height of the porous drainage strip increases. For tube BP with the largest value of p, $\tilde{\phi}_f \simeq 1$ is reached at $l \gtrsim 14$ mm. In contrast, the circumferential variation of δ/h for series E is rather gentle.

The lower half of Fig. 4 shows the circumferential distributions of $\tilde{T}_w$ for seri D, where the local values at 25, 45, 65 and 105 mm distances from the coolant i

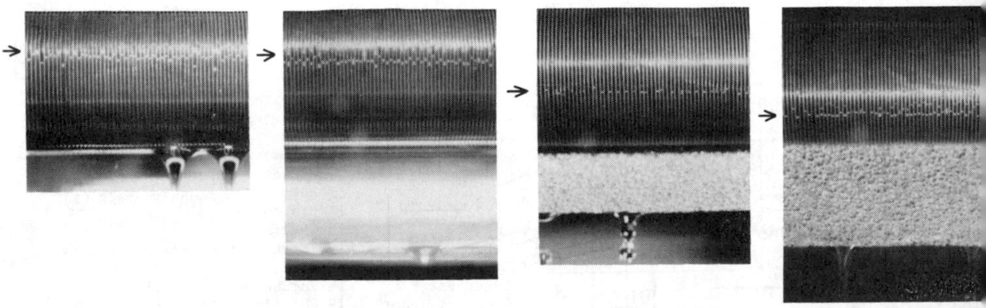

(a) D $\tilde{\phi}_f = 0.38$ (b) DS14 $\tilde{\phi}_f = 0.43$ (c) DP9 $\tilde{\phi}_f = 0.58$ (d) DP14 $\tilde{\phi}_f = 0.68$

FIGURE 3. Side views of Series D tubes during condensation, $(T_s - T_{wm}) \simeq 8$ K.

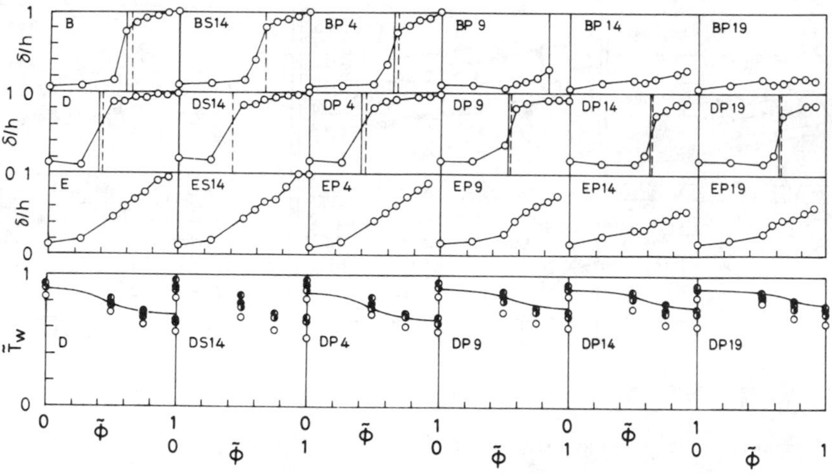

FIGURE 4. Circumferential distributions of δ/h and $\tilde{T}_w$, $(T_s - T_w) = 7 \sim 12$ K.

let are plotted with symbols ○, ◑, ◐ and ●, respectively. Due to the thermal entry length effect of the coolant, the local values at the first section are considerably lower than those at the downstream sections. A comparison of the δ/h and $(1 - \tilde{T}_w)$ distributions reveals a close correlation between the two. This can be attributed to the fact that $\alpha_u \gg \alpha_f$ [13].

Vertical solid lines and solid curves in Fig.4 show the theoretical predictions of $\tilde{\phi}_f$ and $\tilde{T}_w$, respectively (see also the following sections for the calculation method). These predictions are seen to agree well with the measured values.

DETERMINATION OF r_p, x_1 and K VALUES

To calculate $\tilde{\phi}_f$ using equation (9), the values of r_p, x_1 and K are yet to be determined. The value of r_p was determined by the capillary rise method as 0.43 mm. The value of x_1 is related to ΔP_2 by equation (7). Substituting q = 0 into equation (7) yields

$$\Delta P_{20} = \rho g x_1 \tag{10}$$

where ΔP_{20} denotes the limiting value of ΔP_2 at $q \to 0$. Eliminating x_1 from equations (7) and (10) yields

$$\Delta P_2/\Delta P_{20} = 1 - F/K \tag{11}$$

Thus, x_1 and K are obtained from the measured values of ΔP_2 and ΔP_{20} using equations (10) and (11). The values of ΔP_2 and ΔP_{20} were calculated from equation (9) using the measured values of $\tilde{\phi}_f$. These data were plotted on the coordinates of $\Delta P_{20}/\rho g$ versus 1 and $\Delta P_2/\Delta P_{20}$ versus F. From these plots, the values of x_1 and K applicable in the range of $\Delta P_{20} < \Delta P_{2max}$ and $\tilde{\phi}_f < 1$ were determined as $x_1 = 1 - 2.4$ mm and $K = 1.7 \times 10^{-9}$ m^2, respectively.

HEAT TRANSFER

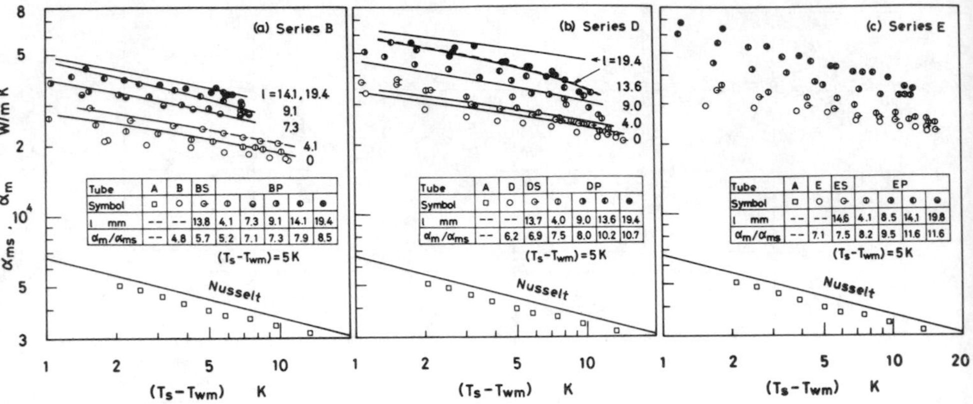

FIGURE 5. Comparison of average heat transfer coefficients for series B, D and E.

Figure 5 shows the average heat transfer results for series B, D and E, where α_m is plotted as a function of $(T_s - T_{wm})$. For comparison, the smooth tube (tube A) results, α_{ms}, the Nusselt equation for a horizontal tube with d_o = 19 mm and the values of the enhancement ratio α_m/α_{ms} at $(T_s - T_{wm})$ = 5 K are also presented. As expected, α_m increases as the height of the porous drainage strip increases. How ever, the difference between the l = 14 and 19 mm results is small as compared with the difference between the l = 9 and 14 mm results. This indicates the effe of capillary limit at large l. The enhancement produced by the solid drainage strip of l = 14 mm is comparable to that produced by the porous one of l = 4 mm.

In Fig. 5 the theoretical predictions of α_m are also shown by solid lines. These predictions are seen to agree well with the measured values except for the case

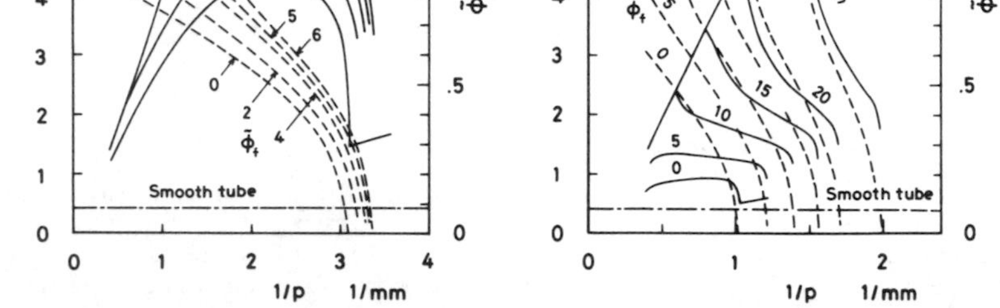

Copper tube, T_s = 35°C, T_c = 25°C, u_c = 2 m/s, d_o = 18.7 mm, d_c = 14.1 mm, h = 1 mm, t_F = 0.2 mm, r_t = 0.02 mm, $\theta = 0°$, t = 2 mm, r_p = 0.4 mm, $K = 1.7 \times 10^{-9}$ m^2

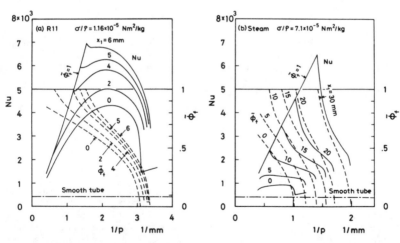

FIGURE 6. Predicted heat transfer performance.

of tube DP19, where the formers are higher than the latters. These results suggest that the assumed values of K and/or x_1 are not applicable to the case of $l = 19$ mm where ΔP_{20} is comparable to ΔP_{2max}. It is assumed here that K is not affected by l. The expression for x_1 at $\Delta P_{20} > \Delta P_{2max}$ is tentatively modified as $x_1 = \Delta P_{2max}/\rho g$. The dotted line in Fig. 5(b) show the theoretical prediction based on the modified expression for x_1. This line is seen to agree well with the measured values. It is mentioned here that the theoretical predictions agree within ±10 % with most of the present and previous experimental data for methanol and R-113.

Figure 6 presents illustrative examples showing the effects of x_1 and p on the predicted heat transfer performance of finned tubes, where Nu and $\tilde{\phi}_f$ are plotted as a function of 1/p with x_1 as a parameter. The Nu value for a smooth tube is also shown by a chain line. As a general trend observed, the Nu and $\tilde{\phi}_f$ values increase as x_1 increases. In the case of R-11 with small σ/ρ, the enhancement ratio is large even for a finned tube without the drainage strip ($x_1 = 0$). In the case of steam with large σ/ρ, on the other hand, the enhancement ratio is small for a finned tube without the drainage strip, and a remarkable heat transfer enhancement is attained by use of the porous drainage strip.

CONCLUSIONS

(1) The average heat transfer coefficient for a finned tube increases as the height of a porous drainage strip increases. The upper limit of the effective height is given by the capillary limit of the porous drainage strip.

(2) The heat transfer enhancement produced by the porous drainage strip is greater than that produced by the solid one. This is due to a greater vapor to condensate pressure difference at the tube bottom produced by the porous drainage strip.

(3) The average heat transfer coefficient for a low integral-fin tube with the porous drainage strip can be predicted by equation (1). The value of $\tilde{\phi}_f$ in equation (1) is obtained from equations (7) and (9). The expressions for Nu_i, η_i and $\tilde{T}_{wi}$ (i = u, f) in equation (1) are basically the same as those for a finned tube without the drainage strip.

(4) Further experimental studies are required to get a better understanding on the mechanism of condensate flow through the porous drainage strip.

NOMENCLATURE

- b = fin spacing at fin tip
- d_c = inner tube diameter
- d_o = tube diameter at fin tip
- G = dimensionless parameter, $(4\sigma\cos\theta/b - 2\Delta P_2)/\rho g d_o - 1$
- g = gravitational acceleration
- h = fin height
- h_{fg} = specific enthalpy of evaporation
- K = permeability
- l = height of drainage strip
- Nu = average Nusselt number, $\alpha_m d_o/\lambda$
- Nu_i = Nusselt number for region i, $\alpha_i d_o/\lambda$
- P = pressure
- ΔP_1 = static pressure difference between bottom and top of retained condensate, equation (4)
- ΔP_2 = pressure rise of condensate flow in porous drainage strip, equation (7)
- ΔP_{2max} = maximum capillary pressure, equation (8)

ΔP_{20} = limiting value of ΔP_2 at $q \to 0$
p = fin pitch
q = average heat flux based on nominal surface area
r_t = radius of curvature at corner of fin tip
r_p = effective pore radius of porous drainage strip
T_c = coolant temperature
T_s = saturation temperature
T_w = wall temperature at fin root
$\tilde{T}_w$ = dimensionless wall temperature, $(T_w - T_c)/(T_s - T_c)$
t = thickness of drainage strip
t_F = average thickness of fin
u_c = coolant velocity
v = condensate velocity in porous drainage strip
x = coordinate measured downward from tube bottom, Fig. 1
x_1 = overflow point on porous drainage strip, Fig. 1
y = condensate flow width in porous drainage strip
z_f = height of retained condensate, Fig. 1
α_i = heat transfer coefficient for region i
α_m = average heat transfer coefficient, $q/(T_s - T_{wm})$
α_{ms} = average heat transfer coefficient for smooth tube
δ = condensate film thickness at fin root, Fig. 1
η_i = fin efficiency for region i
θ = half tip angle of fin
λ = thermal conductivity of condensate
ν = kinematic viscosity of condensate
ρ = density of condensate
σ = surface tension
ϕ = angular coordinate measured from tube top, Fig. 1
$\tilde{\phi}$ = dimensionless angle, ϕ/π

Subscripts

f = flooding point; also flooded region
m = average value
u = unflooded region

REFERENCES

1. Beatty,K.O. and Katz,D.L., Chemical Engineering Progress, 44, 55-70(1948).
2. Pearsons,J.F. and Withers,J.G., ASHRAE Journal, 11, 77-82(1969).
3. Arai,N., Fukushima,T., Arai,A., Nakayama,T. and Fujie,K., Trans. ASHRAE, 83 (2), 58-70(1977).
4. Kisaragi,T, Enya,S., Oshiai,J., Kuwahara,K. and Tanasawa,I., in preprint of Japan Soc. Mech. Engrs., No. 780-1, 1-5, Tokyo(1978).
5. Carnavos,T.C., ASME Paper 80-HT-54, (1980).
6. Rudy,T.M. and Webb,R.L., AIChE Symposium Series, 79(225), 11-18(1983).
7. Honda,H., Nozu,S. and Mitsumori,K., in proceedings of ASME-JSME Thermal Engng. Joint Conference, 3, 289-295, Honolulu(1983).
8. Yau,K.K., Cooper,K.R. and Rose,J.W., Trans. ASME, J. Heat Transfer (in press)
9. Wanniarachchi,A.S., Marto,P.J. and Rose,J.W., ASME HTD, 38, 133-143(1984).
10. Owen,R.G., Sardesai,R.G., Smith,R.A. and Lee,W.C., Inst. Chem. Engrs. Symp. Ser., NO. 75, 415-428(1983).
11. Rudy,T.M. and Webb,R.L., ASME HTD, 18, 35-41(1980).
12. Yau,K.K., Cooper,J.R. and Rose,J.W., ASME HTD, 38, 151-156(1984)
13. Honda,H. and Nozu,S., ASME HTD, 38, 107-114(1984).

An Experimental Study of Potassium Vapor Condensation

R. ISHIGURO, K. SUGIYAMA, and F. TERAYAMA
Hokkaido University
Sapporo, Japan

ABSTRACT

The objective of the present study is to examine the behaviour of metal condensation including intensive condensation far from equilibrium. The condensation coefficient is found to be very close to unity. The general tendency of the present results on intensive condensation is in agreement with the most recent theoretical study.

INTRODUCTION

Over the past 30 years many studies on the condensation phenomenon of metal vapour have been conducted (see for example Ref. [1]). These studies were motivated partly by the need for design data for application in nuclear and space science. Besides these practical purposes, many researchers have been interested in the physical aspects of this problem, because the behaviour of metal condensation is different from that of nonmetal condensation.

In the case of condensation of metal vapour, owing to high thermal conductivity of condensate, an excess temperature drop at the vapour-liquid interface clearly occurs within a few mean free paths in width as a result of net transfer of matter, while for nonmetal vapours the liquid-vapour interface is almost at saturation temperature, except at very low pressures. This temperature drop at the interface can be large compared with that which occurs across the condensate film at temperature ranges used for practical purposes. This means that analyses of the Nusselt type are not applicable to predict the condensation rate of metal. Therefore, to clarify the characteristics of metal condensation, many experimental and analytical studies have been made, but without sufficient success.

The authors formerly reported experimental results of potassium condensation by using a test apparatus in which highly accurate measurements of condensation rate and condenser surface temperature were rendered possible [2, 3]. In those experiments, we observed an unexpected vapour temperature distribution in a duct of a test section which might affect the results to some degree. Another problem of the experiments was that the vapour pressure was directly evaluated from the measured temperature without examining the vapour state, as was done in all other previous works.

The objective of the present study is to examine the behaviour of metal condensation again, including the intensive condensation condition, with a more sophisticated apparatus improved in the two above-stated points. The relationship between condensation mass flux and the difference between vapour and liquid surface temperatures, and the dependence of this relationship on vapour temperature are reported in the present paper.

APPARATUS

The apparatus, in the form of a stainless steel closed loop, is shown in Fig. 1. The potassium vapour generated in the boiler was condensed on the vertical plane surface of the test condenser. The auxiliary condenser, located above the test condenser, can be used to prevent accumulation of non-condensing gas over the test condenser surface. The condensate from both condensers was returned to the boiler by gravity force. The inventory of potassium was 2.5 kgs. The heater capacity of the boiler was 20 kWs. An improvement was made in the configuration of the test section for the apparatus used before [3]. We observed an unexpected vapour temperature distribution in the test section of the previous apparatus. To clarify the cause of this distribution, a numerical analysis on the flow field of vapour was carried out. By this analysis, we concluded that the odd temperature (pressure) distribution was caused by the flow moving towards the downstream region of the test section, which could not be removed perfectly due to the restriction of the configuration in the previous apparatus. This phenomenon might have had some effect on the previous results. In order to correct the problem, an extension duct (150 mm long, 83 mm in diameter) was connected to the previous test section, and the test condenser block was welded onto the end of the extension. The test condenser, which was made of high-purity copper block with 30 mm-square condensing surface, was the same as that used in the previous study. Six thermocouple holes (0.35 mm square), which parallel to the condensing surface accurately determined, ran through the condenser block. Twelve sheathed thermocouples, 0.25 mm in diameter, were inserted into the holes from both sides to measure the surface temperature by extrapolation and the heat flux from the temperature gradient.

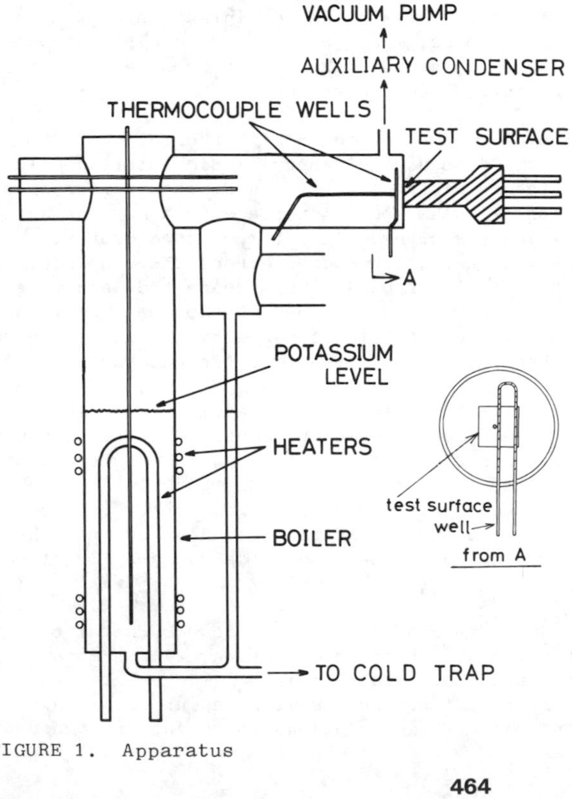

FIGURE 1. Apparatus

The temperature of potassium vapour was measured by the sheathed thermocouples inserted into the thermocouple wells of the thin stainless steel tube normal and parallel to the test surface, as shown in Fig. 1. In the previous studies, without examining the state of vapour, the vapour pressure was evaluated as the saturation state using the measured temperature. To remove this uncertainty, the saturation temperature corresponding to the pressure of the potassium vapour was measured by using the principle of the wet bulb thermometer in the present study. For this measurement a fine-mesh stainless steel net was rapped and spot-welded on the surface of the thermocouple well. This thermocouple well was cooled by flowing gas into it to make a wet condition by potassium condensate on the net. After the gas flow was stopped and a sufficient amout of time had elapsed, the temperature of the condensate on the well surface arrived at the saturation value corresponding to the pressure of the vapour. This temperature was measured by the thermocouples inserted into this wet thermocouple well.

Since a small amount of non-condensable gas in potassium vapour has a significant effect on condensation performance, special care was taken to ensure the vacuum tightness of the apparatus. The apparatus was welded completely by the TIG method, except for the connection between the test condenser block and the stainless steel flange. A helium leak detector with a sensitivity of 10^{-11} Pa m^3/s was used to check the air tightness of all parts of the apparatus. We then baked the apparatus up to 700 K for about 60 hours and obtained a pressure of about 10^{-6} Pa at room temperature. Finally, the apparatus was isolated, and we confirmed that the apparatus maintained a value less than 10^{-3} Pa after 15 hours.

Next potassium was charged from a cold trap vessel into the boiler. The cold trap was operated at a temperature of 350 K for about 6 hours for purification. During the experiment, the auxiliary condenser, located above the test condenser, was operated to prevent any non-condensable gas which might exist in the apparatus from accumulating on the condensing surface. The procedures of the experiment were almost the same as those reported previously [3].

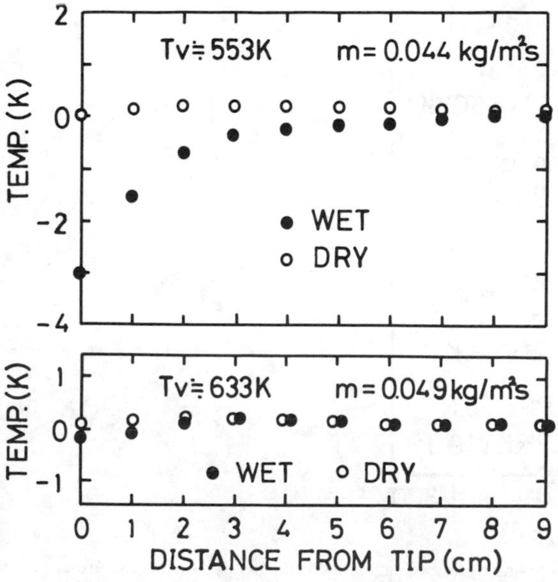

FIGURE 2. Vapour temperature variation along thermocouple well normal to condensing surface

RESULTS AND DISCUSSION

Vapour Temperature

Vapour temperature distributions for the highest T_v and also for the lowest one, measured along the thermocouple well normal to the condensing surface are shown in Fig. 2. It is seen that, for the lowest T_v (553K), the distribution measured by the wet thermocouple well is clearly different from that measured by the dry thermocouple well. This significant difference is explained as follows: The duct diameter of the test section in which the thermocouple wells were installed was 83 mm, while the condensing surface was 30 mm square. Therefore, acceleration of the vapour flow occurred as it approached the condensing surface of a smaller cross section, and this acceleration led to the pressure drop in this region. By using a wet thermocouple well, the temperature distribution corresponding to the local pressure could be measured. However, when a dry well was used as in the previous studies, a relatively higher temperature was obtained, because the vapour became superheated by the pressure decrease.

The difference between wet and dry conditions was small at the highest T_v (633 K). Since the vapour density increases with increasing vapour temperature (pressure), relatively smaller vapour velocity is required for the same condensation rate at high temperatures. Thus the pressure variation associated with the change in velocity which occurred in the vicinity of the test surface was correspondingly small at 633 K. This is the reason why the difference between the wet and the dry conditions at vapour temperature 633 K is relatively small

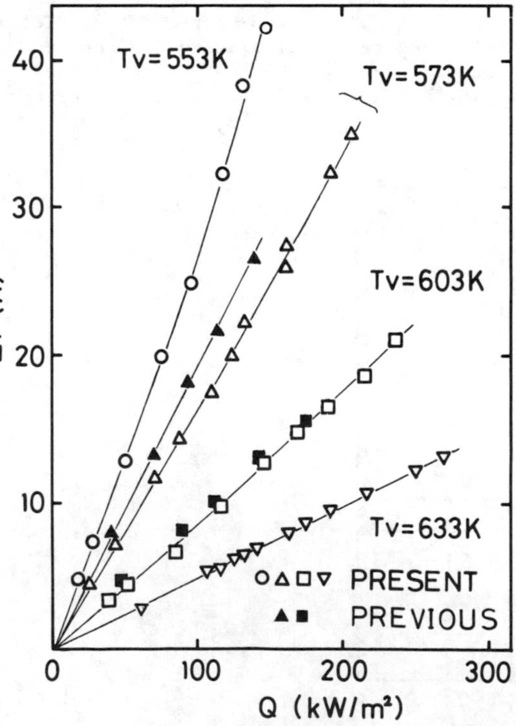

FIGURE 3. Dependence of vapour-to-surface temperature difference on heat flux

compared with the case of 553 K.

Comparison with Our Earlier Work [3]

Fig. 3 shows the relationship between the vapour-to-surface temperature difference and the heat flux at various vapour temperatures. Almost all the experimental results were observed on straight lines through the origin, and the gradient was strongly dependent on the vapour temperature. Some previous results by the present authors are also plotted in this figure. In the case of the vapour temperature of 573 K, a difference between the present results and the previous ones is observed. This difference may be attributed to the configuration of the test section and to the use of the dry thermocouple well in the previous experiment. However, since the difference between them is small at 603 K, it is obvious that large error was not caused at the measurement made at higher temperature ranges. By this comparison, we found that the problems in the previous experiment did not interfere with the observation of the behaviour of the metal condensation.

Discussion

In order to clarify the behaviour of metal condensation, the results in Fig. 3 are rearranged as shown in Fig. 4. The abscissa is the condensation rate and the ordinate on the left side is ξ defined by the following equation:

$$m = \xi \frac{P_v - P_s}{\sqrt{RT_v}} . \tag{1}$$

The solid line is ξ obtained analytically by Labuntsov and Kryukov [4] summarized by the formula

$$m = \frac{1.67}{\sqrt{2\pi}} [1 + 0.515 \ln\{ \frac{P_v}{P_s} (\frac{T_s}{T_v})^{1/2} \}] \frac{P_v - P_s}{\sqrt{RT_v}} . \tag{2}$$

This formula, for σ taken as unity, was obtained allowing a tolerance of 5 % to their numerical result. Labuntsov and Kryukov used a model of one dimension which was divided into a "Knudsen region" and a "gas-dynamic flow region". The former, which exists in the vicinity of the condensing surface, was treated from the viewpoint of the molecular-kinetic theory and the latter on the basis of the continuum equations. In the present study, uncertainties of P_v and T_v arose from the pressure drop due to the converging flow towards the condensing surface especially at a low pressure and high condensation rate. Therefore, the measurement of pressure was made at a position of 4 mm from the condensing surface, where the pressure drop due to the converging flow is negligible. In the analysis of Labuntsov and Kryukov it was concluded that the pressure in the continuum region remained approximately constant. So, the measured pressure corresponds directly to the P_v in Eq. (2). The measured temperature also can be used without essential error, since T_v is included in the form of square root in Eq. (2). From this viewpoint, although exact one dimensional flow was not realized in the present experiment, the comparison of the results of the present experiment with the analysis of Labuntsov and Kryukov is adequate. The values of ξ in the present experiment are larger than those of Labuntsov and Kryukov at lower vapour temperatures, while the values are lower than their results at higher vapour temperatures. Although there is a small discrepancy between the present results and theirs, it can at least be concluded that the value of ξ increases with increasing the condensation rate and this tendency becomes outstanding as the

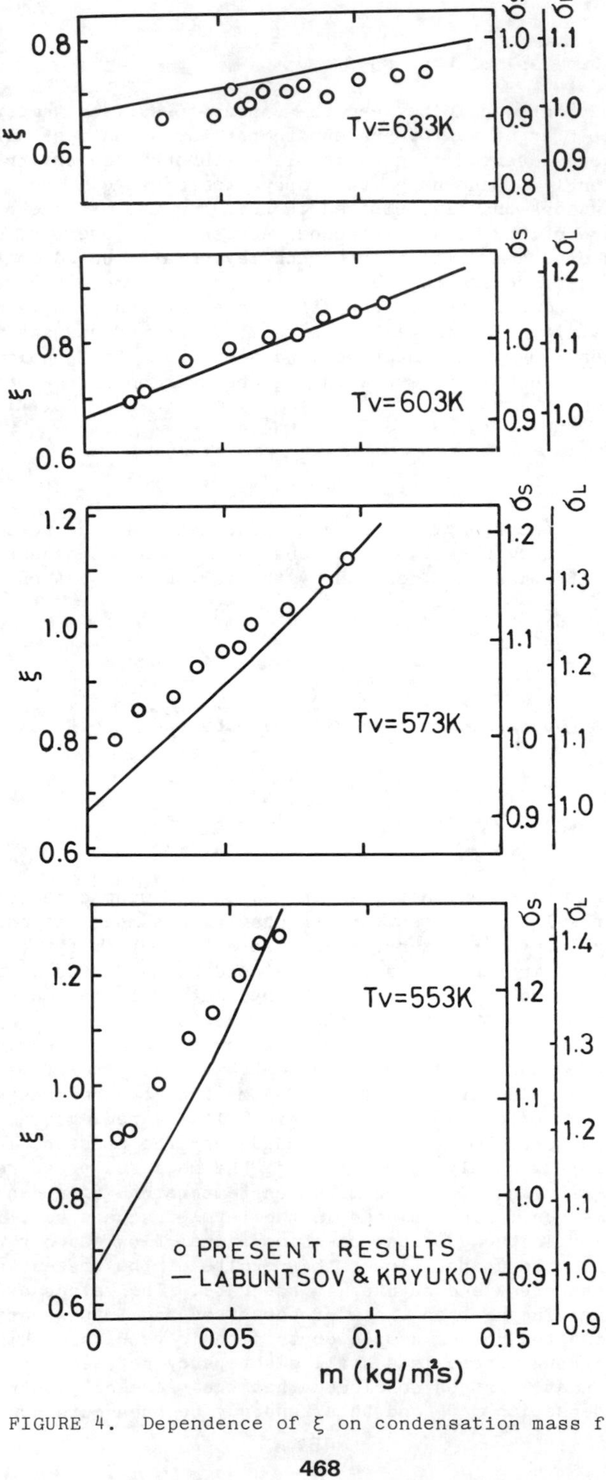

FIGURE 4. Dependence of ξ on condensation mass flux

vapour temperature decreases.

Previous experiments except ones by Rose and coworkers [5, 6] were made near the equilibrium condition, and many of them were compared with Schrage's theory [7] given by Eq. (3)

$$m = \frac{1}{\sqrt{2\pi}} \left(\frac{2\sigma_s}{2 - \sigma_s}\right) \frac{P_v - P_s}{\sqrt{RT_v}} . \tag{3}$$

In those comparisons the condensation coefficient σ_s was evaluated as being around unity at the vapour temperature range of the present experiment. Thus the conformity of the present results with previous ones must be examined. The ordinate of right side in Fig. 4 is the σ_s evaluated by Eq. (3). Extrapolation for $m \rightarrow 0$ of the present results at low vapour temperatures suggests a value of σ_s of around 1, while it indicates around 0.9 for higher T_v. Labuntsov [8] obtained the following equation for the condition of small departure from equilibrium:

$$m = \frac{1}{\sqrt{2\pi}} \left(\frac{2\sigma_\ell}{2 - 0.8\sigma_\ell}\right) \frac{P_v - P_s}{\sqrt{RT_v}} . \tag{4}$$

The term in the parentheses in Eq. (4) becomes 1.67 if σ_ℓ is taken as unity. The term in $\ell n\{P_v/P_s(T_s/T_v)^{1/2}\}$ on the right side in Eq. (2) approaches zero with decreasing condensation rate. So Eq. (4) is the same as Eq. (2) for the condition close to equilibrium. σ_ℓ evaluated by Eq. (4) is indicated for the second ordinate of the right side in Fig. 4. It can be seen that the value of σ_ℓ for $m \rightarrow 0$ approaches to around unity in the experiment of higher vapour temperatures.

Eq. (4) was deduced under the condition of small Knudsen numbers, while Eq. (3) was obtained for large Knudsen numbers. Although previous works were made in a relatively small Knudsen number region, many of them were compared not with Eq. (4) but with Eq. (3), and the results demonstrated that the condensation coefficient σ was close to unity at the temperature range of the present experiment. This point must be examined more closely in future experiments on the effect of non-condensable gas measured quantitatively. Measurements of metal condensation including the region far from equilibrium have been recently reported by Niknejad and Rose [6]. Comparison of the present results with theirs may not be proper because they used mercury in their experiment. However, the phenomenon that the coefficient ξ increases with increasing condensation rate was clearly observed in both studies. The major difference was that at lower vapour temperature the present result showed a higher value than that of Labuntsov and Kryukov, while the result of Niknejad and Rose was always lower than that of Labuntsov and Kryukov. This difference may be attributed to the measurement of vapour temperature. In their experiments, in addition to the uncertainty arising from the pressure drop in the converging flow toward the condensing surface, there is another uncertainty caused by the determination of vapour pressure without examining the vapour state. These problems are removed in our study.

CONCLUSION

An experiment of potassium vapour condensation, in which the problems encountered in the previous measurement are removed has been made for the vapour temperature from 553 K to 633 K. The following conclusions were obtained:

(1) By improving the method to measure the vapour pressure, the present value of ξ, which represents condensation characteristics of metal, becomes larger than that obtained in the previous measurement at lower vapour temperatures, while it

is almost the same value as the previous ones at higher temperatures.
(2) The condensation coefficient σ obtained in the present experiment lies between 0.9 and 1.0 according to Schrage's equation, and between 1.0 and 1.1 based on Labuntsov's equation.
(3) In the present experiment, it is confirmed that ξ values are dependent on vapour temperature and condensation mass flux, as observed in the experiment of mercury by Niknejad and Rose.
(4) The dependence of ξ on vapour temperature and condensation mass flux is generally in agreement with the analytical result of Labuntsov and Kryukov.

NOMENCLATURE

m = net mass flux, kg/km^2s
P_s = saturation pressure corresponding to condensate surface temperature, Pa
P_v = bulk saturation pressure of vapour, Pa
Q = heat flux in test condenser block, kW/m^2
R = specific ideal-gas constant, J/kg K
T = temperature, K
T_s = condensate surface temperature, K
T_v = bulk saturation temperature of vapour, K
$\Delta T = T_v - T_s$, K
ξ = coefficient defined by Eq. (1)
σ = condensation coefficient
σ_e = condensation coefficient defined by Eq. (4)
σ_s = condensation coefficient defined by Eq. (3)

REFERENCES

1. Rohsenow, W. M., Film Condensation of Liquid Metals, in Progress in Heat and Mass Transfer, vol. 7, ed. Dwyer, O. E., pp. 469-484, Pergamon, Oxford, 197[illegible]

2. Ishiguro, R., Sugiyama, K. and Hisamatu, T., A Study of Metal Vapour Condensation, in Heat Transfer in Energy Problems, ed. Mizushina, T. and Yang, W. J., pp. 93-98, Hemisphere, Washington, D. C., 1983.

3. Ishiguro, R., Sugiyama, K. and Hisamatu, T., An Experimental Study of Potassium Vapour Condensation, in Research on Effective Use of Energy, vol. 1, pp 151-158, Min. Educ. Sci. Cult., Tokyo, 1982.

4. Labuntsov, D. A. and Kryukov, A. P., Analysis of Intensive Evaporation and Condensation, Int. J. Heat Mass Transfer, vol. 22 no. 7, pp. 989-1002, 1979.

5. Necmi, S. and Rose, J. W., Film Condensation of Mercury, Int. J. Heat Mass Transfer, vol. 19, no. 11, pp. 1245-1255, 1976.

6. Niknejad, J. and Rose, J. W., Interphase Matter Transfer: An Experimental Study of Condensation of Mercury, Proc. R. Soc. London, vol. 378, no. 1774, pp. 305-327, 1981.

7. Schrage, R. W., A Theoretical Study of Interphase Mass Transfer, pp. 32-43 Columbia Univ., New York, 1953.

8. Labuntsov, D. A., An Analysis of Evaporation and Condensation Processes, Teplofiz. Vysok. Temper., vol. 5, no. 4, pp. 647-654, 1967.

Filmwise Condensation of Vapor Containing Noncondensable Gas in a Horizontal Duct

HIROAKI KUTSUNA
Kobe University of Mercantile Marine
Kobe, Japan

KIYOSHI INOUE
Samson Co., Ltd.
Kagawa, Japan

SHIGEYASU NAKANISHI
Department of Mechanical Engineering
Osaka University
Osaka, Japan

INTRODUCTION

Since classical analyses by Nusselt[1], filmwise condensation has been investigated theoretically by numerous researchers. Most of them were concerned with condensation in unconfined space. In practical condensers vapor condenses in confined spaces. As condensation proceeds, vapor flow decelerates and noncondensable gas in vapor is concentrated more and more. These effects must be included in theoretical analyses. For duct condensation of pure vapor, Shekriladze's theory [2] has been accepted to be useful. On the other hand, condensation under the presence of noncondensable gas has been analyzed by many investigators for unconfined spaces (for example, [3,4,5]). But there is no analysis which consider these two effects simultaneously. Moreover, few experimental data about these effects are available[6,7].

In the first part of this paper, we develop a theoretical analysis for condensation of vapor flow with noncondensable gas in a horizontal two-dimensional parallel duct. In the second part, we describe our experimental results and compare them with our theoretical results. In addition, we propose an approximate expression to facilitate practical calculations.

THEORETICAL ANALYSIS

The model considered in this analysis is shown in Fig. 1. Vapor containing noncondensable gas flows into a horizontal two-dimensional parallel duct with a uniform velocity u_0. The effect of gravity is eliminated by using horizontal arrangement. The bottom wall of the duct is cooled at a constant temperature T_w, while the top wall is adiabatic. Vapor condenses on the bottom wall and forms condensate film. Velocity, temperature and mass boundary layers develop on the film.

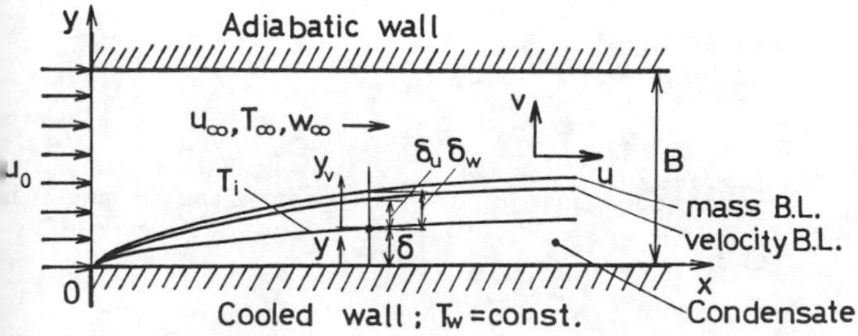

FIGURE 1. Analytical model

The assumptions and simplifications employed in our analysis are as follows:

1. Fluid properties are constant irrespective of temperature and concentration;
2. Boundary layer approximation is valid and the velocity and concentration pro files in gas phase are given by

$$u_v = u_i + (u_\infty - u_i)\left(\frac{2y_v}{\delta_u} - \frac{y_v^2}{\delta_u^2}\right), \quad (1)$$

$$w_{1v} = w_{1vi} + (w_{1v\infty} - w_{1vi})\left(\frac{2y_v}{\delta_w} - \frac{y_v^2}{\delta_w^2}\right), \quad (2)$$

   where u_v is vapor velocity, u_i interface velocity, u_∞ velocity of bulk flow, δ_u velocity boundary layer thickness, w_{1v} vapor concentration, w_{1vi} vapor co centration at interface, $w_{1v\infty}$ vapor concentration of bulk flow, and δ_w mass boundary layer thickness;
3. Condensate film is laminar and the convective term is negligible;
4. The thickness of the condensate film and boundary layers are much smaller th the duct height B, and the ratio of the boundary layer thicknesses $\xi = \delta_w/\delta_u$ constant throughout the duct;
5. The effects of the boundary layers over the adiabatic wall are neglected;
6. The bulk velocity, u_∞, can be replaced by the average vapor velocity, $\overline{u}_v$, an is much larger than the interface velocity u_i;
7. At the interface, the conduction term in the vapor phase is negligible;
8. The interface temperature, T_i, is constant throughout the duct and the conce tration of vapor in the bulk flow, $w_{1v\infty}$, does not change; and
9. The Shekriladze's approximation[2] holds, i.e.,

$$\tau_i = \dot{m}(u_\infty - u_i) \simeq \dot{m}\,\overline{u}_v, \quad (3)$$

where τ_i is the shear stress at the interface.
Under these assumptions, the basic equations commonly used (for example, [8]) ar solved simultaneously. According to assumption 3, the heat balance of the liqu film gives the condensing rate

$$\dot{m} = q_w/h_{fg} = \lambda_L(T_i - T_w)/(h_{fg}\,\delta). \quad (4$$

If T_i=constant along the duct, this equation means that $\dot{m}\delta$ is constant as well as Jakob number H_i, defined as

$$H_i = c_{pL}(T_i - T_w)/h_{fg}\ . \quad (5$$

The mass balance for the vapor phase is expressed as

$$\rho_v(B - \delta)\overline{u}_v = \rho_v B u_o - \int_o^x \dot{m}\,dx\ .$$

Taking account of $\delta \ll B$, we obtain

$$\frac{d\overline{u}_v}{dx} = -\dot{m}/(\rho_v B). \quad (6$$

It can be easily shown that the momentum equation for the vapor flow is given b

$$\frac{dP}{dx} = -\frac{\tau_i}{B} - \frac{d(\rho_v \overline{u}_v^2)}{dx}\ .$$

Substituting Eqs. (3) and (6), this equation becomes

$$\frac{dP}{dx} = \dot{m}\,\bar{u}_v/B\ . \tag{7}$$

The momentum equation for condensate film is given by

$$\mu_L \frac{\partial^2 u_L}{\partial y^2} - \frac{dP}{dx} = 0 \tag{8}$$

with the following boundary conditions:

$$\left.\begin{aligned} &\delta = 0 \quad \text{for} \quad x = 0\ ; \\ &u_L = 0 \quad \text{for} \quad y = 0\ ; \quad \text{and} \quad \mu_L\frac{\partial u_L}{\partial y} = \tau_i = \dot{m}\,\bar{u}_v \quad \text{for} \quad y = \delta. \end{aligned}\right\} \tag{9}$$

Equation (8) is easily integrated giving the solution

$$u_L = \frac{\dot{m}\,\bar{u}_v}{\mu_L}\, y(1 - \frac{\delta}{B} + \frac{y}{2B}).$$

According to assumption 4, this equation can be approximated with

$$u_L = \dot{m}\,\bar{u}_v\, y/\mu_L. \tag{10}$$

This means that the term dP/dx in Eq.(8) is neglected. According to Eq.(7), P varies very little along the duct, which agrees with our observation. The mass balance for the condensate film $\dot{m} = \frac{d}{dx}\int_0^x \rho_L u_L dy$ can be rewritten to the form

$$\frac{d}{dx}(\delta\,\bar{u}_v) = 2\nu_L/\delta. \tag{11}$$

In considering Eqs.(4),(6) and (10), this equation is transformed to

$$\frac{d\delta}{dx} = \frac{1}{\bar{u}_v}(2\nu_L/\delta + \frac{\dot{m}\,\delta}{\rho_v B}). \tag{12}$$

The second term in the parentheses on the R.H.S. of Eq.(12) cannot be neglected because it might be comparable with the first term.
As a result, we obtained a system of ordinary differential equations for $\bar{u}_v$ and δ, Eqs.(6) and (12). Boundary conditions are as follows:

$$\bar{u}_v = u_o \quad \text{and} \quad \delta = 0 \qquad \text{for} \quad x = 0. \tag{13}$$

Now, we turn to the vapor phase. The total mass balance for condensable substance 1 is given by [9]:

$$\mathcal{D}_{12}\left.\frac{\partial w_{1v}}{\partial y}\right|_i = \frac{d}{dx}\int_\delta^\infty u_v(w_{1v\infty}-w_{1v})dy + (w_{1v\infty}-w_{1vi})\frac{\dot{m}}{\rho_v},$$

where $\mathcal{D}_{12}$ is the diffusion coefficient. Substituting Eqs.(1) and (2) into the equation and executing integration of the R.H.S., we obtain

$$\frac{\mu_v}{S_{cv}\rho_v}\,\frac{w_{1v\infty}-w_{1vi}}{2\delta_w} = \frac{d}{dx}\{(w_{1v\infty}-w_{1vi})\Psi(\xi)\bar{u}_v\delta\} + (w_{1v\infty}-w_{1vi})\frac{\dot{m}}{\rho_v}, \tag{14}$$

where

$$\left.\begin{aligned} &\Psi(\xi) = \xi^2(\frac{1}{6} - \frac{1}{30}\xi) \quad \text{for} \quad \xi<1\ ; \\ &\Psi(\xi) = \frac{1}{3}(\xi - 1) + \frac{1}{\xi}(\frac{1}{6} - \frac{1}{30}\frac{1}{\xi^2}) \quad \text{for} \quad \xi>1, \end{aligned}\right\} \tag{15}$$

$S_{cv}=\nu_v/\mathcal{D}_{12}$ is the Schmidt number, μ_v the dynamic viscosity of vapor phase and ξ

the ratio of boundary layer thicknesses. As shown later, the interface concentra w_{1vi} can be taken constant. From this fact and Eqs.(5) and (11), we finally obta

$$4(P_{rL}/H_i)^2\{(\rho_v\mu_v/\rho_L\mu_L)\}\xi\Psi(\xi) + \xi - 1/S_{cv} = 0 \quad (16$$

where $P_{rL} = c_p\mu/\lambda$ is the Prandtl number for condensate.

Next, we find the expression for $\xi=\delta_w/\delta_u$. δ_u can be easily derived from Eqs.(1 and (3) as

$$\delta_u = 2\mu_v/\dot{m}. \quad (17$$

δ_w is derived from the mass balance of the condensate substance at the interfac [10]:

$$\dot{m} = \dot{m}\, w_{1vi} + \mathcal{D}_{12}\left(\frac{\partial w_{1v}}{\partial y}\right)_i,$$

or substituting Eq.(2) and rearranging, we obtain

$$\delta_w = (2\mu_v/S_{cv})(1/\dot{m})(w_{1v\infty} - w_{1vi})/(1 - w_{1vi}). \quad (18$$

Thus,

$$\xi = \delta_w/\delta_u = (1/S_{cv})(w_{1v\infty} - w_{1vi})/(1 - w_{1vi}). \quad (19$$

The concentration of condensable substance at the interface, w_{1vi}, is obtained from the phase equilibrium condition at the interface:

$$w_{1vi} = (P_{1vi}/P)(M_1/M_2)/\{1 - (P_{1vi}/P)(1 - M_1/M_2)\} \quad (20$$

where P is the total pressure, P_{1vi} the saturation pressure of substance 1 corr sponding to interface temperature T_i. M_1 and M_2 are the molecular weights of va and noncondensable gas,respectively. Hence, w_{1vi} is a function of T_i and, in t ξ is so. As H_i is also a function of T_i, we can find T_i by Eq.(16), iterativel Once we get T_i, it is straightforward to find δ and $\overline{u}_v$ by solving simultaneousl Eqs.(6) and (12) with Eq.(13).

Now, we rewrite these equations into nondimensional forms for generalization. Introducing the nondimensional velocity U, the film thickness D and the distanc

$$U = \overline{u}_v/u_o \ ; \ D = u_o\delta/\nu_L \ ; \ Z = u_o x/\nu_L, \quad (21$$

we get

$$dU/dZ = -Q_i/D \quad (22$$

$$dD/dZ = (1/U)(2/D + Q_i) \quad (23$$

with the boundary conditions:

$$U = 1 \text{ and } D = 0 \quad \text{for } Z = 0, \quad (24$$

$$\text{where } Q_i = \mu_L H_i/(P_{rL}\rho_v B u_o). \quad (25$$

Dividing Eq.(22) by Eq.(23), we obtain

$$dU/dD = -U/(D + 2/Q_i), \quad (26$$

which is integrated to

$$Q_i D/2 = 1/U - 1. \quad (27$$

Substituting Eq.(27) into Eq.(22) and integrating it, we get the following equation

$$Q_i^2 Z/2 = U - 1 - \ln U. \quad (28)$$

Equations (27) and (28) give the relationship between D and Z through U. It shou be noted that these equations are valid universally with or without noncondensabl gas. Its effect is included in evaluation of Q_i. The solution of Eqs.(27) and (28) can be approximated with the following expression

$$Q_i D = 2.30(Q_i^2 Z)^{0.5} \exp(0.58\, Q_i^2 Z). \quad (29)$$

The local heat flux, q_w, and heat transfer coefficient, α_x, can be calculated fro

$$q_w = h_{fg}\, \dot{m} = \rho_L u_o h_{fg} H_i/(P_{rL} D), \quad (30)$$

$$\alpha_x = q_w/(T_\infty - T_w) = \rho_L u_o h_{fg} H_i/\{(P_{rL} D)(T_\infty - T_w)\}. \quad (31)$$

By using Eq.(22), the average heat transfer coefficient, $\overline{\alpha}$, is given by

$$\overline{\alpha} = \int_0^x q_w dx/\{x(T_\infty - T_w)\} = h_{fg}\rho_v B u_o(1 - U)/\{x(T_\infty - T_w)\}. \quad (32)$$

To apply these results to the case without noncondensable gas, it is sufficient only to put $T_i = T_\infty$.

RESULTS AND DISCUSSION

Experimental Procedure

Experimental works were conducted on a test rig with a horizontal test section using the steam-air mixture under subatmospheric pressure. The test section was 10 mm in width, 30 mm in height and 500 mm in length and was cooled from the bottom side only. Local heat flux was determined from measurements of temperature distribution in the wall of the cooling surface, and average heat flux was from weighing total condensate in the test section.

Heat Transfer Coefficient

Figure 2 shows the distribution of local heat transfer coefficient along the duct in a typical run of pure steam. In the figure, prediction of our theory as well as those for semi-infinite space by several authors [2,11,12] are shown. Our prediction agrees well with the experimental data. The difference between our theory and the others means the deceleration effect.

The effect of noncondensable gas is shown in Fig. 3 for the inlet velocity $u_o = 6$ m/s. Plotted curves are our theoretical predictions and agree well with the experimental data. Good agreements are also obtained for other experimental condition.

In Fig. 4, we replotted the data in the form of average heat transfer coefficient $\overline{\alpha}$ versus air concentration C. Again the theoretical results agree well with the experimental results calculated from the condensate flow rates.

Nondimensional $Q_i D$ is plotted against nondimensional $Q_i^2 Z$ in Fig. 5 for the theoretical and experimental results, where $Q_i D$ and $Q_i^2 Z$ correspond to both the liquid film thickness and the distance from the leading edge, respectively, as seen from Eq.(21). $Q_i D$ increases exponentially with $Q_i^2 Z$ because the interfacial shear

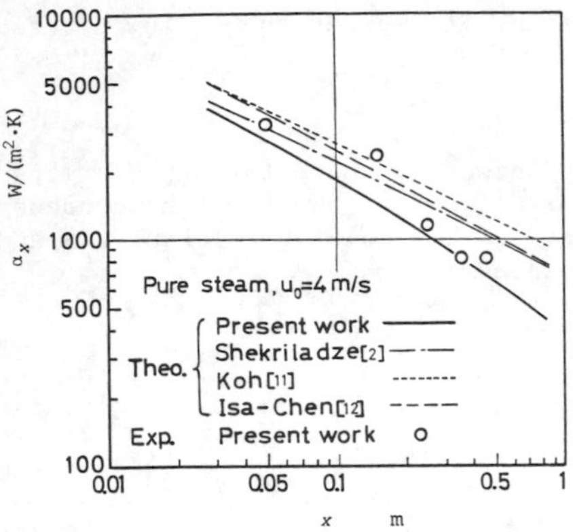

FIGURE 2. Local heat transfer coefficient for pure steam.

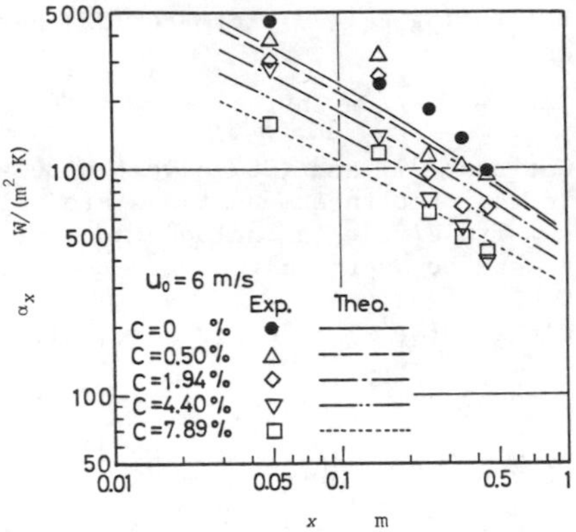

FIGURE 3. Local heat transfer coefficient for steam-air mixture.

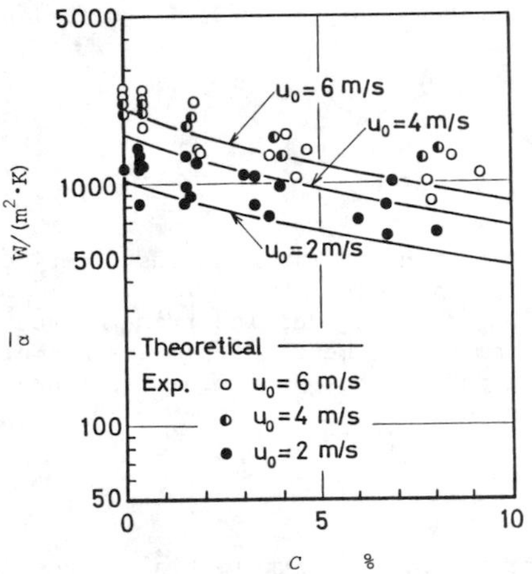

FIGURE 4. Average heat transfer coefficient.

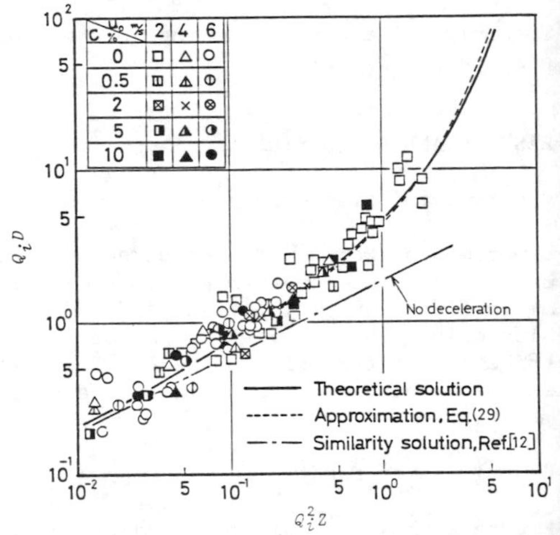

FIGURE 5. Nondimensional $Q_i D$ vs. nondimensional $Q_i{}^2 Z$.

stress decreases with the deceleration of bulk flow. For the filmwise condensation, $Q_i D$ represents the thermal resistance of the condensate film, and its reci rocal corresponds to the heat transfer coefficient. If there is no change in va por flow velocity, the slope in the log-log plot of $Q_i D$ versus $Q_i{}^2 Z$ is equal to 0.5, as shown by a chain line in Fig. 5. Comparison between two curves show tha the heat transfer coefficient with deceleration of the bulk flow becomes smaller than without deceleration, especially in large $Q_i{}^2 Z$ range. Experimental results are in good agreement with our theoretical result. The dotted line in the figur is the approximation, Eq.(29), and agrees well with our theoretical solution.

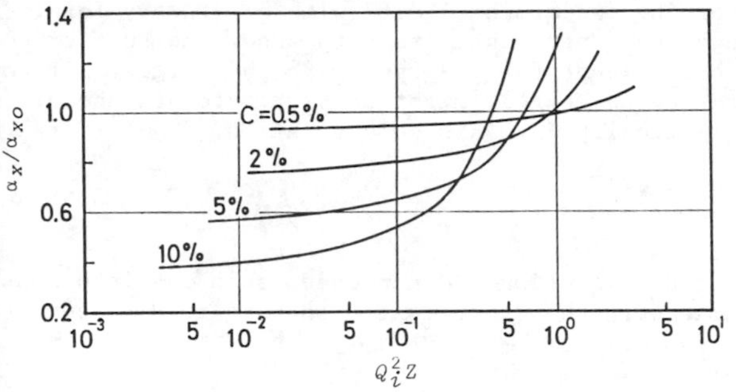

FIGURE 6. Theoretical curves α_x/α_{xo}.

Effect of Noncondensable Gas

The effect of noncondensable gas on the local and the average heat transfer coefficients can be seen from Figs. 3 and 4. Here, to clarify the characteristics of the depression of the heat transfer coefficient, the ratio of α_x to α_{xo} is examined, where α_x and α_{xo} are the local heat transfer coefficients for steam-air mixture and for pure steam, respectively. In Fig. 6, theoretical curves α_x/α_{xo} are shown for various air concentrations. The deceleration effect is eliminated by normalizing by α_{xo}. For the given air concentration, the value of α_x/α_{xo} increases with Q_i^2Z, and the larger the air concentration is, the more rapid the ratio rises. In the range with small values of Q_i^2Z, the ratio decreases with the increasing air concentration as predicted for the case of no deceleration[13]. On the contrary, in the range with large values of Q_i^2Z, the ratio with large air concentration becomes larger than unity and even the order of magnitude is reversed. This anomalous behavior is attributed to the presence of noncondensable gas. In one hand, the noncondensable gas reduces the rate of vapor diffusion, and on the other hand, it decreases the deceleration effect. These two effects work simultaneously. So, for smaller Q_i^2Z the former effect prevails, while for larger Q_i^2Z the latter does.

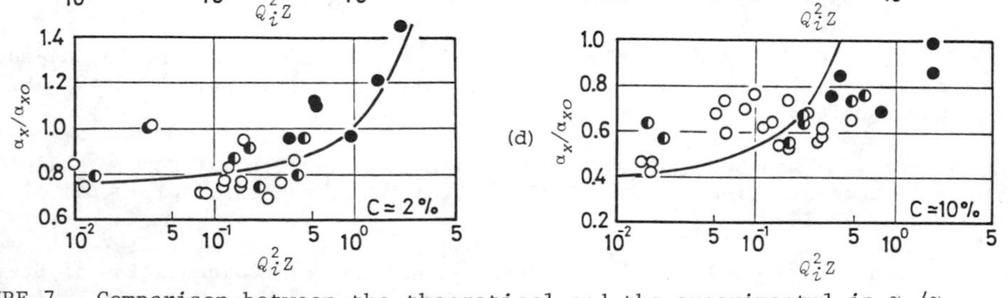

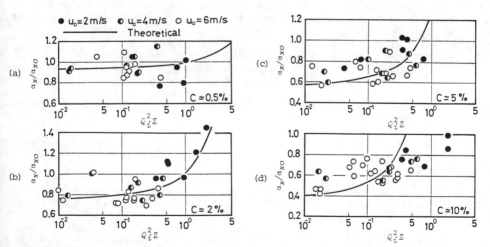

FIGURE 7. Comparison between the theoretical and the experimental in α_x/α_{xo}.
(a) C≃0.5% (b) C≃2% (c) C≃5% (d) C≃10%

Figure 7 shows the comparisons of the experimental data with our theory for various air concentrations. Although the experimental results show some scattering, the theory well represents our data except for the runs with the largest air concentration (C=10%,(d)); the disagreement partly might be attributed to the difficulty of experiments with unrealistically high air concentration.

CONCLUSIONS

An analysis taking the effects of deceleration and noncondensable gas into consideration was developed and compared with the experiments, showing good agreements Results obtained are summarized as follows:

1. The condensation process with deceleration is fully described by two nondimensionals Q_iD and Q_i^2Z irrespective of the presence of noncondensable gas;
2. The effect of noncondensable gas is represented by the nondimensional parameter Q_i;
3. The theoretical relation between Q_iD and Q_i^2Z can be approximated with

$$Q_iD = 2.30(Q_i^2Z)^{0.5} \exp(0.58Q_i^2Z);$$

4. For smaller Q_i^2Z the effect of noncondensable gas prevails, while for larger Q_i^2Z the deceleration effect does.

ACKNOWLEDGEMENT

We would like to thank Professor Koji Akagawa of Kobe University for his valuabl suggestion.

REFERENCES

1. Nusselt, W. Die Oberflächenkondensation des Wasserdampfes, VDI-Z, Band 60, Nr.27, S.541, 1916.

2. Shekriladze, I.G. and Gomelauri, V.I. Theoretical Study of Laminar Film Condensation of Flowing Vapour, Int.J. Heat Mass Transfer, vol.9, no.6, pp.581-591, 1966.

3. Minkowycz, W.J. and Sparrow, E.M. The Effect of Superheating on Condensation Heat Transfer in a Forced Convection Boundary Layer Flow, Int. J. Heat Mass Transfer, vol.12, no.2, pp.147-154, 1969.

4. Rose, J.W. Condensation of a Vapour in the Presence of a Non-Condensing gas, Int. J. Heat Mass Transfer, vol.12, no.2, pp.233-237, 1969.

5. Hijikata, K. and Mori, Y. Gyoshukukitai wo-fukumu Kitai no Heiban ni-sou Kyosei-Tairyu-Gyoshuku, Trans. J.S.M.E., vol.38, no.314, pp.2630-2640, 1972 (in Japanese).

6. Slegers, L. and Seban, R.A. Laminar Film Condensation of Steam Containing Small Concentrations of Air, Int. J. Heat Mass Transfer, vol.13, no.12, pp. 1941-1947, 1970.

7. Al-Diwany, H.K. and Rose, J.W. Free Convection Film Condensation of Steam in the Presence of Non-Condensing Gases, Int. J. Heat Mass Transfer, vol.16, no 7, pp.1359-1369, 1973.

8. Koh, J.C.Y. Laminar Film Condensation of Condensible Gases and Gaseous Mixtures on a Flat Plate, 4th USA Nat. Cong. Appl. Mech., vol.2, pp.1327-1336, 1962.

9. Eckert, E.R.G. and Drake, R.M., Jr., Heat & Mass Transfer, 2nd ed., p.463, McGraw-Hill, New York, 1959.

10. Bird, R.B., Stewart, W.E., and Lightfoot,E,N., Transport Phenomena, p.502, John Wiley & Sons, New York, 1960.

11. Koh, J.C.Y. Film Condensation in a Forced-Convection Boundary-Layer Flow, Int. J. Heat Mass Transfer, vol.5, no.10, pp.941-954, 1962.

12. Isa, I. and Chen C-J. Steady Two-Dimensional Forced Film Condensation with Pressure Gradients for Fluids of Small Prandtl Numbers, Trans. A.S.M.E., Ser. C, vol.94, no.1, pp.99-104, 1972.

13. Sparrow, E.M., Minkowycz, W.J., and Saddy, M. Forced Convection Condensation in the Presence of Noncondensables and Interfacial Resistance, Int. J. Heat Mass Transfer, vol.10, no.12, pp.1829-1845, 1967.

An Experimental Study of Condensation of Refrigerant 113 on Low Integral-Fin Tubes

H. MASUDA
Mitsubishi Electric Co.
Japan

J. W. ROSE
Department of Mechanical Engineering
Queen Mary College
(University of London)
London, UK

ABSTRACT

Vapour-side heat-transfer coefficients have been determined for condensation of refrigerant 113 on low integral-fin tubes. All tubes had a diameter at the fin root of 12,7 mm and had rectangular-section fins with height 1.59 mm and width 0.5 mm. **Fourteen different fin spacings between 0.25 mm and 20 mm were used. Measure**ments were also made using a plain tube having an outside diameter of 12.7 mm. All tests were made at near-atmospheric pressure with vapour flowing vertically downwards with a velocity of 0.24 m/s. The vapour-side heat-transfer coefficient enhancement was found to be only weakly dependent on temperature difference and to increase with decreasing fin spacing (increasing surface area) for tubes with fin spacings down to 0.5 mm. The performance of the tube with fin spacing 0.25 mm was, however, inferior to that with spacing 0.5 mm, despite having around 27% more surface area. For the best tube (fin spacing 0.5 mm) the vapour-side enhancement was around 7.3.

INTRODUCTION

Surface tension forces play an important role in the process of condensation on finned tubes. An early simple model for calculating the vapour-side heat-transfer coefficient [1], which did not include surface tension effects, has been found unsatisfactory as more experimental data have become available covering wider ranges of fluid properties and surface geometries. The deleterious effect of surface tension in giving rise to condensate retention on the lower part of a horizontal tube is well-understood and the extent of the 'flooded' region of the tube can be calculated with good accuracy [2-5]. The beneficial role of surface tension in effectively thinning the condensate film is understood in principle but theoretical approaches are, as yet, imperfect. Relatively simple models in which surface tension effects were partially accounted for [3,6-8] appear to be inadequate. Recent detailed treatments [9,10] are more realistic.

In parallel with the theoretical studies, attention has been devoted to experiments in which the important variables (geometry and fluid properties) have been systematically studied. In [5,11,12] accurate data have been obtained for steam covering, in all, two pressures, two tube diameters, four vapour velocities, two fin heights and thirteen different fin spacings. Some data are available for other fluids eg [1,13-15], but there is not yet a sufficiently comprehensive set for the purpose of proving theoretical models or developing semi-empirical general equations. The present paper reports new data for refrigerant 113 condensing on the same tubes as those used in [5,12] together with an additional tube with smaller fin spacing.

APPARATUS AND PROCEDURE

The apparatus, has been described in [5,12]. Vapour was generated in the stainless steel boiler fitted with four electric immersion heaters. The vapour flowed vertically downwards over the condenser tube. Cooling water was passed through the condenser tube via a float-type flow meter. The condenser tube and the inlet and outlet ducts were well insulated from the body of the test section and from the environment with nylon and ptfe components. Condensate from the tube and uncondensed vapour were led to the auxiliary condenser and the whole of the condensate was returned by the gravity to the boiler. The boiler, vapour supply duct and the test section were thermally well insulated from the surroundings. The temperatures at the test section, condensate return and cooling water inlet and outlet were measured using thermocouples which fitted tightly in closed metal tubes. The test section gauge pressure was measured with an R113 manometer.

The test condenser tubes were of copper with internal diameter 9.78 mm and length exposed to vapour 102 mm. The outside diameter at the root of fins was 12.7 mm, and the fin height and thickness were 1.59 mm and 0.5 mm, respectively. Fin spacings of 0.25, 0.5, 1, 1.5, 2, 4, 6, 8, 10, 12, 14, 16, 18 and 20 mm were used. All tests were conducted at near-atmospheric pressure. The apparatus was first run for around an hour to expel air and to achieve steady operating conditions. The condenser tube was viewed through a pyrex glass window. That the isothermal immersion of all thermocouples was adequate was checked during the operation by withdrawing the junctions from their 'pockets' by 1 or 2 cm. No change in the thermo-emfs was found. For each tube in turn measurements were made on at least two separate occasions.

RESULTS

The heat-transfer rate to the condenser tube was found from the mass flow rate and temperature rise of the cooling water. The vapour mass flow rate and hence velocity were obtained from a steady-flow energy balance between the boiler inlet and the test section. A small correction for the thermal losses from the apparatus was incorporated as indicated in [12]. The non-condensing gas (taken to be air) content was found from measured values of the test section temperature and pressure using the ideal-gas mixture laws and assuming saturation conditions. This was in all cases found to be less than the accuracy with which this quantity could be determined, i.e. the mass fraction of non-condensing gas was less than about 0.0005.

Overall Heat-Transfer Coefficients

Graphs of overall heat-transfer coefficient (based on plain tube area with diameter at fin root) versus coolant velocity are given in Fig.1. Data for higher coolant velocities for which the coolant temperature rise corresponded to a thermo-emf less than 15 μV were judged to be of marginal accuracy (precision of measurement 1 μV) and have been omitted from Fig.1.

The overall coefficient generally increases with decreasing fin spacing. The value of U at the smallest spacing (b = 0.25 mm) was, however, less than that for b = 0.5 mm. The fact that the overall heat-transfer coefficient for b = 8 mm appeared to be slightly lower than that for b = 10 mm is of interest. Although the effect is marginal for R113, similar, but more pronounced behaviour was found for steam [12], where maxima and minima occurred at fin spacings of 16 mm and 14 mm. This was thought to be associated with the fact that the "Taylor instability wavelength" was of similar magnitude to the fin spacing. For R113 under the present experimental conditions the value of this parameter (taken as $\pi(8\sigma/\rho g)^{1/2}$ [16]) is about 9 mm.

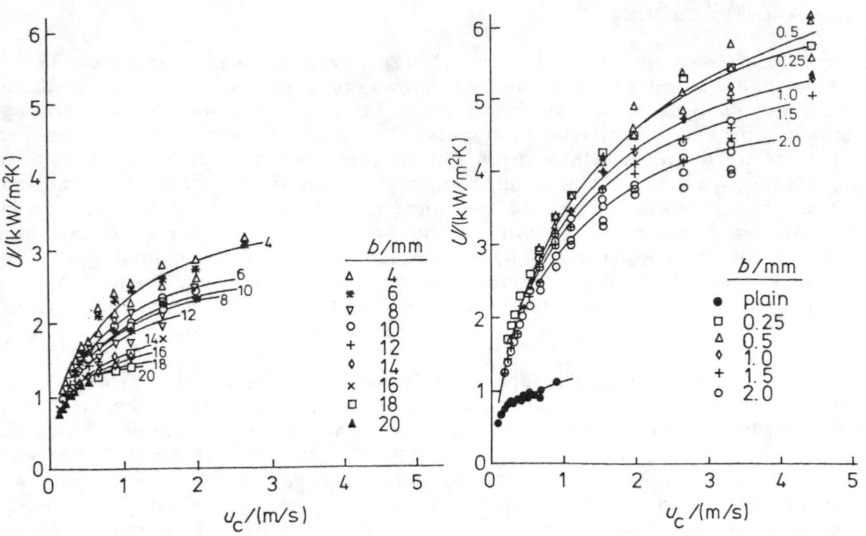

FIGURE 1. Dependence of overall heat-transfer coefficient on coolant velocity.

Vapour-Side Heat-Transfer Coefficients

In [5, 12] the vapour-side coefficient was found by subtracting the coolant-side and wall resistances from the overall resistance. The equation

$$Nu_c = 0.03\, Re_c^{0.8}\, Pr_c^{1/3}\, (\mu_c/\mu_w)^{0.14} \tag{1}$$

used to represent the coolant-side, was derived from earlier experiments in which steam was condensed on an instrumented plain tube. The tube wall was considered to extend from the inside surface to the root diameter of the fins and the tube wall temperature drop calculated on the basis of uniform radial conduction. The vapour-side coefficient thus found is based on the area of a plain tube with diameter equal to that at the root of the fins and includes the effects of both condensate and fins.

When the present data for R113 were treated in this way, the values of the vapour-side temperature difference at small heat flux appear to be somewhat too low, suggesting a finite heat flux at zero temperature difference (see Fig. 2). An alternative "modified Wilson Plot" method of treating the data was then adopted. This is similar to that used for a plain tube in [11]. Equation (2) below, in which *A* is an unknown constant, to be found, was used for the coolant-side and the tube wall again was treated as indicated above. For the vapour-side (including fins) a "Nusselt type" of equation (equation (3) below) was used, in which *B* is an unknown constant to be found. Note that *B* depends on geometry and fluid properties (including surface tension) but is constant for a given tube, fluid and vapour pressure. The use of the index 0.25, as in the Nusselt theory, is suggested by the results for steam [5, 11, 12]. Recent experimental and theoretical results [17] for R113 have also been found by the present authors to be well-fitted by equation (3).

$$Nu_c = A\, Re_c^{0.8}\, Pr_c^{1/3}\, (\mu_c/\mu_w)^{0.14}\ (2) \tag{2}$$

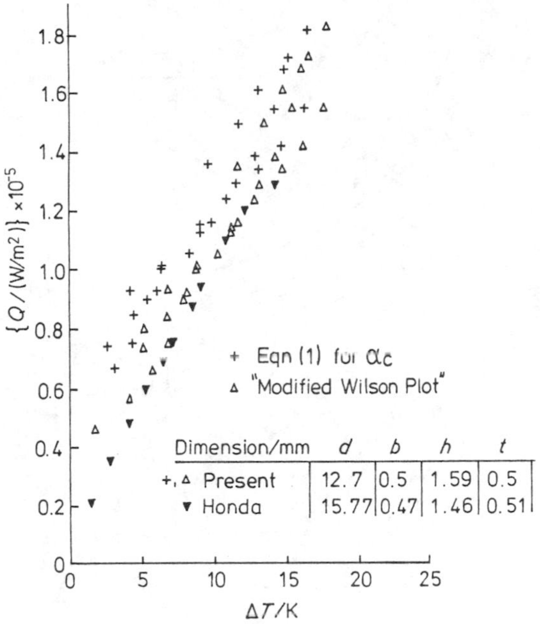

FIGURE 2. Comparison of methods of calculating vapour-side temperature difference and comparison with data of Honda [17] for condensation of R113 with similar tube and fin geometry.

$$Nu = B\left[\frac{\rho^2\, g\, h_{fg} d^3}{\mu\, k\, \Delta T}\right]^{1/4} \tag{3}$$

For each tube, data were obtained for between 12 and 20 coolant flow rates. These data were used to obtain A and B by minimisation of the sum of squares of residuals (measured minus calculated values) of the vapour-to-coolant temperature difference. An iterative procedure was used so as to evaluate the condensate properties at $0.3\,T_v + 0.7\,T_o$ and μ_w at T_w. h_{fg} was taken at T_v and the coolant properties at T_c. This process led to values of A higher than 0.03 (as used in [5, 12]) and generally in the range 0.035 to 0.04, indicating a smaller proportion of the total resistance on the coolant side and consequently higher values of vapour-side temperature difference ΔT. Values of ΔT were then found using the appropriate value of A and treating the data as before. Fig. 2 compares vapour-side results obtained by the two methods. The difference is not large but the "modified Wilson Plot" values appear more reasonable and are closer to data of Honda [17] for similar geometry. This method has been used to obtain the vapour-side data given in this paper. When the earlier steam data [5, 12] were treated in this way, values of A very close to 0.03 were found and the results obtained were virtually identical to those given earlier. It is considered that the effect of temperature variation around the tube (stronger for steam than R113) led to the difference in effective coolant-side resistance for the same coolant flow rate in the two cases.

Since the enhancement was much higher at the higher fin densities, only the data for those tubes with $b \lesssim 2$ mm are considered in more detail. $Q - \Delta T$ relationships determined as indicated above are shown in Fig. 3(a) together with the plain tube

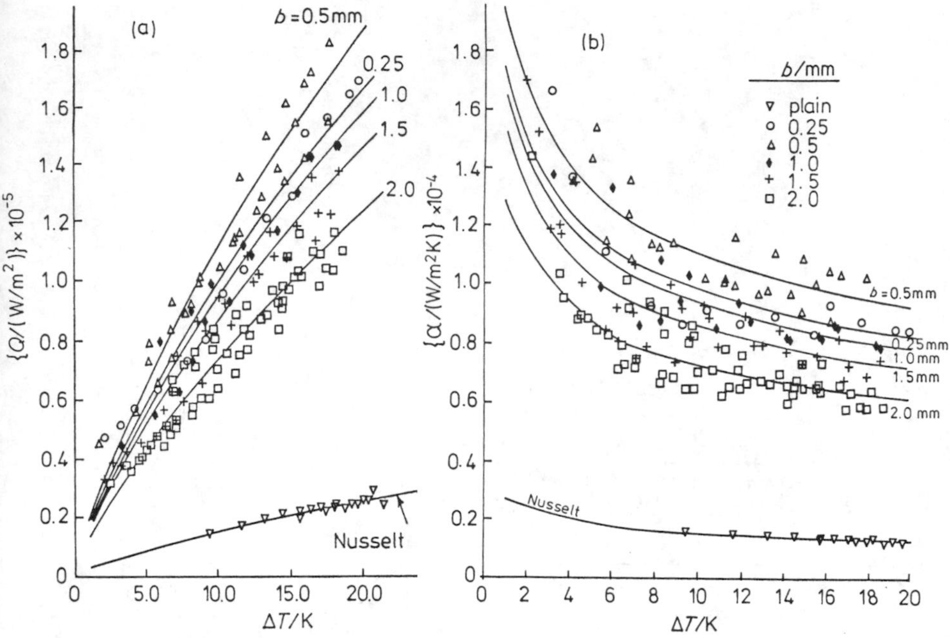

FIGURE 3. Dependence of heat flux and vapour-side heat-transfer coefficient on vapour-side temperature difference.

results. The lines through the data are given by equation (3) with the appropriate values of B. Fig. 3(b) gives the same results on a heat-transfer coefficient basis.

Vapour-side enhancement ratio

Enhancement ratios (vapour-side coefficient for finned tube/vapour-side coefficient for plain tube) may be given either for the same ΔT or for the same Q, and at a specified value of ΔT or Q. The fact that both finned and plain tube data may be represented by equation (3) enables specification of enhancement ratios independent of the values of ΔT and Q. Thus:

$$\epsilon_{\Delta T} = B_{\text{finned tube}}/B_{\text{plain tube}} \qquad \text{(for the same } \Delta T\text{)} \tag{4}$$

$$\epsilon_Q = \{B_{\text{finned tube}}/B_{\text{plain tube}}\}^{4/3} \qquad \text{(for the same } Q\text{)} \tag{5}$$

$$\epsilon_Q = \epsilon_{\Delta T}^{4/3} \tag{6}$$

Fig. 4 shows the dependence of $\epsilon_{\Delta T}$ on the fin spacing. Values of $\epsilon_{\Delta T}$ obtained from the earlier data using the same tubes for steam [5, 12] are also given in Fig. 4. It is seen that the enhancement ratios for R113 are appreciably larger than those for steam and that the peak occurs at a smaller fin spacing for R113. These findings would seem to be associated, in part, with the fact that the extent of condensate retention is significantly less for R113 than for water. The proportions of tube "flooded", indicated on Fig. 4, have been calculated using the analysis of Honda et al. [4] which gives, for parallel-sided fins:

$$\Phi = \cos^{-1}\left\{\frac{2\sigma}{\rho g b R} - 1\right\} \tag{7}$$

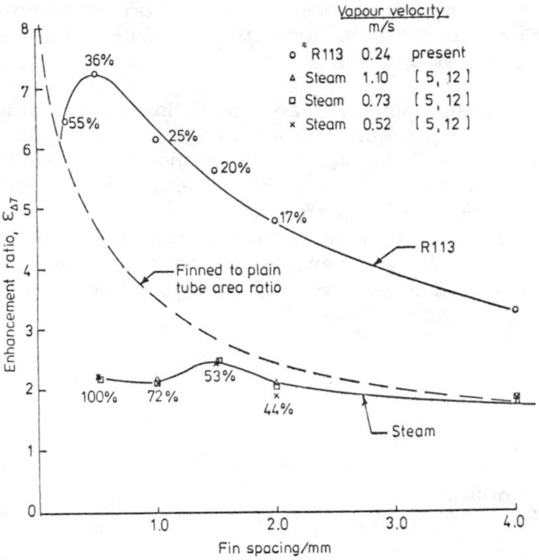

FIGURE 4. Dependence of enhancement on fin spacing for steam and R113. The numbers indicate the proportion of tube "flooded" according to equation (7).

It may be seen that peak enhancement for steam occurs when about half of the tube is "flooded", while this figure is around one third for R113. Though the enhancement curve for steam is rather flat, attention is drawn to the fact that data for three different vapour velocities are in good agreement and all indicate a maximum at a fin spacing of 1.5 mm. It is noteworthy that peak enhancement for steam has also been reported at a fin spacing of 1.5 mm in tests with a fin height and thickness of 1 mm, on a tube with fin root diameter 19 mm [11].

Comparison with theoretical models

The present results have been compared with the simpler models [1, 2, 3]. In all cases there were large discrepancies and systematic dependance on fin spacing; the quantity $\alpha_{cal}/\alpha_{obs}$ increased, for all three models, with decreasing fin spacing. The most satisfactory model was that of Rudy et al, [2] which underestimated the heat-transfer coefficient by about 40% at fin spacing 2 mm and overestimated by up to about 25% at fin spacing 0.25 mm. It may be noted, however, that Honda et al. [10] have indicated that this model gave, for steam, vapour-side coefficients lower than experimental values [5, 12] by a factor of 3.4. Detailed comparisons with the more complex theory of Honda et al. [10] have not yet been made. Provisional calculations supplied by Honda [17] indicate that this gives the correct general dependence on fin spacing and underestimates the present vapour-side coefficients for R113 by between 15% and 35%. Honda et al. [10] also report that their theory agreed with the steam data [5, 12] to within about 35%.

CONCLUDING REMARKS

The work reported above is part of a continuing investigation aimed at providing a satisfactory correlation for the heat-transfer coefficient for condensation on low-finned tubes. A systematic study of the effects of geometry and fluid properties is being conducted. Data have been obtained for steam [5, 12] and R113 covering a range of fin spacing but otherwise fixed geometry. Measurements are currently being made with ethylene glycol, which has surface tension and liquid thermal conductivity intermediate between those of water and R113.

A thorough study of the data has not yet been completed and the following remarks are tentative. Preliminary analysis of the results obtained so far seems to suggest that the higher enhancement for the lower surface tension fluids cannot be explained solely by the lesser degree of condensate retention. It thus appears that, for the 'unflooded' upper part of the tube, enhancement is also higher for lower surface tension fluids. Surface tension forces lead to thinning of the condensate film in some regions and to thickening in others. It has generally been supposed that non-uniformity of film thickness will always lead to an overall beneficial effect. The present tentative conclusions appear not to support this.

NOMENCLATURE

A	constant, see eqn. (2)
B	constant, see eqn. (3)
b	spacing between fins
c_{Pc}	isobaric specific heat-capacity of coolant
d	diameter at fin root, outside diameter of plain tube
d_i	inside diameter
g	specific force of gravity
h	height of fin
h_{fg}	specific enthalpy of evaporation
k	thermal conductivity of condensate
k_c	thermal conductivity of coolant
ℓ	length of condenser tube exposed to vapour
Nu	vapour-side Nusselt number, $\alpha d/k$
Nu_c	coolant-side Nusselt number, $\alpha_c\, d_i/k_c$
Pr_c	coolant Prandtl number, $\mu_c\, c_{Pc}/k_c$
Q	heat flux, $Q_T/\pi\, d\, \ell$
Q_T	Heat-transfer rate to coolant
R	radius of tube to fin tip
Re_c	coolant Reynolds number, $u_c\, \rho_c\, d_i/\mu_c$
T_c	mean coolant temperature, $(T_{in} + T_{out})/2$
T_o	effective outside surface temperature at fin-root diameter
T_v	temperature of vapour
T_w	inside wall temperature
T_{in}	inlet coolant temperature
T_{out}	outlet coolant temperature
U	overall heat-transfer coefficient
u_c	coolant velocity
α	effective vapour-side heat-transfer coefficient $Q/\Delta T$
α_{cal}	calculated vapour-side heat-transfer coefficient
α_{obs}	observed vapour-side heat-transfer coefficient
α_c	coolant-side heat-transfer coefficient, $Q(d/d_i)/\Delta T_c$
ΔT	vapour-side temperature difference, $T_v - T_o$
ΔT_c	coolant-side temperature difference, $T_w - T_c$
$\epsilon_{\Delta T}$	finned-to-plain tube heat-transfer coefficient enhancement ratio for same ΔT, see eqn. (4)
ϵ_Q	finned-to-plain tube heat-transfer coefficient enhancement ratio for same Q, see eqn. (5)
μ_c	coolant viscosity at T_c
μ_w	coolant viscosity at T_w
ρ	condensate density
ρ_c	coolant density
σ	surface tension
Φ	angle from top of tube at which interfin space becomes full of condensate

REFERENCES

1. Beatty, K.O. and Katz, D.L., Condensation of Vapors on Outside of Finned Tubes, Chem. Eng. Prog., Vol. 44, No. 1, pp. 55–70, 1948.

2. Rudy, T.M. and Webb, R.L., An Analytical Model to Predict Condensate Retention on Horizontal Integral-Fin Tubing, Proc. ASME-JSME Thermal Engineering Joint Conference, Honolulu, Vol. 1, pp. 373–378, 1983.

3. Owen, R.G., Sardesai, R.G., Smith, R.A. and Lee, W.C., Gravity Controlled Condensation on Horizontal Low-Fin Tube, Inst. Chem. Engrs Symp. Ser. 75, pp. 415–428, 1983.

4. Honda, H. Nozu, S. and Mitsumori, K., Augmentation of Condensation on Horizontal Finned Tubes by Attaching a Porous Drainage Plate ASME-JSME Thermal Eng. Conference, Vol. 3, pp. 289–296, 1983.

5. Yau, K.K., Cooper, J.R. and Rose, J.W., Effects of Drainage Strips and Fin Spacing on Heat Transfer and Condensate Retention for Horizontal Finned and Plain Condenser Tubes, Fundamentals of Phase Change: Boiling and Condensation, HTD-Vol. 38, Proc. ASME Winter Annual Meeting, New Orleans, pp. 151–156, 1984.

6. Rudy, T.M., Kedzierski, M.A. and Webb, R.L., Investigation of Integral-Fin-Type Condenser Tubes for Process Industry Applications, Inst. Chem. Engrs. Symp. Series 86, pp. 633–647, 1984.

7. Karkhu, V.A. and Borovkov, V.P., Film Condensation of Vapors at Finely Finned Horizontal Tubes, Heat Trans.-Sov. Res, Vol. 3, pp. 183–191, 1971.

8. Borovkov, V.P., Refined Method of Calculating Heat Exchange in Condensation of Stationary Vapor on Horizontal Finned Tubes, Inzhenerno-Fizicheskii Zhurnal, Vol. 39, pp. 597–602, 1980.

9. Adamek, T., Bestimmung der Kondensationgrossen auf feingewellten Oberflächen zur Auslegung optimaler Wandprofile, Wärme- und Stoffübertragung, Vol. 15, pp. 255–270, 1981.

10. Honda, H. and Nozu, S., A Prediction Method for Heat Transfer during Film Condensation on Horizontal Integral-Fin Tubes, Fundamentals of Phase Change: Boiling and Condensation, HTD-Vol. 38, Proc. ASME Winter Annual Meeting, New Orleans, pp. 107–114, 1984.

11. Wanniarachchi, A.S., Marto, P.J. and Rose, J.W., Filmwise Condensation of Steam on Externally-Finned Horizontal Tubes, Fundamentals of Phase Change: Boiling and Condensation, HTD-Vol. 38, Proc. ASME Winter Annual Meeting, New Orleans, pp. 133–141, 1984.

12. Yau, K.K., Cooper J.R. and Rose J.W., Effect of Fin Spacing on the Performance of Horizontal Integral-Fin Condenser Tubes, ASME, J. Heat Transfer, in press, 1985.

13. Pearson, J.F. and Withers, J.G. New Finned Tube Configuration improves Refrigerant Condensing, ASHRA Journal, pp. 77–82, 1969.

14. Katz, D.L. and Geist, J.M., Condensation on Six Tubes in a Vertical Row, Trans ASME, pp. 907–914, 1948.

15. Carnavos, T.C., An Experimental Study: Condensing R-11 on Augmented Tubes, ASME Paper 80-HT-54, 1980.

16. Yung, D., Lorenz, J.J. and Ganic, E.N., Vapor/Liquid Interaction and Entrainment in Falling Film Evaporators, ASME J. Heat Transfer, Vol. 102, pp. 20–25, 1980.

17. Honda, H., Private Communication 1985.

Heat Transfer of Condensation on Vertical Cosine and Semi-circle Type Corrugated Surfaces—Physical Models and Solutions by Finite Element Method

JUEMIN PEI, ZHANFENG CUI, XIUFAN GUO, and JIFANG LIN
Dalian Institute of Technology
Dalian, Liaoning, PRC

ABSTRACT

New physical models of condensation heat transfer on cosine and semi-circle type corrugated surfaces were presented. The numerical solutions to the mathematical models of hydrodynamic behavior have been made with finite element method (FEM), by which the two demensional velocity distribution and the flow rate of condensate have been calculated. Thus the local and average heat transfer coefficients of condensation were obtained.

The condensation heat transfer experiments were carried out on cosine and simi-circle type corrugated surfaces. The comparison of the calculated and experimental results showed that they are well agreed with each other. The theoretical and experimental results also showed that corrugated surfaces offer much higher heat transfer coefficients than the smooth surfaces, and the cosine-type corrugated tube is proved to be the best.

NOMENCLATURE

- a   area of heat transfer   m
- b   pseudo-film-thickness   m
- a_i, b_i, c_i parameters
- D   outer diameter of tube   m
- d   inner diameter of tube   m
- F   condensate flow rate in a single flute m^3/hr
- G   mass flow rate of condensate   kg/hr
- g   gravitational acceleration m/s^2
- J   functional
- H   depth of the groove   m
- Ki   matrix for one element
- K   heat conductivity of condensate   kcal/m·hr·°C
- P   pitch between two adjacent crests m
- Q   matrix for whole region
- Qi   matrix for an element
- q   heat flux $kcal/m^2 \cdot hr$
- Qi   matrix for an element
- T   temperature   °C
- u   velocity of flow   m/s
- X,Y,Z   coordinate system
- x_i, y_i coordinates of nodes
- ∂z local heat transfer coefficient $kcal/m^2 \cdot hr°C$
- ∂m average heat transfer coefficient $kcal/m^2 \cdot hr \cdot °C$
- Δt_m average temperature difference   °C
- m   average film thickness
- λ latent heat   kcal/kg
- ν kinetic viscosity
- ρ density kg/m^3

INTRODUCTION

The utilization of corrugated or longitudinal vertical finned tubes will provide enhancement for heat transfer processes[1,2,3]. The reason of the enhancement is, as Gregorig[4] reported, that the condensate is driven by surface-tension-induced forces from the ridge to the troughs leaving a very thin film on the crest with low thermal resistance.

Several geometrical profiles of corrugated tubes have be employed, such as V-type, rectangular type, cosine type and semi-circle type. These corrugated surfaces are able to enhance condensation heat transfer to varying extents.

Although the Gregorig's theory has been convicted by many author's experiments [5,6,7] and some physical models of several types of corrugated tubes have been presented [8,9,10,], there has not yet been a simple and general calculation method so far, with which condensation heat transfer on corrugated surfaces can be calculated. Paper [11][12] presented a physical model of condensation heat transfer on V-type corrugated tubes, in which the condensate film was divided into two parts: the region of crest for heat transfer and the region of valley for condensate draining, and the concept of pseudo-film-thicknees, b, was introduced. Navier-Stokes equation of the region of valley was solved by Finite Difference Method. The film-thickness-coefficient, $n=b/\delta_m$ was determined by experiments, so that the calculation method and equation for film coefficients was given. The calculated results and the experimental data are well in agreement.

Extending the method to corrugated tubes with cosine-curve and semi-circle configurations(Fig.1) physical models on these corrugated tubes were also presented. Navier-Stokes equation were solved numerically with Finite Element Method. Formulae for calculation of condensation heat transfer were also obtained.

PHYSICAL MODELS

The cross sections of cosine and semi-circle corrugated are showed in Fig.1. The geometry of tubes:

Pitch between two adjacent crests	:	P=1.73 mm
Height of groove	:	H=0.866mm
P/H ratio	:	P/H=2
Tube length	:	L=0.9 m
Number of vertical grooves	:	N=29
Outer diameter of tube	:	D=16 mm
Inner diameter of tube	:	d=9 mm
Tube material	:	copper

Take one single corrugation for analysis. The coordinate system is seen in Fig.2.Z axis is along the tube axis, with positive direction downward. The origion is at the top of the tube. X and Y corrdinates are in the horizontal plane. Curve ABC is the interface between the condensate film and the vapour. The region surrounded by curve ABCDA is condensate film. OB=b is defined as pseudo-film-thickness, b is an increasing function of Z,n is defined as film-thickness coefficient:

$$n = b/\delta_m \qquad (1)$$

where δ_m the average film thickness of heat transfer.

Assumptions:

1. There is only the flow vertically, and the horizontal flow is negligible.
2. Flow in the whole region of condensate film is steadily laminar.
3. The latent heat is transported by conduction through the liquid film.
4. The interface between vapour and condensate is assumed as a cosine-curve coherent with the corrugation of cossine-type. While for semi-circle surface, it may be assumed as a circular arc, with its center on the same axis, OY.
5. The temperature on the outer wall of the tubes are uniform.
6. Ignore the overheating of the vapour, subcooling of the condensate and the existing of shear stress.
7. The physical properties are constant.

SOLUTIONS TO THE MATHEMATICAL MODELS WITH FINITE ELEMENT METHOD

Let u be the velocity at any point in the condensate film, then Navier-Stokes equation may be simplified as the following Poisson equation:

$$\frac{\partial^2 u}{\partial x^2} + \frac{\partial^2 u}{\partial y^2} = -\frac{g}{\nu} \tag{2.a}$$

Because of the symmetry of the corrugation, only half of the corrugation is analysed, that is, the region surrounded by L1, L2, and L3 (Fig.2.)

For cosine curve, the equations of the curves are:

$$L1 : y=\frac{H}{2} - \frac{H}{2} \cos\frac{2\pi x}{p} \tag{3.a}$$

$$L2 : y=y, \quad x = 0 \tag{3.b}$$

$$L3 : y=\frac{H+b}{2} - \frac{H+b}{2} \cos \frac{2\pi x}{p} \tag{3.c}$$

For semi-circle curve, they are

$$L1 : y=\sqrt{2yH-x^2} \tag{4.a}$$

$$L2 : y = y, \ x = 0 \tag{4.b}$$

$$L3 : y=\left\{\left[\frac{2H^2+b^2-2Hb}{2(H-b)}\right]^2 - x^2\right\}^{\frac{1}{2}} + \frac{2H^2-b^2}{2(H-b)} \tag{4.c}$$

The boundary conditions for both the curves are:

$$u\big|_{L1}=0$$

$$\left.\frac{\partial u}{\partial n}\right|_{L_2+L_3} = 0 \tag{2.b}$$

Finite Element Method was used in numerical solution for the equation above due to the complication of the boundary conditions. The velocity distributions and the flow rates of the condensate film for b/H=0.1,0.2,0.3,...1.0 have been computerized with Finite Element Method.

(1). Discretization of the calculated region:

As shown in Fig.3, the region is divided into certain number of triangle elements, with certain number of nodes. Some nodes locate just at the bounderie Number the nodes and the elements. From now on, we just take the cosine curve as an example to explain the calculation process. The coordinates of the nodes for b/H=0.5 can be seen in Table 1.

(2) Derivation of element equation:

Analyse each element by functional analysis. Let the numbers of the nodes o a triangle element be 1,2,3, along counterclock-wise direction for the element itself; and be I,J,M for the over-all region. The area of the element is Ai.

The functional corresponding the element is

$$J_i = \frac{1}{2}\int_{A_i} (\nabla U)^2 dA - g/\nu \int_{A_i} U dA \tag{5}$$

where $\nabla U = \left(\frac{\partial u}{\partial x}, \frac{\partial u}{\partial y}\right)$

Eq.(5) can be changed into the matrix form as follows by linear interpolatio

$$J_i = \frac{1}{2} U_i^T K_i U_i - U_i^T Q_i \tag{6}$$

in which

$$U_i = \begin{pmatrix} u_I \\ u_J \\ u_M \end{pmatrix} = \begin{pmatrix} u_1 \\ u_2 \\ u_3 \end{pmatrix}_i \tag{7.a}$$

is matrix of velocities at nodes, while

$$K_i = \frac{1}{4\Delta} \begin{pmatrix} b_1^2+c_1^2 & b_2b_1+c_2c_1 & b_3b_1+c_3c_1 \\ b_2b_1+c_2c_1 & b_2^2 + c_2^2 & b_3b_2+c_3c_2 \\ b_3b_1+c_3c_1 & b_3b_2+c_3c_2 & b_3^2 + c_3^2 \end{pmatrix} \tag{7.b}$$

is a symmetrical matrix relating only to the geometry of the element.
And

$$Q_i = \begin{pmatrix} Q_1 \\ Q_2 \\ Q_3 \end{pmatrix}_i \tag{7.c}$$

$$(Q_j)_i = g/\nu \int_{A_i} f_j(x,y)\, dA \tag{7.d}$$

where
$$f_j(x,y) = (a_j+b_jx+c_jy)/\, 2\Delta \qquad (j=1,2,3) \tag{7.e}$$

$$\begin{cases} a_1 = x_2y_3 - x_3y_2 \\ b_1 = y_2 - y_3 \\ c_1 = x_3 - x_2 \end{cases} \tag{7.f}$$

Δ is the area of the triangle element (=Ai)

$$\Delta = \frac{1}{2}(x_2y_3 - x_3y_2 + x_3y_1 - x_1y_3 + x_1y_2 - x_2y_1) \tag{7.g}$$

(3). Formation of the assembling element equation:
The functional of the whole region equals to the summations of functionals of all the elements. Then

$$J = \sum_{i=1}^{N} J_i = \sum_{i=1}^{N} (\frac{1}{2} U_i^T K_iU_i - U_i^TQ_i) \tag{8}$$

Where N is number of elements (for the section of b/H=0.5, cosine corrugation, N=25). Eq.(8) can be changed into the matrix form [15] by certain algebric operation:

$$J = \frac{1}{2} U^T KU - U^TQ \tag{9}$$

where $U = (u_1, u_2 \ldots\ldots u_{NN})^T$

where NN is the number of nodes.

K is NN x NN order matrix and Q is NN row matrix (in fact, a vector).
For variational calculus, let $\delta J = KU - Q = 0$ (10)
Then we get an algebric equation group

$$KU = Q \tag{11}$$

(4) Accounting for the boundary condition:
Apply the boundary condition Eq. (2.b) to Eq.(11),

$$K\ U = Q \tag{12}$$

is given in which U is the matrix formed by unknown nodel velocities.

(5). Solution of the resulting system of equations:

After Eq.(12) has been solved [17,18], the values of velocities at the nodes are obtained. See Table 1.

(6). Calculation of flow rate:

The flow rate of the whole region equals the summation of those in all the elements. The flow rates of a single corrugation with various values of b/H are shown in Table 2.

Regress the data with least square method, the corrlation on the relationshi between flow rate, F, and pseudo-film-thickness, b, is given as following:

$$F = -0.177075b+1742.1b^2+4.80748x10^6b^3 \tag{13}$$

Similarly of course for semi-circle corrugation, we have another correlatior

The whole calculation was made by using TRS-80 micro-computer with FORTRAN language program.

The procedure of the program is shown in Fig.5.

BALANCE OF MATERIAL AND HEAT --- DETERMINATION OF BASIC CORRELATION

(1). The relationship between flow rate, F, and the vertical coordinate, Z.

As shown in Fig.6, the heat transfered through the differential area dA (with height dZ) is

$$dQ = \alpha_z \cdot \Delta t_m \cdot dA \tag{14.1}$$

And the condensate on the area

$$dF = \frac{dQ}{\lambda \cdot \rho} = \alpha_z \cdot \Delta t_m \cdot dA/\lambda\rho \tag{14}$$

where λ is the latent heat of the condensate.

ρ is the density of the condensate.

While $\alpha_z = K/\delta_m$ (15.1)

therefore $\alpha_z = \frac{nK}{b}$ (15)

let $dA = P \cdot dZ$ (16)

we will get

$$dF = \frac{nK}{b} \cdot \Delta t_m \frac{P}{\lambda\rho} \cdot dz \tag{17}$$

(2). The relationship between Z and b:

Differentiating Eq. (13), we obtain:

$$dF = (-4.91875x10^{-5}+0.9678333b+4006.2333b^2)db \tag{18}$$

Combining Eqs. (17) and (18), we will get:

Boundary condition

$$Z|_{b=0} = 0 \tag{20}$$

The correlation of Z will be obtained by solving Eqs.(19) and (20), as followin

$$Z = \frac{1}{n \cdot \Delta t_m} (1.854x10^{15}b^4+5.972x10^{11}b^3-4.553x10^7b^2) \tag{21}$$

DETERMINATION OF FILM-THICKNESS-COEFFICIENT, n, BY EXPERIMENTS.

In the experiments, steam condenses on the outside of the tube, that is on the crests of the corrugation. Measuring the flow rate of condensate, wall temperatures, the temperatures of the cooling water at the inlet and outlet, we can calculate average temperature difference, Δt_m, pseudo-film-thickness, b, and film-thickness-coefficient n.

The data are in Table 3. From the values calculated in Table 3, we get the mean value of n, that is, 22.2.

THE COMPARISON BETWEEN CALCULATED AND EXPERIMENTAL DATA

(1). Local heat transfer coefficient, α_z

The relationship between α_z and Z is shown in Table 4, with Δt_m = 8.58°C.

Fig.7 shows the curves of α_m against tube length, Z, α_z decreases from the top downwards. This is because of increase of the film thickness. Because the film thickness is thinner and heat flux is less when Δt_m is smaller than those when Δt_m larger, so the curves with small Δt_m are above the curves with larger Δt_m

(2). Average condensation film coefficients, α_m

Fig.8 is the curve of α_m against q, heat flux. As shown in Fig.8, the calculated values are well agreed with the experimental data.

(3). The comparison of effects of enhancement on different types of corrugated tubes.

Fig.9 shows the curves of the average condensation heat transfer coefficients on several kinds of corrugated tubes with the smooth tube of the same outer and inner diameters, effective length, value of P/H and same material. From the figure, we can see these corrugated tubes can enhance condensation heat transfer significantly, with the cosine-type corrugated tube the best.

CONCLUSION.

(1). The proposed models are recommendable for practical use, since they have been verified by experimental results.

(2). The Finite Element Method is more convenient and more accurate than the Finite Difference Method when it is used for this numerical solution.

(3). The enhancement of condensation heat transfer with the corrugated tubes employed is significant. The coefficients of condensation heat transfer on corrugated tubes with cosine configuration are about 5-10 times higher than that on smooth tube, and about 4-8 times higher with the semi-circle corrugation (Fig.9).

REFERENCES

1. A.E.Bergles; Enhancement of Heat Transfer, Proc. of the 6th Intern. Heat Transfer Conf. Vol.6,P.60 (1978)
2. R.L.Webb; Special Surface Geometries for Heat Transfer Augmentation, Development in Heat Exchanger Technology-1,P.179(1978)
3. W.Nakayama et al; Enhancement of Heat Transfer, Proc. Of the 7th Intern. Heat Transfer Conf., Vol.1,P.223(1982)
4. V.G.Gregorig; Zeit. Angew. Math. Phys., Vol.5,P.36(1954)
5. D.G.Thomas; Prospects for Furture Improvement in Enhanced Heat Transfer Surfaces, Desalination, Vol.12(2) P.189(1978)
6. J.F.Houle and W.T.Buhrig; Proc. of the 4th Intern. Symp. on Fresh Water from the Sea, Vol.1, P.313(1973)
7. M.Beccari and L.Spinosa, Proc. of the 4th Intern. Symp. on Fresh Water from the Sea, Vol.1, P.17(1973)
8. C.Ozgon and G.Somer; Proc.of the 6th Intern. Symp. on Fresh Water from the Sea, Vol,1, P.193(1978)
9. H.Yamamoto and T. Ishibachi, Proc. of the 6th Intern. Symp. on Fresh Water from the Sea, Vol.1, P.203(1978)
10. T.Fujii and H.Honda; Laminar Filmwise Condemsation on a Vertical Single-

Fluted Plate,Proc.of 6th Intern.Heat Transfer Conf. Vol.2,P.419(1978)
11. J.F.Lin,T.Q.Hsu and J,M.Pei; Heat Transfer of Condensation on a Vertical-Type Corrugated Tubes--A New Physical Model,Proc. of the 7th Intern. Heat Transfer Conf.,Vol.5,P.119(1982)

12. J.F.Lin,H.Q.Hsu,J.M.Pei and Yu Hong;Proc.of AIChE/CIESC Joint Meeting,Vol. 2,P.653,Sept. 19-22(1982)
13. S.S.Rao;The Finite Element in Engineering,Pergamon Press (1982)

Table 1. Velocity Distribution in the Film
(b/H=0.5,b=0.433 10 m)

Nodes	x (mm)	y (mm)	u (m/s)
1	0.000	0.000	0.0000
2	0.000	0.100	0.4884
3	0.000	0.250	0.9635
4	0.000	0.433	1.1678
5	0.100	0.028	0.0000
6	0.075	0.175	0.6251
7	0.200	0.250	0.5688
8	0.100	0.447	1.1424
9	0.200	0.109	0.0000
10	0.300	0.549	0.6209
11	0.300	0.350	0.4473
12	0.300	0.232	0.0000
13	0.400	0.624	0.3461
14	0.400	0.500	0.2988
15	0.400	0.381	0.0000
16	0.500	0,650	0.1836
17	0.475	0.499	0.0000
18	0.550	0.739	0.1120
19	0.550	0.611	0.0000
20	0.700	0.828	0.0168
21	0.635	0.723	0.0000
22	0.800	0.860	0.0002
23	0.700	0.790	0.0000
24	0.800	0.854	0.0000
25	0.866	0.866	0.0000

Table 3. Calculation of Pseudo-Thickness-Coefficients,n

Velocities of Cooling Water m/s	1.06	1.69	2.11	2.75	3.38	4.22	5.07	6.34
Ave.Wall Temp. °C	95.90	93.77	91.42	88.55	86.40	84.15	83.07	78.57
Δtm °C	4.10	6.23	8.58	11.45	13.60	15.86	16.93	21.43
G kg/hr	16.25	21.76	25.45	30.39	34.46	39.59	43.62	47.92
$F\times10^4$ m^3/hr	5.83	7.80	9.13	10.90	12.36	14.20	15.64	17.18
$b\times10^3$ m	0.42	0.47	0.49	0.53	0.55	0.58	0.60	0.62
n	24.82	24.26	21.79	20.79	20.78	21.54	23.01	20.66

Table 2. Relationship between F and b in a single flute

b/H	$bx10^3$(m)	$Fx10^4$ (m^3/hr)
0.1	0.0866	0.0684
0.2	0.1732	0.5478
0.3	0.2598	1.6084
0.4	0.3464	3.3944
0.5	0.4330	6.3458
0.6	0.5196	10.0236
0.7	0.6062	16.0236
0.8	0.6928	23.3332
0.9	0.7794	31.9390
1.0	0.866	42.7004

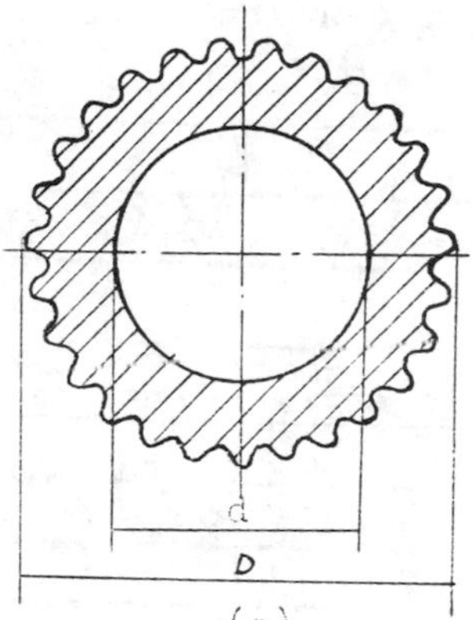

(a)

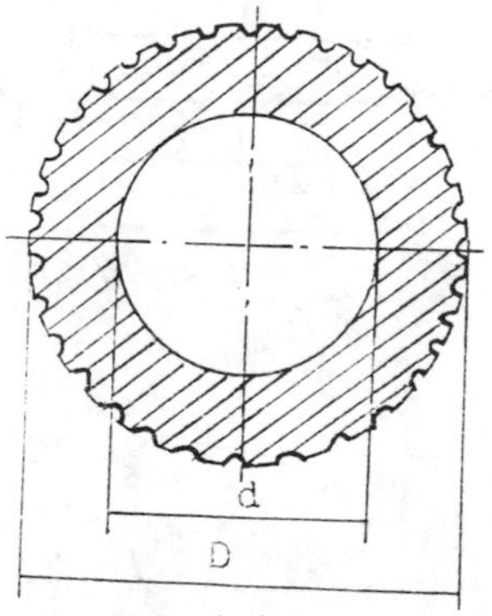

(b)

Fig.1. Cross Section of the Test Tube
(a) cosine type
(b) semi-circle type

Table 4. The variation of with Z (Δt_m = 8.58 C)

Z (m)	$bx10^3$(m)	α_z kcal/$m^2 \cdot hr \cdot {}^oC$
0.1	0.27	47614
0.2	0.33	39498
0.3	0.37	35399
0.4	0.40	32751
0.5	0.42	30832
0.6	0.44	29349
0.7	0.46	28154
0.8	0.48	27154
0.9	0.49	26308

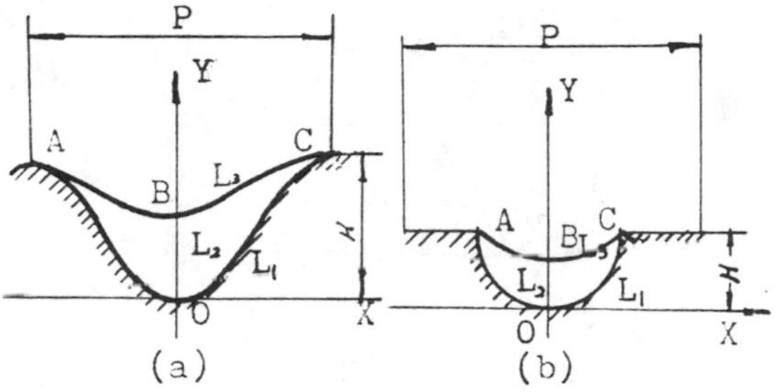

Fig.2. A single groove of the tubes
(a). cosine corrugation
(b). semi-circle corrugation

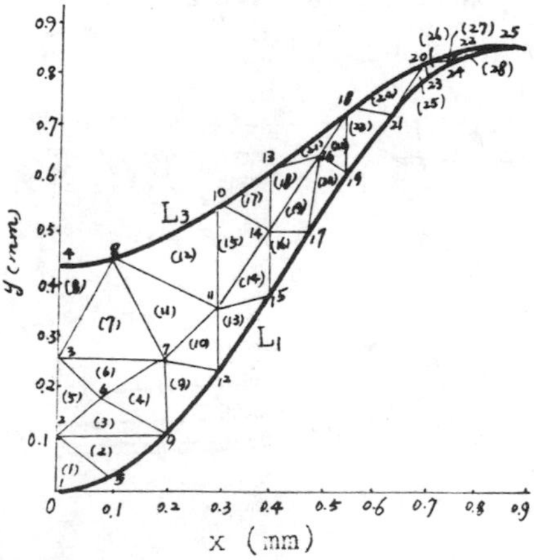

Fig.3. Finite Element Grid of the condensate film in half of the cosine corrugation

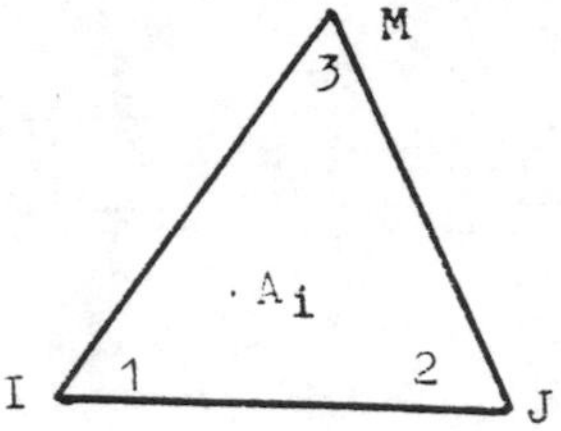

Fig.4. A sketch of a single element

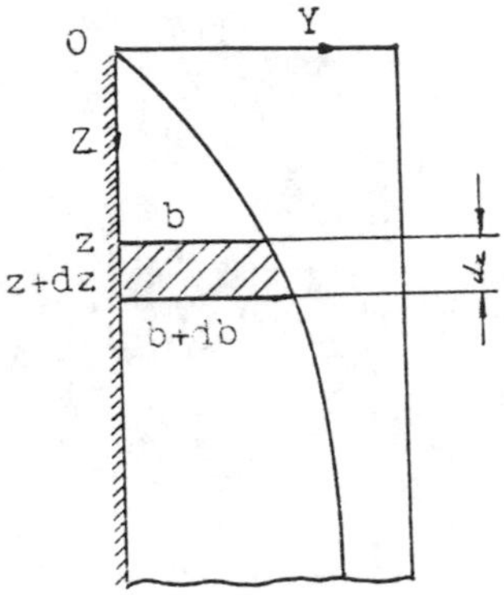

Fig.6. Longitudinal section of a single corrugation

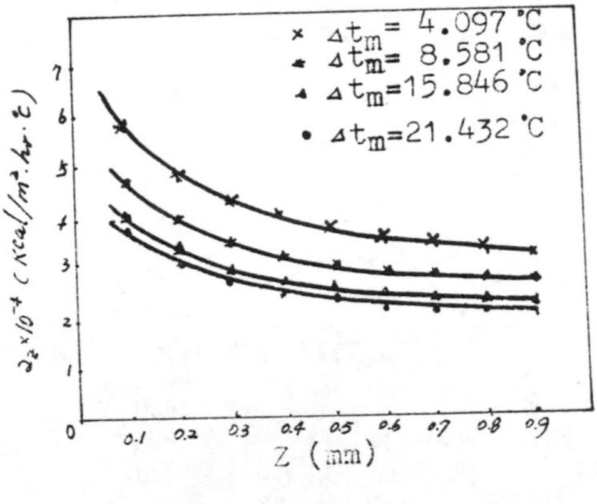

Fig.7. Variation of local heat transfer coefficient with the tube length

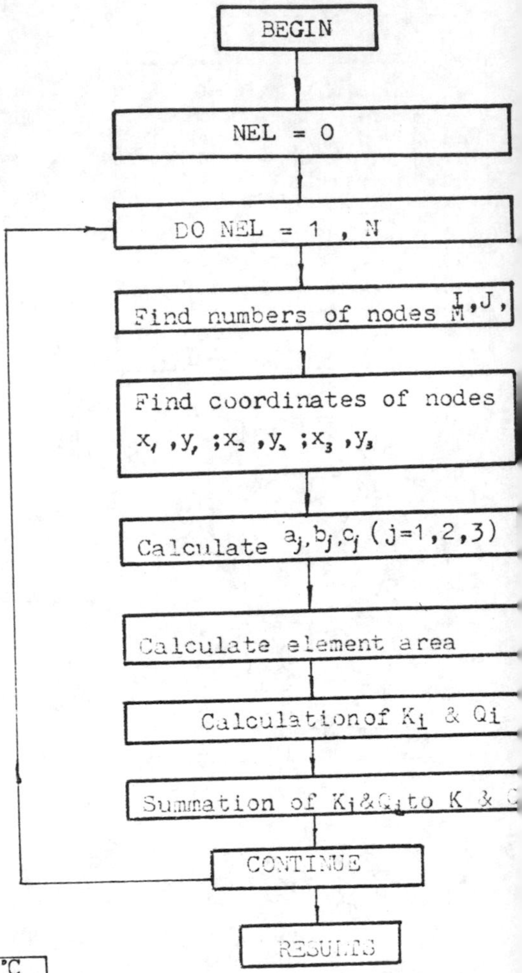

Fig.5. Calculation Process

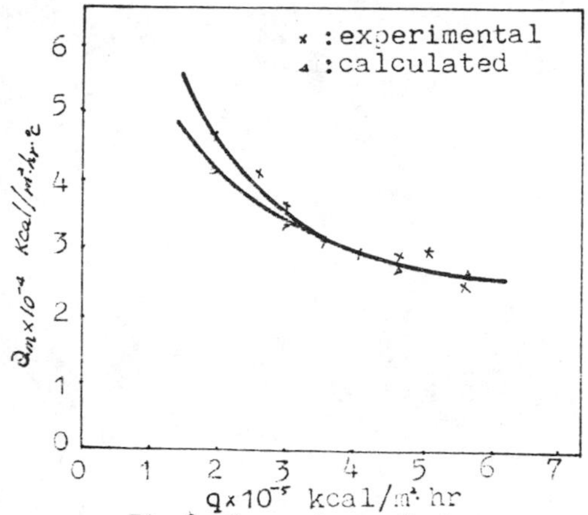

Fig.8. Comparison of the average heat transfer coefficients

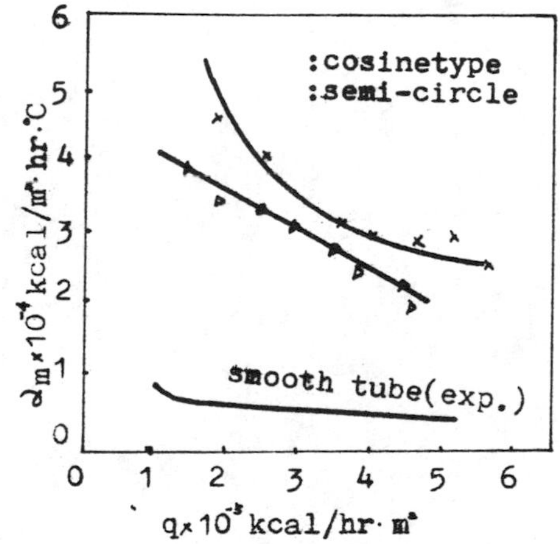

Fig.9. Comparison of heat transfer coefficients on corrugated tubes

THERMAL RADIATION

Radiative Heat Transfer Measurement for a Horizontal Tube Immersed in Small and Large Particle Fluidized Beds

NASSER ALAVIZADEH, ZICHENG FU, RONALD L. ADAMS, JAMES R. WELTY, and ALI GOSHAYESHI
Department of Mechanical Engineering
Oregon State University
Corvallis, Oregon 97331, USA

ABSTRACT

Results of an experimental study of the radiative contribution to heat transfer to a horizontal tube immersed in a gas fluidized bed are presented. The radiative heat transfer was determined using instrumentation designed to separate radiative and convective components. The tests were conducted in a propane gas heated atmospheric pressure facility containing crushed refractory material with mean particle diameters of 0.52 mm, 1.00 mm, 2.14 mm, and 3.23 mm. Data were obtained at temperatures up to 1050 K with gas velocities up to 3.3 times minimum fluidization velocity. The results show that the radiative heat transfer increases with particle diameter while the overall heat transfer decreases. The relative contribution of radiative heat transfer was found to range from 4% for .52 mm diameter particles to 9% for 3.23 mm diameter particles at a bed temperature of 812 K. At a bed temperature of 1050 K and for 3.23 mm particle diameter, 15% of the heat transfer was due to radiation.

NOMENCLATURE

c	specific heat
d_p	mean particle diameter
h_{max}	maximum total heat transfer coefficient
h_r	radiative heat transfer coefficient
h_{rb}	radiative heat transfer coefficient to a black tube wall
h_{to}	total heat transfer coefficient
k	thermal conductivity
q_r	radiative heat flux
q_{to}	total heat flux
T_b	bed temperature
T_w	wall temperature
U	gas velocity
U_{mf}	minimum fluidizing velocity
ε	emissivity
ε_b	effective bed emissivity
ρ	density

Z. Fu is presently with Research Institute of Building Materials, Beijing, PRC.
R. L. Adams is presently with Tektronix Laboratories, Tektronix, Inc., Beaverton, Oregon, USA.

INTRODUCTION

Fluidized beds have a variety of industrial applications as described by Kunii and Levenspiel [1]. Recently, a growing research effort has been concentrated on the study of the fluidized bed as an alternative to the existing conventional coal combustors in electrical power generation. In this case the bed material is predominantly limestone or dolomite. These particles have the ability to adsorb the sulphur dioxide resulting from the combustion of high sulphur content coal so that air quality standards can be met. The inherent higher heat transfer rates from the bed to the immersed tubes result in reduction of the size of the heat exchange area.

In a fluidized bed (Figure 1), gas is directed to the particle chamber through a distributor plate. Minimum fluidization is achieved when the solid particles are sustained by the upward moving gas. The added gas volume produced by increasing the gas flow rate passes through the bed in the form of bubbles. Bubbles move the particles from the core of the bed to the immersed surface and then back to the core. This circulation is responsible for the observed high heat transfer rates. Therefore, heat transfer studies, particularly at the bubbling stage, are of prime importance in designing fluidized bed combustors.

Saxena et al. [2], Zabrodsky [3], and Botterill [4] summarize the heat transfer data for horizontal tubes immersed in a bed at ambient or low temperature. Fewer high temperature measurements have been reported and are contained in Golan and Cherrington [5], Goblirsch et al. [6], Leon et al. [7], Wright et al. [8], Kharchenko and Makhorin [9], and George [10]. Most of these studies, with some variations, involved measurement of the inlet and outlet coolant temperatures along with estimation of the tube outside wall temperature by assuming unidirectional convection and conduction to the tube wall, respectively, to estimate the overall heat transfer coefficient. The data obtained generally shows a sharp increase in heat transfer as a result of transformation from a packed to a fluidized bed. This is followed by a mild increase and then a reduction as the gas velocity is increased.

There have been significantly fewer studies of the radiative heat transfer contribution and in some cases the results are contradictory. Baskakov et al. [11] estimated radiation by comparing the transient total heat transfer to oxidized and silver-plated spheres. Yoshida et al. [12] applied a similar technique to two vertical tubes. Vadivel and Vedamurthy [13], Basu [14], Vedamurthy and Sastri [14], and Ilchenko et al. [16] employed a quartz glass window mounted on top of a cavity within a tube to filter convection and to transmit radiation. A

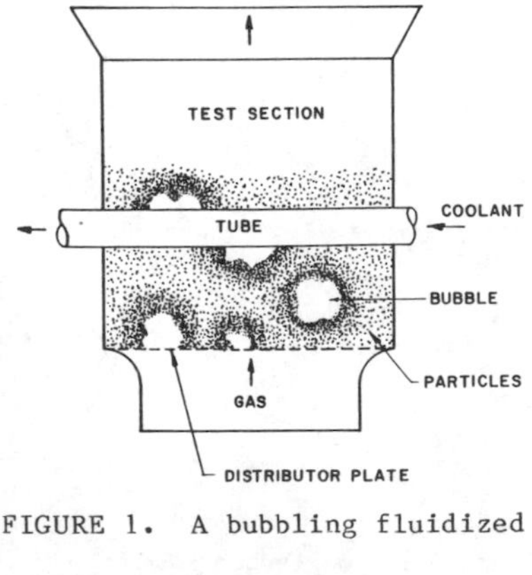

FIGURE 1. A bubbling fluidized bed.

thermoelectric device behind the window detected the transmitted radiation. This design is subject to a "conduction error" produced by the convective component of total heat transfer (see Alavizadeh et al. [17]). Ozkaynak et al. [18] used a two-layer flat window of zinc selenide with a gap between them on one end of a brass tube. Air was passed through the gap to eliminate the conduction error. The results of these studies, along with the operating conditions, are summarized in Table 1. As Table 1 shows, the radiation contribution to total heat transfer based upon these studies varies from 0% to 60% for bed temperatures between 1023 K and 1273 K and different particle sizes.

The present work was directed toward investigation of radiative heat transfer to a horizontal tube for a wide range of particle sizes. Total heat transfer was also measured to assess the relative contribution of radiation. The radiation measurement probe employs a silicon window assembled flush with the tube wall and a thin-film thermopile detector placed behind it. This device was calibrated against a black body cavity. Results obtained in this study can be modified to approximate the radiation contribution to total heat transfer to different tube wall emissivities.

APPARATUS

Figure 2 shows the Oregon State University high-temperature fluidized bed facility. Propane is ignited in the combustion chamber and hot gases are directed to a 30 cm x 60 cm test section. A venturi meter monitored the air flow rate to the combustion chamber. Two type K thermowells, one placed above the distributor plate and one at the top of the bed, measured the bed temperature. Water was used as coolant to control the instrumented tube temperature.

Total and radiative heat transfer measurement probes were mounted side-by-side along the axis of a 51 mm outside diameter bronze tube. Figure 3 illustrates the radiation and total probe instrumentation. The radiation probe makes use of a silicon window and a thin-film thermopile-type heat flow detector (.076 mm x 12.7 mm x 28.5 mm) separated by a 4.8 mm gap. The choice of silicon as the

TABLE 1. Radiative Heat Transfer Studies

Investigation	d_p (mm)	T_b (K)	T_w (K)	h_{to}* (q_{to})†	h_r* (q_r)†	Radiation contribution %
Baskakov et al. [11]	1.25	1123	423	-	30	7 to 9
	0.35			-	25	
Yoshida et al. [12]	0.18	1273	413	343	0	0
Basu [14]	.33-.50	1123	293-303	420	35	8
Ilchenko et al. [16]	.57	1223	333-393	(410,000)	(88,000)	20
	1.75			(354,000)	(83,000)	24
Vedamurthy and Sastri [15]	< 6.3	1173	313	80	50	60
	< 3.15			120	43	35
Vadivel and Vedamurthy [13]	< 6.0	1023	313	121	46	38
Ozkaynak [17]	1.03	1033	303-374	(157,000)	(50,000)	30
Present work	3.23	1050	353	200	45	15

*W/m^2K; †W/m^2

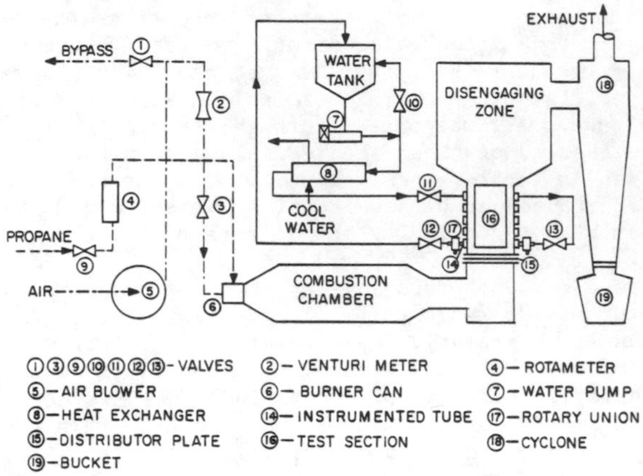

FIGURE 2. Oregon State University high-temperature fluidized bed.

window material was due to its high thermal conductivity (17 times that of quartz) and hence negligible conduction error (less than 3%) and also its wide spectral transmission characteristics (1.3 to 12 μm wavelength). The radiation probe was calibrated by exposing it to a narrow-angle black body cavity. A linear relation was found between the heat flux measured by the thin-film detector and the incident radiation to the window (also measured). The total heat flux probe uses a similar heat flow detector bonded to the tube wall and covered tightly by a stainless steel foil to protect the detector against the bed abrasion. The heat flux was calculated from the emf output and temperature of the detector. A more detailed discussion of the instrumentation, assembly, and calibration is given by Alavizadeh et al. [17].

A high precision, Hewlett-Packard digital voltmeter (HP-3497A), was employed for the measurement of both thermopile and thermocouple signals of the heat flow detectors. Data was collected for a period of about eighty seconds at each gas velocity and then stored as raw voltages using a microcomputer (HP-85) as the control unit.

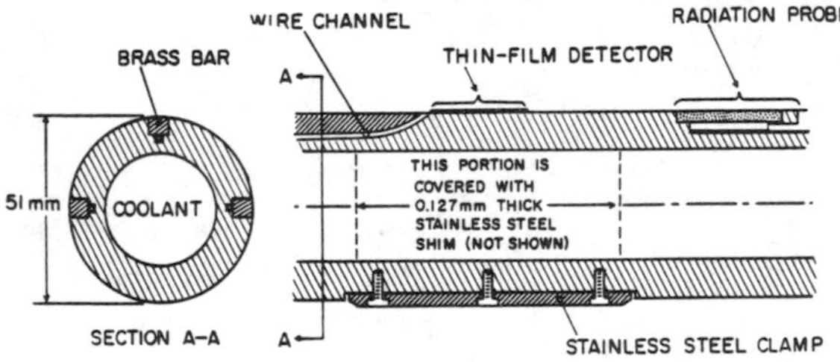

FIGURE 3. Instrumented tube.

TABLE 2. Summary of Test Conditions and Results for Ione Grain* Particles

d_p (mm)	T_b (K)	T_w (K)	U_{mf} (m/sec)	h_{max}†	h_{rb}† (at h_{max})	Radiation contribution % ε = .64	ε = 1	ε_b (at h_{max})
0.52	812	334	0.30	349	18	4	6	.35
1.00	812	331	0.48	275	19	5	8	.38
2.14	812	366	1.62	189	22	8	12	.42
3.23	812	357	2.31	184	24	9	14	.44
1.00	922	319	0.50	291	26	7	10	.39
2.14	1050	349	1.75	226	42	13	20	.43
3.23	1050	355	2.36	204	44	15	22	.45

*53.5% silicon, 43.8% alumina, 2.3% titania, .4% other; ρ = 2700 kg/m^3; k = 1.26 W/mK; C = .22 cal/gK; ε = .86 (estimated)

†W/m^{2}K

EXPERIMENTS

Tests were conducted for ione grain particles with mean diameters of .52 mm, 1.00 mm, 2.14 mm, and 3.23 mm and bed temperatures of 812 K, 922 K, and 1050 K. The tube was positioned at 30 cm above the distributor plate and the packed bed height was 50 cm. The repeatability of the measurements was ±12% and an rms error of -12%, +13% was calculated for the conducted tests. Table 2 shows minimum fluidization velocities (U_{mf}) for different test conditions. Also shown are the ione grain particle properties as measured by Ghafourian [19].

Figure 4 shows the variation in total heat transfer coefficient (h_{to}) with relative gas velocity for different bed temperatures and particle sizes. A sharp increase is noticeable for all the test conditions when the gas velocity exceeds

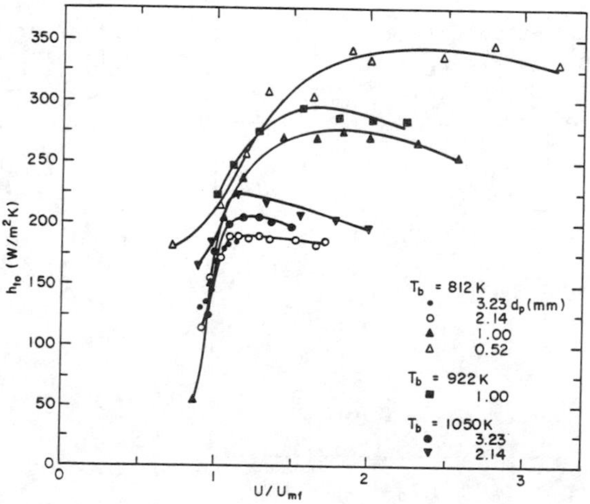

FIGURE 4. Total heat transfer coefficient versus relative gas velocity.

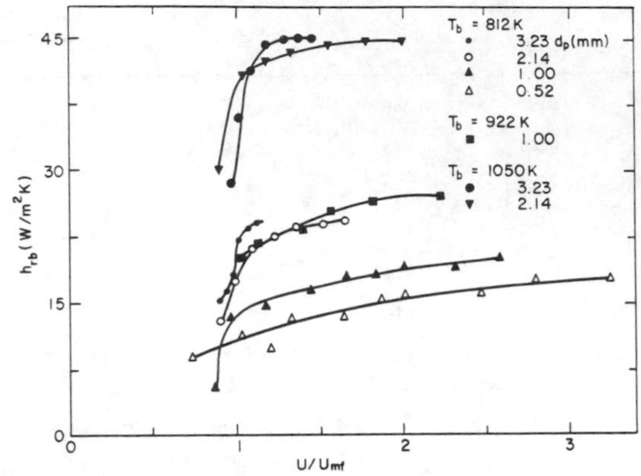

FIGURE 5. Radiative heat transfer coefficient (black tube wall) versus relative gas velocity.

U_{mf}. This is primarily due to the bubble-induced motion which replaces the "cold" particles near the tube surface with "hot" particles. The sharp increase is followed by a mild increase and a slow decay in h_{to}. The latter behavior can be attributed to an increase in the bed voidage (gas volume/total volume) and also possibly greater bubble contact. Similar trends for hot and cold beds have been reported by a number of investigators.

Figure 5 shows the radiative heat transfer coefficient to a black tube wall (h_{rb}). Its behavior is similar to that of h_{to} for gas velocities near the minimum fluidization. But h_{rb} tends to approach a nearly asymptotic value when the gas velocity is increased as also observed by Vedamurthy and Sastri [15]. h_{rb} increases with an increase in both particle size and bed temperature, the latter having a more pronounced effect.

The maximum total heat transfer coefficient (h_{max}) variation with particle diameter is illustrated in Figure 6. h_{max} drops with an increase in particle

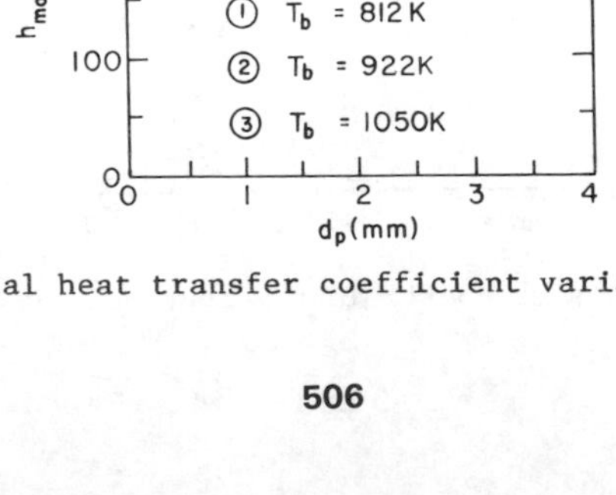

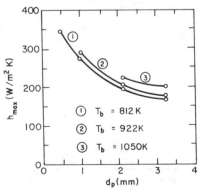

FIGURE 6. Maximum total heat transfer coefficient variation with particle diameter.

size. This is consistent with the findings of Wright et al. [8] and Kharchenko and Makhorin [9]. However, Baskakov et al. [11] reported that h_{max} will start to increase if sufficiently large particles are used. This is due to the larger contribution of gas convection at the expense of particle convection (see Botterill [3]) for larger particle beds. The minimum value of h_{max} can be seen to shift to larger particles when the bed temperature was elevated. Also, h_{max} was observed to decrease when the bed temperature was lowered from 1050 K to 812 K. These behaviors could probably be explained by the change in physical properties of the gas and the change in the radiative heat transfer coefficient as the bed temperature was changed.

The radiative heat transfer coefficient (h_{rb}) at maximum total heat transfer (h_{max}) is shown as a function of bed temperature in Figure 7. Preliminary calculations suggest that h_{rb} from our experiments varies linearly with the particle mean diameter (d_p) and with the third power of the bed temperature (T_b). However, further study is necessary to arrive at any conclusive results.

A summary of the results, along with the test conditions, are furnished in Table 2. The contribution of radiation heat transfer presented in this table is estimated for the wall emissivity of the instrumented tube (ε = .64) and also for a black tube wall (ε = 1.00). The results suggest that the radiation contribution maximum total heat transfer is greater for higher bed temperatures, and also larger particles due to slower cooling of the particles contacting the tube wall. The latter effect is magnified as a result of the reduction in h_{max} when larger particles are used. The radiation contribution varied from 4% to 9% for a bed temperature of 812 K and the particle range of .52 mm to 3.23 mm. Radiation is about 15% of the overall heat transfer at a bed temperature of 1050 K and particle size of 3.23 mm. An effective bed emissivity $\left(\varepsilon_b = q_{rb}/\sigma T_b^4\right)$ of .45 was computed for the largest mean particle diameter and highest bed temperature. This agrees well with ε_b = .40 obtained by extrapolating the data of Baskakov et al. [11]. However, their data shows ε_b > .52 (extrapolation) for a bed temperature of 812 K. At the same bed temperature, as Table 2 indicates, ε_b depends on the particle size and varies between .35 and .44. Botterill and Sealey [20] calculated the effective bed emissivity at a bed temperature of 1073 K to be about .60 for small (.2 mm to .7 mm) and .64 for large (1.0 mm to 1.5 mm) particles.

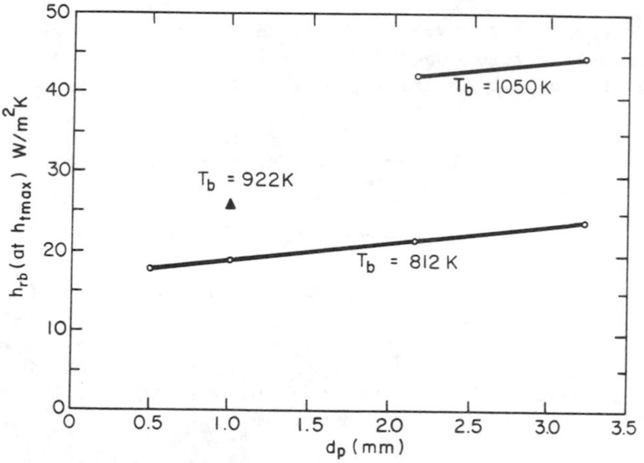

FIGURE 7. Radiative heat transfer coefficient to a black tube wall.

SUMMARY

A tube was instrumented for the measurement of radiative and total heat transfer in a fluidized bed for a variety of particle sizes and bed temperatures. The radiative heat transfer was obtained using a thin-film thermopile-type heat flow detector mounted behind a silicon window. The total heat transfer was measured using a similar detector mounted flush with the tube wall. Spatial average total and radiative heat transfer coefficients were obtained. The total heat transfer coefficient (h_{to}) variation with gas velocity followed the usual pattern observed by a number of investigations and was also found to drop with increasing particle size and decreasing bed temperature. The radiative heat transfer coefficient (h_{rb}) appeared to approach an asymptotic value as the gas velocity was increased. The maximum total heat transfer coefficient (h_{max}) approached a minimum value with increasing particle size. The minimum seems to occur at a larger particle size when the bed temperature was elevated from 812 K to 1050 K. The radiative heat transfer coefficient (h_{rb}) at h_{max} was observed to vary linearly with the particle mean diameter and with the third power of the bed temperature. The effective bed emissivity at h_{max} was also computed. The radiation contribution to total heat transfer dropped with decreasing particle size and bed temperature. The radiation contribution varied from 4% to 15% for the range of particle sizes (.52 mm to 3.23 mm) and bed temperatures (812 K to 1050 K) of these experiments. Also, h_{max} was found to vary from 184 to 349 W/m^2K and h_{rb} (at h_{max}) from 18 to 44 W/m^2K (to a black tube surface) for the same operating conditions.

ACKNOWLEDGEMENT

This work was funded in part by the National Science Foundation under Grant MEA 80-20781. The content of this paper does not reflect the views of the National Science Foundation.

REFERENCES

1. Kunii, D. and Levenspiel, O., Fluidization Engineering, John Wiley and Sons, New York, 1969.

2. Saxena, S.C., Grewal, N.S., Gabor, J.D., Zabrodsky, S.S., and Galershtein, D.M., Heat Transfer between Gas Fluidized Bed and Immersed Tubes, cited in Advances in Heat Transfer by Irvin, Jr., J.F. and Hartnett, J.P., Academic Press, 1978.

3. Zabrodsky, S.S., Hydrodynamics and Heat Transfer in Fluidized Beds, M.I.T. Press, Cambridge, 1966.

4. Botterill, J.S.M., Fluid-Bed Heat Transfer, Academic Press, London, 1975.

5. Golan, L.P. and Cherrington, D.C., Heat Transfer and Unit Response of a Large Fluidized Bed Combustor, AIChE Symp. Ser., no. 208, vol. 27, 1981.

6. Goblirsch, G., Vander Molen, R.H., Wilson, K., and Hajicek, O., Atmospheric Fluidized Bed Combustion Testing of North Dakota Lignite, Combustion Power Co., Inc., Menlo Park, California, May 1980.

7. Leon, A.M., Choksey, P.J., and Bunk, S.A., Design, Construction, Operation and Evaluation of a Prototype Culm Combustion Boiler/Heater Unit; Report FE 3269-9A, Dorr-Oliver, Inc., Stamford, Connecticut, April 1979.

8. Wright, S.J., Hickman, R., and Ketley, H.C., Heat Transfer in Fluidized Beds of Wide Size Spectrum at Elevated Temperature, Brit. Chem. Eng., vol. 15, no. 12, December 1970.

9. Kharchenko, N.V. and Makhorin, K.E., The Rate of Heat Transfer between a Fluidized Bed and an Immersed Body at High Temperatures, Int. Chem. Engr., vol. 4, no. 4, October 1964.

10. George, A., An Experimental Study of Heat Transfer to a Horizontal Tube in a Large Particle Fluidized Bed at Elevated Temperatures, Ph.D. thesis, Oregon State University, Corvallis, Oregon, 1981.

11. Baskakov, A.P., Berg, B.V., Vitt, O.K., Fillippovsky, N.F., Kirakosyan, V.A., Goldobir, J.M., and Maskaev, V.K., Heat Transfer to Objects Immersed in Fluidized Beds, Powder Technology, vol. 8, 1973.

12. Yoshida, K., Ueno, T., and Kunii, D., Mechanism of Bed-Wall Heat Transfer in a Fluidized Bed at High Temperatures, Chem. Eng. Sci., vol. 29, 1974.

13. Vadivel, R. and Vedamurthy, V.N., An Investigation of the Influence of Bed Parameters on the Variation of the Local Radiative and Total Heat Transfer Coefficients around an Embedded Horizontal Tube in a Fluidized Bed Combustor, Proc. of 6th Int. Conf. on Fluidized Bed Combustion, Atlanta, USA, April 1980.

14. Basu, P., Bed to Wall Heat Transfer in a Fluidized Bed Coal Combustor, AIChE Symp. Ser., no. 176, vol. 74, 1978.

15. Vedamurthy, V.N. and Sastri, V.M.K., An Experimental Study of the Influence of Bed Parameters on Heat Transfer in a Fluidized Bed Combustor, Fluidized Bed Library, New York University, pp. 589-597.

16. Ilchenko, A.I., Pikashov, V.P., and Makhorin, K.E., Study of Radiative Heat Transfer in a Fluidized Bed, J. of Engr. Physics, 14, 1968.

17. Alavizadeh, N., Adams, R.L., Welty, J.R., and Goshayeshi, A., An Instrument for Local Radiative Heat Transfer Measurement in a Gas-Fluidized Bed at Elevated Temperatures, 22nd ASME/AIChE Natl. Heat Transfer Conf., Niagara Falls, 1984.

18. Ozkaynak, T.F., Chen, J.C., and Frankfield, T.R., An Experimental Investigation of Radiation Heat Transfer in a High Temperature Fluidized Bed, Int. Conf. on Fluidized Bed, Japan, 1983.

19. Ghafourian, M.R., Determination of Thermal Conductivity, Specific Heat, and Emissivity of Ione Grain, M.S. project, Department of Mechanical Engineering, Oregon State University, 1984.

20. Botterill, J.S.M. and Sealey, C.J., Radiative Heat Transfer between a Gas-Fluidized Bed and an Exchange Surface, Brit. Chem. Eng., vol. 15, no. 9, 1970.

Thermal Analysis and Optimum Design for Radiating Spine of Various Geometries

B. T. F. CHUNG and L. D. NGUYEN
Department of Mechanical Engineering
University of Akron
Akron, Ohio 44325, USA

INTRODUCTION

One of the most important problems of modern space technology is that of thermal control; while the analysis and design of radiating fins plays a significant role in the thermal control of spacecraft and satellites. Since the weight is one of the primary design considerations of heat exchangers, it is highly desirable to obtain the optimum fin designs.

The thermal behavior of radiating longitudinal fins have been studied by numerous investigators [1-7]. The analysis of convecting spine has been presented in detail by Kern and Kraus [8]. However, the optimum results of radiating spines have not been available in the literature. Although Wilkins [9] solved the least material profile for radiating longitudinal fins, the corresponding solution for the radiating spine still does not exist.

The purpose of this study is to solve the general governing differential equations for radiating spines, to formulate the general relationships for spine dimensions, heat transfer characteristics under the optimum condition and to determine a least material profile among radiating spines.

MATHEMATICAL ANALYSIS

The following assumptions are made in the present analysis: steady state holds; the temperature distribution in the fin is one dimensional; the thermal properties at each location along the fin are constant; the fin base surface is isothermal; the radiant interaction with adjacent fins and with the fin base is negligible, and the environmental temperature is at absolute zero.

Optimum Dimensions and Heat Transfer Characteristics

Figure 1 shows the terminology and coordinate system for the general radiating spine profile. The origin of the coordinate system is taken at the center of the base of the spine, and the x-axis is positive in the direction toward the spine edge. The general spine profile may be expressed as [10]

$$f(x) = \frac{\delta}{2}\left(1 - \frac{x}{b}\right)^n \qquad (1)$$

where n equals 0, 0.5, 1, 2, and 37/11 for cylindrical, convex parabolic, conical, concave parabolic and least material profiles (shown later), respectively.

The energy balance along with Equation (1) yields the following non-dimensional differential equation

$$(1 - X)^n \frac{d^2Y}{dX^2} - 2n(1 - X)^{n-1} \frac{dY}{dX} - 2\zeta Y^4 = 0 \qquad (2)$$

where $X = x/b$, $Y = T/T_b$ and $\zeta = 2\sigma\varepsilon b^2 T_b{}^3/k\delta$. The above differential equation is subjected to the following two boundary conditions

$$Y = 0 \quad \text{at} \quad X = 0 \qquad (3)$$

$$q = 0 \quad \text{at} \quad X = 1 \qquad (4)$$

Equations (2) and (4) imply $dY/dX = 0$ for the cases of n equals 0 and 0.5, $dY/dX = -\zeta Y^4$ for $n = 1$, and $Y = 0$ for $n = 2$ and 37/11 at the tip of the fin $(X = 1)$. The subscript b implies the condition at the fin base.

The heat dissipation q, which is the heat flow through the spine base at the steady state, is obtained from Fourier's Law

$$q = \frac{\pi k \delta^2 T_b}{4b} \left(- \left.\frac{dY}{dX}\right|_{X=0}\right) \qquad (5)$$

The spine efficiency is defined as the ratio of actual heat dissipation to the idealized heat transfer which would have been dissipated if the entire spine were to radiate at the base temperature. It can be in terms of n, ζ and temperature gradient at the base.

$$\eta = \frac{n+1}{2\zeta} \left(- \left.\frac{dY}{dX}\right|_{X=0}\right) \qquad (6)$$

The spine width δ and spine height b are written in terms of spine volume V which is obtained from

$$V = \int_0^b \pi f^2(x)dx = \frac{\pi}{4(2n+1)} \delta^2 b \qquad (7)$$

From the definition of ζ and Equation (7), we arrive at fin width and height:

$$\delta = \left\{ \frac{\left[\frac{4(2n+1)}{\pi}\right]^2 (2\sigma\varepsilon) V^2 T_b^3}{k\zeta} \right\}^{\frac{1}{5}} \qquad (8)$$

$$b = \left\{ \frac{\left[\frac{4(2n+1)}{\pi}\right] k^2 V \zeta^2}{(2\sigma\varepsilon)^2 T_b^6} \right\}^{\frac{1}{5}} \qquad (9)$$

The heat dissipation expressed in equation (5) can be rewritten in terms of volume and ζ using Equations (7) - (9).

$$q = \left\{ [\pi^2(2n+1)^3][k(\sigma\varepsilon)^4 V^3 T_b{}^{17}] \right\}^{\frac{1}{5}} \left[\frac{1}{\zeta}\right]^{\frac{4}{5}} \left[- \left.\frac{dY}{dX}\right|_{X=0}\right] \qquad (10)$$

The heat dissipation per unit volume is obtained by combining Equations (6), (7) and (10).

$$q/V = \frac{4\sigma\varepsilon T_b^4}{\delta}\left[\eta\ \frac{(2n+1)}{n+1}\right] \tag{11}$$

Substituting the result of $\left.\frac{dY}{dX}\right|_{X=0}$ from equation (6) into equation (10), the heat dissipation becomes

$$q = \left\{\left[\frac{32\pi^2(2n+1)^3}{(n+1)^5}\right]\left[k(\sigma\varepsilon)^4 V^3 T_b{}^{17}\right]\right\}^{\frac{1}{5}} \eta\zeta^{\frac{1}{5}} \tag{12}$$

The maximum heat dissipation can now be achieved by letting dq = 0, which gives

$$\frac{d\eta}{d\zeta} = -\frac{1}{5}\frac{\eta}{\zeta} \quad \text{or} \quad \frac{\ell n\ \eta}{\ell n\ \zeta} = -\frac{1}{5} \tag{13}$$

If a logarithmic coordinates is used in plotting η vs ζ Equation (13) indicates that the optimum condition is reached when the slope of the curve is -0.2. We therefore have located ζ_o, where the subscript o designates optimum condition.

The Least Material Profile of Radiating Spine

For convenience, the derivation of least material profile is performed by setting the origin of the coordinates at the tip of the spine, i.e.

$$f(x) = \frac{\delta}{2}\left(\frac{x}{b}\right)^n \tag{14}$$

Our objective is to seek the value of n which yields the minimum volume for a given amount of heat dissipation. The heat conduction across any position x is

$$q = k\pi f^2(x)\frac{dT}{dx} \tag{15}$$

while the radiant heat dissipation from the element dx is evaluated from the Stefan-Boltzmann's Law.

$$dq = \sigma\varepsilon[2\pi f(x)]T^4 dx \tag{16}$$

There are two conditions for q and T:

$$q = 0 \quad \text{and} \quad T = T_e \quad \text{at } x = 0 \tag{17}$$

$$q = q_b \quad \text{and} \quad T = T_b \quad \text{at } x = b \tag{18}$$

To determine the radiating spine of least material profile, we need to find q(x), T(x), and f(x) so that they satisfy Equations (15)-(18) and the spine volume given below must be a minimum. (The subscript e implies fin tip condition.)

$$V = \pi\int_o^b [f^2(x)]\ dx \tag{19}$$

To accomplish this, we introduce the following transformations

$$u = (T/T_b)^{17} \quad \text{and} \quad v = (q/q_b)^{5/4} \tag{20}$$

dx, f(x) and V in Equations (15)-(19) can now be in terms of u, v, T_b and q_b as

$$dx = \frac{2\ q_b}{5\pi\sigma\varepsilon T_b^4}\ \frac{1}{f(x)}\ u^{-4/17} v^{-1/5} \frac{dv}{du}\, du \tag{21}$$

$$f(x) = \left(\frac{34q_b^2}{5\pi^2 k\sigma\varepsilon T_b^5}\ u^{12/17} v^{3/5} \frac{dv}{du} \right)^{1/3} \tag{22}$$

$$V = \left[\frac{272q_b^5}{625\pi^2 k(\sigma\varepsilon)^4 T_b^{17}} \right]^{1/3} \int_{u_e}^{1} \left(\frac{dv}{du}\right)^{4/3} du \tag{23}$$

where $u_e = T_e/T_b$. Since the term inside the bracket is a constant, V will have a minimum only if the integral is a minimum. Applying the calculus of variation, we can show that the integral attains a minimum value of unity at v = u. Therefore, Equation (23) gives the minimum spine volume, V_o for a given amount of total heat transfer.

$$V_o = \left[\frac{272q_b^5}{625\pi^2 k(\sigma\varepsilon)^4 T_b^{17}} \right]^{1/3} \tag{24}$$

Alternately, the maximum heat dissipation for any radiating spine profile with a fixed spine volume is

$$q_o = 1.867[k(\sigma\varepsilon)^4 V^3 T_b^{17}]^{1/5} \tag{25}$$

Applying the relationship $u = v = (T/T_b)^{17}$ along with Equations (21), (22) and (25) we arrive at

$$f(x) = \left[\frac{34q_o^2}{5\pi^2 k\sigma\varepsilon T_b^5} \right]^{1/3} \left(\frac{x}{b}\right)^{37/11} \tag{26}$$

Comparing the above expression with Equation (14) we conclude that n = 37/11 yields the design of least material profile. Furthermore, the corresponding optimum fin width becomes

$$\delta_o = 2 \left[\frac{34q_o}{5\pi^2 k\sigma\varepsilon T_b^5} \right]^{1/3} = 2.678 \left[\frac{\sigma\varepsilon V^2 T_b^3}{k} \right]^{1/5} \tag{27}$$

RESULTS AND DISCUSSION

Numerical solutions of Equation (2) are obtained for each given value of ζ by applying the fourth order Runge-Kutta method. This method is designed for initial value problem and requires the value of dY/dX at X = 0. In the present analysis, the value of dY/dX at X = 0 is chosen iteratively until the second boundary condition at X = 1 is satisfied. Once the temperature gradient of the spine base is known, the spine efficiency can be calculated from Equation (6). With the curve η vs ζ available, the optimum design point $\zeta = \zeta_o$ is determined from Equation (13). The optimum efficiency, spine width, height, dissipation per unit volume and total heat transfer are then solved by setting $\zeta = \zeta_o$ in Equations (6), (8), (9), (11), and (12) respectively. Table 1 summarizes the above information for various profiles. As can be seen the dimensionless heat transfer, S_3 reaches the maximum as n increases from 0 to 37/11 and starts to decline very slightly as n continues to increase. The dimensionless spine width, height, heat transfer and spine efficient under optimum conditions are plotted in Figure 2. Similar to convecting spines, the optimum shape of the radiating spines become wider at the base and longer in height. The variation of S_2 with n is found to be linear and may be accurately represented by the following empirical equation

$$S_2 = 0.303\, n + 0.352 \tag{28}$$

The above expression is obtained from least square curve fitting. Similarly S_1, S_3 and η_o can be conveniently calculated from the following working formulae

$$S_1 = \frac{4(2n+1)}{\pi(0.303n+0.352)} \tag{29}$$

$$S_3 = 1.867 - 0.241\, e^{-2.5n} \tag{30}$$

$$\eta_o = \frac{(n+1)(1.867-0.241e^{-2.5n})}{[2\pi(2n+1)(0.303n+0.352)]^{0.5}} \tag{31}$$

As also shown in Table 1, although an increase in both k and ε results in an increase in the amount of heat dissipation, the increase of emissivity has more effect on q_o than that of conductivity as can be observed from Equation (12). Furthermore, the optimum heat transfer rate is proportional to the three-fifths power of the spine volume; to double the optimum heat dissipation of a single spine, its volume would have to increase 3.17 times.

For general radiating spine design consideration, the spine efficiency η as functions of profile number ζ is obtained for each spine profile. The results are presented in Figure 3. The position of maximum heat dissipation of each profile is also marked in this figure. It appears that a short fin is preferred from the heat transfer viewpoint. For any spine profile, as ζ increases, η increases. However, a more realistic comparison should be based on the heat dissipation per unit volume, q/V which is found to be proportional to the product of η and (2n+1)/(n+1) as shown in Equation (11). As n increases, both η and (2n+1)/(n+1) increase; this results in an increase of q/V. The variations of q/V with respect to ζ are presented in Figure 4. It is seen, the spine profile with a larger value of n is always superior in terms of heat dissipation per unit volume or mass.

Figure 5 shows the dimensionless temperature distribution of radiating spines of various profiles. As n increases, the temperature gradient at the spine base increases, and the temperature drops more rapidly. This results in an increase of the optimum heat dissipation. Note that the temperature gradient does not approach zero at the spine tip of conical, concave parabolic, and least material profiles although heat transfer there is zero. Furthermore, the optimum temperature distribution of radiating spine profile of least material is found to be the form of $T/T_b = (1-X)^{5/11}$. This implies that the slope of the temperature curve at the spine tip (at X=1) approaches infinity. However, since the heat flux at the spine tip is found to be proportional to 68(1-X)/11. Therefore, the heat flux is equal to zero at the spine tip.

REFERENCES

1. Lieblein, S., Analysis of Temperature Distribution and Radiant Heat Transfer Along a Rectangular Fin of Constant Thickness, NASA, TN D-196, November 1959.

2. Liu, C.Y., On Minimum Weight Rectangular Radiating Fins, J. of the Aero/Space Science, Vol. 27, pp 871-871, 1960.

3. Shouman, A.R., An Exact General Solution for the Temperature Distribution and the Radiant Heat Transfer Along a Constant Cross-Sectional Area Fin, ASME Paper No. 67-WA/HT-27, 1967.

4. MacKay, D.B. and Bacha, C.P., Space Radiation Design and Design Analysis, North American Aviation, Inc., Report ASD 61-30, October 1961.

5. Nilson, N. and Curry, R., The Minimum Weight Straight Fin of Triangular Profile Radiating to Space, J. of Aero/Space Science, Vol. 27, pp 146-147, 1960.

6. Reynolds, W.C., A Design-Oriented Optimization of Single Tapered Radiating Fins, J. of Heat Transfer, Trans. ASME, Vol. 85, pp 193-202, 1963.

7. Kotan, K. and Arnas, O.A., On the Optimization of the Design Parameters of Parabolic Radiating Fins, ASME-AIChE Heat Transfer Conference and Exhibit, Los Angeles, California, August 8-11, 1965.

8. Kern, D.Q. and Kraus, A.D., Extended Heat Transfer, McGraw-Hill, New York, 1972.

9. Wilkins, J.E., Jr., Minimum Mass Thin Fins Which Transfer Heat only by Radiation to Surroundings at Absolute Zero, J. of the Society for Industrial and Applied Mathematics, Vol. 8, pp 630-640, 1960.

10. Gardner, K.A., Efficiency of Extended Surface, Trans. of ASME, Vol. 67, pp 621-631, 1945.

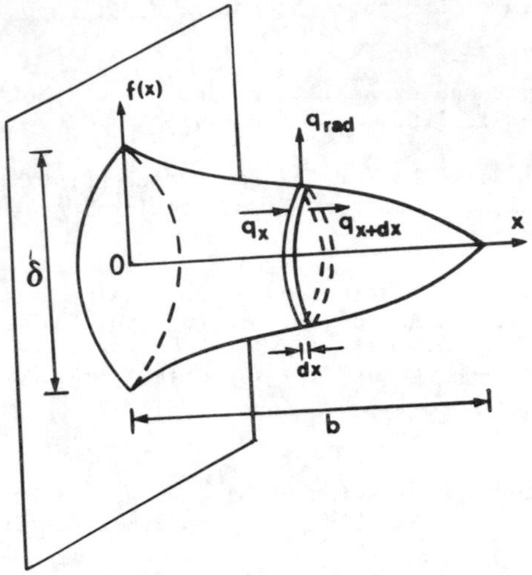

FIGURE 1. Terminology and coordinate systems for general radiating spine profile

TABLE 1. Optimum dimensions, efficiency, and heat transfer characteristics for radiating spines

Radiating Spines	Spine Profile Parameter n	$\delta_o = S_1[\frac{\sigma\epsilon V^2 T_b^3}{k}]^{0.2}$ S_1	$b_o = S_2[\frac{k^2 V}{(\sigma\epsilon)^2 T_b^6}]^{0.2}$ S_2	$q_o = S_3[k(\sigma\epsilon)^4 V^3 T_b^{17}]^{0.2}$ S_3	η_o	$(\frac{T_e}{T_b})_o$	ζ_o	$-\frac{dY}{dX}\Big\vert_{X=0}$
Cylindrical Profile	0	1.891	0.356	1.626	0.769	0.904	0.134	0.206
Convex Parabolic Profile	0.5	2.248	0.504	1.800	0.759	0.868	0.226	0.229
Conical Profile	1	2.428	0.648	1.846	0.747	0.806	0.346	0.258
Concave Parabolic Profile	2	2.587	0.951	1.865	0.723	0	0.700	0.338
Third Order Parabolic	3	2.657	1.263	1.866	0.709	0	1.200	0.425
Least Material Profile	37/11	2.678	1.372	1.867	0.706	0	1.406	0.455
Higher Order Parabolic	3.5	2.680	1.418	1.866	0.704	0	1.500	0.469

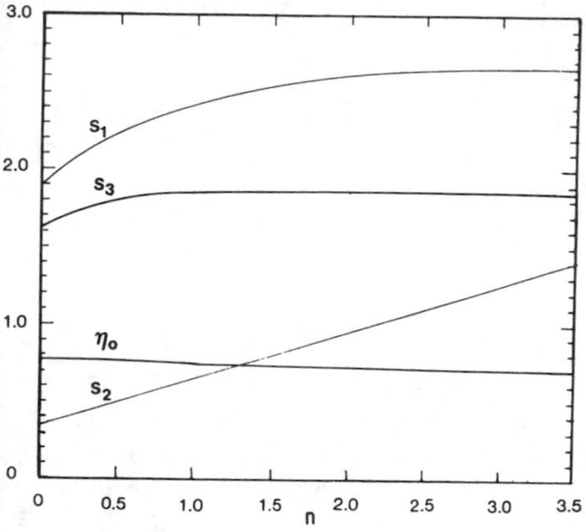

FIGURE 2. Dimensionless optimum width, height, heat transfer, and efficiency for radiating spines of various profiles

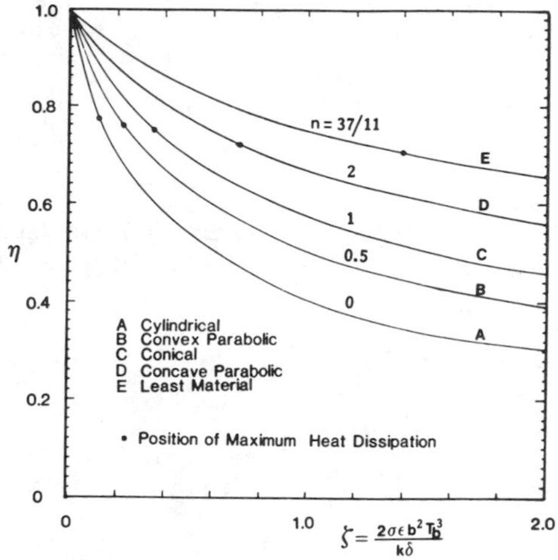

FIGURE 3. Efficiencies of radiating spines as functions of profile number

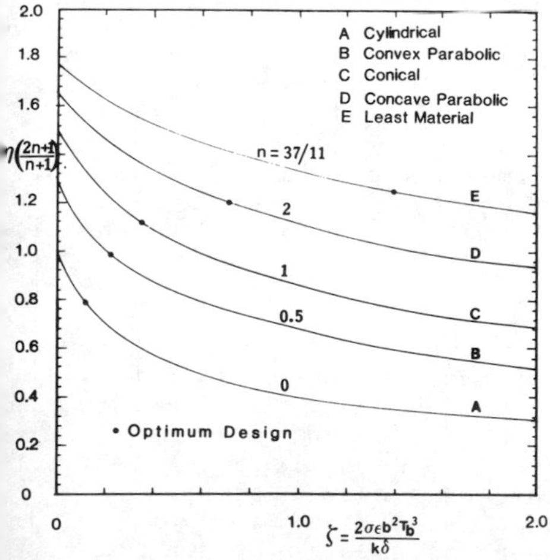

FIGURE 4. Dimensionless heat dissipation per unit volume for radiating spines of various profiles

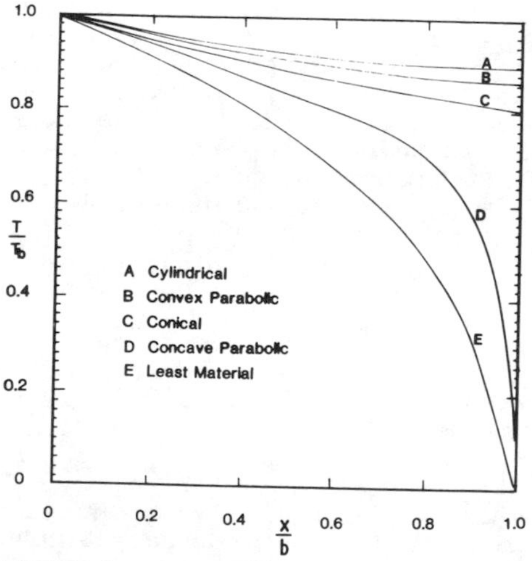

FIGURE 5. Dimensionless temperature profile of optimum radiating spines

Radiative Transfer in a Sodium Mist Layer

K. KAMIUTO
Faculty of Engineering
Ohita University
Ohita, Japan

I. KINOSHITA
Central Research Institute of Electric Power Industry
Tokyo, Japan

1. INTRODUCTION

Prediction of heat transfer through the cover gas space of an Liqui -Metal Fast Breeder Reactor (LMFBR) is indispensable for the therma design of the upper rotating shield plugs of the reactor core. Generally, heat transfer through the cover gas space is due to natural convection, thermal radiation and condensation, but the dominant mechanisms for heat transfer are considered to be natural convection and thermal radiation. Thus it is sufficient for the thermal design of the shield plugs to quantitatively evaluate the amount of heat transferred by both mechanisms.

Calculation of natural convection heat transfer through the cover gas space may be done straightforwardly by utilizing known heat transfer correlations on natural convection, because under the reactor operating condition both the mass flux of sodium vapour an the mass fraction of sodium mist are small and the properties of t cover gas medium can be treated as those of the inert gas. However evaluation of radiative heat transfer through the cover gas space seems to be hard because they call for the knowledge of the radiative properties of sodium mist and time-consuming computations of the equation of transfer. In fact, there has been only a little amount of information about the radiative heat transfer through th cover gas space containing sodium mist.

Furukawa et al.[1] have made an experiment of heat transfer through the cover gas space over a hot sodium pool and have theoretically evaluated the total heat flux at the cold boundary surface, but in their analysis the diffusion approximation to radiative heat flux was utilized, which makes the agreement between the experiment and the theory less satisfactory. Truelove[2] has made a theoretical analysis on radiative transfer through a sodium mist layer, but in his analysis two boundary surfaces were assumed to be gray diffuse emitters and reflectors and moreover only a limited case was discussed.

In view of this situation, the present study aims at evaluating th total radiative heat flux at the cold boundary surface of the cove gas space containing sodium mist by exactly solving the spectral equation of transfer and examining the validities of some approximations, including the diffusion approximation to radiative heat flux Throughout the present analyses the radiative properties of sodium mist required are evaluated rigorously by the Mie theory by taking into account the dependence of the particle size distributions of sodium mist on temperature and the dependence of the complex refractive index of liquid sodium on both temperature and wavelength

2. THEORETICAL ANALYSIS

2.1 Basic assumptions

The physical model and the coordinate system relevant to the present study are shown in Fig.1. The basic assumptions and postulations introduced herein are as follows.

1. The cover gas space is infinite and is bounded by the cold upper and hot lower liquid sodium surfaces.

2. The cover gas medium consists of sodium mist, sodium vapour and an inert gas (argon).

3. Sodium mist consists of a cloud of spherical liquid sodium particles and disperses uniformly in the cover gas space.

4. Sodium mist is continuous for thermal radiation and can emit, absorb and scatter thermal radiation.

5. The extinction coefficient, the albedo and the phase function of sodium mist and the reflectivities (or emissivities) of two boundary surfaces depend on both temperature and wavelength.

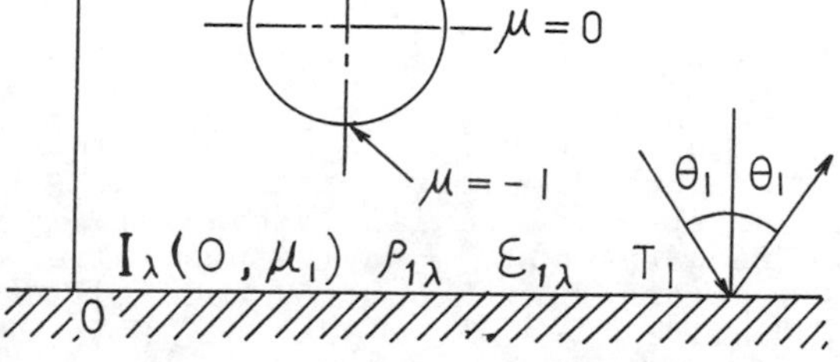

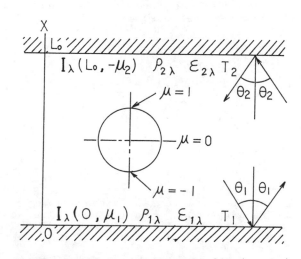

FIGURE 1. Physical model and the coordinate system

6. Two boundary surfaces can be regarded as planar, specular liqui sodium surfaces at constant but different temperatures.

7. Turbulent natural convection takes place in the cover gas mediu and so the cover gas medium is at uniform temperature given by the mean value of temperatures at two boundary surfaces.

8. Effects of thermal and density boundary layers near two boundar surfaces on radiative transfer are negligible because these boundar layers are very thin.

9. The refractive index of the cover gas is unity.

10. Planck's law and Kirchhoff's law are valid.

2.2 Equation of transfer

Under the foregoing assumptions, the spectral equation of transfer governing the radiation field and its relevant boundary conditions can be written as follows:

$$\mu \frac{dI_\lambda(x,\mu)}{dx} + (\sigma_{a,\lambda} + \sigma_{s,\lambda})I_\lambda(x,\mu) = \frac{\sigma_{s,\lambda}}{2}\int_{-1}^{1} P_\lambda(\mu,\mu')I_\lambda(x,\mu')d\mu' + \sigma_{a,\lambda}I_{b,\lambda}(T_m) \quad (1$$

$$I_\lambda(0,\mu_1) = I_{b,\lambda}(T_1)\varepsilon_{1,\lambda}(T_1,\mu_1) + I_\lambda(0,-\mu_1)\rho_{1,\lambda}(T_1,\mu_1) \quad (2$$

$$I_\lambda(L_0,-\mu_2) = I_{b,\lambda}(T_2)\varepsilon_{2,\lambda}(T_2,\mu_2) + I_\lambda(L_0,\mu_2)\rho_{2,\lambda}(T_2,\mu_2) \quad (3$$

where $I_{b,\lambda}$ denotes the spectral black-body intensity. Dividing both sides of eqs.(1)(2) and (3) by the black-body intens $I_{b,\lambda}(T_1)$ yields the dimensionless forms of the spectral equation o transfer and its boundary conditions:

$$\mu \frac{dI^*_\lambda(\eta,\mu)}{\tau_0 d\eta} + I^*_\lambda(\eta,\mu) = \frac{\omega_\lambda}{2}\int_{-1}^{1} P_\lambda(\mu,\mu')I^*_\lambda(\eta,\mu')d\mu' + (1-\omega_\lambda)I^*_{b,\lambda}(\theta_m) \quad (4$$

$$I^*_\lambda(0,\mu_1) = \varepsilon_{1,\lambda}(\theta_1,\mu_1) + I^*_\lambda(0,-\mu_1)\rho_{1,\lambda}(\theta_1,\mu_1) \quad (5$$

$$I^*_\lambda(1,-\mu_2) = I^*_{b,\lambda}(\theta_2)\varepsilon_{2,\lambda}(\theta_2,\mu_2) + I^*_\lambda(1,\mu_2)\rho_{2,\lambda}(\theta_2,\mu_2) \quad (6$$

where $I^*_\lambda(\eta,\mu)$ denotes the dimensionless spectral intensity of radiation and $I^*_{b,\lambda}$ the dimensionless spectral black-body intensity In order to obtain the spectral radiative heat flux at the cold boundary surface defined by eq.(7), eq.(4) was solved by the Barkstrom method together with eqs.(5) and (6) in the waveleng region from 1 to 100 µm.

$$q_{r,2,\lambda} = 2\pi\int_{-1}^{1} I_\lambda(L_0,\mu)\mu d\mu \quad (7$$

Further, to obtain the total radiative heat flux $q_{r,2}$, the spectral radiative heat flux is integrated in the same wavelength region. In addition to the exact analyses, three kinds of approximation respecting to radiative transfer were examined: the isotropic scattering approximation to the phase function, the diffusion approximation to radiative heat flux defined by eq.(8) and the gray approximation to radiative properties.

$$q_{r,\lambda} = \frac{\pi(I_{b,\lambda}(T_1)-I_{b,\lambda}(T_2))}{\frac{3}{4}\tau_{0,\lambda}+\frac{1}{\varepsilon_{1,\lambda}}+\frac{1}{\varepsilon_{2,\lambda}}-1} \qquad (8)$$

2.3 Radiative properties of sodium mist

Knowledge on spectral variations of the radiative properties of sodium mist such as the extinction and scattering coefficients and the phase function is required in order to solve the spectral equation of transfer. These radiative properties may be evaluated rigorously by the Mie theory , provided that the liquid sodium particles are considered spherical particles. From the Mie theory, the extinction, scattering and absorption coefficients are given by

$$\sigma_e = 10^{-6}\int_0^\infty N(r)Q_e(r)\pi r^2 dr \ (m^{-1}) \qquad (9)$$

$$\sigma_s = 10^{-6}\int_0^\infty N(r)Q_s(r)\pi r^2 dr \ (m^{-1}) \qquad (10)$$

$$\sigma_a = \sigma_e - \sigma_s \ (m^{-1}) \qquad (11)$$

where $Q_e(r)$ denotes the extinction efficiency factor, $Q_s(r)$ the scattering efficiency factor and $N(r)$ the number density of sodium mist. Moreover, the phase function of sodium mist is given by

$$P(\mu,\mu') = \sum_{n=1}^{\infty} a_n P_n(\mu)P_n(\mu') \qquad (12)$$

$$a_n = \frac{2n+1}{2}\,\frac{10^{-6}}{\sigma_s}\int_0^\infty N(r)Q_s(r)\pi r^2\left\{\int_{-1}^{1} P(r,\mu)P_n(\mu)d\mu\right\}dr \qquad (13)$$

where $P_n(\mu)$ denotes the Legendre function and $P(r,\mu)$ the phase function of a single particle with radius r.
Since evaluation of these radiative properties calls for the number density of sodium mist N(r), it is necessary to determine the form of the N(r). Hitherto, some experimental work has been devoted to obtain the N(r) of sodium mist , but these previous work was rather limited in respect of the temperatures at the lower liquid sodium surface and the cold upper end unit bounding the cover gas space. Under these circumstances, we have utilized the semi-analytical results for the weight-based particle size distributions of sodium mist obtained by Kudo and Hirata[3]. In their study the dependence of the weight-based particle size distributions of sodium mist on the temperatures at two boundary surfaces was taken into account. However, since the number-based particle size distributions are required for the calculations of radiative properties, their weight-based particle size distributions were converted to the number-based ones. As a result, we have obtained the following forms of

the number-based particle size distributions:

$$N(r)=K_1 bnr^{n-4}\exp(-br^n) \quad (r \leqq r_c) \tag{14}$$

$$=K_2\left(\sum_{i=0}^{\infty}(\ln(r)^{i-1})\exp\left(\sum_{i=0}^{\infty}c_i(\ln(r))^i\right)/r^4\right/$$

$$\exp\left(\exp\left(\sum_{i=0}^{\infty}c_i(\ln(r))^i\right)\right) \quad (r>r_c) \tag{15}$$

where r_c denotes the critical radius where the weight-based particl size distributions begin to deviate from Rosin-Rammler's. The profiles of the number-based particle size distributions thus obtained are shown in Fig.2(a)(b). Once the N(r) is obtained, we can readily get the weight concentration of sodium mist as follows:

$$w_m = \frac{4}{3} \times 10^{-12}\int_0^{\infty}\pi r^3 \rho_s N(r)dr \quad (g/cm^3) \tag{16}$$

where ρ_s denotes the density of liquid sodium and depends on

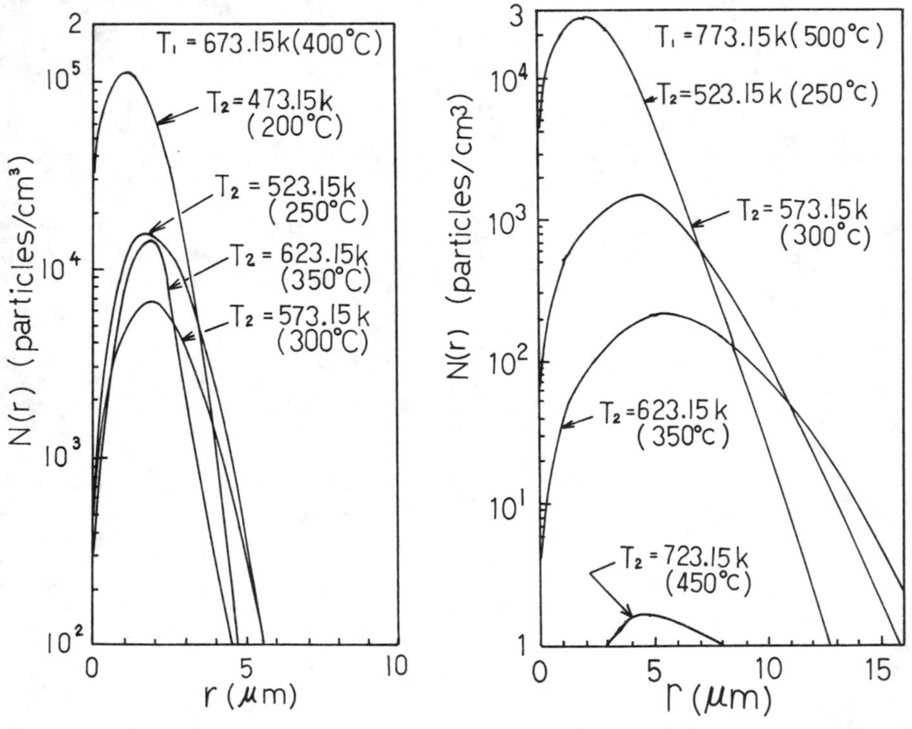

(a) Case of T_1=673.15 K (b) Case of T_1=773.15 K

FIGURE 2(a)(b). Number-based particle size distributions of sodium mist temperature. w_m proves to be a decreasing function of the temperatu: at the cold boundary surface of the cover gas space as far as the temperature at the hot liquid sodium surface is kept at constant[3]. Moreover, since the calculations of the radiative properties of sodium mist also require the complex refractive index of liquid sodium, it was evaluated by utilizing the Drude theory . Further details of the radiative properties of sodium mist in the cover gas over a hot sodium pool will be presented in our separate article[4] .

3. RESULTS AND DISCUSSION

The total radiative heat flux at the cold boundary surface is shown in Fig.3, where the cover gas space is 1 m in thickness and the temperature at the hot boundary surface is varied as a parameter. As seen from this figure, the total radiative heat flux strongly depends on the temperature at the hot boundary surface. Further, as the temperature difference between two boundary surfaces becomes large, the radiative heat flux in the case of ignoring sodium mist increases, while the radiative heat flux in the case of considering sodium mist takes the maximum value at a certain temperature at the cold boundary surface and then decreases with the further decrease in the temperature at the cold boundary surface. This is attributabl to the fact that as the temperature at the cold boundary surface decreases, the concentration of sodium mist in the cover gas space increases, and so the shield effect by sodium mist on the total radiative heat flux becomes evident. However, as the temperature at the cold boundary surface increases, the difference existing between the radiative heat flux in the case of ignoring sodium mist and that

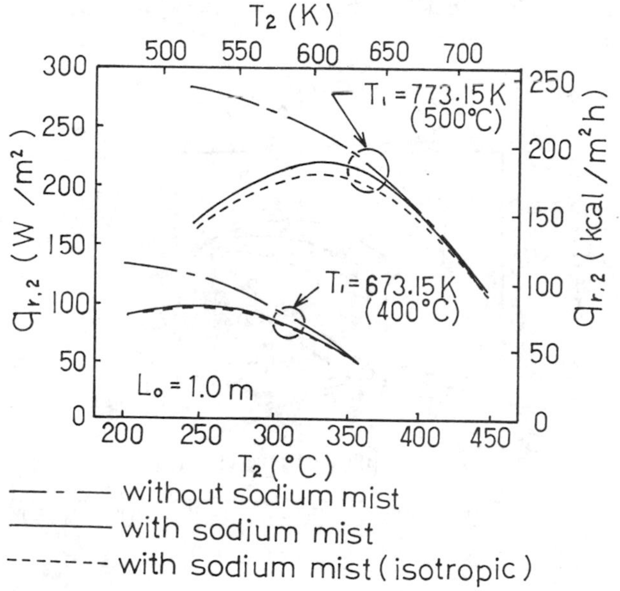

FIGURE 3. Total radiative heat flux at the cold boundary surface (Case of L_0=1.0 m)

in the case of considering sodium mist disappears because the optical thickness of the cover gas space becomes small under such condition. Furthermore, the comparison between the exact calculation of the total radiative heat flux at the cold boundary surface and the approximate one based on the isotropic scattering approximation to the phase function indicates that there is no difference between them in the case of T_1=673.15 K and only a little difference exists in the case of T_1=773.15 K. This is true even in the cases of L_0= 0.25, 0.5 and 1.5 m. Consequently, the isotropic scattering approximation is convinced to be good for evaluating the total radiative heat flux at the cold boundary surface.

The total radiative heat flux at the cold boundary surface is shown in Fig.4, where the thickness of the cover gas space is varied as a parameter. This figure indicates that when the temperature difference between two boundary surfaces is large, the total radiative heat flux decreases as the thickness of the cover gas space increases.

Figure 5 shows the comparison between the total radiative heat flux obtained by the diffusion approximation and that obtained by the exact analysis. As seen from this figure, when the temperature difference between two boundary surfaces is small, the agreement between two results is excellent, but, as the temperature at the cold boundary surface decreases, the diffusion approximation denoted by the dotted line tends to overestimate the total radiative heat flux at the cold boundary surface. This is the reason why the theoretical analysis by Furukawa et al.[1] overestimates the total heat flux in comparison with the experiment.

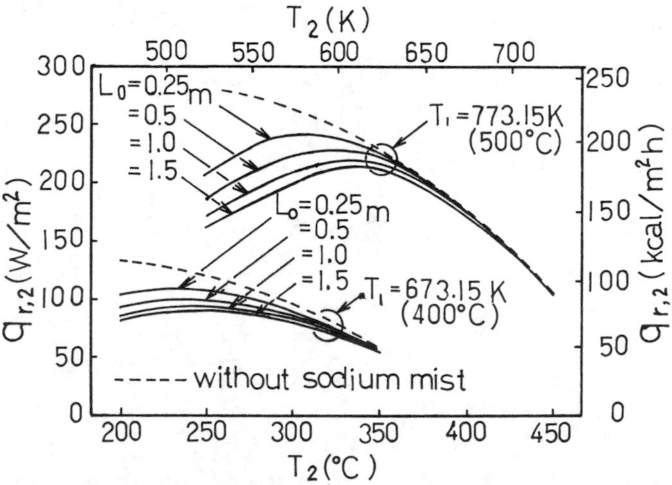

FIGURE 4. Total radiative heat flux at the cold boundary surface (Effect of thickness of the cover gas space)

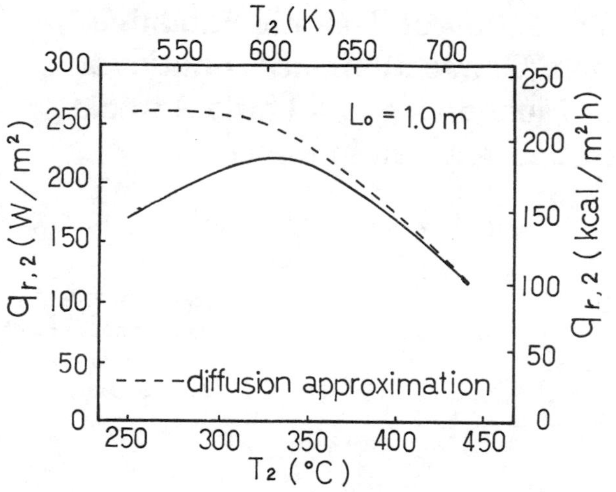

FIGURE 5. Comparison between the total radiative heat flux at the cold boundary surface obtained by the diffusion approximation and that obtained by the exact analysis

Furthermore, the gray and isotropic scattering approximation was also examined, and this approximation proves to be valid for predicting the total radiative heat flux at the cold boundary surface.

4. CONCLUSIONS

Radiative heat transfer through the cover gas space over a hot sodium pool was analyzed theoretically to evaluate the total radiative heat flux at the cold boundary surface of the cover gas space, and some related approximations were examined. As a result, it was found that the radiation shield effect by sodium mist on the total radiative heat flux becomes evident as the temperature difference between two boundary surfaces is large.

5. REFERENCES

1. Furukawa, O., Furutani, A., Hattori, N. and Iguchi, T., Heat Transfer through the Cover Gas Space in an LMFBR, 19-th Japan Heat Transfer Symposium, C112, 1982

2. Truelove, J. S., Radiative Heat Transfer through the Cover Gas of a Sodium-Cooled Fast Reactor, Int. J. Heat Mass Transf., vol. 27, no. 11, pp.2085-2093, 1984

3. Kudo, K. and Hirata, M., Sodium Deposition on the Cold Horizontal Plate over a Hot Sodium Surface, Trans. Japan Soc. Mech. Engrs., vol. 44, no. 379, pp.1025-1033, 1978

4. Kamiuto, K. and I. Kinoshita, Radiative Properties of Sodium Mist, J. Nucl. Sci. & Technol., vol. 22, no. 6, 1985, in press

Theoretical Prediction of Thermal Radiative Interaction with Convection and Conduction Heat Transfer Mechanism in a Large Particle Gas Fluidized Bed with an Immersed Horizontal Tube

BAHRAM MAHBOD
Axel Johnson Engineering Corporation
San Francisco, California 94120, USA

TARANEH TABESH
Department of Mechanical Engineering
University of California,
Berkeley, California 94720, USA

ABSTRACT

A theoretical model is presented to investigate the interaction or coupling of radiation with the con duction and convection mechanism in a large-particle gas fluidized bed. The model is based on the single particle model of Adams and Welty [9] with coupled gas convection and unsteady conduction of Adams [10] with addition of radiative heat transfer. Radiative cooling of the particles is establish by application of net radiative method to an enclosure formed by a gray cylinder surrounding a spheri cal particle. The bubble phase radiative heat flux is approximated by assuming the bubble boundary is isothermal and gray.

A parametric study of the effects of particle thermal and physical properties on heat transfer to the tube is presented. The numerical results are compared with experimental data. The heat transfer coefficients calculated using the model were found to be within the range of experimental results obtained by others. The model is expected to be valid for mean particle diameters greater than 2 mm.

INTRODUCTION

In a bubbling fluidized bed operating at elevated temperatures, the immersed surfaces will be wetted not only by the emulsion phase but also by the bubble phase and energy will be transfered from the cc of the bed to the surface through these two phases. The flow of heat from the core of the bed to the immersed surface through the bubble phase will be by convection and radiation, and that through the emulsion phase by gas convection, conduction and radiation.

When considering the mathematical models, two basic philosophies can be discerned. Both methods use a two-phase description of the system but for each approach the characteristics and the function of the phases are different.

The packet models stem from the work of Mickley and Fairbanks [1]. According to their model, partic move only in small packets, which behave as isotropic substances with a definite thermal property. Kubie and Broughton [2] modified the packet theory to allow for property variations in the packet in the region of the surface. Recently, Chen and Chen [3] modified the transient conduction model of Mickley and Fairbanks [1] to account for the effects of radiative heat transfer.

Gabor [4] developed a model based upon unsteady conduction in alternate slabs of solid and gas which was later refined by Kolar, Grewal and Saxena [5] to account for radiation contribution at high bed temperatures. Boroduly and Kovensky [6] propsed a model of radiative heat transfer in a dispersed medium which allows the calculation of the temperature distribution near a heat transfer surface whe the radiative properties of the particles and the heat exchanger are prescribed. Fatani [7] used th alternate slab model of Kolar, Grewal and Saxena [6] with addition of radiation and modified it so that transmission through the solid slab is considered.

The aim of this paper is to propose a theoretical model of bed tube conductive, convective and radia heat transfer at temperatures likely to be encountered in the fluidized bed combustors. The analysis is based upon single particle model and is expected to be accurate at low Fourier numbers (i.e., large particle and/or short residence time). In this work, the fluid thermal mass is neglected and gas convection effects are included. In the case of emulsion phase, the gas is assumed to be optically thin and in the case of bubble phase, gas radiative absobtivity is considered. The effect of particle roughness in the region of close contact is taken into account by incorporating a thermal contact resistance.

THEORETICAL DEVELOPMENT

The baseline geometry for this analysis consists of a single tube immersed horizontally in a gas fluidized bed made up of large particles with a single bubble contacting the tube as shown in Figure 1. The mechanisms of heat transfer are convection due to flow of gas within bubbles contacting the cylinder and within the interstitial voids bounded by the cylinder wall, as well as particle surfaces and thermal radiation emitted by the hot particles. The gas is expected to be optically thin and negligible heat transfer occurs by radiation from the gas. According to the assumptions of the model, the bubble which is in contact with the tube pushes the emulsion phase aside as it passes the tube so that the time-averaged heat transfer coefficient at a given location on the tube is given according to Botterill [8], as

$$Nu_p = f_b \, Nu_{p_b} + (1 - f_b) \, Nu_{p_e} \tag{1}$$

where $Nu = h \, d_p / k_g^*$, Nu_{p_b} and Nu_{p_e} are bubble and emulsion phase heat transfer coefficients, respectively, and f_b is the bubble contact fraction. The effect of radiative heat transfer has been taken into account also.

The emulsion heat transfer is based on the single particle model of Adams and Welty [9] with refinements of Adams [10] and addition of radiation. The primary element of the model is an interstitial channel adjacent to the cylinder wall. The flow channel is bounded below by the cylinder wall and on the sides by the surfaces approximately defining the circulating gas trapped between adjacent particles

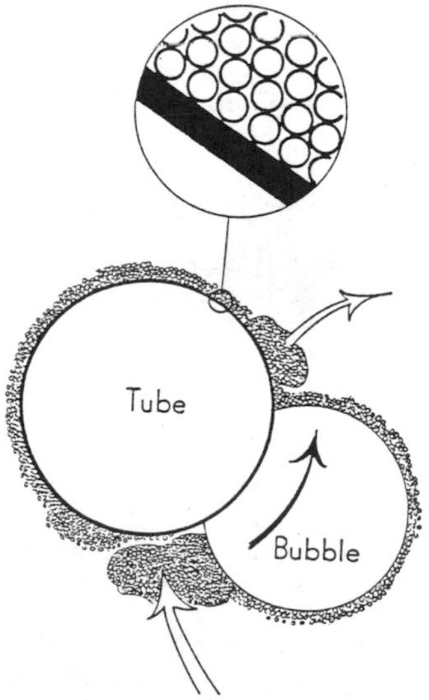

IGURE 1. Horizontal tube with attached bubble in a fluidized bed.

Since the particle surface temperature is influenced by the interstitial gas flow, the particle convective contribution is determined from a coupled analysis of transient conduction within the spheri particle and steady gas conduction within the region near the contact point of the heat transfer sur and the particle (this assumption is valid since the thermal mass of the gas is assumed to be negligible).

The particle surface temperature distribution is determined from an analysis of transient conductior within a spherical particle with convective and radiative boundary conditions. For the purpose of simplification, the temperature distribution is assumed to be symmetric about an axis through the co tact point and is therefore governed by

$$\frac{\delta T''}{\delta Fo} = \frac{1}{r^2}\left[\frac{\delta}{\delta r}\left(r^2\frac{\delta T''}{\delta r}\right) + \frac{1}{\sin\P}\frac{\delta}{\delta\P}\left(\sin\P\frac{\delta T''}{\delta\P}\right)\right] \quad (2$$

where T" is the dimensionless temperature defined as $(T-T_w)/(T_B-T_w)$, T_w and T_B are wall and bed temperatures, respectively, r is the dimensionless radial coordinate, Fo is the Fourier number, and ¶ is the angular position on the surface of the particle measured from the contact point of particle and the tube surface. The initial and boundary conditions are

$$T''(r,\P,Fo)=1 \qquad \text{at } Fo = 0^-$$

$$T''(1,0,Fo)=0 \qquad \text{at } Fo = 0^+ \quad ($$

$$\frac{\delta T''}{\delta r} = \begin{cases} -\left[Bi_c(\P) + Bi_r(\P)\right] T'' & \P \le \P_s \\ -\left[Bi_c(\P) + Bi_r(\P)\right](T'' - 1) & \P > \P_s \end{cases} \quad ($$

The convective Biot number, Bi_c, and parameter $\P_s$ are determined from the coupled gas convection an unsteady conduction model of Adams [10]. The radiative Biot number, Bi_r, is established according to

$$Bi_r = \frac{q_i(\P, Fo)\, r_p}{K_s (T_B - T_w)} \quad ($$

where q_i is the net radiative flux leaving the particle.

To obtain a reasonable estimate for the contact resistance, the surface microstructure must be considered. Due to the microscopic roughnesses and the low contact forces typical of fluidized beds, there are relatively few actual contact points when the particle touches the surface. For the purp of this analysis the model proposed by Decker and Glicksman [11] is used. The dimensions are taken from data for typical rough surfaces given by Bater and King [12].

A numerical solution of eq. (2) with stated initial and boundary conditions is obtained using an Alternate Direction Impilicit (ADI) finite difference scheme. The difference approximation of eq. is given by [10].

Radiation Heat Transfer

The radiation exchange between surfaces of the enclosure shown in Figure 2 is formulated in a conve manner by using the net radiation method of Siegel and Howell [13]. The following system of integr equations is obtained

$$q_{o,i}(d_i) = \epsilon_i T''^4(d_i) + (1-\epsilon_i)\sum_{j=1}^{4}\int_{A_j} q_{o,j}(d_j)\, dF_{di-dj}(d_i,d_j) \quad , i=1,\ldots,4 \quad ($$

where $q_{o,i}$ is the radiosity of surface i at position d_i, nondimensionlized with respect to black be radiative heat flux, i.e. T_B^4, T''_i is the nondimensional temperature of i-th surface and is equal tc T_i/T_B, this parameter is determined from transient conduction analysis of solid particles, and F_{di-} is the angle factor between i-th and j-th surfaces. These angle factors can be found in Mahbod [14 and Mahbod and Adams [15]. The net radiation leaving surface i, $q_i(d_i)$, at position d_i is given by

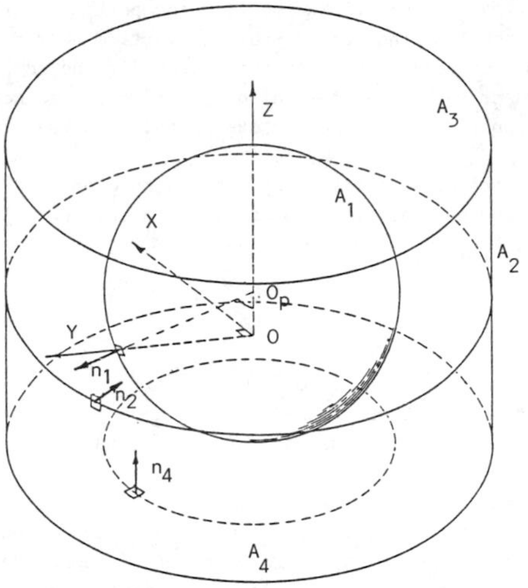

FIGURE 2. Spherical particle and the coaxial cylinder geometry.

$$q_{o,i}(d_i) = T''^4_i(d_i) - \frac{1-\epsilon_i}{\epsilon_i} q_i(d_i) \qquad i=1,\ldots,4 \tag{7}$$

where d_i represents the appropriate coordinate as equations (6) and (7) are applied to a particular surface of the enclosure, Figure 2.

The heating condition of the tube wall and bed particles are such that the particle surface temperature would vary markedly over its surface area. Therefore, to reduce the numerical complications, the surface areas are subdivided into smaller, variable size, and more nearly isothermal portions. As a consequence of this isothermal area assumption, the emitted energy is taken to be uniform over each subsurface of the enclosure and the system of integral equations (6) is reduced to the following set of coupled algebric equations for radiosities, $q_{o,i}$,

$$q_{o,1_i}(d_{1_i}) = \epsilon_1 T''^4_{1_i}(d_{1_i}) + (1-\epsilon_1)\left[\sum_{j=1}^{n} q_{o,2_j} F_{1_i-2_j} + q_{o,3}F_{1_i-3} + q_{o,4}F_{1_i-4}\right] \tag{8}$$

$$q_{o,2_i}(d_{2_i}) = \epsilon_2 T''^4_{2_i}(d_{2_i}) + (1-\epsilon_2)\left[\sum_{k=1}^{2}\sum_{j=1}^{n} q_{o,k_j} F_{2_i-k_j} + q_{o,3}F_{2_i-3} + q_{o,4}F_{2_i-4}\right] \tag{9}$$

$$i=1,\ldots,n$$

$$q_{o,3} = \epsilon_3 T''^4_3 + (1-\epsilon_3)\left[\sum_{k=1}^{2}\sum_{j=1}^{n} q_{o,k_j} F_{3-k_j} + q_{o,4}F_{3-4}\right] \tag{10}$$

$$q_{o,4} = \epsilon_4 T''^4_4 + (1-\epsilon_4)\left[\sum_{k=1}^{2}\sum_{j=1}^{n} q_{o,k_j} F_{4-k_j} + q_{o,3}F_{4-3}\right] \tag{11}$$

where parameter n coresponds to the number of divisions on each surface. The system of equations (8) - (11) is solved iteratively and each equation is updated at every iteration step with the new values of the radiosities. The resulting values of $q_{o,i}$ are substituted into equation (6) to yield the net radiation flux leaving each surface (or subsurface) of the enclosure. When a surface is both diffusely emitting and reflecting, the intensity of all the energy leaving the surface does not vary with angular direction, as a result, the geometric view factors, F_{i-j}, derived for black surfaces can be used for the present model.

Bubble Phase Heat Transfer

The bubble phase contribution to the heat transfer includes the combined effects of gas convection due to flow through the bubble and thermal radiation from the particle surfaces defining the bubble boundary. An approximate model of the bubble phase convective heat transfer to a horizontal tube in a fluidized bed is proposed by Adams [16]. The heat transfer due to particle radiation is approximat by assuming the bubble boundary is an isothermal gray surface. Two models were considered to establi an estimate of lower and upper limits of bubble phase radiative heat transfer to the immersed tube.

The first model considers an enclosure formed by a hemisphere and a flat wall. The net transfer equations are obtained by the net radiation method [13] to this enclosure. Therefore,

$$Nu_{r_b} = Bi_{r_b} \frac{\epsilon_p (1 + T''_w)(1 + T''^2_w)}{1/2 + \epsilon_p (1/\epsilon_w - 1/2)} \qquad (12)$$

where $Bi_{r_b} = d_p T_B^3 / k_g^*$ and $T''_w = T_w / T_B$, ϵ_p and ϵ_w are emissivities of bubble and tube surfaces, respectively. In this model, both bubble boundary and tube surface are assumed to be diffuse and gray, and the gas within the bubble is considered to be optically thin. Further discussion of this model is available in [17].

The second model takes into account the effect of gas absorption in an enclosure formed by the attach bubble and the heat transfer surface. To obtain an order of magnitude estimate, this enclosure is assumed to consist of two infinte parallel plates. The model reported by Heaslet and Warming [18] was adapted and modified for this purpose. Known physical conditions include the temperatures and emissivities of the two walls and absorptivity of the intervening gas. The absoptivity of the medium at the film temperature was determined from the experimental results of Abu-Romia and Tien [19]. Therefore, the bubble phase radiative Nusselt number is given by

$$Nu_{r_b} = Bi_{r_b} \frac{(1 + T''_w)(1 + T''^2_w) B_o (A_1 + B_1)}{1 + (1/\epsilon_w + 1/\epsilon_p - 2) B_o (A_1 + B_1)} \qquad (13)$$

where Bi_{r_b} is the same as before and the term $B_o(A_1 + B_1)$ is interpolate from table 1, and Z_L is the optical thickness given by

$$Z_L = K_p P_a L_b$$

where K_p is the Planck mean absorption coefficint, P_a is the partial pressure of absorbing gas, and L_b is the geometric distance between the two planes.

TABLE 1. Absoption term as a function of optical thickness [18]

Z_L	.1	.2	.3	.4	.6	.7	1.	1.5	2
$B_o(A_1 + B_1)$	.916	.85	.793	.704	.667	.605	.553	.457	.

CALCULATION AND RESULTS

The model described above was used to compute local time-averaged particle convective, radiative, to emulsion phase, and overall average Nusselt numbers for a wide range of parameters. The calculation of the radiative and average heat transfer according to equation (1) requires information regarding thermophysical properties of th solid particles and the hydrodynamic characteristics of the bed. Am the required constants are emulsion phase voidage, mean residence time, the bubble contact fraction, thermal conductivities of solids, specific heat of solid particles, and tube emissivities. The emul phase residence time and the bubble contact fraction must be determined experimentally. The results obtained by Catipovic [20] using the capacitance method are useful for this pupose. The thermophysi properties of the bed materials used in the calculations were obtained from Bater and King [12]. The voidage distribution at the cylinder surface is based in part on the capacitance probe data reported by Canada, McLaughlin and Staub [23].

To carry out the numerical integration according to equation (2), Fourier step, $\triangle$ Fo, was varied from 10^{-4} to 10. Radiative flux, q_i, was computed at each time step basee on the known surface temperature profile of the particle at the previous time step. The model is expected to be accurate if the heat from the heat transfer surface does not penetrate beyond the first layer of particles.

Parametric Analysis

Thermal diffusivity. The effect solid thermal diffusivity, α_s, on the radiative heat transfer is shown in Figure 3. The results indicate that the radiative heat transfer is slightly affected by variations in thermal diffusivity, particularly at higher values of Fourier number. Though particle radiative heat transfer is in no direct dependence on α_s, it may increase the role of α_s indirectly since at the same residence time the first row of particles will be cooled (heated) more intensely. In application of the proposed model to any operating conditions within the fluidized bed, one can allow for such effects by increasing gas conductivity, k_g^*, up to some effective conductivity, thereby yielding higher heat transfer coefficients at small residence times.

Gas absoption in bubble heat transfer. The result of a study of heat transfer to an immersed surface from an attached bubble are shown in Figure 4. This study was conducted for a fluidized bed with a

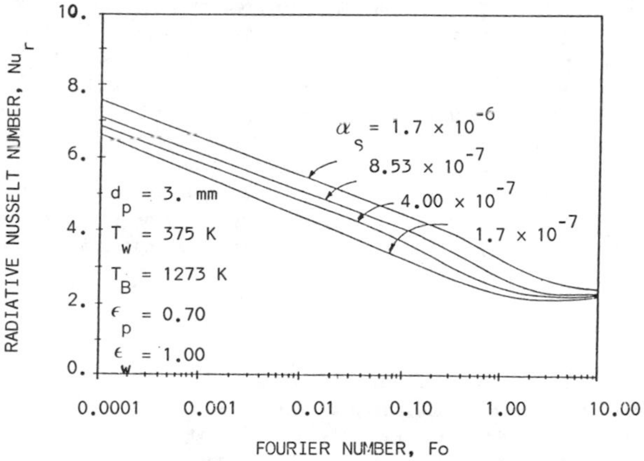

FIGURE 3. Emulsion phase radiative heat transfer vs. Fo, influence of solid thermal diffusivit.

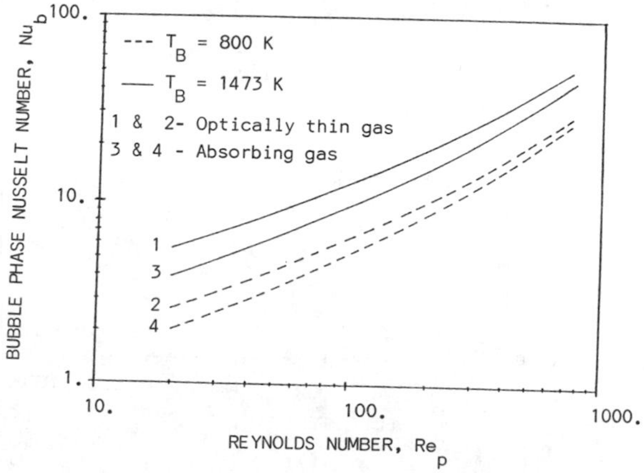

Figure 4. Bubble phase heat transfer to an immersed horizontal tube, effect of gas absorption.

mixture of Carbon dioxide and water vapor as the fluidized gas. Analysis of the illustrated result indicates that at the lower bed temperatures (800 K) the assumption of optically thin gas as sugg by [9] is valid but as the bed temperature rises, the assumption becomes questionable. For this analysis it was assumed that the bubble diameter is in the order of one cylinder diamter. For bubb with larger diameter it is expected that the effect of gas absorption becomes even more significant than that noted in this study.

Relative contribution of radiative heat transfer. The relative contribution of radiation to total h transfer with variation in particle diameter is illustrated in Figure 5. The analysis of this figu indicates that the contribution of radiative heat transfer slightly increases with increase in part diameter but such increase becomes more pronounced as the bed temperature rises. Note that all the curves in Figure 5 tend toward a limiting value as the bed temperature rises beyond 1500 K. In vi of the model discussion, this behavior may be explained by making the following observations. It i shown by [17] that at temperatures in excess of 1200 K particle and gas convective contributions t total heat transfer, while decreasing in magnitudes, are asymptotically reaching a limiting value a the rate at which radiative heat flux increases with temperature, can be seen to decrease.

Comparison with Experiment

The experimental measurements of several investigators, [20],[21] and [22], were used to test the r and verify its validity. The radiative heat transfer coefficient measurements of Alavizadeh [21] a 3.23 mm crushed refractory particle at 812 K with a single 50.8 mm immersed horizontal tube in a dimensional fluidized bed is compared with the calculated values, using the model, in Figure 6. T is shown to predict the local radiative heat transfer coefficient within 10 to 15 pecent of the me values. The experiments of Catepovic [20], Alavizadeh [21] and Baskakov and Suprum [22] were suff to compare the predicted overall time-averaged heat transfer coefficients for a range of particle diameters and bed temperatures, Figure 7. The model is found, generally, to underestimate the val of total heat transfer, particularly at lower values of superficial velocity, U_o. Compensating er in local values result ina good agreement between theory and experiment for large particles and un estimation for smaller particles. The comparison of computed and reported time-averaged radiative heat transfer coefficients was previously reported in [17].

SUMMARY

A theoretical radiative heat transfer model for a high temerature fluidized bed with an immersed h zontal tube is presented. The model is based on a single particle concept and is extension of the convective model of Adams and Welty [9] with the addition of radiative heat transfer. The calcula values of radiative contribution to the total heat transfer agree reasonably well with the experim

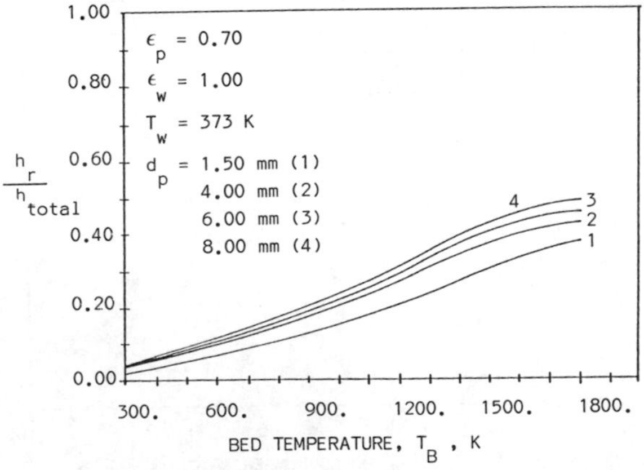

FIGURE 5. Relative contribution of radiation with bed temperature for a range of particle diamete

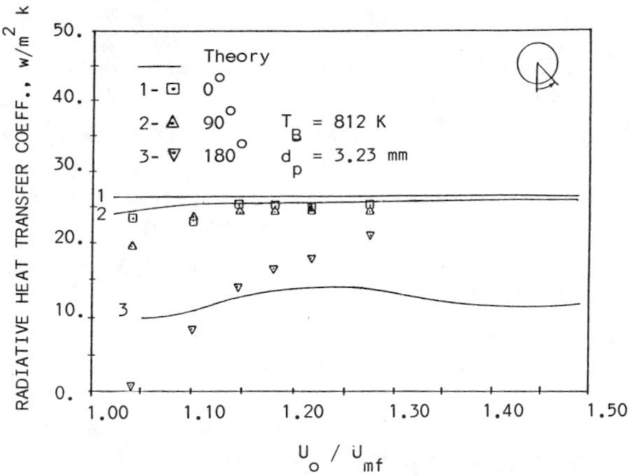

FIGURE 6. Comparison between theory and experiment: local time-averaged radiative heat transfer coefficient vs. U_o / U_{mf} for a 3.23 mm mean diameter particle and bed temperature of 812 K. U_o is the superficial velocity and U_{mf} is the minimum fluidization velocity.

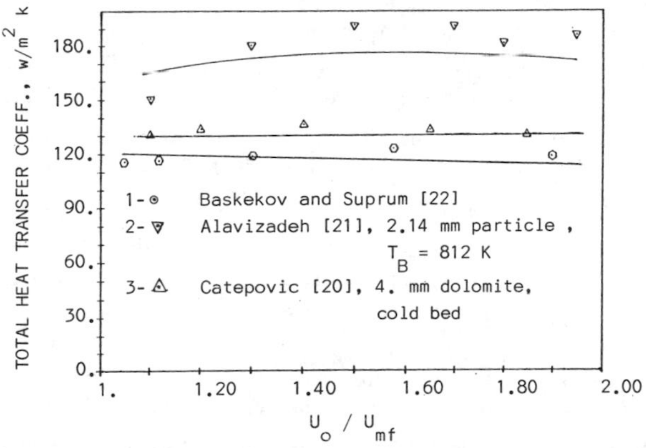

FIGURE 7. Comparison between theoretical prediction of overall time-averaged heat transfer coefficient and experimental values for a range of particle diameter and bed temperature.

values when appropriate hydrodynamic parameters are used. However, local values of heat transfer coefficient differ from experimental values particularly for smaller than 2.00 mm particles. In conclusion, in view of purely theoretical nature of the model and the extreme complexity of the phenomena, it is expected the model to predict the local radiative heat transfer coefficient with reasonable accuracy.

ACKNOWLEDGEMENT

This work was funded by the National Science Foundation under the Grant MEA 80-20781. The content of this paper does not necessarily reflect the view or policies of the foundation.

REFERENCES

1. Mickley, H.S. and Fairbanks, D.F., "Mechanism of Heat Transfer in Fluidized Beds," AICHE J., Vol. 1, pp.374-384, 1955.

2. Kubie, J. and Broughton, J., "A Model of Heat Transfer in Gas Fluidized Beds," Int. J. Heat Mass Transfer, Vol. 18, pp.289-299, 1975.

3. Chen, J.C. and Chen, K.L., "Analysis of Simultaneous Radiative and Conductive Heat Transfer in Fluidized Beds," Chem. Eng. Commun., Vol.9, pp. 255-271, 1981.

4. Gabor, J.D., "Wall to Bed Heat Transfer in Fluidized and Packed Beds," Chem. Eng. Prog. Symp. Se Vol. 66 No. 105, pp. 76-86, 1970.

5. Kolar, A.K., Grewal, N.S. and Saxena, S.C., "Investigation of Radiative Contribution in a High Temperature Fluidized Bed Using the Alternate Slab Model," Int. J. Heat Mass Transfer, Vol. 22, pp. 1695-1703, 1979.

6. Borodulya, V.A. and Kovensky, V.I., "Radiative Heat Transfer Between a Fluidized Bed and a Surfa Int. J. Heat Mass Transfer, Vol. 26, pp. 277-287, 1983.

7. Fatani, A., "An Analytical Model of Radiative and Convective Heat Transfer in Small and Large Particle Gas Fluidized Beds," Ph.D. Thesis, Oregon State University, Corvallis, Oregon, 1984.

8. Botterill, J.S.M., Fluid-Bed Heat Transfer, Academic Press, New York, 1975.

9, Adams, R.L. and Welty, J.R., "A Gas Convection Model of Heat Transfer in Large Particle Fluidize Beds," AICHE J., Vol. 25, No. 3, pp. 395-405, May 1979.

10. Adams, R.L., "Coupled Gas Convection and Unsteady Conduction Effects in Fluid Bed Heat Transfer Based on a Single Particle Model," Int. J. Heat Mass Transfer, Vol. 25, pp. 1819-1828, 1982.

11. Decker, N. and Glicksman, L.R., "Conduction Heat Transfer at the Surface of Bodies Immersed in G Fluidized Beds of Spherical Particles," Chem. Eng. Prog. Ser., Vol. 77,No. 208, pp. 341-349, 198

12. Bater, E.F. and King, H.W., Handbook of Hydraulics, McGraw-Hill, New York, 1976.

13. Siegel, R. and Howel, J.R., Thermal Radiation Heat Transfer, Hemisphere Publ., 1981.

14. Mahbod, B., "A Theoretical Study of the Radiative Contribution to Heat Transfer Between a High Temperature, Large-Particle Gas Fluidized Bed and an Immersed Tube," Ph.D. Thesis, Oregon State University, Corvallis, Oregon, 1984.

15. Mahbod, B. and Adams, R.L., "Radiative View Factors Between Axisymetric Subsurfaces Within a Cyl inder with Spherical Center-body," J. Heat Transfer, Vol. 106, No. 1, pp. 244-248, 1984.

16. Adams, R.L., "An Approximate Model of Bubble Phase Convective Heat Transfer to a Horizontal Tube in a Large Particle Fluid Bed," J. Heat Transfer, Vol. 104, No. 8, pp. 565-567, 1982.

17. Mahbod, B. and Adams, R.L., "A Model of the Radiative Contribution to Heat Transfer in a High Temperature Large-Particle Gas Fluidized Bed," ASME Paper No. 84-HT-112, 1984.

18. Heaslet, M.A. and Warming, R.F., "Radiative Transport and Wall Temperature Slip in an Absorbing Planar medium," Int. J. Heat Mass Transfer, Vol. 8, pp. 979-994, 1965.

19. Abu-Romia, M.M and Tien, C.L., "Appropriate Mean Absorption Coefficients for Infrared Radiation Gases," J. Heat Transfer, Trans. ASME, pp. 321-327, 1967.

20. Catipovic, N.M., "Heat Transfer to Horizontal Tubes in Fluidized Beds: Experiment and Theory," Ph.D. Thesis, Oregon State University, Corvallis, Oregon, 1979.

21. Alavizadeh, N., Private Communication.

22. Baskakov, A.P. and Suprum, V.M., "Determination of the Convective Component of Heat Transfer Coefficient to a Gas Fluidized Bed and a Surface," Int. Chem. Eng., Vol. 12, pp. 324-326, 1972.

23. Canada, G.S., McLaughlin, M.H. and Staub, F.W., "Flow Regimes and Void Fraction Distribution in Gas Fluidization of Large Particles in Beds Without Tube Banks," Chem. Eng. Prog. Series, Vol. 7 No. 176, p. 14, 1978.

A Dynamical Measuring Method on Infrared Emissivity of Solar Coating

RENZHANG QIAN, WEI LIU, WEIHAN CHEN, and HONGJIANG SHENG
Department of Power Engineering
Huazhong University of Science and Technology
Wuhan, Hubei, PRC

ABSTRACT

In this paper, we have presented a method of dynamically measuring thermal emissivity of solar coating and designed a set of experiment devices. In the test, the vacuum in the black cavity is pumped to more than 10^{-6} torr by mechanical and diffused pumps. The sample to be measured is made to be a hollow cylinder, with a heater in its centre and some thermal couples under its coating. Considering the temperature drop in this thin layer, we have adopted a concept of the heat resistance in dynamical measurement to obtain the surface temperature of the sample. For determining the infrared emissivity of solar coating, test period can be shortened and measuring precision can be improved by the present method.

NOMENCLATURE

T	transient temperature of the sample
T_o	initial temperature of the sample
T_p	surface temperature of the coating
T_∞	water temperature in the water bath
T_w	wall temperature of the black cavity
V	volume of the sample
F	surface area of the sample
F_w	inner surface area of the black cavity
α_c	mean heat transfer coefficient of the surface of the bare sample
α_p	mean heat transfer coefficient of the surface of the paint sample
$R, R_{p,cd}$	conductive heat resistance of the coating
$R_{p,cv}$	convective heat resistance of the paint surface
$R_{p,t}$	total heat resistance of the coating sample
$R_{c,cv}$	convection heat resistance of the bare sample
Q	electric heater power
ρ	density of the red copper
τ	the time
c	specific heat capacity of the red copper
c_o	specific heat capacity at initial temperature
c_∞	specific heat capacity at temperature T_∞
c_δ	specific heat capacity at defined temperature T_δ
ε	emissivity to be measured of the paint surface
ε_w	emissivity of the inner wall of the cavity
ε_n	system emissivity
θ	nondimensional temperature

$$\theta = (T - T_o)/(T_\infty - T_o)$$

INTRODUCTION

It is well known that the selective surface, such as black-chromium, black-nickel, black-paint layers and so on, play an important part in solar thermal utilization facilities. The adoption of the selective absorber and the improvement of its optical performance, as solar collectors, are efficacious methods to reduce thermal loss, increase collective energy and enhance collector efficiency. At present, the comparison method or caloric method is used usually in the measurement of the surface performance of solar materials, especially in the measurement of the infrared emissivity. For the former, under the same condition a seneor is taken to detect the thermal radiation emitted by a surface whose emissivity needs to be measured and by a standard surface whose emissivity has been measured, and then both of them are compared so that a normal emissivity is obtained. For the later, the semisphere emissivity of the surface to be measured is determined directly in high vacuum.

It must be noted that there is the temperature drop in the thin selective layer due to the existence of its thermal resistance, no matter what method may be used. Though the physical properties of some coating materials, at present, can not be gotton. The errors will be generated by omiting this factor.

The test rig under investigation consists of a black cavity, a liquid nitrogen jacket, some liquid nitrogen pipes, a vacuum compartment, a outer shell, a foundation frame, a mechanical pump, a diffusion pump, some vacuum meters, a electric heating system, some temperature-measured elements, some controlling instruments and so on (Fig.1). The vacuum in the black cavity and the vacuum compartment is pumped to more than 10^{-6} torr by the mechanical pump and the diffusion pump. The sample to be measured is desigened and made to be a hollow cylinder, with a heater in its centre, some thermal couples under its surface and the uniform black paint on its surface.

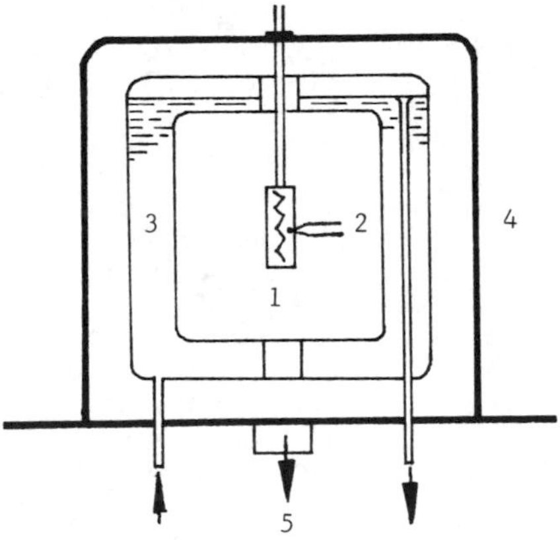

1. test sample
2. thermal couple
3. liquid nitrogen jacket
4. vacuum shell
5. to mechanical pump and diffusion pump

Fig. 1 Experimental apparatus for measuring paint emissivity

The sample material is red copper which has an excellent conductive performance, so that we can assume that its temperature is uniform in the radius direction. Five thermal couples are inlaid in the axis direction. The experimental results show that the temperature different is very small among five points, thus the temperature distribution of the sample can also be regarded as uniform in the axis direction. This leads to the lumped parameter method may be used in theoretical analysis.

METHOD AND PRINCIPLE

In the test, the liquid nitrogen is filled into the jacket which encloses the black cavity and sample. Its evaporation balance temperature is 77 K, and the temperature of the inner wall of the black cavity is able to reach this level, because of good thermal conductivity of the wall metal (stainless steel). When the vacuum in the black cavity, in which the sample is laid, is pumped to more than 10^{-6} torr, the convection and the conduction of the air disappear basically or wholly. Through radiation, the heat flux is transfered from the sample to the cavity wall and the radiation from the cavity wall to the sample can be neglected. Although the black cavity wall is painted with the black coating whose thermal emissivity is closed to one so that the heat flux emitted by the sample is absorbed directly by the black wall without much reflection process, it is difficult for the sample temperature to be steady at a constant value for a constant electric power condition. The results in references [1] and [2] indicate that it will take about 20 hours to heat the sample from heating beginning to heat balance state. If the steady-state caloric method were adopted, it would be necessary continuously to control instrument, record data and complement the liquid nitrogen to maintain the precision. Obviously, if we make experiments, using lots of the samples under different conditions based on balance temperature, a lot of time and money will be waste. So we had devoleped a transient state model, considering both shorter test period and higher measuring precision.

1. Determination of the heat resistance in paint layer

In the usual method, it is generally recognized that both the red copper and the black coating are at the same temperature without thinking the temperature drop in this thin layer. This is because some troubles may be caused by considering the temperature difference of the coating, for example, the determination of physical properties, thickness and surface temperature. In view of this fact, a concept of the conduction heat resietance was established and it can be measured by means of test and presumed to be changeless with the change in temperature. Thereby it is not necessary to pay much attention to the physical properties and the thickness and possible to find out the paint surface temperature.

Taking two same size samples, one with paint surface and other with bare surface, into a special water bath, we can make use of it to determine the coating heat resistance in the controllable temperature water. The water temperature is controlled at certain value under the boiling point, and two samples are heated at the same electric power, then turning off the power, they will cool in constant temperature water.

For the sample with bare surface:

$$\rho c V \frac{dT}{d\tau} = \alpha_c F(T_\infty - T)$$

$$\tau = 0 \qquad T = T_o \tag{1}$$

Let the specific heat capacity be linear

$$c = c_o + a(T - T_o)$$

The solution of formula (1) become

$$\frac{1}{\alpha_c} = \frac{F\tau}{\rho c_o V} \left[(1-\frac{c_\infty}{c_o})\theta - \frac{c_\infty}{c_o}\ln(1-\theta)\right]^{-1} \qquad (2)$$

For the sample with paint surface:

$$\rho c V \frac{dT}{d\tau} = \alpha_p F(T - T_p)$$

$$\tau = o \qquad T = T_o \qquad (3)$$

We define

$R_{p,cd}$ -- the conductive heat resistance of the coating;
$R_{p,cv}$ -- the convective heat resistance of the surface.

It is easy to find

$$T_p = \frac{R_{p,cv}\,T + R_{p,cd}\,T_\infty}{R_{p,cd} + R_{p,cv}} \qquad (4)$$

Substituting (4) into (3) and noting the variation of the specific heat capacity with the change in the temperature, we have the solution of formula (3)

$$\frac{1}{\alpha'_p} = \frac{F\tau}{\rho c_o V} \cdot \left[(1-\frac{c_\infty}{c_o})\theta' - \frac{c_\infty}{c_o}\ln(1-\theta')\right]^{-1} \qquad (5)$$

with

$$\alpha'_p = \alpha_p \cdot \frac{R_{p,cv}}{R_{p,cd} + R_{p,cv}}$$

During the test, the temperature difference between the samples and the water is not more than 40°C by controlling the electric heater power, so that an assumption is allowed to be made, that the two samples are basically identical in the convective condition on their surface. In other wards, the convective heat resistance of the bare surface ($R_{c,cv}=1/\alpha_c$) equals approximately that of the paint surface ($R_{p,cv}=1/\alpha_p$), i.e. $R_{c,cv} \cong R_{p,cv}$.

Evidently, for being at the same water temperature T_∞ and transfering the same heat flux Q, two samples are quite different in the variation of temperature with the time. Due to the effect of the heat resistance of this thin layer, the coating sample needs a longer time than the bare sample to get to heat balance temperature T . If a total heat resistance from the copper surface to the wate is defined as

$$R_{p,t} = 1/\alpha'_p$$

then

$$R_{p,t} = R_{p,cd} + R_{p,cv} \cong R_{p,cd} + R_{c,cv}$$

so that a approximate expression yields

$$R_{p,cd} \cong R_{p,t} - R_{c,cv} \qquad (6)$$

An actual calculating formula on the conductive heat resistance of the coating can be obtained by substituting (2) and (5) into (6). In choosing a constant time interval and making an in-step mersurement of the temperature for two sampl formula (6) becomes the following form and a capital letter R represents the mea value of the conductive heat resistance of the coating.

$$R_{p,cdi}=\frac{F\Delta\tau}{\rho c_o V}\left\{[(1-\frac{c_\infty}{c_o})\theta_i'-\frac{c_\infty}{c_o}\ln(1-\theta_i')]^{-1}-[(1-\frac{c_\infty}{c_o})\theta_i-\frac{c_\infty}{c_o}\ln(1-\theta_i)]^{-1}\right\}$$

$$i = 1,2, \ldots , n \tag{7}$$

$$R = \frac{1}{n}\sum_{i=1}^{n} R_{p,cdi} \tag{8}$$

where n is the total number of times in dynamical measurement; θ_i and θ_i' are the nondimensional temperatures of the bare sample and the paint sample, which are detected in every time step.

2. Determination of the thermal emissivity of the paint surface

In the cavity compartment with the wall temperature Tw, the coating sample irradiated heat energy at the surface temperature T_p to its ambient. Heat energy converted by the electric power is equivalent to an inner heat source with generating heat quantity Q. Omiting the effect of the heat capacity of the coating on the total heat capacity of the sample, the mathematical description of this probl can be written as the following ordinary differential equations.

$$\begin{cases} \rho cV \dfrac{dT}{d\tau} = Q - \varepsilon_n \sigma F(T_p^4 - T_W^4) \\ \tau = o \qquad T = T_o \end{cases} \tag{9}$$

or

$$\begin{cases} \rho cV \dfrac{dT}{d\tau} = Q - \dfrac{T - T_p}{R} \\ \tau = o \qquad T = T_o \end{cases} \tag{10}$$

with

$$\varepsilon_n = [1/\varepsilon + F/F_W(1/\varepsilon_W - 1)]^{-1}$$

where R can be determined by the method introduced in the preceding section. For the present device, the area ratio F/F_W equals approximately 0.02 and the emissivity of wall surface of the black cavity is about 1, and so $\varepsilon_n \cong \varepsilon$.

From (9) and (10), we have

$$T = T_p + \varepsilon\sigma FR(T_p^4 - T_W^4) \tag{11}$$

Differentiate with respect to time

$$\frac{dT}{d\tau} = (1 + 4\varepsilon\sigma FRT_p^3)\frac{dT_p}{d\tau} \tag{12}$$

For the problem of variable specific heat capacity

$$T = T_o + (c - c_o)/a$$

while taking the paint surface temperature as a variable to solve equation (9), some conversion about specific heat capacity must be considered, so we define

(i) the temperature difference between the copper surface and the paint surface is

$$T_\delta = T - T_p$$

(ii) the specific heat capacity corresponding T_δ is

$$c_\delta = c_o + a(T_\delta - T_o)$$

then

$$T_p = T - T_\delta = \frac{c - [c_o + a(T_\delta - T_o)]}{a}$$

i.e.

$$c = c_\delta + aT_p \qquad (13)$$

Accoring to (12) and (13), we rewrite (9) as follows.

$$(c_\delta + aT_p + 4\varepsilon\sigma FRc_\delta T_p^3 + 4\varepsilon\sigma FRaT_p^4)\frac{dT_p}{d\tau}$$

$$= \frac{\varepsilon\sigma F}{\rho V}\left(\frac{Q}{\varepsilon\sigma F} + T_W^4 - T_p^4\right)$$

If let

$$E = 4\varepsilon\sigma FR$$

$$M = \varepsilon\sigma F/(\rho V)$$

$$N = [Q/(\varepsilon\sigma F) + T_w^4]^{\frac{1}{4}}$$

Above formula can be expressed as an ordinary differential equation about variabl T_p

$$\begin{cases} \dfrac{dT_p}{d\tau} = \dfrac{M(N^4 - T_p^4)}{c_\delta + aT_p + c_\delta ET_p^3 + aET_p^4} \\ \tau = o \qquad T_p = T_o \end{cases} \qquad (14)$$

Through generalized integral and deriving integrated constant, the solution of the above equation can be given

$$\tau = \frac{1}{4M}\left(\frac{c_\delta}{N^3} + aEN\right)\ln\frac{T_p+N}{T_p-N} + \frac{1}{2M}\left(\frac{c_\delta}{N^3} + aEN\right)tg^{-1}\frac{T}{N}$$

$$+ \frac{a}{4MN^2}\ln\frac{T_p^2 + N^2}{T_p^2 - N^2} - \frac{c_\delta E}{4M}\ln(T_p^4 - N^4) - \frac{aE}{M}T_p + C \quad (1$$

where the integrated constant C may be gotton by initial condition from (15).

$$C = \frac{1}{M}\left[aET_o + \frac{E(c_o+aT_o)}{4}\ln(T_o^4 - T_W^4) - \frac{a}{4T_W^2}\ln\frac{T_o^2 + T_w^2}{T_o^2 - T_w^2}\right]$$

$$- \frac{1}{2M}\left[\frac{c_o + aT_o}{T_W^3} + aET_w\right]\left[\frac{1}{2}\ln\frac{T_o + T_w}{T_o - T_w} + tg^{-1}\frac{T_o}{T_w}\right]$$

Complex formula (15) can also be described as the simple form of general functio $\tau=f(T_p,\varepsilon)$. Here we know that T_p and ε are unknown numbers, as a constant time interval is chosen. Substituting (11) into (15), the function form may become $\tau=g(T_p,T)$. When time τ is known and the copper temperature is measured, we find that rewritted formula (15) will only include a variable T_p:$F(T_p)=o$. By solving the equation, the surface temperature of the paint layer may be obtained without needing the data about the thickness and physical property. Now,the thermal emissivity of the paint surface may be evaluated by

$$\varepsilon = \frac{T - T_p}{FR(T_p^4 - T_w^4)} \qquad (16)$$

3. Results of experiment and calculation

Through test and evaluation, the relationship among the infrared emissivity, the heat resistance and the sample temperature are indicated in Fig.2a. There R_1,R_2 and R_3 respectively represent three heat resistance values for different coating thicknesses. It is evident that the heat resistance rises as the thickne increases. These curves show that, as solar black coating, decrement of the thermal emissivity is dependent on decrement of heat resistance value for the sam coating material. So it is very important carefully to study spraying and plating technology.

Compared with transient-state model, steady-state caloric method has some advantages too. First, it is not necessary to consider the variation of physical properties in theoretical analysis. Then, it is not imperative to take a dynamica measurement of temperature. Both experiment and calculation, therefore, are very simple. As long as the heat exchange between the sample and the black cavity wall reachs radiative balance, the infrared emissivity can be gotton directly from (9)

$$\varepsilon = \frac{Q}{\sigma F} \left[(T-QR)^4 - T_w^4 \right]^{-1} \tag{17}$$

Formula (17) can ensure a higher precision. Based on this fact, the steady-state test was also carried out to verify the reliability of the transient-state measure ment. In Fig.2b, comparable results, estimated by this two models, have been indicated respectively by smooth curve or broken line and only a small deviation yield between the two.

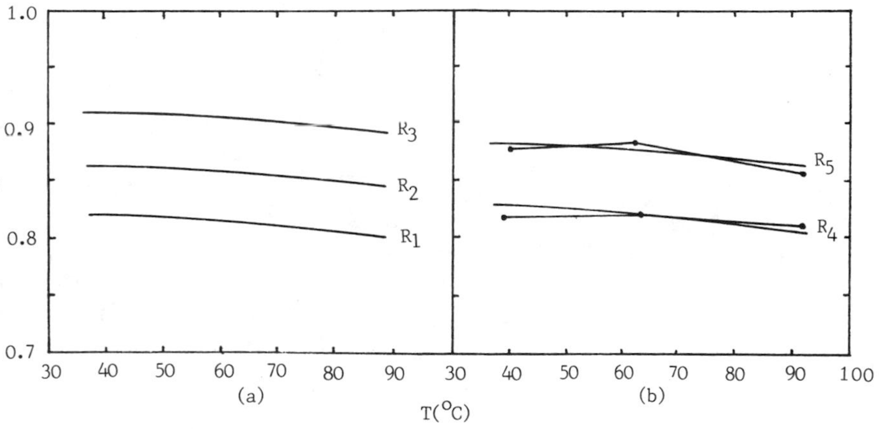

Fig.2 Variation of the thermal emissivity of the paint surface with the temperatur and the heat resistance of the coating

CONCLUSION

In the present paper, we adopt a concept of the conductive heat resistance of the coating and a principle of the lumped parameter method to determine the infrared emissivity of solar black paint. Other materials used as solar selective coating can also be detected by the same method. In analysis, because of conside ing the variation of physical properties of the sample, estimating precision is raised undoubtedly. Not only the results are more reliable, but also the experiment takes a shorter time by the introduced method.

The experimental results show that the thermal emissivity of the paint surfac has a rise tendency, as coating thickness increases and surface temperature decreases. This phenomenon is related to the change in surface state and physical properties of the coating. Due to lack of physical property data of solar selective materials and difficulties of controlling and detecting the thickness of this thin layer, this paper, therefore, use the method of determining thermal resistance to find the surface temperature of the black paint. So the data on physical property and thickness may not be considered and calculating method is simplified.

REFERENCE

[1] Qian Renzhang, "The Measurement of Total Thermal Emissivity of Solid Surface," J. Huazhong University of Science and Technology, Vol. 11, No.3, p.45 1983.

[2] Qian Renzhang, Chen Weihan, Liu Wei and Sheng Hongjiang, "A Method of Measuring Infrared Thermal Emissivity of Solar Selective Coating, "Proceeding of International Symposium on Thermal Application of Solar Energy, 1985.

[3] Nelson K.E., Luedke E.E., "A Device for the Rapid Measurement of Total Emittance." J. Spacecraft and Rocket, Vol.3, 1966.

A Simple Combined-Band Model to Calculate Total Gas Radiative Properties

Q.-Z. YU
Department of Power Engineering
Harbin Institute of Technology
Harbin, PRC

M. A. BROSMER and C. L. TIEN
Department of Mechanical Engineering
University of California
Berkeley, California 94720, USA

ABSTRACT

A combined-band model, derived from Edwards exponential wide-band model, is used to correlate total emissivity and Planck mean absorption coefficient data for radiating gases. Based on the fundamental concepts of gaseous thermal radiation, total gas emissivitics may be specified in terms of four basic parameters. The present work utillizes the combined band model to approximate the total emissivities and Planck mean absorption coefficients for carbon monoxide, carbon dioxide, methane, water vapor and sulfur dioxide. Comparisons of calculations for standard emissivities based on the combined-band model and the wide-band model indicate errors of less than 20 percent over a wide range of temperature and pressure path length.

INTRODUCTION

In engineering applications involving radiative transfer through gaseous media, the radiative properties of the various gas species must be determined. These properties are often presented through figures, charts or sets of algebraic expressions[1-4], which in general are difficult to use. Several models are commonly used when mixtures of gases are present. The first is the empirical gray-gas model for direct calculation of total emissivity[1,2]. It assumes that real gas emissions can be determined from the radiation of several gray gases. This model is not physically realistic and may cuase significant errors when extrapolating to conditions beyond the scope of the original data. It does benefit from its simplicity as the emissivities are easily obtained. A second formulation utilizes the exponential wide-band model to determine band emissivities and, upon combination, total emissivities[3,4]. This method has the advantage of being based on a solid understanding of the spectral nature of gaseous radiation, and has been shown to be applicable to gas mixtures and non-gray, non-homogeneous gases. The primary disadvantage of the method is its rapidly increasing complexity with an increasing number of absorption bands.

The purpose of the present work is to develop a model, based on the fundamental concepts of gaseous radiation, to determine total gas emissivities and yet remains simple to use in engineering analyses.

ANALYSIS

A radiating gas consists of several infrared-active bands and the total gas emissivity is defined as the sum of the emissivities of the various bands. The band emissivity ε_b, based on the total band absorption, A, can be approximated as

$$\varepsilon_b \approx \frac{\pi \, I_b \nu_o}{\sigma T^4} A \qquad (1)$$

where $I_b\nu_o$, the black body intensity evaluated at the band center ν_o, is defined as

$$I_b\nu_o = \frac{2c_1\nu_o^3}{\exp(c_2\nu_o/T)-1} \qquad (2)$$

The evaluation of the total emissivity thus requires a knowledge of the absorption responses of the individual bands to variations in temperature, pressure and path length[3,4]. Edwards exponential wide-band model[5] is based on the assumption that the various lines of an infrared-active band may be rearranged such that the intensities decay exponentially from the band head. This model requires a knowledge of three parameters: the integrated intensity α, the line width parameter β and the exponential decay width ω. It divides the band absorption into three distinct regions, based on the response of the absorption to the changes in pressure path length. The four regions correspond to the linear, square root and logarithmic responses. If it is assumed that the various bands may likewise be rearranged and combined into a single band, the total emissivity may then be directly calculated from this combined-band. Based on the analysis of a number of gases, a combined-band absorption can be defined which exhibits the same responses as the wide-band absorption. A complete and accurate specification of the emissivity, however, requires a fourth parameter in addition to the three required for the wide-band model. This parameter is the wave number associated with the effective band center of the combined-band.

In the optically thin limit, the band absorption is given as

$$A = \alpha X \qquad (3)$$

where the pressure path length X is defined as $X=P_aL$, Pa is the partial pressure of the absorbing gas and L is the physical path length. The integrated intensity α , is defined as

$$\alpha = \int_0^\infty S/\delta \; d\nu \qquad (4)$$

with the spectral line intensity given as S/δ . For real gases with multiple absorption bands, the total integrated intensity for all bands can be specified as the sum of the integrated intensities of the various bands. Then in the limit of zero path length, $X \rightarrow 0$, the total emissivity of a gas consisting of m bands may be expressed as

$$\varepsilon = \sum_{i=1}^{m} \frac{\pi I_b \nu_{oi}}{\sigma T^4} \alpha_i X \qquad (5)$$

for the wide-band analysis, or

$$\varepsilon = \frac{\pi I_b \nu_o}{\sigma T^4} \sum_{i=1}^{m} \alpha_i X \qquad (6)$$

for the combined-band model. Combining these expressions with Eqn. (2), the effective band center of the combined-band, ν_o , can be easily evaluated. For a gas which consists of a single dominant band along with several weaker bands, the band center may be approximated as a constant without incurring significant errors in the emissivity calculations. The optically thin limit may also be used to determine the Planck mean absorption coefficient. In the wide-band approximation, it is given as

$$k \approx \sum_{i=1}^{m} \frac{\pi I_b \nu_{oi}}{\sigma T^4} \alpha_i \tag{7}$$

Based on the combined-band, this may be further approximated as

$$k \approx \frac{\pi I_b \nu_o}{\sigma T^4} \alpha \tag{8}$$

with

$$\alpha = \sum_{i=0}^{m} \alpha_i$$

The two remaining parameters may be obtained from the square-root and logarithmic responses of band absorption. The emissivities of the two regions are given as

$$\varepsilon = \frac{\pi I_b \nu_o}{\sigma T^4} [C_2(XP_e^n)^{\frac{1}{2}} - \omega\beta P_e^n] \tag{9}$$

and

$$\varepsilon = \frac{\pi I_b \nu_o}{\sigma T^4} \omega[\ln(\frac{C_2^2 P_e^n}{4\omega^2} X) + 2 - \beta P_e^n] \tag{10}$$

respectively with $C_2 = (4\alpha\beta\omega)^{\frac{1}{2}}$. The effective pressure P_e is defined as

$$P_e = [1 + (B - 1) P_a/P_t]P_t/P_o \tag{11}$$

where P_t is the total pressure of the gas mixture and the normalizing pressure P_o is specified as one atmosphere. The value of the pressure broadening coefficient B may be obtained through the procedure of Burch et al.[6] and Anderson et al[7]. The value of the pressure broadening exponent n may be obtained from the square root data for various pressure ratios, P_a/P_t. For standard emissivity, the effective pressure is identically one and is obtained in the limit of the partial pressure approaching zero.

A single, continuous band absorption relation for Edwards model is given by Tien and Lowder[8]. Using this approximation for the combined-band absorption, the total emissivity is obtained as

$$\varepsilon = \frac{\pi I_b \nu_o}{\sigma T^4} \omega \ln[\tau f(\beta P_e^n) \frac{\tau + 2}{\tau + 2f(\beta P_e^n)} + 1] \tag{12}$$

where the function $f(\beta P_e^n)$ is defined as

$$f(\beta P_e^n) = 2.94[1 - \exp(2.60\beta P_e^n)] \tag{13}$$

and the dimensionless path length is given as $\tau = \alpha X/\omega$.

This relation may be used over a wide range of pressure path lengths to iteratively determine the remaining parameters of C_2, ω and n.

RESULTS AND DISCUSSION

The total emissivities and Planck mean absorption coefficients have been computed based on the combined-band analysis for carbon monoxide, carbon dioxide, methane, water vapor and sulfur dioxide.

A number of expressions have been published for the emissivities of these gases and, within their ranges of applicability, the discrepancies exceed $\pm$ 20 percent [2-4, 9-14]. This 20 percent error band is thus chosen as a major criterion for acceptability of the combined-band model. The error is defined as

$$\text{Error} = \frac{\varepsilon_c - \varepsilon_r}{\varepsilon_r} \times 100\%$$

where the calculated and reference emissivity values are given as ε_c and ε_r respectively. For all gases, the exponential wide-band model is used to define the reference emissivity values. The wide-band parameters presented by Edwards [3] have been chosen for use with carbon monoxide and carbon dioxide, while the parameters presented by Brosmer and Tien[12], Edwards[13] and Kunitomo et al. [14] have been used for methane, water vapor and sulfur dioxide, respectively.

The effective band centers, ν_o, and the total integrated intensities of the five gases are presented in Table 1. Based on these values, the Planck mean absorption coefficients have been calculated and are presented in Figs. 1 and 2. For comparison, the reference calculations, based on the individual bands, are also presented. The agreement can be seen to be excellent for all gases, with a maximum error of 15 percent for carbon dioxide. Also presented in Table 1 are the pressure broadening parameters, B and n.

The temperature variation of the band width parameter, ω, and modified line width parameter, C_2, for CO, CO_2, and methane may be expressed in the form

$$\omega, C_2 = \left(\frac{T}{T_o}\right)^b \sum_{i=0}^{j} a_i \left(\frac{T}{T_o}\right)^i \tag{14}$$

while they take the form

$$\omega, C_2 = \left(\frac{T}{T_o}\right)^b \exp \sum_{i=0}^{j} a_i \left(\frac{T}{T_o}\right)^i \tag{15}$$

for H_2O, and

$$\omega, C_2 = \exp \sum_{i=0}^{j} a_i \left(\frac{T_o}{T}\right)^i \tag{16}$$

for SO_2. The coefficients of these polynomials are given in Table 2, with T_o = 300 K. These values, when combined with Equs. (12) and (13) are valid for the determination of emissivities for path lengths from zero up to 1000 atm-cm.

For water vapor, care should be taken however at low emissivities, $\varepsilon < 0.015$, where the approximate results deviate from the wide-band computations by greater than 20 percent. For all other gas species, and for $\varepsilon > 0.015$ for water vapor, the emissivity can be calculated to within the 20 percent accuracy levels. The error bands associated with each gas are presented in Table 3 along with the maximum applicable temperature. The minimum temperature in all cases is 300K. The results of the combined-band calculations are presented in Figs.3-7 along with the wide-band calculations which are used for comparison.

Calculations have also been preformed for effective pressures other than unity. For methane and carbon monoxide, the resulting emissivities are accurate to within ± 25 percent for effective pressures between 0.1 and 10. For sulfur dioxide, carbon dioxide and water vapor, an accuracy of ± 30% percent is attained for effective pressures between 1/3 and 3. The combined-band parameters were developed for standard emissivity, and emphasis was placed on the accuracy in this limit. It is therefore not unexpected that larger errors would be incurred at effective pressures much larger or small than unity, even with the pressure broadening effect accounted for.

CONCLUSIONS

A combined band model has been developed which is able to simply and accurately determine total gas emissivities and Planck mean absorption coefficients. The model has been used to compute the total standard emissivities and Planck mean absorption coefficients of five gases with the results compared to values predicted by the wide-band model. In all cases, the emissivities are determined to within ± 20 percent of the reference values. The Planck mean absorption coefficient is accurate, in all cases, to within ± 15 percent.

REFERENCES

[1] Hottel, H. C. and Sarofin, A. F., Radiative Transfer, McGraw-Hill Publishers, New York (1967).

[2] Farag, I. H. and Allen, T. A., AMSE/AICHE National Heat Transfer Conference, Milwaukee, Wisconsin, Aug. 2-5, (1981).

[3] Edwards, D. K., in Advances in Heat Transfer, Vol 5, Academic Press, New York (1972) .

[4] Tien, C. L., in Advances in Heat Transfer, Vol 5, Academic Press New York (1968).

[5] Edwards, D. K. and Menard, W. A., Applied Optics, Vol 3, 621 (1964).

[6] Burch, D. E., Singleton E. B., and Williams, D., Applied Optics, Vol. 1 No. 3 pp. 359-363 (1962).

[7] Anderson, A., Chai, A-T, and Williams, D., J. Optical Society of Am., Vol. 57 No. 2 pp. 240-246 (1967).

[8] Tien, C. L. and Lowder, J. E., Int. J. Heat Mass Transfer, Vol, 8, p 698 (1966).

[9] Leckner, B., Combustion and Flame, Vol. 19,33 (1972).

[10] Sarofim, A. F., Farag, I. M. and Hottel, H. C., AIAA/ASME Thermodynamics and Heat Transfer Conference, Palo Alto, California, May 24-26 (1978).

[11] Chan, S. H. and Tien, C. L., ASME J. Heat Transfer, Vol. 93.

[12] Brosmer, M. A. and Tien, C. L., JQSRT, in press.

[13] Edwards, D. K., Chapter 4 in Handbook of Heat Transfer, in press.

[14] Kunitomo, T., Masuzaki, H., Veoka, S. and Osumi, M., JQSRT, Vol. 25, 345 (1981).

Table 1. Band centers, integrated intensities and pressure broadening parameters for the combined bands of carbon monoxide, carbon dioxide, methane, water vapor and sulfur dioxide.

gas	ν_o cm^{-1}	α $atm^{-1}-cm^{-2}$	B	n
CO	2143	$240\left(\frac{300}{T}\right)$	1.1	0.8
CO_2	$2102\left(\frac{T}{300}\right)exp\left[-0.286\left(\frac{T}{300}\right)\right]$	$2358\left(\frac{300}{T}\right)^{0.98}$	1.3	0.65
CH_4	$805.9+919.8\left(\frac{T}{300}\right)^{0.75}$	$414\left(\frac{300}{T}\right)^{0.98}$	1.3	0.8
H_2O	$56.84+80.25\left(\frac{T}{300}\right)^{1/2}$	$83290\left(\frac{100}{T}\right)^{1.04}exp\left[-2.89\left(\frac{300}{T}\right)^{1/2}\right]$	$8.6\left(\frac{100}{T}\right)^{1/2}+0.5$	1
SO_2	1265	$7.347+1084\left(\frac{300}{T}\right)$	1.9	1

Table 2. Polynomial coefficients for combined band parameters of C_2 and ω

gas		b	a_o	a_1	a_2	a_3	a_4
CO	C_2	-1/2	29.69	5.826	0.1052	-	-
	ω	0	29.77	17.22	1.393	-	-
CO_2	C_2	-1/2	31.34	290.6	-83.84	7.515	-
	ω	-1/2	189.5	-46.82	10.586	1.4929	-
CH_4	C_2	-1/2	248.1	-157.7	65.35	-3.105	-
	ω	1/2	84.69	25.81	3.522	-1.121	-
H_2O	C_2	-1/2	4.426	1.652	-0.2231	0.01770	-0.000612
	ω	3	4.882	0.2888	-0.05821	0.00222	-
SO_2	C_2	-	5.280	-1.144	1.805	-	-
	ω	-	7.757	-7.579	5.316	-	-

Table 3. Temperature limits and error bands for combined-band computations of standard emissivity.

	CO	CO_2	CH_4	H_2O	SO_2
T_{max} (K)	2400	2200	1700	2400	2400
Error Band (%)	-10, +6	-20, +15	-11, +13	-16, +20	-15, +13

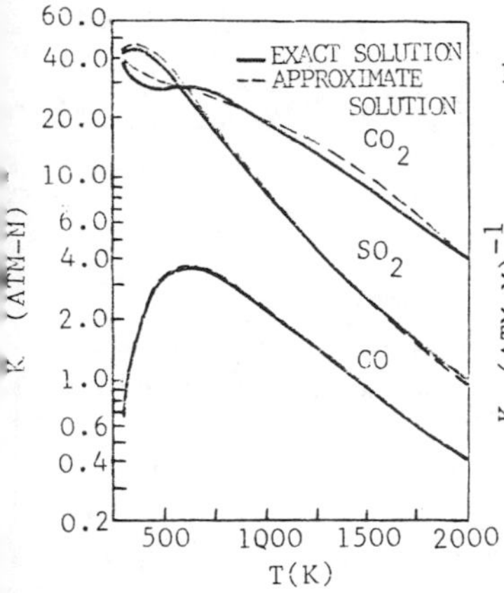

Fig.1 K=f(T)

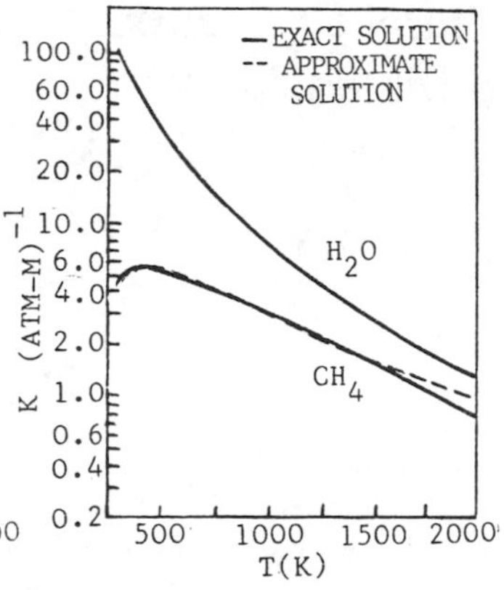

Fig.2 K=f(T)

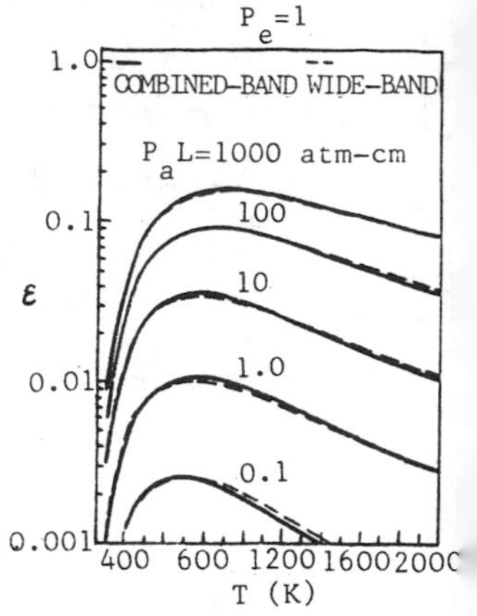

Fig.3 CO

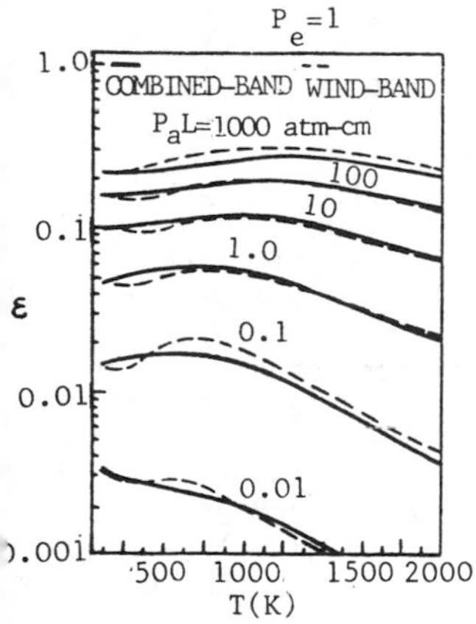

Fig.4 CO_2

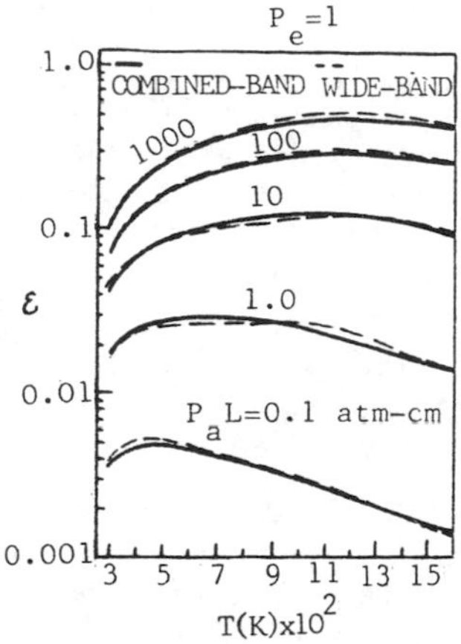

Fig.5 CH_4

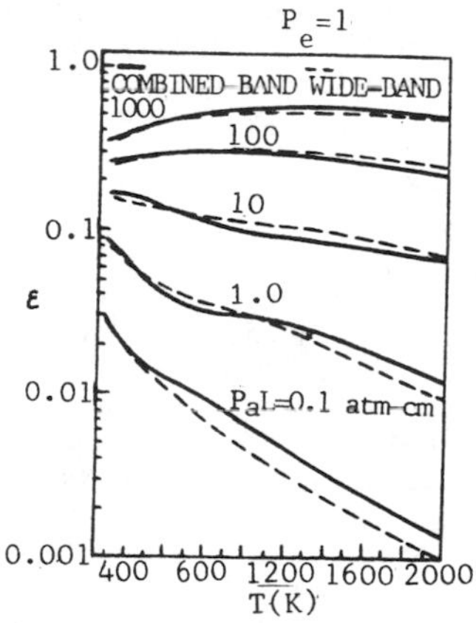

Fig.6 H_2O

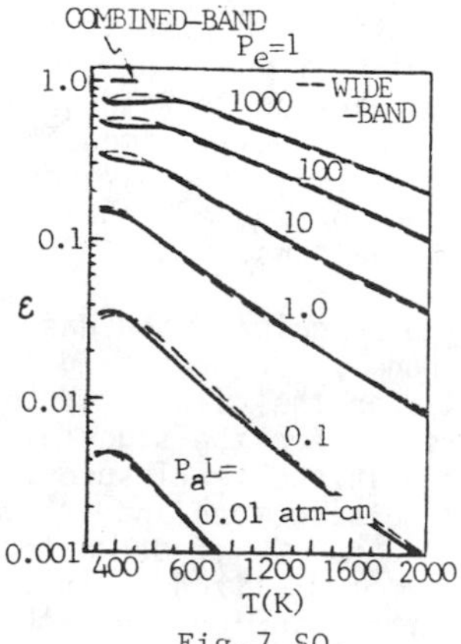

Fig.7 SO_2

A Total Slice Band Absorptance Model for Nongray Gas-Natural Convection Interaction in a Two-Dimensional Square Enclosure

Z. Y. ZHONG
Shanghai Jiaotong University
Shanghai, PRC

J. R. LLOYD
Michigan State University
East Lansing, Michigan 48823, USA

K. T. YANG
University of Notre Dame
Notre Dame, Indiana 46556, USA

INTRODUCTION

In many buoyancy-driven enclosure-flow phenomena involving gases, convection rates are generally relatively low, and consequently thermal radiation effects can become significant even at moderate temperature variations. Good examples are solar collector applications and cooling of electronic components in enclosures. When high temperatures are encountered, effects of thermal radiation become much more important and its interaction with the buoyant flow significantly alters the flow field. Phenomena such as buoyant flows in furnaces and fire and smoke spread in rooms and buildings are good examples in the high-temperature environment. Since radiation participating gases such as carbon dioxide and water vapor are often present in these phenomena, any analysis should realistically include the interaction effects of gas radiation, surface radiation and natural convection. The usual approach is to incorporate radiative transfer resulting from gas absorption and emission into the energy equation for the buoyant flow as a source term. Complexity arises in view of the fact that radiative transfer characteristics depend on the temperature field, which is however at the same time affected by the radiative transfer in the enclosure. Consequently, the radiative transfer equation and the equations for the buoyant enclosure flow must be solved simultaneously. With the recent advances in numerical solutions to the enclosure flow problem and simplified computations of radiative transfer for nongray gases, at least in the two-dimensional situations, quantitative results for enclosure flows with gas radiation-natural convection interactions can be obtained. Reference [1] provides a recent review of the various techniques now available for the computation of radiative transfer in multi-dimensional enclosures containing an absorbing-emitting, but non-scattering, nongray gas or gas mixture. Some emphasis has also been placed on how these techniques can be implemented into the numerical buoyant-flow calculations.

The purpose of this paper is to describe a new gas radiation model, known as the total slice band absorptance model, for nongray gases or gas mixtures. This model is particularly suitable for the computation of two-dimensional radiative transfer in a square enclosure and for the study of nongray gas radiation - natural convection interactions in such enclosures. In the following sections the model will first be described and its validity and accuracy are then demonstrated in three cases for which accurate results are known in the literature. Finally, the model is utilized in an interaction study for a tilted square enclosure containing carbon dioxide and with differentially heated side walls.

A TOTAL SLICE BAND ABSORPTANCE MODEL FOR A NONGRAY GAS IN A TWO-DIMENSIONAL SQUARE ENCLOSURE

Consider a long enclosure with a square cross-sectional area with black side walls as shown in Figure 1(a). For a non-scattering medium in local thermodynamic equilibrium, the radiative transfer along a beam in an arbitrary direction s, say along SA in Figure 1(a), is given by

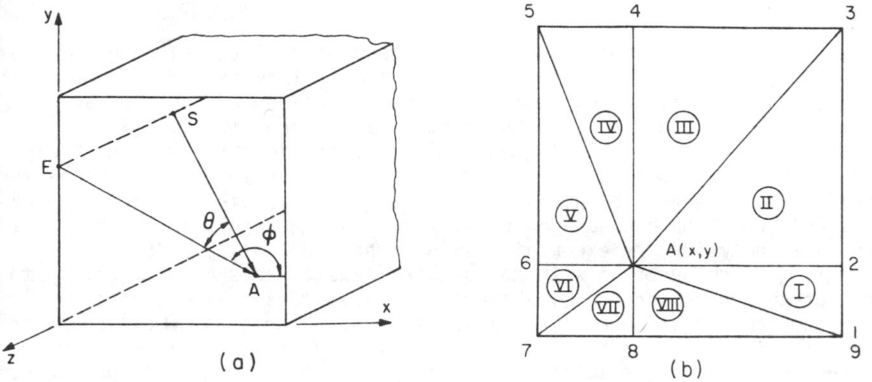

FIGURE 1. (a) Enclosure geometry,(b) Integration limit map for radiation fluxes.

$$\frac{dI_\nu}{ds} = \kappa_\nu (I_{b\nu} - I_\nu) \tag{1}$$

where I_ν is the spectral radiative intensity, κ_ν the spectral absorption coefficient, and $I_{b\nu}$ the Planck function. Equation (1) can be readily integrated along the beam, resulting in

$$I_\nu(\tau) = I_\nu(0)e^{-\tau} + \int_0^\tau I_{b\nu}(\tau-\tau')e^{-\tau'}d\tau' \tag{2}$$

where $I_\nu(0)$ is the intensity leaving the boundary at S and τ is the optical thickness defined by

$$\tau = \int_0^S \kappa_\nu \, ds' \tag{3}$$

where s is again measured from the boundary. Since a two-dimensional enclosure is considered here, equation (2) can be recast in terms of the optical thickness in the plane of z=o containing points A and E as follows:

$$I_\nu(\tau) = I_\nu(0)e^{-\tau/\cos\theta} + \int_0^\tau I_{b\nu}(\tau,\tau')e^{-\tau'/\cos\theta}\frac{d\tau'}{\cos\theta} \tag{4}$$

which, when multiplied by $\cos^2\theta$ and integrated with respect to θ from $\theta = -\pi/2$ to $\theta = +\pi/2$, yields a "slice spectral intensity" $I_{\nu s}$:

$$I_{\nu s} = I_{\nu s}(0)\int_{-\pi/2}^{\pi/2} \cos^2\theta e^{-\tau/\cos\theta}\, d\theta + \int_0^\tau I_{b\nu}[\int_{-\pi/2}^{\pi/2} \cos\theta e^{-\tau'/\cos\theta}\, d\theta]d\tau' \tag{5}$$

By introducing the following integral functions

$$D_n(\tau') = \int_0^1 t^{n-1}(1-t^2)^{-1/2} e^{-\tau'/t} dt \tag{6}$$

$I_{\nu s}$ from equation (5) can now be written in a final form as

$$I_{s\nu} = 2I_{s\nu}(0)D_3(\tau) + 2\int_0^\tau I_{b\nu}(\tau,\tau')D_2(\tau')d\tau' \tag{7}$$

It is evident that D_n are universal functions and therefore D_2 and D_3 can be calculated once for all, and accurate results from direct numerical integration for these functions over a wide range of τ have been obtained and are given in [2].

Before equation (7) can be used to determine the radiant fluxes throughout the enclosure and the subsequent radiant energy source, information on the spectral absorption coefficient is needed, and this coefficient depends on the wave number, temperature and pressure of the participating medium. With a few exceptions, the spectral data on the absorption coefficient for most commonly encountered gases are known [3,4], but must be modeled so that computations for the radiation intensity can be performed. These include both gray and nongray gas models [5], which have also been briefly reviewed recently in Reference [1]. In the present analysis, the exponential wide-band model of Edwards and Menard [6] is utilized in conjunction with the scaling laws of Felske and Tien [7] to scale the optical thickness and line structure parameters to those of an equivalent homogeneous isothermal radiation system, so that nonhomogeneous and nonisothermal conditions can be accommodated. The exponential wide-band model is known to give excellent results in systems at relatively moderate pressures [8,9].

The slice spectral intensity given in equation (7) can now be integrated over the wave number spectrum, yielding the following equation for the total slice intensity:

$$I_s = \frac{1}{2}\sigma T_w^4 - \frac{1}{2}\sum_j \int_0^\tau A_j d\, I_{b\nu}(\tau') \tag{8}$$

where j designates the j-band, and A_j is a total slice band absorptance defined by

$$A_j = \int_{\Delta\nu} [1 - \frac{4}{\pi} D_3(\tau_j)]d\nu \tag{9}$$

In addition σ is the Stefan-Boltzman constant and T_w is the wall temperature. After introducing the exponential wide-band model, A_j in (9) now becomes

$$A_j = \omega_j \int_0^1 [1- \frac{4}{\pi} D_3(\tau_{Hj}\tau_j')] \frac{d\tau_j'}{\tau_j'} \tag{10}$$

where

$$\tau_{Hj} = (\frac{\alpha_j}{\omega_j})\rho s , \qquad \tau_j = \tau_{Hj}\tau_j' , \qquad \tau_j' = \exp[-(\nu_j - \nu)/\omega_j]$$

where ν_j is the wave number at the head of the band, ω_j the band width parameter, and α_j the integrated band intensity parameter. It may be noted tha the quantity A_j/ω_j is similar to the axial band absorptance used by Wassel and Edwards [10] for cylindrical cavities. Before equation (8) can be used to calculate I_s, the band absorptance must further be scaled to account for nonhomogeneous and nonisothermal conditions [7] by using a mass path length. Details of this scaling are given in [2].

Once the total slice intensity I_s is known, the radiative flux components in the two principal directions can then be determined by integrations over the solid angle ϕ (Figure 1(a)) by

$$q_{rx} = -\int_0^{2\pi} \cos\phi \; I_s \; d\phi \quad , \qquad q_{ry} = -\int_0^{2\pi} \sin\phi \; I_s \; d\phi \tag{11}$$

The evaluation of these integrals is expedited by dividing the whole enclosure into eight segments as shown in Figure 1(b), so that the physical path lengths in each segment can be identified, and the corresponding limits of integration are given by

$$\phi_1 = \tan^{-1}[(-y/H)/(1-x/H)], \quad \phi_2 = 0, \quad \phi_3 = \tan^{-1}[(1-y/H)/(1-x/H)],$$
$$\phi_4 = \pi/2, \quad \phi_5 = \tan^{-1}[(1-y/H)/(-x/H)] \, , \quad \phi_6 = \pi, \tag{12}$$
$$\phi_7 = \tan^{-1}[(-y/H)/(-x/H)], \quad \phi_8 = \frac{3}{2}\pi \, , \quad \phi_9 = 2\pi + \phi_1$$

Since the local temperature, density and pressure data are needed along each beam within each segment in the determination of absorption and emission characteritics at A, a bilinear interpolation formula is utilized, based on the respective values from the four neighborhing calculation cells for the enclosure flow calculations. Finally, the energy source term due to gas radiation is simply given by the divergence of the radiative flux vector q_r:

$$\nabla \cdot q_r = \frac{\partial q_{rx}}{\partial x} + \frac{\partial q_{ry}}{\partial y} \tag{13}$$

which can then be evaluated from (11) and incorporated into the energy equation governing the buoyant flow field. It is significant to note that the calculation method based on the total slice band absorptance model just described is completely compatible with the numerical solution to the enclosure flow problem, which is described in [11] and hence will not be repeated here.

MODEL AND CALCULATION PROCEDURE VALIDATION

To determine the validity and accuracy of the band absorptance model and the calculation procedure described in the last section, radiation flux calculations have been carried out in three separate cases for which accurate results are known. The first case treats gas radiation in a square enclosure with four black cold walls and filled with a gray participating medium at a uniform temperature of T_G. Results of the present calculations in terms of the radiation fluxes at the walls for three characteristic optical thicknesses ($\tau_D = \kappa H$) of 0.1, 1.0 and 10.0 are shown in Figures 2(a), (b) and (c), respectively. Here κ is the constant gas absorption coefficient and H is the enclosure height or width. Also shown in these figures are the corresponding exact results [12,13] and the results based on a different calculation procedure [12]. It is clear that the present results are essentially identical to that of the exact solution.

A second case calculated deals with a radiative equilibrium problem in a square enclosure filled with a gray medium. All walls are taken to be black and one wall, y=0, is taken to be at a temperature of T_w, while the other three walls are cold. Since the radiation fluxes depend on local temperatures, an iterative method of calculation is used in this case with an initial estimated temperature field. The final temperature field is determined when radiation equilibrium is achieved to a desired tolerance. The criterion used here is that the sum of

all absolute values of radiation heat gains of all calculation cells in the domain is less than 1.0% of the energy emitted by the hot wall. Figure 3(a) shows the calculated temperature profile along the enclosure center line at X/H = 0.5, and Figure 3(b) gives the radiation flux distribution along the hot wall. Also shown in these figures are the corresponding results based on the Hottel's zone method [14] credited to Larsen and also based on finite-element calculations [14]. It is seen that all results are practically identical.
The above two cases both deal with gray gas only and the excellent results only attest to the proposed calculation procedure. A more definitive case must be sought so that the total band slice absorptance model can be more critically examined. The third case calculated deals with a square enclosure with a temperature field given so that all isotherms are parallel to the y-axis and the temperature decreasese linearly in the x-direction from $T = T_H$ on the hot wall and $T = T_C = 0.5\ T_H$ on the cold wall. The gas medium is a mixture of 96% of air and 4% of CO_2 by volume. Only the optically thin limit is considered here. Since the gas in the enclosure is not isothermal, scaling laws are also used. The results in terms of radiation flux distributions along the center line at X/H = 0.5 and along the side wall y=0 are respectfully shown in Figures 4(a) and (b). Also shown are the corresponding results of Abu-Romia and Tien [15] based on the Plank mean calculations. The two sets of data are within 2% of each other. On the basis of the above three calculated cases, it is felt that the proposed total slice band absorptance model, together with the calculation procedure described previously, is inherently very accurate and can be applied to general radiative transfer calculations in two-dimensional enclosures filled with participating gases or gas mixtures.

GAS RADIATION-NATURAL CONVECTION INTERACTION IN A TILTED SQUARE ENCLOSURE

In the present study, the calculations of pure laminar natural convection in a tilted square enclosure with variable properties [11] has been extended to cases including gas radiation effects [2]. The range of parameters covered in the calculations consists of Rayleigh numbers up to 10^6, temperature difference ratios $\theta_0 = (T_H - T_C)$ up to 2.0, and the whole range of the tilt angle ψ (Figure 5). Both CO_2 and air filled enclosures have been considered. Some details of these results and their physical discussions are given in [16] and will not be repeated here. However, for the purpose of illustration only, two sets of results are shown in Figures 6 and 7. Figure 6 shows the usual average Nusselt number variations for a range of the Rayleigh numbers for a vertical CO_2 enclosure with $\theta_0 = 2.0$. Dimensionless quantities are calculated with properties based on the cold wall temperature. Nu_T accounts for both natural convection and surface and gas radiation and can be adequately correlated by

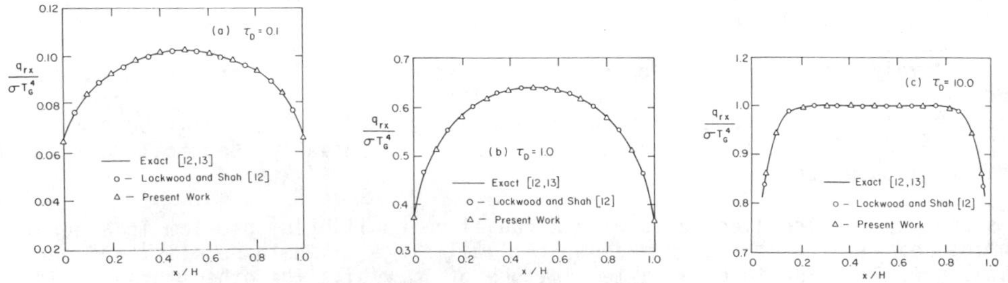

FIGURE 2. Radiation flux distribution along black and cold walls of a square enclosure.

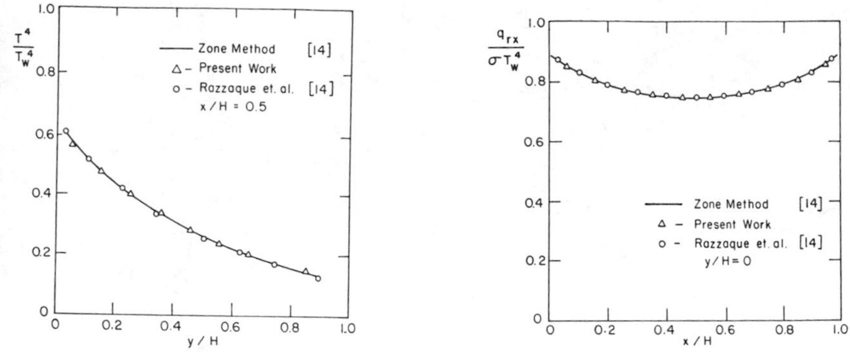

(a) (b)

FIGURE 3. (a) Center line temperature profile, (b) Radiation fluxes along hot wall.

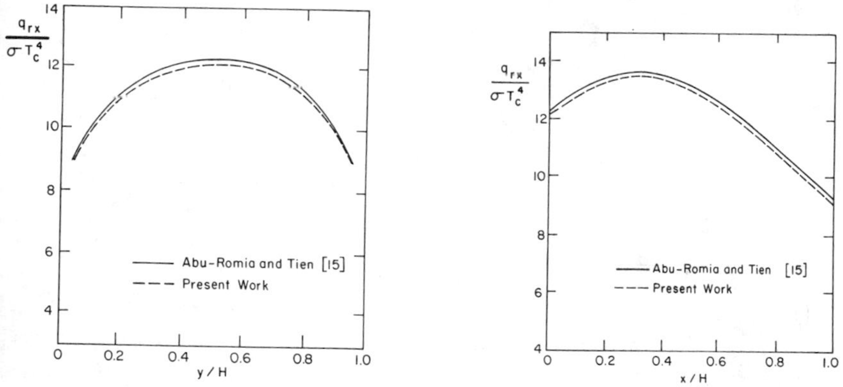

(a) (b)

FIGURE 4. Radiation flux distribution of a square enclosure with linear temperature distribution along (a) center line (b) side wall.

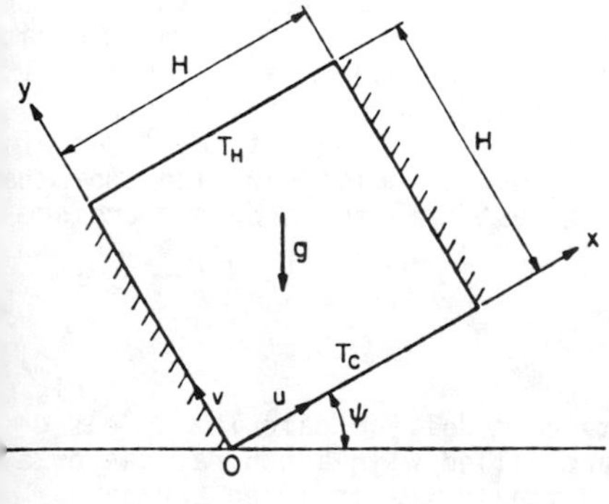

FIGURE 5. Tilted square enclosure geometry

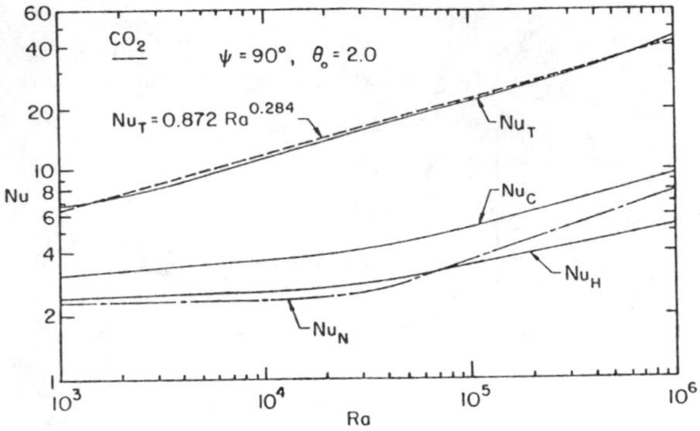

FIGURE 6. Interaction Nusselt numbers for CO_2, ψ = 90° and θ_0 = 2.0.

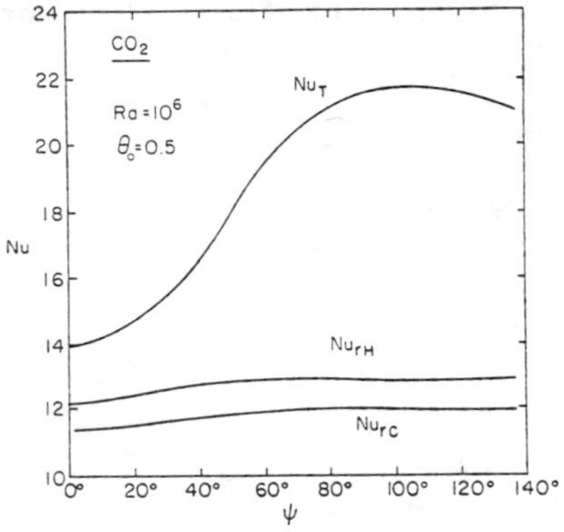

FIGURE 7. Effect of tilt angle on interaction Nusselt numbers for CO_2, Ra=10^6 and θ=0.5.

$Nu_T = 0.872\ Ra^{0.284}$. The corresponding Nusselt numbers without radiation effects, Nu_N are also shown and the effect of radiation for θ_0 = 2.0 is clearly seen. Figure 6 also shows two other Nusselt numbers. Nu_C and Nu_H represent th conductive parts of the total heat transfer at the cold and hot walls, respectively. Their behaviors are attributed to the role played by the insulated walls of the enclosure, as discussed in [16]. Figure 7 shows the sam Nusselt number variations for a CO_2 enclosure at Ra=10^6 and θ_0 = 0.5 as a function of the tilt angle ψ. One interesting result here is that at this value of θ_0, radiation accounts for about 85% of the total heat transfer at ψ = 0°, but decreases to about 50% at ψ = 100°. Our calculations also show tha at θ_0 = 2.0, radiation completely dominates the heat transfer and is therefore not much affected by the tilt angles.

CONCLUDING REMARKS

In this paper a new total slice based absorptance model for calculating gas radiation in a two-dimensional square enclosure filled with a nongray gas or a gas mixture is described and its accuracy is demonstrated in three separate cases for which accurate solutions are known. This model is entirely compatib with numerical solutions to the underlying buoyancy-driven enclosure-flow problem and represents a viable choice among several other models currently available. More research is needed to extend this model to three-dimensional enclosures and radiative transfer including the effects of scattering.

ACKNOWLEDGMENT

The authors gratefully acknowledge the support of this study by the U. S. National Science Foundation under grants CME 79-18682 and MEA 82-19158 to the University of Notre Dame and by the University of Notre Dame Computer Center.

NOMENCLATURE

A_j	total slice band absorptance
D_n	universal functions
g	gravitational acceleration
H	enclosure height or width
I	radiation intensity
Nu	Nusselt number
q_r	radiation flux
Ra	Rayleigh number
s	path length
T	absolute temperature
t	$\cos\theta$
u,v	velocity components
x,y,z	coordinates
α_j	integrated band intensity parameter
θ	angle defined in Figure 1(a)
θ_o	$(T_H - T_C)/T_C$
κ	absorption coefficient
ν_j	wave number at band head
σ	Stefan-Boltzmann constant
τ	optical thickness
τ_D	characteristic optical thickness
τ_H	scaled optical thickness
ϕ	angle defined in Figure 1(a)
ψ	tilt angle, Figure 5
ω_j	band width parameter

Subscripts

b	black wall
C	cold wall
H	hot wall
j	gas band
s	slice
W	wall
x,y	components
ν	spectral

REFERENCES

1. Yang, K. T., Lloyd, J. R., Natural Convection-Radiation Interaction in Enclosures, Natural Convection: Fundamewntals and Applications, eds. W. Aung, S. Kakac and R. Viskanta, Hemisphere Publishing Corp., N.Y., 1985.

2. Zhong, Z. Y., Variable Property Natural Convection with Thermal Radiation Interaction in Square Enclosures, Ph.D. Dissertation, University of Notre Dame, 191 pp., August 1983.

3. Tien, C. L., Thermal Radiative Properties of Gases, Advances in Heat Transfer, eds. T. F. Irvine, Jr., J. P. Hartnett, Academic Press, N.Y., vol. 5, pp. 253-324, 1968.

4. Edwards, D. K., Molecular Gas Radiation, Advances in Heat Transfer, eds. T. F. Irvine, Jr., J. P. Hartnett, Academic Press, N.Y., vol. 12, pp. 115-192, 1976.

5. Siegel, R., Howell, J. R., Thermal Radiation Heat Transfer, Second Edition, McGraw Hill Book Co., N.Y., 1981.

6. Edwards, D. K., Menard, W. A., Comparison of Models for Correlation of Total Band Absorption, Applied Optics, vol. 3, pp. 621-625, 1964.

7. Felske, J. O., Tien, C. L., Infrared Radiation from Nonhomogeneous Gas Mixture Having Overlapping Bands, Journal of Quanti. Spectro. and Radiative Transfer, vol. 14, pp. 35-48, 1974.

8. Tiwari, S. N., Gupta, S. K., Accurate Spectral Modeling for Infrared Radiation, ASME Paper No. 77-HT-69, 1977.

9. Liu, V. K., Lloyd, J. R., Yang, K. T., An Investigation of a Laminar Diffusion Flame Adjacent to a Vertical Flat Plate Burner, Int. Journal of Heat and Mass Transfer, vol. 24, pp. 1959-1970, 1981.

10. Wassel, A. T., Edwards, D. K., Molecular Gas Band Radiation in Cylinders, Journal of Heat Transfer, vol. 95, pp. 21-26, 1974.

11. Zhong, Z. Y., Lloyd, J. R., Yang, K. T., Variable Property Natural Convection in Inclined Square Cavities, Numerical Methods in Thermal Problems, eds. R. W. Lewis, J. A. Johnson, W. R. Smith, vol. 3, Pineridge Press, Swansea, United Kingdom, pp. 968-979, 1983.

12. Lockwood, F. C., Shah, N. G., A New Radiation Solution Method for Incorporation in General Combustion Prediction Procedures, Proc. 18th Symposium (International) on Combustion, pp. 1405-1414, 1981.

13. Fiveland, W. A., Discrete-Ordinates Solutions of the Radiative Transport Equation for Rectangular Enclosures, Journal of Heat Transfer, vol. 106, pp. 699-706, 1984.

14. Razzaque, M. M., Klein, D. E., Howell, J. R., Finite Element Solution of Radiative Heat Transfer in a Two-Dimensional Rectangular Enclosure with Gray Participating Media, Journal of Heat Transfer, vol. 105, pp. 933-936, 1983.

15. Abu-Romia, M. M., Tien, C. L., Appropriate Mean Absorption Coefficients for Infrared Radiation in Gases, Journal of Heat Transfer, vol. 89, pp. 321-327, 1967.

16. Zhong, Z. Y., Yang, K. T., Lloyd, J. R., Variable Property Natural Convection in Tilted Enclosures with Thermal Radiation, Numerical Methods in Heat Transfer, vol. III, John Wiley Ltd., United Kingdom, 1985.

Estimation of the True Temperature of the Surface of a Body by Using a Multi-Spectral Radiance Method

DEZHONG ZHU and YUQIN GU
Tsinghua University
Beijing, PRC

ABSTRACT

A new multi-spectral radiation pyrometry is introduced in this paper. The basic principle, measurement system and the experimental results are presented.

INTRODUCTION

The problem worth to discuss in the radiation pyrometry is to eliminate the effect of emissivity on the measurement results. Many authors have developed a multi-color technique for temperature measuring, that is, the temperature is determined from the radiance signals of three or four wave lengths.[1][2] The multi-spectral radiance pyrometry is a new method based on the principle of the multi-color technique and the principle of equalization of two spectral radiance.[3] As the multi-spectral radiance is the quantity to be measured, it will give a lot of information which can be reduced by many different methods. So its accuracy would be much better and the effect of emissivity on the accuracy will be decreased.

PRINCIPLE

The multi-spectral radiance method as well as the other methods in radiation pyrometry is based on Wien's law. The spectral radiation energy is:

$$E_{\lambda T} = \varepsilon_{\lambda T} C_1 \lambda^{-5} \exp(-C_2/\lambda T) \tag{1}$$

If the measured spectral radiance is shown as in Fig.1, then the following relations can be obtained,

$$R_{\lambda_1 \lambda_4, T} = E_{\lambda_1 T}/E_{\lambda_4 T} = \varepsilon_{\lambda_1 T} \lambda_1^{-5} \exp(-C_1/\lambda_1 T)/\varepsilon_{\lambda_4 T} \lambda_4^{-5} \exp(-C_2/\lambda_4 T) \tag{2}$$

$$R_{\lambda_2 \lambda_3, T} = E_{\lambda_2 T}/E_{\lambda_3 T} = \varepsilon_{\lambda_2 T} \lambda_2^{-5} \exp(-C_2/\lambda_2 T)/\varepsilon_{\lambda_3 T} \lambda_3^{-5} \exp(-C_2/\lambda_3 T) \tag{3}$$

where λ_1, λ_2, λ_3 and λ_4 are four selected wave lengths.
Combining (2) with (3), the true temperature of the irradiant body T will be:

$$T=\frac{C_2[(1/\lambda_1-1/\lambda_2)+(1/\lambda_3-1/\lambda_4)]}{\ln\frac{\varepsilon_{\lambda_1T}\,\varepsilon_{\lambda_3T}}{\varepsilon_{\lambda_2T}\,\varepsilon_{\lambda_4T}}+5\ln\frac{\lambda_2\lambda_4}{\lambda_1\lambda_3}+\ln\frac{E_{\lambda_2T}\,E_{\lambda_4T}}{E_{\lambda_1T}\,E_{\lambda_3T}}} \quad (4)$$

If $R_{\lambda_1\lambda_4,T}=R_{\lambda_2\lambda_3,T}$, $\ln(E_{\lambda_1T}/E_{\lambda_4T})/(E_{\lambda_2T}/E_{\lambda_3T})=0$,

then

$$T=\frac{C_2[(1/\lambda_1-1/\lambda_2)+(1/\lambda_3-1/\lambda_4)]}{\ln(\varepsilon_{\lambda_1T}\,\varepsilon_{\lambda_3T}/\varepsilon_{\lambda_2T}\,\varepsilon_{\lambda_4T})+5\ln(\lambda_2\lambda_4/\lambda_1\lambda_3)} \quad (5)$$

If $\varepsilon_{\lambda_1T}/\varepsilon_{\lambda_4T}=\varepsilon_{\lambda_2T}/\varepsilon_{\lambda_3T}$ is given simultaneously, then

$$\ln(\varepsilon_{\lambda_1T}\,\varepsilon_{\lambda_3T}/\varepsilon_{\lambda_2T}\,\varepsilon_{\lambda_4T})=0.$$

In this case, the true temperature of the surface of the body is:

$$T=\frac{C_2[(1/\lambda_1-1/\lambda_2)+(1/\lambda_3-1/\lambda_4)]}{5\ln(\lambda_2\lambda_4/\lambda_1\lambda_3)} \quad (6)$$

The multi-spectral radiance pyrometry is a method in which four wave lengths are selected and the radiation energy ratio of each two wave lengths are equalized, that is,

$$E_{\lambda_1T}/E_{\lambda_4T}=E_{\lambda_2T}/E_{\lambda_3T}$$

If $\varepsilon_{\lambda_1T}/\varepsilon_{\lambda_4T}=\varepsilon_{\lambda_2T}/\varepsilon_{\lambda_3T}$, then the true temperature of the surface of the irradiant body can be determined by equation (6).
If $\varepsilon_{\lambda_1T}/\varepsilon_{\lambda_4T}\neq\varepsilon_{\lambda_2T}/\varepsilon_{\lambda_3T}$, then ε_{λ_1T}, ε_{λ_2T}, ε_{λ_3T} and ε_{λ_4T} can be determined by the relation between the emissivity, the wave length and the temperature. The true temperature can be calculated by equation (5). Normally, the variation of emissivity with wave length can be expressed in the form.

$$\varepsilon_{\lambda T}=a+b\lambda+c\lambda^2+\ldots$$

where a , b , c , ... are functions of the temperature. In certain range of wave lengths the emissivity variation is linear, that is,

$$\varepsilon_{\lambda T}=a+b\lambda \quad (7)$$

Usually,

$$b\lambda \ll a$$

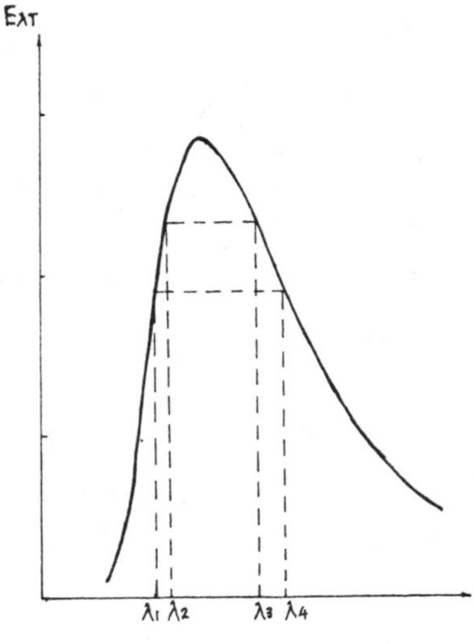

Fig.1 spectral radiance

Approximately,

$$\ln \varepsilon_{\lambda T} = \ln a + b\lambda/a$$

then

$$\ln\left(\varepsilon_{\lambda_1 T}\, \varepsilon_{\lambda_3 T} \,/\, \varepsilon_{\lambda_2 T}\, \varepsilon_{\lambda_4 T} \right) = b\,(\lambda_1-\lambda_2+\lambda_3-\lambda_4)/a$$

$$T = \frac{C_2[(1/\lambda_1-1/\lambda_2) + (1/\lambda_3-1/\lambda_4)}{b(\lambda_1-\lambda_2+\lambda_3-\lambda_4)/a + 5\ln(\lambda_2\lambda_4/\lambda_1\lambda_3)} \tag{8}$$

If we selected such six wave lengths in the spectral radiance, that

$$E_{\lambda_1 T}/E_{\lambda_4 T} = E_{\lambda_2 T}/E_{\lambda_3 T}\ , \quad E_{\lambda_5 T}/E_{\lambda_6 T} = E_{\lambda_2 T}/E_{\lambda_3 T}\ ,$$

the following equations will be given,

$$T = \frac{C_2[(1/\lambda_1-1/\lambda_2) + (1/\lambda_3-1/\lambda_4)]}{b(\lambda_1-\lambda_2+\lambda_3-\lambda_4)/a + 5\ln(\lambda_2\lambda_4/\lambda_1\lambda_3)} \tag{8}$$

$$T = \frac{C_2[(1/\lambda_5-1/\lambda_2) + (1/\lambda_3-1/\lambda_6)]}{b(\lambda_5-\lambda_2+\lambda_3-\lambda_6)/a + 5\ln(\lambda_2\lambda_6/\lambda_5\lambda_3)} \tag{9}$$

Comining (8) with (9), the true temperature is given as follows,

$$T = \frac{C_2(\lambda_1-\lambda_2+\lambda_3-\lambda_4)[(1/\lambda_5-1/\lambda_2 + (1/\lambda_3-1/\lambda_6)] - (\lambda_5-\lambda_2+\lambda_3-\lambda_6)[(1/\lambda_1\ \ 1/\lambda_2) + (1/\lambda_3-1/\lambda_4)]}{5(\lambda_1-\lambda_2+\lambda_3-\lambda_4)\ln(\lambda_2\lambda_6/\lambda_5\lambda_3) - (\lambda_5-\lambda_2+\lambda_3-\lambda_6)\ln(\lambda_2\lambda_4/\lambda_1\lambda_3)} \tag{10}$$

MEASUREMENT SYSTEM

The measurement system consists of a monochrometer, a pyroelectric IR detector, a IR radiation instrument and a recorder. It is shown in Fig.2. Monochrometer is an apparatus which resolves a multi-color light into a set of mono-color lights. Its wave length is ranging from 0.2μ to 12μ. And it is fitted with a wave length scanning system. IR radiation instrument is an instrument which is composed of a chopper, a pyroelectric IR detector and a phase-locked amplifier (Fig.3).

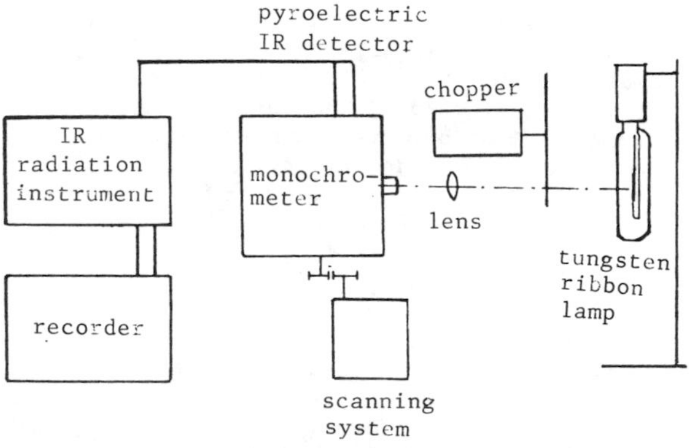

Fig.2 Experimental set-up for a multi-spectral radiation pyrometry

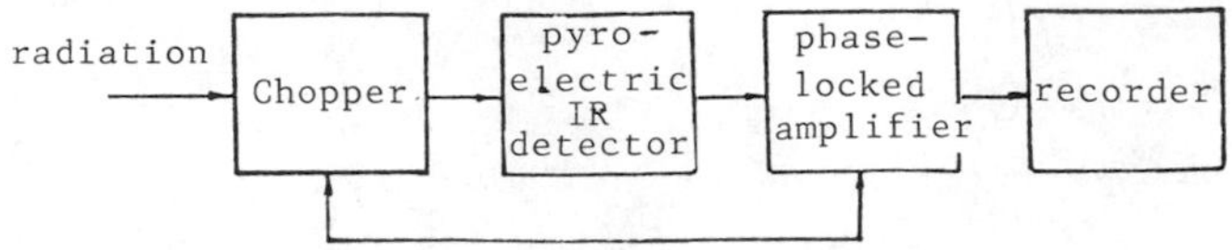

Fig.3 Block of the IR radiation instrument

EXPERIMENTAL RESULTS

It is necessary to calibrate the measurement system before it is used. In calibration the spectral characteristic factor of the selected wave length range must be determined. In our case the spectral characteristic factor is determined with a calibrated standard tungsten ribbon lamp. The relation between the output voltage of the IR radiation instrument $V_{\lambda T}$ and the spectral radiation brightness $L_{\lambda T}$ is:

$$V_{\lambda T} = K_{\lambda} \cdot L_{\lambda T} \tag{11}$$

where K_{λ} is the spectral characteristic factor. In order to examine the reliability of this pyrometry, we have measured the spectral radiance of a standard tungsten ribbon lamp at different temperatures. Meanwhile, the spectral radiance of other irradiant bodies, such as iodine tungsten lamp, a flame, a mercury lamp, a laser beam etc., have also been measured. In Fig.4, there are three curves which are the spectral radiation brightness at different temperatures, 2000°C , 1800°C, 1600°C (i.e. the true temperature are 2491 K, 2250 K, 2015 K, respectively). At first we decided the accuracy of this measurement system with Wien's law. For a black body, when its temperatures are 2491 K, 2250 K, and 2015 K, its maximum radiation wave lengths will be 1.16μ, 1.29μ, and 1.44μ. Correcting the curves in Fig.4 with the emissivity of the tungsten ribbon, the spectral radiance of the black body is obtained. And then we can get the maximum brightness wave length which are 1.15μ, 1.30μ, and 1.40μ for different temperatures. The agreement is satisfied. Later we calculated the true temperature from these curves with the radiation pyrometry of multi-wave length as follows:
(1) selecting λ_1=0.669μ, λ_2=0.7μ, λ_3=0.8μ, λ_4=0.75μ, we get

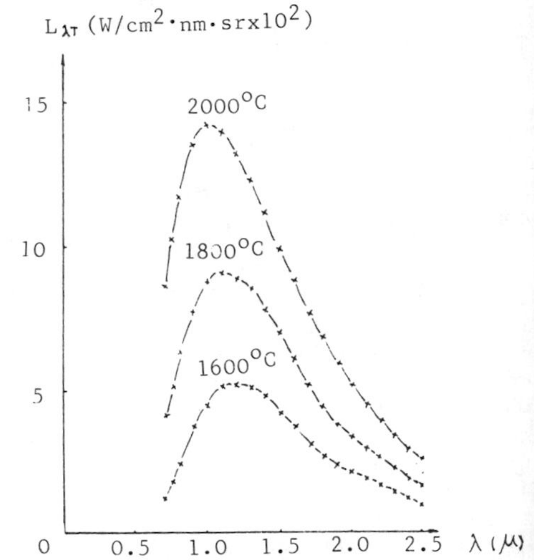

Fig.4 spectral radiation brightness of standard tungsten lamp

$\varepsilon_{\lambda_1 T}=0.43$, $\varepsilon_{\lambda_2 T}=0.425$, $\varepsilon_{\lambda_3 T}=0.41$, $\varepsilon_{\lambda_4 T}=0.415$ by ref.[4] and then the temperatur determined by equation (5) is T=2562 K. Its relative error is 2.9%.

(2) $\lambda_1=0.75^{\mu}$, $\lambda_2=0.96^{\mu}$; $\lambda_3=1.3^{\mu}$, $\lambda_4=1.72^{\mu}$; $\varepsilon_{\lambda_1 T}=0.424$, $\varepsilon_{\lambda_2 T}=0.39$, $\varepsilon_{\lambda_3 T}=0.324$, $\varepsilon_{\lambda_4 T}=0.27$; T=2338 K. Its relative error is 3.9 %.

(3) $\lambda_1=0.72^{\mu}$; $\lambda_2=1.05^{\mu}$; $\lambda_3=1.4^{\mu}$; $\lambda_4=2.4^{\mu}$; $\varepsilon_{\lambda_1 T}=0.434$, $\varepsilon_{\lambda_2 T}=0.37$, $\varepsilon_{\lambda_3 T}=0.306$, $\varepsilon_{\lambda_4 T}=0.214$; T=2084 K. Its relative error is 3.4%.

It is shown that the multi-spectral radiance pyrometry and the measurement system which we used is available. The emissivity of tungsten is not suitable to be expressed by the relation (7). Here we have not calculated the true temperature with equation (10).

REFERENCES

[1] J. F. Babelot, et al., Microsecond and sub-microsecond mutli-wavelength pyrometry for pulsed heating technique diagnostics, 1982 American Institute of Physics, Temperature, Vol. 5, Pt. 1

[2] G. S. Ambrok, Estimation of the true temperature of targets by their thermal radiation based on Planck's law, 1982 American Institute of Physics, Temperature, Vol. 5, Pt. 1

[3] Zhou Peisen Shi Kekuan, Spectro-balance thermometry, Acta Metrologica Sinica Vol.3,No.2 April 1982

[4] R. J. Thorn and G. H. winslow, Radiation of thermal energy from bodies, 1962 American Institute of Physics, Temperature, Vol. 3

HEAT AND MASS TRANSFER IN POROUS MEDIA

A Calculating Method of Heat and Moisture Transfer in Multi-Layered Building Enclosure

FU-QI HUANG and ZHAO-XIAN LI
Chinese Academy of Building Research
PRC

ABSTRACT

A set of eqations is derived according to the transport theory which is based on the non-equilibrium thermodynamics to describe the heat and moisture transfer in multi-layered building enclosure (short as MBE). This set of equations encludes the non-linear equations for heat and moisture transfer in homogeneous medium, the controlling condition at the interface and the third boundary condition. According to this model, a FORTRAN program is worked out to simulate the moisture transfer process in MBE under various outer conditions such as rain, the changes in air temperature and humidity, sun-shining and heating in winter. The analogue analysis of the properties of thermal insulation and moisture proof of the many kinds of commonly used MBE consisted of porous materials in CHINA quite agree with the contrast experiment data. It makes it possible for the first time to predict the changes of the inner moisture distribution of MBE.

NOMENCLATURE

- $\vec{jm}$: moisture flux vector, kg/m^2.h
- r_o: specific weight of dry porous material, kg/m^3
- am: am= am(w,t), moisture diffusivity, m^2/h
- w: moisture content in unit mass, kg/kg
- δ: thermal gradient coefficient, 1/^{o}c
- $\vec{j_q}$: heat flux vector, kcal/m^2.h
- λ: thermal conductivity, kcal/m.h.^{o}c
- L: vaporization latent heat, kcal/kg
- Co: specific heat capacity of dry porous material, kcal/kg.^{o}c
- C: specific heat capacity of water, kcal/kg.^{o}c
- E(T): saturation vapor pressure, mmHg
- αm: mass transfer coefficient, kg/m^2.h
- αn: heat transfer coefficient indoor, kcal/m^2.h.^{o}c
- αw: heat transfer coefficient outdoor, kcal/m^2.h.^{o}c
- t: temperature, ^{o}c
- a: thermal diffusivity, m^2/h
- ε: phase change criterion
- T: absolute temperature, ^{o}k
- τ: time h (hour)
- φ: humidity, %.

INTRODUCTION

The effects of the utilization, the durability and the insulation of a building are strongly influenced by the heat insulation and damp proof properties of the building enclosure(short as BE). For example, the condensation inside surface of outer walls was found in some buildings in the north area of CHINA. As the moisture condition of a porous material has a very great effect on its thermal conductivity λ, the condensation moistened the wall and lowered its insulating effectiveness. Conversely, the decrease in the effectiveness of the insulation of the walls intensified the condensation, forming vicious circle. Therefore, it is very important to design enclosure well.

Up to now, a over simplified calculating method, which is based on the supposition that the moisture in BE was transfered only by the vapor pressure gradient, was still used in the BE design. This method is proper only when the transfer flux of capilary water is very small relatively. Therefore, some new effort was made in our work to improve the situation.

THE EQUATIONS OF HEAT AND MOISTURE TRANSFER IN MBE

The heat and moisture transfer in the MBE in the process of drying and application is composed of three simpler transporting processes: the heat and moisture transfer in homogeneous porous medium, at the interface and between the enclosure surface and the enviroment.

1. The heat and moisture transfer equations in homogeneous porous material:

The moisture movement and the heat flow in porous medium are influenced each other. Under the condition of non-filtering air and water flow, such transport equations have been given by:

$$\text{(2-1)} \quad \begin{cases} \vec{j}_m = -r_o a_m (\nabla w + \delta \nabla t) \\ \vec{j}_q = -\lambda' \nabla \end{cases}$$

or

$$\text{(2-2)} \quad \begin{cases} \dfrac{\partial w}{\partial \tau} = a_m \nabla^2 w + a_m \delta \nabla^2 t \\ \dfrac{\partial t}{\partial \tau} = a' \nabla^2 t + \dfrac{\varepsilon L}{(C_o + Cw)} \dfrac{\partial w}{\partial \tau} \end{cases}$$

Considering the effect of the moisture movement on the heat transfer more precisely, we get

$$\text{(2-3)} \quad \begin{cases} \vec{j}_m = -r_o a_m (\nabla w + \delta \cdot \nabla t) \\ \vec{j}_q = -\lambda \nabla t + c \cdot t \cdot \vec{j}_m \end{cases}$$

or

$$\vec{j}_m = -r_o \cdot a_m \cdot (\nabla w + \delta \cdot \nabla t)$$

$$\text{(2-4)} \quad \begin{cases} \dfrac{\partial w}{\partial \tau} = -\dfrac{1}{r_o} \cdot \nabla \cdot \vec{j}_m \\ \dfrac{\partial t}{\partial \tau} = a \cdot \nabla^2 t - \dfrac{c}{r_o (C_o + C \cdot w)} \vec{j}_m \cdot \nabla t + \dfrac{\varepsilon L}{(C_o + Cw)} \cdot \dfrac{\partial w}{\partial \tau} \end{cases}$$

Equ. (2-2) or (2-4) are non-linear and, up to now, the numerical method is the only way to solve them. But actually, the drying process of BE could last three to five years, even eight years. To reduce the CPU time, the transfer equation (2-4) was further simplified through the careful consideration of the actual process:

$$(2-5) \quad \begin{cases} \iiint_V \frac{\partial w}{\partial \tau} \cdot dV = - \oiint a_m \cdot \left(\frac{\partial w}{\partial x} + \delta \frac{\partial t}{\partial x}\right) \cdot \vec{i} \cdot d\vec{s} \\ \frac{\partial^2 t}{\partial x^2} = 0 \end{cases}$$

Here the following presumption have been made:

a) During the usual drying process, the ε in BE is very small as to be ignored.

b) In MBE (mainly to the various kinds of building boards and prooves), the transfer process is one-dimensional.

c) The temperature gradient has a great effect on moisture transfer, but the moisture flow has a little effect on the heat transfer. The rationality of this assumption can be seen clearly by the following numerical estimation. To most kinds of MBE used in CHINA:
the thickness of the enclosure

$d = 0.1 - 0.4$ (m), taking $[d] = 0.2$;

the greatest temperature diference between any two points in the enclosure

$\Delta t = 0 - 50\ ^{\circ}C$, taking $[\Delta t] = 30$;

the greatest moisture content difference between any two points in the enclosure

$\Delta w = 0 - 50\%$, taking	$[\Delta w] = 30\%$;
$\lambda = 0.1 - 1$ kcal/m.h.$^{\circ}$C,	$[\lambda] = 0.5$;
$r_o = 500 - 2000$ kg/m^3,	$[r_o] - 1500$;
$C = 1$ kcal/kg.$^{\circ}$C,	$[C] = 1$;
$\delta = 10^{-3} - 10^{-1}$ 1/$^{\circ}$C,	$[\delta] = 10^{-2}$;
$am = 10^{-7} - 10^{-5}$ m^2/h,	$[am] = 10^{-6}$;

then by equ. (2-3):

$$\vec{j}_q = -\lambda \nabla t \left(1 + \frac{c t r_o a_m}{\lambda}\left(\frac{\nabla w}{\nabla t} + \delta\right)\right)$$

$$\left[\frac{c t r_o a_m}{\lambda}\left(\frac{\nabla w}{\nabla t} + \delta\right)\right] = \frac{[c][t][r_o][a_m]}{[\lambda]}\left(\frac{[\Delta w]}{[\Delta t]} + [\delta]\right) = \frac{2.7}{100} \ll 1$$

note: these square brackets mean taking numerical estimation.

d) The actual transport process can be assumed to be composed of steady heat transfer and unsteady moisture transfer in each short period.

By equ. (2-5) the CPU time needed by describing a drying process of BE is greatly reduced in contrast with equ. (2-2) or (2-4). For example, to a matrix equation which was established upon a discrete model of N nodes, there will be 2N correlative parameters by equ. (2-2) or (2-4). But by equ. (2-5) only N correlative parameters will come out. And for each step the computing time will reduce from $\frac{1}{3}(2N)^3 + (2N)^2 - \frac{1}{3}(2N)$ for the former to $\frac{1}{3}N^3 + N^2 - \frac{1}{3}N$ for the latter by Gassian method.

2. Controlling conditions at the interface:

Accoring to the moisture transfer theory, which is an analogue theory to thermodynamics, the movement of the moisture depends upon the distribution of moisture transfer potential "θ". A great deal of experiemtns shows that: the value of θ in materials depends not only on the local weight moisture content but also on the local temperature as well as on the properties and structure of the material. For a interface of MBE, there would be:

(2-6)
$$\begin{cases} \theta_{-0} = f_1(w_{-0}, t_{-0}, P_{m-0}) \\ \theta_{+0} = f_1(w_{+0}, t_{+0}, P_{m+0}) \end{cases}$$

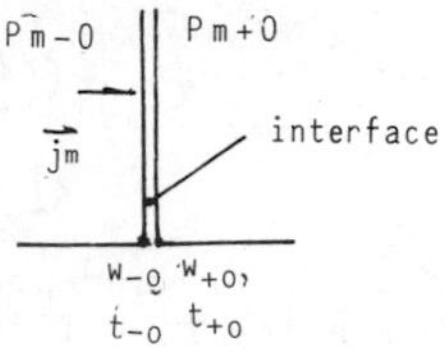

Fig.2.1

Suppose that the interface of MBE is still in the state of quasi-equilibrium during drying process. Then:

$$\theta_{-0} = \theta_{+0} = \theta; \quad t_{-0} = t_{+0} = t$$

(2-7)
$$\begin{cases} w_{-0} = S(\theta, t, P_{m-0}) \\ w_{+0} = S(\theta, t, P_{m+0}) \end{cases}$$

Where Pm indicates material property, S is the anti-function of f_1 to w. When a definite scale of θ is determined, the function form of S is given theoretically. Usually, a kind of test paper is used as a standard material to express indirectly the relationship between θ and w of different kinds of porous materials by setting up equilibrium between the test paper and those materials for a series of temperature and moisture. If the weight moisture content of the test paper is expressed by w_p, then

(2-8)
$$\theta = f_1(w_p, t, P_{m\,paper}) = f_2(w_p, t)$$

(2-9)
$$\begin{cases} w_{-0} = S_2(w_p, t, P_{m-0}) \\ w_{+0} = S_2(w_p, t, P_{m+0}) \end{cases}$$

Equ. (2-9) and continuity equation (see (2-5)) describe the moisture transfer process at the interface. $w = S_2(wp, t, Pm)$ is the function of the potential equilibrium curve given by experiments. To most of the porous materials, this curve will not vary greatly with the change of temperature when its moisture content is not very great. Therefore, as an approximation, we think the potential equilibrium curve is independent of temperature.

(2-10)
$$\begin{cases} w_p = S_2^{-1}(w_{-0}, P_{m-0}) \\ \dfrac{w_{+0}}{w_{-0}} = \dfrac{S_2(w_p, P_{m+0})}{S_2(w_p, P_{m-0})} \\ \vec{j}_{m-0} = \vec{j}_{m+0} \end{cases}$$

3. The third boundary condition:

The transfer driving force is the difference of vapor pressure between the surface of BE and the air:

(2-11)
$$j_m = \pm \alpha_m b\,(P_b(w_b, t_b) - P_a(\varphi_a, t_a))$$

Expanding Pb at wa in Taylor series and taking the first approximation:

(2-12)
$$P_b(w_a, t_b) = P_b(w_a, t_b) + \left.\frac{\partial P_b}{\partial w}\right|_{w_a}(w_b - w_a)$$

Choosing wa to satisfy

(2-13)
$$P_b(w_a, t_b) = P_a(\varphi_a, t_a)$$

then

(2-14)
$$\vec{j}_m = \pm \alpha_m b \left.\frac{\partial P_b}{\partial w}\right|_{w_a}(w_b - w_a)$$

where αmb, defined by equ. (2-11), is moisture exchange coefficient, wa, defined by equ. (2-13), is equivalent weight moisture content of air, the subscripts b and a in other variables indicate wall surface and air respectively.

In terms of the definition of the air humidity:

$$P_a(\varphi_a, t_a) = E(T_a)\varphi_a$$

we can similarly define:

$$P_b(w_a, t_b) = E(T_b)\varphi_b$$

Then yields (see equ. (2-13)):

(2-15) $$\varphi_b = \varphi_a \frac{E(T_a)}{E(T_b)}$$

then

(2-16) $$\left.\frac{\partial P_b}{\partial w}\right|_{w_a} = E(T_a)\left.\frac{\partial \varphi_a}{\partial w}\right|_{w_a}$$

By the isothermal absorbing line:

$$w = g(\varphi, t)$$

when $\varphi_b < 1$, we obtain

(2-17) $w_a = g(\varphi_b, ta)$, (see Fig.2.2)

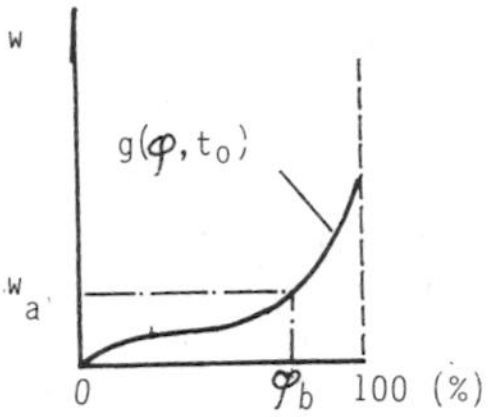

Fig.2.2 the isothermal absorbing line

Experiments show that $g(\varphi,t)$ of most porous building materials do not vary greatly with t, so we can consider $g(\varphi,t) = g(\varphi)$ and get

(2-18) $$\begin{cases} w_a = g\left(\varphi_a \frac{E(T_a)}{E(T_b)}\right) \\ \alpha_m = \alpha_{mb} E(T_a) \left.\frac{\partial g^{-1}}{\partial w}\right|_{wa} \\ \vec{j}_m = \pm \alpha_m (w_b - w_a) \end{cases}$$

When $\varphi_b \geq 1$, it may dew on the surface. Combinating the three sets of equations above, we obtain at last (see Fig.2.3):

(2-19) $$\begin{cases} \iiint \frac{\partial w}{\partial \tau} dv = -\oiint a_m\left(\frac{\partial w}{\partial x} + \delta \frac{\partial t}{\partial x}\right)\vec{i}\cdot d\vec{s} & \tau > 0 \\ \frac{\partial^2 t}{\partial x^2} = 0 & \\ \lambda \frac{\partial t}{\partial x} = \alpha_n (t(0,\tau) - t_n) = \alpha_w (t_w - t(L,\tau)) & \text{on surface} \\ w_{ai} = g_i\left(\varphi_i \frac{E(T_{ai})}{E(T_{bi})}\right) & i = 1,2 \\ \alpha_{mi} = \alpha_{mb} E(T_{ai}) \left.\frac{\partial g_i}{\partial w}\right|_{w_{ai}} & i = 1,2 \\ j_m|_{x=0} = \pm \alpha_{m1} (w(0,\tau) - w_{a1}) & \\ j_m|_{x=L} = \pm \alpha_{m2} (w_{a2} - w(L,\tau)) & \\ w_p = S_2^{-1}(w_{-0}, P_{m-0}) & \\ \frac{w_{+0}}{w_{-0}} = \frac{S_2(w_p, P_{m+0})}{S_2(w_p, P_{m-0})} & \text{at interface} \end{cases}$$

where i = 1 indicates the indoor, and i = 2 the outdoor.

THE SET OF DISCRETE EQUATIONS AND IT'S PROGRAM

The difference equations are obtained by discretizing the set of equations (2-19).

(3-1) $$AW^{n+1} = B$$

where A is a N×N matrix and $A = A(W^n, t^n, t^{n+1})$, B is a N grade vector and $B = B(W^n, t^n, t^{n+1})$, $W^n = (w_1^n, w_2^n, \ldots\ldots, w_N^n)^T$, $t^n = (t_1^n, t_2^n, \ldots\ldots, t_N^n)^T$.

The flowing graph of the computing program is given as Fig. 3.1

THE ANALOGUE RESULTS AND THEIR ANALYSIS

1. The simulation of the drying process of "Jin 80" board under the climatic conditions in Jinzhou district (see Fig.4.1).

At the beginning, there are homogeneous moisture content distributions in layer I and II respectively, that is W^o = 5.47% in layer I and W^o = 42% in layer II. Other parameters are obtained from climatic information.

The simulation stands for the period from October 15, 1980 to April 15, 1983. Fig.4.2, 4.3 and 4.4 show the results. In Fig. 4.4 the calculated results are compared with measured data in the mixture layer of cement and pearlite of "Jin 80" board of the contrast building, which was built in 1980. It can be seen that there are good agreements between them.

The first sample was taking from the test building at January 31, 1981, by boring holes on the wall. After two monthes, the average moisture content w in south wall decreased from 21% to 15%, and from 24% to 16% in north wall. They decreased by 6% and 7.5% respectively. The calculated result in the same period is from 18% to 12%, the moisture content drop is 6%. There is only a small discrepency. The error in absolute value was mainly caused by the differnce of the initial distribution. Again, let's see the lines near peak point 2 and 2' in detail. The measured data of w rose from 6.5% in May to 17% two monthes and a half after, then it fell 10.5% again another 4 monthes later. But the calculated result rose from 4% on July 15 to 16% 2.5 monthes after, and then fell to 8% when another 4 monthes passed. The drying speed of the latter is slightly lower than that of the former. The reason is that the rain season determined by statistical climatic information was two monthes later than the actual rain season in the year and it caused that the calculating period had two monthes which lay in the heating period.

2. The simulation of the drying process of "Yantai board" under the climatic condition in Yantai district (see Fig.4.5).

The simulation stands for the condition of the drying process in the period from October 15, 1981 to April 15, 1984. The results are shown in Fig.4.6. and 4.7.

Besides, it was simulated that the heat and moisture transfer process in other MBE (eg. several multi-layered gas concrete boards) under Beijing's and Shengyang's climatic conditions and come to the following conclusions for the thermal insulation and damp proof properties and drying law of general multi-layered board.

a) The thermal insulation and moisture proof properties of MBE whose thermal insulation layer, which is apt to absorb water, exposes to air, such as "Jin 80" board are affected by rainfall and rain season greatly. They are not suitable to be used in the area where it is rainy in/or near winter. These boards have no definite natural drying state when they are used. Because of the periodic rain season, their moisture content varies up and down periodically. But this kind of boards have good thermal insulating properties in the area where it is dry in winter and have high drying speeds. The thermal insulating effect designed is reached in a short period.

b) To those multi-layered boards that have out-surfaces which are solid and don't easily absorb water, its moisture state and thermal insulating properties are scarecely affected by the rain season. The average moisture content in insulating layer decreaces monotonously with time and tends to a certain stable "equilibrium value". This kind of boards is suitable for various climate. But when it is used in the area where it is very clod in winter and in wet building if the internal surface layer of the plate is not treated well, the locating thermal insulating and moisture proof properties near the outer interface of insulating layer would go into bad condition becauce water will enter the wall through inner surface from the room by virtue of the great temperature difference of indoor and outdoor This case will extend to the whole thermal insulating layer as to damage the structure of the boards. In order to prevent this case, the designed solid layer outside should not be too thick and the solid "vapor insulating" layer inside should not be too thin. We let not only the moisture indoor enter the wall difficultly but also the moisture in the board leaves the surface easily under the condition that the structure has a good "rain-proof" property.

c) The drying speeds of various multi-layered boards depend mainly on their structure. Their drying speeds are independent of orietation. The difference of district climats affects hardly the drying speed of the wall that has a solid "rain-proof" layer.

d) The drying speed of various multi-layered board changes a little under the condition of high moisture content, and there is a "constant speed drying period".

Further, the simulation substantiated that the moisture transfer is mainly caused by temperature gradient when there is a greater temperature difference. The drying speed of the whole multi-layered board depends greatly on the degree of the surface moisture transfer. In winter, there is a greater temperature differnce between indoor and outdoor because of heating, this temperature difference causes the moisture to transport outwards. Therefore, the moisture content at the inner surface falls steeply and sometimes is lower than the equivalent weight moisture content Wa of the air inside the room. In this case, the moisture losing of the wall takes place nearly at the outer surface and the internal surface has a little effect, even though the moisture of the air in the room enter the wall through the internal surface of the board. This is why the drying speed of the wall in heating period becomes slow.

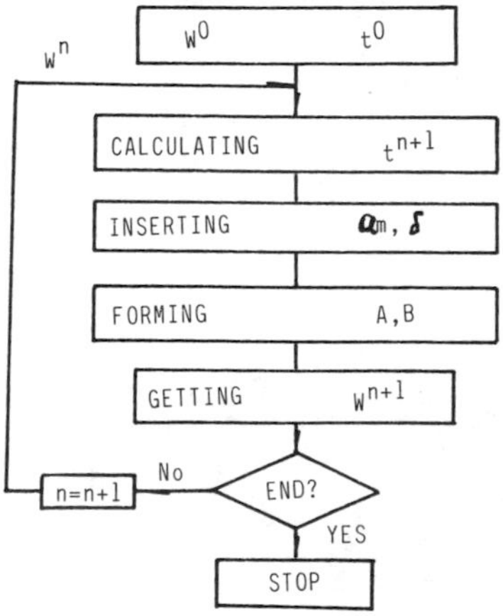

Fig.3.1

The flowing graph of the computing program

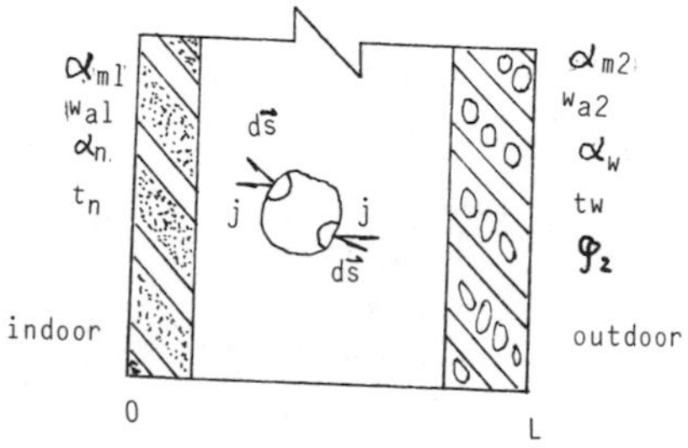

Fig.2.3

Fig.4.1 the sketch of "Jin80" board
I: d_1= 3cm; reinforced concrete: r_o=2250 kg/m^3 w^o=5.47%
II: d_2=14cm; 1:10 mixture of cement and pearlite; r_o=400 kg/m^3; w^o=42%

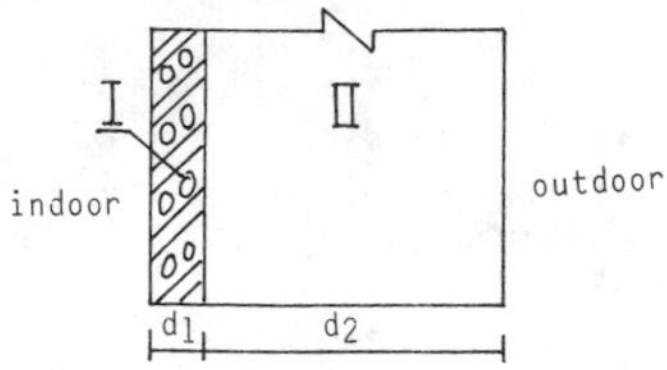

Fig.4.1

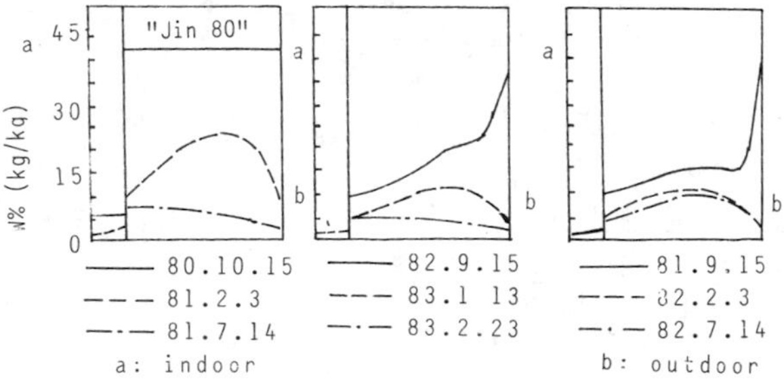

Fig.4.2 moisture distribution

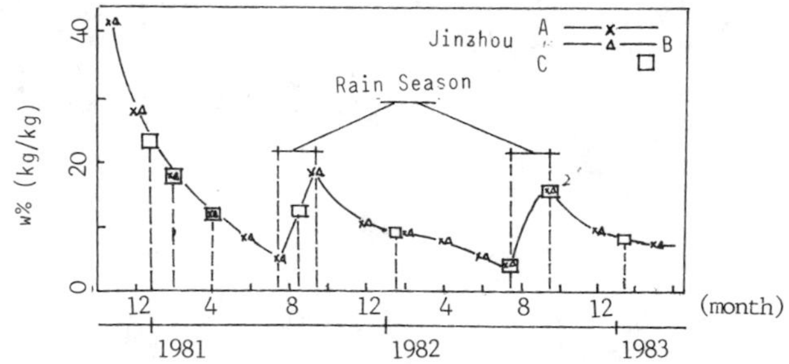

Fig.4.3 average w of insulating layer at different moment. A: south B: north C: calculating points for comparison with measured data

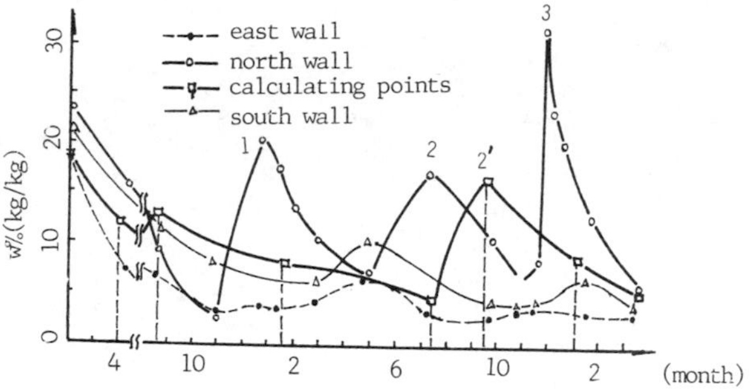

Fig.4.4 Peak point 1 and 3 are measured data after artificial rain peak point 2 is a measured datum after rain and 2' is a calculated datum take from Fig.4.3

Fig.4.5 the sketch of "Yantai" board. I: d_1=2cm; mixture of cement and sand; r_o=2100 kg/m^3, w^o=10%, II: d_2=12.5 cm; 1:10 mixture of cement and pearlite; r_o=400 kg/m^3; w^o=42% III: d_3=2.5 cm; reinforced concrete; r_o=2250 kg/m^3; w^o=5.47%

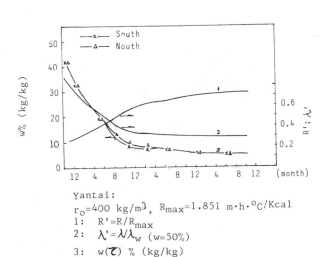

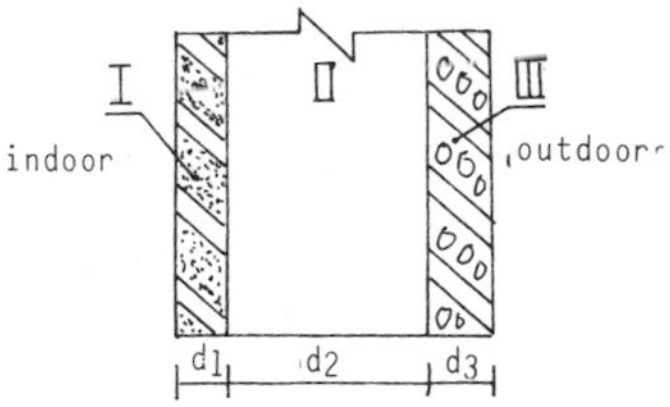

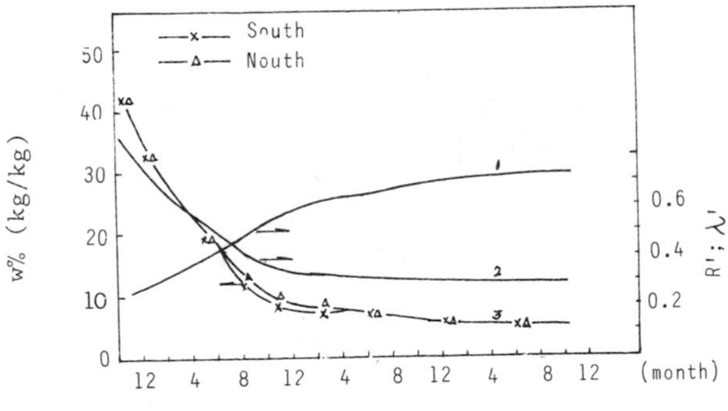

Yantai:
r_o=400 kg/m^3, R_{max}=1.851 m·h·°C/Kcal
1: $R'=R/R_{max}$
2: $\lambda'=\lambda/\lambda_w$ (w=50%)
3: $w(\tau)$ % (kg/kg)
when w=50%, λ_w=1.35 Kcal/m.h.°C

Fig.4.6 The change of thermal insulating property and average moisture content in the layer of mixture of cement and pearlite in "Yantar" board.

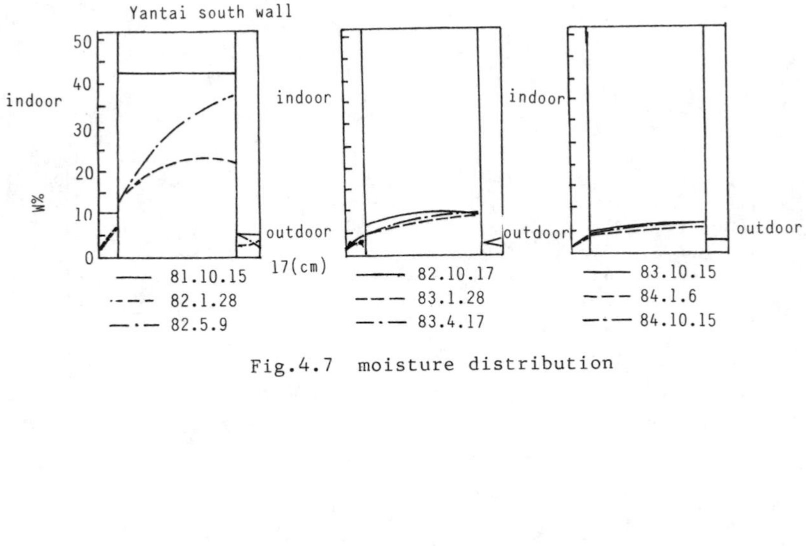

Fig.4.7 moisture distribution

Effects of Porous Layer Thickness of Sintered Screen Surfaces on Pool Nucleate Boiling Heat Transfer and Hysteresis Phenomena

XIN LIU, TONGZE MA, and JIPEI WU
Institute of Engineering Thermophysics
Chinese Academy of Sciences
Beijing, PRC

ABSTRACT

Experimental investgation of the characteristics of nucleate boiling heat transfer was performed on sintered metal screen surfaces in R-113 and water at atmospheric pressure. The results show that the boiling heat transfer coefficient on sintered screen surfaces is dependent on the number of screen layers. The coefficient at moderate heat fluxes is 8-9 times as high as on plain surfases in R-113 and 3-4 times as high in water. It is also found in the experiments tha the boiling heat transfer hysteresis is associated closely with the number of screen layers. In addition, a criterion correlation has been obtained.

NOMENCLATURE

C	specific heat J/(kg.K)	r	characteristic length m
D	diameter of screen pore m	ΔT	wall superheat K
g	gravitational acceleration m/s^2	λ	thermal conductivity W/(m.K)
i	latent heat of vaporization J/kg	μ	dynamic viscosity N.s/m
n	number of screen layers	ν	kinematic viscosity m^2/s
P	system pressure Pa	ρ	density kg/m^3
Q	heat flux density W/m^2	σ	surface tension N/m
R	gas constant J/(kg.K)	α	contact angle

Subscript

l	liquid	v	vapour

INTRODUCTION

Recently much attention is paid to the high transfer performance of enhanced boiling surfaces and various kinds of such surfaces are investigated. One of them is the porous material formed by metal screens. They are of two major types, namely, pressure-stacked screens and sintered screens. Because of the use of pressure-stacked screens in heat pipes, considerable amount of literature has been published which is concerned with boiling heat transfer on these surfaces, such as references (1-3), whereas only few papers are available that deal with boiling heat transfer on the surface of sintered screens.

References (4) and (5) presented the experimental results of boiling heat transfer on surfaces of stainless steel screens, brass screens and copper screens with different mesh and different numbers of layers. In these experiments explosiv boiling phenomena associated with hysteresis were observed.

Although both sintered screen surfaces and porous metallic matrix surfaces formed by sintering particles possess excellent boiling heat transfer characteristics, but they all face the potential start-up problem (6). How to diminish or even eliminate hysteresis is an important question yet to be dealt with.

EXPERIMENTAL APPARATUS

The pool boiling test apparatus used in this work is shown in Fig.1. According to reference (4), whether pool boiling or thin liquid layer boiling, among various kinds of screens used in the experiments copper screens of 185 mesh provide most effective enhancement of heat transfer. Therefore in the experiment the above mentioned screens were used, the parameters of which are shown in Table 1.

Table 1 Parameters of copper screens

Mesh Number	185
Wire Diameter (mm)	0.046
Pore Diameter (mm)	0.0913
Porosity (%)	56
Area Porosity (%)	44

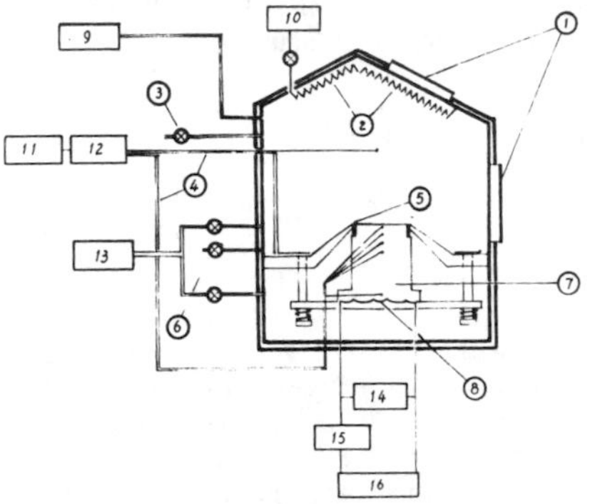

Fig.1 The boiling apparatus
1 observation window; 2 water-cooled condenser; 3 inlet valve; 4 thermocouples; 5 seal; 6 vent valve; 7 test block; 8 electric heater; 9 U-shaped barometer tube; 10 flow counter 11 printer; 12 digit voltmeter 13 vacuum pump; 14 voltmeter; 15 amperemeter; 16 power supply

The electrical heater for heating the test surfaces was tightly pressed on the bottom of the test block and insulated by mica sheet from the block. The thermocouple above the test surface indicated the saturated vapour temperature. In the upper part of the test block, two thermocouples were imbedded in a horizontal plane near the upper surface and the other two also in a horizontal plane, but with a distance of about 11 mm from the surface. Another four thermocouples were provided for monitoring. The heat flux was determined by the temperature gradient near the heating surface.

The screens were sintered on the plain surfaces. The same temperature and sintering time were maintained in each sintering process. The numbers of layers were respectively 1,2,3,4,5,6,7,9, and 13.

Curves of experimental results were coded for convenience. R113 and W represent freon 113 and distilled water respectively, the numbers after a dash standing for the numbers of screen layers. For example, R113-5 represents the boiling curve of R-113 on surface sintered with 5 layers of screens.

RESULTS AND DISCUSSION

In the course of the experiment, records were taken for both increasing and decreasing flux curves. Fig. 2 (a)-(i) show consecutively boiling curves of R-11 on the surfaces with 1 to 13 sintered screen layers. It can be seen from the boiling curve of one-layer screen surface that as the heat flux increased, the wall superheat decreased several times. This effect could be detected up to a heat flux of $1,8\times10^5$ W/m^2. On the increasing boiling curve, we see that when hea flux was 8.5×10^3 W/m^2, the wall superheat was about 11 K; but as heat flux reache 1.8×10^5 W/m^2, the wall superheat was only about 7 K. Thus, for the surface with one-layer sintered screen the boiling curves were characterized by strong hyster-

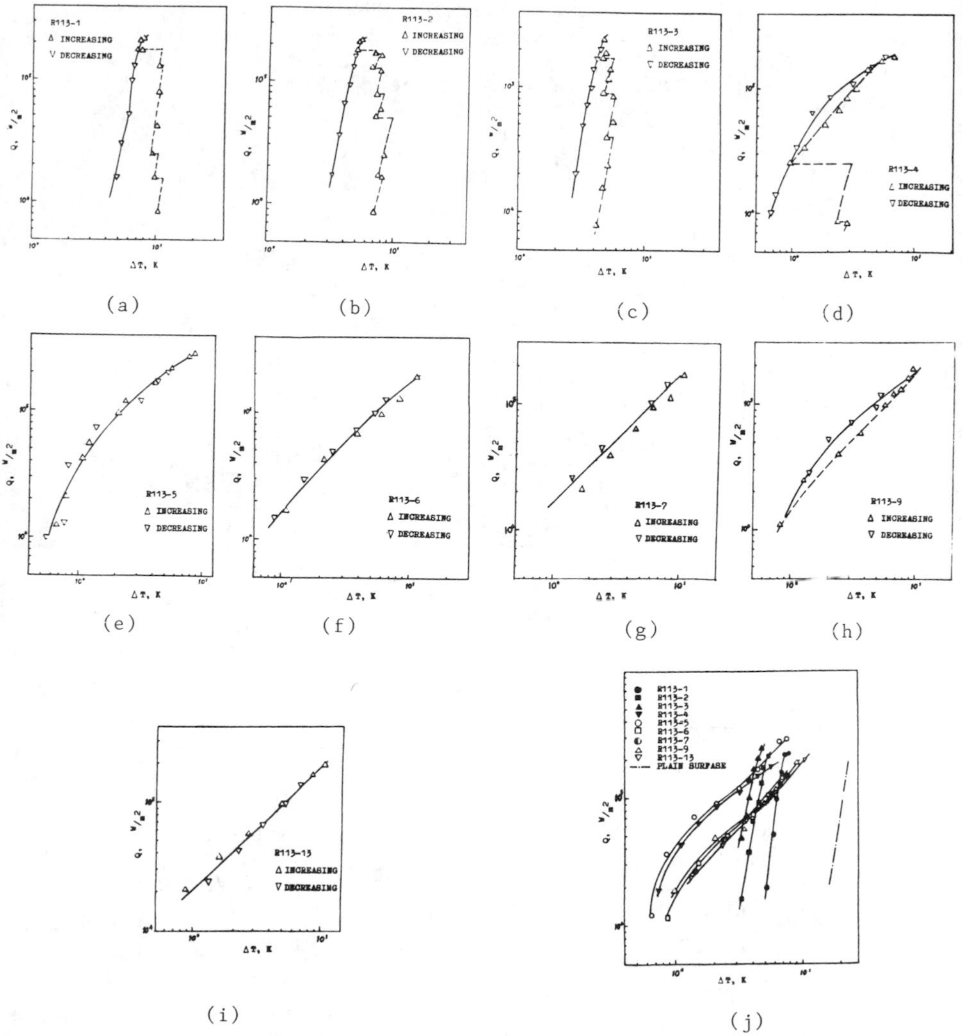

Fig. 2 Boiling curves for R-113

esis. For the two-layer screen surface, the case was somewhat similar, but the wall superheat needed was no more than 8.5 K, as long as the increasing and decreasing curves remain distinct. For the three-layer screen surface, the situation more or less improved. In the non-coinciding portion the highest superheat was less than 6 K. As the number of layers increased to four, the wall temperature decreased with the increase of heat flux only when the heat flux was less than 2.5×10^4 W/m^2 and the wall superheat lower than 3 K. For the number of layers beyond five, there was no apparent hysteresis within the range of the recorded temperature. From the above we see that with the increase of the thickness of porous layer, the heat flux at the point of coincidence of the increasing and decreasing curves goes down gradually. In addition, the start-up superheat declines gradually. These characteristics are of benefit to the application of por-

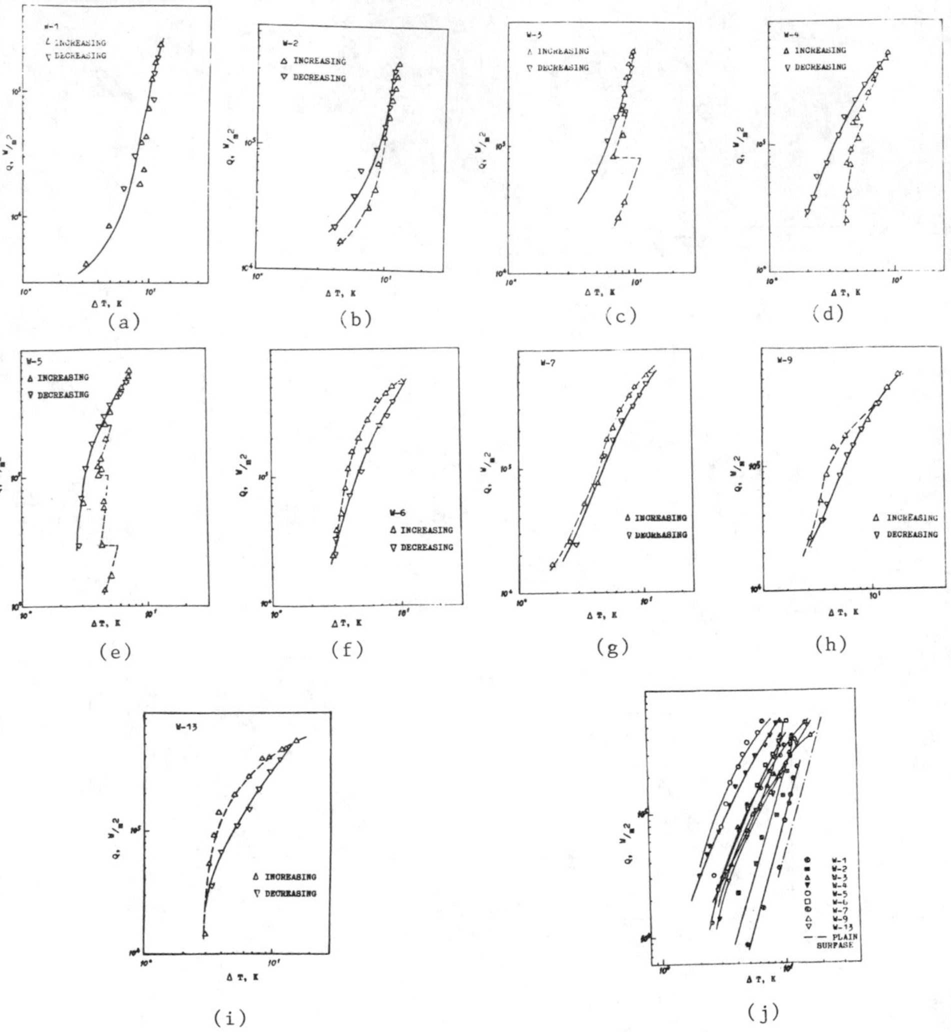

Fig.3 Boiling curves for water

ous surface of sintered screen type. If the start-up superheat decreases with the increase of number of layers, it is then possible to let porous surfaces work in the segment of high effectiveness without the need of high surperheat.

Fig. 2 (j) shows a comparision of the boiling curves of the surfaces with different numbers of screen layers in R-113. It must be kept in mind that these curves are decreasing curves. These curves illustrate that the number of layers is closely connected with the heat transfer coefficient. The following characteristic can be observed from the curves.

1. From the standpoint of boiling enhancement an optimum number of layers is sure to exist. For low and medium heat fluxes four or five layers are desirable, while for high heat fluxes, three layers are the best. The effect of hysteresis, however, should always be kept in mind. The figure does not indicate the increasing curves.

2. When the number of layers is small the shape of the boiling curves change sharply with the number of layers, but for large numbers of layers the change becomes less as the number of layers increases. For the numbers of layers 7,9,13, there are no obvious changes.

The general tendency of the boiling curves for distilled water shown in Fig.3 (a)-(j) is similar to that of the curves for R-113, but for the case of one-layer in water, there is no obvious hysteresis. When the number of layers goes up to 3 4, and 5, hysteresis appears but not so strong as in R-113. When the numbers of layers become greater than 6, hysteresis diminishes quickly.

An explanation (6) of the hysteresis in boiling on porous surfaces was given as follows: It is due to flooding of the porous matrix with liquid so that only relatively small sites are available for nucleation. By the usual theory of boiling nucleation, these sites must have re-entrant characteristic. A similar explanation was given to explosive boiling (4). The authors of this paper tend to suppose that, besides the above explanation, hysteresis is closely connected with the thickness of porous layer due to two reasons: The formation of re-entrant cavities and the interconnection of cavities.

The re-entrant cavities can be activated under low degree of superheat. However, re-entrant cavities are not likely to form all over the surface under the condition of one-layer screen, but more and more cavities will become re-entrant as the number of screen layers increases (Fig.4). Therefore, the increase of number of screen layers brings about the improvement of porous structure. Of course, re-entrant cavities will never increase indefinitely as the thickness of porous layer increases. In this case the increased flow resistence of vapour will play an important role. Therefore we see from the boiling curves that the heat transfer tends to deteriorate when the number of layers goes beyond a certain value.

Fig.4 The increase of porous layers forms new re-entrant cavities

Another important effect of the thickness of porous layers is concerned with the interconnection of cavities . As the thickness of porous layers goes up the originally individual cavities gradually become interconnected throughout to form an integrated surface so that the activation of some re-entrant cavities can induce boiling over the entire surface at relatively low degree of superheat.

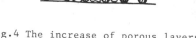

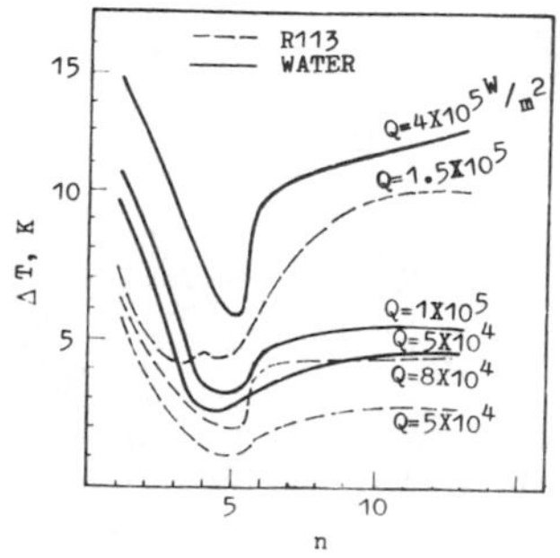

Fig.5 Effect of the numbers of layers on superheat

The curves shwon in Fig.5 are the relationships between superheat and the numbers of layers for different heat fluxes. These curves show that the higher the heat flux, the more obvious is the effect of the number of layers.

It is very difficult, if not impossible, to make a precise mathematical description of the boiling process because too many parameters are involved. Furthermore, porous surfaces themselves introduce new variables, such as the diameter of the pores, the thickness of the porous layer, the number of pores per unit area, the nature of the porous material, the method of manufacture, etc. In order to thoroughly understand the effects of these factors the boiling heat transter mechanism from porous surfaces must be further investigated. For the time being, some of the parameters, which are deemed more important, should be screened from the entirety, and then a relatively simple correlation can be sought among these variables that may be used for design purposes.

A simplification results from the assumption that the vapour in the bubbles is homogeneous. In this way the equations for the liquid and the boundary conditions should be taken into account. Assuming full consistency of size in every layer of sintered screens, only two characteristic values are needed, namely, characteristic length D (diameter of the screen pores) and the number of layers n. Thus, we have:

$$Q = f(\alpha, C_1, \rho_1, \mu_1, \sigma, \lambda_1, D, g, \Delta T, \rho_v, i, n, R, P)$$

By dimensional analysis the following dimensionless numbers have been obtained:

$$\frac{\mu_1^2}{\sigma D \rho_1}, \quad \frac{C_1 \mu_1}{\lambda_1}, \quad \frac{\mu_1^2}{g \rho_1^2 D^3}, \quad \frac{i}{C_1 \Delta T}, \quad \frac{Q\,D}{\mu_1\, i},$$

$$\frac{\mu_1^2}{P D^2 \rho_1}, \; \alpha, \; \frac{\rho_v}{\rho_1}, \; n, \; \frac{R}{C_1}$$

where the first and the sixth dimensionless numbers can be combined to yield Webe number σ/PD. The second is Prandtl number. The third is the reciprocal of Galileo number. Since the Froude number is the ratio of inertia force to gravity force and using the characteristic length r, which is related to the bubble depar ture radius, instead of D, we obtain

$$\nu_1^2/gr^3 \qquad \text{where} \qquad r = \alpha\left[\frac{\sigma}{g(\rho_1-\rho_v)}\right]^{0.5}$$

The fourth is Stanton number, and the fifth Reynolds number. By least square fit to the experimental data the following correlation is obtained:

$$\frac{C_1 \Delta T}{i} = 2.24\text{x}10^{-4}\,\frac{[(4.5-n)^2]^{0.115}}{[(7.5-n)^2]^{0.198}}\left(\frac{Q\,D}{\mu_1 i}\right)^{\frac{1}{1+8.2\text{x}0.55^{n}}}\left(\frac{P\,D}{\sigma}\right)^{1.73}\left(\frac{g\,r^3}{\nu_1^2}\right)^{0.18}$$

$$\alpha^{0.88}\left(\frac{\rho_v}{\rho_1}\right)^{0.744}\left(\frac{\lambda_1}{\mu_1 C_1}\right)^{0.047n}\left(\frac{R}{C_1}\right)^{0.0466n}\left(\frac{D\rho_1\sigma}{\mu_1^2}\right)^{0.00591n}$$

Comparison of the correlation with the experimental results is shown in Fig.6, where the measured values are based on those recorded when the heat flux is decreasing. Fig.6 reveals that scatter of the experimental data is within ±30% from the correlation line.

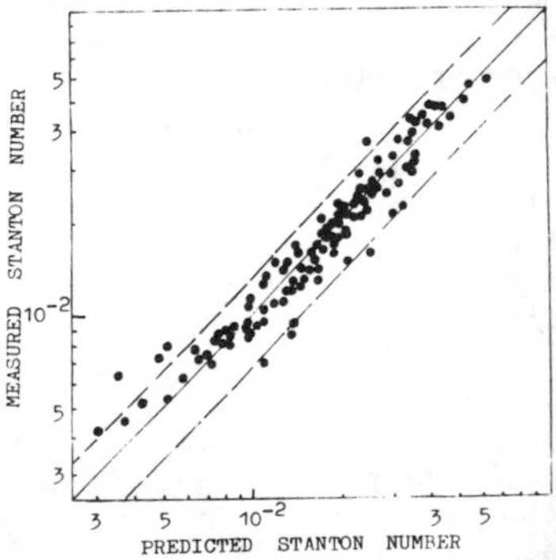

Fig.6 Comparison of predicted and measured Standon numbe

CONCLUSION

Boiling heat transfer coefficient on sintered screen surfaces, in the case of moderate heat fluxes, is 8-9 times as high as on plain surfaces in R-113 and 3-4 times as high as on plain surfaces in water.

Experimental results reveal that heat transfer coefficient increases significantly with the increase of the number of layers n up to 4-5. Further increase in n, however, leads to deterioration of boiling heat transfer, but the change of heat transfer coefficient with n between 7-13 is not as obvious as for $n < 6$.

Boiling heat transfer hysteresis changes with the number of screen layers. When this number is beyond a certain value, hysteresis diminishes significantly.

By means of dimensional analysis a correlation has been obtained for boiling heat transfer from sintered screen surfaces.

REFERENCES

1. Smirnov, G. F. and Afanasiev, B. A., Investigation of Vaporization in Screen Wick-Capillary Structure, Advances in Heat Pipe Technology, pp. 405, Proceedings of Fourth International Heat Pipe Conference, 7-10, September 1981, London, UK.
2. Rannenberg, M. and Beer, H., Heat Transfer by Evaporation in Capillary Porou Wire Mesh Structure, Letters in Heat and Mass Transfer, Vol. 7, pp. 425-436 1980.
3. Xin M. D.,Xie H. D. and Chen Y. G., Boiling Heat Transfer on Heating Surfaces Covered by Metal Screens, The proceedings of 5th Int. Heat Pipe Conf., Part 1, May 14-18, 1984, Tsukuba, Japan.
4. Ma Tongze, Zhang Zhengfang and Li Huiqun, Vaporization Heat Transfer in Capillary Wick Structures Formed by Sintered Screens, The proceedings of 5th Int. Heat Pipe Conf., Part 1, May 14-18, 1984, Tsukuba, Japan.
5. Wu Wen-guang, Ma Tong-ze and Li Hui-qun, Experimental Study of Critical Heat Flux in Pool Boiling at Low Pressure, Proceedings of the Seventh Internationa Heat Transfer Conference, Munchen, Fed. Rep. of Germany, 1982.
6. Bergles, A. E. and Chyn, M. C., Characteristic of Nucleate Pool Boiling from Porous Metallic Coatings, J. Heat Transfer, May 1982.
7. Nakayama, W., Daikoku, T., Kuwahara, H., and Nakajima, T., Boiling Heat Transfer on Porous Surfaces, Part 1, Experimental Investigation, J. Heat Transfer, Vol. 102, Aug., 1980.
8. Arshad, J., and Thome, J.R., Enhanced Boiling Surfaces: Heat Transfer Mechanism Mixture Boiling, ASME JSME Thermal Eng. Joint Conference, Vol. 1, 1983.

Growth Rate of Frost Formation through Sublimation—A Porous Medium Physical Model of Frost Layers

FANJIONG MENG, WANGONG GAO, and YANLING PAN
Dalian Marine College
PRC

ABSTRACT

The paper presents a combined experimental and theoretical study on the phenomenon of frost formation through sublimation of water vapor.

In this paper, taking the porous media as a physical model of frost layer, the laws of heat and mass transfer during the frost formation were analysed, the physical significance of tortuosity was demonstrated and a mathematical formula on frost layer tortuosity has been derived. On this basis, the rate equations for frost formation and equations for finding initial frost density were formulated.

The numerical solutions of frost formation on flat plate under forced convection are in good agreement with the author's experimental data. The maximal error was within 10 percent.

NOMENCLATURE

D	molecular diffusivity
D^*	diffusivity of porous media
D_{ij}	coefficient of mechanical dispersion
E_{ij}	coefficient of heat dispersion
h_H	heat transfer coefficient
h_M	mass transfer coefficient
J	flux
J^*	mass flux
K	thermal conductivity
K_{ij}	permeability
L	heat of sublimation
L_D	Dufour coefficient
L_s	coefficient of thermal diffusion
n	porosity
R	universal gas constant
V^*	mass velocity
V'	volume velocity
ρ	density
τ^*_{ij}	tensor of tortuosity
λ	thermal conductivity
ω	mass fraction

Subscripts

s	frost surface
m	porous matrix
v	water vapor
w	wall
α	component of fluid
a	in the airstream
f	frost
h	heat
i	solid ice
o	reference point

INTRODUCTION

In recent years, because of the advances of the engineering and technology in the field of cryonetics, refrigeration and air conditioning as well as the world's interest in energy resource problems, the investigations of frost formation of water vapor were enhanced and are getting steady progress ever since.

As the phenomenon of frost formation through sublimation of water vapor from the moist air on a cooled surface is a very complicated problem involving the growth of branch ice-crystals, in which heat and mass transfer are accompanied with changes of phases, shifts of boundaries and nonequilibrium, there remains much to be solved in this problem. Attempts in solving these problems in the microscopic level have been proved to be futile, whereas choice of an appropriate physical model appears to be of vital importance when describing these problems in macroscopic level.

In this paper, taking porous media as physical model of frost layers, the mathematical model of frost formation, its rate equations and numerical analyses have been probed to quite an extent. The experimental study on this problem is still in progress.

THE POROUS MEDIUM MODEL OF FROST LAYER

Since the construction of real frost layers is a mixture of air and matrix of branch ice crystals in which the air is filled, in this paper, frost layers play the role as porous medium, the ice crystals as porous matrix of this medium and humid air as the fluid in it.

Conceptual models of porous media of frost layers.

In order to express the principal properties of porous media of frost layers and to suit them to mathematical dealings, simplified conceptual models of porous media of frost layers are introduced in this paper [1].

It is also assumed that void spaces of frost layers are interconnected by stochastically emerged pipes. The length, cross section and direction of these pipes are all mutable and at least, there are three pipes connected to one connector. The main difference between the pipe and connector is that the pipes have slender shapes and their axes can be determined while the connectors have no certain direction in space. Pipes and connectors are evenly distributed in space, and the velocity of the fluid particle is supposed to be parallel to the wall, therefore, the flow of the fluid in void space is in laminar motion. Another assumption is made that the loss of the energy is not caused by connectors but by pipes.

The heat and mass transfer within frost layer

The mathematical formula depicting this physical model has been derived by using the method of continuum and average approach, taking into consideration of the cross and coupled effects of this irreversible process (the summation rule of the tensor analysis has been applied). The conservation equation of mass of water vapor in per unit of void area can be written:

$$\frac{\partial \rho_v}{\partial \tau} + \frac{\partial}{\partial x_i}(\rho_v V_i^{\bullet}) - \frac{\partial}{\partial x_i}\left(D_{ij}\frac{\partial \rho_v}{\partial x_j}\right) + \frac{\partial}{\partial x_i}(J_{vi}^{\bullet}) + \frac{\rho_v V_i'}{n}\frac{\partial n}{\partial x_i} = \frac{\partial \rho_f}{\partial \tau} \qquad (1)$$

The conservation equations of mass, momentum and energy of humid air in per unit of void area can be written:

$$\frac{\partial \rho}{\partial \tau} + \frac{\partial}{\partial x_i}(\rho V_i^{\bullet}) = 0 \qquad (2)$$

$$V_i^{\bullet} + \frac{\overline{B}\rho}{\mu}\frac{\partial V_i^{\bullet}}{\partial \tau} = -\frac{K_{ij}}{n\mu}\left(\frac{\partial P}{\partial x_i} + \rho g \frac{\partial z}{\partial x_j}\right) \qquad (3)$$

$$\rho c\left(\frac{\partial T}{\partial \tau} + V_i \frac{\partial T}{\partial x_i}\right) = -\frac{\partial}{\partial x_i}(J_{hi} + LJ_{vi}) - \frac{\partial}{\partial x_i}\left(E_{ij}\frac{\partial T}{\partial x_j}\right) + h(T_m - T) + \varepsilon \tag{4}$$

where $\overline{B}$ -- conductivity

$$\varepsilon = \rho V_i \frac{\partial}{\partial x_i}\left(\frac{P}{\rho} + gz\right) \tag{5}$$

The conservation equation of ice crystals as porous matrix in per area of ice crystals is given;

$$\rho_m C_m \frac{\partial T_m}{\partial \tau} = -\frac{\partial}{\partial x_i}(J_{hmi}) + h(T - T_m) \tag{6}$$

where

$$J_{hmi} = -(\lambda_m)_{ij}\frac{\partial T_m}{\partial x_j} \tag{7}$$

When heat and mass flux is taken as irreversible coupled processes, the thermal diffusion effect and Dofour effect should be considered.

$$J_{ai} = -\rho(D^*)_{ij}\frac{\partial w_a}{\partial x_j} - (L_S)_{ij}\frac{\partial T}{\partial x_i} \tag{8}$$

$$J_{hi} = -(L_D)_{ij}\frac{\partial w_a}{\partial x_j} - (\lambda)_{ij}\frac{\partial T}{\partial x_j} \tag{9}$$

In per unit of the porous media, the energy equation is written (assuming $T = T_S$)

$$[n\rho c + (1-n)\rho_m C_m]\frac{\partial T}{\partial \tau} + \rho c n V_i \frac{\partial T}{\partial x_i} = \frac{\partial}{\partial x_i}\Big\{[n(\lambda)_{ij} + (1-n)\ (\lambda_m)_{ij}]\frac{\partial T}{\partial x_j} + (L_D)_{ij}\frac{\partial w_a}{\partial x_j} + Ln\rho_f (D^*)_{ij}\frac{\partial w_v}{\partial x_j} + (L_S)_{ij}\frac{\partial T}{\partial x_j}\Big\} - \frac{\partial}{\partial x_i}\left(nE_{ij}\frac{\partial T}{\partial x_j}\right) + n\varepsilon \tag{10}$$

where: $n\rho c + (1-n)\rho_m C_m = C_f \rho_f$; $n\lambda + (1-n)\lambda_m = K_f$

According the Onsager reciprocal relations, the coefficient of irreversible cross coupled effects is $L_D = L_S$. When compared with coefficient for direct effects, their order of magnitude is 10^{-3} and is then neglected [1][2] . As the frost forms on cooled surface, the porous end of the media is closed, therefore, the effect of mechanical and heat dispersion can be neglected. The moist air can be considered as a balanced system, the density of which being unaffected with the elapse of time ($\rho_v \ll \rho$). The viscosity dissipation and heat radiation can be neglected when compared with changes of sublimation phases.

Also assumed that the porosity n is evenly distributed in space and varies with time. From (1)-(10), we obtain the governing differential equations for frost formation through sublimation.

$$\frac{\partial \rho_v}{\partial \tau} = \frac{\partial}{\partial x_i}\left(nD\tau_{ij}\frac{\partial \rho_v}{\partial x_j}\right) - \frac{\partial \rho_f}{\partial \tau} \tag{11}$$

$$C_f \rho_f \frac{\partial T}{\partial \tau} = \frac{\partial}{\partial x_i}\left(K_f \frac{\partial T}{\partial x_j}\right) + L\frac{\partial}{\partial x_i}\left(nD\tau_{ij}\frac{\partial \rho_v}{\partial x_j}\right) \tag{12}$$

When the boundary conditions and initial conditions are added, the universal mathematical model can be obtained.

The porosity and tortuosity of frost layer

The porosity and tortuosity are important macroscopic parameters of porous media, but this problem still remains unsolved as introduced by many literatures [3][4]. The important character of porous media can be proved that volumetric porosity at any point of porous media equals to the mean area porosity of this point [1]. By assuming the frost density is even in space within frost layer, we can prove:

$$n = n_V = n_A = \frac{\rho_t - \rho_f}{\rho_t - \rho_a} \tag{13}$$

The macroscopic parameter τ^*_{ij} emerged in average equations is a second order tensor. τ^*_{ij} is dimentionless operator. It converts any component of the external driving force on an elemental control volume in void space (as ∇C, ∇P) into projective component on axial direction of this point of the pipe [1].

$$\tau^*_{ij} = \cos(I_s, I_{xi}) \cos(I_s, I_{xj}) \tag{14}$$

where I_s – unit vector along the pipe axes,
I_{xi}, I_{xj} – unit vector along the space coordinate axis.
The average parameter $\bar{\tau}^*_{ij}$ can be comprehended as tortuosity. According to conceptual model of porous media of frost layer, the equivalent average results for one-dimentional frost layer can be shown as in fig. 1. The diffusive mass flux along pipe axis direction S of porous media:

$$J_e = -nD\frac{d\rho_v}{dL_e} = -nD\frac{d\rho_v}{dx}\cos\theta \tag{15}$$

The diffusive mass flux along direction x of porous media:

$$J = J_e\cos\theta = -nD\cos^2\theta\,\frac{d\rho_v}{dx} = -nD\left(\frac{L}{L_e}\right)^2\frac{d\rho_v}{dx} \tag{16}$$

therefore, tortuosity will be:

$$\bar{\tau}_{ij} = \left(\frac{L}{L_e}\right)^2 = \cos^2\theta \tag{17}$$

Applying (14), we have the average parameter for one-dimentional frost layer: $\bar{\tau}^*_{ij} = \cos^2\theta$

Under general conditions, the mean length square of the passase L_e^2 is inversely proportional to the area porosity and directionly proportional to the square linearity of the fluid, therefore, L_e^2 can be written:

$$L_e^2 = \frac{c}{n\left(\frac{v}{v_0}\right)^{-\frac{2}{3}}} = \frac{c' L^2}{n\left(\frac{v}{v_0}\right)^{-\frac{2}{3}}} \tag{18}$$

where C, C' is proportion constants. From the equations of ideal gas state we have:

$$\left(\frac{v}{v_0}\right)^{-\frac{2}{3}} = e^{\frac{2}{3}\left(-\frac{P_0}{P} + \frac{T_0}{T}\right)} \qquad \therefore\ \bar{\tau}_{ij} = \left(\frac{L}{L_e}\right)^2 = \frac{n e^{\frac{2}{3}\left(-\frac{P_0}{P} + \frac{T_0}{T}\right)}}{C'}$$

When the initial conditions of frost formation at $\tau = 0$, $T = T_0$, $P = P_0$, $n = 1$,

$(\overline{\tau^*_{ij}}) = 1$ proportion coefficient can be given as: $c' = 1$. Considering the $P = P_0$ during the process of frost formation, the mean tortuosity can be written

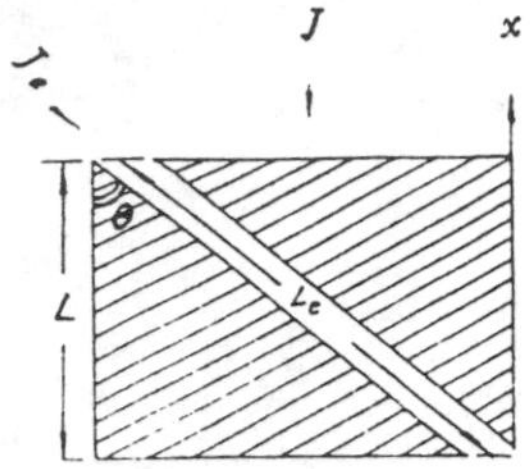

Fig. 1 Eguivalent model

Fig. 2 photograph of the test loop

$$\overline{\tau}_{ij} = \tau^* = n e^{\frac{2}{3}\left(\frac{T_0}{T} - 1\right)} \tag{19}$$

where T_0 is reference temperature. If $T_0 = T_W$ then,

$$\tau^* = n e^{\frac{2}{3}\left(\frac{T_W}{T} - 1\right)} = \frac{\rho_i - \rho_f}{\rho_i - \rho_a} e^{\frac{2}{3}\left(\frac{T_W}{T} - 1\right)} \tag{20}$$

In Equation (20) we find that n reduces and the temperature of frost layer rises as the density of frost layer increases. Flux of the diffusive mass reduces too.

THE RATE EQUATIONS OF FROST FORMATION

The one-dimentional mathematical model of frost formation through sublimation on flat plate under forced convection is derived from the governing equation (11) and (12):

$$\frac{\partial \rho_v}{\partial \tau} = \frac{\partial}{\partial x}\left(nD\tau^* \frac{\partial \rho_v}{\partial x}\right) - \frac{\partial \rho_f}{\partial \tau} \tag{21}$$

$$C_f \rho_f \frac{\partial T}{\partial \tau} = \frac{\partial}{\partial x}\left(K_f \frac{\partial T}{\partial x}\right) + L \frac{\partial}{\partial x}\left(nD\tau^* \frac{\partial \rho_v}{\partial x}\right) \tag{22}$$

The initial and boundary conditions are:

$$T = T_0 = T_w, \quad \rho_f = \rho_{f0}, \quad x = x_0 \qquad (\text{at } \tau = 0) \tag{23}$$

$$T = T_W = \text{const}, \qquad (\text{at } x = 0) \tag{24}$$

$$\left(K_f \frac{dT}{dx}\right)_{x=0} = h_H (T_a - T_S) + \dot{m}_t L \qquad (\text{at } x = 0) \tag{25}$$

The mass flux through cooled surface:

$$-\left(D^{\bullet}\frac{d\rho_v}{dx}\right)_{x=0}=0 \qquad \text{(at } x=0) \tag{26}$$

$$\left(K_f\frac{dT}{dx}\right)_{x=s}=h_H(T_a-T_S)+L\rho_{fs}\frac{dx_s}{d\tau} \qquad \text{(at } x=s) \tag{27}$$

$$\left(D^{\bullet}\frac{d\rho_v}{dx}\right)_{x=s}=\dot{m}_t-\rho_{fs}\frac{dx_s}{d\tau} \qquad \text{(at } x=s) \tag{28}$$

where:

$$D^{\bullet}=nD\tau^{\bullet}=n^2De^{-\frac{2}{3}}\left(\frac{T_W}{T}-1\right) \tag{29}$$

As part of the water vapor, being transported to the frost during the process of frost formation, will be diffused into the existing layer of frost before it sublimates (which serves to increase the frost density), the rest part of the water vapor sublimates outside the frost layer and increase the thickness of the frost layer. The amount of water vapor being transported to the frost:

$$\dot{m}_t=h_M(\rho_{va}-\rho_{vs})=\rho_{fs}\frac{dx_s}{d\tau}+\int_0^S\frac{\partial\rho_f}{\partial\tau}dx \tag{30}$$

The equation (21) is integrated between the interval [0,S] and while applying the equations (26), (28) and (30), then $\partial\rho_v/\partial\tau=0$. Within the frost layer, the temperature is in a quasi-steady state, then $\partial T/\partial\tau=0$ [4].

The equations (21) and (22) can be simplified:

$$\frac{d}{dx}\left(D^{\bullet}\frac{d\rho_v}{dx}\right)-\frac{d\rho_f}{d\tau}=0 \tag{31}$$

$$\frac{d}{dx}\left(K_f\frac{dT}{dx}\right)+L\frac{d}{dx}\left(D^{\bullet}\frac{d\rho_v}{dx}\right)=0 \tag{32}$$

The equation (31) is integrated between the interval [0, S], and when considering the boundary condition (26) and the assumption that the frost density is homogeneous in space, then

$$\left(D^{\bullet}\frac{d\rho_v}{dx}\right)_{x=s}=x_s\frac{d\rho_f}{d\tau} \tag{33}$$

Supposing that the water vapor is an ideal gas and near-thermodynamically equilibrium, we have by Clapeyron equation:

$$\rho_v=\frac{P_0}{RT}exp\left[-\frac{L}{R}\left(\frac{1}{T}-\frac{1}{T_0}\right)\right] \tag{34}$$

$$\frac{d\rho_v}{dx}=\rho_v\left[\frac{L}{RT^2}-\frac{1}{T}\right]\frac{dT}{dx} \tag{35}$$

The growth rate of frost density by substituting ρ_v and $\frac{d\rho_v}{dx}$ into equation (33) as:

$$\frac{d\rho_f}{d\tau}=\frac{D_s^{\bullet}\,\rho_{vs}\left[\frac{L}{RT_s^2}-\frac{1}{T_s}\right]\left(\frac{dT}{dx}\right)_s}{x_s} \tag{36}$$

The growth rate equation of frost thickness can be derived by equation (30):

$$\frac{dx_s}{d\tau} = \frac{\dot{m}_t - x_s \frac{d\rho_f}{d\tau}}{\rho_f} = \frac{h_M(\rho_{va} - \rho_{vs}) - D^{\circ}\rho_{vs}\left(\frac{L}{RT_s^2} - \frac{1}{T_s}\right)\left(\frac{dT}{dx}\right)_s}{\rho_f} \tag{37}$$

In order to solve equations (36) and (37), both $\left(\frac{dT}{dx}\right)_s$ and T_s can be found by integrating the equation (31) and equation (32) when applying the boundary condition equations (25) ~ (28) and assuming that the conductivity K_f of the frost in the space is homogeneous, we have

$$\frac{dT}{dx} = \frac{h_H(T_a - T_s) + h_M(\rho_{va} - \rho_{vs})L}{K_f + LD^{\circ}\rho_v\left[\frac{L}{RT^2} - \frac{1}{T}\right]} \tag{38}$$

$$T_s - T_w + \frac{LD_s^{\circ}\rho_{vs}\left[\frac{L}{RT_s^2} - \frac{1}{T_s}\right]\left(\frac{dT}{dx}\right)_s}{2K_f}x_s = \frac{h_H(T_a - T_s) + h_M(\rho_{va} - \rho_{vs})L}{K_f}x_s \tag{39}$$

The growth rate equations of frost formation through sublimation on flat plate under forced convection can be expressed as:

$$\frac{dx_s}{d\tau} = \frac{h_M(\rho_{va} - \rho_{vs}) - D_s\rho_{vs}\left(1 - \frac{\rho_f}{\rho_i}\right)^2 e^{-\frac{2}{3}\left(\frac{T_w}{T_s} - 1\right)}\left(\frac{L}{RT_s^2} - \frac{1}{T_s}\right)\left(\frac{dT}{dx}\right)_s}{\rho_f} \tag{40}$$

$$\frac{d\rho_f}{d\tau} = \frac{D_s\rho_{vs}\left(1 - \frac{\rho_f}{\rho_i}\right)^2 e^{-\frac{2}{3}\left(\frac{T_w}{T_s} - 1\right)}\left(\frac{L}{RT_s^2} - \frac{1}{T_s}\right)\left(\frac{dT}{dx}\right)_s}{x_s} \tag{41}$$

EXPERIMENTAL RESEARCH AND NUMERICAL ANALYSES

The system used in experimental study of frost formation under forced convection was made up by a polymethyl methacrylate wind tunnel and a closed space with controllable air temperature , air velocity and humidity. Fig. 2 is a photograph of part of the test loop with some of the measure system and refrigeration appliances. The entrance section and exist section of the wind tunnel in the closed space (not shown in Fig. 2) is at the left side.

In Fig. 3 shows the schematic diagram of the equipment. The test section made by brass plate (150/52 mm) is fixed on a thermo-electric refrigeration unit in which the temperature is controlled (0 - 40°C) . The temperature of cooled surface of frost formation is measured by copper-constantan thermocouples, which is evenly buried 1 mm under the cooled surface.

The thermal electric potential is measured by digital millivoltameter with precision rate 1%mv. The air temperature and humidity of wind tunnel is measured by four platinum resistors, which are fixed at the test section (as dry and moist balls), The temperatures are shown by digital centigrade, precision of which is ± 25% °C. The Doppler Laser system was used to measure the air flow velocity with precision rate ± 1%.

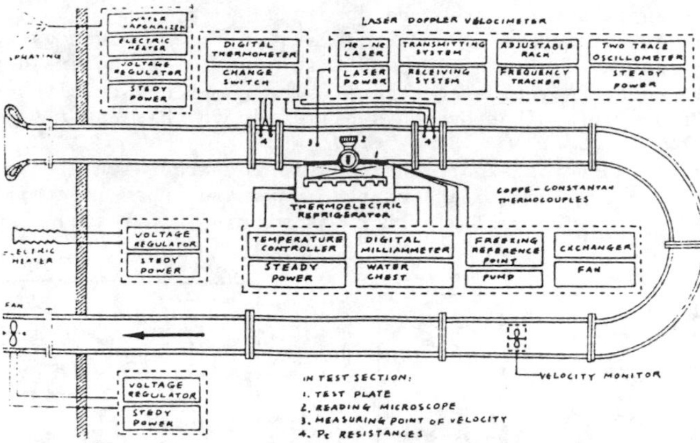

Fig. 3. Schematic diagram of test loop

The whole system can be humidifiled with atomizer by heating water vapor in the closed space while the temperature undulation of the wind tunnel during the expertimental process is less than $\pm$ 0.5 ^{o}C. The growth of the frost thickness with time is shown in Fig. 4.

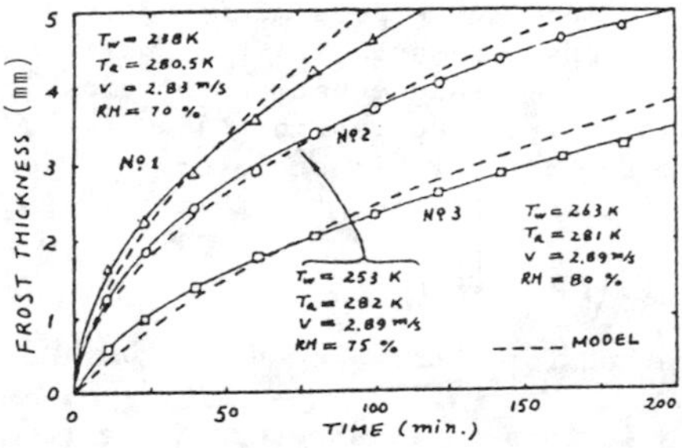

Fig. 4 comparison of model and data for this paper

The rate equations of frost formation (40) and (41) can be solved by numerical method, but first of all, chioce of initial value has to be solved. In literatures correlative to the phenomenon of frost formation, the choice of initial values has different methods [3][4], and has not yet well settled. In this paper, the temperature of initial frost layer is assumed to be linearly distributed and the frost density of initial frost layer is even considered to be homogeneous.

$$T_{so} = T_w + \left(\frac{dT}{dx}\right)_{so} \cdot x_{so} \tag{42}$$

where T_{so} and $\left(\frac{dT}{dx}\right)_{so}$ given by formulae (38) and (39).

In this paper, the initial thickness of frost layer is 2 X 10^{-5}m considering the order of the critial radius of steady ice crystal so that the initial thickness will be no less than the spherular hat length of steady ice crystal. The initial frost denstity can thereby be obtained [5][6].

The choice of parameters in numerical analysis: D_S given by Eckert and Drake [7]; K_f given by Brian Reid and Shan [8]; h_M and h_H are determined by using the Colburn and Reynolds equations. The numerical results using rate equations of frost formation through sublimation in this paper (38) - (41) are shown in Fig. 4, 5, 6.

In Fig. 4 the author's experiment data is compared with theoretical results. In Fig.5,6, the experiment data of the Yonko and Sepsy is compared with theoretical results [9]. The maximal error was within 10 percent.

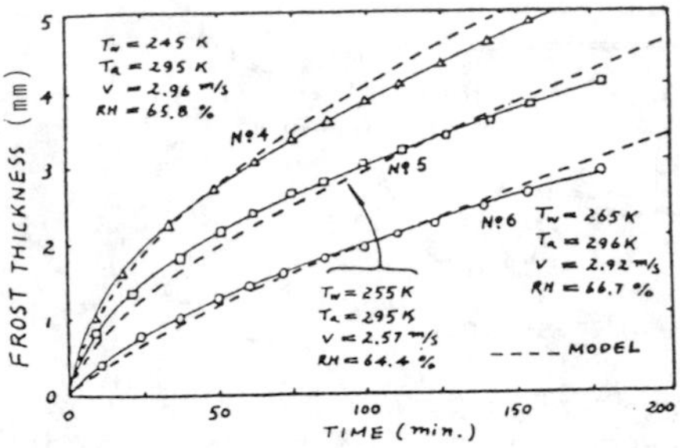

Fig. 5

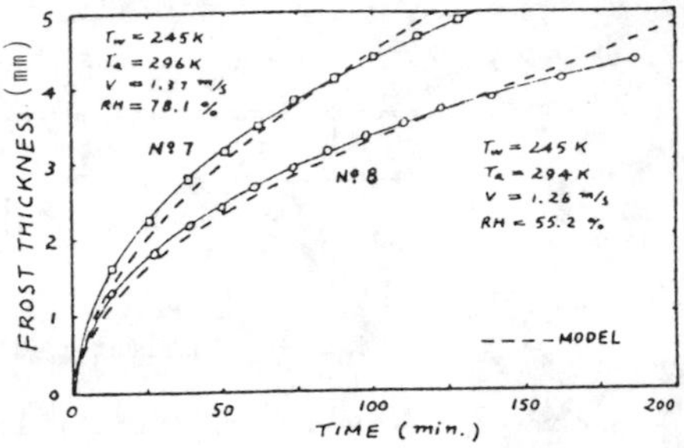

Fig. 6

Comparison of the model and data for Yonko and sepsy

CONCLUSIONS

(1) According to conceptual model of porous media of frost layer, this paper give governing differential equations (11) and (12) for sublimation frost formation.

(2) This paper gives more perfect rate equations of frost formation through sublimation on flat plate under forced convention (40). (41).

(3) The macroscopic parameter τ^*_{ij} in frost layer is a second order tensor, the average parameter $\bar{\bar{\tau}}^*_{ij}$ is tortuosity of frost layer. The one-dimentional frost layer is

$$\tau^* = ne^{\frac{2}{3}}\left(\frac{T_0}{T} - \frac{P_0}{P}\right)$$

(4) This paper gives a method for choice of initial frost density by equations (38) (39) (42).

(5) The experimental data in this paper shows that the rate of frost layer increases with the increase of air flow velocity and humidity and the decrease of temperature in cooled surfaces.

REFERENCES

1. J.Bear, Dynamics of Fluids in Porous Media, American Elsevier Publishing Co., Inc. 1972.
2. S.R. Degroot and P. Mazur, Non-equilibrium Thermodynamics, North-Holland Publishing Co. - Amsterdam, 1962.
3. B.W. Jones and J.D. Parker, Frost Formation with varying Environmental Parameters, Journal of Heat Transfer. Trans. ASME, series C, Vol, 97. No.2, 1976.
4. M.Ditenbeger, P. Kumar and J. Luers, Frost Formation on an Airfoil A Mathematical Model I, NASA Contractor Report 3129, 1979.
5. F.J. Meng and Y.L. Pan, Growth Rate of Frost Formation on Horizontal Surface under Forced Convection and its Numerical Calculation, Journal of Dalian Marine College, No.2. Aug. 1983.
6. Y.B. Li and Y.L. Pan, A Physical Model of Forming Ice Nuclei on Vessel Surface from Vapor in Moist Air through Sublimation,Journal of Engineering Thermophysics, Vol. 3, No. 2 May, 1982.

7. E.R.G. Eckert and R.M. Drake, Jr., Analysis of Heat and Maas Transfer, McGraw-Hill, New York, 1972, P. 787.
8. P.L.T., Brian, R.C. Reid and Y.T. Shan, Frost Deposits on Cold Surface, Ind. Eng. Fundamentals, Vol. 9, No. 3, pp. 375 - 380.
9. J.D. Yonko and C.F. Sepsy, An Investigation of the Thermal conductivity of Frost while Forming on a Flat Horizontal Plate, ASHRAE Trans., Vol. 73, Part I

Analysis on the Moment Method for Determining the Moisture Transport Properties in Porous Media

BU-XUAN WANG
Thermal Engineering Department
Tsinghua University
Beijing, PRC

ZHAO-HONG FANG
Shandong Institute of Civil Engineering
Jinan, PRC

ABSTRACT

A new unsteady-state method is proposed for determining the moisture transport properties in wet porous media. It is based on measurement of the change in moment of gravity caused by the moisture migration. In addition to its high-speed performance, this method may get rid of the difficulty in determination of a changing moisture content or moisture distribution. On this basis, two particular procedures are contrived: a constant heat source method for determining the thermal mass diffusivity and an instantaneous moisture source method for determining the moisture diffusivity.

NOMENCLATURE

- a thermal diffusivity
- A cross-sectional area
- b length
- D_m moisture diffusivity
- D_t thermal mass diffusivity
- E defined by Eq. (11)
- g gravitational acceleration
- G reading of balance
- h thickness of saturated layer
- $\mathbf{i}$ unit vector in x-direction
- J mass flux
- l interval between two bearers
- L length of sample
- m mass of water added
- M moment
- $\mathbf{n}$ unit vector in normal direction
- q heat flux
- t temperature
- w moisture content
- x,y,z coordinates
- η $=h/(2\sqrt{D_m\tau})$, dimensionless
- θ temperature excess
- λ thermal conductivity
- ρ bulk density of dry medium
- τ time

Subscripts

- 0 initial value
- s saturated

INTRODUCTION

The moisture in wet porous media migrates generally by the action of three impetuses, i.e. the total pressure gradient, the moisture content gradient and the temperature gradient in the porous media. Under very small total pressure gradient or in a medium with poor permeability, the migrations caused by the latter two impetuses are prevailing. The moisture transport process in capillary-porous media is much slower as compared with heat transport process. The quick and convenient means is now lacking for measuring the moisture content in a porous medium. Consequently, it becomes a difficult and time-consuming task to determine moisture transport properties in the media. Up to now these data have been not only very deficient but also poor in accuracy, and can therefore meet little need of practical use. The shortage of systematic property data also hampers verification and perfection of the theories on moisture transport mechanisms by experiments.

Since the twenties of this century a number of methods have been developed for measuring the moisture transport properties in porous media. These methods can be classified generally as either steady-state or unsteady-state ones. The steady-state methods are simple in principle and have the formulae in differential forms. It is unnecessary to assume the properties constant. However, the steady-state methods have a common drawback that they spend lots of time because the diffusivities of moisture transport in porous media are usually so small that it takes a long time, often months, for a sample to establish its steady state of moisture distribution. Notice that D_m and D_t are functions of both the moisture content and temperature, so a series of tests would be needed to obtain the moisture transport properties for a given material with different moisture contents and at different temperatures. This is certainly an arduous task. Besides, the procedure of cutting, weighing and drying is rather inconvenient and will introduce some additional errors to the final results.

In order to overcome the obstacle of long test duration, a few unsteady-state methods have been developed since the fifties.[1] They are all based on solutions for the moisture content field under peculiar boundary and initial conditions with the properties involved usually assumed constant. The changing moisture distribution in the samples is ascertained either by direct measurement with the traditional procedure of cutting and weighing, or by indirect measurement, in which the moisture content is transformed into another kind of physical parameters. For example, the electric conductance or electric capacity methods [2], the microwave or γ-ray methods [3] etc. have been reported for this purpose. All the indirect measurements suffer difficulty of having to calibrate before being used. The effect of moisture on indirect measurements depends upon the operating temperature and the nature of the materials as well as the actual moisture content itself. So, most of these unsteady-state methods still remain in their developing stage.

A new unsteady-state method is proposed here for determining the moisture transport properties in wet porous media. It employes a precise balance as the main measuring instrument. The essential idea of "the moment method" is to study the shift of center of gravity of the sample with moisture migration, i. e. the change in the moment of moisture gravity about a reference point. This method attempts to get rid of both cutting the sample and calibrating for indirect measurement and is expected to simplify the operation, improve the accuracy and achieve continuous measurement. It will also be shown that the moment method is capable of being used in tests with different boundary and initial conditions.

PRINCIPLE OF THE MOMENT METHOD

As shown in Fig.1, a cylindrical homogeneous sample, fixed horizontally, is sustained by two bearing points B and B'. The point B' is connected with a precise balance. The cross-sectional area of the sample is A and the bulk density of its dry skeleton is ρ . Suppose the moisture distribution in the sample is one-dimensional and denoted by $w(x,\tau)$. Then the moment of the moisture gravity about B at any instant τ can be written as

$$M = \int_0^L A\rho g(x - b)w \, dx \qquad (1)$$

The balance will accordingly give the reading

$$G = (M + M_d)/(gl), \qquad (2)$$

where M_d denotes the corresponding moment of the dry skeleton of the sample, not changing with time.

At the initial instant the sample has a moisture distribution of $w_o = w(x,o)$, and the balance reading is G_o. Then the increment of the balance reading at any instant τ will be

$$\Delta G = G - G_o$$
$$= \frac{1}{l}\int_o^L A\rho(w-w_o)x\,dx - \frac{b}{l}\int_o^L A\rho(w-w_o)dx, \qquad (3)$$

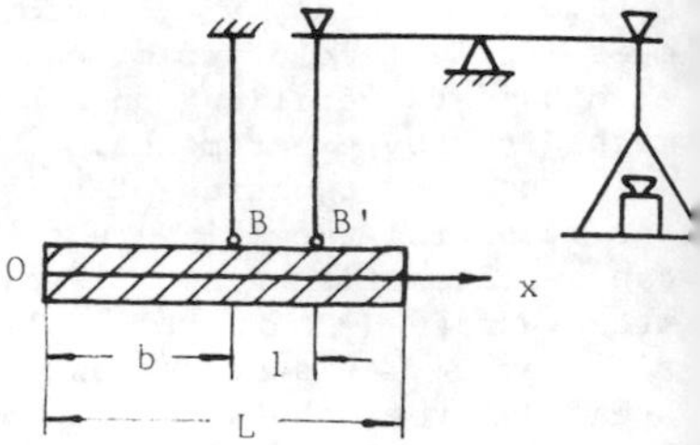

Fig. 1 The schematic diagra
of the moment methoc

where $\int_o^L A\rho(w - w_o)dx$ represents the increase in the total mass of moisture contained in the sample and will be equal to zero if the sample is kept in moisture insulation with its surroundings. In this instance we reduce equation (3) to

$$\Delta G = \frac{1}{l}\int_o^L A\rho(w - w_o)x\,dx. \qquad (4)$$

We can adjust positions of the two bearing points so that only a small load is exerted on B' and ΔG will be measured with a precise balance.

It would be appropriate for determining the transport properties to make the initial moisture distribution uniform in the sample. If a heat or moisture effe is then applied to the end $x = 0$ of the sample, the moisture distribution in the sample will begin to change with time. The sample may be thus regarded as a sem infinite body with respect to the moisture transport process until the test last long enough. On this condition, we can rewrite equation (4) as if the sample we infinitely extended so that

$$\Delta G = \frac{1}{l}\int_o^\infty A\rho(w - w_o)x\,dx. \qquad (5)$$

The equations derived above can also be expressed in differential forms. For example, corresponding to equation (5), we have

$$\frac{dG}{d\tau} = \frac{1}{l}\int_o^\infty A\rho\frac{\partial w}{\partial \tau}x\,dx. \qquad (6)$$

$\frac{dG}{d\tau}$ may be calculated from the measured ΔG at different instants or directly obtained with the aid of a differential circuit.

CONSTANT HEAT SOURCE METHOD FOR DETERMINING D_t

When a constant heat flux q is applied to the end x=o of a sample with uniform initial temperature and moisture distributions, the moisture in the sample will migrate towards the other end. On the assumptions of one-dimensional trar fer and constant properties, the transient temperature and moisture content fie in such a semi-infinite body should be [4]:

$$w-w_o = \frac{2q\,D_t}{\lambda(D_m - a)}\left[\frac{a}{D_m}\sqrt{D_m\tau}\,\mathrm{ierfc}\left(\frac{x}{2\sqrt{D_m\tau}}\right) - \sqrt{a\tau}\,\mathrm{ierfc}\left(\frac{x}{2\sqrt{a\tau}}\right)\right], \qquad (7$$

$$t-t_o = \frac{2q\sqrt{a\tau}}{\lambda}\,\mathrm{ierfc}\left(\frac{x}{2\sqrt{a\tau}}\right), \qquad (8$$

where λ and a are the equivalent thermal conductivity and thermal diffusivity of the wet porous material respectively.

Substitute equation (7) into equation (5) and integrate. Noticing

$\theta_{o,\tau} = t(o,\tau) - t_o = \frac{2q}{\sqrt{\pi}\lambda}\sqrt{a\tau}$ and $\int_o^\infty u\cdot \mathrm{ierfc}(u)du = \frac{1}{6\sqrt{\pi}}$, we have obtained

$$\Delta G = \frac{2}{3} \frac{AD_t \rho\tau\theta_{o,\tau}}{l[1+(D_m/a)^{0.5}]} . \tag{9}$$

ΔG determined by the moment method and $\theta_{o,\tau}$ measured with a thermocouple fixed on the heated surface, the thermal mass diffusivity may be thus evaluated as

$$D_t = \frac{3}{2} \frac{l\,\Delta G}{A\rho\tau\theta_{o,\tau}} [1 + (D_m/a)^{0.5}] \tag{10}$$

Here D_m and a must be determined otherwise. However, their errors will make only minor influence on the measured D_t, because $D_m \ll a$ for most practical materials.

Furthermore, if we measure the temperature at a certain cross-section apart from the heated end, where the constant heat flux is known, the equiavlent thermal conductivity and thermal diffusivity of the sample can be determined simultaneously by the plane heat source method with constant heat rate [6]. The obtained equivalent thermal diffusivity can thence be used for evaluating D_t.

DISCUSSION ON THREE-DIMENSIONAL PLOBLEM WITH VARIABLE PROPERTY

If the moisture migration in a porous medium is caused merely by moisture content gradient and concerned in no temperature variation, the situation will be simpler in nature and can be further discussed. The moisture diffusivity D_m in a homogeneous isothermal medium will be a function of moisture content only. A new variable can be defined as

$$E = \int_{w'}^{w} D_m(w)\, dw, \tag{11}$$

such that $$D_m = \frac{dE}{dw} . \tag{12}$$

Then the mass transfer equation should take the form

$$\frac{\partial w}{\partial \tau} = \nabla\cdot(D_m \nabla w) = \nabla^2 E. \tag{13}$$

Now we consider a homogeneous porous body with an arbitrary shape. The straight line connecting its two bearing points is chosen as the x-axis, with z-axis parallel to the direction of gravity and y-axis normal to both x- and z-axises. The porous body occupies a space with a boundary surface S and has moisture distribution of $w(x,y,z,\tau)$. The moment of the moisture gravity about y-axis is

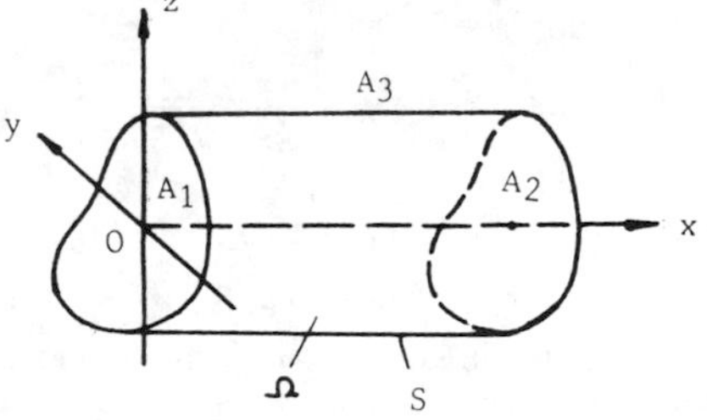

Fig.2 Moisture migration system

$$M = \iiint_\Omega \rho g x w\, dv \tag{14}$$

Taking the derivative of M with respect to τ and using equation (13), we get

$$\frac{dM}{d\tau} = \iiint_{\Omega} \rho g x \, \nabla\cdot(\nabla E)\, dv$$

$$= \rho g \oiint_S (x\nabla E - E\mathbf{i})\cdot\mathbf{n}\, dA \qquad (15)$$

where **i** and **n** denote the unit vectors in the direction of the x-axis and the outward normal direction of the boundary. It is clear that the derivative $dM/d\tau$ depends only upon the values E and ∇E on the boundary. If the body is moisture-insulated from the surroundings, i.e. $\mathbf{n}\cdot\nabla E|_S = 0$, Equation (15) should reduce to

$$\frac{dM}{d\tau} = -\rho g \oiint_S E\,\mathbf{i}\cdot\mathbf{n}\, dA \qquad (16)$$

Now we suppose the body is a column with uniform cross-sectional area A, and that its end surfaces are normal to the x-axis and its generatrix is parallel to the x-axis as sketched in Fig.2. Then,

$$\frac{dM}{d\tau} = \rho g\left[\int_{A_1} E\, dA - \int_{A_2} E\, dA\right]$$

$$= \rho g A\left[\bar{E}\Big|_{A_1} - \bar{E}\Big|_{A_2}\right] \qquad (17)$$

In this case $dM/d\tau$ will depend only upon the averages of E on the two ends and have no concern with the moisture distribution inside the sample. For the special case of one-dimensional transfer, in which the moisture contents are even on every cross-sections, we get

$$\frac{dM}{d\tau} = \rho g A\,(E_{x=o} - E_{x=L}). \qquad (18)$$

If the moisture content on the end x=L keeps unchanged as in the case where a sample with uniform initial moisture content w_o can be considered as a semi-infinite body, it is convenient to define $E = \int_{w_o}^{w} D_m\, dw$ so that $E_{x=L}=0$, and

$$\frac{dM}{d\tau} = \rho g A E_{x=o}\,. \qquad (19)$$

Using the device illustrated in Fig.1, we will be able to determine

$$E_{x=o} = \frac{1}{\rho A}\,\frac{dG}{d\tau} \qquad (20)$$

If a proper method is contrived to ascertain the moisture content on the end x=o changing with time, $D_m = \frac{dE}{dw}$ should be obtained as a function of w. As mentioned above, however, no perfect means are available so far, by which the moisture content can be determined continuously and conveniently. So it still needs some more efforts to realize this measurement of D_m as a variable property.

If the assumption is adopted that D_m is constant during the test, then $E = D_m(w - w_o)$, equation (17) becomes

$$\frac{dM}{d\tau} = \rho g A D_m\left(\bar{w}\Big|_{A_1} - \bar{w}\Big|_{A_2}\right). \qquad (21)$$

For the one-dimensional semi-infinite problem it takes the form

$$\frac{dG}{d\tau} = \frac{A\rho D_m}{l} [w(0,\tau)-w_o] \tag{22}$$

D_m being constant, the moisture content distribution, and then $w(0,\tau)$, might be obtained analytically according to given conditions, and D_m might be determined on the basis of equation (22).

INSTANTANEOUS MOISTURE SOURCE METHOD FOR DETERMINING D_m

The instantaneous moisture source method employs a device illustrated in Fig.1, and the test is carried out isothermally. Before the test, there existed an even moisture content w_o in the sample. A certain amount of water is added to the end x = o of the sample at the beginning of the test, and then it is made to be moisture-insulated again. Assume: the moisture migration is one-dimensional in x-direction; the sample itself may be regarded as a semi-infinite body; and D_m is constant during the test.

As a first approximation, suppose the water added can be considered as an instantaneous plane moisture source, then the moisture content distribution may be found as [6]

$$w - w_o = \frac{m}{A\rho} \frac{1}{\sqrt{\pi D_m \tau}} \exp(- \frac{x^2}{4D_m\tau}) . \tag{23}$$

where m is the mass of the added water.

Setting x=o in equation (23) and using equation (22) we get

$$\frac{dG}{d\tau} = \frac{m}{l}\sqrt{\frac{D_m}{\pi\tau}} \tag{24}$$

With reference to the initial instant, the total increment of the balance reading will be

$$\Delta G = \int_o^\tau \frac{dG}{d\tau} d\tau = \frac{2m}{l}\sqrt{\frac{D_m\tau}{\pi}} \tag{25}$$

If the reference is replaced by the state before the test, which is steady and the balance reading may be taken more accurately, then the total increment becomes

$$\Delta G' = \frac{2m}{l}\sqrt{\frac{D_m\tau}{\pi}} - \frac{mb}{l} \tag{26}$$

D_m can thus be evaluated easily as

$$D_m = \pi\tau(\frac{1}{m}\frac{dG}{d\tau})^2 \tag{27}$$

or

$$D_m = \frac{\pi}{4\tau}(\frac{l\Delta G'}{m} + b)^2 \tag{28}$$

The model discussed above treated the water added as an instantaneous plane source and used a δ-function to describe the initial moisture distribution, which will approach infinity as $x \to o$. This does not agree with the actual condition, because the moisture content of a porous medium should be restricted to a maximum, the saturated moisture content w_s. For this reason we replace the initial condition with a more real one, supposing that at the initial instant there is a thin layer of thickness h at the end x = o of the sample where the moisture is saturated, but outside this layer the moisture content has not yet been affected by the water and remains the original w_o. The thickness of this layer can be calculated by

$$h = \frac{m}{A\rho\,(w_s - w_o)} \tag{29}$$

Solving this problem gives the moisture distribution [6]

$$w - w_o = \frac{w_s - w_o}{2}\left[\operatorname{erf}\left(\frac{x + h}{2\sqrt{D_m\tau}}\right) - \operatorname{erf}\left(\frac{x - h}{2\sqrt{D_m\tau}}\right)\right]. \tag{30}$$

Again, setting x=o and using equation (22), we obtain

$$\frac{dG}{d\tau} = \frac{A\rho D_m}{1}(w_s - w_o)\operatorname{erf}\left(\frac{h}{2\sqrt{D_m\tau}}\right). \tag{31}$$

Let $\eta = \dfrac{h}{2\sqrt{D_m\tau}}$, $R(\eta) = \dfrac{\sqrt{\pi}\operatorname{erf}(\eta)}{2\eta}$, equation (31) can be rewritten as

$$\frac{dG}{d\tau} = \frac{m}{1}\sqrt{\frac{D_m}{\pi\tau}}\;R(\eta). \tag{32}$$

With reference to the state before the water is added, the total increment of the balance reading at any instant τ becomes

$$\Delta G' = \int_o^\tau \frac{dG}{d\tau}\,d\tau - \frac{m}{1}\left(b - \frac{h}{2}\right)$$

$$= \frac{2}{\sqrt{\pi}}\,\frac{m}{1}\sqrt{D_m\tau}\;P(\eta) - \frac{mb}{1} \tag{33}$$

where $P(\eta) = \frac{1}{2}\exp(-\eta^2) + \frac{\sqrt{\pi}}{2}\operatorname{erf}(\eta)\left(\eta + \frac{1}{2\eta}\right)$. Table 1 lists some values of the functions $R(\eta)$ and $P(\eta)$, which approach the limit of 1 as $\eta \to 0$.

Table 1 Functions $R(\eta)$ and $P(\eta)$

η	1	0.5	0.2	0.1
$R(\eta)$	0.7468	0.9226	0.9868	0.9967
$P(\eta)$	1.3042	1.0813	1.0133	1.0033

It can be seen in table 1 that, as η=0.1, the relative difference between the instantaneous plane source model and the modified saturated layer model is merely 0.33% and can be neglected. It is desirable to evaluate D_m through the former model, which is much simpler and clearer. For this purpose the amount of the water added should be appropriate and the test duration should be long enough so as to ensure $\eta \leq 0.1$.

To accord with the semi-infinite body assumption, the length of the sample must be long enough. A calculation on the basis of equation (23) indicates that $w(L,\tau) - w_o < 1.6 \times 10^{-4} m/(A\rho L)$ as $D_m\tau/L^2 \leq 0.025$, which implies the increases in moisture content at the end x = L of the sample will be less then 0.02% of the average increase over the sample and negligible. It is not difficult to ensure $D_m\tau/L^2 \leq 0.025$ since D_m is usually very small.

SUMMARY

The moment method is a new contrivance to determine the moisture transport properties in wet porous media. It involves some mathematical treatment instead of a transform of the physical nature of the measured moisture content, so it can avoid the arduous calibration needed for indirect measurements and is advantageous to improving the accuracy. This method has got rid of the procedure of cutting the sample as well and can record the test processes continuously. By virtue of its unsteady state design, the duration of a single test is estimated at half to ten hours, which is much shorter than traditional steady-state ones.

The constant heat source method proposed for determining D_t and the instantaneous moisture source method for D_m need to be verified through experiments. These two methods have been derived under the assumption of constant properties, which is a common shortcoming of most unsteady-state methods and will certainly introduce some error to the measured results. Nevertheless, if appropriate means are developed for determining continuously the moisture content in porous media, it will be possible to determine D_m as a variable property with w under isothermal condition. Of course, further research on this field remains to be done.

REFERENCES

[1] A.V.Luikov, Heat and Mass Transfer in Capillary-Porous Bodies, Pergamon Press, Oxford, 1966.

[2] G.G.Yan and Y.L.Yan, Principle of instrumental analysis and its application in agricalture (in Chinese), Science Press, Beijing, 1982.

[3] F.Q.Huang, Study on the measurement of the moisture distribution and moisture conductivity in porous bodies by means of the γ-ray, in Nuclear Science and Technology (in Chinese), Science Press, Beijing, 1964.

[4] B.X.Wang and Z.H.Fang, Heat and mass transfer in wet porous media and a method proposed for determination of the moisture transport properties, Heat and Technology, 2(1), Italy, 1984.

[5] B.X.Wang, L.Z.Han, W.C.Wang and Z.L.Jiao, A plane heat source method for simultaneous measurement of the thermal diffusivity and conductivity of insulating materials with constant heat rate, Chinese Journal of Engineering Thermophysics, (in Chinese, with English abstract), 1(1), 1980; English Translation, Engineering Thermophysics in China, 1(2), 255-268, Rumford Pub. Co. Inc., 1980.

[6] B.X.Wang, Engineering Heat and Mass Transfer, (in Chinese), 205-215, Science Press, Beijing, 1982.

On the Heat and Mass Transfer in Moist Porous Media

BU-XUAN WANG and WEI-PING YU
Thermal Engineering Department
Tsinghua University
Beijing, PRC

Abstract---A new synthetic theory is presented for the heat and mass transfer in moist porous media. The phenomenological equations are given. A simplifie energy equation has been derived to measure simultaneously the thermal conductivity and diffusivity of wet sand with transient hot wire method.

NOMENCLATURE

C_p	specific heat	w	moisture content
D_1	molecular diffusivity	α	mass thermal conductance
D_2	coefficient of capillary flux	β	correction factor of area for molecular diffusion
D_t	thermal mass diffusivity	ε	porosity
d	vapor mass fraction in humid air	λ^*	apparent thermal conductivity
h	enthalpy	λ_e	effective thermal conductivity of moist porous media
h_{fg}	latent heat of evaporation	ξ	nonlinear thermal conductivity
I	mass source	ρ	density
J	mass flux	σ	surface tension
m	mass	τ	time
P_c	capillary pressure	φ	relative mass content
t	temperature		
V	volume		

subscripts

a	air	2	liquid
0	solid matrix	w	humid air
1	vapor		

INTRODUCTION

For the research of heat and mass transfer in porous media, there exist wide backgrounds of engineering applications in energy resources, building materials, geological exploration, power, chemical and oil-refining, and so on. Such as industrial drying, soil water controlling, measurement of thermophysical propert of porous insulating and building materials, heat storage in ground aquifers, et are all the practical problems in this field.

The process of heat and moisture migration in porous media without sweeping flow of fluid was to be paid attention to a long time ago. As early as 1915, Bouyoucos[1] found first that moisture in soils moves from higher to lower temperature region. Since then, three classical theories had been formed, based on the transfer mechanism: the theory of moisture diffusion[2], of capillary adsor tion[3], and of vaporization and condensation[4]. However, it was found soon afterwards that the theory of moisture diffusion can adapted only for the moistu migration process under isothermal or near isothermal conditions, and fails to explain the mutual action of temperature and concentration gradients. In the capillary theory, the so-called "capillary potential" in porous media is regarded as the basis to analyze the unsaturated moisture migration. By the theory of vaporization and condensation, the liquid will be evaporated at the hot end while the vapor condensed at the cold end, and the moisture migrates from higher to lower temperature region as the result.

The actual heat and mass transfer in wet porous media is really a very complicated physical process with multi-substances-and-phases. So it is impossible to consider only one single basic transfer mechanism to cover a wide range of moisture content, as being pointed out in the previous papers[5,6]. There appeared also some phenomenological irreversible thermodynamic methods[7-9], which would avoid to assume the transfer mechanism of moisture migration, but it will be still restricted by the difficulty in experimental technicues to determine the coefficients and the complexity of mathematical presentation for the practical problems.

Since 1950's, several synthetic theories have been reported, such as the works of Philip (1957,1958)[10,11], Luikov(1961)[7], Harmathy(1969)[12], Huang(1978)[13], Eckert & Maghri(1980)[14], etc. Recently, Wang & Fang(1984)[6] had considered the contributions of molecular diffusion, vaporization and condensation, and also capillary effect on the heat and mass transfer in wet porous media. In this paper we try to develop further such a theory.

THE SYNTHETIC THEORY OF HEAT AND MASS TRANSFER IN MOIST POROUS MEDIA

1. Essential Thought

The basic assumptions adopted are as follows:

(1) The porous media can be regarded as a macroscopically uniform medium. There is no fluid sweeping across the media. The liquid moisture can migrate only by capillary force while the vapor migrate through diffusion.

(2) The capillary potential is so much greater that the effect of gravitational potential can be neglected [5,7].

(3) The temperature in the media is above the freezing point of moisture containe everywhere, so that there is no freezing or melting boundary inside the media.

(4) The total pressure can be considered constant during the transport process.

(5) The gradients of various physical fields are not great enough that the local thermodynamic quasi-equilibrium state can be reached anywhere.

By conservation of mass on the infinitesimal control volume basis, we get for any arbitrarily differential volume in the interior of the media as:

air:
$$\rho_0 \frac{\partial \varphi_a}{\partial \tau} = -\nabla \cdot J_a \tag{1}$$

vapor:
$$\rho_0 \frac{\partial \varphi_1}{\partial \tau} = -\nabla \cdot J_1 + I_1 \tag{2}$$

liquid:
$$\rho_0 \frac{\partial \varphi_2}{\partial \tau} = -\nabla \cdot J_2 + I_2 \tag{3}$$

where $\varphi = m/m_0$, the relative mass content per unit volume; ρ_0, the apparent density or the mass of solid matrix per unit volume; J, the mass flux; I_1 and I_2, the mass source of vapor and liquid respectively, and $I_1 + I_2 = 0$, or $I_1 = -I_2$ evidently.

The total moisture content w should be the sum of the vapor in humid air and that of liquid, i.e. $w = \varphi_1 + \varphi_2$. So, from Eq.s(2) and (3), the mass equation of moisture migration would be

$$\rho_0 \frac{\partial w}{\partial \tau} = -\nabla \cdot (J_1 + J_2) \tag{4}$$

By conservation of energy on control volume basis,

$$\rho_0 \frac{\partial}{\partial \tau}(h_0 + \varphi_a h_2 + \varphi_1 h_1 + \varphi_2 h_2)$$
$$= -\nabla\cdot(J_a h_a + J_1 h_1 + J_2 h_2) + \nabla\cdot(\lambda^* \nabla t) \qquad (5)$$

where h, the enthalpy; and dh=Cpdt for specific heat at constant pressure, Cp, being considered as constant property; λ^*, the apparent thermal conductivity. Then substituting Eq.s (1),(2) and (3) into Eq.(5), we have

$$\rho_0 (C_{p0} + \varphi_a C_{pa} + \varphi_1 C_{p1} + \varphi_2 C_{p2}) \frac{\partial t}{\partial \tau} + h_{fg} I_1$$
$$= -(J_a\cdot\nabla h_a + J_1\cdot\nabla h_1 + J_2\cdot\nabla h_2) + \nabla\cdot(\lambda^* \nabla t) \qquad (6)$$

where $h_{fg}=(h_1-h_2)$ is the latent heat of evaporation. The transfer of apparent heat $\Sigma J\cdot\nabla h$ in energy equation had been generally neglected in published literatures[6].

We shall use the following characteristic functions of transfer mechanisms for phase change, capillary force and molecular diffusion to study quantitatively the macroscopic behaviours of the porous media. So, it is really a thought of synthetic method.

2. Processing of Transfer Mechanisms

i. Vapor or liquid source The occurrence of vapor source I_1 or liquid source I_2 is caused by the phase change between liquid and vapor, whereas $h_{fg}I_1$ represents the quantity of latent heat removed away from the control volume. Mathematically, we have

$$I_1 = \frac{dm_1}{dt}\frac{\partial t}{\partial \tau} = \rho_0 \frac{d\varphi_1}{dt}\frac{\partial t}{\partial \tau}$$

Noticing that,

$$\varphi_a = \rho_1 V_1/\rho_0 V_0 = [\rho_1 V_1/(\rho_1 V_1 + \rho_a V_a)]\cdot[(\rho_1 V_1 + \rho_a V_a)/\rho_0 V_0] = d\, \rho_w V_w/\rho_0 V_0$$

where d is the vapor mass fraction in humid air, which is function of local temperature, t, only, and the subscript w represents the humid air. Then

$$I_1 = \rho_w V_w/V_0 \cdot \frac{dd}{dt}\frac{\partial t}{\partial \tau} \qquad (7)$$

when the change of $\rho_w V_w/\rho_0 V_0$ with temperature is neglected. Here, dd/dt depends on thermodynamic properties of saturated vapor, V_w/V_0 is the ratio of volume of humid air to total volume, which is dependent upon moisture content, w, and internal geometric characteristics of porous media, while ρ_w is function of temperature at constant pressure. Thereby,

$$\frac{\rho_w V_w}{\rho_0 V_0} = f(t, w) \qquad (8)$$

Hence,

$$I_1 = \rho_0 f(t,w) \frac{dd}{dt}\frac{\partial t}{\partial \tau} = -I_2 \qquad (9)$$

V_w is the volume of pores minus that occupied by liquid. When the porosity of the medium is known as ε, we have

$$\frac{V_w}{V_0} = \frac{(V_0\varepsilon - V_2)}{V_0} = \varepsilon - \frac{V_2}{V_0}$$

or

$$\frac{V_w}{V_0} = \varepsilon - \frac{\rho_0}{\rho_2}\frac{\rho_2 V_2}{\rho_0 V_0} = \varepsilon - \frac{\rho_0}{\rho_2}\varphi_2 \qquad (10)$$

Compared with Eq.(8), it is clear that

$$f(t,w) = \rho_w(\rho_2\varepsilon - \rho_0\varphi_2)/\rho_0\rho_2 \qquad (11)$$

ii. Molecular diffusion between vapor and air On the basis of molecular diffusion theory, the mass flow rate by diffusion is proportional to the gradient of concentration, that is

$$\left.\begin{aligned} J_1 &= -\beta D_1 \rho_w \nabla d \\ J_a &= -\beta D_1 \rho_w \nabla (1-d) = \beta D_1 \rho_w \nabla d = -J_1 \end{aligned}\right\} \tag{12}$$

Here, β, correction factor of the area where the molecular diffusion occurs; D_1, molecular diffusivity, which is function of temperature, i.e. $D_1 = D_1(t)$.

Since d is function of local temperature only, we have $\nabla d = dd/dt . \nabla t$. Therefore,

$$\left.\begin{aligned} J_1 &= -\beta D_1(t) \rho_w \frac{dd}{dt} \cdot \nabla t \\ J_a &= \beta D_1(t) \rho_w \frac{dd}{dt} \cdot \nabla t = -J_1 \end{aligned}\right\} \tag{13}$$

iii. Capillary flow J_2 is the mass flow rate of liquid. From basic assumption(1), liquid migrates by the action of capillary force only, and the liquid flow rate will be proportional to the capillary pressure gradient, in the direction of capillary pressure increasing. That is,

$$J_2 = D_2 \nabla p_c = D_2 \nabla (\frac{\sigma}{r_m}) \tag{14}$$

The proportional coefficient, D_2, is relevant to the shape of flow channel and viscosity of liquid, it is function of moisture content only, $D_2 = D_2(w)$, if the variation of viscosity with temperature is ignored. As the surface tension, σ, is function of temperature, $\sigma = \sigma(t)$, the average radius of curvature for the liquid surface in capillary channels depends only upon the moisture content for a given medium, $r_m = r_m(w)$. It may be a function of temperature and moisture content. So,

$$J_2 = D_2(w) (\frac{\partial p_c}{\partial \tau} \nabla t + \frac{\partial p_c}{\partial w} \nabla w) \tag{15}$$

3. Differential Equations for Heat and Moisture Migration in Porous Media

Substituting Eq.s(13) and (15) into Eq.(4), the differential equation of moisture content becomes

$$\frac{\partial w}{\partial \tau} = \nabla \cdot (D_m \nabla w) + \nabla \cdot (D_t \nabla t) \tag{16}$$

where D_m and D_t are so-called "mass diffusivity" and "thermal mass diffusitity" respectively, defined as:

$$D_m = -\frac{1}{\rho_o} D_2(w) \frac{\partial p_c}{\partial w} ; \tag{17}$$

$$D_t = \frac{1}{\rho_o} [\beta D_1(t) \rho_w \frac{dd}{dt} - D_2(w) \frac{\partial p_c}{\partial t}] . \tag{18}$$

Substituting Eq.s(9), (13) and (15) into Eq.s(12) and (13), we get the mass equations for vapor and liquid respectively as:

$$\frac{\partial \varphi_1}{\partial \tau} = \nabla \cdot (D_{t_1} \nabla t) + W \frac{\partial t}{\partial \tau} \tag{19}$$

$$\frac{\partial \varphi_2}{\partial \tau} = \nabla \cdot (D_m \nabla w) + \nabla \cdot (D_{t_2} \nabla t) - W \frac{\partial t}{\partial \tau} \tag{20}$$

where D_{t1}, D_{t2} and W are defined as:

$$D_{t_1} = \frac{1}{\rho_0} \beta D_1(t) \rho_w \frac{dd}{dt} \; ; \tag{21}$$

$$D_{t_2} = - \frac{1}{\rho_0} D_2(w) \frac{\partial P_c}{\partial t} \; ; \tag{22}$$

$$W = f(t, w) \frac{dd}{dt} \; ; \tag{23}$$

and $D_{t1} + D_{t2} = D_t$. (24)

Substituting Eq.s(9),(13) and (15) into Eq. (6), the energy equation will be

$$\rho_0 C_p^* \frac{\partial t}{\partial \tau} = \nabla \cdot (\lambda^* \nabla t) + \xi (\nabla t)^2 + \alpha \nabla w \cdot \nabla t \tag{25}$$

where the phenomenological coefficients C_p^*, ξ and α are called "nominal specific heat capacity", "nonlinear thermal conductivity" and "mass thermal conductance" respectively, defined as:

$$C_p^* = [C_{p_0} + \varphi_a C_{pa} + \varphi_1 C_{p_1} + \varphi_2 C_{p_2} + h_{fg} f(t, w) \frac{dd}{dt}] / (1 + w) \; ; \tag{26}$$

$$\xi = (C_{p_1} - C_{pa}) \beta D_1(t) \rho_w \frac{dd}{dt} - C_{p_2} D_2(w) \frac{\partial P_c}{\partial t} \; ; \tag{27}$$

$$\alpha = - C_{p_2} D_2(w) \frac{\partial P_c}{\partial w} \; . \tag{28}$$

Comparing various phenomenological coefficients above, it is found that

$$\alpha = \rho_0 C_{p_2} D_m \tag{29}$$

$$\xi = \rho_0 [(C_{p_1} - C_{pa}) D_{t_1} + C_{p_2} D_{t_2}] \tag{30}$$

SIMPLIFIED ENERGY EQUATION AND AN EXPERIMENTAL STUDY ON SIMULTANEOUS MEASUREMENT OF THERMAL CONDUCTIVITY AND DIFFUSIVITY OF WET SAND

If the variation of φ_a, φ_1 and φ_2 with τ are ignored at the left hand of Eq.(5), the resulted energy equation(25) can be rewritten as

$$\rho_0 C_p^* \frac{\partial t}{\partial \tau} = \nabla \cdot (\lambda^* \nabla t) + \nabla \cdot \{ [(h_1 - h_a) \beta D_1 \rho_w \frac{dd}{dt} - h_2 D_2 \frac{\partial P_c}{\partial t}] \nabla t \} + \nabla \cdot (h_2 D_2 \frac{\partial P_c}{\partial w} \cdot \nabla w) \tag{31}$$

Moisture gradient and temperature gradient at any point in the porous media should be relevant to each other. The ratio of moisture gradient to temperature gradient may be expected being a week function of coordinates, and can thus be expressed approximately as

$$\nabla w = g(t, w) \nabla t \tag{32}$$

So, a simplified energy equation can be obtained from Eq.(31),

$$\rho_0 C_p^* \frac{\partial t}{\partial \tau} = \nabla \cdot (\lambda_e \nabla t) \tag{33}$$

where λ_e may be referred to "effective thermal conductivity", and defined as:

$$\lambda_e = \lambda^* + (h_1 - h_a) \beta D_1 \rho_w \frac{dd}{dt} - h_2 D_2 (\frac{\partial P_c}{\partial t} + g \frac{\partial P_c}{\partial w}) \tag{34}$$

where $g=g(t,w)$. It is obvious from Eq.(34) that, the "effective thermal conductivity", λ_e reflects the overall effect of heat conduction, including vapor diffusion and capillary flow of liquid, on the heat transfer.

With the assumed constant properties and basic assumption(5), i.e. the gradients of various physical fields being not great enough, the simplified energy equation(33) can be further reduced to

$$\frac{\partial t}{\partial \tau} = a_e \nabla^2 t, \tag{35}$$

where

$$a_e = \frac{\lambda_e}{\rho_0 C_p^*} \tag{36}$$

is the effective thermal diffusivity of the moist porous media. At the meantime Eq.(16), for moisture migration will be simplified as

$$\frac{\partial w}{\partial \tau} = D_m \nabla^2 w + D_t \nabla^2 t \tag{37}$$

Eq.s(35) and (37) are the same as derived in a previous paper[6], yet reveal the physical nature more logically and rigorously.

Based on the simplified energy equation(35), a transient hot wire method with constant heat rate has been developed to measure simultaneously the effective thermal conductivity and diffusivity of wet sand[15], and the experimental data are plotted here in Fig.s 1 and 2, which are alike qualitatively with that reported previously[5] for moist aerocrete by a plane heat source method[16].

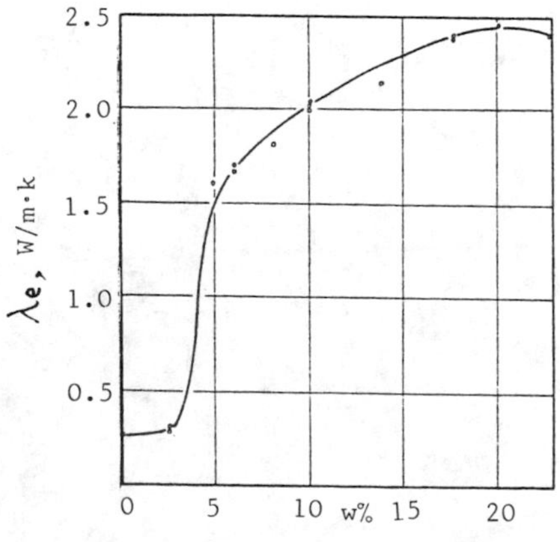

Fig.1 λ_e—w plot of experimental data

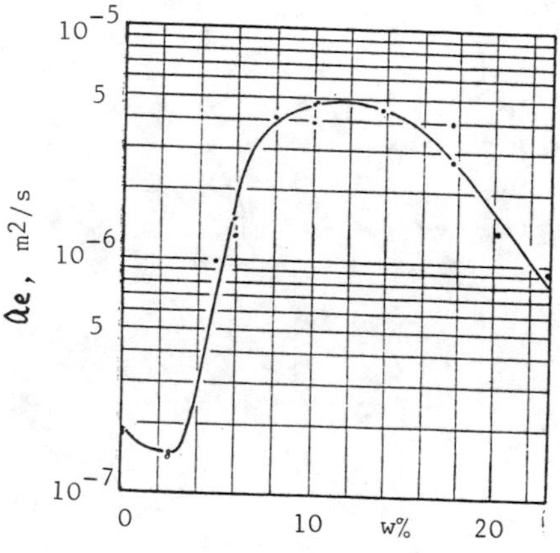

Fig.2 a_e—w plot of experimental data

CONCLUDING REMARKS

A synthetic theory of heat and mass transfer in moist porous media has been establised. The moisture migration equation(16) and energy equation(25) are derived concisely and have a rigid physical background.

A simplified energy equation(35) is obtained conditionally, which is just the same reported previously[6]. On the basis of this equation, a transient hot wire method for measuring simultaneously the thermal conductivity and diffusivity had been developed. The experimental data for sand are alike qualitatively with that reported[5] for moist aerocrete. However, further work needs to check the suitability of the simplified energy equation. In this aspect, it is very important to search for methods to determine moisture transport properties.

REFERENCES

[1] G.J.Bouyoucos, J.Agr.Res. 5:141-172,1915
[2] Y.K.Sherwood, Trans.Am.Inst.Chem.Engrs. 27:190-202,1931
[3] N.H.Ceaglske & O.A.Hougen, Ind.Engng.Chem. 29:805-813,1937
[4] C.G.Gurr et al., Soil Sci. 74:335-345,1952
[5] B.X.Wang & R.Wang, Chinese J.Engr.Thermophysics 4:146-152,1983
[6] B.X.Wang & Z.H.Fang, Heat & Technology 2:29-43,1984
[7] A.V.Luikov, Theoretical basis for achitectural thermophysics, (in Russian), Minsk, USSR, 1961
[8] J.W.Cary, IJHMT, 7:531-538,1964
[9] J.C.Benet & P.Jouanna, IJHMT, 25:1747-1754,1982
[10] J.R.Philip, Soil Sci. 83:345,1957
[11] J.R.Philip, Soil Sci. 85:278,1958
[12] T.Z.Harmathy, I/EC Fundamentals 8:92-103,1969
[13] C.L.Huang, Final report for NSF,1978
[14] E.R.G.Eckert & M.Maghri, IJHMT, 23:1613,1980
[15] W.P.Yu, M.S.Thesis, Tsinghua Univ.Beijing, China,1984
[16] B.X.Wang,L.Z.Han,W.C.Wang & Z.L.Jiao, Engineering Thermophysics in China, 1:255-268, Rumford Pub.Co.,Inc.,1980

NUCLEAR REACTION HEAT TRANSFER

Dispersed Flow Film Boiling at Low Pressure and Low Mass Flow Conditions

HAO-RAN BI, CHI-CHUN HSU, ZHI-CHAO GUO, and ZHI-MING QIAN
Power Machinery Engineering Department
Shanghai Jiao Tong University
Shangshi, PRC

ABSTRACT

A study is made of dispersed flow film boiling of water steam mixture flowir through a 12 mm i.d. uniformly heated tube. Vapour superheats of several hundre degrees (^oC) in the presence of water drops were measured. Experimental data were obtained at the following conditions: Pressure 2 to 7 bar; Mass Flow Rate 40 to 140 Kg/m^2 sec; Inlet Quality 0.025 to 0.834; Heat Flux 32 to 76 Kw/m. A model including heat transfer from wall to vapour and drops and from vapour to drops was presented. A comparison was made between the experimental data and th calculated results with the proposed model.

NOMENCLATURE

a	thermal diffusivity
B	proportionality constant
C_p	specific heat
CHF	critical heat flux
D	test section tube diameter
G	mass flow rate
h	heat transfer coefficient
i	enthalpy
k	thermal conductivity
K	parameter for vapour generation rate
N	droplet number density
P	pressure
Pr	Prandtl number
q	heat flux
Re	Reynolds number
T	temperature
V	velocity
y	distance from inner wall surface
α	vapour void fraction
Γ	vapour generation rate per unit volume
δ	droplet diameter
ρ	density
σ	surface tension
τ	droplet wall contact time

Subcripts

c	critical or convection
d	droplet
dc	direct contact
fg	phase change
H	hot patch
h	homogeneous
o	value at CHF
l	liquid
s	saturation
su	superheat
sv	saturated vapour
t	total
v	vapour
w	wall

INTRODUCTION

Film boiling heat transfer occurs when wall temperature exceeds the minimum film boiling temperature. For flow film boiling conditions it embraces the two distinct regimes: dispersed flow film boiling regime (or liquid deficient regime) and inverted annular flow film boiling regime. Due to poor heat transfer properties of the vapour, especially for vapour of water at low pressure, highly heated surface temperatures are often encountered in film boiling regime. Because of the interest of dispersed flow boiling in various technologies, some experimental studies on water dispersed flow boiling at elevated pressures and mass flow rates had been carried out during 1960's. The typical of those are the experiments by A. W. Bennett [1]. Since Groeneveld [2] developed the hot patch technique to obtain flow film boiling in 1976, it has become much easier to obtain experimental data on dispersed flow boiling heat transfer even for fluid like water.

Parker and Grosh [3] perhaps are the first scientists who discovered the thermodynamic non-equilibrium phenomena between the saturated liquid and the superheated vapour phase, Bennett [1] and Laverty [4] proved the superheated vapour and the saturated liquid coexist in the dispersed flow boiling regime by their experiments respectively and both independently proposed the two-step heat transfer theory almost at the same time of the year 1976. As a model for the high quality, dispersed flow pattern of film boiling, they assumed that the heat transfer process may take place in two steps: (a) from the wall to the superheated vapour (b) from the superheated vapour to the saturated liquid droplets. Because the dispersed saturated droplets tend to quench any temperature sensor and prevent detection of superheated vapour temperature, the measurement of the vapour superheats in dispersed flow boiling is a difficult task. Neither Bennett nor Laverty has measured vapour superheats. Since then a number of scientists have tried to make a thermocouple probe to detect vapour superheats. The vapour probe which was developed by Nijhawan[5]may be the best one at this time. With this kind of vapour probe, Nijhawan [5], Gottula [6], Evans [7] carried out a series of experiments on dispersed flow boiling heat transfer and obtained a range of data including the vapour superheats.

Many models for dispersed flow boiling heat transfer have been proposed. The models can be classified as empirical, semi-theoretical and numerical. In general the empirical models fit the data to the proposed correlations, they are simple to use but have a limited range of validity and should not be extrapolated outside the recommended range. Both semitheoretical and numerical models necessitate all parameters initially evaluated at the CHF location. The heated channel is subdivided axially and the axial gradients in droplet diameter, equilibrium and actual (non-equilibrium) qualities, vapour velocity and superheats and wall temperature are calculated at each node. The conditions at the downstream nodes are found by stepwise integration along the heated channel. For each node the semi-theoretical models decompose the whole heat transfer process in several steps or ways, such as heat transfer from the wall to the vapour and the droplets respectively and from vapour to droplets etc. Each step or way uses existing or modefied correlations. The numerical models solve mass, momentum and energy conservation equation for each node.

EXPERIMENTAL APPARATUS

Test apparatus

A 1.6 m long, vertical test section (inconel tube, 15 mm o.d., 12 mm i.d.) was used in the experiments. The deminerized water was led from pump via rotameter, horizontal preheater and vertical preheater into test section. A hot patch was mounted just upstream of the test section inlet. A vapour probe was provided at 1.4 m downstream of the inlet to measure the vapour superheats. Water steam mixture leaving from the test section was condensed in the cooler then returned to the pump. Test section was instrumented with chromel-alumel thermocouples welded to the outside wall at different intervals described in table 1. Besides, another thermocouple was welded to the hot patch. Some s.s. sheathed thermocouples were used to measure fluid temperatures. Thermocouples' outputs could be read out on either a 20-channel recorder or a precision digital voltmeter. Pressure taps were provided at the inlet and the outlet of the test section. Thus in each run of experiments the following parameters could be measured: pressure, flow rate and temperatures of the fluid, the temperatures of the hot patch and the superheated vapour, the temperature distribution along the test section, the amperes and the volts of power supply to the test section, the preheater and the hot patch. Using the experimental data of heat loss through thermal isolator to the atmosphere, the quantities of heat transferred from the test section, preheaters and hot patch to the fluid could be calculated.

Hot patch

The hot patch is a copper cylinder (15 cm dia., 10 cm length) equipped with twelve 500 W cartridge heaters. The power supply of the heaters can be adjusted by a variac. A hole (12 mm dia.) was drilled in the center of the cylinder. The upper end of this hot patch was connected to the test tube in such a way that the fluid can directly contact the hot patch. The hot patch is an effective one and was used in all runs of the experiments.

Vapour probe for measuring superheats

An exposed hot junction of a 0.5 mm s.s. sheathed thermocouple was placed within the inner of two concentric capillary tubes (5 mm i.d. and 3 mm i.d. respectively). Access holes to the outer and inner tubes were drilled at 90 deg. displacement. The probe assembly was then inserted into the two-phase flow in such a manner that the sampled fluid had to traverse through a 180 deg. and a second 90 deg. change of direction before passing over the thermocouple junction. These directional changes provided the inertial shielding for separation of liquid drops from the vapour. The small fraction of liquid drops which are aspirated into the probe would collect in the outer annulus, to be drawn off by aspiration through the annular space. Essentially liquid-free vapour would then be aspirated through the inner tube past the sensing thermocouple junction. Judicious control of the differential aspiration between the inner tube and the outer annulus permitted decrease of the liquid quenching frequency. Bench test results of the vapour probe show that it can be used effectively in high or moderate vapour quality but failed in vapour quality less than 10%. This kind of vapour probe was developed and first used by Nijhawan [5]

EXPERIMENTAL RESULTS

Ranges of parameters

Groeneveld pointed out in 1977 that experimental data of dispersed flow boiling are non-existent at low flows, low pressures and low qualities. Since then Nijhawan, Chen et al. have published a range of data at these conditions, but they are still thought to be insufficient. Besides, data at low flows and low pressures is particularly useful in predicting fuel sheath temperatures during a LOCA of nuclear reactor. Thus the ranges of parameters in this experiments are as follows

pressure	2 to 7 bar,	mass flow rate	40 to 140 Kg/m^2sec,
quality at CHF	0.025 to 0.834,	heat flux	32 to 76 Kw/m.

Experimental data

Forty five runs were carried out in the experiments. Due to very low quality (or subcooled) inlet conditions, the flow regimes of six runs were thought to be inverted annular flow boiling. The vapour temperatures of nine runs failed to be measured still due to low vapour qualities. The typical experimental data of six runs among the rest are shown in table 1 (L in table 1 represents the distance between the test section inlet and the measuring point). Typical axial variations of local wall temperature are shown in fig. 1 and fig.2. The wall temperatures indicated by open circles, triangles and squares in the figures were obtained from the thermocouple measurements on the outer surface of the test section with a correction for the temperature drop through the tube wall. The temperatures at both ends of the test section were influenced by end conduction effects --- the heating effect by the hot patch at the inlet and the cooling effect by the flanges at outlet and by downstream pipes.

CALCULATING MODEL

A semi-theoretical model is proposed as follows

Letting q_t be the total amount of heat transferred from the wall to the two phase flow; $q_{dc,w-d}$ the heat transferred from the wall to the liquid droplets by direct contact; $q_{c,w-v}$ the heat transferred from the wall to the vapour by convection, the following equation could be obtained by heat balance

$$q_t = q_{dc,w-d} + q_{c,w-v} \tag{1}$$

If $q_{c,v-d}$ represents the heat transferred from the vapour to the droplets by convection, it would be possible to calculate the heat for vapour super-heat (q_{su}) by the following relation

$$q_{su} = q_{c,w-v} - q_{c,v-d} \tag{2}$$

In this model, just like most papers on dispersed flow boiling, radiation heat transfer is neglected.

Heat transfer from wall to superheated vapour by convection

From the data of this experiments such as the heat flux of the test section and temperature difference between the wall at vapour probe location and the superheated vapour, one would find that the following Dittus-Boelter equation holds for the most of this experimental data

$$Nu = 0.023 Re_v^{0.8} Pr_v^{0.4} \tag{3}$$

In view of high void fraction (more than 90% in the most of the experiments), the agreement between the experimental data and the Dittus-Boelter equation seems reasonable. Here the actual vapour velocity should be used for calculating vapour Reynolds Number Re_v.

Heat transfer from wall to droplets by direct contact

Most models for dispersed flow boiling assume that direct wall contact by liquid droplets is negligile. The justification is that the highly superheated wall is above the Leidenfrost temperature. However, recent experiments of Lee et al. [8] showed that transient contacts of liquid on hot surface do occur. Since heat transfer to the contacting liquid would be at very high transient heat flux, a short contact time could make a significant contribution to the overall average heat transfer. The earlier experimental investigations of drop impingement heat transfer by Wachters [9] and Burge [10] had the same results. Nowadays people have recognized that at downstream regions reasonably far from the CHF point (in this experiments, CHF point exists at hot patch location), liquid-soli contact could exist because the liquid film sputtered off the wall at the CHF location would result in droplets with significant transverse initial velocities which are likely to impact on the test section wall. THis 'dynamic Leidenfrost' process would likely promote liquid-solid contact to much higher wall superheats than the classical Leidenfrost static drop phenomenon.

It is difficult to analyse quantitatively heat transfer between the wall and the droplets because it necessitates detailed information about size, density anc velocity of droplets which is difficult to measure and is basically unknown at this time. In order to circumvent these difficulties, a quasi-theoretical model for calculating heat transfer between the wall and the droplets has been proposec In this model the heat transfer between the wall and a single droplet by direct contact is calculated analytically, then it is multified by factor F to obtain the heat transfer between the wall and all droplets $q_{dc,w-d}$. Here F represents the fraction of wall surface area under liquid contact at any instant. F is determined by experimental data.

Some simplifications used in this quasi-theoretical model are as follows

1. The thickness of test section is infinitive.
2. The temperature of wall surface is maintained at the liquid saturation temperature during the period of contact.
3. The axial heat conduction is neglected, thus one dimension heat conductio exists within the wall.

The investigation of Wachters et al.[9] showed that the wall-droplet contact time is very short (about 10^{-2} sec) while the thickness is greater than 1 mm, there- fore it seems reasonable to consider the thickness to be infinitive in view of transient temperature process. Thus the heat conduction equation with its initia and boundary conditions is as follows

$$\frac{\partial T}{\partial t} = a\frac{\partial^2 T}{\partial^2 y} \qquad 0<y<\infty, \tag{4}$$

$$T=T_s \quad \text{at} \quad y=0 \quad \text{for } t>0,$$
$$T=T_w \quad \text{when} \quad t=0 \quad \text{for } 0\leq y\leq\infty$$

The solution of the equation satisfying its initial and boundary conditions [11] is

$$q_d = \frac{\sqrt{(k\rho Cp)_w}\,(Tw-Ts)}{\sqrt{(\pi\tau)}} \tag{5}$$

where τ represents the contact time between the wall and the single droplet.

The experiments of Wachters [9] showed that the contact time τ appeared to be about equal to the first order vibration period of a freely oscillating drop (see Rayleigh [14])

$$\tau = \frac{\pi}{4}\sqrt{\frac{\rho_\ell\delta^3}{\sigma}}$$

Assume that the droplet density N is proportional to the initial droplet den- sity No within the regions not far from the CHF location, i.e.

$$N = \frac{6(1-\alpha)}{\pi\delta^3} = BN_o = B\frac{6(1-\alpha_o)}{\pi\delta_o^3} \tag{6}$$

where B is a proportionality constant.

The heat transferred from the wall to the droplets can be expressed as

$$q_{dc,w-d} = \frac{2}{\pi} BF\sqrt{(k\rho Cp)_w}\,(Tw-Ts)\left[\frac{\sigma(1-\alpha_o)}{\rho_\ell(1-\alpha)}\right]^{\frac{1}{4}}\delta_o^{-3/4} \tag{7}$$

Expression for BF could be obtained from experimental results. The best fit of this experiment is

$$BF = 1.6x10^{-6}\frac{\pi}{2}\left(\frac{Tw-T_L}{Ts+273}\right)^{-3.2} \quad (8)$$

where T_L represents classic Leidenfrost temperature which can be calculated with the equations developed by Spieger [12]. Finally, the heat transferred from the wall to the droplets by direct contact becomes

$$q_{dc,w-d}=1.6x10^{-6}\sqrt{(k\rho Cp)_w}(Tw-Ts)\left[\frac{\sigma(1-\alpha o)}{\rho_l(1-\alpha)}\right]^{\frac{1}{4}}\left(\frac{Tw-T_L}{Ts+273}\right)^{-3.2}\delta_o^{-3/4} \quad (9)$$

Heat transfer from vapour to droplets

For dispersed flow boiling conditions the vapour generation rate model is convenient to calculate the heat transferred from the superheated vapour to the droplets. Webb [13] proposed the vapour generation model in which the vapour generation rate Γ_1 can be formulated to be a direct function of the local thermo-hydraulic conditions and thermodynamic properties as follows

$$\Gamma_1=K\left(\frac{Gx}{h}\right)^2\frac{(1-\alpha_h)^{2/3}(Tv-Ts)k}{\rho_v\,\sigma\,i_{fg}D} \quad (10)$$

where

$$K\propto\frac{h\sigma(1-\alpha_0)^{1/3}}{K_v\rho_vV_v^2(1-\alpha_0^{\frac{1}{2}})} \quad (11)$$

The variables σ, k_v and ρ_v in equation (11) are all thermodynamic properties of the vapour and the variation of all these properties is to decrease the value of K with increase in pressure. Therefore Webb et al. thought that K is at least a function of the reduced pressure (P/Pc) to account for this property variation, and a best fit to the data of Bennett [1] and Nijhawan [5] is

$$K = 1.32\left(\frac{P}{Pc}\right)^{-1.1} \quad (12)$$

A comparison of the Webb's predicting equations with this experimental data shows that at the region near the inlet of test section, the calculated vapour generation rate is underpredicted, while at the region near outlet, it is over-predicted. The anelysis of equation (11) shows that the parameter K depends on a number of variables, Apart from these thermo-dynamic properties, some variables such as heat transfer coefficient h and void fraction α_0, depend on the detailed information of droplets and can not be adequately quantified at present. The remaining variable --- vapour velocity Vv strongly influences the value of K by power of 2 and it seems reasonable that the evaluation of K should include the vapour velocity. A best fit to the data of this experiments is

$$K = \frac{470}{V_v^2}\left(\frac{P}{Pc}\right)^{-1.1} \quad sec^2/m^2 \quad (13)$$

In addition to the variable Vv which is neglected in Webb's prediction, the vapour generation rate Γ comes from two parts. The first part Γ_1 is due to the heat transferred from the superheated vapour to the saturated liquid. The second part Γ_2 is the heat transferred from the wall to the droplets by direct contact. The second part is also omitted in Webb's model. Therefore the following equations are obtained.

$$\Gamma = \Gamma_1 + \Gamma_2 \quad (14)$$

$$\Gamma_2 = \frac{4q_{dc,w-d}}{D \ i_{fg}} \quad (15)$$

$$q_{c,v-d} = [\Gamma_1 i_{fg} + \Gamma(i_v - i_{sv})] \ D/4 \quad (16)$$

COMPARISON OF CALCULATING MODEL AND EXPERIMENTAL DATA

The transition from nuclear boiling to film boiling occurs at the location of hot patch which is made of pure copper providing very good thermal conductive property. Because the heat transfer coefficients of film boiling are approximat of the order of a hundredth of those of the nuclear boiling, it seems reasonalbe to consider the heat transfer from the hot patch to the two phase flow to take place at the pre-dryout condition of the fluid. Since all parameters are initia evaluated at the dryout location, the whole amount of heat inputted to the hot patch is used for increasing the initial vapour quality. With this calculating model and the parameters measured in the experiments, the wall temperatures and vapour superheats at each axial node could be calculated and compared with the experimental data. The standard diviation of predicted wall superheats from experimental data of 30 runs is 8.4%. Fig.3 shows the typical comparison. One could find that the agreement is reasonable except the regions near the inlet an the outlet of the test section where end conduction effects cause wall temperatu to differ greatly from the experimental data.

REFERENCES

1. A. W. Bennett, G. F. Hewitt, H. A. Kearsey and R. K. F. Keeys, Heat Transfe to Steam Water Mixture Flowing in Uniformly Heated Tubes in which the Criti Heat Flux Has Been Exceeded, AERE-R-5373, 1967.
2. D. C. Groeneveld, A Method of Obtaining Flow Film BOiling Data for Subcoole Water, Int. J. Heat Mass Transfer, vol.21, 1978, pp. 664-666.
3. J. D. Parker and R. J. Grosh, Heat Transfer to a Mist Flow, AEC Research an Development Report, ANL-6291, 1961.
4. W. F. Laverty and W. M. Rohsenow, Film Boiling of Saturated Liquid Nitrogen Flowing in a Vertical Tube, vol. 89, 1967, pp. 90-98.
5. A. Nijhawan, J. C. Chen, R. K. Sundaram and E. J. London, Measurement of Vapour Superheat in Post-Critical Heat Flux Boiling, J. of Heat Transfer, vol.102, 1980, pp.465-470.
6. R. C. Gottula, R. A. Nelson, J. C. Chen, S. Neti and R. K. Sundatam, Forced Convective Nonequilibrium Post-CHF Heat Transfer Experiments in a Vertical Tube, ASME-JSME Thermal Engineering Conference, Honolulu, March, 1983.
7. D. Evans, S. W. Webb and J. C. Chen, ExperimentaL Measurement of Axially Varying Vapour Superheat in Convection Film Boiling, Interfacial Transport Phenomena, ASME, 85-92. 1983.
8. L. Lee, J. C. Chen and R. A. Nelson, Surface Probe for Measurement of Liquid Contact in Film transition Boiling on High-Temperature Surfaces, Rev. Sci. Instrum., vol.53, 1982, pp.1472-1476.
9. L. H. J. Wachters and N. A. J. Westerling, The Heat Transfer from a Hot Wall to Impinging Water Drops in the Spheroidal State, Chem. Eng. Sci., vol.21, 1966, pp.1074-1065.
10. H. L. Burge, Chem. Eng. Symp. Series, vol.6, No.59,1966, pp.115-126.
11. E. R. C. Eckert and R. M. Drake, Heat and Mass Transfer, MCGRaq-Hill, 1959, p.91.
12. P. Speigler, J. Hopenfeld, M. Silberberg, C. F. Bumpus and A. Norman, Onset of Stable Film Boiling and the Foam Limit, Int. J. Heat Mass Transfer, Vol.6, 1963, pp.987-989.
13. S. W. Webb, J. C. Chen and R. K. Sundaram, Vapour Generation Rate in Nonequilibrium Convective Film Boiling, Proc. 7th Int. Heat Transfer Conf., Munich, 1982, vol.4, pp.437-442.
14. Lord Rayleigh, Proceeding of Royal Society, vol.29, 1879, p.71.

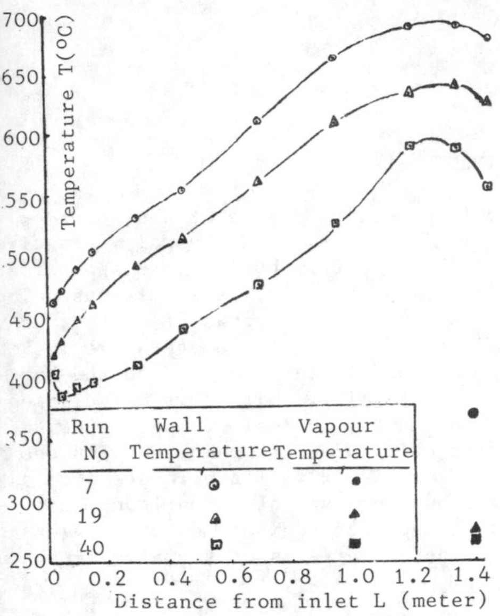

Fig.1. Temperature Viariation along Tube

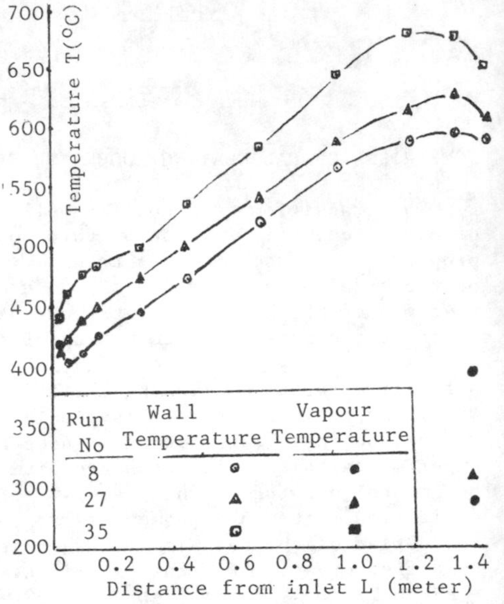

Fig.2. Temperature Variation along Tube

Table 1 Typical experimental data

Run No.		7	8	19	27	35	40
P (bar)		3.41	3.41	4.91	6.91	4.51	2.01
G($Kg/m^2 sec$)		107	105	104	90.4	52.7	41
q (Kw/m^2)		67.1	59	72	63.8	49.8	40.5
x (%)		14.3	23.0	25.4	30.8	24.5	51.6
Ts (°C)		137	137	150	164	147	120
T_H (°C)		514	512	425	425	430	426
L(m)	Tw(°C)						
0.025	T_1	462	420	419	415	442	405
0.05	T_2	472	404	431	424	460	387
0.10	T_3	490	412	448	441	478	395
0.15	T_4	504	427	462	451	487	399
0.30	T_5	530	446	484	471	499	412
0.45	T_6	553	473	514	501	537	442
0.70	T_7	612	520	560	540	583	476
0.95	T_8	663	567	610	591	648	524
1.20	T_9	677	587	634	613	680	581
1.35	T_{10}	676	595	639	626	678	567
1.45	T_{11}	652	582	624	608	653	555
1.60	T_{12}	145	144	159	107	157	135
1.40	T_V(°C)	371	327	275	320	394	268

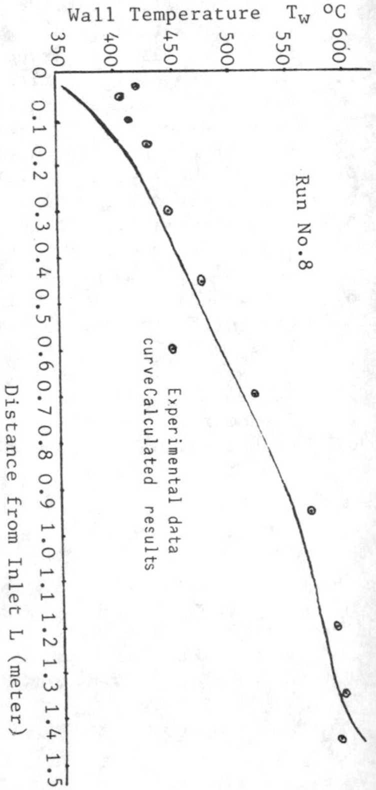

Fig.3. Comparison between Calculated Results and Experimental Data

Safety Investigations on KWU ECC-Systems LOCA Thermohydraulic Phenomena Simulated in the PKL-Facility

BERNHARD BRAND, RAFAEL MANDL, and HEINRICH WATZINGER
Kraftwerk Union AG
Erlangen, FRG

ABSTRACT

The behaviour of a PWR under LOCA conditions was experimentally investigated at KWU in the PKL test-facility. Large break tests, especially under "best estimate" conditions, demonstrate the safety margins of the KWU ECC-System. Small break experiments show that the automatically initiated cooldown procedure for the secondary side together with high pressure injection at the hot side of the reactor pressure vessel (RPV) is an appropriate measure to cover this spectrum of accidents and can avoid pressurized thermal shock problems in the RPV.

INTRODUCTION

With support of the German Federal Ministry of Research and Technology KWU operates the PKL (Primärkreisläufe) Test-Facility which is a scaled down model of a complete primary circuit of a KWU type pressurized water reactor. The system features and scaling principles are given in Figure 1.
The experiments are largely oriented towards the simulation of Loss-of-Coolant-Accidents (LOCA). The aim is to

- understand the physical phenomena occuring during transients
- obtain a data base for assessment of LOCA codes and
- verify the Emergency Core Cooling System (ECCS) safety measures.

Specifically, the following LOCA phases have been investigated so far:

- Core cooling conditions during the Refill and Reflood (R/R)-phase of a large-break LOCA
- Influence of the end phase of depressurization (Blowdown) with the initiation of accumulator injection of the R/R-phase also during large break LOCA
- Energy transport mechanisms occuring during various small break LOCA transients.

LOCA SAFETY MEASURES

Licensing authorities postulate as design basis accident a double ended quilliotine break (200 % cross section) of a recirculation pipe in the primary system. The process engineering measures taken to mitigate the consequences of a LOCA, the emergency core cooling system (ECCS), are shown in Figure 2.
The KWU ECC-system is of modular design and provides separate water injection to each loop (Figure 2 shows one of four subsystems). Each subsystem includes:

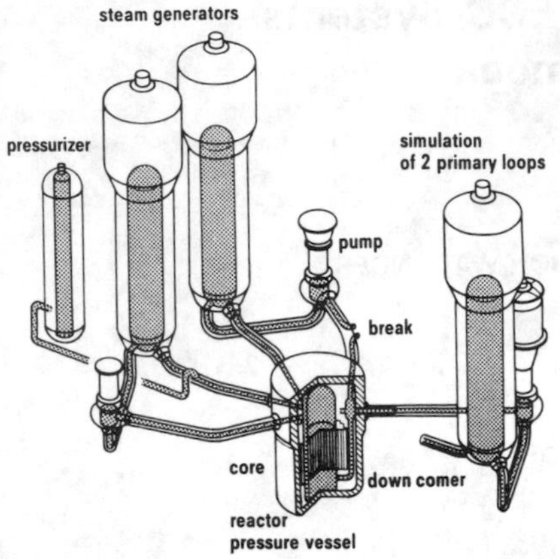

PKL-Systems Effects Tests:

- interplay between components
- code verification

	PWR	PKL
number of rods	45548	314 (+26)
volume scale	145	1
elevation scale	1	1
primary loops	4	2+1 double
ECC-injection	8	4+2 double

FIGURE 1. PKL Simulation of a 4-Loop PWR

- one high-pressure pump injection system
- two accumulators and
- one low pressure pump injection system.

Large effort was expended in order to minimize the fuel rod failure resulting from extended core uncovery following a large break accident. Accumulators and low pressure pumps supply water to both the inlet (cold leg) and exit nozzles (hot leg) of the reactor pressure vessel. The most significant feature of this combined hot and cold leg injection is the precooling of the reactor core by ECC water from the top during the end of blowdown, refill and reflood phase of a large break accident. The hot side injected water penetrates into the core forming a quench front from the top and condenses the steam passing from the core into the upper plenum. This eliminates the possibility of steam binding.

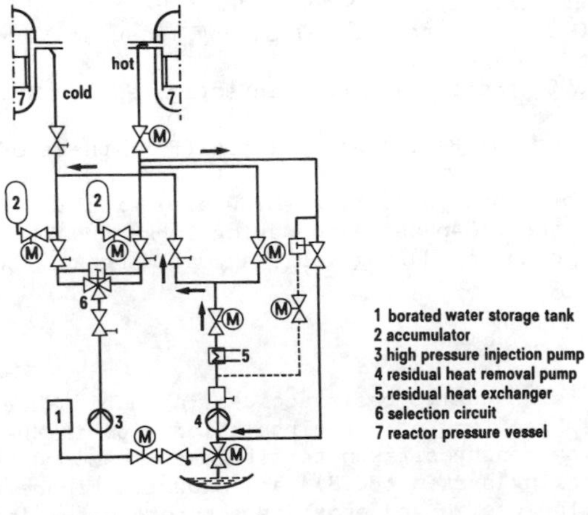

FIGURE 2. KWU-ECCS (1 of 4 subsystems)

Small breaks are characterized by the fact that the break mass flow removes less energy in the form of steam or hot water than is released by decay heat in the core. This results in a relatively slow pressure drop in the primary system and a correspondingly slow reduction of the water inventory, which is counteracted by injecting ECC water using the high pressure pumps. The decrease in coolant inventory is derived from the leakage rate and injection rate, both of which depend on pressure. However in addition to the break flow, an another heat sink is required to remove the decay heat from the primary system. This is obtained in KWU PWRs by automatically cooling down the secondary side of the steam generator (SG) by 100 K/h.

PKL TEST FACILITY AND TEST PROGRAMM

The PKL facility (Figure 3) is scaled down by comparison with the primary system of a commercial KWU 1300 MW PWR four-loop plant. Preserving all vertical dimension full scale (1:1), the power/volume scale is 1:145, so that a 340 rod bundle (314 electrically heated) represents the reactor core. The three loops (one of them of double capacity, simulating two loops in parallel) contain pump simulators and steam generators with original-size tubes. Their secondary sides are also volume scaled and can be operated as required under LOCA condition The max. pressure of the primary side of PKL is 40 bar to investigate for large break LOCA the endphase of blowdown (EOB) with the initiation of the low pressure ECC systems (accumulators $\leq$ 26 bar; pumps $\leq$ 10 bar) and the following Refill- and Reflood phase.
The facility is instrumented with over 600 measuring points. Apart from the conventional measurement of temperatures, pressure (p, Δp) etc. measurement for two phase flow also exists (1).

Several test series have been performed to date. PKL IA and IB series simulated only the Refill-/Reflood phase of a large break LOCA and was focused

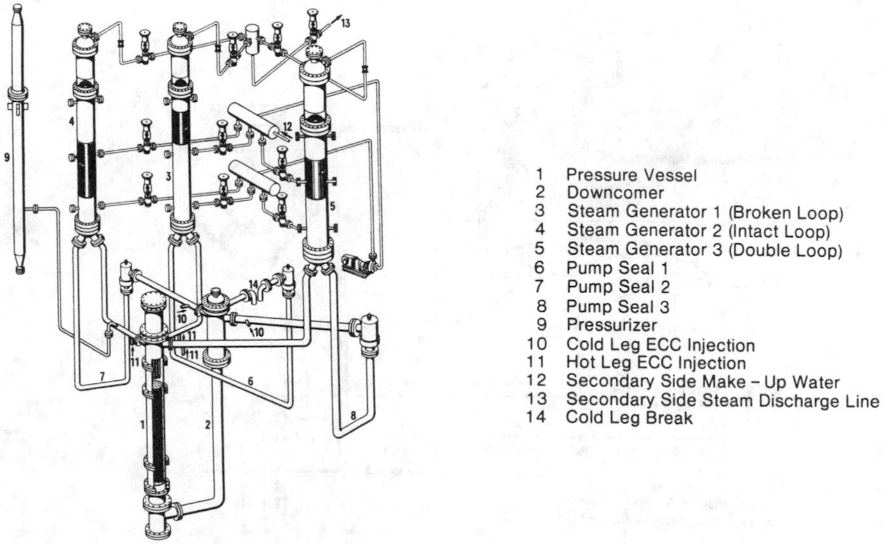

FIGURE 3. PKL Test Facility

on the effects of loop resistance and type on injection (2). In the PKL ID series the thermohydraulic phenomena occuring during small break LOCA were investigated in numberous steady-state and transient test (3). The PKL II experiments were concentrated on the influence of the EOB-phase on Refill/Reflood again for large breaks. The matrix of the latest tests include hot and cold leg break tests and the variation of ECC system availability. Variations of injection rates are carried out to demonstrate the safety margins of an ECC system in case of combined injection.

In the following, some selected results will be discussed, focusing on phenomena which can only be obtained by means of a test in an integral facility.

LARGE BREAK EXPERIMENTS

Figure 4 illustrates the results obtained in experiments simulating cold leg break and combined injection (2 tests): Starting at the beginning of the Refill-phase (4.5 bar) with mid-plane core temperature of about 600°C the heat transfer to the coolant has immediately the same order of magnitude as the decay heat generated and further improvement in heat transfer brings about early quenching. The plot also shows that a change in loop resistance (e.g. by different pump design or pump blockage) does not effect the cooling process significantly due to the fact that most of the steam is condensed in the upper plenum. The situati is different in the case of cold leg injection only (2 curves on the right hand side of the plot). A time delay in the onset of the cooling process is observed during the refill phase. All steam or steam/water mixture generated in the core during the reflooding phase must flow to the break via the steam generators and pumps. It follows that higher loop resistance and the accompanying higher pressu loss reduces the pressure difference (i. e. reflood driving force) between the downcomer and core resulting in longer quench times.

In order to simulate the latter stage of large break LOCA more realistically it is necessary to relax some of the conservative assumptions applied up till now.

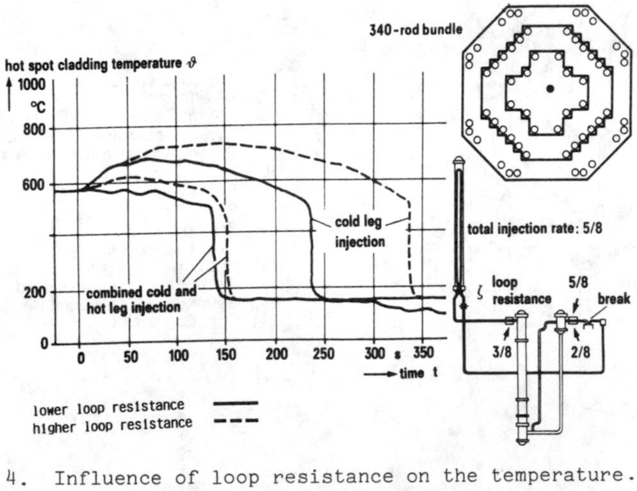

FIGURE 4. Influence of loop resistance on the temperature. Variation with different injection modes.

More representative boundary conditions are achieved by preceding refill/reflood with end-of-blowdown (EOB) and simultaneously applying the so called "Best Estimate" (BE) conditions. Some of the important differences between licensing and Best-Estimate assumptions are the following:

- power level and power density distribution
- decay power
- stored energy in core
- availability of emergency core cooling systems.

In case of a combined injection ECC-system however, the most significant contribution is made by considering the cooling effect of accumulator water in the EOB phase. This water, injected in the latter phase of blowdown and largely ignored in the conservative analysis, assists in early cooling of fuel rods expecially in large sized cold leg breaks. Calculations made using the American Trac PF1 Code, developed for Best-Estimate analysis, showed, for the case of a KWU-line PWR (combined injection), that taking into account the cooling effect of the accumulator water not only reduces the maximum fuel rod temperatures by several hundred degrees K but also prevents the usually expected second temperature peak during the refill and reflood phase.

The results of test series PKL II (Figure 5) show that a significant part of the stored energy is released already during EOB resulting in lower core temperatures and short quench times. These results indicate the safety margins of the present ECC system.

SMALL BREAK EXPERIMENTS

Influenced by risk probability studies such as "WASH 1400" or the "Deutsche Risiko-Studie" the attention has turned from large break LOCA to small breaks. As a result of the TMI-2 accident worldwide experimental activities were started to increase the understanding of the thermohydraulic phenomena during small break LOCA. The first systematic investigation of energy transport as a function

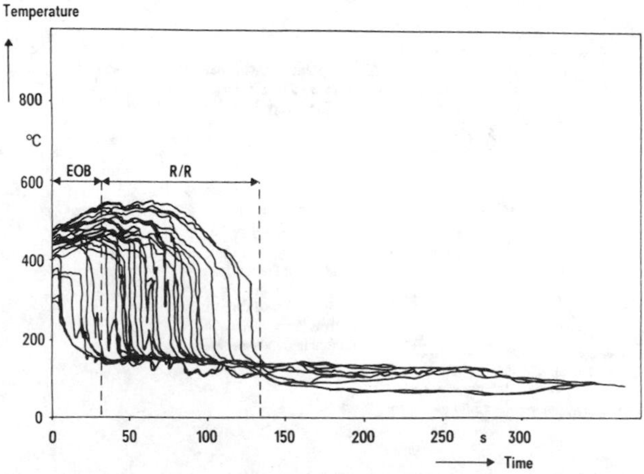

FIGURE 5. Heater rod temperatures for test IIB-1.
Cold leg break conditions with combined ECC-injection

of inventory, influence of noncondensible gases and the time history of transients, although at reduced pressure of 30 bar, were carried out in the PKL test facility.

The most significant results are the relationship between the water inventory versus natural circulation mass flow and the heat transfer in the steam generato as shown in Figure 6. With single-phase natural circulation, the mass flow rate is nearly constant for different degrees of subcooling. When saturation temperature in the core is reached boiling starts. While this vapour is being collected in the upper dome of the reactor pressure vessel, natural circulation remains unchanged. Vapour flowing to the steam generators results in additional driving force for natural circulation and the flow rate increases. With further loss of water inventory phase-separation occurs at the top of the U-tubes and the flow circulation comes to a virtual standstill at about 80 % water inventory At this and lower water inventories the steam generated in the core rises into the steam generators where it is condensed and returns to the core in form a thin water film via both the hot and cold legs (reflux condenser mode). While with single-phase natural circulation a 17 K difference between the primary and secondary side is necessary to transfer 365 kW of heat the required temperature difference is reduced to about 5 K with two-phase natural circulation and to only 2 K with reflux condensation.

These tests show that natural circulation is by no means a necessary prerequisit for energy transport from the core to the steam generators. The cladding temperature in the core does not rise as long as the two-phase coolant mixture covers the core and the secondary side heat sink is available.

The amount of coolant lost during depressurization depends on the relative positions of the break and injection locations. When the break is located in the cold leg and the ECC water is injected into the hot leg the loss of inventory is of the order of 5 to 20 %. In case of cold leg break with cold leg injection a loss of coolant up to 50 % was measured (Figure 7).
By means of these transient tests it could be shown that the phenomena taking place were the same as those observed during steady state tests and that no additional effects took place during the transition from one mode of energy transport to another.

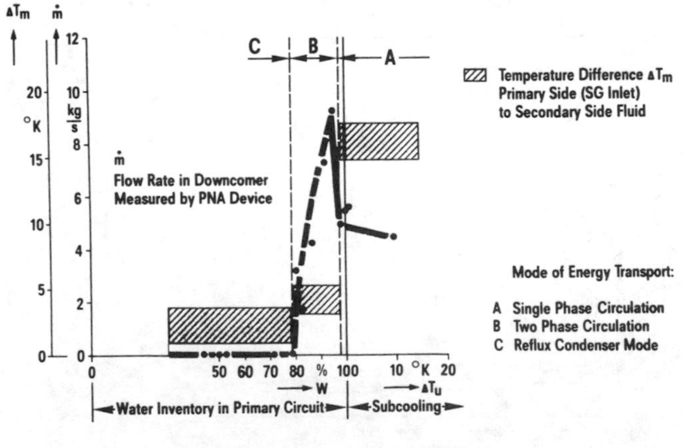

FIGURE 6. PKL-Small Break Tests. Results from test ID1.

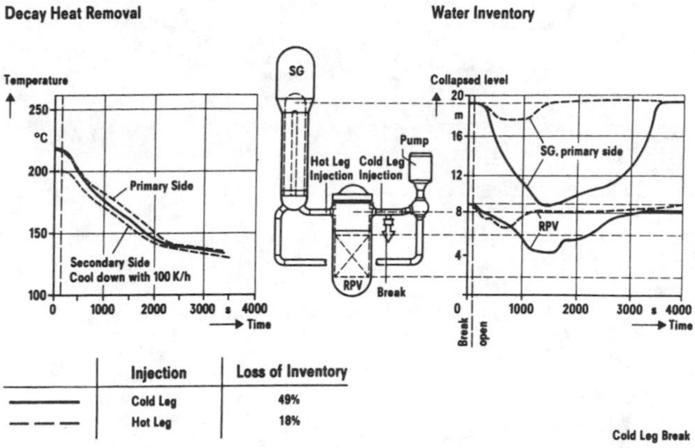

FIGURE 7. PKL tests: Effect of injection location on loss of inventory

At the same time the effectiveness of the special KWU measure of cooling down the steam generators by 100 K/h was demonstrated for all types of loos of coolan accidents.

FUTURE WORK

After completition of the PKL II test series the facility will be modified. Above all the secondary side will be augmented (Figure 8), so a wide range of "transient experiments" can be performed. Several topics for investigation are listed below:

- Supplementary investigations on small break accidents
- simulation of steam generator U-tube rupture
- LOCA on secondary side (steamline or feed water)
- failure of the feed water supply
- various cooldown procedures without break under emergency power conditions and/or partly isolated steam generators.

Although the PKL facility is limited in pressure to 40 bar valuable results are expected for

- verification of advance computer codes including plant analyzer systems
- investigation of optimum operation mode for cooldown procedures
- simplification and standardization of measures to be taken and finally as long term objective
- simplification of reactor plants.

In addition the test facility will be permanently available for experimental investigations at short notice stemming from topical operational problems in nuclear power plants.

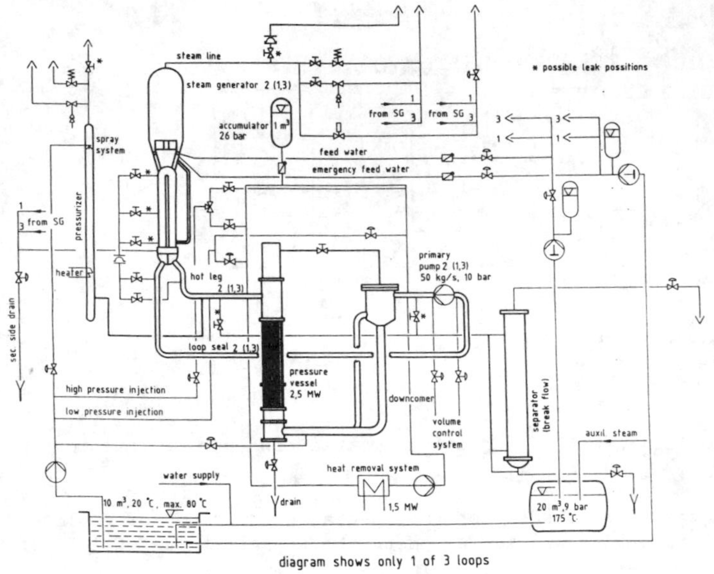

FIGURE 8. PKL III - Loop schematic for transients and small break experiments

REFERENCES

1. Brand, B., Kefer, V., Liebert, H., Mandl, R., Experience with Two Phase Flow Measurement Techniques, Int. Symposium on Heat Transfer, Beijing, Okt. 1985.

2. Hein, D., Riedle, K., Untersuchungen zum Systemverhalten eines Druckwasserreaktors bei Kühlmittelverluststörfällen. Das PKL Experiment, Atomkernenergie Kerntechnik, Vol. 42, 1983, No. 1.

3. Mandl, R., Weiss, P., PKL Test on Energy Transfer Mechanisms during Small-Break LOCA, Nuclear Safety, Vol. 23, No. 2, 1982.

Experimental Study of Inverted Annular Flow Film Boiling Heat Transfer of Water

YUZHOU CHEN
Institute of Atomic Energy
P.O. Box 275
Beijing, PRC

ABSTRACT

Steady state film boiling experiment has been conducted on a tubular test section (7 mm ID) with water flowing upward, using the directly heated hot patch technique. The test covers the following range of parameters: pressure 1.0-5.5ata, inlet subcooling 0-70.7°C, mass flux 130-920kg/m^{2}S. The data indicate a strong increase of heat transfer coefficient with subcooling and mass flux, and a complicated dependence of heat transfer coefficient on pressure, upstream heating power and distance from the dryout point. The effects of the parameters can not be explained well with the previous models.

INTRODUCTION

Inverted annular flow film boiling might take place ahead of a quench front during reflood phase of a loss of coolant accident in nuclear reactor. It plays an important role for the progression of the rewetting front. However, the mechanism of this regime was poorly understood because it was very difficult to establish this regime stably with the techniques used in the normal heat transfer experiments. In recent years, the application of the indirectly heated hot patch technique is gradually improving our knowledge of film boiling.[1-4] In spite of this, there are still insufficient data to generalize the relevant correlation especially for higher subcooling and flow conditions, where it was difficult to reach the high heat flux required with the techniques available.

In the recent computer codes (e.g. the advanced best-estimated codes RELAP5/MOD2 and TRAC-PD2) Bromley correlation or its modified form, in which either or both of effects of subcooling and flow rate of the coolant were not considered, was widely used due to the lack of appropriate correlation to predict the heat transfer of this regime, despite it has been found long ago that the subcooling and velocity of the coolant may greatly affect the heat transfer.[5] In a review of the extensive literature Groeneveld has concluded: "Correlations for the low-quality or subcooled film boiling region should be suspected because of the lack of a reliable data base and the difficulty in accounting for the various physical mechanisms in a single correlation."[6]

The author of this paper has used a directly heated hot patch technique in the experimental study of flow film boiling of water over a wide range of subcooling, and obtained a lot of data at atmospheric pressure.[7] In the present investigation the experiment has been conducted at higher pressure.

EXPERIMENT

Apparatus

The test section, shown in Fig.1, was basically the same as that in the previous work, [7] which consists of a 7 mm ID (10 mm OD) -tainless steel tube with water inside flowing upward during test runs. Sections AB and BC were directly heated respectively by an AC supply connected, adjusting the power levels seperately on the both sections. AB was 10 mm long and BC was 150 mm long. The tube was machined to be thinner locally just upstream of the positions B and C. Thus, heat flux peaks could be achieved there.

There exists an additional pair of clamps O and A at a distance of 225 mm upstream of section AB, which was connected to another AC supply only when the effect of the upstream heating on heat transfer coefficient was studied. (see 3.3) The section BC was thermally insulated.

17 thermacouples were spot-welded on the outside of the section BC at different axial positions to measure the fine axial distribution of the wall temperature.

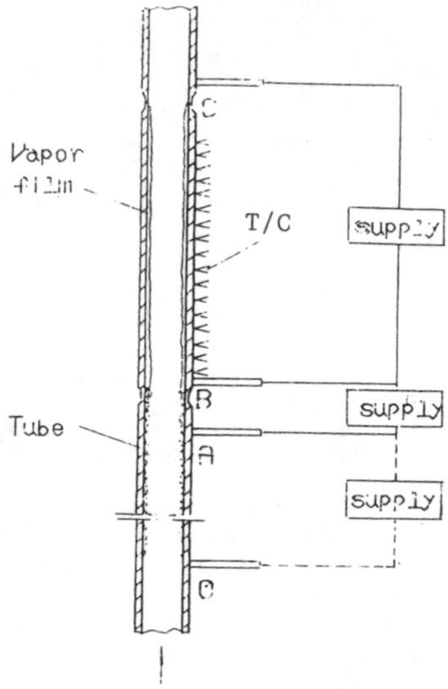

Fig.1 Schematic of test section

Procedure

Initially the desired pressure, flow rate and inlet temperature of water were established in a by-pass. The sections AB and BC were then heated up respectevely. When the wall temperature reached the 500-700°C range, the water flow was slowly diverted from the by-pass into the test section. The tube would start to cool, and the powers supplied to sections AB and BC were appropri ately increased to prevent the propagation of the rewetting front upward through position B and keep the wall temperature of the section BC higher then the min imum film boiling temperature. The heat flux peak near position C prevented the propagation of another rewetting front downward from the outlet as well. And th film boiling could be maintained over the section BC.

When the stable conditions were established, the wall temperatures and other required parameters were recorded. After a test run, some of the parameters wer changed to another desired values with the section BC in film boiling, and next run was started.

It was easy to establish the film boiling regime except at highest subcoolin where the flow exhibited some fluctuation and might cause overheat on the notch, shoretening the life of the test section.

RESULTS AND DISCUSSIONS

The typical results of the film boiling heat transfer coefficients are shown versus length in Figs 2-5, where ΔT_s is the inlet subcooling (at the dryout poin The heat transfer coefficient was evaluated from the heat flux divided by the di ference between the inner surface temperature and the saturated temperature of water, corrected for heat loss and effect of lengthwise variation of temperature on the electrical resistance, neglecting the axial heat conduction. It involves all the convection, conduction and radiation components. At higher subcooling and flow conditions, the radiation component is negligible, whereas at very low subcooling and flow conditions, it would be appreciable.

Effect of mass flux

Fig.2 presents the results obtained with three mass fluxes for different pressures and inlet subcoolings. It can be seen that heat transfer coefficient increases with mass flux increases. This trend is not unexpected. In inverted annular flow film boiling the resistance to heat transfer basically lies on the vapor film covering the heated surface. That is, the heat transfer coefficient dependents on the thickness of the vapor film and the turbulent intensity therein. As the mass flux of the liquid increases, resulting in the increase in the interfacial velocity and, hance, steam velocity, both the thickness and the turbulent intensity in the vapor film would become more favourable to heat transfer.

Effect of inlet subcooling

Fig.3 shows the distributions of the heat transfer coefficients with the inlet subcooling as a parameter. It is evident that the heat transfer coefficient increases markedly with increasing inlet subcooling, except for rather low inlet subcooling, where this effect seems less steep.

Being different from the case at saturated condition, at subcooled condition a part of the heat is transfered from the liquid-vapor interface into the bulk liquid. In some model[8], it was predicted with a convective heat transfer correlation for a single phase flow, and the vapor film was treated as in saturated film boiling. The examination of this model has been made on the basis of the present results. In case of P=2ata, G=880kg/m^2.s and ΔT_s=49.9°C, for example, the heat flux q_l calculated with Dittus-Boelter correlation (R_e=1.48x10^4) is 3.45x10^5 w/m^2. In saturated film boiling with the same pressure, mass flux and similar wall temperature, the datum of heat flux q_v is 1.45x10^5w/m^2. According to such model, at the subcooling of 49.9°C with heat flux of q_l+q_v (4.90x10^5w/m^2), the behavior of the vapor film is the same as that at saturated condition with $q=q_v$. However, at this subcooled condition the heat transfer coefficient is higher than that at saturated condition by about a factor of 4, although the heat flux (8.73x10^5w/m^2) is much higher than q_l+q_v (the increase in heat flux has an unfavourable effect on heat transfer, as shown in the previous work and else).

This discrepancy suggests that in inverted annular flow, the "boundary layer" of the bulk liquid flow is greatly different from that in the single-phase flow with the same average velocity. In addition, the interaction between the subcooled liquid and superheated vapor would largely change the behavior of the vapor film, as demonstrated by Toda and Mori in their visual experiment[9].

Effect of upstream heating power

With upstream heating, the liquid temperature field would be developed. This tends to reduce the effect of subcooling on heat transfer to some extent. When the upstream heating power is high enough, resulting in nucleate boiling, the steam flow coming from its boundary layer would entrain water into the vapor film and enhance the turbulence in it, improving the heat transfer. This effect is illustrated in Fig.4, where q_o is the heat flux on section OA. A limited increase in wall temperature downstream of the dryout point corresponds to lower upstream heat flux (without nucleate boiling), whereas a distinct fall of wall temperature corresponds to high heat flux of nucleate boiling.

In order to study the effects of most important parameters on film boiling heat transfer, in the present experiment the test section with a shorter length of section AB (10 mm) was used compared with the previous work (20mm).

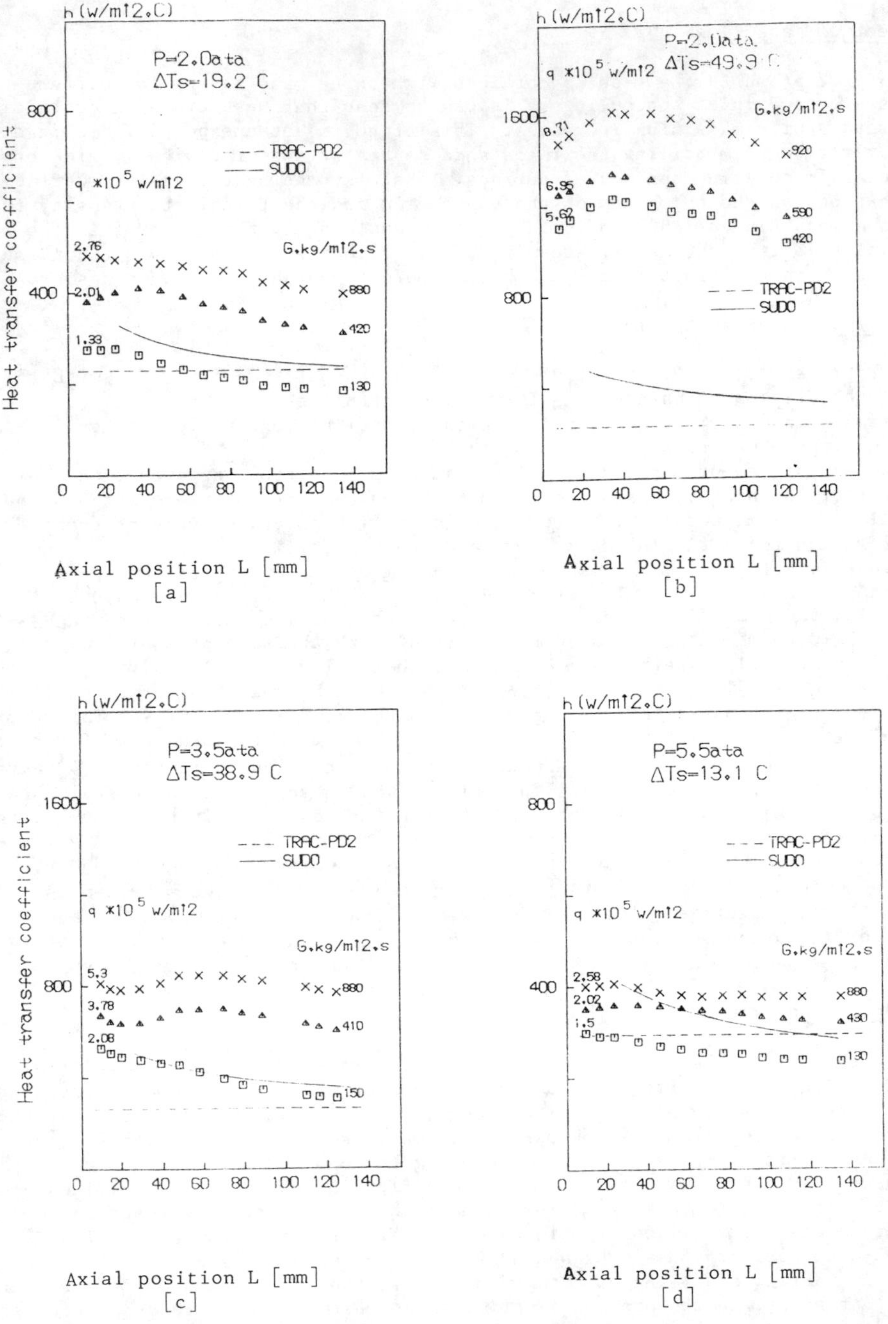

Fig.2. The effect of mass flux on heat transfer coefficient

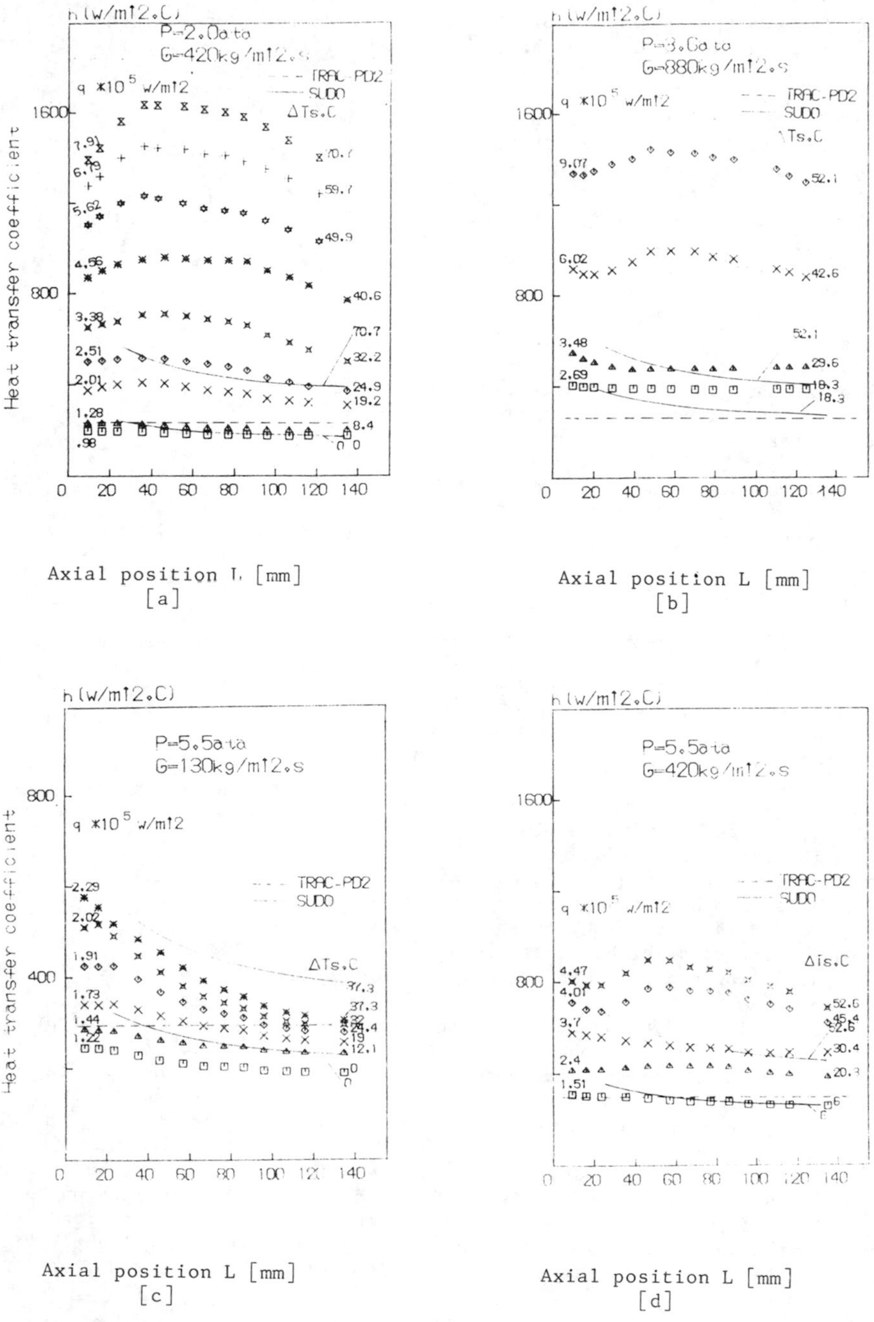

Fig.3. The effect of inlet subcooling on heat transfer coefficient

Effect of pressure

Fig.5 shows the comparison of the results at different pressures. It can be observed that at low subcooling and low flow conditions, the heat transfer coefficient increases with pressure, whereas at higher subcooling and flow conditions it exhibits an opposite trend. This also implies that at lower pressure the efects of subcooling and mass flux are stronger than at higher pressure. This dependence can not be explained with the laminar model of film only. As the pressure increases, the thickness of the vapor film would decrease due to the increase in steam density, improving the heat transfer. On the other hand, at subcooled condition, the heat, mass and momentum transfer at the liquid-vapor interface may promote the turbulence in the vapor film. At lower pressure it would be more drastic due to the greater difference in the properties between two phase. At higher subcooling condition this effect would largely contribute to heat transfer.

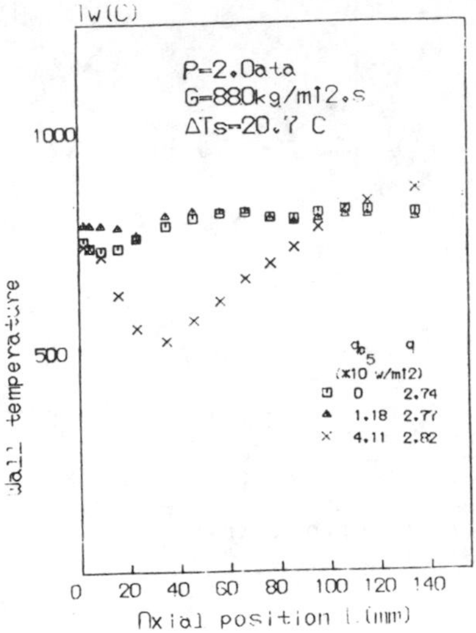

Fig.4 The effect of upstream heating power on heat transfer coefficient

Effect of axial position

The effect of axial position can also be seen from above Figs. The heat transfer coefficients exhibit a complicated variation along the length, except at very low inlet subcooling, where the heat transfer coefficient distribution is rather uniform. At lower mass flux, the heat transfer coefficient decreases along the length, and this trend becomes steeper with the inlet subcooling increases perhaps due to the steeper rise of water temperature. At higher mass flux, the maximum heat transfer coefficient shifts from the dryout point. It is interesting to note that for higher pressure at higher subcooling and flow conditions, a valley is observed downstream of the dryout point.

These complicated trends reflect the combination of the effects of pressure, mass flux and subcooling along the length on heat transfer coefficient.

In Figs 2 and 3 the modified Bromley correlation used in the computer code TRAC-PD2 and SUDO correlation are superimposed for the comparison with the data. As shown, the modified Bromley correlation lies on the lower level of the wide data range apparantly because the effects of both subcooling and mass flux were not included in this correlation. SUDO correlation was based on the experiment at flow rate less than 30cm/s. As can be seen, it essentially lies on the data range for this condition. But it can not predict the complicated trends of the data. At the worst condition (saturated and lowest flow it is slightly higher than the data, whereas at higher flow and high subcooling condition, it is lower than the data.

CONCLUSIONS

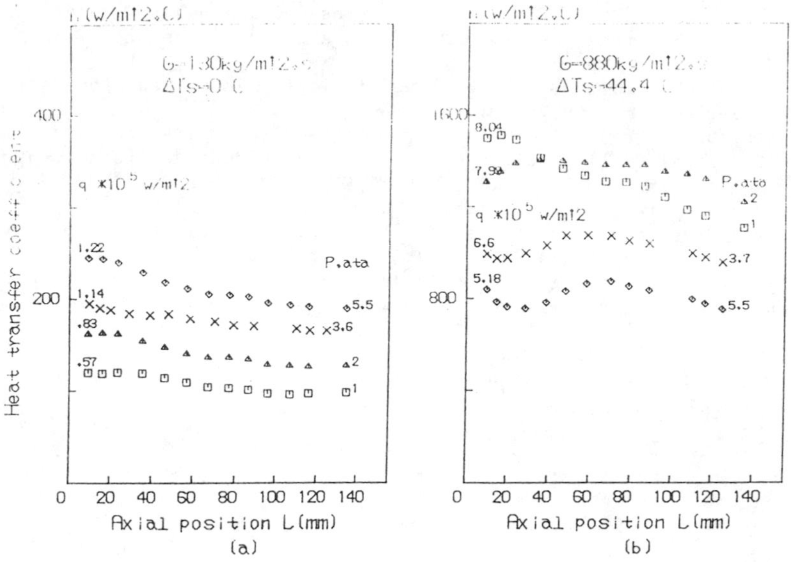

Fig5. The effect of pressure on heat transfer coefficient

At the present experimental conditions, the following conclusions are achieved:

1. Film boiling heat transfer coefficient increases markedly with the subcooling and mass flux of water.
2. At very low subcooling and flow conditions the heat transfer coefficient increases with pressure, and at higher subcooling and flow conditions it exhibits an opposite trend. The variation of heat transfer coefficient distribution along the length is very complicated with various parameters
3. The upstream heating power has an effect on heat transfer coefficient.
4. The interaction of the subcooled bulk liquid and superheated vapor may greatly affect the behavior of vapor film. The correlation and model based on saturated condition are not suitable for the present subcooled condition.

REFERENCES

1. K.K.Fung, S.R.M.Gardiner and D.C.Groeneveld, Subcooled and low Quality Flow Film Boiling of Water at Atmospheric Pressure, Nucl.Eng.Des.55 51-57, 1979
2. Stewart J.C. and Groeneveld D.C., Low Quality and Subcooled Film Boiling of Water at Elevated Pressures, Nucl.Eng.Des. Vol.67 259-272, 1981
3. G.Costigan, A.W.Holmes, J.C.Ralph Steady State Post-Dryout Heat Transfer in a Vertical Tube with Low Inlet Quality, presented at 1st U.K. National Heat Transfer Conference. LEEDS, July 1984
4. K.Johannsen, Observations on Flow Boiling CHF and Post-CHF Heat Transfer

of Water in a Short Vertical Tube at Low Pressure and Quality, presented at the CHINA-U.S seminar on two-phase flows and heat transfer, Xi'an, China, May 1984
5. E.I.Motte and L.A.Bromley, Film Boiling of Flowing Subcooled Liquids, Ind.Eng.Chem,Ind.Edn. 49 1921-1928, 1957
6. Groeneveld, D.C., Prediction Methods for Post-CHF Heat Transfer and Superheated Steam Cooling Suitable for Reactor Accident Analysis, Centre d'Etude Nucleaires de Grenoble, reprot TT/SETRE/82-4-E/DCGr, 1982
7. Chen Yuzhou and Li Jusheng, Subcooled Flow Film Boiling of Water at Atmospheric Pressure, Presented at The First Light Water Reactor Safety Research Workshop on Thermal-Hydraulics, Tokyo, 1984
8. E.K.Kalinin et al, Investigation of Film Boiling in Tubes with Subcooled Nitrogen Flow, Heat Transfer Vol. V B4.5 El evier Pub. Co., Amsterdam, 1970
9. S.Toda, M.Mori, Subcooled Film Boiling and the Behavior of Vapor Film on a Horizontal Wire and a Sphere. 7th Int Heat Transfer Conf.,V4, 173-178, 1982

Velocity and Turbulence Measurements in Water Flowing Axially through a 37-Rod Bundle

D. F. D'ARCY and J. R. SCHENK
Atomic Energy of Canada Limited
Atomic Energy of Canada Research Company
Chalk River Nuclear Laboratories
Chalk River, Ontario, Canada, K0J 1J0

1. EXPERIMENT

Measurements of the axial component of local velocity and turbulence were made at various points in a simulated 37-rod fuel bundle of CANDU design. The test bundle was mounted in the horizontal test section of a water flow loop shown in Figure 1. Two actual fuel bundles were installed in the test section immediately upstream of the test bundle. The latter was one of two simulated bundles of the same geometry, made of hollow rods. The flow tube diameter was 103.6 mm and the bundle string resting on it was 0.72 mm eccentric. All bundles had wear pads on the outer rods and inter-rod spacers as shown in Figures 2 and 3. The flow rate, temperature and pressure, as well as the pressure drop across the test bundle were recorded as in Figure 1.

A typical CANDU fuel bundle is sketched in Figure 2, showing an end plate and the wear pads on which the bundle rests in its flow tube. Rods numbered 9 and 10 are indicated in the outer ring of rods. Rod 9 was made of glass and contained a fibre-optic LDA probe as shown. The probe itself is sketched in Figure 4. The probe slides snugly within the glass rod on teflon rings, and can be rotated and moved axially by a hollow metal rod which passes through the two downstream bundles and exits via an 'O' ring seal.

The fibre-optic probe incorporated a dual beam laser doppler anemometer designed for back-scattering. One to two watts of light from an argon-ion laser were fed to the probe via a 50 μm diameter graded index optical fibre. The scattered light collected by the probe went to an external photomultiplier tube via a similar optical fibre of 200 μm diameter. An adjustable mirror in the end of the probe deflected the laser beams from the axial to the radial direction (Figure 4), so that the beam crossing spot was outside the glass rod. The radial distance of the spot from the glass wall was adjusted by changing the axial location of the mirror in the probe. This was done by pushing the probe upstream to engage a screwdriver mechanism and rotating it to turn the leadscrew supporting the mirror. This also caused an equal movement of the spot in the axial direction. Linear and angular scales were used to set the axial and azimuthal position of the spot. Radial movement was controlled by recording the turns of the 40 threads/inch leadscrew.

The laser doppler signals from the p.m. tube were analyzed by a TSI model 1980B counter type signal processor, which made a velocity-bias correction.

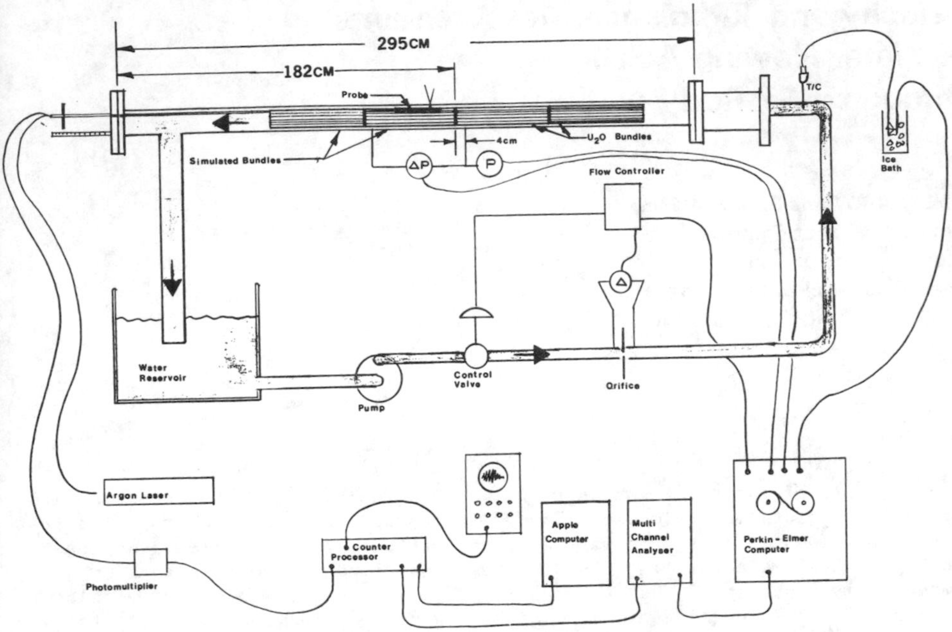

Fig. 1. MR-2 Loop and L.D.A. Data Aquisition System

Batches of 256 or 512 data points were transmitted to an Apple II_e computer, which used TSI software to produce histograms of velocity, and compute the mean and R.M.S. values of the local velocity. With clean tap water it was found that seeding of the water was not necessary.

The water mass flow rate was controlled as shown in Figure 1. The bundle Reynolds number varied over a small range because of small changes of temperature. Pressure in the loop was about 45. kPa gauge.

2. RESULTS

Measurements were made with the bundles in two different configurations. In configuration 'A', all 37 rods were aligned but the adjacent end plates were not, as in Figure 5. In configuration 'B', the upstream bundle was rotated one outer ring pitch (20. deg). This nearly aligned the end plates but misaligned the inner rings of rods (Figure 5).

The bundle Reynolds number was about 2.5×10^4 for all tests.

2.1 Bundle Pressure Drop

For reference, the bundle pressure drop was recorded by the instrument indicated in Figure 1. It could be represented by:

$$CD = A\ Re^{-.101} \tag{1}$$

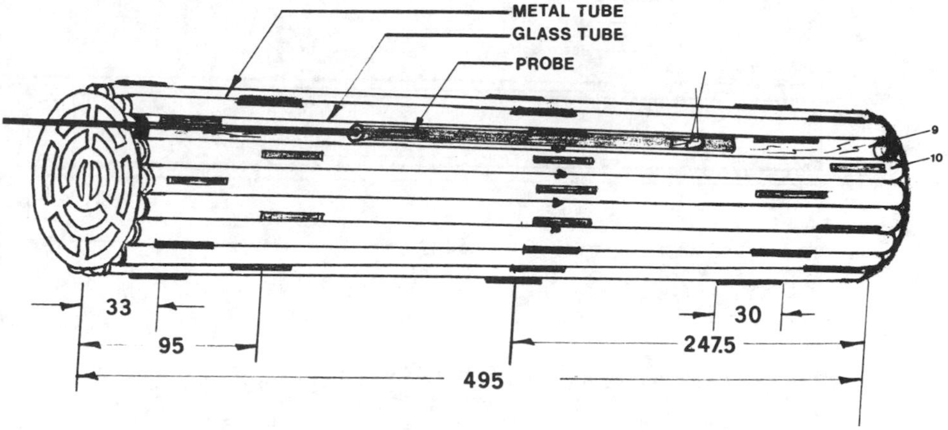

Fig. 2. 37-Element Bundle with LDA Probe

where A = 6.86 For Configuration 'A'
7.08 For Configuration 'B'

The equivalent friction factor is F = CD/(L/DH). With L/DH = 500./7.42 = 67.4, F('A') = 0.037 and F('B') = 0.038 at Re = 2.5 x 10^4. Trupp and Azad [1] give F about 0.03 for a smooth bundle at the same Re. The difference is due to spacing devices and endplates. The exponent -.101 is similar to that for flow in a rough tube, as might be expected.

The pressure drop in configuration 'B' is higher than in 'A' by about 3 pct. This indicates that the misalignment of the inner rods increases the pressure drop more than the near-alignment of the endplates decreases it, i.e., the increase in loss in the interior of the bundle overcomes the decrease in the outer ring (see Figure 5).

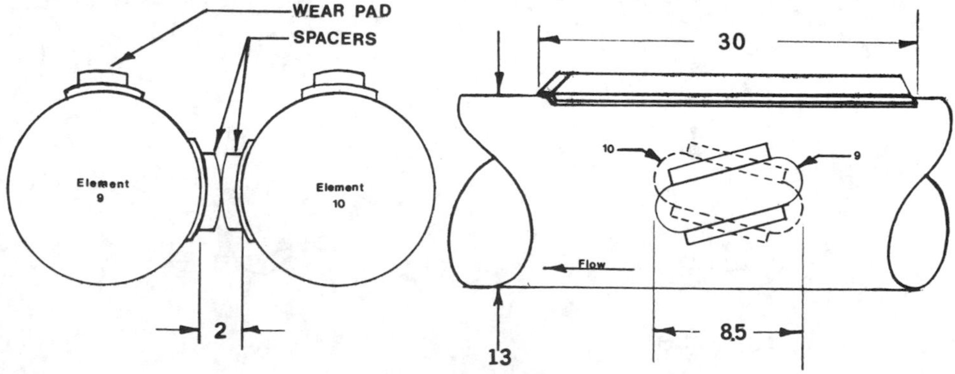

Fig. 3. Spacer and Wear Pads

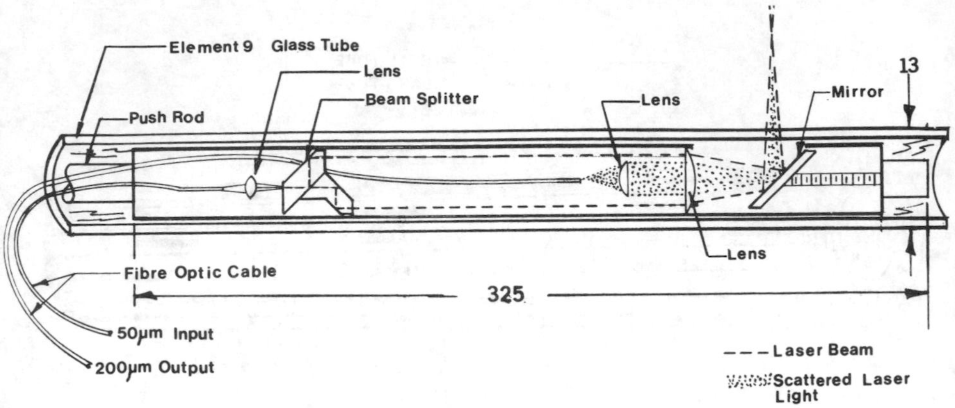

Fig. 4. Fibre Optic Probe

2.2 Velocity and Turbulence Contours

In the first series of tests, the axial velocity and turbulence components in the entire flow field around rod 9 were measured in configurations 'A' and 'B' at 93. mm from the upstream endplates. Contour plots of mean axial velocity u and relative axial turbulence u'/u are given in Figures 6 and 7 for configuration 'A' and in Figures 8 and 9 for configuration 'B'. The contours were drawn from measurements at 1 mm radial, and 5 or 10 deg. azimuthal increments, i.e. a rather coarse grid.

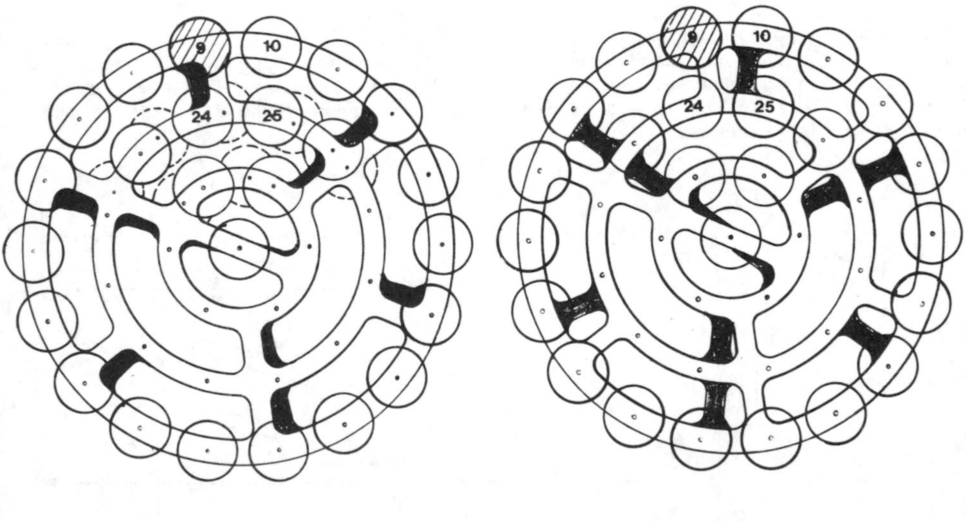

Configuration 'B' Configuration 'A'

Fig. 5. Upstream View of Bundles and Endplates

From Figures 6 and 8, the effect on velocity distribution of the rotation of the upstream bundle is small. In both configurations the velocity distribution in each subchannel has a broad, fairly flat maximum, similar to pipe flow. The increase of velocity in the outer subchannels above rod 9 in configuration 'B' is expected due to the increased resistance of the inner subchannels, as noted above. In the subchannels below rod 9, a weak effect of the endplate web is discernable at this axial location. The velocity distribution became less peaked in the subchannel at right, while in the subchannel at left, some local reduction of velocity can be seen.

In Figures 7 and 9 contours of u'/u = 0.10 (solid line) and u'/u = 0.12 (broken line) are shown. Within the 0.10 contour the values were about = 0.08. Outside the 0.12 contour the values rose appreciably due in part to the decrease in u near the wall.

Crossplots of the data in Figures 6 to 9 are given in Figures 10 and 11. u and u'/u are plotted along the radius midway between rods 9 and 10 (Figure 5). Symbol + is for configuration 'A', X for configuration 'B'. The higher velocity near the outer wall in configuration 'B' noted above is clear in Figure 10. The lower velocity in the gap at ring 2 is likely a shadow of the upstream rod at this location.

In Figure 11 the local axial turbulence intensity is seen to be low at midsubchannel in both configurations. A slight reduction in intensity near the ring 3 gap can also be seen for configuration 'A'.

2.3 Normal Velocity Distribution

Measurements were made at 163 mm from the endplates, at closer spacing along lines perpendicular to the surfaces of two rods, directed into the adjacent subchannels. The spacing was .25 or .5 mm in the radial, or 1 deg. in the azimuthal direction.

The measurements were fitted with a universal velocity distribution:

$$u/u^* = A \, \mathrm{Ln}\, ((Y-YW)\, u^*/\nu) + B \tag{2}$$

With A = 1/.418 and B = 5.45. Y is the measured distance normal to the wall, YW is a small error in Y at the wall due to the finite size of the LDA spot. The value of the friction velocity u* was found by iteration such that a least squares line through the values of YW found from (2) at each point had zero slope, i.e., YW was constant for all points. This overcame the difficulty of determining the exact location of the wall with a finite-sized spot. In each case the value found for YW was within the expected error of the measured wall position. The plot along a radial line at θ = 146 deg is shown in Figure 12. The azimuthal angle θ is defined in Figures 6 to 9, where every 10 deg. is marked around rod 9. The long mark at the top is θ = 0 deg. θ increases

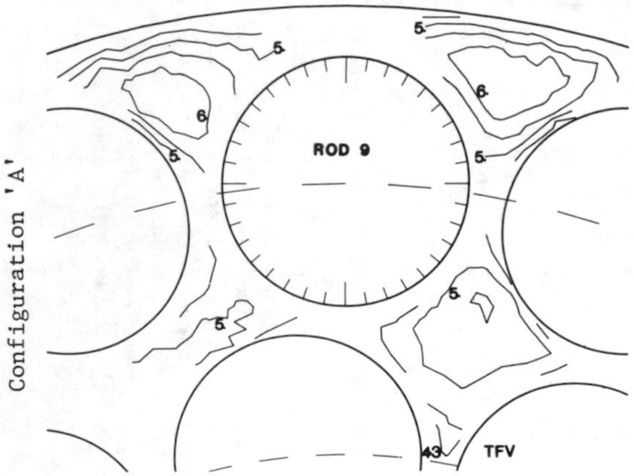

Fig. 6. Mean Axial Velocity Contours (m/s)

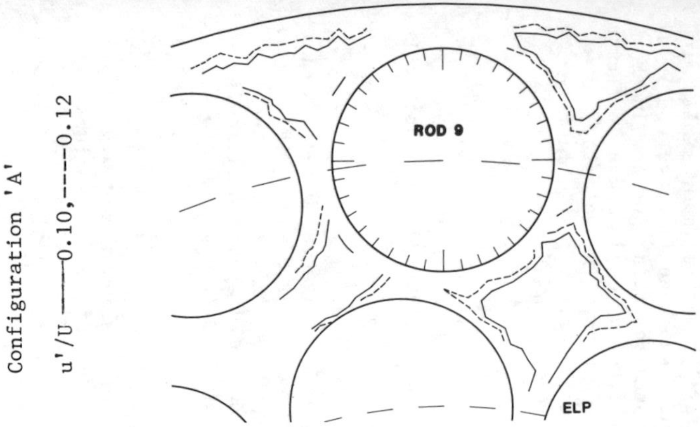

Fig. 7. Axial Relative Turbulence

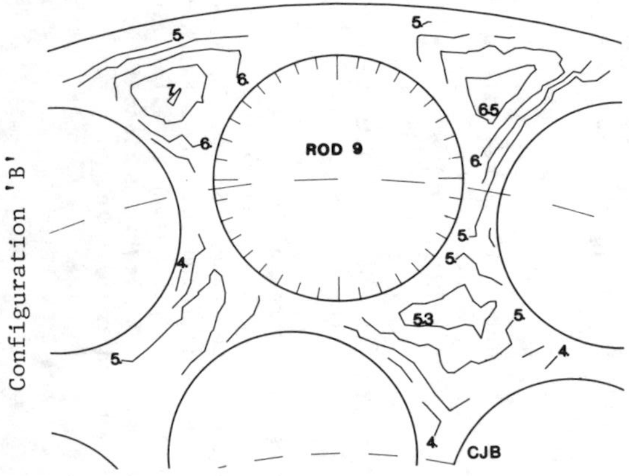

Fig. 8. Mean Axial Velocity Contours

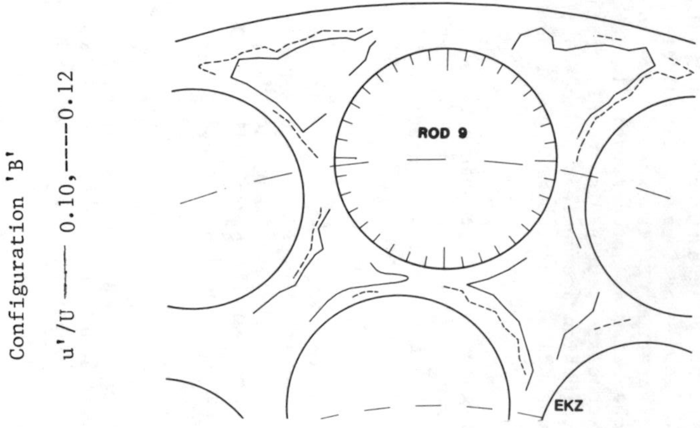

Fig. 9. Axial Relative Turbulence

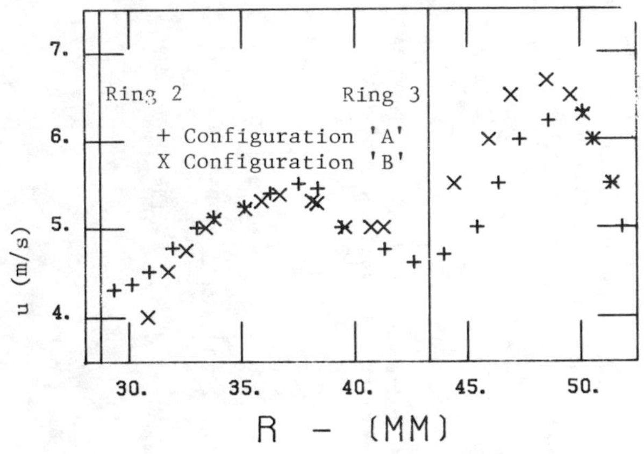

Figure 10: Mean Axial Velocity vs. Radius

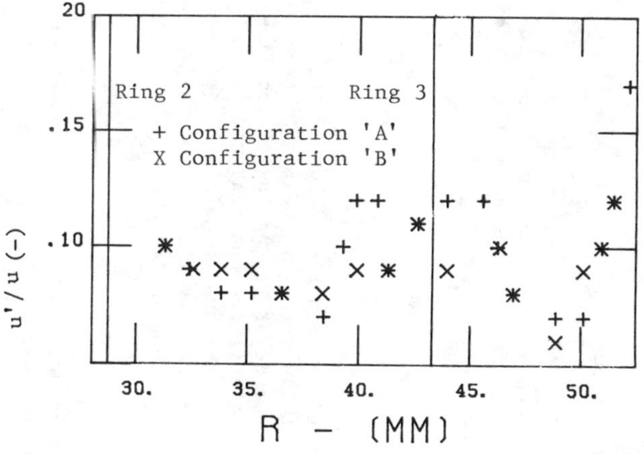

Figure 11: Axial Relative Turbulence vs. Radius

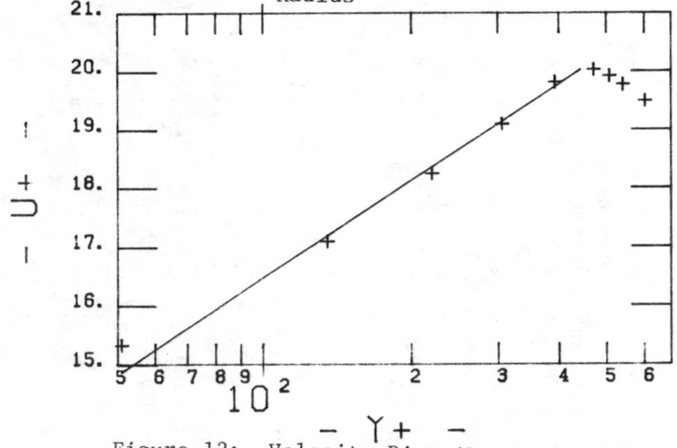

Figure 12: Velocity Distribution at Rod 9 at $\theta = 146^{o}$

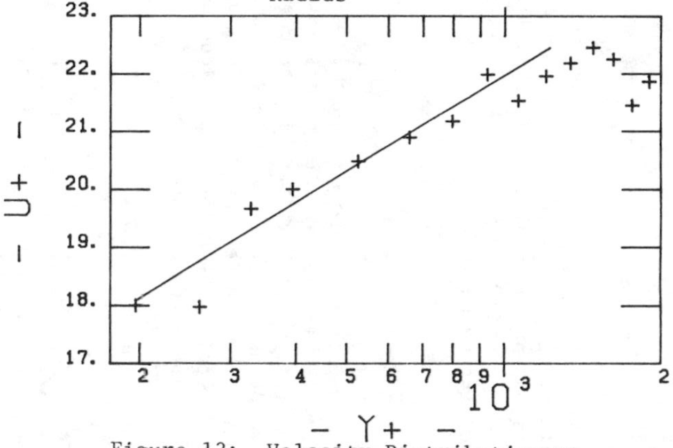

Figure 13: Velocity Distribution at Rod 10, Upper Half, at R=13.54 mm

clockwise. The plot along the radius R = 13.54 mm from the upper edge of rod 10 is shown in Figure 13. The values of u* estimated by this procedure are:

u* = 0.204 for rod 9 at θ = 146 deg.
0.282 for rod 10 at R = 13.54 mm (Upper surface)

The mean value estimated from the equivalent friction factor for the whole bundle is u* = 0.23. This is not an accurate method of measuring u*, but it yields approximate values, and it indicates that the velocity obeys the law of the wall at the points tested. Greater accuracy would require a better traversing mechanism.

2.4 Drag of Rod Spacers

An extra pair of spacers (Figure 3) was mounted between rods 9 and 10 at 161 mm from the endplates. The normal spacers, at 250 mm only, are not in range of the present probe. Velocity traverses were made at several axial locations up and downstream of the spacers with the probe at three constant radii spanning the gap between rods. The form drag of the spacer pair was computed by integrating the momentum loss due to the spacers over the area of their wake. Expressed as a loss coefficient, this was found to be:

CD = 0.71 at Re = 2.5×10^4

CD was based on the frontal area of the spacers and the upstream velocity in the gap. This value is remarkably close to the values found by Arie, et al for square cylinders in turbulent boundary layers, as quoted in [2].

3. CONCLUSIONS

A miniature LDA probe has been used to make local measurements of the axial velocity and turbulence intensity within a fully simulated nuclear fuel bundle at actual scale. Measurements of these quantities and of the drag of a typical rod spacer have been presented.

4. ACKNOWLEDGEMENT

The work reported was funded by AECL-Ontario Hydro CANDEV Program. The LDA probe was developed by TSI Inc. in conjuction with J.R. Nickerson of AECL, Chalk River, Canada.

5. REFERENCES

[1] Trupp, A.C., Azad, R.S., Nucl. Eng. and Des. V32, P47, (1975).

[2] Sakamoto, H., Moriya, M., Taniguchi, S., Arie, M., J. Fluids Engg., V104, P326, (1982).

Reflood Heat Transfer in PWR Fuel Rod Bundles Deformed in a LOCA

F. J. ERBACHER, P. IHLE, K. WIEHR, and U. MÜLLER
Kernforschungszentrum Karlsruhe
Institut für Reaktorbauelemente
Postfach 3640, 7500 Karlsruhe 1, FRG

INTRODUCTION

In the framework of the licensing procedure for pressurized water reactors (PWR) evidence must be produced that the impacts of all pipe ruptures hypothetically occurring in the primary loop and implying loss of coolant can be controlled. The double-ended break of the main coolant line between the main coolant pump and the reactor pressure vessel is presently considered to constitute the design basis accident. Upon rupture of the reactor coolant line the reactor is shut down. However, as the production of decay heat continues, reliable long-term cooling of the reactor core is required.

After depressurization and evacuation of the reactor pressure vessel emergency cooling systems supply the reactor core with the emergency cooling water kept in the accumulators and the flooding tanks. Cooling of the fuel elements is temporarily deteriorated until emergency core cooling becomes fully effective. In this time interval some fuel rod claddings attain temperatures at which they balloon or burst under the impact of the internal overpressure. Figure 1 shows as an example the fuel rod cladding load in a loss-of-coolant accident (LOCA) as predicted by conservative evaluation models for a PWR.

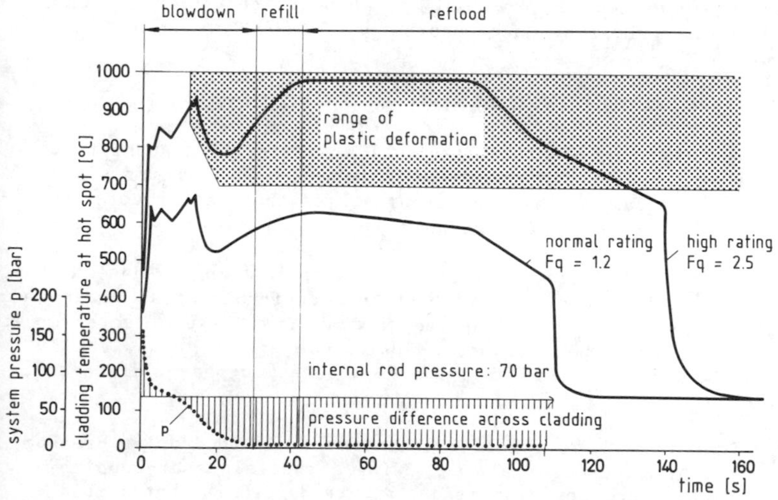

FIGURE 1. Fuel rod cladding load in a double-ended cold leg break LOCA

Ballooned fuel rod claddings lead to local flow blockages in the coolant channels within the fuel elements. Further damage to the fuel elements can be avoided only if the emergency cooling systems, despite of the reduced flow cross sections, ensure reliable cooling of the fuel elements and no further major temperature rise takes place in the fuel elements. Therefore, the prediction of the peak cladding temperature during the reflooding phase of a LOCA and of hot fuel rod quenching are of crucial importance in nuclear reactor safety.

The reflood heat transfer in fuel elements is a complex transient two-phase flow and heat transfer process. It is influenced among others by the non-equilibrium in two-phase flow, the grid spacers, and the flow blockages caused by ballooned and burst fuel rod claddings. These effects of heat transfer have not been modeled sufficiently well in most of the existing computer codes; however, they determine decisively the cladding temperature and the resulting degree of fuel rod failure in a LOCA.

The paper reviews these specific phenomena which were investigated under the FEBA [1] and REBEKA [2] programs performed by the Karlsruhe Nuclear Research Center.

REFLOOD HEAT TRANSFER ABOVE THE QUENCH FRONT

During the reflooding phase in a PWR with cold leg injection the emergency core cooling systems inject the cooling water via the downcomer into the lower plenum of the reactor pressure vessel from where it rises into the fuel elements. However, immediate rewetting of the fuel rod claddings can not be expected, since the cladding temperatures are above the Leidenfrost temperature. They will be cooled initially by a steam flow carrying water droplets.

Heat transfer between the fuel rods and the steam-water droplet mixture takes place by convection from the claddings to the steam and from the steam to the water droplets and by radiation from the claddings to the steam and water droplets. It has been shown that the heat is transfered almost exclusively by convection [3]. Since the heat transfer from the cladding tube wall to the steam is substantially higher than from the steam to the water droplets, a thermodynamic non-equilibrium is established in the two-phase flow during the reflooding phase, i.e. the steam is superheated along the coolant channel. Figure 2 shows as an example that during dispersed flow cooling, when the peak cladding temperature develops, steam superheating up to approximately 500 K was measured.

EFFECT OF GRID SPACERS ON REFLOOD HEAT TRANSFER

Grid spacers are structural components in the fuel elements which support the individual fuel rods at a prescribed rod-to-rod pitch. All fuel elements are equipped with grids at the same axial elevations across the reactor core. Therefore, they represent coplanar blockages without bypass.

In spite of the small flow blockage of about 20 %, depending on the design, the grid spacers produce a turbulence enhancement of the continuous steam phase and cause droplet breakup due to the impact on the grid straps. These effects result in a substantial enhancement of heat transfer downstream of the grids, mainly in the early period of the reflood phase when the flow is a two-phase, dispersed, non-equilibrium flow.

A systematic investigation of heat transfer enhancement due to droplet breakup at grids was performed by S.L. Lee et al. [4] using a special Laser-Doppler anemometry in an air-water droplet dispersion. Figure 3 reveals that a grid spacer intercepts large droplets and transforms them into a large number of

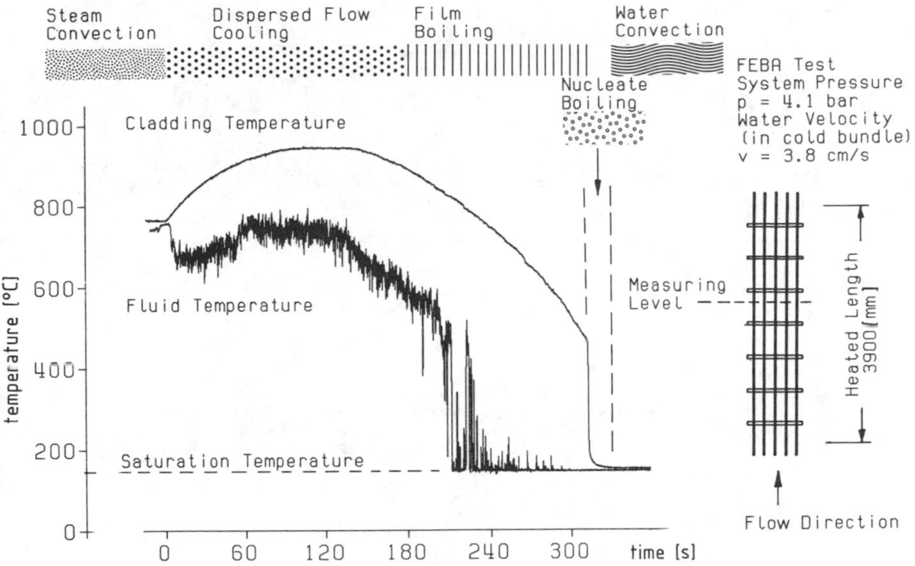

FIGURE 2. Heat transfer regimes, fluid and cladding temperatures during reflooding (FEBA)

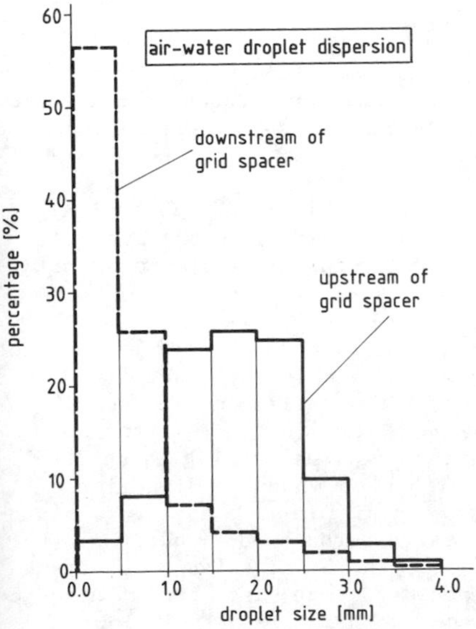

FIGURE 3. Droplet size distribution at grid spacers.

smaller droplets. It has been found that regardless of the initial mean droplet size upstream of the grid, the mean droplet size downstream of the grid takes a stabilized value of well below 0.5 mm. These small droplets, due to their large surface area to mass ratio, serve as a very effective evaporative cooling and as a heat sink for the highly superheated steam. The turbulence enhancing effect of the grid spacers gives rise to intensive mixing of the water droplets with the superheated steam and, consequently, to a reduced degree of steam superheating downstream of each grid spacer. On the way to the next grid spacer in the direction of flow the degree of superheating increases again which leads to the development of an axial temperature profile between two grid spacers.

In the FEBA tests the thermal-hydraulic effect of grid spacers during reflood was investigated. Examples of the axial cladding temperature behavior with and without the midplane grid are shown in Figure 4 for different periods of time after initiation of reflooding. The temperature plots reveal that the presence of the grid results in an enhanced heat transfer downstream of the grid and leads to an axial cladding temperature profile with a local maximum close to the next grid in flow direction.

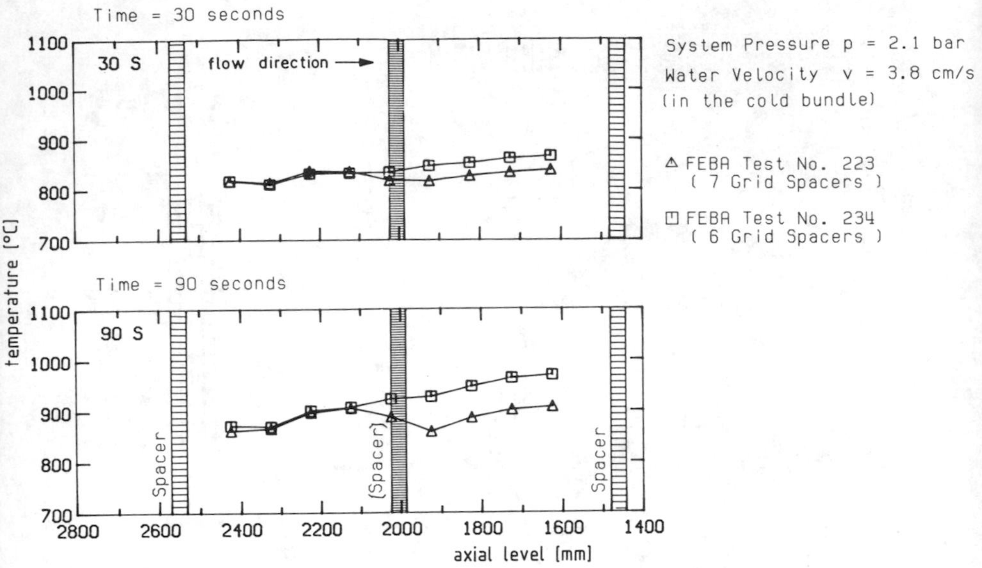

FIGURE 4. Influence of a grid spacer on the axial cladding temperature profile (5x5 FEBA rod bundle)

CLADDING DEFORMATION AND FLOW BLOCKAGE DURING REFLOODING

The REBEKA tests served the primary purpose of investigating the influence of thermal-hydraulics on cladding tube deformation during emergency cooling. The tests were performed under representative emergency cooling conditions with bundles accommodating up to 7x7 rods of full length.

The flow blockage in a fuel element caused by ballooned cladding tubes is determined by the maximum circumferential strain of the ballooned tubes, the axial extension of the ballooning and the axial displacement of the bursts between two grid spacers.

It was proved that heat transfer coefficients greater than 50 W/m^2K, typical of the reflooding phase in a LOCA, cause temperature differences on the circumference of the claddings and limit the mean burst strain to values of approximately 50 % [5].

The axial displacement of the burst points is essentially determined by the fact whether the direction of coolant flow in the fuel element remains unchanged during cladding deformation or whether it undergoes flow reversal during refill and/or reflood. The temperature maximum of the axial cladding temperature profile between the grid spacers as described in the previous section is shifted in flow direction and changes its axial position with time when the direction of flow is reversed. Since the individual rods develop different temperature histories during reflood due to given inhomogeneities in a rod bundle, burst of the claddings occurs at different times. Both effects, the axial shifting of the maximum of the temperature profile with time and different burst times influence the axial displacement of the bursts and together with the relatively low strains prevent high flow blockages.

It has been found in the REBEKA 5 test that reversed flow of steam and water

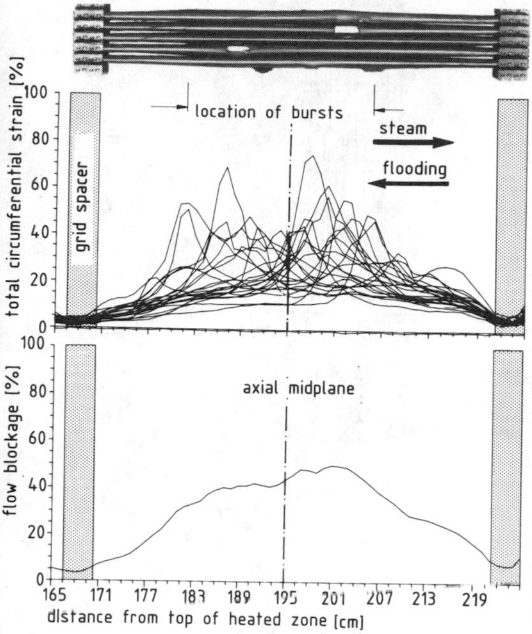

FIGURE 5. Cladding deformation and flow blockage under reversed flow (REBEKA 5)

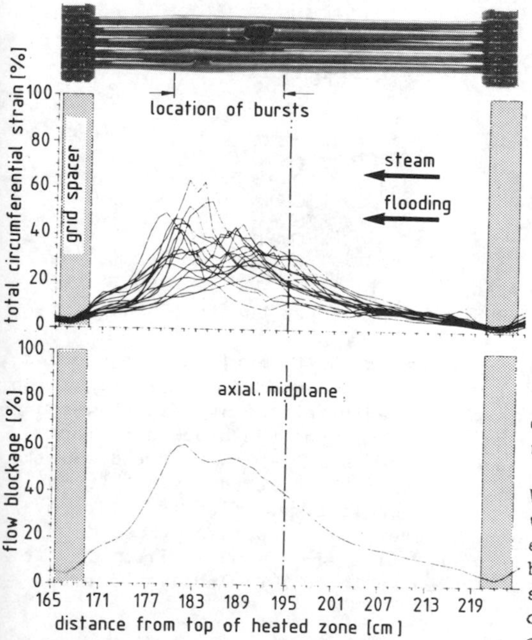

FIGURE 6. Cladding deformation and flow blockage under unidirectional flow (REBEKA 6)

from the refilling to the reflooding phases results in a maximum flow blockage of about 50 %. This situation illustrated in Figure 5 is typical of a PWR with combined injection of emergency cooling water both into the cold leg and into the hot leg [6]. Unidirectional coolant flow during the refilling and reflooding phases results in a higher flow blockage of approx. 60 % which is more typical of a PWR with cold leg injection only. Figure 6 shows the cladding deformation and flow blockage between the inner grid spacers of REBEKA 6 test. The plotted curves make evident the effects of the grid spacers and of the unidirectional flow on cladding deformation.

In the REBEKA 7 test which was performed also under unidirectional flow maximum rod-to-rod interaction and cladding deformation developed. This test resulted in the highest possible flow blockage of 66 %.

EFFECT OF FLOW BLOCKAGES ON REFLOOD HEAT TRANSFER

Flow blockages produced by ballooned fuel rod claddings change the cooling mechanism and cause two counteracting effects on the local reflood heat transfer:

- A flow bypass effect, which reduces the coolant mass flow in the blocked region and, consequently, decreases the local heat transfer.
- A flow blockage effect, which can lead to turbulence enhancement, water droplet dispersion and evaporation, thus increasing local heat transfer.

It depends on a number of boundary conditions which of these effects is more dominant.

Within the FEBA program flooding tests with forced feed were performed using electrical heater rods without gap between filler material and stainless steel claddings. Ballooned fuel rod claddings were simulated by coplanar conical sleeves mounted on the outer surface of the heater rods. Figure 7 shows cladding temperature transients in the blocked and the bypass region

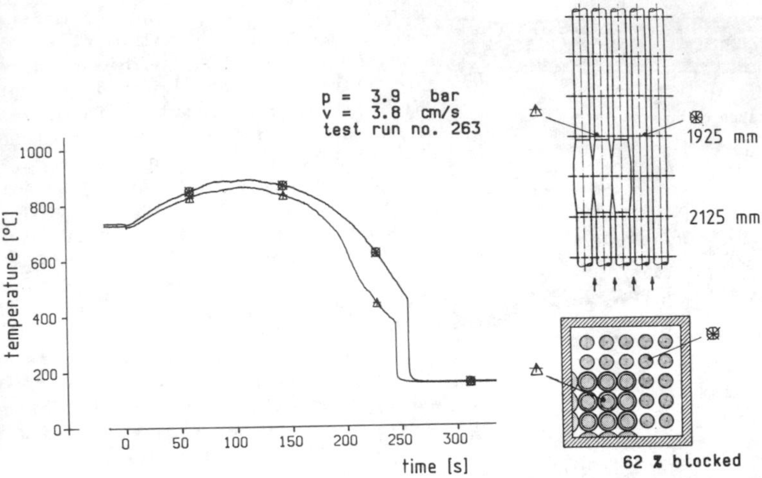

FIGURE 7. Cladding temperatures in a 62 % partly blocked bundle (FEBA)

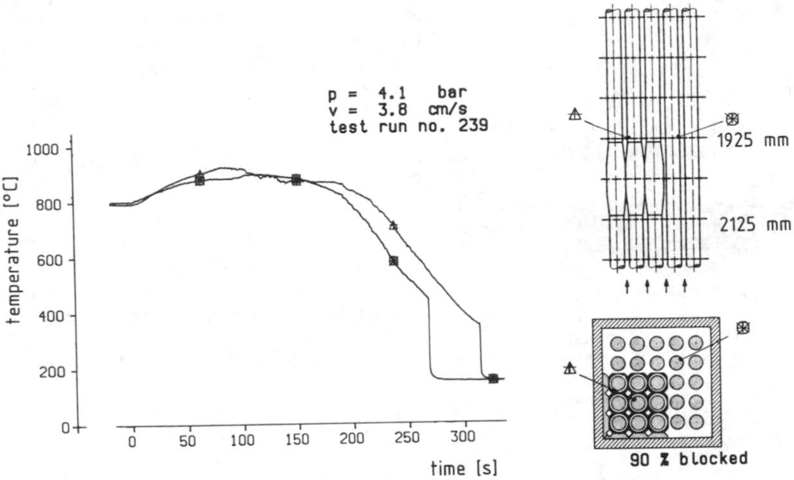

FIGURE 8. Cladding temperatures in a 90 % partly blocked bundle (FEBA)

for a blockage ratio of 62 % in the blocked region. It is evident from the diagram that under the given conditions the effect of water droplet breakup, which improves heat transfer, overcompensates the degrading effect of mass flow reduction resulting in the cladding temperature downstream of the blocked region being somewhat lower compared to that in the unblocked region. Figure 8 shows a corresponding plot for a blockage ratio of 90 % in the blocked region. It makes evident that under these severe conditions the coolant mass flow reduction overshadows the two-phase enhancement effect. However, the temperature rise downstream of the blockage and the delay in quench time are moderate. From this it can be concluded that in fuel elements blocked up to 90 % coolability can be maintained in a LOCA.

In the FEBA tests with blocked bundles solid electrical heater rods were used. Recent reflood tests performed within the SEFLEX-program have shown that the results obtained with solid heater rods are conservative compared to results from tests with fuel rod simulators with a gap between pellets and cladding similar to nuclear fuel rods. The main difference is due to the decoupling effect of the gap between the fuel pellets and the claddings. This effect lowers the maximum cladding temperatures and shortens the quench times significantly [7].

EFFECT OF BURST CLADDING TUBES ON REFLOOD HEAT TRANSFER

The REBEKA tests have shown that ballooned and burst Zircaloy cladding tubes, respectively, improve the coolability compared with non-deformed cladding tubes. In case of burst cladding tubes burst lips penetrate into the coolant channels and the fission gas has escaped from the fuel rod. Steam with its much poorer heat conductivity enters the fuel rod through the burst and decouples the cladding from its heat source. The two-phase coolant flow is now capable of cooling down to quench temperature the cladding at the point of burst very fast and to quench it. A new quench front begins to propagate from the burst point. The regular quench front of the undisturbed geometry may be at a much lower level at that point in time. Two new secondary quench fronts propagate from the burst. One of them moves upwards, the other downwards opposing to the direction of flow. The quench front moving upwards reaches the upper rod end earlier than the regular quench front in a non-deformed fuel element.

For REBEKA 6 with a maximum coolant channel blockage of 60 % Figure 9 shows temperature plots for Zircaloy cladding tubes measured at different axial positions between the two central grid spacers. It can be recognized that the quench front propagates faster in the direction of flow than in the opposed direction. The regular quench front of a non-deformed Zircaloy cladding which is not influenced by premature quenching of adjacent burst rods attains the axial position of 1850 mm only 135 s later than the secondary quench front for the burst rod no. 29.

These results prove that the hot zones in a fuel element accommodating burst claddings are cooled down very quickly.

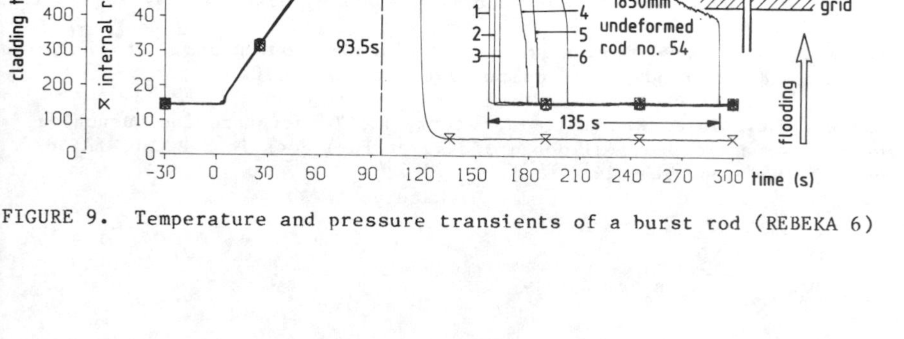

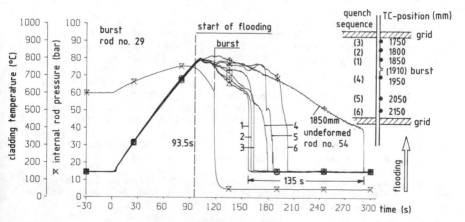

FIGURE 9. Temperature and pressure transients of a burst rod (REBEKA 6)

SUMMARY AND CONCLUSION

The essential results elaborated under the FEBA and REBEKA programs on reflood heat transfer in PWR fuel rod bundles deformed in a LOCA can be summarized as follows:

- Steam superheating in the two-phase flow up to approximately 500 K develops during reflood heat transfer.

- Grid spacers promote heat transfer and reduce maximum cladding temperature.

- Thermal-hydraulics during reflooding limits clad ballooning and the resulting maximum flow blockage to values less than 70 %.

- Coolability in a blocked fuel element can be maintained up to a flow blockage of approx. 90 %.

- Burst cladding tubes result in early quenching and improve coolability.

All these effects of reflood heat transfer indicate that the safety margin for judging the coolability in a LOCA of a PWR is higher than predicted by most of the computer codes.

REFERENCES

1. Ihle, P., Rust, K., FEBA-Flooding Experiments with Blocked Arrays, Evaluation Report, KfK 3657, March 1984.

2. Erbacher, F.J., Neitzel, H.J., Wiehr, K., Brennstabverhalten beim Kühlmittelverluststörfall eines Druckwasserreaktors - Ergebnisse des REBEKA-Programms, KfK-Nachrichten, Jahrgang 16, Heft 2/84, pp. 79-86.

3. Gaballah, I., Ein Beitrag zur theoretischen Untersuchung der Zweiphasenströmung mit Phasenwechsel und Wärmezufuhr in einem Kühlkanal eines LWR-Brennstab-Bündels beim Kühlmittelverluststörfall, KfK 2657, September 1978.

4. Lee, S.L., Rob, K., Cho, S.K., LDA Measurement of Mist Flow across Grid Spacer Plate Important in Loss-of-Coolant Accident Reflood of Pressurized Water Nuclear Reactor, Proceedings of International Symposium on Application of Laser-Doppler Anemometry to Fluid Mechanics, Lisbon, Portugal, 1982.

5. Erbacher, F.J., Neitzel, H.J., Wiehr, K., Effects of Thermohydraulics on Clad Ballooning, Flow Blockage and Coolability in a LOCA, OECD-NEA-CSNI/IAEA Specialists' Meeting on Water Reactor Fuel Safety and Fission Product Release in Off-normal and Accident Conditions, Risø, Denmark, May 16-20, 1983.

6. Erbacher, F.J., Interaction between Fuel Clad Ballooning and Thermal-Hydraulics in a LOCA, KfK 3880/1, December 1984, pp. 299-310.

7. Erbacher, F.J., Ihle, P., Rust, K., Wiehr, K., Temperature and Quenching Behavior of Undeformed, Ballooned and Burst Fuel Rods in a LOCA. KfK 3880/1, December 1984, pp. 516-524.

Some Aspects of Australian Research into Critical Heat Flux and Post Dryout Phenomena

W. J. GREEN
AAEC Research Establishment
Lucas Heights Research Laboratories
Private Mail Bag
Sutherland, Australia

ABSTRACT

This paper provides a brief review of some aspects of the research investigations into boiling crisis and post crisis heat transfer which have been performed at the AAEC Research Laboratories using a pressurised Freon-12 heat transfer facility as the basic experimental tool. The paper also includes information on recent work being performed on rewetting phenomena at low mass fluxes.

INTRODUCTION

More than 15 years ago the Australian Atomic Energy Commission became actively engaged in experimental research using pressurised Freon-12 to investigate the phenomenon of Critical Heat Flux (CHF) under flow conditions. This approach originated on the basis of preliminary investigations conducted by the UKAEA's Atomic Energy Establishment at Winfrith into the feasibility of modelling boiling high pressure water by using Freon-12. Collaboration between the AAEC and the UKAEA led to the construction of a pressurised Freon-12 heat transfer facility at the AAEC's Research Laboratories at Lucas Heights and to a test program to determine CHF data for uniformly heated vertical round tubes, annuli and a rod cluster (1-5). These early experiments enabled information to be gained on the effects of coolant conditions (mass flux, pressure, enthalpy) and on some effects of flow channel geometry, including in particular, spacer elements in annular configurations (6,7). Table I summarises the conditions investigated.

In more recent years, attention has been diverted away from studies of CHF at equivalent reactor operating conditions to ones associated with the low mass fluxes that might occur in loss of coolant accidents (LOCAs).

When the CHF data obtained in the AAEC Freon-12 rig for uniformly heated round tubes were compared with published CHF correlations (8,9), substantial discrepancies were found particularly at low mass fluxes (less than 500 kg $s^{-1}m^{-2}$) (10). Effort was directed therefore toward developing more general, accurate CHF correlations which would describe the experimental results obtained from the AAEC Freon-12 facility and would be suitable for inclusion in computer codes used to calculate thermal transients. These studies, which have been based upon dimensional analysis have led to the formulation of two general correlations which enable CHF predictions to be made for any fluid (including water and Freon-12) at either high or low mass fluxes.

Because of the accuracy and wide range of applicability of the correlations developed, consideration has been given to the significance and nature of the dimensionless groups in these correlations in an effort to better understand the boiling crisis phenomenon. Facets of this work are discussed in the paper.

Table I AAEC CHF experiments with vertical upflow

Geometry Tested	Shroud Tube Bore mm	Inner Tube Outside Diameter mm	Heated Length mm	Coolant Pressure MPa	Exit Quality	Mass Flux kg $s^{-1}m^{-2}$	Heating Distribution	Reference
Round tube	16.1	N/A	2860	0.89 → 1.60	0.035 → 1	200 → 4100	Uniform	[2]
Round tubes	15.3 16.1 21.5	N/A	2850 → 3940	0.96 → 1.05	0.007 → 0.78	450 → 3850	Uniform	[3]
Round tubes	8.5 16.8 21.3	N/A	2870 → 3700	0.90 → 1.32	0.19 → 0.86	380 → 2800	Uniform and axially non-uniform	[10]
Annuli	9.5 → 26.6	8.0 → 18.6	457 → 4280	1.04	0.04 → 1	660 → 4160	(1) Shroud only, (2) Inner tube only, or (3) Both shroud and inner tube; Uniform in each case, also chopped cosine axially for case (2)	[4]
Annuli	21.0 22.1 22.7	14.4 15.9	1830 2740	0.83 → 1.65	0.002 → 0.76	670 → 4200	Heated shroud or inner tube - axially uniform	[5]
19-rod cluster	114	15.9	1250	1.05 → 1.69	0.49 → 1	47 → 260	Axially uniform - shroud not heated	[5]

Apart from CHF studies, in the late 1970s, a sensitivity analysis (11) of the blowdown phase of a hypothetical LOCA in a pressurised water reactor showed that maximum surface temperatures predicted for the cladding of a nuclear fuel element at relatively low coolant mass fluxes are highly dependent on the heat transfer relationships assumed to apply for the onset of dryout (CHF) and during the subsequent post dryout conditions (see figure 1). As a consequence, experimental investigations were commenced using Freon-12 as the coolant with the aim of examining post crisis heat transfer. The first program of tests considered heat transfer under drying out conditions and used electrical power transients applied either to the whole or part length of a heated tube. Later, heat transfer under rewetting conditions was investigated using a high thermal capacity test section situated downstream of a preheater tube. Aspects of this work and the latest findings are also considered in this paper.

2. CRITICAL HEAT FLUX CORRELATIONS

Although there have been many experimental (8,9,12,13) and analytical (14,15, 16) investigations aimed at correlating, understanding and predicting the onset of dryout (or CHF) in a round tube, in general, CHF correlations are only applicable to coolant conditions for which they were developed. An exception is Ahmad (17) who developed his scaling law concept to widen its application to different fluids

In attempts to develop more general CHF correlations, several investigators (18,19,20) have used dimensional analyses. Each of these approaches has assumed that the correlations involve non-dimensional groups related to one another in a simple product form and incorporate the overall heated length and inlet subcooling as independent variables.

At the AAEC, one approach to formulating a general, accurate CHF correlation has been to use a dimensional analysis technique but with attention being paid to the possibility that dimensionless groups may be interrelated with other dimensionless groups, not only in a simple product relationship, but also as power functio of one another, e.g. if ϕ, A, B and C are dimensionless groups then there may be a correlation having the form $\phi \propto A^x B^y C^z$ where the indices x and y could themselve be functions of the dimensionless group C.

A further consideration taken into account when developing such a CHF correla-ation has been that, since computer codes for analysing transients in a water-cooled reactor require correlations based on local conditions, the local quality and boiling length should appear in the final correlation. As quality is a dimen-sionless quantity it was not included directly in the dimensional analysis but was introduced via the mean local density of the coolant. The underlying basic dimensional analysis used to formulate a correlation was as follows:

Considering CHF as a function of the densities, velocities, viscosities, spec-ific heats and thermal conductivities of both the vapour and liquid at satura-tion conditions; together with the diameter of the tube, saturation length, sur-face tension, latent heat of vaporisation, total mass flux, and a temperature diff-erence (as suggested by Griffith (19)); it can be shown, by dimensional analysis, that a critical heat flux number, $\phi D/\mu_v\lambda$, is a function of the following groups:

$$\left\{\frac{\rho VD}{\mu}\right\}_v, \ \left\{\frac{\rho VD}{\mu}\right\}_\ell, \ \left\{\frac{C_p\mu}{k}\right\}_v, \ \frac{\sigma}{\rho_v D\lambda}, \ \frac{\rho_\ell}{\rho_v}, \ \frac{\mu_\ell}{\mu_v}, \ \frac{k_\ell}{k_v}, \ \frac{C_{p\ell}}{C_{p_v}}, \ \frac{GD}{\mu_v}, \ \frac{k_v\Delta T}{\mu_v\lambda}, \ \frac{L_s}{D} .$$

Of these groups, $k_v\Delta T/\mu_v\lambda$ was eliminated as it did not appear to be of prime significance and is difficult to evaluate accurately because of the small values of ΔT and the limited accuracy of wall temperature recording.

If it is assumed that there is thermal equilibrium between the vapour and liquid phases of the coolant and that the flow is homogeneous with the slip ratio qual to unity (i.e. $V_v = V_\ell$) the vapour and liquid Reynolds numbers may be ex-ressed as follows:

$$\left\{\frac{\rho VD}{\mu}\right\}_v = \frac{GD}{\mu_v}\left\{X + (1-X)\frac{\rho_v}{\rho_\ell}\right\}$$

nd

$$\left\{\frac{\rho VD}{\mu}\right\}_\ell = \frac{GD}{\mu_\ell}\left\{\frac{\rho_\ell}{\rho_v}X + 1-X\right\} = \left\{\frac{\rho VD}{\mu}\right\}_v\left(\frac{\mu_v}{\mu_\ell}\right)\left(\frac{\rho_\ell}{\rho_v}\right).$$

Since viscosity and density ratios occur as dimensionless groups, the vapour nd liquid Reynolds numbers defined above are effectively interchangeable, hence nly one need be considered in formulating a correlation. Also, since the vapour eynolds number is a derivative form of the overall Reynolds number, GD/μ_v, it is ot expected that both would appear as prime dimensionless groups in any correla-ion at the same time.

The CHF number can therefore be considered as a function of eight dimension-ess groups, namely vapour Reynolds number, vapour Prandtl number, four coolant roperty ratios, a surface tension number, and a dimensionless saturation length, .e.

$$\frac{\phi D}{\mu_v\lambda} = f\left\{\left\{\frac{\rho VD}{\mu}\right\}_v, \ \left\{\frac{C_p\mu}{k}\right\}_v, \ \frac{\sigma}{\rho_v D\lambda}, \ \frac{\rho_\ell}{\rho_v}, \ \frac{\mu_\ell}{\mu_v}, \ \frac{k_\ell}{k_v}, \ \frac{C_{p\ell}}{C_{pv}}, \ \frac{L_s}{D}\right\}.$$

2.1 Development of Correlations

Originally, a correlation was developed (21) for Freon-12 data which covered a de range of coolant conditions including mass flux as low as 380 kg s^{-1} m^{-2}. The

›rrelation however, was found to be so general in form that only minor modifica- ion was required for it to be applicable to water data (22). Subsequently it was lso compared without modification with nitrogen data (23). The procedure by which e correlation was developed is described in detail in references 21, 23 and 24 id its form is as follows:

$$\frac{\phi D}{\mu_v \lambda} = 9 \times 10^{-5}\, Re_v^{\,n} \left\{\frac{\rho_\ell}{\rho_v}\right\}^m \sigma_N^{\,p}\, Pr_v^{\,w}\, f(L_s/D) \left\{1 + \delta\right\}\left\{1 - \delta_1\right\},$$

where

$$Re_v = \frac{DG}{\mu_v}\left\{x + (1-x)\,\rho_v/\rho_\ell\right\} \qquad \sigma_N = \sigma/(D\lambda\rho_v)$$

$$f(L_s/D) = \exp\left\{3.83\, e^{-0.00396\, L_s/D} - 0.00055\, L_s/D\right\}$$

$$n = 1 - e^{-0.0067\, L_s/D}, \qquad p = -0.5\,(0.15 + e^{-0.007\, L_s D}),$$

$$m = 0.1 + e^{-0.007\, L_s D}, \qquad w = -\,(0.21 + 0.55\, e^{-0.007\, L_s D}),$$

$$\delta = \exp\left\{-0.14 \times 10^8 \sigma_N - 0.02\,(L/D)\, Pr_v\right\}, \qquad \delta_1 = 0.75 \exp\left\{-B \cdot \frac{L}{D} \cdot \frac{G\sigma}{\rho_\ell \mu_\ell \lambda}\right\},$$

$$\text{and } B = 130.5 \exp\left\{5.0 \exp(-0.02\, L/D)\right\}.$$

This correlation was found to be accurate and applicable over a wide range of oolant conditions and tube dimensions. For CHF qualities greater than 0.1 and ass fluxes for Freon-12 and pressurised water greater than ~ 300 kg s^{-1} m^{-2}, the ean ratio of calculated to experimental CHF values was found to be;

(i) 0.992 with a r.m.s. error of 3.3% for 1760 sets of Freon-12 data,

(ii) between 0.97 and 1.03, depending upon the data sets considered, with orresponding r.m.s. errors ranging between 2.0% and 10.9% for over 7000 sets of ressurised water data,

nd (iii) 0.96 with a r.m.s. error of 9.5% for 48 sets of liquid nitrogen data at ass fluxes in the range 220-1800 kg $m^{-2}s^{-1}$.

For mass fluxes less than ~ 300 kg s^{-1} m^{-2} the correlation was found to be less ccurate. It was considered therefore that an intensive examination should be made f low flow data. This led to the formulation of a second correlation (25) which as been verified against the limited experimental data available at low flow ates. This correlation is

$$\frac{\phi D}{\mu_v \lambda} = 0.25\, Re_v \left\{\frac{D}{L_s}\right\}\left\{\frac{1}{1+\delta_2}\right\}$$

here

$$\delta_2 = 0.046\,(L/D)^{1.12}\,(D/L_s)^{1.65} \exp\{25 \times 10^3\, LFN - 0.038\, \rho_\ell/\rho_v\} \qquad LFN = G\sigma/(\rho_\ell \mu_\ell \lambda).$$

When solved iteratively in conjunction with the heat balance equation, this orrelation gave an overall mean ratio of predicted to experimental CHF of 0.986 ith a r.m.s. error of 7.0% for the available low flow data sets examined.

The analysis also found that analogous to the change from turbulent to laminar low in single phase fluids, the boundary between the high and low mass flux egimes is controlled by a complex dimensionless parameter namely δ_1 (specified in he high flow CHF correlation). As can be seen from figure 2, the transition from igh to low flow is marked by values of δ_1 becoming greater than 0.07.

2.2 Physical Significance of the High Flow CHF Correlation

Because of the accuracy and ability of the high flow CHF correlation to des-
:ribe a wide range of experimental data, it was reasoned that closer examination of
it might provide a more fundamental understanding of the boiling crisis phenomenon.
Theoretical work was therefore conducted to examine:

(i) the relative influences of each of the dimensionless groups;
(ii) physical interpretations of the various groups; and
(iii) models which may explain the form of the correlation.

Arising from these studies (26) it was found that (see figure 3) the correla-
ion is consistent with a region in which the relationship between Critical Heat
'lux and critical quality is characterised by quality being apparently independent
f CHF. This phenomenon is sometimes termed boiling crisis of the second kind and
s often postulated as indicating an unusual and separate flow regime which needs a
pecial correlation to describe it. Such assumptions are unnecessary for the
orrelation described in this work.

Consideration of three mechanisms, namely (i) the formation of a bubble; (ii)
he turbulent interchange of fluid to and from a heated surface; and (iii) the
otion that bubbles attached to a heated wall are equivalent to wall roughness,
rovided a theoretical basis for explaining the principal dimensionless groups in
he correlation. These theoretical considerations also indicated that a dimension-
ess group namely $[T_s k_\ell / \mu_v \lambda]$, which is similar to $k_v \Delta T / \mu_v \lambda$, may need to be re-
onsidered for inclusion in correlating formulae.

. POST DRYOUT HEAT TRANSFER

The AAEC Freon-12 experimental program to investigate post-dryout phenomena at
ow flow rates in round tubes has been performed in three stages. Initially, an
xperimental technique was used in which the uniform heat input to the full length
f a stainless steel tube (by direct electrical resistance heating of the wall) was
ncreased until dryout was induced over the exit region of the tube (27).

Later, experiments (28) were performed using a short, independently heated
ownstream section of the same tube, the upstream section of which was maintained
: constant power. In these experiments the power to the short exit section was
icreased in steps until this section reached dryout. For both of these experi-
ental programs the objective was to study crisis and post crisis phenomena during
ryout. Details of the tubular test sections and flow conditions investigated in
iese experiments are given in references (27) and (28).

More recently experiments have been performed to ascertain post crisis heat
ransfer characteristics under slow rewetting conditions. These experiments have
ilised high thermal capacity test sections appended to the exit of a uniformly
eated tube.

3.1 Post Dryout Results Obtained under Dryout Conditions

The experimental data obtained from each of these test programs are in the
rm of wall temperature transients. Such data have been analysed by matching
corded responses with those calculated using the thermal hydraulic transients
de THETRAN (29). To do this it is necessary to formulate a schematic represen-
tion of the boiling curve shown in figure 1. The representation which is used is
own in figure 4. Agreement between calculated and observed tube temperature
sponses was then achieved by adjusting the values of arbitrary factors control-
ng the calculated heat transfer rates at crisis and during the four post crisis
gions shown in figure 4. Five factors were utilised to vary the calculated heat
uxes of these four post crisis regions. They were as follows:

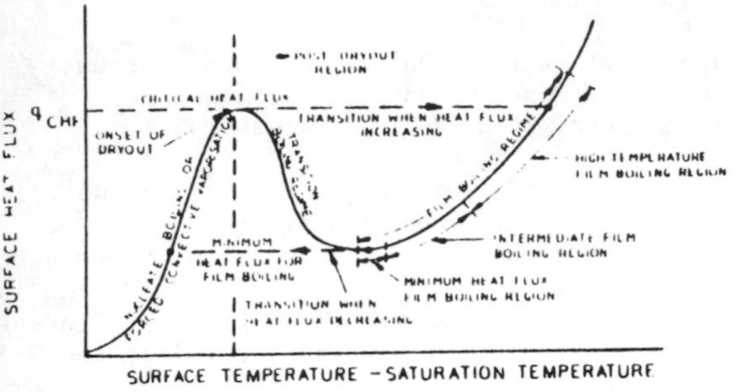

FIG 1 SIMPLIFIED REPRESENTATION OF BOILING CURVE

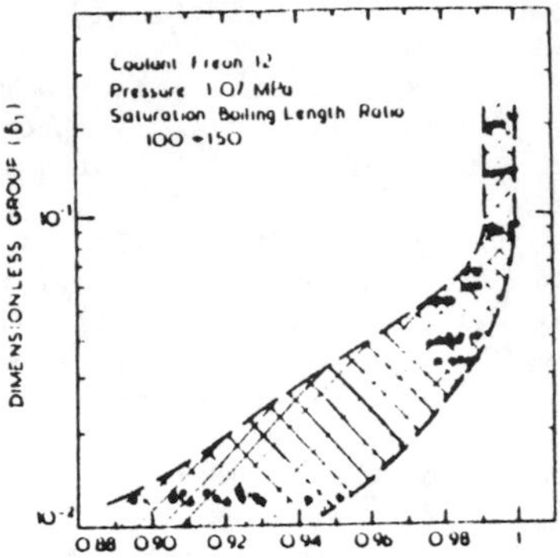

RATIO OF CHF CALCULATED FROM LOW FLOW CORRLN TO EXPTL CHF

FIG 2 TRANSITION FROM LOW TO HIGH FLOW REGIME

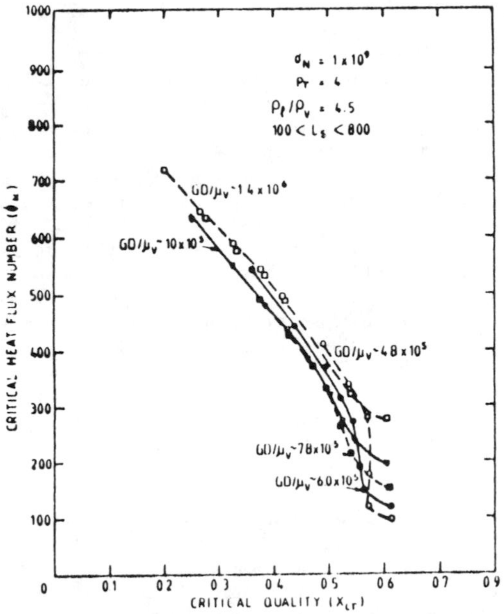

FIG 3 PREDICTED RELATIONSHIPS BETWEEN CHF AND QUALITY FOR WATER AT ~ 17.5 MPa.

. the negative slope of the relationship between q and ΔT in the transition regime
. the minimum-heat-flux level;
. the magnitude of ΔT in the film boiling regime at which the heat flux begins to increase with further increases in wall temperature;
. the positive slope of the relationship between q and ΔT in the intermediate film boiling region (see figure 4); and
. a constant factor applied to the heat transfer coefficient evaluated from one of the film boiling correlation options included in THETRAN.

As a result of this analytic approach it was found that the relative extent of each region can vary substantially; in particular the minimum heat flux region may cover a very wide or a very narrow range of surface temperatures. Other results which have been found for Freon-12 are that:

(1) experimental data for the minimum heat flux region may be correlated by

$$q_{min} = A \left\{DG/\mu_v\right\}^{0.54}$$

where A is a function of the properties of the coolant (see figure 5). This correlation was found to be valid provided that the minimum heat flux is less than 70% of the critical heat flux.

(2) the data for the intermediate film boiling region may be represented by

$$q/q_{fb} = 1 + 3.65 \exp\left[11.0\ (\Delta T/T_s)\ (L_s/L)\right]$$

where q_{fb} is the surface heat flux corresponding to the high temperature film boiling region, ΔT is the temperature difference between the wall and the saturation temperature (T_s), L_s is the boiling length, and L is the total heated length,

and (3) the data for the high temperature film boiling region are correlated by

$$q_{fb} = 0.85 \exp\left(1.33\ D/L_{DO} - 0.15\ L/L_s\right) q_{DR}$$

where L_{DO} is the length of tube in dryout and

$$q_{DR} = 0.023 \frac{k_v}{D} Pr_v^{0.4} \left\{\frac{DG}{\mu_v}\right\}^{0.8} \left[x+(1-x)\rho_v/\rho_\ell\right]^{0.8} (T_w - T_s).$$

3.2 Post Dryout Results under Rewetting Conditions

In order to study the post dryout regimes existing during rewetting, experiments are currently being performed using high thermal capacity (H.T.C.) test sections situated at the exit of the long uniformly heated stainless steel tubes described earlier. Thus instead of measuring the rapid temperature responses which occurred in the experiments using thin-wall tubes, the post dryout heat transfer characteristics are investigated whilst the test sections slowly cool from post dryout to rewetted conditions. Because this experimental program is current and is still in the process of being fully reported, the following paragraphs describe some of the more salient parts of the work.

3.2.1 Experimental equipment and test procedure

Rewetting under slowly decreasing temperature transient conditions with high thermal capacity test sections have been investigated by experimenters in the UK (30,31) using water at close to atmospheric pressure. When designing the test sections shown in figure 6, several factors had to be considered. First, a decision had to be made on whether the heater system of the H.T.C. test section should be within or outside the pressure casing. An external configuration would make heat flows including heat losses extremely difficult to assess and, as a consequence,

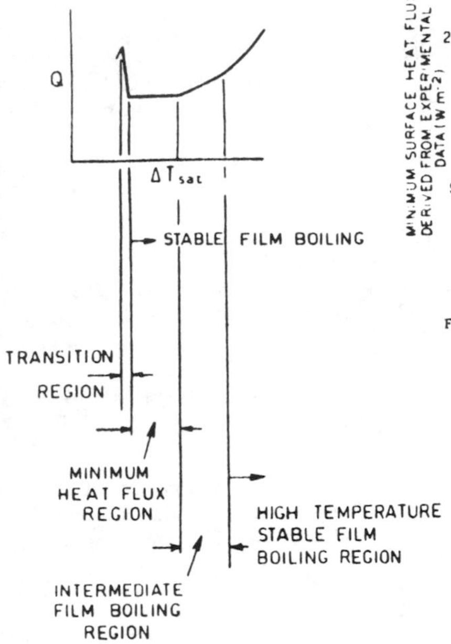

FIG 4 NOTIONAL REPRESENTATION OF POST CRISIS REGIONS OF BOILING CURVE

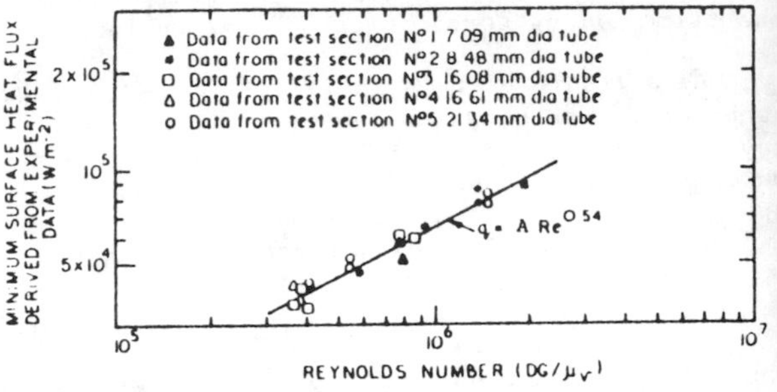

FIG 5 RELATIONSHIP BETWEEN MINIMUM HEAT FLUX OF FILM BOILING AND REYNOLDS NUMBER

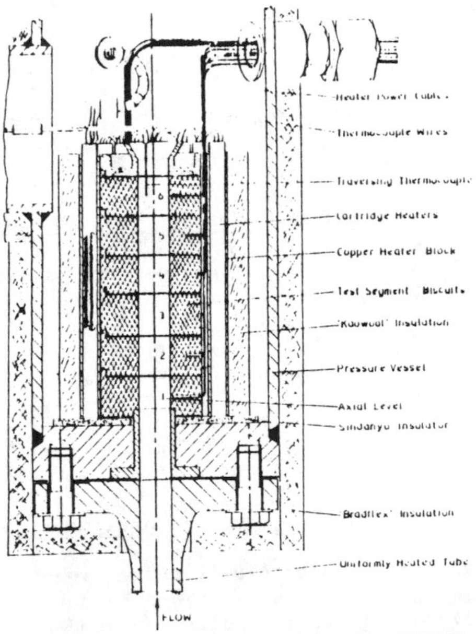

FIG 6 HIGH THERMAL CAPACITY TEST SECTION

the experimental results would be less accurate. If the heater assembly were loca-ted within the pressure casing, the heat loss problems would lessen, but the syste would then function in a hostile environment of liquid and vaporised Freon-12. As can be seen in fig. 6, the heating system was placed inside the pressure vessel bu the test program has suffered from delays arising from a series of heater failures

Although attention was given to minimising external heat losses from the H.T.C test section, nevertheless some losses occurred and these needed to be determined. Each series of flow tests was therefore preceded by static tests in which the test assembly, comprising the H.T.C. test section and the long stainless steel preheate was isolated from the main loop, and the assembly filled with Freon-12 vapour.

Three copper test sections having internal diameters of 8.5, 16.6, 21.3 mm hav been tested. The range of coolant conditions investigated was:

pressure:	0.76 → 1.31 MPa
flow rate:	0.1 → 0.25 m^3/h
flow quality at inlet to the high thermal capacity test section:	0.32 → 0.80.

Before matching calculated temperature responses with those observed during a flow test, information gained from the static tests was analysed to determine:

(1) the external heat losses from the test sections, and
(2) the equivalent thermal capacity of the heater system which was a complex composition of copper, stainless steel, and insulators.

Analysis of the experimental results to date has shown that the minimum film boiling heat flux is strongly dependent upon the local equilibrium quality and mass flux at high coolant qualities and low mass fluxes, but appears to be independent of coolant quality when the coolant quality is low. Overall it would appear that the relationship between the minimum film boiling heat flux, the mass flux and the coolant equilibrium quality is more complex than for the conditions under which a surface is drying out, and that the relationship between surface heat flux and temperature difference $T_w - T_s$ is not the same for dryout and rewetting processes, as is often assumed in analytical codes used to calculate LOCA conditions.

. CONCLUSIONS

Based upon a Freon-12 heat transfer facility, substantial contributions have been possible at the AAEC Research Laboratories in the fields of boiling crisis and post crisis phenomena. These contributions have led to;

(i) the formulation of accurate and wide ranging CHF correlations,
(ii) better understanding of the physical mechanisms that control boiling crisis,
(iii) a unique analytical approach to investigating post dryout regimes,
and (iv) enhancement of knowledge on heat transfer under post dryout conditions.

. REFERENCES

(1) Ilic, V. AAEC/TM632, 1972.
(2) Ilic, V. AAEC/E325, 1974.
(3) Stevens, J.R. and Miles, D.N. AAEC/E506, 1980.
(4) Ilic, V. AAEC/E323, 1974.
(5) Ilic, V. AAEC/E324, 1974.
(6) Ilic, V. and Lawther, K.R. 1st Australasian Conf. on Heat and Mass Transfer, Melbourne, 1973.
(7) Ilic, V. AAEC/E349, 1975.

(8) Bertoletti, S., et al, CISE-99, 1964.
(9) Groeneveld, D.C. AECL-3418, 1969.
(10) Green, W.J. and Stevens, J.R. AAEC/E517, 1981.
(11) Green, W.J. and Lawther, K.R. Nuc. Eng. & Design, Vol. 47, 87-99, 1978.
(12) Merilo, M. and Ahmad, S.Y. AECL-6485 (1979).
(13) Stevens, G.F., Elliott, D.F. and Wood, R.W. AEEW-R321 (1964).
(14) Bowring, R.W. AAEW-R 789, 1972.
(15) Shah, M.M., Int. J. Heat Mass Transfer 22 (1979) 557-568.
(16) Katto, Y., Int. J. Heat Mass Transfer 21 (1978) 1527-1542.
(17) Ahmad, S.Y., AECL-3663, Chalk River, 1971.
(18) Brevi, R. and Cumo, M. CNEN RT/ING(72) 19, 1972.
(19) Griffith, P. WAPD-TM-210 (1959).
(20) Barnett, P.G. AEEW-R134 (1963).
(21) Green, W.J. AAEC/E528 (1981).
(22) Green, W.J. AAEC/E532 and 536 (1982).
(23) Green, W.J. and Lawther, K.R. Proc. 7th Internat. Heat Transfer Conference, Munich, 1982, paper FB18, Vol. 4.
(24) Green, W.J. and Lawther, K.R. Nucl. Engrg. Des. 67 (1981) 13-25.
(25) Green, W.J. Nucl. Engrg. Des. 72 (1982) 381-389.
(26) Green, W.J. and Beattie, D.R.H. Nucl. Engrg. Des. 75 (1982) 33-41.
(27) Green, W.J. and Lawther, K.R. Nucl. Engrg. Des. 55 (1979) 131-144.
(28) Green, W.J. and Lawther, K.R. ANS-ASME Topical Meeting on Nuclear Reactor Thermohydraulics, October 1980, NUREG/CP-0014 p.1092.
(29) Green, W.J. and Jacobs, W.S. THETRAN, AAEC/E507 (1981).
(30) Newbold, F.J., Ralph, J.C. and Ward, J.A. AERE-R-8390 (1976).
(31) Costigan, G., Holmes, A.W. and Ralph, J.C. AERE-R-10579 (1982).

An Analysis of Heat Transfer Based on the Drop Deformation for Nonwetting Impinging Drop

SHIGEAKI INADA and YOSHIKI MIYASAKA
Department of Mechanical Engineering
Gunma University
Gunma 376, Japan

1. INTRODUCTION

Transient heat transfer to a non-wetting impinging drop is a subject of great interest in the fields of nuclear reactor safety, post-dryout dispersion flow and spray cooling. Heat transfer to non-wetting impinging drops was measured and analyzed by Wachters and Westerling [1]. Successful predictions of heat transfer during impact were obtained using a vapor cushion model based on high speed photos of the impinging drop. The disadvantage of Wachters and Westerling's technic is the need for information on drop dynamics which must be obtained from photographs.
Kendall and Rohsenow [2] presented an analysis for saturated drops, in which simple modeling of the drop shape during the impact led to derivation of equations of drop motion using Lagrangian methods and described as a coupling of quasi-steady heat transfer and dynamics.
Heat transfer analysis in this paper is based on this analysis of Kendall and Rohsenow. The temperature distribution between the hot surface and the drop is generally nonlinear because the vertical vapor velocity in the vapor layer exerts influence on heat transfer from the hot surface to the drop. Kendall and Rohsenow considered the nonlinear temperature gradient at the drop bottom as a function of only wall superheat and estimated the vapor generation rate at the bottom. However this temperature gradient must be affected considerably by the vertical velocity of the drop bottom itself since the rate of vapor generation depends on the flow field between the hot surface and the drop. In this analysis,on considering the effect of the gravitational force of the drop, the temperature distribution in the vapor layer was found by taking into account of the time variation of the nonlinear temperature gradient,and furthermore, simple equations adjusted by Weber number, wall superheat and drop diameter were obtained for heat transfer effectiveness and time average heat flux.

2. ANALYSIS OF HEAT TRANSFER BASED ON THE DROP DEFORMATION

2.1 Equation of Motion for the Deforming Drop

When a liquid drop impinges on a very hot solid surface heat is transferred from the surface to the drop and evaporation at the drop bottom can occur fast enough to generate a vapor cushion. The drop deforms to undergo an outward radial liquid film flow on the vapor cushion. The drop restores gradually to the original state by its elasticity due to surface tension and rebounds from

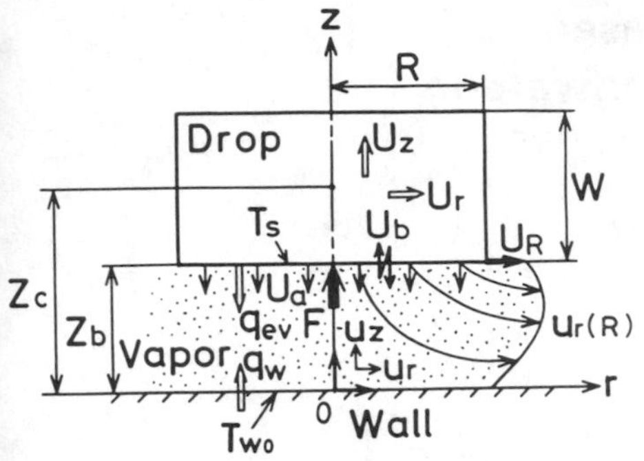

FIGURE 1. Model of deforming drop and vapor flow

the hot wall. In order to derive the equations of drop motion in this deformation process, let the drop geometry throughout impact can be modeled as a right circular cylinder as shown in Fig.1. As the drop deforms, the height W and the radius R change while the drop volume remains constant. If the drop is supposed to have a flat bottom, the location of the bottom of the cylinder Zb can be determined from the coordinate of the center of mass Zc and the cylinder height W, the general equations of motion for two parameter W and Zc are [2],

$$\frac{d^2W}{d\tau^2} = -\frac{6}{\pi D^3\rho_f}\left[\frac{1}{2}F + 2\pi\sigma D\left\{\frac{1}{2}\sqrt{D/6W} - \frac{1}{6}\left(\frac{D}{W}\right)^2\right\} - \frac{\rho_f}{32D}\frac{\pi D^3}{6}\left(\frac{D}{W}\right)^4\left(\frac{dW}{d\tau}\right)^2\right] \Big/ \left\{\frac{1}{12} + \frac{1}{48}\left(\frac{D}{W}\right)^3\right\} \tag{1}$$

$$\frac{d^2z_c}{d\tau^2} = \frac{6}{\pi D^3}\frac{F}{\rho_f} \tag{2.a}$$

Instead of Eq.(2.a),the following Eq.(2), which takes into account of the gravitational effect was used in this analysis, but the effect can be disregarded if the drop diameter is less than 0.5 millimeter.

$$\frac{d^2z_c}{d\tau^2} = \frac{6}{\pi D^3}\frac{F}{\rho_f} - g \tag{2}$$

where g is acceleration of gravity.
Equations (1) and (2) can be solved by 4th order Runge-Kutta methods.

2.2 Estimate of the Force Acted on the Drop Bottom

Consider a quasi-steady vapor flow between the drop and the wall to estimate the pressure rise. The mass continuity equation for the steady uniform density flow between the drop and the wall is:

$$\frac{\partial u_r}{\partial r} + \frac{u_r}{r} + \frac{\partial u_z}{\partial z} = 0 \tag{3}$$

The thickness of the vapor layer under the drop is so small with respect to the radius of the drop bottom that the vertical vapor velocity in the slit can be neglected with respect to the horizontal velocity. Consequently, as for the momentum equation it is sufficient to consider only in the r direction.

$$\rho_g\left(u_r\frac{\partial u_r}{\partial r} + u_z\frac{\partial u_r}{\partial z}\right) = -\frac{\partial P}{\partial r} + \mu_g\frac{\partial^2 u_r}{\partial z^2} \tag{4}$$

If it is assumed that there is no slip either at the hot surface or at the drop bottom, the boundary conditions are:

$u_r=0$, $u_z=0$ at Z=0; u_r=Ur, u_z=-Uo at Z=Zb

Uo is the resultant downward vapor velocity at the drop bottom, that is , the sum of the downward vaporization velocity Ua and the drop bottom velocity Ub in the Z direction. Introducing the deformation velocity Uw, the velocity Ur is replaced as shown by Eq.(5).

$$U_r=-\frac{1}{2}\frac{dW}{d\tau}\left(\frac{r}{W}\right)=-\frac{1}{2}\frac{U_w}{z_b}r \tag{5}$$

$$U_a=\beta_2\cdot\lambda_g\cdot\Delta T_{sat}/(\rho_g\cdot L_g\cdot z_b) \tag{6}$$

$$U_b=dz_c/d\tau-1/2(dW/d\tau) \tag{7}$$

$$U_0=U_a+(-U_b) \tag{8}$$

The external force F that acts on the drop bottom is obtained by integrating Eq.(4).

$$F=\frac{\pi R^4}{4}\left(\frac{\rho_g U_0^2}{z_b^2}I_1-\frac{\mu_g U_0}{2z_b^3}I_2\right) \tag{9}$$

where $I_1=9/10+13/20(U_w/U_0)+1/10(U_w/U_0)^2$, $I_2=-12-6(U_w/U_0)$

Using Eq.(9) and solving Eqs.(1) and (2), W and Zc are determined as functions of time, the initial conditions now being as follows.

W=Wo, dW/dτ=0 ; Zc=Zo, dZc/dτ=-Vo and β_2=1

where Zo is an arbitrary value larger than Wo/2 and Wo can be determined by minimizing the potential energy function, it gives Wo=0.87D.

2.3 Temperature Distribution in the Vapor Layer

Consider a quasi-steady heat flow between the hot surface and the drop. The energy equation is given by

$$u_r\frac{\partial T}{\partial r}+u_z\frac{\partial T}{\partial z}=a_g\frac{\partial^2 T}{\partial z^2} \tag{10}$$

The corresponding boundary conditions are,

$\partial^2T/\partial z^2=0$, $-\lambda_g(\partial T/\partial z)=q_w$, at $z=0$; $T=T_S$, $-\lambda_g(\partial T/\partial z)=q_{ev}$, at $z=z_b$

where q_w and q_{ev} can be expressed by introducing the nonlinearlity distribution factor β ,

$$q_w=\beta_1\lambda_g(T_{wo}-T_S)/z_b \tag{11}$$

$$q_{ev}=\beta_2\lambda_g(T_{wo}-T_S)/z_b \tag{12}$$

Equation (10) is solved by assuming a reasonable shape for the temperature distribution, for instance,

$$T_v=T_{wo}+C_1z+C_2z^2+C_3z^3 \tag{13}$$

where each of the three undetermined coefficients is a function of the variable r, and is determined by applying the boundary conditions. Consequently Eq.(13) can be nondimensionalized by defining θ=(Tv-Ts)/(Two-Ts) and Xv=Z/Zb.

$$\theta=1-1/2(3-\beta_2)X_v+1/2(1-\beta_2)X_v^3 \tag{14}$$

The factor β_2 in Eq(14) is a function of the variable r and is still unknown. The factor β_2 is determined by integrating Eq.(10) with respect to Z. Since the factor β_2 is given in the form of a linear equation of the first order in terms of the independent variable r,it is necessary to integrate the factor β_2

with respect to r. As for the corresponding boundary condition, the factor β_2 at r=0 takes the value which satisfies Eq.(15) because the heat flow at r=0 can sufficiently be assumed to be one-dimensional in the z direction.

$$3/20\ A_1 \beta_2^2+(3/2+7/20\ A_1-3/20\ A_2+1/120\ A_3)\beta_2+(-3/2-7/20\ A_2+3/40\ A_3)=0 \qquad (15)$$

where $A_1=Cpg\cdot\Delta Tsat/Lg$, $A_2=Ub\cdot Zb/\alpha g$, $A_3=Uw\cdot Zb/\alpha g$ and the limiting value of the factor β_2 is given by $0 \leq \beta_2 \leq 3$. The value of the factor β_2 averaged with respect to the variable r was substituted in Eq.(6), but it was proved through our numerical analysis that β_2 was uniform implying that the vapor flow in the r direction was isothermal. When a drop approaches the hot surface initially and the drop-wall separation attains a minimum value, it is assumed in this analysis that a vapor layer, the thickness of which is equivalent to the minimum value, was already formed at the instant. And if the initial separation Zo is not chosen so large, it is considered that the beginning of the fall of the drop and the formation of vapor occur simultaneously at $\tau=0$.

3. RESULTS OF NUMERICAL SOLUTIONS

3.1 The Radius of the Drop as a Function of Time

The numerical calculation based on the drop dynamics can be performed for water drop at atmospheric pressure. All the properties of the superheated vapor are evaluated at a temperature defined by (Two+Ts)/2. R/Rmax are shown against $\tau/\tau r$ by the solid lines in Fig.2 as an example of the numerical calculation for D=0.22 millimeter, ΔTsat=400 K and We=12.3 and 100, where Rmax is the maximum extension radius. According to the results of our numerical analysis in the range of D= 0.22 ~ 4.0 millimeters, ΔTsat=200~600 K and We=12.3~300, Rmax/D can be expressed by a function of Weber number alone, which holds to a fairly good approximation.

$$2\,R_{max}/D=0.87\,(We/6+2)^{0.5} \qquad (16)$$

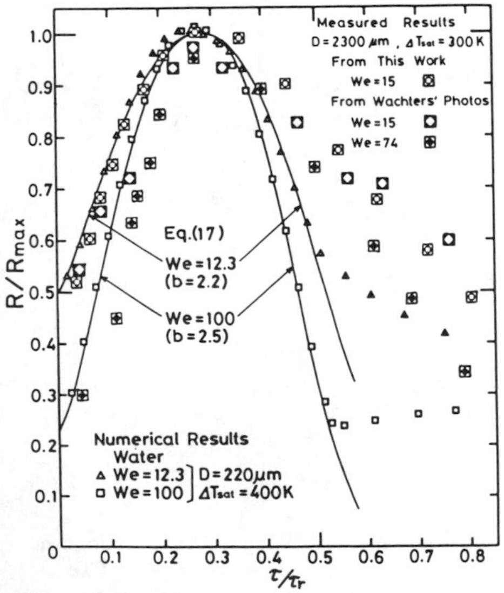

FIGURE 2. Drop bottom radius during impact

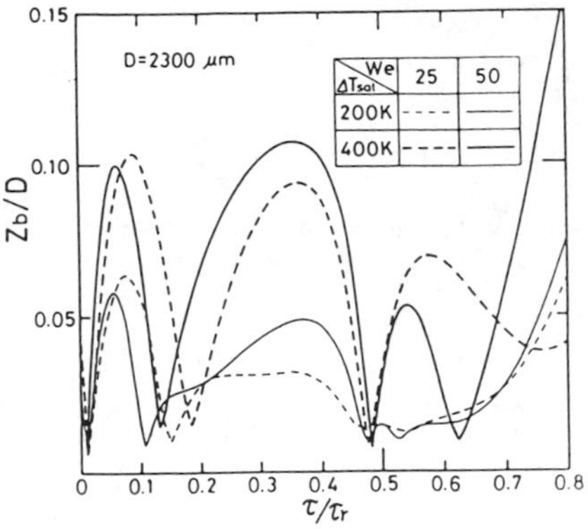

FIGURE 3. Drop-wall separation during impact

This Eq.(16) agrees with the experimental results of Ueda et al.[3]. In Fig.2 R/Rmax is almost independent of D and Δ Tsat, and reaches the maximum extension radius at $\tau/\tau r$=0.28. The time required to reach the initial equilibrium position after the drop deforms in the radial liquid film state is 50~ 60 % of the τr. Data for R as a function of time, taken from our experiment and Wachters' photographs [1] are nondimensionalized by Rmax showed by Eq.(16) and are plotted in Fig.2. The numerical results shows good agreement with data up to the time for completion of the radial spreading, but the difference becomes large after the time. The results of numerical analysis for R/Rmax can be adjusted by the following equation with a sufficient accuracy for $0 \leq \tau/\tau r \leq 0.5$,

$$R/R_{max}=exp\left[\frac{-(|\tau/\tau_r-0.28|)^b}{0.28^b}\sqrt{1-\tau/\tau_r}\cdot \ln\sqrt{We/6+2}\right] \qquad (17)$$

where Weber number is the range of 10~300. Corresponding to Weber number, the power b is 2.2~2.5, but b=2.5 for high Weber number 100~300.

3.2 Drop-Wall Separation and Heat Flux as a Function of Time

The drop-wall separation as a function of time is given in Fig.3 for the case of D=2.3 millimeter. When the drop approaches initially the hot surface, the

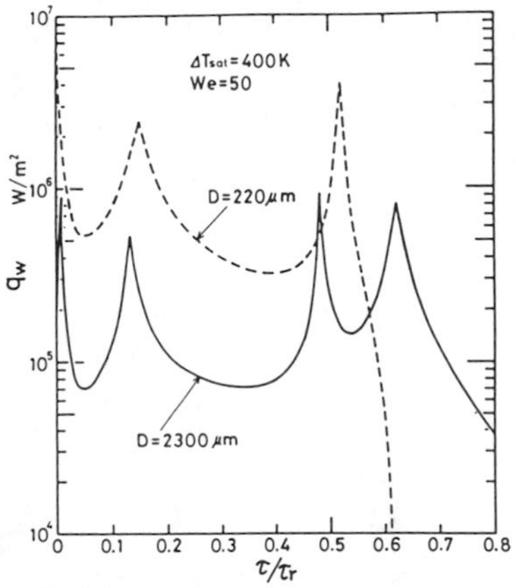

FIGURE 4. Heat flux during impact

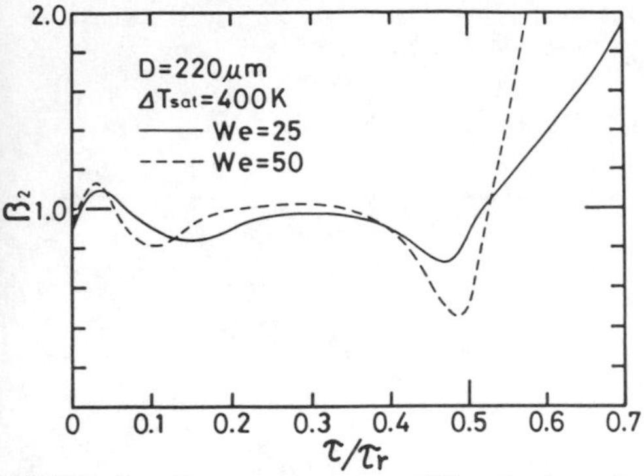

FIGURE 5. Temperature profile factor during impact

drop-wall separation shows the first minimum value. After the time the thickness of the vapor layer oscillates as the time proceeds and then the thickness increases rapidly in the period above $\tau/\tau r$=0.6. Such a variation of the thickness results in the up-and-down motion of the drop on the vapor layer which is formed by evaporation at the bottom of it. If no evaporation occurs, the variation does not occur too. There are mainly four minima in a trace of the variation. In Fig.4, heat flux is shown as a function of time for two drop sizes at constant Weber number and wall superheat. As is evident from Fig.4, the heat flux for small drop diameter is higher than for large. Comparing Fig.3 with Fig.4, q_w shows the maximum value when Zb/D shows the minimum value.

3.3 Temperature Profile in the Vapor Layer

In Fig.5, the factor β_2 is shown as a function of time for two different Weber numbers at constant wall superheat. Picking up some arbitrary times $\tau/\tau r$ from Fig.5, nondimensionalized temperature profiles at the time are shown in Fig.6 for the same Weber numbers and wall superheat as those shown in Fig.5. It becomes clear that the heat flux increases rapidly in the period $\tau/\tau r$=0.47~0.48, where β_2 shows the minimum value because in this period the bottom of the drop moves fast toward the hot surface and the rate of change in enthalpy of vapor flow increases. The temperature profile in this period shows a downward convex-like curve in Fig.6. On the other hand, when $\tau/\tau r$=0.61, especially for We=50,

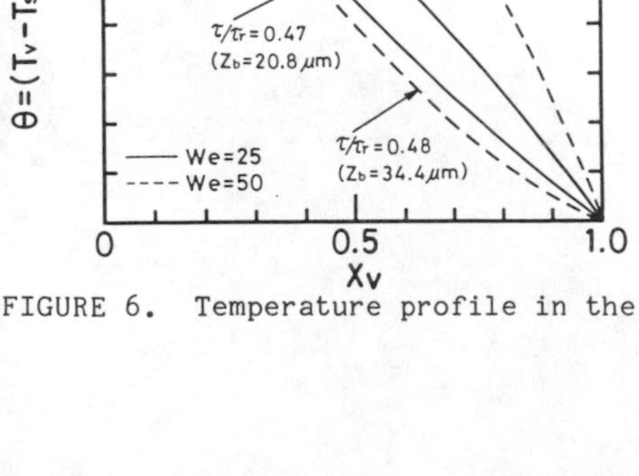

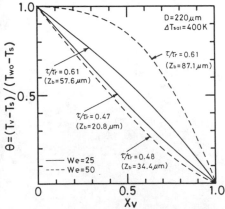

FIGURE 6. Temperature profile in the vapor layer

the bottom of the drop moves so fast in upward direction that the heat flux decreases abruptly (see the dotted line in Fig.4) and the temperature gradient at the hot surface shows an adiabatic condition (see the dotted line for We=50 in Fig.6).

3.4 Heat Transfer Effectiveness and Time Averaged Heat Flux

The drop heat transfer effectiveness ηw at the hot surface is defined as the ratio of the heat transfer rate to the latent heat of the drop,

$$\eta_w=\frac{1}{(\pi D^3/6)\,\rho_f\cdot L_v}\int_0^{\tau_c}\beta_1\lambda_g\pi R^2\frac{(T_{w0}-T_S)}{z_b}\,d\tau \tag{18}$$

where τc is known as the impact period [2]or the effectual heat transfer period [4]. At times greater than τc, it can be considered that the heat transfer to the drop decreases substantially because drop-wall separation increases steadily. The numerical results of the heat transfer effectiveness in the all range of calculation are plotted in Fig.7 against the product of Weber number and the dimensionless conductivity group Th defined by Eq.(19)

$$T_h=(\lambda_g\Delta T_{sat}/L_g)/\sqrt{\rho_g\sigma D} \tag{19}$$

An adjustable equation which is applicable to the all range of numerical calculation was obtained within± 23% in error,

$$\eta_w=1.26\times10^{-2}T_h^{0.5}\left(\frac{We}{6}+2\right)^{0.5} \tag{20}$$

Combining Eqs.(11) and (18) gives

$$\frac{1}{\tau_c}\int_0^{\tau_c}q_w\pi R^2 d\tau=\frac{\pi}{6}D^3\rho_f\,L_v\,\frac{\eta_w}{\tau_c}$$

If the averaged value of R during the impact is known, $\overline{q}w$ must be obtained as function of ΔTsat, WE and D. Then protting $\overline{q}w/(\lambda g\cdot\Delta Tsat/D)$ against Th·(We/6+2), the values of $\overline{q}$w in the all range of calculation fell on a straight line though there was some scattering in high Weber number. The adjustable equation which is applicable to the all range of numerical calculation was obtained within±25% in error except for We=300 ,

$$\overline{q}_w\Big/\left(\frac{\lambda_g\Delta T_{sat}}{D}\right)=1.9\,T_h^{-0.5}\left(\frac{We}{6}+2\right)^{-0.5} \tag{21}$$

Since ηw and $\overline{q}$w have been estimated, the averaged value of R during the impact is given by Eq.(22)

$$\overline{R}=0.7\cdot R_{max} \tag{22}$$

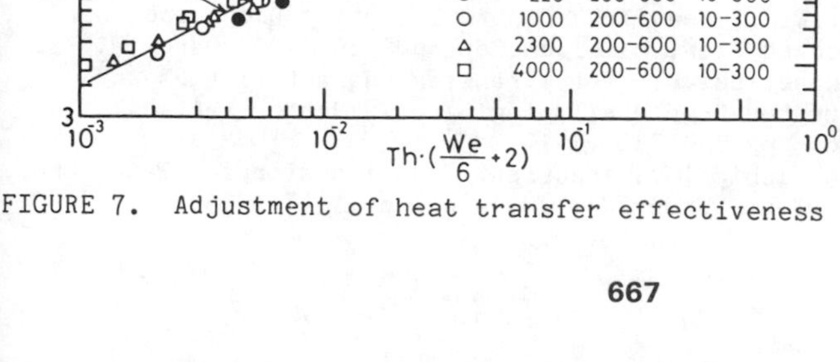

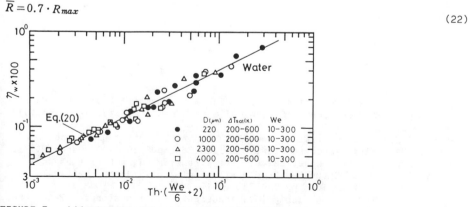

FIGURE 7. Adjustment of heat transfer effectiveness

4. CONCLUSIONS

The numerical calculation based on the drop dynamics can be performed in the range of D=0.22~4.0 millimeters, ΔTsat=200~600 K and We=10~300 for water drops at atmospheric pressure. The results obtained are summarized as follows:

1. The drop bottom radius calculated as a function of time is given by an adjustable equation and shows good agreement with measured data up to the time for completion of the radial spreading.
2. There are three or four minima in the drop-wall separation as a function of time and the drop rebounds away from the hot surface at about half of the characteristic vibration period of the drop.
3. The up-and-down motion of the drop bottom itself exerts strong influence on an instantaneous quasi-steady heat transfer rate.
4. The drop heat transfer effectiveness and the time average heat flux are shown by the following equation respectively,

$$\eta w \propto \Delta Tsat^{0.5} \cdot (We/6+2)^{0.5} \cdot D^{-0.25}, \quad \bar{q}w \propto \Delta Tsat^{0.5} \cdot (We/6+2)^{-0.5} \cdot D^{-0.75}.$$

Nomenclature

a:thermal diffusivity m^2/s
Cp:specific heat kJ/kg K
D:diameter of a drop m
L:specific enthalpy kJ/kg
q_{ev}:heat flux for evaporation W/m^2
q_w:heat flux removed from the wall W/m^2
Ts:saturated temperature ℃
Two:surface temperature of the hot wall ℃
ΔTsat:wall superheat =Two−Ts K
U:internal flow velocity in the drop m/s
u:internal flow velocity in the vapor layer m/s
Vo:initial impinging velocity m/s
We:Weber number $=\rho_f Vo^2 D/\sigma$
τ:time s
r:characteristic vibration period of the drop $=\pi/4\sqrt{\rho_f \cdot D^3/\sigma}$ s
λ:thermal conductivity W/m K
ρ:mass density kg/m^3
σ:surface tension N/m
μ:vapor viscosity Pa s

Subscripts

f:liquid drop
g:superheat vapor
v:saturated vapor
−:time mean value

REFERENCES

1. Watchters,L.H.J.and Westerling,N.A.J., The Heat Transfer from a Hot Wall to Impinging Water Drops in the Spheroidal State,Chem.Eng.Sci.,21-11, pp.1047-1056, 1966.
2. Kendall,G.E.and Rohsenow,W.M., Heat Transfer to Impacting Drops and Post Critical Heat Flux Dispersed Flow, MIT,Tech.Rep.No.85694-100, March. 1978.
3. Ueda,T.,Enomoto,T. and Kanetsuki,M., Heat Transfer Characteristics and Dynamic Behavior of Saturated Droplets Impinging on a Heated Vertical Surface, Bull.JSME,22-167, pp.724-732 ,1979.
4. Inada,S.,Miyasaka,Y. and Nishida,K., Transient Heat Transfer for Water Drop Impinging on a Heated Surface, Trans.JSME, 51-463, pp.1047-1053, 1985.

Prediction of Heat Transfer Coefficient and Pressure Drop in Rifled Tubing at Subcritical and Supercritical Pressure

MAKIO IWABUCHI, TOKUJI MATSUO, and MITSUO KANZAKA
Nagasaki Technical Institute
Technical Headquarters
Mitsubishi Heavy Industries, Ltd.
Nagasaki 850-91, Japan

HISAO HANEDA and KENJIROU YAMAMOTO
Boiler Engineering Department,
Power Systems Headquarters
Mitsubishi Heavy Industries, Ltd.
Tokyo 100, Japan

1. INTRODUCTION

Since 1980, in Japan, in accordance with the increase of the percentage of nucle power and the large change of daily electric power demand, almost all of the new fossil fuel-fired boiler has been designed as the supercritical sliding pressure operation unit suitable for the middle load service. In this type of boiler, th operating pressure and the mass velocity decrease with decrease in a load, which causes the operation below the critical pressure under a certain point in partia loads. Thus, it is impossible to avoid the heat transfer crisis in the evaporat ing tubes in the subcritical pressure region. In order to protect the tubes in this heat transfer crisis, so-called spirally wound type furnace water wall construction has been conventionally adopted, in which the number of the evaporatin tubes is reduced by inclining them for keeping the high mass velocity.

Meanwhile, for the simplification of the furnace water wall construction and the reduction of auxiliary power consumption, the authors planned to develop the supercritical sliding pressure operation boiler with vertical water wall evaporating tubes using rifled tubing, which had been found to have the superior heat transfer characteristics[1][2][3]. The feasibility of this type of boiler has been clarified through the detail experiment on the heat transfer phenomena at the near critical pressure region[4][5]. In the thermal design of the actual boiler, however, it is required to evaluate the heat transfer and hydrodynamic characteristics of the water wall evaporating tubes under all of the operating c ditions from minimum to maximum load.

This paper describes the systematical correlations to be applied to predicting t heat transfer coefficient and pressure drop in each region of subcritical and supercritical pressures, based on the experimental data of rifled tubing under the same condition as that of the actual boiler.

2. TEST PROCEDURE

Figure 1 shows the flow diagram of the test facility. This test facility is a once-through type loop used for the forced convection heat transfer test at a wide pressure range from subcritical to supercritical.

The test section was arranged so as to produce a vertical upward flow and full-circumferential heating (360°-heating) or half-circumferential heating (180°-heating) by means of electric heater. The test section was made of 1.25Cr-0.5Mo steel tube of 28.6 mm in outside diameter, which had a measured inside diameter of 17.7 mm at its groove, a rib-number of four at tube cross section, a rib-heig

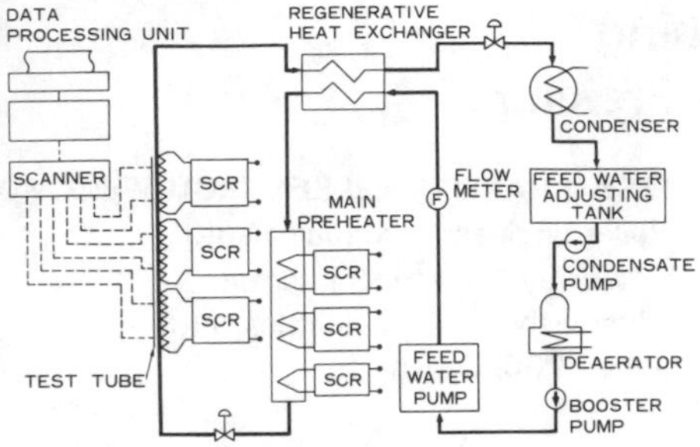

FIGURE 1. Flow diagram of test facility

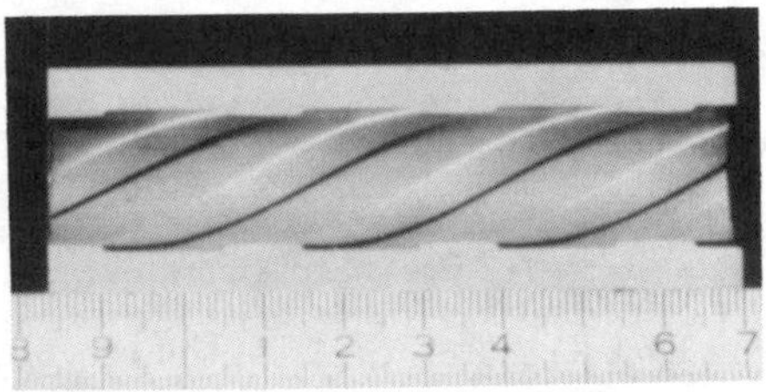

FIGURE 2. Tested rifled tube

of 0.83 mm and spiral angle of 30° against its tube axis as shown in Figure 2. The test section was composed of three subsections. The length and the effective heated length of each subsection are 1,930 mm and 1,730 mm, respectively. The tube wall temperature was measured at intervals of 216 mm.

In this test, the mass velocity was defined as the actual one, that is, the mass flow rate divided by the actual flow area. The heat flux was defined as that on the inner surface at the crown point.

It was verified through the heat conduction analysis that the heat flux distribution on the inner surface at 180°-heated test is quite similar to that in the welded wall panel in an actual boiler.

3. HEAT TRANSFER COEFFICIENT IN SUPERCRITICAL PRESSURE REGION

The validity of Dittus-Boelter's correlation to estimate the heat transfer coefficient at the supercritical pressure liquid region of rifled tubing has been already discussed[2], and it is expected that the present correlations for smooth tubing can be applicable to the higher enthalpy region including pseudo-critical point.

Figure 3 shows the comparisons of measured wall temperatures of 180°-heated rifle tubing with the predicted ones by following correlations.

Dittus-Boelter's correlation: $$Nu_b = 0.023\, Re_b^{0.8}\, Pr_b^{0.4} \quad (1)$$

Styrikovich's correlation[6]: $$Nu_b = 0.023\, Re_b^{0.8}\, Pr_{min} \quad (2)$$

where $$Pr_{min} = Pr_b \ (Pr_b < Pr_w),\ Pr_{min} = Pr_w \ (Pr_b > Pr_w) \quad (3)$$

Bishop's correlation[7]: $$Nu_b = 0.0069\, Re_b^{0.90}\, \overline{Pr}_b^{0.66} \left(\frac{v_b}{v_w}\right)^{0.43} \quad (4)$$

where $$\overline{Pr}_b = \frac{H_w - H_b}{T_w - T_b} \cdot \frac{\mu_b}{k_b} \quad (5)$$

In the above correlations, hydraulic diameter is used in defining Nu_b and Re_b.

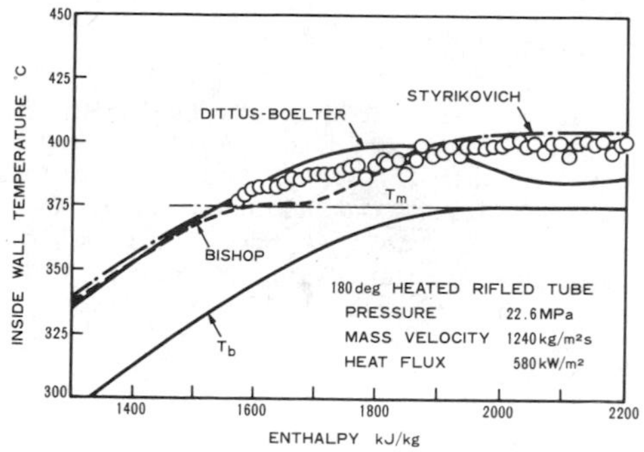

FIGURE 3. Measured and predicted wall temperatures at supercritical pressure

From this figure, it can be said that Styrikovich's correlation agrees with the experimental data at enthalpies below 1,800 kJ/kg and Bishop's agrees above 1,800 kJ/kg. The differences of predicted temperatures seems to be owing to the definitions of the temperature, at which physical properties of fluid are evaluated in the region where the tube wall temperature T_w exceeds the pseudo-critical temperature T_m. This suggests that the further improvement is required in evaluating fluid properties.

Based on the measured data of rifled tubing at the supercritical pressure region we derived the following new correlation (NTI-SC-R-2 correlation):

$$Nu_w = 0.02053\ Re_b^{0.800}\ \overline{Pr}_w^{0.689} \left(\frac{k_b}{k_w}\right)^{0.177} \tag{6}$$

$$\text{where} \quad \overline{Pr}_w = \frac{H_w - H_b}{T_w - T_b} \cdot \frac{\mu_w}{k_w} \tag{7}$$

Figure 4 shows the comparison of this correlation with measured data. RMS (root mean square) error for 622 data is 6.7% and it is regarded to be quite satisfactory.

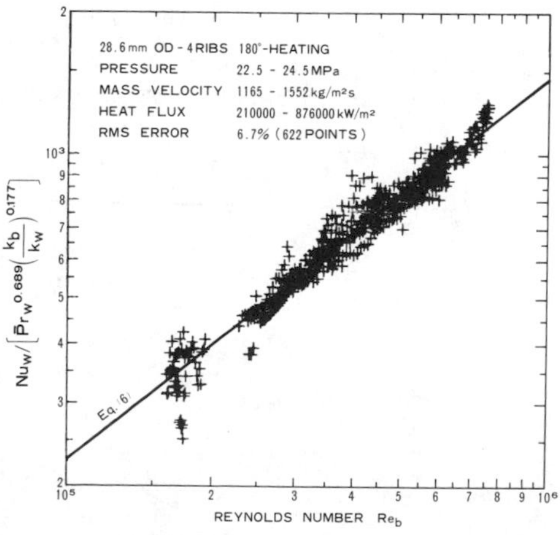

FIGURE 4. Heat transfer coefficient at supercritical pressure

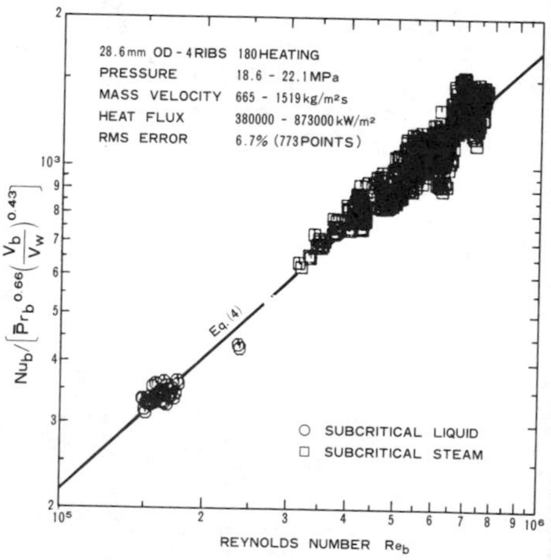

FIGURE 5. Heat transfer coefficient in single phase flow at subcritical pressure

4. HEAT TRANSFER COEFFICIENT IN SUBCRITICAL PRESSURE REGION

4.1 Liquid and Superheated Steam Single Phase Flow Region

Figure 5 shows the comparison of experimental data with Bishop's correlation. The prediction shows good agreement with experimental data both in the liquid and superheated steam regions, and RMS error for 773 data is 6.7%. Thus, Bishop's correlation can be applicable as a design correlation. On the other hand, as for Dittus-Boelter's correlation, if it is limited to the data in the liquid region, the prediction shows good agreement with experimental data with RMS error of 1.1% for 64 data. But in the superheated steam region, it greatly differs from experimental data and can not be used for a design work.

4.2 Subcooled and Saturated Nucleate Boiling Region

The wall temperature in this region is only slightly higher than the saturated temperature. The wall temperature can be predicted by Jens-Lottes' correlation[8] for example, as a design correlation, although it was originally developed to the subcooled nucleate boiling region.

4.3 Post-CHF Region

Figure 6 shows one example of measured wall temperatures at 21.6 MPa near the critical pressure. In this case CHF (critical heat flux) phenomenon occurs at the subcooled region of 1,700 kJ/kg in an enthalpy. The wall temperature reaches its maximum at the subcooled region of 1,870 kJ/kg, and after that, decreases gradually as the enthalpy increases. Since the region after the maximum wall temperature can be regarded as the developed film boiling region, we named the maximum wall temperature point 'OFB' (onset of developed film boiling) point. Figure shows the quality at CHF, x_{CHF}, and that at OFB, x_{OFB}, with reference to the mass velocity. The difference between x_{CHF} and x_{OFB} is not so large at 20.6 MPa, but it becomes considerably large at 21.6 MPa.

As for the wall temperature prediction in the above mentioned high pressure post-CHF region, it is necessary to divide the whole region into the transition region from CHF to OFB and that after OFB. The wall temperature in the former region can be approximately predicted by the linear interpolation between CHF and OFB points.

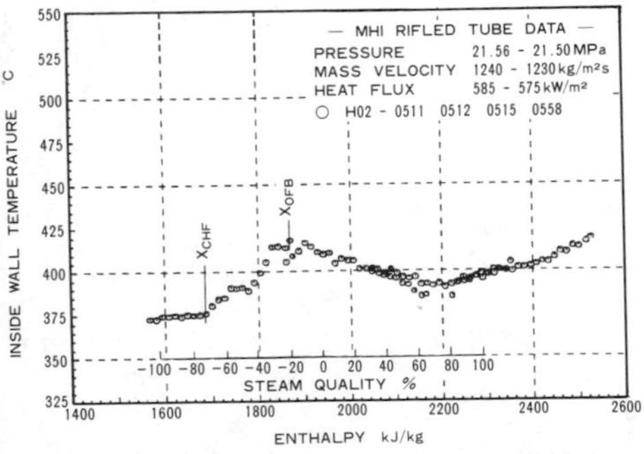

FIGURE 6. Measured wall temperature

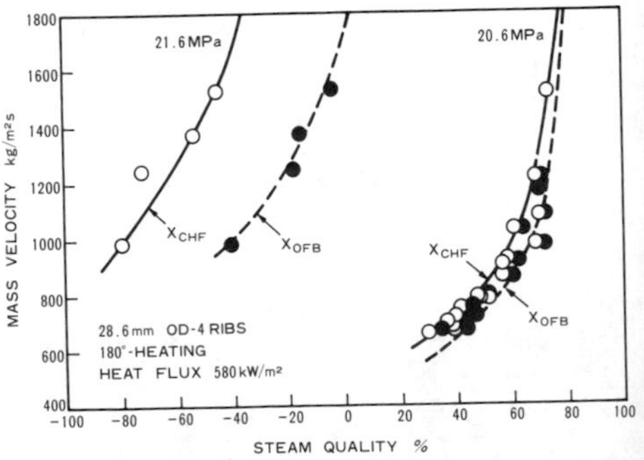

FIGURE 7. Quality at CHF and OFB

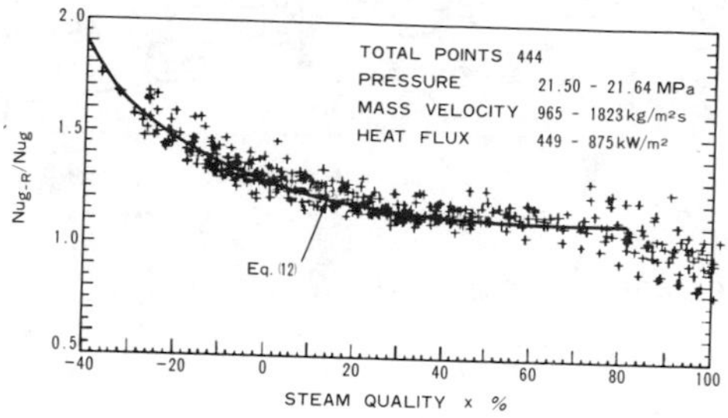

FIGURE 8. Ratio of Nusselt number in post-CHF region vs. steam quality

As a heat transfer correlation in post-CHF region for smooth tubing, Groeneveld[9] proposed the following correlation:

$$Nu_g = C\, Re_x^{n1} Pr_w^{n2} y^{n3} \left(\frac{q}{3.155}\right)^{n4} \quad (8)$$

where Re_x and y are defined by the following equations:

$$Re_x = Re_g \left\{x + \frac{v_\ell}{v_g}(1 - x)\right\} \quad (9)$$

$$y = 1 - 0.1\left(\frac{v_g}{v_\ell} - 1\right)^{0.4}(1 - x)^{0.4} \quad (10)$$

The values of C, n_1, n_2, n_3 and n_4 depend upon the shape and size of flow circuit, and are 1.85×10^4, 1.0, 1.57, -1.12 and 0.131 with 10.1% RMS error for vertical and horizontal tubes. It is thought that the Groeneveld type correlation is also applicable to the case of rifled tubing after OFB.

Figure 8 shows the ratio of Nu_{g-R}/Nu_g vs. quality x, where Nu_g is Nusselt number by Groeneveld correlation and Nu_{g-R} is Nusselt number for rifled tubing defined the following equation:

$$Nu_{g-R} = \frac{q}{T_w - T_{sat}} \cdot \frac{d_h}{k_g} \quad (11)$$

In Figure 8 the ratio of Nu_{g-R}/Nu_g is about 1.1 at x of 0.8, and it increases with the decrease of x, then it becomes about 1.5 at x of -0.2. This means there is the remarkable improvement on the heat transfer characteristics of rifled tubing at the region where the wall temperature reaches its maximum just after OFB.

Nusselt number of rifled tubing in pressures above 20.6 MPa and qualities from -0.4 to 0.8 can be described by modifying equation (8) as follows:

$$Nu_{g-R} = Nu_g\left[0.9 + 0.274/(x + 0.68)\right] \quad (12)$$

In the other pressure range, x_{OFB} becomes large and OFB point can not be observed at the range of x lower than 0.8

5. PRESSURE DROP

5.1 Basic Concept of Pressure Drop of Rifled Tubing

In the pressure drop calculation for non-circular tubes, hydraulic diameter is generally used. For rifled tubing, hydraulic diameter was also used together with the actual mass velocity, and the total pressure drop was presented by the sum of pressure drop components, i.e., friction, static, acceleration pressure drops.

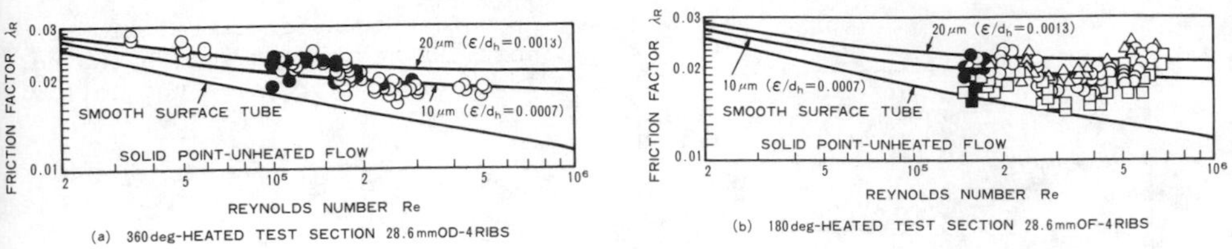

FIGURE 9. Measured friction factor of rifled tubes in single phase flow

5.2 Pressure Drop in Single Phase Flow Region

The static and acceleration pressure drops, ΔP_g and ΔP_a, of rifled tubing in the single phase flow region can be presented by the same equations as those for smooth tubing. The friction pressure drop is described by use of the friction factor λ_R of rifled tubing and hydraulic diameter as follows:

$$\Delta P_{f-R} = \lambda_R \frac{\Delta Z}{d_h} \cdot \frac{G^2}{2} v \tag{13}$$

Figure 9 shows the measured friction factor λ_R, based on the friction pressure drop obtained by subtracting ΔP_g and ΔP_a from the measured total pressure drop in the subcritical and supercritical pressure single phase region. The values of λ_R are distributed within the range from 0.0005 to 0.0015 in relative roughness ε/d_h, and this roughness is equivalent to that of ordinary drawn steel tubes. Accordingly, the friction pressure drop of rifled tubing in the single phase flow region can be predicted by the same expression as that for smooth tubing.

5.3 Pressure Drop in Two-Phase Flow Region

As for the model for the pressure drop prediction of evaporating tube in the two-phase flow region, there are two types of concept, i.e., the homogeneous model and the slip model. If it is limited within the relatively high pressure region, the satisfactory prediction for a design work can be obtained by using the homogeneous model. But, it is necessary to adopt the slip model to predict the pressure drop for the sliding pressure operation boiler whose lower operating pressure becomes 6.9 MPa.

In the analysis, the slip factor S was assumed to be the same as that for smooth tubing by Thom's correlation[10]. Thus, the specific volume of the two-phase flow is presented by the following equation:

$$v = \frac{1 + x(S - 1)}{1 + x(S\, v_\ell / v_g - 1)} v_\ell \tag{14}$$

Then, the static and acceleration pressure drops in the two-phase flow can be obtained by same equations as those in the single phase flow, using the specific volume defined by equation (14).

As for the friction pressure drop for smooth tubing, Thom proposed the following correlation:

$$\Delta P_{f-S} = \lambda_S \frac{\Delta Z}{d_h} \cdot \frac{G^2}{2} v_\ell \, r_{f-S} \tag{15}$$

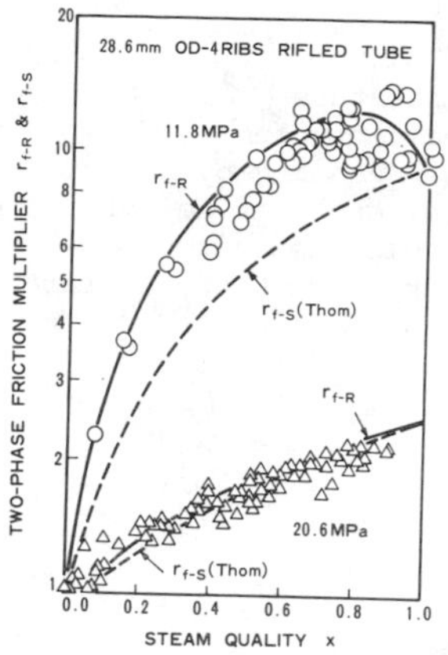

FIGURE 10. Friction multiplier

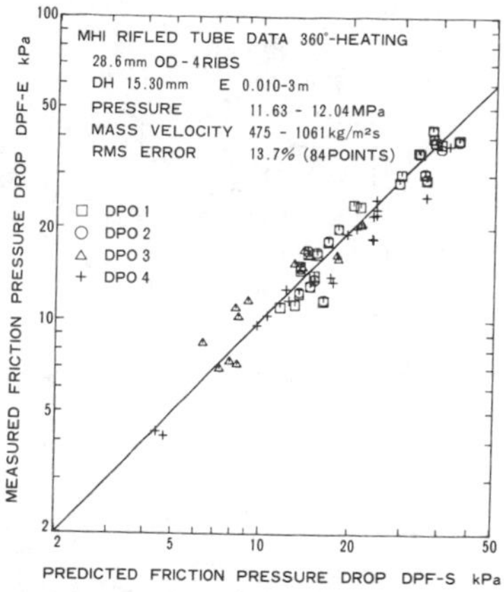

FIGURE 11. Friction pressure drop

where λ_S and r_{f-S} are the friction factor and the two-phase friction multiplier for smooth tubing, respectively. r_{f-S} is given by a function of pressure and quality. For rifled tubing, similar equation can be applied by defining the two-phase friction multiplier r_{f-R} in place of r_{f-S}.

Figure 10 shows the two-phase friction multiplier for rifled tubing obtained by subtracting ΔP_g and ΔP_a from the measured total pressure drop. In the case of 20.6 MPa, the values of r_{f-R} do not differ so much from those of r_{f-S}. On the other hand, in the case of 11.8 MPa, r_{f-R} becomes about one and a half times of r_{f-S} at the middle range of x. Though there is a possibility that the assumption in which the void fraction of rifled tubing is the same as that of smooth tubing does not always hold at low pressures, we take this assumption because we do not have any experimental evidence for judging it so far.

Figure 11 shows the comparison of the predicted friction pressure drop with the experimental data, and it is found that the same prediction method as Thom's correlation for smooth tubing is also applicable to rifled tubing by use of r_{f-R}. In addition, it has been verified that the values of r_{f-R} used in this paper can be applicable to the rifled tubing with different tube diameter and rib numbers.

6. CONCLUSION

The prediction methods were established for the heat transfer coefficient and the pressure drop of rifled tubing for the supercritical sliding pressure operation boiler with vertical water wall evaporating tubes, based on the comprehensive experimental data of heat transfer and hydrodynamic characteristics.

The authors would like to express our sincere gratitude to Professor Emeritus K. Nishikawa and Professor S. Yoshida of Kyushu University for their valuable suggestions on the preparation of this paper.

NOMENCLATURE

Cp	specific heat	kJ/(kg K)	T	temperature		°C
d_h	hydraulic diameter	m	v	specific volume		m^3/kg
G	mass velocity	$kg/(m^2 s)$	x	steam quality by weight		
H	specific enthalpy	kJ/kg	ΔZ	tube length or height		m
h	heat transfer coefficient	$kW/(m^2K)$	μ	dynamic viscosity		Pa s
k	thermal conductivity	kW/(m k)	suffix	b	bulk temperature	
Nu	Nusselt number = hd_h/k			g	saturated steam	
Pr	Prandtl number = $Cp\mu/k$			ℓ	saturated water	
ΔP	pressure drop	Pa		sat	saturation	
q	heat flux	W/m^2		w	tube wall	
Re	Reynolds number = Gd_h/μ					

REFERENCES

[1] Swenson, H. S., Carver, J. R. and Szoeke, G., The Effects of Nucleate Boiling Versus Film Boiling on Heat Transfer in Power Boiler Tubes, Trans. ASME, Ser. A, vol.84, pp. 365-371, 1962.

[2] Nishikawa, K., Fujii, T. and Yoshida, S., Investigation into Burnout in Grooved Evaporator Tubes, Journal of Japan Soc. Mech. Engr., vol.75, pp. 700-707, 1972.

[3] Watson, G. B., Lee, R. A. and Wiener, M., Critical Heat Flux in Inclined and Vertical Smooth and Ribbed Tubes, Proc. 5th Int. Heat Transfer Conf., Tokyo, vol.4, pp. 275-279, 1974.

[4] Kawamura, T., Kunimoto, T., Haneda, H., Sengoku, T., Iwabuchi, M., Tateiwa, M., Muraishi, K. and Fukahori, K., Large Supercritical Sliding Pressure Operation Monotube Boiler of Vertical Water Wall Tube Type, Mitsubishi Heavy Industries, Technical Review, vol.17, No. 3, pp. 213-224, October 1980.

[5] Iwabuchi, M., Tateiwa, M. and Haneda, H., Heat Transfer Characteristics of Rifled Tube in Near Critical Region, Proc. 7th Int. Heat Transfer Conf., München, vol.5, pp. 313-318, 1982.

[6] Styrikovich, M. A., Miropolsky, S. L. und Schitzman, M. E., Wärmeübergang in kritischen Druckgebiet bei erzwangener Strömung des Arbeitsmediums, VGB, Ht 61, pp. 288-294, 1959.

[7] Bishop, A. A., Sandberg, R. O. and Tong, L. S., High Temperature Water Loop, Part 4, Forced Convection Heat Transfer to Water at Near-critical Temperature and Supercritical Pressures, WCAP-2056IV, 1964.

[8] Jens, W. H. and Lottes, P. A., Analysis of Heat Transfer, Burnout, Pressure Drop and Density Data for High-Pressure Water, ANL-4672, 1951.

[9] Groeneveld, D. C., Post-dryout Heat Transfer at Reactor Operating Conditions National Topical Meeting on Water Reactor Safety, 1973.

[10] Thom, J. R. S., Prediction of Pressure Drop during Forced Circulation Boilin of Water, Int. J. Heat and Mass Transfer, vol.7, pp. 709-724, 1964.

Heat Transfer in Bubble Column Reactors

H.-J. KORTE, A. STEIFF, and P.-M. WEINSPACH
Universität Dortmund
P.O.B. 500 500
Dortmund, FRG

INTRODUCTION

Bubble column reactors are commonly used in industrial practice as absorbers, strippers, reactors (for oxidation, nitration, chloration etc.) and bioreactors. The dimensioning of these bubble columns is influenced by fluid dynamic features as well as by heat transport aspects, especially if strong exothermic or endothermic processes should be performed. Not always an indirect heat exchange exclusively via the reactor wall is possible, so that inner heat exchanger-installations, like cross- or lengthwise flow tube bundles, are installed in the bubble column.

EXPERIMENTAL RESULTS ON HEAT TRANSFER IN BUBBLE COLUMNS

Heat Transfer "Reactor Wall/Aerated Liquid"

The experiments on heat transfer "reactor wall/aerated liquid" mainly considered the influence of the gas distributors and the coalescence behaviour of the gas/liquid system.
Figure 1 shows the heat transfer coefficient as a function of the superficial gas velocity. The heat transfer coefficient in the homogeneous bubble flow regime increases more with the gas throughput than the coefficient in the heterogeneous regime. Also it becomes obvious, that the heat transfer in bubble columns depends not significantly on the gas distributor and coalescence behaviour of the gas/liquid system. The bubble sizes are none important parameters for the heat exchange.

Heat Transfer "Single Tube/Aerated Liquid"

Some results on heat transfer "single tube/aerated liquid" are presented in Figure 2. The heat transfer coefficient is shown as a function of the superficial gas velocity for two different systems. The probe, which is used for the measurements of the heat transfer coefficient, was installed in the middle of the column cross section at half the column height. The tested systems show the known curve-slope with a strong increase of the heat transfer coefficient in the homogeneous bubble flow regime; within the heterogeneous regime the dependence of the α-value on the superficial gas velocity is lower. A limiting value for the heat transfer coefficient was not detested for any of the examined systems during variations of the superficial liquid velocity between v_{Lo} = 1.0 cm/s and 38.0 cm/s. With an increasing viscosity η_L the heat transfer gets worse.

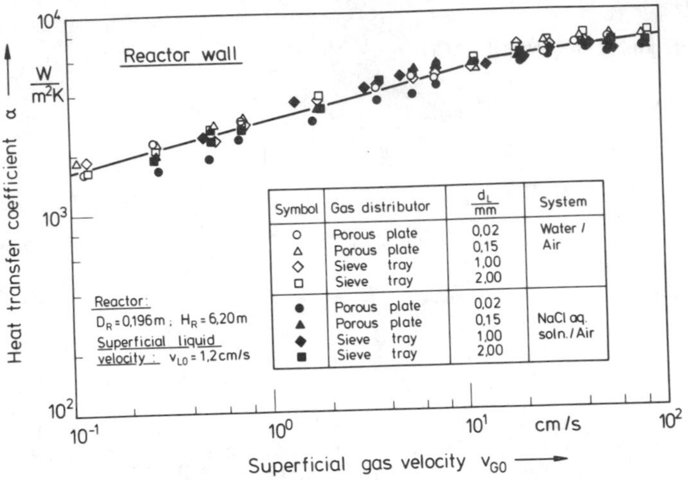

Figure 1: Heat transfer "reactor wall/aerated liquid" in coalescing and non-coalescing systems for different gas distributors

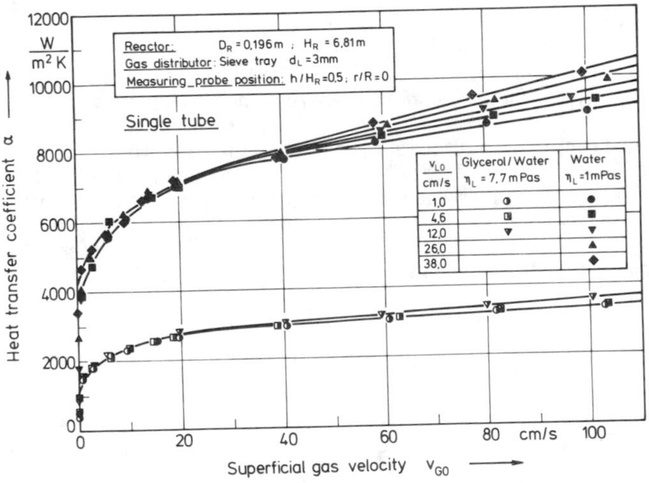

Figure 2: Effect of superficial gas- and liquid velocity on heat transfer "single tube/aerated liquid" in the systems water/air (η_L = 1 mPas) and glycerol-water/air (η_L = 7.7 mPas)

Further measurements of the heat transfer coefficient in different systems with the same liquid viscosity and air as gas phase allow the following conclusion: Not only the liquid viscosity also density, thermal conductivity and specific heat capacity determine the heat transfer. The superficial liquid velocity v_{Lo} is of subordinate importance for heat transfer at the single tube in gas/liquid systems.
An uniform description of the measured values with respect to the influence of operation parameters and physical properties of the liquid on heat transfer is possible with the model equation developed by Kast [1], if a term, which considers the direction of the heat flux is added:

$$St = f\,[(Re_G \cdot Fr_G \cdot Pr_L)^{-1/3} \cdot (\frac{\eta_L}{\eta_{LW}})^{0.30}] \,. \qquad (1)$$

With the help of this relation heat transfer at the single tube in coalescing systems with pure liquids and liquid-mixtures as well as in non-coalescing systems like NaCl water solution/air can be described. The influence of the physical properties is covered by the Prandtl-number of the liquid phase.
Apart from the mentioned influencing parameters the reactor diameter is significant for heat transfer in heterogeneous bubble flow regime. These results were obtained by Wendt [2] for the system water/air at superficial liquid velocities of $v_{Lo} = 0 \div 0.9$ cm/s. The heat transfer coefficient at a single tube installed in the middle of the cross section area is proportional to the reactor diameter:

$$\alpha \sim D_R^{\,0.15} \,. \qquad (2)$$

Measured values on heat transfer "reactor wall/aerated liquid" and "single tube/aerated liquid" can be compared in Figure 1 and 2. It can be realized, that for both the gas throughput has a similar influence. The heat transfer coefficient for the single tube is significantly higher than the coefficient for the reactor wall. With increasing dimensionless column radius r/R the heat transfer coefficient for the single tube gets closer to the one for the reactor wall.

Heat Transfer "Lengthwise Flow Tube Bundle/Aerated Liquid"

This part of the experiments should allow statements on the influence of operation and construction parameters as well as physical properties of the liquid phases on heat transfer at heat exchanger-installations, which are lengthwise flowed by the liquid and gas phase.
Independent of the gas throughputs the superficial liquid velocity has a neglectible influence on heat transfer for each of the tested systems. This means, that a single probe characteristic is valid for the heat transfer at lengthwise flow tube bundles with reference to the influence of the liquid loading.
In Figure 3 the heat transfer coefficient "lengthwise flow tube bundle/aerated liquid" is shown as a function of the superficial gas velocity for the system water/air at different tube arrangements. Up to gas throughputs of $v_{Go} = 20$ cm/s for each presented tube arrangement there is a strong increase of the heat transfer coefficient with this operation parameter. A further increase of the superficial gas velocity leads to a curve-slope depending on the liquid phase. While the heat transfer in aerated water is furtherly raised during an increase of the superficial gas velocities, there is a low increase of the heat transfer for systems with higher viscosities. Generally the heat transfer coefficient at a centrally installed measuring probe and symetrically arranged tubes around the centre is proportional to the tube pitch t_t and decreases with increasing relative

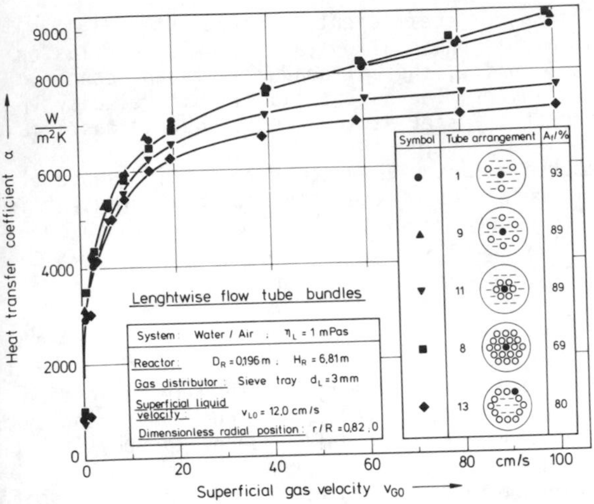

Figure 3: Influence of superficial gas velocity and tube arrangement on heat transfer at lengthwise flow tube bundles in the system water/air

free cross sectional area A_f. Figure 3 also shows the heat transfer coefficients for a concentric ring of tubes near the reactor wall, tube arrangement (13). The α-values conside with the ones for the reactor wall, presented in Figure 1.
A dimensionless description of the results on heat transfer at lengthwise flow tube bundles with centrally arranged measuring probe can be done using relation (1). It determines the influence of the relevant operation parameters and physical properties. This relation is extended by two parameters, which describe the bundle geometry, the relative free cross sectional area A_f and the tube pitch proportion t_t/d_t. For the Stanton-number the following equation was found by a multiple regression:

$$St_{cal} = 0.139 \cdot [(Re_G \cdot Fr_G \cdot Pr_L^{2.26})^{-1/3}]^{0.84} \cdot A_f^{-0.20} \cdot \left(\frac{t_t}{d_t}\right)^{0.14} \cdot \left(\frac{\eta_L}{\eta_{LW}}\right)^{0.30} . \qquad (3)$$

With this equation the measured values are covered with a maximal deviation of ± 15 %. A comparison of the measured and calculated Stanton-numbers is presented in Figure 4. The tube arrangements, their construction data as well as the different systems are indicated in the figure. The given equation (3) is also valid for different liquid throughputs.
Here one aspect should be mentioned. For the tested tube arrangements the tube arrangement (1) is the critical case of the single probe, which heat transfer can also be described with equation (3). For the term (t_t/d_t) the tube pitch t_t changes into the reactor radius R. Therefore it is valid for the single probe:

$$\alpha_{cal} \sim \left(\frac{D_R}{2d_t}\right)^{0.14} . \qquad (4)$$

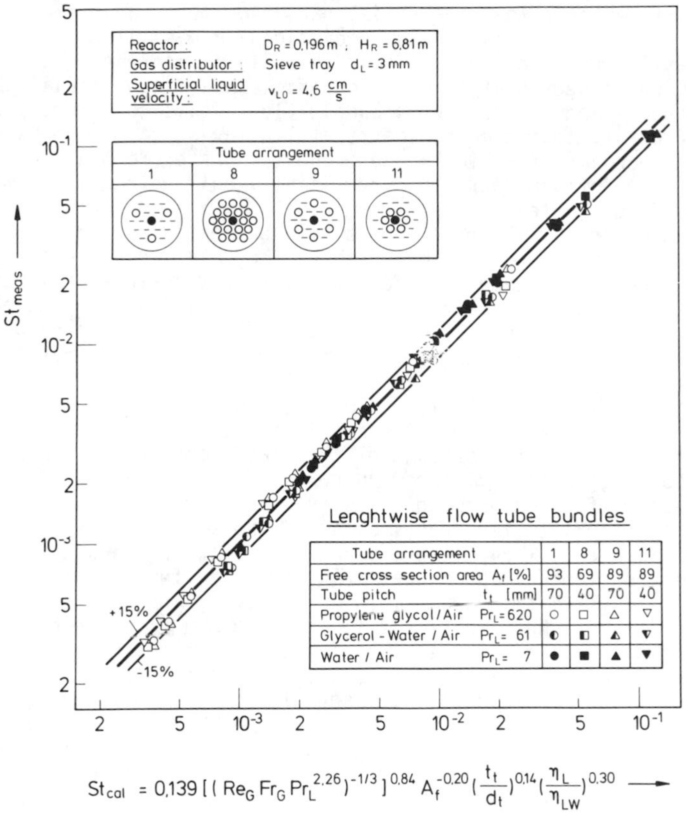

Tube arrangement	1	8	9	11
Free cross section area A_f [%]	93	69	89	89
Tube pitch t_t [mm]	70	40	70	40
Propylene glycol/Air $Pr_L = 620$	○	□	△	▽
Glycerol - Water / Air $Pr_L = 61$	◐	◧	◭	⧩
Water / Air $Pr_L = 7$	●	■	▲	▼

Figure 4: Comparison of measured and calculated Stanton-numbers for different systems and tube arrangements

If the relation is compared with the relation between heat transfer coefficient and reactor diameter ($\alpha \sim D_R^{0.15}$) as presented by Wendt [2] for a single tube, the above relation can also describe the influence of the column diameter on heat transfer at this installation in the bubble column. The conclusion is drawn, that the tubes around the centrally arranged measuring probe have the same effect with reference to the heat transfer as a reactor wall.
The heat transfer coefficient averaged by the reactor cross sectional area is up to 20 % lower than the central heat transfer coefficient in all systems and for all tube arrangements shown in Figure 4.

Heat Transfer "Cross Flow Tube Bundle/Aerated Liquid"

Cross flow tube bundles installed in a bubble column produce a massive disturbance of the flow structure compared with the non-disturbed bubble column. Similar to perforated plate installations the flow profile of the liquid phase becomes flattened above the cross section. With increasing gas and liquid throughputs the reactor is divided into a top and a bottom column part, whereby a cascade bubble column is formed.

With reference to the influence of the superficial gas velocity on heat transfer at cross flow tube bundles two regimes can be distinguished similar to heat transfer at single tubes. If the superficial liquid velocity is increased the difference of the relation between heat transfer coefficient and superficial gas velocity for the homogeneous and heterogeneous bubble flow regime gets more and more insignificantly. Contrary to the measurements at lengthwise flow tube bundles the liquid throughput has an important influence on heat transfer "cross flow tube bundle/aerated liquid". The influence of the superficial liquid velocity becomes dominating for higher viscous systems.
The influence of the geometric parameters of the bundle on heat transfer is mainly determined by the liquid throughput. Figure 5 shows the relation between heat transfer coefficient and number of tube rows for two different systems at a superficial gas velocity of v_{Go} = 40 cm/s. In the range of low liquid throughputs the heat transfer coefficient remains constant or drops slightly setting the number of tube rows from N_R = 0 (single tube) to N_R = 3. These results are an inversion of the relations for the single-phase flow, according to the results of Gnielinsk [3]. With increasing liquid flow the relations are inversed again. From a superficial liquid velocity of v_{Lo} = 12 cm/s onwards, an increase of the number of tube rows leads to a singnificant improvement of the heat transfer, determinded centrally in the tube bundle, up to a tube row-number of N_R = 3.
Another geometric parameter of the tube bundle was examined: the influence of the tube pitch on heat transfer. In-line and staggered tube arrangements with tube pitches of t_t = 32 mm and t_t = 64 mm were used. The following Figure 6 illustrates the influence of tube pitch on heat transfer at different gas- and liquid throughputs for the system propylene glycol/air. During low superficial liquid velocities a doubling of the pitch leads to an increasing of the heat transfer. This contradicts the conditions in the single-phase flow. Higher liquid throughputs inverse this effect and an increase of the tube pitch causes a lower heat transfer coefficient. This is valid for all systems on number of tube rows and tube arrangements.

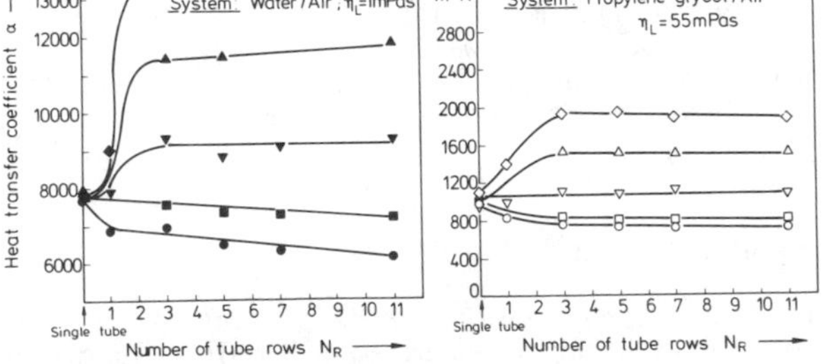

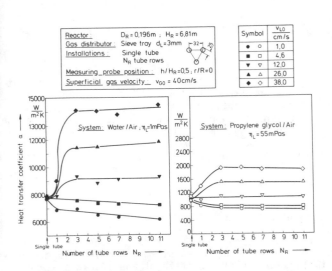

Figure 5: Dependence of heat transfer "cross flow tube bundle/aerated liquid" on number of tube rows in different systems

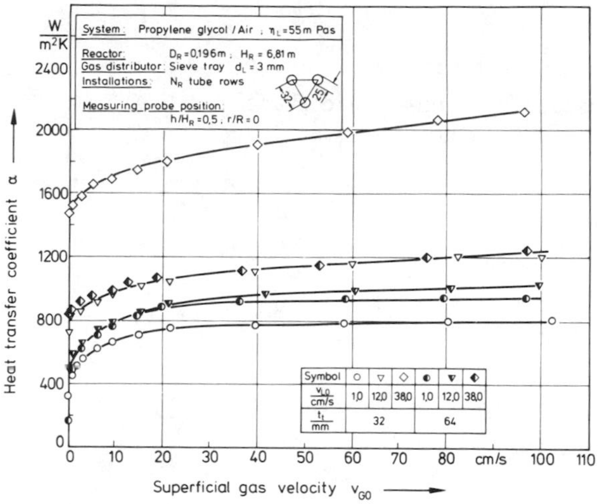

Figure 6: Heat transfer coefficient as function of superficial gas- and liquid velocity as well as tube pitch in the system propylene glycol/air

Summarized the influence of operation and construction parameters on heat transfer in cross flow tube bundles can be characterized as follows:
The higher the flow resistance is, the more the mathematical interrelationships of the single-phase liquid flow determine heat transfer; the influence of gas throughput gets less important. The flow resistance depends on the viscosity of the liquid phase, the superficial liquid velocity, the number of tube rows, the tube pitch and the tube arrangement.

SUMMARY

In this paper results on heat transfer at different heat transfer areas, like reactor wall, single tube, cross- and lengthwise flow tube bundles are presented. The relevant influencing parameters on heat transfer are varied in industrial range.
Independent of gas distributors an equation for an estimation of heat transfer at single tubes and lengthwise flow tube bundles is reported. It is valid in the total range of operation and construction parameters for coalescing and non-coalescing systems with Prandtl-numbers of the liquid phase up to Pr_L = 620. Contradictory to the influence of the liquid throughput on heat transfer at for example lengthwise flow tube bundles the superficial liquid velocity is very important for the relation between the heat transfer and other influencing parameters at cross flow tube bundles.

NOMENCLATURE

A_f	%	relative free cross sectional area
c_{PL}	J/(kg·K)	specific heat capacity of the liquid phase
d_L	m	hole diameter
d_t	m	tube diameter
D_R	m	reactor diameter
g	m²/s	acceleration due to gravity
H_R	m	reactor height
N_R	1	number of tube rows
t_t	m	tube pitch
v_{Go}	m/s	superficial gas velocity
v_{Lo}	m/s	superficial liquid velocity
α	W/(m²·K)	heat transfer coefficient
η_L	Pas	viscosity of the liquid phase
η_{LW}	Pas	viscosity of the liquid phase near the heat transfer area
λ_L	W/(m·K)	heat conductivity of the liquid phase

Dimensionless Numbers

H/H_R	dimensionless column height
r/R	dimensionless column radius
$Fr_G \equiv \frac{v_{Go}^2}{g \cdot D_R}$	Froude-number for the gas phase
$Pr_L \equiv \frac{\eta_L \cdot c_{PL}}{\lambda_L}$	Prandtl-number for the liquid phase
$Re_G \equiv \frac{\rho_L \cdot v_{Go} \cdot D_R}{\eta_L}$	Reynolds-number for the gas phase
$St \equiv \frac{\alpha}{\rho_L \cdot c_{PL} \cdot v_{Go}}$	Stanton-number

REFERENCES

1. Kast, W.:
Chem.-Ing.-Tech. 35 (1963) 11, S. 785/788.

2. Wendt, R.:
Thesis, Dortmund (FRG) 1983.

3. Gnielinski, V.:
Forsch. Ing.-Wes. 44 (1978) 1, S. 15/25 .

Heat Transfer in the Decay Pool of Irradiated Nuclear Fuel

ENRICO LORENZINI, MARCO SPIGA, and DAVIDE ZERBINI
Istituto di Fisica Tecnica
Università di Bologna
Viale Risorgimento 2
40136 Bologna, Italy

INTRODUCTION

The storage of spent nuclear fuel in the decay pool constitutes the stage immediately subsequent its burning in the reactor core and meets some important requirements. The arrangement in pools involves the use of cylindrical tanks (with circular or rectangular cross section) made of reinforced concrete structures, with very thick walls, covered with plates in stainless steel, in order to realize a perfect water seal. At the bottom of the pool the fuel subassemblies are put into small containers under a suitable water head.

In such a way two important requirements are satisfied :

a) the water thickness in the pool gives the essential shielding warranties and the stay in the pool allows the fission products with short half life to decay to a small radioactive level, so that afterwards the spent fuel can be conveyed to the reprocessing plants ;

b) the pool water provides for the removal of the residual heat decaying in the irradiated nuclear fuel.

Therefore the prediction of the thermal behaviour of the water is very important. In fact, one of the most significant features in the pool design consists in the search for its minimum size (in connection with the economic aspects). Obviously the fuel cooling has to be secured not only during routine operations but also in postulated accidents.

To this end the water is kept in forced circulation in order to enhance the cooling of the fuel assemblies, resorting also to external heat exchangers.

The present work aims to analyze the mono-dimensional, time dependent trend of the water temperature, assuming that there are no auxiliary cooling systems; this can simulate one of the worst hypothetical accidents which may occur in the plant. In fact the cooling of the spent fuel is provided only by natural convection; in this situation it is an interesting matter to valuate the time evolution of the water temperature and its maximum value (it is a

fundamental safety requirement the prevention of boiling).
The problem of heat transfer and fluid flow in the storage decay pool is very complicated; it is strictly related to the study of hydrodinamic stability and natural convection in liquid bodies with internal heat source, where the source can be considered as a porous medium. The multidimensional natural convection in porous media has extensively analyzed by many authors, to obtain criteria for the onset of convection and to determine the flow patterns [1 - 5]. The experimental and theoretical investigations have led to the development of numerical techniques for the solution of the conservation equations (mass, momentum, energy) [6 - 10] simplified by the Boussinesq approximation (with constant density for the pool except in the buoyancy term which drives the motion).
In recent papers [11-12] Gay applied this approach to a very accurate investigation of spent nuclear fuel storage pool, including porosity of the assemblies, pressure losses produced by grid spacers, orifices or other planar flow obstructions, heat conduction in the fuel, gap and clad regions.
The great accuracy of this method, unfortunately, requires a very large computer time, and makes it rather heavy and expensive.
In this paper, a very simple method to perform the mono-dimensional transient thermal hydraulic analysis of the storage pool is proposed. The pool is subdivided in horizontal isothermal segments; for every segment an enthalpy balance equation will be deduced, taking account of a vertical mixing flow rate between adjacent segments, due to the convective fluid streams up and down.
The water enthalpy will be considered as a function of pressure and temperature, like all the physical properties (density, specific heat, ..); the pressure is distributed with hydrostatic law.
The lower segment in the pool includes the irradiated fuel heat source, while the upper segment is characterized by heat exchange with the surroundings by evaporation, convection and radiation.
This stratification model allows to get a system of first order non linear differential equations, which can be solved by a numerical procedure.
In particular the Crank Nicolson method will be applied, which, after introducing the linearization hypotheses, will provide an algebraic system.
The mathematical procedure and the computer solution are very handy and simple, and require a short CPU time, making this approach particularly effective for unexpensive thermal hydraulic calculations.

HEAT TRANSFER IN THE STRATIFIED POOL

The approach assumed here consists in a suitable pool water stratification. For each layer, labeled from 1 to N (beginning from the bottom), we consider a thermal balance where the only unknown functions are the enthalpies $h_n = h_n(t)$.

The set of thermal balance equations will give a system whose implicit solution is the temperature at time t, in every stratified layer.

The reasons determining the heat exchange among adjacent layers are of two kinds: conduction and mixing. The former is quite negligible, being the thermal conductivity of water very small; the latter gives the most relevant contribution.

Mixing depends on free convection due to buoyancy effects.

The determination of these gravity driven flows is very difficult; their correct valuation is possible only by having a solution for the complete balance equations (continuity, momentum and energy equations) in the case of free convection. This should require a powerful numerical procedure, very expensive in terms of computer time.

In this paper we use a different point of view which is more practical for engineering applications.

We assume that the velocity of the water, because of the different density, is proportional to the temperature difference between the adjacent layers.

After determining $v_{n \to n+1} = -v_{n+1 \to n} = K(T_n - T_{n+1})$ where K is a known constant, the consequent mass flow rate exchanged between adjacent layers is

$$m_{n \to n+1} = \frac{1}{2} A \varrho_n K(T_n - T_{n+1}) \qquad (1)$$

where A is the transversal cross section of the pool. If m flows upward from the layer n to n+1, then an equal flow rate must go downward in layer n from the adjacent n+1, in order to satisfy the continuity equation; so that

$$m_{n+1 \to n} = m_{n \to n+1} = m_{n,n+1} \qquad n = 1,2, \ldots, N-1. \qquad (2)$$

That is to say A is crossed for a half, by a hot ascending flow, and for the other half, by a discending cold flow.

Particular thermal balance conditions concern the extreme lower and upper layers.

This latter is in conctact with the air of the pool building and the pertinent heat transfer is due to the following three processes (Fig. 1).

a) Convection - this is due to the air motion on the water free surface, and is expressible by means of the usual Newton's law

$$Q_c = \alpha A (T_N - T_a) \qquad (3)$$

where α, the convective heat transfer coefficient, is

$$\alpha = 0.77135 (T_N - T_a)^{0.25} \qquad (W/m^2 °C)$$

Figure 1

Convection, evaporation, radiative heat transfer and gravity driven flows in the spent fuel storage pool.

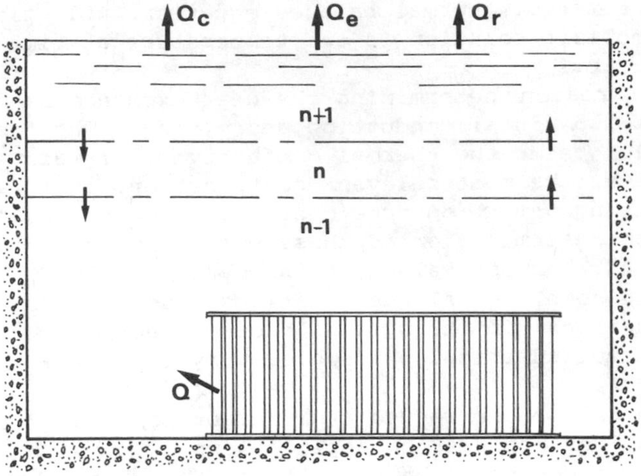

b) Evaporation - this is due to the vapour pressure difference between the water of the N-layer (saturated condition) and the air (whose thermohygrometric parameters are known). The power rate transferred by evaporation from the pool to the air is expressed by the correlation

$$Q_e = J\,A\left[p(T_N) - p(T_a)\right] \tag{4}$$

where $J = 0.0161153\,\alpha \quad (W / Pa\,m^2)$.

The vapour pressure in saturated condition is deduced as a function of the water temperature by means of the polynomial equation ($10°C \leq T \leq 100°C$)

$$p(T) = 630.835 + 40.905\,T + 1.625\,T^2 + 2.216\times10^{-2}T^3 + 3.226\times10^{-4}T^4 + 2.594\times10^{-6}T^5$$

(where $[p] = Pa$ and $[T] = °C$).
The vapour pressure of the moist air in the pool building is simply valued as $p_a = u\,p(T_a)$.

c) Radiative heat transfer - this contribute is only due to the different temperature between the water and the building walls. Assuming that the free surface of the water is quite enclosed in an isothermal convex surface (the walls of the building) and considering water and walls as gray bodies, the radiant power rate is

$$Q_r = \sigma A \frac{T_N^4 - T_w^4}{\frac{1}{a_N} + \varphi\left(\frac{1}{a_w} - 1\right)} \tag{5}$$

For the lower layers, the decay thermal power due to the irradiated fuel must be considered, its expression can be obtained by a polynomial or an exponential interpolation, provided that some experimental data are available (Fig.2 shows a typical trend of the relative power versus time).

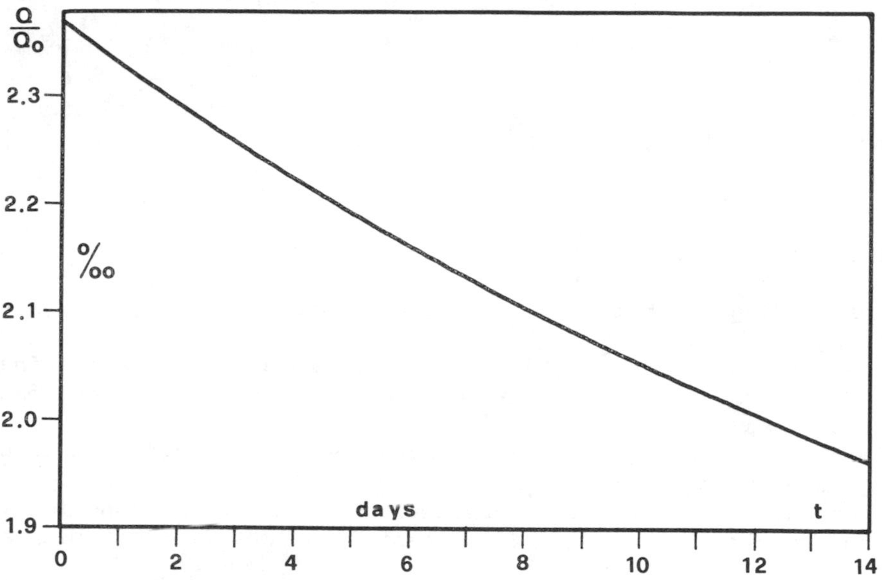

Figure 2 . Relative power decay of the fuel, starrting from sixteen days after shutdown.

Finally, the thermal balance, written for every layer, gives the following system

1) $M_1 \frac{dh_1}{dt} = -m_{1,2} h_1 + m_{1,2} h_2 + Q(t)$

....

n) $M_n \frac{dh_n}{dt} = m_{n-1,n} h_{n-1} - (m_{n-1,n} + m_{n,n+1}) h_n + m_{n+1,n} h_{n+1}$ (6)

....

N) $M_N \frac{dh_N}{dt} = m_{N-1,N} h_{N-1} - m_{N-1,N} h_N - Q_c - Q_e - Q_r$

The enthalpy is a function of the water temperature and pressure, this latter is considered varying with hydrostatic law in the pool. The set of non homogeneous, non linear, first order differential equations (6), can be solved only resorting to numerical techniques in order to evaluate the implicit function T .

NUMERICAL PROCEDURE

Dividing each equation of the system (6) by the corresponding mass layer M_n, the following matricial form can be obtained

$$\frac{d}{dt}\vec{h} = \|C\|\,\vec{h} + \vec{V} \tag{7}$$

Since this is an initial value differential problem depending on time t, the initial condition must be imposed

$$\vec{h}\,[T(t_o)] = \vec{h}_o \tag{8}$$

One can solve the system with a numerical procedure, fixing a suitable time step Δt.
By proceeding in this way, there are 3 large classes of resolutive methods:

a) methods which predict the future functions as linear combination of the past functions, taking account of the present and past derivatives;
b) methods which involve the computation of derivatives with order greater than 1;
c) methods for which the determination of the future functions do not require the memory of the past functions (for example the Runge-Kutta methods).

The procedures b) and c) result too expensive for a numerical treatment, because they require a very long computational time (the water temperature behaviour is studied for a time range greater than a week, by using an integration step whose order of magnitude is approximately the minute).
Hence one of the procedures of the class a) is adopted; this is generally expressed by means of the following equation

$$\vec{h}(t+\Delta t) = \vec{h}(t) + \Delta t\left[\beta\vec{h}'(t) + (1-\beta)\vec{h}'(t+\Delta t)\right] + \vec{E} \tag{9}$$

where $\vec{E}$ is the truncation error.
The method corresponding to $\beta = \frac{1}{2}$ is particularly suitable (Crank-Nicolson method); it is unconditionally stable and the only necessary caution is the choice of a time step smaller than a predetermined value (depending on the maximum eigenvalue of the system), to avoid oscillating solutions.
Now substituting the equation (7) into the equation (9), we have

$$\vec{h}(t+\Delta t) = \vec{h}(t) + \frac{\Delta t}{2}\left[\|C\|_t\,\vec{h}(t) + \vec{V}(t) + \|C\|_{t+\Delta t}\vec{h}(t+\Delta t) + \vec{V}(t+\Delta t)\right] \tag{10}$$

amenable to the form

$$\|C'\|\;\vec{h}(t+\Delta t) = \vec{V}' \tag{11}$$

where

$$\|C'\| = \|I\| - \tfrac{1}{2}\Delta t\;\|C\|_{t+\Delta t}$$

$$\vec{V}' = \vec{h}(t) + \frac{1}{2}\Delta t\left[\|C\|_t \vec{h}(t) + \vec{V}(t+\Delta t) + \vec{V}(t)\right] .$$

The system (11) is not directly resolvable. Difficulties are connected with the non linearity of the terms appearing in $\|C\|$ and $\vec{V}$. This fact occurs directly by virtue of the particular form of the buoyancy mass flow rates and, indirectly by means of the physical water properties and coefficients, which depend on the temperature. The enthalpy vector $\vec{h}$ is calculated step by step resorting to an iterative procedure, until the difference between the predicted and the corrected results is smaller than the pre-established accuracy required in the numerical integration.

RESULTS

This stratified pool model allows to obtain an accurate and fast running estimate licensing tool. It is applied to a hypothetical spent fuel storage pool under operating conditions which approximate the worst accident (loss of cooling). It is assumed that the pool is loaded with one-third core of irradiated fuel assemblies sixteen days after shutdown; the axial power distribution for each assembly is considered homogeneous. The geometrical data and the pertinent parameters of the pool are shown in Table 1.

Table 1. Input data for the thermal hydraulic analysis of the spent fuel storage pool.

Initial water and air temperature	20 °C	Maximum air temperature	30 °C
Building relative humidity	0.7	Air building pressure	1 bar
Nominal power reactor	870 MW	Pool cross section	225 m^2
View factor	0.3	Pool depth	10.5 m

The pool is subdivided in 8 isothermal layers, the first and second segment hold the spent fuel. The equations (7) are solved by the Crank-Nicolson method to predict transient natural circulation flows and heat transfer. Figure 3 illustrates the fluid (demineralized water) temperatures in the stratified pool and the temperature of the air in the building versus time. The temperature increases very slowly (due to the very slow decreasing of the fuel power generation) and reaches its maximum after about 27 days. In every segment the water temperature is smaller than the boiling temperature (120°C for the deepest layer); the temperature difference between the top and the bottom of the pool reaches 24 °C.

At last, the convective, evaporative and radiative power dispersed by the top surface of the pool are shown in Figure 4 versus time. These are the only forms of heat removal in the pool (assuming loss of cooling) and follow the trend of the water temperature; the evaporation is by far the most significant heat transfer mechanism.

A more accurate way to perform this thermal hydraulic analysis could be the use of the large systems codes which are available in the nuclear industry, such as the RETRAN or RELAP codes and the GFLOW code (developed for three-dimensional analysis in porous media). However a severe penalty in computation cost should be paid when such large codes are used for this relatively simple application.

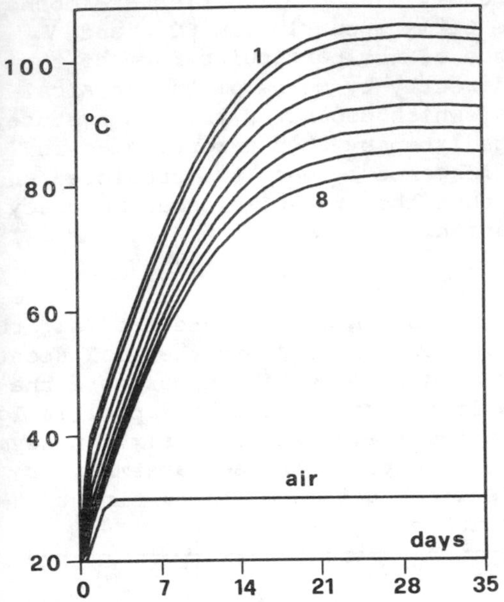

Figure 3. Temperature response of the pool.

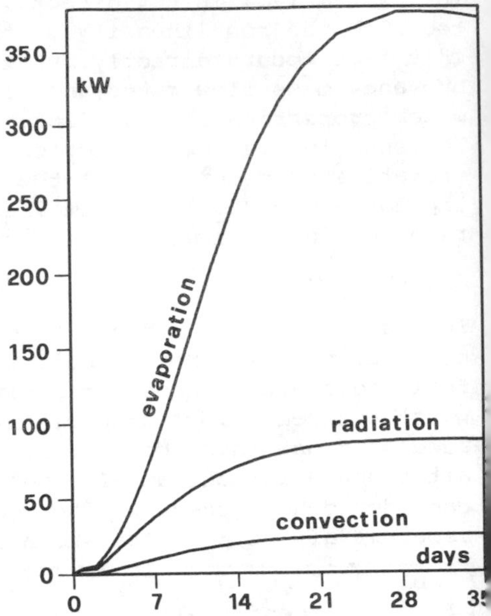

Figure 4. Power losses from the top layer.

ACKNOWLEDGEMENT

This work was financially supported by the CNR.

NOMENCLATURE

a	emissivity
A	cross section of the pool (m^2)
$\|C\|$	matrix of the coefficients
h	water specific enthalpy (J/kg)
J	evaporation coefficient (W/m^2Pa)
m	buoyancy mass flow rate (kg/s)
M	mass of water (kg)
p	vapour pressure (Pa)
Q	thermal power (W)
t	time (s)
T	water temperature (°C)
u	relative humidity
$\vec{V}$	vector of the known terms
α	convective heat transfer coefficient ($W/m^2{}^\circ C$)
φ	view factor
ϱ	water density (kg/m^3)

SUBSCRIPTS

a	air
c	convection
e	evaporation
n	n-th segment
o	initial
r	radiative
w	wall

REFERENCES

1. Malkus, W.V.R. and Veronis, G., Finite amplitude cellular convection, J.Fluid Mech., vol.4, pp.225-260, 1958.

2. Ozoe, H., Yamamoto, K., Sayama, H.and Churchill, S.W., Natural convection patterns in a long inclined rectangular box heated from below, Int.J.Heat Mass Transfer, vol.20,pp.131-139,1977.

3. Davis, S.H., Convection in a box: linear theory, J.Fluid Mech., vol.30, pp.465-478, 1967.

4. Catton, I., The effect of insulating vertical walls on the onset of motion in a fluid heated from below, Int.J.Heat Mass Transfer, vol.15, pp.665-672, 1972.

5. Ozoe, H., Churchill, S.W., Fujii, K.and Lior, N., A theoretically based correlation for natural convection in horizontal rectangular enclosure heated from below with arbitrary aspect ratios, Proc. 7 th Int.Heat Transfer Conf., vol.2,pp.257-262, München, 1982.

6. Chan, A.M.C.and Banerjee, S., Analysis of transient three dimensional natural convection in porous media, J.Heat Transfer , vol.103, pp.242-248, 1981.

7. Chan, A.M.C.and Banerjee, S., Three dimensional numerical analysis of transient natural convection in rectangular enclosures, J.Heat Transfer, vol.101, pp.114-119, 1979.

8. Holst, P.H.and Aziz, K., Transient three dimensional natural convection in confined porous media, Int.J.Heat Mass Transfer, vol.15, pp.73-90, 1972.

9. Exeter, M.K., Hay, N.and Webster, J.J., Natural convection heat transfer in a confined flow geometry, Nucl.Sc.Eng., vol.83 , pp.253-266, 1983.

10. Wu, J.M.and Chuang, C.F., Three dimensional numerical analysis of natural convection of compacted spent fuel, Nucl.Techn. , vol.63, pp.40-49, 1983.

11. Gay, R.R., Gravity driven flow and heat transfer in a spent nuclear fuel storage pool, Nucl.Sc.Eng., vol.83, pp.1-12 , 1983.

12. Gay, R.R., Spent nuclear fuel storage pool thermal hydraulic analysis, Progr.Nucl.Energy, vol.14, pp.199-225, 1984.

Subcooled Forced Convection Film Boiling Heat Transfer

DIMIN YAN
Institute of Atomic Energy
P.O. Box 275
Beijing, PRC

ABSTRACT

In this article the theoretical calculation of film boiling heat transfer to subcooled liquid flowing upwards vertically is presented. Four heat transfer models have been proposed, analysis and comparison have been made in these models We take into account both laminar and turbulent flow patterns for vapor film, different thermal boundary conditions and hydraulic conditions at the liquid-vapo interface, etc. The prediction is compared with ours and Xu Guohua's reflooding experimental data, Fung's steady state film boiling data. The agreement is satis factory. At the same time, the heat transfer in self-modelling region is discussed.

INTRODUCTION

On the basis of Bromleys' film boiling theory, a number of analytical models have been developed for the calculation of heat transfer in film boiling. But most of the work are concerned with the saturation film boiling. Lauer(1) considered liquid subcooling in free convection film boiling. However the application of these models leads to very large discrepancies when compared with experimental data. Therefore, the semiempirical correlations are to be adopted. Based on the experimental data the Bromley correlation (or Ellion correlation) is to be modified to consider the effect of flowrate and subcooling(2).

This paper presents detailed analysis on the subcooled forced convection film boiling heat transfer, describes the adequate physical model and gives the comparison with the experimental data.

ANALYTICAL MODEL

We use two-fluid model to analyse the heat transfer behavior in the inverted annular flow regime. The fluid is considered as divided into a vapor film and subcooled liquid central core. The heat generated in the tube wall is transfered by convection and radiation to the vapor-liquid interface, where on the one hand vapor is produced and on the other hand the heat transfer to subcooled liquid takes place by convection. Radiation heat transfer in inverted annular flow film boiling may be considered as radiation between two flat surfaces:

$$h_{rad}(T_w-T_s) = \frac{\sigma_{SB}}{1/\varepsilon_w + 1/\varepsilon_f - 1}(T_w^4 - T_w^4) \quad (1)$$

where $\varepsilon_w = 0.5$, $\varepsilon_f = 0.95$ (3)

For liquid convection heat transfer under the condition Re>2200, the Dittus-Boelter correlation is used:

$$h_{conv} = \frac{\lambda_f}{d} 0.023 Re_f^{0.8} P_{rf}^{0.4} \tag{2}$$

Under the condition Re<2200, the flow is laminar, but must take into account the effect of free convection. We use Collier correlation

$$h_{conv} = 0.17 \frac{\lambda_f}{d} Re_f^{0.33} P_{rf}^{0.43} \left(\frac{P_{rf}}{P_{rw}}\right)^{0.25} G_{rf}^{0.1} \tag{3}$$

For vapor film we considered both two cases: laminar flow and turbulent flow. For the first case different thermal boundary conditions and hydraulic conditions at the vapor-liquid interface are analyzed.

1. Vapor film laminar flow, zero shear stress at the vapor-liquid interface, uniform wall temperature.

From the momentum conservation equation we get

$$(\rho_f - \rho_g) g (\delta - y) = \mu_g \frac{dU}{dy} \tag{4}$$

where δ = film thickness

U = local velocity

Applying the boundary condition y=0, U=0;

$y=\delta$, $\tau=0$, a integration of above equation is made

$$U = \frac{(\rho_f - \rho_g) g}{\mu_g} \left(\delta y - \frac{y^2}{2}\right) \tag{5}$$

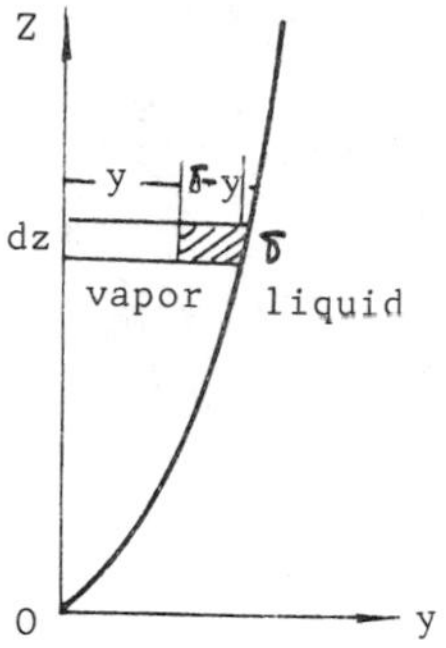

Fig.1 Two-fluid model

From the energy balance we get

$$\left(\frac{\lambda_g}{\delta} + h_{rad}\right)(T_w - T_s) dz$$

$$= \varphi h_{fg} \frac{d\dot{m}_z}{dz} dz + h_{conv}(T_s - T_f) dz \tag{6}$$

where $\varphi = 1 + C_p \frac{T_w - T_s}{2h_{fg}}$

Substituting eq.5 into eq.6 and integrating with the abbreviations

$$A = \frac{\lambda_g \mu_g (T_w - T_s)}{\rho_g (\rho_f - \rho_g) g \varphi h_{fg}}$$

$$B = h_{rad}/\lambda_g$$

$$C = \frac{h_{conv} (T_s - T_f)}{\lambda_g (T_w - T_s)}$$

$$C' = C - B$$

and integrating we obtain

$$\int_0^\delta \frac{\delta^3}{1 - C'\delta} d\delta = AZ \tag{7}$$

or

$$\frac{1}{C'^4}\left\{1.5\,[1-(1-C'\delta)^2] - \frac{1}{3}\,[1-(1-C'\delta)^3]\right.$$

$$\left. -\ln(1-C'\delta) - 3C'\delta\right\} = AZ \qquad (8)$$

Eq. 8 contains the vapor film thickness implicitly. The film boiling heat transfer coefficient can be determined bv

$$h_{FB} = \lambda_g/\delta \qquad (9)$$

2. Vapor film laminar flow, velocity at the vapor-liquid interface U=Uin, uniform wall temperature.

Applying the boundary condition y=0, U=0; $y=\delta$, U=Uin to eq.4 and integrating, we get

$$U = Uin\,\frac{y}{\delta} + \frac{(\rho_f - \rho_g)g}{2\mu_g}\,(\delta y - y^2) \qquad (10)$$

Substituting into eq. 6 we have

$$\frac{1}{4C'^4}\left\{-\ln(1-C'\delta) - 3C'\delta + 1.5\,[1-(1-C'\delta)^2]\right.$$

$$\left. - \frac{1}{3}\,[1-(1-C'\delta)^3]\right\} + \frac{Uin\,\mu_g}{2C'^2(\rho_f-\rho_g)g}\,[\,1-C'\delta - \ln(1-C'\delta)]$$

$$= AZ \qquad (11)$$

3. Vapor film turbulent flow, zero shear stress at the vapor-liquid interface, uniform wall temperature

From momentum balance, we obtain

$$(\delta - y)dz(\rho_f-\rho_g)g = (\mu_g+\rho_f\varepsilon_g)\frac{dU}{dy}\,dz \qquad (12)$$

where ε_g turbulent diffusivity of vapor. Rewrite in dimensionless form

$$(\delta^+-y^+)(\rho_f-\rho_g)g = (U^{*3}/\nu_g^2)\,\mu_g\left(1+\frac{\varepsilon_g}{\nu_g}\right)\frac{dU^+}{dy^+} \qquad (13)$$

where $U^* = \sqrt{\tau_w/\rho_g}$, $\quad \tau_w = \delta(\rho_f-\rho_g)g$

we take

$$\varepsilon_g/\nu_g = b^2(y^+)^2$$

where b=0.091

The velocity distribution of the vapor is (4)

$$U^+ = \frac{-a}{2b^2}\ln(b^2y^2+1) + \frac{a\,\delta^+}{b}\tan^{-1}(by^+) \qquad (14)$$

where $a = \nu_g\,(\rho_f-\rho_g)g/(U^{*3}\rho_g)$

Film heat transfer equation can be written as

$$q_c = -\rho_g C_p (K + \varepsilon_g) \frac{dT}{dy} \tag{15}$$

Express in dimensionless form

$$T^+ = \rho_g C_p U^* (T_w - T)/q_c$$

and a integration is made

$$T^+ = \frac{\sqrt{Pr_g}}{b} \tan^{-1}(b\sqrt{Pr_g} y^+) \tag{16}$$

Since the temperature at the distance δ is saturation temperature, from above equation we obtain

$$q_c = \frac{b \rho_g C_p U (T_w - T_s)}{\sqrt{Pr_g} \tan^{-1}(b\sqrt{Pr_g} \delta^+)} \tag{17}$$

Substituting into energy conservation equation and integrating, we finally get

$$\int_0^{\delta} \frac{f_1 \delta^3 d\delta}{f_2 - C'\delta} = AZ \tag{18}$$

where

$$f_1 = \frac{1.5}{(\delta^+)^3} b^2 [\{ -\frac{1}{b} + b(\delta^+)^2 \} \tan^{-1}(b\delta^+) - \delta^+]$$

$$f_2 = \frac{b\delta^+ \sqrt{Pr_g}}{\tan^{-1}(b\delta^+ \sqrt{Pr_g})}$$

The difference between eq. 18 and eq. 7 is merely attributed to multiplier coefficients f_1, f_2. The film boiling heat transfer coefficient is

$$h_{FB} = f_2 \frac{\lambda_g}{\delta} \tag{19}$$

4. Vapor film laminar flow, zero shear stress at the vapor-liquid interface, uniform heat flux.

In this case from energy balance we obtain

$$\left(\frac{\lambda_g}{\delta} + h_{rad}\right)(T_w - T_s)Z - h_{conv}(T_s - T_f)Z$$

$$= \varphi h_{fg} \frac{\rho_g(\rho_f - \rho_g)g}{\mu_g} \frac{\delta^3}{3} \tag{20}$$

or

$$\frac{\delta^4}{3A(1-C'\delta)} = Z \tag{21}$$

CALCULATION RESULTS

Calculation results of model 1

The detailed calculation have been made for film boiling heat transfer coefficient as a function of distance with different wall temperature, flow rate and fluid subcooling. The heat transfer coefficient is increased with the increasing of the subcooling and flow rate, and is decreased with the increasing of wall temperature. Moreover, the trend of heat transfer coefficient with the distance also depends on these parameters. Generally, the dependence on distance becomes more flat in the high flowrate and high subcooling cases. But during the reflooding of heat element and in steady state thermal experiments the wall temperature is not uniform and always rises with the distance, so that the trend of decreasing heat transfer coefficient with distance is almost identical. The data in literature(5) can confirm this point. It also can be seen from the following comparison. The calculation result is correlated by the form of modified Bromley equation. That is

$$h_{FB} \,/\, \left[\frac{\lambda_g^3 (1 + c_p \frac{T_W - T_s}{2h_{fg}}) h_{fg} \rho_g (\rho_f - \rho_g) g}{\mu_g (T_W - T_s) Z} \right]^{\frac{1}{4}} = C \tag{22}$$

Fig.2 shows the dependence C with $U_{in}\frac{T_s - T_f}{T_W - T_s}$ at the different wall temperature, flowrate and subcooling. Seen, it may be described by a unified curve

$$C = 0.7 + 75(U_{in}\frac{T_s - T_f}{T_W - T_s})^{1.35} \tag{23}$$

The comparison of heat transfer coefficient calculated with four models

Fig.3 shows the comparison of calculation results for above mentioned four models. The heat transfer coefficient obtained under the assumption of vapor film turbulent flow is much higher than other models. Moreover it rises with Z at the distance $Z > 30$mm. These disgree with our experimental data and related foreign literatures. In addition, in the vapor film laminar flow cases the thermal and hydraulic conditions have some effect on the heat transfer, but the difference gradually becomes unimportant with increasing flowrate, which can be seen from Table 1. This is the reason why the discrepancy between the experimental data is great under low flowrate condition.

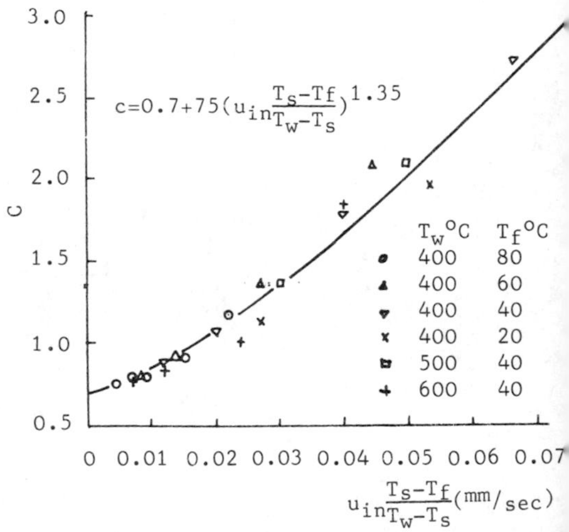

Fig.2 Correlation of calculation results (z = 100 mm)

Comparison with experimental data

As the data in literature (5) (6) are expessed in the form $C/(1+0.025\Delta T_{sub})$, for the convenience we recorre-lated the predicted value C, which can be expressed within ±20% by

$$C = 0.039 U_{in}^{0.55} (1 + 0.025 \Delta T_{sub}) \tag{24}$$

In Fig.4 the data in (3), (5) is compared with above equation. The agreement is satisfactory. Comparison with data in (6) gives the same result. From Fig.5 it is also seen, that concerning the trend of heat transfer coefficient with distance the analysis is very close to the experimental data.

Table 1. The effect of boundary condition on heat transfer coefficient

Uin mm/sec		60	100	200	330
Coef. in model 1	C1	0.880	1.061	1.779	2.710
2	C2	0.704	0.978	1.774	2.708
4	C4	0.970	1.176	1.831	2.723
Effect of hydrau.con.	C2/C1	0.800	0.922	0.997	0.999
Effect of therm.con.	C4/C1	1.102	1.108	1.029	1.005

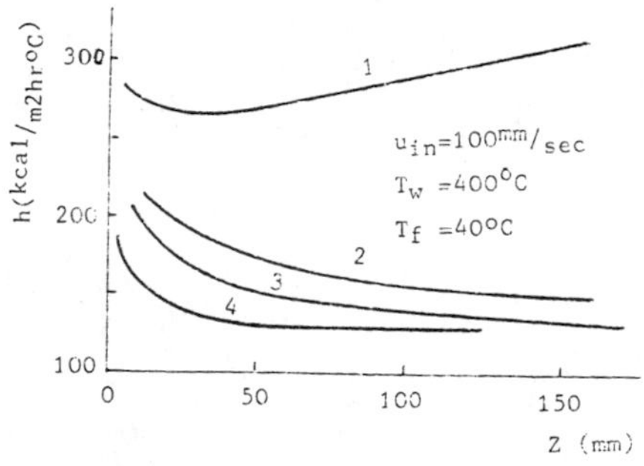

Fig.3 Comparison of different anality cal model

1 turbulent $\tau=0$ at $y=\delta$ T_w=const.
2 laminar $\tau=0$ at $y=\delta$ q_w=const.
3 laminar $\tau=0$ at $y=\delta$ T_w=const.
4 laminar $u_\delta = u_{in}$ T_w=const.

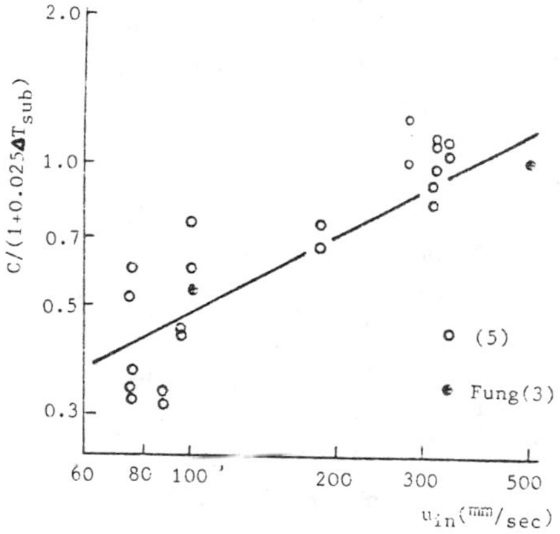

Fig.4 Comparison of theoretical calculation with experimental data

DISCUSSION

The calculation results show that when Z is greater than a certain value, particularly at high flowrate and high subcooling, the heat transfer coefficient always not changes. It is said, the film boiling enters the self-modelling region. In this case, the film thickness is not varied that is $d\delta/dz = 0$

From equation 6, we get

$$Z_{st} = \frac{2.802}{AC'^4} \quad \text{for } 1\%$$

$$Z_{st} = \frac{1.309}{AC'^4} \quad \text{for } 5\%$$

It is seen from fig.6, when Uin> 200 mm/sec, ΔTsub>40°C, the film boiling enters the self-modelling region just as it takes place.

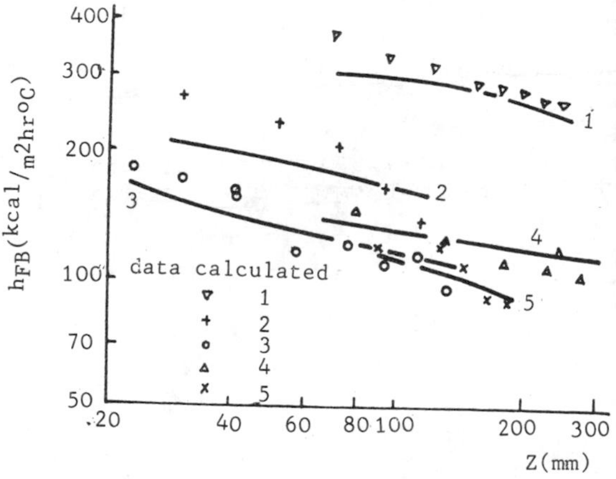

Fig.5 Comparison of calculation results with experimental data

	▼	+	O	▲	×
u_{in}	315	184	74.8	95.1	107.8
ΔT_{sub}	54.6	49.4	55.1	50.5	35.5

CONCLUSION

1. Four heat transfer models are analysed. Under subcooled forced convection condition, film layer is thinner, and turbulent flow doesn't occur in vapor film. For laminar flow vapor film, boundary condition has certain effect on heat transfer, but the effect becomes not important in high flowrate cases.

2. A correlation of subcooled forced convection film boiling heat transfer is suggested, which is essentially consistent with data of (3), (5), (6).

3. The problem of heat transfer of the self-modelling region is discussed.

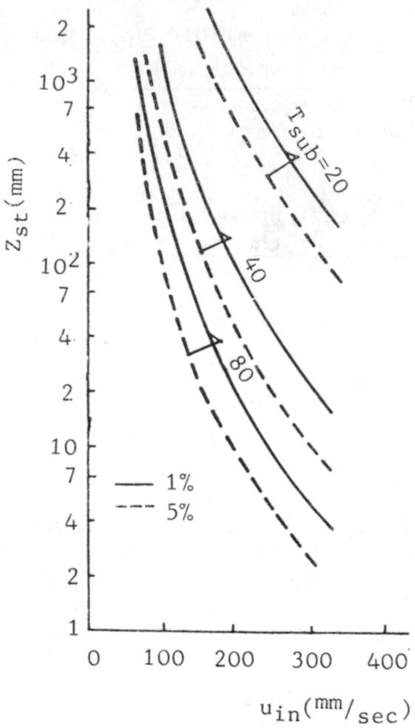

Fig.6 Dependence of Z_{st} on ΔTsub, Uin

Literature

1. Lauer H. & Hufschmit W., Two-Phase Transport and Reactor Safety, V. 3 pp. 1307-1326 (1978)

2. Yan Dimin, Refooding Heat Transfer, Seminar on LWR Safety Heat Transfer (1982)

3. Fung K.K. et al., Nücl. Eng. & Des. Vol.55 No.1 (1979)

4. Osakabi M. & Sudo Y., Jl. of Nucl. Sci. & Techn., Vol.21 No.2 (1984)

5. Yan Dimin, et al., Rewetting Heat Transfer during Bottom Flooding of Tubular Test Section, Journal of Engineering Thermophysics, Vol.6 No.1 (1985)

6. Xu Guohua, et al., Rewetting Heat transfer with Water on Heating Surface, Annual Report of IAE p.165(1983)

COMBUSTION HEAT TRANSFER

Radiative Heat Transfer Enhancement to a Water Tube by Combustion Gases in Porous Media

R. ECHIGO, K. HANAMURA, and Y. YOSHIZAWA
Department of Mechanical Engineering
Tokyo Institute of Technology
Meguro-ku, Tokyo, 152, Japan

T. TOMIMURA
Research Institute of Industrial Science
Kyushu University
Kasuga City
Fukuoka Prefecture, 816, Japan

INTRODUCTION

Although flame radiation plays an important role in conventional furnaces and heat transfer facilities, it requires a capacious volume for the heating chamber to enlarge the dimension of flame and the resultant optical thickness for heat transfer augmentation. Moreover the heating object must be installed away from the reaction zone to prevent the emission of unburnt species when its surface temperature is low. These disadvantages make it difficult to reduce the size of the heating chamber appreciably. One of the authors has shown in a previous work [1] that the porous medium of an appropriate optical thickness placed in a duct was very effective for energy conversion from flowing gas enthalpy to thermal radiation directing toward the higher temperature side, and the successful applications to an industrial furnace [1] and to a combustor for extremely low calorific gas [2] have been reported. Recently, K.Y. Wang and C.L. Tien [3] have reported the influence of scattering on the effectiveness of the energy conversion in a similar flow system. The thermal structures in the porous media with internal heat generation have also been studied analytically and experimentally [4], and it has been shown that it provided a uniform space of high temperature and extremely high radiant energy density in it.

A high performance heat transfer facility has been developed in this study, in which a water tube has been directly inserted into the high radiant energy density space. The temperature profiles and the radiation field in the porous media have been clarified by experiments and numerical analyses based on a proposed model, the results of which demonstrate the outstanding heat transfer characteristics by as much as 3 to 5 times higher compared with the smooth tube. It is also shown that combustion reaction is completed in a very thin space in the porous medium. These results are promising to reduce the size of heat transfer facilities with combustion.

EXPERIMENTAL APPARATUS

A cross sectional view of the experimental apparatus is schematically sketched in Fig.1. It consists of stainless steel duct of the cross section 200 x 50 mm^2, in which Porous Medium I, II and III are installed. A water tube of the diameter 14 mm is directly inserted in the combustion space packed with Porous Medium II. The dense porous media, Porous Medium I and III, are respectively made of 20 sheets of a stainless steel wire net of 50 meshes and their packing fractions are 26 %. On the other hand, a pile of several sheets of a wavy net with very large meshes is used as Porous Medium II. It is very sparse and its packing fraction is 5.18 % at most. City gas is used as the fuel whose lower heating value is 1.87×10^4 kJ/Nm^3. A mixture of city gas and air is supplied from the perforated pipe, which works as a line burner when the porous media are removed, and flows into the porous media after rectified by ceramic porous plate.

Experiments were also executed using a line burner for comparison, in which the porous media and the bottom cover ① were removed and the backet ② was attached instead. The secondary air was supplied from the bottom of the backet. The equivalence ratio of a mixture of the primary air and city gas is kept constant 2.4, and the total equivalence ratio is controlled by varying the secondary air.

Eleven chromel-alumel thermocouples, whose diameters are 0.2 mm, are embeded along the center line of the apparatus. Since both diameters of the thermocouples and the elemental wires of the porous media are the same, they are able to be regarded as the elements of the porous media and the measured temperatures are considered to be those of the solid phase. The entrance temperature of water is about 293 K, the flow rate of which is 16.7 cm^3/s.

EXPERIMENTAL RESULTS AND DISCUSSION

Fig.2 shows a typical result of the temperature profiles along the flow direction for combustion load L_c = 2.5 kW and equivalence ratio ϕ = 0.45. The packing fraction of Porous Medium II, f_p, was taken as the parameter and varied from 0 to 2.59 %. f_p = 0 corresponds to experiments without Porous Medium II. x coordinate is measured from the boundary between Porous Medium I and II toward the flow direction. The position of the water tube, l_p, is expressed using the x coordinate of its center. It is installed, in this case, at x = 25 mm. Since the maximum temperature appears at the downstream end of Porous Medium I except for f_p = 0, flame is stabilized there. The stabilization mechanism is considered to be the same with that of the conventional one-dimensional flat flame burner. The temperatures in the upstream of the water tube are high when f_p is large because the energy recirculation is formed to some extent within the upstream region. On the other hand, the temperatures for f_p = 0 are higher than the others in the downstream. This evidence implies that the energy recircu-

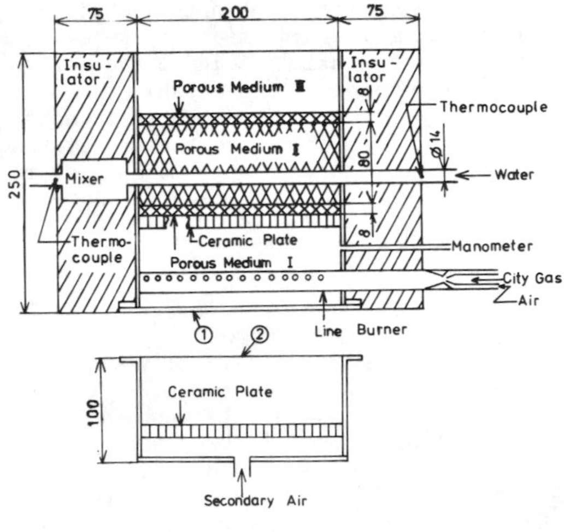

FIGURE 1. Experimental apparatus

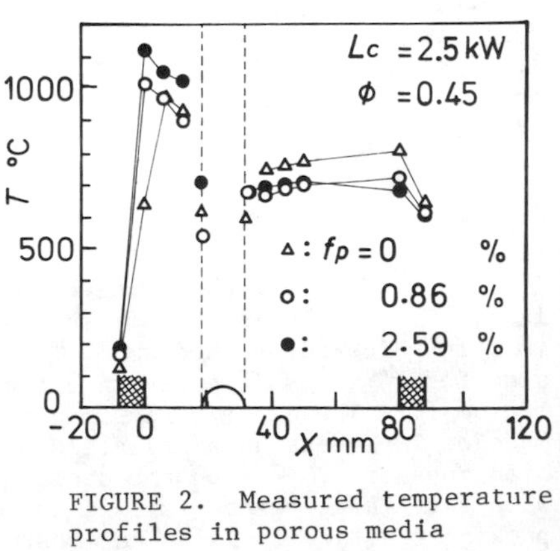

FIGURE 2. Measured temperature profiles in porous media

lation is formed only between Porous Medium I and III. Porous media also have a prominent effect in radiation shielding, which is well demonstrated by the temperature differences across Porous Medium I and across III.

Mean Nusselt numbers, Nu_m which are shown in Fig.3, were evaluated from the heat fluxes at the water tube surface calculated from water temperature rises, ΔT, and the water tube surface area. Here, the adiabatic flame temperature based on the ambient temperature was employed as the higher representative temperature. On the other hand, the surface temperature of the water tube was estimated to be 373 K from a simple calculation of mean heat transfer coefficient of the inner surface (about 1 kW/m^2K). The curve denoted by Nu_{me} represents the mean Nusselt numbers around a cylinder given by an empirical formula for forced convective heat transfer. The water tube touches Porous Medium I, when l_p = 7 mm in Fig.3-(b). The results for the line burner heating type are about twice as large as Nu_{me}, which is considered to be attributable to the effects of the blockage of the duct by the water tube and of radiative heat transfer between the duct wall and the water tube. Moreover when porous media are installed, Nu_m's are about three times as large as that for the line burner heating type. The increment of Nu_m in Fig.3-(a) with increasing f_p corresponds to the temperature rise in the upstream of the water tube. Nu_m decreases abruptly in Fig.3-(b) with decreasing ϕ when l_p = 7 mm, that is, the water tube touches the downstream surface of Porous Medium I. It corresponds to the observation that the flame was partially quenched and moved to the downstream of the water tube.

ANALYTICAL MODEL

A one-dimensional model was adopted for numerical analysis. The analytical model is shown in Fig.4 with the side view of the experimental apparatus, where x and τ denote geometrical and optical coordinate systems, respectively. The working gas flows in across the section at $x = -x_1$ at the ambient temperature T_{m1} and flows out across the section at $x = x_e$ at the temperature T_{me}. It gains, en route, Q_1 in the flame zone of the thickness δ and loses Q_7 through the side wall. There are incident radiative heat fluxes J_1^+ and J_e^- at the both ends of Porous Medium I and III from full radiators at the ambient temperature. It is assumed that the water tube is square of the side d and that the upper and lower surfaces of it are heated only by radiative heat transfer and the side surfaces are heated only by convective heat transfer.

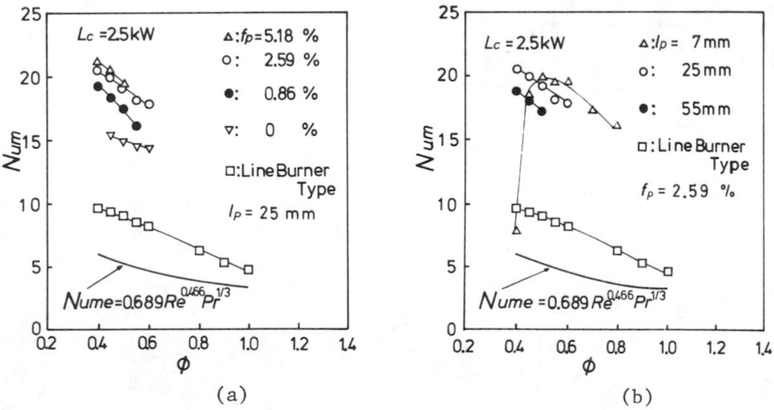

FIGURE 3. Nusselt numbers (measured) (a) Effects of f_p (b) Effects of l_p

The assumptions employed for radiative heat transfer are almost the same as the previous work [1], among of which the principal ones are as follows; (i) The working gas is non-radiating. (ii) The porous media consist respectively of fine solid particles dispersed homogeneously in the layers. (iii) The porous media are able to emit and absorb thermal radiation in local thermal equilibrium and the scattering is not taken into account. (iv) The one dimensional approximation of radiative propagation along x-direction is to be valid. The other assumptions introduced here are; (v) The flow is one dimensional along x-direction in steady state. (vi) There is the exothermic zone at the downstream end of Porous Medium I. (vii) The volumetric heating rate is uniform throughout this zone. (viii) The overall heat transfer coefficient of the duct wall, K, is constant. (ix) The heat transfer coefficient of the water tube surface, K_w, is constant. (x) The absorptivity of the water tube surface is unity, and its temperature, T_w, is constant to be 373 K. (xi) The physical properties are constant.

BASIC EQUATIONS AND BOUNDARY CONDITIONS

Following the foregoing assumptions, the energy equations for gas- and solid-phases are formulated, respectively, as follows,

$$\rho_m \cdot c_p \cdot u_m \cdot b \cdot (dT_m/dx) + 2 \cdot K \cdot (T_m - T_{m1})$$

$$+ b \cdot h \cdot n \cdot A \cdot (T_m - T_p) - b \cdot q + 2 \cdot K_w \cdot (T_m - T_w) = 0 \quad (1)$$

$$b \cdot h \cdot n \cdot A \cdot (T_m - T_p) - b \cdot (dq_R/dx) = 0 \quad (2)$$

The fourth and the fifth terms in eq.(1) express the exothermicity in the flame zone and the heat transmission to the water tube, respectively, and are to be taken into account only in the specified zones. The third term in eq.(1) and the first and the second terms in eq.(2) have to be evaluated using the physical properties of the solid-phase corresponding to the porous medium on the spot. Dimensionless variables and parameters are defined as follows,

$Re = 2 \cdot b \cdot \rho_m \cdot u_m / \mu$, $P = (4 \cdot K \cdot b / k_m)/(Re \cdot Pr)$, $P_w = (4 \cdot K_w \cdot b / k_m)/(Re \cdot Pr)$,

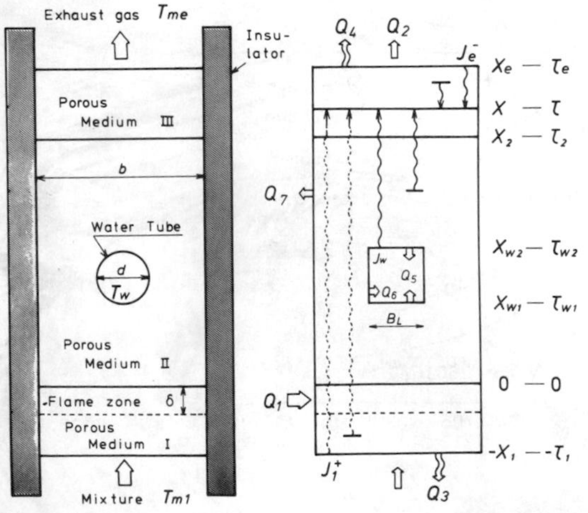

FIGURE 4. Analytical model and side view of the apparatus

$$Nu_p = h \cdot d_p/k_m, \quad Jw = \sigma \cdot T_w{}^4/(\sigma \cdot T_{m1}{}^4), \quad J_1{}^+ = J_e{}^- = \sigma \cdot T_{m1}{}^4/(\sigma \cdot T_{m1}{}^4),$$

$$Pr = \mu \cdot c_p/k_m, \quad M_p = 2 \cdot Nu_p \cdot n \cdot A \cdot b^2/d_p, \quad N_{Rp} = \kappa_p \cdot k_m/(4 \cdot \sigma \cdot T_{m1}{}^4),$$

$$Q_p = M_p/(Re \cdot Pr), \quad H(\tau) = q_R(x)/(4 \cdot \sigma \cdot T_{m1}{}^4), \quad q_c = q \cdot \delta/(\rho_m \cdot c_p \cdot u_m \cdot T_{m1}),$$

$$\tau = \kappa_p \cdot x, \; \tau_o = \kappa_p \cdot b, \; \Delta = \delta/b, \; X = x/b, \; \theta = T/T_{m1} \quad (3)$$

Substitutions of these parameters into eqs.(1) and (2) yield

$$\frac{d\theta_m}{dX} + P \cdot (\theta_m - \theta_{m1}) + Q_p \cdot (\theta_m - \theta_p) - \frac{q_c}{\Delta} + P_w \cdot (\theta_m - \theta_w) = 0 \quad (4)$$

$$Q_p \cdot (\theta_m - \theta_p) - \frac{2 \cdot \tau_o{}^2}{Re \cdot Pr \cdot N_{Rp}} \cdot \frac{dH(\tau)}{d\tau} = 0 \quad (5)$$

Taking optical properties along the path and the blockage by the water tube into account, radiative heat fluxes from every part of the porous media have to be integrated at each section in the porous media to evaluate $H(\tau)$ in eq.(5). For example, it is written as follows at a section in Porous Medium III (i.e. $X_2 < X < X_e$ in Fig.(5)) when the properties of Porous Medium I and III are the same and are different from those of Porous Medium II.

$$\begin{aligned} H(\tau) = \frac{1}{2} \cdot [\ & (1-B_L) \cdot J_1{}^+ \cdot E_3(\tau+\tau_1) - J_e{}^- \cdot E_3(\tau_e-\tau) + B_L \cdot J_w \cdot E_3(\tau-\tau_{w2}) \\ & + (1-B_L) \cdot \int_{-\tau_1}^{0} \theta_p{}^4 \cdot E_2(\tau-\tau') \cdot d\tau' + (1-B_L) \cdot \int_{0}^{\tau_{w2}} \frac{\kappa_{pI}}{\kappa_{pII}} \cdot \theta_p{}^4 \cdot E_2(\tau-\tau') \cdot d\tau' \\ & + \int_{\tau_{w2}}^{\tau_2} \frac{\kappa_{pI}}{\kappa_{pII}} \cdot \theta_p{}^4 \cdot E_2(\tau-\tau') \cdot d\tau' + \int_{\tau_2}^{\tau} \theta_p{}^4 \cdot E_2(\tau-\tau') \cdot d\tau' \\ & - \int_{\tau}^{\tau_e} \theta_p{}^4 \cdot E_2(\tau'-\tau) \cdot d\tau'\ ] \end{aligned} \quad (6)$$

where, $E_n(\tau)$ is exponential integral function of n'th order.

$$E_n(\tau) = \int_0^1 \tilde{\mu}^{\,n-2} \cdot \exp(-\tau / \tilde{\mu}) \cdot d\tilde{\mu} \quad (7)$$

The boundary conditions for eq.(1) reduces to

$$\theta_m = \theta_{m1} = T_{m1}/T_{m1} = 1 \quad (\text{ at } X = X_1) \quad (8)$$

The parameters for the calculation are determined based on the experimental conditions and observations as much as possible. The overall heat transfer coefficient around the tube is estimated using the velocity at the minimum duct width. Here, B_L is 0.28 (14 mm / 50 mm).

The overall energy balance of the system is expressed as the balance between the combustion load Q_1 and the sum of the out-flow enthalpy of the working gas Q_2, radiation from the upstream end of Porous Medium I and from the downstream end of Porous Medium III Q_3 and Q_4, respectively, the amounts of heat transfer to the water tube surface by radiation Q_5 and convection Q_6, respectively, and heat loss throughout the duct wall Q_7.

$$Q_1 = Q_2 - Q_3 + Q_4 + Q_5 + Q_6 + Q_7 \quad (9)$$

here Q_5 is described as follows,

$$Q_5 = B_L \cdot [\ H^+(\tau_{w1}) - H^-(\tau_{w2}) - 2 \cdot J_w\ ] \tag{10}$$

ANALYTICAL RESULTS AND DISCUSSION

Fig.5 shows typical results of profiles of temperature and radiative heat flux along the flow direction, in which the positions of the water tube are sketched by the broken line. They are described in terms of the optical length from the downstream end of Porous Medium I to the water tube, τ_d, since the thermal structures in porous media are determined principally by the optical thickness and the geometrical length can be changed arbitrarily by using a porous medium of appropriate optical properties. The optical thickness of Porous Medium I and II are 10 and 20 respectively, and III is not installed. Heat loss Q_7 through the duct wall is neglected. The larger optical thickness, τ_d, is preferable to raise the maximum temperature in the porous media. On the other hand, the optical thickness between the water tube and the downstream end of porous media is also important for the radiation shielding. The radiative heat fluxes toward up- and downstream change discontinuously across the water tube, and the absolute values of the jumps are proportional to the radiative heat transfer to the water tube. They show the maximum in Fig.5 when τ_d = 9.8 because the temperature in the porous media is low when τ_d = 1.5 and the radiation shielding is insufficient when τ_d = 17.7.

COMPARISON BETWEEN ANALYSES AND EXPERIMENTS

Fig.6 shows the analytical and experimental results for f_p = 2.59 %, ϕ = 0.45 at various combustion loads. The analytical results show reasonable agreement with the experimental ones except for the points close to the water tube in the downstream. This difference is attributable to the gap between the one dimensional model and the experiments, since the thermocouples were embeded along the center line and some of them submerged in the wake of the water tube. The

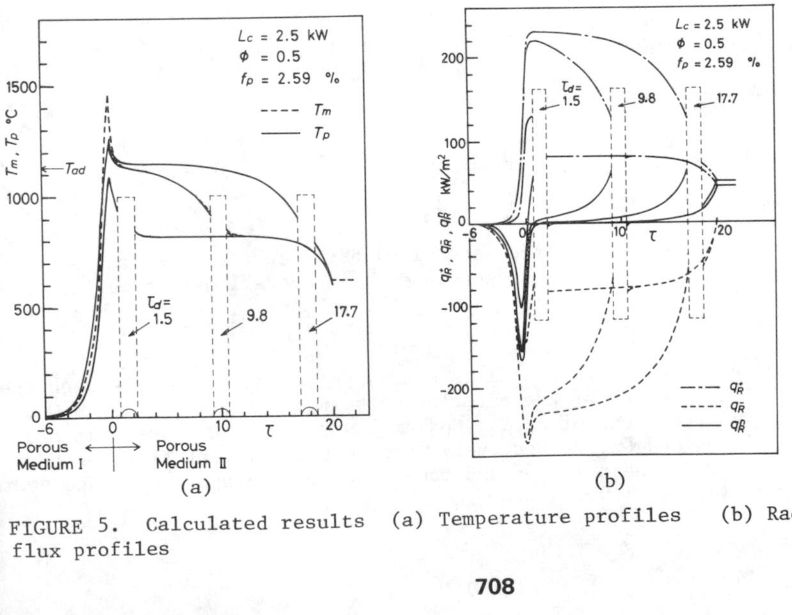

(a) (b)

FIGURE 5. Calculated results (a) Temperature profiles (b) Radiative heat flux profiles characteristic of this apparatus is also assessed in term of the efficiency defined as the ratio between the amount of heat transfer to the water tube and the combustion load ($\eta = (Q_5 + Q_6) / Q_1$) which are shown in Fig.7. Here, the conditions f_p = 0.86, 2.59 and 5.18 % correspond to the optical lengthes of Porous Medium II, τ = 2.7, 8.1 and 16.2, respectively, when the geometrical thickness of Porous medium II is 80 mm. On the other hand, l_p = 7, 25 and 55 mm correspond to τ_d = 0.7, 2.5 and 5.5, respectively, when f_p = 2.59 %. It is noteworthy that around 30 to 40 % of the combustion load can be withdrawn by only one smooth water tube. Both the experimental and analytical results show the same trends, but the experimental results of η show somewhat stronger dependencies on f_p and l_p than analytical ones. This discrepancy is partially attributable to the position of the flame zone. It is fixed in the analysis, but it moves around the boundary following to the condition of flame stabilization in the experiment.

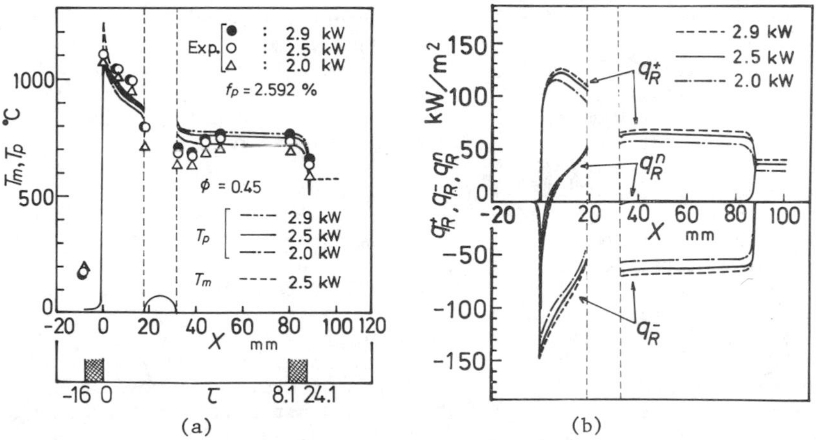

FIGURE 6. Comparison between analyses and experiments; effects of combustion loads

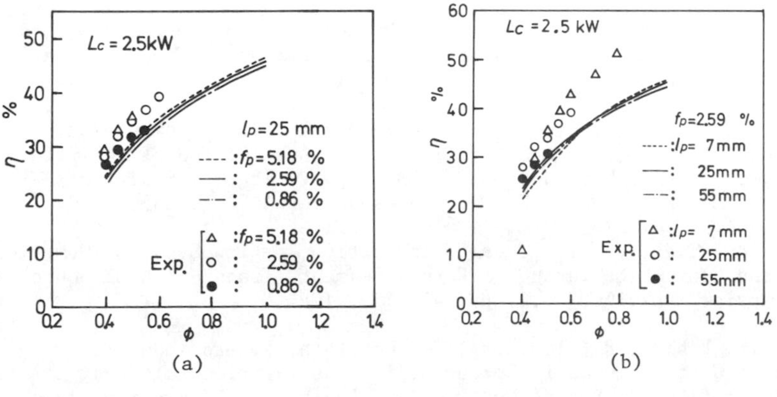

FIGURE 7. Efficiency (a) Effects of f_p (b) Effects of l_p

CONCLUSIONS

(1) A high performance heat transfer facility has been developed in which a water tube has been directly inserted into the high radiant energy density space in a porous medium combustor, and its outstanding characteristic of heat transfer has been demonstrated.
(2) The radiation field in the heat transfer facility and its basic characteristics have been clarified by numerical analysis based on a proposed one dimensional model.

NOMENCLATURE

A : surface area of a particle (m^2)
B_L : blockage ratio
b : width of the duct (m)
c_p : specific heat at constant pressure ($J \cdot kg^{-1} \cdot K^{-1}$)
d : diameter of the water tube (m)
d_p : diameter of a particle (m)
f_p : packing fraction of Porous Medium II (%)
h : heat transfer coefficient around the particles ($W \cdot m^{-2} \cdot K^{-1}$)
K : overall heat transfer coefficient ($W \cdot m^{-2} \cdot K^{-1}$)
k : thermal conductivity ($W \cdot m^{-1} \cdot K^{-1}$)
L_c : combustion load (kW)
l_p : height of the water tube (mm)
n : number density of particles (m^{-3})
Nu_m : mean Nusselt number
q : volumetric heating rate ($W \cdot m^{-3}$)
q_R : radiative heat flux ($W \cdot m^{-2}$)
Q_i : dimensionless energy (Fig.1)
T : temperature (K)
ΔT : water temperature rise (K)
u : velocity (m/s^{-1})
x : geometrical coordinate system

Greek Symbols
δ : width of reaction zone (m)
η : thermal efficiency
κ_p : absorption coefficient (m^{-1})
  κ_{pI} for Porous Medium I,III
  κ_{pII} for Porous Medium II

μ : viscosity ($Pa \cdot s$)
ρ : density ($kg \cdot m^{-3}$)
σ : Stefan Boltzmann constant (5.67×10^{-8} $W \cdot m^{-2} \cdot K^{-4}$)
τ : optical coordinate system
ϕ : equivalence ratio

Subscripts
1,0,2,e : see coordinate system(Fig.1)
m : gas-phase
p : solid-phase
w : wall surface of the water tube

Superscripts
$+,-,n$: radiative heat fluxes toward positive, negative directions and net radiative heat flux
' : integral variable
$H, J, M_p, N_{Rp}, P, Pr, P_w, q_c, Q_p, Re, X, \Delta, \theta, \tau$: dimensionless quantities, see eqn.(3)

REFERENCES

1. Echigo, R. Effective Energy Conversion Method Between Gas Enthalpy and Thermal Radiation and Application to Industrial Furnaces, Proc. 7th Int'l Heat Transfer Conference, vol.VI (1982), p.361.

2. Echigo, R., Kurusu, M., Ichimiya, K. and Yoshizawa, Y. Combustion Augmentation of Extremely Low Calorific Gases (Application of the Effective Energy Conversion Method from Gas Enthalpy to Thermal Radiation), Proc. 1983 ASME-JASME Thermal Engng. Conference, vol.IV (1983), p.99.

3. Wang, K. Y. and Tien, C. L. Thermal Insulation in Flow Systems: Combined Radiation and Convection Through a Porous Segment, Trans. of ASME Journal of Heat Transfer, vol.106, No.2, pp. 453-459, 1984.

4. Yoshizawa, Y., Echigo, R., Tomimura, T., Hanamura, K. and Koda, M. Combustion of Gaseous Fuels in Porous Media (An Experimental Study on Temperature Profiles in Porous Media and an Analysis on Radiation Field), Trans. of JSME (in Japanese), vol.B-51, No.466 (1985), forthcoming.

Calculation of Three Dimensional Heat Transfer in Utility Boiler Combustors

ZHAOXING SUN and XIUGUO MA
Beijing Graduate School
North China Institute of Electric Power
Beijing, PRC

ABSTRACT

In this paper, Monte Carlo Solution is presented in calculating three dimensional heat transfer in combustors with inclined walls. In order to improve accuracy and to save time in calculation, random emission model and DPE(Differential Emissive Power Emission Method) iterative algorithm are employed. This method was applied to calculating heat transfer in a 400 T/H boiler manufactured by Shanghai Boiler Plant. Results obtained have been proved to be satisfactory in comparing with experimental data.

NOMENCLATURE

$A, \delta A$	Area and element area.
K	Gas absorption coefficient.
L	Energy bundle free path.
m, n	Gas volume and wall element number divided a combustor into for calculation.
P	Probability of an element absorbing radiant energy given out by another.
Q, q	Quantity of heat.
R	Uniform random number within [0,1] and emissive directional cosine.
T	Temperature.
$V, \delta V$	Gas volume of combustor and gas volume element.
x_j, y_j, z_j	Coordinates of reference point of element j.
$\Delta X, \Delta Y, \Delta Z$	Measures of element.
ε	Waterwall surface emissivity.
η, θ	Angle.
σ	Stefan-Boltzmann Constant.

SUBSCRIPTS

a	absorption;	conv	convection;	g	gas;
r	radiation;	w	waterwall		

INTRODUCTION

Till now, studies on three dimensional heat transfer in utility boilers have been made for more than two decades. The common practice in such calculation nowadays is to divide the combustor into numerous elements transforming continuous quantities into discrete ones; then the discrete values of temperature distribution and heat flux distribution for each element could be obtained by solving energy equations stand for respective elements.

The three dimensional energy equations are:

$$q_f = q_{cond}+q_{r.v\to\delta v}+q_{r.A\to\delta v}+q_{comb}-q_{r.g} \quad \text{(for gases)}$$
$$q_{conv}+q_{r.v\to\delta A}+q_{r.A\to\delta A}=q_{r.w}+q_{w.a} \quad \text{(for wall)} \qquad (1)$$

In equation (1), q_f, the term of convection and q_{cond}, the term of turbulent heat conduction can be solved by difference method. q_{comb}, combustion heat, is a given value. Hence the calculation of heat transfer in a combustor turns out to be that of radiant heat transfer in non-black system.

Various methods are availabe in calculating radiant heat transfer. Through comparing and analysing, Monte Carlo Method is preferred[1]. In Monte Carlo Method, the physical process of radiant heat transfer is simulated by mathematical principles of probability and statistics and through which the radiant heat transfer is solved. Studies on applying Monte Carlo Method in calculating radiant heat transfer in boiler combustors have been carried out as beginning from the sixties, calculation on utility boiler of large capacity has been made and some achievements have been obtained[2]~[4].

However , the above mentioned calculations were, without exception, made by simplifying boiler combustors into a box of regular shape. Practically, boiler combustors are not necessarily regular in shape the overall lay-out of a boiler is full of variety. By simplification, calculations may be made more convenient, but at the same time additional errors may result. Hence, it is necessary to develop an acceptable method in calculating three dimensional heat transfer in combustors by adopting original, or in close proximity to original, shapes of combustros.

For this purpose, we through years of study, developed a solution of calculating three dimensional heat transfer in combustors with inclined walls using Monte Carlo Method. Here, random emission mode[5] and DPE iterative algorithm [4] [5] are employed. Results obtained by applying this method to the calculation of three dimensional heat transfer in an SG-400 T/H superpressure reheating boiler have been proved to be satisfactory while comparing with the experimental data.

BASIC PRINCIPLE OF CALCULATION OF THREE DIMENSIONAL HEAT TRANSFER

The calculation of three dimensional heat transfer in a utility boiler combustor is to solve the three dimensional energy equation (1), and the key point is to solve those terms related to radiant heat transfer.

In Monte Carlo Method, the probability simulating formulae for radiant heat transfer are:

$$Q_{r.v\to\delta v_j} + Q_{r.A\to\delta v_j} = \sum_{i=1}^{m} Q_{r.\delta v_i} P_{\delta v_i\to\delta v_j} + \sum_{i=1}^{n} Q_{r.\delta A_i} P_{\delta A_i\to\delta v_j} \qquad (2)$$

$$Q_{r.v\to\delta A_j} + Q_{r.A\to\delta A_j} = \sum_{i=1}^{m} Q_{r.\delta v_i} P_{\delta v_i\to\delta A_j} + \sum_{i=1}^{n} Q_{r.\delta A_i} P_{\delta A_i\to\delta A_j} \qquad (3)$$

where $Q_{r.\delta v_i} = 4K_i\sigma T_i^4 \delta V_i$; $Q_{r.\delta A_i} = \varepsilon_i \sigma T_i^4 \delta A_i$

Random samplying formulae should be used in each procedure in probability simulation, yielding random variables with specific probability distribution, in order that the direction with which the emitted energy bundle in an element of gas would be in compliance with isotropic emissive characteristies, the energy bundle in a wall element would be in compliance with Lambert's Law, and the free path of energy bundle would be in compliance with Bouguer's Law. Sampling formul of random simulation in the process of radiation were given in references [1] and

[5].

The sampling formulae mentioned above cannot be applied directly to areas at boundary region of a combustor with arbitrary shapes, because they are usuall irregular. Hence, the point of emission of a bundle, the direction of emission and the position it arrives at must be re-defined according to the geometry of these areas. Accordingly, these sampling formulae are deducted as follows:

Determination of the Point of Emission

In combustor, typical irregular wall elements and volume elements are shown in Fig.1. When energy bundle is emitted from these elements, the point of emission may be determined as follows.

For wall element'a', random emission point is sampled first by assuming in to be a rectangle (Fig.2), we have

$$x = x_j$$
$$y = y_j + R_1 . \Delta Y \qquad (4)$$
$$z = z_j + R_2 . \Delta Z$$

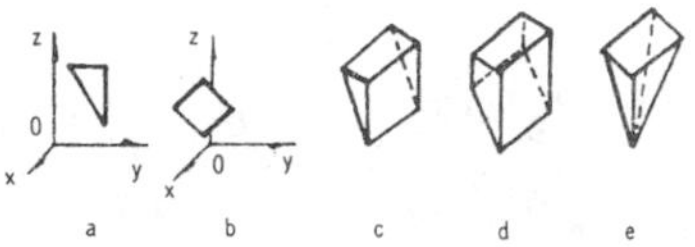

Fig.1

Then, judge whether z is greater than f(y), if z⩾f(y), the sampling has been proved to be right. Otherwise repeat the process and determine y and z, compare them with f(y), until it is satisfactory. It is clear from the sampling process that the determination of point of emission is made at random, so the same is true for subareas.

As for a volume elemnt, say 'c', sampling can be done by first assuming it to be a cuboid, we have

$$x = x_j + R_1 \cdot \Delta X$$
$$y = y_j + R_2 \cdot \Delta Y \qquad (5)$$
$$z = z_j + R_3 \cdot \Delta Z$$

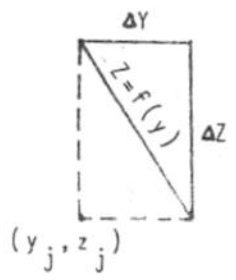

Fig.2

Then compare z sampled with boundary equation Z=f(x,y), if z falls within the extent meaningful to calculation, then the sampling is successful; otherwise repeat the process until the goal is reached.

For other forms of typical elements (or elements of more complicated geometr random emission point can be determined in ways described above; only that the more complicated the geometry is, the more restrictions there will be. Other th that nothing will be different, so it will be unnecessary to go on in details.

Sampling of Direction of Emission of Energy Bundles at Arbitrary Surfaces

For the sake of convenience in discussion and disposal, an auxiliary frame of reference is used.

Suppose there is a surface A at ordinary position in xyzo system, and an auxiliary frame of reference XYZO is set up on surface A with its axis OY coinci ing with the mormal of surface A (Fig.3).

With this auxiliary framework, direction of emission of energy bundle at arbitrary surface could be determined by methods used where surfaces are specified.

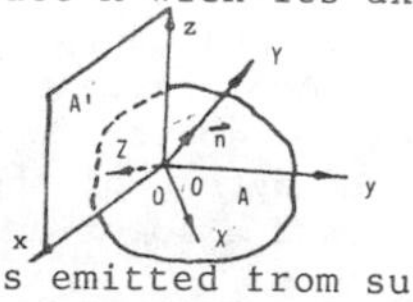

Fig.3

At first we assume that the energy bundle $\vec{I}'$ is emitted from surface A' which coincides with xz plane. Then, rotate surface A' to a surface that coincic with plane A, and the energy bundle $\vec{I}'$ emitted from surface A' is rotated togethe with plane A' in order to ensure the relative positions between $\vec{I}'$ and A' remai unchanged. The energy bundle after rotation is designated by $\vec{1}$. The relation in positions between plane A and energy bundle $\vec{1}$ in auxiliary system XYZO, are:

$$R_Y' = R_y' = \sqrt{1 - R_\eta}$$
$$R_Z' = R_z' = \sqrt{1 - R_y'^2}\cos(2\pi R_\theta) \qquad (6)$$
$$R_X' = R_x' = \sqrt{1 - R_y'^2}\sin(2\pi R_\theta)$$

Sampling formulae for determining the direction of emission of energy bundle emitted from arbitrary inclined surface can be derived through a series of deductions using the theory of coordinate transformation. We have

$$R_x = I_1 R_x' + I_2 R_y' + I_3 R_z'$$
$$R_y = m_1 R_x' + m_2 R_y' + m_3 R_z' \qquad (7)$$
$$R_z = n_1 R_x' + n_2 R_y' + n_3 R_z'$$

where $\{I_1, m_1, n_1\}$, $\{I_2, m_2, n_2\}$ and $\{I_3, m_3, n_3\}$ are unit directional vectors of axes OX, OY and OZ of the auxiliary system respectively, their values can be determined by geometric relations or by Euler's angles.

General Method in Calculating the Position at Which the Energy Bundle Arrives

In order to make the discussion feasible for general application, assume that the system is composed of arbitrary limited number of walls (n) and absorbent gase within the space enclosed by these walls. Wall equations described these walls can be expressed as follow. (It is understood that in following discussions the system referred to is xyzo system if not specified.)

$$A_1 x + B_1 y + C_1 z + D_1 = 0$$
$$A_2 x + B_2 y + C_2 z + D_2 = 0 \qquad (8)$$
$$\vdots$$
$$A_n x + B_n y + C_n z + D_n = 0$$

where $\{A_i, B_i, C_i\}$ is the vector normal of No. i wall, and the direction of the inner normal is considered to be positive.

Within the length of free path, energy bundle could be regarded as a 'half-line' and is expressed by

$$x = x_o + R_x t$$
$$y = y_o + R_y t \qquad (9)$$
$$z = z_o + R_z t$$

where (x,y,z) is the coordinate of an arbitrary point in the energy bundle, and (x_o, y_o, z_o) the coordinate of emitting point (or reflecting point) of the energy bundle, t is a parameter.

As the free path L of the energy bundle is known, it is possible to trace the energy bundle and a solution of the point at which the energy bundle arrives could be obtained.

(1) When the energy bundle is moving toward wall No.i, the intersecting point could be determined by equations (8) and (9).

Thus the parameter of the energy bundle related to the wall No.i is

$$t_i = -\frac{A_i x_o + B_i y_o + C_i z_o + D_i}{A_i R_x + B_i R_y + C_i R_z} \qquad (i=1,2,\ldots\ldots,n) \tag{10}$$

By substituting t_i into equation (9), we have the coordinate of the intersecting point of energy bundle and wall No. i.

Then the distance travelled by the energy bundle when it reaches the wall No. i is:

$$L_i = \left|\frac{x_i - x_o}{R_x}\right| \quad \text{or} \quad \left|\frac{y_i - y_o}{R_y}\right| \quad \text{or} \quad \left|\frac{z_i - z_o}{R_z}\right| \tag{11}$$

$$(R_x \neq 0;\ R_y \neq 0;\ R_z \neq o;\ i = 1,2,\ldots,n)$$

(2) When the energy bundle is moving away from or in parallel with wall No.i, there will be no intersection of the bundle and wall, the distance travelled could be considered as infinity ($L_{away,\ parallel} = \infty$).

(3) The minimum length L_{min} is:

$$L_{min} = \mathrm{Min}(L_1, L_2, \ldots, L_n) \tag{12}$$

Then the coordinate of the point that the energy bundle reaches can be obtained by replacing t using L_{min} in equation (9).

And finally the remainder length of free path is

$$L_{rem} = L - L_{min} \tag{13}$$

CALCULATION OF THREE DIMENSIONAL HEAT TRANSFER IN SG-400 T/H BOILER

Following is an example of applying the method to an SG-400 T/H boiler

This is a boiler of superpressure and reheating type, burning pulverized coal. The burners are in three rows along the front wall.

The temperature distribution and heat flux distribution of the boiler have been surveyed by the Research Institute of Shanghai Boiler Plant in 1976. Operation conditions during experimentation are shown in table 1.

Fig.4 shows the division of combustor and choice of coordinate system.

The combustor is divided into 1,506 elements, of them 880 volume elements of gas and 626 wall elements.

Distribution of mass flow in combustor is calculated from theories of free jet and viscous flow, the results are shown in Fig.5. Distribution of heat generated during combustion is taken from experimental dat.

Tab. 1

item	unit	data
electric load	MW	102
excess air factorat the exit of the chamber		1.095
main steam temperature	°C	528/526
main steam pressure	kgf/cm	107.7
feedwater temperature	°C	214.8
boiler efficiency	%	93.45

The results of calculation are shown in Fig. 6-9. Fig.6 is the temperature distribution of gas on certral section of the chamber (K), Fig.7 is the drstribution of heat flux projected to each wall of the chamber (kcal/m^2·hr), Fig.8 and Fig.9 are un-even coefficient of the distribution of heat flux projected to each wall over height and over width.

Comparing the results with the experimental data, we find that the results are mainly identical with the experimental data, the differences between corrresponding values are mostly below 10%. For those spots located at front and rear rows on side walls, these differences are somewhat larger, this is probably due to the nearness of the spots to the chamber corners, where temperature gradient is quite large; however, elements for calculation at that part of the chamber is about 1 meter apart from the corners. Besides, measuring at charmber corner is liable to cause greater errors arising from heat flux.

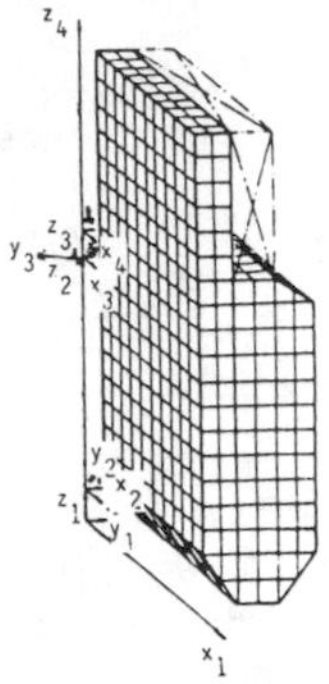

Fig.4

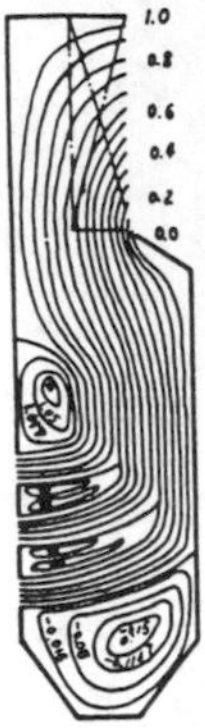

Fig.5

CONCLUSION

We have been tackling the problem concerning three dimensional heat transfer in combustors composed of arbitrary inclined walls, and putting forward a solution to the radiant heat transfer with Monte Carlo Method by determining emitting point, sampling the direction of energy bundle emission and supplying a general method for determination of the points where energy bundles reach. This leads to laying the foundation for radiant heat transfer calculation on practical combustros.

Temperature distribution and heat flux distribution are obtained by applying this method to the calculation of heat transfer to combustor of an SG-400 T/H boiler. The results are proved to be satisfactory when compared with experimental data.

The realization of this method illustrates the adaptability and flexibility of Monte Carlo Method in solving complex multi-dimensional problems. Hence we suggest that in calculating three dimensional heat transfer in a complexshaped combustor of large capacity boiler, original combustor should not be simplified, especially over-simplified.

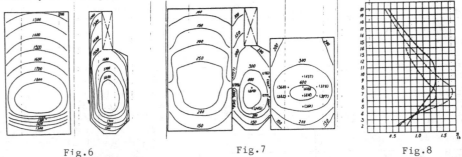

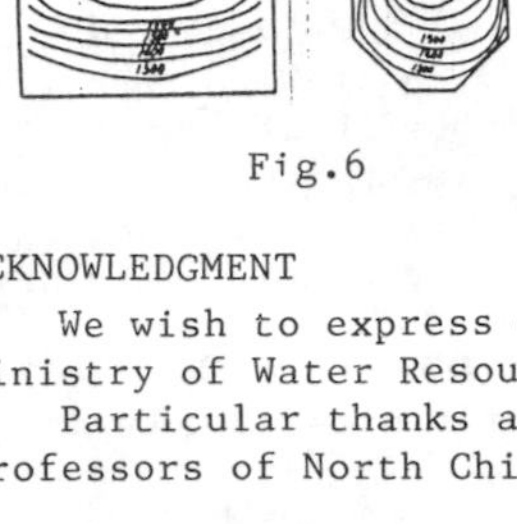

Fig.6

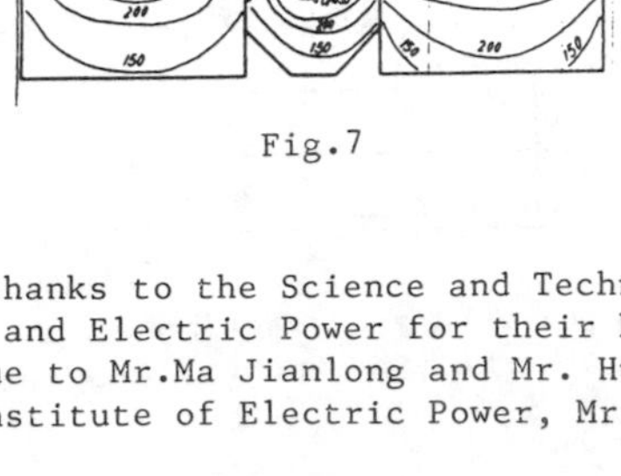

Fig.7

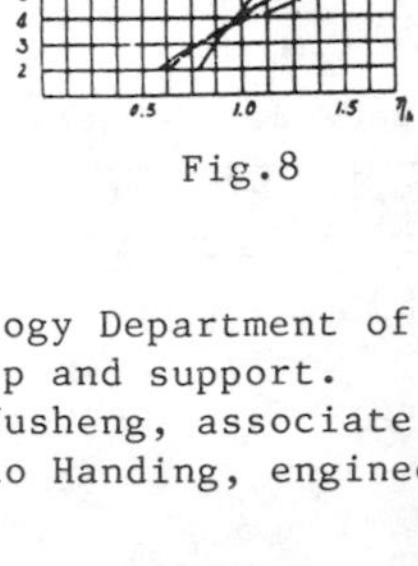

Fig.8

ACKNOWLEDGMENT

We wish to express our thanks to the Science and Technology Department of Ministry of Water Resources and Electric Power for their help and support.

Particular thanks are due to Mr.Ma Jianlong and Mr. Hu Yusheng, associate professors of North China Institute of Electric Power, Mr.Cao Handing, engineer of Shanghai Power Plant Equipment Research Institute for thier suggestions and helpful comments.

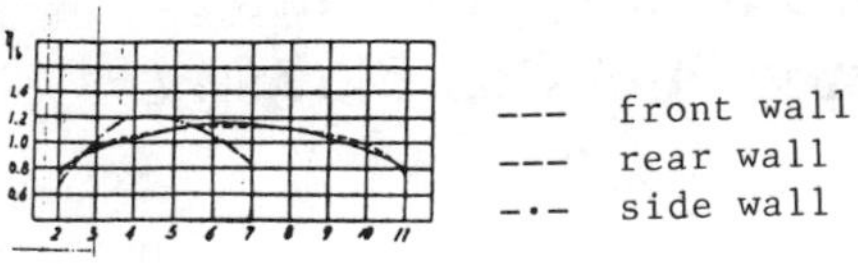

Fig.9

REFERENCES

[1] J.R.Howell, M.Perlmutter, Monte Carlo solution of thermal transfer through radiant media between gray walls, Trans. ASME, Ser.C, 86-1(1964-2), 116.

[2] Hiroshi Taniguchi and Masayuki Funazu, The numerical analysis of temperature distributions in a three dimensional furnace, Bulletin of the JSME, No.66, Vol.13, 1970.

[3] Xü-Chang Xü, Mathematical modelling of three-dimensional heat transfer from the flame in combustion chambers 18th Symposium (International) on Combustion, The Combustion Institute, 1981.

[4] Sun Zhaoxing and Hu Naiyi, Monte-Carlo Method for radiative heat transfer in a furnace, Journal of Electrical Engineering, No.4, Vol.4, 1984(in Chinese).

[5] Sun Zhaoxing, Application of Monte Carlo Method to the calculation of heat transfer in furnace, Journal of North China Institute of Electric Power, No.2, 1983(in Chinese).

Heat and Mass Transfer and Combustion Velocity Profile over a Flat Plate with Distributed Fuel Gas Blowing

X. Q. WANG
Anshan Iron and Steel Complex
Anshan, PRC

K. KUDO, H. TANIGUCHI, and M. OGUMA
Hokkaido University
Sapporo 060, Japan

K. F. CEN
Zhejiang University
Hangzhou, PRC

1. INTRODUCTION

Recently the coal became one of primary energy sources. The combustion characteristics have not been deeply evaluated to be compared with gas and liquid fuels. As solid-fuel ordinarily consists of two burning components of fixed carbon and volatile substances, the characteristics of solid-fuel itself under heating condition had to be studied before combustion analysis.

Essenhigh [1] and Juntgen and Van Heek [2] developed the methods to determine the release rate and composition of volatile substances as a function of temperature and heating rate. Krazinski, et al. [3] developed an analytical model for coal-dust air combustion. These methods and model for uniform velocity field are unfit to determine the ignition and extinction limits. In order to understand the mechanisms of solid-fuel combustion, it is essential to treat the flow field over the fuel surface as non-uniform. Another important problem in solid-fuel combustion is the role of radiation, flame structure and its holding. In usual theoretical studies of combustion phenomena, the radiation effect has been neglected. Since the release rate and composition of volatile substances are temperature-dependent and combustion velocity is an exponential function of temperature, the temperature analysis may be significantly required to study the combustion model.

The present study is to investigate the temperature distribution of fuel surface with radiation heating and the stability of a roll-up flame observed in coal combustion. The roll-up flame is analyzed by use of a flame holding model which has the solid-fuel plate with inflammable gas blowing from upstream surface and flammable gas from downstream. These results were confirmed by using a combustion wind tunnel. The objective of this study is to estimate the temperature distribution of the fuel surface before ignition, the complex heat transfer and the combustion after the ignition.

2. FORCED CONVECTIVE HEAT TRANSFER FROM A RADIATION HEATED SOLID-FUEL PLATE

2.1 Numerical Analysis

Fig.1 shows a solid-fuel plate (subscript p) of length X_p and a heater surface (subscript hp) placed in parallel to the fuel plate at a distance Y_L. A gas flows over the plates at an inlet velocity (subscript in) of u_{in} and forms a turbulent boundary layer. The following assumptions are imposed on this analysis

1. The flow is steady and two-dimensional, and the properties are variable.
2. Conduction in the fuel plate is negligible.
3. Frictional heating is negligible.
4. The gas (air, volatile substances and their mixture) is ideal and transparent to radiation.
5. All wall surfaces are isothermal except the fuel plate whose surface temperature is x-dependent.

Under the above assumptions, the governing equations [4,5,6] lead the followings:

$$\frac{\partial}{\partial x}(\rho u) + \frac{\partial}{\partial y}(\rho v) = 0 \tag{1}$$

$$\rho u\frac{\partial u}{\partial x} + \rho v\frac{\partial u}{\partial y} = -\frac{\partial p}{\partial x} + \frac{\partial}{\partial x}(\mu_t\frac{\partial u}{\partial x}) + \frac{\partial}{\partial y}(\mu_t\frac{\partial u}{\partial y}) \tag{2}$$

$$\rho u\frac{\partial v}{\partial x} + \rho v\frac{\partial v}{\partial y} = -\frac{\partial p}{\partial y} + \frac{\partial}{\partial x}(\mu_t\frac{\partial v}{\partial x}) + \frac{\partial}{\partial y}(\mu_t\frac{\partial v}{\partial y}) - \rho g \tag{3}$$

$$\rho u\frac{\partial h}{\partial x} + \rho v\frac{\partial h}{\partial y} = \frac{\partial}{\partial x}(\frac{\mu_t}{Pr_t}\frac{\partial h}{\partial x}) + \frac{\partial}{\partial y}(\frac{\mu_t}{Pr_t}\frac{\partial h}{\partial y}) \tag{4}$$

$$\rho u\frac{\partial Ci}{\partial x} + \rho v\frac{\partial Ci}{\partial y} = \frac{\partial}{\partial x}(\frac{\mu_t}{Sc_t}\frac{\partial Ci}{\partial x}) + \frac{\partial}{\partial y}(\frac{\mu_t}{Sc_t}\frac{\partial Ci}{\partial y}) \tag{5}$$

$$\rho = \frac{p}{RT} \tag{6}$$

$$\dot{q}_{Ri} = \varepsilon_i E_i - \frac{1}{A_i}\int_A \varepsilon_j E_j S_{ij}\, dA \tag{7}$$

$$\mu_t = \mu + \rho\varepsilon_m \tag{8}$$

$$\varepsilon_m = l_m^2\left|\frac{\partial u}{\partial y}\right| \tag{9}$$

$$l_m = ky\{1-\exp(-\frac{y\sqrt{\tau/\rho}}{26\nu})\} \tag{10}$$

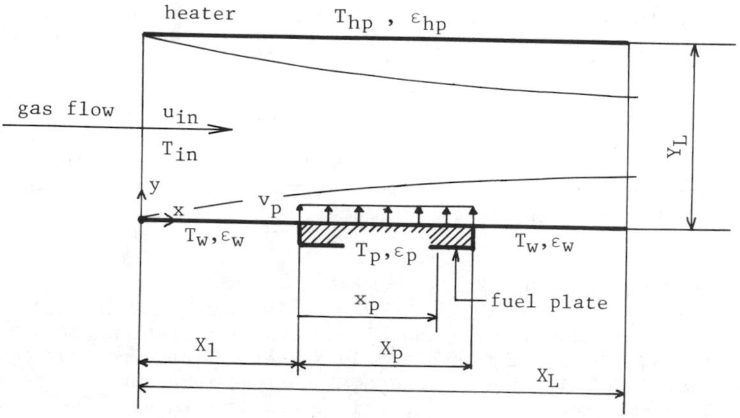

FIGURE 1. A schematic drawing of physical system.

The eddy diffusivity ε_m is evaluated from Prandtl's transport theory in which the mixing length l_m by Van Driest [5] is utilized and the value of k is taken from Nikuradse [6] as 0.4. Furthermore the turbulent Prandtl number Pr_t and Schmidt number Sc_t are assumed unity in this present analysis. The fuel plate is assumed to be insulated except its surface (exposed to the air stream) and the wall temperature T_w is equal to the inlet air temperature Tin.

The flux terms were converted into a finite-difference form by the hybrid scheme and the central or upwind difference was used to depend on the cell Peclet number. The SIMPLER algorithm [7] was employed for the determination of pressure which derives the required condition by coupling the continuity equation Eq.(1) to the momentum equations Eqs.(2) and (3). Then, radiative heat transfer was treated by a Monte Carlo method [8]. A HITAC M-200H digital computer was employed in numerical computations over the range of input variables listed in Table 1.

3.2 Calculated Results

Fig.2 shows an influence of the heater surface temperature T_{hp} (solid line) on the temperature distribution of the fuel surface T_p^4/T_{hp}^4 under T_w and T_{in} of 300 K. A convection effect is also illustrated. With the neglect of convection, the T_p^4/T_{hp}^4 profiles exhibited a concave symmetrical shape due to radiation loss to both the inlet and exit sections. In the presence of convective heat transfer, T_p^4/T_{hp}^4 was low but increased rapidly in the upstream half because thin boundary layer thickness and large temperature gradient near the fuel plate induced a high convection heat loss to the main stream, resulting in lower surface temperature. Fig.2 also shows another influence of the inlet air temperature Tin and T_w (broken line) on the temperature distributions T_p^4/T_{hp}^4 under the condition of some heater temperatures in which the temperature of central fuel surface is nearly equal to that under the other condition of T_{hp}=700 K and T_w=T_{in}=300 K. As the increase in inlet air temperature raised the fuel surface temperature, T_p^4/T_{hp}^4 profile approached to that without convection.

The effect of gas blowing on T_p^4/T_{hp}^4 is shown in Fig.3. An increase in v_p introduced the higher temperature gas penetration into the boundary layer which resulted in a growth of this boundary layer, a reduction in the convective heat transfer and a raise in the fuel surface temperature (solid line). When v_p was assumed as a function of the fuel surface temperature, the convective heat transfer rate under the constant v_p was less than that of the temperature-dependent v_p. So the mean temperature gradient on the fuel surface T_p^4/T_{hp}^4 (broken line) under temperature-dependent v_p is greater than that under the constant v_p (solid line).

Fig.4 illustrates an influence of system geometry on the temperature distribution of fuel surface. It was observed that the tendency of concave temperature profile (solid line) was amplified with a decrease in X_L/Y_L. The simplification of radiative heat transfer by an one-dimensional model (broken line) may lead to a significant error in the temperature profile as shown in Fig.4.

3. EXPERIMENT FOR COAL COMBUSTION

Fig.5 shows the combustion wind tunnel and its test section. By use of an air heater by propane burner and an electric heater on the upper surface of the test section (max. 20 kW/m^2), the air and coal surface temperatures are able to be controlled respectively. The coal particle bed is used as a fuel plate which is put in a shallow pan (200×300×5 mm) set on a lower surface of the wind tunnel.

Five windows are equipped on each side of the test section to observe the combustion. The fuel surface temperature was increased by radiation after inserting the plate at its place (Fig.5) in the combustion wind tunnel with an air velocity of 1.4 m/s, an inlet air temperature of 520 K and a total radiation heat flux of 16 kW/m^2. Then the fuel surface began to burn from its upstream edge where oxygen diffused rapidly. At this time, volatile substances were released from the fuel surface as visible grey smoke. Next, when ignited by the

TABLE 1. Range of input variables.

T_{hp} , K	500~900
T_{in} , K	300~600
u_{in} , m/s	2.0
X_1 , m	0~1.0
X_L , m	0.5~1.0
X_p , m	0.5
Y_L , m	0.2
ε_{hp}	0.8
ε_w	0.8
ε_p	1.0

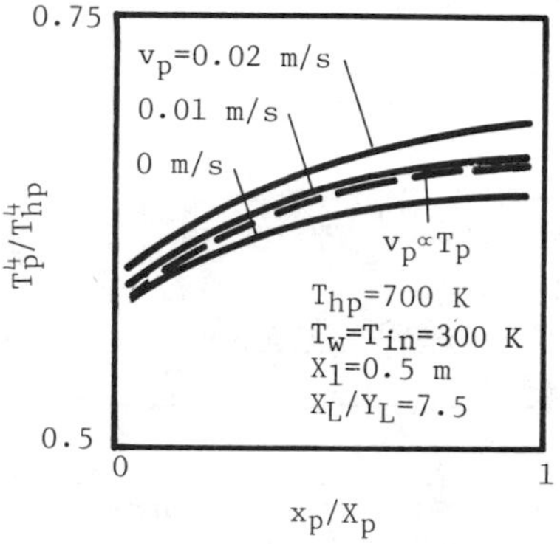

FIGURE 3. Effect of v_p on the temperature distribution of fuel surface.

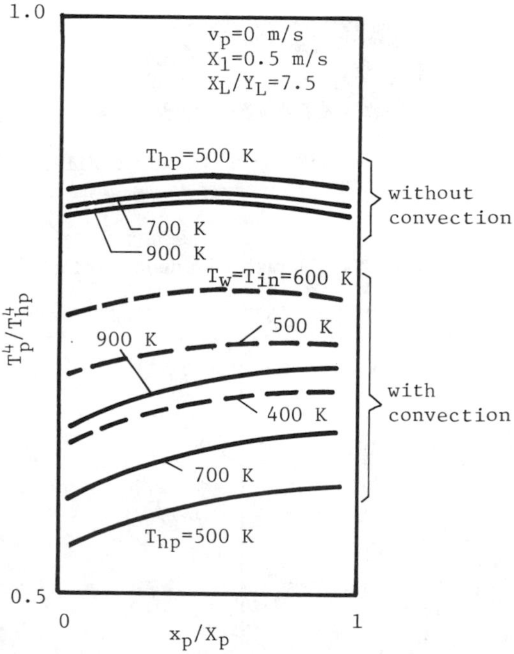

FIGURE 2. Effect of T_{hp} and Tin on the temperature distribution of fuel surface.

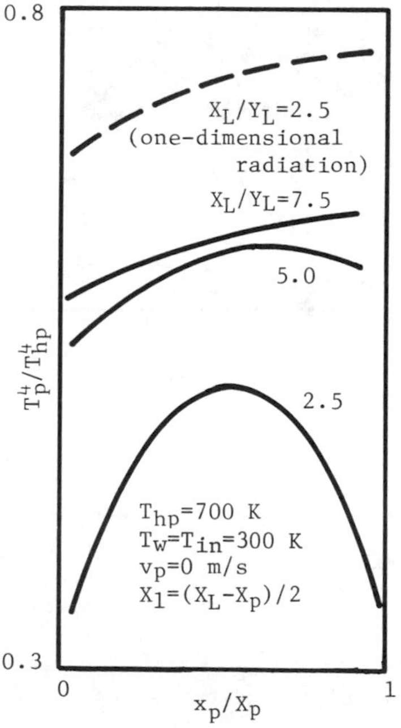

FIGURE 4. Effect of system geometry on the temperature distribution of fuel surface.

ignition stick, the bed was wholly covered by diffusion flame (Fig.6). After several minutes since the ignition, the flame moved gradually toward downstream with the surface combustion region and the front end of this flame rolled up. This roll-up flame was a premixed flame because of the flame profile and its color (Fig.7). The upstream part of the flame may be holded by the propagation characteristics.

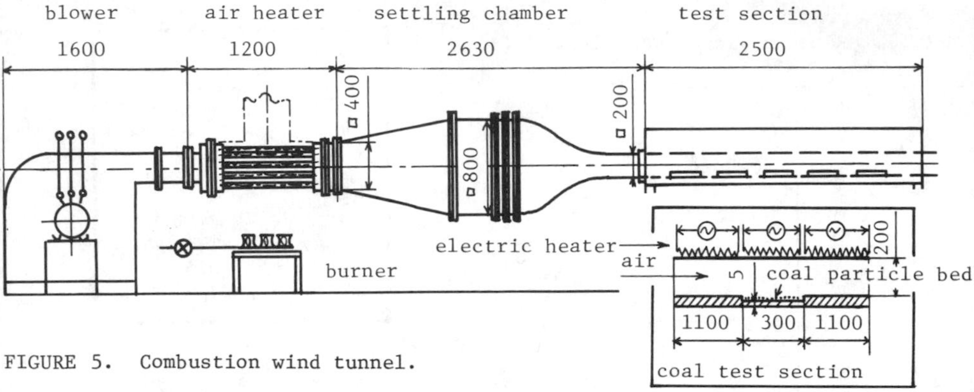

FIGURE 5. Combustion wind tunnel.

FIGURE 6. Diffusion flame profile.

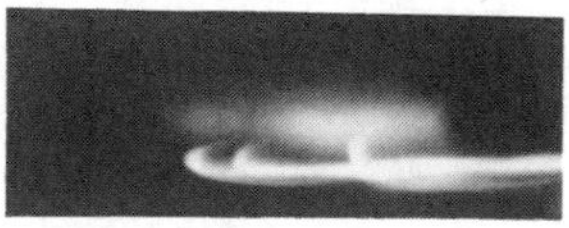

FIGURE 7. Roll-up flame profile.

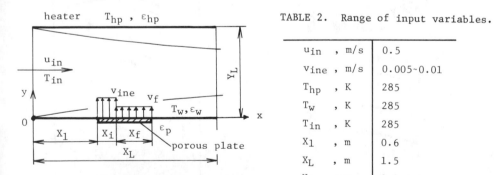

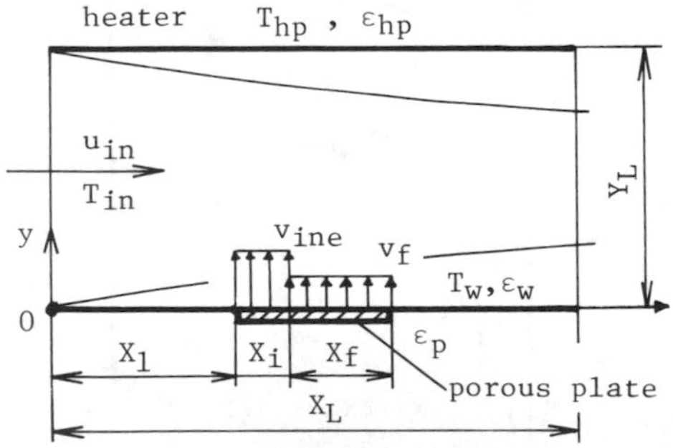

FIGURE 8. Holding model test section of combustion wind tunnel.

TABLE 2. Range of input variables.

u_{in} , m/s	0.5
v_{ine} , m/s	0.005~0.01
T_{hp} , K	285
T_w , K	285
T_{in} , K	285
X_1 , m	0.6
X_L , m	1.5
X_i , m	0.1
X_f , m	0.2
Y_L , m	0.2

4. NUMERICAL ANALYSIS AND EXPERIMENT FOR THE ROLL-UP FLAME

4.1 Flame Holding Model

Fig.8 shows the test section of combustion wind tunnel for a proof of the flame holding model. A porous plate is imitated as the solid-fuel and placed at $y=0$ and $x \geq X_1$. The inflammable gas blowing velocity v_{ine} at $X_1 \leq x \leq X_1+X_i$ and flammable gas blowing velocity v_f at $X_1+X_i \leq x \leq X_1+X_i+X_f$ are applied for the above section. The inflammable and flammable gases are imitated as the surface combustion products and volatile substances respectively. A main stream is assumed to be laminar.

4.2 Caluculated Results

Eqs.(1) through (7) are utilized at this analysis and the range of input variables for computation is shown in Table 2. The inlet air temperature is equal to the room temperature, and T_{hp}, T_w and T_p are assumed to be equal to the room temperature. In this analysis, inflammable and flammable gases are nitrogen and propane respectively, and Kuehl's combustion velocity [9] of propane and air mixture is applied for the calculation.

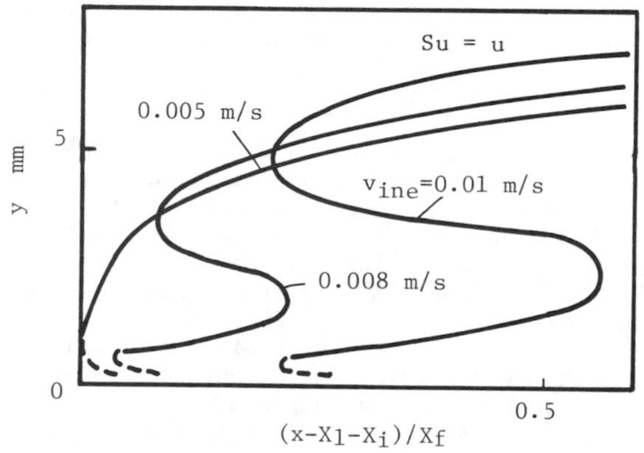

FIGURE 9. Effect of inflammable gas blowing velocity v_{ine} on Su=u line.

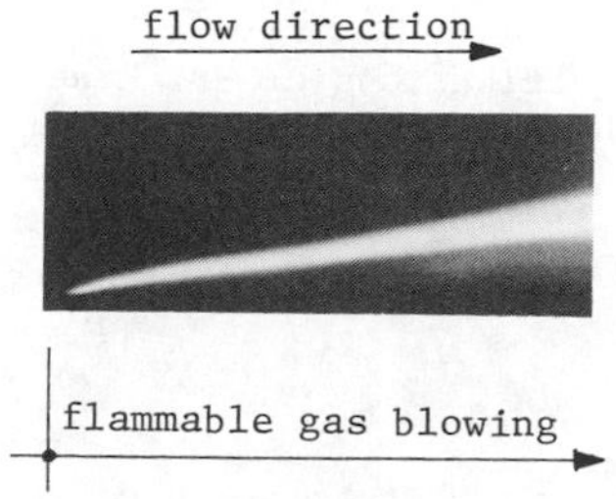

FIGURE 10. Diffusion flame profile without inflammable gas blowing.

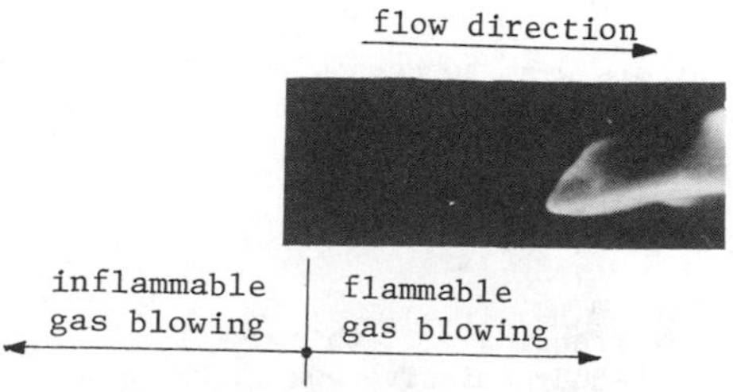

FIGURE 11. Roll-up flame profile with inflammable gas blowing.

Fig.9 shows an effect of inflammable gas blowing velocity v_{ine} on Su=u line. When a flame front was vertical to flow direction, flame was stabilized on the Su=u line. The stabilized flame front may exist on a top of convex to the upstream on the Su=u line and the flame front profile may be similar to the roll-up flame which is observed in the coal combustion experiment. So this analysis showed that the roll-up flame profile could be stabilized at the upstream end in a boundary layer over the porous plate which had the inflammable gas blowing from upstream surface and flammable gas from downstream. A combustion velocity near the porous plate may be less than Kuehl's value because an extinction effect is existed, so Su=u lines near the porous plate may be shown in Fig.9 (broken line). When v_{ine} is little (v_{ine}=0.005 m/s) or zero, the stability point of the premixed flame may be set at the upstream end of the porous plate and the ordinary diffusion flame will be settled.

4.3 Experimental Results

Fig.10 shows a diffusion flame over the porous plate with flammable gas blowing only. When the inflammable gas of nitrogen and flammable gas of propane were blowing from the upstream and downstream surfaces of the porous plate respectively, the roll-up flame profile was observed (Fig.11).

5. CONCLUSIONS

The temperature distributions of a radiation heated fuel surface and the stability of a roll-up flame profile were analyzed for the coal combustion and its imitated model. The stabilized mechanism was confirmed by the experimental result by use of a combustion wind tunnel. It is concluded that

1. The blowing (resulting from the release of volatile substances) from the fuel plate causes a reduction in the convective heat transfer. The effect of radiation heating from the heater surface controls the fuel surface temperature and increases in the release rate of volatile substances.
2. The radiation heat loss from the fuel surface to the ambient of inlet and exit sections is so significant that the one-dimensional simplification of radiative heat transfer should be avoided.
3. The heater surface temperature, the inlet air temperature, the blowing velocity and the system geometry are important factors which affect the temperature distribution of the fuel surface.
4. The roll-up flame profile can be stabilized in the boundary layer over the plate by means of inflammable gas blowing from upstream surface and flammable gas from the downstream.

ACKNOWLEDGEMENT

The authors wish to acknowledge the invaluable contributions of Mr. Kanefumi Suda and Mr. Koichi Watanabe in the conduct and help of the experiments.

NOMENCLATURE

- A area, m^2
- C_i mass fraction of i-th species
- E black-body emissive power, kW/m^2
- F_{ij} view factor between i and j surfaces
- h enthalpy, kJ/kg
- p pressure, Pa

$\dot{q}_R$ radiation heat flux, kW/m^2
R gas constant, kJ/(kg K)
S heat exchange coefficient for radiation
ε emissivity, ε_m eddy diffusivity, m^2/s
μ absolute viscosity, kg/(m s)
ν kinematic viscosity, m^2/s
ρ gas density, kg/m^3
σ Stefan-Boltzmann constant, $W/(m^2\ K^4)$
τ wall shear stress, N/m^2

REFERENCES

1. Essenhigh, R.H., Combustion and Flame Propagation in Coal Systems, *16th Symposium (International) on Combustion*, Academic Press, pp.353, 1976.

2. Juntgen, H. and Van Heek, K.H., Gas Release from Coal as a Function of the Rate of Heating, *Fuel*, vol.5, pp.31, 1979.

3. Krazinski, J.L., et al., Coal Dust Flame, *Prog. Ener. Comb. Sci.*, vol.5, pp.31, 1979.

4. Libby, P.A. and Williams, F.A., Fundamental Aspects, in *Turbulent Reacting Flows*, ed. P. A. Libby and F. A. Williams, pp.1, Springer, New York, 1980.

5. Van Driest, E.R., On Turbulent Flow near a Wall, *J. Aeronaut. Sci.*, vol.23, pp.1007, 1956.

6. Nikuradse, J., Gesetzmassigkeiten der Turbulenten Stromung in Glatten Rohren, *VDI-Forschungsh.*, vol.353, 1932.

7. Patankar, S.V., *Numerical Heat Transfer and Fluid Flow*, Hemisphere, Washington, D.C., 1980.

8. Siegel, R. and Howell, J.R., *Thermal Radiation Heat Transfer*, McGraw-Hill, New York, 1972.

9. Kuehl, D.K., Laminar-Burning Velocities of Propane-Air Mixtures, *8th Symposium (Internatinal) on Combustion*, The Williams & Wilkins Co., Baltimore, pp.510, 1964.

Improvement and Industrial Application of Numerical Computation of Flame Heat Transfer with Monte-Carlo Method

XUCHANG XU and YUNSHAN WANG
Tsinghua University
Beijing 10084 PRC

YONGFU ZHAO
Beijing Institute of Iron and Steel Technology
Beijing, PRC

ABSTRACT

This paper deals with the improvement of numerical computation of flame heat transfer with Monte-Carlo method used previously by the authors and others. It is considered that, in the probability simulation of pulverized coal flame, in which a lot of solid particles are entrained by the gases, radiation energy is not only absorbed but also scattered by them. The simulation method of uniform random distribution of emission positions of the energy bundles in any element is discussed.

In this article the practical problems of industrial application of numerical computation of flame heat transfer with Monte-Carlo method is also studied. The influence of heat transfer in the furnace by the flame in a 50MW oil-fired boiler is discussed. Comparing the results of numerical computation of flame heat transfer in a 100MW pulverized coal fired boiler furnace with the experimental measurements, the authors found that the calculation conforms well with the test data. The method worked out by the authors for the computation of flame heat transfer with Monte-Carlo method of a 300MW pulverized coal fired boiler furnace with platen superheaters is discribed and its results of computation are discussed.

NOMENCLATURE

Notation:
A-Area
C_p-Specific heat
K-Coefficient of absorption or scattering of medium
N-Number of energy bundles in probability simulation
p-Probability
Q-Radiation heat transfer rate, heat realeasing rate by Chemical reaction
R-Uniformly distributed random Number
r-Radius, distance
S-Source term in transport equations
T-Temperature
V-Volume
u-Velocity
y-Coefficient of diviation of heat absorption
Γ_T-Turbulent thermal conductivity
ρ-Density
η-Angle between the energy bundle and the normal of element surface
σ-Stefan-Boltzmann constant

Subscripts:
a-absorption
g-gas, medium
s-scattering, surface
$\delta V_i, \delta V_j$ -element volume
$\delta A_i, \delta A_j$ -element wall surface
$\delta V_i \rightarrow \delta V_j$ -from element volume δV_i to element volume δV_j
$\delta A_i \rightarrow \delta V_j$ -from element wall surface δA_i to element volume δV_j
V-entire volume of furnace
A-entire wall area of furnace

INTRODUCTION

In order to solve the heat flux distribution on the walls of a furnace, to find the working condition of the heating surface material, to analyze correctly the hydrodynamic characteristics of the water screen tubes, and to know the temperature distribution of the gases in the three-dimensional space of the furnace, the following three-dimensional energy equation must be solved.

$$\nabla\cdot(\rho u c_p T) = \nabla\cdot(\Gamma_T \nabla T) + S_Q \quad (1)$$

In which, S_Q is the source term of the energy equation, that is the sum of the combustion heat release and the radiative heat transfer in an element volume in unit time.

The common methods used to solve energy equation (1) are heat flux method [1 4], zone method[5-10], and Monte-Carlo method[10-18]. By Monte-Carlo method, in the simultaneous solution of the equations (1) of the three dimensional space, probability simulation is used to calculate the multi-integration terms in the very complicated three-dimensional radiative heat transfer, the amount of the computation is relatively reduced and the results are close to the real data. For an element volume, we may write the source term of Eq. (1) as[12,18]

$$S_Q = Q_R + Q_j = \frac{Q_{V\to\delta V_j}}{\delta V_j} + \frac{Q_{A\to\delta V_j}}{\delta V_j} - \frac{Q_{\delta V_j}}{\delta V_j} + Q_j \quad (2)$$

in which
$$Q_{V\to\delta v_j} = \delta V_j \iiint_V \frac{K_{ai}K_{aj}}{\pi r^2} e^{-\int_0^r K_a dr}\sigma T_{gi}^4 \, dV_j \quad (3)$$

$$Q_{A\to\delta v_j} = \delta V_j \iint_A \frac{K_{aj}}{\pi r^2} e^{-\int_0^r K_a dr}\varepsilon_{si}\sigma T_{si}^4 \cos\eta \, dA_i \quad (4)$$

$$Q_{\delta v_j} = 4K_{aj}\sigma T_{gj}^4 \delta V_j \quad (5)$$

$$Q_{\delta Ai} = \varepsilon_{si}\sigma T_{si}^4 \delta A_i \quad (6)$$

and then the calculation by probability simulation is performed, in which the energy $Q_{\delta vi}$ radiated by the element volume δV_i is divided into N_i energy bundles consequently, the radiative energy absorbed by each element volume δV_j is calculaed by statiotic method with the probabilities of the absorption of every energy bundle by the surrounding element volumes to $P_{\delta vi\to\delta vj}$:

$$Q_{v\to\delta vj} = \sum_{N_g} \frac{Q_{\delta vi}}{Ni} P_{\delta vi\to\delta vj} \quad (7)$$

$$Q_{A\to\delta vj} = \sum_{N_A} \frac{Q_{\delta Ai}}{Ni} P_{\delta Ai\to\delta vj} \quad (8)$$

Using Monte-Carlo method, we can make the calculation approach the real physical process quite well. The absorption coefficient of the gases should be taken as a variable value in the furnace space but not a constant. The reflection of the incidental radiative energy by the wall surfaces can be considered with ease, and the influence of the fouling layer of the water wall can also be accounted for. Even if we divide the furnace volume into thousands element volumes, the amount of the calculation would still not be very bulky. Therefore, this method is suitable to be used to calculate the heat transfer in the furnace of utility boilers with complicated configurations.

To calculate the real heat transfer in the furnace of a utility boiler by Monte-Carlo method, several problems will be confronted which must be solved one by one.

THE IMPROVEMENT OF NUMERICAL COMPUTATION OF FLAME HEAT TRANSFER WITH MONTE-CARLO METHOD

Numerical computation of flame heat transfer with Monte-Carlo method has been developed and improved for several times. Now, it can be used to calculate the radiative heat transfer in furnaces with medium of non-homogeneous coefficient of absorption[11,12]. It can take into account the number of energy bundles in the simulated computation assumed in accordance with the intensity of the radiation of each element volume and the influence of the fouling layer of the water walls of the furnace. It can be also used for the furnace of various configurations. The numerical computation of flame heat transfer with Monte-Carlo method may be greatly simplified by using emissive energy difference method for iteration[21,22]. To approach closely the real physical process of heat transfer by the flame, the method of calculation may still be improved by the following ways.

In the flame of natural gas or oil, the scattering effect of soot particles is negligibly small, and the calculation can be simplified by taking into account only the absorption of radiative energy by the medium in the furnace, However, in our country, majority of the utility boilers burn pulverized fuels, ash particles and coke particles of size much greater than that of the soot particles present in the flame. The scattering coefficient K_s of these ash and coke particles is always as big as 1/6 of the value of the absorption coefficient K_a or even greater[16,19,20], therefore the influence of the scattering of the radiative energy by the medium should be put into account in pulverized fuel fired flames.

Under this condition, all the element volumes δV_j , beside the volumetric radiative energy $4K_{aj}\sigma T_{gj}^4\delta V_j$ emitted by themselves, they scatter the incidental radiative energy to their surroundings. This part of energy is $K_{sj}(Q'_{V\to\delta vj}+Q'_{A\to\delta vj})$.

Therefore, Eq.(5) should be written as

$$Q'_{\delta vj} = 4K_{aj}\sigma T_{gj}^4\delta V_j + K_{sj}(Q'_{V\to\delta vj} + Q'_{A\to\delta vj}) \qquad (9)$$

and Eq. (7) showed be written correspondingly as

$$Q'_{V\to\delta v_j} = \sum_{N_g}\frac{Q'_{\delta vi}}{Ni}P'_{\delta vi\to\delta vj} = \sum_{N_g}\frac{4K_{ai}\sigma T_{gi}^4\delta V_i + K_{si}(Q'_{V\to\delta v_i}+Q'_{A\to\delta vi})}{N_i}P'_{\delta vi\to\delta vj} \qquad (10)$$

where the upper right note of the synbols "$'$" means that the scattering effect is considered in the probability simulation[16]. It differs from the Eq. (5), (7) in which the influence of scattering is not accounted for. The probability $P'_{\delta vi\to\delta v_j}$ in Eq. (10) takes into account the absorption and scattering of energy by the medium, through which the energy bundles passed. To consider the influence of the scattering of the medium on the range of penetration of the energy bundles, it is necessary to change the equation to calculate the probability simulation of the absorption of the energy bundle into the following form

$$\sum_m (K_a+K_s)_m\delta r_m = -\ln(1-Rr) \qquad (11)$$

In which, the note m is number of element volumes passed through by the energy bundle after it is emitted. The calculation of probability simulation is the same as the previous method[11-18].

Although every element volume is divided into a number of bundles of emissive energy in the computation of radiative heat transfer by probability simulation, the center of the element volume is taken as the common origin of these bundles of emissive energy. When the furnace of a boiler with great capacity is calculated, although the furnace volume is subdivided into thousands of element volume the dimension of these element volumes generally is still at the order of one meter. If all the bundles of the emissive energy of an element volume is assumed to be emitted from the center of the volume, and if the sum of the coefficients of absorption and scattering $K_a+K_s=0.8$, 40% of the energy bundles emitted will be absorbed or scattered by the volume itself, and only 60% of them are emitted to the surroundings of this element volume. This is different from the real physical phenomenon.

In the calculation by simulation which is closer to the real process, the emission of energy bundles should be distributed uniformly in the space of the volume, therefore, the origin of the emission of energy bundles should also be random and distributed uniformly in the space of the volume [16]. The origins of the emission of the energy bundles are not the geometric centers of the element volumes (x_i, y_i,z_i,), they can be obtained by using the following equatio

$$X'_i = X_i + (0.5-R_{\delta x})\delta X \quad (12)$$

$$y'_i = y_i + (0.5-R_{\delta y})\delta y \quad (13)$$

$$z'_i = z_i + (0.5-R_{\delta z})\delta z \quad (14)$$

In which, $R_{\delta x}$, $R_{\delta y}$, $R_{\delta z}$ are uniformly distributed random number in domain 0-1. After the above improvement, the radiative heat transfer is then closer to the real process. Under the same conditions, only 30% of the emitted energy is absorbed and scattered by that element volume itself. Thus the energy emitted to the surroundings by the element volume will be increased to 70%, and an increase of 16.7% of emitted energy happens in comparison with the case when the origins of the energy bundles are assumed to be the geometric centers[21-23]. Therefore, by different simulation methods, the calculated absorpt heat flux distributions on the walls will also be different.

For cylindrical combustion chambers, the random origin of the emission of energy bundles should also be taken as uniformly distributed in the element annular volumes[16]. At first, we may take uniformly distributed random number $R_{\delta A}$, $R_{\delta z}$ in domain 0-1, then we may obtain the coordinates of the random origins in the annular volume:

$$r'_i = \sqrt{\frac{\delta A_i}{\pi} R_{\delta A} + [(i-1)\delta r]^2} \quad (15)$$

$$z'_i = z_i + (0.5-R_{\delta z})\delta z \quad (14)$$

For two-dimensional coordinates, the azimuth direction of energy bundle emission and the way to follow the tracks of energy bundles in a cylindrical combustion chamber, the relevant literature[14,16] may be referred to.

The authors have already used the improvements stated above for different requirments in their calculation of the flame heat transfer in real boiler furnaces.

THE APPLICATION OF NUMERICAL COMPUTATIONS OF FLAME HEAT TRANSFER WITH MONTE-CARLO METHOD

The 50MWe Oil-fired boiler of the SG-200/100-Y type of the Beijing 2nd District Heating Power Station has troubles of overheating of superheated steam. In order to find the reason of these troubles, a calculation of the flame heat transfer was carried out by the authors. Through the calculation, the effect of the suggestion to alter the circular turbulent burners on the front wall was studied. The alteration of the burners would lower the center of combustion by 0.75 m. To find out whether this measure would augment the flame heat transfer to the water walls and lower the temperature of the gases at the exist of the furnace and thus the overheating of the superheated steam would disappear[23]. Therefore, a comparison of the flame heat transfer calculations under the heat releasing conditions of the both cases was performed. In Fig.1a the distribution of heat releasing before the alteration is shown, and in Fig.1b the heat releasing condition of the postulated alteration of the burners is shown. The curves are the contours of equal heat releasing. The calculated absorptive heat flux on the water walls is shown in Fig.2, temperature distribution of the gases in the furnace is shown in Fig.3. It is obvious that, after the alteration of the burners, the position of the isopleth absorbed heat flux is lowered; the position of the climax of the temperature of the gases is also lowered somewhat. From the calculations, the temperatures of the gases at the exist before and after the alteration are 1201°C and 1189°C respectively, i.e., the alteration can lower the temperature by only 12°C, and the heat absorption increases only the value of 2.2%. Before the alteration, the gas temperature at the exist was measured to be 1206°C, and there was no significant change of it by measurements after real alteration. All these were well predicted by the calculations. The ratio of the absorbed heat flux of the crossection of the exit of the furnace to the average absorbed heat flux of the entire furnace walls is 0.536, i.e. the coefficient of deviation of heat absorption y_P=0.536; while the coefficient of deviation of heat absorption of the ceiling of the furnace is greater in value, and y_T=0.696.

The upper left corner of Fig.4 is the schematic diagram of a 100MWe pulverized coal fired boiler of the Gaojing Power Station in Beijing of the HG410/100-1 type Twelve circular turbulent burners are located at the front wall of the furnace. In the operation of this boiler, the flames are skewed to the right wall. In order to understand the distributions of the gas temperature and the heat flux of the walls, a calculation of the flame heat transfer was performed according to the real pulverized coal supply and heat release of the furners. In the calculation, the absorption coefficient of the betuminous coal flames was assumed to be $0.97m^{-1}$, and that of the gases to be $0.3m^{-1}$, and the heat resistance of the fouling layer on the water walls was assumed to be $0.005m^2h{\cdot}C/kcal$. The curves of the distribution of the incidental radiation heat flux on the walls are shown in Fig.4. 58 openings for heat flux probe were anranged on the furnace walls, and the values of the measurements from these openings are shown in Fig.4. The curves of regression of these measurements are shown in Fig.5. From Fig.4 it is evident that calculated values coincide well with the measured values. The calculated ratioes of the heat flux of the exit window of the furnace and the ceiling of the furnace to the average value (i.e., the coefficients of deviation of absorption heat) are both close to 0.7. The calculated gas temperature distribution is shown in Fig.6. The loads of the boiler for operation and calculation were both 420t/h. The measured average gas temperature at the furnace exit was 1096°C, that for the calculated value was 1139°C, i.e., 43°C higher than the measured value. If the coefficient of absorption of the flames was assumed to be $0.77m^{-1}$, while that of the gas zone still to be $0.3m^{-1}$, then the calculated average gas temperature at the exit would be 61°C higher than the measured value.

To calculate the flame heat transfer in furnaces of boilers of greater capacities, the method of numerical computation of furnace heat transfer with platen superheaters has been developed[15]. By this method, the authors calculat the heat transfer in the furnace of a 300MWe once through boiler of Shanghai Boiler Works in design (Type SG-1025/170.5-M310)(see Fig.7), for different fuels, at various loading conditions.

Because there are platen heating surfaces in the furnace, two-grid system has to be adopted in the numerical computation of probability simulation of the positions of absorption of the energy bundles at the zone of the platen heating

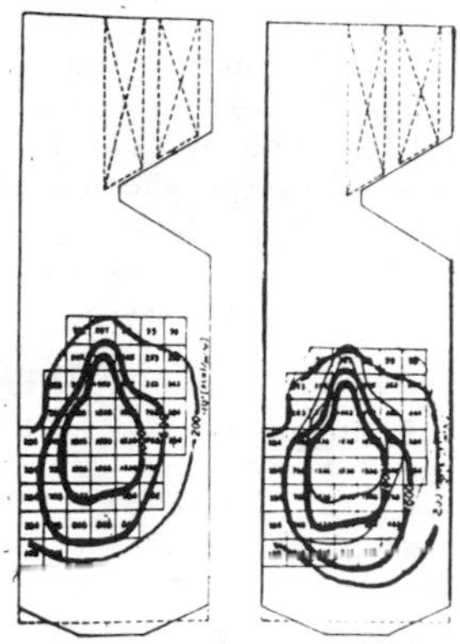

Fig.1 The distribution of heat releasing in the calculation of flame heat transfer in a 50MW oil-fired boiler furnace

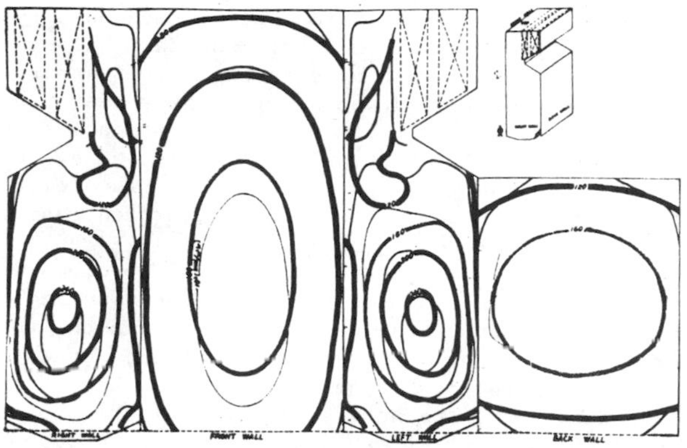

Fig.2 The calculated absorbed heat flux distrib tion on the water walls of 50MW boiler.

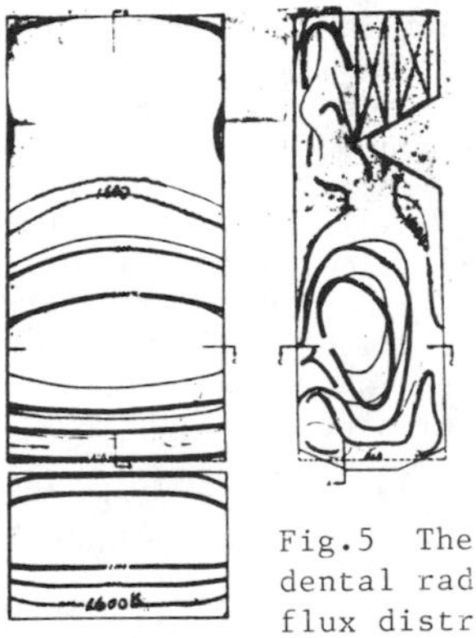

Fig.3 The gas temperature distribution in the furnace of 50MW boiler

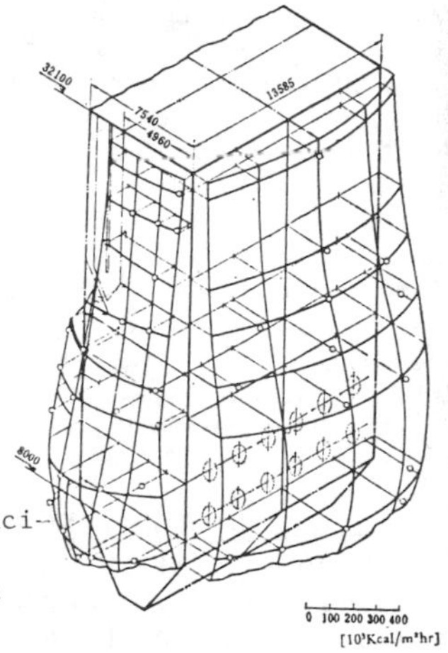

Fig.5 The measured incidental radiation heat flux distribution in a 100MW boiler furnace

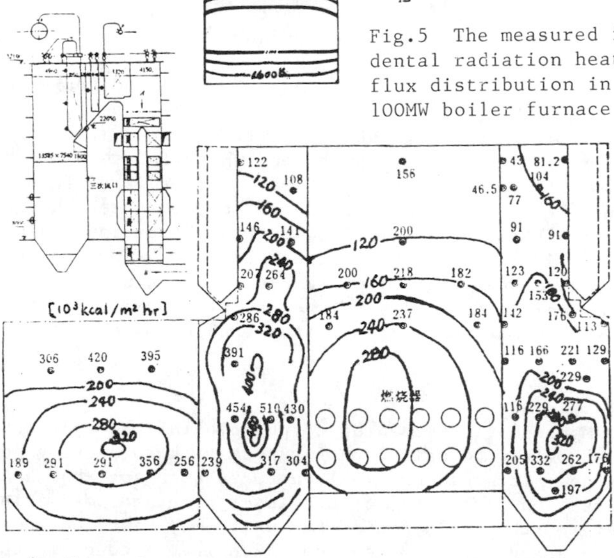

Fig.4 The calculated and measured incidental radiation heat flux distribution on the walls in a 100MW boiler furnce.

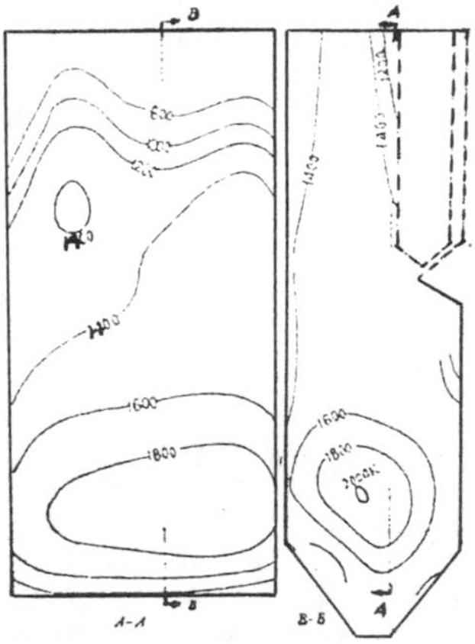

Fig.6 The calculated gas temperature distribution in the 100MW boiler furnace.

surfaces. The energy bundle, which is being tracked at the zone of the platen heating surfaces, when it reached to the boundary of every element volume, it must be judged that whether this boundary belongs to that of a platen heating surface, if not, it can pass through the boundary of this element volume, and the tracking of it continues; if it is a platen heating surface, the evaluation of its probability simulation of absorption must be done, to determine whether this energy bundle is absorbed or reflected by this platen heating surface, then the absorbed energy will be stored in this element surface of this grid system which represents the platen heating surface, and the tracking of this energy bundle will be ended.

The calculated distribution of the absorbed heat flux in the furnace of this 300MWe boiler is shown in Fig. 8 when pulverized thin-bituminous coal is fired. The distribution of the gas temperature in the furnace is shown in Fig.9. The burners of this boiler are located at the corners of the furnace. The coefficient of absorption of the flame zone is assumed to be $0.73m^{-1}$, that for the gas zone $0.3m^{-1}$, the heat resistance of the fouling layer on water walls and the platen heating surfaces are 0.005 and $0.0055m^2h.C/kcal$ respectively. The calculated gas temperature at the exit of the platen heating surface is 1124°C.

For the heating surface below the platen, the variation of the ratio of the calculated absorbed heat flux averaged at the height of the walls to the average heat flux of the entire furnace (i.e. coefficient of deviation of the heat absorption) is shown in Fig.10a as solid line. The dotted line is the value of the coef. of deviation of heat absorption recommended by the method used widely in boiler works of our country, it is obvious that these two lines are relatively close to each other. However the later cannot be extrapolated to all the heating surfaces in the furnace, including the platen heating surfaces. In Fig.10b, the dotted line is the extropolated curve without considering the inadequacy of the extropolation, and the solid line is the calculated result. Since the absorbed heat flux in the zone of platen heating surface decreases abruptly near the ceiling, the coefficient of deviation of the heat absorption may approach 0.3, thus the average absorbed heat flux of the whole furnace is decreased, the coefficient of deviation of heat absorption near the burners may attain the values greater than 1.8. Without considering of this real condition of the absorbed heat flux distribution in the furnace, great error will be resulted in the design of furnace and in the calculation of the hydrodynamic characteristics in the water wall tubes.

The coefficient of deviation of the heat absorption at various height along the breadth of the furnace is shown in Fig.11. It is evident that the distribution of this coefficient is different at different heights. The variation of the average absorbed heat flux in the zone of platen heating surfaces is shown in Fig.12. It is obvious that the heat absorption of each platen is different for different location of the platen, besides, for one platen, the heat absorptions of the surface facing the flame and the surface backward from the flame are different. If the calculated values of the absorbed heat flux for the surface facing the flame and backward from the flame are used, the values of the wall metal temperatures of the heating surfaces may be calculated with greater accuracy, thus the designer may choose the grade of steel tubing more reasonably.

CONCLUSION

1. By improvements, the Monte-Carlo Method of calculation of the radiative heat transfer by the flame can be use to calculate furnaces with relatively complicated configurations, besides, the real physical factors, such as: the heat resistance of the fouling layer, the influence of the absorption and scattering by the medium in the furnace, and the influence of the uniformly distributed origins of the emission of the energy bundles, can be taken into account in the calculation.

2. The real examples given in this article show that the numerical computa-

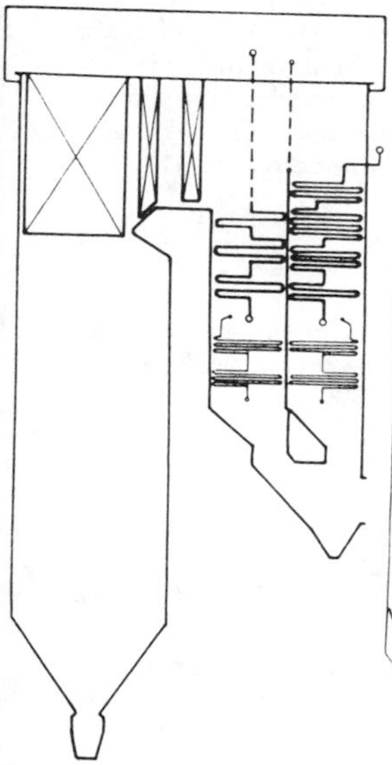

Fig.7 The 300MW once through boiler of SG1025/170.5-M310 type

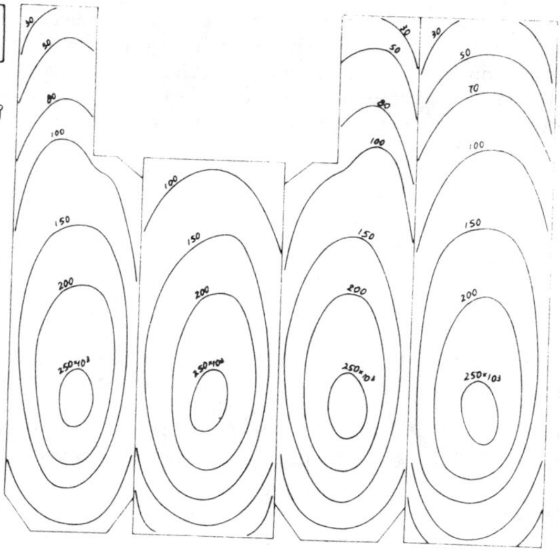

Fig.8 The calculated absorbed heat flux distribution on the water walls of 300MW boiler.

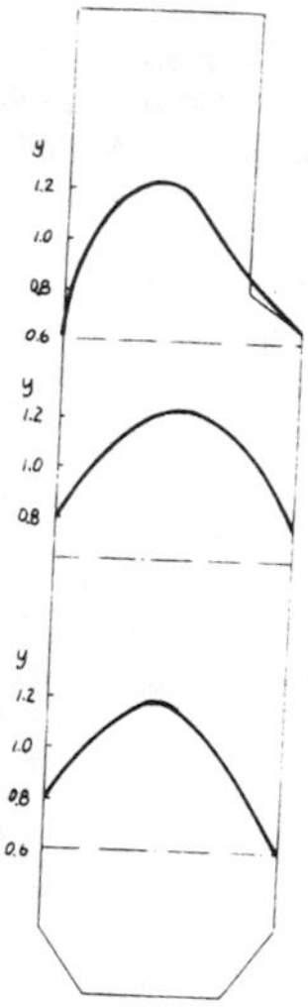

Fig.11 The coefficient of deviation of heat absorption at various height along the breadth of the 300MW boiler furnace.

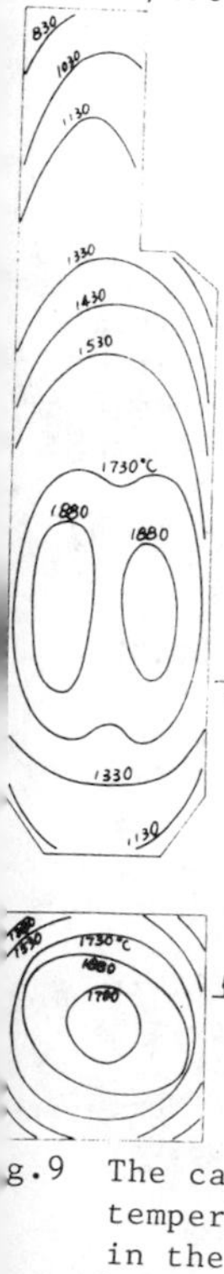

g.9 The calculated gas temperature distribution in the furnace of 300MW boiler.

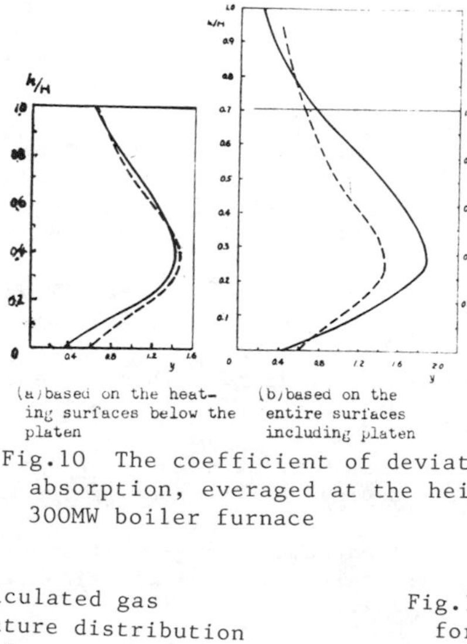

(a) based on the heating surfaces below the platen

(b) based on the entire surfaces including platen

Fig.10 The coefficient of deviation of heat absorption, everaged at the height of the 300MW boiler furnace

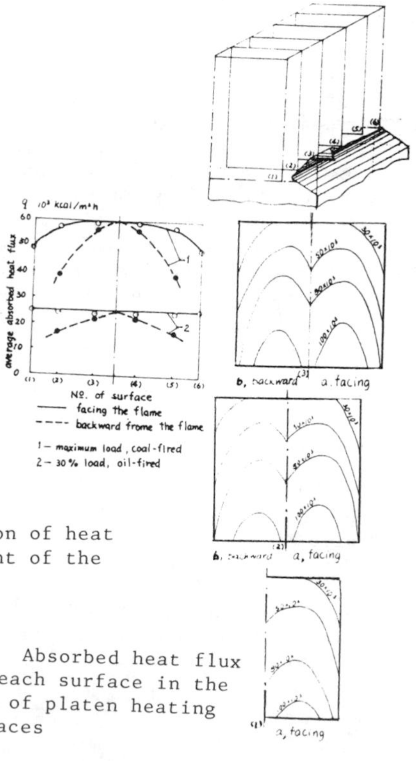

Fig.12 Absorbed heat flux for each surface in the zone of platen heating surfaces

tion adopted by the authors can reflect and solve problems of operation and design of steam boilers.

3. In order to improve further the method of numerical calculation of the flame heat transfer in furnaces, the flow process, the chemical reaction of the heat releasing combustion process and the radiative characteristics of the medium in the furnaces should be considered jointly as a whole, besides, experimental measurement data relevant to these processes and phenomena should be supplemented.

ACKNOWLEDGEMENT

The authors wish to express their hearty thanks to Gaojing Power Station in Beijing, the Second District Heating Power Station in Beijing, Dongfang Boiler Works, Shanghai Boiler Works, Science Foundation of Academia Sinica, who give financial supports and helps to this research project.

REFERENCES

1 N.Selçuk, R.C.Siddall, J.M.Beér, "A Comparison of Mathematical Models of the Radiative Behavior of An Industrial Heater," Chemical Engineering Science, 30, (1975), 871-876.
2 N.Selçuk, R.G.Siddall, J.M.Beér, "Two-flux Modelling of Two-dimensional Radiative Transfer in A Large-scale Experimental Furnace", J.Institute of Fuel, 49, 400, (1976).
3 S.V.Patankar, D.B.Spalding, "Mathematical Models of Fluid Flow and Heat Transfer in Furnaces: A Review", J.Institute of Fuel, 46, 388, (1973), 279.
4 W.Richter, R.Quack, "A Mathematical Model of A Low-volatile Pulverised Fuel Flame", Heat Transfer in Flames, p.95-109,(1974).
5 H.C.Hottel, E.S.Cohen, "Radiant Heat Exchange in A Gas-filled Enclosure, Allowance of Nonuniformity of Gas Temperature", A.I.CH.E.J., 4,1, (1958), 3-14.
6 H.C.Hottel, A.F.Sarofilm, "Radiative Transfer", (1967).
7 A.Lowe, T.F.Wall, I.McC.Stewart, "A Zoned Heat Transfer Model of A Large Tangentially Fired Pulverized Coal Boiler", 15th Symp. (Intern.) on Combustion, p.1261-1270. (1974).
8 Z.H.Yu, C.D.Shen et al., "The Mathematical Model of Zone Method to Calculate Radiative Heat Transfer and Application in Box or Cylinder Furnace", (in Chinese), Journal of Chemical Engineering, No.2, p.143-164. (1980).
9 Tatuzo Hirose, Akikaru Mitunaga, "An Investigation of Radiant Heat Exchange in Boiler", Bulletin of JSME, 14,74, (1971), 829.
10 F.R.Steward, H.K.Guruz, "Mathematical Simulation of An Industrial Boiler by the Zone Method of Analysis", Heat Transfer in Flames, p.47, (1974).
11 Haku Taniguchi et al., "Three-dimensional Analysis of Temperature Distribution in Combustion Chamber", Transaction of JSME, NO.284, p.610, (1970); NO.324, p.2473, (1973).
12 Xuchang Xu, "Mathematical Modelling of Three-dimensional Heat Transfer of Flame in Combustion Chambers", 18th Symp. (Inter.) on Combustion, p.1919, (1980).
13 Xuchang Xu, "Application of Mathematical Modelling of Three-dimensional Flame Heat Transfer in Utility Boilers", (in Chinese), Journal of Engineering Thermophysics, 2, p.161-168, (1982).
14 Yongfu Zhao, Xuchang Xu, "The Mathematical Simulation of Heat Transfer in A Cylindrical Furnace", (in Chinese), Journal of Engineering Thermophysics, 3,p.275-280, (1983).
15 Yongfu Zhao, Xuchang Xu, "The Method of Mathematical Simulation of A Pulverize Coal Boiler Furnace with Platen Superheaters", Journal of Tsinghua University, No.2, p.23-33, (1984).

16 Xuchang Xu, "Numerical Computation of Flame Heat Transfer", in Numerical Computation of Combustion Processes, Chapter 4. (in Chinese), China Science Press, (in the press 1985).
17 Yoshi Hayasaka, Haku Taniguchi et al., Journal of Thermal and Nucleal Power, (in Japanese), No.2, (1983).
18 W.Richter, "Anwendung von Berechnungsmodellen fur Feuerraume", VGB, 10, (1982
19 A.Lowe, I.McC. Stewat, T.F.Wall, "The Measurement and Interpretation of Radiation from Fly Ash Particles in Large Pulverized Coal Flames", 17th Symp. (Intern.) on Combustion, p.105-114, (1978).
20 J.L.Krazinski, R.O.Buckius, H.Krier, "Coal Dust Flames: A Review and Development of a Model for Flame Propagation", Prog. Energy Comb. Sci.,vol.5, p. 31-71, (1979).
21 Haku Taniguchi et al., 11th Japanese Symp. on Heat Transfer, (1974).
22 Zhaoxing Sun, Naiyi Hu, "Monte-Carlo Method for Radiative Heat Transfer in a Furnace", Journal of Electrical Engineering, No.1, p.41-50, (1984).
23 Xuchang Xu, "Numerical Computation of Flame Heat Transfer in The Furnace of Boiler No.1 in The 2nd Station of Beijing Thermal Power Plant", Research report (in Chinese). (1981).

HIGH–TEMPERATURE HEAT TRANSFER

Dynamics and Heat Transfer of Drops Impacting on a Hot Surface at Small Angles

KANG YUAN CAI
Southwestern Reactor Design and Research Institute
P.O. Box 291 (205)
Chengdu, PRC

SHI-CHUNE YAO
Department of Mechanical Engineering
Carnegie-Mellon University
Pittsburgh, Pennsylvania 15213, USA

INTRODUCTION

Heat transfer of impacting spray to hot surfaces is of great concern in mar metallurgical processes. The dilute sprays have been used frequently in the cooling alloys at very high temperatures. When a dilute spray impinges normally onto a soli the small drops may follow the air stream and divert from the solid surface such th the drop, in fact, impacts to solids at a small angle. In some other cases the drops with the gas stream on to the solid surface at small angles. The solid may also moving at such high velocity that the spray impacts on the surface at a very hi relative velocity and, therefore, at very small angles.

Substantial information has been reported on the normal impaction of a drop to hot solid surface. The mechanism of the impaction dynamics and the accompanyir heat transfer at above and below Leidenfrost temperature have been described b Pederson [1] for the condition of normal impactions. He reported that at the no wetting region, where the surface is at beyond the Leidenfrost temperature, the surfac temperature has little effect on the impaction dynamics of drops. If the velocity low, the drop does not break up into smaller drops. At the wetted region, where th surface is below the Leidenfrost temperature, the drop may wet a considerable area the surface. Nucleate boiling may occur and vapor forms beneath the liquid film. Th vapor sometimes breaks through the film and expels small drops away from the surfac The normal impaction dynamics and the heat transfer of drops at different surfac temperatures has been reviewed by Hall [2]. An entrainment model has been used t fit the data of high pressure spray cooling.

In many practical applications the drops impact to a hot surface at small angle The mechanism of this type of impaction is quite different from that of the norm impaction. The tangential relative velocity gives a rotational force to the drop durin impaction. As a result, the drop deforms in a non-symmetric form. Before th impaction, the drop travels through the hydrodynamic boundary layer of the gas near th wall that the drop decelerates and deforms before hitting to the wall. Furthermore, du to the large relative tangential velocity between the drop and the solid wall a gas laye may be entrained between the drop and the wall that it enhances the physical separatio of the drop from the wall.

Although the small angle impaction of a drop to surface contains variou unconventional mechanisms, there is little information reported on this problem. Bigo and Cumo et al. [3] observed the effect of surface temperature on the impaction hea transfer of 1 mm drops at large angle of collisions and for a very limited number o cases. Takeuchi et al. [4] reported the heat transfer effectiveness at beyond Leidenfrost temperature for the impaction angle of drops at 90, 45, and 30 degrees to the surface. The dynamics of large water drops falling onto a polished gold surface at angles has been investigated by Wachters and Westerling [5]. The angle of impaction is about 60 degree. Therefore, as a result, they can use the Weber number of the normal velocity component to interpret the date. All these studies are for the impaction of isolated drops. No consideration of the effect of the gas boundary layer has been made, and which is likely to be important for small angle impaction between drop flow and hot surfaces.

The only work closely relevant to this study is the experiment performed by Shalnev et al. [6] in simulating the impaction of a drop to steam turbine blades. A disk of 0.5 m diameter rotating at various speeds is used to study the impact dynamics of vertically falling water drops. The air boundary layer on the rotating disk could be either laminar or turbulent and the impacting angle of drop to disk could be very small. The disk is not heated; however, when the disk speed is high enough the drop may bounce away without any contact to the disk surface due to the cushion effect of entrained air beneath the drop.

The experiment of [6] did not look into the heat transfer of drops at small angle impaction because the disk was kept at the room temperature. In practical applications the heat transfer mechanism of drop flow interacting at small angle to a hot surface is determined by the criterions of whether the drops break up during the impact and whether the surface is beyond or below the Leidenfrost temperature. The existence of the tangential velocity components of the drop and the entrained gas will affect the criterion of drop break up and the Leidenfrost temperature as compared with the cases of normal impaction. It is of great interest to know whether the tangential velocity component will stabilize or de-stabilize the impacting drops and whether it will increase or decrease the surface Leidenfrost temperature. These are the objectives of the present study.

EXPERIMENTAL APPARATUS AND PROCEDURE

The schematic of the experimental apparatus is shown in Fig. 1 where monosize drops are directed vertically downwards on the hot rotating disk. The water drops are generated by piezoelectric pulses to produce small drops, and by dripping from a needle to produce large drops. The generated drops are of uniform size and velocity that the initial conditions of the impactions are precisely controlled.

The piezoelectric drop generator is shown schematically in Fig. 2. The drops are produced with a hypodermic needle which is attached to the end of a 4 mm O.D. thin-wall stainless steel tubing. The middle section of the tubing is pressed flat for a length of 7 cm, on which two strips of thin piezoelectric crystals with electric wires are epoxied. To generate the drops, slight pressure is applied to the water feed line and electric pulses of 20 to 100 volts with duration of about 20 μs are imposed on the piezoelectric crystals. The rapid deflection of the piezoelectric crystals provide a pressure pulse to the fluids in the tubing. By assigning a regular frequency of the electric pulses a regular disturbance is applied to the flowing liquid jet out of the needle. As a result, the liquid jet is disintegrated into drops of regular sizes. The jet diameter is determined by the needle; the drop velocity is controlled by the pressure of water feed line; and the drop size is adjusted by varying the electric pulse frequency. When very large drops are required the electric pulses are stopped and water pressure is reduced that the large drops drip from the needle at a low frequency.

The hot surface is made in the form of a rotating disk with different rotating speeds correspond to various relative tangential velocities and impacting angles. Using this system the drop phenomena may be examined easily by a stationary observer. The disk is made of brass with 76 mm radius and is rotated by an A.C. motor at speed of 0 to 800 rpm. The surface of the disk is of mirror finish and is oxidized uniformly in a furnace to increase its surface emissivity.

In conducting the experiment, distilled water was used and the disk was heated by a propane torch up to 400°C. The drops fell on the disk at a location about 64 mm from the disk center and the instantaneous surface temperature was measured at the location of 15 degree downstream of the impacting location. The surface temperature was detected with a MIKRON 80 Series digital infrared thermometer. The calibration was performed when the disk is stationary.

Calibration of the drop size and velocity was performed by photographing method. Back lighting with two flashes during the camera shutter opening time were employed to give drops with double images. Then the photograph was analyzed. The velocity profiles in the air boundary layer of the cold disk was measured with a hot wire anemometer. The dynamic behavior of the impinging drop was recorded by photographing with a strobe light which was placed at the same side of the camera.

In the study of the Leidenfrost temperature of the impacting drops the torch was removed and allowed the disk to cool down in a very slow transient. Through all the experiments no significant contamination of the surface was observed. The output of the infrared thermometer was recorded on a chart recorder. When the surface temperature fell below the Leidenfrost temperature the slope of the temperature trace indicated a 50 ~ 100 percent increase if the rotating speed of disk was not very high. When the rotating speed was increased the change of slope was less significant because at this moment the air convective cooling became more dominant. Due to the non-uniformity of the surface emissivity, slight oscillation of the sensed temperature was observed while the disk was rotating. However, this did not affect the data reduction. When the rotating speed was increased the oscillation diminished due to the averaging effect of the reading.

RESULTS AND DISCUSSION

Velocity profiles in boundary layer

In most of the practical applications the drops are carried with the flow in making impacts at small angles. Before hitting to the surface, the drops pass through the hydrodynamic boundary layer and experience deceleration and distortion. As a result, the impaction dynamics and the heat transfer behavior of the drops are influenced by the behavior of the boundary layer. Although the practical boundary layers may have various velocity profiles and may be of a turbulent flow, the present study considers the laminar boundary layer on a rotating disk as a typical situation.

The laminar boundary layer on a rotating disk is a classical problem and has been solved analytically in detail [7]. The velocity profiles in the present experiment have been measured using a hot wire anemometer at two different rotating speeds and room temperature condition. The measurements were performed in the boundary layer at various distances till 1 mm from the surface. The comparison of data and theoretical results are shown in Fig. 3 in terms of the non-dimensional distance ζ, radial velocity $u/\omega r$, and tangential velocity $v/\omega r$ where ω is the rotating angular velocity. Very good confirmation is obtained for both sets of data. Therefore the boundary layer, which is the boundary condition of the present study, is clearly described.

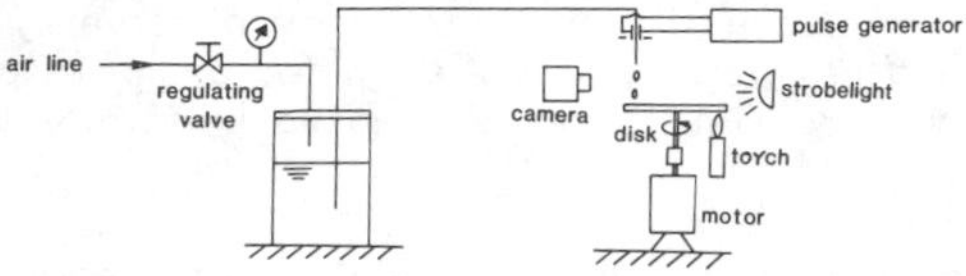

1. Schematic Diagram of Experimental Apparatus

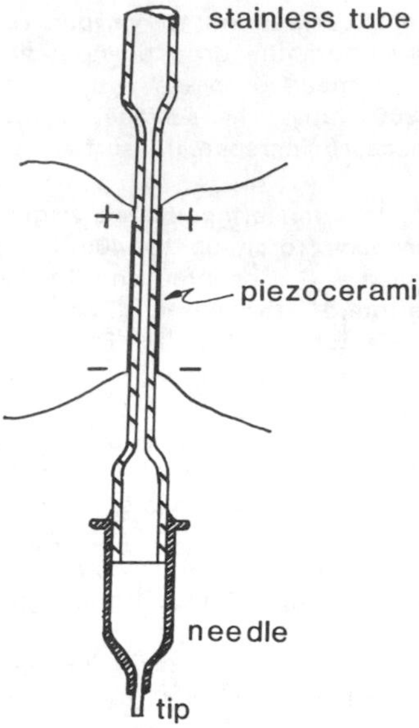

2. Piezoelectric Drop Generator

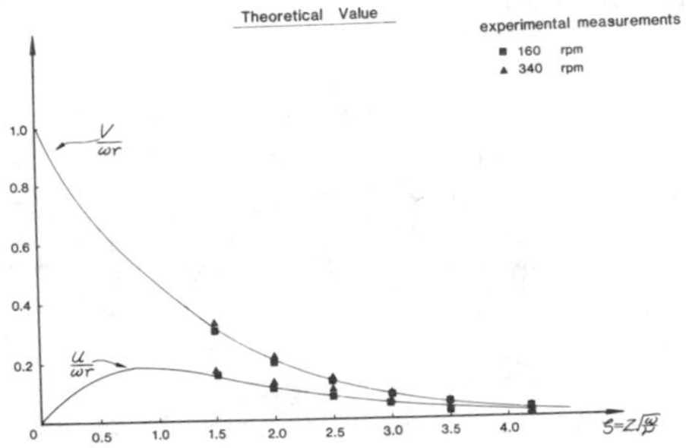

3. Comparison of Experimental Boundary Layer Velocity Profile With Theoretical Predictions

Drop impaction dynamics

One of the most important information of drop impaction dynamics is the breakup of the drop. When the surface temperature is beyond the Leidenfrost temperature a drop usually evaporates little during an impaction. However, if the drop breaks into smaller ones the effective surface area of the drops in the flow will increase significantly at downstream, and subsequently enhance heat transfer of the drop flow.

The impaction dynamics of water drop is primarily dependent upon the drop size, the normal velocity, surface temperature, and the relative tangential velocity. Observation has been made on the variation of each parameter. First of all, the effect of surface temperature has been investigated with drops of 3 mm diameter, 0.8 m/s normal velocity and at zero tangential velocity when the disk sitting stationarily. If the surface temperature is in the range of 150 to 200°C an impinging drop adheres to the surface and spreads as a film with nucleation occurring beneath the film and bubbles breaking through the film. When the temperature is in the range of 200 to 260° a large amount of mist forms immediately after the impaction. At higher temperatures, beyond the Leidenfrost temperature, a liquid film appears after the impaction but it rebounds into a drop without any breakup. At this condition, the behavior is rather independent of the surface temperature.

Keeping the same condition as the previous case at beyond Leidenfrost temperature but increasing the normal velocity to 1.5 m/s the drop shatters into several smaller ones in all directions. Maintaining this same normal velocity but reducing the drop size to 1 mm diameter the drops rebound without breakup. As it is well known, the effects of these parametric variations can be characterized by the Weber number of the incoming drop.

It is interesting to observe the effect of the relative tangential velocity. The disk rotates at a low speed and is at beyond Leidenfrost temperature. After impacting to the surface the drop rebounds with rolling motion and gains tangential linear velocity. By increasing the disk rotation speed to a high value the drop breaks into many smaller ones after the impaction. Fig. 4 and 5 present typical phenomena at these conditions. These Figures are traced from photographs with multiple exposure of a single drop.

Experiments of the drops impaction dynamics at various conditions have been conducted. Table 1 summarizes the conditions which cover the size 0.6 to 3.5 mm diameter, normal velocity 0.8 to 1.63 m/s and wide ranges of rotating speed and surface temperature. The relative tangential velocity shows definite effect to the breakup phenomena and could be referenced to the conventional normal impaction behavior. The Weber number of the incoming drop, based upon its normal velocity component, has been used generally as one of the criterions of breakup. With this incoming Weber number used, the effects of the relative tangential velocity can be expressed in terms of either an horizontal Weber number or the angle of impaction θ which is measured with respect to the surface. In the present study the impaction angle θ is chosen as the second parameter.

The data of the present study are presented in Fig. 6 in terms of the We_v and the angle θ. For any initial condition corresponding to a point below the curve in the Figure the drop will not break up after the impaction. For the initial condition at above this curve the breakup occurs. At a normal impaction that θ is 90 degrees, the critical We_v is about 45. This is quite consistent with published information in literature [1]. As the impaction angle decreases a drop with smaller incoming We_v may also break up. In other words, with the same vertical Weber number the existence of the relative tangential velocity destabilizes the drop in an impaction. The existence of the tangential velocity gives two effects. The drop will experience a tangential gas velocity in the boundary layer such that some distortion and stretching occurs before the impaction to the surface. Additionally, the surface exerts a horizontal force suddenly to the drop to cause further distortion and stretching. As a result, the drop is destabilzed significantly at the condition of small angle impaction.

This curve determining the drop breakup can be correlated by

$$We_v = 12.89 + 0.85\,\theta - 0.0053\,\theta^2 \qquad (1)$$

where the θ is expressed in degree. In Fig. 6, the data of a chrome plated disk are also presented. No apparent effects to the impaction dynamics is observed as long as the surface temperature is beyond the Leidenfrost temperature.

Drop impaction heat transfer

The heat transfer of an impacting drop is very much dependent upon whether the surface temperature is beyond or below the Leidenfrost temperature. The Leidenfrost temperature, however, is a function of the complicated dynamic conditions of the impaction. For the reference condition of the normal impaction, the Leidenfrost temperature of water drops can be deduced from the papers of Pederson [1] and Kendall [8]. It was observed that the higher the incoming Weber number of drops the higher the Leidenfrost temperature because the gas film between the drop and the hot surface can be squeezed out easily. Their results and the present results of the Leidenfrost temperature are plotted against the incoming drop Weber number in Fig. 7. The trend is consistent and the data can be correlated in the form of

$$Tq_v - Ts = 135.6\, We_v^{0.09} \qquad (2)$$

where the Ts is the saturation temperature of water and the temperatures are expressed in degree C.

The effect of the tangential velocity to the Leidenfrost temperature has been observed. When the disk was at very low speed of rotation and the surface temperature was cooled to slightly below the Leidenfrost temperature, the drops impacted on the disk and produced mist due to the partial boiling as shown in Fig. 8. By increasing the rotating speed of the disk the mist disappeared and the drops rebounded after impaction. The Leidenfrost temperature decreases with the increasing of the tangential relative velocity.

Considering that the Leidenfrost temperature of normal impaction can be predicted by the equation (2), it would be interesting to know the change of Leidenfrost temperature with respect to that of the normal impaction when the tangential velocity exists. A non-dimensional Leidenfrost temperature, $(Tq - Ts)/(Tq_v - Ts)$, can be used. The data of non-dimensional Leidenfrost temperature in the present study can be correlated with respect to the angle θ of the incoming drop. As shown in Fig. 9 at several different vertical Weber numbers, all the data fall onto the same curve. The correlation of least square fit is

$$\frac{Tq - Ts}{Tq_v - Ts} = 0.028\,\theta - 0.00019\,\theta^2 \qquad (3)$$

where the θ is expressed in degree. When the θ is at 90 degree the Leidenfrost temperature will be the same value as that of normal impaction. When the θ reduces the Leidenfrost temperature reduces. The present study covers the angle down to 15 degree. But the experiment of Shalnev et. al. [6] indicated that at very small angle, say 3 degree, the drop may not touch to cold disk due to the aerodynamic effects Therefore, it could be suggested that the curve in Fig. 9 goes toward zero when the angle is reduced toward zero.

It appears that if the angle of impaction is large, for example between 45 to 90 degree, the Leidenfrost temperature is not affected significantly. That is the reason why Wachters and Westerling [5] were able to correlate the data simply using the vertical component of the incoming drop velocity. If the angle of impaction is less than 45 degree the tangential velocity component becomes important. When the drop slides down to the surface a layer of gas is entrained beneath the drop. According to the Lubrication Theory [7] this gas layer cannot be squeezed out easily that the drop may not be able to have physical contact with the hot surface easily. As a result, the Leidenfrost temperature is reduced.

It should be pointed out that although the present experiment is performed on rotating disk the results may be applicable with minor modifications for many practical conditions. The effects of the overall profile of the boundary layer may not be as important as the velocity gradient near the wall where the impaction occurs. Due to the same reason the present results may show more deviation when a turbulent boundary layer is considered. Further study will be required to resolve this difference.

CONCLUSION

1. The impaction of drops in flow to a hot surface is affected by the angle of impaction. At small angle impactions, the drop will experience (a) the gas

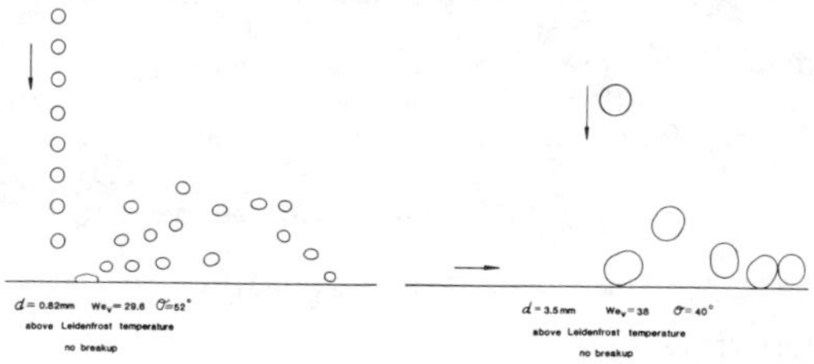

4. No Breakup After Impaction

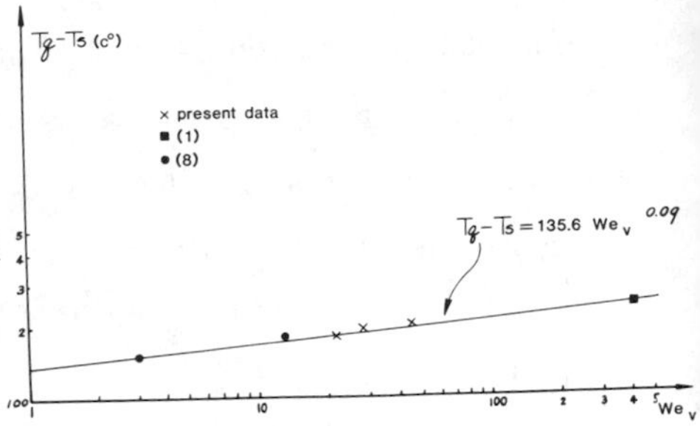

7. Effect Of Vertical Weber Number on Surface Quench Temperature

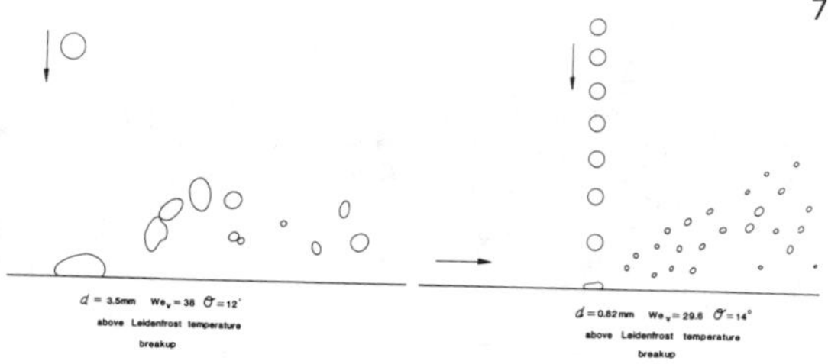

5. Breakup After Impaction

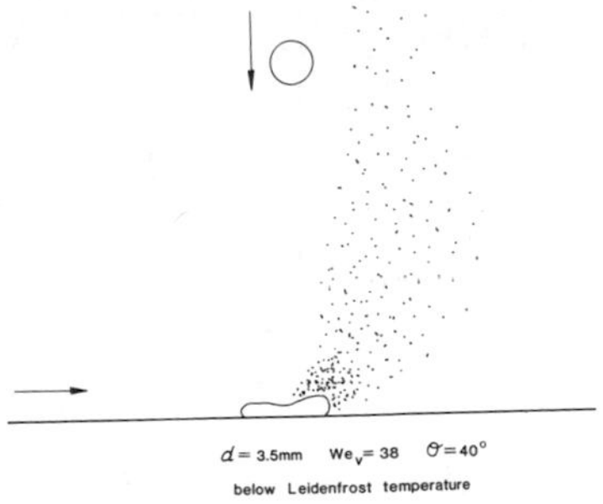

8. Partial Boiling Phenomena

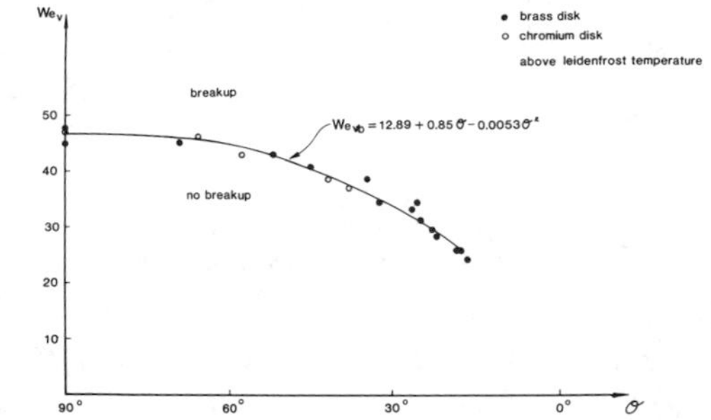

6. Effect Of Impact Angle On Breakup Of Drop

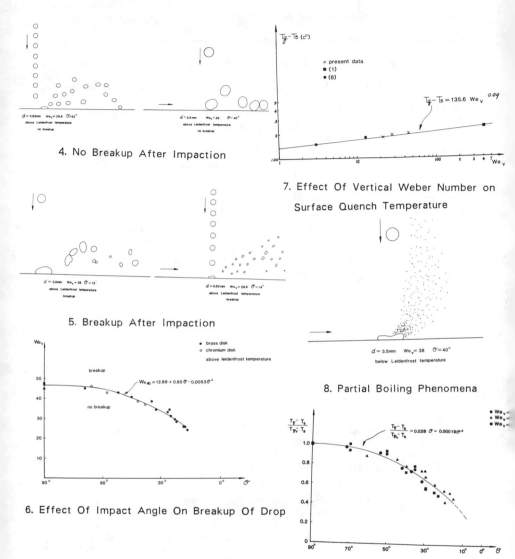

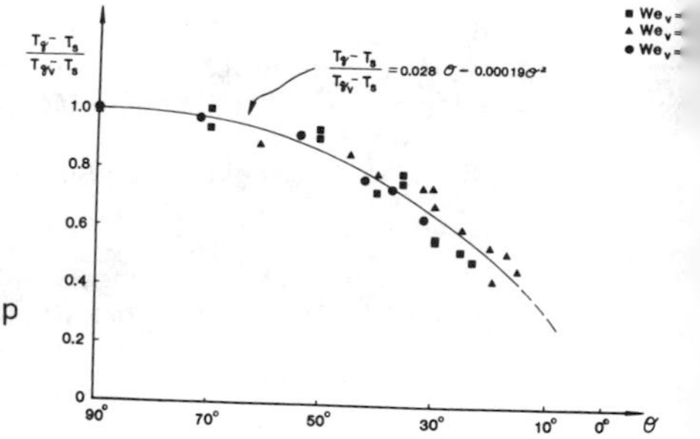

9. Effect Of Impact Angle On Surface Quench Temperature

boundary layer at significant relative velocity, (b) the entrained gas layer beneath the drop during the impaction, and (c) the strong shearing force of the surface at the instance of impact. The overall result of the tangential velocity component is to destabilize the drop. The onset condition of the drop breakup can be correlated with the equation (1).

2. The existence of the tangential relative velocity tends to stabilize the drop in the film boiling condition. During an impact, the entrained gas layer beneath the drop increases the resistance of drop-surface contact. The Leidenfrost temperature decreases with the decreasing of the impact angle with respect to the surface. The Leidenfrost temperature can be correlated by equation (3)

TABLE 1. CONDITIONS OF PRESENT EXPERIMENT

Drop Dia. mm	Drop Velocity m/s	Disk Speed rpm	Disk Temp. c
0.6	1.63		
0.68			
0.82			
0.86			
0.91	1.45	0-800	100-380
1.0			
3.0	0.8 1.15		
	0.9 1.40		
	1.10 1.45		
3.5	0.85 0.97		
	0.90 1.13		

REFERENCES

1. Pederson C O, "The Dynamics And Heat Transfer Characteristics Of Water Droplets Impinging Upon A Heated Surface", Ph.D. Thesis, Mechanical Engineering Department Carnegie-Mellon University 1967.

2. Hall P C, "The Cooling Of Hot Surfaces By Water Sprays", CEGB RD/B/N3361 1975.

3. Bigiom A, Cumo M et al., "Observation On The Impingement Of Droplets On Heated Walls In Emergency Core Cooling", CNEN-RT/ING (82)8.

4. Takeuchi K et al., "Heat Transfer Characteristics And The Breakup Behavior Of Small Droplets Impinging Upon A Hot Surface", ASME JSME Thermal Engineering Joint Conference, Vol. 1, pp.165-172 1983.

5. Wachters L and Westerling N, "The Heat Transfer From A Hot Wall To Impinging Water Drops In The Spheroidal State", Chemical Engineering Science, Vol. 21, pp.1047-1056 1966.

6. Shalner F K, Povarov O A, Nazarov O I, and Shalobasov I A, Proceedings of the Fifth Conference of Fluid Machinery, Vol. 2, pp.1011-1019 1975.

7. Schlichting H, "Boundary-Layer Theory", 6th ed. New York, McGraw-Hill 1968.

8. Kendall G E and Rohsenow W M, "Heat Transfer To Impacting Drops And Post Critical Heat Flux Dispersed Flow", Report of Heat Transfer Laboratory, No. 85694-100 MIT 1978.

Nomenclatures

d	diameter of incoming drop
r	disk radial
T	disk surface temperature
u	radial linear velocity of air boundary air
V	incoming velocity of drop
v	tangential linear velocity of air boundary air
We	Weber number of the incoming drop $We = \frac{\rho V^2}{(\sigma/d)}$
z	distance from disk

Greek symbols

θ	impaction angle of water drop with respect to surface, degree.
ζ	non-dimensional distance from disk
ω	rotating angular velocity of disk
ν	kinematic viscosity of air
ρ	density of water
σ	surface tension of water to air

Subscripts

q	quench temperature of disk surface
s	saturation temperature of water
v	vertical impaction

A Proposed Expression for Heat Transfer between a Thermal Plasma Flow and a Particle

XI CHEN
Engineering Thermophysics Division
Department of Engineering Mechanics
Tsinghua University
Beijing 100084, PRC

ABSTRACT

This paper is concerned with the heat transfer between a thermal plasma flow and an immersed particle without evaporation and without accounting for the Knudsen effect. A new heat-transfer expression is proposed based on an appropriate summation of the exact solution for the case of pure conduction and the approximate analysis for the case with high Reynolds numbers. The new expression is identical to the exact solution as the Reynolds number equal to zero and gives results which are consistent with the experimental finding for the case with high Reynolds numbers. In addition, within the range of low and intermediate Reynolds numbers typical for applications in plasma chemistry and plasma processing, the new expression predicts heat fluxes which are in good agreement with corresponding computational ones obtained by directly solving the simultaneous set of continuity, momentum and energy equations including actual properties of argon and nitrogen plasmas.

NOMENCLATURE

C_p	specific heat
D	particle diameter
h	specific enthalpy
k	thermal conductivity
Nu	Nusselt number
Nu_s	Nusselt number based on S
Nu_{s1}	Nusselt number for high Re
Pr	Prandtl number
q	specific heat flux
Re	Reynolds number
r_w	particle radius
S	heat conduction potential
T	temperature

Greek Letters

α	correction factor
μ	viscosity
ρ	density

Subscripts

f	film temperature
w	particle surface
∞	oncoming plasma flow

INTRODUCTION

Powder processing represents a key link in many industrial applications of thermal plasmas such as plasma spheroidizing, plasma spraying, decomposition and synthesis of refractory materials, ect. There is consensus among the researcher in this field that the heating and movement of particles injected into a thermal plasma flow are important aspects to be clarified for the further development of the thermal plasma processing. In recent years many publications have appeared /1——21/.

As a continuation of previous work/10,13,16,17/, this paper is concerned wit the heat transfer to a particle without evaporation exposed to a thermal plasma flow, ignoring the Knudsen effect/13/. Although in practical calculations of particle heating the Knudsen effect as well as evaporation may be important, a heat transfer expression under the above mentioned conditions is a must for the modeling of the powder processing. It is because that computational studies have shown/10,16/ that the ratio of the heat flux with accounting for Knudsen effect or evaporation to that without these effects remains almost the same no matter if convection exists. Hence, practical calculations for the case with convection, evaporation and Knudsen effects can be simply performed by using the heat flux expression presented in the present paper and the calculated results concerning the heat flux ratios obtained simply for the case without convection/10,13/.

ASSESSMENTS OF AVAILABLE EXPRESSIONS

There are many expressions available in the literature concerning the heat transfer to a spherical particle exposed to a thermal plasma flow. A widely employed expression is the well known Ranz-Marshall formula with the film temperature as the reference temperature/3,7/:

$$Nu_f = 2 + 0.6Re_f^{1/2}Pr_f^{1/3} \qquad (1)$$

This formula was considered to be applicable mainly to the case with lower plasma temperatures. Recognizing this limitation of Eq.(1), many authors presented suggestions about how to correct Eq.(1) in order to account for the presence of the great temperature difference between the plasma flow and the particle surface. Examples of such corrected expressions are as follows/1,2,4,5,9,18/:

$$Nu_f = (2 + 0.6Re_f^{1/2}Pr_f^{1/3}) \left(\frac{\rho_\infty \mu_f}{\rho_f \mu_\infty}\right)^{0.15} \qquad (2)$$

$$Nu_f = (2 + 0.6Re_f^{1/2}Pr_f^{1/3}) \left(\frac{\rho_\infty \mu_\infty}{\rho_w \mu_w}\right)^{0.6} \qquad (3)$$

$$Nu_f = (2 + 0.6Re_f^{1/2}Pr_f^{1/3}) \left(\frac{\rho_\infty \mu_\infty}{\rho_w \mu_w}\right)^{0.6} (C_{p\infty}/C_{pw})^{0.38} \qquad (4)$$

Ref./21/ suggested to use the following expression for calculating the heat flux to the particle:

$$Nu^* = \frac{DC_{pw}q}{k_\infty(h_\infty - h_w)} = 2(k_w/k_\infty)+0.5Re_\infty^{0.5}Pr_\infty^{0.4}\left(\frac{\rho_\infty \mu_\infty}{\rho_w \mu_w}\right)^{0.2} \qquad (5)$$

Sayegh and Gauvin/6/ used the following expression to correlate their computational results of the heat flux under plasma conditions:

$$Nu_w = 2f_o + 0.473Pr_w^m Re_{0.19}^{0.552} \qquad (6)$$

where

$$m = 0.78Re_{0.19}^{0.145}$$

the subscript 0.19 in the Reynods number means using $[T_w + 0.19(T_\infty - T_w)]$ as the reference temperature. f_o in Eq.(6) was the analytical result, and $f_o = [1-(T_w/T\ )^{1+x}]/\{(1-x)[1-(T_w/T_\infty)](T_w/T_\infty)^x\}$ which was obtained under pure conduction conditions based on the assumptions $\rho \sim 1/T, \mu \sim T^x$ and $k \sim T^x$. For argon plasma, x=0.8.

Since there exist pronounced differences between these suggested expressions for the heat transfer to a particle, it is highly desirable to have knowledge about how well they can be used.

Fortunately, an exact solution for the case of pure heat conduction under thermal plasma conditions has been obtained/10/:

$$q_0 = (S_\infty - S_w)/r_w \tag{7}$$

where

$$S = \int_{T_0}^{T} k\,dT$$

is the so-called heat conduction potential, and T_0 is a reference temperature (e.g. T_0=300K). The heat conduction potential S can be treated as one of the thermophysical properties of plasma. Hence, we can use the exact solution, Eq. (7), to check the accuracy of the calculated heat flux by using each expression among Eqs.(1)-(6) for the limit case with zero Reynolds numers. Fig.1 shows the calculated variations with the plasma temperature of the ratios of the predicted heat flux by each expression among Eqs.(1)—(6) to the exact value given by Eq.(7). It can be found from Fig.1 that no expression among Eqs.(1)—(6) can be considered as a satisfactory one within the wide range of plasma temperature (up to 20000K), although Eq.(1) predicts correct values of the heat flux as $T_\infty \leq$ 7000K. They all deviate substantially from the exact results at higher plasma temperatures.

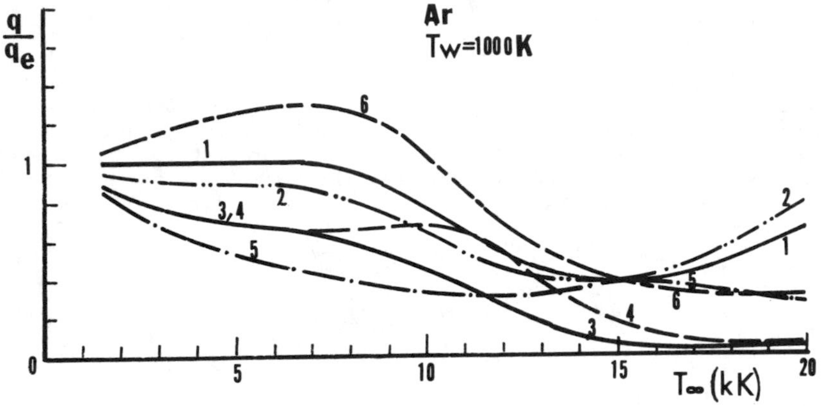

Fig.1. Variations with plasma temperatures of ratios of the calculated heat fluxes by Eqs.(1)-(6) to that of the exact solution (Re_∞ = 0). Curves 1—6 correspond Eqs.(1)—(6).

Deviation of Eq.(1) from the exact value for the case with higher plasma temperatures is not a surprising result, because Eq.(1) was built for ordinary temperature fluid conditions and it did not account for some new phenomena under plasma conditions (e.g. gas ionization, dissociation, etc.). Deviations of Eqs. (2)—(6) from the exact solution are somewhat surprising. One of the possible reasons for explaining these deviations is that Eqs.(2)—(6) were all proposed based on correlating some computational or experimental results within certain ranges of plasma/particle parameters. In addition, the authors of Ref./21/ took $q=(2/D)(k_w/Cp_w)(h_\infty - h_w)$ as the exact result for the case of pure conduction, but it is not a correct judgement as the comparison of their result with the exact solution, Eq.(7), shows. The error assumed by Eq.(6) can be explained by that the assumptions concerning the variations of ρ, μ and k with the plasma temperature employed in deriving Eq.(6) do not represent accurately the actual variations of argon plasma properties.

Since the typical value of the Reynolds number met in plasma processing of powder is in general small(<10), heat conduction would contribute a large fraction to the total heat flux under convection conditions. Hence, the errors of Eqs.(1)—(6) exhibited in Fig.1 reflect the disadvantage of these equations, although the comparisons in Fig.1 are only for the case of pure heat conduction.

As indicated in Refs./10,14,15/, a more suitable definition of the Nusselt number under thermal plasma conditions is

$$Nu_S=qD/(S_\infty - S_W)$$

because it always gives an value $Nu_S=2$ from Eq.(7) as Re=0. This result is similar to that obtained for the ordinary temperature fluid.

Based on this finding, Ref./14/ suggested to use the following expression to calculate the heat flux to a particle exposed to a thermal plasma flow:

$$Nu_S = 2 + 0.514Re_{av}^{0.5} \qquad (8)$$

where the subscript av in the Reynolds number means Re is based on the integral mean values of plasma properties between T_W and T_∞. The disadvantage of Eq.(8) is that it does not include Prandtl number effect on heat transfer as Re$\neq$0. And the comparisons of the predicted results by Eq.(8) with corresponding computational values demonstrate that Eq.(8) is not a satisfactory expression, as we see later on in Figs.2—4.

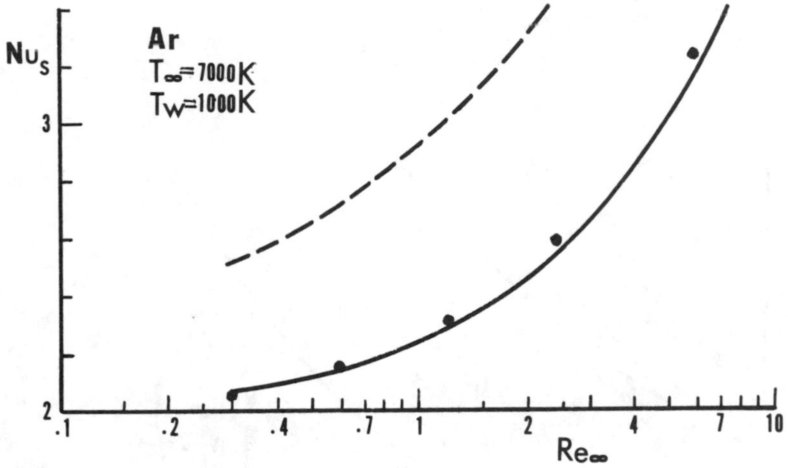

Fig.2. Variations of the calculated Nusselt number with the Reynolds number. ———Eq.(19); ----Eq.(8);•computation (u_∞ =200m/s, r_W=2.5, 5, 10, 20 and 50μm, T_∞=7000K)

THE NEW PROPOSED EXPRESSION

For the case in which the Reynolds number is so large that the boundary layer approximation is applicable, the local or circumferentially average heat flux over a spherical particle can be obtained by using the method similar to that employed for the cylinder immersed in a thermal plasma cross flow/22,23/. Based on the assumptions concerning plasma properties

$$\rho h=\rho_\infty h_\infty \qquad (9)$$

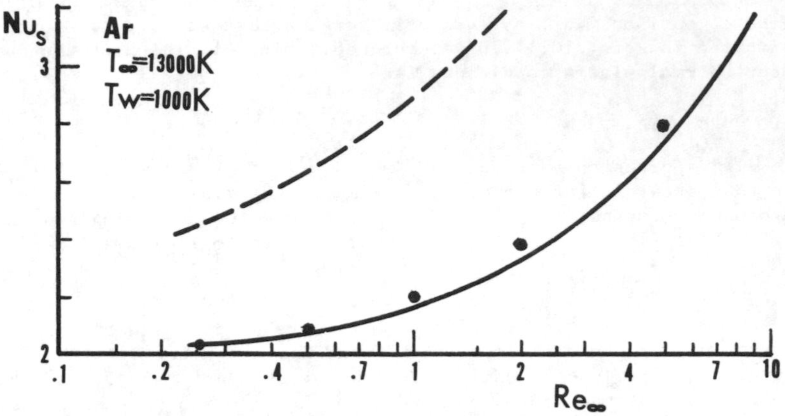

Fig.3. Variations of the calculated Nusselt number with the Reynolds number. ——— Eq.(19); ---- Eq.(8); • computation (u_∞=200 m/s, r_w=2.5, 5, 10, 20 and 50 μm, T_∞=13000 K)

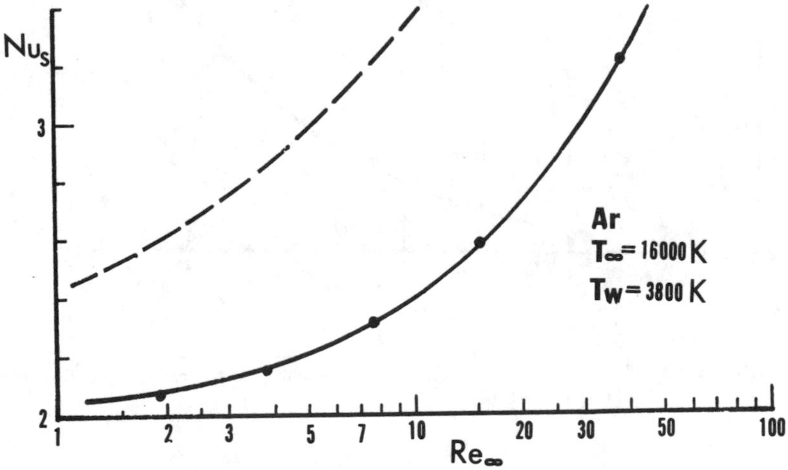

Fig.4. Variations of the calculated Nusselt number with the Reynolds number. ——— Eq.(19); ---- Eq.(8); • computation (T_∞=16000K, u_∞=200 and 500 m/s, r_w=12.5, 25, 50 and 100 μm)

$$\rho\mu = \rho_\infty \mu_\infty \tag{10}$$

$$\rho k/Cp = \rho_\infty k_\infty / Cp_\infty \tag{11}$$

and considering the feature of the spherical symmetry in the present problem, we can obtain the following average heat flux expression over the frontal semisphere by using the Merk-Chao transformation/24,25/:

$$\bar{q}_{0-90^\circ} = 0.63\left(\frac{k_\infty}{Cp_\infty}\right)\left(\frac{h_\infty}{D}\right)Re_\infty^{0.5}\ Pr_\infty^{0.4}[1-(h_w/h_\infty)^{1.14}] \tag{12}$$

In order to obtain the average heat flux over the whole sphere, the concentration coefficient of the heat flux on the frontal semisphere (Q_{0-90°}/Q_{0-180°}) is determined by solving numerically the simultaneous governing equations including actual properties of argon plasma. It is found that this coefficient is approximately equal to 0.68 within the parameter range studied. Hence, the average heat flux over the whole sphere is

$$q = 0.46(k_\infty/Cp_\infty)(h_\infty/D)Re_\infty^{0.5}\ Pr_\infty^{0.4}[1-(h_w/h_\infty)^{1.14}] \tag{13}$$

However, since the assumptions employed in the derivation of Eqs.(12) and (13) (i.e. Eqs.(9)-(11)) do not represent accurately the actual variations of practical plasma properties, a correction factor is needed to be introduced into these expressions in order to obtain correct results.

Now let us discuss this correction. For the case of pure conduction, the approximate solution of the heat flux as the assumptions (9) and (11) are employed is/10/

$$q = (k_\infty/Cp_\infty)(h_\infty/D)[1-(h_w/h_\infty)^2] \tag{14}$$

Comparison of this approximate solution with the exact result (7) gives the correction factor due to the assumptions (9) and (11):

$$2(S_\infty - S_w)/[(k_\infty/Cp_\infty)h_\infty(1-(h_w/h_\infty)^2)]$$

If we assume this factor is also applicable to the case with convection, Eq.(13) can be rewritten as

$$q = 0.92\alpha(S_\infty - S_w)(1/D)Re_\infty^{0.5}Pr_\infty^{0.4}[1-(h_w/h_\infty)^{1.14}]/[1-(h_w/h_\infty)^2] \tag{15}$$

or

$$Nu_{s1} = 0.92\alpha Re_\infty^{0.5}\ Pr_\infty^{0.4}[1-(h_w/h_\infty)^{1.14}]/[1-(h_w/h_\infty)^2] \tag{16}$$

Here a new factor α is introduced in order to correct the error due to the third assumption Eq.(10) employed for the case with convection and due to accounting for the difference between the case with convection and pure conduction.

Eq.(15) or (16) shows that the heat flux or Nusselt number for the case with higher Reynolds numbers is proportional to the square root of the Reynolds number. This predicted result is identical to the experimental finding reported in the literature. However, Eq.(16) predicts $Nu_{s1}=0$ as $Re_\infty=0$, instead of the exact result $Nu_{so}=2$.

An usual practice for the ordinary temperature fluid is to add simply Eq.(16) with $Nu_{so}=2$ and use the summation expression for the calculation of heat fluxes under low or intermediate Reynolds numbers. However, such a simple summation is not a satisfactory approach even for the ordinary temperature fluid for it would predict an overestimated heat flux under low Reynolds numbers/6/.

A more general approach, which can always satisfy the requirements that $Nu_s \sim \sqrt{Re_\infty}$ as Re_∞ is large and $Nu_s=2$ as $Re_\infty=0$, is as follows:

$$Nu_s = (2^n + Nu_{s1}{}^n)^{1/n} \quad (17)$$

where Nu_{s1} is given by Eq.(16).

A computational study shows for the parameters (e.g. particle size, plasma temperature, plasma velocity, etc.) typical in plasma chemistry and plasma processing, the value of "n" in Eq.(17) can be taken as 2 and the factor α can be expressed as

$$\alpha = 1.73(Pr_w/Pr_\infty)^{0.21}(\rho_\infty\mu_\infty/\rho_w\mu_w)^{0.26} \quad (18)$$

Hence, the final form of the new proposed expression for the heat transfer to a particle exposed to a thermal plasma flow is as follows

$$Nu_s = 2\left\{1 + 0.63Re_\infty Pr_\infty^{0.8}\left(\frac{Pr_w}{Pr_\infty}\right)^{0.42}\left(\frac{\rho_\infty\mu_\infty}{\rho_w\mu_w}\right)^{0.52}\left[\frac{1-(h_w/h_\infty)^{1.14}}{1-(h_w/h_\infty)^2}\right]^2\right\}^{0.5} \quad (19)$$

COMPARISONS WITH COMPUTATIONAL RESULTS

For small particles exposed to an argon plasma with typical parameters for the applications of plasma chemistry and plasma processing, Figs.2-4 compare the predicted results by Eq.(19) and those by computation, respectively, for plasma temperatures T_∞=7000, 13000 and 16000K.

The computational results are obtained by using the equations and solving technique similar to those described in Refs./10,16/, but evaporation and the Knudsen effect are not included in the present study as mentioned above. Briefly the set of two dimensional (spherically symmetrical), elliptic equations (continuity, momentum and energy equations) including actual plasma properties is solved by using the finite-difference method/26/. 20(uniform in θ direction)X24 (non-uniform in radial direction) grid points are employed in the computation.

It is seen from Figs.2-4 that the agreement between the predicted results by the proposed expression (19) and corresponding computational ones is fairly good, while Eq.(8) suggested in /14/ assumes great errors.

In Table I, additional comparisons for argon plasma are presented for a wider range of the plasma temperature (r_w=50μm, T_w=3800K and u_∞=200m/s). Deviations are seldom more than 5%.

In Table II, we compare the predicted results by the proposed expression (19) with corresponding ones by computation for nitrogen plasma. Fairly good agreement in general is also observed except for the case with lower plasma temperatures, where 5-7% errors probably appear.

Table I. Comparisons of Eq.(19) with computation (argon plasma, u_∞=200m/s, r_w =50μm, T_w=3800K)

T_∞ (K)	5000	6000	8000	9000	10000	12000	14000	16000
Nu_s,Eq.(19)	3.41	3.17	2.84	2.76	2.72	2.67	2.51	2.32
Nu_s,computation	3.37	3.15	2.80	2.70	2.60	2.54	2.50	2.34

Table II. Comparisons of Eq.(19) with computation (nitrogen plasma, u_∞=200 m/s, r_w=50μm, T_w=3800 K)

T_∞ (K)	5000	7000	9000	10000	12000	14000	16000
Nu_s, Eq.(19)	3.23	2.78	2.40	2.36	2.29	2.21	2.14
Nu_s,computation	3.01	2.63	2.42	2.37	2.32	2.30	2.23

CONCLUSIONS

Since none of the available expressions in the literature concerning the heat transfers to a particle exposed to a thermal plasma flow is satisfactory, a new expression (i.e. Eq.(19)) is proposed in this paper based on the power-law summa tion of the exact solution for the pure conduction (i.e. $Nu_s=2$) and the simplifi boundary-layer solution for the case with high Reynolds numbers. Within the range of parameters typical for many applications in plasma chemistry and plasma processing, comparisons with corresponding computational results show that the new proposed expression can predict fairly well the heat flux to a particle in thermal plasma flow without taking evaporation and the Knudsen effect into account.

ACKNOWLEDGMENT

This work was supported by the Science Foundation of the Chinese Academy of Sciences under grant TS-83-758.

REFERENCES

(1) Lewis,J.A. and W.H.Gauvin, AIChE J.,19,982(1973).
(2) Yoshida,T. and K.Akashi, J. Appl. Phys.,48,2252(1977).
(3) Harvey,F.J. and T.N.Meyer, Metall. Trans.,9B,615(1978).
(4) Boulos,M.I., IEEE Trans. Plasma Sci.,4,93(1978).
(5) Fiszdon,J.K., Int. J. Heat Mass Transfer,22,749(1979).
(6) Sayegh,N.N. and Gauvin,W.H., AIChE J.,25,522(1979).
(7) Gal-Or,B., ASME J. Eng. of Power,102,589(1980).
(8) Rykalin,N.N. et al., High Temperature,19,404(1981).
(9) Lee,Y.C.,K.C.Hsu and E.Pfender, Fifth Int. Symp. Plasma Chem., Vol.2,795(1981).
(10) Chen,Xi and E.Pfender, Plasma Chemistry and Plasma Processing, 2,185(1982).
(11) Chen, Xi and E.Pfender, Plasma Chemistry and Plasma Processing,2,293(1982).
(12) Vardelle,A, M.Vardelle and P.Fauchais, Plasma Chemistry and Plasma Processir 2,255(1982).
(13) Chen,Xi and E.Pfender, Plasma Chemistry and Plasma Processing,3,97(1983).
(14) Vardelle,M.,A.Vardelle,P.Fauchais and M.I.Boulos, AIChE J.,29,236(1983).
(15) Bourdin,E.,P.Fauchais and M.I.Boulos, Int. J. Heat Mass Transfer,26,567(1983
(16) Chen,Xi and E.Pfender, Plasma Chemistry and Plasma Processing,3,351(1983).
(17) Chen,Xi,Y.C.Lee and E.Pfender, 6th Int. Symp. Plasma Chem.,Vol.1,59(1983).
(18) Lee,Y.C., Modeling Work in Thermal Plasma Processing, Ph.D. Dissertation, University of Minnesota(July,1984).
(19) Proulx,P.,J.Mostaghimi and M.I.Boulos, 6th Int. Symp. Plasma Chem., Vol.1,59 (1983).
(20) Chen,Xi and E.Pfender, ASME Paper No.84-GT-287(1984).
(21) Kalganova,I.V. and V.S.Klubnikin, High Temperature,14,408(1976).
(22) Chen,Xi,ASME Paper No.82-HT-30(1982).
(23) Chen,Xi, ASME Journal of Heat Transfer,105,418(1983).
(24) Merk,H.J.,J. Fluid Mechanics, 5,460(1959).
(25) Chao,B.T. and R.O.Fagbenle, Int. J. Heat Mass Transfer,17,223(1974).
(26) Patankar,S.V., Numerical Heat Transfer and Fluid Flow, Hemisphere,N.Y.(1980)

Thermal Radiation Properties of Refractory Metals and Electrically-Conductive Ceramics at High Temperatures

TOSHIRO MAKINO, HIROFUMI KINOSHITA, YOSHINAO KOBAYASHI, and TAKESHI KUNITOMO
Department of Engineering Science, Faculty of Engineering
Kyoto University
Kyoto 606, Japan

ABSTRACT

Near-infrared and infrared spectra of reflectivity and emissivity are measured on refractory metals (Mo, Ta and W) and electrically-conductive ceramics (TiC and TiN) at temperatures from room temperature to 1 300 K. The spectra are analyzed by using the Kramers-Kronig method to obtain the optical constants. These metals and ceramics are found to have a common mechanism of absorption and emission of radiation by electrons : strong interband transition in the near-infrared region and conduction absorption by electrons on an anisotropic Fermi surface in the infrared region. A dispersion formula of optical constants is used and the electronic parameters are determined, in order to estimate the radiation heat transfer in high temperature systems.

1. INTRODUCTION

Refractory metals are structural materials for high temperature systems where the temperature is so high that heat resisting alloys can not be used. Refractory metals are protected by ceramic coating layers, and/or used in pure atmosphere such as vacuum and rarefied hydrogen gas. Radiation properties of the metals have been investigated mostly in the visible or near-infrared region for illumination engineering and/or optical techniques for temperature measurement. As coating materials for refractory metals, new ceramics with high electric conductivities are good choices, but the radiation properties have scarcely been investigated. The properties in the wide spectral regions are needed for the thermal engineering applications.

The authors have studied radiation properties of heat resisting alloys [1] and dielectric ceramics [2] by spectroscopic methods. In the present study, the same methods are applied and the properties of refractory metals (Mo, Ta and W and electrically-conductive ceramics (TiC and TiN) are investigated. These ceramics are used in the same engineering systems as those of refractory metals, and have similar physical properties as those of refractory metals [3]. An electronic theory of metals is examined to find a common expression for radiation properties of the metals and the ceramics. Macroscopic properties for radiation heat transfer are calculated systematically by using the method.

2. EXPERIMENTAL PROCEDURE

Experimental specimens of the refractory metals are bulk plates of 50 mm square

TABLE 1. Chemical compositions of refractory metals (ppm)

	Mo	Ta	W	Mg	Al	Si	Cr	Mn	Fe	Ni	Cu	Sn
Mo	Bal.			<3	<15	<15	<40	<50	<100	<50	<3	<10
Ta	<20	Bal.	<20		<10	<10	<10		<10	<10		
W	20		Bal.	<1	<1	<1	1	<1	3	5	<1	<5

TABLE. 2 Specifications of specimens of conductive ceramics

	Chemical composition	Sintering condition
TiC (Sintered)	$TiC_{0.977}$> 96 wt. %, TiN, TiO, W	2 073 K, 1×10^7 Pa
(CVD)	$TiC_{0.97}$	
TiN (Sintered)	$TiN_{0.959}$> 99 wt. %, TiC, TiO	2 023 K, 1×10^7 Pa
(CVD)	$TiN_{0.98}$	

and 1 mm in thickness. Table 1 shows the chemical compositions. The purities are higher than 99.9 %. The surfaces are lapped and/or buffed to realize an optical smoothness of a maximum roughness less than 30 nm. For the conductive ceramics, two kinds of specimens are used. The first ones are thin layers of three kinds of thickness of 3-20 µm, which are coated on the optically smooth substrates of Mo by a CVD (chemical vapor deposition) technique. The second kind of specimens are bulk plates of 50 mm square and 2 mm in thickness, which are sintered at high temperatures and a high pressure. Table 2 shows the specifications of the specimens. The chemical compositions are deviated a little from the stoichiometric values of TiC and TiN. The surfaces of both kinds of specimens are polished by the same means as those for the refractory metals.

At room temperature, three kinds of spectroscopic measurements are carried out. Firstly, the normal reflectivity R_N is measured at wavelengths of 0.34-22.2 µm. Secondly, the reflectivity R_{70p} for the p-polarized component of radiation incident at 70° is measured at four wavelengths in the visible region. Thirdly, for the CVD ceramics, the diffuse component R_D of the normal-incident-hemispherical reflectance is measured at five wavelengths in the visible to the infrared regions. At high temperatures from 900 to 1 300 K, the normal emissivity ε_N is measured on the refractory metals and the CVD ceramics. The wavelength region is from 0.80 to 3.8 µm. The ε_N measurement is carried out at a reduced pressure less than 1×10^{-3} Pa. The experimental apparatuses and procedures of the above four measurements are the same as those in the previous studies [2, 4].

3. EXPERIMENTAL RESULTS AND DISCUSSION

3.1 Kramers-Kronig Analysis

In order to clarify the mechanism of absorption and emission of radiation in refractory metals and conductive ceramics, the spectra of R_N are analyzed by the Kramers-Kronig method [5] to obtain the spectra of optical constants, $\hat{n} = n - i k$. Since the Kramers-Kronig equations include the integral over the whole wavelength region, the R_N spectra should be extrapolated out of the measured region. The determined values of the optical constants depend on the choice of the method of extrapolation, but the qualitative behavior of the spectrum is not affected so much. Thus, when the optical constants are preliminarily known at a specified wavelength, accurate values of the constants can be obtained in the whole wavelength region of the R_N measurement.

In the present analysis the following procedure is adopted. For the extrapolation to the longer wavelengths, a Hagen-Rubens-type equation is used. For the extrapolation to the shorter wavelengths, R_N and R_{70p} at the four visible wavelengths are analyzed firstly. Optical constants at the wavelengths are determined, so that the values of the reflectivities calculated by the Fresnel's formulae can fit the measured values. Secondly, extrapolating equations,

$$R_N(\lambda)/R_N(\lambda_{min}) = (\lambda/\lambda_{min})^h \qquad (\lambda \leqq \lambda_p) \tag{1}$$

and

$$R_N(\lambda) = R_N(\lambda_{min}) \qquad (\lambda_p \leqq \lambda \leqq \lambda_{min}) \quad , \tag{2}$$

are presumed, where λ is the wavelength of radiation in vacuum and λ_{min} is the shortest wavelength of the R_N measurement. The constants h and λ_p are determined, so that the optical constants at the four wavelengths can be expressed well. There is not any physical meaning in Eqs. (1) and (2).

The results of the above analysis are presented in the following in the form of $(2nk/\lambda)$. The energy of radiation absorbed by solid materials (per unit volume, unit time and unit wavelength interval) is proportional to this quantity.

3.2 Spectra of Refractory Metals and Electrically-conductive Ceramics

Figures 1 and 2 show the spectra of ε_N and $(2nk/\lambda)$ of the refractory metals, respectively. The ε_N at room temperature is calculated by $(1-R_N)$. The figures include the calculated values based on the discussion in Section 3.3 and also the values of other authors [6-9].

In every spectrum in Figure 1, a strong absorption band is observed in the visible to near-infrared region. The ε_N spectra do not depend on temperature in this region. This region corresponds to the longer wavelength tail of the interband transition region in Figure 2, where $(2nk/\lambda)$ decreases with an increase of wavelength. The values of Nestell and Christy [9] in the figure, which are measured on evaporated films, agree well with the present results on bulk specimens. In the wavelength region of 2–5 μm adjacent to the interband region, ε_N decreases and $(2nk/\lambda)$ increases with an increase of wavelength. It is due to the conduction absorption by electrons. But the gradient of the $(2nk/\lambda)$ spectrum is of the order of 1 in the logarithmic coordinate of Figure 2, and smaller than 2 for the Drude's free electron model. With an increase of temperature ε_N increases mainly at the longer wavelengths. The temperature dependence is weak at the shorter wavelengths. These behaviors suggest that the Fermi surfaces of refractory metals are anisotropic, as in the case of other transition metals [4]. In the wavelength region of 10–20 μm, the gradient of the ε_N spectrum tends to (−1/2) even at room temperature, which is the value for the Hagen-Rubens approximation. The gradient of the $(2nk/\lambda)$ spectrum begins to decrease. These behaviors suggest that the relaxation wavelengths of refractory metals are smaller than those of other transition metals.

Most of previous measurements of ε_N at high temperatures were made on foils or filaments heated by electric current, and the surface conditions were not described sufficiently. By comparing the spectra of Price [6] and Reithof et al. [8] with the present ones in Figure 1, differences are larger in the longer wavelength region in the infrared rather than in the shorter wavelength region where the effect of surface roughness appears sensitively. In the present measurements on optically smooth surfaces of bulk specimens, the reproducibility of ε_N in the longer wavelength region was not good at temperatures higher than 1 000 K for Ta, and higher than 1 300 K for W. It is considered to be related to the crystal growth at the surface, but it is not clear.

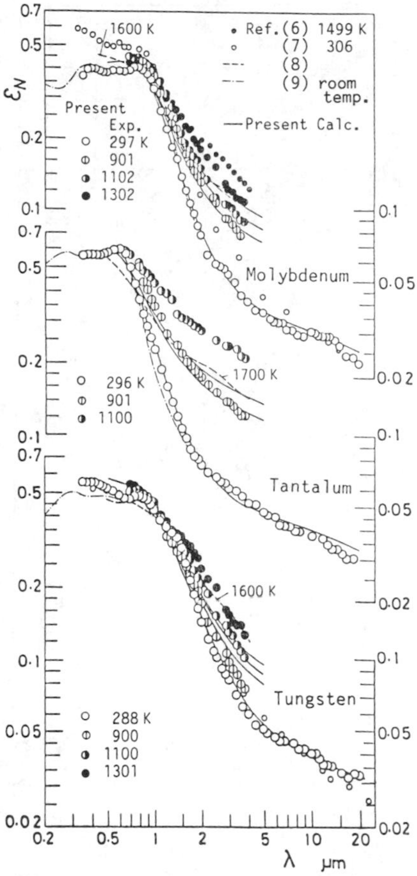

FIGURE 1. Emissivity spectra of refractory metals

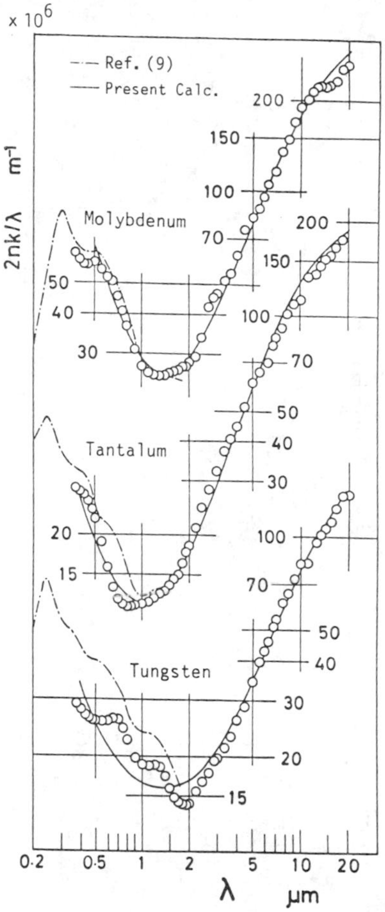

FIGURE 2. Spectra of (2nk/λ) of refractory metals at room temperature

In the R_N and R_D measurements on the CVD specimens of three kinds of thickness, R_N does not depend on thickness at every wavelength, and R_D is zero at every wavelength and at every thickness. The semi-transparent and scattering properties, which characterize the visible and near-infrared spectra of dielectric ceramics, do not appear. The absorption bands of lattice vibration, which characterize the far-infrared spectra of dielectrics, also do not appear. Figures 3 and 4 show the spectra of ε_N and $(2nk/\lambda)$ together with those of other authors [10-12]. The spectra of the CVD specimens can be regarded as those of infinitely thick layers. They are a little different from those of the sintered ones, but it is considered to be caused by the fact that the cystal grains at the surfaces of CVD layers have particular orientations such as (200) or (111), while the grains in the sintered ceramics are oriented randomly. In Figure 4, $(2nk/\lambda)$ is larger than 10 in the infrared region. The penetration depth of radiation $(\lambda/8\pi k)$ is less than 20 nm. The large values of $(2nk/\lambda)$ suggest that absorption and emission of radiation in the ceramics are caused by the conduction electrons. The Fermi level lies in the conduction band of electrons [10]. The

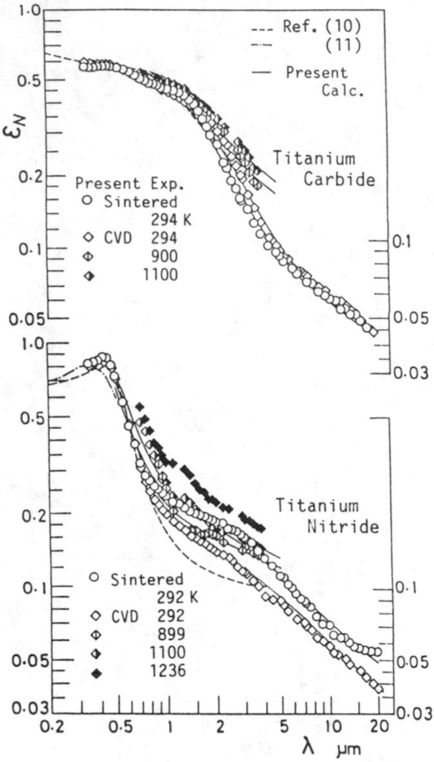

FIGURE 3. Emissivity spectra of conductive ceramics

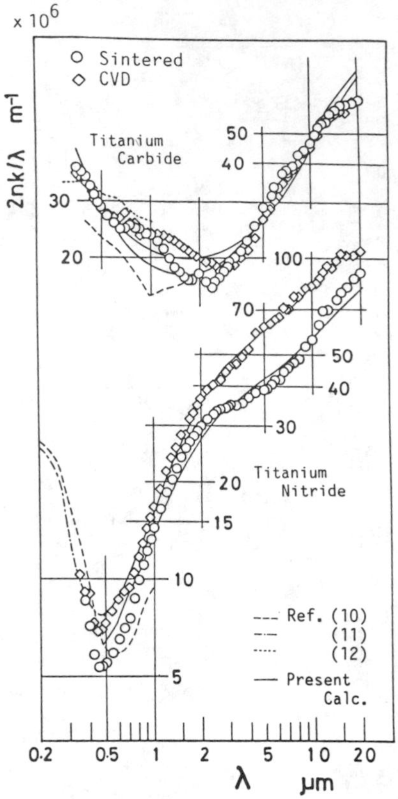

FIGURE 4. Spectra of (2nk/λ) of conductive ceramics at room temperature

electric conductivities are comparable to those of stainless steels [3]. Radiation properties of the ceramics can be discussed on a basis of the electronic theory of metals.

In Figures 3 and 4, the spectral behavior and the temperature dependence are similar to those of refractory metals, in every ceramic and in every wavelength region. The interband transition is predominant at the shorter wavelengths, and the temperature dependence is weak. The absorption band of TiN lies at the shorter wavelengths in the visible region, corresponding to the gold-metallic color of the ceramic. In the infrared region, conduction absorption of the electrons on an anisotropic Fermi surface is predominant, and the temperature dependence is strong at the longer wavelengths. The spectra of previous works on RF-sputtered films [10, 11] (Figures 3 and 4) agree qualitatively with the present ones for CVD ceramics and sintered ones.

3.3 Dispersion Formula of Optical Constants

On the basis of the above discussion, an electronic theory of transition metals is used to describe the spectral behavior of refractory metals and conductive ceramics. A dispersion formula of optical constants $\hat{n}$ [13],

$$\hat{n}^2 = 1 + (S - i\frac{S\delta\lambda_0}{\lambda}) - \frac{\lambda^2}{2\pi c\varepsilon_0}\sum_{k=1}^{2}\frac{\sigma_k}{\lambda_k - i\lambda} \quad , \qquad (3)$$

is used. Unity in the first term of the right hand side of the formula is the relative dielectric constant of vacuum. The second term corresponds to the contribution of the interband transition. It is described by a wide-band model, in which the absorption is represented by a broadened band and the effect at the longer wavelength tail is involved. The parameter λ_0 is the central wavelength of the band, and S and δ are the parameters for the strength and the broadening of the band, respectively. The third term corresponds to the contribution of the conduction absorption. The effect of the anisotropic Fermi surface is represented by the simplest form, where c is the speed of light in vacuum and ε_0 is the dielectric constant of vacuum. The parameter σ_k is the k-th component of the optical dc conductivity σ_0, and λ_k is the corresponding relaxation wavelength. The conductivity σ_0 is related to the optical constants by

$$\sigma_0 = \lim_{\lambda\to\infty}\{-2\pi c\varepsilon_0 \mathrm{Im}(\hat{n}^2)/\lambda\} = \sum_{k=1}^{2}\sigma_k \quad . \qquad (4)$$

Dispersion formula (3) includes six independent parameters : S, $(S\delta\lambda_0)$, σ_0, (σ_1/λ_1), σ_2 and λ_2. As to the absolute values and the temperature dependencies of these parameters, the following facts have been known. For the interband transition, the experimental results show the weak temperature dependence. The parameters S and $(S\delta\lambda_0)$ are regarded as independent of temperature. For the conduction absorption it is predicted that (σ_k/λ_k)'s (k=1, 2), which are related to the curvature of the energy curve of electrons, are substantially independent of temperature. Also, it is known experimentally that σ_2 does not depend on temperature [12]. Accordingly, the following assumption is adopted in the whole temperature range and for each metal and ceramic.

$$S,\ (S\delta\lambda_0),\ (\sigma_1/\lambda_1),\ \sigma_2,\ \lambda_2 = \mathrm{const.} \qquad (5)$$

The effect of temperature T is represented by the conductivity $\sigma_0(T)$. The temperature dependence of σ_0 is known to be similar to that of the reciprocal of the resistivity [1, 4]. Since the resistivity increases linearly with an increase of temperature [3], an empirical equation,

$$\sigma_0 = 1/(A+BT) \quad , \qquad (6)$$

is presumed. Additionally, the value of relaxation wavelength λ_2 can be presumed to be zero, except for the case of TiN, since the effect of λ_2, which is predominant in the visible and near-infrared regions, is obscured by the effect of the strong interband transition in these regions.

Firstly, the ε_N spectra at room temperature are analyzed to determine the above six parameters. Secondly, the ε_N spectra at high temperatures are analyzed to determine the values of A and B in Eq. (6). The spectra at the maximum temperatures of the experiments are excluded in the analysis. Table 3 shows the results. For the sintered ceramics, on which experiments were made only at room temperature, the values of σ_0 are shown. By using these values in Eqs. (3)-(6), optical constants can be calculated as a function of wavelength and temperature. The solid lines in Figures 1-4 are the calculated values of ε_N and $(2nk/\lambda)$ by this method. In Figures 2 and 4, the results of the present analysis agree well with those of the Kramers-Kronig analysis (circles and squares), although the two analyses were carried out independently.

TABLE 3. Values of the electronic parameters in Eqs. (3)-(6)

	S	$S\delta\lambda_0$ m	A Ω m	B Ω m/K	σ_1/λ_1 S/m^2	σ_2 S/m	λ_2 m
		$\times10^{-6}$	$\times10^{-8}$	$\times10^{-10}$	$\times10^{11}$	$\times10^{5}$	$\times10^{-6}$
Mo	19.5	12.9	5.88	3.51	5.66	2.18	0
Ta	6.51	3.22	-1.90	9.47	4.06	1.07	0
W	6.44	3.47	17.7	3.63	2.27	2.06	0
TiC (Sintered)	7.85	3.59	σ_0=1.82x10^6 S/m		1.15	2.52	0
(CVD)	8.45	2.90	4.64	15.4	0.990	2.99	0
TiN (Sintered)	5.12	0.514	σ_0=1.92x10^6 S/m		0.698	6.56	1.45
(CVD)	8.13	0.783	8.42	7.24	1.04	9.72	1.71

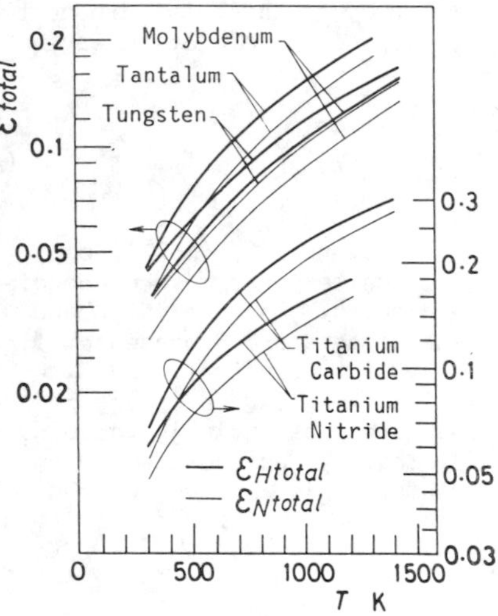

FIGURE 5. Temperature dependencies of total emissivities [Calculated]

The results of the present analysis are applicable to the estimation of various properties for radiation transfer. Figure 5 shows calculated values of the total hemispherical emissivity $\varepsilon_{H\ total}$ and the total normal emissivity $\varepsilon_{N\ total}$. The values for ceramics are those for CVD ceramics. Since refractory metals and conductive ceramics have steeply inclined spectra of emissivity, the total emissivities depend on temperature sensitively.

4. CONCLUSIONS

An experimental study was performed on thermal radiation properties of refractory metals and electrically-conductive ceramics at high temperatures. Results are summarized as follows :

(1) Radiation properties of refractory metals are dominated by strong interband transition of electrons which extends to the near-infrared region, and by conduction absorption of electrons on an anisotropic Fermi surface which makes a steeply inclined spectrum of emissivity in the most important wavelength region for heat transfer.

(2) Radiation properties of conductive ceramics are caused by a behavior of electrons. The behavior is similar to that in refractory metals, and it can be investigated on a common basis of an electronic theory of transition metals.

(3) Optical constants of refractory metals and conductive ceramics are described by a dispersion formula (3) and Eqs. (4)-(6) for the temperature dependencies of the electronic parameters. Various properties for radiation heat transfer can be calculated by using the equations.

This work was partially supported by the Grant-in-Aid for Scientific Research of the Ministry of Education, Science and Culture of Japan.

REFERENCES

1. Makino, T., Kunitomo, T. and Mori, T., Study on Radiative Properties of Heat Resisting Metals and Alloys (2nd Report), *Bull. JSME*, vol. 27, no. 223, pp. 57-63, 1984.

2. Makino, T., Sakai, I., Kinoshita, H. and Kunitomo, T., Thermal Radiation Properties of Ceramic Materials, *Trans. JSME*, ser. B, vol. 50, no. 452, pp. 1045-1053, 1984.

3. Touloukian, Y. S., *Thermophysical Properties of High Temperature Solid Materials*, vols. 1 and 5, Macmillan, New York, 1967.

4. Makino, T., Kawasaki, H. and Kunitomo, T., Study on Radiative Properties of Heat Resisting Metals and Alloys (1st Report), *Bull. JSME*, vol. 25, no. 203, pp. 804-811, 1982.

5. Moss, T. S., *Optical Properties of Semi-Conductors*, Chap. 2, Butterworths, London, 1959.

6. Price, D. J., The Emissivity of Hot Metals in the Infra-red, *Proc. Phys. Soc. Lond.*, vol. 59, no. 331, pp. 118-131, 1947.

7. Edwards, D. K. and de Bolo, N. B., Useful Approximations for the Spectral and Total Emissivity of Smooth Bare Metals, *Advances in Thermophysical Properties at Extreme High Temperatures and Pressures*, pp. 174-188, ASME, 1965.

8. Reithof, T., Acchione, B. D. and Branyan, E. R., High-Temperature Spectral Emissivity Studies on Some Metals and Carbides, *Temperature : Its Measurement and Control in Science and Industry*, vol. 3, part 2, pp. 515-522, 1962.

9. Nestell, Jr., J. E. and Christy, R. W., Optical Conductivity of bcc Transition Metals : V, Nb, Ta, Cr, Mo, W, *Phys. Rev.*, ser. 3, vol. 21, no. 8 B, pp. 3173-3179, 1980.

10. Karlsson, B., Sundgren, J. and Johansson, B., Optical Constants and Spectral Selectivity of Titanium Carbides, *Thin Solid Films*, vol. 87, pp. 181-187, 1982.

11. Schlegel, A., Wachter, P., Nickl, J. J. and Lingg, H., Optical Properties of TiN and ZrN, *J. Phys. Chem.*, vol. 10, pp. 4889-4896, 1977.

12. Lye, R. G. and Logothetis, E. M., Optical Properties and Band Structure of Titanium Carbide, *Phys. Rev.*, ser. 2, vol. 147, no. 2, pp. 622-635, 1966.

13. Makino, T., Hasegawa, H., Narumiya, Y., Matsuda, S. and Kunitomo, T., Thermal Radiation Properties of Transition Metals and Alloys in the Liquid State, *Trans. JSME*, ser. B, vol. 50, no. 459, pp. 2655-2660, 1984.

Numerical Studies of the Multiple-Jet Impingement and Internal Cooling of Small Confined Space with Exit Slots

JINGZHONG XU and SHAOYEN KO
Institute of Engineering Thermophysics
Chinese Academy of Sciences
P.O. Box 2706, PRC

FU-KANG TSOU
Drexel University
Philadelphia, Pennsylvania 19104, USA

ABSTRACT

The heat transfer and pressure loss in the multiple-jet impingement and internal cooling of small confined space with exit slots are studied in this paper. Flow visualization studies are also carried out for such flow fields. Eleven cases are studied for the stream function, internal flow pressure distribution, the local and the average coefficients of heat transfer on the upper and lower surfaces of the small confined space. Geometric parameters are varied in the following ranges, H/b=1,2 and 4, L/b=2,4 and 8. The Reynolds numbers used in the analysis are 200, 377 and 650, respectively.

The maximum values of the local flow resistances are found in the vicinity of the exit slots where the pressure drop is quite large. The total flow resistance increases by a factor of 3 when the H/b value decreases from 2 to 1. The L/b value seems to have stronger effect on flow resistance in certain range of flow parameters. The average coefficient of heat transfer and the stagnation point heat transfer coefficient are evaluated at different H/b and L/b values. An equation of average Nusselt number is suggested for design purpose.

INTRODUCTION

Impingement is one of the most effective ways of local cooling of high temperature components. In the recently developed multiple-mode-cooling walls such as Lamilloy and Transply[Ref.1], the cooling air is first admitted into the wall by impingement, and then internal cooling takes place before the cooling air leaves as a full-coverage film. Such composite cooling configuration takes the full advantage of effective cooling by impingement, internal cooling and film cooling. It has been used recently in gas turbine combustors. And it is promising for use in air cooled gas turbine blade and other high temperature enviroment applications.

Generally speaking, the local heat flux of the impinged surface is quite large and the flow must be organized in such a way that the cooling effect should be high and the flow resistance should be low. Therefore, a detailed information on the flow field and heat transfer should be desirable for optimization designs.

Many papers on impingement flow heat transfer have been published in recent years [Ref. 2,4,5]. Only few dealt with the heat transfer on the permeable impinged surfaces. The objective of this paper is to study numerically the flow and heat transfer characteristics of the 2-D laminar multiple-jet impinging on the surface with exit slots in a small confined space.

MATHEMATICAL SIMULATION

The flow system is shown in Fig.1 . The cooling air is first admitted from the slots of width b at the upper surface, impinging on the lower surface at a distance H from the upper surface, and then turnning in a horizontal direction over a distance L before leaving from the exit slots on the lower surface. The width of the exit slots is the same as that of the entrance ones. L/b, H/b and Reynolds number are varied over a given range. The chosen calculation domain is shown in Fig.1. For the reason of simplicity, the fluid is considered as incompressible and constant physical properties. The governing equations are as following:

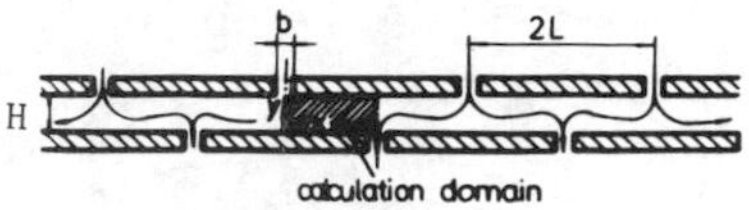

Fig.1 Geometry of the flow system

$$\frac{\partial}{\partial x}(\rho u^2) + \frac{\partial}{\partial y}(\rho vu) = -\frac{\partial p}{\partial x} + \mu\left(\frac{\partial^2 u}{\partial x^2} + \frac{\partial^2 u}{\partial y^2}\right) \quad (1)$$

$$\frac{\partial}{\partial x}(\rho uv) + \frac{\partial}{\partial y}(\rho v^2) = -\frac{\partial p}{\partial y} + \mu\left(\frac{\partial^2 v}{\partial x^2} + \frac{\partial^2 v}{\partial y^2}\right) \quad (2)$$

$$\frac{\partial}{\partial x}(\rho u) + \frac{\partial}{\partial y}(\rho v) = 0 \quad (3)$$

$$\frac{\partial}{\partial x}(\rho uT) + \frac{\partial}{\partial y}(\rho vT) = \frac{k}{c_p}\left(\frac{\partial^2 T}{\partial x^2} + \frac{\partial^2 T}{\partial y^2}\right) \quad (4)$$

The boundary conditions for the dependent variables are given in table 1.

The SIMPLE computational program for 2-D elliptic flows have been adopted with some modification. The details of the computational procedure are given in Ref.3. A non-uniform grid was used. Five grid networks of 52x26, 52x16, 52x12, 32x12 and 22x12 were used to correspond to five different schemes where H/b took the values of 4,2,1, and L/b took the values of 8,4,2. The average velocity V_o was chosen as 4.16, 7.84 and 13.5 m/s. It was assumed that the convergence of the numerical results was achieved if the local absolute additional mass resources were every-where within 10^{-6}. The relax factors used in the iterations were changeable in the range of 0.5 to 0.005.

Table 1

	U	V	T	P
jet entrance	0	V_o	350°K	–
the upper surface	0	0	350°K	–
the lower surface	0	0	300°K	–
jet exit	0	$\frac{\partial V}{\partial y}=0$	$\partial T/\partial y=0$	–
symmetric plan	0	$\partial v/\partial x=0$	$\partial t/\partial x=o$	–

RESULTS AND DISCUSSIONS

Eleven cases were computed in this paper. The typical distribution of stream function is shown in Fig.2. The curves represent the stream function contours. The upper and the lower thin straight lines indicate the confinement plate and the impingement plate respectively. The dashed lines at both sides represent the symmetric planes. There is a primary vortex in the counterclockwise direction near the jet entrance. Its strength increases with the jet Reynolds number. It will be seen that this primary vortex is located at the place where a low pressure zone occurred. Meanwhile, a secondary vortex is found at the upper right hand corner where the pressure is relatively high due to the secondary stagnation of the fluid at this symmetric plan. Fig.3 is a photograph

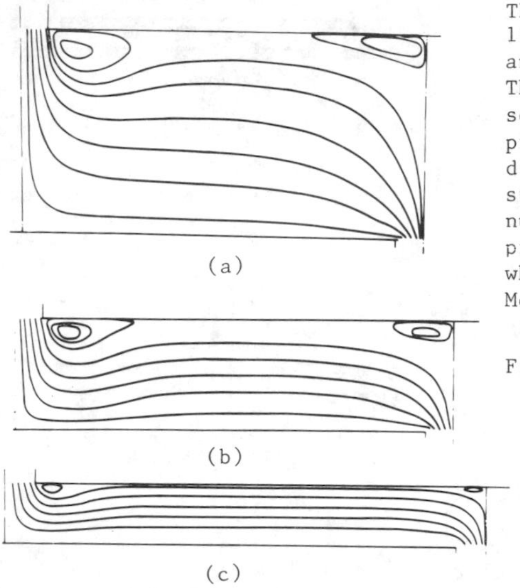

Fig.2 Contours of stream function

(a) Re_b=377, H/b=4, L/b=8
(b) Re_b=377, H/b=2, L/b=8
(c) Re_b 377, H/b=1, L/b=8

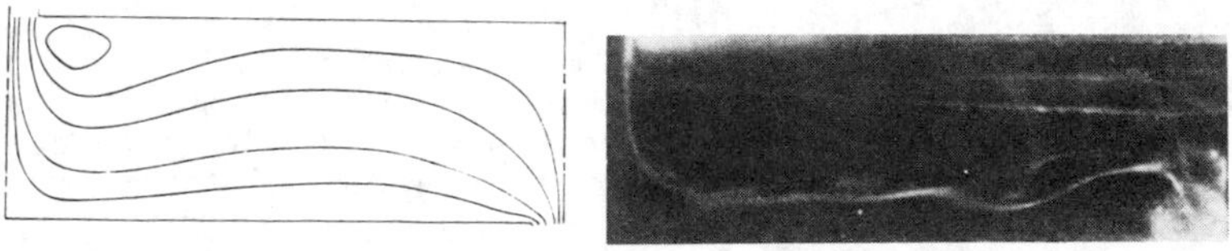

Fig.3 Comparasion between visualized and calculated flow field, Re_b=650, H/b=3, L/b=8

of the visualization of the flow field where H/b=2, L/b=8 and Re_b=800. The streamlines of the jet were traced by the hydrogen bubble method. To compare with the numerical results, the stream function was also computed with the same geometrical parameters and fluid properties. It must be emphasized that under the condition of low Reynolds number the strength of the vortex is relatively weak and could hardly be visualized with the tracer of hydrogen bubble. Fig.4 shows the typical pressure distribution. The maximun value of the pressure takes place at the stagnation point of the impinged plate. After impingement, the flow turns to the x direction and the local pressure gradually decreases. At the exit, however, because of the secondary stagnation and because of the existence of a throat at the exit of the flow, the pressure

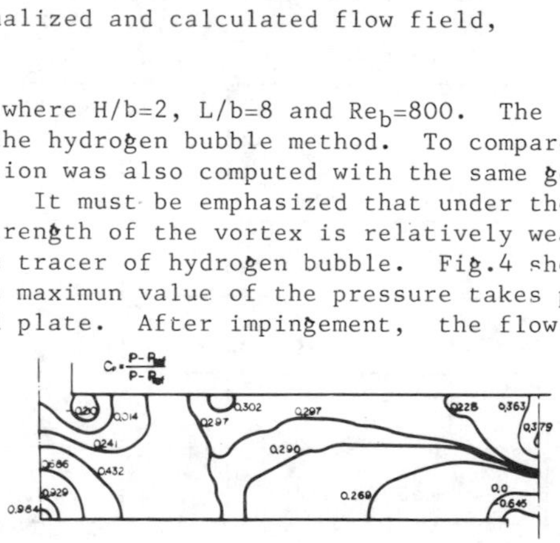

Fig.4 Pressure distribution, Re_b=649, H/b=2, L/b=8

drop and its gradient are quite large. So, the local flow resistance is also very large. Therefore, the geometry in the vicinity of the exit is important f decreasing the flow resistance of the entire flow system. Fig.5 shows the ave age flow resistance varying with the jet Reynolds number. The slope and the in- tercept of the given straight lines indicate the flow resistance characteristic of the corresponding configurations.

The distribution of the local Nusselt number is given in Fig. 6. The local heat transfer coefficient h(x) can be calculated from equation

$$k[\Delta T_w(x)/\Delta y_w] = h(x)(T_i - T_w) \quad (5)$$

where $\Delta T_w(x)$ is the temperature difference between the local impinged surface and the grid close to it, Δy_w is the corresponding distance. When the impinged surface with exit slots is considered, the distribution of the local heat transfer coefficients have the following features:

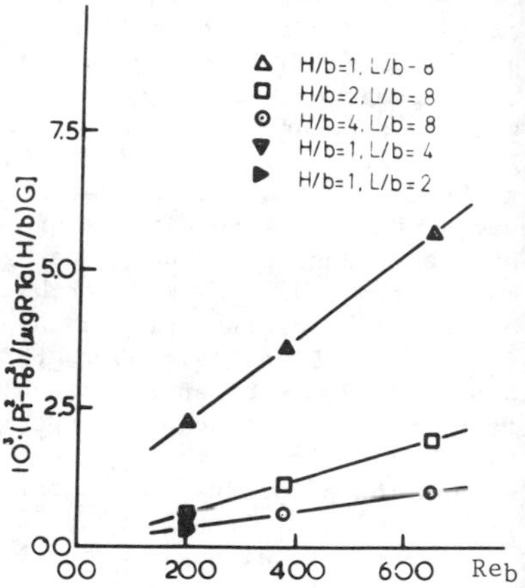

Fig.5 Typical flow resistance data

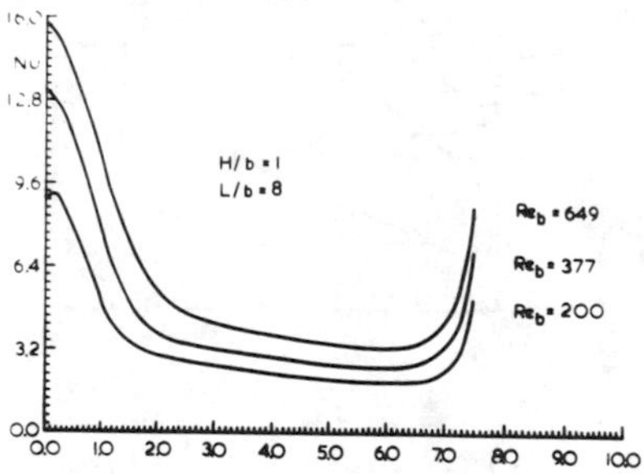

(a)

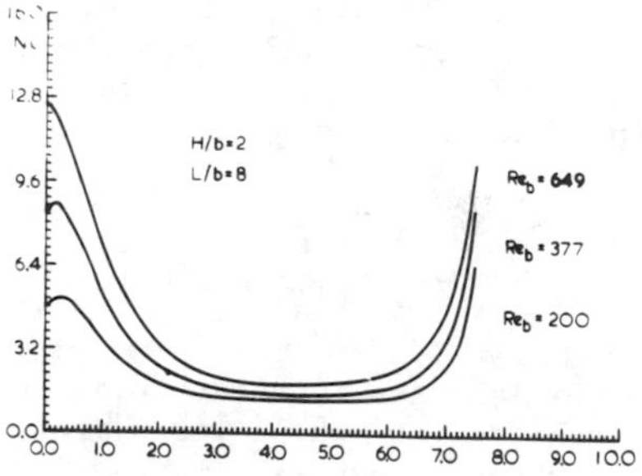

(b)

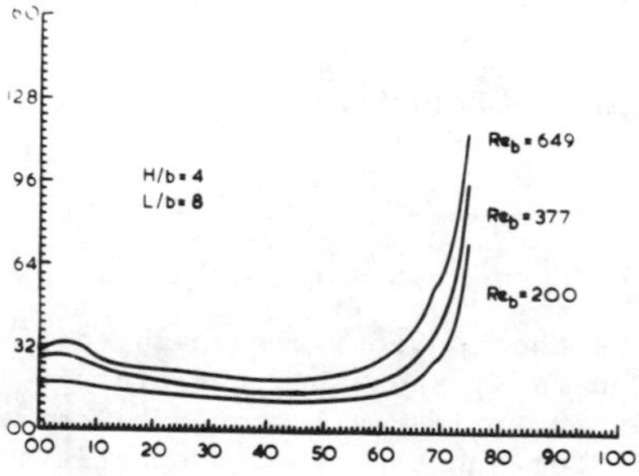

(c)

Fig.6 Distribution of local Nusselt number

(1) At the exit slot, the pressure drop and the pressure gradient are quite large due to the effect of secondary stagnation and due to the throat effect, the mechanism of local convective heat transfer is thus enhanced. There appears to have a secondary peak in the distribution of the local heat transfer coefficients, and its amplitude could reach or even surpass the main peak.

(2) With small H/L values, the horizontal distance between the entrance and the exit is relatively long, the influence of outflows on the entering jets is small and the high local heat transfer coefficient at the stagnation point can be expected. But for relatively large H/L values, the low pressure distribution at the exit slot influence the entering jet to a great extent, and the jet partly deviates from the stagnation point before impinging on the surface, so the impingement is weakened and the local heat transfer coeffcient is thus greatly decreased.

(3) For small H/b values and given range of Reynolds number the local heat transfer coefficients at the stagnation zone appear to be double-peak distributions. Under the conditions of this paper, this can be obtained when the values of the dimensionless parameter group $Re_b/(H/b)$ are within 50 to 300. Beyond this range, the double-peak distributions will disappear.

Fig.7 shows the relationship between the stagnation point Nusselt number and the jet Reynolds number. For the reason of comparison, the stagnation point Nusselt numbers of an impermeable wall in the reference 4 are also presented in the same figure. Compared with impermeable walls, the stagnation point Nusselt number at the surface with exit slots is much lower.

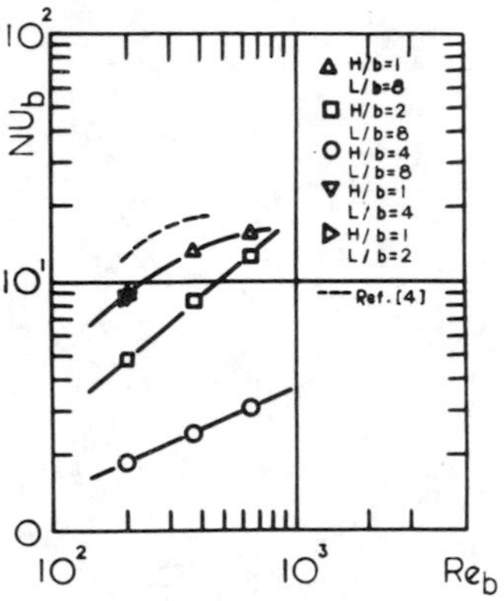

Fig.7 Stagnation point Nusselt number

For practical engineering evaluations, the average heat transfer coefficient of the whole internal surfaces is often needed. Imagine that the thicknesses of both plates are quite thin and the temperature of each plate is uniform, then the heat equivalance equation can be obtained under the specified heat boundary conditions

$$\rho_i V_j (b/2)\, C_p (T_i - \overline{T}_o) = 2(L - b/2)\, \overline{h}(T_i - T_a) \qquad (6)$$

where h is the average heat transfer coefficient of the whole internal surfaces and can be expressed as

$$\overline{h} = \frac{1}{2(L-b/2)} \int_{\text{internal surface}} h(x)dx \qquad (7)$$

T_a is the average temperature of the whole internal surfaces

$$T_a = \frac{1}{2(L-b/2)} \sum_j (T_j \Delta L_j) \qquad (8)$$

The left hand side of equation (6) represents the convective enthalpy difference or the cooling air from entrance to exit. Generally speaking, the average Nusselt number of the similar flow system as shown in Fig.1 should be correlated with jet Reynolds number, geometry of the system and the fluid properties, i.e.,

$$\overline{Nu} = f(Re_b,\ H/b,\ L/b,\ r,\ Pr) \qquad (9)$$

where the $\overline{Nu}$ is defined as

$$\overline{Nu} = \frac{h.2(L-\frac{b}{2})}{k} \tag{10}$$

and r is the area ratio of the entrance and exit. It is difficult to express the $\overline{Nu}$ with a relatively simple form of equation. However, for the case of this paper, the parameters are varied only in a limited range, so it is possible to obtain a simple equation. Within the given range, the variation of Prandtl number is very small and can be absorbed into the coefficient of the equation. The parameter r equals one and thus disappeared in the equation. The regression equation has the form as

$$\overline{Nu} = 0.996Re_b^{\ 0.50}(H/b)^{-0.45}(L/b)^{0.59} \tag{11}$$

$$200 \leqslant Re_b \leqslant 650 \qquad 1 \leqslant H/b \leqslant 4 \qquad 2 \leqslant L/b \leqslant 8$$

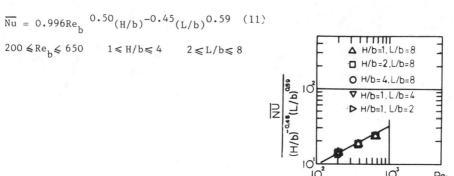

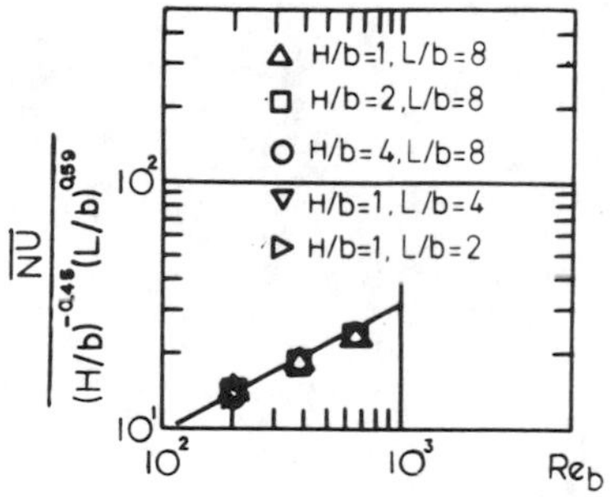

Fig.8 Fitness of equation (11)

The agreement between numerical results and the equation (11) is shown in Fig.8. The maximun error is within 5.1%.

CONCLUSION

(1) For the multiple-jet impingement in a small confined space with exit slots, as shown in Fig.1, the maximun value of the pressure is located at the stagnation point. Near the jet entrance at the confinement plate, there is a low pressure region corresponding to a main vortex. This vortex region is so called "death zone". At the symmetric planes between entering jets, the pressure is retatively higher due to the secondary stagnation, and a pair of secondary vortices is found near the confinement plate. The size and strength of the vortex are decreased with the jet Reynolds number. In the vicinity of the exit slots, because of the secondary stagnation and being an outflow throat, the pressure gradient is quite large and the flow resistance is also very large. So,it is important to select the proper exit geometry so that the total flow resistance can be reduced.

(2) The maximun value of the local heat transfer coefficient over the impinged surface generally takes place at the stagnation point and decay rapidly in x direction. At the exit slots, however, because of the enhancement of local convective mechanism, the heat transfer coefficient distribution has a secondary peak, its amplitude may reach or even surpass the main peak.

(3) The average heat transfer coefficients in a small confined space usually recieve more attention in practice. A simple correlation equation has been obtained in this paper.

REFERENCES

1. Nealy, D.A. and Reider, S.B., "Evaluation of Laminated Porous Wall Materials for Combustor Liner Cooling," ASME 79-GT-100, 1979.

2. Martin, H., "Heat and Mass Transfer Between Impinging Gas Jets and Solid Surfaces." Advances in Heat Transfer, Vol.13, Academic Press, New York, pp-60, 1977.

3. Patankar, S.V.,"A Calculation Procedure for Two-Dimensional Elliptic Situations." Numerical Heat Transfer, Vol.4, No.4, Oct-Dec., 1981.

4. Hin-Sum Law and Masliyah, J.H., "Mass Transfer Due to a Confined Laminar Impinging Two-Dimensional Jet," Int. J. Heat and Mass Transfer, Vol.27, No.4, April, 1984.

5. Hwang, J.C. and Tsou, F.K., "Numerical Calculation of Jet-Induced Ground Effects in VTOL," AIAA-81-0015, 1981.

ENHANCED HEAT TRANSFER

Heat Transfer Enhancement for Gravity Controlled Condensation on a Horizontal Tube by Coiling Wires

T. FUJII
Research Institute of Industrial Science
Kasuga-shi, Fukuoka, 816, Japan

W. C. WANG
Thermal Engineering Department
Tsinghua University
Beijing, PRC

S. KOYAMA and Y. SHIMIZU
Research Institute of Industrial Science
Kasuga-shi, Fukuoka, 816, Japan

INTRODUCTION

High performance and low capital cost of condensers are among the vital needs the economic consideration of the development of systems for waste heat a natural energy utilization. For example, more than half of capital cost of electric power plant utilizing waste heat at 100 °C temperature level is spent f the condenser.

A device for enhancing condensation heat transfer coefficient by means protrusions on a tube surface was proposed by Gregorig[1], where the fi thickness at the top of the protrusion is thinned by the effect of surfa tension. As extensions of his idea, various kinds of high performance condens tubes have been developed. Though the development is made possible by t progress of manufacturing technique, the price is still much higher than that smooth tubes.

One of the directions of the research and development of new condenser tubes is pursue a maximum performance. The other direction is to minimize the overall co by combining moderate performance with moderate manufacturing cost. Thomas[measured the condensation heat-transfer coefficient for a vertical tube on whi fine wires were attached vertically, and proved the wires' enhancement effect. The idea of Gregorig is to thin the film thickness at the top parts of t protrusions, while that of Thomas is to thin the film thickness at the ar between protrusions.

The use of wire coiled tubes is also effective for horizontal tube bundles. Thoma et al.[3] reported that the heat transfer coefficient was about two times t value predicted by the Nusselt's expression for a smooth tube [4] in an ammoni condenser. Marto and Wanniarachchi[5] found significant reduction of the effec of condensate inundation in a steam condenser with large tube bundles. However the selection of the most effective wire diameter and pitch is still uncertai even for a single condenser tube. To clarify this problem is the objective of th present paper.

THEORETICAL ANALYSIS

Figure 1 shows the physical model and coordinate system. A horizontal tube o diameter D is coiled with a fine wire of diameter D_w in pitch p. When the tube i cooled from inside, vapor condenses on the tube. Since the condensate is drawn t the wires by surface tension as well as to the circumferential direction b gravity, the condensate film thickness between the fine wires decreases and th average heat transfer coefficient increases.

Under the assumptions that (i) the condensate film is much thinner than the tube diameter, (ii) the inertia term in the momentum equation and the convection term in the energy equation are negligible, (iii) shear stress at the vapor-condensate interface is negligible, (iv) tube surface temperature is uniform and (v) physical properties of the condensate are independent of temperature, the governing equations are simplified as follows;

$$\mu(\partial^2 u/\partial y^2) + \rho g \sin\phi = 0 \qquad \text{(momentum equation in } \phi\text{-direction)} \tag{1}$$

$$\mu(\partial^2 w/\partial y^2) - (\partial P/\partial z) = 0 \qquad \text{(momentum equation in z-direction)} \tag{2}$$

$$\frac{2}{D}\frac{\partial}{\partial\phi}\int_0^\delta u\,dy + \frac{\partial}{\partial z}\int_0^\delta w\,dy = \frac{\lambda(T_s - T_w)}{\rho L \delta} \qquad \text{(heat balance)} \tag{3}$$

together with the boundary conditions

$$u = 0 \quad \text{and} \quad w = 0 \qquad \text{at } y=0 \tag{4),(5}$$

$$\partial u/\partial y = 0 \quad \text{and} \quad \partial w/\partial y = 0 \qquad \text{at } y=\delta \tag{6),(7}$$

By assuming that $\partial P/\partial z$ is independent of y because of a thin condensate film, equations (1) and (2) subjected to the conditions (4),(5),(6) and (7) can be solved as follows;

$$u = (g\sin\phi/\nu)(\delta y - y^2/2) \quad , \qquad w = (-1/\mu)(\partial P/\partial z)(\delta y - y^2/2) \tag{8),(9}$$

By substituting equations (8) and (9) into equation (3) and introducing dimensionless numbers, the following equation is obtained.

$$\frac{\partial}{\partial\phi}(\delta^3\sin\phi) - \frac{D}{2\rho g}\frac{\partial}{\partial z}\left(\delta^3\frac{\partial P}{\partial z}\right) = \frac{3D\nu^2 Ph}{2gPr\delta} \tag{10}$$

Then, a model for the distributions of film thickness δ and pressure P is introduced as shown in Fig.2. $\delta(\phi)$ is constant in the region of $0<z<s/2$ and abruptly changes at z= s/2 as shown in Fig.2(b). Similarly, P changes abruptly from 0 to a constant value of $-(\sigma/r_s)$ at z=s/2 as shown in Fig.2(c), where s and r_s are unknown at present. The term $-(\partial P/\partial z)$ changes like the delta-function as shown in Fig.2(d). The portion between s/2 and p/2 in z coordinate is assumed to be adiabatic because of the existence of wire.

Under the conditions derived from Fig. 2(d)

$$\partial P/\partial z = 0 \quad \text{at } z=0 \quad \text{and} \quad \partial P/\partial z \neq 0 \quad \text{at } z=z \tag{11),(12}$$

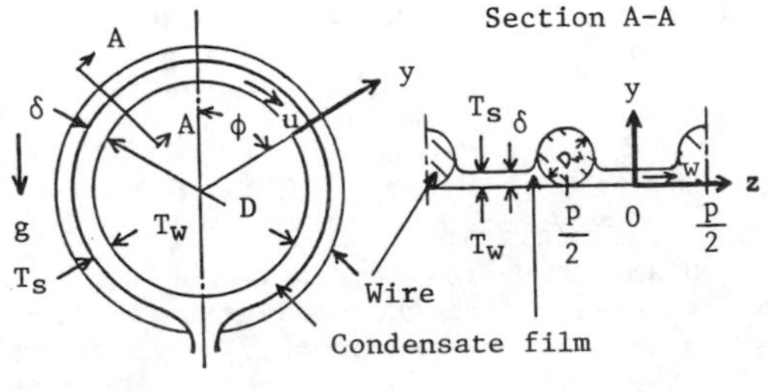

FIGURE 1. Physical model and coordinate.

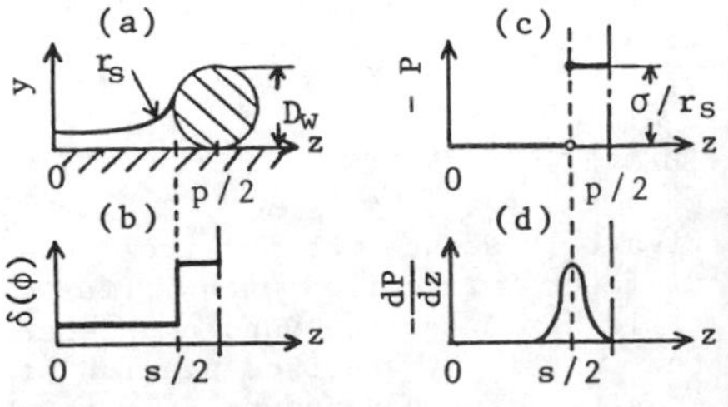

FIGURE 2. Simplified film model.

equation (10) is integrated with respect to z as

$$z\frac{\partial}{\partial\phi}(\delta^3\sin\phi) - \frac{D}{2\rho g}\left(\delta^3\frac{\partial P}{\partial z}\right) = \frac{3D\nu^2 Phz}{2gPr\delta} \quad (13)$$

Further, under the conditions derived from Fig.2(c)

$$P=0 \text{ at } z=0 \quad\text{and}\quad P=-\sigma/r_s \text{ at } z=s/2 \quad (14),(15)$$

equation (13) is integrated from 0 to s/2 with respect to z and changed by introducing dimensionless numbers as

$$\bar{\delta}\frac{d}{d\phi}(\bar{\delta}^3\sin\phi) + \frac{4\sigma D}{\rho g s^2 r_s}\bar{\delta}^4 = \frac{3Ph}{2GaPr} \quad (16)$$

The analytical solution of equation (16) is given as

$$\frac{\bar{\delta}^4}{B} = \frac{4\int_0^{\phi}(\tan(\phi/2))^{4A/3}(\sin\phi)^{1/3}d\phi}{3(\tan(\phi/2))^{4A/3}(\sin\phi)^{4/3}} \quad (17)$$

where

$$A = 4\sigma D/(\rho g s^2 r_s) \quad\text{and}\quad B = 3Ph/(2GaPr) \quad (18),(19)$$

Therefore, the local and average Nusselt numbers are expressed as

$$Nu_\phi = \alpha_\phi D/\lambda = s/(p\bar{\delta}) = (s/p)\{3/(4B)\}^{1/4}F_1(\phi,A) \quad (20)$$

$$Nu = \alpha D/\lambda = (s/p)\{GaPr/(2Ph)\}^{1/4}F_2(A) \quad (21)$$

where

$$F_1(\phi,A) = \frac{(\tan(\phi/2))^{A/3}(\sin\phi)^{1/3}}{\{\int_0^{\phi}(\tan(\phi/2))^{4A/3}(\sin\phi)^{1/3}d\phi\}^{1/4}} \quad (22)$$

$$F_2(A) = \int_0^{\pi}F_1(\phi,A)d\phi/\pi \quad (23)$$

Equations (20) and (21) coincide with the Nusselt's equation[4] for a smooth horizontal tube when $A\rightarrow 0$ and $s\rightarrow p$.

Figures 3 and 4 show the values of $F_1(\phi,A)$ and $F_2(A)$, respectively. In Fig.3 it is seen that the film thickness distribution in ϕ-direction becomes flat as the value of A increases. The value of $F_1(0,A)$ at $\phi=0$ can be derived from equation (22) as

$$F_1(0,A) = \{4(1+A)/3\}^{1/4} \quad (24)$$

As seen in Fig.4, equation (24) represented by a chain line coincides with equation (23) represented by a solid line for $A>15$, that is, the average Nusselt number can be represented by the local Nusselt number at $\phi=0$. The heat-transfer enhancement ratio is expressed by

$$Nu/Nu_n = (s/p)F_2(A)/F_2(0) \quad (25)$$

where Nu_n is the Nusselt's equation for a smooth tube. Unknown parameters s and r_s in the above analysis will be determined experimentally.

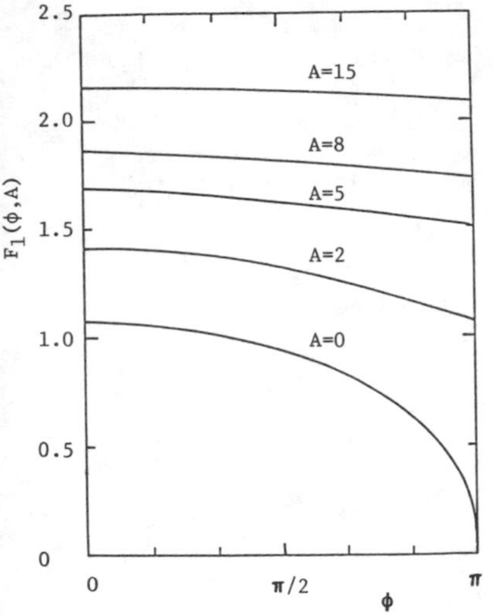

FIGURE 3. Values of $F_1(\phi,A)$, eq. (22).

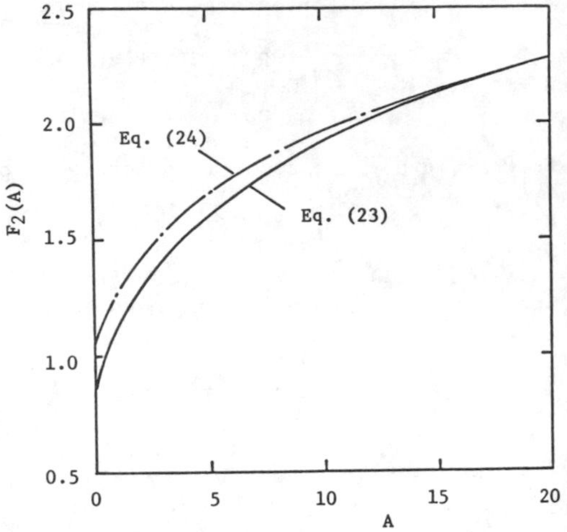

FIGURE 4. Values of $F_2(A)$ and $F_1(0,A)$, eqs. (23) and (24).

EXPERIMENTAL APPARATUS AND MEASUREMENT

The experimental apparatus is shown schematically in Fig.5. A vessel consisting of glass cylinder of 159 mm i.d. and 393 mm length and two flanges made of stainless steel contains the test fluid. The test liquid at the bottom of the cylinder is heated and evaporated by using two 500 watt sheath heaters. A copper test tube of 18 mm o.d., 16 mm i.d. and 385 mm effective length is cooled by water from inside. Vapor temperature is regulated by both electric input to the heaters and flow rate of the cooling water. On the outer surface of the tube a fine wire is coiled. The diameter and pitch of the wire were changed as D_w=0.1, 0.2, 0.3 mm and p=0.5, 1.0, 2.0 mm, respectively.

Vapor temperature was measured by a copper-constantan thermocouple at a position 50 mm horizontally apart from the center of the test tube. The saturation state of the vapor was confirmed by the reading of a Bourdon-tube pressure gauge. Since the test tube was electrically insulated, the volume averaged temperature of the tube was measured by means of the variation of the electrical resistance, whose calibration was made in a constant temperature bath prior to the experiment. Constant current type D-C electric source was used for the measurement and calibration. The difference between the measured average temperature and surface temperature was estimated to be about 1% of the temperature difference between the tube and vapor. The e.m.f. of the thermocouple, voltage drop through a standard resistor for electric current measurement and voltage drop through test tube were measured by a digital voltmeter of 0.1 μV sensitivity.

The temperature rise of the cooling water was measured by two copper-constantan thermocouples installed in the mixing chambers at the tube inlet and exit, and the flow rate was measured by a rotameter and a weight-scale. From these values the heat transfer rate was reckoned. The condensate was collected by a tray beneath the test tube and flowed down to a filling-cup. By summing up the filling time the condensation rate was measured. The heat transfer rate reckoned from the condensation rate agreed with that measured from the enthalpy rise of cooling water within an accuracy of 10% . For the data processing the latter was used.

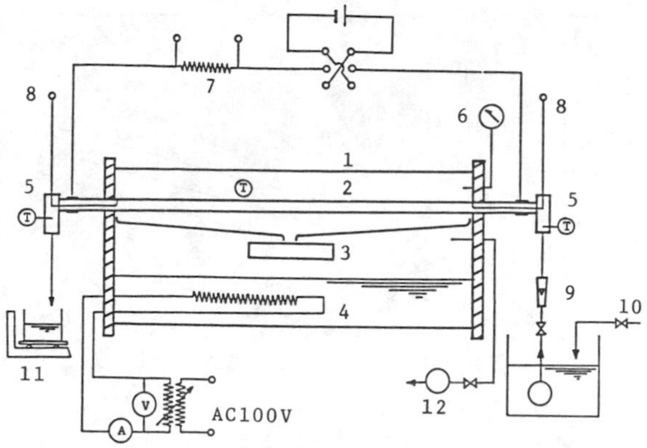

1. Glass cylinder
2. Test tube
3. Measuring cup
4. Heater
5. Mixing chamber
6. Pressure gauge
7. Standard resistor
8. Voltage tap for test tube
9. Rotameter
10. Cooling water
11. Weight-scale
12. Vacuum pump

(T) Thermocouple

FIGURE 5. Schematic of experimental apparatus.

EXPERIMENTAL RESULTS AND CONSIDERATION

Condensations of refrigerant 11 (R11) vapor are shown in Fig.6(a) for a smoot tube and in Fig.6(b) for a tube around which a fine wire of 0.3 mm dia and 1 mm pitch is coiled, both at $T_s - T_w = 5$ K. Different features observe in the falling condensate are due to the difference of condensation rate.

Heat transfer results are plotted in Fig.7(a) for R11 and in Fig.7(b) for ethano with coordinates Nu versus GaPr/Ph. The data for the smooth tube in both figure can be correlated by the chain lines, which represent the following Nusselt' equation

$$Nu_n = 0.725(GaPr/Ph)^{1/4} \qquad (26$$

This fact exhibits high accuracy of the present experiment.

The data for the same coiled tubes can be correlated by respective solid lines which are drawn parallel to the chain lines. These solid lines have almost th same tendency of equation (21). Now, the values of s and r_s, unknown parameters i the theory above-mentioned, will be determined. Observations of the condensat film allow us to estimate the relation

$$s = p - D_w \qquad (27$$

Since r_s is the radius of curvature of the condensate film surface near the fin wire, we can assume that r_s is a function of capillary constant $a=\sqrt{2\sigma/\rho g}$ and D_w.

(a) Smooth tube (b) Wire coiled tube

FIGURE 6. Photographic illustration of condensation phenomena for R11.

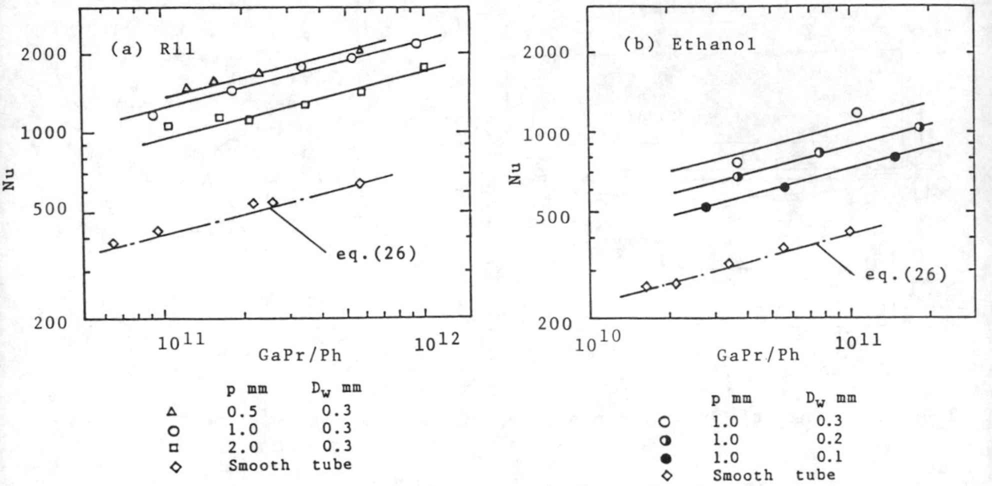

FIGURE 7. Correlations between Nu and GaPr/Ph for R11 and ethanol.

One of the simplest functional form is that r_s is proportional to $a^{n+1}/D_w^{\,n}$. Here n=1 is meaningless, because surface tension is diminished when n=1 is applied t equation (18). For n=2, equation (25) becomes a function of p/D_w and D/a.

Figures 8(a), (b) show the relation between Nu/Nu_n and p/D_w. The data in th figures are obtained from the values on solid and chain lines in Fig.7. Th chain, solid and broken lines in Figs. 8(a),(b) show the values of equation (21 with k = 0.02, 0.03 and 0.04, respectively, in the equation

$$r_s = ka^3/D_w^{\,2} \qquad (28$$

The solid lines for k=0.03 correlate the experimental data fairly well.

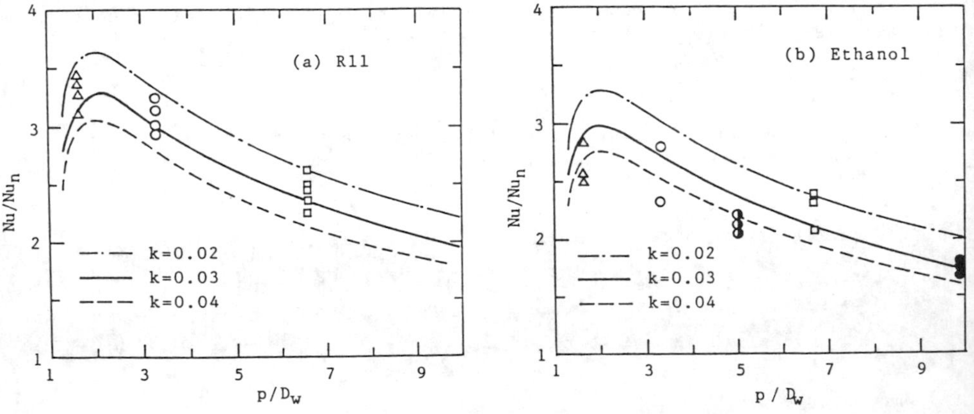

FIGURE 8. Correlations between Nu/Nu_n and p/D_w for R11 and ethanol.

Figure 9 shows equation (25) calculated under the assumption that equation (28) along with k=0.03 is applicable. Each curve with different D/a-value has the maximum at $p/D_w=2$, which is expressed by

$$(Nu/Nu_n)_{max} = 1.78(D/a)^{1/4} \qquad (29)$$

This equation can also be derived analytically from equations (25) and (28) for large value of D/a.

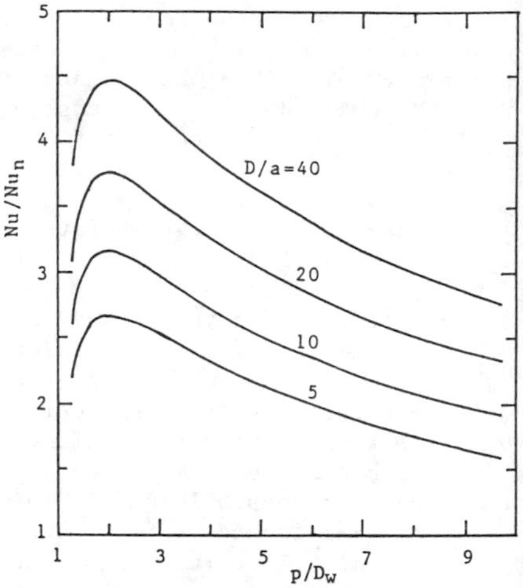

FIGURE 9. Relation between Nu/Nu_n and p/D_w as parameter D/a.

Finally, the practical importance of the wire coiled tube will be discussed. Consider the case, in which a quiescent R11 vapor of 35°C condenses on a titanium tube of 30.4 mm i.d. and 31.8 mm o.d., which is cooled by sea water of 20°C and 1.5 m/s. For the same heat exchange rate, the ratios of the predicted tube length to that of smooth tube are about 0.48 for a wire coiled tube, for which the wire diameter D_w is 0.3 mm and the pitch is 0.6 mm, about 0.5 for a circumferentially grooved tube[6] and about 0.38 for Thermoexcel tube[7]. These values exhibit that the heat transf enhancement by coiling fine wires deserves the practical application consideri the low manufacturing cost.

CONCLUSION

The effect of fine wire coiling in enhancing condensation heat transfer coeff cient for gravity controlled condensation on a horizontal tube has been clarifi theoretically and experimentally. The increase of the heat transfer coefficient comparison with that for a smooth tube is maximum at a wire diameter to pit ratio equal to 1/2 , and the maximum value is expressed by equation (29). Th relation has been confirmed by the experiment for the condensation of R11 a ethanol vapors on a copper tube of diameter D=18 mm in the range of fine wi diameters D_w=0.1∿0.3 mm. The present method for condensation heat transf enhancement will be effective for organic vapor condensers.

NOMENCLATURE

A = dimensionless number defined by equation (18)
a = $\sqrt{2\sigma/\rho g}$ = capillary constant, m
B = dimensionless number defined by equation (19)
c_p = specific heat of condensate, J/kg K
D = outer diameter of the tube, m
D_w = outer diameter of the fine wire, m
$F_1(\phi,A)$= equation (22)
$F_2(A)$= equation (23)
g = gravitational acceleration, m/s^2
Ga = D^3g/ν^2 = Galileo number
Δh_v = specific latent heat of condensation, J/kg
k = constant in equation (28)
Nu = $\alpha D/\lambda$ = average Nusselt number
Nu_n = Nusselt's equation, equation (26)

Nu_ϕ = $\alpha_\phi D/\lambda$ = local Nusselt number
P = pressure in the condensate film, Pa
p = wire pitch, m
Ph = $c_p(T_s - T_w)/\Delta h_v$ = phase change number
Pr = Prandtl number of condensate
r_s = radius of curvature of film surface shown in Fig.2 and equation (28)
s^s = z-coordinate limiting effective condensing part shown in Fig.2 and equation (27), m
T_s = saturation vapor temperature, K
T_w = outer surface temperature of the tube, K
u,w = velocity components in ϕ-and z-directions, respectively, m/s
y = coordinate in the outward normal direction measured from tube surface, m
z = coordinate in the direction of tube axis measured from a mid-point between wires on the tube surface, m
α = average heat transfer coefficient, W/m^2K
α_ϕ = local heat transfer coefficient, W/m^2K
$\delta(\phi)$ = film thickness, m
$\bar{\delta}$ = δ/D = dimensionless film thickness
λ = thermal conductivity of condensate, W/m K
μ = dynamic viscosity of condensate, Pa s
ν = kinematic viscosity of condensate, m^2/s
ρ = density of condensate, kg/m^3
σ = surface tension of condensate, N/m
ϕ = angular coordinate measured from the top of the tube, rad

REFERENCES

1. Gregorig, R., Hautkondensation an Feingewellten Oberflächen bei Berücksichtigung der Oberflächenspannungen, Z. Angew. Math. Phys., Vol.5, No. 1, pp. 36-49, 1954.

2. Thomas, D. G., Enhancement of Film Condensation Heat Transfer Rates on Vertical Tubes by Vertical Wires, I & EC Fundamentals, Vol.6, No.1, pp. 97-103, 1967.

3. Thomas, A., Lorenz, J. J., Hillis, D. A., Young, D. T. and Sather, N. F., Performance Tests of 1MWt Shell-and-Tube and Compact Heat Exchangers for OTEC, Proc. 6th OTEC Conf., V-2, Paper 11.1, pp. 1-12, 1979.

4. Nusselt, W., Die Oberflächenkondensation des Wasserdampfes, VDI-Z, Band 60, Nr.27, pp. 541-546, Nr.28, pp. 569-580, 1916.

5. Marto, P. J. and Wanniarachchi, A. S., The Use of Wire-wrapped Tubing to Enhance Steam Condensation, ASME-HTD-Vol.37,pp. 9-16, 1984.

6. Fujii,T. and Honda,H., Laminar Filmwise Condensation on a Horizontal Tube with Circumferential Groove (in Japanese), Trans. Jap. Soc. Mech. Engrs, Vol.45(B) No.393, pp. 740-748, 1979.

7. Ito, Y., Noguchi, H., Tatsumi, A. and Ohizumi, K., Newly Developed Tubes —— It's Application to Heat Exchangers (in Japanese), Refrigeration and Air-conditioning, No.253, pp. 12-19, 1982.

An Analysis on the Enhanced Heat Transfer Induced by Square-Ribbed Surface Roughness

Y. LEE
Department of Mechanical Engineering
University of Ottawa
Ottawa, Canada, K1N 6N5

INTRODUCTION:

Asymmetric flow and heat transfer are encountered in many practical situations where engineering requirements impose dissimilar boundary conditions, as in artificially roughened or finned annuli in heat exchangers. Many artificial roughness configurations are employed in various heat transfer devices to enhance the convective heat transfer rate.

Since the roughness of the surface not only increases the heat transfer rate, as a consequence of the higher turbulence induced in the flow, but also produces additional pressure losses, the heat transfer per unit pumping power expended may not be improved. Therefore, it is desirable to obtain optimum or advantageous geometrical shapes and arrangements of the surface roughness elements.

Surface roughness has an important effect on the fluid flow, but their patterns and effects on turbulent flow cannot be effectively described by any single parameter, such as the average roughness size, k. Nevertheless many studies proceeded as if k alone were sufficient to describe the flow induced by the surface roughness.

Schlichting [1] introduced the concept of equivalent sand-grain roughness, k_s, as means of characterizing different types of surface roughness by referring to the equivalent net effect produced by Nikuradse's experiments [2], which were carried out in pipes that were artificially roughened with uniform grains of sand.

Some other attempts have been made to set up models of turbulent flow over rough surfaces (e.g. Allan [3], Halls [4], Musker and Lewkowicz [5], etc.). Unfortunately, all these models require a priori knowledge of the function to describe a particular set of shapes and arrangements of surface roughness elements.

In the present analysis, a different approach from these models was used to indirectly describe a particular surface roughness for the prediction of pressure loss and heat transfer rate in an asymmetric flow induced by the given roughness elements (i.e. square-ribbed). This approach was taken because, with asymmetric boundaries, the flow exhibits some characteristic aspects by virtue of symmetry that have been hidden.

It is shown in the present study that certain artificial roughness elements may be used to enhance heat transfer rates with advantages from the overall efficiency point of view. Some characteristics of asymmetric flows are also identified in the present paper.

ANALYSIS:

Since no satisfactory surface roughness description exists, especially for those artificial roughness elements used in heat exchangers, a simple modified mixing-length model for flow turbulence is used for the analysis.

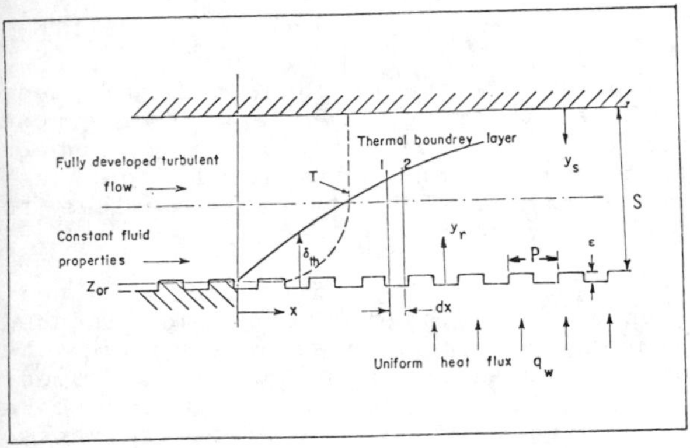

Fig. 1 Idealized Model

For simplicity, the energy integral method based on a boundary-layer model is applied to the present case of the thermal entrance region heat transfer in an asymmetric flow induced by square-ribbed surface roughness elements on the wider side of a large aspect ratio rectangular duct as shown in Fig. 1.

Velocity Profiles.

The basic equation governing the transport of momentum with a turbulence model, based on a modified mixing length theory, would result in a logarithmic velocity profile. Therefore it is assumed that the velocity distribution for the rough wall region can be expressed as:

$$\frac{u_r}{u_{\tau_r}} = \frac{1}{K} \ln \frac{y_r}{Z_{or}} \qquad (1)$$

where K = 0.4, von Karman's constant and Z_{or} is often taken as m.

A number of different values of m are reported in the literature for different shape and density of roughness, ranging from 1/3.58 [10] to 1/9400 [11], indicating that there is no apparent agreement in describing the roughness in terms of m.

For smooth wall regions, a number of expressions for the velocity profile are available. The expression due to Reichardt [6] was chosen for its simple and continuous form valid for y>0 as:

$$u_s^+ = \frac{1}{K} \ln(1 + Ky^+) + 7.8[1 - \exp(- \frac{y^+}{11}) - \frac{y^+}{11} \exp(- \frac{y^+}{3})] \tag{2}$$

For axi-symmetric turbulent fluid flow the turbulent stress at the center of a channel is zero, not necessarily because there is zero velocity gradient but for reasons of symmetry. This implies that this may not be true in asymmetric fluid flow. However, the locations for zero velocity gradient, y_{mr}, and zero total shear stress, y_{or}, must be known to close the governing equations.

To obtain these values experimental studies were made by Alp [7] and Bhuiyan [8]. Asymmetry was introduced into the air flow in rectangular ducts of very large aspect ratio by roughening one of the wider sides with equilateral triangular [7] and square ribs [8] of different pitches.

The empirical correlations for the values of y_{or} and y_{mr}, deduced from the dimensional argument for the case of square-ribbed surface roughness elements are given as [8]:

$$\frac{y_{mr}}{S} = 0.299\, Re^{0.066} \left(\frac{S}{\varepsilon}\right)^{0.140} \left(\frac{p}{\varepsilon}\right)^{0.201} \tag{3}$$

and

$$\frac{y_{or}}{S} = 0.523\, Re^{0.033} \left(\frac{S}{\varepsilon}\right)^{0.157} \left(\frac{p}{\varepsilon}\right)^{0.158} - 0.25 \tag{4}$$

Having the values of y_{or} and y_{mr}, Z_{or} can be shown to be:

$$Z_{or}^{++} = y_{mr}^{++} \exp(- D\{\ln [(S^{++} - y_{mr}^{++})D] + c.K\}) \tag{5}$$

where

$$D = \frac{u_{\tau_s}}{u_{\tau_r}} = \sqrt{\frac{S^{++}}{y_{mr}^{++} - 1}}$$

Temperature Profiles and Energy Integral Equation.

Knowing the velocity distributions, Eqs. (1) and (2), together with the assumed turbulent Prandtl number of unity and a linear heat flux distribution given as:

$$q\,(y_r^{++}) = q_w\left(1 - \frac{y_r^{++}}{\delta_{th}^{++}}\right) \tag{6}$$

the temperature profiles across the flow channels is then obtained as:

$$T^{++} = \int_{Z_{or}^{++}}^{S^{++}} \frac{1}{(\varepsilon_H/\nu + 1./Pr)} dy_r^{++} - \frac{1}{S^{++}} \int_{Z_{or}^{++}}^{S^{++}} \frac{y_r^{++}}{(\varepsilon_H/\nu + 1/Pr)} dy_r^{++} \quad (7)$$

where $(\varepsilon_H/\nu)_r = \{1/(du_r^{++}/dy_r^{++})\} - 1 \quad \text{for } Z_{or}^{++} < y_r^{++} < y_{or}^{++}$

and $(\varepsilon_H/\nu)_s = \{(1/d\,u_s^{++}/dy_r^{++})\} - 1 \quad \text{for } y_{or}^{++} < y_r^{++} < S^{++}$

The thermal entrance lengths are obtained from the derived relations based on the idealized model shown in Fig. 1 as:

$$\frac{x^{++}}{S^{++}} = \frac{1}{S^{++}} \int_{Z_{or}^{++}}^{\delta_{th}^{++}} u^{++} (T_{\delta_{th}}^{++} - T^{++}) dy_r^{++} \quad (8)$$

and the local Nusselt number and the flow Reynolds number are obtained from the definitions as:

$$Nu_x = De^{++} \cdot Pr/(T_b^{++})_x \quad (9)$$

and

$$Re = 2S^{++} u_b^{++} \quad (10)$$

The friction factor can be stated in dimensionless parameters as:

$$f = 2/u_b^{++} \quad (11)$$

RESULTS AND DISCUSSION:

An example of predicted velocity profiles is compared with the experimental measurement of Bhuiyan [8] in Fig. 2. In general the logarithmic velocity profiles given by Eqs. (1) and (2) agree reasonably well with the experimental data, except near the rough wall. This deviation is attributed to the extremely complex flow structures in the region due to the roughness elements. It indicates the deficiency of the turbulence model used in the analysis.

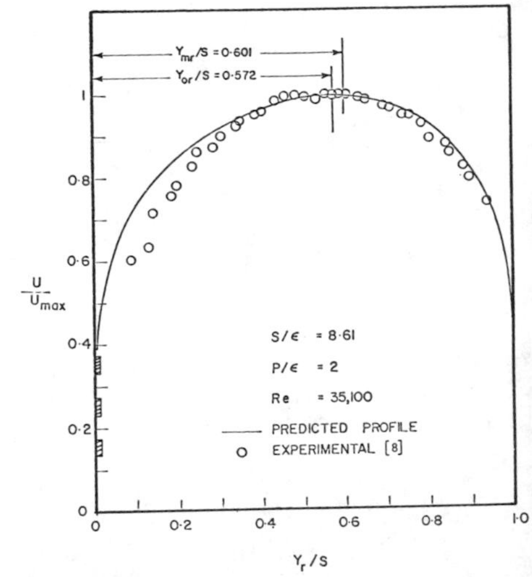

Fig. 2 Velocity Profile

However, this did not affect the friction factors which were predicted with these velocity profiles as they agreed fairly well with the experimental data. The figure shows clearly that the locations of zero shear stress and zero velocity gradient do not coincide as previously discussed. The variation in the positions of these points seems to depend on the values of Reynolds number, S/ε and P/ε. They become proportionately closer to the smooth wall values with increasing Reynolds number and decreasing S/ε.

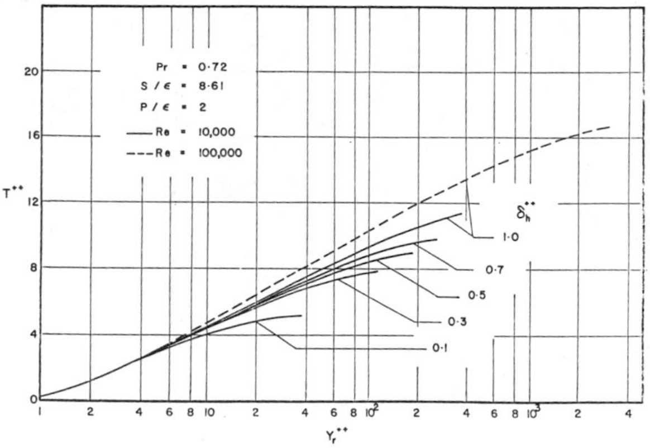

Fig. 3 Temperature Profiles

The development of typical temperature profiles in the thermal entrance region is shown in Fig. 3. The analysis showed that the effect of the relative height of roughness, S/ε, seemed to be less than that of the roughness density, P/ε, for the ranges of the variable studied. The effect of Reynolds number is clearly shown in the figure.

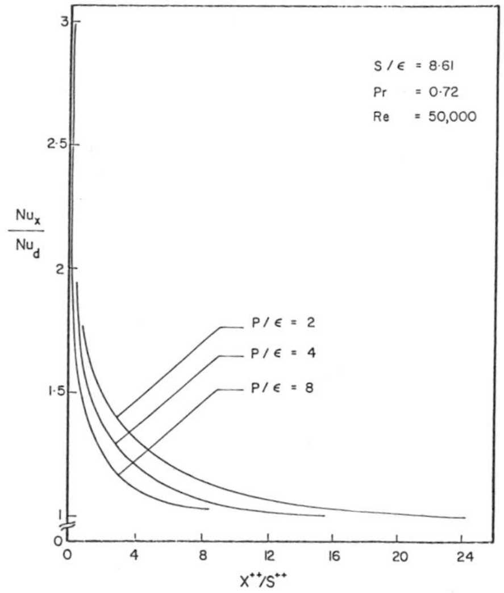

Fig. 4 Thermal Entrance Regions

Representative Nusselt number distributions in the entrance region for a given set of variables are presented in Fig. 4. Nusselt number increases with increasing S/ε and decreases with increasing P/ε. The effect of P/ε on the variation of Nusselt numbers and the entrance lengths can be seen in Fig. 4.

The resultant effect of artificial roughness is determined from a comparison of rough and smooth surfaces with respect to the heat transfer increase relative to the increase in pressure losses. This is expressed in terms of a non-dimensional parameter, F/H, which is defined as:

$$F/H = (f_r/f_s)/(Nu_r/Nu_s) \qquad (12)$$

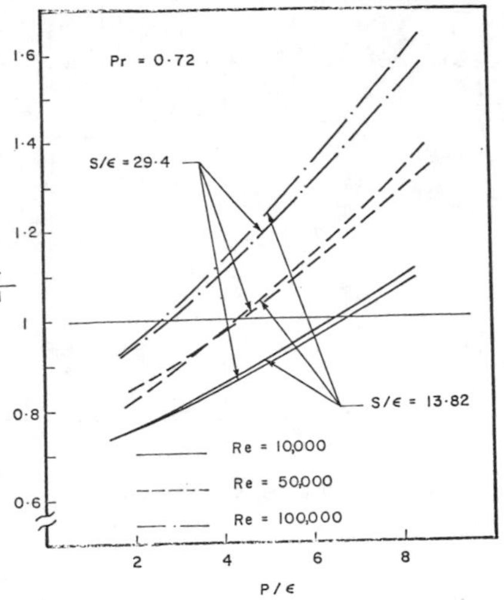

Fig. 5 Effect of Re on F/H

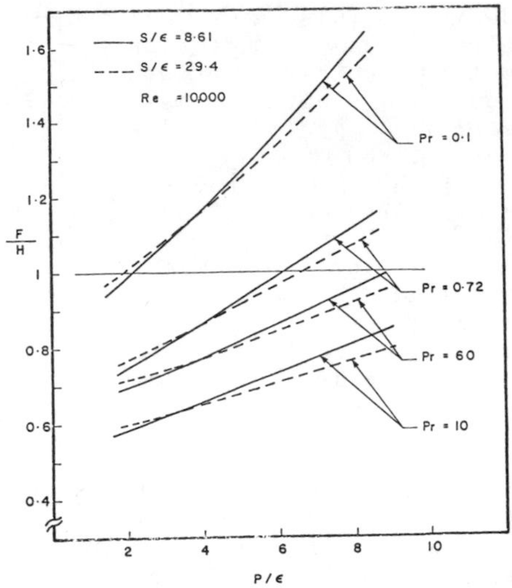

Fig. 6 Effect of Pr on F/H

Fig. 5 shows the effect of Reynolds number on F/H for Prandtl numbers of 0.72 while Fig. 6 shows that of Prandtl numbers for Reynolds number of 10,000. The ratio F/H increases with increasing P/ε while there was a small increase in F/H with decreasing S/ε.

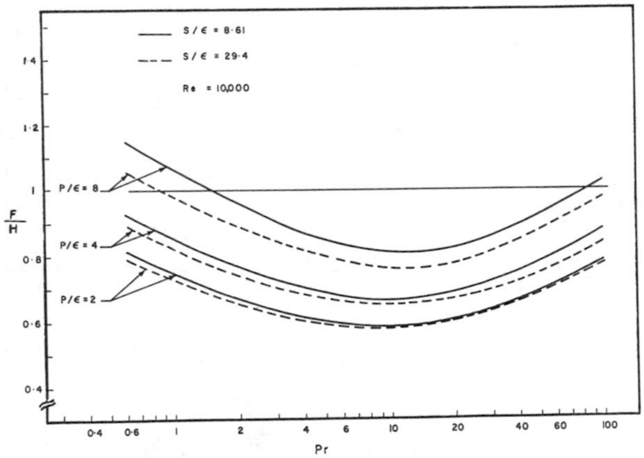

Fig. 7 Effects of S/ε, P/ε and Pr on Re

The effect of increasing Reynolds number is to increase the ratio F/H but the effect of Prandtl number on the ratio is not simple. As seen in Fig. 7, the ratio F/H decreases with increasing Prandtl number up to Prandtl numbers of about 10. However, further increase in Prandtl number would bring an increase in F/H. It seems that there exists a minimum value of Prandtl number beyond which friction forces become dominant.

CONCLUDING REMARKS:

The effects of roughness density, relative roughness height, Reynolds and Prandtl numbers on the ratio F/H were identified for an asymmetric flow induced by square-ribbed surface roughness elements on the wider side of a large aspect ratio rectangular.

The analysis demonstrates that certain artificial roughness elements can be used to enhance heat transfer rates with advantages from the overall efficiency point of view and that the required pumping power per unit heat transfer rate decreases with decreasing values of the roughness density, the relative roughness height and Reynolds number, and increasing value of Prandtl number up to about 30, respectively.

ACKNOWLEDGMENT:

The present study is financed by the Natural Science and Engineering Research Council of Canada under grant number A5175.

NOMENCLATURE: (see also Fig. 1)

c	constant, 5.52
De	equivalent diameter
F	f_r/f_s ratio
f	friction factor
H	Nu_r/Nu_s ratio
Nu	Nusselt number
Pr	Prandtl number
q	heat flux
Re	Reynolds number
T^{++}	$(T_w - T)c_p\tau/(q_w u_{\tau r})$
u	velocity in x-direction
u_τ	shear velocity
Z_{or}	parameter defined by Eq. (1), hydrodynamic roughness

Subscripts:

b	bulk
m	corresponding to the maximum velocity point
r	rough wall
s	smooth wall
o	corresponding to the zero shear stress point
w	wall

Superscripts:

+	quantity non-dimensionalized with $u_{\tau s}$
++	quantity non-dimensionalized with $u_{\tau r}$

REFERENCES:

1 Schlichting, H., Boundary Layer Theory, 6th ed., pp. 578-589, McGraw Hill Book Co., Inc., 1968.

2 Nikuradse, J., Stromungsgesetze in rauhen Rohren, Forsch. Arb. Ing.-Wes. No. 361, 1933.

3 Allan, W.K. and Sharma, V., An Investigation of Low Turbulent Flows over Smooth and Rough Surface, Jour. Mech. Eng. Sci. vol. 16, pp. 71-78, 1974

4 Hall, W.B., Heat Transfer in Channel having Rough and Smooth Surfaces, Jou[illegible]h. Eng. Sci. vol.4, No. 2, 1962

5 Musker, A.J. and Lewkowicz, A.K., The Effect of Ship Hull Roughness on the Development of Turbulent Boundary Layers, Int. Symposium on Ship Viscous Resistance, SSPA, Goteborg, 1978.

6 Reichardt, H. Vollstandige Darstellung der Turbulenten Geschwindigkeits-verteilung in Glatten Leitungen, vol. 31, pp. 208-219, 1951.

7 Alp, E., Asymmetric Turbulent Flow: Analysis and Experiment, M.A.Sc. Thesis, Dept. of Mech. Eng., University of Ottawa, 1974.

8 Bhuiyan, A., An Asymmetric Turbulent Fluid Flow Induced by Rectangular Ribbed Surface Roughness, M.A.Sc. Thesis, Dept. of Mech. Eng., University of Ottawa, 1977.

9 Hanjalic, K., Two-Dimensional Asymmetric Turbulent Flow in Ducts, Ph.D. Thesis, University of London, 1970.

10 Simpson, R.L., A Generalized Correlation of Roughness Density Effects on the Turbulent Boundary Layer, AIAA Jour. vol. 2, pp.242-244, 1973.

A New Method of Heat Transfer Augmentation by Means of Foreign Gas Jet Impingement in Liquid Bath

CHONG-FANG MA, YONG-PING GAN, and FU-JING TANG
Institute of Thermal Sciences and Energy Conservation
Beijing Polytechnic University
Beijing, PRC

A. E. BERGLES
Department of Mechanical Engineering
Iowa State University
Ames, Iowa, USA

ABSTRACT

A two-phase two-component experimental system was developed to study enhancement of heat transfer from a vertical chip size heater to surrounding liquid (Freon 113 or Ethanol) due to air jet impingement. Heat transfer coefficient was measured as function of jet velocity, heat flux and the distance between exit of jet tube and heated surface. The injecter was placed very close to the hot surface.

It was found that the heat transfer was notably enhanced with foreign gas impingement, especially in the cases of small temperature differences between wall and coolant. In fact for higher jet velocities the heat transfer coefficient tended to infinite while the temperature differences between wall and coolant were zero. The heat transfer coefficients decreased with the increacing of heat flux. High-speed camera was used to study the behaviour of the gas bubbles which play very important role in the heat transfer process. A physical model of simultaneous heat and mass transfer was presented to explain the special heat transfer characteristics in this system. It was suggested by the experimental and analytical investigations that the heat transmission was enhanced not only by agitation of the liquid by the gas bubbles but also by evaporative cooling of the liquid film beneath the gas bubble. Based on this model a semiempirical method was developed and can be used to correlate the non-boiling data.

NOMENCLATURE

- C concentrarion
- C_p specific heat of liquid at constant pressure
- d jet diametor
- g gravititational acceleration
- H_{fg} Heat of vaporization
- h heat transfer coefficient
- h_D mass transfer coefficient
- k thermal conductivity
- M molecular weight
- P pressure
- q'' heat flux
- R_o general gas constant
- T temperature
- U_j exit velocity of jet
- x vertical distance measured from leading edge of heater
- Z distance between exit of jet nozzle and heated surface
- β isobaric thermal expansion coefficient
- μ viscosity of fluid
- ν kinematic viscosity
- G^*_{rx} modified Grashof number, $G^*_{rx} = g\beta q''x^4/k$
- N_{ux} Nusselt number, $N_{ux} = xh/k$
- P_r Prandtl number, $P_r = C_p\mu/k$

Subscripts

- C convection or coolant
- D,m mass transfer
- l liquid
- t total
- w wall
- ∞ incoming flow

INTRODUCTION

Heat transfer has become a critical consideration in the design of microelectronic devices and computers. Efforts to improve the performance of cooling systems in microelectronic equepments continoe to be stimulated by increasing demands for energy dissipation and temperature control. A number of advanced heat removal techniques have been developed for cooling microelectronic components or systems with high heat flux. Liquid immersion cooling has attracted much interest since 1970's and has been adopted by industry for "supercomputer."[1] Two of the authors of the present paper suggested that boiling jet impingement cooling might be used in testing of arrays of microelectronic chips.[2]

Foreign gas injection in liquid pool is a potential method of microelectronic cooling. Baker investigated enhancement of heat transfer from a small heat source to R-113 or silicone coolants by bubble induced mixing.[3] It was found that the heat transfer coefficients with bubble agitation, which did not depend on heat flux according Baker's experimental result, were higher than those with boiling for R-113 and forced convection for silicone. Keening and Kao reported that upstream injection of Nitrogen gas caused increases of up to 50% in heat transfer coeffecent for water flowing in a channel of rectangular cross section [4]. They concluded that the enhancement did not depend on heat flux and suggested that a possible machanism for augmentation was secondery flow production by the interaction of bubbles with the shear flow near the wall. Resently Brosmer and Incropera presented their experimental results for enhancement of heat transfer from a cylinder to water with low velocities due to gas impingement [5]. The average Nusselt number with gas injection was as much as seventeen times larger than results associated with single phase free and mixed convection. The authors also found the heat transfer enhancement was independent of heat flux. In contrast to almost all the other studies, the investigation of Tamari and Nishikawa showed that average heat transfer coefficient of vertical flat heater to ethyl alcohol or water with air bubble injection were proportional to $q''^{-1/8}$ [6]. Because of the complication of the heat transfer process with bubble injection and the lack of available literature , there are many problems to be solved.

In present work a experimental system was developed for studing augmentation of heat transfer from a vertical heater of chip size to surrouding liquid with air jet impingement. The injector was placed very close to the heater surface. Local heat transfer coefficient at stagnation point was measured as an function of jet velocity, heat flux and distance between the exit of jet tube and the heated surface. It was noted that the heat transfer coefficient was very stronly affected by heat flux. In fact it tended to infinite while the temperature difference between wall and coolant was zero. Heat flow from wall to liquid, as high as $3.3 \times 10^4 w/m^2$ for R-113 and $1.3 \times 10^4 w/m^2$ for Ethanol, was recorded in the cases of negative temperature potential ($T_w < T_c$). This phenomenon suggested that the mechanism of heat transfer enhancement in present investigation must be very different from that in previous studies, in which bubble agitation was considered to be the primary contributor for enhancement. High-speed photography was employed to disclose the heat transfer mechanism Based on the experimental data and the photographic study, a possible modele was presented, which revealed that the evaporation of the liquid film beneath the gas bubble made very important contribution to the enhanced heat transfer. A semiempirecal correlation method was developed. All the non-boiling data were well correlated with this method using a electronic computer. The present report is a follow-up paper of literature[7] finished in 1983 at ISU Heat Transfer Laboratory.

EXPERIMENTAL APPARATUS AND PROCEDURE

The experimental system is shown in Fig.1 and 2. The working fluids were R-113 and Ethanol. Dry air was supplied from a high pressed clumn. The jet temperature was always adjusted to equal to pool temperature. All the experiments were conducted at atmospheric pressure. Further details of experimental apparatus and procedure were described in[2]. The local heat transfer coefficie at the center of the heater can be calculated by the following formular

$$h_t = q''_t / (T_w - T_c) \qquad (1)$$

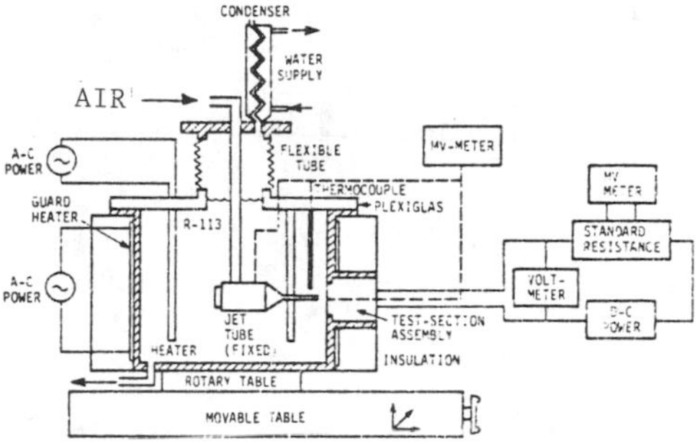

Fig.1 Test chamber and instrumentation.

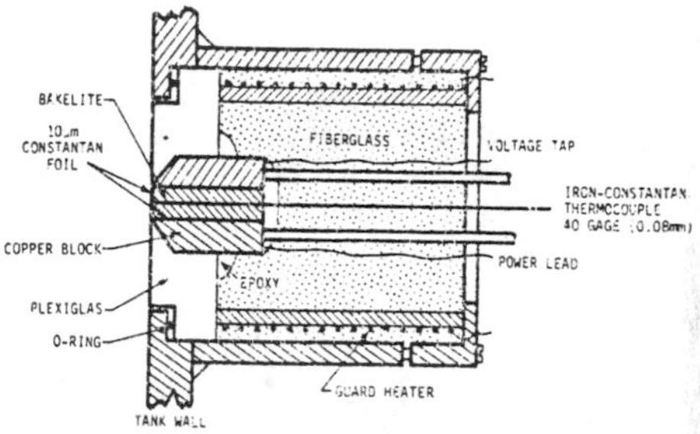

Fig.2 Details of test section.

MEASUREMENT RESULTS

Natural Convection Without Gas Impingement

Single phase free convection data were taken first both for R-113 and Ethano as shown in Fig.3(a) and (b) respectively. In the same figures the calculated results from Fujii's formular[8] for vertical constant heat flux are also presented. The data of present study are much higher than the calculated results, about 100% for Ethanal and over 200% for R-113. This discrepancy may be explaine by the three-dimensional boundary layer effects, which can not be neglected for samll heat sources like those used in present investigation. This trend has been confirmed by the previous study of Baker with a 4.56 x 2.22mm heater[3] as seen in the Fig.3 (a).

Enhanced Heat Transfer for Ethanol with Low Gas Velocity

The exprimental result of heat transfer augmentation for Ethanol with low air jet velocity is shown in Fig.3(b). In the logarithmic plot the data may be represented by correspondent stright lines, which run almost parallet to that for zero gas velocity. The slight departures from the slope of single phase free convection curve imply the decrease of enhancement of heat transfer with increasing of heat flux. The improvement in heat transfer of up to 200% was obtained in low heat flux as illustrated in Fig.3(b).

Enhanced Heat Transfer with Higher Velocity of Gas Jet

Measurements were made at four jet velocities for R-113 and six velocities for Ethanol with Z/d = 1.0, 1.5 or 2.0. Heat flux and measured temperature difference between wall and coolant are given in semilogarithmic plots for dif-

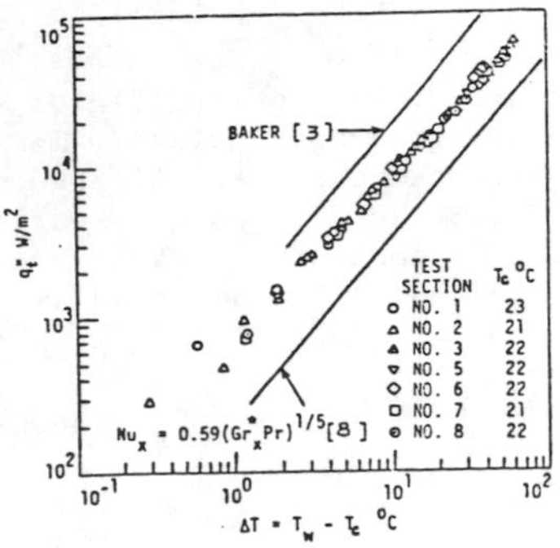

(a)

Fig.3 Natural convection heat transfer for R-113 (a) and Ethanol (b).

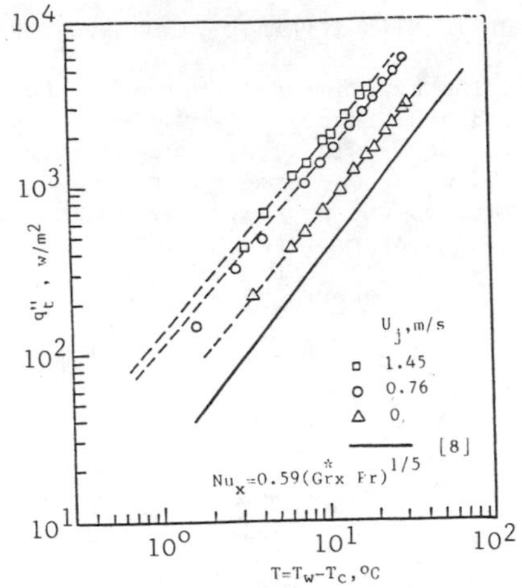

(b)

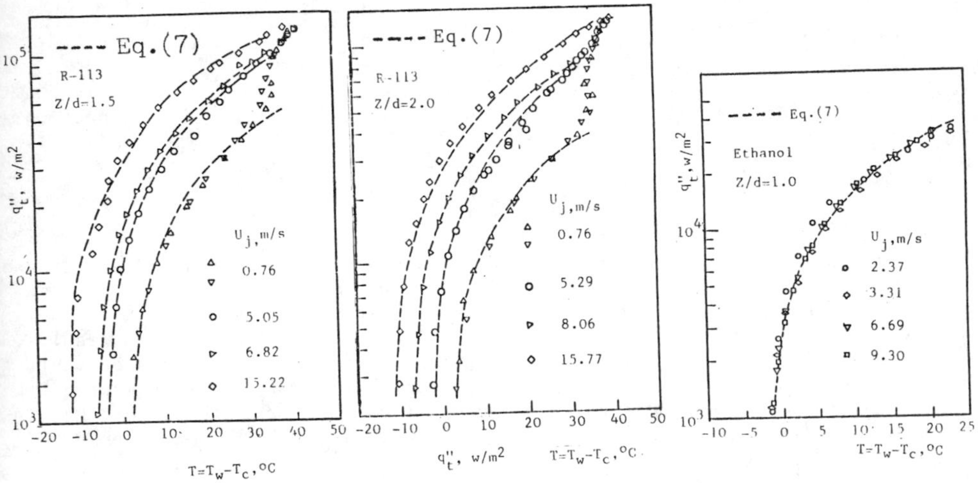

(a) (b) (c)

Fig.4 Enhanced Heat Transfer with air jet impingement for R-113 at Z/d = 1.5 (a) and 2.0 (b) and for Ethanol at Z/d = 1.0 (c).

ferent constant jet velocities (Fig.4). As seen in the figures gas jet impingement was most effective in inproving heat transfer at lower heat flux. In fact the heat flux still maintained at considerably high level (over 10^4w/m^2) with zero temperature potential (T_W=T_C). Heat flux from wall to fluid was also recorded in the cases of "negative temperature difference" (T_W<T_C) as shown in the figures. The concept of heat transfer coefficient defined by Eq. (1) may be used in the cases of negative temperature potential only with negative values. It tended to infinite with higher velocities of gas jet impingement while temperature difference being zero as illustrated in Fig.5. It is of interest to compare the effect of heat flux on heat transfer coefficient in present investigation with those for boiling or forced convection. It has been established that heat transfer coefficient considerably increases with increasing of heat flux for boiling and essentially keeps constant with variation of heat flux for single phase forced convection. In our study the heat transfer coefficient was stronly affected by heat flux and decreased fast with increasing of heat flux until a horizental asymptote was closed as illustrated in Fig.5. The heat transfer characteristics reaveled in present experiment suggested that the enhancement mechanism must be very different from those in previous studies.

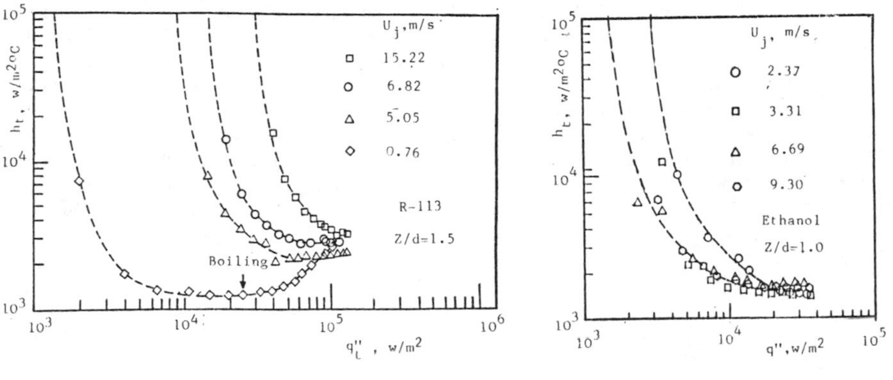

(a) (b)

Fig.5 Variation of Heat Transfer Coefficients with air jet impingement for R-113 (a,b) and for Ethanol (c).

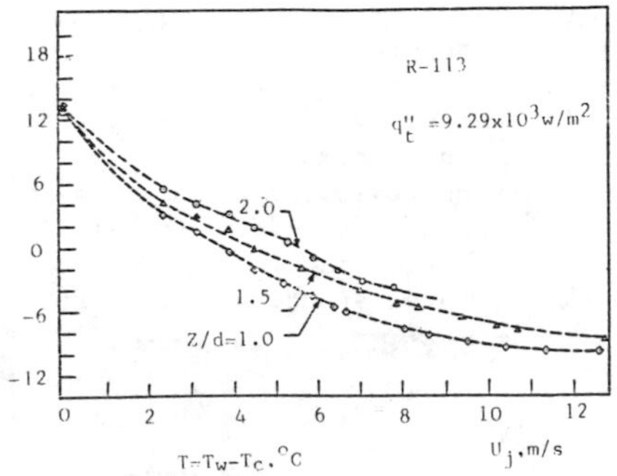

Fig.6 Effect of gas jet velocity and nozzle-to-plate spasing on Heat Transfer for R-113.

Effect of Gas Jet Velocity and

Nozzle-to-Plate Spasing

Experiment was conducted for studing the effect of gas jet velocity and nozzle-to-plate spasing on heat transfer with constant heat flux ($q''=9.29x10^3 w/m^2$). The result is given in Fig.6. As most data hade negative temperature difference we did not use heat transfer coefficient to present the experimental result. The air jet velocity ranged from zero to 13m/s, at Z/d=1.0, 1.5 and 2.0. Heat transfer enhancement was improved with increasing of gas jet velocity. The temperature potential required for the given heat duty decreased with the jet velocity increase. After the jet velocity excessing a necessary value, which depended on the nozzle-to-plate spasing, the temperature difference between wall and coolant became negative. As shown in the figure wall temperature might be lower than that of coolant as high as 10°C although high heat flux ($9.29x10^3 w/m^2$) transfered from wall to fluid. Nozzle-to-plate spasing has important influence on heat transfer. When the injector was placed very close to the heated surface the heat transmission was extramerly enhanced. For Ethanol heat transfer was essencially unaffected by the gas jet velocity as shown in Fig.4(c).

ANALYSIS AND DISCUSSION

Photographical Study

In order to disclose the mechanism of the enhanced heat transfer in present investigation photographical study was carried out. First the flow patten was photographed using a single-flash light source of very short duration. Then high-speed photography was employed with a camera of 3000 frames per second. All the photographical experiments were conducted with Ethanol at zero heat flux. Some typical results are presented in Fig.7. As show in the pictures, the growth of an air bubble initiated at the exit of the jet tube without touching the solid wall, then the bubble expanded and got in touch with the wall removing the liquid adjacent to the wall. Finally the air bubble grew to a sufficient volume (diameter about 3 mm) and rose to the free liquid level, liquid refilled the space where the leaving gas bubble existed just before. Some bubble parameters were determined with the high-speed motion pictures. The bubble departure frequency increased with air jet velocity, for example f=1578 1/s for U_j = 7.06 m/s and f=1910 1/s for U_j=14.12 m/s. But the average free rose velocity of the bubbles was not affected by the gas jet velocity significantly. It was 0.32 m/s and 0.39 m/s respectively at U_j=7.06 m/s and 14.12 m/s.

Possible Heat Transfer Model and Data Correlation

From the photogaphical results we may postulate the heat transfer augmentation was caused by the following factors:

(1) On the base of the growing air bubble a liquid film formed and existed. The heat resistance of this liquid film might be negligible as its thickness was so small as seen in the high-speed motion pictures. The evaporotive cooling of the liquid film played very important role in the heat transfer process.

(2) Gas bubble formation, growth and departure caused quick exchange of liquid in the region between the wall and injector. This mechanism enhanced heat transfer just like that happened in nucleate boiling.

(3) Secondary flow in the wake region of the moving bubble.

In order to simplify the anylisis of this simultaneous heat and mass transfer problem, the following assumptions may be further made:

(1) The heat transfer process is quasi-steady.

(2) The temperature of the liquid film may be considered as that of the wall as the heat resistance of the very thin film is negligable.

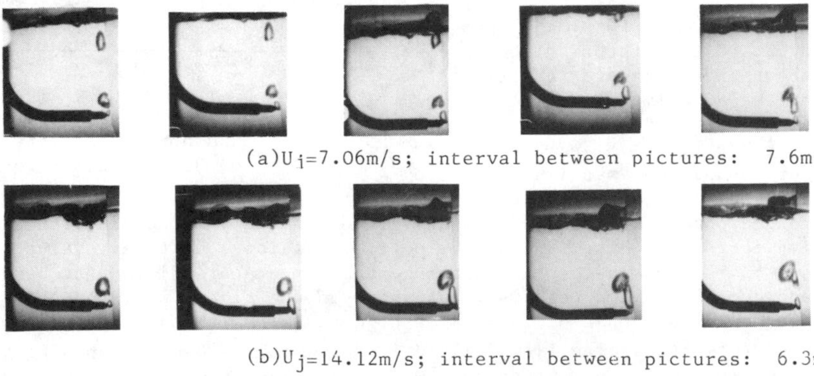

(a)U_j=7.06m/s; interval between pictures: 7.6ms.

(b)U_j=14.12m/s; interval between pictures: 6.3ms.

Fig.7 High-speed motion pictures for Ethanol with air jet impingement at zero heat flux (Z/d=1.5).

(3) All the effects of convection may be combined with a effective convective heat transfer coefficient.

Based on these assumptions the total heat flux can be expressed by the equation

$$q_t'' = q_c'' + q_m'' \tag{2}$$

Convective heat flux q_c'' may be calculated by the formula

$$q_c'' = h_c(T_w - T_c) \tag{3}$$

Heat flux due to evaporation q_m'' can be expressed as

$$q_m'' = h_D H_{fg}(C_w - C_\infty) \tag{4}$$

Using the ideal-gas equation of state and the assumption (2), the vapor concentration at wall C_w may be calculated:

$$C_w = \frac{P_{1,w}\ M}{R_o\ \ T_w} \tag{5}$$

Noting $C_\infty = 0$, Eq. (4) may be rewritten as

$$q_m'' = h_D H_{fg} \frac{P_{1,w}\ M}{R_o\ \ T_w} \tag{6}$$

Substituting Eqs.(3) and (5) into Eq. (2), we obtain

$$q_t'' = h_c(T_w - T_c) + h_D H_{fg} \frac{P_{1,w}\ M}{R_o\ \ T_w} \tag{7}$$

The validity of the analysis based on the simplified model will depend upon whether the non-boiling data may be correlated by Eq. (7) with constants h_c and h_D for given jet velocity. By means of a electronic computer the coefficents h_c and h_D were determined for different velocities. Then the Eq. (7) can be used with the constants h_c and h_D for correlating the experimental data. As shown in Fig.4 (a)(b), all the data for R-113 are well correlated by this methed, with

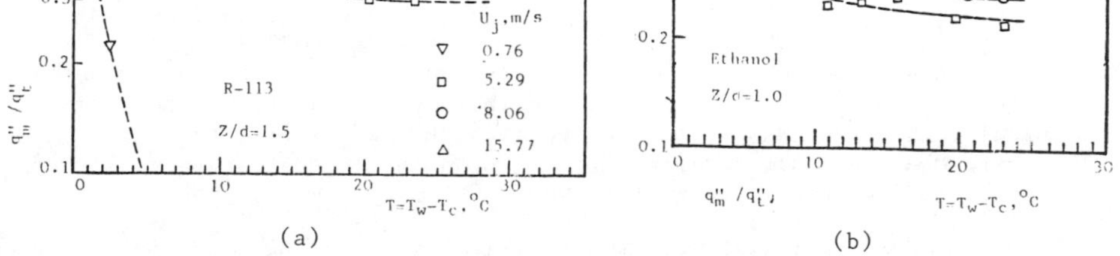

(a) (b)

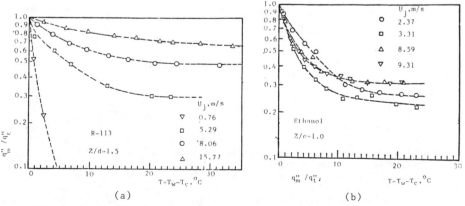

Fig.8 Variation of ratio q_m''/q_t'' for R-113 (a) and Ethanol (b).

exception of the data associated with boiling. For Ethanol all the data with high air jet velocities are correlated by a single curve, indicating the negligable influence of jet velocity on heat transmission as seen in Fig. 4 (c). The good agreements between experimental data and correlation curves with Eq. (7) testify that the simplified model of heat transfer may be employed for the enhanced heat transfer in present investigation. The contributions of evaporative cooling of the liquid film and convection were estimated with the calculated values of h_C and h_D. The results are presented in Fig. 8 with the ratio q''_m/q''_t . It is found that the evaporation of the liquid film is the main contributor in heat transmission for both R-113 and Ethanol in the cases of lower heat flux, it still makes very important contribution in the heat transfer process while the heat flux is increased, especially for high jet velocities as illustrated in the figures.

CONCLUSIONS

(1) A jet impingement system was developed to study the augmentation of heat transfer from a vertical heater of chip size to surrouding liquid due to air jet impingement. Local heat transfer characteristics at stagnation point were studied experimentally.

(2) Heat transfer was notably enhanced with foreign gas impingement, especially in the cases of lower heat flux. The heat transfer coefficient tended to infinite with zero temperature difference between wall and coolant. Heat flux from wall to liquid, as high as $3.3 \times 10^4 w/sm^2$ for R-113 and $1.3 \times 10^4 w/m^2$ for Ethanol, was recorded while the wall temperature was lower than the temperature of coolant.

(3) Photographical investigation with high-speed camera was carried out for studing the bubble behaviour. A physical model of simultaneous heat and mass transfer was presented. It was found that the evaporative cooling of the liquid film made very important contribution in the heat transfer process.

(4) Based on the simplified model a semiempirecal method was developed, with which all non-boiling data were well correlated using a electronic computer.

ACKNOWLEDGMENT

This work was portially supported by Science Foundation of Chinese Academy of Sciences.

REFERENCES

[1] Bar-cohen, A., Kraus, A.D. and Davidson, S.F., "Thermal Frontiers in the Design and Packaging of Microelectronic Equipment," Mechanical Engineering, Vol. 105, No. 6, PP. 53-59, 1983.

[2] Ma, C.-F. and Bergles, A.E., "Boiling Jet Impingement Cooling of Simulated Microelectronic Chips," ASME Publication HTD-Vol. 28, "Heat Transfer in Electronic Equipment -- 1983," PP. 5-12.

[3] Baker, E., "Liquid Immersion Cooling of Small Electronic Devices," Microelectronics and Reliability. Vol. 12, PP. 163-173, 1973.

[4] Kenning, D.B.R. and Kao, Y.S., "Convective Heat Transfer to Water Containing Bubbles: Enhancement Not Dependent on Thermaocapillarity," International Journal of Heat and Mass Transfer, Vol. 15, PP. 1709-1717, 1972.

[5] Brosmer, M.A. and Incropera, F.P., "Augmentation of Heat Transfer from A Cylinder to A Liquid In Crossflow Due to Gas Impingement," ASME Paper 83-HT-41, 1983.

[6] Tamari, M. and Nishikawa, K., "The Stirring Effect of Bubbles Upon the Heat Transfer," Heat Transfer-Japanese Research, Vol. 5, PP. 31-44, 1976.

[7] Ma, C.-F. and Bergles, A.E., "Augmentation of Heat Transfer from A Vertical Heated Surface to Liquid By Foreign Gas Impingement," to be published.
[8] Fujii, T. and Fujii, M., "The Dependence of Local Nusselt Number on Prandtl Number in the Case of Free Convection Along a Vertical Surface with Uniform Heat Flux," International Journal of Heat and Mass Transfer. Vol. 12 PP.121-122. 1976.

An Investigation on the Heat Transfer Augmentation and Friction Loss Performances of Plate-Perforated Fin Surfaces

JIARUI SHEN, WEIZAO GU, and YUMING ZHANG
Institute of Engineering Thermophysics
Chinese Academy of Sciences
Beijing, PRC

ABSTRACT

In this paper, heat transfer augmentation and friction loss performances of plate-perforated rectangular fin surfaces in a conductive-convective heat transfer system are studied. The effects of the effectiveness of the fins having different perforation geometries on the heat exchanger performance are determined, and performance comparison of the plate-perforated fin heat exchangers relative to a reference exchanger having plate-monperforated fin surfaces is included. It is found that under certain circumstances the perforation will produce substantial improvement in heat transfer without introducing pronounced form drag. The slotted fins are seen to be the high-performance extended surfaces.

NOMENCLATURE

- a slot spacing mm
- Ac flow area in flow channels mm^2
- b plate spacing mm
- C_p specific heat of air at constant pressure $kcal/kg^oC$
- d_s slot width mm
- d_t hole diameter of orificemeter mm
- G mass flow rate of air kg/h
- F_f fin surface area mm^2
- F_o total heat transfer surface area mm^2
- h heat transfer coefficient $kcal/m^2h^oc$
- L length of test core in flow direction m
- p_l longitudinal pitch mm
- p_t transverse pitch mm
- S perimeter of plate-fin mm
- t_δ plate thickness mm
- W flow channel width mm
- W_s length of slot mm
- γ_a specific weight of air kg/m^3
- μ viscosity kg s/m^2
- σ_F frontal porosity
- σ_s plate porosity
- α orifice meter coefficient

INTRODUCTION

In order to develop new heat transfer augmentaion techniques for the design of more efficient plate-fin heat exchanger equipment such as air-cooled condensers an investigation of heat transfer and friction loss performance of plate-perforate fin surfaces in a conductive-convective heat transfer system (not in simple convective system) has been performed. Previous reports indicated that the perforation in fin surfaces resulted in a substantial increase in heat transfer without introducing a pronounced form drag 1-5. However, these studies were carried out in simple convective heat transfer systems. In other words, the effects of fin efficiencies on heat exchanger performance were not considered. In the present paper, this problem is tackled in order to determine the effects of efficiencies of the fins having different hole geometries on the size of finned exchanger. The direct goal of this work is to add one more study on perforated fins to the list of investigation in this field so that the spectrum would be more complete. Specifically speaking, this work contains performance comparison between plate-perforated fin heat exchangers and a reference exchanger having plate-nonperfora fins for a specified heat transfer rate. This comparison serves the purpose of heat exchanger design.

EXPERIMENTAL APPARATUS AND DATA REDUCTION

A schematic diagram of the experimental facility is presented in Fig.1. The air from the compresser passes through the cleaner, orifices, and stabilizer, and enters the test section, and finally flows out through the exhaust valve. The test core consisted of a number of carbon-steel sheets welded to the two side walls of a rectangular stainless steel tube to form parallel flow channels as shown in Fig.2. The electric heaters were set on the side walls with heated effective length of 0.5 m. As the air flowed through the test section, the side walls and fins of the test core were cooled. To measure the temperatures over the side walls and fin surfaces, 12 copper-constantan thermocouples, 0.12 mm dia., were laid on three cross sections of both side walls and fins of the test core, respectively. The inlet and outlet temperatures of the air in the test core were also measured by thermocouples. All the temperature measurements were recorded by PF15 digital voltmeter. Air flow rate through the test core was measured by orificemeter and U-tube manometer, and the pressure loss in the test core was evaluated from U-tube manometer readings. The test fins were perforated with slotted holes and round holes, respectively, arranged in staggered patterns. The geometric properties of some typical test cores are listed in Table 1.

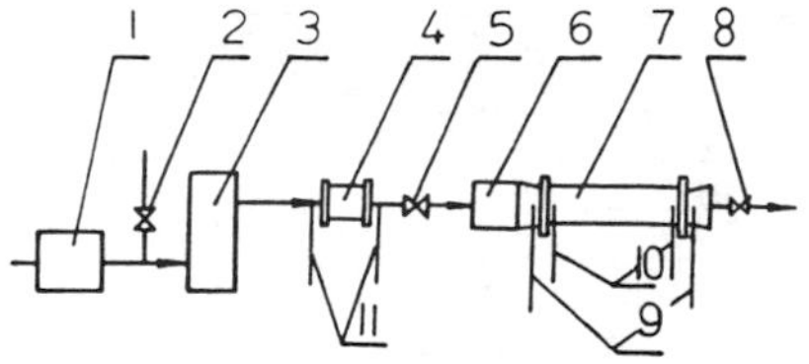

1. compresser 2. valve 3. cleaner
4. orifices 5. valve 6. stabilizer
7. test section 8. exhaust valve
9. air temperature measurements
10. 11. pressure measurements

Fig.1 A schematic diagram of experimental apparatus

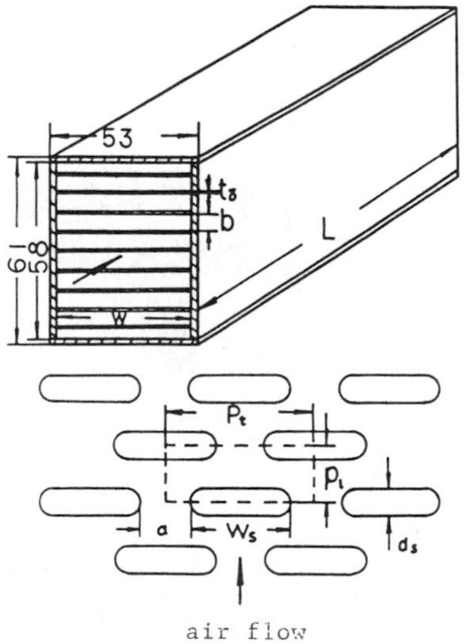

air flow

Fig.2 Test core and plate perforation pattern

Table 1 Geometric properties of typical test cores

definition		slotted holes	round holes	nonperforated
t_δ	mm	0.8	0.8	0.8
d_s	mm	4	3	
d_s/t_δ		5	3.75	
W_s	mm	10	3	
P_l	mm	17	8	
P_t	mm	15	10	
a	mm	5	7	
b	mm	5.08	5.08	5.08
W	mm	50	50	50
L	mm	485	500	500
de	mm	9.2	9.2	9.2
L/de	mm	52.6	54.2	54.2
Ac	mm^2	2540	2540	2540
σ_F		0.864	0.864	0.864
σ_s		0.1434	0.0883	

The heat transfer and friction loss performance of the test fins was interpreted in terms of the Colburn j factor and Fanning f factor:

$$j = \frac{hA_c}{GC_p} P_r^{2/3} \qquad (1) \qquad\qquad f = \frac{d_e}{4L} \frac{2g\, \gamma_a \Delta P_c A_c^2}{G^2} \qquad (2)$$

With the pressure difference Δh across the orificemeter and the pressure P in the inlet of the orificemeter, the air flow rate G can be calculated by the following equations:

$$G = 1.252\alpha d_t^2 \left[(\gamma_{H_2O} - \gamma_a)\, \gamma_a \Delta h \right]^{\frac{1}{2}} \qquad (3)$$

where

$$\gamma_a = 1.252P \frac{273}{273+Ta}, \qquad (4)$$

The Reynolds number is defined on the basis of the hydraulic diameter d_e of the air flow channel in the test core as

$$R_e = \frac{Gd_e}{A_c \mu} \qquad (5)$$

where

$$d_e = \frac{2Wb}{W+b} \qquad (6)$$

In the heat transfer measurements, the air temperature distribution along the heated length of the test core was considered linear since the two side walls had equal and uniform heat flux. Because the fins extended from one side wall to the other, the effective fin height l was half the wall spacing. i.e. $l = W/2$. As $L \gg t_\delta$ and $l \gg t_\delta$, heat conduction along the fin may be considered one dimensional. In the heat transfer system as shown in Fig.3, the general solution to the differential equation

$$\frac{d^2T}{dx^2} - \frac{hs}{\lambda A}(T - T_a) = 0 \qquad (7)$$

is

$$\theta = C_1 e^{mx} + C_2 e^{-mx} \qquad (8)$$

where

$$\theta = T - T_a$$

$$m = \left(\frac{hs}{\lambda A}\right)^{\frac{1}{2}}$$

$$S = 2(L + t_\delta)$$

$$A = Lt_\delta$$

C_1 and C_2 are integration constants, and λ is thermal conductivity of the fin material.

In the tests, two values of x were selected on cross section i of the test fin, one B unit from the fin base and other precisely 2B units from the fin base. Three temperature measurements, θ_o at $x = 0$, θ_B at $x = B$, and θ_{2B} at $x = 2B$ can be taken, and from equation (8)

$$\theta_o = C_1 + C_2$$

$$\theta_B = C_1 e^{mB} + C_2 e^{-mB} \qquad (9)$$

$$\theta_{2B} = C_1 e^{2mB} + C_2 e^{-2mB}$$

are obtained. Defining a temperature ratio

$$R = \frac{\theta_o + \theta_{2B}}{\theta_B} \qquad (10)$$

which can be evaluated experimentally by utilizing the temperature measurements. We have, using equation (9)

$$R = 2\cosh(mB)$$

and we may write

$$mB = \text{arc cosh}\,(R/2) \qquad (11)$$

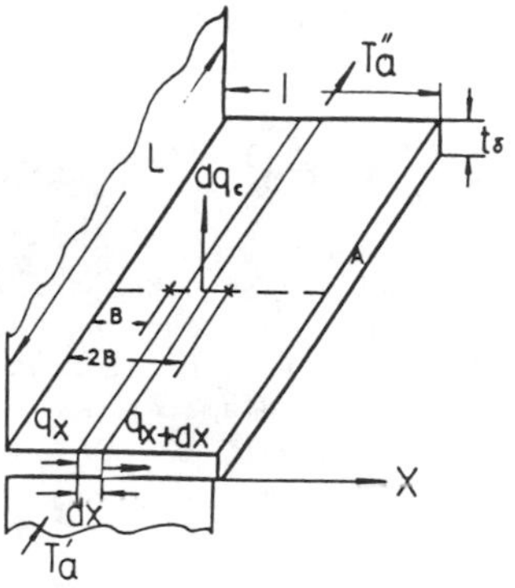

Fig.3 heat conduction and convection of a plate-fin

Then, the heat transfer coefficient can be obtained:

$$hi = \frac{\lambda A}{S}\left[\frac{\text{arc cosh}\,(R/2)}{B}\right]_i^2 \qquad (12)$$

The fin effectiveness from 6 can be writen as

$$\eta_f = \frac{\tanh\left(\frac{\text{arc cosh}(R/2)}{B}\,1\right)}{\frac{\text{arc cosh}\,(R/2)}{B}\,1} \qquad (13)$$

and the effectiveness of the total heat transfer surface is expressed as

$$\eta_o = 1 - (1 - \eta_f)\,\frac{F_f}{F_o} \qquad (14)$$

In the friction loss measurements, all data were taken under isothermal conditions so as to eliminate the effect of temperature difference between the plate-fins and the air stream.

TEST RESULTS AND ANALYSIS

The test results are correlated in the forms of j and f factors as functions of Re. Heat transfer and friction loss results of the representative cores are presented in Fig.4. For performance comparison, the test results are plotted as $(1/j)$, $(f/j)^{\frac{1}{2}}$, and $(f/j^3)^{\frac{1}{2}}$ vs. $Re(f/j)^{\frac{1}{2}}$, respectively, as shown in Figs. 5-7, and are also shown as diagrams of $\eta_o h$ and $\eta_o h\beta$ against $E/\eta_o h$, respectively, in Fig.8.

It is a simple matter to develop the following equations:

$$Re(f/j)^{1/2} = [(2g\gamma_a \Delta P_c/G^2)(GC_p/hF_o)(1/P_r^{2/3})]^{1/2}(d_e G/\mu) \quad (15)$$

$$1/j = (4GC_p/d_e hF_o P_r^{2/3})\ L \quad (16)$$

$$(f/j)^{1/2} = [(2g\gamma_a \Delta P_c/G^2)(GC_p/hF_o)(1/P_r^{2/3})]^{1/2} A_c \quad (17)$$

$$(f/j^3)^{1/2} = [(2g\gamma_a \Delta P_c/G^2)(GC_p/hF_o)(1/P_r^{2})]^{1/2} F_o \quad (18)$$

Thus it can be seen that the test cores which lie lower on the plots of Figs. 5-7 represent heat exchangers which will need shorter length L, smaller flow area A_c, and less total heat transfer surface area F_o, respectively, for the same hydraulic diameter d_e, friction less ΔP_c, air flow rate G, and number of heat transfer units $\dfrac{hF_o}{GC_p}$

The plots of $\eta_o h$ and $\eta_o h\beta$ vs. $E/\eta_o h$ indicate the ratio of the transported energy to the expended energy for the test surfaces, and from these the required heat transfer surface area and volume of exchangers for a fixed heat flux and fluid pumping power can be directly compared. Evidently, the test surface whose data points lie higher on the plots of Fig.8 will require smaller heat transfer surface area and volume for an exchanger using the geometry of the test surface. The fluid pumping power per unit heat transfer surface area on one side of an exchanger is given by

$$E = \Delta P_c G/102\gamma_a F_o \quad (19)$$

and the heat transfer surface area per unit volume is defined as

$$\beta = F_o/V \quad (20)$$

In this simplified method of comparison. it is reasonable to consider the performance of only one side of an exchanger such as air-cooled condenser where the controlling thermal resistance is on the air side.

The test results are summarised as follows.

Fig.4 indicates that the flow regime in which $R_e < 6000$ is characterized by a continuous decrease in f and j with an increase in Re for a nonperforated fin, but the flow regime in which $Re > 6000$ is marked by an upturn of f and j as Re continues to increase. It can be seen that the former should be laminar flow regime and the latter transition flow regime. However, the perforation in fin surfaces prompts early transition from laminar to turbulent flow, and the critical value of Re is reduced from 6000 for nonperforated fins to about 3000 for perforated fins. For larger d_s/t_δ ratio and the value of σ_s critical Re will be smaller. The perforation does not produce significant change in heat transfer and friction loss in the laminar regime, but it does yield higher j factor in transition regime, and the corresponding rise of f is lower. The values of j and f for the fin with slotted holes are about 87% and 17% higher than those for nonperforated fin, respectively. Heat transfer and friction loss performances of slotted fin are better than of fin with round holes.

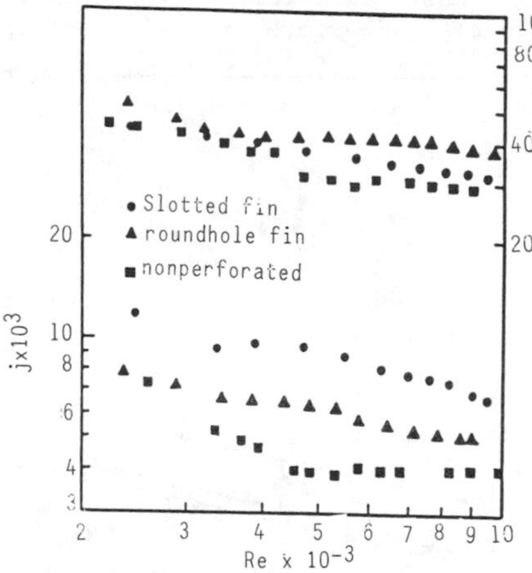

Fig.4 Variations in j and f with Re

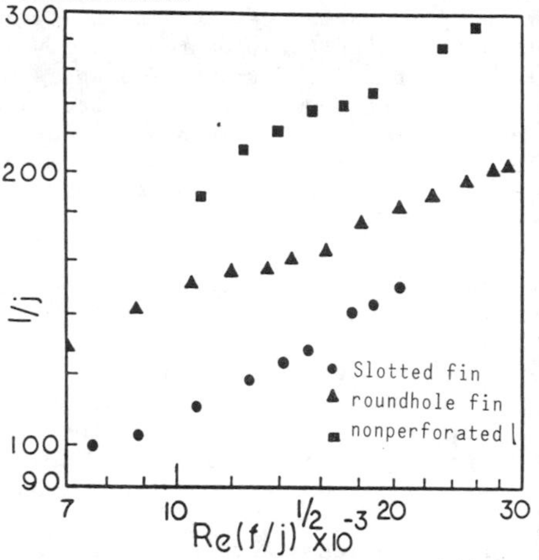

Fig.5 1/j vs. $Re(f/j)^{1/2}$

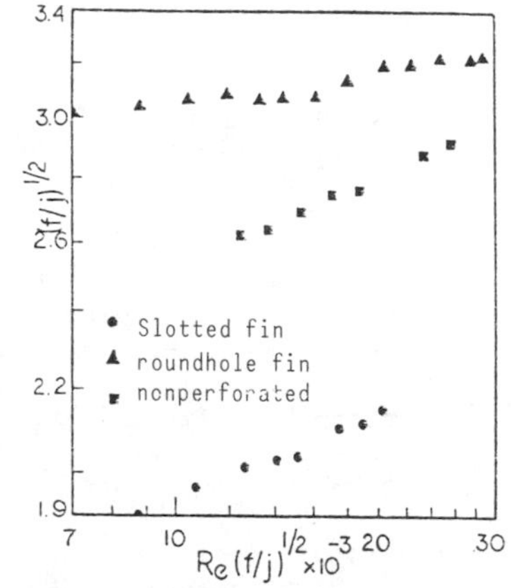

Fig.6 $(f/j)^{1/2}$ vs. $Re(f/j)^{1/2}$

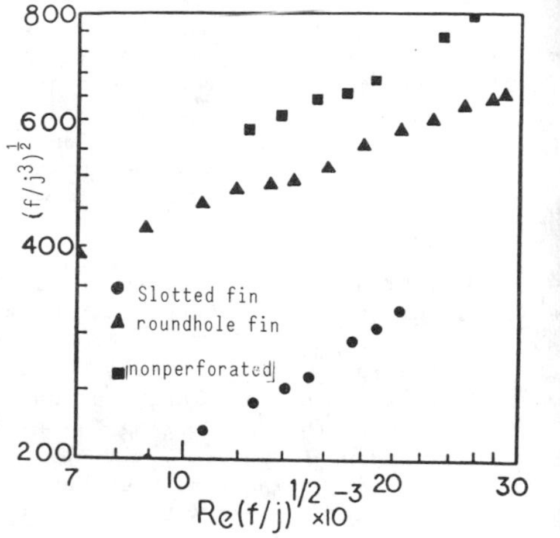

Fig.7 $(f/j^3)^{1/2}$ vs. $Re(f/j)^{1/2}$

Fig. 5 shows that the required length of the plate-slotted fin heat exchanger is about 40% shorter than that of the plate-nonperforated fin exchanger. The exchanger with plate-round hole fins is only about 21% shorter than the plate-nonperforated fin exchanger.

From Fig.6 it can be ovserved that the flow channel area of the plate-slotted fin exchanger is about 32% smaller than that of the reference exchanger. Unfortunately, since the round hole fin surface has greater friction loss, the flow channel area of the plate-round hole fin exchanger is about 16% larger than of the reference exchanger.

Fig.7 shows that the total heat transfer area of the plate-slotted fin and the plate-round hole fin exchangers is about 57% and 20% less than of the reference exchanger, respectively.

Fig.8 depicts the effect of effectiveness of the total heat transfer surface on the performances of the exchangers. The heat transfer augmentation produceed by the perforation results in the overall performance improvement of the exchanger although the perforation reduces fin effectiveness. The core with slotted fins gives the best performance for the same heat flux and pumping power needed. The required area and volume of the plate-slotted fin exchanger are about 57% and 50% less than for the nonperforated case, respectively.

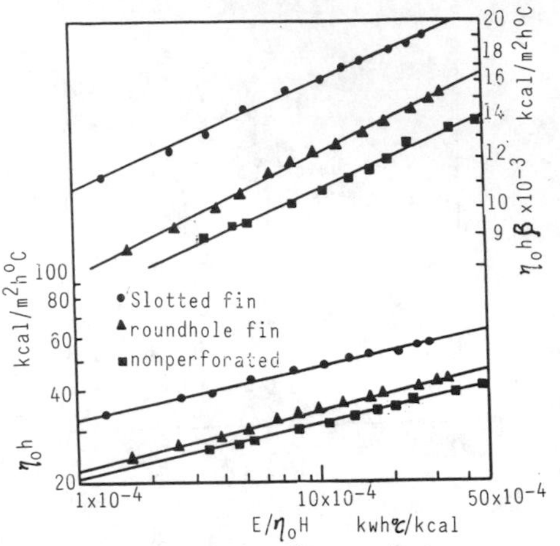

Fig.8 $\eta_o h$ and $\eta_o h\beta$ vs. E/$\eta_o h$

CONCLUSIONS

1. Perforations do not produce significant change in heat transfer and friction loss performance of heat transfer surface in the laminar regime, but they do result in substantial improvement in heat transfer without introducing pronounced friction loss increase in the transition and turbulent regimes.
2. Perforations induce early transition from laminar to turbulent flow. The higher the values of d_s/t_δ and σ_s, the earlier will transition take place.
3. Perforations can improve the overall exchanger performance. The plate-slotted fin exchangers are characterized by high-performance and high-compactedness, although perforations reduce fin effectiveness. The plate-slotted fin heat exchanger should be designed to operate in the transition flow regime.

REFERENCES

(1) C.Y. Liang and Wen-Jei Yang: Heat Transfer and Friction Loss performance of Perforated Heat Exchanger Surfaces. J. Heat Transfer Vol.97, 1975, p.9-15

(2) C.Y. Liang and Wii-Jei Yang: Modified Single-Blow Technique for Performance Evaluation on Heat Transfer Surfaces. J.Heat Transfer Vol.97, 1975, P. 16-21.

(3) C.Y. Liang. Wei-Jei Yang and Johb A. Clark: Slotted-Fin Tubular Heat Exchangers for Dry Cooling Towers. AIAA/ASME 1974 Thermophysics and Heat Transfer Conference. Vol.1, 1974, p.1-10.

(4) Forgo. L.: Some Extra High Capacity Heat Exchangers of Special Design. Proceedings of 1972 International Seminar on Recent Developments in Heat Exchangers. 1972.

(5) Pucci, P.F., Howard, C.P., Piersall, C.H.: The Single Blow Transient Test Technique for Compact Heat Exchanger Surfaces. J. Engineering for Power, Series A, Vol.89, 1967.

(6) J.P. Holman: Heat Transfer. McGraw-Hill Book Company. 1976.

The Enhancement of Condensation Heat Transfer for Stratified Flow in a Horizontal Tube with Inserted Coil

WEICHENG WANG
Thermal Engineering Department
Tsinghua University
Beijing, PRC

ABSTRACT

Theoretical anlysis of condensation for stratified flow in a horizontal tube with a inserted coil is presentet by using a simplified physical model. The experimental data of R-11 shows that heat transfer coefficient can be increased about 40% by inserted coil. Therefore, it is a practical method of heat transfer enhancement for condensers with organic condensation in tube.

Key words: Phase change, Stratified-flow condensation, Heat transfer enhancement, Horizontal tube, Theory, Experiment

NOMENCLATURE

- a $=(2\sigma/\rho g)^{\frac{1}{2}}$ capillary constant of condensate, m
- D Inside diameter of test tube, m
- G mass velocity, kg/m^2s
- g gravitational acceleration, m/s^2
- Ga $=D^3\ g/\nu^2$ Galileo number
- H $=c_p(T_s-T_w)/L$ phase change number
- L specific latent heat of condensation, J/kg
- Nu, Nu_s average Nusselt number of tube with a inserted coil and smooth tube
- P pressure, bar
- Pr Prandtl number of condensate
- q_o heat flux at inner surface of tube, W/m^2
- r_s curvature radius of surface of condensate film, m
- Re_l $=G(1-X)D/\mu$ Reynolds number of condensate
- s $=P-D_w$ effective length between wires, m
- T_s saturated temperature, K
- T_w inner surface temperature of tube, K
- u, w velocity components in φ-and z-directions, m/s
- X quality of vapor
- y coordinate in the inward normal direction measured from tube surface, m
- z coordinate in the direction of tube axis measured from a mid-point between wires on tube surface, m
- α, α_s heat transfer coefficient of tube with a inserted coil and smooth tube W/m^2K
- $\bar{\delta}$ $=\delta/D$ dimensionless film thickness
- λ thermal conductivity of condensate, W/mK
- μ dynamic viscosity of condensate, kg/ms
- ν kinematic viscosity of condensate, m^2/s
- ρ density of condensate, kg/m^3
- ρ_v density of vapor, kg/m^3
- σ surface tension of condensate, N/m

INTRODUCTION

The condensers with organic condensation in tube are widely used in refrigeratory plant, low potential waste heat recovery powerplant, geothermal powerplant and petrochemical industry. With the application of high-performence on-tube fin the enhancement of in-tube condensation becomes more significant for economy and compactification of these condensers. Today, the main enhancement methods applied in practice are of two types: mechanically extended surface (finned tubes) and twisted tape inserts. The former is more superior, its average heat transfer coefficient can be increased by 150% , the pressure drop increases slightly[1]. But the cost is expensive, especially with tubes other than of copper. And by the latter, the average heat transfer coefficient is increased by 30%, the pressure drop is increased over 250%[2]. In this paper the effect of inserted coil closely contacted with the wall of the tube on condensation heat transfer augmentation for stratified flow in a horizontal tube is investigated. Because in stratified flow vapor velocity is very low and condensate depth is large, the condensation heat transfer coefficient is small. Therefore it is necessary to enhance condensation heat transfer in this region. Besides, the condensation in this region is mainly gravity controlled, for this reason it is possible to enhance heat transfer by the surface tension of condensate.

PHYSICAL MODEL AND ANALYSIS

The physical model and coordinates are shown in Fig.1. Wire with diameter D_w is wound into coil with pitch p. It is assembled in tube with a special tool. Due to the effect of surface tension the condensate in the gaps between wires is drawn up to the wire, so the liquid film there becomes much thinner than usual, the average thermal resistance decreases. During analysis the following factors are assumed: (1) the condensate film is laminar; (2) the inertia term in dynamic equation and the convection term in energy equation are negligible; (3) shear stress at the vapor-condensate interface is negligible; (4) the temperature profile of cooled surface is even, properties of condensate keep constant. On the basis of the aforesaid assumption, the fundamental equations are simplified as follows:

$$\mu\,(\partial^2 u/\partial y^2) + \rho g \sin\varphi = 0 \qquad (1)$$

$$\mu\,(\partial^2 w/\partial y^2) - (\partial P/\partial z) = 0 \qquad (2)$$

$$\frac{2\partial}{D\partial\varphi}\int_0^{\delta} u\,dy + \frac{\partial}{\partial z}\int_0^{\delta} w\,dy = -\frac{\lambda(T_s-T_w)}{\rho L \delta} \qquad (3)$$

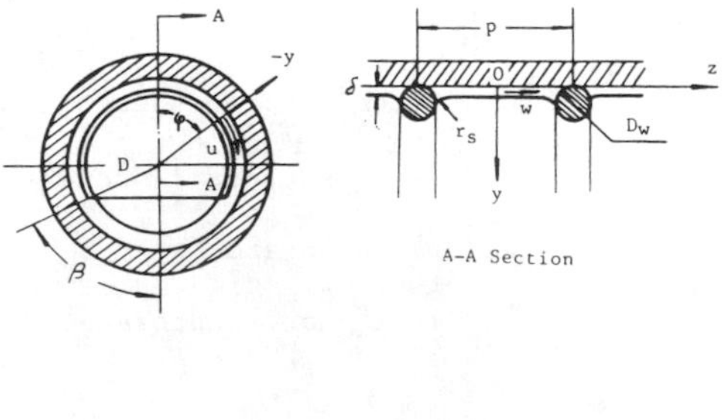

Fig.1. Physical model and coordinate

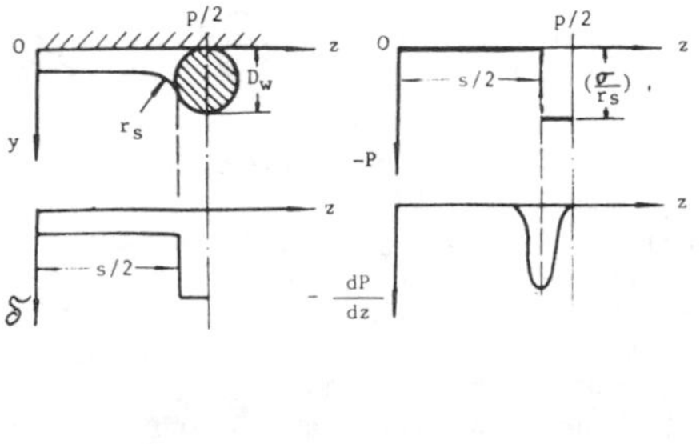

Fig.2. Simple model of film

corresponding boundary conditions are

at $y=0$ $u=w=0$ (4) (5)

at $y=\delta$ $\partial u/\partial y=\partial w/\partial y=0$ (6) (7)

Because the condensate film is very thin, we do not deem that $(\partial P/\partial z)$ is the function of y. Applying boundary conditions (4)(5), the solutions of Eqs. (1) (2) are as follows:

$$u = (-g \sin\varphi/\nu)(y^2/2 - \delta y) \tag{8}$$

$$w = (1/\mu)(\partial P/\partial z)(y^2/2 - \delta y) \tag{9}$$

Compare Eq. (3) with Eqs. (8),(9), we can write down the following equation:

$$\frac{\partial}{\partial\varphi}(\delta^3\sin\varphi) - \frac{D}{2\rho g}\frac{\partial}{\partial z}\left(\delta^3\frac{\partial P}{\partial z}\right) = \frac{3D\nu^2 H}{2gPr\delta} \tag{10}$$

Since rigorous analysis is very difficult, a simple model is used for thickness $\delta(z)$, pressure-distribution $P(z)$ and pressure-gradient (dP/dz) of the condensate film, which is shown in Fig.2. Corresponding boundary conditions are

at $z = 0$ $dP/dz = 0$ and $P = 0$ (11) (12)

at $z = s/2$ $dP/dz \neq 0$ and $P = -(\sigma/r_s)$ (13) (14)

Integrate Eq.(10), with boundary conditions (11),(13), we can obtain

$$z\frac{\partial}{\partial\varphi}(\delta^3\sin\varphi) - \frac{D}{2\rho g}\left(\delta^3\frac{\partial P}{\partial z}\right) = \frac{3D\nu^2 H z}{2gPr\delta} \tag{15}$$

Eq.(15) is integrated with boundary conditions (12),(14), the following dimensionless equation is obtained

$$\bar{\delta}\left(\bar{\delta}^3\cos\varphi + \frac{3}{2}\bar{\delta}^2\sin\varphi\right) + (4\sigma D/\rho g s^2 r_s)\,\bar{\delta}^4 = (3H/2GaPr) \tag{16}$$

then the solution for the top of the tube is

$$\bar{\delta}_{\varphi=0} = (3H/2GaPr)^{\frac{1}{4}}(1+A)^{-\frac{1}{4}} \tag{17}$$

where $A = 4\sigma D/\rho g s^2 r_s$ (18)

As a result, Nusselt number at the top of the tube can be expressed as

$$Nu_{\varphi=0} = 0.904\left(\frac{Ga\ Pr}{H}\right)^{\frac{1}{4}}(1+A)^{\frac{1}{4}}\frac{(p-D_w)}{p} \tag{19}$$

The contact heat resistance between wire and tube wall is very large, the condensate film around the wire is thick, the main function of this region is to drain off the condensate. Therefore the projection surface of the wires must be deducted from heat transfer surface. So the right side of Eq.(19) is multiplied by $(p-D_w)/p$.

The average Nusselt number over circumference ought to relate with the depth of condensate stored up. For stratified flow in smooth tube this Nusselt number is given in references[3] and [4] as following semiempirical equations:

$$Nu_s = 0.725\left(\frac{Ga\ Pr}{H}\right)^{\frac{1}{4}} / (1+5B)^{\frac{1}{4}} \tag{20}$$

where $$B = \frac{[1+1.6\times10^{11}(H/Pr)^5]^{\frac{1}{4}}}{(\rho/\rho_v)^{\frac{1}{2}}}\left[\frac{4(GaPr/H)^{\frac{1}{4}}}{\pi Re_l\ X/(1-X)}\right]^{1.8} \tag{21}$$

and $$Nu_s = 0.725\left(\frac{Ga\ Pr}{H}\right)^{\frac{1}{4}}(1.158-0.054Re_l^{0.34}) \tag{22}$$

In Eqs.(20) and (22) the influence of depth of condensate stored up is reflected by dimensionless number B and Re_l, however in the space of condensation $\varphi = 0$--$(\pi-\beta)$ the Nusselt Eq. $Nu=0.725(Ga\ Pr/H)^{\frac{1}{4}}$ is applied. After insertion of the coil, the circumferential distribution of film thickness becomes different due to horizontal drag of the surface tension. The analytical result of a horizontal tube with coiled wire on the outside indicated that when $A>15$ $Nu_{\varphi=0}$ can represent the average Nusselt number over circumference approximately [5]. The physical model of stratified flow condensation in a tube with inserted coil is similar in region $\varphi=0$ -- $(\pi-\beta)$. Therefore the average Nusselt number in this space can also be represented by $Nu_{\varphi=0}$ (when $A>15$). Compare Eq.(19) with Eq.(20) and Eq.(22), the enhancement rate of heat transfer by coil can be expressed as follows:

$$Nu/Nu_s = 1.25\,\eta\,(1+A)^{\frac{1}{4}}(p-D_w)/p \tag{23}$$

where the correction constant η represents the influence of axial vapor velocity and the change of condensate-depth due to the existence of coil. η is determined experimentally. The curvature radius $r_s=0.03a^3/D_w^2$ [5].

For stratified flow the condensation Nusselt number in horizontal tube with inserted coil can be expressed as follows:

$$Nu=0.904\,\eta\left[\frac{(GaPr/H)(1+A)}{(1+5B)}\right]^{\frac{1}{4}}\frac{(p-D_w)}{p} \tag{24}$$

EXPERIMENTAL APPARATUS AND METHOD

The experimental system is shown in Fig.3. R-11 saturated vapor produced in boiler 1, passes through superheater 2, mixing chamber 3, sight glass 4, gets

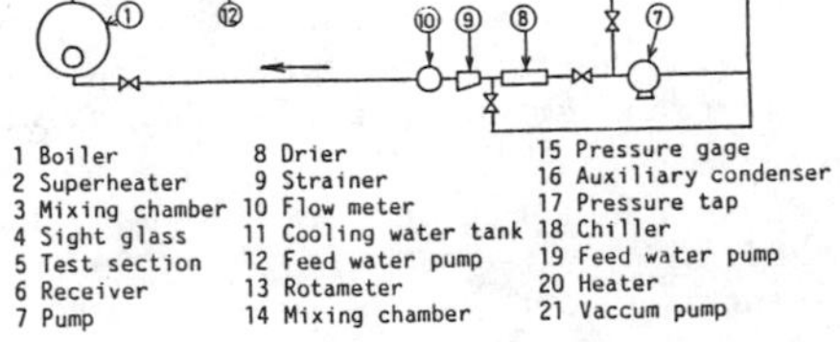

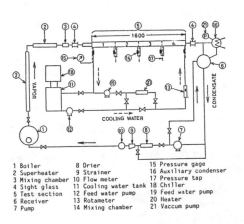

Fig.3 experimental system

into test section 5. The condensate from test section passes through sight glass and auxiliary condenser 16, turns into subcooled condensate. Then it passes through pump 7, dryer 8, strainer 9, flow meter 10, and returns to the boiler. The cooling water from tank 11 passes through pump 12, rotameter 13, mixing chamber 14, gets into each region of test section in order. Temperature of cooling water is controlled by chiller 18 and heater 20. The function of vaccum pump 21 is to remove noncondensable gases.

The test section is a structure of concentric tube. The annular passage for cooling water is divided into four segments, the length of each segment is 400mm The temprature rise of cooling water is measured by two copper-constantan thermo couples installed in the mixing chambers at the inlet and exit of each segment. For each segment the tempratures of test tube wall are measured at two cross-section by six copper-constantan thermocouples buried in tube wall (at one cross-section-four and another--two). The static pressure at the inlet of test section is measured by pressure gauge, and the pressure drops of each segment are measured by inverted U pipes.

The size of test tube and coil and test parameters are collected in table 1.

EXPERIMENTAL RESULTS AND DISCUSSION

This paper discusses the heat transfer enhancement by inserted coil for condensation in stratified flow only, and for annular flow it will be discussed in another paper. For stratified flow the photographs of flow appearance in tube with inserted coil are shown in Fig.4. Due to surface tension condensate is

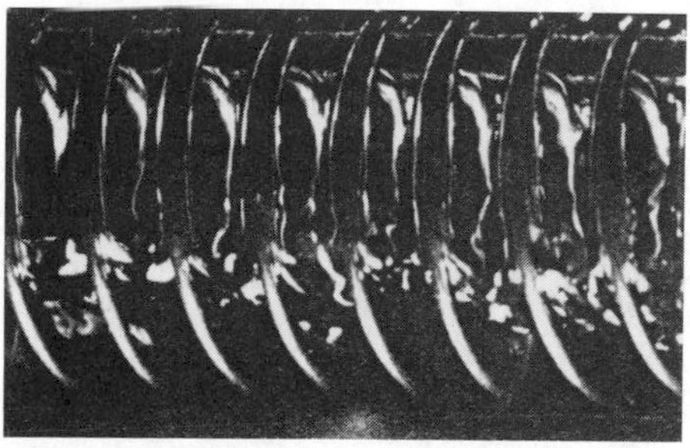

(a) G=93 kg/m^2s X=0.27

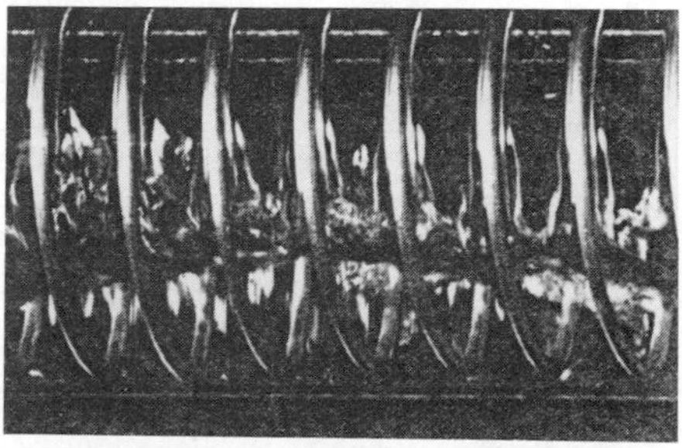

(b) G=99 kg/m^2s X=0.17

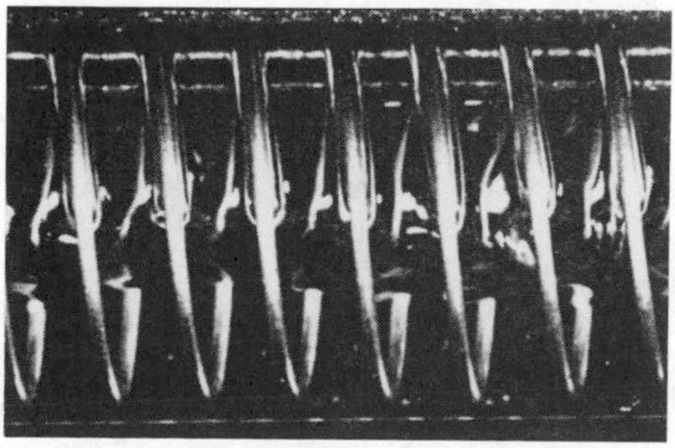

(c) G=99 kg/m^2s X=0.11

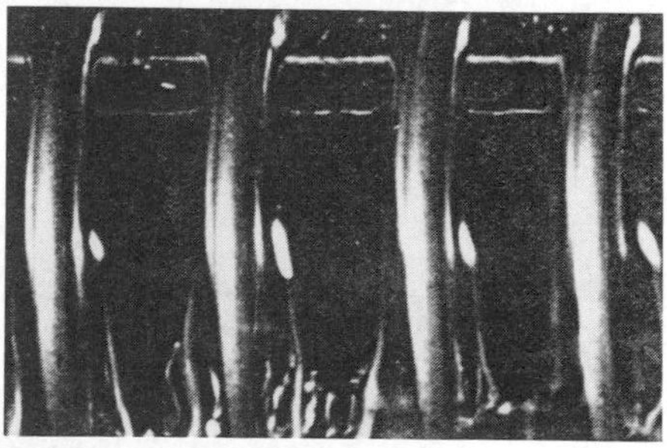

(d) G=99 kg/m^2s X=0.11

Fig.4 photographs of flow appearance in tube with inserted coil

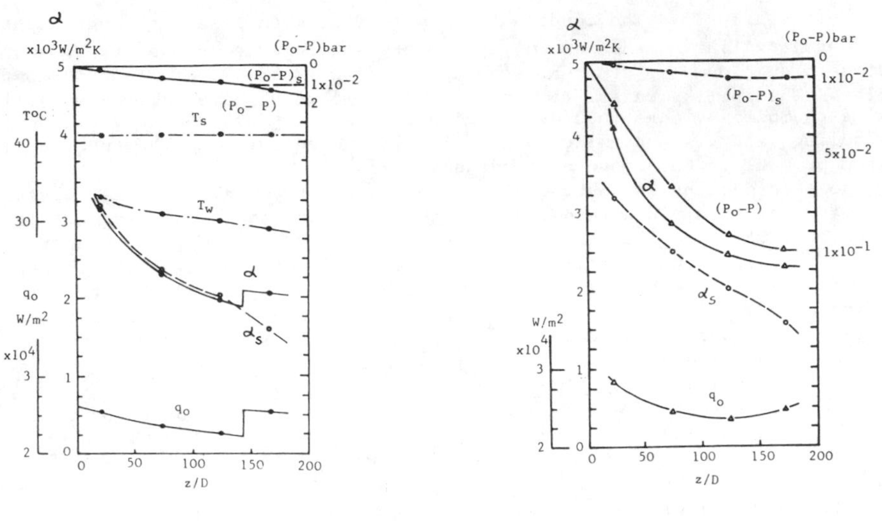

(a) $G = 117 kg/m^2 s$ $P_o = 1.794 bar$ (b) $G = 116 kg/m^2 s$ $P_o = 1.749 bar$

Fig.5 Experimental result

drawn up to the wires and then flows downward. The condensate film in gaps between wires turns into thinner, as can be observed. This is the main mechanism of heat transfer enhancement by coil for in-tube condensation in startified flow. It is consistent on the whole with the assumed physical model.

The experimental data for tube partially (only in the fourth segment) with inserted coil and for smooth tube are shown in Fig.5(a). Since in the fourth segment vapor velocety rapidly drops and the depth of stored up condensate increases, α_s rapidly dicreases. After the insertion of coil heat transfer coefficient is increased by 40%, heat flux raises correspondingly. In addition, since vapor velocity is very low, the pressure drop is not remarkable, and is acceptable for practical application. As shown in Fig.5(b), when coil is inserted in whole tube, heat transfer coefficient in annular flow is enhanced remarkably, but the pressure drop is about 10 times larger than smooth tube,

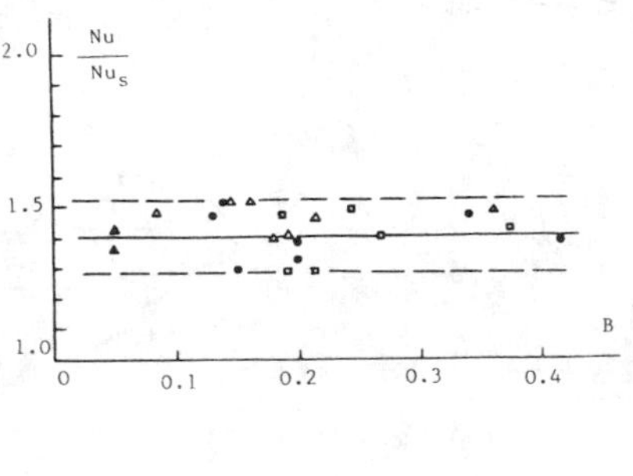

Fig.6 Relation of Nu/Nu_s and B

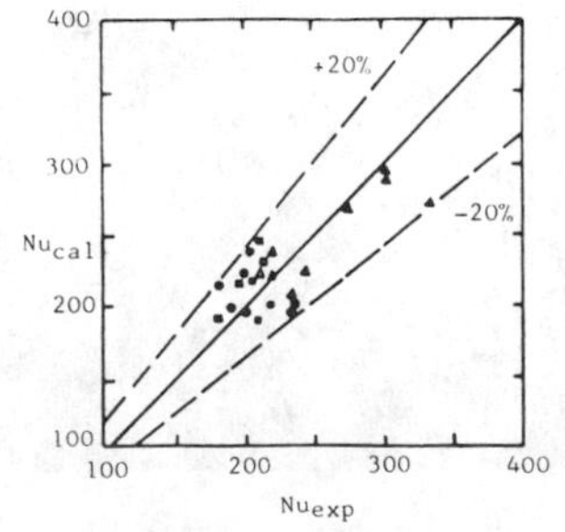

Fig.7 Eq.(24) is compared with experimental result

which limits the practical application.

The experimental results of heat transfer enhancement rate in startified flow are shown in Fig.6. It indicates that Nu/Nu_s is independent of dimensionless number B reflecting the influence of condensate depth. According to Eq.(23) Nu/Nu_s is determined only by dimensionless number A reflecting the ratio of surface tension to gravity and geometry of coil. For coil of D_w=0.5 mm, p=2.6 mm, in test condition A=20.9--21.2; for coil of D_w=0.3 mm, p=1.5 mm, A=22.8--24.0. Its heat transfer enhancement rate Nu/Nu_s=1.28--1.52, the scatter of data is mainly due to the influence of axial vapor velocity.

The comparison between Nusselt numbers which come from experiment and calculation by Eq.(24) is shown in Fig.7. When correction constant η=0.6, the deviation is within ± 20%.

table 1

	smooth tube	coil in whole tube	coil in whole tube	coil in fourth segment only
Tube diameter D mm	8.4	8.4	8.4	8.4
Wire diameter D_w mm	---	0.5	0.3	0.3
Coil pitch p mm	---	2.6	1.5	1.5
Inlet pressure P_o bar	1.56--2.00	1.59--2.03	1.71--1.96	1.65--2.11
Mass velocity G kg/m^2s	65--270	65--210	65--210	65--290
Exit quality X	0.03--0.06	0--0.09	0.01--0.07	0--0.09
mark	○	□	△	●

CONCLUSION

(1) For enhancing condensation heat transfer in stratified flow inserting coil is effective. Heat transfer coefficient is increased by about 40%, However, the absolute value of flow resistance increased slightly.
(2) Eq.(24) obtained on the basis of simple physical model and its analytical solution is compared with experimental result, the deviation is within ± 20%.

The experiment of this paper have been done during my visit in Okayama University of Japan. Express heartfelt thanks for the help of professor H. HONDA, lecture S. NOZU and H. KATAYAMA (MS).

REFERENCES

[1] Johnh.R and Arthure.B., Experimental Study of The Augmentation of Horizontal In-tube Condensation,ASHRAE Trans Vol 82-1 (1976), 925
[2] Luu.M.,Enhancement of Horizontal In-tube Condensation of R-113, ASHRAE Trans Vol 86-1 (1980), 300
[3] Tetsu FUJII, Hiroshi HONDA, Shigeru NOZU, Condensation of Fluorocarbon Refrigerants Inside A Horizontal Tube, REFERIGEARTION Vol. 55, No.627 (1980), 7
[4] Myers.J.A. and Rosson.H.F., ibib, 57-32,(1961),150
[5] T.FUJII,W.C.WANG,Sh.KOYAM,Y.SHIMIZU, Heat Transfer Enhancement For Gravity Controlled Condensation on A Horizontal Tube by Coiling Wires, Proc.ISHT, Beijing (1985)

Augmentation of Condensation Heat Transfer around Vertical Cooled Tubes Provided with Helical Wire Electrodes by Applying Nonuniform Electric Fields

A. YABE, T. TAKETANI, and K. KIKUCHI
Mechanical Engineering Laboratory
MITI, Tsukuba, Japan

Y. MORI
The University of Electro-Communications
Chofu, Tokyo, Japan

K. HIJIKATA
Tokyo Institute of Technology
Meguro, Tokyo, Japan

1. INTRODUCTION

Promotion of energy conservation has been achieved recently by three methods. The first method is to reduce energy consumption in power plants. The second is to effectively utilize waste heat by applying conventional technologies. The third one is to improve production processes not to use much energy. However, the further promotion of energy conservation by introduction of newly developed technologies continues to be of major concern. The utilization of large amount of waste hot water in factories is of particularly important interest. However, if an energy conservation equipment is made by conventional technologies, they would need prohibitive pay-back periods. One of main reasons for high cost equipments is the small temperature difference between heat transferring fluids in heat exchangers. In case of generating electricity utilizing waste hot water, the small temperature difference in the condenser and evaporator necessitated by the small temperature difference between the waste hot water and the cooling water, makes heat exchangers larger and more expensive. This problem is also important for making geothermal binary power plants, ocean thermal energy conversion, solar energy electric power conversion and large scale heat pump systems economically feasible. Therefore, to manufacture compact heat exchangers of high performance is one of important breakthroughs common to small temperature difference energy conversion problems.

Meanwhile, organic heating mediums such as R-113(Freon 113) and R-114 have come to be extensively utilized in energy conversion cycles of small temperature difference because of their low boiling temperature. In such cases, by utilizing advantages of the small electric conductivity of organic medium, heat transfer enhancement techniques by making use of electric field with negligible electric power consumption have come to be feasible for practical applications. Effects of electrostatic fields on condensation heat transfer have been studied a little, mainly by demonstration experiments. According to Asakawa[1], heat exchange between condensation of steam vapor inside a tube and forced convection of air outside the tube is enhanced by applying an electric field. In that case, the corona wind caused by the corona discharge[2] is supposed to play the main role of heat or mass transfer. Recently, Didkovsky-Bologa[3] investigated condensation enhancement of diethylether as the main medium on a vertical smooth plate by use of a plate electrode or slotted electrode. Then, the condensate spouted out like a spray to realize a very thin liquid film and consequently over ten times higher heat transfer coefficient could be attained. As for R-113, they could realize two times higher heat transfer coefficient in the same condition. However, the mechanism of these phenomena have not yet been made clear and a theoretical study has not been conducted either.

Fundamental studies on heat transfer augmentation of the filmwise condensation by applying a non-uniform electric field have been carried out by the authors from the electro-hydro-dynamical(EHD) viewpoint to realize high performance condensers for small temperature difference energy conversion cycles. In the previous report[4], liquid extraction phenomena by applying non-uniform electric fields to make the liquid column between the electrodes were studied and reported and, by a theoretical study, this EHD liquid extraction phenomenon was explained as a gas-liquid surface instability due to non-uniform electric fields. Then, by applying this phenomenon to condensation along a vertical cooled plate, the EHD augmentation phenomenon of removing some amount of condensate from the plate was observed by use of wire electrodes stretched horizontally and parallel to the plate. Consequently, condensation heat transfer coefficients was found to increase over 2.24 times higher. In this paper, the EHD augmentation method has been applied to condensation heat transfer outside of a vertical cooled tube to examine the augmentation effect as a next forward step to apply this method to practical condensers. As for a heat transfer surface, the vertical circular tube was selected since vertical type condensers need smaller area than horizontal type condensers and also since the EHD augmentation method is easily applicable to make the thin condensate film all along the vertical tube, especially at the bottom part which is usually covered by thick condensate film and has disadvantage against vertical type condensers. In experiments, combination of single smooth surface tube and R-113 was selected since they were proper for the basic research. Smooth surface tubes were used as they had several economical advantages.

2. EXPERIMENTAL APPARATUS

The experimental apparatus for condensation heat transfer is shown in Fig. 1(a). The experimental apparatus is composed of an evaporation part, heated by an electric sheath heater, and a condensation part cooled by water. R-113 is used as a working fluid and the condensation heat transfer surface is a smooth vertical circular tube made of brass. The condensation tube is 18 mm in outer diameter, 6 mm in inner diameter, 540 mm in length and its roughness is smaller than 10 μm. The vessel of the condensation part is made of acrylic resin. Before experiments were started, the condensation part and the evaporation part were

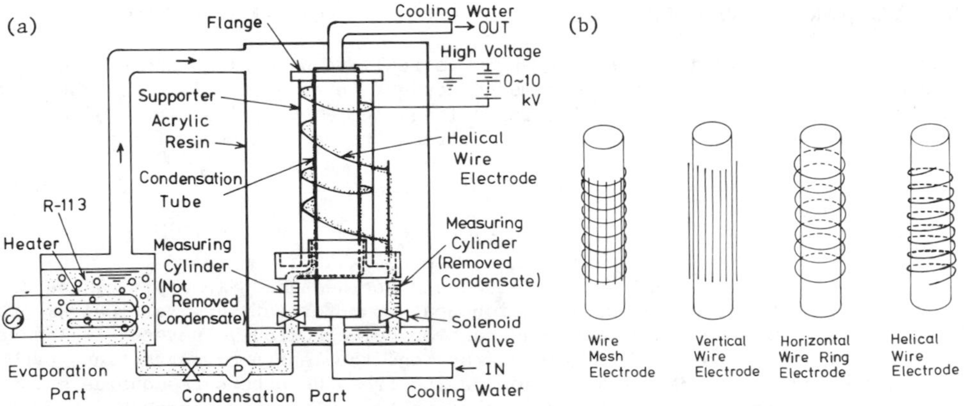

FIGURE 1. Experimental apparatus:(a) equipment for the removal of condensate and heat transfer,(b) electrodes

evacuated by a water-sealed vacuum pump to eject noncondensable gases from the vessel so that the pressure inside of the vessel was maintained at the vapor pressure of R-113. Experiments were conducted mainly at pressure of about 108 kPa (1.1atm) and vapor temperature of about 50°C. The condensate accumulated at the bottom of the condensation part was pumped back to the evaporation part. Three C-A electrically sheathed thermocouples of 0.5 mm in diameter were immersed along a radial direction in the brass tube at the point 150 mm upwards from the bottom. The local heat flux was measured by the temperature gradient in the tube. The flow rate and its temperatures of the cooling water at the inlet and the outlet of the condensation tube were measured. The temperature of the cooling water was about 20°C and its flow velocity was set at 5 m/s so as not to make a large contribution of its forced convection heat transfer coefficient to the total thermal resistance around the condensation tube. Copper was used for the material of wire electrodes since it was easy to machine.

As for the polarity of the electrodes, the EHD extraction phenomenon can be realized at both alternating and direct current conditions. In order to carry out research from a basic viewpoint, the tube was grounded and wire electrodes were applied by positive high potential. The electric high voltage power supply consisted of a transformer, regulating condensers and a small capacity high electric resistor for safety. In order to maintain a constant distance between the wire electrodes and the condensation tube surface, flanges made of electrically insulating synthetic resin were used at the top and the bottom of the tube and the wire electrodes were fixed to these flanges by use of supporting metal rods. The amount of the condensate and the amount of the removed condensate were measured by use of measuring cylinders provided with solenoid valve. The amount of the condensate on the tube surface at the wire electrode position was estimated by the Nusselt theory. The total amount of the condensate not removed from the tube was obtained by subtracting the estimated amount of the condensate from the amount of condensate measured by the cylinder. Initially, local condensation heat transfer coefficients were measured without electric fields. There was a good agreement within an error of 20% between those measured by the total amount of condensate and measured by the temperature difference between the inlet and the outlet of the cooling water. Therefore, the experimental data were mainly calculated from the local heat transfer coefficients.

3. EXPERIMENTS OF REMOVING CONDENSATE BY USE OF HELICAL WIRE ELECTRODES

Enhancement of condensation heat transfer is realized by making condensate film thinner, maximizing the amount of removed condensate so as to minimize the film thickness of the condensate falling along the tube. Therefore, electrode arrangements to maximize the amount of condensate removed were first investigated.

3.1. Selection of Helical Wire Electrodes

Several types of electrode arrangement could be designed in applying the EHD liquid extraction method to condensation heat transfer outside vertical smooth tubes. Concerning the shape and construction of electrode, several important factors such as the point of liquid extraction, the length of extraction part, the removal method of the extracted condensate from the tube were considered. In a preliminary experiment four types of electrodes such as wire mesh electrodes, vertical wire electrodes, horizontal wire ring electrodes and helical wire electrodes as shown in Fig. 1(b) were manufactured and experimentally investigated. As for the wire mesh electrodes, the coarse mesh and the fine mesh electrodes were both experimented on. In case of these mesh electrodes the EHD extraction phenomenon occurred at many points. But, the extracted condensate column fell mainly along the vertical wires. Therefore, the liquid film on the tube was thin at several parts of the tube but thick at several other parts. Consequently, a large amount of augmentation could not be realized by these wire mesh electrodes. As for the vertical wire electrodes, condensate was extracted to the vertical wire and fell downwards with a high speed by gravity, but the removal mechanism from the tube was not clear. In case of the horizontal wire ring electrodes, wire rings were fixed horizontally around the vertical tube and then the condensate liquid was extracted to the ring electrode forming a horizontal sheet. However, since gravity was not effective, the extracted condensate could not be removed from the tube so much. It was additionally noted that in case of helical wire electrodes, the extracted condensate could fall down as column or sheet with a high speed along the helical wire by gravity. Then the liquid film became thinner in the wide region between the successive helical wires over the whole circumference. Furthermore, the centrifugal force caused by the falling down motion of the condensate along the helical wire could make removal motion of the liquid from the tube more easy. Therefore the helical wire electrodes were selected as the basic and appropriate shape of the electrodes and used in the following experiments.

3.2. Removal Methods of the Condensate from the Heat Transfer Tube

When the helical wire electrode had a uniform helix angle, the condensate columns or sheets imparted between the helical wire and the condensation tube moved downward along the helical wire and could not be removed away from the heat transfer tube surface. However, when the helical wire was cut at several points and the tips were bent to keep a little distance away from the tube so as to make the radius of the helix a little larger at several cut ends of the helical wire as shown in Fig. 2(a), the condensate column or sheet was transferred to the helical wire at the tip of the helical wire and also was removed from the condensation tube by the effect of the centrifugal force. An initial observation of this phenomenon was made and the flow along a conduit separated from the condensation tube was seen. In this way, the condensate could be removed at any point by cutting the helical wire. Additionally, the mechanism of this EHD liquid removal phenomenon can be estimated as follows. When the tip of the helical wire was kept a little distance away from the tube, the electric Maxwell stress force to hold the liquid column decreased, because of the longer distance between the tube and electrode. Therefore, the centrifugal force caused by the falling down motion of the liquid along the helical wire became greater than the electric force and then the liquid removal from the tube could be realized.

3.3. Optimizing the Shape of the Helical Wire Electrodes

The parameters considered for optimizing the shape of helical wire electrode were; the helix angle, the cross section structure of the wire, the applied voltage and the distance between the tube and the electrode. In the following, these parameters were investigated.

<u>Optimizing the helix angle</u>. The helical wire electrode of constant helix angle have been researched at first by changing the angles. From experiments of changing the angle, small helix angle at the top part of the electrode was found to be important for extracting enough condensate. That is why the low flow velocity of the liquid along the helical wire was necessary for extracting the enough amount of the condensate from the tube surface nearly perpendicular to

(a)

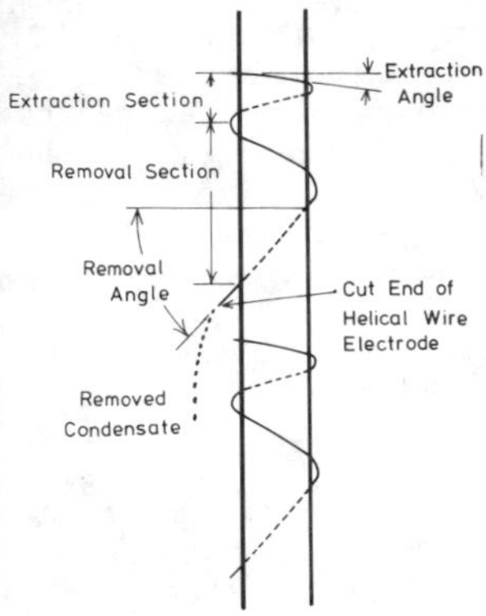

(b)

Extraction Angle	Removed Amount Ratio
>20°	<50%
20°	64%
10°	72%
4°~10°	85%

Unit Length of Wire: 450 mm
Distance between Electrode: 1.1 mm
Cross Section Structure: 2 Wires

(c)

Cross Section (Number of Wires)	Removal Angle	Removed Amount Ratio
2	30°	<70%
2	30°~35°	78%
4	30°~45°	95%
4	45°	90%

(d)

Cross Section Structure

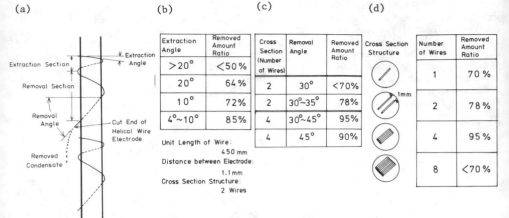

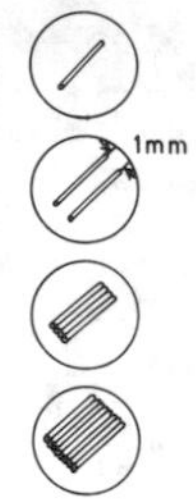

Number of Wires	Removed Amount Ratio
1	70%
2	78%
4	95%
8	<70%

FIGURE 2. Optimization of helical wire electrodes:(a) sketch of the helical wire,(b) extraction angle vs. removed amount ratio,(c) removal angle vs. removed amount ratio,(d) cross section structure vs. removed amount ratio

the helix. After removing the large amount of the condensate extracted from the condensation tube, the large helix angle was also important to make the falling velocity of the liquid large enough so as to make the centrifugal force larger than the electrical holding force. Therefore, the combination of the small helix angle of the extraction section and the large helix angle of the removal section were preferably selected as shown in Fig. 2(a) instead of choosing the constant helix angle. Furthermore, since one helix unit was composed of an extraction section and a removal section, the several units could be placed in line along the condensation tube. The optimum helix angle for the extraction section, that was called the extraction angle, existed between 4° and 10° as shown in Fig. 2(b). The optimum helix angle for the removal section, that was called the removal angle, existed between 30° and 45° as shown in Fig. 2(c). The helix length for both sections was fixed constant over about one pitch, since even if the total helix length was set over one pitch, the removal amount of the condensate obviously did not increase. On the other hand, in case of the helix length shorter than one pitch, there remained the thick film region in the circumference of the tube.

Optimizing the cross section structure of the helical wire. Two important factors were necessary to be explained here for the cross section structure of the helical wire to effectively realize the EHD extraction phenomenon. The first one was that the surface of the electrode was smooth and should not have sharp-pointed edges in order to prevent corona discharge. The second one was that non-uniform electric fields had to be produced to extract the condensate to the helical wire by the EHD extraction phenomenon. In experiments, the copper fine wire of 0.5 mm diameter was used for the wire electrode to prevent the corona discharge, and by changing the number of this fine wires composing the helical wire, the non-uniformity of the electric fields was also varied. Four types of the cross sectional structures as shown in Fig. 2(d) were investigated and the ratio of the amount of the removed condensate to that of the total one was taken as the evaluating function. As can be seen from the Fig. 2(d), the belt-shaped cross section composed of the four fine wires shows the maximum performance. Therefore, the belt-shaped cross section which was formed by four wires is the optimum construction for the above-mentioned type of the helical wire electrode. By broadening the width of the belt-shaped cross section, the degree of the non-uniform electric fields decreases. Therefore, the absolute amount of extracted condensate decreases. However, at the same time, the centrifugal force caused by the falling down motion of condensate becomes greater than the electric force holding condensate between the wire and the tube due to the decrease of the non-uniform electric fields. Consequently, the removal phenomenon of extracting liquid from the tube to the helical wire is realized more easily. These trends are considered to explain the existence of the optimum cross section. Furthermore, by use of the belt-shaped cross section composed of four fine wires or eight fine wires as shown in the lower picture of Fig. 2(d), the drag force worked on the falling liquid could be decreased when the centrifugal force is big enough. This was caused by the fact that the supporting rods for a single wire or two wires consisting of the separated two wires at a distance of 1 mm disturbed the flow of the falling liquid and would cause a larger drag force. The amount of removed condensate is chosen as the evaluating function in the above discussion. But if the strength of wire electrodes is taken as another evaluating function, that of the belt-shaped cross section would be evaluated highly, since they are stronger in the transformation and vibration condition.

Optimizing the applied voltage and the distance between electrodes. The relation between the applied voltage of the helical wire electrode and the removed amount ratio of condensate is shown in Fig. 3. As can be seen from the figure, the maximum removal ratio is realized at the voltage of about 4.2 kV. The following reasons are considered to explain the existence of the optimum voltage. As the voltage becomes larger in the lower voltage region, the volume of the extracted liquid would increase and the removal amount would become too larger. On the other hand, as the voltage becomes higher in the higher voltage region, the electric Maxwell stress holding liquid columns or sheets between the electrodes would increase and would become stronger than the centrifugal force that removes condensate from the tube to the helical wire electrode so that the removed amount would decrease.

As for the optimum distance between the electrodes, the cases of 1.1 mm and 1.6 mm distance were investigated. As the distance between the electrodes becomes larger, the applied voltage necessary for extracting the liquid becomes higher in almost linear proportion. However, the removed amount of condensate remains almost constant. Therefore, as for the distance between the electrodes, there

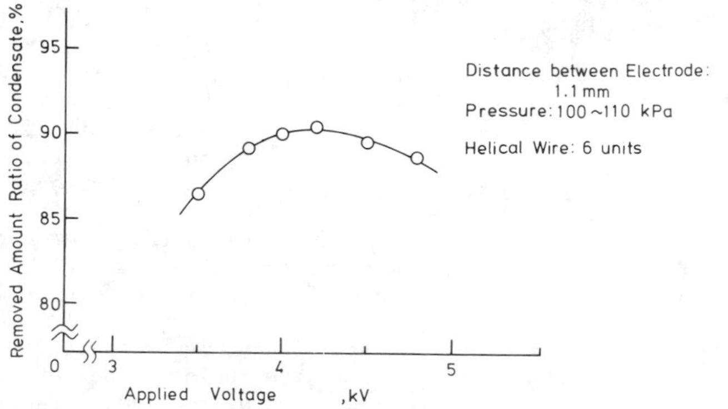

FIGURE 3. Relation between the applied voltage and the removed amount ratio of condensate

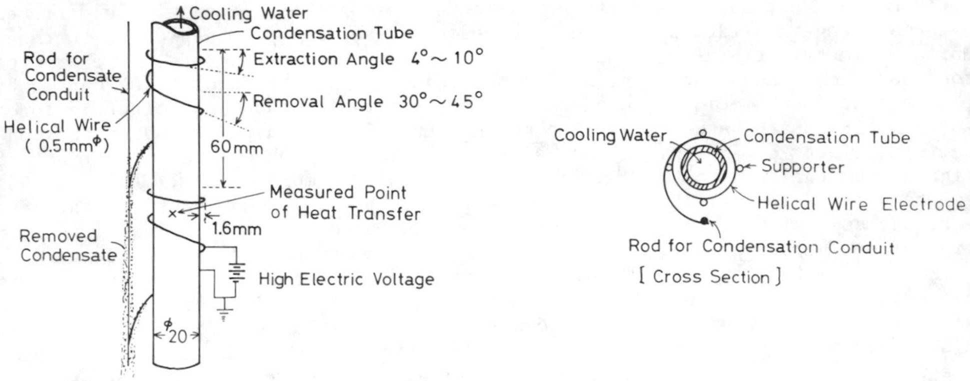

FIGURE 4. Helical wire electrodes optimized experimentally

exist no maximum in the region from 1.1 mm to 1.6 mm. In addition to these experiments, a double helix structure of wire electrode was also investigated. Since the removed amount increased but not drastically, this configuration was not preferable. From the experimental research described above, the construction and condition of the helical wire electrode were optimized for the helix angles, the cross section structure and the applied voltage. Under the condition of the optimized shape of the helical wire electrode shown in Fig. 4., the maximum removed amount ratio of condensate was achieved up to 94.6 %.

4. AUGMENTATION OF CONDENSATION HEAT TRANSFER BY USE OF HELICAL WIRE ELECTRODES

The helical wire electrodes obtained by the experimental optimization described in the previous section were used for experiments of heat transfer augmentation. At the voltage of 4.65 kV, heat transfer coefficients increased from 0.506 to 1.45 kW/m^2 K. This heat transfer coefficient was 2.8 times higher than that without the EHD effect and then the removed amount ratio of condensate was 94.6 %. The point where heat transfer measurements were performed was 10 mm downwards of the top of the helical wire between the extraction helical wire and the removal helical wire as shown in Fig. 4. In order to explain the heat transfer enhancement quantitatively, the thickness of the condensate film needs to be discussed. Even if the electric field was applied, the film Reynolds number was less than 100 and was small enough compared to the transition Reynolds number of 1400. Therefore, the flow would be laminar. In case of the laminar film flow, the film thickness was proportional to the cubic root of the flow rate, and accordingly the decrease of the condensate film thickness by removal of condensate from the tube surface could also be calculated. Consequently, for the condensate flow rate of 5.4 % of the original value which was obtained for the removed amount ratio of 94.6 %, the film thickness was considered to reduce by 62% of the original film thickness. This reduction of film thickness causes 2.6 times higher heat transfer coefficient than that without electric field. This result was in good agreement with the heat transfer enhancement of 2.8 times which was obtained in the experiments. Accordingly, the relation between the removal value of the condensate and the augmentation value of the heat transfer coefficient can be well explained by considering the reduction of condensate film thickness due to the electric field.

A typical applied voltage needed for the heat transfer enhancement was 4.4 kV, and the current was about 0.3 μA. Then the electric power consumption was about 1.3 mW. This consumed electric power was 0.0004 % of the total transferred heat and was negligibly small, even if the difference in the quality of heat and electricity is taken into account. This feature of negligibly small electric power consumption is one of the important merits of EHD heat transfer augmentation method. The removed amount of condensate by EHD effect is supposed to be dependent upon the total amount of condensate. According to our experimental results, in the wide range of the heat flux from 3.7 to 17 kW/m^2, the removed amount of 89 %-95 % could be achieved by use of helical wire electrodes.

5. CONCLUSION

In this paper, a study on augmentation of condensation heat transfer outside of a smooth vertical cooled tube by utilizing non-uniform electric field has been experimentally made. Investigation of the performance of removal of condensate from the cooled tube by utilizing the EHD liquid extraction phenomenon and of the EHD augmentation effects on condensation heat transfer has taken out the following conclusions:

(1). By applying a high electric voltage between the helical wire and the electrically grounded cooled tube, and additionally, by making use of the helical wire electrode set apart from the tube with adequately varying helix angle at the lower end of the helical wire, it has been proven experimentally that the condensate could be extracted and removed from the condensation tube.

(2). Concerning the shape and construction of the helical wire electrode, the helix angles of the extraction and removal sections, the detail structure of the helical wire, and the applied voltage were shown to have the optimum values for the EHD removal purpose. By optimizing these parameters, about 95 % of the total condensate could be removed from the tube surface.

(3). The relation between the removed condensate amount and heat transfer augmentation value is explained quantitatively by considering decrease of the film thickness due to condensate removal under condition of laminar falling film flow. Heat transfer coefficients 2.8 times larger than those without electric field were achieved in conjunction with the 95 % removed amount. It should be stressed that the electric power consumption is extremely negligible compared with transferred heat in condensation.

REFERENCES

1. Asakawa, Y., Promotion of Combustion Vaporization and Heat Transfer by Means of Application of Electric Field, IEEE-IAS Annual Meeting, Vol.1, pp.120-127, 1978.

2. Yabe, A., Mori, Y. and Hijikata, K., EHD Study of the Corona Wind between Wire and Plate Electrodes, AIAA J. Vol.16, no.4, pp.340-345, 1978.

3. Didkovsky, A. B. and Bologa, M. K., Vapour Film Condensation Heat Tranfer and Hydrodynamics under the Influence of an Electric Field, Int. J. Heat Mass Transfer, Vol.24, no.5, pp.811-819, 1982.

4. Yabe, A., Kikuchi, K., Tatetani, T., Mori, Y. and Hijikata,K , Augmentation of Condensation Heat Transfer by Applying Non-uniform Electric Fields, Proceedings of 7th International Heat Transfer Conference, Vol.5, pp.189-194, 1982.

HEAT EXCHANGER

Effect of Inlet Fluid Temperature Maldistributions on Both Sides and the Induced Flow Nonuniformities on the Performance of Crossflow Heat Exchanger

J. P. CHIOU
Mechanical Engineering
University of Detroit
Detroit, Michigan 48221, USA

INTRODUCTION

In the thermal analysis of heat exchanger, it is normally assumed that both the flow and inlet fluid temperature are uniformly distributed [1, 2, 3]*. The arithmatic mean value of the mass velocity distribution and the arithmatic mean value of the inlet fluid temperature distribution are then used in the analysis. Study on the effect of flow maldistribution alone, or of the inlet fluid temperature maldistribution alone had been reported [4 to 14]. Report on the study on the combined effect of inlet fluid temperature maldistributions and the temperature induced flow distortions of both fluid sides on the thermal performance of crossflow heat exchanger does not seem to be available. It is the subject discussed in this paper.

ANALYSIS

A single pass crossflow heat exchanger may be visualized as having a wall separating the two fluid streams flowing at right angles. The idealizations of this analysis are:

1. Steady state and steady flow conditions.
2. Both fluids pass in crossflow patterns on both sides of the exchanger wall. Each fluid flow is assumed to be unmixed in their respective side.
3. The fluid pressure distributions at both inlet and exit sections of the core are uniform. In other words, the fluid pressure drops across all the flow passages on either fluid side of the core are identical on that side.
4. The inlet fluid temperature is not uniformly distributed on either side. The distortions of the mass velocity distributions over the core face area on both sides are accounted for. The friction factor f is assumed to be directly proportional to the Reynolds Number, or $f \sim Re^{-n}$. The convection heat transfer coefficient between the fluid and the heat transfer surface area is governed by this equation, $Nu = \text{constant} \times Re^{\beta}\, Pr^{m}$.
5. Longitudinal heat conduction in either fluid is neglected.
6. Thermal resistances through the exchanger wall in the directions perpendicular and parallel to the fluid flows are neglected.
7. No heat generates within the exchanger. No phase change occurs in

---

*Numbers in brackets designate references at the end of the paper.

the fluid streams flowing through the exchanger. Heat transfer between the exchanger and the surrounding is neglected.

Based on these idealizations, the governing differential equations of heat transfer for the subject system are:

$$\frac{-C_x \Phi}{y_o z_o} \frac{\partial T'}{\partial x'} = (\eta' h' a')_x (T' - \theta') \quad (1); \qquad \frac{C_y \phi}{x_o z_o} \frac{\partial t'}{\partial y'} = (\eta' h' a')_y (\theta' - t') \quad (2)$$

$$(\eta' h' a')_x (T' - \theta') = (\eta' h' a')_y (\theta' - t') \quad (3)$$

The boundary conditions are:

$$T' (0,y',k) = T'_{in} (0,y',k) \quad (4); \qquad t' (x',0,k) = t'_{in} (x',0,k) \quad (5)$$

Since the thermo-physical properties of fluid can be expressed as functions of its temperature as shown in Table 5 for a specified temperature range, therefore, Eqs. (1) to (5) can be transformed into;

$$\frac{\partial T}{\partial x} = NTU_1 (T - \theta) = 0 \quad (6); \quad \frac{\partial t}{\partial y} + NTU_2 (t - \theta) = 0 \quad (7)$$

$$\text{and } C^* \Phi \, NTU_1 (T - \theta) + NTU_2 \, \phi \, (t - \theta) = 0 \quad (8)$$

The boundary conditions are:

T (0,y,k) = given inlet conditions; t (x,0,k) = given inlet conditions (9)

Since,

$$\Delta p = \frac{G^2}{2g_c \rho} f \frac{L^*}{r_h} \quad (10); \quad \text{and} \quad f \sim Re^{-n} \quad (11) \quad \text{then, } \Delta p \sim (\mu^n G^{2-n}/\rho) \quad (12)$$

or,

$$\Delta p_x \sim T'^{dn} G_x^{2-n} / T'^{b} \quad (13) \quad ; \quad \Delta p_y \sim t'^{d'n} G_y^{2-n} / t'^{b'} \quad (13a)$$

The local flow nonuniformity parameters Φ and ϕ can be expressed as,

$$\Phi = \frac{G_{x,\text{ a flow passage}}}{G_{x},\text{ all flow passages}} = \frac{T^*}{\Sigma T^*,\text{ all flow passages}} \quad (14)$$

$$\text{Here, } T^* = \Big(\sum_{\text{all subdivisions in a flow passages}} T'^{dn-b} \Big)^{\frac{1}{n-2}} \quad (15)$$

and

$$\phi = \frac{G_{y,\text{ a flow passage}}}{G_{y,\text{ all flow passage}}} = \frac{t^*}{\Sigma t^*,\text{ all flow passages}} \quad (16)$$

Here, $t^* = \left(\sum_{\text{all subdivisions in a flow passages}} t'^{d'n-b'} \right)^{\frac{1}{n-2}}$ (17)

The crossflow heat exchanger is divided into M x N x L subdivisions as shown in Fig. 1. For each x'y' or each kth plane, the arrangement of the subdivision is shown in Fig. 2. As shown previously [8, 9], the following three equations can be obtained by solving Eqs. (6) to (8) by using the Successive Substitution Technique,

$$T_{i,j,k} = \frac{1}{D}\left[\theta_{i,j-1,k} + \sigma\theta_{i,j-2,k} + \sigma^2\theta_{i,j-3,k} + \ldots + \sigma^r\theta_{i,j-r-1,k} + \ldots + \sigma^{j-2}\theta_{i,1,k}\right] + \sigma^{j-1}T_{i,1,k} \quad (18)$$

$$t_{i,j,k} = \frac{1}{F}\left[\theta_{i-1,j,k} + \delta\theta_{i-2,j,k} + \delta^2\theta_{i-3,j,k} + \ldots + \delta^r\theta_{i-r-1,j,k} + \ldots + \delta^{i-2}\theta_{1,j,k}\right] + \delta^{i-1}t_{1,j,k} \quad (19)$$

and

$$W\theta_{i,j,k} + P_1\,\theta_{i,j-1,k} + P_2\,\theta_{i-1,j,k} + P_3\left[\theta_{i,j-2,k} + \sigma\theta_{i,j-3,k} + \sigma^2\theta_{i,j-4,k} + \ldots + \sigma^{j-3}\theta_{i,1,k}\right] + P_4\left[\theta_{i-2,j,k} + \delta\theta_{i-3,j,k} + \delta^2\theta_{i-4,j,k} + \ldots + \delta^{i-3}\theta_{1,j,k}\right] = -P_1D\,\sigma^{j-1}T_{i,1,k} - P_2F\,\delta^{i-1}t_{1,j,k} \quad (20)$$

For the reason of simplicity, the two-dimensional array $\theta_{i,j,k}$ on any kth plane is transformed into an one-dimensional array $\theta^*_{I,k}$ on that kth plane as shown in Fig. 3. After this transformation, Eq. (20) may be written as,

$$W\theta^*_{I,k} + P_1\,\theta^*_{I-1,k} + P_2\,\theta^*_{I-N,k} + P_3\left[\theta^*_{I-2,k} + \sigma\theta^*_{I-3,k} + \sigma^2\theta^*_{I-4,k} + \ldots + \sigma^{j-3}\,\theta^*_{I-j+1,k}\right] + P_4\left[\theta^*_{I-2N,k} + \delta\theta^*_{I-3N,k} + \delta^2\,\theta^*_{I-4N,k} + \ldots + \delta^{i-3}\,\theta^*_{j,k}\right] = -P_1D\,\sigma^{j-1}\,T_{i,1,k} - P_2F\delta^{i-1}\,t_{1,j,k} \quad (21)$$

First of all, mass velocity distributions on both sides are assumed. For every kth plane, M x N simultaneous equations of the type of Eq. (21) are solved. After the θ^*'s are determined, they are transformed back to the two-dimensional array $\theta_{i,j,k}$ on the kth plane concerned. The temperature distributions of both fluids throughout the exchanger can then be calculated. The local flow nonuniformity parameters Φ and ϕ or the mass velocity distributions on both sides are determined by Eqs. (14) and (16). The assumed and the calculated mass velocity distributions are then compared. Iteration scheme may be needed to obtain a reasonable accurate mass velocity distribution. Finally the arithmetic average temperatures of both fluids at their respective inlet and exit sections and the exchanger heat transfer effectiveness ε, can then be determined. The deterioration factor for the heat transfer effectiveness of the heat exchanger τ can be expressed as;

$$\tau = \frac{\Delta\varepsilon}{\varepsilon} = \left[\varepsilon_{\text{uniform inlet fluid temperature and mass velocity distributions}} - \varepsilon_{\text{nonuniform inlet fluid temperature distribution and together with the distorted mass velocity distribution}} \right] \Big/ \left[\varepsilon_{\text{uniform inlet fluid temperature and mass velocity distributions}} \right] \quad (22)$$

RESULTS AND DISSCUSSIONS

In this study, air-to-air, water-to-water and water-to-air heat exchangers are considered. It had been observed that there are many possible models of fluid temperature distributions over the inlet sections of exchanger cores. Since the primary objective of this paper is to suggest a mathematical method of solving the subject problem, a presentation of the comprehensive compilation of data of the effects of various conceivable inlet fluid temperature distribution models on the thermal performances of heat exchangers is out of the scope of this paper. In this study, two typical models of inlet water temperature distributions and two typical models of inlet air temperature distributions as shown in Figs. 4 to 7 are used. The local dimensionless temperatures of these models are presented in Table 1 to Table 4.

Fig. 8 presents results of this study. The τ - NTU relationships of two different model arrangements for each exchanger are shown. The average inlet temperatures of the hot and cold sides used are 400°K and 300°K respectively. The relationships of the fluid properties and the temperature are shown in Table 5. The values of the exponents β, m and n used in this discussion are 0.8, 0.333 and 0.25 respectively. Thus $Nu \sim Re^{0.8}\ Pr^{0.333}$ and $f \sim Re^{-0.25}$ The method developed in this paper, of course, is not limited to the fluids or limited to the heat transfer and flow friction characteristics or to the temperature levels as mentioned above.

As shown in Fig. 8, the value of τ is positive; and it increases when NTU increases. This indicates that the nonuniformities of the inlet fluid temperature distributions on both sides and their induced mass velocity nonuniformities have detrimental effect on the thermal performance of the exchanger. The degree of this detrimental effect increases when NTU increases. It is also shown in Fig. 8 that the pattern of the maldistribution of the inlet fluid temperatures has significant effect on the degree of the deterioration of the exchanger performance.

CONCLUSIONS

It is found in this study that the deterioration effect on the exchanger performance can reach a significant level and it should not be ignored. The designer can apply the method presented in this paper to estimate the degree of deterioration for the unit concerned; and then to make necessary modification for his design.

Using the method developed in this paper, the effort of testing and modification of the prototype of the heat exchanger before its design is finalized can be reduced to the minimum. Savings of manpowers and resources in production of exchangers using this method as a tool are believed to be significant.

NOMENCLATURE

a'	= heat transfer area per unit core volume, m^2/m^3
a	= $a'x_o y_o z_o$ = total heat transfer area, m^2
C	= total fluid heat capacity rate, J/s °K
c_p, c'_p	= specific heat of fluid at constant pressure based on the average inlet or local fluid temperature, J/Kg.°K
G	= mass flow velocity, kg/m^2 s
g_c	= conversion constant, kg m/KN s^2
h, h'	= convection heat transfer coefficient based on the average inlet or the local fluid temperature, w/m^2 °K
k, k'	= thermal conductivity of the fluid based on the average inlet or local fluid temperature, w/m °K
L^*	= length of the flow path in each subdivision, m
Δp	= fluid pressure drop, Kpa
r_h	= hydraulic radius of the flow passage, m
T', t'	= fluid temperature in x' and y' directions respectively, °K
x'	= coordinate in x' direction, m
x_o	= overall dimension of exchanger core in x' direction, m
y'	= coordinate in y' direction, m
y_o	= overall dimension of exchanger core in y' direction, m
z'	= coordinate in z' direction (non-flow direction), m
z_o	= overall dimension of exchanger core in z' direction, m
θ'	= exchanger wall temperature, °K
μ, μ'	= absolute viscosity of the fluid based on average inlet or local fluid temperature, Kg/m s
ρ, ρ'	= fluid density based on average inlet or local fluid temperature Kg/m^3

Dimensionless Quantities

$B = NTU_1\ \Delta x$; $B = (1 - 0.5\ B')/B'$; b,b' = see Table 5; $C^* = C_x/C_y$

$D = (1 + 0.5\ B')/B'$; d,d' = see Table 5; $E' = NTU_2\ \Delta y$

$E = (1 - 0.5\ E')/E'$; e,e' = see Table 5; $F = (1 + 0.5\ E')/E'$

f = friction factor defined by $f \sim Re^{-n}$

L,M,N, = number of subdivisions in z', y' and x' directions of the core respectively

m = an exponent defined by $Nu \sim Re^{\beta}\ Pr^{m}$

NTU, NTU_x, NTU_y = number of transfer units for the entire heat exchanger and for the x' or y' flow side respectively (based on the average inlet). fluid temperature and without distortion of mass velocity distribution).

$NTU_1 = NTU_x \phi^{\beta-1}\ (T'/T'_{avg})^{d(m - \beta) + (1 - m)\ (e - q)}$

$NTU_2 = NTU_y \phi^{\beta-1}\ (t'/t'_{avg})^{d'(m - \beta) + (1 - m)\ (e' - q')}$

Nu = Nusselt Number; n = an exponent defined by $f \sim Re^{-n}$

Pr = Prandtl Number; $P_1 = 0.5\ NTU_1\ (1 + \sigma)/D$, $P_2 = 0.5\ NTU_2\ (1 + \delta)/(C^*\ F)$

$P_3 = \sigma P_1$; $P_4 = \delta P_2$; q,q' = see Table 5; Re = Reynolds Number; r = an integer

$S^* = (\eta h a)_x/(\eta h a)_y$; $T = (T' - t'_{in,avg})/(T'_{in,avg} - t'_{in,avg})$

$t = (t' - t'_{in,avg})/(T'_{in,avg} - t'_{in,avg})$

$W = - NTU_1 \ \Phi - NTU_2\phi/C^* + 0.5 \ NTU_1\Phi/D + 0.5 \ NTU_2\phi/(C^* F)$; $x = x'/x_o$;

$y = y'/y_o$; $z = z'/z_o$; β = an exponent defined by $Nu \sim Re^{\beta} \ Pr^m$; $\delta = E/F$

$\varepsilon = [C_x \ (T'_{in,avg} - T'_{exit,avg})]/[C_{min} \ (T'_{in,avg} - t'_{in,avg})]$

η, η' = overall surface efficiency based on average inlet or local fluid temperature

$\theta = (\theta' - t'_{in,avg})/(T'_{in,avg} - t'_{in,avg})$, three subscripted dimensionless temperature of the exchanger wall

θ^* = two subscripted dimensionless temperature of the exchanger wall; $\sigma = B/D$

τ = deterioration factor of the heat transfer effectiveness of the exchanger due to the nonuniformity of inlet air temperature distribution and the induced flow nonuniformity as defined by Eq. (22)

Φ, ϕ = local flow nonuniformity parameters for x' and y' - directions respectively. It is defined as the ratio of the actual local mass flow rate to the average mass flow rate if the flow is uniformly distributed as expressed by Eqs. (14) and (16).

SUBSCRIPTS

avg = arithmetic average temperature at either inlet or exit section on either fluid side

exit = exit section of exchanger

i,j,k = ith row, jth column on the kth x'y' plane

I = (i - 1)N + j ; in = inlet section of exchanger

min = minimum magnitude; x = for flow in x' direction

y = for flow in y' direction

REFERENCES

1. Kays, W. and London, A. L., Compact Heat Exchangers, 2nd ed., McGraw-Hill, 1964.

2. Schlünder, E. U., Heat Exchanger Design Handbook, Hemisphere, 1983.

3. Taborek, J., Hewitt, G. F. and Afgan, N., Heat Exchanger: Theory and Practice, Hemisphere, 1983.

4. Kutchey, J. A. and Jullien, H. L., "The Measured Influence of Flow Distribution on Regenerator Performance", SAE TRANS., Vol. 83, Paper No. 740164, 1974

5. Shah, R. K and London, A. L., "Influence of Brazing on Very Compact Heat Exchanger Surfaces", ASME Paper No. 71-HT-29, 1971.

6. Mondt, J. R., "Effects of Nonuniform Passages on Deepfold Heat Exchanger Performance", ASME Paper No. 76-WA/HT-32, 1976.

7. Muller, A. C., "An Inquiry of Selected Topics on Heat Exchanger Design", Donald Q. Kern Award Lecture, 16th National Heat Transfer Conference, August, 1976.

8. Chiou, J. P., "Thermal Performance Deterioration in Crossflow Heat Exchanger due to the Flow Nonuniformity", J. of Heat Transfer, ASME TRANS., Vol. 100, November, 1978, pp. 580-587.

9. Chiou, J. P., "Thermal Performance Deterioration in Crossflow Heat Exchanger due to Longitudinal Heat Conduction and Flow Nonuniformity on both Fluid Sides", ASME Paper No. 79-WA/HT-59, 1979.

10. Dobryakov, B. A., Gagarina, M. V., Efremov, A. S., Mokin, V. N., Morgulis-Yakushev, V. Y. and Rostovtsev, V. I., "The Calculation of the Heat Exchanger Equipment with Crossflow of the Heat Transfer Agents", International Chemical Engineering, Vol. 13, No. 1, 1973, pp. 81-84.

11. Bentwich, M., "Multistream Countercurrent Heat Exchangers", J. of Hea Transfer, ASME TRANS., Vol. 95, Nov., 1973, pp. 458-463.

12. Gaddis, E. S. and Schlünder, E. U., "Temperature Distribution and Hea Exchange in Multipass Shell-and-Tube Exchangers with Baffles", Heat Transfer Engineering, Vol. 1, No. 1, July, 1979, pp. 43-52.

13. Chiou, J. P., "The Effect of Nonuniformities of Inlet Temperatures of Both Fluids on the Thermal Performance of a Crossflow Heat Exchanger", ASME Paper No. 82-WA/HT-42, 1982.

14. Chiou, J. P., "Effect of Longitudinal Heat Conduction and Inlet Fluid Temperature Nonuniformities of Both Sides on the Thermal Performance of Cross-flow Heat Exchanger", ASME Paper No. 83-WA/HT-84, 1983.

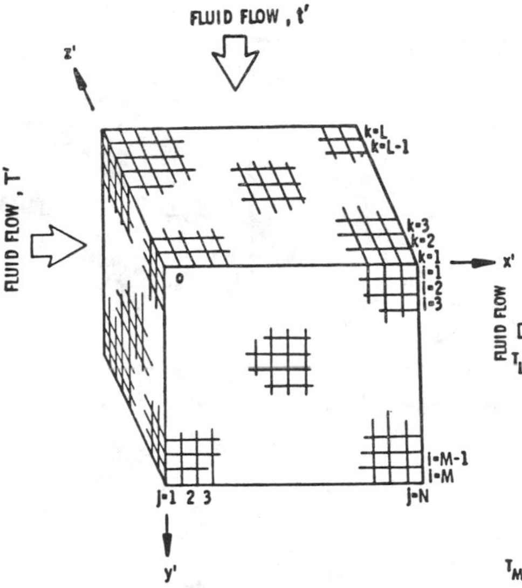

FIG. 1 SINGLE PASS CROSSFLOW HEAT EXCHANGER

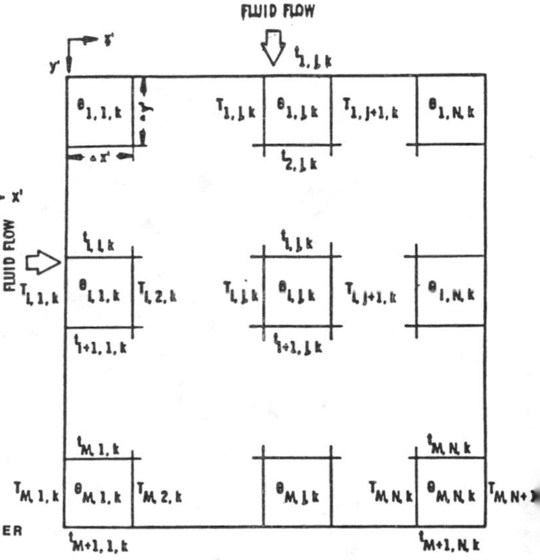

FIG. 2 SUBDIVISIONS ON Kth x'y' PLANE

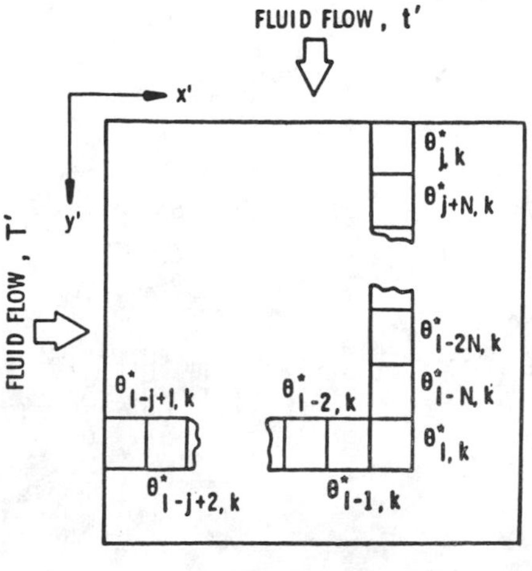

FIG. 3 RELATED NODES ON ANY Kth PLANE

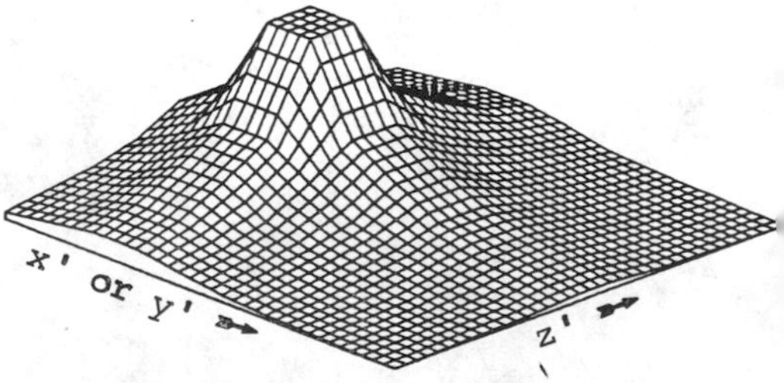

FIG. 4 INLET FLUID TEMPERATURE DISTRIBUTION MODEL A

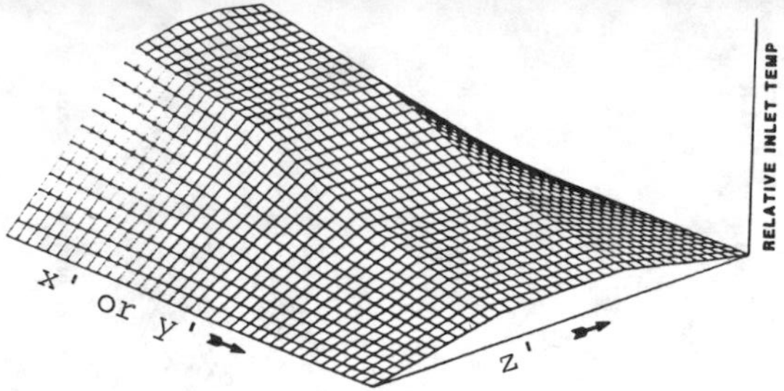

FIG 5 INLET FLUID TEMPERATURE DISTRIBUTION MODEL B

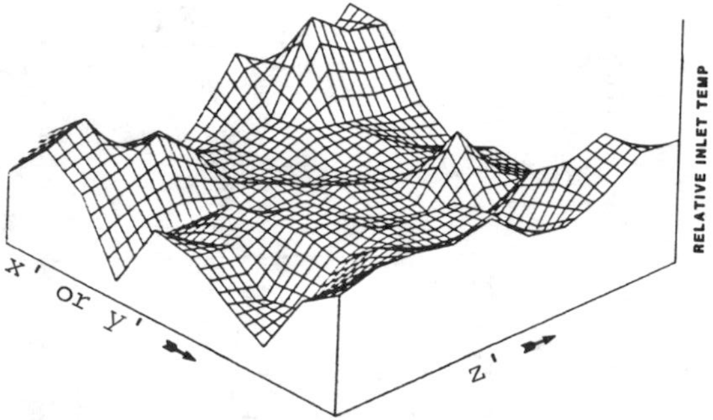

FIG. 6 INLET FLUID TEMPERATURE DISTRIBUTION MODEL C

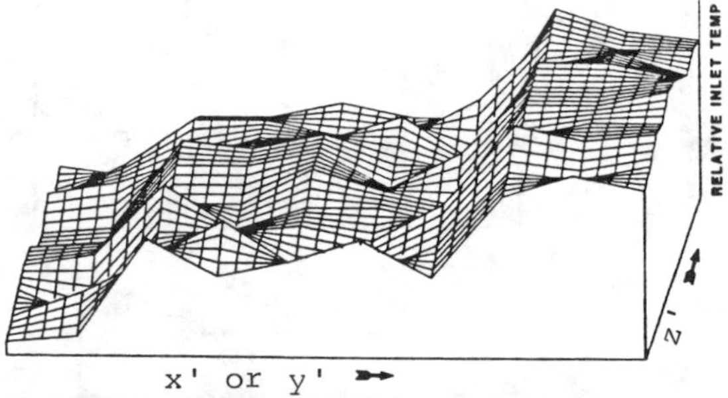

FIG. 7 INLET FLUID TEMPERATURE DISTRIBUTION MODEL D

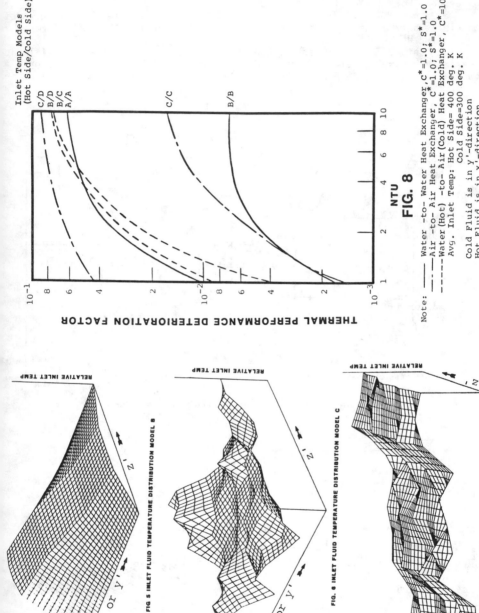

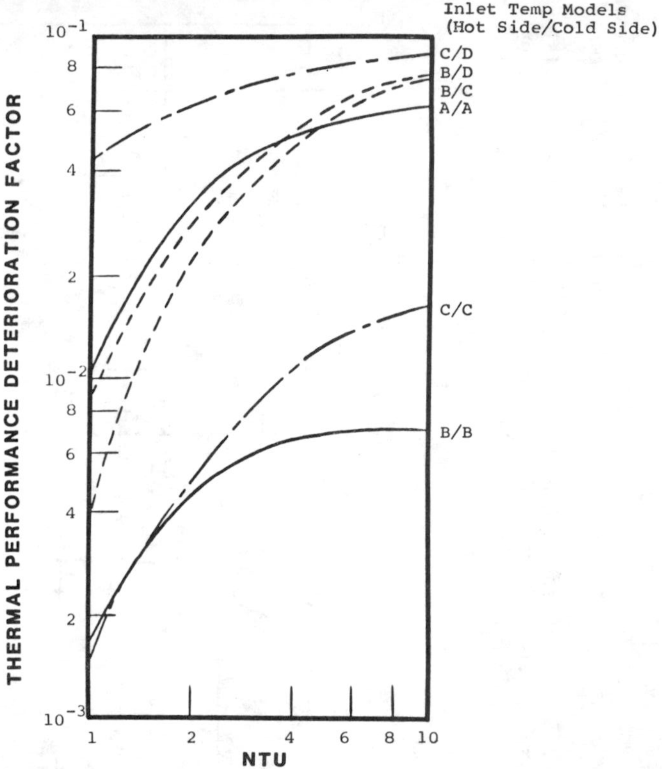

FIG. 8

Note: —— Water -to- Water Heat Exchanger, $C^*=1.0$; $S^*=1.0$
—·—Air -to- Air Heat Exchanger, $C^*=1.0$; $S^*=1.0$
-----Water (Hot) -to- Air (Cold) Heat Exchanger, $C^*=10.$; $S^*=5.0$
Avg. Inlet Temp: Hot Side= 400 deg. K
Cold Side=300 deg. K
Cold Fluid is in y'-direction
Hot Fluid is in x'-direction

TABLE 1A RELATIVE INLET FLUID TEMPERATURE T', DISTRIBUTION MODEL A (HOT SIDE)

i =	1	2	3	4	5	6	7	8	9	10
K=1,10	.300	.500	.700	.700	.500	.300	.200	.160	.120	.100
K=2, 9	.500	.700	1.040	1.040	.700	.500	.300	.200	.160	.120
K=3, 8	.700	1.040	1.700	1.700	1.040	.700	.500	.300	.200	.160
K=4, 7	1.040	1.700	2.500	2.500	1.700	1.040	.700	.500	.300	.200
K=5, 6	1.700	2.500	5.000	5.000	2.500	1.700	1.040	.700	.500	.300

TABLE 1B RELATIVE INLET FLUID TEMPERATURE t', DISTRIBUTION MODEL A (COLD SIDE)

i =	1	2	3	4	5	6	7	8	9	10
K=1,10	-.700	-.500	-.300	-.300	-.500	-.700	-.800	-.840	-.880	-.900
K=2, 9	-.500	-.300	.040	.040	-.300	-.500	-.700	-.800	-.840	-.880
K=3, 8	-.300	.040	.700	.700	.040	-.300	-.500	-.700	-.800	-.840
K=4, 7	.040	.700	1.500	1.500	.700	.040	-.300	-.500	-.700	-.800
K=5, 6	.700	1.500	4.000	4.000	1.500	.700	.040	-.300	-.500	-.700

TABLE 2A RELATIVE INLET FLUID TEMPERATURE T', DISTRIBUTION MODEL B (HOT SIDE)

i =	1	2	3	4	5	6	7	8	9	10
K=1,10	.886	.886	.886	.886	.826	.886	.826	.886	.886	.886
K=2, 9	.991	.984	.977	.970	.956	.942	.928	.921	.913	.906
K=3, 8	1.096	1.082	1.068	1.055	1.026	.938	.970	.956	.942	.928
K=4, 7	1.203	1.181	1.160	1.139	1.096	1.055	1.012	.991	.970	.949
K=5, 6	1.222	1.181	1.160	1.139	1.096	1.055	1.012	.991	.970	.949

TABLE 2B RELATIVE INLET FLUID TEMPERATURE t', DISTRIBUTION MODEL B (COLD SIDE)

i =	1	2	3	4	5	6	7	8	9	10
K=1,10	-.114	-.114	-.114	-.114	-.114	-.114	-.114	-.114	-.114	-.114
K=2, 9	-.009	-.016	-.023	-.030	-.044	-.058	-.072	-.079	-.087	-.094
K=3, 8	.096	.082	.068	.055	.026	-.002	-.030	-.044	-.058	-.072
K=4, 7	.203	.181	.160	.139	.096	.055	.012	-.009	-.030	-.051
K=5, 6	.222	.181	.160	.139	.096	.055	.012	-.009	-.030	-.051

TABLE 3A RELATIVE INLET FLUID TEMPERATURE T', DISTRIBUTION MODEL C (HOT SIDE)

i =	1	2	3	4	5	6	7	8	9	10
K= 1	1.072	1.580	1.254	.434	1.281	1.181	.844	.553	1.341	1.654
K= 2	1.341	1.756	1.341	.747	.939	1.107	.696	.791	1.313	1.788
K= 3	1.027	1.072	1.580	.733	1.181	1.254	1.281	1.163	1.580	1.756
K= 4	.739	1.181	1.163	.642	.733	1.254	1.281	1.341	1.378	1.656
K= 5	.193	.687	.584	.791	.584	.818	.730	1.299	1.254	1.756
K= 6	.618	.584	.604	.584	.799	.642	.765	1.281	1.379	1.341
K= 7	1.281	.733	.799	.604	.642	.604	.434	.799	.756	1.251
K= 8	.948	1.756	.696	.730	.584	1.281	.584	1.254	1.593	1.437
K= 9	.696	1.379	.584	.434	.553	.652	.678	.584	1.756	1.756
K=10	1.281	1.254	.553	.223	.443	.390	.443	.295	1.281	1.698

TABLE 3B RELATIVE INLET FLUID TEMPERATURE t', DISTRIBUTION MODEL C (COLD SIDE)

i =	1	2	3	4	5	6	7	8	9	10
K= 1	.072	.580	.254	-.566	.281	.181	-.156	-.447	.341	.654
K= 2	.341	.756	.341	-.253	-.061	.107	-.304	-.209	.313	.788
K= 3	.027	.072	.580	-.267	.181	.254	.281	.163	.580	.756
K= 4	-.261	.181	.163	-.358	-.267	.254	.281	.341	.378	.656
K= 5	-.807	-.313	-.416	-.209	-.416	-.182	-.270	.299	.254	.756
K= 6	-.382	-.416	-.396	-.416	-.201	-.358	-.235	.281	.379	.341
K= 7	.281	-.267	-.201	-.396	-.358	-.396	-.566	-.201	-.244	.251
K= 8	-.052	.756	-.304	-.270	-.416	.281	-.416	.254	.593	.437
K= 9	-.304	.379	-.416	-.566	-.447	-.348	-.322	-.416	.756	.756
K=10	.281	.254	-.447	-.777	-.557	-.610	-.557	-.705	.281	.698

TABLE 4A RELATIVE INLET FLUID TEMPERATURE T', DISTRIBUTION MODEL D (HOT SIDE)

i =	1	2	3	4	5	6	7	8	9	10
K= 1	.243	.289	1.094	.789	.911	1.094	.789	1.460	1.643	1.521
K= 2	.302	.500	.789	1.094	.729	.656	.911	1.521	1.460	1.583
K= 3	.363	.132	.850	.850	.911	.789	1.094	1.763	1.643	1.702
K= 4	.423	.515	1.094	.729	.911	.911	.789	1.521	1.643	1.643
K= 5	.423	.455	.789	.789	1.094	.911	.789	1.643	1.643	1.763
K= 6	.243	.289	.971	.971	1.094	.789	1.094	1.643	1.643	1.702
K= 7	.243	.289	1.032	.971	1.094	.911	.781	1.702	1.702	1.583
K= 8	.423	.526	.729	.911	.911	1.094	.729	1.521	1.643	1.460
K= 9	.243	.529	.911	.911	.789	.789	1.094	1.643	1.521	1.521
K=10	.363	.289	.789	.789	.911	.789	.789	1.702	1.521	1.460

TABLE 4B RELATIVE INLET FLUID TEMPERATURE t', DISTRIBUTION MODEL D (COLD SIDE)

i =	1	2	3	4	5	6	7	8	9	10
K= 1	-.757	-.711	.094	-.211	-.089	.094	-.211	.460	.643	.521
K= 2	-.698	-.500	-.211	.094	-.271	-.344	-.089	.521	.460	.583
K= 3	-.637	-.868	-.150	-.150	-.089	-.211	.094	.763	.643	.702
K= 4	-.577	-.485	.094	-.271	-.089	-.089	-.211	.521	.643	.643
K= 5	-.577	-.545	-.211	-.211	.094	-.089	-.211	.643	.643	.763
K= 6	-.757	-.711	-.029	-.029	.094	-.211	.094	.643	.643	.702
K= 7	-.757	-.711	.032	-.029	.094	-.089	-.219	.702	.702	.583
K= 8	-.577	-.474	-.271	-.089	-.089	.094	-.271	.521	.643	.460
K= 9	-.757	-.471	-.089	-.089	-.211	-.211	.094	.643	.521	.521
K=10	-.637	-.711	-.211	-.211	-.089	-.211	-.211	.702	.521	.460

TABLE 5

	b	d	e	q
AIR	-1.0	0.694	0.805	0.000
WATER	0.0	-5.000	0.403	0.053

NOTE: $\rho' \sim T'^{b}$; $\mu' \sim T'^{d}$; $k' \sim T'^{e}$; $c_p' \sim T'^{q}$

An Apparatus for Measuring the Heat Transfer Coefficients of Finned Heat Exchangers by Use of a Transient Method

K. FUJIKAKE, H. AOKI, and H. MITSUI
Toyota Central Research and Development Laboratories, Inc.
Nagakute, Aichi, Japan

ABSTRACT

For development of high performance finned heat exchangers, an experimental apparatus which can measure the heat transfer coefficients for various fin shapes is essential. This paper describes the transient testing technique by use of a modified curve matching method in comparison with the water steady-state method and the maximum slope method for measurement of heat transfer coefficients. For the experimental accuracy, fine wires for a heater, 20 fine thermocouples to measure the temperature-time history of the air flowing out of a fin core, and a desk-top computer for data-processing are used. The experimental error of the measured heat transfer coefficients was within ±3%.

INTRODUCTION

Along with a recent increase in heat loads to automobiles, the heat transfer performance of the heat exchangers is required to be further improved. On the other hand, light weight and small size are required for such exchangers. However, finned heat exchangers for automobiles already have high performance. For optimum design of fin dimensions, the fin performance should be accurately measured in relation to the fin dimensions(Fig. 1).

Generally, the performance of heat exchangers for automobiles is measured by utilizing a steady-state heat transfer process, in which hot water or steam is flowed in the tubes and air is flowed in the duct. Heat transfer coefficients are determined by the heat flux for the fluid or the air. Kays and London [1] measured the performances of various heat transfer surfaces. Disadvantages of

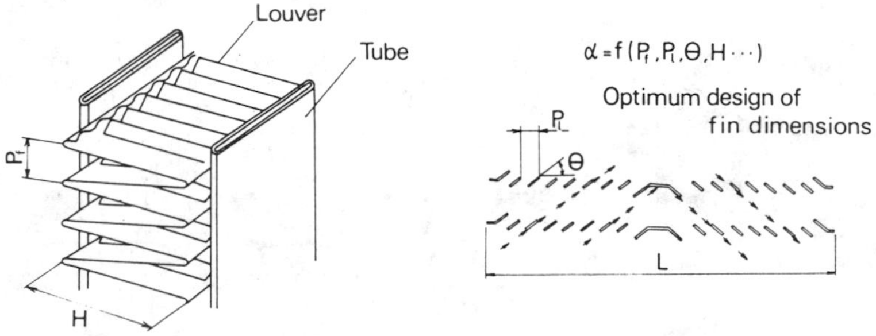

FIGURE 1. Optimum design of corrugated louvered fins

the method are as follows; long measurement time, difficulty in obtaining an accurate temperature difference between inlet and outlet, and heat transfer coefficients including the effects of fin efficiency, thermal contact resistance between fin and tube, and irregular fin spacing. So, the steady-state method is not suitable for optimum design of the fins.

On the other hand, according to a transient testing technique, heat transfer coefficients are determined with a temperature-time history of the air flowing out of a fin core. It needs a steplike temperature change of the air flowing into it and highly responsive temperature measurement of the air flowing out of it. But it does not need hot water or steam to heat a heat exchanger and the heat transfer coefficient is easily measured in a short time. Pucci, et al.[2] measured the perfomances of plate-fin type heat transfer surfaces with large flow length/hydraulic diameter ratio L/d_H by a maximum slope method [3, 4]; one of the transient testing techniques. Senshu et al.[5] proposed an apparatus to measure the heat transfer coefficients of finned tube surfaces for air conditioners by a curve matching method; one of the transient testing techniques.

Since louvered fins, widely used for recent automotive heat exchangers, are very thin and the ratio L/d_H is small, the conventional transient testing techniques can not be used for accurate measurement of the coefficients owing to rapid temperature change of the out-flowing air. The purpose of this paper is to develop an apparatus by which the coefficients of various heat transfer surfaces with low numbers of transfer units N_{tu} and low heat capacities such as automotive radiator fins can be accurately measured in an easy and simple way.

PRINCIPLE OF THE METHOD

Transient testing technique for measuring heat transfer coefficients is divided into two groups, curve matching and maximum slope methods. In the latter, the heat transfer coefficients are determined from the maximum slope of temperature-time history of the air flowing out of a test-core. Since this method has a disadvantage of low calculation accuracy for automotive heat transfer fins with low N_{tu} and disconnections of longitudinal heat conduction by louvers, it is not suitable for the purpose of this study. In contrast with it, heat transfer coefficients of low N_{tu}-fins are accurately measured by the curve matching method. Fig. 2 shows a conceptual diagram of the measurement. The time sequential change of air temperature flowing out a test-core after a step change of the in-flowing air temperature can be uniquely fixed by only the heat transfer coefficient, if fin dimensions, physical properties, and inlet air velocity are fixed. The change curve is obtained by two different processes, calculation executed for an assumed coefficient α_f and experimental measurement. Computation is repeated by modifying the value of the coefficient with the calculated temperature change coincides with the experimental one within a certain small limit.

The computation is executed under the following assumptions in order to facilitate a theoretical handling of heat transfer between air and a fin.

1. Physical properties of air and a fin are temperature independent;
2. A heat transfer coefficient is uniform in a fin core;
3. Thermal conductivity of air and a fin is infinite perpendicular to the flow direction;
4. The existence of louvers is ignored except for adiabatic conditions at louver edges.

Assumption 1. is granted by keeping the temperature change within several degrees. Assumption 2., 3., and 4. strictly differ from the actual conditions,

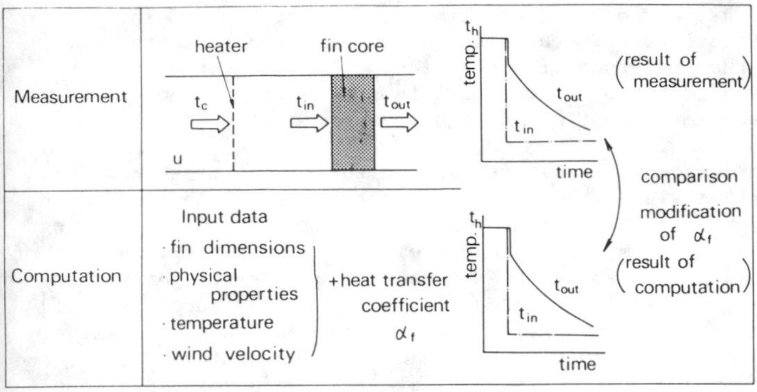

FIGURE 2. Conceptual diagram of transient testing technique

but they do not matter from a practical point of view. Under these assumptions, the basic equations for air and a fin are as follows:

$$\frac{\partial t_a}{\partial t}=-u\frac{\partial t_a}{\partial x}+\frac{\alpha_f A_f}{C_p \gamma_a V_c}(t_f-t_a) \tag{1}$$

$$\frac{\partial t_f}{\partial x}=\frac{\lambda_f}{C_f \gamma_f}\frac{\partial^2 t_f}{\partial x^2}+\frac{2\alpha_f}{C_f \gamma_f \delta_f}(t_a-t_f) \tag{2}$$

EXPERIMENTAL APPARATUS

An experimental apparatus must be designed to meet the above mentioned assumptions imposed by the idealization of the theoretical analysis. Fig. 3 show the apparatus, which consists of two parts; a wind tunnel and a data processing unit. The tunnel is a suction-type one having 2m length, 0.4×0.44m frontal dimensions, and 0.07×0.11m test section dimensions. The wind velocity is controled by rotating speed of a fan and a throttle at the exit and metered by a pitot tube and a displacement micro manometer. Except for the vicinity of duct walls, the flow uniformity in the test section was within 1%.

It is desired that the temperature of the air entering a fin core is made to change from high to low as quickly as possible. In the conventional method, the temperature change is simulated by extracting a heater out of the wind tunnel. But, since it takes several tens of milliseconds or more than to extract the heater and the temperature distribution of the air entering a fin core is disturbed by the extraction, the extraction-method can not be employed for automotive heat transfer fins. The measurement of heat transfer coefficients is finished with in 1s. In order to achieve a step change in air temperature by quickly switching off the heater, the heat is made of 110 fine tungsten wires, 15μm in diameter, spacing at intervals of 1mm, and the heat capacity is very small. The temperature change after switching off the heater is expressed by the time-lag of first order. The time constant is about 5ms at wind velocity u=1m/s. It is sufficiently small in comparison with a measuring period. The temperature-time hystory curves of the air entering a fin core are measured by preliminary experiments and expressed with the wind velocity u. They are employed in a computation instead of a step change. Fig. 4 shows the air temperature distributions in the rear of the heater-wires, X=4, 14, and 24mm. The

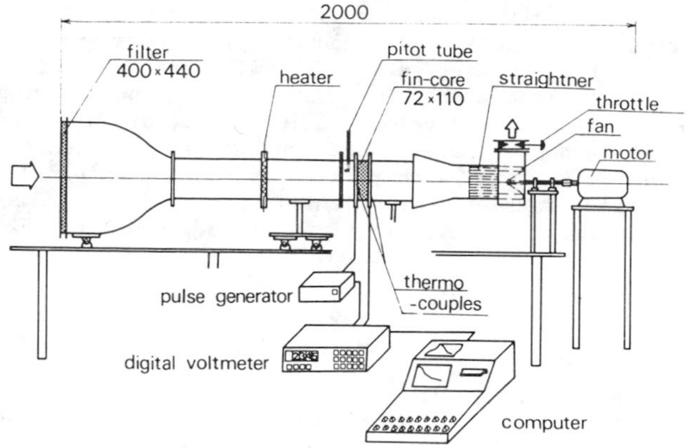

FIGURE 3. Schematic diagram of the apparatus

difference is decreased less than 0.2°C at X=24mm, whereas it is about 8°C at X=4mm. Since the heater is located at more than 60mm upstream of the test section, the distributions are sufficiently unified.

Since the measuring period as stated above is within 1s, the temperature change of the air flowing out of a fin core must be measured with high responsive thermocouples. Besides, the measurement at multipoints is necessary for high accuracy owing to the small temperature changes and spatial temperature distributions of the out-flowing air. Three thermocouples whose diameters are 10, 15, and 25μm were estimated about the response to measure the temperature changes, and the estimations about the number of thermocouples were also employed. Consequently, in the present apparatus, 20 fine chromel-alumel thermocouples of 15μm diameter are connected in series and installed in front and in the rear of a fin core.

The data-processing unit consists of a pulse generator, a digital voltmeter, and a desk-top computer. The pulse generator is motivated by a electromotive force change of the thermocouples located immediately upstream of a fin core. Then, it generates a trigger signal every 50ms for a fixed time to operate the digital voltmeter, which takes in and stores the electromotive forces of the ther-

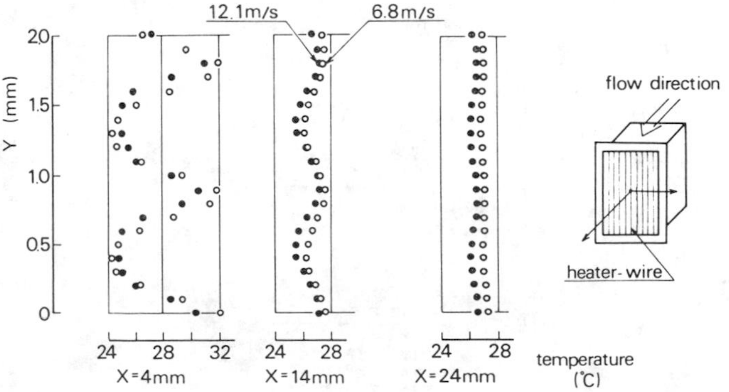

FIGURE 4. Temperature distributions of air in the rear of the heater-wires

mocouples in the rear of the core at every signal input. After taking in the data for a fixed time, they are transmitted to the computer. The above mentioned operations are controled by the computer except for switching off the heater. The computer calculates the temperature-time hystory of the out-flowing air under the input data; fin dimensions, wind velocity, initial conditions, and assumed heat transfer coefficient, and a comparison between the measured result and the calculated one is performed by it. The comparison is performed for 10 data at one measurement and the heat transfer coefficient is computed by automatically modifying it until the difference between both results becomes less than a certain small limit.

COMPARISON WITH OTHER METHODS

Fin cores to be measured consisted of corrugated louvered fins and partition sheets which were made of copper. Both thicknesses were 46 µm. They were framed at fixed fin-pitches p_f with a fixture and fixed with a very small amount of adhesive. The results measured by the present apparatus were compared with the ones measured by the steady-state method with hot water and the maximum slope method.

Comparison with the Steady-state Method

The measurements by the steady-state method were performed with test cores; 0.2×0.2m frontal dimensions and 0.032m air flow length. The test apparatus consisted of a suction-type duct in which air flow rate could be accurately controled and metered, and in which air temperature could be accurately measured. The temperature of hot water flowed in the water tubes was controled at fixed temperature within a deviation of ±0.1°C.

Since the heat transfer coefficients measured by the steady-state method were the apparent coefficients α_a which were affected by the fin efficiency, contact resistance, irregular fin spacings, etc., they can not be directly compared with the heat transfer coefficients α_f measured by the transient testing method. So, the heat transfer coefficients α_f were converted into apparent ones α'_a by calculation of the fin efficiency. Fig. 5 shows the conversion. The fin and water tubes were divided into square segments for use of the finite difference method, and the temperature distributions of them, the air temperature outflowing a test-core, and total heat transfer rate Q were calculated under input

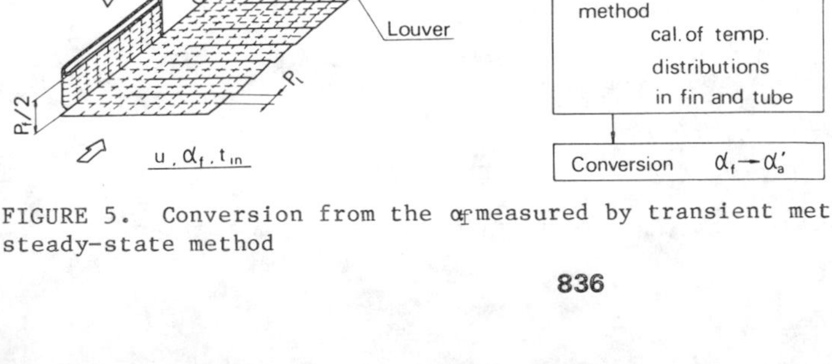

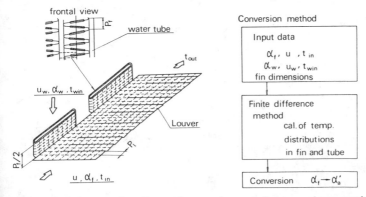

FIGURE 5. Conversion from the α_f measured by transient method into the α_a by steady-state method

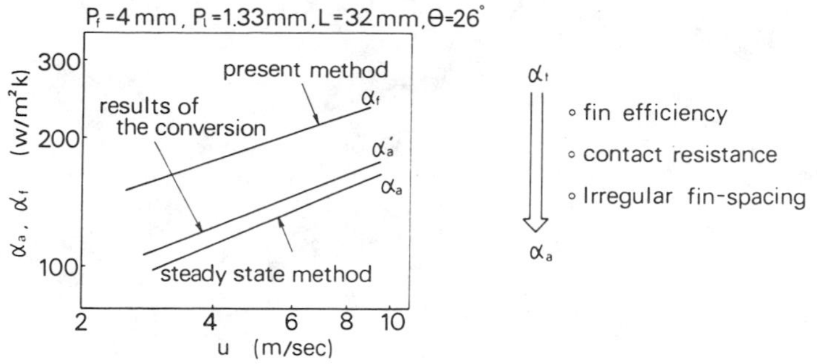

FIGURE 6. Comparison with results of the steady-state method

data of the heat transfer coefficient α_f measured by the transient method, air velocity u, heat transfer coefficient in water tube α_w, etc.. Finally, the measured heat transfer coefficient α_f was converted into the apparent one α'_a affected by the temperature distributions of the fin and tubes by eq. (4):

$$Q=\dot{m}c_p(t_{out}-t_{in}) \tag{3}$$

$$\frac{1}{\alpha'_a}=\frac{A_t(t_w-t_a)_{mean}}{Q}-\frac{A_t/A_w}{\alpha_w} \tag{4}$$

Fig. 6 shows the result of the conversion. The calculated coefficients α'_a were greater than the measured coefficients α_a by the steady-state method by 10% or so. These differences are mainly due to the contact resistance, irregular fin spacings of the core used for the steady-state method, and the difference of air-flow-pattern in the cores for both methods.

Comparison with the Maximum Slope Method

Numbers of transfer units N_{tu} are theoretically calculated as a function of the maximum slope of the exit air temperature curve with the longitudinal conduction parameter of fin Λ(Fig. 7). The heat transfer coefficients are determined from the value of N_{tu} obtained by the experimentally measured maximum slope. Since the inclination of the curve, N_{tu} vs. maximum slope, is steep in the low N_{tu}-range for fins with the low Λ, the measurement should be carefully carried out.

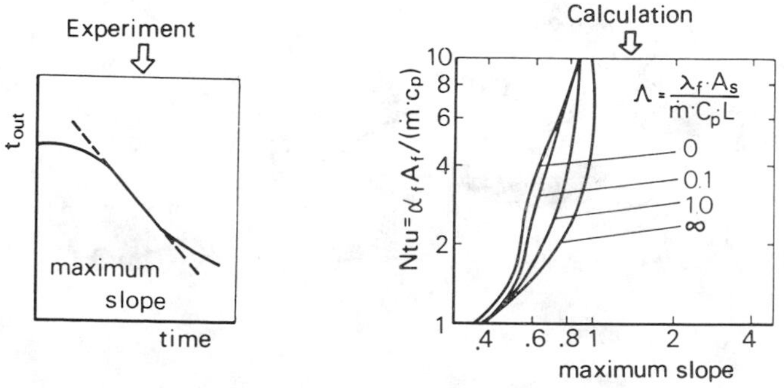

FIGURE 7. Maximum slope method

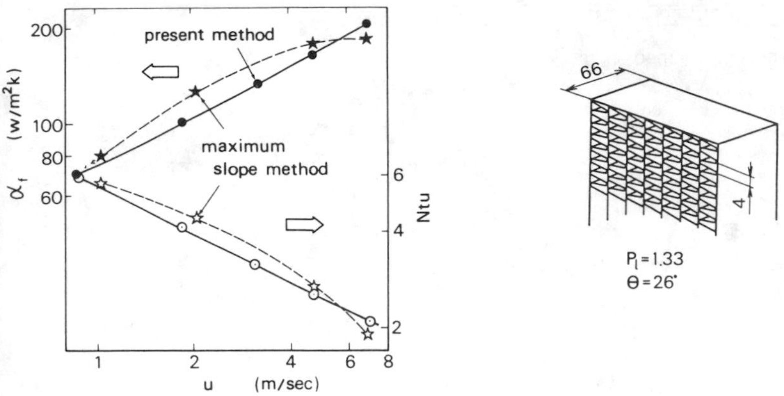

FIGURE 8. Comparison between the present method and the maximum slope method

Almost all fins for use as automotive heat exchangers fall into this range. So, fin cores with a long flow length L were constructed for the comparison. The temperature change of the air flowing out of the core was recorded by a high speed data recorder with the amplifying electromotive forces of thermocouples, and the value of maximum slope was obtained from it. Fig 8 shows the comparison of the heat transfer coefficients measured by the present method and the maximum slope method. Considering the relation between the measured coefficients and the wind velocity, it was experimentally confirmed that the present method is superior to the maximum slope method for estimation of fins used for automotive heat exchangers.

EXPERIMENTAL RESULTS

Heat transfer coefficients and pressure drops of 80 corrugated louvered fin cores were measured by the present apparatus. Fig. 9 shows the examples of the measured results which were estimated with colburn j-factor and Fanning friction factor. It is proved that fin performances affected by fin dimensions can be quantitatively estimated and this apparatus is very useful for determining the optimum dimensions of fin cores. The experimental error of measured heat transfer coefficients is within ±3%.

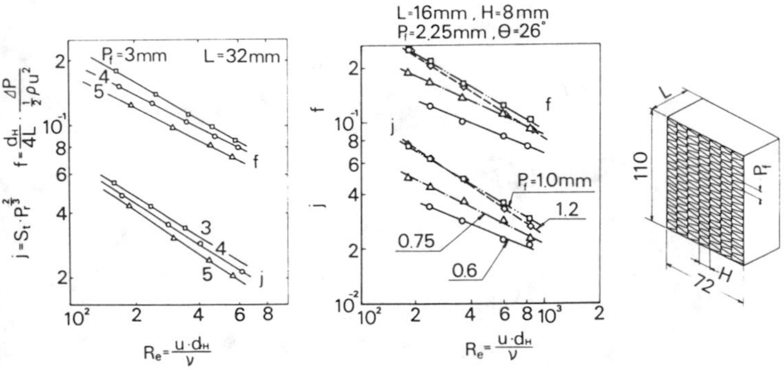

FIGURE 9. Heat transfer and friction data

CONCLUSION

It has been found that the heat transfer coefficients of fins for automotive heat exchangers were accurately measured with the apparatus by the transient testing technique, which consisted of a very fine wire heater, very fine 20 thermocouples connected in series, and a on-line data processing unit. Developments of high performance heat transfer fins can be efficiently performed by accurately measuring the heat transfer coefficients with this apparatus.

NOMENCLATURE

A_f=frontal area of air flow passage, m^2
A_s=cross-sectional area of fin, m^2
A_t=heat transfer surface area including tube, m^2
A_w=heat transfer surface area of tube, m^2
c_f=specific heat of fin, J/Kg°C
c_p=specific heat of air, J/Kg°C
d_H=hydraulic diameter, m
H =fin height, m
L =flow length through fin core, m
$\dot{m}$ =mas flow rate, Kg/s
P_f=fin pitch, m
P_l=louver length, m
ΔP=pressure drop of fin core, Kg/m^2
Vc=volume of air flow passage in core, m^3
α_a=apparent heat transfer coefficient measured by steady-state method, w/m^2K
α_f=heat transfer coefficient measured by transient method, w/m^2K
γ_a=density of air, Kg/m^3
γ_f=density of fin, Kg/m^3
δ_f=thickness of fin, m
Λ =longitudinal conduction parameter, $\lambda_f A_s/mc_p L$
λ_f=thermal conductivity of fin, w/mK
θ =louver-angle, degree
τ =time, sec

REFERENCES

1. Kays, W. M. and London, A. L., Compact Heat exchangers, McGraw-Hill, New York, 1964.

2. Pucci, P. F., Howard, C. P. and Piersall, C. H., Jr., The Single-Blow Transient Testing Technique for Compact Heat Exchanger Surfaces, ASME Paper N. 66-GT-93, 1966.

3. Locke, G. L., Heat Transfer and Flow Friction Characteristics of Porous Solids, TR No. 10, Department of Mechanical Engineering, Stanford University, Stanford, California, 1950

4. Kohlmayr, G. F., Extension of the Maximum Slope Method to Arbitrary Upstream Fluid Temperature Changes, Journal of Heat Transfer, Trans. ASME, Series C, Vol. 90, No. 1, pp. 130-134, 1968.

5. Senshu, T., Hatada, T. and Ishibane, K., Surface Heat Transfer Coefficient of Fins Utilized in Air-Cooled Heat Exchangers, Journal of Japan Soc. Refrigeration, Vol. 54. No. 615, pp. 11-17, 1979.

Unsteady Heat Transfer in a Packed Bed-Type Heat Regenerator by Use of a $CaO/Ca(OH)_2$ Reversible Thermochemical Reaction

MASANOBU HASATANI, HITOKI MATSUDA, and TAKASHI ISHIZU
Department of Chemical Engineering
Nagoya University
Nagoya 464 Japan

1. INTRODUCTION

For the requirements of an effective use of thermal energy, various kinds of heat exchangers are designed and developed in order to recover available heat from a lot of exhaust heat sources. As is well known, heat exchangers are, in general, classified broadly into two types of a direct and an indirect contact ones. Direc contact type heat exchangers like regenerative heat exchangers have a high heat exchange capacity and a high heat exchange rate, and are available when heat carriers are to be heated up to a relatively high temperature level. As to the fundamental problems of the conventional regenerative heat exchangers, heat transfer and heat exchange characteristics as we-l as optimum operation conditions like the time for a switchover from the heat-release to the heat-storage operations have recently been studied on the basis of the simulated results of the dispersion model[7] or the two-phase model[9] by taking an example of a packed bed-type regenerative heat exchanger.

In addition to the conventional regenerative heat exchangers by means of sensible heat of heat storing materials, new types of ones such as by means of the phase change of some organic compounds, and by means of reversible thermochemical reactions have been proposed. In these regenerative heat exchangers, there is a fund mental common advantage that the stored heat can be released at the nearly consta temperature where the phase change or the chemical reaction occurs. In view of this respect, the heat storage devices by means of latent heat have been investigated, and the heat-release as well as the heat-storage characteristics of the device are being studied by using such as naphtalene[5] as a heat storing materia However, in the heat storage devices by means of latent heat as well as sensible heat, the thermal insulation is required in order to reserve the stored heat for long period without leakage.

While, in the heat storage by means of thermochemical reactions, the thermal insu lation is not required, since in this heat storage system the thermal energy is converted into the chemical substances and is stored in the form of the chemical energy by the reaction. Furthermore, this heat storage system has a higher heat capacity per unit volume, compared with those by means of latent heat and sensibl heat, because the reaction heats, in general, are much larger than latent heats and sensible heats. Therefore, many kinds of catalytic or non-catalytic reaction have recently been studied for the purpose of applying these chemical reactions t the heat storage system[1,2,4,10].

In the heat storage by means of thermochemical reactions, the fundamental characteristics of the heat storage device are considered to be affected by the complicated behaviours caused by the chemical reaction. Nevertheless, any systematic

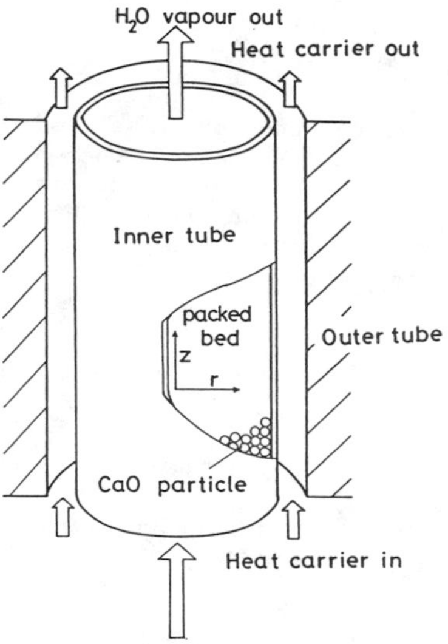

FIGURE 1. Conceptual drawing of the double-pipe regenerative heat exchanger by use of a $CaO/Ca(OH)_2$ reaction.

studies concerning these problems so far has not been undertaken. In view of the present situation, in this study, the following exo-/endothermic reversible reaction was employed as a reaction candidate and was applied to a packed bed-type regenerative heat exchanger.

$$CaO + H_2O \rightleftharpoons Ca(OH)_2 + 104.2 \text{ kJ/mol}$$

This reaction system may be available to a hi temperature heat exchange system, taking its high reaction-temperature(about 773 K) into consideration. Further, this reaction system has the following advantages; i) a high rate of the reaction, ii) a high reversibility of the reaction, iii) a relatively large quantit of the reaction heat, iv) non-toxic and non-corrosive reactant materials, v) reasonable material cost.

The heat-release characteristics of the proposed heat exchanger, which are considered as one of the most important key factors in discussing the thermal performance of the heat exchanger, was studied both experimentally and theoretically as a first step of a chain of research works.

2. DERIVATION OF FUNDAMENTAL EQUATIONS

As a heat exchange system, such a double-pipe regenerative heat exchanger as show conceptually in FIGURE 1 was dealt with in this study. In this heat exchanger, CaO particles in which the reaction heat had already been stored in the form of chemical energy, was packed in an inner pipe, and through which steam was passed to react with CaO particles. Then, the reaction heat was transferred directly to the reactant gas of steam and indirectly to a heat carrier flowing through an annular part.

At the derivation of the fundamental equations with respect to this heat exchange system, the following assumptions were made;

- i) CaO packed particles are so fine that intraparticle resistances in both heat and mass transfers in a CaO particle could be neglected.
- ii) The fluid flow through the bed is of the piston flow type, and the heat and mass transfer rate due to diffusion and mixing in the axial direction are negligible compared to the transfer rate due to the flow.
- iii) The values of physical properties such as void fraction, particle diamete and so on do not undergo any change in the course of reaction.

On the basis of these assumptions, fundamental differential equations concerning both heat and mass transfer in a packed bed were derived through Equations(1) to Equation(9) as follows;

(1) Heat Transfer Equations for a Packed Bed

$$c_p \rho_p (1-\varepsilon) \frac{\partial T_p}{\partial \theta} = \lambda_e^0 \left(\frac{\partial^2 T_p}{\partial r^2} + \frac{1}{r} \frac{\partial T_p}{\partial r} \right) - h_p a_p (T_p - T_g) + H \rho_p (1-\varepsilon) \frac{\partial X}{\partial \theta} \tag{1}$$

where T_p and T_g show the temperature of the particles bed and that of steam, respectively. c_p denotes the specific heat, ρ_p the density, H the reaction heat of the hydration of CaO with steam, X the conversion of the CaO reactant, θ the time, h_p the heat transfer coefficient between the CaO particles and the steam, a_p the specific surface area of a particle, ε the void fraction of the bed, and λ_e^0 the effective thermal conductivity of the CaO packed bed.
The initial and the boundary conditions are written as;

$$\theta=0,\ 0<r<R_0\ ;\ T_p=T_0,\ X=0 \tag{2}$$

$$\theta>0,\ r=0\ ;\ \partial T_p/\partial r=0 \tag{3}$$

$$\theta>0,\ r=R_0\ ;\ -\lambda_e^0(\partial T_p/\partial r)=h_w(T_{av}-T_w) \tag{4}$$

where, the subscript, 0 shows the initial condition, h_w is the heat transfer coefficient between the packed bed and the wall, T_w the wall temperature, and T_{av} is represented as $T_{av}=(T_p + T_g)/2$.

(2) Heat and Mass Transfer Equations for Steam

$$c_g \rho_g \varepsilon \frac{\partial T_g}{\partial \theta} + c_g \rho_g u_g \frac{\partial T_g}{\partial z} = \lambda_g \left(\frac{\partial^2 T_g}{\partial r^2} + \frac{1}{r} \frac{\partial T_g}{\partial r} \right) + h_p a_p (T_p - T_g) \tag{5}$$

$$\varepsilon \frac{\partial H_g}{\partial \theta} + u_g \frac{\partial H_g}{\partial z} = D_{er} \left(\frac{\partial^2 H_g}{\partial r^2} + \frac{1}{r} \frac{\partial H_g}{\partial r} \right) - \frac{\rho_p}{\rho_{N_2}} \frac{M_{H_2O}}{M_{CaO}} (1-\varepsilon) \frac{\partial X}{\partial \theta} \tag{6}$$

where u_g is the superficial velocity of steam, H_g the absolute humidity of steam, D_{er} the effective diffusivity of steam through the bed, and M the molecular weight. c_g and ρ_g denote the specific heat and the density of steam, respectively.
The initial and the boundary conditions are;

$$\theta=0,\ 0<r<R_0,\ z=0;\ T_g=T_0,\ H_g=H_{g0} \tag{7}$$

$$\theta>0,\ r=0,\ 0<z<L;\ \partial T_g/\partial r=0,\ \partial H_g/\partial r=0 \tag{8}$$

$$\theta>0,\ r=R_0,\ 0<z<L;\ -\lambda_g(\partial T_g/\partial r)=h_w(T_{av}-T_w),\ \partial H_g/\partial r=0 \tag{9}$$

where, L is the bed height, and λ_g is the thermal conductivity of steam.

As for a heat carrier flowing through an annular part, Equations(10)-(12) may hold.

(3) Heat Transfer Equations for a Heat Carrier

$$c_f \rho_f \frac{\partial T_f}{\partial \theta} + c_f \rho_f u_f \frac{\partial T_f}{\partial z} = \frac{2R_1}{R_2^2-R_1^2} h_{c1}(T_{w1}-T_f) - \frac{2R_2}{R_2^2-R_1^2} h_{c2}(T_f-T_{w2}) \tag{10}$$

where, T_f is the temperature of a heat carrier, c_f the specific heat, ρ_f the density, and u_f the flow rate of a heat carrier. R_1 is the outer diameter of the inner reaction tube, R_2 the inner diameter of the outer reaction tube. h_{c1} and h_{c2} represent the convective heat transfer coefficient between the wall and the heat carrier. The subscripts, 1 and 2 show the inner tube and the outer tube, respectively.
The initial and the boundary conditions are;

$\theta=0,\ 0<z<L;\ T_f=T_0$ (11

$\theta>0,\ 0<z<L,\ r=R_1;\ h_w(T_{av}-T_w)=h_{c1}(T_{w1}-T_f)+h_r(T_{w1}-T_{w2})$ (12

where, h_r is the radiative heat transfer coefficient between the walls.

The reaction rate of the hydration of CaO, $\partial X/\partial\theta$ which appears in the right-hand side of Equation(6) was determined by the present authors[8], and was expressed as follows;

(4) Reaction Rate Expression

$$\frac{\partial X}{\partial\theta}=\frac{51.1\ \exp(-1.096\mathrm{x}10^4/R_gT)}{1+5.25\mathrm{x}10^4\exp(-4.81\mathrm{x}10^4/R_gT)}(1-X)^{2/3}(P-P_e)\quad (13$$

where P denotes the partial pressure of steam, and P_e the pressure at the reacti equilibrium. R_g is the gas constant.

The unsteady temperature changes in this heat exchanger and the concentration changes of steam flowing through the packed bed can be obtained theoreticlally b solving these differential equations simultaneously, and the thermal performance of the heat exchanger can be estimated.

In the theoreticla analysis, the differential equations of Equations(1),(5),(6), (10) and (13) were transformed into difference equations with regard to the boun ary conditions of Equations(2)-(4), (7)-(9), (11)-(12). These simultaneous diff ence equations were calculated numerically by an explicit method with successive integration to obtain the unsteady, two-dimensional temperature and concentratio fields in a packed bed and the unsteady temperature rise of the heat carrier. I the numerical calculations of the fundamental equations, the value of λ_e^0 and h_p were determined by the Yagi-Kunii[11] and the Kunii-Suzuki[6] equations, respec- tively. The value of the reaction heat of the hydration of CaO with steam was employed as that reported by the Halstead and Moore[3].

3. EXPERIMENTS AND PROCEDURES

In view of the preliminary examination of the

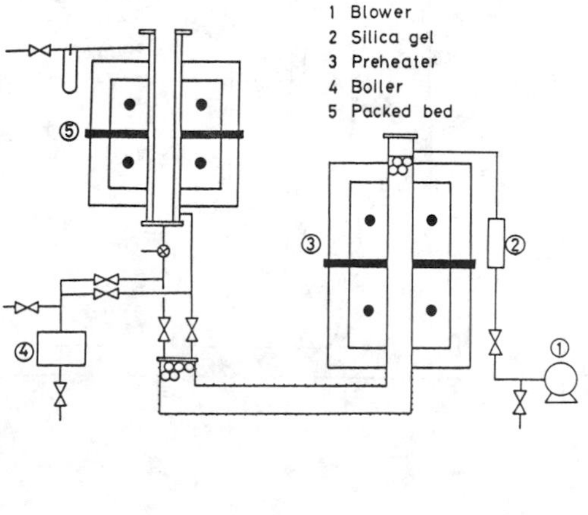

FIGURE 2. Schematic drawing of the experimental apparatus.

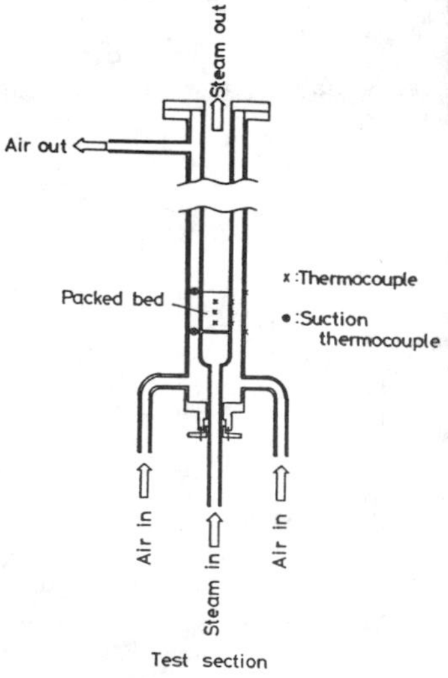

FIGURE 3. Detail of the reaction tube.

proposed regenerative heat exchanger which was discussed on the basis of the theoretical analysis in Section 2, this heat exchanger was expected more favorable to be used as steam superheaters. That is, it was found in the theoretical study that the amount of heat transferred to the heat carrier was at most 10% of the total amount of heat generated in the course of the hydration of CaO with steam, and that the rest of the reaction heat was used to heat up the packed bed and the steam itself. Therefore, in this study, the experiment concerning direct superheating of steam was carried out by taking this point into consideration.

The outline of the regenerative heat exchanger employed in this study is schematically shown in FIGURE 2. The apparatus consists mainly of the packed bed reactor, the steam generator and the air preheater. As shown in FIGURE 3, the reaction tube is composed of a stainless steel(SUS 304) double pipe and is 625 mm in height. The inner and the outer diameter of the inner tube is 56.5 mm and 60.5 mm, respectively. A 20 mesh stainless wire net is attached to the bottom of the tube in order that the CaO particles bed may be formed successfully within the tube. The inner tube is connected to the outer tube(96.7 mm in inner, 101.6 mm in outer diameter) with flanges, and the inner and the outer tubes can easily be separated each other.

The CaO particles specimen could be charged or taken out by drawing the inner tube out from the reactor. Steam or preheated air could be introduced into the inner tube and the annular part by switching the control valves. Thermocouples of 0.3 mm in diameter were set in each position within the CaO particles bed, at the wall and in the annular part so that the temperature changes of each position could be measured continuously in the course of reaction. Suction thermocouples were used to measure the fluid temperature.

The CaO particles of which the average diameter was 1.0 mm, were prepared by calcining the limestone from Okayama(purity over 99%) at 1253.2 K. Then, the CaO particles were packed in the inner tube(the bed height; 7-15cm), and the CaO particles bed was heated up to a constant desired temperature(373.2-573.2 K) by introducing the preheated air into both the packed bed and the annular part and by controlling the electric power to the electric furnace.

After each position of the reaction tube attained the constant desired temperature the steam at the constant partial pressure

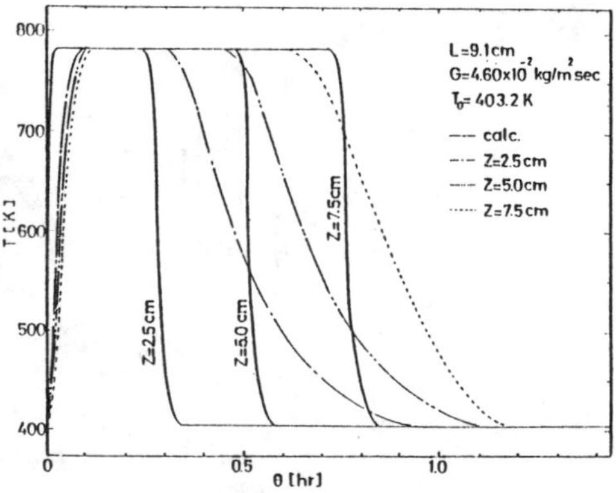

FIGURE 4. A typical example of the temperature changes in the packed bed.

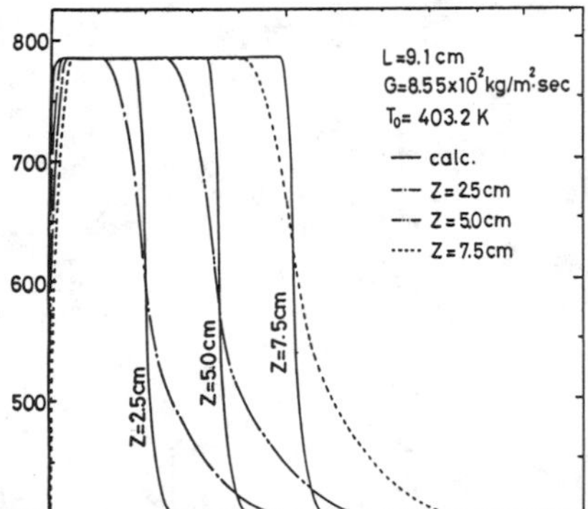

FIGURE 5. A typical example of the temperature changes in the packed bed.

was introduced into the CaO packed bed, and the heat-release experiment was init: ated. The temperature changes at each measuring point were measured continuousl by an automatic high speed temperature recorder(Yokogawa-Hokushin Electric Co., Ltd. Model 3088-23E) until the exothermic process of the hydration of CaO with steam was completed. The partial pressure of steam was determined by the weight change of the adsorbent accompanied by the adsorption of steam.

4. RESULTS AND DISCUSSION

FIGURE 4,5,6 show typical examples of the unsteady temperature changes in the ax: direction of the CaO packed bed during the exothermic process of the hydration o CaO with steam. These experimental data were obtained by packing CaO particles : 9.1 cm height and by introducing pure steam into the bed. FIGURE 4 and 5 are the results of the initial bed temperature of 403.2 K, and the mass flow rate of ste G are 4.6×10^{-2} and 8.55×10^{-2} kg/m^2 s for the case of FIGURE 4 and FIGURE 5, resp tively. FIGURE 6 is the result of the initial bed temperature of 503.2 K and the steam flow rate of 8.55×10^{-2} kg/m^2 s. The solid lines in these figures show the calculated results.

It is seen from these figures that the temperatures throughout the CaO packed be attain about 783.2 K which corresponds to the temperature at the pseudo-equilibr of $CaO/Ca(OH)_2$ reaction under an atmospheric pressure of 101.3 kPa, immediately after the steam was introduced into the bed, and that the bed temperature is kept almost constant at 783.2 K for a relatively long period during the reaction. The bed temperature dropped to return to the initial bed temperature successively fr the inlet part to the outlet part of the bed with the progress of the reaction.

Through FIGURE 4 to 6, there seems a fairly good agreement between the experiment and the calculated results on the whole in terms of the temperature profiles. Then, it is considered that the present theoretical analysis may be valid, although several assumptions are included in the fundamental differential equations. In these figures, however, at the beginning and at the end of the reaction, a slight differences of the temperatures are observed between in the experimental and in the calculated results. Among several assump-

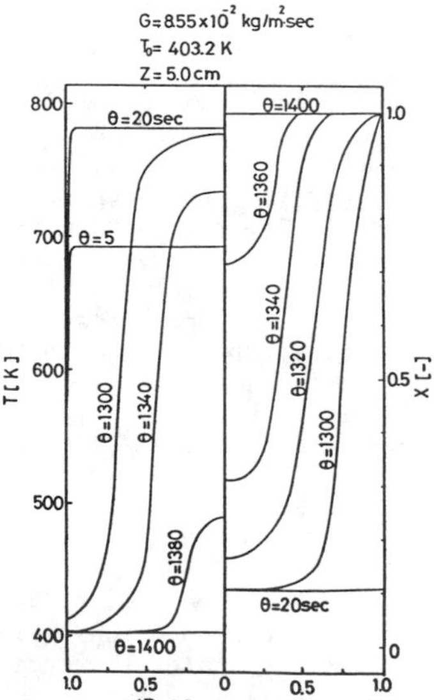

FIGURE 7. Radial distributions of both the temperature and the conversion of reaction in the bed

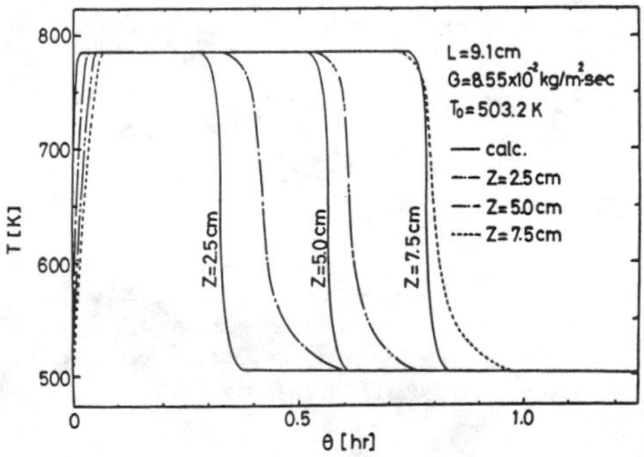

FIGURE 6. A typical example of the temperature changes in the packed bed.

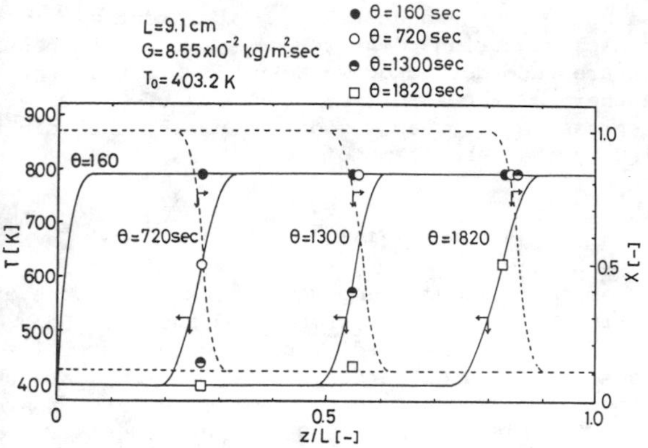

FIGURE 8. Axial distributions of both the temperature and the conversion of reaction in the bed.

tions these differences may be caused by neglecting the sensible heat of the reaction tube.

In order to discuss the packed bed behaviours during the heat-release process, typical examples of the distribution of the temperature and the conversion of reaction within the bed are shown in FIGURES 7 and 8, the experimental conditions corresponding to those in FIGURE 5. FIGURE 7 ahows the theoretical results of the radial distribution of both the bed temperature and the conversion of reaction at the bed height of 5.0 cm. It is seen from this figure that the bed temperature at this position rises up to about 783.2 K in an extremely short time of about 20 seconds after the initiation of the reaction, and that this reaction temperature is kept constant throughout the cross section of the bed except for the vicinity of the wall until 1300 seconds. The reaction rate is extremely slow at the pseudo-equilibrium state and the reaction is apparently at a stop at this bed height throughout the cross section of the bed, the conversion of reaction, X remaining at X≒0.11 in this case. Subsequently after 1300 seconds, the bed temperature begins to drop from the wall side, because the generated heat is taken out of the bed into the steam or the heat carrier. Then, the reaction is completed gradually from the low-temperature side near the tube wall to the center.

FIGURE 8 shows the axial distributions of both the bed temperature and the conversion of reaction at the center. The keys in the figure signify the measured experimental data of the axial bed temperature. The solid lines and the broken lines represent the calculated results of the temperature and the conversion of reaction, respectively. It is seen from the figure that at 160 seconds after the initiation of reaction the exothermic reaction of the hydration of CaO with steam is apparently stopped in almost all the region within the bed with the conversion remained at X≒0.11 under the present operational condition. While, after 720 seconds, the bed temepratur dropped to the initial value within the bed height of 2.5 cm, when the reaction was completed there. Above 2.5 cm of the bed height, the reaction still remained apparently stopped indicating X≒0.11, and so the bed temperature was kept at about 783.2 K.

5. CONCLUSION

A packed bed-type regenerative heat exchanger by use of the $CaO/Ca(OH)_2$ reaction system was designed and operated. The experimental results were compared with the calculated ones in terms of the unseady temperature changes of the packed bed and the steam, and the heat-release characteristics of this heat exchanger were discussed.

As a result, it was found that the CaO packed bed attained the pseudo-equilibrium reaction temperature of about 783.2 K throughout the entire part of the bed, immediately after the steam was introduced. In this heat-release period, the reaction was found to proceed successively from the inlet to the outlet of the bed with both the temperature and the conversion of reaction kept almost constant through the cross section of the bed and with the temperature of steam at the outlet of bed kept at the almost constant high temperature independently of the initial bed temperature and the steam flow rate, until the reaction was completed. The experimental results agreed fairly well with the calculated ones in terms of the temperature profiles in the packed bed.

6. REFERENCES

1. Fahim, M. A. and Ford, J, D., Energy Storage Using the BaO_2-BaO Reaction Cycle, Chem. Eng. Journal, vol. 27, pp. 21-28, 1983.

2. Fujii, I., Heat Energy Storage Based on Reversible Chemical Reactions, Part -Principle and the Base Experiments, Kuuki-Chowa Eisei-Kogaku Ronbunshu, No. pp. 21-27, 1977.

3. Halstead, P. E. and Moore, A. E., The Thermal Dissociation of Calcium Hydroxide, J. Chem., Soc., pp. 3873-3875, 1957.

4. Kanzawa, A. and Arai, Y., Thermal Energy Storage by the Chemical Reaction (Augmentation of Heat Transfer and Thermal Decomposition in the $CaO/Ca(OH)_2$ Powder), Solar Energy, vol. 27, pp. 289-294, 1981.

5. Katayama, K., Saito, A., Utaka, Y., Saito, A., Matsui, H.,Maekawa, H. and Saifullah, A. Z. A., Heat Transfer Characteristics of the Latent Heat Therma Energy Storage Capsule, Solar Energy, vol. 27, pp. 91-97, 1981.

6. Kunii, D. and Suzuki, M., Particle-To-Fluid Heat and Mass Transfer in Packed Beds of Fine Particles, Int. J. Heat Mass Transfer vol. 10, pp. 845-852, 196

7. Levenspiel, O., Design of Long Heat Regenerator by use of the Dispersion Model, Chem. Eng. Sci., vol. 38, No. 12, pp. 2035-2045, 1983.

8. Matsuda, H., Ishizu, T., Lee, S. K. and Hasatani, M., Kinetic Studies on $Ca(OH)_2/CaO$ Reversible Thermochemical Reaction for Thermal Energy Storage, Kagaku Kogaku Ronbunshu, unpublished.

9. Spiga, G. and Spiga, M., Analytical Simulation in Heat Storage Systems, Wärm und Stoffübertrangung, vol. 16, pp. 191-198, 1982.

10. Wentworth, W. E. and Chen, E., Simple Thermal Decomposition Reactions for Storage of Solar Thermal Energy, Solar Energy, vol. 18, pp. 205-214, 1976.

11. Yagi, S. and Kunii, D., Studies on Effective Thermal Conductivities in Packe Beds, AIChE J., vol. 3, No. 3, pp. 373-381, 1957.

A Theoretical and Experimental Investigation of Heat Transfer Performance of Rotary Regenerative Heat Exchanger

ZEPEI REN and SIYONG WANG
Tsinghua University
Beijing, PRC

ABSTRACT

In this paper, a theoretical analysis and calculation of the transient and periodic steady-state temperature distribution of the rotor material and fluids in rotary regenerator are reported. The transient temperature distribution of the fluids was measured in a test regenerator, and the results obtained by the numerical method agree with those of the experimental measurement. The effect of longitudinal heat conduction of solid matrix, temperature dependence of the thermal properties of the fluids and the effect of the heat storage of fluids on regenerator performance are discussed and presented graphically. In addition, the mathematical model presented in this paper is completely general, it provides a theoretical basis for the optimum designing of this type of regenerator.

NOMENCLATURE

- A convective heat transfer area
- A_s or A_f cross sectional area
- c specific heat
- G mass flow rate
- h heat transfer coefficient
- k thermal conductivity
- l total length of the solid
- t temperature
- U circumference of channels
- u average velocity of fluids in channels
- x flow length coordinate measured from hot gas inlet

Greek symbols

- η effectiveness of the regenerator
- ρ density
- τ time

Subscripts

- c cold gas
- f fluids, hot gas or cold gas
- h hot gas
- p the end of a period
- s solid matrix
- c,in cold gas inlet
- h,in hot gas inlet
- f,n fluids in normal state

Dimensionless parameters

$$\Theta = \frac{t - t_{c,in}}{t_{h,in} - t_{c,in}}$$ dimensionless temperature

$$X = \frac{x}{l}$$ dimensionless coordinate

$$T = \frac{\tau}{\tau_p}$$ dimensionless time

$$\rho_f^* = \frac{\rho_f}{\rho_{f,n}}$$ density ratio of fluids

$$C_f^* = \frac{C_f}{C_{f,n}}$$ specific heat ratio of fluids

$Bi^* = \frac{hlA}{K_s A_s}$ Biot number

$St^* = \frac{hA}{G_f C_{f,n} A_f}$ Stanton number

$Fo^* = \frac{hU}{\rho_s c_s A_s}$ ratio between convective heat transfer and capacity of solid matrix

$Ho = \frac{u \tau_p}{l}$ time ratio between the length of a period and residence time of fluids

INTRODUCTION

In view of the widespread application of the periodic-flow rotary heat exchanger for regeneration of exhaust gas thermal energy from the furnace of metallurgical industry, there has been much interest in evaluating the performance of this type of heat exchanger. In the analysis of the performance of the periodic-flow heat exchangers, the differential equations and boundary conditions considering the most general case are sufficiently complex. Thus the results of work in the published literatures[1]-[8] are only obtained with certain simplifying assumptions. In fact, some assumptions are unacceptable in practice. In this paper, therefore a mathematical model, in which the effect of the longitudinal heat conduction of solid matrix, temperature dependence of the thermal properties of fluids and the effect of heat storage of fluids and reversals are considered, is set up to describe regenerator performance. Using the finite difference method, a numerical solution of the mathematical model is obtained over a wide range of possible application in heat recovery. Fig.1 is a representation of such an axial flow rotary regenerator, in which the fluids pass in counterflow direction. The rotor is made of glass ceramic CERCOR 9475 produced by Corning Co., it is a solid matrix with many small size parallel square channels. There are two stages of operating heat regenerator, namely, transient and steady-state periodic stages. The transient temperature distribution of fluids was measured in a test regenerator for proving the numerical calculation. The influence of the various factors on the rotary regenerator effectiveness are discussed. It is found that the effectiveness of the regenerator mainly depends on a good matching between the convective heat transfer rate, the heat capacity of fluids and the heat storage capacity of the solid matrix. The results shown graphically here are useful for designing of this type of the regenerator.

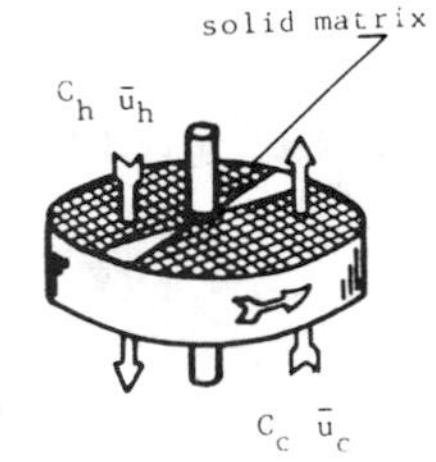

Fig.1 Illustrative rotor arrangement for axial flow

MATHEMATICAL MODEL

The cross section of the small size channels is $1.1 \times 1.1 mm^2$, and the thickness of its wall is 0.3mm. If the fluids, hot gas or cold gas, passing through every channel is uniform, the attention can be focused on one of the channels with half wall thickness. For fluids passing through a channel, as indicated schematically in Fig.2, the energy equation in dimensionless form can be expressed as follows:

$$\frac{1}{St^* \, Ho} \frac{\partial}{\partial T} \left(\rho_f^* C_f^* \Theta_f + \rho_f^* C_f^* \frac{t_{c,in}}{t_{h,in} - t_{c,in}}\right) +$$

$$+\frac{1}{St^*}\frac{\partial}{\partial X}\left(C_f^*\Theta_f+\rho_f^*\frac{t_{c,in}}{t_{h,in}-t_{c,in}}\right)=\Theta_s-\Theta_f \qquad (1)$$

heating period cooling period

Fig.2 Schematic representation of mathematical model

For the wall of a channel, the heat conduction equation in dimensionless form can be written the following:

$$\frac{1}{Bi^*}\frac{\partial^2\Theta_s}{\partial X^2}-\frac{1}{Fo^*}\frac{\partial\Theta_s}{\partial T}=\Theta_s-\Theta_f \qquad (2)$$

In the heating period, the initial conditions are as follows:

for fluid $\Theta_{f,h}(X, 0) = \Theta_{f,h,o}$ (starting or transient) (3a)

$\Theta_{f,h}(X, 0) = \Theta_{f,c}(1-X, 1)$ (periodic steady-state) (3b)

and for wall $\Theta_{s,h}(X, 0) = \Theta_{s,h,o}$ (starting or transient) (4a)

$\Theta_{s,h}(X, 0) = \Theta_{s,c}(1-X, 1)$ (periodic steady-state) (4b)

The boundary conditions will be

$$\Theta_{f,h}(0, T) = 1 \qquad (5)$$

$$\frac{\partial\Theta_{s,h}(0, T)}{\partial X}=0, \qquad \frac{\partial\Theta_{s,h}(1, T)}{\partial X}=0 \qquad (6)$$

In cooling period, corresponding initial and boundary conditions can be written in a similar way to those in heating, only subscript h is replaced by c except for starting or transient behavior, because in any case the solid and fluid temperature at the begining of cooling are the same as those at the end of the previous heating period.

In comparison with published models in literatures, some improvements on consideration following are made in the above mentioned model:

1. Temperature dependence of the thermal properties of fluids which are assumed to be constant in all published literatures.

2. The effect of heat conduction of solid matrix which is neglected in most of literatures. In literatures[1][8], the effect of heat conduction is analysed, but the effect of temperature dependence of thermal properties is ignored.

3. The effect of heat storage of fluids in channels, which is considered[4][7] under conditions of constant thermal properties of fluids and zero thermal conductivity of solid matrix.

4. The effect of reversals associating with the residual fluid in the regenerator.

EXPERIMENTAL STUDY

The test apparatus scheme is shown in Fig.3. 100mm in diameter and 150mm in length solid matrix is fixed in a test section. The hot gas and the cold gas alternatively pass through the channels in solid matrix in counterflow directions. Thus the heat transfer process in rotary regenerator was realized in the test regenerator.

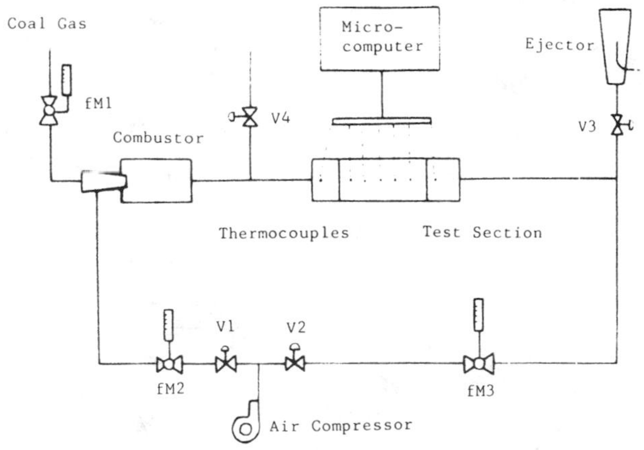

Fig.3 Schematic drawing of test apparatus

In heating period, coal gas and air passed through flowmeters fM1 and fM2, mixed and burnt in a combustor to produce hot gas. Then the hot gas flowed through the test section and was exhausted by an ejector. At the end of heating, a reversal occurred at which valves V1, V3 were closed and the flow of hot gas was shut off. Then, during the cooling period, the valves V2, V4 were opened and air of lower temperature flowed through the test section in the opposite direction of flow to that in the heating period. The experiments were carried out under various mass flowing rates of hot gas and cold gas. The gas temperatures were measured by seven thermocuples which were equidistently located along a channel in solid matrix, and the measured results were recorded by the microcomputer. At the same time, the pressure drop in the test section was measured.

RESULTS AND DISCUSSION

Mathematical model has been solved by use of numerical method. As shown in

Fig.4, the results obtained by the numerical method agree with those of the experimental measuements. Therefore, the mathematical model mentioned above is acceptable.

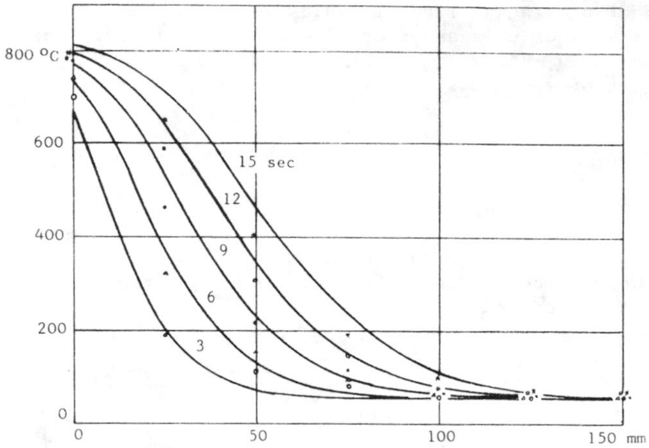

Fig.4 Numerical and experimental results of hot gas temperature

After obtaining the temperature distribution of the solid matrix and fluids, the effectiveness of rotary regenerator can be evaluated which is shown in Fig.5, in the range of 10 St^* 1000 and 1 Fo^* 1000.

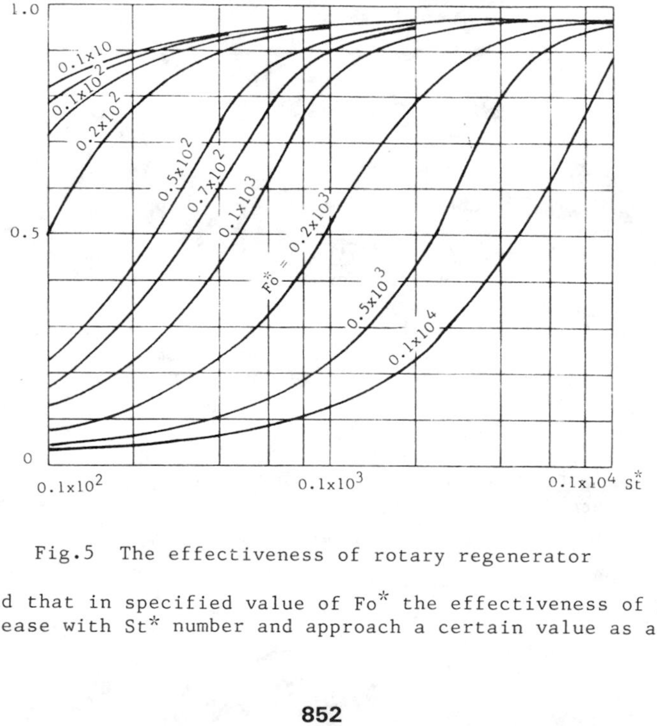

Fig.5 The effectiveness of rotary regenerator

It is found that in specified value of Fo^* the effectiveness of rotary regenerator increase with St^* number and approach a certain value as a limit.

St^* number is representative of the ratio between the convective heat transfer rate and heat capacity rate of the fluids. A small St^* number means that the heat capacity rate is greater than convective heat transfer rate. Because of insufficient convective heat transfer rate, the heat can not be absorbed from or released to fluids by the solid matrix. As a result, the effectiveness of rotary regenerator brings down. Of course, a very large St^* number is not necessary, because it can not obviously improve the effectiveness of regenerator.

In specified value of St^*, the effectiveness of regenerator decrease with increasing Fo^* number. Fo^* number is representative of the ratio between the convective heat transfer in a period and heat capacity of solid matrix. A large Fo^* number means that the heat storage capacity of solid matrix is not enough to store the heat from fluid. In sum, it is found that the effectiveness of the regenerator mainly depends on a good matching between the convective heat transfer rate, heat capacity of fluids and heat storage capacity of solid matrix. In other words, the dimensionless parameters, St^* and Fo^*, must be appropriately selected in designing the regenerator for successful operation.

To best illustrate the influence of the conduction on regenerator performance, the conduction effect is defined as the ratio of the difference between the effectiveness with no conduction and that with conduction to the no conduction effectiveness

$$\frac{\Delta \eta}{\eta} = \frac{\eta_{\text{no Cond}} - \eta_{\text{Cond}}}{\eta_{\text{no Cond}}}$$

and correlated with the Bi^* number which is representative of the ratio of the longitudinal conduction resistance of matrix and the convective thermal resistance. The values of the conduction effect is shown in Fig.6.

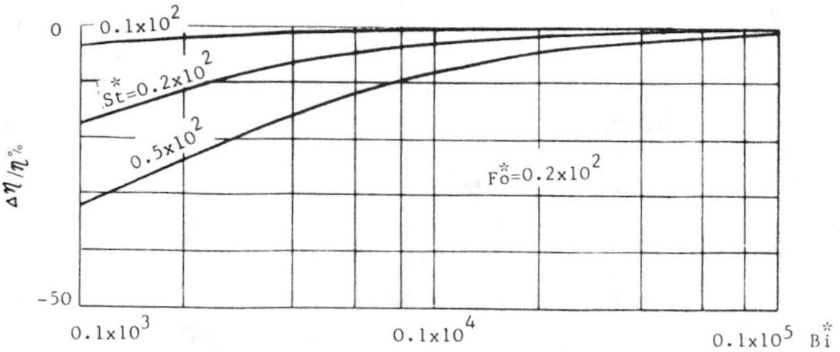

Fig.6 The influence of the longitudinal conduction on regenerator performance

Obviously the influence of the longitudinal heat conduction of the matrix results in a decrease of the effectiveness of the regenerator. But when the Bi^* number is greater than 0.2×10^4, the conduction effect can be neglected. Because of the heat transferred to or from the fluids is restricted in large St^* number, so in this case the conduction effect is somewhat obvious.

From the energy equation of the fluids it will be known, the influence of heat storage of the fluids is related to the product St^*Ho, as shown in Fig.7. In. Fig.7 $\Delta \eta$ is the difference between the effectiveness with no effect and that with effect; Ho is the ratio between the length of a period depended on the rotational velocity of the regenerator and the residence time of fluids in channels. When the Ho number is large, in comparison with the length of a period, the residence time is relatively short, the influence of this effect can be neglected. The influence of Ho varies inversely as St^* and directly as Fo^*. In large value of number St^* and small value of number Fo^*, the convective heat transfer rate is sufficiently large to reduce the effect of heat storage of the fluids and it can be neglected.

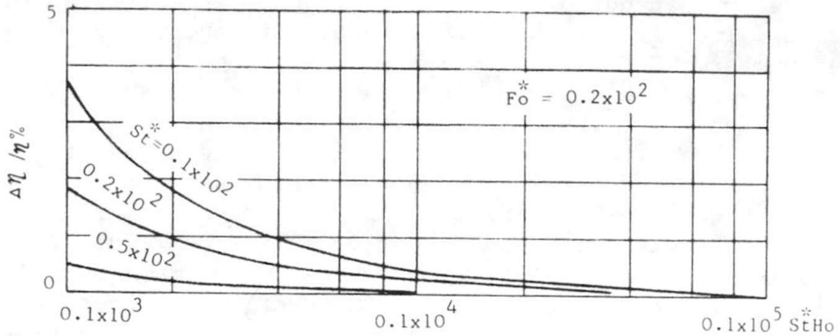

Fig.7 The influence of heat storage of the fluids

In designing a regenerator working in temperature ranging from 800°C to 25°C, the assumption of constant thermal properties of fluids is unacceptable. To illustrate the influence of temperature dependence of thermal properties on performance of regenerator, the variable thermal properties effect is defined as follows:

$$\frac{\Delta\eta}{\eta} = \frac{\eta_{const} - \eta_{var}}{\eta_{const}}$$

As shown in Fig.8 and Fig.9, this effect correlates with the temperature difference $t_{h,in} - t_{c,in}$ and temperature $t_{c,in}$. The effect of temperature dependence of thermal properties causes the effectiveness of the regenerator to decrease by 5%, and becomes important as increasing St^* and Fo^*.

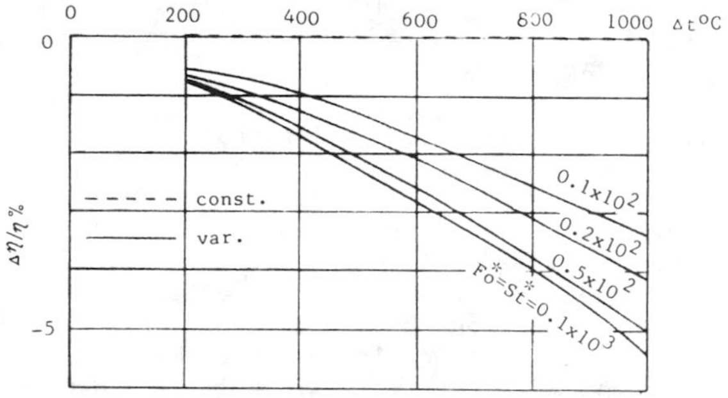

Fig.8 The influence of temperature dependence of thermal properties on regenerator performance

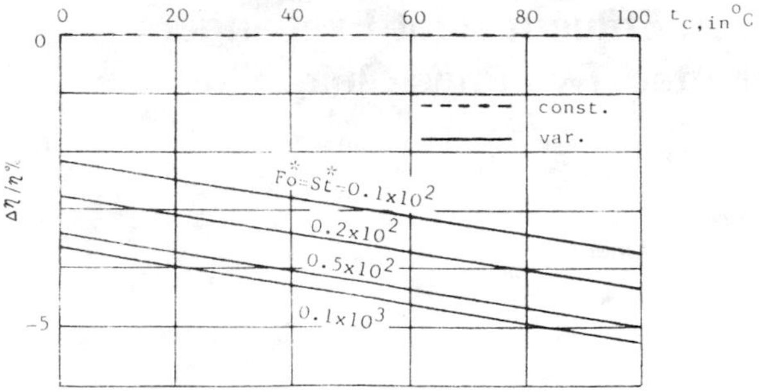

Fig.9 The influence of temperature dependence of thermal properties on regenerator performance

CONCLUSION

1. For successful operating and designing rotary regenerator, the dimensionless parameters St^* and Fo^* should be appropriately selected and matched.

2. In general, the longitudinal heat conduction effect causes the effectiveness of the regenerator to decrease by 5% and it can be neglected in case of $Bi^* \geqslant 0.2 \times 10^4$.

3. The effect of heat storage of fluids results in increasing the performance of the regenerator. In the range of possible applications in industry, it is less than 5%.

4. The effectiveness of the regenerator under variable thermal properties condition is lower 5% than that under constant thermal properties.

REFERENCE

[1] F.W.Schmidt and A.J.Willmott, Thermal Energy Storage and Regeneration, Hemisphere pub. Co., 1981.

[2] H.Hausen, Naherungsverfahren Zur Berechnung des Warmeaustausches in Regeneratoren, Z. angew. Math. 11, 1931, pp. 105-114.

[3] C.E.Iliffe, Thermal Analysis of the Counterflow Regenerative Heat Exchanger, Proc. Inst. Mech. Engrs 159, 1948, pp. 363-372.

[4] A.N.Nahavandi and A.S.Weinstein, A Solution to the Periodic Flow Regenerative Heat Exchanger Problem, Appl. Sci. Res. Section A, 10, 1961, pp. 335-348.

[5] T.J.Lambertson, Performance Factors of a Periodic-Flow Heat Exchanger, Trans. ASME, 80, 1958, pp. 586-592.

[6] A.J.Willmott, Digital Computer Simulation of a Thermal Regenerator, Int. J. Heat Mass Transfer, 7, 1964, pp. 1291-1302.

[7] A.J.Willmott and C.Hinchcliffe, The Effect of Gas Heat Storage Regenerator, Int. j. heat Mass Transfer, 19, 1976, pp. 821-826.

[8] G.D.Bahnke and C.P.Howard, The Effect of Longitudinal Heat Conduction on Periodic-Flow Heat Exchanger Performance, Trans. ASME, 86, 1964, pp. 105-120.

Analysis of Feed-Effluent Heat Exchangers with Maldistributed Two-Phase Inlet Flow

ZHANWU SHEN
Department of Chemical Machinery
Dalian Institute of Technology
Dalian, PRC

KENNETH J. BELL
Chemical Engineering
Oklahoma State University
Stillwater, Oklahoma 74078, USA

ABSTRACT

This paper examines the effect of a maldistributed two-phase (gas/vapor-liquid) feed on the performance of a 1-1 feed-effluent heat exchanger. Equations are developed to predict the performance of a specified unit for cases in which the maldistribution of the feed can be characterized by N classes of tubes, each having a given vapor-liquid feed ratio. A computer program has been developed to solve the sets of equations.

INTRODUCTION

One means of conserving energy in the process industries is to use a feed-effluent heat exchanger in connection with a chemical reactor. The purpose of this heat exchanger is to heat the feed to the reactor (in which an exothermic reaction takes place) by exchanging heat from the hot effluent stream from the reactor. One common configuration is a vertical 1-1 shell and tube heat exchanger with the feed stream flowing vertically upwards through the tubes and the effluent flowing downwards on the shell side (see Fig. 1). The feed stream is commonly a mixture of liquid and gas/vapor which is sprayed or blown upwards into the tubes. The liquid phase is supposed to evaporate entirely before reaching the top of the tubes. The shell side fluid usually begins to condense near the end of the cooling process.

These heat exchangers are generally designed on the assumption that the feed stream is uniformly distributed, each tube receiving the same amounts of liquid and vapor. It is also commonly assumed that the phases in each tube are in thermodynamic equilibrium at each cross-section. Fig. 2 shows a typical set of curves of stream temperature vs. heat transferred for a case satisfying these assumptions, where both the feed and effluent streams are multicomponent.

These are easier assumptions to make than to realize in practice. There is no known way in which to uniformly distribute each phase of a two-phase flow among a number of parallel flow channels (other than the obviously impractical one of metering each fluid independently into each tube). Further, the vaporization process at high quality is usually not at equilibrium, being a mist flow with a superheated vapor in contact with the wall and transferring heat to the entrained droplets, which slowly evaporate. Finally, the vaporization heat transfer coefficient in particular can not generally be assumed to be constant, especially at high qualities.

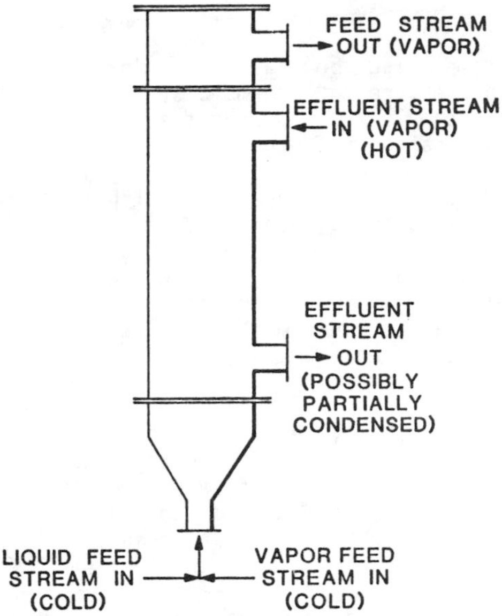

Figure 1. Typical feed-effluent heat exchanger

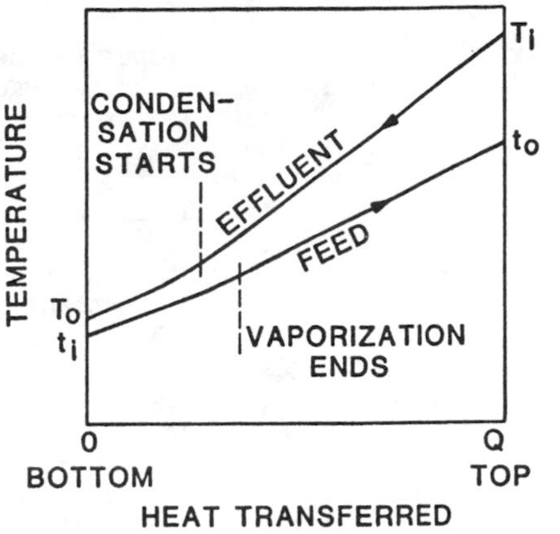

Figure 2. Temperature profiles for a typical feed-effluent exchanger with perfect distribution and multi multi component streams

Failure to uniformly distribute the phases imposes an operational penalty against the heat exchanger. For example, a tube receiving an excessive amount of liquid may not be able to transfer enough heat under the existing temperature profiles and heat transfer coefficients to completely vaporize that liquid by the end of the tube. Correspondingly, a tube receiving a lesser amount of liquid may superheat the vapor to a higher temperature than expected. Usually the amount of excess superheat provided in tubes with insufficient liquid flow is not as great as the shortfall of heat provided to the tubes with an excessive amount of liquid, resulting in an overall failure of the equipment to meet the nominal design conditions.

If the sole concern of the engineer were designability, i.e., the ability to predict within usual engineering precision the outcome of the exchanger operation, he would solve the problem by separating the phases and heating each one individually (the liquid undergoing a phase change) to the desired outlet temperature. Such a practice, while conservative from the heat exchanger design point of view, will generally impose an excessive penalty against the plant in that it will normally require substantial extra heat exchanger surface in order to achieve the desired outcome. It also requires two heat exchangers instead of one and additional piping, instrumentation and control systems.

It is the purpose of this paper to suggest a very simple model for the analysis of this kind of heat exchanger allowing for a certain degree of maldistribution of the liquid and vapor among the tubes. By estimating the degree of maldistribution that may occur, the designer may gauge the effect upon the outcome and therefore have some basis upon which to make decisions concerning safety factors. This paper is only the first of several which the authors have in progress and in which the simplifications are successively eliminated.

DESCRIPTION OF THE MODEL

The model employed in this paper has been simplified about as far as is possible while still retaining the essential features of the maldistribution problem. The feed stream is assumed to be a single pure component entering as a two-phase (vapor-liquid) mixture at its saturation temperature. This stream is assumed to vaporize isothermally and in equilibrium with the vapor until the liquid is totally vaporized, after which point the vapor superheats with a constant specific heat. The shell-side flow is assumed to enter as a superheated single component vapor (in general not the same component as the feed). The vapor cools at a constant specific heat by exchanging heat with the tube side fluid. After reaching saturation temperature, it condenses isothermally; it is assumed that a saturated vapor-liquid mixture exits the exchanger from the shell side. It is also assumed that the shell-side fluid is well-mixed and isothermal at any cross section.

Depending upon the amount of liquid entering each tube, the amount of heat required to vaporize the liquid will vary from tube to tube. Once the liquid is totally vaporized, and now depending upon the total vapor flow in each tube, the rate of temperature rise in each tube will also vary. The mixed mean temperature of the feed stream exiting the exchanger is the mass-averaged outlet temperature of the vapor from each of the tubes in the exchanger.

We will assume that all of the tubes may be classified into one of N classes. In any class n, each tube receives given vapor and liquid mass flow rates $\dot{m}_{v,n}$ and $\dot{m}_{\ell,n}$. Thus for each class, it will be possible to calculate the heat required to vaporize isothermally the liquid in each tube and the superheating of the total vapor flow in that tube that will result from a given further heat addition.

Fig. 3 shows a possible set of temperature profiles for N=2. Here for the n = 1 class of tubes, it is assumed that less liquid enters each tube compared to each of the n = 2 tubes, whereas more vapor enters each of the n = 2 tubes. It will also be assumed that $(\dot{m}_{\ell,1} + \dot{m}_{v,1}) > (\dot{m}_{\ell,2} + \dot{m}_{v,2})$. Thus, vaporization

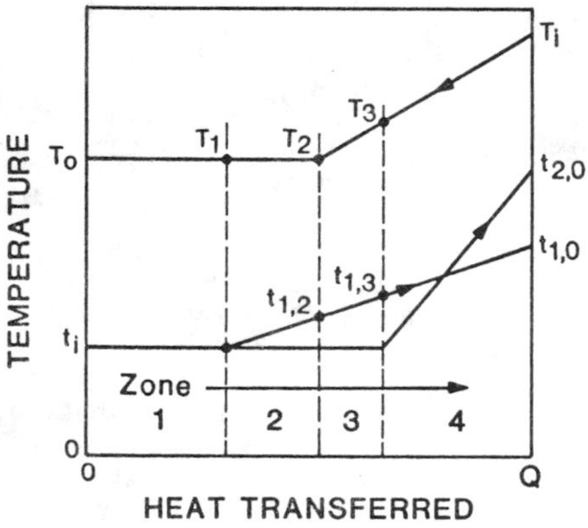

Figure 3. Temperature profiles in an exchanger with N=2

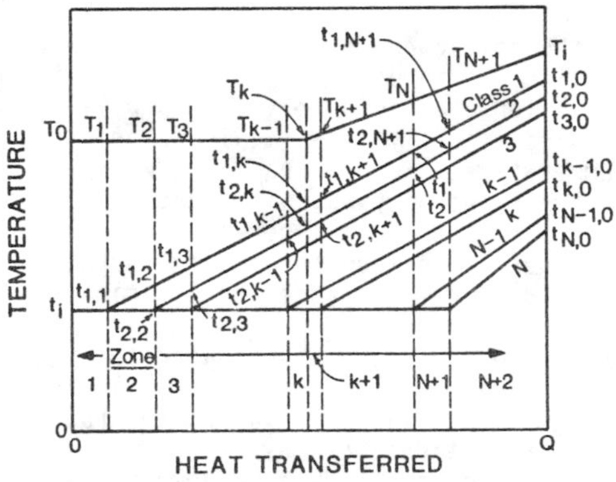

Figure 4. Possible temperature profiles in an exchanger with N classes

is completed more quickly in the n = 1 tubes than in the n = 2 tubes. Since, however, the total mass flow rate in the n = 1 tubes is greater than in the n = 2 tubes, the rate of temperature rise in the vapor sensible heating range is less in the n = 1 tubes than in the n = 2 tubes.

The shell-side fluid may reach its saturation temperature and begin isothermally condensing at a point in the exchanger different from either the point at which vaporization is complete in the n = 1 tubes or the point at which it is complete in the n = 2 tubes. So, as one possibility, we could have the following four zones in the heat exchanger:

1. Liquid is being vaporized in both n = 1 and n = 2 tubes, while the heat transfer process on the shell side is isothermal condensation of the shell side fluid. This zone (1) ends when all of the liquid in the n = 1 tubes is vaporized.

2. The all-vapor flow in n = 1 tubes is being superheated while isothermal vaporization continues in n = 2 tubes and isothermal condensation continues on the shell side. This zone (2) ends where the condensation process initiates on the shell side.

3. In this zone (3), isothermal vaporization continues for the n = 2 tubes and vapor superheating for the n = 1 tubes while the shell side heat transfer process is sensible cooling of the vapor. This zone ends when the vaporization in the n = 2 tubes is completed.

4. In zone (4), the vapor in all tubes is being superheated while the vapor on the shell side is being sensibly cooled.

Assuming that flow rates, heat transfer coefficients, and enthalpy data are given for each of the classes of tubes and the heat transfer processes, there are found to be nine unknown quantities for the N = 2 (or four zone) case. For the case shown in Fig. 3, these are the heat transfer areas of each zone A_1, A_2, A_3, A_4, the outlet tube side temperatures of each zone in which vapor sensible heat transfer is occurring on the shell side, $t_{1,2}$, $t_{1,3}$, $t_{1,0}$, $t_{2,0}$, and the shell side outlet temperature from the zone in which the desuperheating of the vapor is incomplete, T_3. Nine independent equations can be written in order to get the outlet temperatures and the transfer areas required. If there are N classes of tubes, there are (N + 2) zones and $[2 + 3N + N(\frac{N-1}{2})]$ unknown values. We can write a sufficient set of equations, using heat balances and rate equations for each zone, plus the overall heat balance.

BASIC EQUATIONS

We will illustrate the derivation of the equations for the case shown in Fig. 4, where $\dot{m}_{\ell,N} > \ldots \dot{m}_{\ell,n} > \ldots \dot{m}_{\ell,1}$.

Overall heat transfer coefficients are denoted by $U_{n,i}$, where the subscripts denote the class of the tube and the zone, respectively. In the first zone, all classes of tubes have condensing vapor on the shell side and vaporizing liquid on the tube side; therefore all $U_{n,1}$ are equal and have a value characteristic of that combination of processes. (This paper does not attempt to consider the differences in U due to differences in flow rate.) In the last zone, all classes of tubes have sensibly cooling vapor on the shell side and sensibly heating vapor on the tube side and again all U's are equal, with a value characteristic of this combination of processes. For the intermediate zones, there will be various processes present.

Overall Heat Balance:

By an overall heat balance between the N cold streams and the hot stream,

$$Q = \sum_{n=1}^{N} Q_{cv,n} + \sum_{n=1}^{N} Q_{cs,n} = Q_{cv} + Q_{cs} = Q_{hs} + Q_{hc} \quad (1)$$

The total heat required to vaporize the cold stream is:

$$Q_{cv} = \lambda_c \sum_{n=1}^{N} \dot{m}_{\ell,n} \cdot N_n \quad (2)$$

The total superheat of the cold stream is:

$$Q_{cs} = \sum_{n=1}^{N} (\dot{m}_{\ell,n} + \dot{m}_{v,n})\, c_p\, (t_o - t_i)\, N_n \quad (3)$$

The sensible heat duty of the hot stream is

$$Q_{hs} = \dot{M}\, C_p\, (T_i - T_o) \quad (4)$$

The heat of condensation given up by the hot stream is:

$$Q_{hc} = Q - Q_{hs} = \dot{M}_{\ell,o} \lambda_h \quad (5)$$

1st Zone In the first zone, the cold stream is a saturated two phase flow in all tubes. All heat absorbed vaporizes liquid, until the first class of tubes reaches the dryout point. Total heat transfer rate in the first zone is

$$Q_1 = \dot{m}_{\ell,1}\, N_T\, \lambda_c \quad (6)$$

Total heat transfer area in 1st zone is

$$A_1 = \int_0^{Q_1} \frac{dQ_1}{U_{n,1}\,(T_o - t_i)} = \frac{\dot{m}_{\ell,1}\, N_T\, \lambda_c}{U_{n,1}\,(T_o - t_i)} \quad (7)$$

2nd Zone In this zone, the 1st class of tubes is full of vapor and the heat absorbed in this class heats the vapor from $t_{1,1}$ to $t_{1,2}$. The heat balance for one tube of the 1st class in 2nd zone:

$$q_{1,2} = U_{1,2} \frac{A_2}{N_T} \frac{(T_1 - t_{1,2}) - (T_2 - t_{2,1})}{\ln \frac{(T_1 - t_{1,1})}{(T_2 - t_{1,2})}} = (\dot{m}_{v,1} + \dot{m}_{\ell,1})\, c_p\, (t_{1,2} - t_{1,1}) \quad (8)$$

For other classes of tubes in this zone, the heat absorbed will continue to vaporize liquid, until all of the liquid entering the class 2 tubes $(\dot{m}_{\ell,2} - \dot{m}_{\ell,1})$ is vaporized. The heat needed for vaporizing liquid in zone 2 is

$$q_{2,2} = U_{2,2} \frac{A_2}{N_T} (T_2 - t_{2,2}) = (\dot{m}_{\ell,2} - \dot{m}_{\ell,1})\, \lambda_c \quad (9)$$

From Equations (8) and (9), we find A_2 and $t_{1,2}$. Similar arguments apply to subsequent zones of the same type.

kth Zone The kth zone is the one in which the shell side fluid begins to condense. In this zone, there are k-1 classes of tubes full of vapor, so there are k-1 heat balance equations for each class. The amount of liquid vaporization in this zone is unknown, but the amount of vapor condensing from the stream in this zone is:

$$Q_k = Q - \dot{M}\, C_p\, (T_i - T_k) - \sum_{i=1}^{k-1} Q_i = \dot{M}_{\ell,k}\, \lambda_h \tag{10}$$

We can write another independent equation

$$Q_k = \sum_{i=1}^{k-1} q_{i,k} \tag{11}$$

From the above equations, we find A_k, $t_{1,k} \ldots t_{k-1,k}$

N + 2th Zone All tubes are full of vapor in this zone, so we get N equations of heat transfer:

$$q_{i,N+2} = U_{i,N+2} \frac{A_{N+2}}{N_T} \frac{[(T_i - T_{i,o})-(T_{N+1} - t_{i,N+1})]}{\ln \dfrac{(T_i - t_{i,o})}{(T_{N+1} - t_{i,N+1})}} \tag{12}$$

$$= (\dot{m}_{v,i} + \dot{m}_{\ell,i})\, c_p\, (t_{i,o} - t_{i,N+1})$$

The heat balance for this zone is

$$Q_{N+2} = \dot{M}\, C_p\, (T_i - T_{N+1}) = \sum_{i=1}^{N} q_{i,N+2} \tag{13}$$

From the above N + 1 equations, we get the area, A_{N+2}, and the outlet temperatures of the N classes, $t_{i,o}$.

EXAMPLE

The example case is a vertical 1-1 shell and tube heat exchanger with n-heptane flowing upwards through the tubes and n-nonane flowing downwards on the shell side. The n-heptane is at its saturation temperature (98.4°C) at the inlet and is to be superheated to 160°C at the outlet. The total inlet mass flow rates of vapor and liquid are 1.25 kg/s and 0.75 kg/s respectively. The n-nonane is at 220°C at the inlet and at its saturated temperature (150.5°C) at the outlet; its mass flow rate is 2.0 kg/s. Assume the overall heat transfer coefficients for the vapor-vapor U_{vv}, vapor-boiling U_{vb}, condensing-vapor U_{cv}, and condensing-boiling U_{cb} zones are 300, 500, 500, and 1000 w/m^{2}k respectively.

Case I If the feed steam is uniformly distributed among all the tubes, the total area required is 12.445 m^2. This area would correspond to a heat exchanger with 104 tubes, each 19.0 mm OD x 2.0 m long.

Case II The feed stream is maldistributed among 4 classes of tubes with a total of 104 tubes, and the flow rate of each tube class is:

No. of Tubes	17	35	35	17
Class	1	2	3	4
$\dot{m}_{vi}$ kg/s per tube	0.0194	0.0166	0.00714	0.00529
$\dot{m}_{\ell i}$ kg/s per tube	0.00235	0.00457	0.0100	0.0118

If we fix the outlet temperature of the feed stream and the operating conditions as in Case I, the following results are obtained from the computer program.

	A_1	A_2	A_3	A_4	A_5	A_6	$\sum A$
Heat transfer area, m^2	1.481	1.481	2.152	2.281	1.214	4.356	12.965

	$t_{1,o}$	$t_{2,o}$	$t_{3,0}$	$t_{4,0}$	t_o
Outlet temperature of each tube class, °C	166.43	163.14	156.49	151.06	160.00

This shows that, for the same operating conditions, the heat transfer area must be increased and the outlet temperatures of each tube class are different, due to malidistribution.

Case III In an actual exchanger, the area is fixed. Then the outlet temperature of the feed stream will be lower due to maldistribution and the heat recovery decreased. If the feed stream is maldistributed as in Case II, and the exchanger area is fixed as in (uniform) Case I, the following results are obtained.

	A_1	A_2	A_3	A_4	A_5	A_6	$\sum A$
Heat transfer area, m^2	1.481	1.481	2.152	2.281	1.214	3.836	12.445

	$t_{1,o}$	$t_{2,o}$	$t_{3,0}$	$t_{4,0}$	t_o
Outlet temperature of each tube class, °C	163.90	160.40	152.36	146.22	156.6

CONCLUSION

A very simplified model of a feed-effluent heat exchanger with feed stream maldistribution has been proposed, and the computational structure for its analysis developed. This can serve for the present to identify the general magnitude of the effects of flow maldistribution. More realistic models are required, and are being developed.

NOMENCLATURE

A	- Total heat transfer area of the exchanger, m^2
A_1, A_2, ...	- Heat transfer area of each zone in the exchanger, m^2

C_p, c_p	-Specific heats of hot stream and cold streams, J/kg K
$\dot{M}$	- Total mass flow rate of hot stream, kg/s
$\dot{M}_{\ell,o}$	- Mass flow rate of hot stream condensed, kg/s
$\dot{M}_{\ell,k}$	- Mass of hot fluid condensing in the kth zone, kg/s
$\dot{m}_{\ell}$, $\dot{m}_v$	- Total liquid and vapor flow rates of cold stream, kg/s
$\dot{m}_{\ell,n}$, $\dot{m}_{v,n}$	- Liquid and vapor mass flow rates per tube in class n, kg/s
N	- Number of classes of maldistributed tubes
N_n	- Number of tubes in the nth class
N_T	- Total number of tubes in the exchanger, $N_T = \sum_{i=1}^{N} N_n$
Q	- Total heat transfer rate of the exchanger, J/s
Q_1, Q_2, ...	- Heat transfer rate of each zone in the exchanger respectively, J/s
Q_{hc}	- Heat transfer rate for condensing of the hot stream, J/s
Q_{hs}	- Heat transfer rate for sensible cooling of the hot stream, J/s
Q_{cs}	- Heat transfer rate for sensible heating of the cold stream, J/s
$Q_{cs,n}$	- Heat transfer rate for sensible heating of cold stream in the nth class of tubes, J/s
Q_{cv}	- Heat transfer rate for vaporization of the cold stream, J/s
$Q_{cv,n}$	- Heat transfer rate for vaporization of the cold stream in the nth class of tubes, J/s
Q_k	- Heat transfer rate in the kth zone, J/s
$q_{n,i}$	- Heat transfer rate of one tube of the nth class in ith zone, J/s
T_i, T_o	- Inlet and outlet temperatures of hot stream, °C
T_1, T_2, ...	- Outlet temperature of hot stream from each zone, °C
t_i, t_o	- Inlet and outlet mixed mean temperatures of cold stream, °C;
$t_{n,i}$	- Outlet cold fluid temperature of nth class of tubes in ith zone, °C
$U_{n,i}$	- Overall heat transfer coefficient for nth class of tubes in ith zone, w/m^2K
λ_c, λ_h	- Latent heat of vaporization of the cold and hot streams, J/kg

A Further Study of the Mean Temperature Difference of Crossflow Heat Exchangers with One Fluid Mixed and the Other Unmixed

YOUCHUN WANG, HUANZHUO CHEN, SHUQIN JIANG, and SUPING CHEN
Institute of Engineering Thermophysics
Academia Sinica
Beijing, PRC

ABSTRACT

The mean temperature difference (MTD) characteristics of crossflow heat exchangers with one fluid mixed and other unmixed were obtained by D.M. Smith in 1934 and are still in use in sizing the core of this type of exchanger. However, his analysis was based on the assumption that there was no temperature gradient in the shell fluid normal to its direction of flow because of fluid mixing in t this direction. Experiments show that the fluid mixing effect is slight. Improved analysis further demonstrates that mixing effect in the shell fluid is also affected by the plan form of the exchanger. So in some cases the actual MTD characteristics of the exchanger in question are very close to those of crossflow exchangers in which both fluids are unmixed and Smith's over conservative assumption should no longer be followed. In short, much better use can be made of this type of exchanger.

NOMENCLATURE

A	heat transfer surface
A_i	heat transfer surface based on inner diameter of tube
A_o	heat transfer surface based on outer diameter of tube
a	number of heat transfer units based on tube fluid capacity rate=UA/C_1
b	number of heat transfer units based on shell fluid capacity rate=UA/C_2
c	mixing coefficient = $\varepsilon Y/vX^2$, see Eq.(15)
C_1	tube fluid capacity rate=G_1c_{p1}
C_2	shell fluid capacity rate= G_2c_{p2}
c_{p1}	tube fluid specific heat
c_{p2}	shell fluid specific heat
d_o	exchanger tube outer diameter
F	mean temperature difference correction factor
F_1	tube fluid flow area
F_2	shell fluid mean flow area
F_{min}	minimum flow area of shell fluid
G_1	tube fluid flow rate
G_2	shell fluid flow rate
k	thermal conductivity of exchanger tube material
Pe	Peclet number of shell fluid $=Re_oPr$
Pr	Prandtl number of shell fluid taking into account the combined molecular and turbulent thermal conductivity
Q	heat exchange of exchanger
R	ratio of tube fluid capacity rate to shell fluid capacity rate=C_1/C_2
Re_o	shell fluid Reynolds number
t_1	temperature of tube fluid
t_2	temperature of shell fluid
ΔT_m	mean temperature difference of the exchanger under consideration
ΔT_{ln}	log mean temperature difference of a counterflow exchanger with terminal temperatures same as those of the exchanger under considera

tion
ΔT_{max} shell air maximum temperature difference across the length of the exchanger
Δt_m mean temperature drop of tube air from inlet to outlet of the exchanger
U overall heat transfer coefficient of exchanger
v,w velocities of fluid 2 (shell fluid) and 1 (tube fluid) in the y,x directions respectively
v_{max} velocity of fluid 2 at the minimum flow area
x,y space corrdinates

Greek Letters:

α convective heat transfer coefficient
δ exchanger tube wall thickness
δ_T see Table 1
ε combined molecular and turbulent thermal diffusivity, see Eq.(10)
λ_T combined molecular and turbulent thermal conductivity
μ_2 viscosity of shell fluid
ρ_2 density of shell fluid
ς_I dimensionless temperature of fluid 1, see Eq. (11)
ς_{II} dimensionless temperature of fluid 2, see Eq. (11)
ξ,η dimensionless space coordinates, see Eq. (12)

Subscripts:

i inlet or inner
e outlet
c cold
h hot
m space mean value

1. INTRODUCTION

The mean temperature difference (MTD) characteristics of crossflow heat exchangers with one fluid mixed and the other unmixed were obtained by D.M. Smith [1] in 1934. His analysis was based on the assumption that there was no temperature difference in the shell fluid normal to its direction of flow because of fluid mixing in this direction. This assumption is questionable as it postulated infinitely strong mixing in the shell fluid which is obviously impossible. Half a century has elapsed since the publication of Smith's paper and no experimental study of the problem has been found in open literature. So in order to get a true picture of the MTD characteristics of the type of exchanger in question, experiments were conducted using air as heat transfer medium which gave measured values of MTD of the exchanger and measured temperature profiles of the shell air at the exit of the exchanger. The exchanger used in the tests was a crossflo recuperator, 1.09 m long and 0.345 m x 0.345 m in cross section with 390 tubes of size 12 x 1 arranged in staggered pattern, as shown in Fig.1. Three movable baffles were installed on the exchanger which could be slided along the tubes. When they are placed in positions such that the shell space is separated into several partitions, tests would give information as to whether or not the baffles have effects on MTD between the two fluids. This is so because the baffl prevent gross mixing of the shell air between partitions and, if Smith's assump- tion were true, the measured MTD with baf- fles in position would be higher than if they are placed at one end of the exchang leaving the shell space fully open to fluid mixing. Test results show no chang in MTD with and without the baffles in position to prevent fluid mixing, thus invalidating Smith's assumption. Alongsi with this, an improved mathematical model of the exchanger has been developed that takes into account mixing effect in the

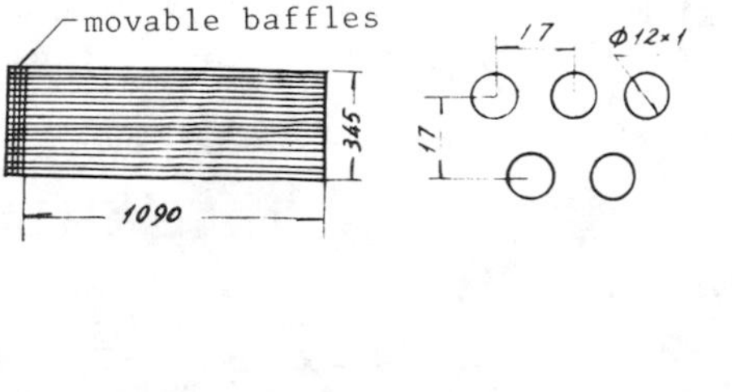

Fig.1 Model Exchanger with Movable Baffles in Shell Space (all dimensions in mm)

shell fluid but does not assume no temperature difference in this fluid normal to its direction of flow. Experiments and model analysis indicate that mixing effect in the shell fluid is slight and can be controlled by adjusting the length and width of the exchanger. Consequently, in some cases, the MTD characteristics of the exchanger in question are very close to those of crossflow exhangers with both fluids unmixed. This is true particularly when the exchanger has a slender plan form, i.e., short flow length of shell fluid and long flow length of tube fluid.

2. EXPERIMENTAL INVESTIGATION OF MTD OF A MODEL EXCHANGER

As is commonly known, the heat exchange of an exchanger is

$$Q = UA\Delta T_m = UAF\Delta T_{ln}$$

or

$$F = \frac{Q}{UA\Delta T_{ln}} \tag{1}$$

So the MTD correction factor F of an exchanger may be obtained by experiment if Q,U, A, and ΔT_{ln} are measured or based on measured quantities. To do this a test loop was constructed as shown shematically in Fig.2 . The tube air (hot) and shell air (cold) flows were supplied by their respective blowers and the former was heated by an electric heater before entering the exchanger. Air flow rates wered measured by 'Flute' type flowmeters and the mean temperatures of air at the inlets and outlets of the exchanger were measured by mercury thermometers. Since the air temperatures at the exits of the exchanger were not uniform, thermometers for measuring the average outlet air temperatures were placed near the mouths of pipings (extending from the exchanger exits) open to the ambient where the air temperatures were well equalized. In addition, ten copper-constantan thermocouples were placed along the mid-line of the exit plane of the exchanger shell space in order to obtain temperature profiles of the shell air normal to its flow direction.

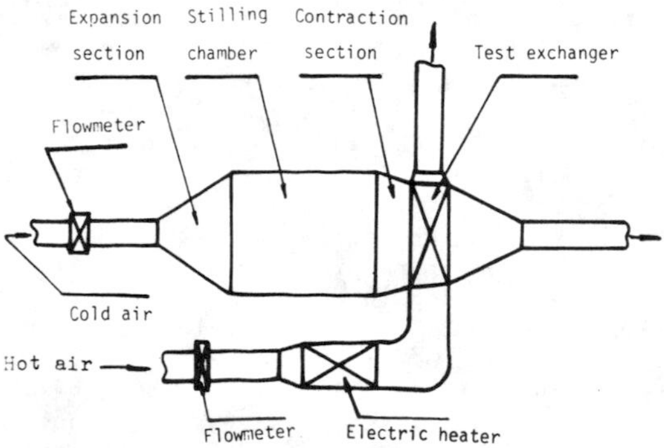

Fig. 2 Schematic of the Test Loop

The reliability of the test data was ensured by the heat balance between the tube and shell air streams. All the data points that were judged useful deviated within 4.5% from the 100% balance line as seen in Fig.3 where some of the data points are shown.

Q and ΔT_{ln} in Eq.(1) were calculated by

$$Q=G_1c_{p1}(t_{1,i}-t_{1,em})=G_2c_{p2}(t_{2,em}-t_{2,1}) \tag{2}$$

$$T_{ln}= \frac{(t_{1,i}-t_{2,em})-(t_{1,em}-t_{2,i})}{\ln \frac{t_{1,i}-t_{2,em}}{t_{1,em}-t_{2,i}}} \tag{3}$$

In these two equations the temperatures $t_{1,i}$, $t_{2,em}$, $t_{1,em}$, and $t_{2,i}$ were all measured quantities. The heat transfer surface area, A, of the test exchanger was known, equal to 16 m^2.

The overall heat transfer coefficient, U, was calculated by the equation

$$U = \frac{1}{\frac{A_o}{A_i}\frac{1}{\alpha_i} + \frac{A_o}{A_i}\frac{\delta}{k} + \frac{1}{\alpha_o}} \approx \frac{1}{\frac{A_o}{A_i}\frac{1}{\alpha_i} + \frac{1}{\alpha_o}} \quad (4)$$

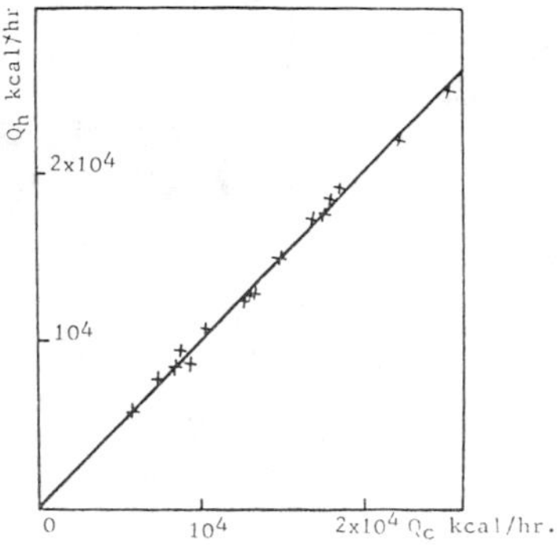

Fig.3 Heat Balance between Tube and Shell Air

because the tube wall (1 mm thick) thermal resistance, $(A_o/A_i)(\delta/k)$, is negligible compared to the other two resistances. α_i and α_o were evaluated by the following correlations respectively

$$Nu_i = \frac{\alpha_i d_i}{k_i} = 0.018\ Re_i^{0.8} \quad (5)$$

$$Nu_o = \frac{\alpha_o d_o}{k_o} = 0.31\ Re_o^{0.6} \quad (6)$$

where Re_i and Re_o were determined by the flow conditions of the tube and shell air.

With Q, ΔT_{ln}, and U obtained respectively from Eqs.(2), (3), and (4), F was readily calculated by (1). The results are plotted in Figs. 4-6 as F vs. P curves with R as parameter. From these figures we see:

(1) When R=1 the measured F is basically in agreement with Smith's theory which is plotted according to the equation

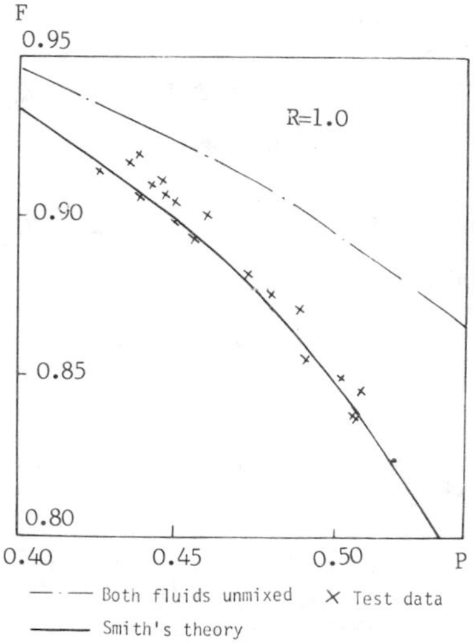

Fig.4 F vs. P Curves for R=1.0

$$F = \frac{1}{\ln\dfrac{1}{1-\frac{1}{R}\ln\frac{1}{1-PR}}} \Bigg/ \frac{R-1}{\ln\dfrac{1-P}{1-PR}} \quad (7)$$

(2) When R=0.8 and 0.6 the measured F deviates from Smith's theory. This deviation becomes more and more pronounced as R is decreasing from unity and, in this process, F is approaching the value for crossflow exchangers with both fluids unmixed as obtained by Nusselt [2].

The measured shell air temperatures normal to its direction of flow are far from being uniform as postulated by Smith. In order to have a quantitative impression of the non-uniformity of the shell air temperature, a dimensionless ratio δ_T is defined as $\delta_T = \Delta T_{max}/\Delta T_{av}$, where ΔT_{max} is the maximum temperature difference of shell air across the exchanger length and ΔT_{av} is the average temperature difference of shell air between the inlet and outlet of the exchanger. Tests show that δ_T increases with decreasing R as indicated in Table 1. Experi-

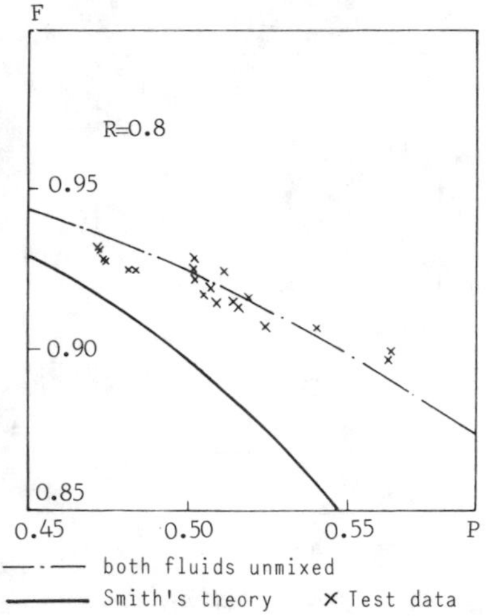

Fig.5 F vs. P Curves for R=0.8

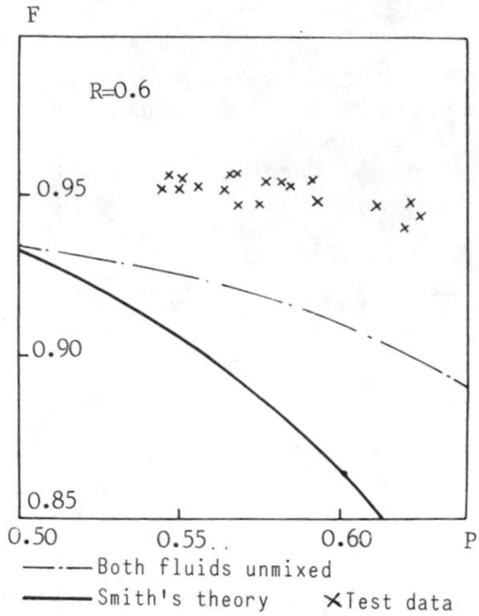

Fig.6 F vs. P Curves for R=0.6

ments were also conducted in which the tube air mean temperature difference between the inlet and outlet of the exchanger was varied while keeping the shell air Reynolds number constant. Three test runs were made with shell air Reynolds numbers equal to 3000, 5000, and 8000, respectively. The results are shown in Fig.7. It is seen that the maximum temperature difference of shell air in the transverse direction, characterizing the shell air temperature non-uniformity, is directly proportional to the tube air mean temperature difference. So we can conclude that the shell air temperature distribution in the transverse direction depends mainly on the temperature non-uniformity of the tube air in the longitudinal direction, and to a much less degree on fluid mixing in the shell air itself.

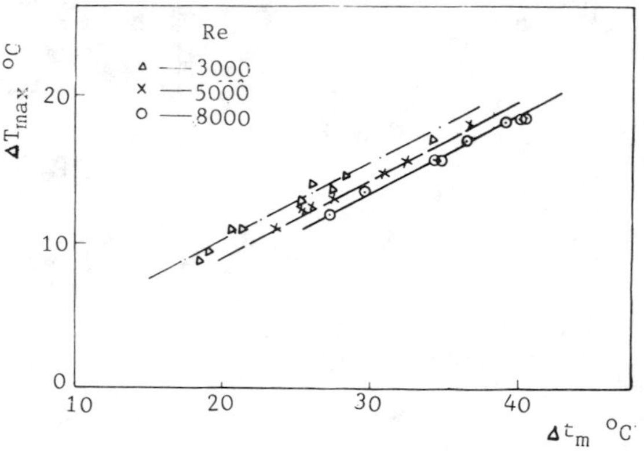

Fig.7 Maximum Temperature Difference of Shell Air across the Exchanger Length vs. Tube Air Mean Temperature Difference

Table 1

Change of δ_T with R

R	1	0.8	0.6
δ_T	0.6	0.69	0.81

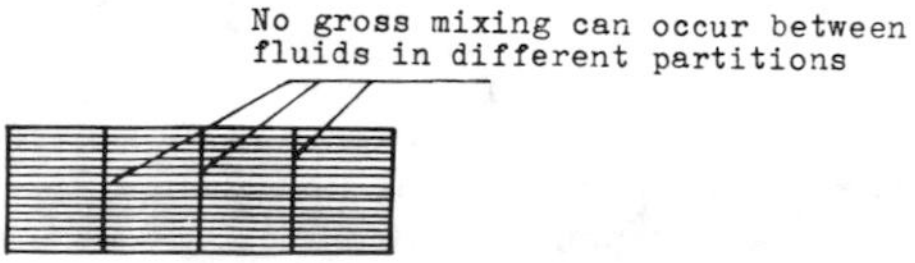

Fig.8 Baffles Separating the Shell Space into Four Equal Partitions

In order to further know the effect of mixing on exchanger MTD, additional tests were performed with the baffles in position such that the shell space was separated into four equal partitions as shown in Fig.8, and gross mixing of fluid between partitions could not take place. So, had Smith's assumption been true, the measured MTD correction factor would have been higher with baffles in position than without. But tests indicated no change in MTD under the two conditions as shown in Fig.9. This is further evidence that fluid mixing in the shell air is slight and Smith's assumption is over conservative.

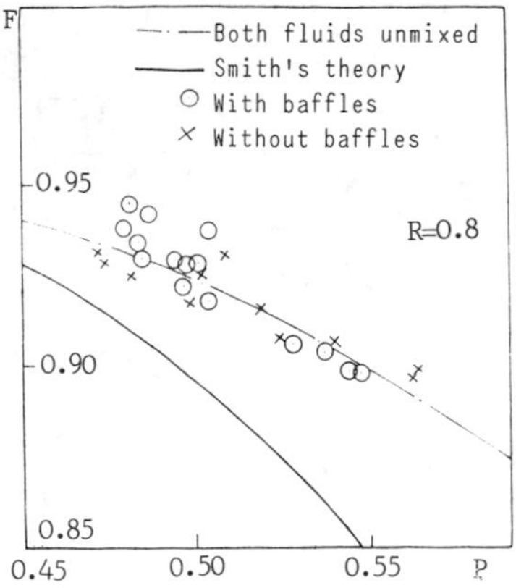

Fig.9. F vs. P Curves with and without Baffles in Position

IMPROVED MATHEMATICAL MODEL OF THE EXCHANGER AND ITS COMPARISON WITH EXPERIMENT

In view of the over conservativeness of Smith's theory an improved mathematic model of the exchanger has been developed that takes mixing effect in the shell fluid into account, but does not assume no temperature difference in this fluid normal to its flow direction. This model is based on the following idealizations

(1) In the "mixed fluid" (shell fluid) mixing is important only in the direction perpendicular to fluid flow, and in the flow direction mixing effect can be ignored.

(2) The "mixed fluid" has no "macroscopic" motion normal to its flow direction (neglecting natural convection in this direction).

(3) The problem is two-dimensional.

(4) Physical properties of the fluids are constant.

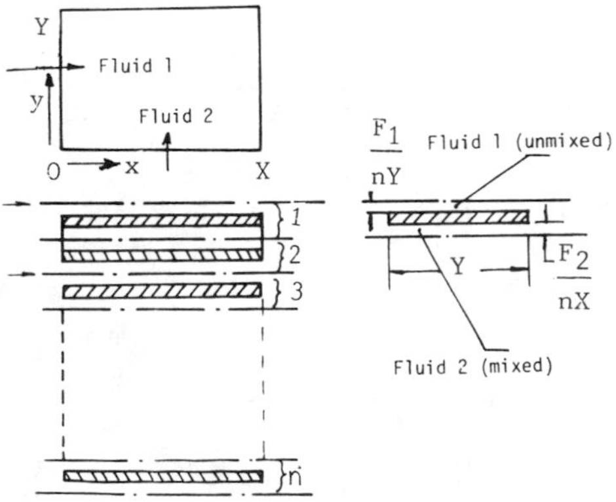

Fig.10. Schematic Crossflow Exchanger Taking into Account Mixing Effect in Shell Space

As depicted in Fig.10, the exchanger is reduced to a schematic structure formed by several parallel flat plates, on one side of each being the unmixed fluid, and on the other the mixed fluid The former is designated fluid 1 and the latter fluid 2, F_1/nY and F_2/nX are the thicknesses of fluid 1 and 2 respective on each side of one plate.

Referring to Fig.11, energy balance of the two fluid elements give the energ equations of the two fluids as follows:

For fluid 1

$$\frac{\partial t_1}{\partial x} = \frac{UnY}{C_1}(t_2 - t_1) \qquad (9)$$

For fluid 2

$$\frac{\partial t_2}{\partial y} = \frac{UnX}{C_2}(t_1 - t_2) + \frac{\varepsilon}{v}\frac{\partial^2 t_2}{\partial x^2} \qquad (10)$$

Introducing two dimensionless temperatu

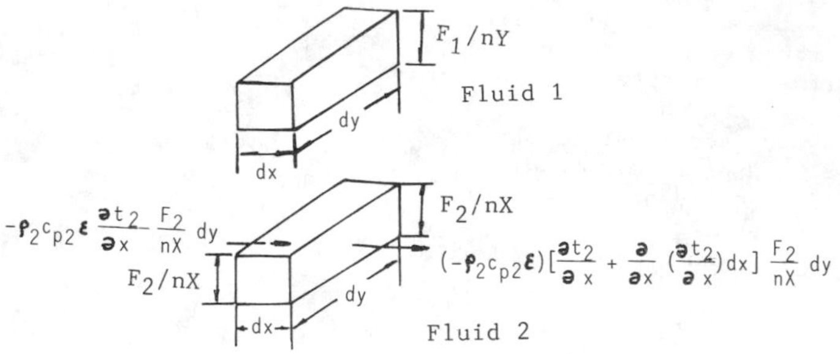

Fig.11 Fluid Elements for Energy Balance Analysis

$$\zeta_I = \frac{t_1 - t_{2,i}}{t_{1,i} - t_{2,i}} \quad ; \qquad \zeta_{II} = \frac{t_2 - t_{2,i}}{t_{1,i} - t_{2,i}} \tag{11}$$

and two dimensionless space coordinates

$$\xi = \frac{x}{X} \; ; \quad \eta = \frac{y}{Y} \tag{12}$$

Eqs. (9) and (10) become

$$a(\zeta_I - \zeta_{II}) = -\frac{\partial \zeta_I}{\partial \xi} \tag{13}$$

$$b(\zeta_I - \zeta_{II}) + c\frac{\partial^2 \zeta_{II}}{\partial \xi^2} = \frac{\partial \zeta_{II}}{\partial \eta} \tag{14}$$

where $c = \varepsilon Y/v^2$ (15)

$$a = UA/C_1 \tag{16}$$

$$b = UA/C_2 \tag{17}$$

Now write ε of Eq.(10) in the form

$$\varepsilon = \frac{\lambda_T}{\rho_2 c_{p2}} \tag{18}$$

where λ_T is the combined molecular and turbulent thermal conductivity of fluid 2.

The shell fluid Reynolds number is defined as

$$Re_o = \rho_2 v_{max} d_o / \mu_2 \tag{19}$$

Define the Prandtl number of fluid 2 as

$$Pr = \mu_2 c_{p2} / \lambda_T \tag{20}$$

and the Peclet number is defined as

$$Pe = Re_o Pr = \rho_2 v_{max} d_o c_{p2} / \lambda_T \tag{21}$$

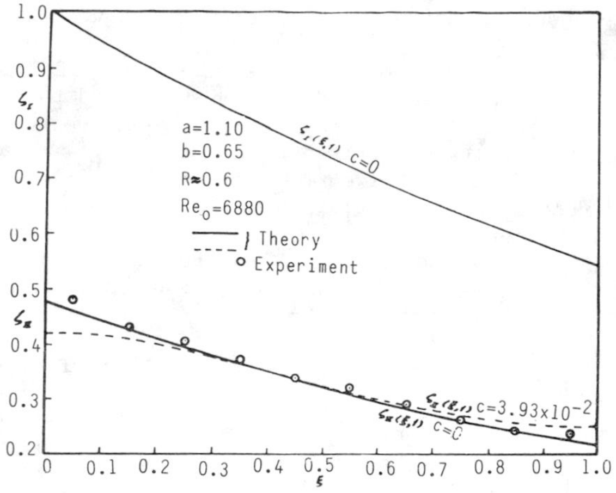

Fig.12 Theoretical Tube and Shell Air temperatures and Measured Shell Air temperatures for R=0.6, a=1.10, b=0.65, Re_o=6880

When (18)-(21) are used in (15) c becomes

$$c = \frac{Y}{\frac{F_{min}}{F_2} Pe\, X^2/d_o} \qquad (22)$$

This equation shows that the mixing coefficient, c, depends not only on the Peclet number, characterizing turbulent mixing, but also on the dimensions X and Y of the exchanger.

Eqs. (13) and (14) are similar to the energy equations of crossflow exchangers with both fluids unmixed [2] except that a mixing term is included in the shell fluid energy equation. So this mathematical model is more general than Nusselt's.

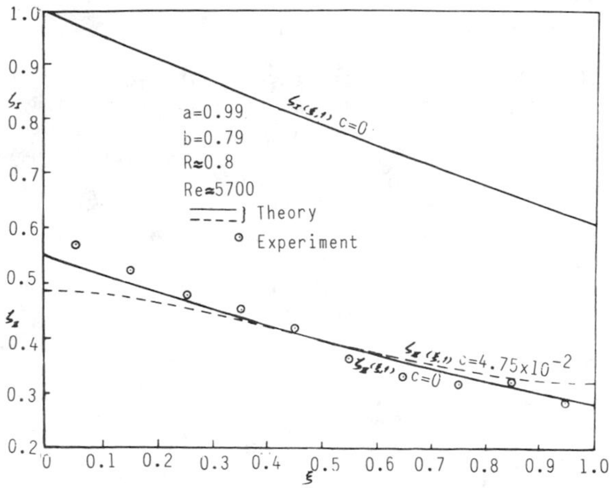

Fig.13. Theoretical Tube and Shell Air Temperatures and Measured Shell Air Temperatures for R=0.8, a=0.99, b=0.08, Re_o=5700

The boundary conditions for (13) and (14) are

$$\left.\begin{aligned} \zeta_I &= 1 \quad \text{when } \xi = 0 \\ \zeta_{II} &= 0 \quad \text{when } \eta = 0 \\ \frac{\partial \zeta_{II}}{\partial \xi} &= 0 \quad \text{when } \xi = 0,1 \end{aligned}\right\} \qquad (23)$$

For the test exchanger, X=1.09m, Y=0.345m, d_o=0.012m, and F_{min}/F_2=0.658, and then c=0.005296/Pe. This means that for the test exchanger c is a small quantity unless Pe is also very small.

A computer program was designed to solve (13) and (14) numerically. Since the heat transfer surface area, A, is known, the constants a and b in (13) and (14) are readily determined for given capacity rates of the two air streams. The mixing coefficient c is a function of $Pe=\rho_2 v_{max} d_o c_{p2}/\lambda_T$ in which the only unknown is λ_T. Logically λ_T is a function of Re_o. But no known relation between λ_T and Re_o is available. So the only way to establish such a relation is by experiment. That is, with given a,b, and Re_o, assume a λ_T by which to determine c. Then the numerical solutions for ζ_I and ζ_{II} may be found. When the calculated ζ_{II} (ξ, 1) is compared with measured data, the agreement between the two indicates that the assumed λ_T suits the given Re_o. A series of tests covering a range of Re_o will yield a useful functional relation between λ_T and Re_o. But this procedure turned out to be unnecessary, for the test exchanger had a slender plan form (small Y, large X) which made the mixing coefficient, c, be significant only when λ_T was very large. For instance, if c is to be of the order of magnitu 10^{-2}, λ_T must exceed 350 w/m-°C below which the second term on the left of (14) will be negligible. Therefore, for ease of comparison, based on each group of te data (a group of known a,b, and Re_o) two ζ_{II} (ξ, 1) curves were computed, one wit c=0 and the other with c>0. All the curves with c>0 correspond to λ_T=978 w/m-°C which is more than twice the thermal conductivity of silver. Even so, the shell air temperature profile at the exit of exchanger is not completely flattened. The curve ζ_{II} (ξ, 1) with c=0 is the shell air dimensionless outlet temperatures when both air streams are unmixed.

CONCLUSIONS

(1) Smith's analysis for crossflow heat exchangers with one fluid mixed and the other unmixed is over conservative. When $R<1$ the MTD characteristics of this type of exchanger are almost the same as for crossflow exchangers with both fluids unmixed. Hence for $R<1$ the MTD correction factor of the exchanger in question may be calculated by Nusselt's method.

(2) The plan form of the exchanger has important effect on mixing. If the exchanger has a slender plan form, short in shell fluid flow direction and long in tube fluid flow direction, the mixing coefficient c of Eq.(22) can be very small and the MTD characteristics of the exchanger will be much better than calculated by Smith's theory.

ACKNOWLEDGEMENT

The experiments of this work were conducted at the Capital Iron and Steel Co., Beijing. Two engineers at this enterprise, Messrs. Zhang Bopeng and Chen Xiaoquan took an active part in the design and construction of the test loop and in the calibration of the flowmeters. Mr. Zhu Rongguo at the Institute of Engineering Thermophysics designed the computer program and performed sample calculations and Mrs. Xu Hongkun prepared the drawings. Their contributions are gratefully acknowledged.

REFERENCES

1. D.M. Smith: "Meam Temperature Difference in Cross Flow," Engineering, Vol.138, 1934, pp 479-481, 606-607.
2. Max Jakob: Heat Transfer, Vol. II, John Wiley and Sons, 1957, pp 219-229. Eqs.(34-76), (34-77).

Analysis of 2-N Split Flow Heat Exchangers

ZHENWAN ZHUANG
Department of Chemical Engineering
Nanjing Institute of Chemical Technology,
Nanjing, PRC

ABSTRACT

Governing equations for heat transfer in 2-n split flow heat exchangers are derived. Using these equations, the formulas and curves for temperature profile logarithmic mean temperature difference correction factor, and effectiveness of heat exchangers are presented.

NOMENCLATURE

C_1--C_{20}	defined by equation (4)
D_1--D_{15}	defined by equation (4)
F	LMTD correction factor
J_1--J_4	defined by equation (4)
K_1--K_3	defined by equation (4)
L	one half of tube pass length
(NTU)	number of heat transfer units
P	temperature efficiency
R	heat capacity rate ratio
T	shell-side fluid temperature
t	tube-side fluid temperature
$\bar{x}$	dimensionless coordinate, $\bar{x} = x/L$
$\bar{x}'$	dimensionless coordinate, $\bar{x}' = x/L$
ε	effectiveness of the exchanger
θ	dimensionless temperature, $\theta = (t - t_{oi})/(T_{oi} - t_{oi})$
θ'	dimensionless temperature, $\theta' = (T - t_{oi})/(T_{oi} - t_{oi})$

Subscripts

e	outlet
i	i-th section
oi	inlet

INTRODUCTION

Due to their ability to produce "temperature correction factors" comparable to those in parallel flow exchangers but a smaller shellside pressure loss, split flow heat exchangers are often used. However, the complete analytical solutions of the temperature profiles of the shell - and tubeside streams have not yet been presented in the open literature even for the simple 1-n split flow heat exchangers. Recently, considering the knowledge of temperature profil is conductive to rational design of heat exchangers, Murty [1] tried to make sufficient analysis for the relative simple case of 1-2 split flow heat exchanger. In his work, however, formulas used for calulating temperature profi have not been derived, the assumption and calculation on value of N (the ratio of heat transfer rates in the two halves of the split flow exchanger) is not appropriate, his result is incorrect as discussed in [2]. As for 2-n split flow heat exchangers, there are no analytical solutions except the numerical result of 2-4 split flow heat exchanger [3].

This paper makes penetrating analysis for heat transfer process in 2-n (n = 4,6,8) split flow heat exchangers as shown in Fig.1.

GOVERNING EQUATIONS AND SOLUTION

As an analytical example, the analysis of 2-8 split flow heat exchanger is given here. It is apprpriate to outline the major assumptions made in this analysis. These are:

a. Constant heat transfer coefficient and specific heat of both heat exchanging fluids.
b. Steady state conditions.
c. Negligible heat losses from the system.

The heat transfer region is divided into four subregions as shown diagrammatically in Fig.2. The heat balances on each section can be written:

$$\frac{d\theta_9}{d\bar{x}} = \frac{(NTU)}{16}(\theta_1' - \theta_9), \qquad \frac{d\theta_{10}}{d\bar{x}} = -\frac{(NTU)}{16}(\theta_1' - \theta_{10}),$$

$$\frac{d\theta_{11}}{d\bar{x}} = \frac{(NTU)}{16}(\theta_1' - \theta_{11}), \qquad \frac{d\theta_{12}}{d\bar{x}} = -\frac{(NTU)}{16}(\theta_1' - \theta_{12}),$$

$$\frac{d\theta_1'}{d\bar{x}} = \frac{R(NTU)}{8}(4\theta_1' - \theta_9 - \theta_{10} - \theta_{11} - \theta_{12}),$$

$$\frac{d\theta_{13}}{d\bar{x}} = \frac{(NTU)}{16}(\theta_2' - \theta_{13}), \qquad \frac{d\theta_{14}}{d\bar{x}'} = -\frac{(NTU)}{16}(\theta_2' - \theta_{14}),$$

$$\frac{d\theta_{15}}{d\bar{x}'} = \frac{(NTU)}{16}(\theta_2' - \theta_{15}), \qquad \frac{d\theta_{16}}{d\bar{x}'} = -\frac{(NTU)}{16}(\theta_2' - \theta_{16}),$$

$$\frac{d\theta'_2}{d\bar{x}'} = -\frac{R(NTU)}{8}(4\theta_2' - \theta_{13} - \theta_{14} - \theta_{15} - \theta_{16}),$$

$$\frac{d\theta_1}{d\bar{x}} = \frac{(NTU)}{16}(\theta_3' - \theta_1), \qquad \frac{d\theta_2}{d\bar{x}} = -\frac{(NTU)}{16}(\theta_3' - \theta_2),$$

$$\frac{d\theta_3}{d\bar{x}} = \frac{(NTU)}{16}(\theta_3' - \theta_3), \qquad \frac{d\theta_4}{d\bar{x}} = -\frac{(NTU)}{16}(\theta_3' - \theta_4),$$

$$\frac{d\theta_3'}{d\bar{x}} = -\frac{R(NTU)}{8}(4\theta_3' - \theta_1 - \theta_2 - \theta_3 - \theta_4),$$

$$\frac{d\theta_5}{d\bar{x}'} = \frac{(NTU)}{16}(\theta_4' - \theta_5), \qquad \frac{d\theta_6}{d\bar{x}'} = -\frac{(NTU)}{16}(\theta_4' - \theta_6),$$

$$\frac{d\theta_7}{d\bar{x}'} = \frac{(NTU)}{16}(\theta_4' - \theta_7), \qquad \frac{d\theta_8}{d\bar{x}'} = -\frac{(NTU)}{16}(\theta_4' - \theta_8),$$

$$\frac{d\theta_4'}{d\bar{x}'} = \frac{R(NTU)}{8}\,(4\theta_4' - \theta_5 - \theta_6 - \theta_7 - \theta_8) \tag{1}$$

The appropriate boundary and interface conditions are:

$$\theta_9(0) = \theta_4(0),\ \theta_{10}(0) = \theta_{11}(0),\ \theta_9(1) = \theta_{13}(0),\ \theta_{10}(1) = \theta_{14}(0), \theta_{11}(1)=\theta_{15}(0),$$

$$\theta_{12}'(1) = \theta_{16}(0),\ \theta_{13}(1) = \theta_{14}(1),\ \theta_{15}(1) = \theta_{16}(1),\ \theta_5(1) = \theta_6(1), \theta_7(1)=\theta_8(1),$$

$$\theta_2(0) = \theta_3(0),\ \theta_1(1) = \theta_5(0),\ \theta_2(1) = \theta_6(0),\ \theta_3(1) = \theta_7(0), \theta_{11}(1)=\theta_8(0),$$

$$\theta_1(0) = 0, \theta_1'(1)=1,\ \theta_2'(0)=1,\ \theta_3'(0)=\theta_1'(0),\ \theta_4'(1) = \theta_2'(1) \tag{2}$$

The following solution of Eq.(1) with the conditions (2) can be obtained:

$$\begin{bmatrix}\theta_9\\ \theta_{10}\\ \theta_{11}\\ \theta_{12}\\ \theta_1'\end{bmatrix} = C_1\begin{bmatrix}1\\1\\1\\1\\1\end{bmatrix} + C_2\begin{bmatrix}0\\1\\0\\-1\\0\end{bmatrix} K_1^{\bar{x}} + C_3\begin{bmatrix}1\\0\\-1\\0\\0\end{bmatrix} K_1^{-\bar{x}} + C_4\begin{bmatrix}J_1\\J_2\\J_1\\J_2\\1\end{bmatrix} K_2^{\bar{x}} + C_5\begin{bmatrix}J_3\\J_4\\J_3\\J_4\\1\end{bmatrix} K_3^{\bar{x}}$$

$$\begin{bmatrix}\theta_{13}\\ \theta_{14}\\ \theta_{15}\\ \theta_{16}\\ \theta_2'\end{bmatrix} = C_6\begin{bmatrix}1\\1\\1\\1\\1\end{bmatrix} + C_7\begin{bmatrix}0\\1\\0\\-1\\0\end{bmatrix} K_1^{\bar{x}'} + C_8\begin{bmatrix}1\\0\\-1\\0\\0\end{bmatrix} K_1^{-\bar{x}'} + C_9\begin{bmatrix}J_4\\J_3\\J_4\\J_3\\1\end{bmatrix} K_3^{-\bar{x}'} + C_{10}\begin{bmatrix}J_2\\J_1\\J_2\\J_1\\1\end{bmatrix} K_2^{-\bar{x}'}$$

$$\begin{bmatrix}\theta_1\\ \theta_2\\ \theta_3\\ \theta_4\\ \theta_3'\end{bmatrix} = C_{11}\begin{bmatrix}1\\1\\1\\1\\1\end{bmatrix} + C_{12}\begin{bmatrix}0\\1\\0\\-1\\0\end{bmatrix} K_1^{\bar{x}} + C_{13}\begin{bmatrix}1\\0\\-1\\0\\0\end{bmatrix} K_1^{-\bar{x}} + C_{14}\begin{bmatrix}J_4\\J_3\\J_4\\J_3\\1\end{bmatrix} K_3^{-\bar{x}} + C_{15}\begin{bmatrix}J_2\\J_1\\J_2\\J_1\\1\end{bmatrix} K_2^{-\bar{x}}$$

$$\begin{bmatrix}\theta_5\\ \theta_6\\ \theta_7\\ \theta_8\\ \theta_4'\end{bmatrix} = C_{16}\begin{bmatrix}1\\1\\1\\1\\1\end{bmatrix} + C_{17}\begin{bmatrix}0\\1\\0\\-1\\0\end{bmatrix} K_1^{\bar{x}'} + C_{18}\begin{bmatrix}1\\0\\-1\\0\\0\end{bmatrix} K_1^{-\bar{x}'} + C_{19}\begin{bmatrix}J_1\\J_2\\J_1\\J_2\\1\end{bmatrix} K_2^{\bar{x}'} + C_{20}\begin{bmatrix}J_3\\J_4\\J_3\\J_4\\1\end{bmatrix} K_3^{\bar{x}'} \quad (3$$

where $J_1 = \dfrac{1}{1+4R+\sqrt{1+16R^2}}$, $J_2 = \dfrac{1}{1-4R-\sqrt{1+16R^2}}$, $J_3 = \dfrac{1}{1+4R-\sqrt{1+16R^2}}$,

$$J_4=\frac{1}{1-4R+\sqrt{1+16R^2}}, \quad K_1=\exp[\frac{(NTU)}{16}], \quad K_2=\exp[\frac{(NTU)}{4}(R+\frac{\sqrt{1+16R^2}}{4})],$$

$$K_3=\exp[\frac{(NTU)}{4}(R-\frac{\sqrt{1+16R^2}}{4})], \quad C_1=1+(1-\frac{K_3}{K_2})\frac{1}{D_{12}D_8}, \quad C_2=(D_{14}+D_{15})\frac{1}{D_{12}D_8},$$

$$C_3=K_1^4(D_{14}+D_{15})\frac{1}{D_{12}D_8}, \quad C_4=[\frac{1}{K_2^2}(K_3-K_2)-\frac{4RK_3(1+K_1^4)D_{15}}{K_2-K_3}]\frac{1}{D_{12}D_8},$$

$$C_5=[\frac{4RK_2(1+K_1^4)D_{15}}{K_2-K_3}]\frac{1}{D_{12}D_8}, \quad C_6=1+(\frac{K_3-K_2}{K_2})\frac{1}{D_{12}D_8}, \quad C_7=K_1(D_{14}+D_{15})\frac{1}{D_{12}D_8},$$

$$C_8=K_1^3(D_{14}+D_{15})\frac{1}{D_{12}D_8}, \quad C_9=-(\frac{K_3}{K_2})\frac{1}{D_{12}D_8}, \quad C_{10}=\frac{1}{D_{12}D_8},$$

$$C_{11}=D_7+(\frac{K_3}{K_2}-1)C_{15}+(K_3-\frac{K_3^2}{K_2})C_{20}, \quad C_{12}=\frac{(K_3-K_2)C_{15}}{4RK_2(1+K_1^4)}+\frac{(K_3K_2-K_3^2)C_{20}}{4RK_2(1+K_1^4)}.$$

$$C_{13}=\frac{K_1^4(K_3-K_2)C_{15}}{4RK_2(1+K_1^4)}+\frac{K_1^4(K_3K_2-K_3^2)C_{20}}{4RK_2(1+K_1^4)}, \quad C_{14}=-\frac{K_3C_{15}}{K_2}+K_3(\frac{K_3}{K_2}-1)C_{20},$$

$$C_{15}=\frac{D_7}{D_{13}}+\frac{D_2D_5}{D_{12}D_{13}D_8D_9}, \quad C_{16}=1+(\frac{K_3}{K_2}-1)\frac{1}{D_{12}D_8},$$

$$C_{17}=K_1K_2D_{14}C_{15}+\frac{K_1(K_3K_2-K_3^2)C_{20}}{4RK_2(1+K_1^4)},$$

$$C_{18}=K_1^3K_2D_{14}C_{15}+\frac{K_1^3(K_3K_2-K_3^2)C_{20}}{4RK_2(1+K_1^4)}, \quad C_{19}=-\frac{K_3C_{20}}{K_2}, \quad C_{20}=\frac{D_2}{D_{12}D_8D_9}+\frac{D_{10}C_{15}}{D_9}.$$

$$D_1=(\frac{K_3}{K_2}-1)(\frac{1}{K_2}+\frac{2-\frac{K_3}{K_2}-J_1-J_2}{K_3\sqrt{1+1/16R^2}}), \quad D_2=(1-\frac{K_3}{K_2})\left(2-\frac{1}{K_2}+\frac{\frac{K_3}{K_1}+J_1+J_2-2}{K_3\sqrt{1+1/16R^2}}\right),$$

$$D_3=\frac{\frac{K_3^2}{K_2}-\frac{K_3}{K_2}-K_3+J_4+J_3}{K_3(\frac{K_3}{K_2}-1)}, \quad D_4=\frac{1}{K_3}(J_4+J_3)-\frac{1}{K_2}(J_4-J_2)-\frac{1}{K_2}-2$$

$$D_5=K_3-\frac{K_3^2}{K_2}-K_3J_4+\frac{K_3^2J_4}{K_2}+\frac{K_1^4(K_3K_2-K_3^2)}{4RK_2(1+K_1^4)}$$

$$D_6=1-\frac{K_3}{K_2}+\frac{K_3J_4}{K_2}-J_2-\frac{K_1^4(K_3-K_2)}{4RK_2(1+K_1^4)}, \quad D_7=1-\frac{1-\frac{K_3}{K_2}-D_2}{D_{12}D_8}$$

$$D_8 = \frac{2RD_3D_{13}(1+D_{11})}{D_4 - 2RD_3D_{13}} , \quad D_9 = \frac{K_3^2}{K_2} - \frac{K_3}{K_2} - K_3+J_4+J_3$$

$$D_{10} = \frac{K_3}{K_2} - 1 - \frac{1}{K_2}(J_4-J_2) , \quad D_{11} = \frac{D_4}{2RD_{12}D_{13}D_3}(1-\frac{K_3}{K_2}-D_2) - \frac{D_2}{2RD_{12}D_3}(1+\frac{D_4D_5}{D_{12}D_9})$$

$$D_{12} = (1-\frac{K_3}{K_2}) + \frac{J_1}{K_2}(\frac{K_3}{K_2}-1) + \frac{K_1^4(K_3-K_2)}{4RK_2^2(1+K_1^4)} + \frac{\frac{K_3}{K_2}+J_1+J_2-2}{K_3\sqrt{1+1/16R^2}}[\frac{K_1^4(K_2-K_3)}{4RK_2(1+K_1^4)}$$

$$+J_3 - \frac{J_1K_3}{K_2}], \quad D_{13} = D_6 - \frac{D_5D_{10}}{D_9} , \quad D_{14} = \frac{K_3 - K_2}{4RK_2^2(1+K_1^4)} ,$$

$$D_{15} = \frac{(K_2-K_3)(K_3/K_2+J_1+J_2-2)}{4RK_2K_3(1+K_1^4)\sqrt{1+1/16R^2}} \quad (4)$$

From Eq.(3) the temperature efficiency or effectiveness can be determined as

$$p = \mathcal{E} = \theta_{12}(0)$$

$$= 1 + \frac{1}{D_{12}D_8}\left\{1 - \frac{K_3}{K_2} - \frac{J_2}{K_2} + \frac{J_4}{K_2} - D_1[\frac{1}{4R(1+K_1^4)} - \frac{J_2K_3 - J_4K_2}{K_3 - K_2}]\right\} \quad (5)$$

In the similar way, formulas used for calculating the temperature profile, temperature efficiency or effectiveness of 2-4 and 2-6 split flow heat exchangers can also be obtained. Only the calculated results on F and $\mathcal{E}$ values of 2-4 and 2-6 split flow heat exchangers are listed in the next section and the detailed results are not presented here for simplicity.

RESULTS AND DISCUSSIONS

1. Considering that logarithmic mean themperature difference correction factor can be determined by

$$F = \begin{cases} \frac{1}{(NTU)} \cdot \frac{\ln[(1-p)/(1-RP)]}{R - 1} , & R \neq 1 \\ \frac{1}{(NTU)} \cdot \frac{P}{1 - P} , & R = 1 \end{cases} \quad (6)$$

thus, using Eq.(5) and (6) the logarithmic mean temperature difference correction factor and the temperature efficiency or effectiveness of 2-8 split flow heat exchanger can be calculated. The results are shown in Fig.3 and 4. The corresponding results for 2-4 and 2-6 split flow heat exchangers are also shown in Fig.3 and 4.

2. The complete analytical expressions including the temperature profiles, LMTD correction factor, and effectiveness of 1-n and 2-n split flow heat exchangers can be obtained. As an calculated example, the temperature profiles of the shell-and tubeside streams by Eq.(3) are shown in Fig.5. It is observed that

in some portions of the heat transfer surface "reverse heat transfer" takes place (hatched region in the figure). It is obviously undesirable.

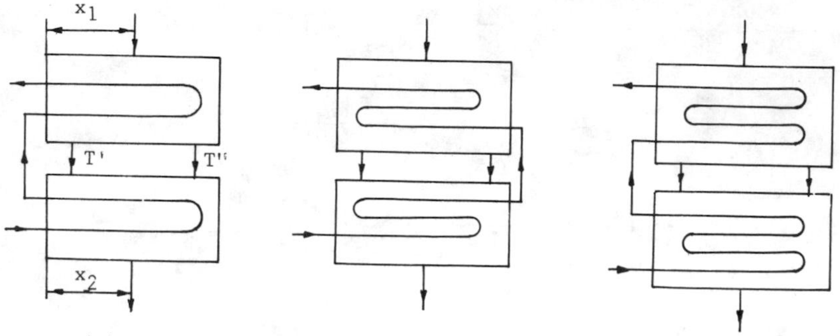

(a) 2-4 split (b) 2-6 split (c) 2-8 split

Fig.1 Schematic diagram of 2-n split flow heat exchangers.

3. It is shown that for different values of P (or (NTU)) the factor F (or P and ε) for 2-4 split flow heat exchanger is slightly different from that for 2-8 split flow heat exchanger, and the factor F (or P and ε) for 2-6 split

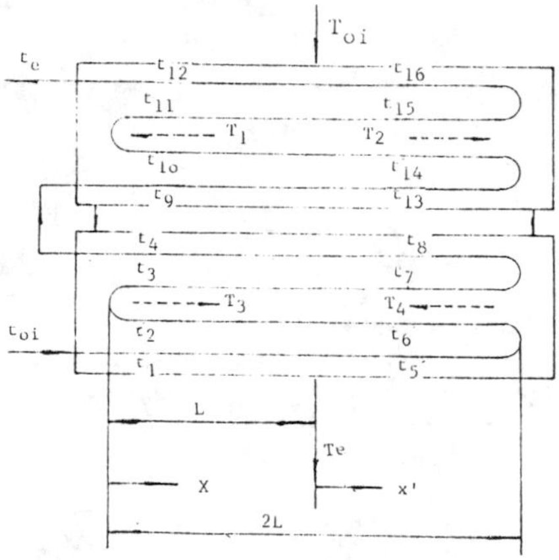

Fig.2. Thermal subregions for 2-8 split flow exchanger geometry.

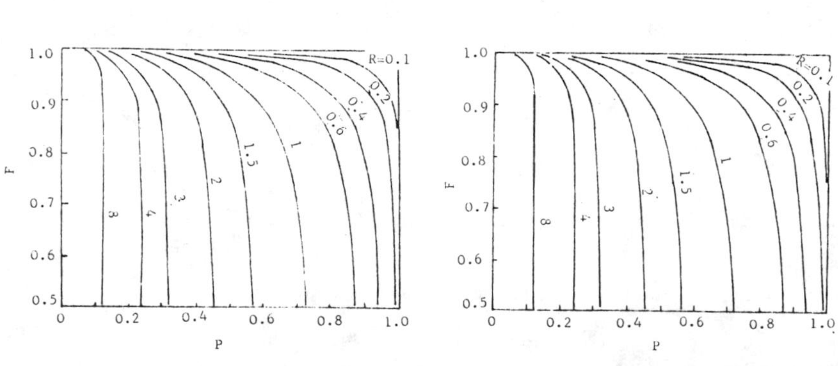

(a) 2-4 split (c) 2-8 split

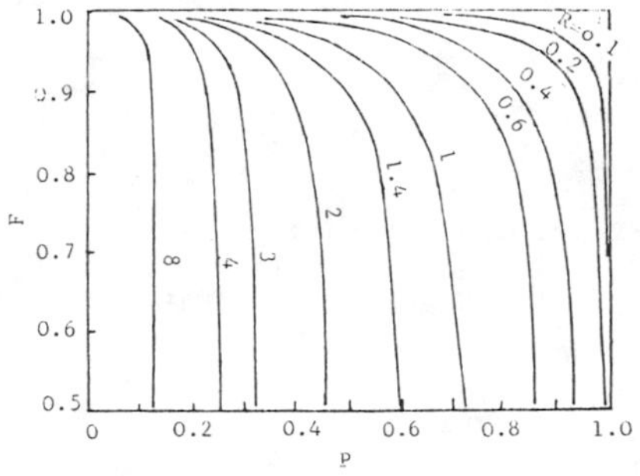

Fig.3. Temperature correction factors for 2-n split flow heat exchangers.

(b) 2-6 split

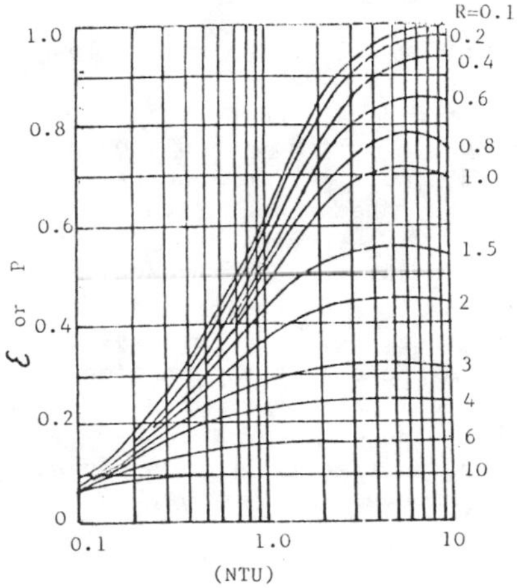

(a) 2–4 split

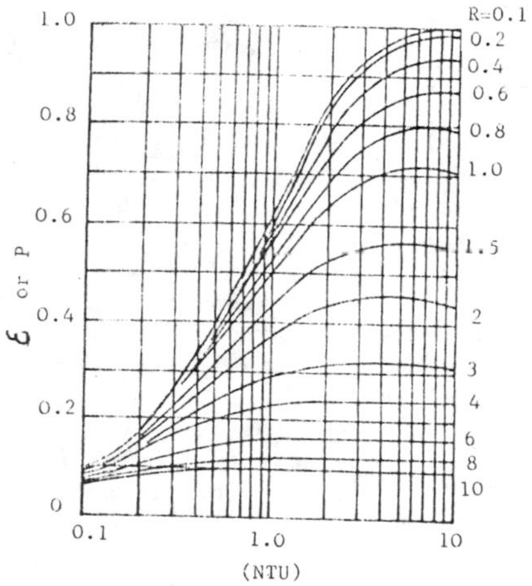

(c) 2–8 split

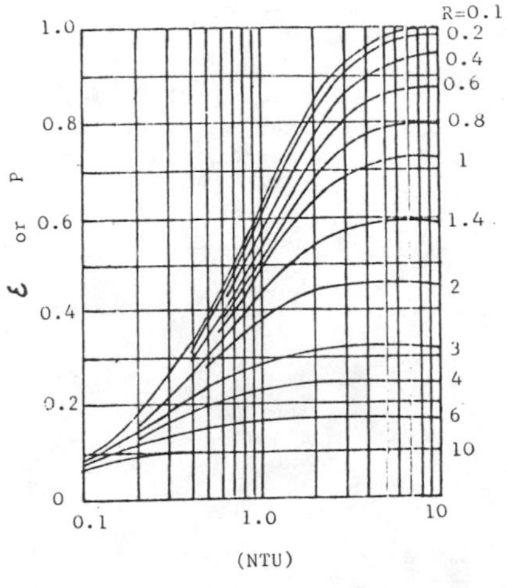

Fig.4. Temperature efficiency curves for 2-n split flow heat exchangers.

(b) 2-6 split

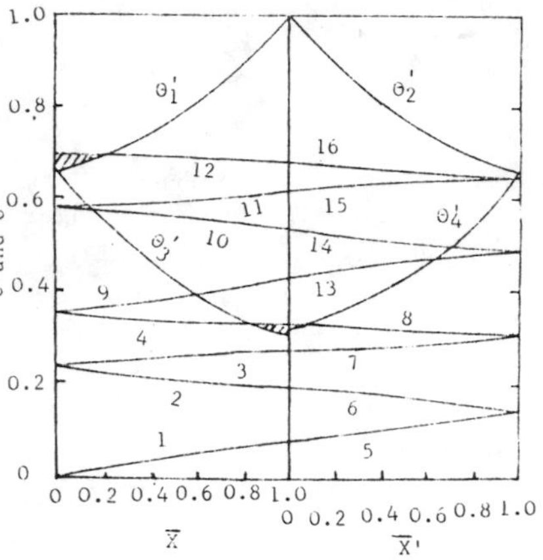

Fig.5. Temperature profile in 2-8 split flow heat exchanger. (R=1,(NTU)=2)

flow heat exchanger is somewhat different from that for 2-4 or 2-8 split flow heat exchagers (see Fig.3 and 4).

4. The calculated results show that the shell fluid exit temperature of the upper hemishell $T'' > T'$ (see Fig.1). In reference [2], it has been proved rigorously for the case in Fig.1(a), and so Murty's result [1] is incorrect.

5. To reduce or even eliminate the effect of the reverse heat transfer segment, the asymmetric nozzle location was suggested in discussion on 2-4 split flow heat exchanger by Singh and Holtz [3]. Based on the method proposed in this paper, the analytical expressions of temperature profile and effectiveness can be obtained for variable nozzle location x_1 and x_2 (see Fig.1, let $x_1 \neq x_2$). Then let the independent variables be x_1 and x_2, and the objective function to be optimized be effectiveness ε . The optimun nozzle locations x_1 and x_2 to maximize the effectiveness for certain R and (NTU) value can be determined by simplex method. The result shows that the increase in effectiveness is generally less than 1% or so in the meaningful design range, however, and so Singh and Holtz's proposal may not be desirable.

ACKNOWLEDGMENTS

The author thanks prof. Shi Jun for his encouragement. Thanks are also due to Mr. Liu Zhuhai for the preparation of Fig.3(c) and 4(c).

REFERENCES

[1] Murty, K.N., Int.J. Heat Mass Transfer, Vol. 26, p.1571 (1983).
[2] Zhuang Zhenwan, Analysis of 1-2 Split Flow Heat Exchanger, Presented at National Conference On Heat And Mass Transfer, Wuhan, China, (1984)
[3] Singh, K.P., A.I.Ch.E. Symposium Series, Vol. 75, No. 189, p.219 (1979)

INDUSTRIAL HEAT TRANSFER

Heat Flux Measurements in Experimental Thermophysics

O. A. GERASHCHENKO
USSR

Measurements of temperature and heat flux density are of great importance in experimental studies and technology. There is a well developed traditional equipment and metrology for temperature measurements.

The second trend of the same importance was "a virgin land" for a long time. That is why 30 years ago the Institute of Engineering Thermophysics has initiated the research into heat flux measurements.

Heat flux measurements imply the combination of means and methods of obtainin experimental data on heat flux density under various conditions[I].

Due to the "virgin" nature of the studies carried out they cover theory, technology, metrology and application.

Theory is associated with creating mathematical models of thermal and electrical conductivity in the body of primary sensing elements allowing for thermoelectric effects in isotropic and anisotropic fields. Analytical and numerical solutions obtained relate the electrical signal being generated to the heat flux density to be measured which changes arbitrarily across the sensing element in space and time. The direct problem, i.e. determination of the electrical signal variation at the given variation and distribution of the heat flux density can be solved completely for all possible cases. In practice one deals sometimes with the reverse formulation. Incorectness of the reverse problem can be avoided by some experimental means. Due to the sensing elements small size it is possible in the most cases to neglect variation in the heat flux density being measured. This simplifies the problem considerably. In such a formulation the problem can be solved completely.

Technology plays an ever growing role in all branches of life. Heat flux measurements are also closely connected with the technology. The modern state is characterized by density of packing which makes it possible to reach 2.000 of single elements per square cm at the battery transducer thickness about Imm. The modern methods make it possible to increase the density of packing by two orders of magnitude, i.e. up to $2.IO^5$ but there is no practical need in it as yet. The commercial transducers allow to record reliably heat fluxes beginning from the density of $IO^{-3}Wm^{-2}$. While studying the phenomena with such low density the main difficulties are connected with the insulating the process under study from thermal disturbances rather than with the transducer sensitivity. Therefore the value of heat flux fluctuations in laboratories or lecture halls is $IOWm^{-2}$ which is four orders of magnitude higher than the transducer sensitivity. The upper limit of heat fluxes being measured constitutes several units of MWm^{-2}. Thus the range of values covered by measurements is about nine orders of magnitude. Naturally, such a wide range cannot be covered by a single-type transducer There are almost I5 types of transducers, this number being conditioned by peculiarities of measurements. In practice, the whole range of nine orders of magnitude is covered by 3 types of transducers.

Metrology can and should be the object of an individual report. It demands not only a profound understanding of thermophysical properties of the processes but also meeting all the requirements and rules accepted in the system of State Standards. Thermophysical grounds of metrological certification are being developed at the Institute of Engineering Thermophysics. A number of scientific establishments take part in the development of formal (in the best sense of the word) satisfying the requirements of the State Standards Committee. Thanks to the combined efforts, the calibration error for a reference means of the I-st class was reduced to 2% of the value being measured. While passing through a reference means of the 2-nd class to the working ones the error increases to 8%. This increase is essentially formal. Our goal is to decrease the calibration error not formally but essentially. For this purpose at the Institute of Engineering Thermophysics a heat flux measuring comparator has been created with an error of comparison less than 0.I%. At the institutes of the State Standards Committee laser standards of heat flux density are being developed with an error less than 0.5%. Thus in the nearest future transducers will be available with an error of measurements down to I%.

Application. The "virgin" nature of heat flux measurements development has cleared the way to their wide application. In the interest of science and technology, more than I40 types of new instruments have been developed. The sensing elements with unique characteristics have acquired good reputation which promoted great demand from various scientific institutes and industrial plants. To satisfy this demand, a small manufacturing division was created with the output of 2.000 pieces a year. Under the influence of our Institute the State Design Office for Thermophysical Instruments making has initiated the production of semiconducting heat flux measuring transducers at the beginning of the 70-s. These instruments are characterized by high energy factor at lower constancy of properties and more narrow temperature range than the metallic ones. Institute of Engineering Thermophysics alone has manufactured for other organizations, those in other countries including, more than IO thousand transducers. Obviously, just these tnes of thousands of new sources of primary information contributed most of all into the development of modern thermophysics. Due to substantial increase of the thermophysical experiment culture induced by heat flux measurements approach the search for the higher orders non-linearity, i.e. dependence of the material thermal conductivity on the temperature gradient value and direction have ceased completely by the beginning of the 70-s[2,3]. The search has lasted about one century and more than once led to false discoveries in semiconducting heat conductionwhich as if corresponded to the electrical rectifying effect. The value of the scientific research is determined by a number of old problems being solved rather than by a number of new problems to be studied.

The possibility to measure heat flux density has allowed to measure directly heat losses in various branches of technology, power engineering etc. Heat loss meters of ITP series are widely known now. There arised a necessity to control heat losses through the building enclosure due to changes in energy-economic balance and the fuel becoming more expensive.

By combined efforts of the Institute of Engineering Thermophysics and some building research institutes two standards were developed and approved by the USSR State Standard Committee - GOST-25380-82 and GOST-26254-84. They regulate the methods of determining heat losses in commercial and residential buildings. To provide these measurements, special instruments were developed and produced. Nowadays it is not only possible but also obligatory to control heat losses from the buildings. The importance of this work can be judged by the fact that through the building enclosure almost 30% of the fuel energy is lost into the atmosphere. Decrease of heat losses by 3% is equivalent to I% economy of the total fuel produced in the country. The instrument full-scale tests revealed some design and constructional drawbacks of some buildings after which certain measures were developed to create comfort conditions inside the buildings. New instruments of our Institute make it possible to set up radical measures on rational use of thermal energy.

Certain thermodynamic relations allow to conclude that heat inputs in cryogenics are not less actual than heat losses in heat engineering which is confirmed not only theoretically but also practically.

New sensing elements have broadened the boundaries of the experimental determination of thermophysical properties. The material thermal conductivity is determined as the ratio of the heat flux density to the temperature gradient. Direct precision measurements of the heat flux density reduce the difficulties by halves. On the basis of these measurements, a series of instruments IT has been developed for measuring thermal conductivity of non-metals within the temperature range from IOO to 360K. One of the modifications has successfully passed State acceptance tests and then control tests and the Institute got the right to manufacture these instruments. At present more than two hundred of such instruments are operated in the USSR and abroad. This allows to think that in our country more than a half of non-metals thermal conductivity measurements are carried out with our Institute's instruments. They possess the highest sensitivity and the results reproducibility.

Closed heat measuring shells used in calorimétry have some advantages against other traditional schemes which consist in the possibility to adopt the calorimetric cell to specific (technological) conditions of the processes under study. In this connection new successful solutions were obtained for control of the processes of products cultivation and drying in ferment and microbiological industry. Here the intensity of heat release specifies in many cases the stages of the culture growth. Such calorimetry becomes the source of information for control and automatization of technological processes in such a perspective branch as microbiological industry.

Measurements of radiant heat flux density has great advantages because an error at calculation of fluxes into temperatures decreases 4 times according to Stefan-Boltzmann law. Depending on various conditions of heat flux measurements more than 30 types of contactless thermometers have been manufactured. In Fig.I RAPP-3 is shown - an instrument developed in cooperation with our colleagues - machine builders. The prototypes of these instruments were tested successfully at thermal power plants during a period of more than 5.000 hours. The experimental results obtained allow to fulfill the effective guidance of combustion processes in furnaces of steam generators at electric power plants. Up to recently, these processes were carried out blindly.

Radiometer "Quantum-RT" developed in cooperation with the All-Union Research Institute of mine-rescue work facilitates prognoses of endogenic fires in mines coal beds 2 months before on the basis of contactless measurements of the rock. The old truth that it is easier to prevent the fire than to quench it has been confirmed in practice. In april-may I984 "Quantum-RT" underwent State acceptance tests at numerous mines of Donbass and was recommended to production with the highest degree of quality. Batch production of "Quantum-RT" was initiated at "Torch" plant in Makejevka for coal mining industry.

For needs of medicine and veterinary sciences a series of PSI instruments has been developed which is presented in the poster as PSI-II. The period of a single measurement is several seconds.

New possibilities opened the ways to new solutions. One of them is connected with direct measurement of pumps and hydraulic turbines efficiency. The signal panel of the instrument displays the value of the throughflow machine efficiency. The process of pumping is presented by -diagram for water in Fig.2. This approach is taken from the theory of gas and steam turbines. The value of the pump efficiency is determined by the ratio of differences of heat content in the ideal and the real processes; the differences of heat content are found by temperature gradients in the process of throttling a small amount of liquid which

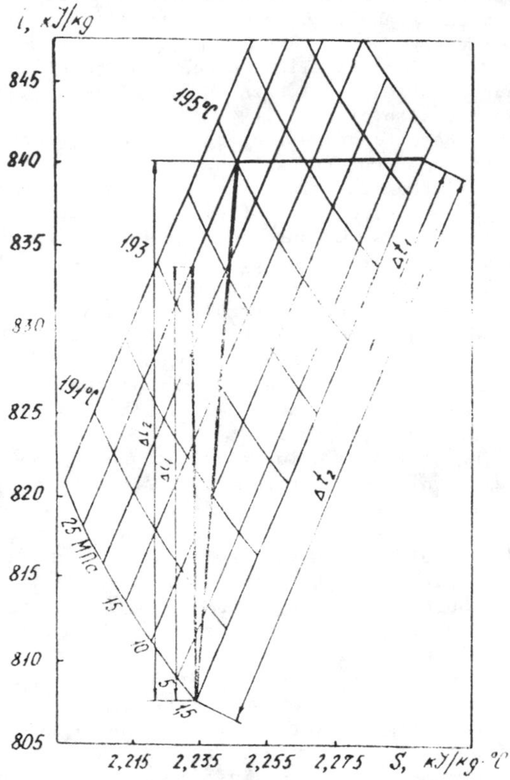

$$\eta_{oi} = \frac{\Delta i_1}{\Delta i_2} = \frac{\left(\frac{\partial i}{\partial t}\right)_S}{\left(\frac{\partial P}{\partial t}\right)_S \left(\frac{\partial i}{\partial t}\right)_P} \frac{\Delta t_1}{\Delta t_2}$$

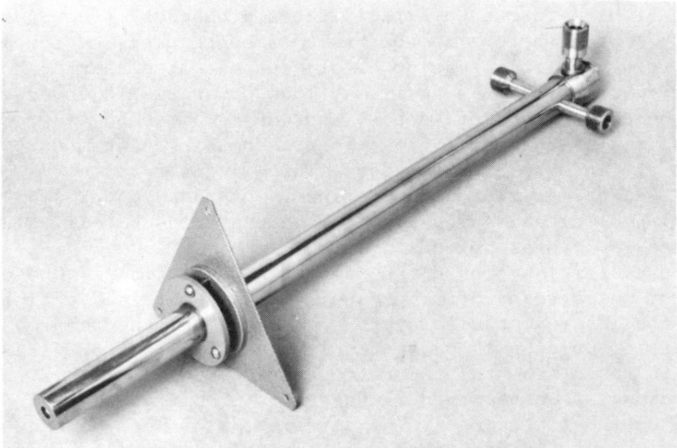

is bypassed from the discharge nozzle to the suction one.

During the whole time of heat flux measurements development, medicine was paid the greatest attention of the scientists. It should be noted that energy release is a characteristic feature of physiological processes both the normal and the pathological ones. Heat flux measurements provided the hardware in medical diagnostics by energy release. For example, the energy release of the sick gingiva differs 7 times from the normal one. In the diagnostics of the liver, the lungs, the muscles etc. by heat flux measurements it is an effective aid in addition to conventional means such as roentgen, ultrasonics, infrared vision. For diagnostics by heat flux measurements, complete absence of adverse action on the human body is typical. In our archives we have a photo of a baby-patient who was erraneously diagnosed the thymus pathology. Thanks to heat flux measurements the diagnosis was argued. Now the patient is a school-boy.

REFERENCES

I. Геращенко О.А. Основы теплометрии. - Киев: Наукова думка, I97I. - I82 с.

2. Г.Карслоу и Д.Егер. Теплопроводность твердых тел. - М.: Наука, I964. - 487 с.

3. Грищин В.А. Тепловые измерения методом текущей компенсации. - М. Энергия, I97I. - 96 с.

4. **Höfflinger W. Thermodynamische Wirkungsgradmessung an hydrostatischen Verdrängungsmaschinen nach dem Drassel-Drucktopf-Verfahren-Ölhydraulik und Pneumatik. - I976, vol.20, N 6, p.426-428.**

5. Witt K. Thermodynamisches Messen in der Ölhydraulik. Ölhydraulik und Pneumatik, I977, v.2I, N 3, p.I6I-I68.

6. Макаров Р.А., Шолом А.М. Термодинамический метод диагностирования составных частей гидропровода. Строительные и дорожные машины, I97[illegible]

Heat Transfer and Thermal Storage Characteristics of Optically Semitransparent Material Packed-Bed Solar Air Heater

MASANOBU HASATANI, YOSHINORI ITAYA, KOJI ADACHI, and HITOKI MATSUDA
Department of Chemical Engineering
Nagoya University
Furo-cho, Chikusa-ku, Nagoya 464, Japan

1. INTRODUCTION

Solar air heaters begin to be increasingly paid vigorous attention since they would be available for low-grade heating systems of air at temperature up to about 100 °C, air conditioning, drying of grains, foods, coals and so on, the regeneration of desiccant and others. In flat plate type of solar air heaters, however, since the heat transfer rate between the absorber on which solar radiation can be transformed into heat and the air is usually much lower than the case of water heating, it is difficult to obtain the high efficiency of the energy collection. In order to enhance the heat transfer between the absorber and the air, for instance, the following devices have been proposed and those heat transfer characteristics have been reported since Close [1]; setting of the absorber under the air channel [2], the use of the V-shaped absorber [3], the multi-pass mode of air channel [4], and packing of porous materials in the air channel [6] etc. Further, it is to be desired that a solar energy utilization system should also have a heat storage effect in order to compensate for an irregular supply of solar radiation. But the air itself can not be used for heat storage because of low sensible heat of the air and then a separate heat storage system such as a rock bed is substantially necessary. This heat storage system would also decrease the overall thermal efficiency of the air heater. Few researches including the heat storage system have been seen with the exception of the report by Mishra and Sharma [5], in which they tested the performance of a packed-bed solar air heater with iron-chips, aluminum-chips and pebbles.

In this work, an optically semitransparent materials packed-bed solar air heater, which may have two advantages to the flat plate solar air heaters was proposed; (1) it has accelerative effects of heat transfer rate due to increases in the optical depth of the air layer and the heat transfer area, and (2) the packed material itself can be used as a heat storage material. As one of basic steps of the R&D on the solar air heater of the proposed type, the effects of the thermal and optical properties of packed materials and the air flow rate on the overall efficiency of the energy collection and thermal storage were investigated theoretically and experimentally. Then the applicability of this type of solar air heater to a practical solar system was discussed comparing with the flat plate type.

2. THEORY

2.1 Basic Equation

Consider the optically semitransparent material packed-bed solar air heater

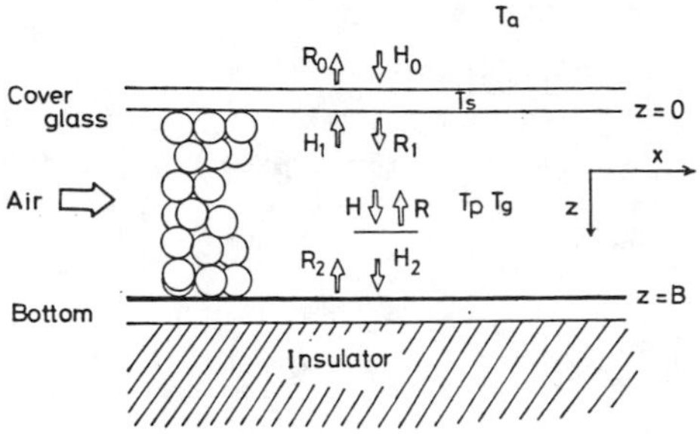

FIGURE 1. Radiative heat transfer model

shown in FIGURE 1. The channel with a cover glass of width D, depth B and length L, of which an aspect ratio is comparatively small is uniformly heated from the upper side by radiation. In the channel, an air is flowing with a constant and uniform velocity u in the direction x. Assuming that the temperature distributions within each packed material and the cover glass are uniform, conductive heat transfer in the flow direction being negligibly small, natural convection not being generated within the channel, and that the physical properties of the packed material are independent of the temperature, each basic equation is given for the unsteady heat transfer in the channel as follows:
(packed material)

$$C_p\rho_p(1 - \varepsilon_p)\frac{\partial T_p}{\partial \theta} = \frac{\partial}{\partial z}\left(\lambda_e \frac{\partial T_b}{\partial z} - Q_r\right) - h_p a_p (T_p - T_g) \tag{1}$$

The initial and boundary conditions for Eq.(1) are:

$$\theta = 0 \ , \ 0 \leqq z \leqq B \ ; \ T_p = T_0 \tag{2}$$

$$\theta \geqq 0 \ , \ z = 0 \qquad ; \ \lambda_e \frac{dT_b}{dz} + h_w(T_p - T_s) = 0 \tag{3}$$

$$\theta \geqq 0 \ , \ z = B \qquad ; \ \lambda_e \frac{dT_b}{dz} - Q_r = 0 \tag{4}$$

(air)

$$C_g\rho_g\varepsilon_p\frac{\partial T_g}{\partial \theta} = - h_p a_p (T_g - T_p) - C_g\rho_g u\frac{\partial T_g}{\partial x} \tag{5}$$

For Eq.(5) the initial and boundary conditions become

$$\theta = 0 \ , \ 0 \leqq x \leqq L \ ; \ T_g = T_0 \tag{6}$$

$$\theta \geqq 0 \ , \ x = 0 \qquad ; \ T_g = T_0 \tag{7}$$

(cover glass)

$$C_s\rho_s d_s\frac{dT_s}{d\theta} = - h_g(T_s - T_a) - h_w(T_s - T_p) + Q_{r0} - Q_{r1} \tag{8}$$

and initial condition is given

$$\theta = 0 \ ; \ T_s = T_0 \tag{9}$$

where T_b is a temperature of the packed-bed and could be approximated by the average temperature of the packing and the air as:

$$T_b = \frac{T_p + T_g}{2} \tag{10}$$

2.2 Radiative Heat Transfer Equations

We make the following assumptions for the radiative heat transfer:

1) the radiative heat transfer is one-dimensional in the direction z.
2) the packed-bed is an isotropically homogeneous semitransparent layer for incident radiation.
3) the effect of scattering is negligibly small.
4) the incident radiation from the heat source is diffuse.
5) the emission in the layer is involved in the effective thermal conductivity expression of the semitransparent material packed-bed.
6) the boundary surface and the packed-bed are gray materials, and the optical properties are independent of the wavelength.

Under these conditions, the relations between the radiosity R and the irradiation H shown in FIGURE 1 are given as below.

$$\left.\begin{aligned} H_0 &= Q_1 + \sigma T_a^4 & H_2 &= 2R_1E_3(\tau_0) \\ R_0 &= r_sH_0 + \varepsilon_s\sigma T_a^4 + (1 - r_s - \alpha_s)H_1 & R_2 &= (1 - \varepsilon_w)H_2 \\ R_1 &= r_sH_1 + (1 - r_s - \alpha_s)H_0 & H &= 2R_1E_3(\tau) \\ H_1 &= 2R_2E_3(\tau_0) & R &= 2R_2E_3(\tau_0 - \tau) \end{aligned}\right\} \quad (11)$$

Then the radiative heat flux Q_r is given by:

$$Q_r = H - R \quad (12)$$

2.3 Effective Thermal Conductivity and Wall Heat Transfer Coefficient

As described in the above Assumption (5), in this work, the effect of the emission in the bed has been tentatively involved in the effective thermal conductivity and the wall heat transfer coefficient. At first, according to the previous paper [7], the stagnant effective thermal conductivity of the semitransparent material packed-bed has been calculated.

$$\lambda_e^0 = \frac{1 + \varepsilon_p^{1.3}}{\dfrac{1}{\lambda_p + h_{rv}d_p\varepsilon_p^{1.3}/(1 - \varepsilon_p^{1.3})} + \dfrac{1}{(\lambda_g/\phi) + h_{rs}d_p}}$$

$$+ \frac{4h_{rv}\varepsilon_p^{1.3}d_p\lambda_p}{\lambda_p + h_{rv}d_p\varepsilon_p^{1.3}/(1 - \varepsilon_p^{1.3}) + (\lambda_g/\phi) + h_{rs}d_p} + \frac{16n^2\sigma}{3\alpha}T^3 \quad (13)$$

where h_{rv} and h_{rs} are radiative heat transfer coefficients [8] and are given by

$$\left.\begin{aligned} h_{rv} &= [0.1952/\{1 + \frac{\varepsilon_p}{2(1 - \varepsilon_p)}\frac{1 - \varepsilon}{\varepsilon}\}](\frac{T_b}{100})^3 \\ h_{rs} &= 0.1952\{\varepsilon/(2 - \varepsilon)\}(\frac{T_b}{100})^3 \end{aligned}\right\} \quad (14)$$

Then substituting the λ_e^0 obtained by Eq.(13) into Yagi and Kunii's equation [8], the effective thermal conductivity in the normal direction z to the airflow could be estimated. The wall heat transfer coefficient h_w has been also estimated as according to the similar idea [9] as

$$\frac{1}{h_wd_p/\lambda_g} = \frac{1}{2}\{\frac{1}{\lambda_{ew}/\lambda_g} - \frac{1}{\lambda_e/\lambda_g}\} \quad (15)$$

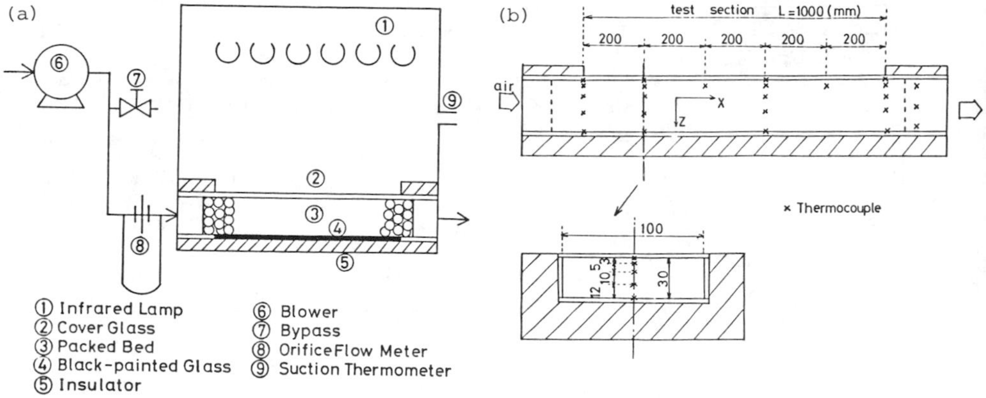

FIGURE 2. Experimental apparatus

where λ_{ew}/λ_g could be seen in the literature [9].

In order to solve the basic equations, the similar dimensionless variables to those defined in the previous papaer [10] have been introduced.Then the basic equations have been solved by the numerical method [10] to obtain the theoretical time-changes of the temperature distribution in the packed-bed.

3. EXPERIMENTAL APPARATUS AND PROCEDURES

A schematic diagram of the experimental apparatus employed is shown in FIGURE 2. An air channel was made of glass with the test section of 1000 mm in length, 100 mm in width and 30 mm in depth. The side wall, the bottom and the other top of the channel than the test section were thermally insulated with Styrofoam. The bottom surface of the test section was painted black and the top surface was covered with a glass plate of 3 mm in thickness. Glass beads of 3.4 mm, 5.2 mm and 10.0 mm in diameter and glass tubes of 5 mm in outer-diameter and 4 mm in inner-diameter as semitransparent packings and porcelain beads of 3 mm in diameter as opaque packings were employed, respectively. As a radiative heat source model, eight infrared lamps (100V-125W) were used and were arranged in two lines to obtain a uniform heat flux along the flow direction of the channel. The time-change of temperature distribution in the packed-bed was measured by 100 μmϕ CA-thermocouples placed at the position shown in FIGURE 2(b). The pressure drop between the inlet and the outlet of the bed was also measured.

The experimental procedures are as follows:After it was confirmed that the air flow and the bed temperature were steadily uniform and constant, the channel was stepwisely heated by the lamps and then the time-change of the temperature distribution in the bed was measured. When the steady state of the temperature was confirmed under heating conditions, the lamps heating was cut off and then transient cooling temperature distribution was measured. Further, the same measurement was conducted in the case of a usual flat plate air heater, no packing in the channel, for the purpose of comparison.

The entire measuring apparatus was surrounded with an aluminum foil and a vinyl sheet to ensure diffuse irradiation- and natural convection-conditions on the cover glass of the channel.

4. RESULTS AND DISCUSSION

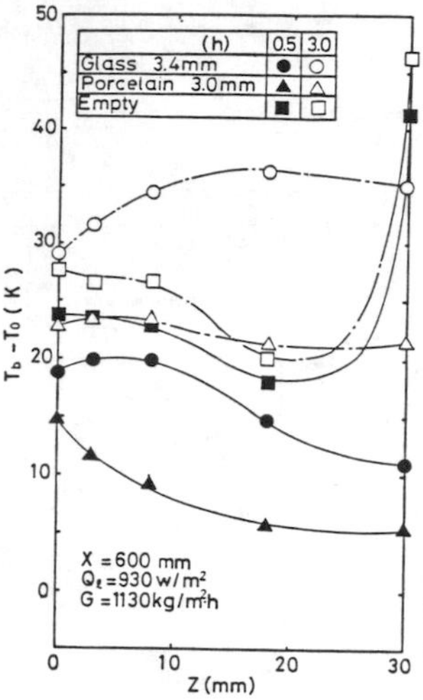

FIGURE 3. Effect of optical properties of packed material on temperature distribution

TABLE 1. Packed materials used in this experiment

Packed Materials	d_p [mm]	ε_p [-]	a_p [m^2/m^3]	α [1/m]
glass beads	3.4	0.41	1140	70
	5.2	0.43	430	60
	10.0	0.46	360	50
porcelain beads	3.0	0.43	1120	-
glass tubes	5.0	0.76	960	-

4.1 Characteristic Values Used in Theoretical Calculations

The other characteristic values than those described in 2.3, which were used in solving theoretically the basic equations, were estimated or measured in the following manners:
The absorption coefficient of the packed-bed was experimentally determined from the results of a stagnant experiment by a trial-and-error procedure. It was to find the absorption coefficient satisfying that the theoretical, steady state temperature distribution at the air flow velocity u=0 showed a good agreement with the experimental data obtained by radiative heating of the stagnant packed-bed. The absorption coefficients thus determined are listed in TABLE 1.

For the optical properties of the cover glass, α_s=0.25, r_s=0.09 and ε_s=0.90 were used, which were calculated from the refractive index n=1.5 and the absorption coefficient given by Neuroth [11]. ε_w=0.8 was used as the emissivity of bottom wall. The void fraction of the packed bed ε_p was measured and the specific surface area a_p was calculated from the ε_p and the diameter of the packing. These values are also summarized in Table 1. The other thermal properties were quoted from the literature. The heat transfer coefficient h_g on the surface of the cover glass was calculated from the authors' empirical equation [12].

4.2 Heat-Trap Characteristics

Temperature distributions. FIGURE 3 shows an example of the experimental results for the time-change of the temperature distribution in the bed at x=600 mm, which is compared with the results of the porcelain beads packed-bed and the empty channel. In the case of the empty, the temperatures of the cover glass and the bottom surface were remarkably higher than that of the air layer. In this case, the incident radiation is absorbed only by the cover glass and on the bottom surface and the air is heated only by convective heat transfer from the glass and the bottom. The temperature distribution in the porcelain beads packed-bed showed high temperature on the top surface at the first period of heating and became gradually uniform to the steady state (θ=3.0 h) since a large portion of the incident radiation is absorbed around the top of the bed and the heat absorbed there is transferred into the bed by thermal conduction and convection. While, in the case of the glass beads packing, the temperature distribution was remarkably different from the results of those two cases, that

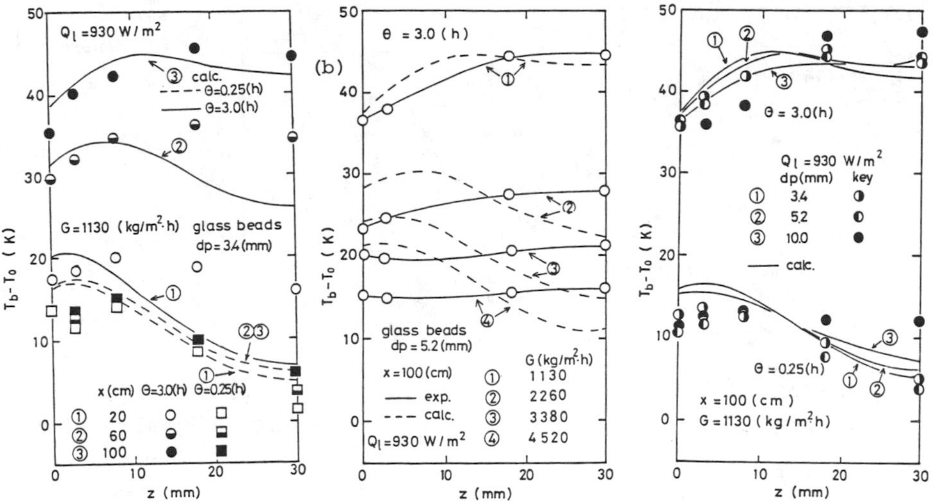

FIGURE 4. Comparison of the experimental data with the calculated results for unsteady temperature distribution within the packed bed

is, the temperature peak was observed within the bed.

FIGURE 4 (a) shows the temperature distribution change in the air flow direction x. The position of the temperature peak in the bed shifted from the top to the bottom with increase in x or θ. This tendency would be desirable from the viewpoint of improvement in heat collection efficiency since the top surface temperature is the lower, the heat loss from the top the lower. FIGURE 4 (b) and (c) show the effects of the air flow velocity and the packed glass beads diameter on the temperature distribution, respectively. The experimental temperature distribution is seen from FIGURE 4 (b) to become more uniform with increase in the air flow velocity. Any remarkable effect of the packed glass beads diameter on the temperature distribution was not observed from FIGURE 4 (c) except that the absorption coefficient of the packed-bed slightly decreases and the bottom temperature becomes slightly higher with increase in packed glass beads diameter.

Through FIGURE 4 (a) to (c), the experimental data are compared with the theoretical calculations. As is seen from the Figure, the agreement between the theoretical results and the experimental data was fairly good in the case of low air flow velocity ($G \lt \sim 1500$ kg/m²h). However, fairly large differences between them appeared with increase in the air velocity. This disagreement may be due to the following reason; the air flow is not really a plug flow but is nonuniformly distributed in the packed-bed because of biased packing conditions or end effects though the plug flow is assumed in the theoretical model.

The temperature distributions in FIGURE 4 show similar configuration to that of the semitransparent liquid heated by radiation [13], and then the glass beads packing is considered effective to increase the optical depth of the air layer for enhancement of the heat trap.

Heat trap efficiency. FIGURE 5 shows the effects of experimental parameters on the heat trap efficiency defined by the following equation:

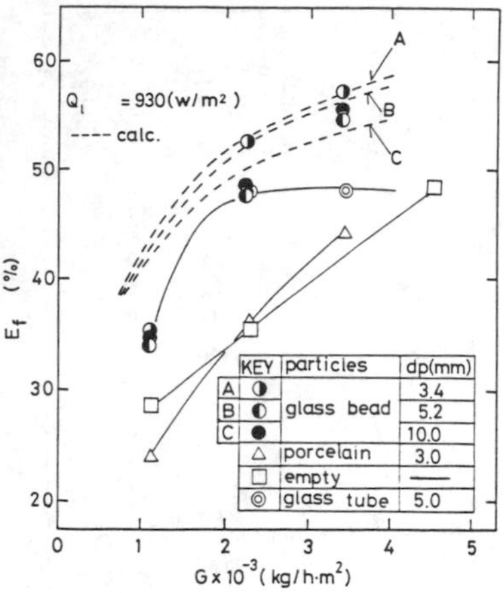

FIGURE 5. Comparison of efficiency of heat collection in various experimental conditions

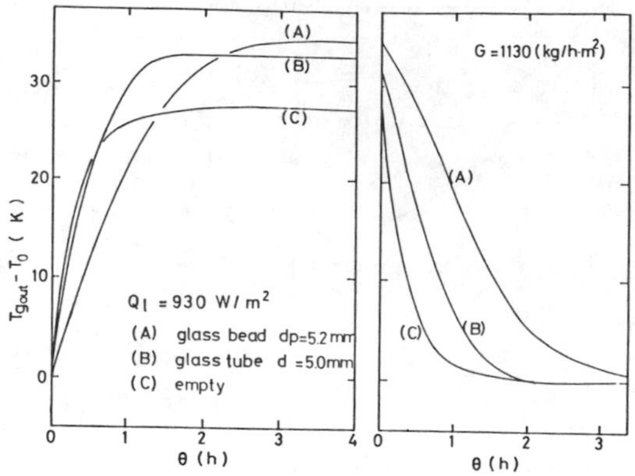

FIGURE 6. Time-change of outlet air temperature

$$E_f = \frac{C_g GB(T_{gout} - T_0)}{Q_1 L} \times 100 \quad (\%) \tag{16}$$

In the range of the experimental conditions employed, the glass beads packing show 5 to 15 % higher heat trap efficiency in an absolute value than those of the empty or the porcelain beads packings. Only little effect of the glass beads diameter was observed in the range of d_p 3.4 to 10.0 mm. In the Figure, the result in the case of glass tubes packing is also indicated. When G<∿2000 kg/m^2h, the efficiency was almost equal to those in the case of the glass beads packings.

4.3 Heat Storage Characteristic

FIGURE 6 indicates an example of the experimental results for the time-change of the outlet air mean temperature in the heating period as well as that in the cooling period. As already described, in the cooling experiment, the lamps heating was stepwisely cut off keeping the other conditions the same. The steady outlet air temperature in the case of the glass beads packing was about 8 °C higher than that of no packing. The hot air of higher temperature over 10 °C than the inlet was obtained for 0.3 h in no packing and for 1.7 h in the glass beads packing after cutting off the lamps. This means that the glass beads packing has a considerable heat storage effect by its heat capacity. However, in the case of glass tubes packing, it was obtained only for 0.9 h. This may be due to larger void fraction, that is, the heat capacity decreases with increase in void fraction and then a heat storage effect decreases, too.

5. CONCLUSION

An optically semitransparent materials packed-bed solar air heater was proposed and the heat collection and storage characteristics were investigated theoretically and experimentally.

It was observed from the experimental data analysis that the solar air heater in which semitransparent materials like glass beads or a glass tubes were used for a heat collection- and storage-material had higher efficiency of the energy collection and the thermal storage than a usual flat plate collector or the collector in which opaque packings like porcelain beads were packed. Those experimental tendencies could be predicted fairly well by the numerical calculation based on a formulated model. From these results, it is likely that the proposed solar air heater has enough heat transfer and thermal storage characteristics to enable the development of a new type of solar air heater with high efficiency.

NOMENCLATURE

a_p = specific surface area
C_g, C_p, C_s = specific heats of air, packed material and cover glass, respectively
d_p = diameter of packed material
d_s = thickness of cover glass
E_3 = exponential integral function
G = mass flow rate of air
h_g = convective heat transfer coefficient between cover glass and surrounding
h_p = convective heat transfer coefficient between packed material and air
Q_l = radiative heat flux from lamp
Q_r, Q_{r0}, Q_{rl} = radiative heat fluxes
r_s = reflectivity of cover glass
T_a, T_g, T_p, T_s = temperature of surrounding, air, packed material and cover glass, respectively
T_0 = inlet and initial temperature
α = absorption coefficient
α_s = absorptivity
ε_p = void fraction of packed bed
$\varepsilon_s, \varepsilon_w$ = emissivities of cover glass and bottom wall, respectively
θ = time
λ_e = effective thermal conductivity
λ_g, λ_p = thermal conductivities of air and packed material, respectively
ρ_g, ρ_p, ρ_s = densities of air, packed material and cover glass, respectively
σ = Stefan-Boltzmann constant
τ = optical distance

Subscript

out = outlet

REFERENCES

1. Close,D.J. Solar Air Heaters, Solar Energy, vol.7, no.3, pp.117-124, 1963.
2. Bhargava,A.K., Garg,H.P. and Sharma,V.K. Evaluation of the Performance of Air Heaters of Conventional Designs, Solar Energy, vol.29, no.6, pp.523-533, 1982
3. Parker,B.F. Derivation of Efficiency and Loss Factors for Solar Air Heaters, Solar Energy, vol.26, no.1, pp.27-32, 1981.
4. Wijieysundera,N.E., Ah,L.L. and Tjioe,L.E. Thermal Performance Study of Two-Pass Solar Air Heaters, Solar Energy, vol.28, no.5, pp.363-370, 1982.
5. Mishra,C.B. and Sharma,S.P. Performance Study of Air-Heated Packed-Bed Solar-Energy Collectors, Energy, vol.6, pp.153-157, 1981.
6. Lalude,O. and Buchberg,H. Design and Application of Honeycomb Porous-Bed Solar-Air Heaters, Solar Energy, vol.13, pp.223-224, 1971.
7. Sugiyama,S., Hasatani,M. and Yada,A. Effective Thermal Conductivity of a Packed Bed of Glass Sphere, Kagaku Kogaku, vol.34, no.5, pp.545-548, 1970.
8. Yagi,S. Kunii,D. Studies on Effective Thermal Conductivities in Packed Beds, AIChE J., vol.3, no.3, pp.373-381, 1957.
9. Kunii,D., Suzuki,M. and Ono,N. Heat Transfer from Wall Surface to Packed Beds at High Reynolds Number, J. Chem. Eng. Japan, vol.1, no.1, pp.21-26, 1968.
10. Hasatani,M. and Arai,N. Unsteady, Two-Dimensional Heat and Mass Transfer in a Packed Bed of Fine Particles with an Endothermic Process, Chem. Eng. Commun., vol.10, pp.223-242, 1981.
11. Neuroth,N. Der Einfluß der Temperature auf die Spektrale Absorption von Glasern im Ultraroten, Glastechn. Ber., vol.25, pp.242-249, 1952.
12. Arai,N., Takahashi,S. and Sugiyama,S. Natural-Convection Heat and Mass Transfer from a Horizontal Upward- Facing Plane Surface to the Air, vol.5, no.5, pp.471-475, 1979.
13. Arai,N., Itaya,Y. and Hasatani,M. Development of a "Volume Heat-Trap" Type Solar Collector Using a Fine-Particle Semitransparent Liquid Suspension (FPSS as a Heat Vehicle and Heat Storage Medium, Solar Energy, vol.32, no.1, pp.49-56, 1984.

Latent Heat Recovery from Gas Flow with Water Vapor

Q. L. HWANG
Electric Power Administration of Northeast China
Shengyang, PRC

H. TANIGUCHI and K. KUDO
Hokkaido University
Sapporo 060, Japan

1. INTRODUCTION

Reduction of exhaust gas loss in boiler is effective for an increase in boiler efficiency. Low temperature corrosion has limited its wide application. Recent progress in the field of anti-corrosive materials [1,2] and in the combusion technique of wet fuels like Coal Water Slurry or Oil Water Mixture [3] makes it possible to recover even latent heat from boiler flue gas. Boiler flue gas contains 10~20 wt% of water vapor. Many investigators have studied condensation heat transfer in humid air [4,5,6,7] (water vapor content of 0~2 wt%) or steam with slight air [8,9] (water vapor content of 70~100 wt%). But the other region between these two extreme conditions has not been studied.

Condensation heat transfer is limited by the heat resistances of a condensate film and another outer boundary layer. When the main heat resistance is in the outer boundary layer, the total heat transfer is obtained by adding condensation to convection heat transfer. Small water vapor concentration is observed in the humid air case. Under such condition, the mass transfer in the outer boundary layer can be calculated by using the analogy between heat and mass transfer, since the mass flux perpendicular to a wall (Stephan flow) may be neglected. When the condensing rate is increased, the heat and mass transfer coefficients will be thought to be changed. Convection heat transfer will be neglected when water vapor including slight air is condensed and the heat resistance in the condensate film is dominant. The water vapor concentration in the boiler flue gas falls between these two extreme conditions. The objective of this study is to investigate the applicability of the heat and mass transfer equations which are obtained for humid air, water vapor with slight air and the medium water vapor content conditions. By using these results, analyses were also made on the performance of a latent heat recovery heat exchanger and on the effects of its installation on a thermal efficiency of thermoelectric power station.

2. EXPERIMENT

2.1 Experimental Apparatus

Convection and condensation heat transfer were measured on a surface of cylindrical tube in a high temperature and high humid air flow. Experimental apparatus is shown in Fig.1. Air from a blower is mixed with steam and led to a test section after heated by a gas fired air heater. Test section has a rectangular cross section of 300×120 mm as shown in Fig.2. In the test section, three horizontal copper tubes of 20 mm outer diameter, 17 mm inner diameter and 300 mm length are arranged in parallel with a pitch of 40 mm and connected in series. Two mixing chambers are attached to the both ends of inlet and outlet of the tube series. Cooling water is kept on a constant temperature and fed to the inlet mixing chamber. Inlet and outlet average temperatures of cooling water and the temperature rise between them are measured by copper-constantan thermocouples which are connected to a microvolt meter with 0.2 μV accuracy. Surface temperature of the tube is also measured by copper-constantan thermocouples of 0.32 mm diameter which are embeded in each tube wall at three points of the surface facing upstream, downstream and side wall. The water vapor concentration in a main flow can be measured by sucking the main flow gas through a gas sampling tube by a vacuum pump as shown in Fig.3. The sucked gas is cooled by a water cooler and its condensate is collected in a measuring cylinder. Ordinary sampling method applied to falling condensate drops is to underestimate the condensing rate because the drops will evaporate during these falling period in a high temperature gas conditions. So a newly developed condensate sampling device is set just beneath the tubes as shown in Fig.2 and Fig.3. This device is composed of stainless steel tubes 0.5 mm inner diameter and 30 mm length arranged 1 mm beneath the test tubes with 10 mm pitch, and a copper tube header is connected to the stainless steel tubes. All condensate on the test tubes is sucked through the stainless steel tubes by the vacuum pump just before it drops. A volume of the sucked condensate is examined by the measuring cylinder. As this device sucks not only the condensate but also adjacent humid air, so the water vapor contained in the air is subtracted from the measured condensate volume. As the sucking treatment may change a thickness of the condensate film, heat transfered to the cooling water in the test tubes should be checked under each case with or without sucking , which has showed that there is no effect of the condensate sucking on the condensation heat transfer. The condensate film on the test tube was confirmed with the eye to be filmwise. Then experimental conditions are shown in Table 1.

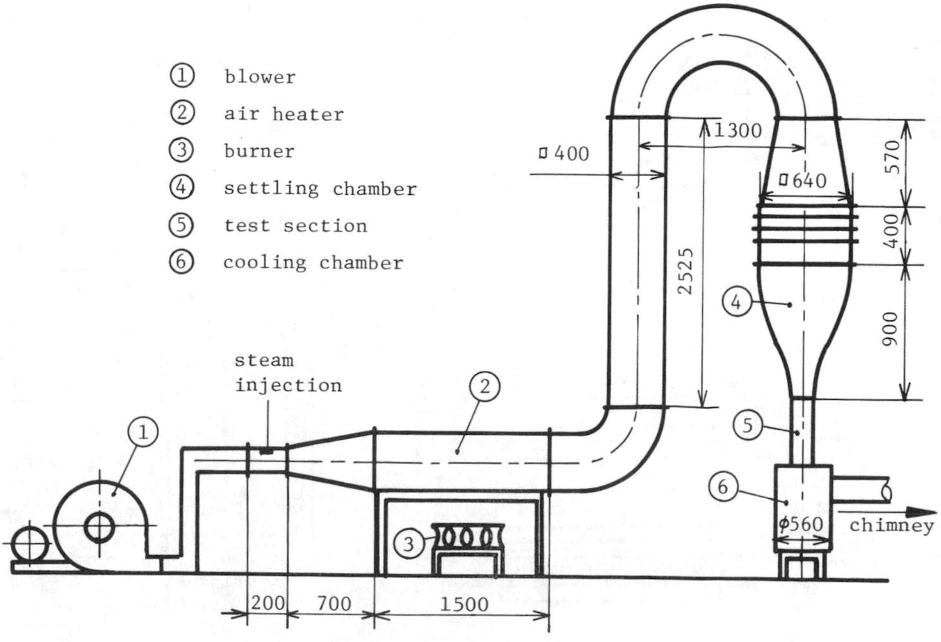

FIGURE 1. Test apparatus

2.2 Experimental Results

Property values used for each variable are obtained by the followings. Diffusion coefficient of water vapor on air can be calculated by Londolt-Bärnstein equation [8]. In this case, the film temperature of the condensate and the gas temperature in a surrounding boundary layer [4] are

$$t_f = t_w + 0.3(t_i - t_w) \tag{1}$$

$$t_r = (t_\infty + t_i)/2 \tag{2}$$

<u>Heat transfer without condensation.</u> To check the accuracy of this experiment, Nu-Re correlation is measured under a condition without condensation. This result is shown in Fig.4. The measured values represented by open circle are higher than the preceding heat transfer equations for the first row of in-line arrangement tubes [10] and staggered-arrangement tubes [11]. The difference is thought to be due to a thermal radiation from the wind tunnel wall, whose temperature is as high 378 K. So radiation corrections should be made by subtracting the estimated radiation heat transfer. This estimation was done by assuming an emissivity of the copper test tubes 0.5 (a value for oxidized copper surface) and a factor of surrounding wall to the tube unity. The compensated values are represented by solid circles in Fig.4 and the following equation:

$$Nu = 0.19Re^{0.60} \quad (Re > 4.5\times10^3) \tag{3}$$

The compensated values coincide with the preceding equations, which shows an accuracy of this experiment.

TABLE 1. Experimental conditions.

main flow				cooling water			test tube	
temp. K	vel. m/s	water vapor conc.	Reynolds number Re	inlet temp. K	temp. rise K	vel. m/s	Reynolds number	temp. K
401.9 ~ 405.2	1.45 ~ 4.66	0 ~ 0.13	2.8×10^3 ~ 9.0×10^3	280.8 ~ 289.4	0.70 ~ 3.35	0.41 ~ 0.43	4.9×10^3 ~ 5.5×10^3	292.0 ~ 303.6

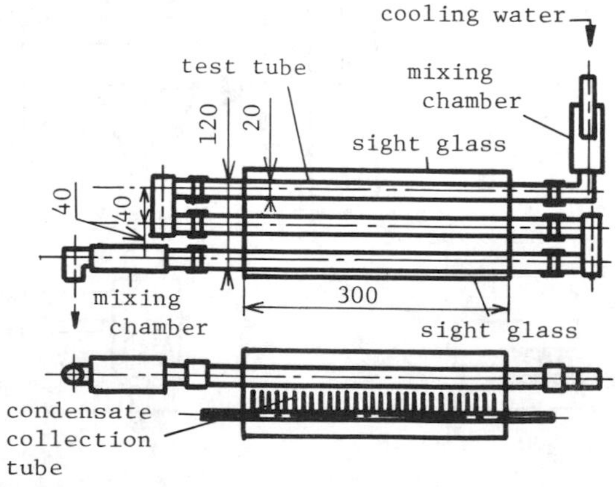

FIGURE 2. Test section.

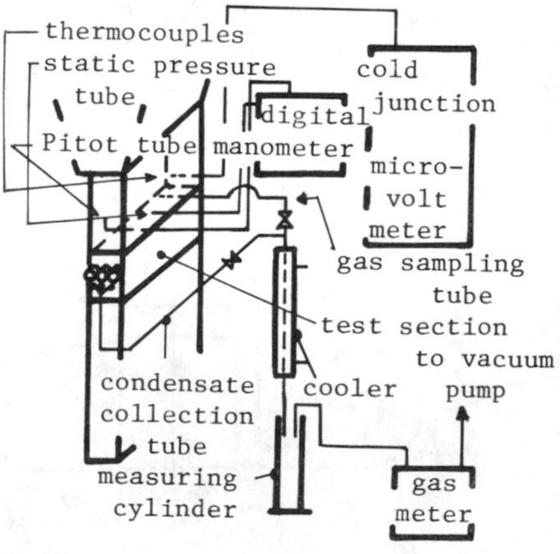

FIGURE 3. Measuring devices.

Heat and mass transfer with condensation. A result of condensation mass transfer is shown in Fig.5. The result coincides with a correction derived by Rose for the mass transfer in multicomponent boundary layer with Stephan flow [5] and a correlation for the condensation in humid air [10] and it is represented by the following equation:

$$Sh = 0.40Re^{0.49} \tag{4}$$

This equation is valid when the water vapor content $W_{v,\infty}$ is changed from 0.10 to 0.13 as shown in Fig.5. Then Fig.6 shows the convection heat transfer Nus with condensation effect. The value Nus is calculated by subtracting condensation heat transfer represented by Eq.(4) from the measured total transfered heat with radiation compensation. Under a condition of $W_{v,\infty}=0.10$, this result is represented by the following:

$$Nus = 0.31Re^{0.60} \tag{5}$$

This Nus value is higher than the Nu value in Fig.4 which is obtained without condensation. Rather high condensing rate of this experiment causes Stephan flow towards the heat transfer surface of test tube and makes a concentration gradient near the surface steeper, which leads to an increase in Nus. When the water vapor concentration in the main flow $W_{v,\infty}$ is increased from 0.10 to 0.13, the Nus value is increased as shown in Fig.6, but Sh value does not change. These results indicate the inter-correlation between the condensing rate and the convective heat transfer coefficient. Such tendency is also shown in the condensation from humid air ($W_{v,\infty}=0.023$) [9] and an increase in the Nusselt number is reported by 15 to 20 %.

3. PERFORMANCE ANALYSIS ON LATENT HEAT RECOVERY HEAT EXCHANGER

3.1 Analytical Model of Heat Exchanger

In this section, the performance of a latent heat recovery heat exchanger is

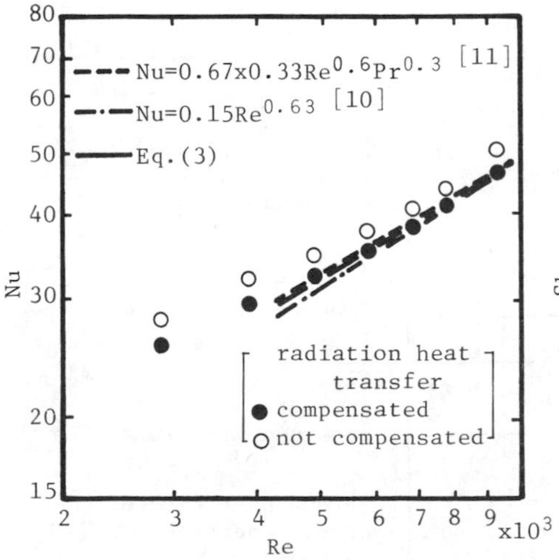

FIGURE 4. Convection heat transfer (without condensation).

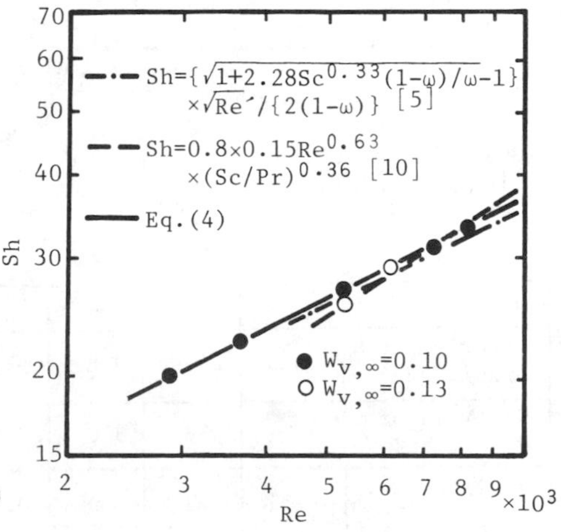

FIGURE 5. Mass transfer. during condensation.

analyzed by utilizing the results obtained in the preceding section. The analysis is made for a coal fired thermoelectric power station of 350 MW output. A schematic model of the heat exchanger is shown in Fig.7. Analytical conditions are listed in Table 2.

3.2 Analytical Results of Heat Exchanger

Fig.8 shows the analytical result of recovered enthalpy change in accordance with a variation of the cooling water inlet temperature. This figure shows that a decrease in the cooling water inlet temperature causes an increase in the recovered enthalpy and another increase from 40 to 110 % for cooling water inlet of 313~293 K is calculated in the total recovered enthalpy compared with this total enthalpy under the condition without condensation. The rightward deviation of broken curves B and D from A and C in Fig.8 indicates a latent heat release by sulfuric acid condensation. This deviation shows that the recovered heat by sulfuric acid condensation is negligibly small compared with the total recovered heat. By applying these results for water inlet 293 K, boiler efficiency increase can be calculated as shown in Table 3. The boiler flue gas loss is assumed to be 6 % before its installation. The boiler efficiency increase is 7.0 % for case A and B of higher water vapor concentration and 5.6 % for case C and D of lower water vapor concentration.

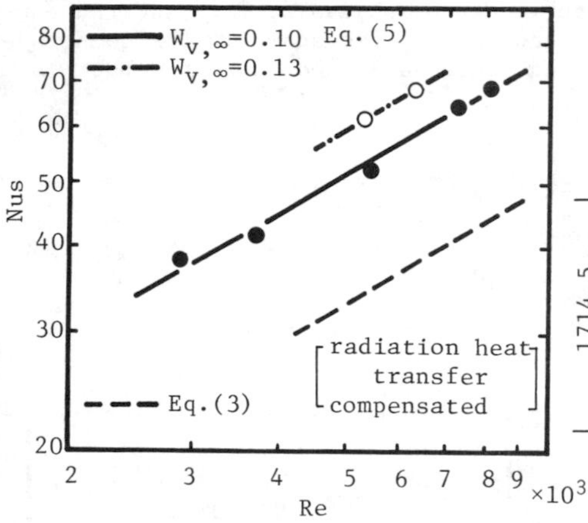

FIGURE 6. Convection heat transfer during condensation.

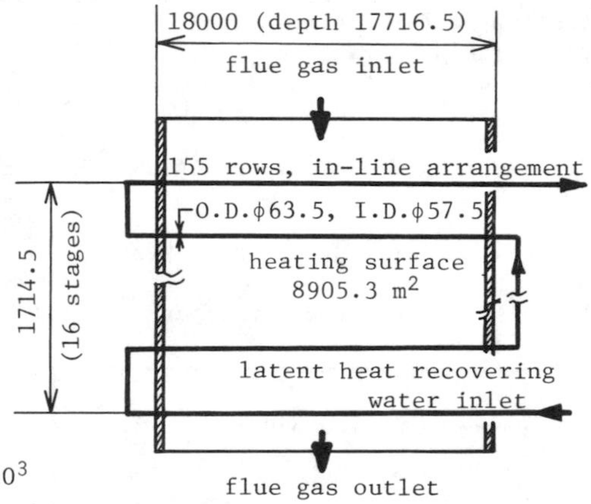

FIGURE 7. Model of latent heat recovery heat exchanger.

TABLE 2. Analytical conditions.

Case	A	B	C	D	E
Water vapor press. kPa	24.9	24.9	9.8	9.8	21.7
Water vapor dew pt. K	338	338	318	318	335
H_2SO_4 press. Pa	0	0.25	0	0.3	0
Dew pt. K	338	413	318	396	335

flue gas condition for each case

flow rate	273.6 m_N^3/s
inlet temp.	423 K
pressure	101.3 kPa

latent heat recovering water condition for each case

flow rate	277.8 kg/s
inlet temp.	278~413 K

3.3 Thermal Efficiency Increase of Thermoelectric Power Station

Increase in the thermal efficiency is analyzed for a reheat and regeneration thermoelectric power plant cycle with a latent heat recovery heat exchanger as a feed water heater or an air heater for its boiler.

Application to feed water heater. Since the recovered enthalpy by sulfuric acid condensation is negligibly small, an analysis is limited for case A, C and E in Table 2 where $SO_3=0$. A size of the heat exchanger is the same as in Fig.7 and analytical conditions are the same as in Table 2. Boiler flue gas of 423 K is fed to the heat exchanger in which the feed water is heated from inlet of 313 K. Table 4 shows the analytical result of the plant thermal efficiency and its boiler efficiency. Since this efficiency increase (values in parentheses in Table 4) in the regeneration cycle is decreased due to a reduction of bleeding rate which is caused by the latent heat recovery, the thermal efficiency increase by the heat recovery is higher in the reheat cycle than the other reheat and regeneration cycle.

Application to air heater. An analysis is made for the air heater type latent heat recovery heat exchanger with 5728.5 m^2 heating surface. Air is heated from 278 K to 405 K, while the boiler flue gas is cooled from 423 K to 313 K in the air heater. Boiler efficiency increase of 6.2 % for case A is calculated by its installation. Since the installation of the air heater has no bad effect on a cycle thermal efficiency, the plant thermal efficiency is increased as much as the boiler efficiency increase of 6.2 %.

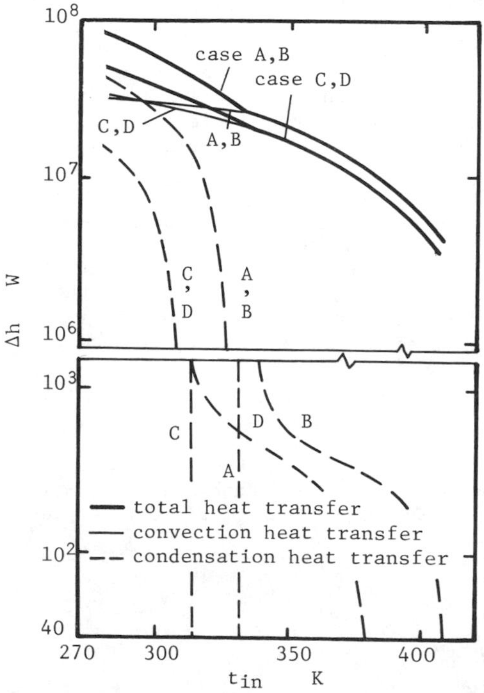

FIGURE 8. Recovered enthalpy change with inlet water temperature.

TABLE 3. Boiler efficiency increase.

Case	A,B	C,D
$P_{v,\infty}$ kPa	24.9	9.8
$t_{d,v}$ K	338	318
$\Delta\eta_B$ %	7.0	5.6

TABLE 4. Effect of latent heat recovery on plant thermal efficiency.

case	A		C		E/A	E	A
type of cycle	reheat		reheat and regeneration				
latent heat recovery	×	o	×	o	×	o	o
$p_{v,\infty}$ kPa	24.9	24.9	9.8	9.8	21.7/24.9	21.7	24.9
plant thermal eff. % (eff. increase by latent heat recovery)	38.06	40.27 (2.21)	42.89	43.33 (0.44)	42.83	43.28 (0.45)	43.31 (0.48)
boiler eff. η_B % (eff. increase by latent heat recovery)	93.61	96.82 (3.21)	93.74	96.45 (2.71)	93.61	96.33 (2.72)	96.22 (2.61)

4. CONCLUSIONS

Convection and condensation heat transfer coefficients were measured on horizontal tubes in a gas flow with water vapor content $W_{v,\infty}$ of 0.1~0.13. By using the above results analyses were made on the performance of a latent heat recovery heat exchanger and the effect of its installation in a thermoelectric power station. Following results were obtained:

1. Radiation heat transfer should be considered when the applied temperature is over than 400 K.
2. Condensation mass transfer coefficient in humid air with a water vapor content $W_{v,\infty}$ of 0.1~0.13 can be represented by the Rose's equation [5].
3. Convection heat transfer coefficient is increased by the co-existence of condensation. Its effect is changed by the water vapor content.
4. In the latent heat recovery heat exchanger, total transfered heat is increased by 40~110 % after condensation starts. But latent heat release of sulfuric acid vapor can be neglected.
5. Boiler efficiency increase of 5~7 % is obtained by using the latent heat recovery heat exchanger as the economizer. When the heat exchanger is used as the feed water heater in a reheat and regeneration cycle, the thermal efficiency increase is reduced due to the reduction of bleeding. So the thermal efficiency increase is only 0.3~0.5 % in this case. When it is used as the air heater in these cycle, the bleeding effect is not affected and the thermal efficiency increase of 6 % is obtained.

ACKNOWLEDGEMENT

The authors would like to express our appreciation to Mr. Kanefumi Suda, Mr. Akihiro Fujii, Mr. Noriyoshi Machida and Mr. Yoichiro Asano for their help and support with the experiments.

NOMENCLATURE

Nu Nusselt number
Nus Nusselt number with condensation effect
Pr Prandtl number

p pressure, Pa
Re Reynolds number for u_m
Re´ Reynolds number for u_∞
Sh Sherwood number
t temperature, K
t_d dew point temperature, K
t_{in} inlet water temperature of heat exchanger, K
u velocity, m/s
u_m average velocity of main flow at minimum area, m/s
W mass fraction
Δh recovered enthalpy change, W
$\Delta \eta_B$ boiler efficiency increase, %
(subscript)
a air
f condensate film
i condensate film surface
v water vapor
r boundary layer
w test tube surface
∞ main flow

REFERENCES

1. *R&D(KOBE STEEL Technical Review)*, vol.19, no.3, pp.2, 1969.

2. Barkley, J.F., et al., *Bureau of Mines Report of Investigation*,pp.4996,1953.

3. Liu, C.K., et al., Experimental Study on Combustion of Single Drops of Coal-Water Slurry and Washery Tailings, *Bulletin of Zhejiang University*, pp.43, 1984.

4. Fujii, T., et al., Forced Convection Condensation from a Low Pressure Steam-Air Mixture on a Horizontal Tube, *Trans. JSME(B)*,vol.47,no.417,pp.836,1984.

5. Rose, J.W., Approximate Equations for Forced-Convection Condensation in the Presence of a Non-Condensing Gas on a Flat Plate and Horizontal Tube, *Int. J. Heat and Mass Transfer*, vol.23, no.4, pp.539, 1980.

6. Yang, S.M., *Heat Transfer*, People's Educational Publishing Co.,Beijing,1981.

7. Fujii, T. and Oda, K., Effect of Air on Condensation of Steam Flowing through Tube Banks, *Trans. JSME(B)*, vol.50, no.449, pp.107, 1984.

8. Fujii, T., et al., Condensation of Water Vapor and Heat Transfer from Humid Air to Horizontal Tubes in a Bank, *Refrigeration*,vol.52,no.602,pp.1059,1977.

9. Fujii, T., et al., Condensation of Water Vapor and Heat Transfer from Humid Air to Horizontal Tubes in a Bank (2nd Report: In-Line Arrangement), *Refrigeration*, vol.57, no.658, pp.787, 1982.

10. Fujii, T., et al., Experimental Study of Heat and Mass Transfer for Humid Air to a Horizontal Tube in False Tube Bank, *Trans. JSME(B)*, vol.50, no.455, pp.1716, 1984.

11. *Engineering Data Book for Heat Transfer Calculation*, 3d ed., pp.40, JSME, 1975.

Study of Thermal Loading on the Piston of a Petrol Engine Measurement and Calculation

ZHI-MIN JIANG
Shanghai Jiao Tong University
Shanghai, PRC

ABSTRACT

Investigations of thermal loading on the piston of spark-ignition petrol engine have been carried out much less than that on the compression-ignition engine whether in P.R C. or abroad. In recent years, the rapid advance of technolohy in the car industry calls for the petrol engine to be further boosted for power, by either raising the R.P.M. or M.E.P. or both. And, as its consequence, the problem of thermal loading on the piston becomes more and more serious.

The experiments of temperature measurement at the characteristic points of the piston by using fusible plug method as well as thermocouple method are presented in this paper. The engine used by the author for this study is a 4-stroke cycle, water cooled, 6-cylinder high speed petrol engine with 100mm bore and 115mm stroke. Study of the temperature field by means of [Finite Element Method] was performed and experimental versus calculated results compared.

EXPERIMENTAL INVESTIGATION

There exists quite a number of methods, old and new, for the measurement of temperature of thermally loaded components of the engine, amongst them the thermocouple method, the fusible plug method and the tempered-hardness-plug method are perhaps the most popular ones. The author and her colleagues have carried out tests on diesel engine piston and liner with all these methods in the past and their experience showes that the fusible plug method which yields reasonable accuracy is the simpliest and quickest.

The fusible plugs used by author were produced by vacuum-method and muffle method. In the latter method melted metal are shielded by activated carbon over the liquid surface to avoid contact with air. 41 types of fusible alloys were developed to cover a temperature range from 46.7°C to 38.2°C. The majority of the alloys are eutectic. The rest are eutectic-peritectics and pure metals. The composition of our fusible alloys please see Table 1. The melting points of all the alloys have been checked to give errors $\leq$ 1%. Those which are ductile were drawn into wires of 1.5mm in diameter while the brittle ones were directly cast into shape in graphite moulds.

The test have been carried out in two stages. In the first stage, 46 measureing points were allocated for each of the piston of No. 2,3,5, & 6 cylinders in order to find out which cylinder or cylinders are thermally loaded to the utmost. Results showed that 5th and 6th cylinders gave the hightest temperatures, a fact which agrees with both the bed-test and the road-test records.

Accordingly a decision was made that the piston temperature fields of the 5th and 6th cylinder must be under investigation.

In the second stage of the test, about 200 plugs at 52 measuring points were inserted on the piston of the 6th cylinder, Fig. 1 showes their measuring point locations and numbers. Three or four plugs about 3 mm apart were inserted at the vicinity of one measuring point along the same diameter. The plug pits on the piston wall were formed by flat bottomed drilling of diameter 1.5 mm and depths 2mm in the skirt and 4mm elsewhere. The equipment was warmed up by running at low load until the temperatures of the cooling water and lubricating oil were raised to their normal levels and then brought to full load (130 BHP, 3000 RPM, 7.14 BMEP) for at least twenty minutes. Such a test was repeated four times, so as to get the final values given in Table2.

Table 1

melting point	composition of alloys %	melting point	composition of alloys %
46.7	Bi44.7Pb22.6 Sn8.3 Cd5.3 In19.1	215	Pb84.7Au15.3
58	Bi49.4Pb18.0 Sn11.6In21.0	221	Sn96.5Ag3.5
70	Bi50 Pb26.3 Sn13.3Cd10	227	Sn99.1Cu0.9
78	Bi57.5Sn17.3 Cd25.2	232	Sn100
93	Sn42 Cd14 In44	240	Pb85 Sn3.5 Sb11.5
102.5	Bi53.9Sn25.9 Cd20.2	245	Pb81.7Cd17.3Zn1
108	Sn46 In52.2 Zn1.8	248	PB82.5Cd17.5
117	Sn48 In52	254.5	Bi97.3Zn2.7
124	Bi55.5Pb44.5	266	Cd82.6Zn17.4
130	Bi56 Sn40	271	Bi100
138	Pb28.6Sn52.45Cd16.7In2.25	280	Sn20 Au80
145	Pb32 Sn49.8 Cd18.2	290	Pb95 Pt5
156.4	In100	304	Pb97.5Ag2.5
163	Sn66.5Cd31 Zn2.5	309	Cd87 Au13
170	Sn57 Ti43	318.2	Pb99.5Zn0.5
177	Pb24 Sn71 Zn5	321	Cd100
183	Pb37.7Sn62.3	327	Pb100
189	Pb40 Sn57.5 Sb2.5	356	Au88 Ge12
198	Sn91.1Zn8.9	363	Si2.85Au97.15
203.5	Cd83 Ti17	377	Zn89 Al7 Cu4
		382	Zn95 Al5

The first three tests have revealed that temperatures at the top ring groove (locations 17,20,30 & 36) were all comparatively high. It is thought that these might be attributed to the fact that the combustion gas blew directly on the fusible plug.

Therefore in the fourth test a locating pin was fitted in the groove with only 1 mm clearance to fix the ring from rotation practically. As a result, the plug with melting point 266°C at the gap melted while the other two with melting points 254°C and 270°C which were 3 mm to its left and right remained intact.

Table 2

	Temperature °C		Temperature °C		Temperature °C
1	< 304	19	163-171	37	203-221
2	290-304	20	177-189	38	221-232
3	-290	21	124-130	39	203-215
4	-280	22	271-280	40	189-203
5	280-290	23	254-266	41	189-198
6	271-280	24	227-240	42	-189
7	-280	25	-156	43	138-145
8	271-280	26	138-145	44	130-145
9	-271	27	138-145	45	-138
10	271-280	28	138-145	46	-138
11	-266	29	254-266	47	232-240
12	254-266	30	232-240	48	198-215
13	-280	31	215-227	49	198-215
14	266-271	32	-170	50	189-198
15	254-266	33	177-189	51	189-203
16	266-271	34	124-130	52	
17	248-254	35	254-266	53	
18	221-227	36	254-266	54	

The thermocouple test was carried out on the piston of the 5th cylinder*. Thermocouples were of intermittent contact. The thermo-electric potential of the intermittent working circuit was measured by means of a capacitance accumulator and the corresponding temperature evaluated therefrom. Twelve thermocouples were fitted on the crown and ring groove, the results of which are shown in Fig. 2. Comparison of these results obtained by fusible plug and by thermocouple tests reveals that the patterns of the temperature field are similar while the values given by the thermocouple tests are always lower. This is due to the fact that touching of the moving and stationary contacts of the circuit occurs only in the period near the B.D.C.(in this instance ±14° crankangle) when the curve of the temperature wave is at its lowest. The output of the thermocouple yields therefore always the value of thermo-electric potential corresponding to the valley of the wave. This of course, differs from the time-mean value of the temperature of the measured point.

CALCULATING METHOD OF TEMPERATURE FIELD AND NUMERICAL SCHEME

The purpose of an investigation into the piston temperature field is to master the temperature value at some characteristic points, such as the maximum temperature of the piston crown and the temperature of the top ring groove, to

*This test was carried out by Mr. Cao, an engineer of the Second Automobile Factory of China.

get the axial and radial temperature distribution of the piston, and to get the heat flux distribution. These are not only the important basis for designing and improving a piston, but the necessary information in studying the heat transfer of internal combustion engine cylinder.

Using the finite element method, the piston temperature field of 6100 petrol engine is calculated according to a two-dimensional steady problem. The piston crown is of flat plate type in this engine. Because of the geometry is not entirely axi-symmetry, the influence of pin foot on heat transfer, and spark plug situated toward the non-thrust side is equivalent to increase a local external heat source for that side, as a result, circumferential temperature distribution of the piston can not be uniform exactly, the temperature will be higher at the minor thrust side. Ths non-uniformity of the circumferencial temperatures by experimental obtainment are: around the circumference of piston crown ......15 deg.C(5.8%); around the top ring groove ......8 dey.C(3%); around the piston shaft for upper region······15 deg.C(10.%); for lower region...... 3 deg.C(2.1%). It may be seen that although the differences are quite obvious, simplification to an axi-symmetry problem with known extent of non-uniformity is allowed in engineering calculation.

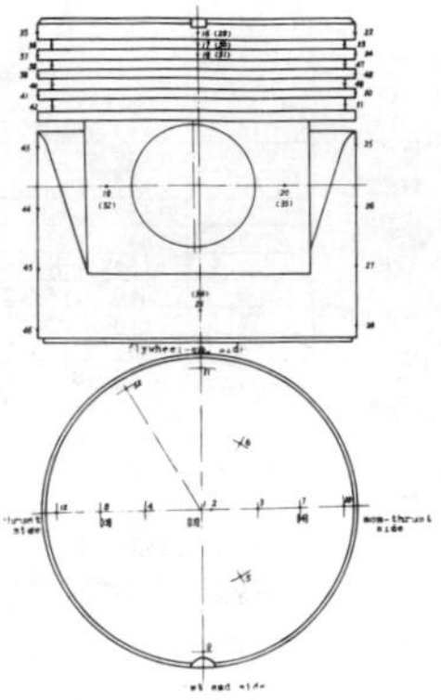

Fig.1 Location of Measuring points and numbers.

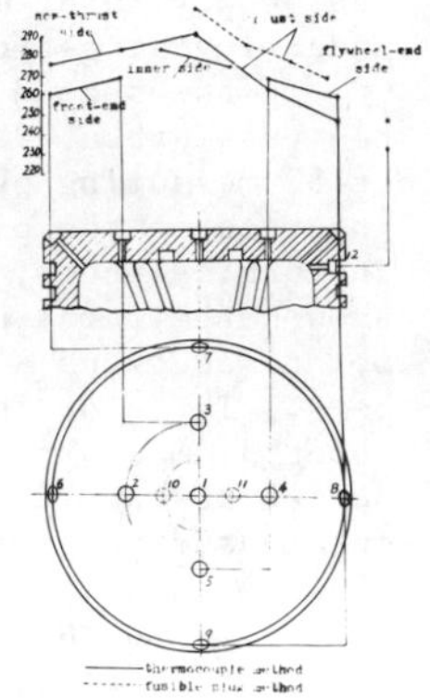

Fig.2 Piston temperature of the 5th cylinder.

Heat conduction equation and boundary equation

Considering a meridion plane (x,r) through symmetry axis, the harmonic function T=T(x,r) in calculating region G should satisfy the following harmonic equation for steady axi-symmetry heat conduction without internal heat source:

$$\frac{\partial^2 T}{\partial x^2} + \frac{\partial^2 T}{\partial r^2} + \frac{1}{r}\frac{\partial T}{\partial r} = 0 \qquad (1)$$

The third kind of heat boundary condition is applied in this paper, i.e. given data are the heat transfer coefficients between the surface and the ambient fluid. Fourier's Law of heat conduction and Newton's Law of convective heat transfer are as follows:

$$dQ = -\lambda \frac{\partial T}{\partial n} dF \quad , \qquad dQ = \alpha (T - T_o) dF$$

where
λ ---- thermal conductivity of piston material;
α ---- convective heat transfer coefficient between piston surface and ambient fluid;
T_o---- temperature of the neighbouring fluid near piston surface;
$\frac{\partial T}{\partial n}$---- partial derivative of piston surface temperature along the external normal of the region boundary (temperature gradient).

Therefore on the boundary: $\lambda \frac{\partial T}{\partial n} = \alpha (T_o - T)$ Defining $\alpha T_o = f$, the heat transfer boundary condition would be:

$$\left(\lambda\frac{\partial T}{\partial n}+\alpha T\right)_{r}=f \qquad (2)$$

Expression of function analysis

Applying finite element analysis to solve this problem,the mathematical basis is variation principle and separate interpolation. It can be show that the calculus of variation to solve the extreme value function T(x,r) is equivalent to solve a boundary value problem of second order partial differential equation about T(x,r). According to variation principle, the temperature function T=T(x,r) which satisfies differential equation (1) and boundary condition (2) should enable the functional analysis

$$U(T)=\int_G\int \frac{\lambda r}{2}\left[\left(\frac{\partial T}{\partial x}\right)^2+\left(\frac{\partial T}{\partial r}\right)^2\right]dxdr+\int_r\left(\frac{1}{2}\alpha T^2-fT\right)rds \qquad (3)$$

to take its minimum value, i.e. to set $\partial u/\partial T = 0$.

Discretization of the region

To transform the temperature distribution problem of the considered region to an approximate temperature value problem of discrete points in the same region discretization is thus needs. The specific scheme is to divide the whole region into a large amount of amsll triangular elements e, the three vertices of a triangle are written as $i(x_i,r_i)$ $j(x_j,r_j)$ and $m(x_m,r_m)$. Specify the line at a boundary element as $\overline{jm}$. The functional expression of an element can be written as follows

$$U_e(T)=\int_e\int \frac{\lambda r}{2}\left[\left(\frac{\partial T}{\partial x}\right)^2+\left(\frac{\partial T}{\partial r}\right)^2\right]dxdr+\int_{\overline{jm}}\left(\frac{1}{2}\alpha T^2-fT\right)rds \qquad (4)$$

The functional expression of the whole region can be summarized from all of these elements:

$$U=\sum_{e=1}^{E} U_e \qquad (5)$$

Then, to find the extreme value of the functional expression can be transformed into a task of finding the extreme values of the discrete nodal temperature.

$$\frac{\partial U}{\partial T_1}=\sum_{e=1}^{E}\frac{\partial U_e}{\partial T_1}=0 \qquad (1=1,2,3......Lo) \qquad (6)$$

Construction of an interpolation function

It is necessary to know the temperature distribution inside an element to solve Ue(T). Such distribution is called on "interpolation function". An linear distribution is used in this paper:

$$T=\frac{1}{2\Delta_e}\left[\sum_{ijm}(a_i+b_ix+c_ir)T_i\right] \qquad (7)$$

where Δ_e is the area of element e, $\Delta_e=\frac{1}{2}(b_ic_j-b_jc_i)$ (8)

$$\begin{cases} a_i=x_jr_m-x_mr_j, & a_j=x_mr_i-x_ir_m, & a_m=x_ir_j-x_jr_i, \\ b_j=r_j-r_m\ , & b_j=r_m-r_i\ , & b_m=r_i-r_j\ . \\ c_i=x_m-x_j\ , & c_j=x_i-x_m\ , & c_m=x_j-x_i\ , \end{cases} \qquad (9)$$

Thus, a set of algebraic equations is formed:

$$[K]^e[T]^e = [P]^e \tag{10}$$

Stiffness matrix of the temperature

Differentiating Eq. (7), gives

$$\frac{\partial T}{\partial x} = \frac{1}{2\Delta_e}\Big[\sum_{ijm} b_i T_i\Big]$$

$$\frac{\partial T}{\partial r} = \frac{1}{2\Delta_e}\Big[\sum_{ijm} c_i T_i\Big] \tag{11}$$

If e is a boundary element, the length of an element line situated at the boundary would be

$$S_i = \sqrt{(x_j-x_m)^2-(r_j-r_m)^2} \tag{12}$$

The temperature value of any point at $\overline{jm}$ line is between T_j and T_m, it satisfies the following linear relationship, r value at $\overline{jm}$ line variates according to a linear relationship, too.

$$t=s/s_i$$

$$T\Big|_{\overline{jm}} = (1-t)T_j+tT_m, \qquad r\Big|_{\overline{jm}} = (1-t)r_j+tr_m \quad 0 \le t \le 1 \tag{13) (14}$$

Substitute Eqs. (11)-(14) into Eq. (4), consider

$$\frac{1}{2\Delta_e}\iint_e r\,dx\,dr = \frac{1}{b}(r_j+r_i+r_m)$$

and

$$\frac{1}{12\Delta_e}\lambda(r_i+r_j+r_m) = \phi$$

Ue takes the form

$$U_e = \frac{\phi}{2}\Big[\Big(\sum_{ijm} b_i T_i\Big)^2+\Big(\sum_{ijm} c_i T_i\Big)^2\Big]+S_i\int_0^1\Big\{\frac{1}{2}\alpha_i[(1-t)T_j+tT_m]^2-f[(1-t)T_j+tT_m]\Big\}$$

$$[(1-t)r_j+tr_m]dt \tag{15}$$

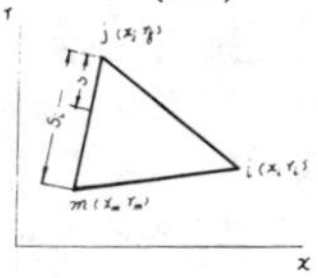

Fig.3

The second term of the right hand side is integrated for the boundary only. If element e is an internal one, then this term does not exist, since $\alpha_i=f_i=0$. It can thus be concluded that the stiffness of the temperature $[K]^e$ is a symmetric and regular matrix:

$$[K]^e = \begin{bmatrix} \phi(b_i^2+c_i^2) & & \\ \phi(b_jb_i+c_jc_i) & \phi(b_y^2+c_j^2)+\frac{\alpha_i S_i}{4}\left(r_j+\frac{rm}{3}\right) & \\ \phi(b_mb_i+c_mc_i) & \phi(b_mb_j+c_mc_j)+\frac{{}_iS_i}{4}\left(\frac{r_m+r_3}{3}\right) & \phi(b_m^2+c_m^2)+\frac{{}_iS_i}{4}\left(r_m+\frac{r_3}{3}\right) \end{bmatrix} \tag{16}$$

The equation of right hand side

$$\{P\}^e = \Big[0 \quad \frac{f_iS_i}{3}\Big(r_j+\frac{r_m}{2}\Big) \quad \frac{f_iS_i}{3}\Big(r_m+\frac{r_i}{2}\Big)\Big]^T \tag{17}$$

Every element inside the considered region G will be dealt with in the same way. After summing up and rearrangement, gives the following flobal matric equation

$$[K]\{T\} = \{P\} \tag{18}$$

Solving equation set (18), nodal temperature T_1, T_2,......T_{Lo} can be got. Based on this scheme, source program is compiled. 6100 piston is divided into 589 elements. There are totally 379 nodes, where one-dimensional total length of stiffness matrix is 3243.

DETERMINATION OF BOUNDARY CONDITION

Defining boundary condition preasely is always a decisive factor for the accuracy of calculating piston temperature field. It has to determine the heat transfer coefficient between the piston surface and the ambient fluid (combustion gas, cooling water, air in the crank case), and the fluid temperature for the third kind boundary condition. The former are calculated from the basic equations of heat transfer and some semi-empirical formulae. The accuracy of the boundary condition can be examined by experimental values. By try-and-error method, the boundary condition suited for this piston can be determined.

The heat transfer coefficient from the combustion gas to the crown surface and equivalent mean temperature

Due to reciprocating motion of the piston, the volume, pressure, temperature and velocity of the gas in each cycle have changed with different crank angle, so the heat transfer coefficient from the combustion gas to the crown surface at each transient is not constant, it is a transient heat transfer process. With the experimental indicator diagram, at hand the transient temperatures versus crank angles can be easily calculated through Clapeyron's equation PV=mRT, our $T-\phi$ curve is shown in Fig. 4(a). The average heat flux of every cycle can be written as

$$q = \frac{1}{\tau_o}\int_0^{\tau_o} \alpha(T-T_w)d\tau \tag{19}$$

where τ_o is the period of a cycle. The working process of an engine under steady state can be treated as a periodic steady process. If τ_o is a timing unit, then it would be a quasi-stable process. The average heat flux has the form

$$q = \alpha_m(T_{res}-T_w) \quad (20) \qquad \alpha_m = \frac{1}{\tau_o}\int_0^{\tau_o}\alpha d\tau$$

Since the oscillate amptitude of piston surface temperature is a small value (several degrees) in each cycle, for example, far an air-cooled 2-stroke petrol engine with diameter 80 mm, temperature variation of piston surface is only $\pm 1.25^oC$[3], the depth penetrated is not larger than 0.5 mm. On the other hand, gas temperature in a cycle is within the range of 300 to 2500 K. Therefore, the variation of piston surface temperature is negligiblely small. By comparing Eqs. (19) and (20) it gives the equivalent mean temperature of the combustion gas[4]

$$T_{res} = \frac{\int_0^{\tau_o}\alpha T d\tau}{\int_0^{\tau_o}\alpha d\tau} \tag{21}$$

Taking every 10 degrees of crank angle as a calculating point interval, it corresponds i = 0,1,2......71 when =0,10,20......710, then the calculating formulae are

$$m = \frac{\sum_0^{71}\alpha_i}{72} \quad , \qquad T_{res} = \frac{\sum_0^{71}(\ T)_i}{\sum_0^{71}\alpha_i}$$

Temperature variation of gas inside the internal combustion engine cylinder is so complicated, that the velocity, pressure and temperature in every local position have changed rapidly, the variation pattern of local parameters in a cycle differs from each other[5]. Petrol engine corresponds approximately to a constant-volume combustion, exothermic rate is quite large. The shape of the combustion chamber is complicated, too, and the deflection to one side of offset spark plug would cause local vortex. In addition, the study on radiation heat transfer inside the cylinder is not yet sufficient. As a result of all these factors, it would be difficult to predict the transient convective coefficient α accurately from the theory of heat transfer. Researchers from various countries have recommended different formulae, nearly all of them are in accordance with a specific type of engine, and some are formulated in non-dimensional form. The experimental study in this area is quite limited, especially for petrol engine. We have collected some formulae, part of them are chosen for calculation, the results are listed in Table 3.

It can be seen from these results that the values of α_m and T_{res} from different formulae differ from each other significantly. According to the heat flux from the combustion gas to piston crown and the experimental temperature values, the different expressions of boundary condition near the combustion gas side are checked by try-and-error method. Finally, the expression by G Woschni published in 1967[6] from a high speed diesel engine is adopted for this piston. Where α_m 245.4 kcal/($m^2 \cdot h^oC$), T_{res} 1222.6K. Fig.4(b) shwon the α-ϕ curve.

Having considered that the disturbance of gas near the combustion chamber is very strong, and that the top land is handled as an insulator, the heat transfer coefficient along the radial direction should be corrected properly, at out edge increases 10%.

Table 3

No.	author & the year published	formula	α_m kcal/($m^2 \cdot h \cdot {}^oC$)	T_{res} K
1	G.Woschni 1967	$\alpha = 0.035 \lambda/D \, Re^{0.8}$ exhaust $w = 6.18C_m$, compression $w = 2.28C_m$, combustion and expansion, $w = 2.28C_m + 6.22 \times 10^{-3} \frac{V_s T_1}{P_1 V_1} \times 10^4 (P - P_{comp})$	245.38	1222.59
2	G.Woschni 1965	$\alpha = 0.045 \lambda/D \, Re^{0.786}$, $w = 5.5 C_m$	392.15	1309.54
3	Eichelberg.G ·1939	$\alpha = 2.1 \sqrt[3]{C_m} \sqrt{PT}$	293.93	1570.78
4	W.Pflaum 1961	$\alpha = f_1(P_L) f_2(C_m) \sqrt{PT}$ $f_1(P_L) = 1.10 + 0.366 (\frac{P_L - P_0}{P_0})$ $f_2(C_m) = 3.00 + 2.57[1 - e^{(1.5 - 0.416 C_m)}]$	371.72	1570.78
5	W.Pflaum 1962	$\alpha = f_1(P_L) f_2(C_m) \sqrt{PT}$ $f_1(P_L) = 2.3 P_L^{0.25}$ $f_2(C_m) = 6.9 - 5.9 \times 4.5^{-(0.008 C_m)^{1.8}}$	854.54	1570.78
6	Nusselt.W 1923	$\alpha = 0.99 \sqrt[3]{P^2 T} (1 + 1.24 C_m) + 0.362 \frac{(\frac{T}{100})^4 - (\frac{T_w}{100})^4}{T - T_w}$, $T_w = 100 + 12 P_e$	403.55	1641.52
7	БРИЛИНГ 1931	$\alpha = 0.99 \sqrt[3]{P^2 T} (1 + a + b C_m)$ $a = 3.5$ (prechamber) 4.2 (vortex chamber) $b = 0.185$	171.29	1635.08
8	Van.Tijen 1959	$\alpha = (3.19 + 0.885 C_m) \sqrt[3]{P^2 T}$	348.98	1635.08
9	Van.Tijen 1959	$\alpha = 4.04 \sqrt[5]{C_m} \sqrt[3]{P^2 T}$	238.07	1635.08
10	Jaklitsch 1929	$\alpha = 0.0224 (1 + 1.24 C_m)(288.3 P)^m T^{(1-m)}$, $m = 0.394 + 0.1685 \times 10^{-3} T$	417.84	1683.13
11	H.Brosinsky 1953	$\alpha = 0.305 C_m (489 - T_n) \sqrt[3]{P}$	635.90	1186.69
12	W.J.D.Annand 1963	$\alpha = 0.49 \lambda/D \, Re^{0.7} + C \frac{(\frac{T}{100})^4 - (\frac{T_w}{100})^4}{T - T_w}$ suction and compression c = 0, combustion and expansion c = 2.8, $\lambda = 0.000314 T^{0.746}$ $\nu = 0.528 \times 10^{-6} \times 9.81 RT^{1.62/P}$	505.30	1395.23
13	G.Sitkei 1972	$\alpha = 25.2 (1 + B_s) \frac{(P C_m)^{0.7}}{D_e^{0.3} T^{0.4}}$ $B_s = 0 - 0.03$ $D_e = \frac{4 V_c}{\pi D (S_c + \frac{1}{2} D) + \frac{4 V_c}{D}}$	217.96	1437.99
14	Hassan 1971	$\alpha = 0.023 \lambda/D_e \, Re^{0.8}$, $\lambda = [2.56 + 7.3(T - 54)/1000] \times 0.0086$ $\nu = 1.966 \times 10^{-5} \frac{(T/273)^{1.8}}{T + 117} \cdot \frac{T}{P}$, $D_e = 2D$, $w = 2C_m$	128.72	1318.61
15	Tatsu Oguri 1960	$Nu = 1.75 (1 + \Delta S/c_p)[2 + \cos(\theta - 20)] \sqrt{P_e}$	836.7	1394.0

Equivalent convective coefficient along piston skirt

For an uncooled piston, the heat flux from combustion gas to piston is transfered mainly through piston skirt, lubricating oil film, and cylinder wall to the cooling water. The equivalent convective coefficient of different parts including piston skirt can be calculated based on the principle that the total thermal resistance is equal to the summation of heat-conduction resistance and convective resistance, then considering the pecific construction, gives some adequate corrections. The average measured temperature of cylinder coolant is 71°C. In this paper, the convective coefficient between the cooling water and inner surface of cooling channel is 1300 kcal/$m^2 \cdot h \cdot {}^oC$.

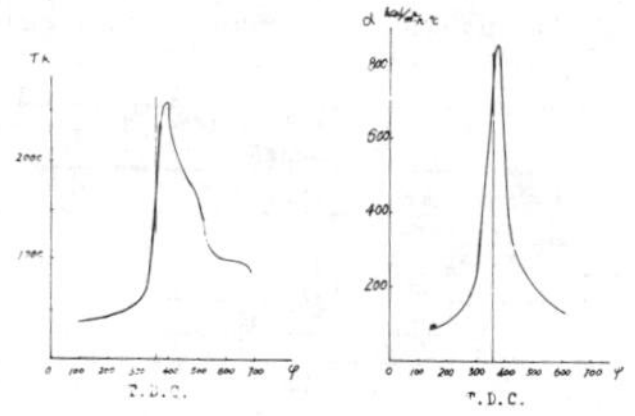

Fig. 4 (a) A T-ϕ diagram
(b) A α-ϕ diagram

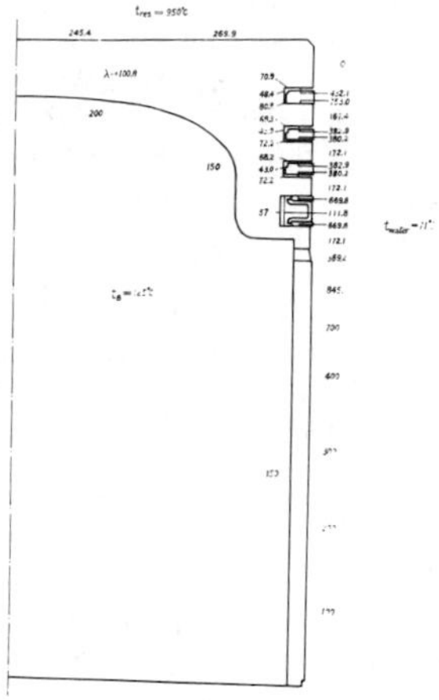

Fig. 5 Boundary conditions of calculating piston temperature field

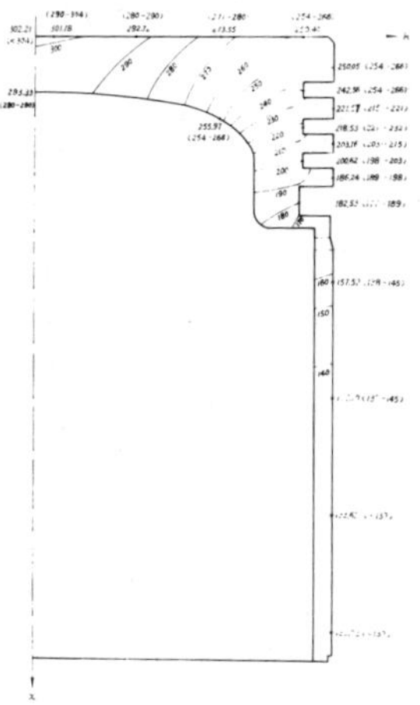

Fig. 6 The distribution of piston temperature

Inner side of the piston

There is no oil-spray cooling for this piston, the inner surface of it is cooled only by the oil mist splashed, and its convective coefficient is small. Its value is 150 and 200 kcal/($m^2\cdot h\cdot {}^oC$) for two sections according to our experience. The temperature of oil-gas mixture inside crank-case is within the range of 114-113°C by measurements. It is higher than that of diesel engine or other petrol engine. This may be one of the reasons to explain why its piston thermal loading is a little high. Average value of 125°C is taken for calculation.

The piston material is ZL8. Thermal conductivity of this material is 100.8 kcal/($m\cdot h\cdot {}^oC$) at mean temperature 200°C. All the boundary condition of the thrust side is shown in Fig. 5.

RESULTS AND ANALYSIS

The calculating results of the piston temperature field are shown in Fig. 6. Temperatures at some characteristic points are listed in this figure, the values inside the parenthesis are measuring ones. It can be seen that the calculating values agree well with the measuring ones. The error is not larger than 2%. Therefore the boundary condition determined above is proved suitable for this piston.

The maximum temperature of the piston crown and the top ring groove can be considered as the evaluation of piston thermal loading. (1) The highest temperature is at the crown center, about 300°C, some reference recommend that the maximum temperature on the piston crown made fromaluminum alloy material should not exceed 350-360°C. Average radial temperature gradient is 0.96^{o}C/mm, and axial temperature gradient of its central section is 1.0^{o}C/mm; (2) Constant temperature lines incline to the piston crown center. Since the heat flux is perpenticular to the constant temperature line, therefore the heat flow rate from the combustio gas is largely transfered to piston skirt side, mainly through the top ring to the cooling water for this kind of uncooled piston. The heat flow rate of piston crown from the combustion gas is 1315.71 kcal/h, the distribution of heat dissipa tion, see Tab. 4; (3) The temperature of the top ring groove seems too high, it reaches 243°C, the measurement value at non-thrust side exceeds 254°C. No doubt, this is the main problem of this piston. To assure that the lubrication is in good condition and to avoid deterioration of oil and increasing of carbon deposit, it can be maintained at a level not exceed 220-240[7].

Table 4

			heat loss rate kcal/h	Percentage %
Percentage into water 85.9%	Percentage into ring zone 58.9%	first ring zone	252.57	19.2
		land	27.44	2.1
		second ring zone	132.97	10.1
		land	21.37	1.6
		third ring zone	122.64	9.3
		land	18.54	1.4
		oil groove	200.56	15.2
	thermal insulating groove		10.28	0.8
	piston skirt		344.15	26.2
inner side			185.30	14.1

The following method are recommended to lower the temperature of the top ring groove: (1) To decrease the thickness of upper part of cylinder liner. When piston is at T.D.C., the first and second ring do not locate at the cylinder where cooling water can be flow freely, as shown in Fig.7. Since the temperature in combustion process is the highest, but the cooling condition is worsen with a large value of convective thermal resistance. This will cause overheat of ring zone; (2) To increase the water velocity, it is an effective measure to decrease the temperature of the top ring groove of uncooled piston. Average velocity of cooling water in this engine is about 0.12 m/s. According to statistical data, it is generally in the range of 0.1-0.2 m/s[8], and the higher value should be used for high speed engine with small cylinder diameter. (3) To reduce the clearance between piston top land and cylinder wall reasonably, it will reduce the temperature of the top ring groove. Reference[7] has introduced that in a non-supercharged 5-cylinder passenger car diesel engine, the top ring groove temperature raises 2-3°C per 0.1 mm of clearance increase. THe phenomenon that the combustion gas goes directly to fusible plug through the gap of the top ring which influence the temperature measurement exists during experiments. The piston top land clearance is about 0.6-0.745 mm. It seems the clearance can be improved; (4) To increase top land length it can also decrease the temperature of the topring groove. Usually, it can reduce 5-7°C for every 1 mm increased.

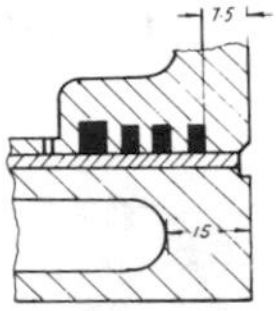

Fig.7

CONCLUSION

Using fusible plug method to measure the temperature at some specific points on the piston of a high speed petrol engine, the boundary condition can be checked and the piston temperature field can be calculated accurately. This is an available way to improve the design of a piston. How to determine the boundary condition is still a subject to be studied. By measuring the local heat flux from combustion gas to piston crown and using the second kind of boundary condition, the temperature field can be calculated, this problem is worth to explore.

REFERENCES

1. Ricardo Consulting Engineers "Piston Temperature Measurement by Fusible Plugs" Ricardo Report (1971)
2. Li Da-qian "Selected Topics on Finite Element Method" Shanghai Fu-Dan Univ.
3. Otto Kruggel "Calculation and Measuring of Piston Temperature of Air-Cooled Two-Stroke Gasoline Engine" SAE Paper 710578
4. G.Eichelberg "Some New Investigations on Old Combustion Engine Problems" Engineering, Vol. 148 (1939)
5. G. Woschni "Prediction of Thermal Loading of Supercharged Diesel Engine" SAE Paper 790821
6. G. Woschni "A Universally Applicable Equation for the Instantaneous Heat Transfer Coefficient in the Internal Combustion Engine" SAE Paper 670931

7. M.D.Roehrle "Thermal Effects on Diesel Engine Pistons" SAE Paper 780781
8. Jiang Zhi-Min "Calculation of Piston Temperature Field of Diesel Engine" Selected Paper in Science and Technology of Shanghai Jiao-Tong University (1978)

Dynamic Characteristics of the Surface-Installed Heat Flux Meter

BU-XUAN WANG and LI-ZHONG HAN
Thermal Engineering Department
Tsinghua University
Beijing, PRC

ZHAO-HONG FANG
Shandong Institute of Civil Engineering
Jinan, PRC

ABSTRACT

This paper analyses the dynamic characteristics of a heat flux meter probe installed on the surface of a wall for which heat flux is to be measured. A double-plate model is presented and the temperature response in the probe and wall is obtained analytically by Laplace's transformation. It agrees reasonably with the experimental results. The transition time, needed for the probe and wall to establish a new steady state, is evaluated and discussed in detail. The figures plotted may provide practical uses for quickly estimating the transition time of surface-installed heat flux meters.

NOMENCLATURE

- a thermal diffusivity
- q heat flux
- r_c thermal contact resistance
- t temperature
- x coordinate
- α heat transfer coefficient
- μ eigenvalue
- δ thickness
- λ thermal conductivity
- τ time

Subscripts

- 1 measured wall
- 0 probe
- e output of probe
- f surrounding fluid
- s steadiness
- w wall surface

Dimensionless Variables

$Bi_0 = \alpha\delta_0/\lambda_0$

$Bi_1 = \alpha\delta_1/\lambda_1$

$E = q_e/q_s$

$Fo = a_0\tau/\delta_0^2$

$K_a = a_1/a_0$

$K_s = K_\lambda/\sqrt{K_a}$

$K_\delta = \delta_1/\delta_0$

$K_\lambda = \lambda_1/\lambda_0$

$M = K_\delta/\sqrt{K_a}$

$Q = q/q_s$

$R = r_c\lambda_0/\delta_0$

$T = (t - t_f)/(t_w - t_f)$

$X = x/\delta_0$

INTRODUCTION

The heat flux meter is a device for direct measurement of local heat transfer, usually composed of a probe and a reading instrument. Of all the present kinds of heat flux meters, the thermal resistance type proposed first by E. Schmidt [1] has been matured most and used in various branches of industry and scientific researches [2], especially as monitors in energy conservation techniques. The probes can be burried in or stuck on a body, and the surface installation is more convenient for applications.

The heat flux meters are usually employed to measure steady heat flux. Having been installed onto a surface, the probe, as well as the body to be measured, undergoes a transient heat conduction process. If the surrounding conditions keep unchanged, they will approach a new steady state through a certain period of time known as the "transition time". However, such a steady state can hardly be reache in practical applications because of the slight fluctuations of surrounding conditions. The dynamic characteristics of the probe will be even of greater significance for its proper use if a transient heat flux is to be measured. In addition, the calibration of the probes involves certainly a transition process, and a rational design of the calibrating device can shorten the time needed for calibration of the probes [3].

An analysis on the dynamic characteristics of the probe was reported recently [4], assuming a constant heat flux at the inner surface of the probe during the transient process. Such an assumption surely simplified the solution but departed from the practical situation.

In this paper, we try to discuss this problem more properly with the probe and the body connected together as a whole, and consider also the thermal contact resistance existing on the interface. In order to discuss the problem analytically, we assume in this paper that the probe is stuck to a plane wall consisting of a homogeneous material, and that the heat flows through the wall and the probe one-dimensionally.

The temperature field in the probe and wall approaches the new steady state gradually. For practical purpose, we define in following discussion that the prob reaches its "steady" state if the heat flux on its outer surface has risen to 0.9 of the steady heat flux. And the duration from its installation to this moment is taken as the transition time of the probe.

THEORETICAL ANALYSIS

Referring to Fig.1, we consider a heat conduction system composed of two plane plates — the probe and the measured wall. At the initial instant, i.e. the probe is just installed to the wall surface, there is a linear temperature distribution in the wall and a uniform temperature same as that of the surroundings in the probe. We assume that the temperature on the inner side of the wall, t_w, remains unchanged during the whole transition process. The heat conduction in both the probe and the measured wall interact each other on the interface, which is known as the fourth boundary condition [5]. The heat fluxes on both sides of the interface remain equal, but a temperature difference, $\Delta t = r_c q$, will be brought about due to the existance of a thermal contact resistance, r_c, on the interface.

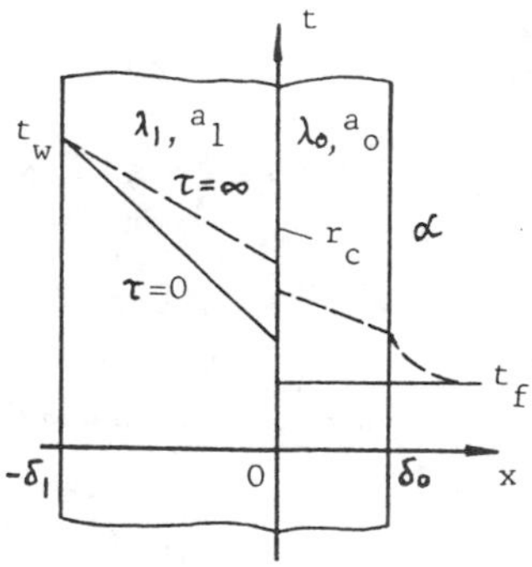

Fig.1 The analytical double-plate model

The transient heat conduction in the double-plate system can thus be formulated as follows:

For the wall

$$-\delta_1 \leq x \leq 0,\ 0 \leq \tau < \infty: \quad \frac{\partial t_1}{\partial \tau} = a_1 \frac{\partial^2 t_1}{\partial x^2}, \tag{1}$$

$$-\delta_1 \leq x < 0 \ , \quad \tau = 0 : \qquad t_1(x,0) = t_w - \frac{\alpha(t_w - t_f)}{\delta_1\alpha + \lambda_1}(x + \delta_1), \tag{2}$$

$$x = -\delta_1, \ 0 \leq \tau < \infty \quad : \qquad t_1(-\delta_1,\tau) = t_w; \tag{3}$$

For the probe

$$0 \leq x \leq \delta_o, \ 0 \leq \tau < \infty \ : \qquad \frac{\partial t_o}{\partial \tau} = a_o \frac{\partial^2 t_o}{\partial x^2} \ , \tag{4}$$

$$0 < x \leq \delta_o, \ \tau = 0 \quad : \qquad t_o(x,0) = t_f \ , \tag{5}$$

$$x = \delta_o, \ 0 \leq \tau < \infty \quad : \qquad -\lambda_o \frac{\partial t_o}{\partial x}(\delta_o,\tau) = \alpha[t(\delta_o,\tau) - t_f]; \tag{6}$$

For the interface

$$x = 0, \ 0 < \tau < \infty \quad : \qquad t_1(0,\tau) - t_o(0,\tau) = -r_c \lambda_o \frac{\partial t_o}{\partial x}(0,\tau), \tag{7}$$

$$x = 0, \ 0 < \tau < \infty \quad : \qquad \lambda_1 \frac{\partial t_1}{\partial x}(0,\tau) = \lambda_o \frac{\partial t_o}{\partial x}(0,\tau). \tag{8}$$

With the dimensionless variables listed in the nomenclature, the above formulation can be nondimensionalized as follows:

$$-K_\delta \leq X \leq 0, \ 0 \leq Fo < \infty: \qquad \frac{\partial T_1}{\partial Fo} = K_a \frac{\partial^2 T_1}{\partial X^2} \ , \tag{1'}$$

$$-K_\delta \leq X < 0, \quad Fo = 0 : \qquad T_1(X,0) = \frac{1}{1 + Bi_1}\left(1 - \frac{Bi_1}{K_\delta} X\right) \ , \tag{2'}$$

$$X = -K_\delta \ , \ 0 \leq Fo < \infty \ : \qquad T_1(-K_\delta, Fo) = 1 \ , \tag{3'}$$

$$0 \leq X \leq 1 \ , \ 0 \leq Fo < \infty : \qquad \frac{\partial T_o}{\partial Fo} = \frac{\partial^2 T_o}{\partial X^2} \ , \tag{4'}$$

$$0 < X \leq 1 \ , \ Fo = 0 \quad : \qquad T_o(X,0) = 0 \ , \tag{5'}$$

$$X = 1 \ , \ 0 \leq Fo < \infty \quad : \qquad \frac{\partial T_o}{\partial X}(1,Fo) = -Bi_o \cdot T_o(1,Fo) \ , \tag{6'}$$

$$X = 0 \ , \ 0 < Fo < \infty \quad : \qquad T_1(0,Fo) - T_o(0,Fo) = -R \frac{\partial T_o}{\partial X}(0,Fo) \ , \tag{7'}$$

$$X = 0 \ , \ 0 < Fo < \infty \quad : \qquad K_\lambda \frac{\partial T_1}{\partial X}(0,Fo) = \frac{\partial T_o}{\partial X}(0,Fo) \ . \tag{8'}$$

Applying Laplace's transformation to equations (1') and (4') together with their initial conditions (2') and (5'), the following images of the solutions are obtained:

$$-K_\delta \leq X \leq 0, \quad \overline{T}_1 = A_1 ch\left(\sqrt{\frac{p}{K_a}} X\right) + A_2 sh\left(\sqrt{\frac{p}{K_a}} X\right) + \frac{1}{p(1+Bi_1)}\left(1 - \frac{Bi_1}{K_\delta} X\right) \ , \tag{9}$$

$$0 \leq X \leq 1 \ , \quad \overline{T}_o = A_3 ch(\sqrt{p}\, X) + A_4 sh(\sqrt{p}\, X) \ , \tag{10}$$

where p denotes the transform of Fo.

From the boundary conditions (3'),(6'),(7') and (8'), the constants A_1, A_2, A_3 and A_4 can be determined as

$$A_1 = sh(M\sqrt{p})[Bi_o(\sqrt{p}\ ch\sqrt{p} + Bi_o sh\sqrt{p}) + (Bi_o R-1)\sqrt{p}\ (\sqrt{p}\ sh\sqrt{p} + Bi_o ch\sqrt{p})]/F(p), \quad (11)$$

$$A_2 = ch(M\sqrt{p})[Bi_o(\sqrt{p}\ ch\sqrt{p} + Bi_o sh\sqrt{p}) + (Bi_o R-1)\sqrt{p}\ (\sqrt{p}\ sh\sqrt{p} + Bi_o ch\sqrt{p})]/F(p), \quad (12)$$

$$A_3 = (\sqrt{p}\ ch\sqrt{p} + Bi_o sh\sqrt{p})[Bi_o sh(M\sqrt{p}) + K_s\sqrt{p}\ ch(M\sqrt{p})]/F(p), \quad (13)$$

$$A_4 = -(\sqrt{p}\ sh\sqrt{p} + Bi_o ch\sqrt{p})[Bi_o sh(M\sqrt{p}) + K_s\sqrt{p}\ ch(M\sqrt{p})]/F(p), \quad (14)$$

where
$$F(p) = (1 + Bi_1)p\sqrt{p}\left\{K_s ch(M\sqrt{p})(\sqrt{p}\ ch\sqrt{p} + Bi_o sh\sqrt{p}) + [sh(M\sqrt{p}) + K_s R\sqrt{p}\ ch(M\sqrt{p})](\sqrt{p}\ sh\sqrt{p} + Bi_o ch\sqrt{p})\right\}. \quad (15)$$

The inverse Laplace transformation of equations (9) and (10) may be carried out by defining

$$T = \frac{1}{2\pi i}\int_{b-\infty}^{b+\infty} \bar{T}\ e^{p\cdot Fo}\ dp \quad (16)$$

It is found that p = 0 and $p_n = -\mu_n^2$ are the polar points of the functions $\bar{T}_o e^{p.Fo}$ and $\bar{T}_1 e^{p.Fo}$ and that μ_n is the n-th positive root of the eigenequation

$$tg(M\mu) + K_s R\mu = \frac{K_s(\mu + Bi_o tg\mu)}{\mu\, tg\mu - Bi_o} \quad . \quad (17)$$

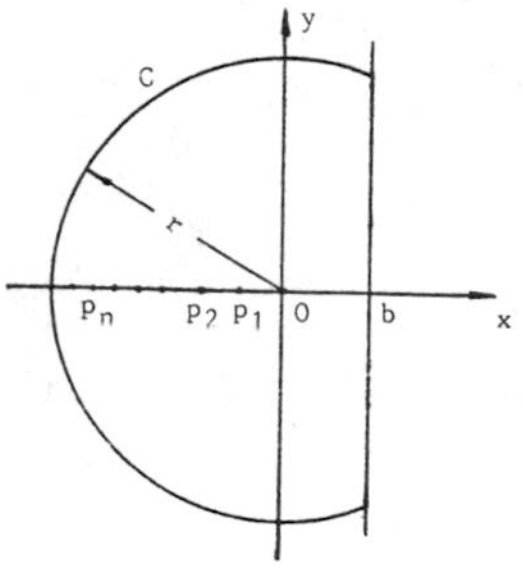

Fig.2 Diagram of the inverse Laplace transformation

Then the inverse Laplace transformation can be done by means of the residue theorem, or

$$T = Res(\bar{T}\ e^{p.Fo},0) + \sum_{n=1}^{\infty} Res(\bar{T}\ e^{p.Fo},p_n) - \frac{1}{2\pi i}\int_c \bar{T}\ e^{p.Fo}\ dp\ . \quad (18)$$

It can be proved that $\int_c \bar{T}\ e^{p.Fo}\ dp = 0$ due to the symmetry of the integrand about the real axis. Calculating the residues, we have finally obtained the solution:

$-K\delta \leq X \leq 0,\ 0 \leq Fo < \infty$:

$$T_1 = \frac{1+Bi_o+Bi_oR-Bi_1X/K\delta}{1+Bi_o+Bi_1+Bi_oR} - \frac{2}{1+Bi_1}\sum_{n=1}^{\infty}\frac{1}{D_n}e^{-\mu_n^2 Fo}$$

$$\cdot[Bi_o(\mu_n cos\mu_n + Bi_o sin\mu_n) + (1-Bi_oR)\,\mu_n\,(\mu_n sin\mu_n - Bi_o cos\mu_n)]$$

$$\cdot[sin(M\mu_n)\ cos\ (\frac{\mu_n}{Ka}X) + cos\ (M\mu_n)\ sin\ (\frac{\mu_n}{Ka}X)], \quad (19)$$

$0 \leq X \leq 1\ ,\ 0 \leq Fo < \infty$:

$$T_o = \frac{1+Bi_o(1-X)}{1+Bi_o+Bi_1+Bi_oR} - \frac{2}{1+Bi_1} \sum_{n=1}^{\infty} \frac{Bi_o \sin(M\mu_n)+K_s\mu_n\cos(M\mu_n)}{D_n} e^{-\mu_n^2 Fo}$$

$$\cdot [(\mu_n\cos\mu_n+Bi_o\sin\mu_n)\cos(\mu_nX)+(\mu_n\sin\mu_n-Bi_o\cos\mu_n)\sin(\mu_nX)] \quad (20)$$

where D_n is defined as

$$D_n = \mu_n\sin(M\mu_n)\,[-K_sRM\mu_n^3\sin\mu_n+(1+K_sM+K_sMBi_oR)\mu_n^2\cos\mu_n+(4+Bi_o+K_sMBi_o)\mu_n\sin\mu_n$$

$$-3Bi_o\cos\mu_n]+\mu_n\cos(M\mu_n)\,[K_sR\mu_n^3\cos\mu_n+(K_s+M+5K_sR+K_sRBi_o)\mu_n^2\sin\mu_n$$

$$-(4K_s+K_sBi_o+MBi_o+4K_sBi_oR)\mu_n\cos\mu_n-3K_sBi_o\sin\mu_n] \;. \quad (21)$$

The normalized temperature response in the probe, equation (20), interests us much more as to predict the dynamic characteristics of the surface-installed heat flux meter.

RESULTS OF CALCULATION AND EXPERIMENTS

The heat flux in the probe can be obtained as

$$q_o(x,\tau) = -\lambda_o\frac{\partial t_o}{\partial x} \;. \quad (22)$$

The new steady state heat flux in the probe will be uniform and given by

$$q_s = (t_w-t_f)/(\frac{1}{\alpha} + \frac{\delta_o}{\lambda_o} + \frac{\delta_1}{\lambda_1} + r_c) \;. \quad (23)$$

From equations (23) and (20) the dimensionless heat flux in the probe, defined as $Q_o(X,Fo)=q_o/q_s$, can be calculated thereby.

As shown in figure 3, the heat flux meter probe of the thermal resistance type is composed of a thermopile wound round a base plate. A steady heat flux passing through the probe should be proportional to the temperature difference across the base plate. When a transient process is involved, the output of the probe expresses in fact the temperature difference across the base plate instead of the transient heat flux. Converting the output of the probe into a relevant heat flux

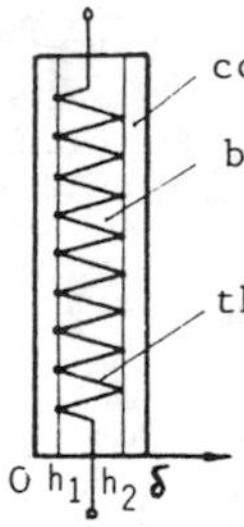

Fig. 3 The composition of the probe

$$q_e = \lambda_o(t_{h_1}-t_{h_2})/(h_2-h_1), \quad (24)$$

We can calculate the normalized output of the probe $E=q_e/q_s$ from equations (20) and (23).

The device for experimental check of the theoretical consideration is sketched in figure 4. A plane plate was fixed on a bulk of aluminium, which is placed in an oil-bath thermostat. The plate was made of cellular concrete and its thermophysical properties were determined in advance. After the temperature field in the plate reached its steady state, a heat flux meter probe was put on it, and the temperature on the outer surface of the probe, t', the temperatures on the interfaces between the probe and the plate, t'', and between the plate and the aluminium base, t''' , and the temperature of the surrounding air, t_f, were recorded continuously. The records showed that the change in temperature on the interface between the plate and the aluminium base did not exceed 0.1°C during the experiment and could be regarded as constant.

Figures 5 and 6 show the records of a typical experiment and its corresponding results through theoretical calculations. Some discrepances in the early stage may be accounted for by disagreement of conditions of the experiment with those assumed in the calculation in addition to errors in the parameters measured. For example, the heat conduction in the wall and probe was not exactly one-dimensional because of the rather finite area of the probe compared with the thickness of the wall; the heat transfer coefficient on the surface is hardly constant during the whole experiment; and the thermal contact resistance, r_c, was not measured but assumed R=0.35 in the calculation.

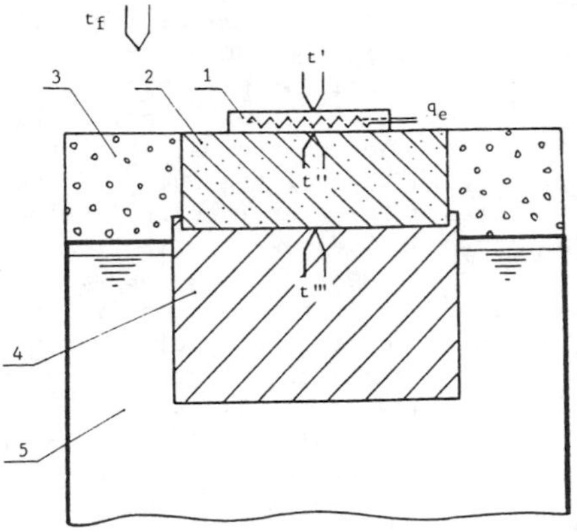

1. probe 2. plate 3. insulatio
4. aluminium base 5. oil-bath

Fig.4 The schematic diagram of the experiment device

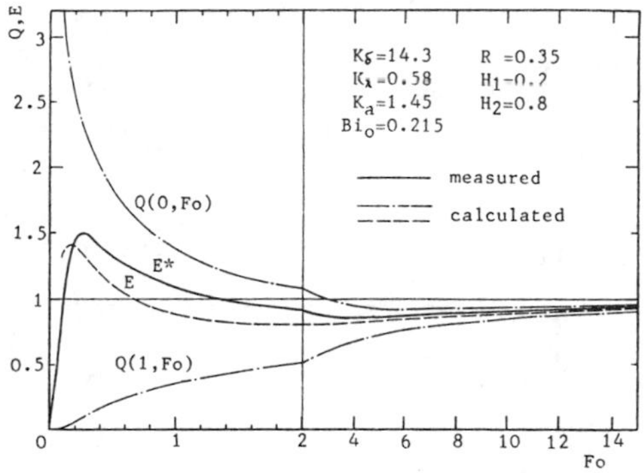

Fig.5 The heat flux and output of the probe

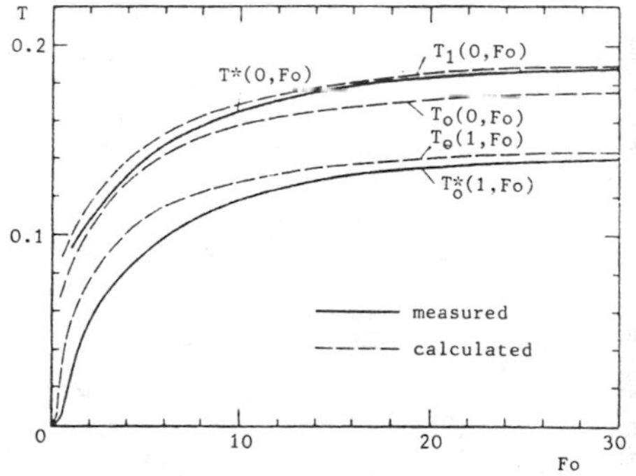

Fig.6 The temperatures on both sides of the probe

TRANSITION TIME OF SURFACE-INSTALLED PROBE

In spite of its intricacy the solution on the double-plate model indicates that the normalized temperature and heat flux in the probe and wall are functions of seven independent dimensionless variables, i.e. X, Fo, Bi_o, K_δ, K_λ, K_a, R. With this analytical solution the transition time of the probe, which is determin by the criterion $Q_o(1,Fo)=0.98$, can also be calculated. Therefore the dimensionless transition time, Fo_s, is a function of five independent variables

$$Fo_s = f(Bi_o, K_\delta, K_\lambda, K_a, R). \qquad (25)$$

Here, we discuss only in the ranges of the parameters which are come across commonly when the heat flux meter is used.

It is known from dimension analysis that the characteristic time of the transien heat conduction of the measured wall may be expressed as δ_1^2/a_1 while tha of the probe as δ_o^2/a_o. If $\delta_1 \gg \delta_o$, which is common in practice, or if $a_1 \ll a_o$, the characteristic time of the wall will be much greater than that of the probe, and the characteristic time of the whole system will be affected much more by the measured wall. On the other hand, the thermal contact resistance should be controlled as small as possible, and under proper use of the probes, R<0.2 may be estimated. It is clear that, as $Bi_o \ll 1$ or $Bi_o \ll Bi_1$, the thermal resistance of the probe and the thermal contact resistance will be almost negligible as compared with that of the wall and the heat transfer resistance on the surface, and the steady temperature distributions in the wall before and after the probe is attached will differ little from each other. This is the desirable situation to measure the heat flux in the wall with a heat flux meter. If the temperature distribution in the wall changes little after it is disturbed by the probe, the time duration for the whole system to establish the new steady state would shorten.

So, a remarkable factor which influences on the dynamic characteristics of the double-plate system is the thickness of the wall. The thicker the wall, the longer the transition time is. Howerer, with the increase in wall thickness, its thermal resistance increases and the disturbance of the additional thermal resistance of the probe abates, which is advantageous for shortening the transition time. Such an effect is especially notable when K_λ is small as shown in figure 7.

It can be seen in Fig.8 that Fo_s-K_a curves relative to greater K_δ seem to be straight lines with inclination of 45°, which means Fo_s is approximately inversely propotional to K_a. This can be easily explained by the fact that the transition time of the system mainly depends upon δ_1^2/a_1 as $K_\delta \gg 1$. Since thermal diffusivities of most non-metalic materials are quite close to one another, K_a does not change much in practical applications and is usually of the order of one.

The effect of Bi_o, or of the heat transfer coefficient, α, is illustrated in Fig.9. A small α hinders the heat transfer on the surface but reduces the differences between the two steady-state temperature distributions in the wall before and after the probe is attached to it. As $Bi_o \to \infty$, the temperature on the outer surface of the probe becomes constant, and Fo_s approaches a limit with other parameters fixed.

Like α, the thermal conductivity of the wall, or the dimensionless variable K_λ, influences on Fo_s in opposite ways as well. As the result, some peaks appear

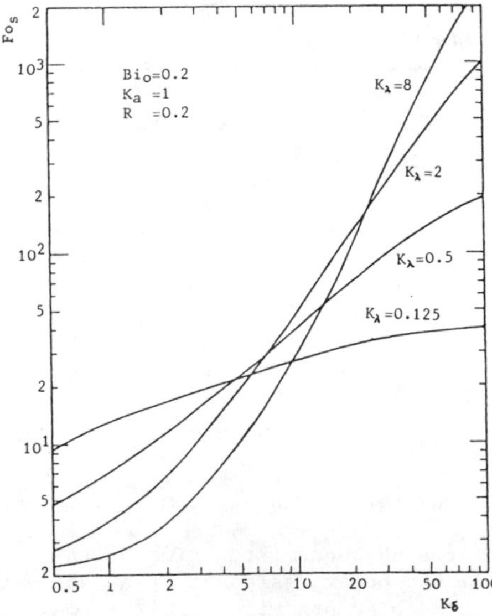

Fig.7 Influence of K_δ on FO_s

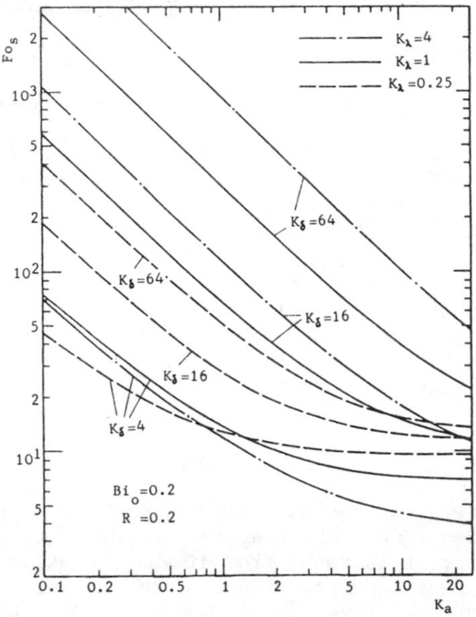

Fig.8 Influence of K_a on FO_s

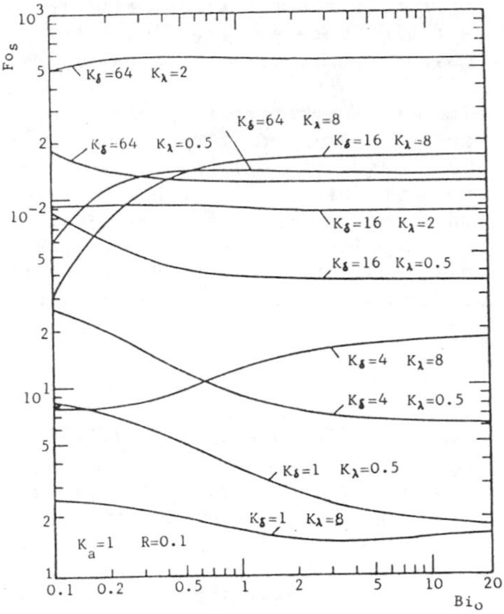

Fig. 9 Influence of Bi_o on FO_s

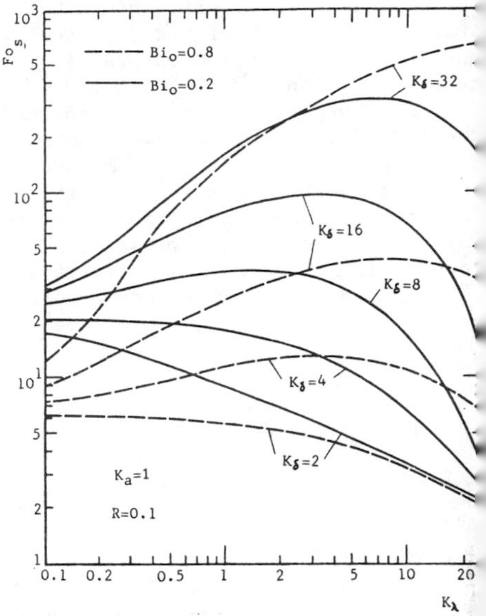

Fig.10 Influence of K_λ on Fo_s

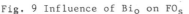

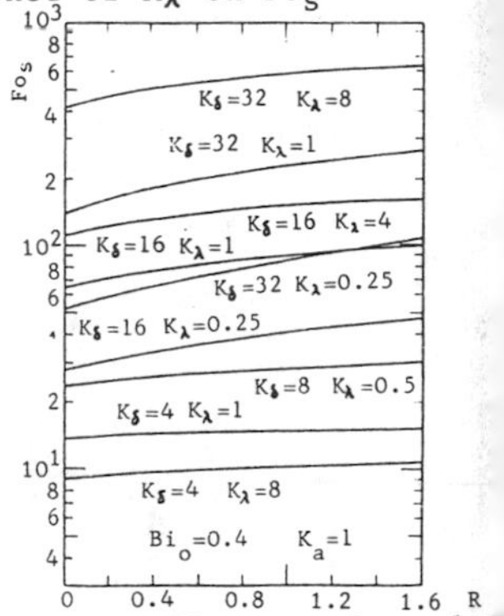

Fig.11 Influence of R on F

at Fo_s-K_λ curves with larger K_δ, which can be seen in figure 10.

Calculation indicates that the thermal contact resistance influences considerably on the response of the probe output at the early stage of the transient process, and that the transition time of the system will be prolonged with the increase in this contact resistance. Under normal conditions, however, R is estimated to be less than 0.2, and within this range the increase in the transition time is about under 10 percent compared with the ideal situation of R=0. So, this is a minor effect.

CONCLUDING COMMENTS

The discussions above indicate that the transition time of the system increa with increase in wall thickness and the thermal contact resistance, but increase with decrease in the thermal diffusivity of the wall. The transition time does not change monotonically with the thermal conductivity of the wall and the heat transfer coefficient on the surface. The typical results presented in figures 7-11 show that the thickness of the measured wall affects more prominently on the transition time, and that Fo_s can range from about 2-15 as $K_\delta=1$ to about 30-3000 as $K_\delta=100$. For illustration, when the heat flux in a 240 mm thick brick wall is measured with a 4 mm thick probe ($a_o=1.5x10^{-3}m^2/h$), the output of the probe reaches 0.9 of its steady one after about 26 minutes, while rises to 0.95 after 2.4 hours, and 0.98 after 7.5 hours (Fo=700). As shown in figure 5, apart from the early stage, the probe output changes more and more slowly and will be often concealed by the disturbance resulting from small fluctuations of the surrounding conditions. Hence, the operator may get a false impression that the steady state has been reached before it really has, which causes error in measu ment. So, it is significant to estimate appropriately the transition time for proper application of the surface-installed heat flux meters. The curves and qualitative analysis presented in this paper may provide a scientific basis for such estimation.

The boundary condition of constant heat flux or of constant temperature on the inner side of the wall also makes difference to the prediction of the dynamic characteristics of the system. However, for most cases of applying the surface-installed heat flux meter, the thermal resistance of the wall is much greater than that of the probe, the actual boundary condition on the inner side of the wall influences slightly on the process. The assumption of one-dimensional heat conduction has been adopted in this paper. This is helpful to obtain an analytical solution and to discuss the effects of various factors. Numerical calculations on two- or three-dimensional model will be discussed later.

REFERENCES

[1] E.Schmidt, Device for the measurement of heat, U.S.A.Patent 1528383,(1925).

[2] F.C.Stemple, D.L.Rall, Application and advancement in the field of direct heat transfer measurements, Annu. ISA Conf. Proc. No 8-1-63, pp1-10, (1963).

[3] B.X.Wang, L.Z.Han, Z.H.Fang, Heat transfer analysis of a high speed calibration device for heat flux meters, Chinese Journal of Metrology, 5(3), (1984).

[4] G.Wang, Y.L. Qin, An analysis on dynamic characteristics of the portable surface heat flow meter, Chinese Journal of Metrology, 4(3), (1983).

[5] B.X.Wang, Engineering Heat and Mass Transfer, § 2-2, Science Press, Beijing, (1982).

Prediction of Fluid Flow in a Tundish Using the Effective Viscosity Formula Method

LIYA WANG and YING QU
Beijing University of Iron and Steel Technology
Beijing, PRC

1. INTRODUCTION

The metallurgists are getting more and more interested in researches on flow characteristics of melts in various kinds of metallurgical reactors because they have great influences on heat transfer, melt temperature distribution and melting processes. It is quite difficult to measure velocity distribution in a metallurgical vessel since the melt temperature is fairly high. As a result, modelling methods have got a widespread application in recent years. (1) (2) (3) (4) Water model experiments are often carried out and have proved useful. (2) (5) Nevertheless, it is difficult to make water models similar to prototypes rigrously so mathematical modelling method is also required to get a good knowledge of flow fields.

A tundish in continuous casting system is the last refractory vessel in steelmaking technological process. The flow pattern in it has great effects on both casting process and slab quality. The temperature of the melt entering the water-cooled crystallizer should be controlled precisely to normal solidification, which depends on the flow pattern of melt in the tundish to some extend; besides, flow pattern affects also the removal of large nonmetallic inclusions and inclusion distribution in slabs. (5) (6) Therefore, it is of significance to make an investigation into flow characteristics in the tundish,

The tundish studied here is a boat-like vessel. The melt is poured into the tundish at one end and flows out through a nozzle into the crystallizer at the other end. For simplicity of numerical solution, it is simplified as a two-dimensional flow across both jet inlet and the nozzle.

2. MATHEMATICAL MODEL

2.1 The Fundamental Equations

The fundamental equations for liquid flow are continuity equation and Navier-Stokes' equation which can be solved numerically under some conditions. For the purpose of easy computation, the governing equations are written in terms of the vorticity ξ and the stream function ψ defined as:

$$\xi = \frac{\partial v}{\partial x} - \frac{\partial u}{\partial y} \tag{1}$$

$$u = \frac{1}{\rho}\frac{\partial \psi}{\partial y} \qquad v = -\frac{1}{\rho}\frac{\partial \psi}{\partial x} \tag{2}$$

The vorticity transport equation is written as:

$$\frac{\partial}{\partial x}\left(\xi \frac{\partial \psi}{\partial y}\right) - \frac{\partial}{\partial y}\left(\xi \frac{\partial \psi}{\partial x}\right) - \frac{\partial}{\partial x}\left(\mu_{eff} \frac{\partial \xi}{\partial x}\right) - \frac{\partial}{\partial y}\left(\mu_{eff} \frac{\partial \xi}{\partial y}\right) = 0 \quad (3)$$

The relationship between the stream function and the vorticity gets the following form:

$$\xi + \frac{\partial}{\partial x}\left(\frac{1}{\rho} \frac{\partial \psi}{\partial x}\right) + \frac{\partial}{\partial y}\left(\frac{1}{\rho} \frac{\partial \psi}{\partial y}\right) = 0 \quad (4)$$

2.2 The effective viscosity formula

The effective viscosity μ_{eff} in equation (3) can be obtained with various kinds of models such as the laminar model and the turbulent model. Most of the flows in metallurgical reactors are of turbulence which is composed of a lot of eddies of which the variations in sizes and directions are favourable for momentum and heat transfer. Therefore, the turbulent models, being able to describe such nature, are more avaliable for the calculation of flow fields in most of the metallurgical reactors.

Of all turbulent models, two-equation models such as k $-\varepsilon$ model and k $-\omega$ model are often applied to get the effective viscosity values. (2)(4)(1) They do be able to present us the local turbulent level of a flow well, however, they need a lot of CPU time and computer storages because two more partial differential equations have to be solved and they often show to be more difficult in convergence than the others. To simplify calculations, some researchers have developed and applied another kind of model ---- effective viscosity formula model ---- in recent years. (7) (8) This method estimates the general turbulent level of the whole flow field with an empirical formula. It can save a lot of CPU time and computer storages, especially in the three-dimensional flow field calculations, in addition to its being easy applied. Of course, the method is not so exact as the two-equation models and it can only be applied to the system on which the effective viscosity formula has been derived. As there was no such formula for vessels like the tundish before, one is to be established here.

The factors affecting the effective viscosity in the tundish are shown in Table 1. Suppose:

$$\mu_{eff} = f\,(V_{in}, \rho, d_o, h, H) \quad (5)$$

According to the similarity theory, there should be three dimensionless groups. Here we take:

$$\pi_1 = \frac{\mu_{eff}}{\rho V_{in} d_o} \quad (6)$$

$$\pi_2 = \frac{d_o}{H} \quad (7)$$

$$\pi_3 = \frac{h}{H} \quad (8)$$

Where π_1, π_2, π_3 are dimensionless groups.

Without the consideration of the contraction of the jet inlet, π_3 can therefore be neglected. Suppose:

$$\pi_1 = C'\pi_2^{\alpha} \quad (9)$$

We can get
$$\mu_{eff} = C' d_o^{\alpha+1} H^{-\alpha} \rho V_{in} \tag{10}$$

or
$$\mu_{eff} = C\ d_o^{\alpha-1} H^{-\alpha} \rho Q \tag{11}$$

where Q is the volumatric flow rate of the liquid and C', C coefficients.

$$Q = \frac{\pi}{4} d_o^2 V_{in} \tag{12}$$

Strictly speaking, C and α in (11) should be determined through experiments. Here, for simplicity, α would be determined with the method similar to that used by Guthrie in establishing an effective viscosity formula for ladels, (8) i.e. by comparing the present formula with that of Pun's.(9)

Pun's formula for a combustion chamber can be written as:

$$\mu_{eff} = k' D^{2/3} L^{-1/3} \rho U_o \tag{13}$$

where L, D are respectively the length and the diameter of the chamber, ρ the density of the gas, U_o the velocity of the gas inlet and K' coefficient.

It can be found that both (10) and (13) have the same form. So far as it is concerned that in a tundish the deeper the melt, the lower the turbulent level, it can be affirmed H in (10) has the same effect on the effective viscosity as L in Pun's formula. So we take $\alpha = 1/3$. Therefore (11) can be written as:

$$\mu_{eff} = C\ d_o^{-2/3} H^{-1/3} \rho Q \tag{14}$$

The coefficient C was obtained through comparison of values of μ_{eff} from (14) and average spatial values of μ_{eff} obtained with $k - \varepsilon$ model.

2.3 Boundary conditions

To establish the mathematical model, following assumption have been made:

(1) The flow was steady, viscous, impressible and constant in temperature.

(2) The influence of top slag on the flow was neglected and the melt surface was taken as a free one, therefore the voticity was zero at the surface.

(3) The jet inlet was uniform.

(4) The tundish was regarded as a rectangle vessel.

The boundary conditions are shown in Fig.1.

3.EXPERIMENTAL WORK

Water model experiments were carried out in order to investigate the flow patterns and confirm the general adequacy of the mathematical model developed. The model was a geometrically scaled one-third one. Because the flow was dominated by gravity, Froude number was taken as the similarity criterion.

The velocity distribution was measured with Laser Doppler Anemometry (LDA-10). The flow patterns were visualized with the photographic technique which was carried out by adding Al-powder, interfused with ethyl alchol, into the water which was partially illuminated by a slit of light and then taking photos

4. RESULTS AND DISCUSSION

The fundamental equations, together with the boundary conditions, were changed into finite difference forms and then translated into FORTRAN language. The computation was carried out on a M - 150 computer at BUIST by means of iteration method. To promote convergence, relaxation method was used, with the vorticity being under-relaxed and the stream function over-relaxed.

$$\frac{\sum|\phi^{(n)} - \phi^{(n-1)}|}{\sum|\phi^{(n)}|} \leq \varepsilon$$

was taken as the criterion of convergence and $\varepsilon = 0.005$.

K - ε model was also applied to calculate the velocity distributions in the tundish with various kinds of dam setting. The results were verified by measurement. From the average effective viscosity obtained with this model, C in (14) was obtained by C = $8.8\text{x}10^{-3}$. Thus, the effective viscosity formula for the flow in the tundish would be

$$\mu_{eff} = 8.8\text{x}10^{-3} d_o^{-2/3} H^{-1/3} \rho Q \tag{15}$$

By use of the formula, the velocity distributions were also calculated. One of them (double-1-a type of dam setting) was shown in Fig.2. Fig.3 and Fig.4 are respectively the measured result and the flow field photograph. Inspection of Fig.2,3,4 reveals a close geometric similarity between the predicted and the measured flow fields. The predicted result with the effective viscosity formula method was also in good agreement with that got by k - ε model(shown in Fig.5). The relative errors between the velocity values obtained with two methods at corresponding nodes were generally under 10%. Therefore it is avaliable for the effective viscosity formula method to be applied in the researches on the flow characteristics in the tundish. Under this case, CPU time can greatly reduced. For example, the CPU time for double-1-a type of dam setting was 605 s by k - ε model while 65 s by the effective viscosity formula method, being saved by nearly 90%.

The comparison between the effective viscosity values obtained with the two methods under conditions of different types of tundish dam setting was shwon in Table 2. It can be found that the present prediction for the effective viscosity by the formula was reliable. The reason was that the local effective viscosity values in the tundish do not vary much except in the jet inlet zone (shown in Fig. 6), which made the application of the formula more reasonable.

5.CONCLUSION

The mathematical modelling method has been applied to calculate the velocity distribution in a continuous casting tundish. For simplicity, an effective viscosity formula for the tundish has been established on the basis of analogy analysis as well as the k - ε model calculation. The formula got the form:

$$\mu_{eff} = 8.8\text{x}10^{-3}\ d_o^{-2/3}\ H^{-1/3} \rho\ Q$$

It was found that the formula was reliable in estimating the general effective viscosity of the flow in the tundish. The predicted velocity distributions were in good agreement with those got with k - ε model and the measured results. Comparing with k - ε model, the present method could save CPU time about 90%. As a method for the investigation on flow characteristics in industrial vessels, effective viscosity formula method can be widespreadly applied.

ACKNOWLEDGEMENT

The authors are grateful to the Fund Council of Chinese Academy of Sciences for the financial support to this work which is a part of the subject No. 82-426.

REFERENCES

1. E.D.Tarapore et al: Met. Trans. B, 7B (1976), p.343
2. R.I.L.Guthrie et al: Met. Trans. B, 9B (1978), p.673
3. M.Nakata et al: Tetsu-to-Hagane, vol.68(1983), s212
4. J. Szekely et al: Met. Trans. B, 5B (1974), p.463
5. Y. Habu et al: Tetsu-to-Hagane, vol.62 (1976), p.1803
6. K.Yamaguchi et al: Tetsu-to-Hagane, vol.67 (1981), s 145
7. Qu Ying et al: SCANINJECT III, June, 1983, Luleå, SWEDEN p.21:1-21:16
8. R.I.L.Guthrie et al: Met. Trans. B, 13B (1982), p.125
9. W.M.Pun et al: XVIII International Astronatical Congress, Proceedings, Pergamon Press/PWN-polish Scientific Publisher, 1967, vol.3, p.269

Table 1. Factors Affecting the Flow Effective Viscosity and Their Dimensions

Factor	Jet velocity	Density	Depth	Jet height	Jet diameter
Symbol	V_{in}	ρ	H	h	d_o
Dimension	Lt^{-1}	mL^{-3}	L	L	L
*	m - mass ; L - lengh ; t - time				

Table 2. Comparison between Effective Viscosity Values Obtained with the Two Methods

Dam	Q (1/min)	H (cm)	d_θ (cm)	liquid	ρ (g/cm^3)	μ_{eff} Eq. (15)	g/cms k-ε
D-1-a	18	25	1.8	Water	1.0	0.610	0.613
D-1-b	18	25	1.8	Water	1.0	0.610	0.630
D-1-c	18	20	1.8	Water	1.0	0.665	0.669
D-1-a	280.5	75	5.4	Steel	7.0	22.0	19.0
D-1-b	280.5	75	5.4	Steel	7.0	22.0	18.3
D-2	280.5	75	5.4	Steel	7.0	22.0	27.3

* D ——— double

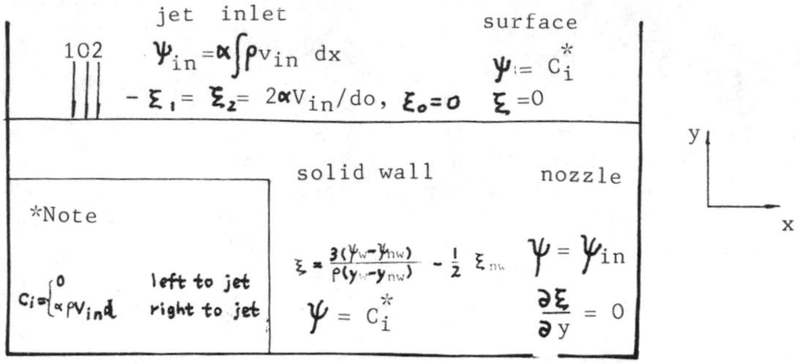

Fig.1 Schematic representation of the system and boundary conditions

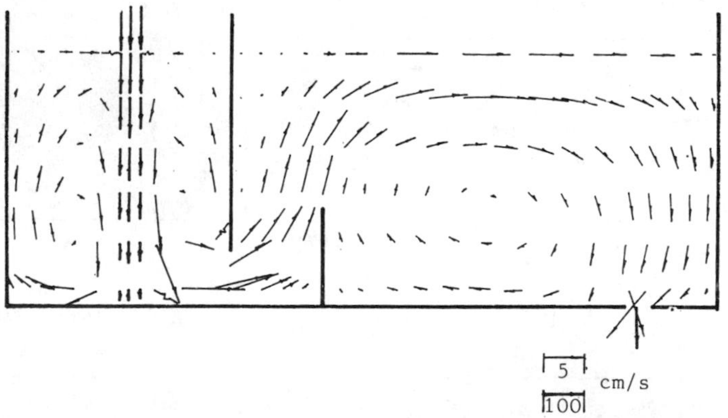

Fig.2 Predicted velocity distribution using μ_{eff} formula (Q=18 1/min, H=25 cm, Water model)

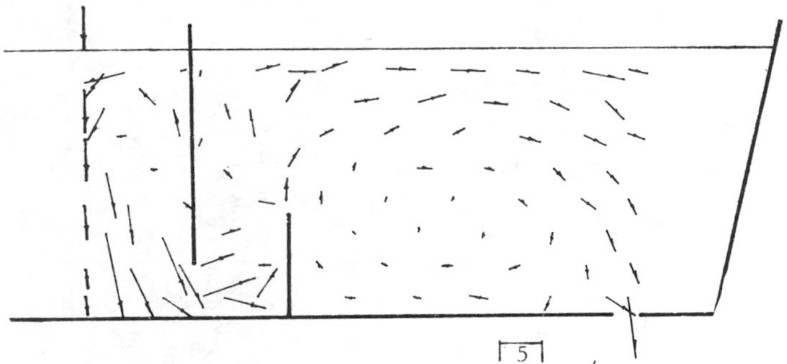

Fig. 3 Measured velocity distribution
(Q=18 1/min, H=25 cm, Water)

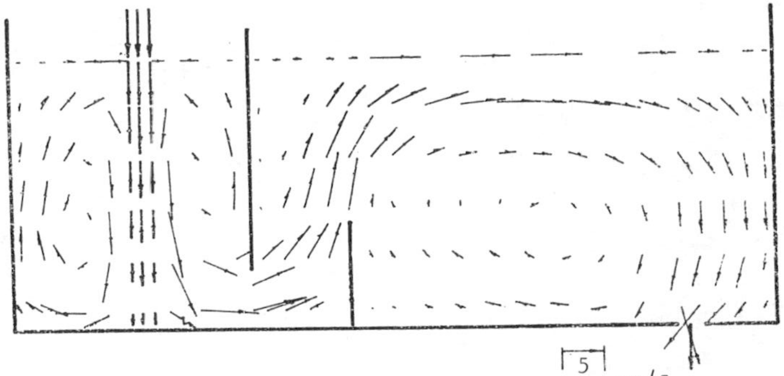

Fig.5 Predicted velocity distribution using
K - ε model (Q=18 1/min, H=25cm, Water model)

Fig.4 Photograph of the flow pattern
(Q=18 1/min, H=25 cm, Water)

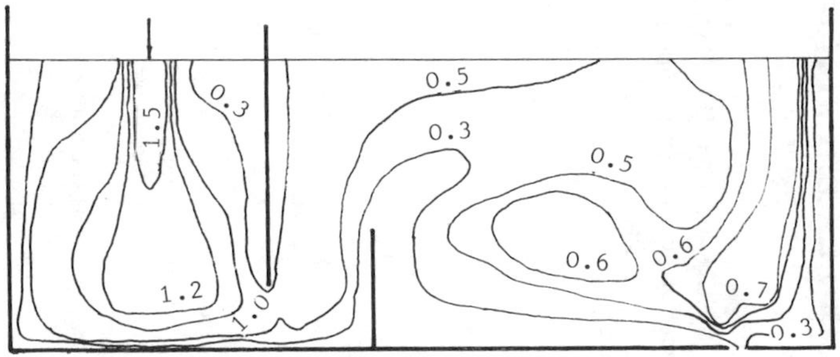

Fig.6 The effective viscosity profile
predicted by k - ε model (g/cms)

Simulation of Time-Temperature History and Cure Extent in Rubber Articles

ZHANG NENGLI and YING DANYANG
Department of Thermal Engineering
Tsinghua University
Beijing, PRC

ABSTRACT

A computer simulation program based on the finite element analysis has been developed to calculate numerically two dimensional time-dependent temperature distribution and the extent of cure in rubber articles with irregular geometries and different material sandwiches. In the program variable thermal properties, internal heat generations which depend on the temperature, cure state during the molding vulcanization as well as during their cooling in air have been considered. As a practical applied example the temperature distribution history and the state of cure at any point in rubber tire during both molding cure cycle and cooldown stages were calculated and compared with results of experiment. A satisfactory agreement was obtained. The program can successfully simulate the cure cycle process of the rubber tire and any rubber articles. It is obvious that the program can be used more widely and effectively to solve any two-dimensional, unsteady heat conduction problems.

NOMENCLATURE

- A, B coefficient matrices
- ko_i, $k1_i$, $k2_i$ constants in eq.(2)
- co_i, $c1_i$, $c2_i$ constants in eq.(3)
- c_p specific heat of the materials
- E activation energy of cure reaction
- G intensity of heat generation
- G_o, G_1 constants in equation (5)
- h convective heat transfer coefficient
- k thermal conductivity
- K, K_t overall conductance and overall capacitance matrices
- K_t^e, K_t^e, K_t^e matrices of conduction, convection coefficient and capacitance in element e
- n normal direction
- P overall heat load vector
- P time-weighted overall heat load vector
- P_Q^e, P_{qs}^e, P_{hs}^e heat load vectors of internal heat generation, specified surface heating and surface convection
- Q internal heat generation
- r_1 time-relaxation factor
- R gas constant
- SOC state of cure
- T, T_a, T_e, T_R local, ambient, effective ambient and reference temperatures
- t, t' time and dimensionless time interval
- w_i weight function
- x, y horizontal and vertical coordinates
- ρ density of materials
- τ time variable

Subscripts

- s_1, s_2, s_3, s_4 surface b-c, d-a, a-b and c-d (Fig.2)

Ying Danyang is a visiting Scholar on leave from Chemical Machinary Institute, Lanzhou, Pople's Republic of China.

INTRODUCTION

The curing of rubber articles presents two basic problems: the simulation of real processes of curing and the prediction of curing cycle of new developed products. Both are in order to optimize the cure cycle and thus reduce energy requirement of the process which is the goal of engineers in rubber industry for a long time. The word "optimize" means reducing the cure time cycle without detriment to quality and a compromise between the over curing portion and the under curing portion which rise from the poor thermal conductivity of rubber. It is obvious that a detail knowledge of vulcanization which will enable us to determine the proper time carrying out the process is necessary to the engineer. On the other hand, as Vandoren point out, nobody has an accurate idea of a good cycle and it is not economically reasonable to make a great number of experiments on real pieces to establish the optimal cure cycle[1]. For technical products the optimum is usually defined by some technical oriteria which may be quite different from product to product. Sometimes the emphasis is put on pure economical criteria. For these cases a good simulation program is needed for saving a great amount of money. Obviously, it is the most important to obtain a sufficient knowledge of the temperature distribution history in rubber articles. Accetta and Vergnaud[2-4] solved the problem for rubber sheets, the one-dimensional problem, by using an explicit numerical method with finite differences. Schlanger[5] described the development and application of an one-dimensional numerical model for a tire vulcanization. Hubbard and Simpson[6] used finite element method to try solve the same problem in two-dimensions, but they did not consider the cure of rubber products during their cooldown stage in which the rubber products were extracted from the molds and contacted with motionless air. As we will show in this paper that this so-called "residual curing" take considerable portion of the total cure extent. Prentice and Williams[7] calculated the time-dependent temperature distribution in a vulcanizing tire, using the finite difference method. They assumed that the tire consists of uniform material.

Actually, to solve the problem of simulating vulcanization process of a real tire or other rubber articles the following factors must be considered: (1) rubber article may be a multi-layer construction consisted of different materials, for example, the rubber tire, (2) the thermophysical properties of the article vary with temperature and position, (3) the internal heat generation due to cure reaction should take into account and a function of temperature and local state of cure, (4) the geometry of the rubber articles may be irregular (5) the boundary conditions usually are time-varying and have two quite different stages -- the heating cycle and the cooldown. In the pressent work all of the factors mentioned above are considered and as a practical example we calculate the temperature distribution history in a real tire and its extent of cure.

GOVERNING EQUATIONS

Most of the calculating problems of temperature distribution history in vulcanizing articles can be summarized into a transient two-dimensional heat conduction with time-dependent internal heat generation within the article. The basic differential equation governing the problem is

$$\frac{\partial}{\partial x}\left(k\frac{\partial T}{\partial x}\right)+\frac{\partial}{\partial y}\left(k\frac{\partial T}{\partial y}\right)+Q=\rho c_p\frac{\partial T}{\partial t} \qquad (1)$$

where Q is the internal heat generation rate due to the vulcanization reaction. The thermal conductivities and the heat capacities of different kind of rubber can be all expressed as following

$$k_i = ko_i + k1_i T + k2_i T^2 \quad (2)$$

$$c_{pi} = co_i + c1_i T + c2_i T^2 \quad (3)$$

where, the subscript i denotes different material. But the accurate data for the temperature dependence of heat capcity and bulk density are not available usually. Fortunately, the density decreases and the heat capacity increases with increasing temperature, so theproduct ρc_p is relatively insensitive to temperature.

It should be noted that the internal heat generation rate Q depends on both the temperature and the local state of cure (SOC). According to Prentice and Williams[7] we can use an approximation

$$Q=G(1-0.02(SOC)) \quad (4)$$

where, G is intensity of heat generation which depends on temperature and can be arranged as a linear function, based on experiment data. Q should be set to zero when the expression in brackets of Eq. (4) is negative. It should be also noted that the experiment shows no heat released from the vulcanization reaction when the temperature is lower than a certain value. For example, the experiment of the materials of a rubber tire gives the approximation

$$G=\begin{cases} G_o+G_1T & \text{when } 120^oC \leq T \leq 180^oC \\ 0 & \text{when } T < 120^oC \end{cases} \quad (5)$$

where $G_o= -0.11083\ W/cm^3$; $G_1= 0.9069x10^{-3}\ W/cm^3 \cdot {}^oC$
The SOC is defined as following

$$SOC=\int_0^t \exp(\frac{E}{R}(\frac{1}{T_R} - \frac{1}{T}))d\tau \quad (6)$$

Here, E is activation energy of the reaction; R, gas constant; T_R, reference temperature; T, the temperature of cure reaction.

MODEL

The time-temperature history and cure extent in rubber articles can be simulated by numerical solution of Eq. (1-6) with realistic mulation of the entire cure cycle of a vulcanizing tire as a example.

Fig.1 shows the scheme of the vulcanization set up of the rubber tire. A realistic cure cycle consists of molding, molding vulcanization, extraction and cooldown stage. The molding and extraction process are much short relative to others and their effects on the entire cure process can be negligible. So we can take the cure cycle as molding vulcanization and cooldown processes and the initial condition is

$$T(x,y,t)_{t=0} = T(x,y,0) \quad (7)$$

During molding vulcanization the tire is heated by steel mold with high pressure steam outside and bladder with superheated water inside. The experiment shows that there are only negligible differences of temperature along the boundaries both outside and inside individually. The thermocouples were arranged at the interfaces of tire with the mold and with bladder and got two time-dependent temperature boundary conditions:

$$\left.\begin{aligned} T_{s1}(x,y,t)&=T_1(t)\\ T_{s2}(x,y,t)&=T_2(t)\end{aligned}\right\} \quad \text{when} \quad 0 < t \leq t_1 \tag{8}$$

Here, subscripts s_1 and s_2 denote the outer-boundary and innerboundary of the tire, surface b-c and d-a in Fig.1, respectively. The time t_1 is the duration of molding vulcanization. After extracting from the mold the tire is cooled by natural convection in air. The heat transfer through rubber-air interface is very complex. It strongly depends on the position of tire and the surroundings. In present paper we use the convective boundary conditions given in reference[7] and take them as

$$\left.\begin{aligned} -k_{s1}\frac{\partial T_{s1}}{\partial n_1} &= h_{s1}(T_{s1}-T_a)\\ -k_{s2}\frac{\partial T_{s2}}{\partial n_2} &= h_{s2}(T_{s2}-T_e)\end{aligned}\right\} \quad \text{when} \quad t_1 < t < t_\infty \tag{9}$$

Here, T_a is ambient temperature; $T_e=(T_{s2}-T_a)/2$, is effective ambient temperature to correct the influence of trapping of air.

The interested points of cure locate at the thickest portion of the shoulder area. Therefore the attention was concentrate on this area, the region a-b-c-d in Fig.1 and in Fig.2 for detail. Because the surface a-b is a line of symmetry it can be taken as adiabatical. The heat transfer across surface c-d where the tire section is out can be neglected since the thermal surroundings and its thin section. Thus the boundary conditions at surfaces a-b and c-d are

$$\left.\begin{aligned} \left(\frac{\partial T}{\partial y}\right)_{s3} &= 0\\ \left(\frac{\partial T}{\partial x}\right)_{s4} &= 0\end{aligned}\right\} \quad \text{when} \quad 0 < t < t_\infty \tag{10}$$

where subscripts s3 and s4 denote the section a-b and c-d respectively. The boundary conditions given by Eq. (8) and used in the calculation of the example are detailed in Fig.3.

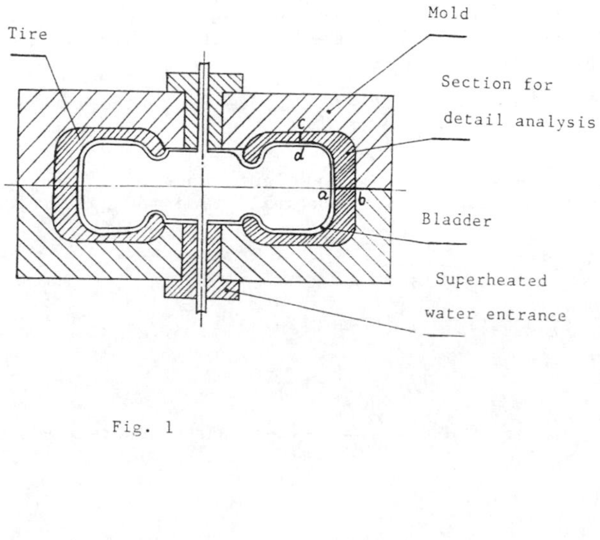

Fig. 1

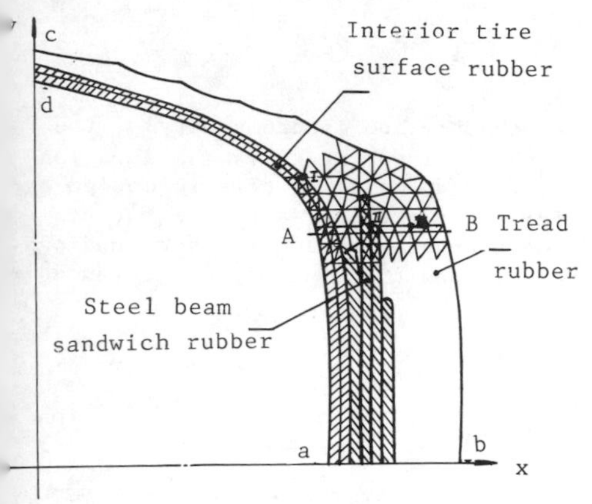

Fig.2

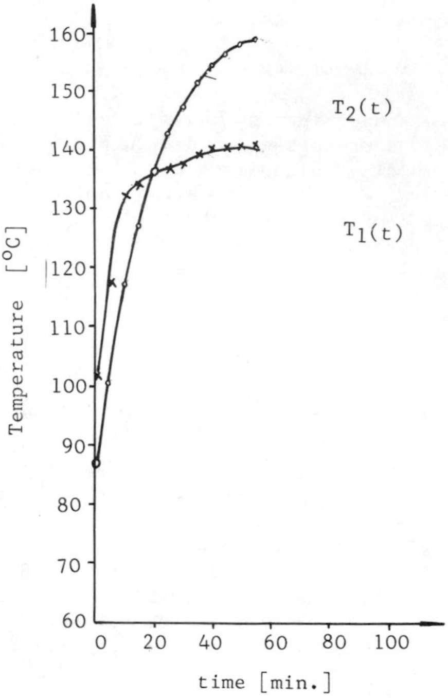

Fig.3

NUMERICAL METHOD

We use the finite element in the space domain and finite difference in the time domain to solve the unsteady state heat transfer problem. The overall finite element equation corresponding to the Eq. (1) is

$$[K_t]\frac{dT}{dt} + [K]T = P \tag{11}$$

where the overall capacitance matrix $[K_t]=\sum_{e=1}^{n}[K_t^e]$; the overall conductance matrix $[K]=\sum_{e=1}^{n}\{[K_k^e] + [K_s^e]\}$ which consists of condution and convection coefficient matrices; $P = \sum_{e=1}^{n}(P_Q^e + P_{qs}^e + P_{hs}^e)$, the heat load vector arising from internal generation, specified surface heating and surface convection. In these relations the subscript e denotes the parameters belong to element e.

Weighted residual technique with finite difference scheme in the time domain can be applied to Eq.(11) and we obtain

$$\int_0^1\left\{[K_t]\frac{T^{i+1} - T^i}{\Delta t_i} + [K]\,(T^i(1-t')+T^{i+1}\,t') - P(t_i+t'\,t_i)\right\}w_i dt' = 0 \tag{12}$$

(i=0,1,2, ...)

where $t'=(t-t_i)/\Delta t_i$, $\Delta t_i = t_{i+1} - t_i$; w_i is weight function. The $[K_t]$ and $[K]$ can be considered to remain constant over a small time interval Δt_i and then

$$\left(\frac{[K_t]}{\Delta t_i} + r_i[K]\right) T^{i+1} + \left(\frac{-[K_t]}{\Delta t_i} + (1-r_i)[K]\right)T^i = \bar{P} \tag{13}$$

here, $r_i=(\int_0^1 w_i t' dt')/(\int_0^1 w_i dt')$; $\bar{P} = (1-r_i)P^i+r_iP^{i+1}$; i.e.

$$[A]T^{i+1} + [B]T^i = \bar{P} \tag{14}$$

where $[A]=\frac{[K_t]}{\Delta t_i} + r_i[K]$; $[B]= -\frac{[K_t]}{\Delta t_i} + (1-r_i)[K]$; $\bar{P}=(1-r_i)P^i+ r_iP^{i+1}$

By choosing different forms of the weighting function w_i a variety of solution schemes can be available. In the present paper, we choose $r_1=0.60$.

Once the time-temperature history at points of interest is calculated the state of cure at the points at any time can be determined very easily in the following

$$SOC^{i+1} = SOC^{i} + \int_{t}^{t+\Delta t} \exp\left(\frac{E}{R}\left(\frac{1}{T_R} - \frac{1}{T}\right)\right) d\tau \qquad (15)$$

where SOC^{i} is the state of cure at the instent of $t=t_i$; SOC^{i+1}, the value at $t=t_{i+1}$.

RESULTS AND DISCUSSION

Three typical points, I,II,III in Fig.2 were chosen to calculate their time-temperature history and compare with the results of measurement which were provided by Beijing Rubber Institute. Fig.4-6 are the comparisons. The results show that the satisfactory agreements are reached. Both cases corresponding to varying thermophysical properties (case 1 in Fig.4-6) and constant

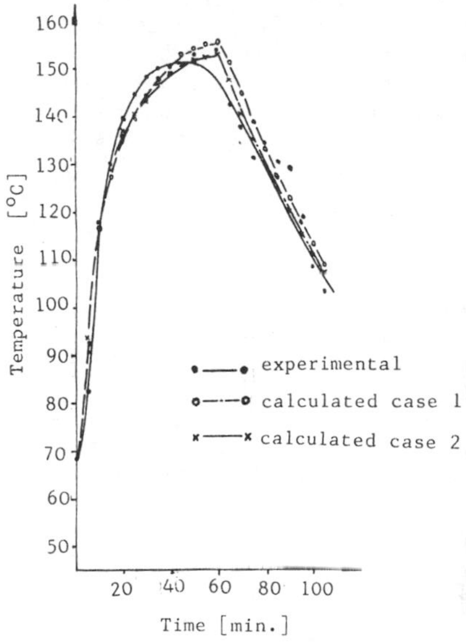

point I in Fig.3

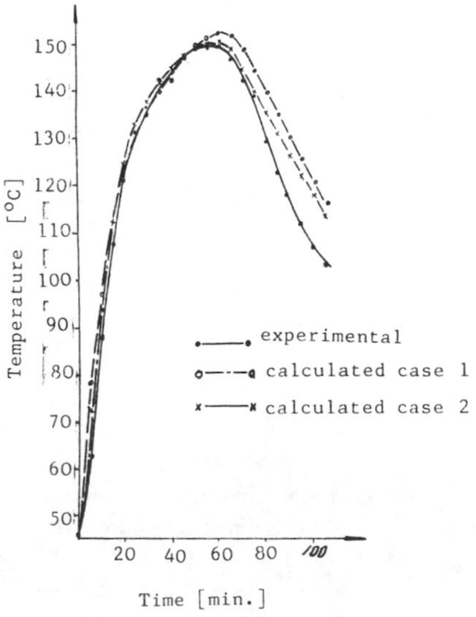

Fig.5 Temperature history at point II in Fig.3

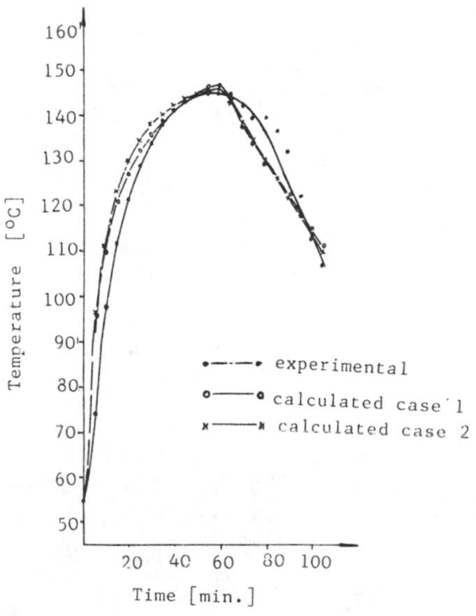

Fig.6 Temperature history at point III in Fig.3

roperties (case 2 in Fig.4-6) ere calculated. They do not eviate from the results of experiment more than 12% which is ot beyond the error of the measurement. Considering the elasicoviscosity of the rubber the esults of calculation should be onsidered to be sufficient satisfactory.

The distribution of temperaure along a section, for exmple, A-B section in Fig.2 and ts time history can be calcuated. Fig.7 shows the temperaure distributions along A-B ection at several moments. 'ig.8 shows the distribution of he state of cure along the secion at the moment of the end of eating cycle and at a moment vithin cooldown stage. The reult shows that the cure process ot only exists in heating cycle ut also in cooldown stage and he latter takes a considerable ortion of the total cure extent. So one has to take it into account. From Fig.8 the nonuniform state of cure across A-B section was found. It means that the components of rubber tire or the control temperature of mold and the bladder in this example should be regulated to improve the overall state of cure of the tire. In other vords, the results of the simulation show that the present operation using in some rubber tire manufacturer which is cited in this paper would not produce qualified tire and some new technical regulation have to made.

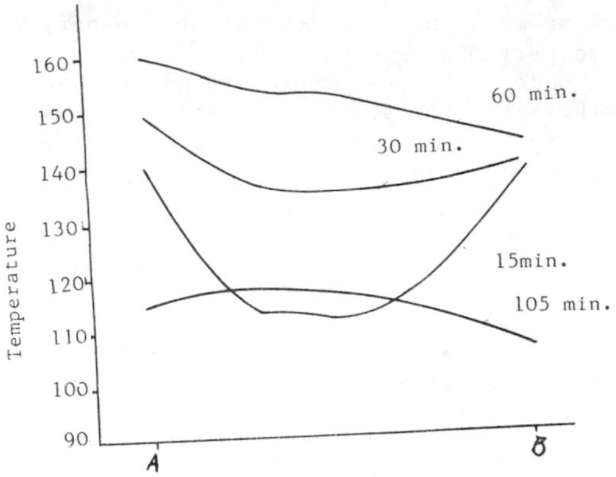

Fig.7

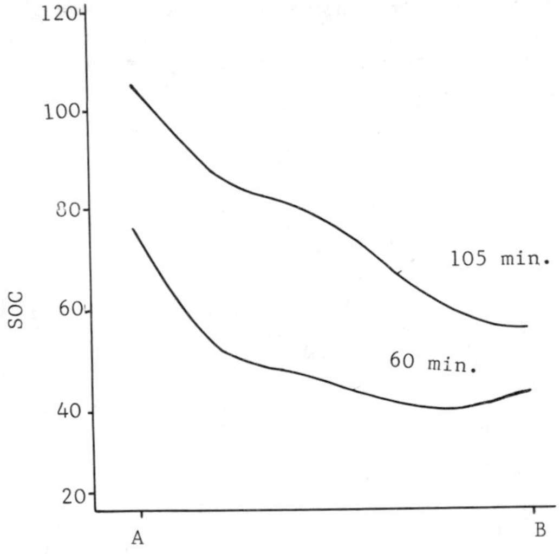

Fig.8

The effects of the key parameters, like conductivities $k_i(T)$, Q(T) and the control temperature of the mold and bladder (the boundary conditions), were examined. The simulations show that these parameters have very important influence to the overall SOC. The detail discussion will be presented in a separate paper later.

CONCLUSIONS

The accurate two-dimensional time-dependent temperature distribution and extent of cure in rubber articles with irregular geometries and different material sandwiches can be simulated based on the finite element analysis. In the simulations the variable thermophysical properties and variable internal heat generations which are also functions of the state of cure have been considered. As a practical application example, the real process of cure of the tire in a manufacturer has been simulated and some directional results have been got. The program can be applied directly any two-dimensional unsteady heat conduction problem and easily extented to cover three-dimensional analysis.

ACKNOWLEDGMENTS

This work was supported in part by Beijing Rubber Institute. The cooperation of Mr. Hailin Wang at Beijing Rubber Institute is greatly acknowledged.

REFERENCES

[1] Vandoren, P., "Minicomputer Technique for Simulation of Heat Transfer and Cure Level in the Thick Rubber Elements", Kautschuk and Gummi Kunststoffe 37, Jahrgang, Nr. 5/84 p. 398

[2] Accetta, A. and Vergnaud, J. M., "Calculation of the Temperature and Extent of Reaction During the Vulcanization of Powdered Rubber", Rubber Chem. Technol. 56, 4, (1983) p. 689

[3] Accetta, A. and Vergnaud, J. M., "Kinetic Parameters of the Overall Reactic of Scrap Rubber Vulcanization by 2% Sulfur", Thermochim. Acta, 59, (1982) p. 149

[4] Vergnaud, J. M,,"Process Monitoring and Simulation of Production Process", Internat. Rubber Conf.,Moscow (1984)

[5] Schlanger, H. P., "A One-Dimensional Numerical Model of Heat Transfer in the Process of Tire Vulcanization", Rubber Chem. Technol..56, 2, (1983) p. 304

[6] Hubbard, G. D. and Simpson, G. M., "Application of Finite Element Theory to Heat Flow in Rubber Products", Proc. Int. Rubber Conf., Venice, (1979)

[7] Prentice, G. A. and Williams, M. C., "Numerical Evaluation of the State of Cure in a Vulcanizing Rubber Article", Rubber Chem. Technol., 53, 4, (1980) p. 1023

[8] Rao, S. S., "The Finite Element Method in Engineering" Pergamon Press Ltd., (1982)

Author Index

Subject Index